# Springer Texts in Business and Economics

Springer Texts in Business and Economics (STBE) delivers high-quality instructional content for undergraduates and graduates in all areas of Business/Management Science and Economics. The series is comprised of self-contained books with a broad and comprehensive coverage that are suitable for class as well as for individual self-study. All texts are authored by established experts in their fields and offer a solid methodological background, often accompanied by problems and exercises.

Patrice Poncet • Roland Portait

# Capital Market Finance

An Introduction to Primitive Assets, Derivatives, Portfolio Management and Risk

Volume 1

With Contributions by Igor Toder

Patrice Poncet
ESSEC Business School
Cergy Pontoise, France

Roland Portait
ESSEC Business School
Cergy Pontoise, France

ISSN 2192-4333   ISSN 2192-4341 (electronic)
Springer Texts in Business and Economics
ISBN 978-3-030-84598-8    ISBN 978-3-030-84600-8 (eBook)
https://doi.org/10.1007/978-3-030-84600-8

Mathematics Subject Classification: 91G10; 91G15; 91G20; 91G30; 91G40; 91G60; 91G70

Translation from the French language edition: Finance de marché by Patrice Poncet, et al., © Éditions Dalloz 2008. Published by Éditions Dalloz. All Rights Reserved.

© Springer Nature Switzerland AG 2022
This work is subject to copyright. All rights are solely and exclusively licensed by the Publisher, whether the whole or part of the material is concerned, specifically the rights of reprinting, reuse of illustrations, recitation, broadcasting, reproduction on microfilms or in any other physical way, and transmission or information storage and retrieval, electronic adaptation, computer software, or by similar or dissimilar methodology now known or hereafter developed.
The use of general descriptive names, registered names, trademarks, service marks, etc. in this publication does not imply, even in the absence of a specific statement, that such names are exempt from the relevant protective laws and regulations and therefore free for general use.
The publisher, the authors, and the editors are safe to assume that the advice and information in this book are believed to be true and accurate at the date of publication. Neither the publisher nor the authors or the editors give a warranty, expressed or implied, with respect to the material contained herein or for any errors or omissions that may have been made. The publisher remains neutral with regard to jurisdictional claims in published maps and institutional affiliations.

This Springer imprint is published by the registered company Springer Nature Switzerland AG
The registered company address is: Gewerbestrasse 11, 6330 Cham, Switzerland

*This book is dedicated to the dear memory of Roland Portait (1943–2021), my long-time friend, wonderful colleague and inspiring co-author, whose untimely death leaves a deep and painful void among his family and friends.*

# Preface

This textbook is primarily aimed at graduate students of market finance (MBA, MiM, and Ms in business schools and engineering schools at universities, first-year PhD programs in finance or economics). It is also aimed at market finance practitioners: trading rooms, asset management firms (using quantitative tools), actuaries in banks and insurance companies, risk management and control outlets, and research teams in market finance.

The objective is to make up for the relative inadequacy of most traditional finance books and mathematical finance books. The former do not offer in general the advanced mathematical techniques currently used by expert professionals. And the latter often focus on mathematical refinements at the expense of the economics behind financial contracts and products, the practical use of instruments, and the financial logics of markets; in addition, some of them are of a level inaccessible to non-mathematicians, even engineers.

This book is the result of many years of teaching market finance in the specialized programs of ESSEC (market finance track, master's in finance), CNAM (master's in market finance), University of Paris 1-Panthéon-Sorbonne (master's and PhD in finance) and within the trading rooms of Société Générale, a major French bank, and other banking institutions.

It offers a comprehensive and consistent presentation (from the point of view of analysis and notation) of the whole of market finance. In particular, it covers all primitive assets (equities, interest and exchange rates, indices, bank loans) except real estate, most vanilla and exotic derivatives (swaps, futures, options, hybrids, and credit derivatives), portfolio theory and management, and risk appreciation and hedging of individual positions as well as portfolios and firms' balance sheets. It emphasizes the methodological aspects of the analysis of financial instruments and of risk assessment and management. In particular, it devotes an important space to the probabilistic foundations of asset valuation and to credit and default risks, the poor understanding of which aggravated the 2007–2008 financial crisis originated in the subprime credit market.

The introductory chapter (Chap. 1) is devoted to the economic role played by the financial and banking markets and their organization and functioning. It is followed by four parts.

The first part, consisting of Chaps. 2 to 8, deals with primitive assets (debt securities, bonds, equities), the term structure of interest rates, floating rate instruments, and vanilla swaps. It does not use complicated mathematical tools and is accessible to undergraduate students. It is aimed at readers who are new to market finance, the most advanced of which can access the second part almost directly.

The second part, consisting of Chaps. 9 to 20, is devoted to derivatives (options and futures) and presents the main models, stochastic calculus tools, and probabilistic theories on which modern methods of valuing contingent assets or claims and financial risks are based. This part, much more technical than the previous one, is aimed at graduate students and practitioners operating in financial markets. The mathematical background required to read part of the material is that acquired during the first two years of scientific studies and quantitative economics or management studies (analysis, differential and integral calculus, linear algebra, and probability theory). These developments are signaled by one star (*) or, rarely, two stars (**). This is also the case for books referenced at the end of each chapter. Readers equipped with this mathematical background will find in the book all the complements concerning stochastic calculus and probabilistic theories, necessary and sufficient for an in-depth understanding of modern market finance.

The third part, including Chaps. 21 to 25, is devoted to portfolio theory and management. After a presentation of the standard portfolio theory (Markowitz, capital asset pricing model, and arbitrage pricing theory), various techniques of strategic and tactical asset allocation (benchmarking, portfolio insurance, alternative investment, etc.) are discussed.

Finally, the fourth part, consisting of Chaps. 26 to 30, deals with risk management with particular attention to analytical methods (simulations, value-at-risk, expected shortfall, value adjustments), credit risk (theoretical and empirical analysis of counterparty and default risks, credit VaR, credit derivatives), and new regulation regarding financial institutions and banks.

Cergy Pontoise, France

Patrice Poncet
Roland Portait

# Main Abbreviations and Notations

Despite our efforts, because of the variety of themes covered in this book and its length and technicality, we could not prevent certain symbols from having distinct meanings in different chapters. Although we always define our notations before using them and the context in principle removes any ambiguity, the following list of key notations and abbreviations may prove useful.

General conventions: an underlined variable ($\underline{x}$) denotes a vector (or sequence), matrices are written in bold ($\mathbf{X}$), a ' denotes the derivative of a function or the transpose of a vector or matrix, and $[X]^+$ means $\max[X, 0]$. The reading of more technical paragraphs equipped with one or two stars can be omitted or postponed.

| | |
|---|---|
| AAO | Absence of arbitrage opportunity, no arbitrage |
| ABS | Asset-backed security |
| $\alpha$ | "abnormal" return rate (in asset pricing and portfolio management), or recovery rate of a debt instrument in the event of default (in credit risk analysis) |
| BS | Black–Scholes (formula of, or model of) |
| BSM | Black–Scholes–Merton (model of). Alternative name: Gaussian evaluation model |
| $B_T(t)$ | Price on date $t$ of a zero-coupon bond delivering \$1 or €1 on date $T$ (duration: $T$-$t$) |
| $b_\theta(t)$ | Discount factor or price on date $t$ of a zero-coupon bond delivering \$1 or €1 on date $t + \theta$ (duration: $\theta$). We have: $B_T(t) = b_{T-t}(t)$. |
| $\beta$ | Beta of a stock (sensitivity to a stock market index) or, more generally, the slope of a linear regression |
| $C$ | Usually the price of a call, sometimes a coupon, sometimes a Cap |
| CDO | Collateralized debt obligation |
| CDS | Credit default swap |
| CIR | Cox–Ingersoll–Ross (model of) |
| CRR | Cox–Ross–Rubinstein (model of). Alternative name: binomial model |
| $D, d$ | In general, time to maturity (duration), or Macaulay duration |
| $\delta$ | The delta (sensitivity) of an option |
| DD | distance to default |

| | |
|---|---|
| E | mathematical expectation ($E^Q$, $E^{QT}$, $E^*$, $E^{RN}$, if the probability measure is specified) |
| ES | Expected shortfall (alternatively: conditional VaR or tail VaR) |
| F | A cash flow; the price of a floor; a forward or futures price; a distribution function |
| $\underline{F}$ | A cash flow sequence (vector): $F_{\theta \mid \theta = t1, \dots, tn}$ |
| $F_T(t)$ | Forward or futures price on date $t$ for maturity $T$ (less precise notation: $F$ $(t)$) |
| $\Phi_T(t)$ | Forward price when it is distinguished from the futures price denoted by $F_T(t)$ |
| $f_{T,D}(t)$ | Forward rate prevailing at $t$ relative to the period $(T, T+D)$ |
| $f_T(t)$ | Instantaneous forward rate (limit of the previous one when $D$ tends to 0) |
| $\gamma(t)$ | Probability of survival at $t$ (in case of risk of default) |
| $\Gamma, \theta, \rho, \nu$ | The various other "Greek" parameters of an option (gamma, theta, rho, vega, respectively) |
| $H(t)$ | Value at $t$ of the optimal growth (logarithmic) portfolio |
| $k$ | Often a fixed interest rate (nominal rate of a fixed income instrument, fixed rate of a swap,...) |
| $K$ | Exercise price (strike) of an option |
| $\lambda$ | Market price of risk, Lagrange multiplier, intensity of a jump process, hazard rate, scalar |
| $L$ | Loss in default risk analysis, Lagrangian |
| $\mu$ | Often an expectation or a mean |
| $N(\mu, \sigma^2)$ | Normal (Gaussian) distribution of mean $\mu$ and variance $\sigma^2$ |
| $N(x)$ | Distribution function of a reduced centered (standard) normal variable |
| P | Historical probability |
| $P$ | Usually the price of a put |
| $p_d$ | Probability of default |
| PDE | Partial differential equation |
| Q | Risk-neutral probability measure (denoted also by RN and sometimes by *) |
| $Q_T$ | Forward-neutral probability measure relative to maturity $T$ (denoted also by FN-$T$) |
| $r$ | Generically means a rate. In general, our notation does not distinguish between proportional, compounded, discrete, and continuous rates. From Chap. 6 on, it is a zero-coupon rate, with yields to maturity of bullet bonds being denoted by $y$. Also, $r$ represents a risk-free rate and $r$ $(t)$ the instantaneous risk-free rate on date $t$. |
| $r_\theta$ | In general a zero-coupon rate of duration $\theta$ |
| $R, R_i$ | Often a random return (chapters on portfolio theory) |
| $\rho$ | Often a correlation coefficient; also the sensitivity of an option to interest rate variations; exceptionally an interest rate |
| $S$ | Sometimes the sensitivity of an interest rate instrument to interest rate variations |
| $s, s_\theta$ | Often a spread (difference between rates, margin,...) |

# Main Abbreviations and Notations

| | |
|---|---|
| $S(t)$, $S_t$ | The price of a spot asset on date t |
| $S_{k\text{-}r}$ | Swap fixed-rate $k$ receiver (floating rate $r$ payer) or lender swap |
| $S_{r\text{-}k}$ | Swap fixed-rate $k$ payer (floating rate $r$ receiver) or borrower swap |
| $\Sigma$ | Return rate diffusion matrix (dynamics of a portfolio in continuous time) |
| $\sigma$ | Volatility, or standard deviation |
| $\sigma_{XY}$ | Covariance $(X, Y)$ |
| SDE | Stochastic differential equation |
| $T$ | Maturity date (distant from current date $t$ by duration $\theta = T\text{-}t$) |
| $\tau$ | A date, a duration, a first-default date |
| $U$ | Often a reduced centered (standard) normal variable |
| UA | Underlying asset (of a derivative instrument) |
| $V$ | In general a market value or price |
| $\mathbf{V}$ | A variance–covariance matrix |
| VaR | Value at risk |
| $var$ | Variance (alternatively $\sigma^2$) |
| $W(t)$, $W_t$ | Standard Brownian motion or Wiener process (value at $t$) |
| $y$, $y_\theta$ | Often a yield to maturity, to be distinguished from a zero-coupon rate $r_\theta$ |
| $Z$ | Standardized profitability (in credit risk analysis) |
| ZC | Zero-coupon bond |
| $\underline{1}$ | Unit vector |
| $\mathbf{1}_E$ | Indicator of event E ($= 1$ if E is true, and $= 0$ if not) |

# Acknowledgments

The authors acknowledge the participation, assistance, and support of Igor Toder who contributed to five chapters and is the main author of half of the last one.

The authors thank Sami Attaoui, Riadh Belhaj, Frédéric Bompaire, Thierry Charpentier, Minh Chau, Guillaume Coqueret, Andras Fulop, Vincent Lacoste, Didier Maillard, Jacques Olivier, and Bertrand Tavin for their many valuable comments over the long years spent to write this book.

They also thank the students of ESSEC Business School majoring in finance and students of the master's in finance at ESSEC, the CNAM, and the Sorbonne, as well as the participants in the Société Générale (a major French global bank) training seminars, for their support and many relevant questions and remarks over the same long years. They acknowledge special support from Chloé Baraille, Fei Wang, and, particularly, Alexis Marty.

They also wish to express their gratitude to Springer's Catriona Byrne, Joerg Sixt, and Rémi Lodh for their unfailing trust, as well as for editing the book.

Of course, they are responsible for any errors and omissions.

# Contents

**1 Introduction: Economics and Organization of Financial Markets** . . .  1
  1.1  The Role of Financial Markets . . . . . . . . . . . . . . . . . . . . .  1
      1.1.1  The Allocation of Cash Resources Over Time . . . . . . .  1
      1.1.2  Risk Allocation . . . . . . . . . . . . . . . . . . . . . . . . .  3
      1.1.3  The Market as a Supplier of Information . . . . . . . . . .  5
  1.2  Securities as Sequences of Cash Flows . . . . . . . . . . . . . . . .  5
      1.2.1  Definition of a Security (or Financial Asset) . . . . . . . .  5
      1.2.2  Characterizing the Cash Flow Sequence . . . . . . . . . .  6
  1.3  Equilibrium, Absence of Arbitrage Opportunity, Market
      Efficiency and Liquidity . . . . . . . . . . . . . . . . . . . . . . . . .  8
      1.3.1  Equilibrium and Price Setting . . . . . . . . . . . . . . . .  8
      1.3.2  Absence of Arbitrage Opportunity (AAO) and the
            Notion of Redundant Assets . . . . . . . . . . . . . . . . .  9
      1.3.3  Efficiency . . . . . . . . . . . . . . . . . . . . . . . . . . . .  11
      1.3.4  Liquidity . . . . . . . . . . . . . . . . . . . . . . . . . . . .  16
      1.3.5  Perfect Markets . . . . . . . . . . . . . . . . . . . . . . . .  18
  1.4  Organization, a Typology of Markets, and Listing . . . . . . . . . .  19
      1.4.1  The Banking System and Financial Markets . . . . . . . .  19
      1.4.2  A Simple Typology of Financial Markets . . . . . . . . . .  19
      1.4.3  Market Organization . . . . . . . . . . . . . . . . . . . . . .  24
  1.5  Summary . . . . . . . . . . . . . . . . . . . . . . . . . . . . . . . . .  32
  Appendix: The World's Principal Financial Markets . . . . . . . . . . . .  33
      Stock markets, market indexes and interest rate instruments . . .  33
      Organized Derivative Markets (Futures and Options, Unless
      Otherwise Indicated) . . . . . . . . . . . . . . . . . . . . . . . . . .  34
  Suggestion for Further Reading . . . . . . . . . . . . . . . . . . . . . . . .  35
      *Books* . . . . . . . . . . . . . . . . . . . . . . . . . . . . . . . . . .  35
      *Articles* . . . . . . . . . . . . . . . . . . . . . . . . . . . . . . . . .  35

**Part I  Basic Financial Instruments**

**2  Basic Finance: Interest Rates, Discounting, Investments, Loans** . . .  39
  2.1  Cash Flow Sequences . . . . . . . . . . . . . . . . . . . . . . . . . .  39

xv

| | | | |
|---|---|---|---|
| 2.2 | Transactions Involving Two Cash Flows | | 40 |
| | 2.2.1 | Transactions of Lending and Borrowing Giving Rise to Two Cash Flows over One Period | 41 |
| | 2.2.2 | Transactions with Two Cash Flows over Several Periods | 42 |
| | 2.2.3 | Comparison of Simple and Compound Interest | 47 |
| | 2.2.4 | Two "Complications" in Practice | 49 |
| | 2.2.5 | Continuous Rates | 53 |
| | 2.2.6 | General Equivalence Formulas for Rates Differing in Convention and the Length of the Reference Period | 53 |
| 2.3 | Transactions Involving an Arbitrary Number of Cash Flows: Discounting and the Analysis of Investments | | 55 |
| | 2.3.1 | Discounting | 57 |
| | 2.3.2 | Yield to Maturity (YTM), Discount Rate and Internal Rate of Return (IRR) | 63 |
| | 2.3.3 | Application to Investment Selection: The Criteria of the NPV and the IRR | 66 |
| | 2.3.4 | Interaction Between Investing and Financing, and Financial Leverage | 69 |
| | 2.3.5 | Some Guidelines for the Choice of an Appropriate Discount Rate | 72 |
| | 2.3.6 | Inflation, Real and Nominal Cash Flows and Rates | 74 |
| 2.4 | Analysis of Long-Term Loans | | 76 |
| | 2.4.1 | General Considerations and Definitions: YTM and Interest Rates | 76 |
| | 2.4.2 | Amortization Schedule for a Loan | 79 |
| 2.5 | Summary | | 84 |
| Appendix 1: Geometric Series and Discounting | | | 85 |
| Appendix 2: Using Financial Tables and Spreadsheets for Discount Computations | | | 87 |
| | 1. Financial Tables | | 87 |
| Suggested Reading | | | 92 |

| **3** | **The Money Market and Its Interbank Segment** | | **93** |
|---|---|---|---|
| 3.1 | Interest Rate Practices and the Valuation of Securities | | 94 |
| | 3.1.1 | Interest Rate Practices on the Euro-Zone's Money Market | 95 |
| | 3.1.2 | Alternative Practices and Conventions | 99 |
| 3.2 | Money Market Instruments and Operations | | 100 |
| | 3.2.1 | The Short-Term Securities of the Money Markets | 100 |
| | 3.2.2 | Repos, Carry Trades, and Temporary Transfers of Claims | 101 |
| | 3.2.3 | Other Trades | 103 |
| 3.3 | Participants and Orders of Magnitude of Trades | | 105 |
| | 3.3.1 | The Participants | 105 |

| | | | |
|---|---|---|---|
| | 3.3.2 | Orders of Magnitude | 106 |
| 3.4 | | Role of the Interbank Market and Central Bank Intervention | 107 |
| | 3.4.1 | Central Bank Money and the Interbank Market | 107 |
| | 3.4.2 | Central Bank Interventions and Their Influence on Interest Rates | 110 |
| 3.5 | | The Main Monetary Indices | 113 |
| | 3.5.1 | Indices Reflecting the Value of a Money-Market Rate on a Given Date | 113 |
| | 3.5.2 | Indices Reflecting the *Average* Value of a Money-Market Rate During a Given Period | 115 |
| 3.6 | | Summary | 117 |
| | | Suggestions for Further Reading | 118 |

**4 The Bond Markets** ..................................................... 119

| | | | |
|---|---|---|---|
| 4.1 | | Fixed-Rate Bonds | 121 |
| | 4.1.1 | Financial Characteristics and Yield to Maturity at the Date of Issue | 121 |
| | 4.1.2 | The Market Bond Value at an Arbitrary Date; the Influence of Market Rates and of the Issuer's Rating | 126 |
| | 4.1.3 | The Quotation of Bonds | 131 |
| | 4.1.4 | Bond Yield References and Bond Indices | 135 |
| 4.2 | | Floating-Rate Bonds, Indexed Bonds, and Bonds with Covenants | 135 |
| | 4.2.1 | Floating-Rate Bonds and Notes | 136 |
| | 4.2.2 | Indexed Bonds | 137 |
| | 4.2.3 | Bonds with Covenants (Optional Clauses) | 137 |
| 4.3 | | Issuing and Trading Bonds | 138 |
| | 4.3.1 | Primary and Secondary Markets | 138 |
| | 4.3.2 | Treasury Bonds and Treasury Notes Issues: Reopening and STRIPS | 139 |
| 4.4 | | International and Institutional Aspects; the Order of Magnitude of the Volume of Transactions | 141 |
| | 4.4.1 | Brief Presentation of the International Bond Markets | 142 |
| | 4.4.2 | The Main National Markets | 143 |
| 4.5 | | Summary | 146 |
| | | Suggested Readings | 148 |

**5 Introduction to the Analysis of Interest Rate and Credit Risks** ... 149

| | | | |
|---|---|---|---|
| 5.1 | | Interest Rate Risk | 149 |
| | 5.1.1 | Introductory Examples: The Influence of the Maturity of a Security on Its Sensitivity to Interest Rates | 150 |
| | 5.1.2 | Variation, Sensitivity and Duration of a Fixed-Income Security | 152 |
| | 5.1.3 | Alternative Expressions for the Variation, Sensitivity and Duration | 156 |

| | | | |
|---|---|---|---|
| | 5.1.4 | Some Properties of Sensitivity and Duration | 158 |
| | 5.1.5 | The Sensitivity of a Portfolio of Assets and Liabilities or of a Balance Sheet: Sensitivity and Gaps | 161 |
| | 5.1.6 | A More Accurate Estimate of Interest Rate Risk: Convexity | 167 |
| 5.2 | Introduction to Credit Risk | | 170 |
| | 5.2.1 | Analysis of the Determinants of the Credit Spread | 170 |
| | 5.2.2 | Simplified Modeling of the Credit Spread; the Credit Triangle | 172 |
| 5.3 | Summary | | 174 |
| | Appendix 1 | | 175 |
| | Default Probability, Recovery Rate and Credit Spread | | 175 |
| | Suggested Reading | | 176 |

**6 The Term Structure of Interest Rates** .......... 177

| | | | |
|---|---|---|---|
| 6.1 | Spot Rates and Forward Rates | | 177 |
| | 6.1.1 | The Yield Curve | 177 |
| | 6.1.2 | Yields to Maturity and Zero-Coupon Rates | 179 |
| | 6.1.3 | Forward Interest Rates Implicit in the Spot Rate Curve | 184 |
| 6.2 | Factors Determining the Shape of the Curve | | 187 |
| | 6.2.1 | The Curve Shape | 187 |
| | 6.2.2 | Expectations Hypothesis with Term Premiums | 188 |
| | 6.2.3 | Influence of the Credit Spread on Yield Curves | 191 |
| 6.3 | Analysis of Interest Rate Risk: Impact of Changes in the Slope and Shape of the Yield Curve | | 192 |
| | 6.3.1 | The Risk of a Change in the Slope of the Yield Curve | 192 |
| | 6.3.2 | Multifactor Variation and Sensitivity and Models of Yield Curves | 194 |
| 6.4 | Summary | | 203 |
| | Suggested Readings | | 204 |

**7 Vanilla Floating Rate Instruments and Swaps** .......... 205

| | | | |
|---|---|---|---|
| 7.1 | Floating Rate Instruments | | 206 |
| | 7.1.1 | General Discussion and Notation | 206 |
| | 7.1.2 | "Replicable" Assets: Valuation and Interest Rate and Spread Risks | 212 |
| 7.2 | Vanilla Swaps | | 228 |
| | 7.2.1 | Definitions and Generalities About Swaps | 228 |
| | 7.2.2 | Replication and Valuation of an Interest Rate Swap | 232 |
| | 7.2.3 | Interest Rate, Counterparty and Credit Risks for an Interest Rate Swap | 241 |
| | 7.2.4 | Summary of the Various Types of Swaps | 247 |
| 7.3 | Summary | | 251 |

# Contents

xix

Appendix . . . . . . . . . . . . . . . . . . . . . . . . . . . . . . . . . . . . . . . . 252
    Proof of the Equivalence Between Eq. (7.2') and
    Proposition 1 . . . . . . . . . . . . . . . . . . . . . . . . . . . . . . . . . 252
Suggested Reading . . . . . . . . . . . . . . . . . . . . . . . . . . . . . . . . . 253
    Books . . . . . . . . . . . . . . . . . . . . . . . . . . . . . . . . . . . . . . . 253
    Articles . . . . . . . . . . . . . . . . . . . . . . . . . . . . . . . . . . . . . . 253

**8 Stocks, Stock Markets, and Stock Indices** . . . . . . . . . . . . . . . . . . 255
  8.1    Stocks . . . . . . . . . . . . . . . . . . . . . . . . . . . . . . . . . . . . . 255
      8.1.1   Basic Notions: Equity, Stock Market Capitalization,
              and Share Issuing . . . . . . . . . . . . . . . . . . . . . . . 256
      8.1.2   Analysis of Stock Issues, Dilution, and Subscription
              Rights . . . . . . . . . . . . . . . . . . . . . . . . . . . . . . 261
      8.1.3   Market Performance of a Share and Adjusted Share
              Price . . . . . . . . . . . . . . . . . . . . . . . . . . . . . . . 263
      8.1.4   Introduction to the Valuation of Firms and Shares;
              Interpretation and Use of the PER . . . . . . . . . . . . . . 266
  8.2    Return Probability Distributions and the Evolution of Stock
      Market Prices . . . . . . . . . . . . . . . . . . . . . . . . . . . . . . . . 273
      8.2.1   Stock Price on a Future Date, Stock Return, and Its
              Probability Distribution: Static Analysis . . . . . . . . . . 273
      8.2.2   Modeling a Stock Price Evolution with a Stochastic
              Process: Dynamic Analysis . . . . . . . . . . . . . . . . . . 276
  8.3    Placing and Executing Orders and the Functioning of Stock
      Markets . . . . . . . . . . . . . . . . . . . . . . . . . . . . . . . . . . . . 284
      8.3.1   Types of Orders . . . . . . . . . . . . . . . . . . . . . . . . . 284
      8.3.2   The Clearing and Settlement System . . . . . . . . . . . . 286
      8.3.3   Investment Management . . . . . . . . . . . . . . . . . . . . 288
      8.3.4   The Main Stock Markets . . . . . . . . . . . . . . . . . . . . 293
  8.4    Stock Market Indices . . . . . . . . . . . . . . . . . . . . . . . . . . . . 294
      8.4.1   Composition and Calculation . . . . . . . . . . . . . . . . . 295
      8.4.2   The Main Indices . . . . . . . . . . . . . . . . . . . . . . . . 300
  8.5    Summary . . . . . . . . . . . . . . . . . . . . . . . . . . . . . . . . . . . 302
Appendix 1 . . . . . . . . . . . . . . . . . . . . . . . . . . . . . . . . . . . . . . 304
    Skewness and Kurtosis of Log-Returns . . . . . . . . . . . . . . . . . 304
Appendix 2 . . . . . . . . . . . . . . . . . . . . . . . . . . . . . . . . . . . . . . 305
    Modeling Volatility with ARCH and GARCH . . . . . . . . . . . . 305
Suggestions for Further Reading . . . . . . . . . . . . . . . . . . . . . . . . 308
    Book Chapters . . . . . . . . . . . . . . . . . . . . . . . . . . . . . . . . 308
    Articles . . . . . . . . . . . . . . . . . . . . . . . . . . . . . . . . . . . . . 308
    For an Online Comparative Description of Investment Funds
    from Different Countries . . . . . . . . . . . . . . . . . . . . . . . . . . 308
    For an Online Description and Analysis of the Asset
    Management Industry . . . . . . . . . . . . . . . . . . . . . . . . . . . 308

**Part II  Futures and Options**

**9  Futures and Forwards** ................................. 311
  9.1   General Analysis of Forward and Futures Contracts ......... 312
      9.1.1   Definition of a Forward Contract: Terminology and
              Notation ................................. 312
      9.1.2   Futures Contracts: Comparison of Futures and Forward
              Contracts ................................. 314
      9.1.3   Unwinding a Position Before Expiration ........... 317
      9.1.4   The Value of Forward and Futures Contracts ........ 318
  9.2   Cash-and-carry and the Relation Between Spot and Forward
     Prices ......................................... 319
      9.2.1   Arbitrage, Cash-and-Carry, and Spot-Forward Parity . . 319
      9.2.2   Forward Prices, Expected Spot Prices, and Risk
              Premiums ................................. 324
  9.3   Maximum and Optimal Hedging with Forward and Futures
     Contracts ...................................... 325
      9.3.1   Perfect or Maximum Hedging .................. 326
      9.3.2   Optimal Hedging and Speculation ............... 333
  9.4   The Main Forward and Futures Contracts .............. 334
      9.4.1   Contracts on Commodities .................... 335
      9.4.2   Contracts on Currencies (Foreign Exchanges) ....... 338
      9.4.3   Forward and Futures Contracts on Financial Securities
              (Stocks, Bonds, Negotiable Debt Securities), FRA,
              and Contracts on Market Indices ................ 340
  9.5   Summary ...................................... 348
  Appendix ........................................... 349
     The Relationship Between Forward and Futures Prices ....... 349
  Suggestions for Further Reading ........................... 352
     Books .......................................... 352
     Articles ......................................... 352

**10  Options (I): General Description, Parity Relations, Basic
Concepts, and Valuation Using the Binomial Model** ........... 353
  10.1   Basic Concepts, Call-Put Parity, and Other Restrictions from
     No Arbitrage .................................... 354
      10.1.1   Definitions, Value at Maturity, Intrinsic Value,
              and Time Value ........................... 354
      10.1.2   The Standard Call-Put Parity .................. 358
      10.1.3   Other Parity Relations ....................... 361
      10.1.4   Other Arbitrage Restrictions ................... 364
  10.2   A Pricing Model for One Period and Two States
     of the World .................................... 367
      10.2.1   Two Markets, Two States ..................... 368
      10.2.2   Hedging Strategy and Option Value in the Absence of
              Arbitrage ................................. 369

| | | | |
|---|---|---|---|
| | 10.2.3 | The "Risk-Neutral" Probability | 372 |
| | 10.2.4 | The Risk Premium and the Market Price of Risk | 374 |
| 10.3 | The Multi-period Binomial Model | | 377 |
| | 10.3.1 | The Model Framework and the Dynamics of the Underlying's Price | 377 |
| | 10.3.2 | Risk-Neutral Probability and Martingale Processes | 379 |
| | 10.3.3 | Valuation of an Option Using the Cox-Ross-Rubinstein Binomial Model | 381 |
| 10.4 | Calibration of the Binomial Model and Convergence to the Black-Scholes Formula | | 386 |
| | 10.4.1 | An Interpretation of Premiums in Terms of Probabilities of Exercise | 387 |
| | 10.4.2 | Calibration and Convergence | 389 |
| 10.5 | Summary | | 391 |
| Appendix 1 | | | 393 |
| | Calibration of the Binomial Model | | 393 |
| *Appendix 2 | | | 395 |
| Suggestions for Further Reading | | | 397 |
| | Books | | 397 |
| | Articles | | 397 |

**11  Options (II): Continuous-Time Models, Black–Scholes and Extensions** ............................................................ 399

| | | | |
|---|---|---|---|
| 11.1 | The Standard Black-Scholes Model | | 399 |
| | 11.1.1 | The Analytical Framework and BS Model's Assumptions | 400 |
| | 11.1.2 | Self-Financing Dynamic Strategies | 401 |
| | 11.1.3 | Pricing Using a Partial Differential Equation and the Black–Scholes Formula | 402 |
| | 11.1.4 | Probabilistic Interpretation | 406 |
| 11.2 | Extensions of the Black–Scholes Formula | | 412 |
| | 11.2.1 | Underlying Assets That Pay Out (Dividends, Coupons, etc.) | 412 |
| | 11.2.2 | Options on Commodities | 421 |
| | 11.2.3 | Options on Exchange Rates | 422 |
| | 11.2.4 | Options on Futures and Forwards | 424 |
| | 11.2.5 | Variable But Deterministic Volatility | 427 |
| | 11.2.6 | Stochastic Interest Rates: The Black–Scholes–Merton (BSM) Model | 429 |
| | 11.2.7 | Exchange Options (Margrabe) | 434 |
| | 11.2.8 | Stochastic Volatility (*) | 437 |
| 11.3 | Summary | | 443 |
| Appendix 1 | | | 444 |
| | Historical and Risk-Neutral Probabilities and Changes in Probability | | 444 |

| | | | |
|---|---|---|---|
| Appendix 2 | | | 446 |
| | Changing the Probability Measure and the Numeraire | | 446 |
| Appendix 3 | | | 449 |
| | Alternative Interpretations of the Black–Scholes Formula | | 449 |
| Suggested Reading | | | 451 |
| | Books | | 451 |
| | Articles | | 451 |

**12 Option Portfolio Strategies: Tools and Methods** .............. 453

| | | | |
|---|---|---|---|
| 12.1 | Basic Static Strategies | | 454 |
| | 12.1.1 | The General P&L Profile at Maturity | 454 |
| | 12.1.2 | The Main Static Strategies | 455 |
| | 12.1.3 | Replication of an Arbitrary Payoff by a Static Option Portfolio (*) | 457 |
| 12.2 | Historical and Implied Volatilities, Smile, Skew and Term Structure | | 457 |
| | 12.2.1 | Historical Volatility | 458 |
| | 12.2.2 | The Implied Volatility | 459 |
| | 12.2.3 | Smile, Skew, Term Structure, and Volatility Surface | 461 |
| 12.3 | Option Sensitivities (Greek Parameters) | | 463 |
| | 12.3.1 | The Delta ($\delta$) | 464 |
| | 12.3.2 | The Gamma ($\Gamma$) | 467 |
| | 12.3.3 | The Vega ($\upsilon$) | 468 |
| | 12.3.4 | The Theta ($\theta$) | 470 |
| | 12.3.5 | The Rho ($\rho$) | 471 |
| | 12.3.6 | Sensitivity to the Dividend Rate | 472 |
| | 12.3.7 | Elasticity and Risk-Expected Return Tradeoff | 473 |
| 12.4 | Dynamic Management of an Option Portfolio Using Greek Parameters | | 475 |
| | 12.4.1 | Variation in the Value of a Position in the Short Term and General Considerations | 475 |
| | 12.4.2 | Delta-Neutral Management | 476 |
| | 12.4.3 | A Tool for Risk Management: The P&L Matrix | 485 |
| 12.5 | Summary | | 486 |
| Appendix 1 | | | 488 |
| | Computing Partial Derivatives (Greeks) | | 488 |
| Appendix 2 | | | 493 |
| | Option Prices and the Underlying Price Probability Distribution | | 493 |
| Appendix 3 | | | 496 |
| | Replication of an Arbitrary Payoff with a Static Option Portfolio | | 496 |
| Suggestions for Further Reading | | | 498 |
| | Books | | 498 |
| | Articles | | 498 |

# Contents

xxiii

**13 American Options and Numerical Methods** . . . . . . . . . . . . . . . . . 501
    13.1   Early Exercise and Call-Put Parity for American Options . . . . . 502
          13.1.1   Early Exercise of American Options . . . . . . . . . . . . . 502
          13.1.2   Call-Put "Parity" for American Options . . . . . . . . . . 515
    13.2   Pricing American Options: Analytical Approaches . . . . . . . . . 516
          13.2.1   Pricing an American Call on a Spot Asset Paying
                 a Single Discrete Dividend or Coupon . . . . . . . . . . . 517
          13.2.2   Pricing an American Option (Call and Put) on a Spot
                 Asset Paying a Continuous Dividend or Coupon . . . . . 520
          13.2.3   Prices of American and European Options: Orders of
                 Magnitude . . . . . . . . . . . . . . . . . . . . . . . . . . . . 528
    13.3   Pricing American Options with the Binomial Model . . . . . . . . 529
          13.3.1   Binomial Dynamics of Price $S$: The Case of a Discrete
                 Dividend . . . . . . . . . . . . . . . . . . . . . . . . . . . . . 529
          13.3.2   Binomial Dynamics of Price $S$: The Continuous
                 Dividend Case . . . . . . . . . . . . . . . . . . . . . . . . . . 531
          13.3.3   Pricing an American Option Using the Binomial
                 Model . . . . . . . . . . . . . . . . . . . . . . . . . . . . . . . 531
          13.3.4   Improving the Procedure with a Control Variate . . . . . 533
    13.4   Numerical Methods: Finite Differences, Trinomial and Three-
         Dimensional Trees . . . . . . . . . . . . . . . . . . . . . . . . . . . . . 533
          13.4.1   Finite Difference Methods (*) . . . . . . . . . . . . . . . . 534
          13.4.2   Trinomial Trees . . . . . . . . . . . . . . . . . . . . . . . . . 539
          13.4.3   Three-Dimensional Trees Representing Two
                 Correlated Processes . . . . . . . . . . . . . . . . . . . . . . 541
    13.5   Summary . . . . . . . . . . . . . . . . . . . . . . . . . . . . . . . . . . . 543
    Appendix 1 . . . . . . . . . . . . . . . . . . . . . . . . . . . . . . . . . . . . . 545
         Proof of the Smooth Pasting (Tangency) Condition (13.5b) . . . 545
    Appendix 2 . . . . . . . . . . . . . . . . . . . . . . . . . . . . . . . . . . . . . 546
         Orthogonalization of the Processes $\ln S_1$ and $\ln S_2$ and
         Construction of a Three-Dimensional Tree . . . . . . . . . . . . . . 546
    Suggestion for Further Reading . . . . . . . . . . . . . . . . . . . . . . . . 547
         Books . . . . . . . . . . . . . . . . . . . . . . . . . . . . . . . . . . . . . . 547
         Articles . . . . . . . . . . . . . . . . . . . . . . . . . . . . . . . . . . . . . 548

**14 \*Exotic Options** . . . . . . . . . . . . . . . . . . . . . . . . . . . . . . . . . 549
    14.1   Path-Independent Options . . . . . . . . . . . . . . . . . . . . . . . . 550
          14.1.1   The Forward Start Option (with Deferred Start) . . . . . 550
          14.1.2   Digital and Double Digital Options . . . . . . . . . . . . . 551
          14.1.3   Multi-underlying (Rainbow) Options (*) . . . . . . . . . . 555
          14.1.4   Options on Options or "Compounds" . . . . . . . . . . . . 559
          14.1.5   Quantos and Compos . . . . . . . . . . . . . . . . . . . . . . 560
    14.2   Path-Dependent Options . . . . . . . . . . . . . . . . . . . . . . . . . 567
          14.2.1   Barrier Options . . . . . . . . . . . . . . . . . . . . . . . . . 567
          14.2.2   Digital Barriers . . . . . . . . . . . . . . . . . . . . . . . . . 573

|  |  |  |  |
|---|---|---|---|
| | 14.2.3 | Lookback Options (*) | 576 |
| | 14.2.4 | Options on Averages (Asians) | 578 |
| | 14.2.5 | Chooser Options (*) | 586 |
| 14.3 | Summary | | 587 |

Appendix 1 ... 589
**Value of a Compo Call ... 589

Appendix 2 ... 591
**Lemmas on Hitting Probabilities for a Drifted Brownian Motion ... 591

Appendix 3 ... 594
**Proof of the "Inverses" Relation for Barrier Options ... 594

Appendix 4 ... 595
**Valuing a Call Up-and-Out with L (Barrier) > K (Strike) ... 595

Appendix 5 ... 597
**Valuing Rebates ... 597

Appendix 6 ... 598
**Proof of the Price of a Lookback Call ... 598

Appendix 7 ... 601
**Options on an Average Price ... 601

Appendix 8 ... 602
**Options with an Average Strike ... 602

Suggestions for Further Reading ... 604
Books ... 604
Articles ... 604

## 15 Futures Markets (2): Contracts on Interest Rates ... 607

|  |  |  |  |
|---|---|---|---|
| 15.1 | Notional Contracts | | 608 |
| | 15.1.1 | Basket of Deliverable Securities (DS) and Notional Security | 608 |
| | 15.1.2 | The Euro-Bund Contract | 609 |
| | 15.1.3 | Settlement and Conversion Factors | 611 |
| | 15.1.4 | Cheapest to Deliver and Quoting Futures at Expiration | 613 |
| | 15.1.5 | Arbitrage and Cash-Futures Relationship | 618 |
| | 15.1.6 | Interest Rate Sensitivity of Futures Prices | 624 |
| | 15.1.7 | Hedging Interest Rate Risk Using Notional Bond Contracts | 629 |
| | 15.1.8 | The Main Notional Contracts | 637 |
| 15.2 | Short-Term Interest Rate Contracts (STIR) (3-Month Forward-Looking Rates and Backward-Looking Overnight Averages) | | 641 |
| | 15.2.1 | STIR 3-Month Contracts (LIBOR Type, Forward-Looking) | 641 |
| | 15.2.2 | Futures Contracts on an Average Overnight Rate | 647 |
| | 15.2.3 | Hedging Interest Rate Risk with STIR Contracts | 656 |
| 15.3 | Summary | | 659 |

Contents xxv

Appendices . . . . . . . . . . . . . . . . . . . . . . . . . . . . . . . . . . . . . . . . . . . 661
1 Valuation of the Delivery Option . . . . . . . . . . . . . . . . . . . . . . . . . 661
2 Relationship Between Forward and Futures Prices . . . . . . . . . . . . . 662
Suggestions for Further Reading . . . . . . . . . . . . . . . . . . . . . . . . . . . 666
    Books . . . . . . . . . . . . . . . . . . . . . . . . . . . . . . . . . . . . . . . . . . . 666
    Articles . . . . . . . . . . . . . . . . . . . . . . . . . . . . . . . . . . . . . . . . . . 666
    Internet Sites . . . . . . . . . . . . . . . . . . . . . . . . . . . . . . . . . . . . . . 666

**16   Interest Rate Instruments: Valuation with the BSM Model, Hybrids, and Structured Products** . . . . . . . . . . . . . . . . . . . . . . . . . 667
  16.1   Valuation of Interest Rate Instruments Using Standard Models . . . 667
      16.1.1   Principles of Valuation and the Black-Scholes-Merton Model Generalized to Stochastic Interest Rates . . . . . . 668
      16.1.2   Valuation of a Bond Option Using the BSM-Price Model . . . . . . . . . . . . . . . . . . . . . . . . . . . . . . . . . . . . 671
      16.1.3   Valuation of the Right to a Cash Flow Expressed as a Function of a Rate and the BSM-Rate Model . . . . . . 672
      16.1.4   Convexity Adjustments for Non-vanilla Cash Flows (*) . . . . . . . . . . . . . . . . . . . . . . . . . . . . . . . . . 676
  16.2   Nonstandard Swaps and Swaptions . . . . . . . . . . . . . . . . . . . . . 680
      16.2.1   Review of Swaps and Notation . . . . . . . . . . . . . . . . . 680
      16.2.2   Some Nonstandard Swaps . . . . . . . . . . . . . . . . . . . . 682
      16.2.3   Swap Options (or Swaptions) . . . . . . . . . . . . . . . . . 686
  16.3   Caps and Floors . . . . . . . . . . . . . . . . . . . . . . . . . . . . . . . . . . 688
      16.3.1   Vanilla Caps . . . . . . . . . . . . . . . . . . . . . . . . . . . . . . 689
      16.3.2   A Vanilla Floor . . . . . . . . . . . . . . . . . . . . . . . . . . . . 692
  16.4   Static Replications and Combinations; Structured Contracts . . . 693
      16.4.1   Basic Instruments: Notation and General Remarks . . . 693
      16.4.2   Replication of a Capped or Floored Floating-Rate Instrument Using a Standard Asset Associated with a Cap or a Floor . . . . . . . . . . . . . . . . . . . . . . . . . . . . 696
      16.4.3   Collars . . . . . . . . . . . . . . . . . . . . . . . . . . . . . . . . . . 699
      16.4.4   Non-standard Caps and Floors . . . . . . . . . . . . . . . . . 702
      16.4.5   Other Static Combinations; Structured Products; Contracts on Interest Rates with Profit-Sharing . . . . . . 705
  16.5   Bonds with Optional Features and Hybrid Products . . . . . . . . . 707
      16.5.1   Convertible Bonds . . . . . . . . . . . . . . . . . . . . . . . . . 708
      16.5.2   Other Bonds with Optional Features . . . . . . . . . . . . . 711
  16.6   Summary . . . . . . . . . . . . . . . . . . . . . . . . . . . . . . . . . . . . . . . 713
  Appendix . . . . . . . . . . . . . . . . . . . . . . . . . . . . . . . . . . . . . . . . . . . 715
    The $Q_a$-Martingale Measure . . . . . . . . . . . . . . . . . . . . . . . . . . . 715
  Suggestions for Further Reading . . . . . . . . . . . . . . . . . . . . . . . . . . . 716
    Books . . . . . . . . . . . . . . . . . . . . . . . . . . . . . . . . . . . . . . . . . . . 716
    Articles . . . . . . . . . . . . . . . . . . . . . . . . . . . . . . . . . . . . . . . . . . 716

xxvi | Contents

**17 Modeling Interest Rates and Options on Interest Rates** .......... 719
   17.1  Models Based on the Dynamics of Spot Rates ............... 720
       17.1.1  One-Factor Models (Vasicek, and Cox, Ingersoll and
             Ross) ....................................... 721
       17.1.2  Fitting the Initial Yield Curve; the Hull and White
             Model ....................................... 726
       17.1.3  Multifactor Structures .......................... 728
   17.2  Models Grounded on the Dynamics of Forward Rates ........ 729
       17.2.1  The Heath–Jarrow–Morton Model (1992) ........... 730
       17.2.2  The Libor (LMM) and Swap (SMM) Market
             Models ...................................... 737
   17.3  Summary ......................................... 749
   Appendix 1 ............................................ 751
      *The Vasicek Model ................................. 751
   Appendix 2 ............................................ 754
      *The LMM and SMM Models .......................... 754
   Suggestions for Further Reading ............................ 762
      Books ............................................ 762
      Articles .......................................... 762

**18 Elements of Stochastic Calculus** .......................... 765
   18.1  Definitions, Notation, and General Considerations About
       Stochastic Processes ................................ 766
       18.1.1  Notation .................................... 766
       18.1.2  Stochastic Processes: Definitions, Notation, and
             General Framework ........................... 766
   18.2  Brownian Motion ................................... 769
       18.2.1  The One-Dimensional Brownian Motion ............ 769
       18.2.2  Calculus Rules Relative to Brownian Motions ....... 775
       18.2.3  Multi-dimensional Arithmetic Brownian Motions .... 777
   18.3  More General Processes Derived from the Brownian Motion;
       One-Dimensional Itô and Diffusion Processes ............. 779
       18.3.1  One-Dimensional Itô Processes ................... 779
       18.3.2  One-Dimensional Diffusion Processes ............. 780
       18.3.3  Stochastic Integrals (*) ........................ 782
   18.4  Differentiation of a Function of an Itô Process: Itô's Lemma ... 785
       18.4.1  Itô's Lemma ................................. 785
       18.4.2  Examples of Application ....................... 787
   18.5  Multi-dimensional Itô and Diffusion Processes (*) .......... 789
       18.5.1  Multivariate Itô and Diffusion Processes ........... 790
       18.5.2  Itô's Lemma (Differentiation of a Function of an
             $n$-Dimensional Itô Process) ................... 791
   18.6  Jump Processes .................................... 793
       18.6.1  Description of Jump Processes ................... 793
       18.6.2  Modeling Jump Processes ...................... 793
   18.7  Summary ......................................... 794

Contents xxvii

Suggestions for Further Reading . . . . . . . . . . . . . . . . . . . . . . . . . . 796
Books . . . . . . . . . . . . . . . . . . . . . . . . . . . . . . . . . . . . . . . . . . . . . 796

**19 *The Mathematical Framework of Financial Markets Theory** . . . . 797
19.1 General Framework and Basic Concepts . . . . . . . . . . . . . . . . 798
  19.1.1 The Probabilistic Framework . . . . . . . . . . . . . . . . . . 798
  19.1.2 The Market, Securities, and Portfolio Strategies . . . . . 799
  19.1.3 Portfolio Strategies . . . . . . . . . . . . . . . . . . . . . . . . . 800
  19.1.4 Contingent Claims, AAO, and Complete Markets . . . . 802
  19.1.5 Price Systems . . . . . . . . . . . . . . . . . . . . . . . . . . . . . 804
19.2 Price Dynamics as Itô Processes, Arbitrage Pricing Theory and
  the Market Price of Risk . . . . . . . . . . . . . . . . . . . . . . . . . . . . 805
  19.2.1 Price Dynamics as Itô Processes . . . . . . . . . . . . . . . . 806
  19.2.2 Arbitrage Pricing Theory in Continuous Time . . . . . . . 806
  19.2.3 Redundant Securities and Characterizing the Base of
    Primitive Securities . . . . . . . . . . . . . . . . . . . . . . . . . 808
19.3 The Risk-Neutral Universe and Transforming Prices into
  Martingales . . . . . . . . . . . . . . . . . . . . . . . . . . . . . . . . . . . . . . 809
  19.3.1 Martingales, Driftless Processes, and Exponential
    Martingales . . . . . . . . . . . . . . . . . . . . . . . . . . . . . . . 809
  19.3.2 Price and Return Dynamics in the Risk-Neutral
    Universe, Transforming Prices into martingales
    and Pricing Contingent Claims . . . . . . . . . . . . . . . . . 813
  19.3.3 Characterizing a Complete market and Market Prices
    of Risk . . . . . . . . . . . . . . . . . . . . . . . . . . . . . . . . . . . 816
19.4 Change of Probability Measure, Radon-Nikodym derivative
  and Girsanov's Theorem . . . . . . . . . . . . . . . . . . . . . . . . . . . . 818
  19.4.1 Changing Probabilities and the Radon-Nikodym
    Derivative . . . . . . . . . . . . . . . . . . . . . . . . . . . . . . . . . 818
  19.4.2 Changing Probabilities and Brownian Motions:
    Girsanov's Theorem . . . . . . . . . . . . . . . . . . . . . . . . . 820
  19.4.3 Formal Definition of RN Probabilities . . . . . . . . . . . . 822
  19.4.4 Relations between Viable Price Systems, RN
    Probabilities, and MPR . . . . . . . . . . . . . . . . . . . . . . . 823
19.5 Changing the Numeraire . . . . . . . . . . . . . . . . . . . . . . . . . . . . 825
  19.5.1 Numeraires . . . . . . . . . . . . . . . . . . . . . . . . . . . . . . . . 825
  19.5.2 Numeraires and Probabilities that yield martingale
    Prices . . . . . . . . . . . . . . . . . . . . . . . . . . . . . . . . . . . . 826
19.6 The P-Numeraire (Optimal Growth or Logarithmic Portfolio) . . . 832
  19.6.1 Definition of the Portfolio $(\underline{h}, H)$ as the $P$-Numeraire . . . 832
  19.6.2 Characterization and Composition of the P-Numeraire
    Portfolio $(\underline{h}, H)$ . . . . . . . . . . . . . . . . . . . . . . . . . . . . 833
19.7 ** Incomplete Markets . . . . . . . . . . . . . . . . . . . . . . . . . . . . . 836
  19.7.1 MPR and the Kernel of the Diffusion Matrix $\sum(t)$ . . . . 837
  19.7.2 Deflators . . . . . . . . . . . . . . . . . . . . . . . . . . . . . . . . . . 840

| | | | |
|---|---|---|---|
| 19.8 | Summary | | 842 |
| Appendix | | | 844 |
| | Construction of a One-to-one Correspondence between $\mathbb{Q}$ and $\Pi$ | | 844 |
| Suggestions for further reading | | | 845 |
| | Books | | 845 |
| | Articles | | 846 |

**20 The State Variables Model and the Valuation Partial Differential Equation** ........... 847

| | | | |
|---|---|---|---|
| 20.1 | Analytical Framework and Notation | | 847 |
| | 20.1.1 | Dynamics of State Variables | 847 |
| | 20.1.2 | The Asset Pricing Problem | 848 |
| 20.2 | Factor Decomposition of Returns | | 849 |
| | 20.2.1 | Expressing the Return $dR$ as a Function of the $dX_j$ | 849 |
| | 20.2.2 | Expressing the Return $dR$ as a Function of the $dW_k$ | 850 |
| 20.3 | Expected Asset Returns and Arbitrage Pricing Theory (APT) in Continuous Time | | 850 |
| | 20.3.1 | First Formula for Expected Returns | 851 |
| | 20.3.2 | Continuous Time APT in a State variables Model | 851 |
| 20.4 | The General valuation PDE | | 853 |
| | 20.4.1 | Derivation of the General valuation PDE | 853 |
| | 20.4.2 | Market Prices of Risk and Risk Premia | 854 |
| | 20.4.3 | The Relation between MPR and Excess Returns on Primitive Securities and the Condition for Market Completeness | 855 |
| 20.5 | Applications to the Term Structure of Interest Rates | | 857 |
| | 20.5.1 | Models with One State Variable | 857 |
| | 20.5.2 | Multi-Factor models and valuation of Fixed-Income Securities | 859 |
| 20.6 | Pricing in the Risk-Neutral Universe | | 862 |
| | 20.6.1 | Dynamics of Returns, of Brownian Motions and of State Variables in the Risk-Neutral Universe | 862 |
| | 20.6.2 | The Valuation PDE | 863 |
| 20.7 | Discounting under Uncertainty and the Feynman–Kac Theorem | | 864 |
| | 20.7.1 | The Cauchy-Dirichlet PDE and the Feynman-Kac Theorem | 864 |
| | 20.7.2 | Financial Interpretation of the Feynman–Kac Theorem and Discounting under Uncertainty | 865 |
| 20.8 | Summary | | 866 |
| Appendix | | | 868 |
| Suggestions for Further Reading | | | 869 |
| | Books | | 869 |
| | Articles | | 869 |

# Contents for Volume 2

**Part III   Portfolio Theory and Portfolio Management**

**21   Choice Under Uncertainty and Portfolio Optimization in a Static Framework: The Markowitz Model** ............................ 873

21.1   Rational Choices Under Uncertainty: The Criteria of the Expected Utility and Mean-Variance ..................... 874

21.1.1   The Expected Utility Criterion ................. 874

21.1.2   Some Features of Utility Functions ............. 876

21.1.3   Risk Aversion and Concavity of the Utility Function .. 877

21.1.4   Some Standard Utility Functions .............. 879

21.1.5   The Mean-Variance Criterion ................. 881

21.2   Intuitive and Graphic Presentation of the Main Concepts of Portfolio Theory ............................... 882

21.2.1   Assumptions, General Framework and Efficient Portfolios .............................. 883

21.2.2   Two-Asset Portfolios ...................... 884

21.2.3   Portfolios with $N$ Securities .................. 887

21.2.4   Portfolio Diversification .................... 890

21.3   Mathematical Analysis of Efficient Portfolio Choices ........ 894

21.3.1   General Framework and Notations .............. 894

21.3.2   Efficient Portfolios and Portfolio Choice in the Absence of a Risk-Free Asset and of Portfolio Constraints ............................. 898

21.3.3   Efficient Portfolios in the Presence of a Risk-Free Asset, with Allowed Short Positions; Tobin's Two-Fund Separation ...................... 904

21.4   Some Extensions of the Standard Model and Alternatives ..... 906

21.4.1   Problems Implementing the Markowitz Model; The Black-Litterman Procedure .................. 906

21.4.2   Ban on Short Positions ..................... 908

21.4.3   Separation Results When Investors Maximize Expected Utility But Do Not Follow the Mean-Variance Criterion (Cass and Stiglitz) ........ 910

xxix

| | | | |
|---|---|---|---|
| | 21.4.4 | Loss Aversion and Introduction to Behavioral Finance | 911 |
| 21.5 | Summary | | 915 |

Appendix 1: The Axiomatic of Von Neuman and Morgenstern and Expected Utility ........................................................ 916

| | A1.1 | The Objects of Choice | 917 |
|---|---|---|---|
| | A1.2 | The Axioms Concerning Preferences | 918 |
| | A1.3 | The Expected Utility Criterion | 919 |
| | A1.4 | Notes and Complements | 921 |

Appendix 2: A Reminder of Quadratic Forms and the Calculation of Gradients ........................................................ 922

Appendix 3: Expectations, Variances and Covariances—Definitions and Calculation Rules ........................................................ 923

| | A3.1 | Definitions and Reminder | 923 |
|---|---|---|---|
| | A3.2 | Calculation Rules | 924 |

Appendix 4: Reminder on Optimization Methods Under Constraints ... 925

| | A4.1 | Optimization When the Constraints Take the Form of Equalities | 925 |
|---|---|---|---|
| | A4.2 | Optimization Under Inequality Constraints | 926 |

Suggestions for Further Reading ........................................ 927

| | Books | 927 |
|---|---|---|
| | Articles | 927 |

**22 The Capital Asset Pricing Model** ........................................ 929

| 22.1 | Derivation of the CAPM | | 929 |
|---|---|---|---|
| | 22.1.1 | Hypotheses | 930 |
| | 22.1.2 | Intermediate Results in the Presence of a Risk-Free Asset | 931 |
| | 22.1.3 | The CAPM | 933 |
| 22.2 | Applications of the CAPM | | 943 |
| | 22.2.1 | Use of the CAPM for Financial Investment Purposes | 943 |
| | 22.2.2 | Physical Investments by Firms | 946 |
| | 22.2.3 | Standard Performance Measures | 947 |
| 22.3 | Extensions of the CAPM | | 953 |
| | 22.3.1 | Merton's Intertemporal CAPM | 953 |
| | 22.3.2 | International CAPM | 954 |
| 22.4 | Limits of the CAPM | | 955 |
| | 22.4.1 | Efficiency of the Market Portfolio and Roll's Criticism | 955 |
| | 22.4.2 | Stability of Betas | 956 |
| 22.5 | Tests of the CAPM | | 957 |
| 22.6 | Summary | | 960 |

Suggestions for Further Reading ........................................ 961

| | Books | 961 |
|---|---|---|
| | Articles | 962 |

# Contents for Volume 2

**23 Arbitrage Pricing Theory and Multi-factor Models** . . . . . . . . . . . 963
    23.1  Multi-factor Models . . . . . . . . . . . . . . . . . . . . . . . . . . 964
        23.1.1  Presentation of Models . . . . . . . . . . . . . . . . . . 964
        23.1.2  Portfolio Management Models in Practice . . . . . . . . . 966
    23.2  Arbitrage Pricing Theory . . . . . . . . . . . . . . . . . . . . . . . 966
        23.2.1  Assumptions and Notations . . . . . . . . . . . . . . . . . 967
        23.2.2  The APT . . . . . . . . . . . . . . . . . . . . . . . . . . . 968
        23.2.3  Relationship with the CAPM . . . . . . . . . . . . . . . . 976
    23.3  APT Applications and the Fama-French Model . . . . . . . . . . . 976
        23.3.1  Implementation of Multi-factor Models and APT . . . . 977
        23.3.2  Portfolio Selection . . . . . . . . . . . . . . . . . . . . . 979
        23.3.3  The Three-Factor Model of Fama and French . . . . . . . 980
    23.4  Econometric Tests and Comparison of Models . . . . . . . . . . . 982
        23.4.1  Tests of the APT . . . . . . . . . . . . . . . . . . . . . . 982
        23.4.2  Empirical and Practical CAPM-APT Comparison . . . . 983
        23.4.3  Comparison of Factor Models . . . . . . . . . . . . . . . 984
    23.5  Summary . . . . . . . . . . . . . . . . . . . . . . . . . . . . . . . 985
    Appendix 1: Orthogonalization of Common Factors . . . . . . . . . . . 987
    Appendix 2: Compatibility of CAPM and APT . . . . . . . . . . . . . . 988
    Suggestions for Further Reading . . . . . . . . . . . . . . . . . . . . . . 989
        Books . . . . . . . . . . . . . . . . . . . . . . . . . . . . . . . . . 989
        Articles . . . . . . . . . . . . . . . . . . . . . . . . . . . . . . . . 990

**24 Strategic Portfolio Allocation** . . . . . . . . . . . . . . . . . . . . . . 991
    24.1  Strategic Asset Allocation Based on Common Sense Rules . . . . 992
        24.1.1  Common Sense Rules . . . . . . . . . . . . . . . . . . . . 993
        24.1.2  Reactions to the Evolution of Market Conditions and
                of the Portfolio: Convex and Concave Strategies . . . . . 996
    24.2  Portfolio Insurance . . . . . . . . . . . . . . . . . . . . . . . . . . 997
        24.2.1  The *Stop Loss* Method . . . . . . . . . . . . . . . . . . . 998
        24.2.2  Option-Based Portfolio Insurance . . . . . . . . . . . . . 999
        24.2.3  CPPI Method . . . . . . . . . . . . . . . . . . . . . . . . 1005
        24.2.4  Variants and Extensions of the Basic Methods . . . . . . 1011
        24.2.5  Portfolio Insurance, Financial Markets Volatility and
                Stability . . . . . . . . . . . . . . . . . . . . . . . . . . . 1013
    24.3  Dynamic Portfolio Optimization Models . . . . . . . . . . . . . . 1014
        24.3.1  Dynamic Strategies: General Presentation and
                Optimization Models . . . . . . . . . . . . . . . . . . . . 1014
        24.3.2  The Case of a Logarithmic Utility Function and the
                Optimal Growth Portfolio . . . . . . . . . . . . . . . . . 1017
        24.3.3  The Merton Model . . . . . . . . . . . . . . . . . . . . . 1020
        24.3.4  The Model of Cox-Huang and Karatzas-Lehoczky-
                Shreve . . . . . . . . . . . . . . . . . . . . . . . . . . . . 1022

| | | | | |
|---|---|---|---|---|
| 24.4 | | Summary | | 1026 |
| | | Suggestions for Further Reading | | 1028 |
| | | Books | | 1028 |
| | | Articles | | 1029 |

## 25 Benchmarking and Tactical Asset Allocation ........ 1031

| | | | | |
|---|---|---|---|---|
| 25.1 | | Benchmarking | | 1031 |
| | 25.1.1 | Definitions and Classification According to the Tracking Error | | 1032 |
| | 25.1.2 | Pure Index Funds and Trackers | | 1033 |
| | 25.1.3 | Replication Methods | | 1033 |
| | 25.1.4 | Trackers or ETFs | | 1034 |
| 25.2 | | Active Tactical Asset Allocation | | 1035 |
| | 25.2.1 | Modeling and Solution to the Problem of an Active Manager Competing with a Benchmark | | 1035 |
| | 25.2.2 | Analysis of the Performance of Active Portfolio Management: Empirical Information Ratio, Market Timing, and Security Picking | | 1038 |
| | 25.2.3 | Beta Coefficient Equal to 1 | | 1038 |
| | 25.2.4 | Beta Coefficient Different from 1 | | 1040 |
| | 25.2.5 | Information Ratios, Sharpe Ratio, and Active Portfolio Management Theory | | 1043 |
| | 25.2.6 | The Construction of a Maximum IR Portfolio from a Limited Number of Securities | | 1043 |
| | 25.2.7 | The Construction of a Portfolio That Dominates the Benchmark (Higher Sharpe Ratio) | | 1045 |
| | 25.2.8 | Synthesis, Interpretation and Application to Portfolio Management | | 1047 |
| 25.3 | | Alternative Investment Management and Hedge Funds | | 1047 |
| | 25.3.1 | General Description of Hedge Funds and Alternative Investment | | 1048 |
| | 25.3.2 | Definition of the Main Alternative Investment Styles | | 1049 |
| | 25.3.3 | The Interest of Alternative Investment | | 1051 |
| | 25.3.4 | The Particular Difficulties of Measuring Performance in Alternative Investment | | 1052 |
| 25.4 | | Summary | | 1054 |
| | | Appendix | | 1056 |
| | | Breakdown of the Tracking Error and Performance Attribution | | 1056 |
| | | Suggestion for Reading | | 1059 |
| | | Books | | 1059 |
| | | Articles | | 1059 |

# Contents for Volume 2

**Part IV   Risk Management, Credit Risk, and Credit Derivatives**

**26   Monte Carlo Simulations** .................................... 1063
  26.1   Generation of a Sample from a Given Distribution Law ...... 1064
    26.1.1   Sample Generation from a Given Probability
             Distribution ............................... 1064
    26.1.2   Construction of a Sample Taken from a Normal
             Distribution ............................... 1065
  26.2   Monte Carlo Simulations for a Single Risk Factor .......... 1065
    26.2.1   Dynamic Paths Simulation of $Y(t)$ and $V(t, Y(t))$ in
             the Interval $(0, T)$ ........................ 1065
    26.2.2   Simulations of $Y(T)$ and $V(T, Y(T))$ at Time $T$ (Static
             Simulations) .............................. 1068
    26.2.3   Applications ............................... 1070
  26.3   Monte Carlo Simulations for Several Risk Factors: Choleski
         Decomposition and Copulas ......................... 1073
    26.3.1   Simulation of a Multi-variate Normal Variable:
             Choleski Decomposition ..................... 1073
    26.3.2   Representation and Simulation of a Non-Gaussian
             Vector with Correlated Components Through the
             Use of a Copula ........................... 1075
    26.3.3   General Definition of a Copula, and Student
             Copulas (*) ............................... 1080
    26.3.4   Simulation of Trajectories ................... 1081
  26.4   Accuracy, Computation Time, and Some Variance Reduction
         Techniques ....................................... 1085
    26.4.1   Antithetic Variables ........................ 1086
    26.4.2   Control Variate ............................ 1086
    26.4.3   Importance Sampling ........................ 1088
    26.4.4   Stratified Sampling ......................... 1089
  26.5   Monte Carlo and American Options ................... 1089
    26.5.1   General Description of the Problem and
             Methodology .............................. 1090
    26.5.2   Estimation of the Continuation Value by Regression
             (Carrière, Longstaff and Schwartz) ............ 1091
    26.5.3   Overview of the Carrière Approach ............ 1093
    26.5.4   Introduction to Longstaff and Schwartz Approach .... 1094
  26.6   Summary ........................................ 1099
  Suggestion for Further Reading ........................... 1101
    Books .......................................... 1101
    Articles ........................................ 1101

**27  Value at Risk, Expected Shortfall, and Other Risk Measures** .... 1103

27.1  Analytic Study of Value at Risk .......................... 1105

    27.1.1  The Problem of a Synthetic Risk Measure and
Introduction to VaR .......................... 1105

    27.1.2  Definition of the VaR, Interpretations, and Calculation
Rules ..................................... 1108

    27.1.3  Analytic Expressions for the VaR in the Gaussian
Case ..................................... 1112

    27.1.4  The Influence of Horizon $h$ on the VaR of a Portfolio
in the Absence or Presence of Serial Autocorrelation ... 1116

27.2  Estimating the VaR .................................. 1120

    27.2.1  Preliminary Analysis and Modeling of a Complex
Position ................................... 1121

    27.2.2  Estimating the VaR Through Simulations Based
on Historical Data ........................... 1123

    27.2.3  Partial Valuation: Linear and Quadratic
Approximations (the Delta-Normal and Delta-Gamma
Methods) .................................. 1130

    27.2.4  Calculating the VaR Using Monte Carlo
Simulations ................................ 1138

    27.2.5  Comparison Between the Different Methods ........ 1141

27.3  Limitations and Drawbacks of the VaR, Expected Shortfall,
Coherent Measures of Risk, and Portfolio Risks ........... 1141

    27.3.1  The Drawbacks of VaR Measures .............. 1142

    27.3.2  An Improvement on the VaR: *Expected Shortfall*
(or Tail-VaR, or C-VaR) ..................... 1145

    27.3.3  Coherent Risk Measures ...................... 1148

    27.3.4  Portfolio Risk Measures: Global, Marginal, and
Incremental Risk ............................ 1150

27.4  Consequences of Non-normality and Analysis of Extreme
Conditions ....................................... 1155

    27.4.1  Non-normal Distributions with Fat Tails and
Correlation at the Extremes .................... 1155

    27.4.2  Distributions of Extreme Values ................ 1159

    27.4.3  Stress Tests and Scenario Analysis .............. 1164

27.5  Summary ........................................ 1166

Suggestions for Further Reading ............................ 1168

    Books .......................................... 1168

    Articles ......................................... 1168

**28  Modeling Credit Risk (1): Credit Risk Assessment and Empirical
Analysis** ............................................ 1171

28.1  Empirical Tools for Credit Risk Analysis ................ 1172

    28.1.1  Reminder of Basic Concepts, Empirical Observations,
and Notations ............................... 1172

Contents for Volume 2 xxxv

28.1.2 Historical (Empirical) Default Probabilities and
Transition Matrix . . . . . . . . . . . . . . . . . . . . . . . . . . 1176
28.1.3 Risk-Neutral Default Probabilities Implicit in the
Spread Curve and Discounting Methods in the
Presence of Credit Risk . . . . . . . . . . . . . . . . . . . . . . 1179
28.2 Modeling Default Events and Valuation of Securities . . . . . . . 1189
28.2.1 Reduced-Form Approach (Intensity Models) . . . . . . . . 1189
28.2.2 Structural Approach: Merton's Model and Barrier
Models . . . . . . . . . . . . . . . . . . . . . . . . . . . . . . . 1197
28.2.3 A Practical Application: the Valuation of Convertible
Bonds . . . . . . . . . . . . . . . . . . . . . . . . . . . . . . . . 1207
28.3 Summary . . . . . . . . . . . . . . . . . . . . . . . . . . . . . . . . . . 1216
Appendix . . . . . . . . . . . . . . . . . . . . . . . . . . . . . . . . . . . . . . 1219
Suggestions for Further Reading . . . . . . . . . . . . . . . . . . . . . . . 1219
Books . . . . . . . . . . . . . . . . . . . . . . . . . . . . . . . . . . . . . . . 1219
Articles . . . . . . . . . . . . . . . . . . . . . . . . . . . . . . . . . . . . . . 1220
Website . . . . . . . . . . . . . . . . . . . . . . . . . . . . . . . . . . . . . . 1220

**29 Modeling Credit Risk (2): Credit-VaR and Operational Methods
for Credit Risk Management** . . . . . . . . . . . . . . . . . . . . . . . 1221
29.1 Determining the Credit-VaR of an Asset: Overview and
General Principles . . . . . . . . . . . . . . . . . . . . . . . . . . . . . 1223
29.2 Empirical Credit-VaR of an Asset Based on the Migration
Matrix . . . . . . . . . . . . . . . . . . . . . . . . . . . . . . . . . . . . 1224
29.2.1 Computation of the Credit-VaR of an Individual
Asset . . . . . . . . . . . . . . . . . . . . . . . . . . . . . . . . . 1224
29.2.2 Limitations of the Empirical Approach . . . . . . . . . . . 1227
29.3 Credit-VaR of an Individual Asset: Analytical Approaches
Based on Asset Price Dynamics (MKMV...) and on Structural
Models . . . . . . . . . . . . . . . . . . . . . . . . . . . . . . . . . . . . 1228
29.3.1 Asset Dynamics, Standardized Return, Default
Probabilities, and Distance to Default . . . . . . . . . . . . 1229
29.3.2 Derivation of the Rating Migration Quantiles
Associated with the Standardized Return . . . . . . . . . . 1230
29.3.3 Computation of the Distance to Default and Expected
Default Frequency (MKMV-Moody's Analytics
Method) . . . . . . . . . . . . . . . . . . . . . . . . . . . . . . . 1233
29.3.4 Comparing the Two Approaches . . . . . . . . . . . . . . . 1236
29.3.5 Estimation of the Credit-VaR of an Asset Using EDF
and a Valuation Model Based on RN-FN
Probabilities . . . . . . . . . . . . . . . . . . . . . . . . . . . . 1236
29.3.6 Relationship between Historical and RN Default
Probabilities . . . . . . . . . . . . . . . . . . . . . . . . . . . . 1238
29.4 Credit-VaR of an Entire Portfolio (Step 3) and Factor
Models . . . . . . . . . . . . . . . . . . . . . . . . . . . . . . . . . . . . 1239

| | | | |
|---|---|---|---|
| | 29.4.1 | Marked-to-Market (MTM) Models Involving Simulations | 1240 |
| | 29.4.2 | A Single-Factor DM Model of the Credit Risk of a Perfectly Diversified Portfolio (The Asymptotic Granular Vasicek-Gordy One-Factor Model) | 1242 |
| | 29.4.3 | Extensions of the Asymptotic Single-Factor Granular Model | 1247 |
| | 29.4.4 | Alternative Approach: Modeling the Default Dependence Structure with a Copula | 1250 |
| | 29.4.5 | Probability Distribution of the Default Dates Affecting a Portfolio | 1251 |
| | 29.4.6 | Portfolio Comprising Several Positions on the Same Obligor: Netting | 1253 |
| 29.5 | Credit-VaR, Unexpected Loss and Economic Capital | | 1254 |
| | 29.5.1 | Definition of Unexpected Loss (UL) | 1254 |
| | 29.5.2 | Probability Threshold and Rating | 1257 |
| 29.6 | Control and Regulation of Banking Risks | | 1257 |
| | 29.6.1 | Regulators and the Basel Committee: General Presentation | 1258 |
| | 29.6.2 | Capital and liquidity Rules under Basel 3 | 1260 |
| | 29.6.3 | Pillar 1 Capital Requirements under Basel 3 | 1261 |
| | 29.6.4 | Details on Pillar 1 Liquidity Requirements | 1265 |
| | 29.6.5 | Additional Basel 3 Reflections and Reforms | 1266 |
| 29.7 | Summary | | 1269 |

Appendix 1. Correlation of Defaults in a Portfolio of Debt Assets . . . . 1271
Appendix 2. Regulatory Capital, Market VaR, and Backtesting . . . . . . 1273
Appendix 3. Calculation of Regulatory Capital under the IRB
Approach: Adjustment to the Infinitely Grained One-Factor Model . . . 1274
Suggestion for Further Reading . . . . . . . . . . . . . . . . . . . . . . . . . . 1276

| | | |
|---|---|---|
| | Books | 1276 |
| | Articles and Documentation | 1276 |
| | Websites | 1277 |

**30 Credit Derivatives, Securitization, and Introduction to xVA** . . . . . 1279

| | | | |
|---|---|---|---|
| 30.1 | Credit Derivatives | | 1280 |
| | 30.1.1 | General Principles and Description of Credit Default Swaps | 1281 |
| | 30.1.2 | Single-Name CDS Valuation Techniques | 1285 |
| 30.2 | Securitization | | 1306 |
| | 30.2.1 | Introduction to Securitization and ABS | 1308 |
| | 30.2.2 | ABS Tranching Structuration | 1311 |
| 30.3 | The "xVA" Framework | | 1313 |
| | 30.3.1 | Counterparty Risk Exposure Measurement and Risk Mitigation Techniques | 1314 |
| | 30.3.2 | Counterparty Risk Exposure Modeling Techniques | 1319 |

|  |  | 30.3.3 | Collateralized vs Non-collateralized Trades: Some Statistics | 1324 |
|---|---|---|---|---|

30.3.3  Collateralized vs Non-collateralized Trades: Some
Statistics ................................... 1324
30.3.4  Introduction to CVA ......................... 1327
30.3.5  Introduction to DVA ........................ 1332
30.3.6  The FVA Puzzle ........................... 1334
30.4  Summary ....................................... 1339
Appendix 1 .......................................... 1342
Asset Swap Analysis ............................... 1342
Suggestion for Further Reading .......................... 1345
Books ........................................ 1345
Articles ....................................... 1346
Website: defaultrisk.com ........................... 1346

**Index** ............................................. 1347

# Introduction: Economics and Organization of Financial Markets

**1**

Anything more than a superficial understanding of the nature of market finance requires, for a start, a good familiarity with the economic role played by the banking and financial markets and with the exact nature of the objects traded there. Thus Sect. 1.1 is concerned with showing how these markets are useful for an effective allocation of cash as well as for the allocation of risks among economic agents to the benefit of *each* of the participants. The objects traded, financial assets, are precisely defined in Sect. 1.2 as sequences of future cash flows. Section 1.3 begins to address the essential, and more difficult, notions of financial market equilibrium, of the absence of arbitrage opportunities, of informational efficiency, and of liquidity. Finally, Sect. 1.4 offers a typology of financial instruments, describes the different types of organization encountered, and outlines the various ways of listing securities.

## 1.1 The Role of Financial Markets

The primary role of financial markets, the one which motivated their creation and which remains one of the most important roles, is to facilitate the transfer of cash from savers to borrowers. This allocation of cash is the business of so-called "cash" markets (Sect. 1.1.1). To this primary role is added the function of allocating risks, which has taken on considerable importance in the past decades with the development of derivative products (Sect. 1.1.2). Finally, the financial market contributes to informing economic agents, notably through prices that reflect the positions of enterprises and of the market, insofar as the latter is efficient (Sect. 1.1.3).

### 1.1.1 The Allocation of Cash Resources Over Time

At any given instant savers and borrowers coexist. The savers are willing to invest in financial assets; these take the form of lending or buying of financial instruments of

© The Author(s), under exclusive license to Springer Nature Switzerland AG 2022
P. Poncet, R. Portait, *Capital Market Finance*, Springer Texts in Business and
Economics, https://doi.org/10.1007/978-3-030-84600-8_1

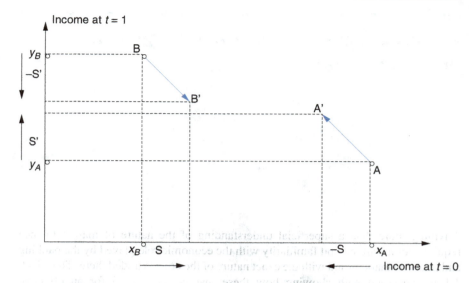

**Fig. 1.1** Allocation of cash over time

all sorts (shares, debt securities, cash, etc.). Symmetrically, borrowing needs may be covered by bank loans and by issuing securities. It is on the cash financial markets that these offers of funds (demands for securities) and needs for funds (offers of securities) meet, and transactions are realized. Outside these transactions, the flows of cash are initially transferred from an agent $a$ to an agent $b$ and then from $b$ to $a$ at subsequent times. For example, $a$ lends a sum $S$ to $b$ at time 0 and, as a counterpart, $b$ will reimburse $S' = S(1 + r)$ at $t = 1$ ($r$ denoting the interest rate).

Such a transaction is described in Fig. 1.1 where the position of each individual is represented as a point, with disposable income at time 0 (today) as its abscissa and disposable income at time 1 as the ordinate. For example, the initial situation of agent $a$ is represented by point A, which means that $a$ has at his disposal wealth $x_A$ today and that he will have $y_A$ at $t = 1$. Agent $b$ has at his disposal $x_B$ today and will have $y_B$ at $t = 1$. In this example, $a$ is rich at time 0 and poor at 1 while $b$ finds himself in the opposite position, which is presumably the reason that motivates the transaction represented in Fig. 1.1. This transaction may be interpreted as the loan today by $a$ of an agreed sum $S$ to $b$, against payment by the latter of a sum $S' = S(1 + r)$ at time $t = 1$.

The movement of their point representatives, A to A′ and B to B′, respectively, characterizes the change in the situations of the two agents that results.

This same operation can also be interpreted as the purchase by $a$ and the sale (or issuing) by $b$ of a security. This security creates a debt (for $b$ here), in an amount $S'$ and with maturity $t = 1$. Remark that such an operation is by nature something that benefits the "utility" (satisfaction) *of the two agents*, since each one of them assents

## 1.1 The Role of Financial Markets

to it, while in and of itself it does not create any apparent value, since the algebraic sum of the cash flows it generates is zero, whether at time $t = 0$ or at time $t = 1$.

### 1.1.2 Risk Allocation

Some financial operations are not intended to cover borrowing needs. They result in weak flows of cash, even zero, at the moment they begin (instant 0) and in future cash flows that depend on the situation pertaining in the future, under some rules initially fixed. As instruments these operations are called *contingent claims* and, generally speaking, are derivative products.

Let us consider an example of a contingent claim signed at 0 (today) by two agents $a$ and $b$. Under the terms of the contract, at time 1 (in the future) $a$ will give a sum $S$ to $b$ if a certain event $e$ happens, and as a counterpart, $b$ will give to $a$ the sum $S'$ if a different event, denoted by $e'$, occurs. At time $t = 0$, which event will occur at time $t = 1$ is not known; it is at best given with a probability. In the sequel, it will be referred to either as the "environment" at $t = 1$ or the "state of the world". In Fig. 1.2, the position of each individual at time $t = 1$ is represented by a point whose abscissa is the income (or wealth) of the individual if $e$ occurs, and whose ordinate is the income if $e'$ occurs.

For example, the situation of agent $a$ at $t = 1$, before any contingent transaction, is represented by point A, which signifies that he has wealth $x_A$ if $e$ happens and $y_A$ if $e'$ occurs. Agent $b$ has $x_B$ in environment $e$ and $y_B$ in environment $e'$. In this example $a$ will be rich in eventuality $e$ and poor in eventuality $e'$ (he thus bears a risk) while $b$ runs the opposite risk. But agents are (save for some exceptions) allergic to risk, that is, for a given expected wealth they prefer situations without risk. Such scenarios, in which the future income is independent of the environment, are

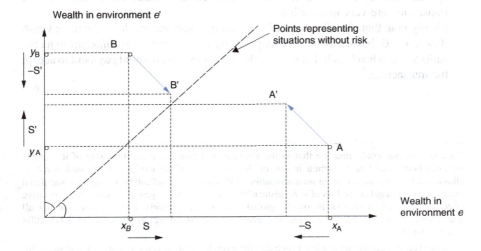

**Fig. 1.2** Risk allocation

characterized by points on the main bisector (the dashed line at 45 degrees). It is because the two agents run risks with opposite signs that *the contingent contract described in the preceding decreases the risk for each of the two agents.*

The consequences of the transaction are depicted in Fig. 1.2. The operation allows $a$ to pass from a situation characterized by A to a situation A' and $b$ to go from B to B'. *Both* operators find themselves, thanks to this contingent contract, in a less risky situation (since A' and B' are both "closer" to the bisector than, respectively, A and B). To make these ideas more concrete, one could imagine that $e$ represents the event that "the \$/€ exchange rate is high" while $e'$ means that "the \$/€ exchange rate is low"; $a$ is a European exporter who has some receivables payable in \$ from American clients while $b$ is an American exporter with receivables payable in €. The event $e$ enriches $a$ and impoverishes $b$, while $e'$ has the reverse effect. The contract in question could consist of a forward sale by $a$ of dollars, bought by $b$, which reduces the risks run *by both agents.*

The analysis just carried out leads to the following remarks:

- First, one notes that in the case of allocation of risks, just as for the allocation of cash, the transaction permits the welfare of the two agents to increase (because it reduces their risk) although the algebraic sum of the cash flows it generates directly is zero, for whatever period and in any state of the world.[1]
- One also notes the essential identity of Figs. 1.1 and 1.2. In fact, the only difference is in the interpretation of the coordinates: in Fig. 1.1 the horizontal axis represents wealth at time 0 while, in Fig. 1.2, it represents the wealth under the circumstance $e$ (at time 1); in Fig. 1.1, the vertical axis represents wealth at time 1 while, in Fig. 1.2, it represents wealth under circumstance $e'$. This graphical similarity comes from a conceptual similarity: in the first case, cash flows indexed by the date are traded while, in the second, it is cash flows indexed by the state of the world that are traded. In this sense, these two types of transaction are very much alike.
- Finally note that, for allocation of risks, no mention was made of the traded cash flow at $t = 0$. Indeed this flow may perfectly well be zero: *it is sufficient to fix the ratio $S'/S$* at a level such that none of the parties exacts an initial payment to accept the transaction.[2]

---

[1] In addition, one could imagine that in the absence of a contingent contract, one of the agents (or even both) could be in default in one of the two states of the world and that such a risk is eliminated by this transaction. In such a situation, with the costs of default (winding up at sacrificial prices of assets, legal costs, loss of synergistic effects, etc.) the contingent transaction has a positive effect on the cash flow, that is, on the global wealth of the agents. Remark also that not all transactions on a contingent basis are as "ideal" as described here, as the crisis in 2007–2008 cruelly revealed.

[2] In a market, this ratio is set at a level such that the overall supply is equal to the overall demand for a given initial cash flow, which could be zero.

### 1.1.3 The Market as a Supplier of Information

The market is not just the "place" where cash flows are traded, whether over time or contingent, but it is equally a vehicle for information. Notably, agents acquire there the information contained in prices. This incorporation becomes operational as a result of the transactions effected by well-informed agents and communicated by the reactions of less informed agents faced with the informed ones' transactions. According to how the prices incorporate, to a greater or lesser degree, this available information, the market will be more or less efficient in the informational sense.

The mechanism by which the prices incorporate information as well as the notion of efficiency will be the subject of Paragraph 3 of Sect. 1.3.

## 1.2 Securities as Sequences of Cash Flows

The objects traded in the market of interest to us here are financial assets in the strict sense (as opposed to derivative products, such as options, futures, or swaps, which will be analyzed in subsequent chapters). We shall give a general definition for these financial assets and then we will analyze all their elements in detail.

### 1.2.1 Definition of a Security (or Financial Asset)

A cash flow is a sum of money transferred from one agent to another. Let us take up again the example of a basic financial operation: lending and borrowing. Let us represent a cash flow received (cash inflow) by a positive number and a cash outflow (disbursement) by a negative number. Any positive flow for the borrower is negative for the lender and *vice versa*. Two symmetrical points of view can thus be adopted: that of the lender or investor (who loans, places or invests a certain sum of money) and that of the borrower (who signs for or contracts the loan and receives this sum). Providing a loan represents an *investment* (or a placement) for the lender: indeed it implies the transmission of initial funds (an outflow of money equal to the amount of the loan) as a counterpart to revenues (receipts of money) hoped for in the future in the form of successive reimbursements of this capital increased by interests. For the borrower, the operation is also called *funding*. The interest rate on the loan corresponds to the yield (return) on the investment for the lender and to the cost of credit for the borrower.

This operation of lending/borrowing in fact is only a special case. A more general concept is that of a security.

**Definition** A security (financial asset) is a right (a claim) to a sequence (or a series, or a vector) of future cash flows.

The purchaser of a security gives to the seller a sum of money corresponding to the value of the security, in exchange for which she acquires the right to receive one or more cash flows spaced over time. For example, the buyer of a *bond* will receive a sequence of interest payments (called coupons) and finally the reimbursement of the capital initially invested. The buyer of an *equity share* will receive a sequence of dividends followed by the (future) value of the share at the date of its resale. The seller of a security receives from the buyer the sum of money corresponding to the value of the security, in exchange for which he will pay out one or several cash flows in the future. As a result, we will analyze lending as equivalent to the *purchase* of a security (or securities), and borrowing as equivalent to the *sale* of (or issuing) a security (or securities). These securities are traded on the financial market in the broad sense, which we consider includes the lending and borrowing that takes place in the banking sector.[3]

In professional jargon, a newly created financial security is traded on the *primary market*. Thus, it provides a relationship between an *issuer* (the seller of the security, the borrower) and one or more buyers (the lenders, or investors). This is the "new (first hand) securities" market. For example, companies issue shares or bonds on the primary market to finance their physical investments (factories, materials, research, etc.). An existing financial security is traded on the *secondary market*. Thus, it provides a relationship between a *seller* and a buyer. This is the "resale" market. For example, an investor having bought a share resells it to another investor because of a need for cash.

## 1.2.2 Characterizing the Cash Flow Sequence

Let us now be more precise about the elements involved in the definition of a security: "cash flow" and "sequence." Four "dimensions" must be specified to characterize cash flows: the discrete or continuous character of their timing (which influences the mathematical way we model it); their number (in the discrete case); their certainty or randomness; and the currency unit (or numeraire) in which they are expressed.

– **Time.** The representation of the cash flow sequence generated by a security makes it necessary to have a definition of the way in which time "flows" and the unit in which it is to be reckoned. The choice considered here is between a discrete and a continuous time. Almost always, *flows are discrete and are generated punctually* on dates bounding the period of interest (e.g., a year for a bond, whose coupon is paid, on April 22 each year, a trimester for a bank loan which is partially paid back on the 7th of March, June, September, December, etc.), and nothing happens between these dates. A cash flow is therefore denoted

---

[3]The definition of a security is very general. It allows us, for example, to analyze a tangible (or physical) investment by a firm (a factory, equipment, research, etc.) as the purchase of a security: a disbursement of cash followed by a sequence of cash flows resulting from the investment. One can therefore apply the theory of selection of financial securities to the analysis of choices of tangible investments (see Chap. 22, Sect. 22.2).

## 1.2 Securities as Sequences of Cash Flows

by $F_t$ where $t$ is the date on which it occurs. So the mathematics used will be that of discrete time (see Chap. 2). However, it is sometimes preferable to represent the passage of time as continuous in the same way as for physical time; the reference period then is an infinitesimal "$dt$." The security gives rise to a *flow density* per unit time, denoted $F(t)dt$, between the instants $t$ and $t + dt$. An example of this is given by a savings account that receives interest continuously. Such a representation can be required especially when the cash flows are daily or even more frequent. Indeed, as we shall see in later chapters, a continuous model of time often facilitates calculations. In the rest of this chapter, we shall consider only *discrete* cash flows.

- **Number**. This notion obviously makes sense only for discrete flows. The sequence of future flows defined by a security is completely general: it can be only one element (repayment of capital and payment of interest all at once), or $N$ elements where $N$ is finite and known (e.g., a bond issued by the State, paying $N$ coupons and the reimbursement of the capital along with the $N$th coupon), or even an infinity of elements (the case of a so-called "perpetuity" (or "consol"), whose capital is never paid back and which thus pays off only in coupons, usually annual and going on forever), or finally could be $N$ elements with $N$ non-negative but random. If we consider a share, for example, the buyer is not sure of the number of dividends he will receive, since the issuer might in certain years not proceed with a distribution and could even go bankrupt; similarly, the payouts for a bond issued by a private enterprise can be interrupted temporarily or conclusively by financial difficulties which it experiences.

- **Certain and random flows**. If the security gives the right to *predetermined* cash flows, and if, in addition, the issuer of the security (the debtor) can be considered immune to all risks of default, for example, because it is an economically robust sovereign State, future cash flows can be considered certain or sure. In this case, estimating the elements in the cash flow sequence is trivial and valuing the security is very simple, using only the mathematics of certainty (see Chap. 2). The security is then without risk or *risk-free*. On the other hand, when the cash flows are fraught with uncertainty, whether it is because the issuer could default, partly or completely, or because they are not fixed from the start (the dividends paid by a share, for instance) and depend on economic circumstances, they are analyzed as random variables and valuing a security becomes much more complicated. The security is then termed *risky*; this is usually the case. It could then be necessary in order to make computations to specify the characteristics of the distribution of each flow, such as its expectation (mean), its variance, its covariance with another random variable, etc. A large portion of this book is devoted to the risky case.

- **Currency unit (or numeraire)**. Finally, we need to specify in what currency unit (numeraire) the cash flows to which the buyer of the security is entitled are to be expressed. Mostly, they are expressed in the currency (or money) of the country where the security is traded (or issued). If the cash is in foreign currency, it is useful to exchange it for domestic currency using the exchange rate current at the date of receipt of the cash flow.[4] In these two cases, the cash flows are "present dollars (euros, sterlings, yens, yuans, . . .)" and one speaks of "nominal" flows. Very exceptionally, the flows can be expressed in "constant dollars," a way of saying protected from inflation, which is just the continual loss of purchasing power in domestic currency. Therefore, one speaks then of "real flows" (for a more detailed analysis, see Chap. 2).

To conclude, the series (or vector) of (discrete) flows defining a security will generally be denoted $F_\theta]_{\theta = t_1, \ldots t_N}$, where the $t_i$ indicate the dates of occurrence (not necessarily equidistant) and the $F_\theta$ are possibly random.

## 1.3 Equilibrium, Absence of Arbitrage Opportunity, Market Efficiency and Liquidity

The role of the financial markets is to ensure the efficiency of the distribution of financial resources and risks amongst different agents. This section succinctly presents the essential characteristics of a market that govern its economic efficiency: equilibrium (Sect. 1.3.1), absence of arbitrage opportunity (Sect. 1.3.2), informational efficiency (Sect. 1.3.3), and liquidity (Sect. 1.3.4).

### 1.3.1 Equilibrium and Price Setting

Let us recall several definitions concerning economic equilibrium. A market is said to be *in equilibrium* if supply is equal to demand. This equality holds thanks to the adjustment of prices. Because of this, in a sufficiently liquid financial market in equilibrium, *every agent will find a counterparty* willing to trade at the market price. The equilibrium is said to be *"competitive"* if no agent has sufficient relative weight to influence the prices by his demand or supply (bid or offer, also ask), in financial market jargon). It follows that in a competitive financial market in equilibrium, any participant may buy or sell arbitrary quantities at the market price while he will find neither a seller accepting a lower price nor any buyer at a greater price. Each agent thus considers the price as an external given: one says he is a *"price taker."* The theory of "general equilibrium" shows that, under certain conditions, a competitive equilibrium for all markets displays a sort of optimality (in regard to the satisfaction or "utility" of the agents). The concept of equilibrium is fundamental in economic and financial theories. In this book, financial markets will, in general, be implicitly assumed to be in competitive equilibrium.

---

[4] It is easy to see that, even if a cash flow in foreign currency were risk-free (as, e.g., in the case of a European Treasury Bond which is paid out in euros), it becomes risky once it is translated (for example, into dollars) since future exchange rates (here euro/dollar) are random, as viewed from the date of purchase of the security.

## 1.3 Equilibrium, Absence of Arbitrage Opportunity, Market Efficiency and Liquidity

To explain how prices are set it is necessary to have a model for the process of trading. Equilibrium is called "Walrasian" if the trading mechanism and setting of prices is in accordance with the paradigm considered by the economist Léon Walras.[5] According to this theoretical model, trading and setting of prices happens *as if there were* an "auctioneer" trying a series of "iterations." The auctioneer announces an initial price taking into account all bids and offers the buyers and sellers have made; he collects these, consolidates them, and compares them (with no trading yet having taken place); according to whether the total demand is greater or less than the total supply, he posts a second price, larger or smaller than the first, and so on. This series of iterations ("tâtonnements") is continued until a price that equalizes supply and demand is reached. The trading transactions take place at this equilibrium price at the end of the process. We emphasize that *"no trades are carried out at other than the equilibrium price"* and that the iteration process is considered quasi-instantaneous. As we shall see in the following paragraph 4 (Sect. 1.3.3), the quotes offered and the trades carried out on organized financial markets do take place according to a procedure that does approximate, as nearly as it can, the Walrasian paradigm.

### 1.3.2 Absence of Arbitrage Opportunity (AAO) and the Notion of Redundant Assets

In financial theory, arbitrage is an operation that involves no investment of funds (no cash outflow) and which only generates positive or zero cash flows (positive in certain cases but never negative). The agent who carries out arbitrage is called an *arbitrageur*. An opportunity for arbitrage may occur, for example, when an asset A can be "synthesized" by combining two assets B and C. One says that A is synthesized or "replicated" by a combination of B and C if the cash flows generated (possibly randomly) in the future by A are equal, *with certainty at any instant*, to a linear combination of the cash flows generated by the assets B and C. As a result, if we denote respectively by $\underline{A}$, $\underline{B}$, and $\underline{C}$ the cash flow sequences generated by the assets A, B, and C, there exist two scalars $\lambda_1$ and $\lambda_2$ such that:

$$\underline{A} = \lambda_1 \underline{B} + \lambda_2 \underline{C}, \quad \text{with certainty.}$$

In this situation a portfolio made up of $\lambda_1$ units of B and $\lambda_2$ units of C is a clone of asset A since it surely generates at each instant exactly the same revenues; we say that this portfolio replicates or synthesizes A and that the assets A, B, and C, taken together, are redundant (one may obtain the revenue of any one of the three starting from the other two). An arbitrage opportunity would then present itself if today the prices $P_A$, $P_B$, and $P_C$ of A, B, and C did not satisfy the following condition for

---

[5] French economist, professor at the University of Lausanne, whose major contribution *Elements of Pure Economics* (*Éléments d'économie pure*) appeared in 1874.

coherence (or parity) called the *"principle of linearity in the values"* (or *"linearity principle"* for short):

$$P_A = \lambda_1 P_B + \lambda_2 P_C$$

Indeed, if $P_A > \lambda_1 P_B + \lambda_2 P_C$, the arbitrage consists in buying $\lambda_1$ units of B and $\lambda_2$ units of C while selling A short,[6] which gives rise today to the positive cash flow $f = (P_A - (\lambda_1 P_B + \lambda_2 P_C))$. Later on, the cash flows paid to the buyer of the asset A will exactly cover the cash flows received from the purchase of B and C. The cash flow series is thus $\{f, 0, \ldots, 0\}$, which is obviously a "free lunch" for the arbitrageur. If we have $P_A < \lambda_1 P_B + \lambda_2 P_C$, the arbitrage amounts to the reverse position (purchase of A and short sales of B and C). For example, in the absence of arbitrage opportunities (AAO in the sequel), that is with no free lunches, the value $v$ of a share of a fund whose assets are made up of different securities $S_1, S_2, \ldots, S_N$ will be given by $v = \frac{1}{M} \sum_{i=1}^{N} n_i P_i$ where $n_i$ is the number of securities $S_i$ in the fund's assets, $P_i$ is the price of security $S_i$ and $M$ the number of shares making up the capital of the fund. We will meet later on in this work numerous examples of redundant assets of the type just described which lead to parity conditions when AAO prevails. We will study replication and the conditions that make it possible more rigorously especially in Chaps. 7 and 15.

The existence of arbitrage opportunities in a market is incompatible with its equilibrium. Indeed, arbitrage opportunity triggers a supply (of over-priced instruments) and a demand (for under-valued instruments) of a theoretically infinite magnitude that is incompatible with equilibrium.[7] In contrast, a market free of arbitrage opportunities is not necessarily in equilibrium. The condition for equilibrium is, therefore, stronger than just the absence of arbitrage opportunities although the latter does lead, in numerous situations, to very productive models[8] and is the keystone of modern financial theory. In the sequel, financial markets will mostly be supposed free from arbitrage opportunities (the AAO situation). It is in any case clear that, when an arbitrage opportunity turns up, to seize it is preferable to *any other* investment strategy since it is a free lunch.

---

[6] A *short sale* is the sale of a security which one does not own but that one borrows temporarily from a third party, either in the market or over-the-counter (OTC) market, i.e., "privately". When the arbitrage is unwound, the arbitrageur repurchases the security that was sold short to give it back to the third-party lender. It is implicitly assumed that this provisional loan of the security takes place at no cost. In practice this is not entirely true.

[7] See Sect. 23.2 of Chap. 23 for examples.

[8] Such as discounting, arbitrage pricing models, option pricing, and so on.

### 1.3.3 Efficiency

The equilibrium of a market may result from offers and bids coming from irrational and/or ill-informed actors, or, more precisely, from the fact that the influence of rational and informed actors is not decisive. In such conditions, the equilibrium price does not necessarily incorporate all the potentially available information and does not play its role in allocating resources effectively. The concept of market efficiency goes back to the work of Arthur Cowles in the 1930s. After the Second World War, economists started to doubt the validity of methods of predicting stock prices. They contrasted the *random walk* model of market prices with the idea that the prices are characterized by tendencies to increase or decrease. According to this model, the relative variation of a price during a given period is a random variable independent of all the information available at the start of the period. It was only in 1970 that a more precise formulation due to Paul Samuelson and later to Eugene Fama[9] gave the concept of market efficiency its present form.

#### 1.3.3.1 The Notion of Efficiency

**Definition** *A market is said to be efficient when, at any instant, its prices take into account all the available relevant information.*

In order for efficiency to be possible, it is required that all new information should be instantaneously and entirely reflected in the prices. The incorporation of information in the prices results essentially from the actions of well-informed agents and from the conclusions drawn by less well-informed, but rational, participants from their observation of the markets. Furthermore, if some information is correctly integrated into the prices, it cannot be used to make an abnormal profit, that is, to buy an underpriced asset or to sell an over-priced one. This last property serves furthermore as the basis for an alternative definition: a market is efficient if, at every instant, the available information does not permit the realization of "abnormal" profits. A very simple model will illustrate these ideas.

Let us consider a security giving the right to a cash flow $X + x$ in the future (at $t = 1$) and nothing thereafter, today's date being $t = 0$. Two types of economic agents, called respectively $i$ and $j$, coexist in this economy. At $t = 0$, the agents $i$ are perfectly informed (or insiders) and know the exact values of $X$ and $x$ while the agents $j$ are, a priori, imperfectly informed and only know $X$ ($x$ being for them a priori random). Furthermore, we assume that any agent may lend or borrow for one period at the interest rate $r$. Such an economy, which is very simplified, works according to the following principles:

---

[9]Fama, E., 1970, "Efficient Capital Markets: a review of Theory and Empirical Work," *Journal of Finance*, p. 383-417. The history of market efficiency as well as the development of the rest of modern finance is recounted by Peter L. Bernstein in *Capital Ideas: The improbable origins of modern Wall Street*, Wiley, 2005.

12 1 Introduction: Economics and Organization of Financial Markets

- The behavior of the informed agents determines the price $P$ at 0.
- The observation of $P$ reveals to the less informed agents $j$ the actual value of $x$.
- The level of the market price perfectly reflects the information on the true value of $x$ and because of this prevents the insider agents $i$ from using their inside information to obtain "abnormal" returns.

**Proof**

- Let us first show that, because of the operations carried out by informed agents, the price of the security is established at $t = 0$ at the level $P = \frac{X+x}{1+r}$. Indeed, an agent $i$ is willing to buy the security as soon as its price is less than $\frac{X+x}{1+r}$ and to sell it as soon as it is greater than this value, thus fixing the equilibrium price at $\frac{X+x}{1+r}$. Let us analyze the mechanism in more detail. Assume to start with that $P < \frac{X+x}{1+r}$. In such a case, $i$ will contract a loan of amount $P$ allowing him to purchase the security with no net outlay at $t = 0$; at $t = 1$, the cash flow $X + x$ returned by the security (which is certain for $i$) will allow him to repay the loan and its associated interest, i.e. $P(1 + r)$, and will leave him with a certain positive cash flow $X + x - P(1 + r)$. Such an operation, which generates a sure positive cash flow without any capital outlay is an arbitrage (*see* the preceding paragraph). These arbitrages, which could be of theoretically infinite amounts, put pressure on the price of the security in such a way as to prevent its being at a level below $\frac{X+x}{1+r}$. Arbitrages with the opposite sign (sale of the security and lending the proceeds) also prevent the price $P$ from exceeding $\frac{X+x}{1+r}$. The equilibrium price $P$ and the AAO price thus equal $\frac{X+x}{1+r}$.
- The agents $j$, ill-informed but not irrational and aware of the mechanism previously described, will deduce the true value of $x$ from the observation of price $P$: $x = P(1 + r) - X$.
- In a market at Walrasian equilibrium where no trade can take place except at the equilibrium price $P$, informed agents cannot exploit inside information (knowledge of the exact value of $x$) to obtain an abnormal return (that is differing from the interest rate $r$). A return differing from $r$ would only be possible if $P$ "did not incorporate" the information about the true value of $x$, that is if it were to differ from its "fundamental" value $\frac{X+x}{1+r}$.

The market we have just imagined, where the prices incorporate all available information, even that held by insiders, is said to possess the "strong form of efficiency." We remark that our model is simplistic and that, in a more complex and realistic situation, asset prices only *partially reveal* insider information.[10] In

---

[10] One can, for example, imagine that the insiders only have partial information on $x$; for example $x = x_1 + x_2$ where $x_1$ is unknown to $j$ but known to $i$, and $x_2$ is random for both $i$ and $j$. If the participants $i$ are risk averse and the number of them is not known to $j$, the observation of the equilibrium price will not be enough for $j$ to deduce the exact value of $x_1$; he will simply be able to reduce the variance of $x_1$.

## 1.3 Equilibrium, Absence of Arbitrage Opportunity, Market Efficiency and Liquidity

contrast, it is more reasonable to suppose that only the information made public is incorporated in the prices as it is in theory known to everyone, so one is led to consider less restrictive hypotheses than strong efficiency (which presumes that *all* information is incorporated in asset prices, including that held by insiders).

The considerations that this introductory example raises concerning the notion of efficiency and its "relative" nature can be expressed in greater generality and more formally. Let us denote by $I_t$ the set of relevant and potentially available information at time $t$. $I_t$ encompasses not only the pertinent factual information but also the collection of methods and models that allow processing of this information. Let us denote $P^*_t$ the prices of the different assets *"which would prevail"* at $t$ if all agents were to use $I_t$ to determine their offers and bids, and denote $P_t$ those which effectively do prevail. The market is efficient if $P^*_t = P_t$. Remark that, in order to obtain such a result, it is not required that all agents be well informed (that is know $I_t$ *ex ante*) but the informed agents must have sufficient weight and influence that they determine the process of setting the price (*see* the introductory example).

Furthermore, using the information set $I_t$ assumed available at $t$ and used as a "benchmark" for efficiency, we define (following E. Fama) three forms of efficiency: *weak efficiency* is that of a market whose prices incorporate at every instant the unique history of past prices ($I_t =$ history of the prices at $t-1$, $t-2$, $\ldots$). In a weakly efficient market, the participants cannot take advantage of the knowledge of the history of past prices to realize systematic "abnormal" profits (for example, with the help of technical or chartist analysis). *Semi-strong efficiency* is that of a market whose prices at each instant incorporate not just the history of past prices but also all the relevant information (profits, intended investments or financing, dividends, etc.), which has been made public ($I_t$ includes all public information). In a market displaying this form of efficiency, it is useless to try and utilize public information to realize "abnormal" profits because the prices will adjust (quasi-) instantaneously as soon as relevant information is announced. *Strong efficiency* is that which characterizes a market where the prices incorporate all available information, whether it has been made public or not: ($I_t =$ all information). In such a market, inside information is very rapidly (quasi-instantaneously) incorporated in the equilibrium prices and insider profits are practically non-existent.

To visualize the difference between the functioning of an efficient market and of an inefficient market, let us graph the possible evolution of a price in an efficient market (Fig. 1.3a) and in an inefficient market (Fig. 1.3b). In both cases, new information, favorable (+) and unfavorable (−), hits the market. In the efficient case, it is instantaneously and completely incorporated in the prices (Fig. 1.3a), while in the inefficient market this happens only gradually (Fig. 1.3b). In an efficient market no transaction takes place except at the price that reflects all information, while in the contrary case the informed and quick-witted agents can take advantage of their information (during the periods enclosed in braces on Fig. 1.3b). Let us remark, to complete the previous explanation, that the instantaneous integration of the information into the prices is a necessary but not a sufficient condition for efficiency. Indeed, the information must also be "processed" in a rational manner (with good analytical tools), so that the market does not under- or over-react.

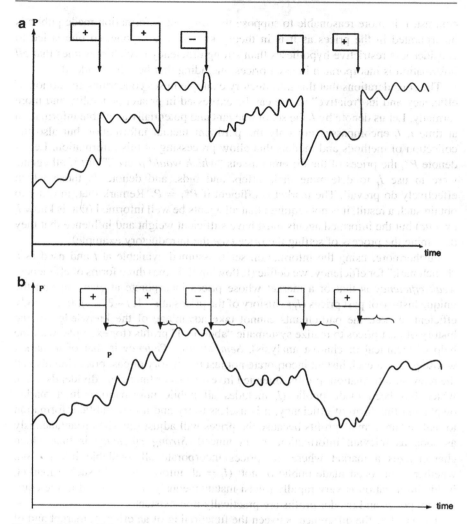

**Fig. 1.3** (a) An efficient market. (b) An inefficient market

### 1.3.3.2 Theoretical and Empirical Considerations

The three hypotheses of efficiency have been the objects of many studies both theoretical and empirical. It is obviously the strong form of efficiency that is the most controversial: not only does the number of insider affairs that are reported in the newspapers directly contradict it, but it also faces theoretical arguments. The first of these is known as the "Grossman–Stiglitz paradox": if the prices were to reflect at any time all insider information (which one could then not use to any advantage) it would not be profitable to search for or develop any costly information. In the limit, since in fact, any information does have a cost, no equilibrium could even be attained! However, this argument due to Grossman and Stiglitz is at least partly

## 1.3 Equilibrium, Absence of Arbitrage Opportunity, Market Efficiency and Liquidity 15

questionable because it does not follow the logic to its conclusion: An agent who has insider information has no interest in immediately and completely revealing it with massive orders which would have the effect of immediately fixing the equilibrium price at the full information level, thus preventing any possible abnormal return. The informed agent thus should behave "with moderation," and because of this the prices only reveal at best a fraction of the insider information: thus, logically, strong efficiency can never hold. Let us furthermore remark that Grossman and Stiglitz's argument also makes the hypotheses of semi-strong and weak efficiency fragile. Indeed, it is because the agents do research, take account of the available information, and then determine their bids and asks as a result of this, that the information is incorporated into the prices: this mechanism does not operate if the participants are not motivated to do research and develop information. An alternative definition of efficiency does allow us to avoid this logical nit-picking: a market is efficient if the use of information does not allow abnormal returns *when the costs of developing and implementing it are taken into account*. In addition, such a definition, which is more satisfactory than the basic definition, does not exclude a useful (and compensated) role for forecasters and analysts while such people have no place in an efficient market in the sense of the basic definition.

On an empirical basis, the three hypotheses of efficiency have been the subjects of experiments based on the following principles:

- **Tests of weak efficiency** attempt to detect a possible *auto-correlation* over time of the successive asset returns; such an auto-correlation would invalidate, under certain conditions, the weak efficiency hypothesis since it would permit taking advantage of knowledge of past prices to reap abnormal returns.
- **Tests of semi-strong efficiency** try to detect the presence of possible abnormally high returns following good news or abnormally low returns following bad news. Such "anomalies" are incompatible with semi-strong efficiency.
- **Tests of strong efficiency** try to detect abnormally large returns after important decisions concerning the relevant security have been taken but not yet made public (e.g., prior to the announcement of a takeover but after it has been decided upon). The precise description and the conclusions to be drawn from these tests are beyond the scope of this introductory chapter. We shall be satisfied here with just mentioning the existence of numerous papers that cast doubts on the efficiency of markets and sometimes reveal the possibility of abnormal profits. It seems, however, excessive to simply reject the hypothesis of efficiency because, even if anomalies have been detected,[11] they are difficult to exploit. A reasonable assumption is that, even though perfect efficiency does not exist, *most financial markets are near to semi-strong efficiency; it is therefore very difficult to take advantage of information that everyone knows to gain abnormal returns, taking*

---

[11] Without any claim of completeness we mention: the "value" effect (stocks with a large [book value/market value] ratio are the most profitable), the "size" effect (small capitalizations are the most profitable), "the Friday the 13th" and the "momentum" effect.

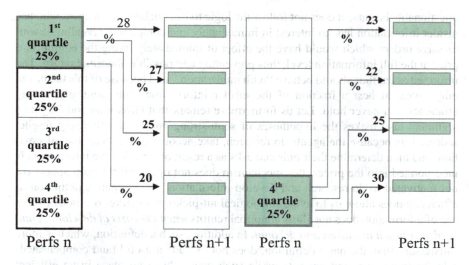

**Fig. 1.4** Typical distribution of performances in two consecutive periods

into account the costs of handling the information and the "normal" compensation for the risk incurred.[12]

Another notion related to the notion of efficiency is that of the absence of persistent performances. The simplest empirical study is to rank a sufficient number of funds into categories according to their performances in the period $n$ and then to see how these categories perform in the succeeding period $n + 1$. Up to some differences from the funds studied and the period chosen (1 year, 2 years or more) one will typically obtain a result conforming to Fig. 1.4 which shows that the persistence of performance from one year to another is limited: The first quartile, made up of the best performing 25% of the funds in period $n$ are distributed fairly uniformly over the four quartiles for period $n + 1$, the persistence of bad performance being slightly more obvious in some studies.

The careful interpretation of these results leads one to think that, in most cases, good (or bad) performances result from luck (good or bad) without excluding the notion that (good or bad) performance of a small number of funds could be linked to the quality of their management.

### 1.3.4 Liquidity

The liquidity of an asset is linked to the ease with which agents can open or close positions in the asset under fair conditions, that is, at the market price. In particular,

---

[12] We will see later, notably in Chaps. 21 and 22 devoted to portfolio theory, what one means by "normal" compensation for the risk *incurred*.

## 1.3 Equilibrium, Absence of Arbitrage Opportunity, Market Efficiency and Liquidity

**Table 1.1** Depth and width of a market

| Bid price | Size of the existing demand (Current price = 50) | | | |
|---|---|---|---|---|
| | (1) Shallow Narrow | (2) Deep Narrow | (3) Shallow Wide | (4) Deep Wide |
| 50 | 100 | 100 | 500 | 500 |
| 49 | 200 | 200 | 500 | 500 |
| 48 | 0 | 320 | 0 | 700 |
| 47 | 0 | 280 | 0 | 600 |
| 46 | 0 | 250 | 0 | 1400 |

when an agent places a sizeable quantity of a liquid asset on the market there should result in no significant drop in its value. In addition, an asset will be less illiquid the more rapidly the shift in its price caused by an offer or a bid is absorbed and the "long-term equilibrium price" is promptly re-established.

Liquidity depends on the *depth, width* (see Table 1.1, for bids), *elasticity,* and *resilience* of the market. A market is deep if supply and demand exist at prices that are far from the current price (smaller for demand, greater for supply). A market is wide if at each price level (around the equilibrium price) both the supply and demand are large. Width and depth can be assessed on the basis of the order book (see Sect. 1.4.3.4 *infra*) of which Table 1.1, for four markets (1), (2), (3), and (4), constitutes a very simplified example.

The *elasticity* of a market depends on its depth and width and is defined as the ratio $\left|\frac{\Delta Q/Q}{\Delta P/P}\right|$ (or $\left|\frac{\Delta Q}{\Delta P}\right|$) where $\Delta P$ denotes the variation in the price produced by a variation $\Delta Q$ in the quantity supplied or demanded. The market is thus elastic if putting a great number of securities in the market does not cause a noticeable variation of the price.[13] It is *resilient if, after some negotiation involving a large number of securities has caused a "jump" in the price, up or down, the price returns rapidly to the value judged to be the equilibrium price.*

The liquidity of an asset surely depends on its nature and, for primitive securities, on the "stock" made available to the market by the issuer and therefore likely to be traded (the *"float"* in financial jargon). It also depends on the transaction costs (taxes, brokers' fees, the bid-ask or bid-offer spread, etc.) which affect each trade. More generally, liquidity also depends on the way exchanges are organized. In particular, the organization has an influence on transaction costs both direct (access to the market, handling, and execution of orders) and indirect (interaction between agents, spread between the bid and offer prices) and on information costs (who holds what,

---

[13] This is the idea behind one of the measures of liquidity due to Amihud (2002): one considers $N$ successive returns $r_j = \Delta P_j/P_j$ ($j = 1, \ldots, N$) associated with transactions in the amounts $v_j$ of an asset, and one is interested in the ratios $\frac{|r_j|}{v_j}$ ($|r_j|$ is the absolute value of $r_j$). The liquidity of this asset becomes greater as the mean value $\frac{1}{N}\sum_{j=1}^{N}\frac{|r_j|}{v_j}$ becomes less (for a given transaction amount, the variation in price is less).

18          1 Introduction: Economics and Organization of Financial Markets

what quantities are being bid or asked, and at what prices). The different ways of organizing markets will be described in Sect. 1.4, and their effects on liquidity, and so on efficiency, revisited there.

Finally, it is important to establish the link between auto-correlation of successive returns, liquidity, and efficiency. We have already mentioned above that a serial auto-correlation of returns, in general, invalidates even the hypothesis of weak efficiency because it allows taking advantage of just knowing past prices. Moreover, let us remark that not incorporating the information immediately in prices (which is the *very definition* of inefficiency) leads to positive auto-correlations. To visualize this it is enough to return to Fig. 1.3b: in the face of good news, the stock will only rise progressively and one will see a *succession* of returns with the same sign (positive); conversely, in case of bad news, the stock will only fall slowly and one will see a *succession* of returns of the same sign (negative). Furthermore, a serial correlation of the returns can be generated by the illiquidity of a security. Work from the "market microstructure" literature suggests that auto-correlation induced by illiquidity can be either positive or negative (successive returns then display alternating signs in a manner that is more systematic, thus more predictable, than when there is no serial correlation). Illiquidity can in this way be the source of a positive auto-correlation if it encourages big participants to distribute their large buy or sell orders over several successive periods. Meanwhile, if the market is efficient, these successive orders will be "compensated for" by those from agents assuring efficiency and the auto-correlation is thus eliminated. Therefore, it is necessary not to confuse illiquidity and inefficiency, which are two different concepts. Illiquidity can equally well lead to a negative auto-correlation, notably because of the share of "market makers" (specialists in a security responsible for its listing; see Sect. 1.4): Following an increase in prices leading to selling on the part of investors, the positions of market makers, who are to buy these securities, become positive; to rub out undesired positions (on average, the position on each security should be zero), and thus to encourage the investors to buy, they lower their prices in the succeeding period.[14] So after prices go up they go down. The reverse effect prevails when an initial price reduction leads to buying and so to undesirable short positions for the market makers, with a consequent increase in the price to encourage investors to sell.

### 1.3.5 Perfect Markets

It is conventional to qualify as perfect those markets in competitive equilibrium (therefore without arbitrage opportunities), satisfying at least the condition of semi-weak efficiency, and which are also perfectly liquid and so exempt from transaction costs, fees discriminating between different components of the returns and

---

[14]The existence of a spread, bid price – offer price, can also lead to negative auto-correlation by a "rebound" effect, but exploring such questions is outside the scope of this book.

constraints due to regulation or techniques preventing certain positions (ratios, short sales, etc.). Such "ideal" markets are suitable for modeling and have proved useful benchmarks with respect to which real markets may be compared and the consequences of their "imperfections" appreciated. Because of a more intensive international competition, of the new types of instruments that are traded, of closer monitoring, and of weaker transaction costs, one may claim that, except during periods of crisis like that of 2007–2008 and ignoring the issue of financial bubbles not discussed in this book, it is the financial markets that are nearest to this "ideal situation" (while never attaining it).

## 1.4  Organization, a Typology of Markets, and Listing

This section first points out the distinction between "intermediated" and "disintermediated" (Sect. 1.4.1), then a typology of financial markets and the instruments traded on them (Sect. 1.4.2), and finally an analysis of the different types of organization and their influence on the efficiency and liquidity of the market (Sect. 1.4.3).

### 1.4.1  The Banking System and Financial Markets

The connection between economic agents with capacities for investment and those who need financing is realized through two distinct channels or circuits. The historically older channel is that of the *banks,* who in fact act as *intermediaries*: the banks "insert themselves," in a sense, between the agents who lend their surpluses (notably in the form of cash or term deposits) and the agents who borrow funds from them (in the form of bank loans, permitted by retail deposits), according to the following diagram (Circuit 1) in Fig. 1.5.

The second channel, the financial market, is more direct. In this scenario, the economic agent who has investment capacity lends directly (without any intermediary, except simply for a broker who is just concerned with the physical execution of buy and sell orders; *see infra* Sect. 1.4.3.2) to someone who needs financing (Circuit 2 in the diagram).

Setting up financing through the market is faster in developed economies than through bank financing (intermediated). This tendency toward dominance of the markets is "disintermediation."

### 1.4.2  A Simple Typology of Financial Markets

Although many classifications are conceivable, we shall retain the distinction between markets in "primitive" assets and markets in "derivative" products. For the first, the securities are quoted on the spot (immediate payment to the seller by the buyer) and their essential role is the reallocation of cash between lenders and

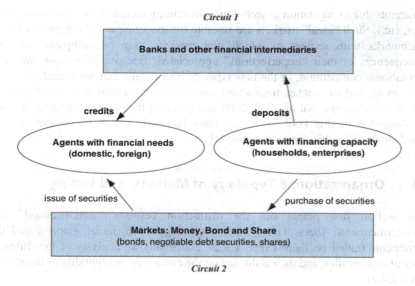

**Fig. 1.5** The two circuits of financing (the direction of the arrow indicates the flow of funds)

borrowers. In the second, the instruments are quoted either spot (for most options and swaps), or at maturity, the payment of the seller by the buyer being delayed to a date settled in advance (forward contracts, futures, certain options, and swaps), and their essential role is the reallocation of risks between agents. This section succinctly describes the principal products that will be analyzed in detail in the course of the book. The appendix to this chapter gives a list of the main markets in the world with their internet addresses.

### 1.4.2.1 Primitive Spot Assets: Allocation of Cash

One should distinguish here interest-rate markets (bonds, loans, fixed-income securities, receivables and payables) from stock markets (capital or equity, or property rights).

(a) *Interest-rate markets*

What is traded on these markets are securities providing coupons and reimbursement of capital. We may further distinguish short-term markets (initial maturity of the asset less than one year), from medium-term markets (initial maturity of less than 5 years) and long-term markets, because the conventions used and their practices differ (see Chaps. 2–4).

It is on the m*oney market* (see Chap. 3) that short-term interest rate products are traded. Issuers are governments, banks and financial establishments and companies, both domestic and otherwise. The investors are banks and financial establishments, insurance companies, UCITS (Undertakings for Collective Investments in Transferable Securities, investment trusts, and mutual funds), companies, and individuals. The most important portion of the money market is the *interbank market, where*

## 1.4 Organization, a Typology of Markets, and Listing

*trading restricted to banks and credit institutions* is carried out. It is on this market that the latter manage their needs and cash surpluses and that the Central Bank intervenes to exercise its monetary policy. It should be remarked that on the money market, the instruments are mostly at *fixed rates* (interest) but that there do also exist products with *floating rates* (forward-looking and backward-looking) for which neither the issuer nor the investor knows in advance the interest that will in fact have to be paid, since this depends on future interest rates in the economy.

The interest-rate products with long terms are traded on the *bond market* (see Chap. 4): bonds and long-term notes issued by the State, corporate bonds, bonds convertible into shares, and all the types of bonds with more or less complicated option clauses. The goal of the issuers is to find stable debt resources to finance long-term investments (in practice from 7 to 20 years).

### (b) *Stock markets*

Shares of stocks give the right to some securitized property listed on an official market (e.g., NYSE, Euronext, or the London Stock Exchange), representing the capital of the issuing firm, and entitling one to future dividends and to vote in the annual general meetings of shareholders (ordinary and special). The NYSE is the largest exchange in the world.

### 1.4.2.2 Derivative Product Markets: Risk Allocation

Derivative products are so-called because their value depends in an essential way on the value of the underlying asset (also called simply the underlying), which is often a cash asset: a share, a stock exchange index, a bond, a merchandise, an interest, or an exchange rate. The two other essential characteristics of these products are the following. On the one hand, their initial value is low compared to that of the underlying, even *nil* (no immediate payment takes place between the buyer and the seller). On the other hand, these are zero-sum games: The final gain to the buyer is a loss to the seller, and the reverse is true as well.[15] We distinguish four principal products.

The two first, forward contacts and swaps, are *unconditional instruments* (like cash instruments) in the sense that the buyer and the seller are both definitely obligated by their transaction (they could obviously unwind their position by going on to a later sale or purchase, respectively, on the secondary market). In contrast, the two others, options and hybrid products or products structured with option clause(s), are *conditional instruments* (in whole or in part), in the sense that the *buyer* (and only the buyer) can decide, if it is worthwhile, not to exercise the option offered by the *seller* to buy or sell the underlying. It is this possibility that makes options fundamentally different from the other products.[16]

---

[15] More generally, a zero-sum game between several players is such that each dollar gained by one of the players is a loss for the *set* of all the others.

[16] To such an extent that another classification than that adopted here consists in distinguishing unconditional instruments from conditional ones, then in subdividing each of these two large groups

## (a) *Forwards and Futures*

A forward (contract) is an agreement between two parties to exchange at a predetermined later date (the maturity) a specified quantity of a specific underlying (asset) at a price (called the *forward price*) also determined at the start. At the moment the contract is concluded no exchange has taken place. At maturity, the buyer pays the price agreed in advance to the seller, and the latter as his counterparty obligation hands over the underlying. In fact, there are two types of contract, called respectively *forwards and futures*. The technical distinction between these two types will be made precise in Chap. 9. It is here enough to understand that the buyer at maturity gains if the value of the underlying goes up, and the seller gains at maturity if it goes down. Let us suppose that indeed on January 2nd the two counterparties exchange at maturity (March 31) 1000 ounces of gold at a price of $1400 the ounce. If on March 31 gold is priced at $1530 (or $1310), the buyer will pay the seller $1400 the ounce and immediately resell it at the market price of $1530 (or $1310), for a total gain (or loss) of $130,000 (or $90,000).

## (b) *Swaps*

A swap is the trade of two cash flow series between two parties, denominated in the same currency (the typical example is an interest rate swap; for a detailed analysis, see Chap. 7), or denominated in different currencies (a currency swap).

The contract specifies the following main elements: the total duration of the swap (most often from 7 days to 10 years), the notional amount (usually not exchanged) on which depends the computation of the flows exchanged, the periodicity of flows received and the (possibly different) periodicity of flows paid, and the (*reset*) dates on which the prices and interest rates are reported which are the concern of the trade. For example, two parties may trade, at the end of each trimester over 2 years (8 flows), on a notional amount of 10 million £ sterling, a fixed interest rate (annualized) of 3.6% against the variable IOS rate (annualized) recorded in the market on the last day of a "trimester" month (March, June, ...). In such a swap, the party who pays the other fixed flows (90,000 £ at the close of each trimester) wins if the future IOS rates rise and loses if they fall. It is the reverse for the party receiving the fixed cash flows.

## (c) *Options*

Numerous types of options are traded on the markets (several chapters will be devoted to them), but we will content ourselves here with defining the simplest standard options, which is enough to understand the basic mechanism at work. From the point of view of the buyer (who initially paid the seller the value of the option,

---

(e.g., in the first case, to distinguish cash assets from forward assets (futures), and in the second standard options from exotic options, etc.).

## 1.4 Organization, a Typology of Markets, and Listing

often called the *premium*), an option (called a "European option") is the right—but not the obligation—to buy (in the case of a buy option or *call*) or to sell (in the case of a sell option or *put*) a definite quantity of a specified underlying[17] at a price fixed in advance (called the exercise price, or *strike* price) on a definite date (*maturity or expiry*). For example, the purchase of a call written on asset XXX with strike $100 and maturity of April 30 allows obtaining that security on that date by paying the seller $100, and that regardless of the then-current cash price of XXX. It is clear that the buyer will not exercise his option unless XXX is priced at more than $100 on April 30, and otherwise, he will give it up and do nothing. If the security is priced at $107, he will obtain for $100 a security that he will immediately resell on the market for $107, for a gain of $7. The seller loses $7. However, since the buyer did pay the seller an initial premium, say $p_0$, his *net gain* will only be $(7 - p_0)$, which could in fact be *negative* (of course the seller has to profit some of the time, otherwise there would be no sellers, thus no options). The purchase of a call is thus a bet on an increase in the price of the underlying, just as the purchase of a put (the right to sell at a fixed strike price) is a speculation on the price falling.

### (d) *Hybrids with optional clause(s)*

Here again, given the fertile imagination of financiers and the search for specialized niche mechanisms for saving, a full list of hybrid instruments containing one or more optional clauses and of products termed "structured" is impossible to make. We shall give here only one example to illustrate the concept of a hybrid security quote on the stock market, the convertible bond. This is a classical bond (paying a series of coupons and reimbursement of the principal), however, with an additional clause allowing the holder, often under certain conditions defined in advance, to convert it into a share: if such is the case the debt is transformed into equity capital (whence the notion of "hybrid"). For example, say a bond is convertible into 11 shares and will reach maturity in several days and be reimbursed, if it is not converted with $880. If the shares of the issuing company are quoted at less than $80, then the investor will give up his option and be paid $880. But if, on the other hand, a share is worth $91 he will convert to obtain shares worth $1001 (and so realizes a conversion profit of $121). As far as structured products are concerned, the simplest example is that of a product with guaranteed capital which, other than a series of (modest) coupons, pays the subscriber on maturity a capital amount which while indexed partially by one of the indexes (e.g., the S&P500) is at least equal to the capital that he put up in the first place. Such a product can be analyzed as a classical bond (with a coupon that is weaker than that which the market would require) plus a call written against an underlying such as the S&P500.

---

[17] Note that this underlying could be anything at all, for example, itself a derivative product such as a swap, a future or even an option.

## 1.4.3 Market Organization

Financial markets, also designated as markets in financial instruments, are systems whereby natural persons (individuals) or legal persons (corporations, banks, States) trade financial instruments. To put it differently with an additional layer of complexity, financial markets are systems that bring together or facilitate the bringing together of multiple third parties buying and selling financial instruments. We use indistinctly market participants to designate parties, persons, or counterparties engaging in the trading of financial instruments.

Financial markets are variously organized to meet the various expectations and needs of participants related to the complexity, liquidity, or price transparency of financial instruments. As those expectations and needs cannot be met all at once, different markets organizations coexist.

We examine below the most significant distinctions between financial market organizations and how they address the above expectations.

### 1.4.3.1 Over-the-Counter/Exchange Traded Markets

*Over the counter* (OTC) trading designates any system where two parties engage in the trading of financial instruments pursuant to the provisions laid down in a private bilateral contract reflecting the right and duties of the parties among which the quantity, agreed price, type of instrument, maturity. Pre-trade negotiation on price, quantity, and other contractual provisions is conducted bilaterally without the intervention of any third party. OTC trading, therefore, provides maximum flexibility regarding the characteristics of the financial instrument to be traded, as well as pre-trade discussion. In that regard, we may speak of a "bespoke" trade, since it exactly meets the needs of the counterparties.

However, as regards price transparency, OTC trading may allow one party to benefit from an asymmetry of information over its counterpart, especially when it comes to trading complex products whose price cannot be compared from an alternative and reliable source.[18] Although we address OTC trading in the section dedicated to financial markets organization, speaking of "OTC markets" is an abuse of language as OTC trades are purely bilateral and "decentralized" as they do not require the use of any intermediary or third-party. "OTC markets" is used by practitioners to emphasize the contrast between purely bilateral trades with centralized systems that we examine below.

Market organization has undergone significant changes over the recent years. Centralized systems bringing together multiple parties engaged in trading financial instruments are now designated as *trading venues*.[19] These may be divided into *regulated markets* and *multilateral trading facilities ("MTF") in Europe or*

---

[18] From an asymmetry information perspective, trading complex instruments is exposed to the same pitfalls as those described by Akerloff (1970).

[19] In the European Union trading venues operate under the Markets In Financial Instruments Regulation and Directive (MIFIR/MIFID) regulatory environment.

## 1.4 Organization, a Typology of Markets, and Listing

*Alternative Trading Systems ("ATF") in the US. Regulated markets* are subject to the most stringent rules imposed by the regulator to onboard market participants allowed to trade. Regulated markets are also subject to a continuous scrutiny of market supervisors (e.g., CFTC in the US) as regards surveillance of unlawful trading behaviors denoted as "market abuse."[20] Market abuse encompasses market manipulation and insider dealing and any violation of those rules has led over the recent years to significant sanctions pronounced by the supervisors.[21] Among practitioners, *regulated markets* designate the most famous "exchanges" (such as ICE, LIFFE, Euronext) where various instruments are traded (stocks, bonds, and derivatives such as Futures and options). Financial instruments traded on "exchanges" are commonly denoted as "listed instruments or listed products."

Regulators were willing to increase price and volume transparency as well as liquidity and limit counterparty risk on OTC trading and hence recast the regulatory applicable framework to foster the entry of MTF/ATF. These newcomers disrupt traditional OTC trading by proposing a centralized system, usually electronic, bringing more price transparency, increasing volumes, and a higher level of regulation whereby they encourage market participants to switch from OTC trading to join them. MTF/ATF offer to trade on a wide range of vanilla financial derivatives (e.g., interest rate swaps or interest rate options) similar to those traded OTC. It must be noted that MTF/ATF are private companies who compete to propose, on a similar set of instruments, the largest volumes, the best price transparency, and the lowest trading costs. By contrast, *exchanges* do not directly compete with MTF/ATF as they do not offer trading facilities on those vanilla derivatives.

Most provisions recited below apply similarly to both regulated markets and MTF/ATF except Sect. 1.4.3.4 even though some MTF are currently implementing proprietary order books designated as "price streaming". For the ease of exposition, we stick to the use of "exchange" or "regulated market" and examine below the exchanges that boast the highest degree of price transparency.

*Trading venues* and especially *exchanges* offer trading financial instruments with standardized characteristics (with regards to their unitary nominal size, maturity, asset class). Trading is possible only during market opening hours. Market participants buy and sell financial instruments on the exchange and the equilibrium price is obtained through a mechanism described in Sect. 1.4.3.4. To use a sartorial analogy, *trading venues* propose *ready-to-wear* products, and market participants admitted to trade[22] on those venues accept the standardization to benefit from a larger liquidity (i.e., ability to buy and sell large volumes on the market with little impact on quoted prices and minimal trading costs).

---

[20] In Europe, Market Abuse Regulation ("MAR" applicable since 2016) sets out the enforceable rules as regards market manipulation. MAR applies to all participants and to all financial instruments including trades carried out in OTC. The latter are obviously more difficult to monitor by the supervisors due to their decentralized nature.

[21] Penalty fines over 1 billion dollars and imprisonment.

[22] It is on *exchanges* that admittance rules are the most stringent.

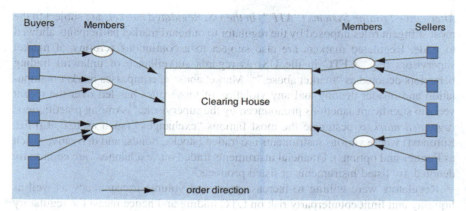

**Fig. 1.6** Clearing house mechanism attached to an exchange

Trading venues do not only offer price transparency related to pre-trade execution (as set out in Sect. 1.4.3.4) to the expense of a greater standardization but also do not compromise on post-trade conditions. Both parties are *committed* once the trade is concluded. The uncertainty attached to the failure to deliver payment(s) or the instrument from one party is called *counterparty risk*. It is mitigated by using *central clearing counterparties* (CCP) also known as *clearing houses*.[23]

Historically, *the use of a clearing house* was mandatory for all market participants admitted to trade on an *exchange*. In the aftermath of the 2007–2009 crisis, regulatory requirements became more stringent and *mandatory clearing* has been gradually extended to a various range of financial instruments traded on MTF/ATF and also to OTC trades where a certain degree of standardization can be achieved.[24] Therefore, CCP have built capacities to be in a position to clear an increasing number of trades and are also more closely scrutinized by the supervisors as they bear a systemic risk. We examine below the clearing mechanisms for CCP in the context of an exchange. Those mechanisms are also applicable to a large extent for any CCP.

The *clearing house* on an exchange is solely accessible to designated and accredited market participants called clearing members or clearing brokers (both being "members"). For each financial instrument listed on the exchange, the net position held by the clearing house is zero at each instant, which means that it sells exactly as many instruments as it purchases. To achieve this result, the price is fixed at its equilibrium level.

In any transaction carried out on an exchange, any person contemplating to buy or sell a financial instrument uses the services offered by a member, according to Fig. 1.6. The exchange's clearing house holds the accounts of its members, and

---

[23] Nowadays, CCP is the most popular denomination among practitioners.

[24] The European Market Infrastructure Regulation ("EMIR") applicable in the EU since 2016) requires mandatory clearing on a wide range of vanilla derivatives and repos and regular trade reporting to the supervisory bodies.

## 1.4 Organization, a Typology of Markets, and Listing

**Table 1.2** Comparison between market organizations

| | Price transparency | Liquidity | Degree of product standardization | Counterparty Risk |
|---|---|---|---|---|
| Regulated markets (*exchanges*) | High | High[a] | High | Low |
| MTF/ATF | High | Medium/high | Medium/high | Low |
| OTC trading | Low for the most complex instruments– Medium/High for similar instruments traded on MTF | Low for the most complex instruments– Medium/High for similar instruments traded on MTF | Low for complex instruments. Medium/High for vanilla instruments | Low (for instruments subject to mandatory clearing). High for non-cleared instruments |

[a]In Europe, the liquidity of bonds remains extremely low on exchanges and most trading is performed through MTF or OTC

the latter oversee the management of their clients' accounts. The overall framework mitigates to a very large extent the counterparty risk since the clearing house is the legal counterparty of all trades, its net position is nil on all instruments and it is subject to regulations (and an appropriate level of equity) so as to ensure that its risk of default is close to zero even under a stressed situation. Besides, the standardization of the trades as well as the legal framework ensure that all trades borne by any ailing member are reallocated in a timely manner to the other members (hence the phrasing "portability") to ensure seamless continuation.

Recall, as a conclusion to this section, that in response to the "subprime" crisis, and under the pressure of regulators, pre-crisis OTC trading on vanilla derivatives instruments has been increasingly streamlined, which paved the way to MTF/ATF offering similar price transparence, volumes, counterparty risk and supervisory surveillance as the exchanges. OTC trading on more complex instruments is also required to use a CCP (or a bilateral "CCP-like mechanism") to mitigate counterparty risk although full price transparency is not achieved. These evolutions blur the traditional difference between OTC trades and exchanges.

Table 1.2 approximately sketches the stages attained by the various organizations as regards the four main properties of a trade.

### 1.4.3.2 Intermediation

The markets differ in the nature and role of the intermediaries who are entitled to quote bids and ask prices. We distinguish two types of intermediaries with very different roles: brokers and dealers.

**Brokers** (who are not dealers) simply bring together buyers and sellers. This is, for example, the case for a significant part of the activity carried out by some investment firms holding a license to solely operate as a broker. They do not deal on their own account and only receive and transmit orders without being a legal counterparty.[25] Therefore, they are never in the position of a buyer or a seller. They do not participate directly in the liquidity of the market, but they do contribute to it, sometimes significantly, by their efforts to bring together potential participants willing to engage in buying or selling financial instruments. Their compensation relies exclusively on a brokerage fee (fixed and/or variable) paid by each of the parties. Brokerage reduces both transaction and information costs for the parties by limiting the number of connections required[26]: If there are $n$ potential market participants willing to engage bilaterally (pair-wise) who need to know the volumes and prices offered to buy or sell, the number of connections would amount to $(n(n-1)/2)$, a number that increases much more rapidly than $n$ (3 for $n = 2$, for example, but 4950 for $n = 100$). If there a broker aggregates all the bids and asks, then only one connection is needed by each market participant (including the broker), i.e. only $n$ in total.

**Dealers**, by contrast, trade for their own account: they legally act as a counterparty and buy from the seller to resell to a buyer, or sell to a buyer and then buy back from another participant. Their trades usually end up with a net buyer or seller remaining position (on a short-term horizon, say a day) which is accounted accordingly as an asset or liability on their balance sheet. Dealers thereby directly influence the liquidity of the market, but in general to a marginal extent. Their compensation derives from the fact that, on average, the prices at which they buy are less than those at which they sell. Note that usually dealers are *also* entitled to act as brokers.

In certain markets, notably for futures and options on exchanges, dealers play the role of *market makers* (also called *specialists* or sometimes *liquidity providers*). Market makers contribute in an essential way to market liquidity. Indeed, they must execute any purchase or sale of any participant, at the prices and volumes they have posted publicly. Thus, they provide liquidity to all other market participants.

Concretely, market makers agree to act, permanently, as a counterparty (for a purchase at a bid price $P_B(t)$ (bid) and for a sale at a higher asking or offer price of $P_A(t)$) for a certain minimum volume fixed by the rules set out in public documents ("exchange rule book"). They are therefore potentially going "against" the market, which is a risky position to be in: If the market is globally buying the instrument, anticipating (for more or less informed reasons) an increase in its price, then the specialists are net sellers and thus exposed to the price increase actually happening. The difference $[P_A(t)- P_B(t)]$ is the *bid-ask spread,* which compensates market makers for their activity and the risk they bear.[27] The *bid–ask spread* is posted for

---

[25] In the economic literature brokers are said to act as "agents" whilst dealers act as "principals."

[26] Therefore, if the value generated by brokers regarding information and transaction costs is not entirely compensated by fees, brokers may even bring positive externalities to the market.

[27] The literature on market microstructure, which is out of this book's scope, distinguishes three justifications for this bid-ask markup: *(i)* the cost of *inventory* connected with the market maker's holdings of undesirable positions (inventory) resulting from the possible delays in the execution of different operations (notably arbitrage or hedging) which it would be optimal to realize simultaneously; *(ii)* the cost connected with the asymmetry of information due to the presence of clients who are better informed than the market maker (anti-selection), about which the latter may have a

## 1.4 Organization, a Typology of Markets, and Listing

a time $t$ and varies according to the fluctuation of bids and asks from participants. It constitutes a measure of the liquidity of the market, narrowing as liquidity increases.

The distinction between pure brokers and dealers also occurs between *agency* markets and *counterparty* markets:

- **Agency markets involve "intermediaries" who are purely brokers** (like real estate brokers) and not counterparties: the market is said to be *order driven* because no price is posted *ex ante*, it depends on the bids and asks of the final participants (excluding the intermediaries) and varies with fluctuations of demand and supply. This is the case, essentially, for the large exchanges such as Euronext, the New York Stock Exchange and its 500 brokers, and the London stock market.
- **On counterparty markets**, dealers, notably those who are market makers, play an essential role by quoting bid-ask spreads (whether purely indicative or binding) on an *ex ante* basis: The market is said to be *price driven* (or quote driven) by the specialists' quotes. These market makers are somehow guessing where the bids and asks from the participants may end up. We can mention here the NASDAQ, the American technology market, and the SEAQ, its equivalent London peer.

Each type of organization has its merits and drawbacks and the choice to trade on one of these markets depends in part on the "natural" liquidity of the instruments traded: dealer markets often have the advantage of liquidity and flexibility since market makers comply with strict rules (to ensure orderly market trading) set by the market supervisory bodies. Besides, the depth of liquidity allows negotiating and trading significant blocks of securities or a large volume of derivatives. In broker markets, information is often of a better quality as the brokers may have a comprehensive view regarding the total amount available for bid and ask, while the bid-ask spreads quoted by market makers are often only indicative and hence non-binding.

In practice, there are also *mixed* markets in which market makers and brokers co-exist, none of them having the upper hand but both being useful as liquidity providers. This is, for example, the case in some stock markets in which specialists are responsible for ensuring a minimum degree of liquidity for small and medium capitalizations.

### 1.4.3.3 Centralized and Decentralized Markets

Another distinction is made between *centralized* and *fragmented* markets. In the first, orders are centralized in a single "location" (physical or electronic; see Sect. 1.4.3.4) and financial instruments are only admitted for trading on this single location and thus have each a unique price at a given instant. Any trading outside the market is prohibited, all orders being transmitted to brokers or specialists who carry out their centralization. In the second case, transactions can be carried out in

---

suspicion but be unable to identify among those issuing orders; *(iii)* the cost called order handling, connected with a negative self-correlation of the prices induced by the very existence of the bid–ask price range.

30    1 Introduction: Economics and Organization of Financial Markets

different "locations," with the risk (or the opportunity) that the prices of an instrument, at a given instant, may differ and so subject to arbitrage (simultaneously buying low here and selling higher there).

### 1.4.3.4 Quotation on an Exchange; Order Book, Fixing and Clearinghouse

As far as transaction organization and quotation are concerned, three systems exist in exchanges: fixing, continuous quotation, and exchange of blocks.

(a) *Fixing order books: general principles*

Buy and sell orders are entered and recorded in an *order book*.

Periodically (once or twice a day) the equilibrium price is determined from the order book according to the following rules:

- The sellers making tenders are dealt with in increasing order of price, and the buyers in decreasing order.
- No participant is required to trade more than he wishes, to buy at more than a (possible) maximum price he has fixed or to sell at less than a required minimum.
- The equilibrium price is the one that maximizes the quantity traded.

It is, as we have seen, the *clearinghouse* of an Exchange that records the transactions (quantity and price) and recalculates each evening the net positions of each participant. In addition, it is the market authorities that fix the tick of the quotes, i.e., the minimum price variation allowed. The "tick" depends on the level of a security's price. For example, it could be $0.01 for market prices less than $50, $0.05 for prices between $50.05 and $100, etc. It should be remarked that these rules are as near to the Walrasian paradigm as possible.

---

**Example of Order Book and Fixing**

A wishes to buy 200 shares of X at a price less than a maximum $41, B to buy 300 shares at less than a maximum of $40, C to buy 150 shares at $39... A′ wants to sell 100 shares at a minimum price of $35, B′ to sell 200 shares at $36... The asks are ordered by decreasing price (A, B, C, ...), the bids by increasing price (A′, B′, ...) so that the order book will look like Table 1.3.

The price of shares of X is fixed at $39 because that is the price at which the volume of trading is maximized,[28] and all transactions are carried out at that price. 550 shares are traded. On the buyers' side, A and B are completely served, as are A′, B′, and C′ on the sellers' side; buyer C is only partially served (he gets 50 shares of the 150 desired). D, E, E′, and F′ are not served.

---

[28] For example, at a price of $40, only 500 shares are wanted while 570 are on offer. The transaction could thus only be carried out on 500 shares; seller D' would only be partially served although he is

## 1.4 Organization, a Typology of Markets, and Listing 31

**Table 1.3** Order book and fixing

| Ask | | | | Bid | | | |
|---|---|---|---|---|---|---|---|
| Quantity | Cumulation | Price | Buyer | Quantity | Cumulation | Price | Seller |
| 200 | 200 | 41 | A | 100 | 100 | 35 | A' |
| 300 | 500 | 40 | B | 200 | 300 | 36 | B' |
| 150 | 650 | 39 | C | 50 | 350 | 37 | C' |
| 50 | 700 | 38 | D | 200 | 550 | 39 | D' |
| 10 | 710 | 37 | E | 20 | 570 | 40 | E' |
| | | | | 100 | 670 | 41 | F' |

### (b) *Continuous quotation*

Continuous quotation, which is relevant only for the most liquid of assets, is made up of three different phases: Opening, properly continuous trading, and closing:

- In the opening phase (e.g., 8:30–9:00), the order book is made up, no transaction is carried out and the orders pile up. A fixing closes the opening phase.
- During continuous quotation (e.g., 9:00–17:00) the order book is kept up to date and the orders carried out continuously, as a function of the orders remaining, insofar as new orders are entered in the order book (as soon as a new order allows a transaction).
- In the few minutes of the closing phase (e.g., 17:30) an order book is made up, transactions stop and orders accumulate; a fixing closes the day at 17:35.

It should be noted that the recent evolution of financial markets is towards *electronic quotation* (remote computer networks connected to the central market authorities) to the detriment of *open outcry*. In this system, employees of the establishments allowed to trade all meet physically at the same time and at the same location and "cry" the bids and asks placed in their order books. Even though some option and futures markets have initially continued this old tradition, this form of price quotation essentially disappeared for lack of efficiency. Remark in this connection the remarkable takeoff of markets based on ECN (Electronic Communication Networks) like Island and Instinet which have no market authorities, with the advantages (reduced access costs and transaction costs) and the risks (manipulation, fraud, non-execution of orders) that such a state of affairs brings with it.

### (c) *Block trading*

Here trades are sizable, carried out *outside* an order book, with the price a matter of mutual agreement by the parties, and usually completed by a broker or specialist.

---

willing to sell at $39. At a price of $38, only 350 shares would be on offer and available to trade against a demand of 700 shares.

### 1.4.3.5 Primary Markets, Secondary Markets and Over-the-Counter (OTC)

We have already met (in Sect. 1.2) the distinction between the primary market (issuing of new securities) and the secondary market (trading of existing securities), the first allowing the financing of enterprises, financial institutions and local authorities and the second ensuring the liquidity of securities. In some cases, there may exist a *grey market*, reserved for institutional investors (banks and financial establishments, insurance and reinsurance companies, savings banks, retirement, and pension funds, etc.). This is a market that *precedes* the primary market and on which a security may be traded for several days before its official listing. How a security performs on the grey market is a good measure of the market's opinion on the issuer and the demand likely to spring from investors for the security about to be issued.

## 1.5  Summary

- Financial markets in which are traded securities, or financial assets, allow for inter-temporal re-allocation of cash resources between borrowers and savers (either through standard borrowing-lending or through issues and purchases of securities) and the re-allocation of risks (through derivatives). They diffuse, more or less efficiently, the information embedded in prices, which is welfare improving for all participants.
- A financial asset is a claim to future cash flow(s) which can be certain or random. These assets are issued in the primary market, the secondary market ensuring their liquidity.
- The stream of cash flows generated by a security are represented either by a series $F_1, F_2, \ldots, F_N$ in discrete time or by a density $F(t)dt$ in continuous time. When the cash flows are random, information is given by their probability distribution.
- The desirable properties of a financial market are equilibrium, absence of arbitrage opportunities, information efficiency and liquidity.
- At equilibrium between supply and demand, *every participant finds a counterparty* willing to trade at the equilibrium market price. Under competitive equilibrium, no agent has sufficient weight to influence prices, so that all agents are "*price takers*." Under a Walrasian equilibrium, all transactions are made at the equilibrium price set through virtual and instantaneous iterations.
- An arbitrage is a trade that generates only positive or zero cash flows (at least one positive). An arbitrage opportunity *may* appear in presence of redundant assets. An asset is redundant when it can be replicated by a combination of other assets which yields identical cash flows *with certainty at any date*. In absence of arbitrage opportunities (AAO), the price of the replicated asset is equal to the cost of the replicating portfolio.
- A market is said to be efficient when, at any date, its prices embeds all the available relevant information (which, therefore, cannot be used to obtain "abnormal" profits). Depending on how large the information embedded in prices is, the

Appendix: The World's Principal Financial Markets

weak, semi-strong, or strong forms of efficiency obtain. This topic has elicited many controversial studies both at the theoretical and empirical levels.

- The liquidity of an asset is linked to the ease with which agents can open or close positions on the asset at a market price and with low transaction costs. In particular, placing a sizeable quantity of a liquid asset should result in no significant drop in its value.
- Lending and borrowing are realized through two distinct channels: the *banks,* acting as intermediaries *on their own account,* and the *financial markets* connecting investors and issuers of securities with brokers or dealers as only intermediaries.
- Fixed-income securities and stocks are cash instruments generally issued by firms or institutions to finance their investments.
- Fixed-income securities yield interests and reimbursement of capital. Short-term ones are traded in *money markets* (where central banks are major operators) and long-term debts are traded in *bond markets.*
- *Derivative assets,* such as options, swaps, forwards, and futures, are used to re-allocate risks. As their name indicates, their value crucially depends on the value of an underlying asset.
- Financial assets can be traded *over-the-counter* (OTC) or on trading venues (*exchanges or MTF/ATF*) equipped with a clearing house that settles all trades. OTC markets are increasingly organized and regulated. Brokers and dealers are intermediaries who compare bids and offers. *Brokers* just make buyers and sellers meet. *Dealers* buy from sellers and sell to buyers, their operations being (in the short term) asynchronous. Some dealers play the role of *market makers* by agreeing to act as counterparties.
- There are three types of trade organization and quotation in Exchanges: *fixing,* *continuous* quotation, and exchange of *blocks.*

## Appendix: The World's Principal Financial Markets

Below we list the principal world markets (with their internet addresses), distinguishing ordinary shares from derivative products.

### Stock markets, market indexes and interest rate instruments

*Euronext* (Paris, Brussels, Amsterdam, and Lisbon) [www.euronext.com]
*London Stock Exchange* [www.londonstockexchange.com]
*Frankfurt* [www.deutsche-boerse.com]
*Suisse* [www.swx.com]
*Milan* [www.borse.it]
*Madrid* [www.bolsamadrid.es]
*SEAQ* [www.lseg.com/areas-expertise/our-markets/london-stock-exchange/equities-markets/trading-services/domestic-trading-services/setsqx]

*New York Stock Exchange* [www.nyse.com]
*American Stock Exchange* [www.amex.com]
*NASDAQ* [www.nasdaq.com]
*Toronto Stock Exchange* [www.tse.com]
*Sao Paolo Stock Exchange* [www.bovespa.com]
*Mexico* [www.bmv.com]
*Australian Stock Exchange* [www.asx.com.au]
*Singapore Stock Exchange* [www.sgx.com]
*Hong Kong Stock Exchange* [www.sehk.com.hk]
*Shanghai* [www.sse.com.cn]
*Shenzhen* [www.szse.cn]
*Tokyo Stock Exchange* [www.tse.org.jp]

## Organized Derivative Markets (Futures and Options, Unless Otherwise Indicated)

*Chicago Board of Trade* (CBOT): Futures on long and medium-term U.S. Treasury Bonds and agricultural products [www.cbot.com]. Part of the *CME Group*.

*Chicago Board of Option Exchange* (CBOE): Options on interest rates, stocks, and stock indices [www.cboe.com]. Part of the *CME Group*.

*Chicago Mercantile Exchange* (CME): S&P 500 stock index, dollar interest rates, currencies, and livestock [www.cme.com]. Part of the *CME Group*.

*New York Mercantile Exchange* (NYMEX): Metals, petroleum, and natural gas [www.nymex.com].

*EUREX* (Frankfurt, Zurich, Geneva): European state bonds (e.g., Euro-Bobl and Euro-Bund), EONIA, European stocks, and European stock indices (for example, Dax, Dow Jones Euro-Stoxx and Dow Jones Stoxx) [www.eurexchange.com].

*LIFFE* (London): European stock indices (e.g., the FTSE 100),

*PHLX* (Philadelphia): currencies [www.phlx.com].

*SIMEX* (Singapore): Asiatic interest rates (long-, medium-, and short-term) and stocks [www.simex.com.sg].

*TIFFE* (Tokyo): Currency and interest rates [www.tiffe.org.jp].

*BM&F* (Brazil): Gold, stock indices, interest rates, and currencies [www.bmf.com.br].

*SFE* (Sydney): Stocks, stock indices, interest rates, and commodities, [www.sfe.com.au].

## Suggestion for Further Reading

### *Books*

Bodie, Z., Kane, A. & Marcus, A. (2014). *Investments* (10th ed.). Irwin.
Brealey, R., Myers, S., & Allen, F. (2019) *Principles of corporate finance* (13th ed.). McGraw Hill.
Campbell, J., Lo, A., & Craig MacKinlay, A. (1997). *The econometrics of financial markets*. Princeton University Press (first chapters).

### *Articles*

Akerloff, G. (1970). The market for "lemons": Quality uncertainty and the market mechanism. *Quarterly Journal of Economics, 84*(3), 488–500 (the classic article on the quality risk borne by the buyer).
Fama, E. (1970). Efficient capital markets: A review of theory and empirical work. *Journal of Finance, 25,* 383–317 (the classic article on market efficiency).
Fama, E. (1991). Efficient capital markets II. *Journal of Finance,* 1574–1617 (a revision and extension of the preceding item).
Grossman, S., & Stiglitz, J. (1980). On the impossibility of informationally efficient markets. *American Economic Review, 70,* 393–408.
Malkiel, B. G. (1995). Returns from investing in equity mutual funds 1971 to 1991. *Journal of Finance, 50,* 549–572.

# Part I

# Basic Financial Instruments

This first part is made up of Chaps. 2–8 and deals with primitive financial assets (stocks, bonds, and money market instruments) and vanilla swaps that are underlying the derivative products studied in the following parts. These "basic instruments" are analyzed in detail but without resorting to complicated mathematical tools. This part is accessible to undergraduate students and is aimed at students and professionals new to market finance. The more advanced will just explore a few concepts in more depth and quickly move on to the following parts:

Chapter 2, in a very classic way, is devoted to the mathematics of interest rates.

Chapters 3 and 4 present short- and long-term fixed income securities (negotiable debt securities, bonds) and their valuation in the money and bond markets.

Chapter 5 provides a detailed introduction to interest rate and credit risks, the more advanced analysis of which will be covered in depth in Parts II and mainly III.

Chapter 6 studies the term structure of interest rates and generalizes the interest rate risk measurement and management tools introduced previously.

Chapter 7 describes floating rate products (forward- and backward-looking floaters) then vanilla swaps and analyzes their valuation by arbitrage.

Chapter 8 is devoted to stocks and stock indices.

# Basic Finance: Interest Rates, Discounting, Investments, Loans

## 2

This chapter presents the elementary notions and methods of basic finance. These are interest rates, as well as the assessment of present values that allow estimating the value today of cash flows expected in the future and to define and calculate the returns on investments and the cost of financing. These basic and elementary mathematical tools will be used throughout this work. In this chapter, they are applied to the study of investments and the transactions of lending and borrowing. In addition, certain fundamental concepts of finance such as financial leverage or the distinction between real and nominal cash flows and interest rates are also discussed.

A first introductory section explains how we represent lending and borrowing and, more generally, investments and financing through cash flow sequences. The second section is about transactions with two cash flows that allow a simple presentation of interest rates and the different conventions they give rise to. Transactions with several cash flows, methods of estimating present values, and their applications to the analysis of investments are treated in Sect. 2.3. Section 2.4 is devoted to the study of long-term credits.

## 2.1 Cash Flow Sequences

From a mathematical point of view, an investment or financing transaction is defined by the sequence of cash flows that it generates. A cash flow is a sum of money transferred from one agent to another. A cash flow received (cash inflow) is represented by a positive number while a cash outflow (disbursement) is quantified with a negative number. Any positive cash flow for the borrower is negative for the lender, and conversely. Two symmetrical points of view can be adopted: that of the lender or investor (who loans or invests a certain sum of money) or that of the borrower (who signs or contracts the loan and receives the money).

The loan represents an *investment* (or an asset placement) for the lender; indeed it involves an initial transfer of funds (cash outflow in the amount of the loan) as the counterpart to expected revenues (cash inflows) to follow in the form of successive

---

© The Author(s), under exclusive license to Springer Nature Switzerland AG 2022

P. Poncet, R. Portait, *Capital Market Finance*, Springer Texts in Business and Economics, https://doi.org/10.1007/978-3-030-84600-8_2

39

reimbursements of this capital plus interest payments. For the borrower, the transaction is a *financing*. The interest rate corresponds to the return on the investment for the lender and the cost of financing for the borrower.

Before trying to represent cash flow sequences generated by financial transactions, it is convenient to make precise the unit in which time is counted. We call reference period the period over which the interest for the transaction is defined. Thus, if the interest is annual, the reference period is a year; if the interest is monthly, the reference period can be a month, etc. In general, the reference period chosen is either a year or the period separating two coupons. The term of the loan is thus expressed as a number of reference periods and is not necessarily an integer.

The evolution over time of a sequence of cash flows can be represented either by a *cash flow diagram* or by a *vector* (or *sequence* or *series* of cash flows).

A cash flow diagram involves a horizontal axis representing time, on which each cash flow is placed at the date of its occurrence; the cash inflows are represented by arrows pointing up, the cash outflows by arrows pointing down.

### Example 1
The diagram in Fig. 2.1 represents the borrowing of 100 contracted at $t = 0$, which gives rise to reimbursements (of capital and interest) of 0, 20, 30, and 115 at the ends of the first, second, third, and fourth years, respectively. The year is the reference period here. The diagram in Fig. 2.1b represents the same transaction from the lender's viewpoint.

Cash flow diagrams allow one to visualize financial transactions as a whole, and we will use them in this chapter and, occasionally, in the remainder of the book.

The lender's cash flow diagram of the loan is symmetrical to the borrower's:

More succinctly, the cash flow sequence can be written as a sequence (or series) or as a vector.

Thus the lending in Example 1 is represented by $\{-100, 0, 20, 30, 115\}$ and the corresponding borrowing by $\{+100, 0, -20, -30, -115\}$.

## 2.2 Transactions Involving Two Cash Flows

Most of the fundamental concepts concerning interest rates can be explained simply from transactions involving two cash flows. In addition, in practice, most short-term transactions (less than a year) only give rise to two cash flows. That is why we devote the whole section to the study of such transactions. Those that last one period are examined before those involving several periods which lead to different ways of computing the interests.

## 2.2 Transactions Involving Two Cash Flows

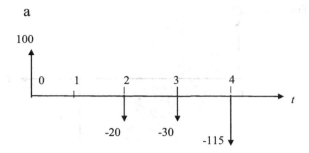

The lender's cash flow diagram of the loan is symmetrical to the borrower's:

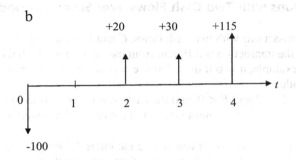

**Fig. 2.1** Cash flow diagrams. (**a**) borrower. (**b**) lender

### 2.2.1 Transactions of Lending and Borrowing Giving Rise to Two Cash Flows over One Period

First, we consider financial transactions with the timeframe of a single period running from date $t = 0$ to date $t = 1$ (the reference period is then the time interval separating the two cash flows). Such transactions thus involve two cash flows, $C$ (capital lent at 0) and $F$ (final cash flow, at $t = 1$, which includes reimbursement of the capital and interest $I$, with $I = F - C$.

For a capital $C$ borrowed at the interest rate $r$ between dates 0 and 1, the cash flow diagram is as shown in Fig. 2.2:

The borrower receives at 0 the capital amount $C$. At $t = 1$, she pays back the capital and the interest, in an amount equal to the interest rate $r$ multiplied by the capital borrowed $C$. At $t = 1$, the total cash flow $F$ is therefore equal to $-C(1 + r)$. More simply, the loan is given by the sequence: $\{+C, -C(1 + r)\}$.

The lending diagram is the reflection with respect to the time axis of the borrowing diagram; the loan gives rise to the sequence $\{-C, +C(1 + r)\}$.

**Fig. 2.2** Two cash flows

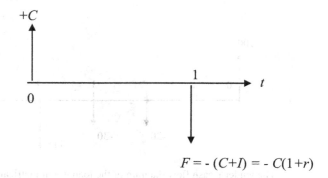

$$F = -(C+I) = -C(1+r)$$

### 2.2.2 Transactions with Two Cash Flows over Several Periods

Now we consider transactions with two cash flows, $C$ and $F$, separated by $n$ periods; here the duration of the transaction is different from the period for which the interest rate is defined. For example, $n = 6$ if the reference period is a month while the term of the loan is 6 months.

For the borrower, $C > 0$ and $F < 0$, and the cash flow diagram is as in Fig. 2.3a.

The final cash flow $F$ includes reimbursement of the capital borrowed, $C$, and the interest, $F - C$.

There are two possible methods for calculating the interest: "simple interest" and "compound interest." We study them in turn and then compare them.

(a) *Simple interest (proportional or linear)*

The transaction takes place over $n$ periods; we analyze it from the borrower's point of view (the lender's is its mirror image).

If capital $C$ is borrowed at an interest rate $r$ between dates $t = 0$ and $t = n$, with reimbursement of capital and interest at the same time on the final date $t = n$, the "simple" or "proportional" interest is proportional to the rate $r$ and the number of periods $n$:

$$I = C\,r\,n$$

The final cash flow is, therefore, $F = -(C + C\,r\,n)$ and the cash flow diagram is in Fig. 2.3b.

Often the number $n$ of periods is not an integer. Such is the case, notably, if the rate is annual and the term of the loan is less than one year. Calling $T$ the term of the loan (as a fraction of the period), the simple interest and the final cash flow are equal to:

$$I = C\,r\,T$$

## 2.2 Transactions Involving Two Cash Flows

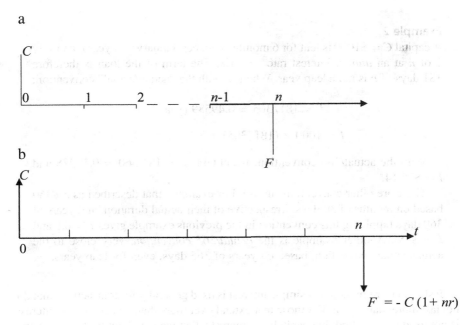

**Fig. 2.3** (a) Two cash flows, borrowing general case. (b) Two cash flows, borrowing at simple interest

$$F = -C(1 + rT). \qquad (2.1)$$

Most often formula (2.1) is applied for an annual interest rate (i.e. a reference period of 1 year) and a term $T < 1$. Several conventions (or "bases") are in use for calculating $T$. In this book we mainly adopt:

– The "*actual/actual*" convention for which:

$$T = Nd/Na$$

where $Nd$ is the duration of the transaction expressed as a number of days, and $Na$ equals the actual number of days in the year (365 or 366 for leap years). To clarify, note that $Nd$ is the number of daily *intervals* between dates 0 and $n$, or, equivalently, the number of nights between the beginning and the end of the transaction. For example, $Nd$ is 47 between April 11 ($t = 0$) and May 28 ($t = n$).

– The "*actual/360*" convention for which $T = Nd/360$; the length of the year is thus expressed on a *360-day basis*, which artificially increases the duration $T$, and so the interest due, by the ratio 365/360 (or 366/360). We will return to this point in subsection 4.

> **Example 2**
>
> A capital $C = \$1000$ is lent for 6 months, between January 1 in year $n$ and July 1 of $n$ at an *annual* interest rate $r = 5\%$. The term of the loan is therefore 181 days, if $n$ is not a leap year. Whence, with the "actual/actual" convention:
>
> $$T = 181/365 = 0.49589 \text{ (year)},$$
>
> $$I = 1000 \times (181/365) \times 5\% = \$24.79.$$
>
> With the actual/360 convention, we obtain: $T = 181/360 = 0.50278$ and $I = \$25.14$.
>
> There are other conventions in use. For example, that described as *30/360* based on months of 30 days, irrespective of their actual duration, and years of 360 days (applying this convention in the previous example gives $T = 0.5$ and $I = \$25$). Another example is the *actual/365 convention*, very close to the actual/actual convention, based on years of 365 days, even for leap years.

It should be remarked that simple interest is used generally for transactions that do not last more than 1 year. For those that extend over more than 1 year, simple interest is not used (the method that capitalizes interests, that we are about to discuss, is the rule).

### (b) *Compound (or capitalized) interest*

The computation principle is as follows: assume capital $C$ is lent at an interest rate $r$ for $n$ periods between $t = 0$ and $t = n$ denoted by 0–1, 1–2, ..., $(n - 1)$–$n$; interest is calculated at the end of each period and added to the capital (it is "capitalized") and the accrued interest produces, in turn, interest at the same rate $r$ in later periods.

If the transaction only lasts one period ($n = 1$) it is represented by the sequence: $\{-C, + C(1 + r)\}$.

In fact, when $n > 1$, no payment occurs at time $t = 1$ and the interest $Cr$ is added to the capital so that the capital due from the borrower at the start of period 1–2 is equal to $C(1 + r)$; everything is as if the investor "re-invests" at $t = 1$ the amount $C(1 + r)$ in the same investment at interest rate $r$. We emphasize that there is no transfer of cash at $t = 1$: the interest is simply *calculated* at the end of the period and *added to the capital*, thus re-invested and carrying interest.

At the end of the second period, that is on date $t = 2$, the amount owed by the borrower is $[C(1 + r)] (1 + r) = C (1 + r)^2$. The cash flow sequence for a transaction that lasts two periods (0–1 and 1–2) thus is: $\{-C, 0, +C(1 + r)^2\}$. When $n > 2$, the amount $C(1 + r)^2$ is not paid by the borrower but added to the capital owed (therefore provided by the lender during the period 2–3) and so on until $t = n - 1$. The capital owed is therefore multiplied by $(1 + r)$ at the end of each period. At time $n$, which is the date of maturity of the loan, the borrower pays back $C(1 + r)^n$. This loan thus

## 2.2 Transactions Involving Two Cash Flows

**Fig. 2.4** Two cash flows, lending at compound interest

leads to the sequence: $\{-C, 0, \ldots, 0, +C(1 + r)^n\}$, represented by the diagram in Fig. 2.4.

Note that the amount of the interests paid is:

$$I = F - C = C(1+r)^n - C, \text{ that is } C\left[(1+r)^n - 1\right].$$

In the general case, the number of periods is not an integer. Then $T$, the duration of the loan, is in general a fraction (of a year); if $Nd$ is the term of the transaction expressed in days, $T$ is calculated for example as the ratio $Nd/365$. For a capital $C$ lent at $t = 0$, the lender receives at $T$:

$$F = C(1+r)^T. \tag{2.2}$$

Therefore, the interests paid by the borrower are:

$$I = C(1+r)^T - C = C\left[(1+r)^T - 1\right].$$

Lastly, we note that if the interest rate $r$ is part of a compound interest calculation it is called a *compound* (interest) rate; if it produces proportional interest, it is called a *proportional* (simple, or linear) rate.

**Remark**

Formula (2.2) is completely general and only requires consistency between $r$ and $T$: $r$ is the *periodic rate* of compounding, and $T$ the number (generally a fraction) of periods of compounding, however long the period is. If $r$ is an *annual* rate, $T$ is a number of years (generally fractional), but if $r$ is a monthly rate, $T$ is a number of months, etc.

> **Example 3**
> A capital of $1000 lent at an annual rate of 7% produces after 10.25 years a cash flow of $1000(1.07)^{10.25} = \$2000.71$, which amounts to a compounded interest of $1000.71 (the capital has doubled). Lent at 1.75% (7/4%) per quarter over 41 quarters (i.e., the same maturity), it yields a compounded
>
> (continued)

> interest of \$1036.63. Compounding annually at 7% and quarterly at 1.75% are not equivalent for they do not lead to the same terminal value.

### (c) *Future value, present value and internal rate of return*

The *future value* (sometimes called *acquired value*) is equal to the final amount (capital and interest) recovered by the lender at maturity. For a capital $C$ lent for a term $T$ at rate $r$, the future value is given by Eqs. (2.1) or (2.2), according to whether rate $r$ is simple or compound:

$$V_{fut} = F = C\,(1 + rT) \text{ if } r \text{ is proportional.}$$

$$V_{fut} = F' = C\,(1 + r)^T \text{ if } r \text{ is compound.}$$

Conversely, a cash flow $F$ at $T$ can be obtained (or acquired) through lending at interest rate $r$, at $t = 0$, an amount equal to.

$$C = \frac{F}{1 + rT} \quad \text{if } r \text{ is proportional;}$$

$$C' = \frac{F}{(1 + r)^T} \quad \text{if } r \text{ is compound.}$$

**Definition** The expression $\frac{F}{1+rT}$ (if $r$ is proportional) or $\frac{F}{(1+r)^T}$ (if $r$ is compound) is called the *present value* (at 0), or the *discounted value* (at 0), of the cash flow $F$ available at $T$.

Now let us consider investing a capital $C$ giving rise to a final cash flow (future value) $F$ at the end of term $T$. By definition, *the Internal Rate of Return (IRR)* is the interest rate that would allow obtaining the cash flow $F$ from a loan of $C$ for a maturity $T$. In this problem, $F$, $C$, and $T$ are given, and the interest rate is the unknown. The result is different according to whether the rate sought is simple or compound.

For proportional interest, we denote the rate by $r_p$, called a proportional internal rate of return, and we have $F = C\,(1 + T\,r_p)$, whence

$$r_p = \frac{F - C}{CT}$$

If the interest is compound, we denote by $y_c$ the compound rate (or *yield*) to be found and we have $F = C\,(1 + y_c)^T$, whence

$$y_c = \left(\frac{F}{C}\right)^{1/T} - 1.$$

## 2.2 Transactions Involving Two Cash Flows

### 2.2.3 Comparison of Simple and Compound Interest

Since a compound rate and a proportional (simple) rate lead to different results, two questions arise: For given rate $r$ and term $T$, are simple interests larger or smaller than compounded interests? When are a proportional rate and a compound rate "equivalent"?

(a) *Comparison of simple and compound interest, for a given rate*

The following proposition answers the question about the amount of interests due according to whether the rate is simple or compound.

**Proposition**
Assuming the rates are positive, $C(1 + rT) < C(1 + r)^T$ if and only if $T > 1$: therefore, for a rate $r$, a capital $C$ and a term $T$, simple interests *due are smaller than compounded interests* if and only if $T > 1$. If $T = 1$, simple interests are *equal* to compounded interests.

**Graphical proof**

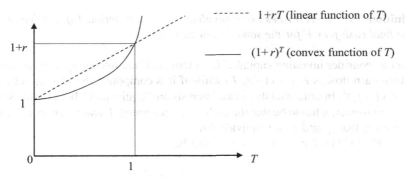

The graph above shows $f(T) = 1 + rT$ (dashed line) and $g(T) = (1 + r)^T$ (solid convex curve). Since the two functions coincide for $T = 0$ ($f(0) = g(0) = 1$) and for $T = 1$ ($f(1) = g(1) = 1 + r$), the curve $g$ is below the line $f$ for $T < 1$ and above for $T > 1$. Therefore, $(1 + r)^T < 1 + rT$ if and only if $T < 1$.

---

**Example 4**
Let us consider $100 invested at a rate of 10% for different terms.

- If they are lent for 2 years ($T > 1$):

    According to the simple interest method, their future value is $100 \times (1 + 2 \times 0.1) = \$120$.
    According to the compound interest method, their future value is $100 \times 1.1^2 = \$121$ ($>120$).

(continued)

- If they are invested for 9 months (the first 9 months of a non-leap year):

According to simple interest, the future value is $100 \times \left(1 + \frac{273}{365} \times 0.1\right) = \$107.48$.

According to compound interest, the future value is $100 \times 1.1^{273/365} = \$107.39 < \$107.48$.

The reader will recall that, usually, the year is the reference period (because official rates are in general annual), compound interest is applied for periods longer than a year (medium or long term) and simple interest for periods shorter than a year (short term). The actual market practice, therefore, favors lenders and penalizes borrowers.

(b) *Calculating "equivalent" rates*

We use in the sequel the notion of "equivalence" for interest rates.

**Definition** *Two interest rates are equivalent for a given period T if they lead to the same final cash flow F for the same initial capital C.*

Let us consider investing capital $C$ for a term $T$. If the interest $r_p$ is proportional, the final cash flow is $F = C(1 + r_p T)$ while if it is compound the final cash flow is $F' = C(1 + y_c)^T$. In order that these two interests are "equivalent", both for the lender and the borrower, it has to be that the cash flows generated, $F$ and $F'$, are equal. From this, we see that $r_p$ and $y_c$ are equivalent if:
$F = F' = C(1 + T r_p) = C(1 + y_c)^T$, that is:

$$\left(1 + T r_p\right) = \left(1 + y_c\right)^T. \tag{2.3}$$

Equation (2.3) can be used both ways: to find a simple interest rate equivalent to a compound interest rate or to find a compound interest equivalent to a simple one.

Note that two rates $y_c$ and $r_p$ equivalent over the period $T$ are *not* equivalent over a period $T'$ different from $T$.

**Example 5**

Consider an investment of 100 for 2 years at a simple interest of 10%. The future value is:

$$V_{\text{fut}} = 100 \, (1 + 2 \times 10\%) = 120.$$

To obtain the same future value with compound interest, the interest rate $y_c$ must be such that:
$120 = 100 \, (1 + y_c)^2$, i.e. $y_c = (120/100)^{1/2} - 1 = 9.54\%$ only.

## 2.2.4 Two "Complications" in Practice

What banks and markets actually do often makes the calculation of interests more complicated. These complications arise, on the one hand, from how the length of the period is calculated, and from how interest is calculated and paid, on the other.

(a) *Calculating the period length T: the different bases*

In most countries and in most cases, the reference period is the year (that is, unless the contrary is made explicit, interest rates are annual). However, in a great number of cases, the term of the transaction is not an integer number of years. So what are the conventions in use to calculate the length of the transaction?

Two cases must be distinguished according to whether $T$ is greater or less than a year, as mentioned in Sect. 2.2.2.

*If the term of the transaction is shorter than a year* in the money market (see. Chap. 3), the money markets in the Euro and American zones use: $T = Nd/360$ (actual/360 basis).[1]

*If the term of the transaction is longer than a year,* one generally uses: $T = Nd/Na$ (actual/actual basis).[2] $Na$ is the actual number of days in a year (365 or 366 for a leap year).

Using the 360 basis rather than the actual value obviously increases the amount of interests, and so the rate, by a ratio of 365/360 or 366/360.

---

**Example 6**

Consider a loan of $1000 for 67 days, from February 1 to April 8 of the *leap* year $n$, at a rate of 5%. The interest $I$ is calculated as simple interest since $T < 1$ year.

On a 360 basis: $I = 1000 \times 5\% \times 67/360 = \$9.31$.

On an actual basis (here, 366): $I = \$1000 \times 5\% \times 67/366 = \$9.15$. The interest is thus higher on a 360 basis, by 366/360.

---

(b) *Interest payment modalities*

Interest can be paid at the end of a period: one says that it is paid *after accrual* or *in arrears*. Interest may equally well be paid at the start of a period: one says that it is paid *in advance*, or *before accrual* (or *up front*).

---

[1] There are exceptions, notably the British money market that uses the actual/365 convention ($T = Nd/365$).

[2] But in some countries the basis 30/360 is used, making the assumption that all months have 30 days.

**Fig. 2.5** Lending with interests in arrears

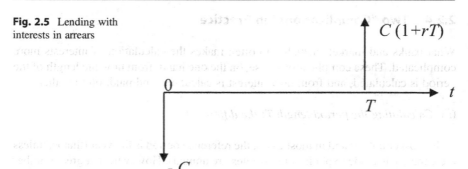

– *Interest accrued (in arrears)*

The interest rate $r$ in this case is called the *"money market rate."*

For example, Fig. 2.5 displays the cash flows of a loan with accruing interest of a capital $C$ at the simple rate $r$, for a term $T$:

> **Example 7**
> Consider, in the money market borrowing, with interests in arrears, for a term of 60 days, an amount $1000, at a simple annual rate of 3.5% using the 360 basis.
> The final cash flow equals $-\$1,000 \left(1 + 3.5\% \times \frac{60}{360}\right) = -1,005.83$.

– *Interest up front.*

Interest is paid up at the start of the transaction. The initial cash flow $F_0$ is in an amount equal to the capital less the interest $I$; the capital (or nominal value, or face value of the transaction) is paid out by the borrower at the end of the investment. This is implicitly the case for a security issued at a discount relative to par, i.e. at a price smaller than the par value $C$. The cash flow diagram for the lender/investor is therefore as in Fig. 2.6.

In the case of interest paid up front, we distinguish two ways of computing the cash flow $F_0 = C - I$, according to whether the calculations are based on a *"bank discount rate,"* or on a *"money market rate."*

**First method: using a bank discount rate (or simple discount rate)**
This method consists in calculating the interests, in the standard way, by applying the rate pro rata (more precisely, pro rata *temporis*) to the capital $C$, and subtracting the resulting amount from capital C to obtain the initial cash flow $F_0$. We then have, denoting the discount rate by $r_d$:

## 2.2 Transactions Involving Two Cash Flows

**Fig. 2.6** Lending with up front interests

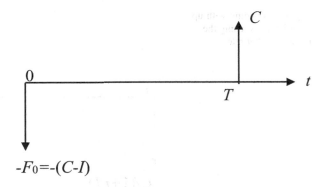

$$F_0 = C - r_d\, C\, T = C\,(1 - r_d\, T).$$

> **Example 8 (Bank Discount)**
> Discounted transfers of bills of exchange (drafts) to a bank give rise to up-front interest computed with a discount rate.
>
> Consider a bill of exchange in the amount of $1000 issued by one of firm X's customer, with a 60-day maturity. Needing funds immediately, X discounts this bill at its bank, at an annual discount rate of 3.5% on a 365-day basis. The cash flows for the bank (the lender) thus are:
> At $t = 0$, the interest is $1000 \times 3.5\% \times 60/365 = \$5.75$, therefore: $-F_0 = -(1000{-}5.75) = -\$994.25$.
> At $t = 1$, the bank recovers $C = +\$1000$.[3]

Note that, for a given amount of interests, a borrower clearly prefers interests in arrears to up-font interests since the payment is due later.

**Second method: using a money market rate**
Since the immediate payment of interests is a disadvantage to the borrower if one uses the same rate as for interest in arrears, in some transactions with up-front interest (notably on the money market), the up-front interest is reduced to compensate for the fact that it is paid in advance, and the actual rate comparable to the money market rate is $r$ such that: $F_0 = C/(1 + rT)$.

We still obtain an initial cash flow ($F_0$) equal to the present value of the final cash flow ($C$). Interests then are equal to $C - C/(1 + rT) = CrT/(1 + rT)$.

---

[3] In practice, the bank recovers the discounted sum ($1000) directly from the customer of its borrower X who issued the bill at $t = 0$: it is with this recovered amount that the bank recoups its loan and earns an interest. From X's point of view, the financial transaction can be seen as borrowing {+ $994.25; −$1000}, the last cash flow not being a real cash transfer, but the *non-collection* of the debt (of $1000) owed it by its customer (since the bank collected the debt in X's place).

**Fig. 2.7** Lending with up front interests, using the money market rate

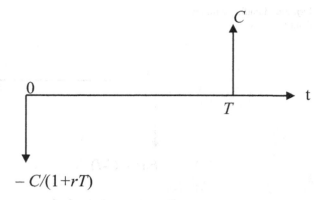

### Remarks

- This reduction is materialized by the fact that the interests $CrT$ (which would be paid in arrears) are discounted by the factor $(1 + rT)$ since they are paid in advance.
- $r$ is indeed the money market rate of the series $(-F_0, C)$ since $C = F_0(1 + rT)$. In Fig. 2.7 represents such a transaction.
- This shows that the practice of asking up-front interest is purely *conventional* (arbitrary) and fundamentally changes nothing, the rate of the transaction always being the money market rate $r$ (*in arrears*): it is enough to say that the capital lent is the sum $C/(1 + rT)$ to make it obvious that the final cash flow to be reimbursed is $C$.

### Example 9

Consider the investment of a capital $C$, of amount \$1000 (it might, for example, be the purchase of a short-term security in the money market). The interests are paid up front but the rate is the money market rate. The final cash flow received by the lender is $+C = +\$1000$ since the interest is paid on the start date. The annual money market rate is 3.5%, the maturity of the security is 60 days and the calculation basis is 360 days.

The initial cash flow then is $-F_0 = -\$1000/(1 + 3.5\% \times 60/360) = -\$994.20$.

(c) *Tabular summary of practices*

Applying different conventions to the same instruments in different places, and to different instruments in the same places, is a source of confusion and complexity. The following table exhibits, without being exhaustive, the different practices in most countries.

## 2.2 Transactions Involving Two Cash Flows

| Term | Transaction type | Interest basis | Term basis |
|---|---|---|---|
| Short term (<1 year at transaction start) | Banking transactions | Simple *Bank discount rate* or *money market rate* | $T = $ actual/360 or actual/actual |
| | Market transactions | Simple *Money market rate* or *Bank discount rate* | $T = $ actual/360 or actual/365 (UK and Commonwealth) |
| Medium and long term (> 1 year) | All transactions | Compound *Compound rate* | $T = $ actual/actual or actual/365 or actual/360 |

### 2.2.5 Continuous Rates

After the conventions used for interest calculation relative to simple interest and compound interest, a third computational convention concerns capitalizing/discounting in continuous time using a *continuous compound rate* $r$ (or briefly, continuous rate). Interests are calculated over every interval $(t, t + dt)$. These interests, which are proportional to rate $r$, to the acquired value $V(t)$ at $t$ and to the maturity $dt$, are added to the acquired value, which thus is incremented by the amount $dV(t) = r V(t) dt$.

This differential equation expresses the fact that $V(t)$ increases exponentially at a constant rate $\frac{1}{V} \frac{dV}{dt} = r$ and admits the solution: $V(t) = V(0) e^{rt} = C e^{rt}$.

As a result, a capital $C$ invested between 0 and $T$ with a continuous capitalization of the interest, at rate $r$, yields a final cash flow $F = C e^{rT}$, and the cash flow series writes: $\{-C, C e^{rT}\}$.

Continuous rates are not used in standard market practice. They are nonetheless very convenient because they often lead to simpler formulas than discrete rates and do approximate reality sufficiently when cash flows are daily or intra-day. We will in fact use continuous rates repeatedly in the remainder of this book.

### 2.2.6 General Equivalence Formulas for Rates Differing in Convention and the Length of the Reference Period

(a) *Equivalence of rates defined on different bases*

The co-existence of several calculation bases often necessitates comparing rates using different conventions. This comparison requires calculating a rate basis $x$ equivalent to some other rate basis $y$. Recall that two rates are equivalent over a given period if, for the same sum invested, they yield the same final cash flow. Here we denote by $c$ a continuous rate (on an actual/actual basis), $y_c$ a compound rate (actual/actual basis), $r_{p1}$ for a proportional rate *in arrears* 365 days, $r_{p2}$ for a proportional rate *in arrears* 360 days and $r_d$ for a *bank discount* rate (up front and 365 days).

The following Eq. (2.4) expresses the condition under which these different rates are equivalent over a given horizon. It is obtained by setting equal the final cash flows $F$ obtained from investing \$1 over $T$ periods according to the different methods of computing interests:

$$e^{cT} = (1 + y_c)^T = 1 + r_{p1}T = 1 + r_{p2}T\frac{365}{360} = \frac{1}{1 - r_dT} \qquad (2.4)$$

where $T = Nd/Na$ (actual/actual).

This equation allows us to transform a rate or yield on any basis into an equivalent one relative to a different basis. In particular, it implies that the continuous rate $c$ equivalent to a compound rate $y_c$ is given by $c = \log(1 + y_c)$, independently of the duration of the transaction[4]; it also implies, as we have seen, that, for any $T$, a market rate over 365 days ($r_{p2}$) is equivalent to a market rate over 360 days ($r_{p1}$) multiplied by 365/360.

Note that if two simple rates $r_{p1}$ and $r_{p2}$ are equivalent over a term $T$, they are also equivalent over any other term $T'$. It is the same for any two equivalent compound rates such as $y_c$ and $c$. By contrast, we recall that if a proportional rate such as $r_{p1}$ and a compound rate such as $y_c$ are equivalent for a particular term $T$, they will not be for any term $T'$ different from $T$.

---

**Example 10**

- A continuous rate of 5% is equivalent to a compound rate of $e^{0.05} - 1 = 5.13\%$ (whatever the term $T$ of the transaction).
- A simple rate over 360 days of 5% is equivalent to a simple rate over 365 days equal to $5\% \times 365/360 = 5.07\%$ (regardless of the term of the transaction).
- A compounded rate of 5% is equivalent to a continuous rate of log $(1.05) = 4.88\%$ (regardless of the term of the transaction).
- For a transaction of term 90 days ($T = 90/365 = 0.246$ years), a compound rate of 5% is equivalent to a simple market rate (basis 365) equal to $(1.05^{0.246} - 1)/0.246 = 4.91\%$. The result does depend on the term of the transaction considered (here 90 days).

---

(b) *Annualizing a periodic rate*

The reference period can be different from one year. It is then often convenient to "annualize" the periodic rate, i.e. to convert it to an equivalent annual rate.

---

[4]Here, as everywhere in the book, log denotes Napier's natural logarithm ($\log e = 1$).

We denote by $m$ the number of periods contained in one year (e.g., $m = 12$ for the month). The problem is then to convert any periodic rate into an annual rate and the other way around.

Here we denote by $r_m$ (or $y_m$) a periodic rate and by $r_a$ (or $y_a$) the corresponding annualized rate.

The periodic rate can be *compound* ($y$) or *proportional* ($r$). The formulas for converting a periodic rate to an annual rate and the other way around follow from what we have already seen:

For *proportional rates*, the conversion formula is:

$$r_a = m\, r_m. \tag{2.5a}$$

This annualized rate is sometimes called Annual Proportional Rate (APR).

For *compound rates*, the conversion formula is:

$$y_a = (1 + y_m)^m - 1. \tag{2.5b}$$

This annual rate is also called annual effective yield, annual compound yield or Annual Percentage Yield (APY).

---

**Example 11**
- A monthly rate of 1.0% is equivalent to an annual compound rate of 12.68% since $(1.01)^{12} - 1 = 0.1268$.
- The annual compound rate of 10.25% is equivalent to a semi-annual compound rate of 5% since $1.1025^{1/2} - 1 = 0.05$.
- The proportional quarterly rate equivalent to a proportional annual rate of 5.2% is 1.3% since $0.052/4 = 0.013$.

---

In certain countries coupons are paid several times a year and computed on the basis of an annual rate $r_a$ *which is divided on a proportional basis* to obtain the periodic coupon: if $m$ denotes the number of coupons paid out in a year, the coupon is thus computed with the rate $r_a/m$. The resulting annual percentage yield (APY) therefore exceeds $r_a$ and is equal to $y_a = \left(1 + \frac{r_a}{m}\right)^m - 1$. As we will see in Chap. 4, this is notably the case for American, British and Japanese bonds that distribute a coupon each semester ($m = 2$).

Finally recall that $m$ *is not necessarily an integer*. In general, if $D$ denotes the length of the reference period as a fraction of a year, we have: $m = 1/D$. The conversion formulas (2.5a and 2.5b) continue to hold replacing $m$ by $1/D$.

---

## 2.3 Transactions Involving an Arbitrary Number of Cash Flows: Discounting and the Analysis of Investments

We now analyze transactions that give rise to an arbitrary number of cash flows, in any amounts and payable at any time.

We consider a transaction that generates the cash flows $F_0, F_{t_1}, F_{t_2}, \ldots, F_{t_N}$ at the respective times $0, t_1, t_2, \ldots, t_N$. The series can be denoted by $\{F_\theta\}_{\theta=0, t_1, \ldots, t_N}$, or, in a simpler but ambiguous way, by $\underline{F}$. If the cash flows are regularly spaced at times $0, 1, 2, \ldots, N$, the sequence will be written $\{F_0, F_1, F_2, \ldots, F_N\}$.

*Recall that, by convention, negative numbers represent cash outflows and positive numbers cash inflows.* We distinguish investments from financing:

*Investments* are characterized by a stream of cash flows that includes, at first, one (or more) negative flows, corresponding to the placement of the funds, followed by positive flow(s).

One can distinguish financial investments such as loans, bank accounts, or securities (stocks, bonds, etc.) from "real" investments such as the acquisition of machinery for production, buying a company, buying real estate, etc. For example, the sequence: $-100{,}000; 30{,}000; 30{,}000; 30{,}000; 30{,}000$ could represent a loan of \$100,000 leading to 4 annual constant reimbursements (capital + interests) of \$30,000 or the purchase of a security for the price of \$100,000 that leads to 4 cash flows of \$30,000, or else the purchase of a machine costing \$100,000 which allows production over 4 years that leads to an annual cash inflow of \$30,000.

– *Financing* involves, for the person who contracts it, an initial cash inflow followed by disbursements (outflows).

Generally, one has:

$F_0, F_{t_1}, F_{t_2}, \ldots, F_{t_N}$ for an investment;
$< 0, > 0, > 0, \ldots, > 0$
$-F_0, -F_{t_1}, -F_{t_2}, \ldots, -F_{t_N}$ for the corresponding financing
$> 0, < 0, < 0, \ldots\ldots, < 0$.

The homology of these two series leads us to analyze decisions about investment and financing with the same mathematical methods.

In this section, in general, we consider the investor's viewpoint. The upcoming Sect. 2.4 takes the matter up again from the borrower's perspective and analyzes lending and borrowing involving several cash flows using the tools introduced in this section.

The fundamentals of discounting and the concept of present value are dealt with in Sect. 2.3.1. The Yield to Maturity (YTM, which is called the Internal Rate of Return (IRR) in an investment context) is presented in Sect. 2.3.2. In Sect. 2.3.3 these concepts are applied to the analysis of investments. The interactions between an investment and its financing (by debt and by own capital or equity) as well as the financial leverage effect are discussed in Sect. 2.3.4. Some useful guidelines for the choice of the discount rate are provided in Sect. 2.3.5. The important distinction, between *real* rates and flows and *nominal* rates and flows is made in Sect. 2.3.6.

## 2.3 Transactions Involving an Arbitrary Number of Cash Flows: Discounting...

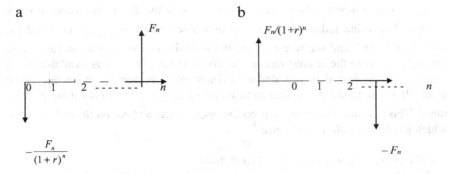

**Fig. 2.8** Present value of a future cash flow

### 2.3.1 Discounting

The considerations that follow assume the existence of a market for lending and borrowing with capitalization of interest, where a rate, denoted by $r$, prevails. To begin with, $r$ is a discrete compound rate and is assumed unique for all maturities.

(a) *Fundamentals of discounting, and valuing the right to a cash flow using its present value*

Lending and borrowing transactions with two cash flows as studied in the preceding section allow us to assert:

- At the cost of initially giving up a sum $\frac{F_n}{(1+r)^n}$, one can generate a unique cash flow $F_n$ at date $n$. The transaction that permits the transfer of cash over time is lending, as represented in Fig. 2.8a following, which leads to the series $\left\{-\frac{F_n}{(1+r)^n}, +F_n\right\}$:

   Transaction (Fig. 2.8a) can be interpreted as the "purchase" of a future cash flow $F_n$ at the "cost" of $\frac{F_n}{(1+r)^n}$ disbursed today.

- Conversely, one can have immediately available the amount $\frac{F_n}{(1+r)^n}$, at the cost of giving up a cash flow $F_n$ on date $n$, by contracting a loan of $\frac{F_n}{(1+r)^n}$ characterized by diagram Fig. 2.8b and the sequence $\left\{+\frac{F_n}{(1+r)^n}, -F_n\right\}$. In operation (Fig. 2.8b), one "sells" the future cash flow $F_n$ at the "price" of $\frac{F_n}{(1+r)^n}$ received today. The possibility of "exchanging" cash flows taking place at different dates, which is the primary reason for the existence of a financial market, justifies the method of *discounting* which attributes the value $\frac{F_n}{(1+r)^n}$ to the claim on (a right to) a cash flow $F_n$ available in $n$ periods.

$\frac{F_n}{(1+r)^n}$ is the *present value* of cash flow $F_n$, or else the *discounted value at $t = 0$* (today). The value today of one dollar available at date $n$, $\frac{1}{(1+r)^n}$, is called the *"discount factor"* and the rate $r$ which is the basis for the calculation (and is, in principle, equal to the interest rate) is the *discount rate*. One dollar available in the future is worth less than one dollar available now, because to have one dollar available in the future it is sufficient to invest today $\frac{1}{(1+r)^n} < \$1$ (assuming positive rates). This discount factor depends on the interest rate, and *not on the inflation rate which has here no direct relevance.*[5]

(b) *The present value of a series of cash flows*

The idea of discounting can easily be extended to any date and to the case of sequences with several cash flows $\{F_\theta\}_{\theta=0,\, t_1,\, ...,\, t_N}$.

To discount at time $t$ a sequence of cash flows consists in determining the unique cash flow available at time $t$ that is equivalent to the whole sequence.

Most often the problem is to calculate the value of a sequence of cash flows at the present date ($t = 0$). In the sequel, except when the contrary is explicitly mentioned, discounting takes place on today's date $t = 0$.

Since a sequence of cash flows $\underline{F} = \{F_\theta\}_{\theta=0,\, t_1,\, ...,\, t_N}$ is the right to $F_0$ at 0, $F_{t_1}$ at $t_1, \ldots$ and $F_{t_N}$ at $t_N$, its present value is:

$$PV_{\underline{F}}(r) = F_0 + \frac{F_{t_1}}{(1+r)^{t_1}} + \frac{F_{t_2}}{(1+r)^{t_2}} + \cdots + \frac{F_{t_n}}{(1+r)^{t_n}} \equiv F_0 + \sum_{\theta=t_1}^{t_n} \frac{F_\theta}{(1+r)^\theta}$$

**Definition** *The sum*

$$F_0 + \sum_{\theta=t_1}^{t_n} \frac{F_\theta}{(1+r)^\theta} \equiv PV_{\underline{F}}(r) \qquad (2.6)$$

*is called the present value (PV) of the series of cash flows $\underline{F}$, discounted at rate $r$.*

The present value $PV_{\underline{F}}(r)$ represents the sum that must be invested today to have the right to the future cash flows generated by $\underline{F}$; it is, therefore, an objective estimate of the value today of the cash flows represented by the sequence $\underline{F}$, when the market interest rate is $r$.

Among the cash flow sequences representing investment and financing, those we call "standard" are of special interest.

**Definition** *A cash flow sequence is called "standard" if $F_0$ has a sign opposite to $F_{ti}$ for every $t_i > 0$.*

---

[5] In fact, the inflation rate does affect the discount rate indirectly because of its influence on the nominal interest rate $r$ (see Sect. 2.3.6).

## 2.3 Transactions Involving an Arbitrary Number of Cash Flows: Discounting...

If $F_0 < 0$ and $F_{t_i} \geq 0$ for $t_i = t_1, \ldots, t_n$, this is a standard investment. In the opposite case, this is a standard financing.

Since the present value of a series of cash flows $\underline{F}$ depends on the discount rate $r$ used, we analyze it as a function denoted by $PV_{\underline{F}}(r)$.

Let us first examine the case of the standard investment whose present value writes:

$$PV_{\underline{F}}(r) = F_0 + \sum_{\theta=t_1}^{t_n} \frac{F_\theta}{(1+r)^\theta} \quad \text{with } F_0 < 0 \text{ and } F_{t_i} \geq 0 \text{ for } t_i \geq 0.$$

We study the behavior of $PV_{\underline{F}}(r)$ as $r$ varies from $-1$ to $+$ infinity:

- Note that as a sum of decreasing hyperbolic functions of $r$, $PV_{\underline{F}}(r)$ is also a decreasing hyperbolic function of $r$.
- Note also that: $\lim_{r \to -1} PV_{\underline{F}}(r) = +\infty$; $\lim_{r \to +\infty} PV_{\underline{F}}(r) = F_0 < 0$.

$PV_{\underline{F}}(r)$ may thus be represented as in Fig. 2.9a. Its curve is decreasing, convex, and tends asymptotically to the vertical axis with abscissa $-1$ as $r$ goes to $-1$ (from above) and to the horizontal with ordinate $F_0$ as $r$ tends to infinity.

**Important Remarks**

- The expression $PV_{\underline{F}}(r) = F_0 + \sum_{\theta=t_1}^{t_n} \frac{F_\theta}{(1+r)^\theta}$ is called the "*net present value*" (NPV), to distinguish it from $\sum_{\theta=t_1}^{t_n} \frac{F_\theta}{(1+r)^\theta}$, the present value (PV) of the (strictly) future cash flows. In the sequel, we use the NPV phrasing when we want to stress that the initial cash flow $F_0$ is taken into account.
- We note that the curve representing $PV_{\underline{F}}(r)$ intersects the $x$-axis in the interval $[-1, +\infty$ [at a single point, denoted by $r^*$: the NPV thus is nil for a unique value $r^*$ of $r$ between $-1$ and $+\infty$. The existence and uniqueness of $r^*$, obvious from the graphical illustration, relies mathematically on the monotonicity and continuity of the function $PV_{\underline{F}}(r)$. The existence of a single discount rate for which the NPV is nil will be important in the definition of the YTM (Yield to Maturity) or of the IRR (Internal Rate of Return).

(c) *Calculating present values*

- In the general case, the calculation of the PV of a cash flow series requires first calculating all the *discount factors* $1/(1+r)^{t_i}$. This can be done directly with a calculator or a computer, or can be read from financial tables. Using such tables or the financial functions present in spreadsheet softwares is briefly described in Appendix 2.

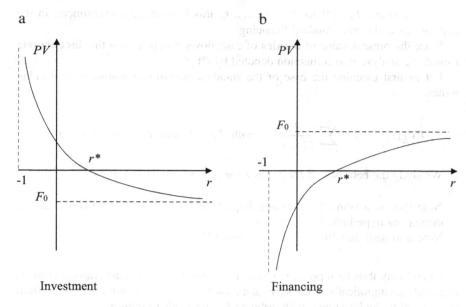

**Fig. 2.9** Present value as a function of interest rate

- If the future cash flows in a series ($F$) are all equal, the sequence writes: $\{F_0, F, F, \ldots, F\}$ and the calculation is much simpler.

$$PV_F(r) = F_0 + \sum_{t=1}^{n} \frac{F}{(1+r)^t} = F_0 + F\left(\sum_{t=1}^{n} \frac{1}{(1+r)^t}\right).$$

The calculation thus reduces essentially to the sum: $\left(\sum_{t=1}^{n} \frac{1}{(1+r)^t}\right)$.

This is the sum of the first $n$ terms of a geometric sequence with ratio $1/(1 + r)$, and first term $1/(1 + r)$. Calculating this sum is done in Appendix 1. We obtain:

$$\sum_{t=1}^{n} \frac{1}{(1+r)^t} = \frac{1-(1+r)^{-n}}{r}. \qquad (2.7)$$

The result in Eq. (2.7) can be read from financial tables A2 (see Appendix 2) or obtained by using a computer or a calculator.

## 2.3 Transactions Involving an Arbitrary Number of Cash Flows: Discounting... 61

> **Example 12**
>
> Consider an initial investment of \$1200 followed by withdrawals of \$360 each year for 10 years. Suppose the prevailing market rate is 10%. The cash flow sequence $(F)$ is:
>
> $$\{-1200; +360; +360; \ldots; +360\}.$$
>
> The net present value of this sequence, discounted at 10%, is therefore:
>
> $$NPV = -1200 + 360 \sum_{t=1}^{10} \frac{1}{(1,1)^t}, \text{ with } \sum_{t=1}^{10} \frac{1}{(1,1)^t} = 6.14457 \text{ (see Table A2 or a}$$
>
> calculator), whence:
>
> $$NPV = -1200 + 360 \times 6.14457 = \$1012.05.$$
>
> This means that the net gain resulting from the transaction which consists in investing \$1200 today to receive \$360 a year for 10 years is \$1012.05. One can also note that the series of \$360 per year paid for 10 years is worth today $360 \times 6.14457 = \$2212.05$ or \$1012.05 more than the initial disbursement of \$1200.

– Discounting may also be done with a proportional rate $r_p$ or a continuous rate $c$.

Taking account of the conversion relation (2.4), the present value of \$1 available at time $t$ writes:
$\frac{1}{1+r_p T}$ if the discounting is done with a proportional rate $r_p$ (often used if $T < 1$ year), and $\exp.(-r_c T)$ if the continuous rate $r_c$ is used. In the latter case, the cash flows are often represented in continuous time by a *density* $F(t)]_{t \in (0,T)}$. The cash flow in the interval $(t, t + dt)$ being $F(t)dt$, its present value is $\exp.(-r_c T)F(t)\,dt$, and that of the whole series over $(0, T)$ is:

$$PV_F(r_c) = \int_0^T F(t)e^{-r_c t} dt. \tag{2.8}$$

> **Example 13**
>
> Consider an investment yielding a cash flow of \$10 per unit of time for 5 years, discounted using a discrete annual rate of $r = 10\%$.
>
> The continuous rate equals $\log(1 + 0.1) = 9.53\%$ and the present value of the cash flow generated by the investment is:
>
> (continued)

$$PV = 10 \int_0^5 e^{-0.0953t} dt = 10 \times \left[ -\frac{1}{0.0953} e^{-0.0953t} \right]_0^5$$

$$= [10/0.953] \left[ 1 - e^{-5 \times 0.0953} \right] = 39.77.$$

Remark that if the investment was a perpetuity, its value would be $10/0.0953 = \$104.92$, while an annuity yielding $10 each year discounted with a discrete annual rate of 10% would be worth $10/0.10 = \$100$, which is less than \$104.92, because the interest payments are capitalized once a year only instead of continuously.

(d) *Applying a discount rate specific to the cash flow maturity*

Up to now, we have assumed that $r$ is unique and independent of the maturity of the cash flows. But the annual discount rate at which one may borrow or lend in general depends on the *time length* of the transaction. We denote by $r_\theta$ the rate at which one may lend or borrow for a time length $\theta$, for a transaction involving two cash flows only (one lends \$1 at date 0 and receives $\$(1 + r)^\theta$ at date $\theta$). $r_\theta$ is called the *zero-coupon rate* since between 0 and $\theta$ no interest is paid (interests are capitalized over the entire life of the transaction). The set of the $r_\theta$, for all maturities $\theta$ traded in the market is called the Term Structure of Interest Rates (TSIR) and its graphical representation ($r$ as a function of $\theta$) is called, somewhat loosely, the *yield curve*. We devote all of Chap. 5 to studying the TSIR and revisit on that occasion the zero-coupon rates.

Under these conditions:

- The present value at 0 of a cash flow $F_\theta$ available at time $\theta$ writes: $\frac{F_\theta}{(1+r_\theta)^\theta}$.
- The present value of the series $\underline{F} = \{F_\theta\}_{\theta=0,\,t_1,\,...,\,t_N}$, discounted using the rates $\underline{r} = \{r_\theta\}_{\theta=0,\,t_1,\,...,\,t_N}$, is equal to:

$$PV_F(\underline{r}) = \sum_{\theta=t_1}^{t_n} \frac{F_\theta}{(1+r_\theta)^\theta}. \tag{2.9}$$

- We remark that the only difference between this expression and that for the PV calculated with a single rate $r$ (Eq. 2.6) is the use in Eq. (2.9) of discount rates $r_\theta$ specific to the different maturities $\theta$.

## 2.3.2 Yield to Maturity (YTM), Discount Rate and Internal Rate of Return (IRR)

We have already defined the IRR, in the case of a loan/investment over a period $T$ generating two cash flows $\{-C, F\}$. It is the interest rate that, for a capital $C$ invested, gives a cash flow $F$ at $T$. It is, therefore, $r*$ such that $F = C\,(1 + r*)^T$ or equivalently:

$$-C + \frac{F}{(1 + r*)^T} = 0.$$

The IRR $r*$ thus is the special discount rate which makes the net present value of the two-cash-flow sequence vanish. The term IRR is used for investments. The term *Yield To Maturity* (*YTM*) corresponds to exactly the same concept but is used in the more general context of a sequence of cash flows that can represent an investment, a financing, a security, ... The YTM of the sequence $\{-C, F\}$ (or $\{C, -F\}$) is thus the discount rate $r*$ which makes the net present value of a two-payment schedule equal to zero. We now generalize this definition to a transaction characterized by an arbitrary sequence $\underline{F} = \{F_\theta\}_{\theta=0,\,t_1,\,...,\,t_N}$.

(a) *Definition and interpretation of the YTM*

**Definition** *We call the Yield To Maturity (YTM) of a sequence of cash flows* $\underline{F} = \{F_\theta\}_{\theta=0,\,t_1,\,...,\,t_N}$ *a special value* $r*$ *of the discount rate which makes the **net** present value of the sequence $\underline{F}$ equal to zero.*
*From this we see that the YTM $r*$ is such that:*

$$PV_{\underline{F}}(r*) \equiv F_0 + \sum_{\theta=t_1}^{t_n} \frac{F_\theta}{(1 + r*)^\theta} = 0. \tag{2.10}$$

So the YTM is the abscissa $r*$ of the intersection of the curve representing $PV_{\underline{F}}(r)$ with the horizontal axis (see Fig. 2.9a or b).

When the sequence $\underline{F}$ represents an investment, the YTM $r*$ is called the *internal rate of return* (IRR) of the sequence $\underline{F}$. When $\underline{F}$ characterizes a funding, the YTM equals the interest rate.[6]

The definition of the YTM, as expressed in Eq. (2.10), raises problems regarding the existence and uniqueness of such a rate. Let us consider the special case where $t_1$, $t_2$, ..., $t_N$ are integers. The calculation thus reduces to solving the polynomial equation of the $n$th degree:

---

[6]Up to the fees and commissions that contribute in the calculation of a compound yield but not in that of an interest rate (see Sect. 2.5).

$$F_0 + \sum_{i=1}^{n} F_i x^i = 0 \quad \text{where } x = 1/(1+r).$$

We know that a polynomial of degree $n$ has $n$ real or complex roots, that in certain cases there are no real roots at all, and that in others there may be several. However, only real values for the YTM (or IRR) make any financial sense. As a result, there is a problem with the existence and uniqueness of the YTM.

**Proposition**

For a sequence of standard cash flows ($F_0$ with the opposite sign to $F_\theta$ for any $\theta > 0$), the YTM exists and is unique in the interval $[-1, +\infty[$.

This proposition follows directly from the second remark made at the end of Sect. 2.3.1(b): *in the interval* $[-1, +\infty[$, *the curve representing* $PV_E(r)$ *intersects the vertical axis at a single point* $r^*$ (see Fig. 2.9a, b), which therefore corresponds to the unique YTM, between $-100\%$ and $+\infty$.

Existence and uniqueness of the YTM, shown here for the special case of standard sequences, also holds in the more general case of cash flows with signs that do not alternate (all the $F_\theta$ for $\theta < \theta^*$ have the same sign which is opposite to that of the $F_\theta$ for $\theta >, \theta^*$).

For sequences with alternating signs, there may exist several values of the discount rate that cancel the net present value of the sequence, or none (according to Descartes' rule of signs). The case of alternating-sign cash flows occurs in the real world for certain hybrid savings and credit products (such as some homebuyer savings plans).

The YTM can be interpreted using the idea of interest rates. Let us consider a standard sequence representing an investment: $\{F_\theta\}_{\theta=0, t_1, \ldots, t_N}$ with $F_0 < 0$ and $F_{t_i} \geq 0$ for $t_i > 0$. The YTM of this sequence is the interest rate on a capital $C = -F_0$ which would give rise to a payment schedule of reimbursements (capital + interest) made up of $F_{t_1}, F_{t_2} \ldots F_{t_N}$.

(b) *Calculating the YTM*

*In the general case*, calculating the YTM requires solving Eq. (2.10):

$$F_0 + \sum_{\theta=t_1}^{t_n} \frac{F_\theta}{(1+r^*)^\theta} = 0; \quad \text{the } F_\theta \text{ are given and the unknown is } r^*.$$

In general, this equation has no analytical solution. As a result, in most cases, the solution is done numerically. The computation requires successive iterations, getting closer and closer to the solution(s) if they exist, up to a given precision. A calculator or a computer provides a direct solution (using iterative algorithms); their use is briefly described in Appendix 2.

In what follows we limit ourselves to the study of several "sequence types" with equally spaced cash flows ($t_i = i$ for $i = 1, \ldots, n$).

## 2.3 Transactions Involving an Arbitrary Number of Cash Flows: Discounting... 65

- Sequence of type 1: *a sequence of constant cash flows* $\{-x, y, y,\ldots, y\}$ with $x$ and $y$ of the same sign.

Equation (2.10) then simplifies to $-x + \sum\limits_{\theta=1}^{n} \frac{y}{(1+r*)^{\theta}} = 0$, or else:

$$\sum_{\theta=1}^{n} \frac{1}{(1+r*)^{\theta}} = \frac{1-(1+r*)^{-n}}{r} = x/y.$$

The value of $r*$ can therefore be deduced by a simple examination of Tables A2.

- Sequence of type 2: *a sequence with two cash flows* $\{-x, y\}$ where $x$ and $y$ have the same sign. Its YTM is $(y-x)/x$. Indeed, a capital loan of $x$, for 1 year, at an interest rate of $\frac{y-x}{x}$, leads to interest payments of $y-x$ which add to the capital $x$ to give the final cash flow $(y-x) + x = y$. This loan indeed generates the sequence $\{-x, y\}$.

We check that the solution to (2.10) confirms the intuitive result. Indeed (2.10) can be written:

$$-x + y/(1+r*) = 0, \quad \text{whence } r* = (y-x)/x.$$

- Sequence of type 3: *a bullet sequence* $\{-x, y, \ldots, y, y+x\}$, where $x$ and $y$ have the same sign.

Its YTM is $y/x$. Indeed, a loan of $x$ on the basis of an interest rate $= y/x$, which is repaid in full at the end of $n$ years (thus generating, throughout the $n$ years, an amount $y$ in interest) does generate a sequence of type 3.

The mathematical proof justifies the intuition. Equation (2.10) may be written:

$$0 = -x + y\sum_{\theta=1}^{n-1} \frac{1}{(1+r*)^{\theta}} + \frac{y+x}{(1+r*)^{n}} = -x + y\sum_{\theta=1}^{n} \frac{1}{(1+r*)^{\theta}} + \frac{x}{(1+r*)^{n}}$$

$$= -x[1-(1+r*)^{-n}] + y\left[\frac{1-(1+r*)^{-n}}{r*}\right].$$

So, $0 = -x + \frac{y}{r*}$, which proves the result.

- Sequence of type 4: *a perpetuity* generating $\{-x, y, y, \ldots, y, \ldots$ to infinity$\}$. Its YTM is again $y/x$.

In fact, the loan of a capital $x$ which is never repaid, with an interest rate $= y/x$, (thus generating every year $y$ in interest forever) does generate a sequence of type 4.

We can verify that the YTM of sequence 4 coincides with this rate by solving the equation:

$$0 = -x + \sum_{i=1}^{\infty} \frac{y}{(1+r*)^i} = -x + \frac{y}{r*}, \quad \text{whence} \quad r* = \frac{y}{x}.$$

(We use the fact that $\sum_{i=1}^{\infty} \frac{1}{(1+r*)^i} = \frac{1}{r*}$; see relation (2.19) in Appendix 1).

For this *perpetuity* (or *consol*, in the case of a perpetual bond), $y$ is the amount of interests paid in each period, up to infinity, for an initial investment $x$.

### 2.3.3 Application to Investment Selection: The Criteria of the NPV and the IRR

The concepts and methods developed in the two preceding paragraphs are the foundation for the study of investments (either financial or physical such as a factory, a piece of land or of art, etc.; the selection of physical investments by firms is referred to as *capital budgeting*) and the criteria of Net Present Value and IRR.

(a) *Net Present Value (NPV)*

An investment, as we have seen, is characterized by the sequence of cash flows it generates:

$$F_0, F_{t_1}, F_{t_2}, \ldots, F_{t_N}$$

$$< 0, \quad > 0, \quad \ldots\ldots, \quad > 0.$$

The present value of this series, discounted today at $t = 0$, is the NPV of the investment (see formula (2.6)):

$$NPV = PV_{\underline{F}}(r) = F_0 + \sum_{\theta=t_1}^{t_n} \frac{F_\theta}{(1+r)^\theta}.$$

Recall that "*net* present value" indicates that we subtract the initial disbursement $F_0 < 0$.

The NPV of an investment calculated using formula (2.6) therefore represents the "net gain" or net increase in value resulting from the investment. As a result, an investment should only be undertaken if it produces a positive NPV, and when choosing between mutually exclusive investments one should prefer that with the largest NPV. This allows us to state the following two rules that constitute the NPV criterion.

*Rule* 1: *An investment should only be adopted if its NPV is positive.*
*Rule* 2: *Among several mutually exclusive and non-renewable investments, one should adopt that with the largest NPV (if it is positive).*

The NPV can be positive or negative according to the discount rate, as shown in Fig. 2.9a. For the NPV to give the algebraic value of the "net gain" or "value creation" generated by the sequence $\underline{F}$, it is necessary that the discount rate be chosen correctly. When the investment is riskless (the cash flows $F_\theta$ are certain), which we assume in this chapter, the discount rate should be equal to the market rate. This fundamental principle is, however, not enough in practice to choose the right discount rate. We have already seen the adjustments made necessary by the fact that the interest rate depends on the length of a loan: the NPV must be computed using formula (2.9) which contains a different rate for each payment. More serious methodological and theoretical difficulties appear when the cash flows cannot be considered risk-free, and when financing is partly debt and partly equity (own resources). Paragraph 5 below will provide some hints in this respect and the dependence of market rates on risk will lie at the heart of this book.

(b) *Internal Rate of Return (IRR)*

**Definition** *The IRR (internal rate of return) of an investment is the YTM of the cash flow sequence $(F_0, F_{t_1}, F_{t_2}, \ldots, F_{t_n})$ that it generates. It is thus the rate $r^*$ given by Eq. (2.10) repeated below:*

$$PV_{\underline{F}}(r*) = F_0 + \sum_{\theta=t_1}^{t_n} \frac{F_\theta}{(1+r*)^\theta} = 0.$$

The IRR of an investment is therefore the particular discount rate that cancels its NPV.

Graphically, it corresponds to the intersection point of the curve of the $PV_{\underline{F}}(r)$ (=NPV) as a function of $r$ with the horizontal axis (see Fig. 2.10).

The IRR criterion may be formulated in this way: *An investment should be accepted if and only if the IRR of the cash flow sequence that it generates exceeds the minimum rate of return required by the investor.* This rate is in general greater than or equal to the interest rate that the investor can obtain by simply placing the funds in the market.

The justification of the IRR criterion is simple in the case depicted in Fig. 2.10 (the NPV decreases as $r$ increases) since, in such a situation, the IRR criterion is strictly equivalent to that of the NPV (see Rule 1 stated in the previous paragraph): It is easy to check in Fig. 2.10 that the NPV is positive if and only if the IRR exceeds $r$, and the investment should be made only in this case.

In contrast, the IRR should not be used in choosing among mutually exclusive investments, i.e. in the case that one needs to select at most one investment among several. The point is that one may be in the situation depicted in Fig. 2.11, which exhibits the variation of the NPV of two investments A and B as a function of the discount rate $r$.

In this example, $IRR_B > IRR_A$; nonetheless, investment B should only be favored over A for discount rates above $r_c$ since, for lower values, the NPV of A is greater

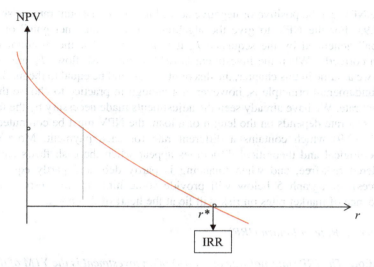

**Fig. 2.10** Internal rate of return

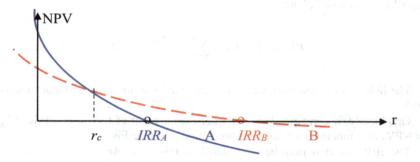

**Fig. 2.11** Comparing two investments

than that of B. Simply knowing the IRR is therefore not enough when comparing two investments.

In addition, using the IRR may elicit other problems, such as the non-existence of the IRR or a multiplicity of values for the IRR, which we will not study.[7]

Although the IRR is a selection criterion both less general and less rigorous than the NPV from which it is derived, it is often used because it apparently does not require choosing a discount rate: the IRR only depends on the cash flows generated by the investment, while to calculate the NPV it is necessary to choose, sometimes rather arbitrarily, a discount rate (see Sect. 2.3.5). However, we do remark that this

---

[7]These problems may occur when the financial instrument generates a cash flow sequence with alternating signs (such as <0, >0, <0), for example in the case of a house-buyer savings plan that is a hybrid product of savings (investment) and borrowing.

## 2.3 Transactions Involving an Arbitrary Number of Cash Flows: Discounting...

argument is deceptive, since the IRR must be compared with the minimum required rate $r$, which in any case has to be determined.

### 2.3.4 Interaction Between Investing and Financing, and Financial Leverage

We establish here the distinction and the connection between the rate of return on an investment and that of the equity committed by shareholders to finance it. The two yields are different if all or part of the investment is financed through borrowing. We begin the analysis starting with an example; general conclusions will be drawn from it in a second step.

(a) *Introductory example of the interaction between investment and financing*

A company acquires a factory for $24 million; the annual production of this factory gives rise to an annual net margin of $3 million over 10 years. This net margin is computed as the difference between the sales, on the one hand, and the different costs of production and marketing and taxes on the other. Taxes are computed under the assumption of zero borrowing, hence zero financing costs, so that the net margin is independent of the financing of the investment.[8]

At the end of 10 years of operation, the factory will be resold, net of taxes (on the capital gain), for $24 million.

The cash flow sequence $I$ generated by this investment thus writes:

$$I = \{-24, 3, 3, \ldots, 3, 3 + 24 = 27\}.$$

The IRR of this sequence, denoted by $r_i$, can be interpreted as the "intrinsic" rate of return (or yield) on the investment since the cash flows $I$, and thus $r_i$ are independent of the financing. We obtain (see the type 3 sequence in Sect. 2.3.2(b)):

$$r_i = 3/24 = 12.5\%.$$

---

[8] An example of computing this net margin is as follows:

Let us assume annual sales of $10 million and costs of materials, production and marketing of $6.484 million yearly, with no financing costs. Suppose provisions for depreciation and amortization are $2 million each year for 10 years, and the tax rate is 34%.

Earnings Before Interests, Taxes, Depreciation and Amortization (EBITDA) $= 10 - 6.484 = \$3.516$ million per year.

Profit before tax $=$ EBITDA $-$ provisions for depreciation and amortization $= 3.516 - 2 = \$1.516$ M per year.

Taxes $= 0.34 \times 1.516 = \$0.516$ million per year.

Net margin (which generates the cash flow) $= 3.516 - 0.516 = \$3$ million per year.

70          2 Basic Finance: Interest Rates, Discounting, Investments, Loans

- Let us first suppose that the *investment is financed exclusively with equity* (the company's own capital), which means that the initial payment ($24 million) is completely financed by the shareholders and that all later positive cash flows are paid out to them. The sequence $\underline{I}$ of cash flows related to the physical investment can then be viewed as the sequence $\underline{S}$ of cash flows disbursed or received by the shareholders (financial investments): the rate of return (or yield) $r_s$ of funds invested by the shareholders is, therefore, equal to $r_i$, i.e., to the IRR of the physical investment. Here, $r_s = r_i = 12.5\%$.
- Now let us consider the more general case where *part of the physical investment (the factory, in the preceding example) is financed by issuing debt (for instance, bonds) and the rest with equity.* The *physical* investment characterized by $\underline{I}$ here gives rise to two distinct *financial* investments, $\underline{S}$ and $\underline{B}$, which correspond, respectively, to the transactions carried out by the shareholders and the bondholders.

The sequence $\underline{S}$ representing the investment of the shareholders will therefore *differ* from $\underline{I}$ and this means that $r_s$ (the YTM of $\underline{S}$) differs from the "intrinsic" yield $r_i$ (the YTM of $\underline{I}$).

> **Example 14**
> Let us suppose that a third of the preceding investment is financed by borrowing $8 million over 10 years, payable in arrears at an annual rate of 10%; the cash flow sequence for the company that results from the debt will be:
>
> $$-\underline{B} = \{+8, -0.8, -0.8, \ldots, -0.8, -8.8\}.$$
>
> ($\underline{B}$ is the sequence for the bondholders and $-\underline{B}$ the corresponding sequence for the company).

Furthermore, since the firm makes profits and is subject to a corporate tax of 34%, it saves $0.80 \times 0.34 = \$0.272$ million in taxes each year over the situation where it does not pay financing costs (since these are deductible from the taxable amount). So financing costs lead to the sequence $\{0, 0.272, 0.272, \ldots, 0.272, 0.272\}$ of *tax savings (tax shelters, tax benefits)* and the firm's net cost of financing then is $(1-0.34) \times 10\% = 6.6\%$.

The cash flow sequence for the shareholders that results is:

- Immediately (date 0) = $-(24-8) = -\$16$ million
- In years 1 to 9 = $3-0.8 + 0.272 = \$2.472$ million
- In year 10 = $24 + 3-8.8 + 0.272 = \$18.472$ million

Or, in other words: $\underline{S} = \{-16, 2.472, \ldots, 2.472, 18.472\}$.

## 2.3 Transactions Involving an Arbitrary Number of Cash Flows: Discounting... 71

The YTM of the cash flow sequence $\underline{S}$ represents the return $r_s$ on equity. Here, $r_s = \frac{2,472}{16} = 15.45\%$ (this is again a sequence of type 3). In this example, it exceeds the investment's "intrinsic" yield $r_i$ (12.5%).

This result can also be obtained by considering that the investment generates an "intrinsic" yield of 12.5% which allows, on the one hand, to pay on the debt, which financed a third of it, an interest rate after taxes of $(1–0.34) \times 10\% = 6.6\%$ and, on the other hand, to grant the shareholders, who financed two-thirds of it, a return (yield) rate of 15.45%. In effect, we have: $12.5\% = 1/3 \times 6.6\% + 2/3 \times 15.45\%$.

### (b) *Generalization and leverage*

The preceding equality can be generalized to[9]:

$$r_i = d(1 - \tau) \, r_d + (1 - d) \, r_s. \tag{2.11}$$

where:

$r_i = $ internal rate of return on the investment (IRR of $\underline{I}$);
$\tau = $ company tax (34% for example);
$r_d = $ cost of debt before taxes (YTM of $\underline{B}$);
$d = $ proportion of debt ($d \in [0, 1]$)[10];
$(1 - d) = $ proportion of equity ($(1 - d) \in [0, 1]$);
$r_s = $ return (yield) rate for the equity invested (YTM of $\underline{S}$).

In Example 14, we had:
$r_i = 12.5\%$; $r_d = 10\%$; $d = 1/3$; $\tau = 0.34$; $r_s = 15.45\%$ and we check again that:

$$12.5\% = \frac{1}{3} \times 0.66 \times 10\% + \frac{2}{3} \times 15.45\%.$$

Furthermore, formula (2.11) implies:

$$r_s = r_i + \frac{d}{1 - d} (r_i - (1 - \tau) r_d). \tag{2.12}$$

Equation (2.12) can be written as an ex-post or ex-ante relationship. In the latter case, $r_i$ and $r_s$ are, in general, random.

This equation leads directly to the proposition which expresses what is conventionally called financial leverage.

---

[9]One can show that the relationship (2.11) is exact if the proportion of debt remains constant throughout the life of the investment, which is true in our example (a loan with the same "time profile" as the investment), but only approximate for sequences $\underline{B}$ and $\underline{I}$ with different profiles.

[10]For a physical investment made by an existing firm, the debt proportion may reach 100%. For the entire firm, which legally must have some equity capital, $d$ may take a value within [0, 1].

## Proposition: financial leverage

(i) *The rate of return on equity increases with the debt-financed proportion of the investment if and only if $r_i > (1 - \tau)\, r_d$.*

(ii) *When $r_i$ is random, $r_s$ is random and we have:*

- *$E(r_s)$ increases with debt if and only if $E(r_i) > (1 - \tau)\, E(r_d)$ where $E$ denotes an expectation.*
- *Variance($r_s$), assumed to represent the risk borne by the shareholders, increases with the debt proportion d.*

## Proof

(i) and the first part of (ii) result directly from Eq. (2.12) since $\frac{d}{1-d}$ increases with $d$. The second part of (ii) also follows from Eq. (2.12) since it implies:

$$\text{Variance } (r_s) = \left(1 + \tfrac{d}{1-d}\right)^2 \text{Variance } (r_i), \text{ which increases with } d.$$

## Remarks

- Under the assumption of zero debt, $\underline{B}$, $d$ and the financial costs are zero; $\underline{I} = \underline{S}$ and $r_s = r_i$: the cash flows of the "physical investment" are entirely payable to the shareholders.

The "primary" or "physical" investment described by $\underline{I}$ is not necessarily an industrial or commercial investment as in the numerical example above. It could be, amongst other possibilities, a real estate investment or even a financial one (acquiring shares of another company) partly financed with debt and partly with equity. If $\underline{I}$ itself is financial, the use of the adjective "physical" for the investment is not appropriate and it would be better to call the investment "primary."

## 2.3.5 Some Guidelines for the Choice of an Appropriate Discount Rate

In general, the discount rate used for a given investment is the minimum rate of return, after taxes, required from the investment. This rate should allow a "normal return" on the funds that finance the investment. The problem of determining such a rate, called the "cost of capital" (for reasons that will become clear in the sequel) is complex and we provide some guidelines only.

(a) First let us assume that we can neglect the randomness affecting the cash flows from the investment under consideration. Two cases may then occur:
- In the first, the investment project is completely financed by debt: we then use as the discount rate the cost, net of taxes, of the contracted debt, that is **(1 − T) × interest rate**; the interest rate is that on bank loans or bonds whose term and profile are analogous to the required financing induced by the investment. It is important to note that the rate to be used is equal to the cost of

## 2.3 Transactions Involving an Arbitrary Number of Cash Flows: Discounting...

financing *net of taxes* since the financial costs may be deducted from taxable income: the tax saving is in this way accounted for in the interest rate (say, for a tax rate of 34%, $0.66\times$ the interest rate).

- In the second, equity is available, but its assignment to the investment project prevents the firm from realizing other investments. Therefore, for the considered project to be worth making, its after-tax yield must exceed that of the most profitable alternative investment, which is called *the project's opportunity cost*. This is the appropriate discount rate for the project under scrutiny.

(b) The previous results hold if the cash flows are known with certainty. This assumption being, in most cases, unrealistic, the preceding decision rules should be adjusted for uncertainty.

We suggest some guiding principles for the choice of an appropriate discount rate when the randomness of future cash flows cannot be ignored.

- Calculate the present value of the expected cash flows using a higher discount rate when the hazards that affect the cash flows are larger. This discount rate can be viewed as an interest rate plus a risk premium.
- Furthermore, some proportion of the investment has to be financed by equity. The role of equity is to cover most of the risk, so that, as a compensation, the expectation of its rate of return must be larger than what is paid to the debt. Indeed, contrary to a wrong idea that is based on an erroneous accounting conception unsuitable for finance, equity is not free but is in fact more costly than debt.

Let us assume that the investment is financed by debt and equity in a proportion of $d$ and $(1-d)$ respectively. Denote by $i$ the interest rate on the debt, so that $(1-\tau)i$ is its cost net of tax, and let $r_e$ be the cost of equity, i.e. the minimum rate of return *required* by the shareholders.

The investment project should then not be adopted unless its IRR allows, *at least*, paying the debt that finances it the rate $(1-\tau)i$, and the equity the rate $r_e$. Taking into account the respective weights of debt and equity, the minimum rate of return that the investment must provide, i.e. the discount rate to be used, is the "weighted average cost of capital" or *wacc*, which writes:

$$wacc = d(1-\tau)i + (1-d)r_e.$$

---

**Example 15**

Let us consider a firm that maintains a ratio "equity to assets" of 2/3. Taking into account the random hazards that affect its results, its shareholders require a return on equity of 15%. The average long-term credit rate is 9.1%, and the

(continued)

corporate income tax is 34%. The discount rate used by this firm to select its investment projects is therefore:

$$wacc = (2/3) \times 0.15 + (1/3) \times 0.66 \times 0.091 \approx 12\%.$$

Let us add two important observations intended to prevent confusion:

- The cost of equity $r_e$ rises as the risk which affects it grows larger. Since this risk increases with debt $d$ (see the proposition about leverage), $r_e$ is an increasing function of $d$. This dependence is too often forgotten when the *wacc* formula is applied mechanically.[11]

The equity cost $r_e$, which is a minimum requirement, should not be confused with the return on equity $r_s$ which appears in formula (2.11) for leverage (with $i = r_d$):

$$\text{IRR} = r_i = d\,(1-\tau)\,i \ + \ (1-d)\,r_s.$$

In this formula, $r_s$ *results* from the IRR. Indeed, it is *because $r_s$ must exceed $r_e$* that the IRR of the investment project must exceed the *wacc* (so that its NPV computed with the *wacc* is positive), and thus that the pertinent discount rate is the *wacc*!

### 2.3.6 Inflation, Real and Nominal Cash Flows and Rates

After defining the basic notions of real cash flow and real rate, we will examine the discounting methods to make them more easily applicable in the context of currency depreciation (i.e., inflation, the growth rate of the global Consumption Price Index).

(a) *Cash flows in constant dollars and real rates*

Up to now, we have considered sequences $F_0, F_1, \ldots, F_n$ of cash flows in current dollars (or nominal ones) and have discounted them with "the nominal rate" denoted by $r$, for example, a market rate: $F_0 + \frac{F_1}{1+r} + \frac{F_2}{(1+r)^2} + \ldots + \frac{F_n}{(1+r)^n}$.

Let us now suppose that the future inflation rate is equal to $\pi$; then:

- $F_1$ dollars in 1 period will have the same purchasing power as $f_1 = \frac{F_1}{1+\pi}$ dollars today.
- $F_2$ dollars in 2 periods will have the same purchasing power as $f_2 = \frac{F_2}{(1+\pi)^2}$ dollars today.
- - - - -

---

[11] Specifying the form of the function $r_e(d)$, however, is outside the scope of this book.

## 2.3 Transactions Involving an Arbitrary Number of Cash Flows: Discounting... 75

- $F_n$ dollars in $n$ periods will have the same purchasing power as $f_n = \frac{F_n}{(1+\pi)^n}$ dollars today.

We are therefore led to define a cash flow sequence "adjusted for currency depreciation":

**Definition** *Let us start with the nominal sequence $F_0, F_1, \ldots, F_n$ and the inflation rate $\pi$ (assumed constant over time) and let us define a new sequence $f_0, f_1, \ldots, f_t, \ldots, f_n$ by the following formulas:*

$$f_0 \equiv F_0; f_1 \equiv \frac{F_1}{1+\pi}; \ldots; f_t \equiv \frac{F_t}{(1+\pi)^t}; \ldots; f_n \equiv \frac{F_n}{(1+\pi)^n}$$

*The sequence $f_0, f_1, \ldots, f_t, \ldots, f_n$ defined in this way is called the sequence of constant dollars (or real cash flows) corresponding to the sequence of current dollars (or nominal cash flows) $F_0, F_1, \ldots, F_n$ and the inflation rate $\pi$.*

Furthermore, just as one adjusts the nominal cash flows to obtain real cash flows, one adjusts the nominal rate to obtain a real rate unaffected by the inflation rate.

Let us consider an investment or a loan of \$1 for one period undertaken at the (nominal) rate $r$ and characterized by two current dollar cash flows $(-1, 1 + r)$, i.e. $(-1, \frac{1+r}{1+\pi})$ in real (constant dollar) terms.

The "real" yield of this investment is the IRR of the real sequence (and not the nominal one), that is $\frac{1+r}{1+\pi} - 1$; thus we are led to the definition:

**Definition** *The real rate $k$ is defined from the nominal rate $r$ and the inflation rate $\pi$ by the relation:*

$$1 + k = \frac{1+r}{1+\pi} \tag{2.13}$$

This equation implies, for small values of $\pi$ and of $r$, that $k$ is approximately equal to $r - \pi$; the rate $k$ is called the *real* interest rate because it measures the interest rate *net* of currency depreciation. For example, if $r = 5\%$ and $\pi = 3\%$, $1 + k = 1.05/1.03$, thus $k = 1.94\% \approx 2\%$.

(b) *Discounting in an inflationary context*

Consider the present value of an arbitrary sequence of nominal cash flows ($F_0, F_1, \ldots, F_n$):

$$VP_F(r) = F_0 + \frac{F_1}{1+r} + \frac{F_2}{(1+r)^2} + \ldots + \frac{F_n}{(1+r)^n}$$

or, equivalently:

$$VP_F(r) = F_0 + \frac{F_1/(1+\pi)}{(1+r)/(1+\pi)} + \frac{F_2/(1+\pi)^2}{(1+r)^2/(1+\pi)^2} + \cdots + \frac{F^n/(1+\pi)^n}{(1+r)^n/(1+\pi)^n}$$

$$= f_0 + \frac{f_1}{\frac{1+r}{1+\pi}} + \cdots + \frac{f_n}{(\frac{1+r}{1+\pi})^n} = f_0 + \frac{f_1}{1+k} + \cdots + \frac{f_n}{(1+k)^n} = VP_f(k).$$

Consequently, the present value of $(F_0, F_1, \ldots, F_n)$ discounted at the nominal rate $r$ is also the present value of $(f_0, f_1, \ldots, f_n)$ discounted at the real rate $k = \frac{1+r}{1+\pi} - 1$.

The preceding considerations are summarized in the following proposition:

**Proposition**
- *For an inflation rate equal to $\pi$, the present value of a sequence of current dollar cash flows discounted at the nominal rate $r$ equals that of the corresponding constant dollar cash flows discounted at the corresponding real rate $k = \frac{1+r}{1+\pi} - 1$.*
- *Let $r$ be the nominal IRR of a current dollar sequence $(F_0, F_1, \ldots, F_n)$; the real IRR $k$ of the corresponding constant dollar sequence $(f_0, f_1, \ldots, f_n)$ is $k = \frac{1+r}{1+\pi} - 1$.*

In any case, the reader should avoid the error (which is relatively common) of discounting real cash flows[12] with a nominal interest rate or to compare a real IRR and a nominal interest rate.

## 2.4 Analysis of Long-Term Loans

After several definitions about rates, we show how to construct amortization schedules for loans.

### 2.4.1 General Considerations and Definitions: YTM and Interest Rates

(a) *Some definitions*

The borrower cancels her debt by making payments (interest and capital) to the lender according to a predetermined schedule. As the case may be, payments may be monthly, quarterly, semi-annual, or annual. The reference period corresponds to the periodicity of the cash transfers: 3 months if the payments are quarterly, a month if they are monthly, etc. The rate that is used (or needed) in the calculations will consequently be so defined: quarterly rate for quarters, monthly rate for months, etc. We use the term "periodic rate" to avoid any confusion with an annual rate.

---

[12] That is, estimated with an implicitly zero inflation rate.

## 2.4 Analysis of Long-Term Loans

Furthermore, for instruments longer than a year, interest is capitalized, so the rate is compound.

We will consider a loan that takes place over $n$ periods. The borrower receives the capital $C$ at the start time $t = 0$ and installments are paid on the dates $t = 1, 2, \ldots, n$. $F_t$ represents the cash outflow (disbursement by the borrower corresponding to reimbursement of the capital and interest payment) on the date $t$. We will denote by $r$ the periodic interest rate on the loan and $C_t$ the remaining capital owed on the date $t$, after the payment on date $t$.

Excluding administrative fees, possible commissions, and insurance costs, the cash flow sequence resulting from this credit can be written:

$$\underline{F} = \{-C, F_1, F_2, \ldots, F_n\}, \text{for the lender};$$

$$-\underline{F} = \{+C, -F_1, -F_2, \ldots, -F_n\}, \text{for the borrower}.$$

Most often, the bank does levy administrative fees and commissions. Calling $c$ the total amount of these fees assumed levied up front on the capital lent, and $c_i$ the fees, commissions, insurance and other charges due along with the $i$th installment, the borrower' bank account is credited, at time 0, with $C- c$ and not $C$, and is debited by $F_i + c_i$ and not $F_i$, so that the actual cash flow sequence resulting from the transaction is:

$$\underline{F}' = \{-(C - c), F_1 + c_1, F_2 + c_2, \ldots, F_n + c_n\}, \text{for the lender};$$

$$-\underline{F}' = \{C - c, -F_1 - c_1, -F_2 - c_2, \ldots, -F_n - c_n\}, \text{for the borrower}.$$

---

### Example 16

A loan of \$13,010, paid back over 2 years in semi-annual payments of \$3500, with administrative costs of \$100, generates semi-annual cash flows that, for the lender, write:

$$F = \{-13,010; 3500; 3500; 3500; 3500\}, \text{without commissions and fees}.$$

$$F' = \{-12,910; 3500; 3500; 3500; 3500\}, \text{net of commissions and fees}.$$

---

(b) *Periodic interest rate, periodic YTM and IRR*

We saw in the previous section that, in the absence of commissions and administrative fees, the *IRR of the investment constituted by a loan is equal to its interest rate*. In effect, in the case of a capital $C$ to be repaid in annual installments $F_t$, the periodic interest rate $r$ satisfies:

$$VP_F(r) \equiv -C + \frac{F_1}{1+r} + \frac{F_2}{(1+r)^2} + \cdots + \frac{F_n}{(1+r)^n} = 0.$$

It is therefore equal to the IRR of the sequence $F = \{-C, F_1, F_2, \ldots, F_n\}$.

When commissions, administrative fees and insurance costs are present, the sequence resulting from the transaction is $F'$ rather than $F$ and the rate that represents the effective cost of credit is the IRR or YTM of the sequence $F'$, that is the rate $r*$ for which:

$$VP_{F'}(r*) \equiv -(C - c) + \frac{F_1 + c_1}{1+r*} + \frac{F_2 + c_2}{(1+r*)^2} + \cdots + \frac{F_n + c_n}{(1+r*)^n} = 0.$$

*The rate $r*$ incorporates all the cost elements in contrast with the interest rate $r$.* As $r$ is the IRR of the sequence $F$ free of the fees, commissions and insurance costs that load the effective cost of financing, $r*$ is always larger than $r$.

Both $r$ and $r*$ are periodic rates, and they can be annualized using formula (2.5b) of Sect. 2.2.

---

### Example 17

In the example of a loan of \$13,010 repaid over 2 years in semi-annual payments of \$3500 and administrative costs of \$100, the semi-annual interest rate is such that:

$$-13010 + \frac{3500}{(1+r)} + \frac{3500}{(1+r)^2} + \frac{3500}{(1+r)^3} + \frac{3500}{(1+r)^4} = 0, \text{ i.e. } r = 3\%.$$

The semi-annual YTM $r*$ is such that:

$$-12910 + \frac{3500}{(1+r*)} + \frac{3500}{(1+r*)^2} + \frac{3500}{(1+r*)^3} + \frac{3500}{(1+r*)^4} = 0, \text{ i.e., } r*$$

$$= 3.32\%.$$

The corresponding *annual* rates are $(1.03)^2 - 1 = 6.09\%$ for the interest rate, and $(1.0332)^2 - 1 = 6.75\%$ for the compound rate (YTM).

---

Finally, let us remark that the definitions we have just presented are relative to the payment schedules in arrears (the usual case) and not to those which result from the advance payment of interests. These will be analyzed in Sect. 2.2(d).

(c) *Annualization of the periodic rate*

The calculation carried out using the schedule of cash flow yields a periodic rate corresponding to the period of capitalization of the interests (in example 17, a semi-annual rate).

## 2.4 Analysis of Long-Term Loans

Mostly, the rate posted is annualized and the problem is to convert it into a periodic rate. However, the problem may be the opposite: one starts from the cash flow sequence from which is computed the periodic rate (as in the preceding example), then one annualizes the periodic rate. The conversion of annual rates to periodic rates and the annualization of periodic rates are done, according to the context, using Eq. (2.5a) or (2.5b). In accordance with what has already been explained, the annual rate obtained from Eq. (2.5a) (sometimes called the APR) is a lower bound on the rate, while Eq. (2.5b) leads to a "truer" rate (sometimes called the APY).

### 2.4.2 Amortization Schedule for a Loan

The amortization schedule for a loan is made up of the following elements, for each repayment date $t$:

- The amount $F_t$ of the installment
- The amount $I_t$ of interests included in the installment
- The amount $R_t$ of capital reimbursement included in the installment
- The capital still due (outstanding) to the lender *after* the payment $t$, denoted by $C_t$. The interests $I_{t+1}$ of installment $t+1$ are based on this capital (or principal) outstanding due at $t+1$.
- The cost of death and disability insurance, if applicable.

(a) *Detailed description of the amortization schedule of the loan*

Here we assume that there are no fees or insurance premiums connected with the loan.

At $t = 0$, the borrower receives the capital $C$ thus $F_0 = C$ and the initial capital still due, $C_0$, is therefore equal to $C$.

For each following maturity date $t$, the installment $F_t$ can be split into two parts: $F_t = I_t + R_t$:

- The interests for the period $(t-1, t)$: $I_t = r\, C_{t-1}$
- The reimbursement $R_t$, (partial or total) of the principal outstanding:

$$R_t = F_t - I_t = F_t - r\, C_{t-1}.$$

Furthermore, $C_t = C_{t-1} - R_t$ (capital outstanding on the date $t$ = capital outstanding on the date $(t-1)$ – reimbursement made at $t$).

Recall that $r$ denotes a periodic interest rate and remark that the capital outstanding between dates $t-1$ and $t$ equals $C_{t-1}$ and is smaller than $C$ since a portion has

# 80      2   Basic Finance: Interest Rates, Discounting, Investments, Loans

already been refunded with the installments paid on dates between 1 and $t - 1$. Thus it is on the amount $C_{t-1}$ and not on $C$ that the interests for the period $(t-1, t)$, payable at $t$, must be calculated.

Let us also note that this "algorithm" also allows computing all the $I_t$, $R_t$, and $C_t$ step by step, as the table below illustrates (constructed from $t = 0$ then proceeding forward up to $t = n$ and filling out each line from left to right).

| Date | Installment $F_t$ | Interests $I_t$ | Capital reimbursed for the period $R_t$ | Capital outstanding $C_t$ |
|---|---|---|---|---|
| $t = 0$ | 0 | 0 | 0 | $C_0 = C$ |
| $t = 1$ | $F_1$ | $I_1 = r\,C_0$ | $R_1 = F_1 - I_1$ | $C_1 = C_0 - R_1$ |
| $t = 2$ | $F_2$ | $I_2 = r\,C_1$ | $R_2 = F_2 - I_2$ | $C_2 = C_1 - R_2$ |
| .... | .... | .... | .... | .... |
| $t$ | $F_t$ | $I_t = r\,C_{t-1}$ | $R_t = F_t - I_t$ | $C_t = C_{t-1} - R_t$ |
| ... | ... | ... | ... | ... |
| $t = n$ | $F_n$ | $I_n = r\,C_{n-1}$ | $R_n = F_n - I_n$ | $C_n = C_{n-1} - R_n = 0!$ |

---

**Example 18**

Let a loan of \$13,010 at a semi-annual rate of 3% be reimbursed in four semi-annual payments of \$3500:

| Date | Installment | Interest: $I_t$ | Capital reimbursed: $R_t$ | Capital outstanding: $C_t$ |
|---|---|---|---|---|
| $t = 0$ | 0 | | | 13,010 |
| $t = 1$ | 3500 | $13{,}010 \times 3\% = 390.30$ | $3500 - 390.30 = 3109.70$ | $13{,}010 - 3109.70 = 9900.30$ |
| $t = 2$ | 3500 | $9900.30 \times 3\% = 297.00$ | $3500 - 297.00 = 3203.00$ | $9900.30 - 3203 = 6697.30$ |
| $t = 3$ | 3500 | $6697.30 \times 3\% = 200.92$ | $3500 - 200.92 = 3299.08$ | $6697.30 - 3299.08 = 3398.22$ |
| $t = 4$ | 3500 | $3398.22 \times 3\% = 101.95$ | $3500 - 101.95 = 3398.05$ | $3{,}398.05 - 3{,}398.22 \approx 0$[a] |

[a]One does not actually obtain exactly zero at the end due to rounding errors. In practice, the last payment is adjusted so that the result is zero

---

One could have remarked that at the end of the last installment, the remaining capital of $C_n$, calculated step by step with the preceding algorithm, is zero. This result, clearly necessary for the whole procedure to be consistent, is not obvious on a strictly mathematical basis. It follows from a more general property of the capital outstanding, $C_t$, on any payment date $t$. We state and then prove this property after presenting several general formulas about the capital outstanding.

Let us consider a loan resulting in the cash flow sequence $\{- C, F_1, F_2, \ldots, F_n\}$.

## 2.4 Analysis of Long-Term Loans

Recall that its interest rate $r$, by definition, is such that: $C = C_0 = \sum_{t=1}^{n} \frac{F_t}{(1+r)^t}$.

Furthermore, $C_0 = C$ and the repayment of the principal in the $t^{th}$ installment is: $R_t = F_t - r\,C_{t-1}$, so that:

$$C_t = C_{t-1} - R_t = C_{t-1} - (F_t - r\,C_{t-1}), \text{ and so :}$$

$$C_t = (1+r)\,C_{t-1} - F_t. \tag{2.14}$$

If we know the installments $F_t$, this last equation allows us to calculate the $C_t$ step by step (starting from $C_0 = C$). So we can state the following proposition.

**Proposition**
*The capital outstanding just after an installment $t$ ($0 \le t \le n$) equals the present value at $t$ of the payments still to be made, discounted with the interest rate $r$ on the loan:*

$$C_t = \frac{F_{t+1}}{1+r} + \frac{F_{t+2}}{(1+r)^2} + \ldots + \frac{F_n}{(1+r)^{n-t}} \tag{2.15}$$

**Proof**
Let us proceed by iteration. Equation (2.15) holds for $t = 0$, by definition of $r$. Let us assume it holds at $t - 1$:

$$C_{t-1} = \frac{F_t}{1+r} + \frac{F_{t+1}}{(1+r)^2} + \ldots + \frac{F_n}{(1+r)^{n-t+1}} = 0.$$

Now, from (Eq. 2.15):

$C_t = (1+r)\,C_{t-1} - F_t$, or: $C_t = (1+r)\left(\frac{F_t}{1+r} + \frac{F_{t+1}}{(1+r)^2} + \ldots + \frac{F_n}{(1+r)^{n-t+1}}\right) - F_t$;

therefore

$C_t = \frac{F_{t+1}}{1+r} + \frac{F_{t+2}}{(1+r)^2} + \ldots + \frac{F_n}{(1+r)^{n-t}}$; Eq. (2.15) is true up to date $t$, which proves the proposition. $\square$

Note that after repayment of the last installment no further payment is owed, so that $C_n = 0$.

(b) *Common profiles of repayment*

- Bullet repayment

The borrowed capital is repaid all at one time on the final date. However, interests are paid at each period.

The cash flow diagram for the borrower is as in Fig. 2.12.

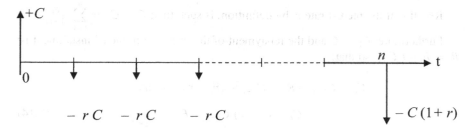

**Fig. 2.12** Bullet loan

This corresponds to a cash flow sequence: $\{+C, -rC, \ldots, -rC, -C(1+r)\}$.

We will see in Chap. 4 that bonds are generally *bullet* loans with interest payable in arrears, which corresponds to the case of Fig. 2.12.

- Loans with constant capital repayments (decreasing installments)

The capital outstanding is repaid in constant fractions: on each date $t$, a fraction $(C/n)$ is thus repaid. As a result, the values of the capital outstanding are:

| $t=0$ | $t=1$ | $t=2$ | $t$ | $t=n$ |
|---|---|---|---|---|
| $C_0 = C$ | $C_1 = C - C/n$ | $C_2 = C - 2C/n$ | $C_t = C - t\, C/n$ | $C - nC/n = 0$ |

The successive interest payments are easily deduced: $I_t = rC(1-(t-1)/n)$; they get smaller as time goes by. The resulting (decreasing) installments are equal to: $F_t = I_t + C/n$.

- Constant installments

The amortization schedule is calculated so that the installments $F_t$ are all equal to a constant $F$. As a result, the constant amount of the payments (capital reimbursed + interest) is linked to the capital borrowed $C$, to the period rate $r$, and to the number of periods $n$, by the relation: $F = \frac{rC}{1-(1+r)^{-n}}$.

Indeed, as we have previously seen, the present value at rate $r$ of the payment sequence equals the capital borrowed: $C = F \sum_{t=1}^{n} \frac{1}{(1+r)^t} = F \frac{1-(1+r)^{-n}}{r}$.

It follows that: $F = \frac{rC}{1-(1+r)^{-n}}$.

> **Example 19**
> Consider a loan of $200,000 with constant semi-annual payments at a semi-annual interest rate $r$ of 5.35%. Repayment is made in 10 semi-annual installments. The cash flow diagram is the following:

(continued)

## 2.4 Analysis of Long-Term Loans

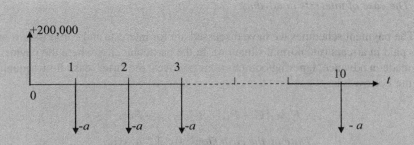

The present value discounted with the rate $r$ of the sequence of semi-annual payments equals the capital borrowed. Therefore, we have: $200,000 = \sum_{t=1}^{n} \times \frac{1}{(1+5,35\%)^t} a = 7.59 \, a$ from which $a = 200,000/7.59 = \$26,342.98$.

The amortization schedule thus is:

| Date | Total to pay | Interest | Amortization of capital | Capital outstanding |
|---|---|---|---|---|
| 0 | | | | 200,000.00 |
| 1 | 26,342.98 | 10,700.00 | 15,642.98 | 184,357.02 |
| 2 | 26,342.98 | 9863.10 | 16,479.88 | 167,877.14 |
| 3 | 26,342.98 | 8981.43 | 17,361.55 | 150,515.59 |
| 4 | 26,342.98 | 8052.58 | 18,290.40 | 132,225.19 |
| 5 | 26,342.98 | 7074.05 | 19,268.93 | 112,956.26 |
| 6 | 26,342.98 | 6043.16 | 20,299.82 | 92,656.44 |
| 7 | 26,342.98 | 4957.12 | 21,385.86 | 71,270.58 |
| 8 | 26,342.98 | 3812.98 | 22,530.00 | 48,740.57 |
| 9 | 26,342.98 | 2607.62 | 23,735.36 | 25,005.22 |
| 10 | 26,342.98 | 1337.78 | 25,005.20 | 0.00 |

(c) *Taking account of insurance and other costs proportional to the capital outstanding*

When the borrower signs up for death and disability insurance at a rate $i$, in each period, an insurance premium equal to $i$ times the capital outstanding is added to the interests. This amounts to increasing the loan rate $r$ by $i$. We obtain a loan rate "insurance included" $(r + i)$.

84  2 Basic Finance: Interest Rates, Discounting, Investments, Loans

(d) *The case of interests in advance*

The payment schedules we have discussed are for interests and commissions and fees paid in arrears (the normal situation). In the particular case where the payments are made in advance, ignoring commissions and fees, the "true" cash flow sequence for the borrower is:

$$\underline{F}' = \{C - F_1, -F_2, \ldots, -F_n\}.$$

*Date of the cash flow* : 0  1  n − 1.

The interest rate $r$ is always calculated from the sequence $\underline{F} = \{C, -F_1, -F_2, \ldots, -F_n\}$:

$$VP_{\underline{F}}(r) = -C + \frac{F_1}{1+r} + \frac{F_2}{(1+r)^2} + \ldots + \frac{F_n}{(1+r)^n} = 0$$

while the YTM $r^*$ calculated from the "true" sequence $\underline{F}'$ expresses the true yield:

$$VP_{\underline{F}'}(r*) \equiv -(C - F_1) + \frac{F_2}{1+r*} + \frac{F_3}{(1+r*)^2} + \ldots + \frac{F_n}{(1+r*)^{n-1}} = 0.$$

The amortization schedule for the loan with advance payments is set using the interest rate $r$, just as if it were for payments in arrears, and is similar to the latter.

## 2.5  Summary

- An interest rate $r$ may be *simple* (or proportional), *compound*, or *continuous compound*. Lending \$1 over $T$ periods yields $1 + rT$ ($rT$ being the interests) if $r$ is proportional, $(1 + r)^T$ if compound and $e^{rT}$ if continuous compound. Simple interest is mainly used for short-term instruments in the money market.
- To compute the time length $T$, the convention "*actual/360*" is used in the money market and the convention "*actual/actual*" elsewhere, where "actual" is the exact number of days.
- Two interest rates are *equivalent* for a given period, if they lead to the same final capital for the same initial capital. Simple $r$, compound $r'$ and continuous compound $r''$ are thus equivalent if: $1 + rT = (1 + r')^T = e^{r''T}$.
- There also exist formulas *to annualize periodic rates:* if 1 year contains $p$ periods: $1 + r_a = 1 + pr_p = (1 + r_p')^p = e^{rp''T}$, $r_a$ denoting annual and $r_p$ periodic rates.
- Interests can be paid *in arrears* (end of period) or *in advance* (beginning of period).
- The present value (PV) of a cash-flow $F$ available at future date $T$ is the amount that should be lent *now* in order to get $F$ on $T$. Depending on whether the *discount rate* is proportional, compound, or continuous, PV($F$) is given by: $F/(1 + rT)$, $F/$

Appendix 1: Geometric Series and Discounting                                    85

$(1 + r)^T$ or $Fe^{-rT}$. The PV of a *stream* is the *sum of its components PV's*.
*Discounting* is to find a present value while *capitalizing* is to find the future,
terminal, value of a (stream of) cash-flow(s).

- The NPV (*Net Present Value*) of an investment includes the initial outflow and
  represents the net gain or net increase in value resulting from the investment, as of
  its date of inception.
- The IRR (*Internal Rate of Return*) of an investment is the *special discount rate
  that makes its net present value equal to zero*. The *Yield to Maturity* (YTM) is
  exactly the same concept but is used in general for a financial instrument. In
  absence of commissions and administrative fees, the *IRR* of the investment
  constituted by a *loan* is equal to *its (periodic) interest rate*.
- A *zero*-coupon rate, or *pure discount rate*, is a special, but very important case of
  YTM when there are no intermediate cash-flows.
- An investment should be accepted, and any financial asset bought, if and only if
  their NPV is *non-negative*.
- *Leverage* is when an investment is partly financed by debt, the cost of which is
  smaller than the cost of equity. Leverage usually results in a higher expected rate
  of return for the investor, but at the cost of a higher risk as in general the future
  cash flows generated by the investment are random.
- The *wacc* (weighted average cost of capital) is the discount rate a firm should use
  to compute the NPV of its investments, when it issues both equity and debt.
- Nominal cash-flows must be distinguished from real cash-flows in that the latter
  are discounted by the rate of inflation. A *real* and a *nominal* interest rate differ
  (roughly) by the rate of inflation.
- In a standard credit, the borrower cancels their debt by making predetermined
  payments $F_t$ divided in interest and capital reimbursements according to the
  following step by step *algorithm*:
  Initially, the borrower receives a capital $C_0$. For each following maturity date $t$,
  the installment $F_t$ is split into two parts: the interests for the period $(t-1, t)$: $I_t = r$
  $C_{t-1}$; and the (partial) reimbursement $R_t = F_t - I_t$ of the capital outstanding so that
  $C_t = C_{t-1} - R_t$. The periodic interest rate $r$ is the IRR of the sequence $(-C_0, F_1, \ldots,$
  $F_T)$ but the *effective cost* of the credit is obtained from the sequence *including*
  commissions and fees.
- Repayment schemes for a loan are typically: *zero*-coupon (no intermediate cash-
  flows), *bullet* (no intermediate capital reimbursements), *constant installments*, or
  *constant capital repayments*.

## Appendix 1: Geometric Series and Discounting

Financial calculations with cash flows over many periods often employ the
properties of geometric sequences, also called geometric progressions.

In a geometric progression, each element is derived from the previous one
through multiplication by a factor called the ratio of the sequence. Thus a geometric
sequence with first term $a$ and ratio $q$ is written: $a, aq, aq^2, \ldots, aq^{n-1}, aq^n, \ldots$ the

$(n + 1)$st term $(aq^n)$ being obtained from $n$th $(aq^{n-1})$ by multiplication of it by the constant ratio $q$.

The sum $S_n$ of the first $n$ terms of a geometric sequence, whose initial term is $a$ and whose ratio is $q$, is given by:

$$S_n \equiv \underbrace{a + aq + aq^2 \ldots + aq^{n-1}}_{n \quad terms} = a\frac{1 - q^n}{1 - q}. \qquad (2.16)$$

Indeed

$$S_n = a + aq + \ldots + aq^{n-1};$$

therefore

$$qS_n = aq + aq^{2+} + \ldots + aq^n.$$

Subtracting the second equality from the first we get:

$$S_n(1 - q) = a - aq^n; \text{ from which}: S_n = a\frac{1 - q^n}{1 - q}.$$

Let us consider, in particular, the present value of a cash flow sequence of $1 for $n$ years: $\frac{1}{(1+r)} + \frac{1}{(1+r)^2} + \ldots + \frac{1}{(1+r)^n}$.

We have a sum of the first $n$ terms of a geometric progression whose initial term and ratio are both equal to $\frac{1}{(1+r)}$. (2.16) thus gives us:

$$\sum_{i=1}^{n} \frac{1}{(1 + r)^i} = \frac{1 - (1 + r)^{-n}}{r}. \qquad (2.17)$$

---

**Example 20**

$$1000 + 1000 \times 1.01 + 1000 \times (1.01)^2 + \cdots + 1000 \times (1.01)^{11}$$
$$= 1000[1 - (1.01)^{12}]/[1 - (1.01)] = 12{,}683.$$

The capital resulting at the end of 1 year (earned value) in the case of a savings plan with monthly interest of 1% or where the payments are $1000 *at the end* of each month is therefore $12,683.

---

Equation (2.16) giving the sum of the first $n$ terms of a geometric progression allows us, by taking the limit, to calculate the sum when the number of terms is infinite, if the ratio is smaller than one; indeed if $|q| < 1$, $q^n$ tends to zero as $n$ tends to infinity, and the sum $S_n = a\frac{1-q^n}{1-q}$ tends to $\frac{a}{1-q}$; i.e.:

Appendix 2: Using Financial Tables and Spreadsheets for Discount Computations      87

$$\text{Lim}_{n\to\infty}(S_n) \equiv S_\infty = \frac{a}{1-q}.$$ (2.18)

Such a computation comes up when one considers a perpetual annuity that provides a constant annual coupon (interest) indefinitely. For a discount rate $r$, its value is:

$$V(r) = \frac{a}{1+r} + \frac{a}{(1+r)^2} + \ldots + \frac{a}{(1+r)^n} + \ldots = \frac{\frac{a}{1+r}}{1 - \frac{1}{1+r}} = \frac{a}{r}$$

In particular:

$$\sum_{i=1}^{\infty} \frac{1}{(1+r)^i} = \frac{1}{r}.$$ (2.19)

---

**Example 21**

$$S_\infty = \frac{1000}{1,05} + \frac{1000}{(1,05)^2} + \frac{1000}{(1,05)^3} + \ldots + \frac{1000}{(1,05)^n} + \ldots = \frac{1000}{0,05} = 20,000.$$

If the interest rate is 5%, the value of a perpetual annuity giving a coupon of $1000 each year thus equals $S_\infty = \$20,000$.

---

## Appendix 2: Using Financial Tables and Spreadsheets for Discount Computations

### 1. Financial Tables

(a) *Computation of present value.*

Tables A1 and A2, at the end of the book, give present value factors for different values of $r$ and different maturities $n$. In Table A1, each line corresponds to a period $n$, and each column to an interest rate $r$. The value of $\frac{1}{(1+r)^n}$ is entered at the intersection of a line and a column. The terms of the sum of formula (2.6) can then be calculated one by one.

For example, to determine the present value of $\{-50, 6, 6, 6, 56\}$ discounted at a rate of 6%, we read, from the intersection of the sixth column (rate $= 6\%$) and the lines 1 to 4, respectively, the coefficients $1/(1.06)^1$, $1/(1.06)^2$, $1/(1.06)^3$ and $1/(1.06)^4$. Thus we obtain a net present value of:

**Financial Table A1** Present value of 1 currency unit: $1/(1 + r)^n$

| $n \backslash r$ | 1% | 2% | 3% | 4% | 5% | 6% | 7% | 8% | 9% | 10% | 11% | 12% | 13% | 14% | 15% | 16% | 17% | 18% | 19% | 20% | 22% | 24% | 26% | 28% | 30% | 32% |
|---|---|---|---|---|---|---|---|---|---|---|---|---|---|---|---|---|---|---|---|---|---|---|---|---|---|---|
| 1 | 0.9901 | 0.9804 | 0.9709 | 0.9615 | 0.9524 | 0.9434 | 0.9346 | 0.9259 | 0.9174 | 0.9091 | 0.9009 | 0.8929 | 0.8850 | 0.8772 | 0.8696 | 0.8621 | 0.8547 | 0.8475 | 0.8403 | 0.8333 | 0.8197 | 0.8065 | 0.7937 | 0.7813 | 0.7692 | 0.7576 |
| 2 | 0.9803 | 0.9612 | 0.9426 | 0.9246 | 0.9070 | 0.8900 | 0.8734 | 0.8573 | 0.8417 | 0.8264 | 0.8116 | 0.7972 | 0.7831 | 0.7695 | 0.7561 | 0.7432 | 0.7305 | 0.7182 | 0.7062 | 0.6944 | 0.6719 | 0.6504 | 0.6299 | 0.6104 | 0.5917 | 0.5739 |
| 3 | 0.9706 | 0.9423 | 0.9151 | 0.8890 | 0.8638 | 0.8396 | 0.8163 | 0.7938 | 0.7722 | 0.7513 | 0.7312 | 0.7118 | 0.6931 | 0.6750 | 0.6575 | 0.6407 | 0.6244 | 0.6086 | 0.5934 | 0.5787 | 0.5507 | 0.5245 | 0.4999 | 0.4768 | 0.4552 | 0.4348 |
| 4 | 0.9610 | 0.9238 | 0.8885 | 0.8548 | 0.8227 | 0.7921 | 0.7629 | 0.7350 | 0.7084 | 0.6830 | 0.6587 | 0.6355 | 0.6133 | 0.5921 | 0.5718 | 0.5523 | 0.5337 | 0.5158 | 0.4987 | 0.4823 | 0.4514 | 0.4230 | 0.3968 | 0.3725 | 0.3501 | 0.3294 |
| 5 | 0.9515 | 0.9057 | 0.8626 | 0.8219 | 0.7835 | 0.7473 | 0.7130 | 0.6806 | 0.6499 | 0.6209 | 0.5935 | 0.5674 | 0.5428 | 0.5194 | 0.4972 | 0.4761 | 0.4561 | 0.4371 | 0.4190 | 0.4019 | 0.3700 | 0.3411 | 0.3149 | 0.2910 | 0.2693 | 0.2495 |
| 6 | 0.9420 | 0.8880 | 0.8375 | 0.7903 | 0.7462 | 0.7050 | 0.6663 | 0.6302 | 0.5963 | 0.5645 | 0.5346 | 0.5066 | 0.4803 | 0.4556 | 0.4323 | 0.4104 | 0.3898 | 0.3704 | 0.3521 | 0.3349 | 0.3033 | 0.2751 | 0.2499 | 0.2274 | 0.2072 | 0.1890 |
| 7 | 0.9327 | 0.8706 | 0.8131 | 0.7599 | 0.7107 | 0.6651 | 0.6227 | 0.5835 | 0.5470 | 0.5132 | 0.4817 | 0.4523 | 0.4251 | 0.3996 | 0.3759 | 0.3538 | 0.3332 | 0.3139 | 0.2959 | 0.2791 | 0.2486 | 0.2218 | 0.1983 | 0.1776 | 0.1594 | 0.1432 |
| 8 | 0.9235 | 0.8535 | 0.7894 | 0.7307 | 0.6768 | 0.6274 | 0.5820 | 0.5403 | 0.5019 | 0.4665 | 0.4339 | 0.4039 | 0.3762 | 0.3506 | 0.3269 | 0.3050 | 0.2848 | 0.2660 | 0.2487 | 0.2326 | 0.2038 | 0.1789 | 0.1574 | 0.1388 | 0.1226 | 0.1085 |
| 9 | 0.9143 | 0.8368 | 0.7664 | 0.7026 | 0.6446 | 0.5919 | 0.5439 | 0.5002 | 0.4604 | 0.4241 | 0.3909 | 0.3606 | 0.3329 | 0.3075 | 0.2843 | 0.2630 | 0.2434 | 0.2255 | 0.2090 | 0.1938 | 0.1670 | 0.1443 | 0.1249 | 0.1084 | 0.0943 | 0.0822 |
| 10 | 0.9053 | 0.8203 | 0.7441 | 0.6756 | 0.6139 | 0.5584 | 0.5083 | 0.4632 | 0.4224 | 0.3855 | 0.3522 | 0.3220 | 0.2946 | 0.2697 | 0.2472 | 0.2267 | 0.2080 | 0.1911 | 0.1756 | 0.1615 | 0.1369 | 0.1164 | 0.0992 | 0.0847 | 0.0725 | 0.0623 |
| 11 | 0.8963 | 0.8043 | 0.7224 | 0.6496 | 0.5847 | 0.5268 | 0.4751 | 0.4289 | 0.3875 | 0.3505 | 0.3173 | 0.2875 | 0.2607 | 0.2366 | 0.2149 | 0.1954 | 0.1778 | 0.1619 | 0.1476 | 0.1346 | 0.1122 | 0.0938 | 0.0787 | 0.0662 | 0.0558 | 0.0472 |
| 12 | 0.8874 | 0.7885 | 0.7014 | 0.6246 | 0.5568 | 0.4970 | 0.4440 | 0.3971 | 0.3555 | 0.3186 | 0.2858 | 0.2567 | 0.2307 | 0.2076 | 0.1869 | 0.1685 | 0.1520 | 0.1372 | 0.1240 | 0.1122 | 0.0920 | 0.0757 | 0.0625 | 0.0517 | 0.0429 | 0.0357 |
| 13 | 0.8787 | 0.7730 | 0.6810 | 0.6006 | 0.5303 | 0.4688 | 0.4150 | 0.3677 | 0.3262 | 0.2897 | 0.2575 | 0.2292 | 0.2042 | 0.1821 | 0.1625 | 0.1452 | 0.1299 | 0.1163 | 0.1042 | 0.0935 | 0.0754 | 0.0610 | 0.0496 | 0.0404 | 0.0330 | 0.0271 |
| 14 | 0.8700 | 0.7579 | 0.6611 | 0.5775 | 0.5051 | 0.4423 | 0.3878 | 0.3405 | 0.2992 | 0.2633 | 0.2320 | 0.2046 | 0.1807 | 0.1597 | 0.1413 | 0.1252 | 0.1110 | 0.0985 | 0.0876 | 0.0779 | 0.0618 | 0.0492 | 0.0393 | 0.0316 | 0.0254 | 0.0205 |
| 15 | 0.8613 | 0.7430 | 0.6419 | 0.5553 | 0.4810 | 0.4173 | 0.3624 | 0.3152 | 0.2745 | 0.2394 | 0.2090 | 0.1827 | 0.1599 | 0.1401 | 0.1229 | 0.1079 | 0.0949 | 0.0835 | 0.0736 | 0.0649 | 0.0507 | 0.0397 | 0.0312 | 0.0247 | 0.0195 | 0.0155 |
| 16 | 0.8528 | 0.7284 | 0.6232 | 0.5339 | 0.4581 | 0.3936 | 0.3387 | 0.2919 | 0.2519 | 0.2176 | 0.1883 | 0.1631 | 0.1415 | 0.1229 | 0.1069 | 0.0930 | 0.0811 | 0.0708 | 0.0618 | 0.0541 | 0.0415 | 0.0320 | 0.0248 | 0.0193 | 0.0150 | 0.0118 |
| 17 | 0.8444 | 0.7142 | 0.6050 | 0.5134 | 0.4363 | 0.3714 | 0.3166 | 0.2703 | 0.2311 | 0.1978 | 0.1696 | 0.1456 | 0.1252 | 0.1078 | 0.0929 | 0.0802 | 0.0693 | 0.0600 | 0.0520 | 0.0451 | 0.0340 | 0.0258 | 0.0197 | 0.0150 | 0.0116 | 0.0089 |
| 18 | 0.8360 | 0.7002 | 0.5874 | 0.4936 | 0.4155 | 0.3503 | 0.2959 | 0.2502 | 0.2120 | 0.1799 | 0.1528 | 0.1300 | 0.1108 | 0.0946 | 0.0808 | 0.0691 | 0.0592 | 0.0508 | 0.0437 | 0.0376 | 0.0279 | 0.0208 | 0.0156 | 0.0118 | 0.0089 | 0.0068 |
| 19 | 0.8277 | 0.6864 | 0.5703 | 0.4746 | 0.3957 | 0.3305 | 0.2765 | 0.2317 | 0.1945 | 0.1635 | 0.1377 | 0.1161 | 0.0981 | 0.0829 | 0.0703 | 0.0596 | 0.0506 | 0.0431 | 0.0367 | 0.0313 | 0.0229 | 0.0168 | 0.0124 | 0.0092 | 0.0068 | 0.0051 |
| 20 | 0.8195 | 0.6730 | 0.5537 | 0.4564 | 0.3769 | 0.3118 | 0.2584 | 0.2145 | 0.1784 | 0.1486 | 0.1240 | 0.1037 | 0.0868 | 0.0728 | 0.0611 | 0.0514 | 0.0433 | 0.0365 | 0.0308 | 0.0261 | 0.0187 | 0.0135 | 0.0098 | 0.0072 | 0.0053 | 0.0039 |
| 25 | 0.7798 | 0.6095 | 0.4776 | 0.3751 | 0.2953 | 0.2330 | 0.1842 | 0.1460 | 0.1160 | 0.0923 | 0.0736 | 0.0588 | 0.0471 | 0.0378 | 0.0304 | 0.0245 | 0.0197 | 0.0160 | 0.0129 | 0.0105 | 0.0069 | 0.0046 | 0.0031 | 0.0021 | 0.0014 | 0.0010 |
| 30 | 0.7419 | 0.5521 | 0.4120 | 0.3083 | 0.2314 | 0.1741 | 0.1314 | 0.0994 | 0.0754 | 0.0573 | 0.0437 | 0.0334 | 0.0256 | 0.0196 | 0.0151 | 0.0116 | 0.0090 | 0.0070 | 0.0054 | 0.0042 | 0.0026 | 0.0016 | 0.0010 | 0.0006 | 0.0004 | 0.0002 |
| 35 | 0.7059 | 0.5000 | 0.3554 | 0.2534 | 0.1813 | 0.1301 | 0.0937 | 0.0676 | 0.0490 | 0.0356 | 0.0259 | 0.0189 | 0.0139 | 0.0102 | 0.0075 | 0.0055 | 0.0041 | 0.0030 | 0.0023 | 0.0017 | 0.0009 | 0.0005 | 0.0003 | 0.0002 | 0.0001 | 0.0001 |
| 40 | 0.6717 | 0.4529 | 0.3066 | 0.2083 | 0.1420 | 0.0972 | 0.0668 | 0.0460 | 0.0318 | 0.0221 | 0.0154 | 0.0107 | 0.0075 | 0.0053 | 0.0037 | 0.0026 | 0.0019 | 0.0013 | 0.0010 | 0.0007 | 0.0004 | 0.0002 | 0.0001 | 0.0001 | 0.0000 | 0.0000 |
| 45 | 0.6391 | 0.4102 | 0.2644 | 0.1712 | 0.1113 | 0.0727 | 0.0476 | 0.0313 | 0.0207 | 0.0137 | 0.0091 | 0.0061 | 0.0041 | 0.0027 | 0.0019 | 0.0013 | 0.0009 | 0.0006 | 0.0004 | 0.0003 | 0.0001 | 0.0001 | 0.0000 | 0.0000 | 0.0000 | 0.0000 |
| 50 | 0.6080 | 0.3715 | 0.2281 | 0.1407 | 0.0872 | 0.0543 | 0.0339 | 0.0213 | 0.0134 | 0.0085 | 0.0054 | 0.0035 | 0.0022 | 0.0014 | 0.0009 | 0.0006 | 0.0004 | 0.0003 | 0.0002 | 0.0001 | 0.0000 | 0.0000 | 0.0000 | 0.0000 | 0.0000 | 0.0000 |

**Financial Table A2**  Present value of a sequence of 1 currency unit per period during n periods $1/(1+r) + 1/(1+r)^2 + \ldots + 1/(1+r)^n$

| $n \backslash r$ | 1% | 2% | 3% | 4% | 5% | 6% | 7% | 8% | 9% | 10% | 11% | 12% | 13% | 14% | 15% | 16% | 17% | 18% | 19% | 20% | 22% | 24% | 26% | 28% | 30% | 32% |
|---|---|---|---|---|---|---|---|---|---|---|---|---|---|---|---|---|---|---|---|---|---|---|---|---|---|---|
| 1 | 0.9901 | 0.9804 | 0.9709 | 0.9615 | 0.9524 | 0.9434 | 0.9346 | 0.9259 | 0.9174 | 0.9091 | 0.9009 | 0.8929 | 0.8850 | 0.8772 | 0.8696 | 0.8621 | 0.8547 | 0.8475 | 0.8403 | 0.8333 | 0.8197 | 0.8065 | 0.7937 | 0.7813 | 0.7692 | 0.7576 |
| 2 | 1.9704 | 1.9416 | 1.9135 | 1.8861 | 1.8594 | 1.8334 | 1.8080 | 1.7833 | 1.7591 | 1.7355 | 1.7125 | 1.6901 | 1.6681 | 1.6467 | 1.6257 | 1.6052 | 1.5852 | 1.5656 | 1.5465 | 1.5278 | 1.4915 | 1.4568 | 1.4235 | 1.3916 | 1.3609 | 1.3315 |
| 3 | 2.9410 | 2.8839 | 2.8286 | 2.7751 | 2.7232 | 2.6730 | 2.6243 | 2.5771 | 2.5313 | 2.4869 | 2.4437 | 2.4018 | 2.3612 | 2.3216 | 2.2832 | 2.2459 | 2.2096 | 2.1743 | 2.1399 | 2.1065 | 2.0422 | 1.9813 | 1.9234 | 1.8684 | 1.8161 | 1.7663 |
| 4 | 3.9020 | 3.8077 | 3.7171 | 3.6299 | 3.5460 | 3.4651 | 3.3872 | 3.3121 | 3.2397 | 3.1699 | 3.1024 | 3.0373 | 2.9745 | 2.9137 | 2.8550 | 2.7982 | 2.7432 | 2.6901 | 2.6386 | 2.5887 | 2.4936 | 2.4043 | 2.3202 | 2.2410 | 2.1662 | 2.0957 |
| 5 | 4.8534 | 4.7135 | 4.5797 | 4.4518 | 4.3295 | 4.2124 | 4.1002 | 3.9927 | 3.8897 | 3.7908 | 3.6959 | 3.6048 | 3.5172 | 3.4331 | 3.3522 | 3.2743 | 3.1993 | 3.1272 | 3.0576 | 2.9906 | 2.8636 | 2.7454 | 2.6351 | 2.5320 | 2.4356 | 2.3452 |
| 6 | 5.7955 | 5.6014 | 5.4172 | 5.2421 | 5.0757 | 4.9173 | 4.7665 | 4.6229 | 4.4859 | 4.3553 | 4.2305 | 4.1114 | 3.9975 | 3.8887 | 3.7845 | 3.6847 | 3.5892 | 3.4976 | 3.4098 | 3.3255 | 3.1669 | 3.0205 | 2.8850 | 2.7594 | 2.6427 | 2.5342 |
| 7 | 6.7282 | 6.4720 | 6.2303 | 6.0021 | 5.7864 | 5.5824 | 5.3893 | 5.2064 | 5.0330 | 4.8684 | 4.7122 | 4.5638 | 4.4226 | 4.2883 | 4.1604 | 4.0386 | 3.9224 | 3.8115 | 3.7057 | 3.6046 | 3.4155 | 3.2423 | 3.0833 | 2.9370 | 2.8021 | 2.6775 |
| 8 | 7.6517 | 7.3255 | 7.0197 | 6.7327 | 6.4632 | 6.2098 | 5.9713 | 5.7466 | 5.5348 | 5.3349 | 5.1461 | 4.9676 | 4.7988 | 4.6389 | 4.4873 | 4.3436 | 4.2072 | 4.0776 | 3.9544 | 3.8372 | 3.6193 | 3.4212 | 3.2407 | 3.0758 | 2.9247 | 2.7860 |
| 9 | 8.5660 | 8.1622 | 7.7861 | 7.4353 | 7.1078 | 6.8017 | 6.5152 | 6.2469 | 5.9952 | 5.7590 | 5.5370 | 5.3282 | 5.1317 | 4.9464 | 4.7716 | 4.6065 | 4.4506 | 4.3030 | 4.1633 | 4.0310 | 3.7863 | 3.5655 | 3.3657 | 3.1842 | 3.0190 | 2.8681 |
| 10 | 9.4713 | 8.9826 | 8.5302 | 8.1109 | 7.7217 | 7.3601 | 7.0236 | 6.7101 | 6.4177 | 6.1446 | 5.8892 | 5.6502 | 5.4262 | 5.2161 | 5.0188 | 4.8332 | 4.6586 | 4.4941 | 4.3389 | 4.1925 | 3.9232 | 3.6819 | 3.4648 | 3.2689 | 3.0915 | 2.9304 |
| 11 | 10.3676 | 9.7868 | 9.2526 | 8.7605 | 8.3064 | 7.8869 | 7.4987 | 7.1390 | 6.8052 | 6.4951 | 6.2065 | 5.9377 | 5.6869 | 5.4527 | 5.2337 | 5.0286 | 4.8364 | 4.6560 | 4.4865 | 4.3271 | 4.0354 | 3.7757 | 3.5435 | 3.3351 | 3.1473 | 2.9776 |
| 12 | 11.2551 | 10.5753 | 9.9540 | 9.3851 | 8.8633 | 8.3838 | 7.9427 | 7.5361 | 7.1607 | 6.8137 | 6.4924 | 6.1944 | 5.9176 | 5.6603 | 5.4206 | 5.1971 | 4.9884 | 4.7932 | 4.6105 | 4.4392 | 4.1274 | 3.8514 | 3.6059 | 3.3868 | 3.1903 | 3.0133 |
| 13 | 12.1337 | 11.3484 | 10.6350 | 9.9856 | 9.3936 | 8.8527 | 8.3577 | 7.9038 | 7.4869 | 7.1034 | 6.7499 | 6.4235 | 6.1218 | 5.8424 | 5.5831 | 5.3423 | 5.1183 | 4.9095 | 4.7147 | 4.5327 | 4.2028 | 3.9124 | 3.6555 | 3.4272 | 3.2233 | 3.0404 |
| 14 | 13.0037 | 12.1062 | 11.2961 | 10.5631 | 9.8986 | 9.2950 | 8.7455 | 8.2442 | 7.7862 | 7.3667 | 6.9819 | 6.6282 | 6.3025 | 6.0021 | 5.7245 | 5.4675 | 5.2293 | 5.0081 | 4.8023 | 4.6106 | 4.2646 | 3.9616 | 3.6949 | 3.4587 | 3.2487 | 3.0609 |
| 15 | 13.8651 | 12.8493 | 11.9379 | 11.1184 | 10.3797 | 9.7122 | 9.1079 | 8.5595 | 8.0607 | 7.6061 | 7.1909 | 6.8109 | 6.4624 | 6.1422 | 5.8474 | 5.5755 | 5.3242 | 5.0916 | 4.8759 | 4.6755 | 4.3152 | 4.0013 | 3.7261 | 3.4834 | 3.2682 | 3.0764 |
| 16 | 14.7179 | 13.5777 | 12.5611 | 11.6523 | 10.8378 | 10.1059 | 9.4466 | 8.8514 | 8.3126 | 7.8237 | 7.3792 | 6.9740 | 6.6039 | 6.2651 | 5.9542 | 5.6685 | 5.4053 | 5.1624 | 4.9377 | 4.7296 | 4.3567 | 4.0333 | 3.7509 | 3.5026 | 3.2832 | 3.0882 |
| 17 | 15.5623 | 14.2919 | 13.1661 | 12.1657 | 11.2741 | 10.4773 | 9.7632 | 9.1216 | 8.5436 | 8.0216 | 7.5488 | 7.1196 | 6.7291 | 6.3729 | 6.0472 | 5.7487 | 5.4746 | 5.2223 | 4.9897 | 4.7746 | 4.3908 | 4.0591 | 3.7705 | 3.5177 | 3.2948 | 3.0971 |
| 18 | 16.3983 | 14.9920 | 13.7535 | 12.6593 | 11.6896 | 10.8276 | 10.0591 | 9.3719 | 8.7556 | 8.2014 | 7.7016 | 7.2497 | 6.8399 | 6.4674 | 6.1280 | 5.8178 | 5.5339 | 5.2732 | 5.0333 | 4.8122 | 4.4187 | 4.0799 | 3.7861 | 3.5294 | 3.3037 | 3.1039 |
| 19 | 17.2260 | 15.6785 | 14.3238 | 13.1339 | 12.0853 | 11.1581 | 10.3356 | 9.6036 | 8.9501 | 8.3649 | 7.8393 | 7.3658 | 6.9380 | 6.5504 | 6.1982 | 5.8775 | 5.5845 | 5.3162 | 5.0700 | 4.8435 | 4.4415 | 4.0967 | 3.7985 | 3.5386 | 3.3105 | 3.1090 |
| 20 | 18.0456 | 16.3514 | 14.8775 | 13.5903 | 12.4622 | 11.4699 | 10.5940 | 9.8181 | 9.1285 | 8.5136 | 7.9633 | 7.4694 | 7.0248 | 6.6231 | 6.2593 | 5.9288 | 5.6278 | 5.3527 | 5.1009 | 4.8696 | 4.4603 | 4.1103 | 3.8083 | 3.5458 | 3.3158 | 3.1129 |
| 25 | 22.0232 | 19.5235 | 17.4131 | 15.6221 | 14.0939 | 12.7834 | 11.6536 | 10.6748 | 9.8226 | 9.0770 | 8.4217 | 7.8431 | 7.3300 | 6.8729 | 6.4641 | 6.0971 | 5.7662 | 5.4669 | 5.1951 | 4.9476 | 4.5139 | 4.1474 | 3.8342 | 3.5640 | 3.3286 | 3.1220 |
| 30 | 25.8077 | 22.3965 | 19.6004 | 17.2920 | 15.3725 | 13.7648 | 12.4090 | 11.2578 | 10.2737 | 9.4269 | 8.6938 | 8.0552 | 7.4957 | 7.0027 | 6.5660 | 6.1772 | 5.8294 | 5.5168 | 5.2347 | 4.9789 | 4.5338 | 4.1601 | 3.8424 | 3.5693 | 3.3321 | 3.1242 |
| 35 | 29.4086 | 24.9986 | 21.4872 | 18.6646 | 16.3742 | 14.4982 | 12.9477 | 11.6546 | 10.5668 | 9.6442 | 8.8552 | 8.1755 | 7.5856 | 7.0700 | 6.6166 | 6.2153 | 5.8582 | 5.5386 | 5.2512 | 4.9915 | 4.5411 | 4.1644 | 3.8450 | 3.5708 | 3.3330 | 3.1248 |
| 40 | 32.8347 | 27.3555 | 23.1148 | 19.7928 | 17.1591 | 15.0463 | 13.3317 | 11.9246 | 10.7574 | 9.7791 | 8.9511 | 8.2438 | 7.6344 | 7.1050 | 6.6418 | 6.2335 | 5.8713 | 5.5482 | 5.2582 | 4.9966 | 4.5439 | 4.1659 | 3.8458 | 3.5712 | 3.3332 | 3.1250 |
| 45 | 36.0945 | 29.4902 | 24.5187 | 20.7200 | 17.7741 | 15.4558 | 13.6055 | 12.1084 | 10.8812 | 9.8628 | 9.0079 | 8.2825 | 7.6609 | 7.1232 | 6.6543 | 6.2421 | 5.8773 | 5.5523 | 5.2611 | 4.9986 | 4.5449 | 4.1664 | 3.8460 | 3.5714 | 3.3333 | 3.1250 |
| 50 | 39.1961 | 31.4236 | 25.7298 | 21.4822 | 18.2559 | 15.7619 | 13.8007 | 12.2335 | 10.9617 | 9.9148 | 9.0417 | 8.3045 | 7.6752 | 7.1327 | 6.6605 | 6.2463 | 5.8801 | 5.5541 | 5.2623 | 4.9995 | 4.5452 | 4.1666 | 3.8461 | 3.5714 | 3.3333 | 3.1250 |

$$-50 + 6 \times 0.9434 + 6 \times 0.89 + 6 \times 0.8396 + 56 \times 0.7921 = 10.396.$$

For constant cash flows, it is preferable to use Tables A2 which give, at the intersection of line $n$ and column $r$, the value of $\sum_{i=1}^{n} \frac{1}{(1+r)^i} = \frac{1-(1+r)^{-n}}{r}$, see Eq. (2.17), in the preceding appendix.

For example, Tables A2 indicate that $\sum_{i=1}^{10} \frac{1}{(1,1)^i} = 6.1446$. As a result, the present value of the cash flow sequence: $\{-1200; 360; \ldots; 360\}$ discounted at a rate of 10%, equals:

$$-1200 + 360 \times 6.1446 = 1012.06.$$

(b) *Computation of a YTM by interpolation, using tables.*

If one has neither a calculator nor a spreadsheet, computing the YTM may be done using financial tables, by successive approximation: one brackets the IRR between two near values, one too large (NPV $<0$) and the other too small (NPV $>0$), then one makes a linear interpolation.

---

### Example 22

Let us consider the cash flow sequence: $\left\{ -1000; \underbrace{40; \ldots; 40}_{9 \text{ times}}; 1090 \right\}$

$NPV(r) = -1000 + 40 \sum_{i=1}^{9} \frac{1}{(1+r)^i} + 1090 \frac{1}{(1+r)^{10}}$; the curve of this function of $r$ is drawn in Fig. 2.13.

We wish to determine $r^*$ so that $NPV(r^*) = 0$ (abscissa of the point a in the figure, intersection of the curve of $NPV(r)$ with the horizontal axis). We proceed by increments, finishing with a linear interpolation.

For $r = 4\%$, we find $NPV = + 33.78$, which means the $r^*$ sought is $> 4\%$. For $r = 5\%$, we obtain $NPV = -46.52$, which means the rate $r^*$ sought is $< 5\%$.

By linear interpolation, we find $NPV = 0$ for $r = 4.42\%$:
$(4\% + \frac{33,78}{33,78+46,52} \times 1\% = 4,42\%$; abscissa of point b).

The actual IRR (4.409%, abscissa of the point a) is in fact slightly smaller than the result of linear interpolation, due to the convexity of the function $NPV(r)$ (see the figure).

# Appendix 2: Using Financial Tables and Spreadsheets for Discount Computations

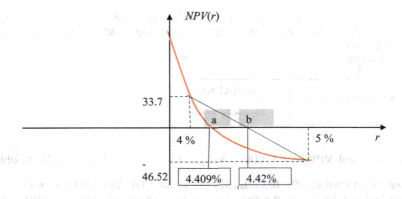

**Fig. 2.13** Computing a yield-to-maturity

Remark that Table A2 may allow us to find, relatively simply, the YTM of a sequence of constant cash flows $F$ following an initial disbursement of $F_0$. As a matter of fact, in this case,

$r*$ is simply the solution of the equation: $\sum_{i=1}^{9} \frac{1}{(1+r*)^i} = -F_0/F$.

To find the value of $r*$, it suffices to consider the line corresponding to $n$ periods, and to run along this line until one finds the closest value to $-F_0/F$; the corresponding column gives the approximate value of the period YTM.[13]

(c) *Using spreadsheets*

Spreadsheets allow calculating the present value or the IRR very easily. However, reading the manual is absolutely necessary to check the mathematical formula used and the assumptions that are made for its use (particularly the reference time) which differ from one piece of software to another.

In Excel, for example, the *NPV* of a sequence is computed using the function *NPV*(discount rate; sequence). However, *this function discounts the first flow of cash*. Let us consider for example the sequence $\{-1000; 40; \ldots; 40; 1090\}$ whose terms are written in the cell range A1:K1 (see the Excel table below):

$NPV(0.04; A1:K1) = -1000 \frac{1}{1,04} + 40 \sum_{i=1}^{9} \frac{1}{(1,04)^{i+1}} + 1090 \frac{1}{(1,04)^{11}} = 32.48$.

---

[13] In general, a linear interpolation is also needed here.

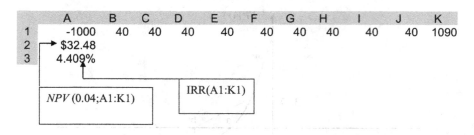

But the usual *NPV* is: $-1\,000 + 40 \sum_{i=1}^{9} \frac{1}{(1{,}04)^i} + 1090 \frac{1}{(1{,}04)^{10}} = 33.78$; to obtain this we have to write: *NPV* (0.04;B1:K1) + A1 (or 1.04\**NPV* (0.04;A1:K1)).

Continuing with Excel, the financial function which calculates the IRR is IRR (sequence). For example, IRR(A1:K1) will give the IRR of the cash flow sequence entered in the cell range A1:K1, which is 4.409% in our example.

## Suggested Reading[14]

Brealey, R., Myers, S., & Allen, F. (2019). *Principles of corporate finance* (13th ed.). McGraw Hill, Pearson Education.

Fabozzi, F., Modigliani, F., & Jones, F. (2014). *Foundations of financial markets and institutions* (4th ed.). Pearson Education.

---

[14] The mathematics of interest rates is addressed in all the books about the basics of finance. The reader may supplement the study of this chapter by reading:

# The Money Market and Its Interbank Segment

**3**

Borrowing on the interest-rate markets (money and bond markets) can be carried out by companies, banks, sovereign states, and local collectives. The differences that distinguish the bond market from the money market arise from the terms of the transactions at inception (long and medium term on the bond market, short term on the money market) as well as from regulations and certain practices, notably regarding quotation. We will use the following convention: Short-term (less than 1 year at their inception) securities and transactions belong in the money market and longer-term instruments belong in the bond market. Together, they are often referred to as *fixed-income* instruments, as the (implicit or explicit) coupon(s) they generate is (are) usually predetermined and constant.[1]

Loans issued on interest-rate markets are materialized by *negotiable securities*. This "incorporation" of a debt into a security negotiable on a market (namely, a bond), has been called "securitization"[2] and has important consequences.

– First, in contrast with a debt not securitized, the bond can, after being issued, easily be traded in the secondary market (to be distinguished from the primary market which concerns debts at issuance). Trading the debt on a secondary market provides it with a *liquidity* that is a definite advantage. This explains why, other things being equal, interest rates are in general lower for debts that have been securitized.

– A second difference between bank financing and market financing arises from the fact that securitized debts traded on a market are generally priced at their market values and not on the (accounting) basis of their capital outstanding.

---

[1] Even if there are some instruments using variable rates and indexed bonds, as we will see in Chap. 4.

[2] The term "securitization" is above all used to describe more complex arrangements involving the issuing of new securities by an ad hoc entity (a Special Purpose Vehicle) backed by pre-existing assets. Such arrangements are studied in detail in Chap. 29.

© The Author(s), under exclusive license to Springer Nature Switzerland AG 2022
P. Poncet, R. Portait, *Capital Market Finance*, Springer Texts in Business and Economics, https://doi.org/10.1007/978-3-030-84600-8_3

- A final difference should also be mentioned: recall that, for bank credit, the banks "intermediate" between economic agents who have a lending capacity and those in need of financing. The first agents lend their excess savings to banks (e.g., in the form of bank deposits or certificates of deposit or bonds) who in turn lend the collected funds to their clients who need funding, in accordance with the scheme shown in Fig. 1.4 of Chap. 1. By contrast, investors lend directly to borrowers on a financial market. The development of market financing has been faster than that of bank financing in the last three decades. This trend is called "disintermediation."

This chapter deals with the money market. It is on this market that short-term interest-rate products are traded. The issuers (borrowers) are sovereign states, banks, savings and loan associations, insurance companies, investment and pension funds, companies, and individuals. The transactions between two banking organizations are carried out in the *interbank market*, which is by far the main compartment of the money market. It is on the interbank market that the banks manage their cash positions and that Central Banks intervene to influence interest rates and monetary aggregates. For instance, following the introduction of the Euro, the European interbank markets became relatively integrated (about half of the transactions are "transnational") and concentrated (only several hundred financial institutions in the Euro-zone are active on this market). In contrast with the interbank market, the other compartments of the money market in Europe remain local, notably in London and Paris. Indeed, two markets make up the majority of the outstanding European short-term banking securities: that on Titres de Créances Négociables (TCN, negotiable debt certificates) based in Paris and that on "Euro Commercial Paper" based in London. The largest money market in the world remains the US one.

The first section is devoted to market practices regarding interest rates and the valuation of money market instruments. The second section describes the principal transactions and instruments on the money market: short-term marketable securities, "repos" (repurchase agreements), and other transactions. The third section covers the market participants and provides the orders of magnitude regarding the volumes of transactions. The role of the interbank market, the Central Bank interventions as well as their role in setting interest rates, are studied in the fourth section. We define the main references (or indices) for money-market rates in the last section.

## 3.1    Interest Rate Practices and the Valuation of Securities

Two general rules hold in the Euro-zone's money market:

- Securities are *listed and traded on an interest-rate basis* and not on a price basis. Thus, it is from the interest rate that one determines a security's price and not the reverse as in the bond market (see the following chapter).

## 3.1 Interest Rate Practices and the Valuation of Securities

- The rates used for short-term transactions ($\leq 1$ year) are *simple* rates, in general money market actual/360, rates, while those used for securities with medium and long terms (bonds and notes) are *compound*, actual/actual, rates.

This second rule has its roots in the Napoleonic Code, which has influenced several countries and forbids the capitalization of interest over a period of less than a year.

We describe the practices for interest rates and valuing short-term securities in the Euro-zone in Sect. 3.1.1 and, briefly in Sect. 3.1.2, the Anglo-Saxon practices which are slightly different.

### 3.1.1 Interest Rate Practices on the Euro-Zone's Money Market

In the Euro-zone, the elements that determine the cash flows effectively generated by a fixed income instrument depend on two types of conventions: those which concern how time is counted and those which concern the definition of the interest rate used.

Securities whose duration at issue is less than or equal to one year are traded on the *actual/360 day* basis[3] and with a *simple (proportional) interest rate*, called the *money market rate or yield*. This implies that the calculation of interest and cash flows obeys the following rules:

- *Actual/360 day basis*:
  The interest that accrues over a number $Nd$ of days (duration of the security), are calculated *pro rata temporis*, as though it accrued over a fraction $T$ of a year with $T = Nd/360$ (and not $Nd/365$ or $Nd/366$). As the denominator has been artificially reduced, the period during which the interest accrues is increased, and as a consequence the interest is increased, by the factor 365/360 (or 366/360 for leap years). It is important to note that, in calculating $Nd$, it is the number of *intervals* (or of nights) separating the start day from the end day of the transaction which counts (e.g., there is 1 day between April 1 and April 2 and there are 12 days between April 1 and April 13).
- *Money market rate (yield)*
  Interests may be calculated ex post (that is, payable in arrears, in accordance with Fig. 3.1a) or ex ante (that is, payable by the issuer in advance, as in Fig. 3.1b). In this last category belong the pure discount (zero-coupon) instruments which pay the nominal (face) value at maturity and are issued at a discount.

Then, in general, at the date of issue: (a) if the interests are in arrears, the first cash flow is $F_0 =$ Face value and the final cash flow $F_T =$ Face value + Interests; (b) if the interests are in advance, or for pure discount instruments, the first cash flow is $F_0 =$ Face value − Interests (or discount) and the final cash flow $F_T =$ Face value.

---

[3] Belgium is an exception, where a basis of 365 is in use.

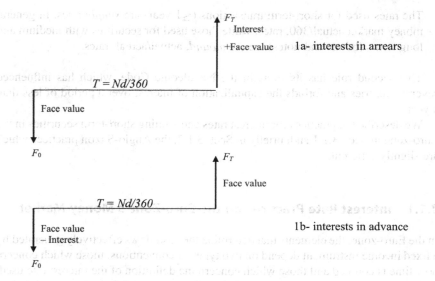

**Fig. 3.1** Interest payment scheme

For the investor, buying at the price $F_0$ and holding on until maturity induces the cash flow sequence: $-F_0$, $F_T$ (opposite signs for the issuer).

The effective proportional yield or *money market yield* is based on the effective cash flows $F_0$ and $F_T$, irrespective of the convention, denomination, or composition used: If the posted rate is $r$, in *both* cases *(interest in advance or in arrears)* the initial cash flow $F_0$ and the final $F_T$, separated by a period of length $T$, are related by:

$$F_0(1 + rT) = F_T \text{ with } T = Nd/360. \tag{3.1}$$

So: $F_0 = \frac{F_T}{1+rT}$ (the discounting formula for simple interest).

Furthermore[4]:

$$\text{Interests} = F_T - F_0 = F_T - \frac{F_T}{1+rT} = F_T(1 - \frac{1}{1+rT}) = F_T \frac{r}{1+rT} T.$$

Therefore, when interests are paid in advance, $F_T$ is the face value, and applying a money market rate $r$ is equivalent to applying a simple discount rate equal to $r/(1 + rT)$.

---

[4]Note that for interests in advance, $F_T$ equals the face value: using the money market yield to calculate the amount of advance interests therefore comes down to calculating this amount with a bank discount rate (see Chap. 2) equal to $\frac{r}{1+rT}$ (less than $r$). The practice of using money market yields is common on the European money market and differs from using a bank discount rate (see Chap. 2).

# 3.1 Interest Rate Practices and the Valuation of Securities

**Example 1 (Interests in arrears)**

Let us consider a bill with a face value $100,000 paying after 60 days interests (in arrears) in addition to its face value. Interests are calculated on the basis of a 6% rate.

The interests amount to: $0.06 \times 100,000 \times 60/360 = \$1000$. The security can thus be viewed as the right to (a claim on) a cash flow of $F_T = \$101,000$ due in 60 days ($T = 60/360$); it is issued at the price $100,000 (= F_0$).

**Example 2 (Interests in advance)**

Let us consider a pure discount bill with a face value 100,000 € (*interests implicitly in advance*), effective yield of 6%, and maturity 60 days. The final cash flow $F_T$ on which the security is a claim = 100,000 € since this cash flow is made up exclusively of the face value (interests are paid at the time of issue). The initial cash flow $F_0 = \frac{F_T}{1+rT} = \frac{100,000}{1+0.06 \times \frac{60}{360}} = 99,009.90$ €.

The interests (in advance) thus amount to: $100,000 - 99,009.90 = 990.10$ € (an amount less than the 1000 € interests in the previous example).[5]

Consider now an alternative financial practice for the same bill. Following this practice used for some instruments outside the Euro-zone money markets, called *simple* discounting, the initial interest will be: $0.06 \times 100,000 \times 60/360 = 1000$ €.

The initial cash flow would then be $F_0 = 99,000$ €, instead of 99,009.90 €, which is favorable for the lender and detrimental to the borrower (exactly as with a bank discount rate, see Chap. 2).

The calculations and examples which have just been presented refer to issuing a security on the primary market ($t = 0$, by convention). The rate $r$ used to calculate the cash flow depends on the market rate at date 0. More precisely, $r$ equals the market rate prevailing at time 0 for transactions of maturity $T$, augmented by a possible "risk premium," which is a function of the quality of the issuer's signature. $F_0$ can be interpreted as the market price (at $t = 0$) of the security or indeed as the present value of the cash flow $F_T$ available at $T$.

These relationships can be extended to the purchase or sale of a short-term security on the secondary market at some time $t$ after its issuance; at $t$, the security has a remaining duration $d = T-t$ (see Fig. 3.2). In this context, $F_T$ (the final cash flow on which the security is a claim) is fixed: according to the case, it equals the face value (for interest in advance) or the face value augmented by interests (for interest

---

[5]It should be noted that $990.10 \text{ €} = \frac{006}{1+0.06 \times \frac{60}{360}} 100,00 \times \frac{60}{360}$, so that the bank discount rate is $\frac{006}{1+0.06 \times \frac{60}{360}} = 5.94\%$.

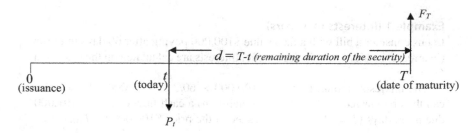

**Fig. 3.2** Transaction involving a monetary security at any instant $t$

in arrears). The question that arises then is how to determine the price $P_t$ of the security at time $t$.

Let us call $r_d(t)$ the simple money market interest prevailing on the market at time $t$ for transactions with duration $d$ (for example, the 2-month EURIBOR if at $t$ the remaining duration $d$ of a security is 2 months).

The formula to be applied then to determine the value of the security at any date $t$ (which generalizes Eq. (3.1)), can be written as:

$$P_t(1 + r_d(t)d) = F_T P_t(1 + r_d(t)d) = F_T; \text{ or } P_t = \frac{F_T}{1 + r_d(t)d} \quad (3.2)$$

with $d$ = number of days remaining before maturity/360.

Both Eqs. (3.1) and (3.2) mean that the value at time $t$ of the cash flow $F_T$ is obtained by discounting with the (proportional) money market rate. Note that this rate depends on the conditions prevailing in the market at the time of valuation.

> **Example 3**
> Bank B issues a pure discount security (*interests payable in advance*) with a maturity of 90 days, a face value of $1,000,000, using a *money market yield* of 8%.
> 
> The terminal cash flow $F_T$ is $1,000,000, and the initial flow is:
> 
> $$F_0 = \frac{F_T}{1 + rT} = \frac{1,000,000}{1 + 0.08\frac{90}{360}} = \$980,392.$$
> 
> The interests paid in advance are $1,000,000-980,392 = \$19,608$.
> 
> In this context, $F_T$ is the given face (nominal) value $N$ and $F_0$ is calculated using a money market rate. Interests equal $F_T - F_0$.
> 
> We note that the amount of interest is less than 8% ×1,000,000×90/360 = $20,000, which would be applied to a commercial bill of exchange for a duration of 90 days at a bank discount rate (see Chap. 2) of 8%.

## 3.1 Interest Rate Practices and the Valuation of Securities

**Example 4**

On June 1 of year $n$, corporation X issues a note that commits to paying its holder 2,000,000 € in 3 months (91 days). The EURIBOR-3 month is 3.70% but the note was written on the basis of a money market yield of 4.00%: the *yield spread* (an increase intended to compensate the lender for the risk of default by the issuer) is therefore 0.3%, or 30 *basis points* (*bips*). Interests are paid in advance.

Corporation Y, which wants to invest its cash surplus, buys the note (on June 1 of year $n$) at the price $P_0$ with $P_0\left(1 + 0.04\frac{91}{360}\right) = 2,000,000$, so that $P_0 = 1,980,002$ €. The interests paid in advance are equal to: $2,000,000 - 1,980,002 = 19,998$ €.

Now, on July 1 of year $n$, corporation Y sells the note in the market, 61 days before its maturity. The EURIBOR-2 month is at the time 4.30%, and the sale of the note takes place at a money market rate of 4.60% (again, with a 30 bips spread). As a result, at the end of this holding period of 1 month, the transfer price $P_1$ is

$$P_1 = \frac{2,000,000}{1 + 0.046\frac{61}{360}} = 1,984,532 \ \text{€}.$$

For corporation Y the monthly yield of this transaction is thus $\frac{P_1 - P_0}{P_0} = 0.229\%$ (2.748% annualized). This mediocre yield results from the decrease in the price of the note caused by the increase in the market rate (issued at a rate of 4% and transferred on the basis of a rate of 4.60%). This interest rate risk, which incidentally will be seen to affect more long-term securities than short-term ones, will be studied in detail in Chap. 5.

## 3.1.2 Alternative Practices and Conventions

The two main day-counting methods are actual/360 and actual/365 (fixed). The basis for computing interest payments is therefore a year of either 360 days or 365 days.[6] Roughly, the UK and the old Commonwealth or Empire countries (Australia, New Zealand, Canada) use the 365-day basis, the rest of the world, with few exceptions, the 360-day.

The second important element is the type of interest rate computation, which can be based on a money market yield as previously explained or on simple discounting (see Example 2). Using the simple discounting (or bank discount) method, the price at time 0 of a bill paying its nominal (face) value $N$ at $T$ ($F_T = N$) is $F_0 = N -$

---

[6]The 365-day basis assumes that the year is 365 days long, even when it is a leap year and, in this respect, is slightly different from the actual basis used in most bond markets.

$NrT = N(1 - rT)$ where $r$ is a bank discount rate and $T$ = number of days/360 or number of days/365. We notice that, for the same rate $r$, $N(1 - rT)$ is smaller than $N/(1 + rT)$, which would be the price obtained by using the money market yield method. In Example 2, we showed that the same bill can be bought for 99,000 € when priced through simple discounting instead of 99,009.90 € when priced by the effective yield method.

In the USA, UK, and other countries outside the Euro-zone, some money market instruments are priced according to the bank discount method (such as Treasury bills, commercial paper, bank acceptances, or bills) while others (typically Certificates of Deposit or CD) are priced using money market yields.

## 3.2 Money Market Instruments and Operations

The transactions on the money market may take different forms and be undertaken for different purposes. Some of them are standard lending and borrowing carried out through the issuance of a short-term security by the borrower bought by the lender. In other cases, the financial technique is less standard and/or the purpose of the market participants may be different from simply investing or financing (e.g., risk hedging).

We describe the main short-term securities traded in the money market in Sect. 3.2.1, repurchase agreements (repos) in Sect. 3.2.2 and other types of operations in Sect. 3.2.3.

### 3.2.1 The Short-Term Securities of the Money Markets

Different short-term securities are traded, mainly OTC (over the counter), on the primary and secondary money markets.

The following properties are common to these various securities:

- Maturity at issuance less than 1 year.
- Money market or zero-coupon structures.
- Trading (pricing) based on interest and not on price, in contrast with the case of bonds.
- Existence of a secondary market allowing the holder of a security to transfer it at any time, at least for certain maturities.
- The default risk they usually bear is small, as they are issued by governments or big and highly rated corporations or institutions.

Most of these securities are traded in large denominations and hence may be out of reach of individual investors. However, even small investors have indirect access to money market assets through money market mutual funds.

Depending on the issuer, money market securities are called:

## 3.2 Money Market Instruments and Operations 101

- *Certificate of deposit (CD)* when issued by a bank or a financial institution. They are usually considered to be fairly safe.[7] Interest may be in arrears or in advance; the interest rate is a money market yield, in general, fixed; maturity is usually between 10 days and 3 months. Short-term CDs are usually highly liquid in the secondary market.
- *Commercial paper* when issued by a nonfinancial corporation. The rate is either a money market yield or a simple discount rate, in general fixed and the maturity is mostly between 10 days and 3 months. Many firms roll over their commercial paper at maturity, thereby generating a stable source of financing.
- *Treasury bills* when issued by a government. Issued on the primary market, short-term Treasury Bills, tradable on the secondary market, are useful for managing government cash (and sovereign debt) positions used by all developed countries. They usually are highly marketable and considered as fairly riskless (except for the sovereign risk they may bear).
- A *banker's acceptance* is an order issued by a firm to its bank to pay a given amount at a future date (as in a post-dated check); when the bank agrees by writing "accepted", the draft becomes a bank's IOU and a negotiable security. Bankers' acceptances, typically of 6-month maturity or less, are very safe and widely used in foreign trade.[8]

Finally, let us note that the practice of issuing *ABCP* (Asset Backed Commercial Paper) backed by previously securitized assets has strongly developed.

### 3.2.2 Repos, Carry Trades, and Temporary Transfers of Claims

These transactions, usually of short duration, are based on the principle of a "repurchase agreement," called a "repo." By definition, a repo involves two economic agents L and B who agree at $t = 0$ on the following transactions:

- L (the lender) buys a security $x$ with cash (at date 0) with the promise to resell it (at $T$) to the counterparty B.
- B (the borrower) sells this security at date 0 to L with the promise to repurchase it (at $T$).
- The price $P_1$ of the transaction to be carried out at $T$ is fixed at date 0.

In general, the underlying security $x$ is a claim on a third party with a very good rating (e.g., the Treasury).

The repo can therefore be interpreted in three ways:

---

[7] In particular because in many countries they are considered as bank deposits and insured as such.

[8] The issuer pays a commission to the bank that guarantees the note, which benefits from a smaller interest rate because of the guarantee that decreases its risk.

(i) As a *cash purchase* of the security (at a price $P_0$) *combined with a forward sale* (at a price $P_1$) for L (and the symmetric for B).

(ii) As a *temporary transfer of a claim* (between 0 and $T$) by the seller B to the buyer L who "carries" the claim between 0 and $T$: this transaction, therefore, is known as a "carry" trade.

(iii) As a *guaranteed loan/borrowing*.

In effect, from the buyer's viewpoint (L), the cash flow sequence produced by this transaction can be written:

$$-P_0 \text{ (purchase price); } +P_1 \text{ (resale price fixed at 0)}.$$

Such a sequence (involving an initial negative cash flow followed by a positive flow) is a characteristic of a loan (lending).

From the seller's point of view (B), the cash flow sequence produced by the transaction is:

$$+P_0 \text{ (sale price); } -P_1 \text{ (buy back price)}.$$

This is equivalent to borrowing.

In the event of a default by B (borrower), the lender (L) keeps the underlying security $x$ (which is worth $P(T)$): the amount at risk is thus limited to $P_1 - P(T)$ (and not the entire $P_1$); the underlying security is therefore a "collateral" which markedly decreases the counterparty risk and thus affords the borrower the most advantageous conditions (even possibly an interest rate slightly less than the LIBOR or EURIBOR used for transactions without collateral).

Furthermore, the simple interest $r$ (money market yield) of this loan is such that:

$$P_1 = P_0(1 + rT), \text{ i.e., } r = \frac{P_1 - P_0}{P_0 T}.$$

In fact, $P_1$ is specified at date 0 so that the interest $r$ of this borrowing/lending is equal to the money market yield of maturity $T$:

$$r = r_T(0).$$

$P_1$ is a function of $r$ by the condition: $P_1 = P_0(1 + rT)$.

In professional jargon, the *repo* often refers to the transaction made by the seller or borrower (B), the transaction made by the buyer or lender (L) being called an *inverse repo*. Moreover, overnight repos are distinguished from longer-term transactions called "term repos."

## 3.2 Money Market Instruments and Operations

**Example 5**

A repo on a security $x$ with a value of 1,000,000 €, for 6 months ($T = 1/2$), interest rate $= 6\%$ (6 month-EURIBOR + small margin).

The buyer (lender, investor) pays the seller (borrower): $P_0 = 1,000,000$ €. She commits to resell $x$ at $T$ at the price $P_T = 1,030,000$ € since:

$$P_T = 1000,000 \times (1 + 0.06 \times 0.5) = 1,030,000 \text{ €.}$$

The sequence for the investor then is: $- 1000,000$; $+1,030,000$.

If the borrower defaults (at $T$), the lender retains the security $x$ which she owns (the transaction is not settled). In this way, the security $x$ constitutes a guarantee (collateral) for the lender.

Repo transactions are usually of much shorter duration (typically a week or less) than in Example 5. They are carried out in several forms: in particular, the collateral may be held in custody by the borrower or actually delivered to the lender.

In the first form, the collateral pledged by the borrower is not actually delivered to the lender but is kept as a segregated asset ("held in custody") by the borrower for the duration of the trade. Widely used several years ago, this form of repo has become rarer after the financial crisis of 2007–2008 because the lender still bears a default risk.

The second form requires the physical delivery of the collateral, which provides more safety to the lender. It can be analyzed as a temporary transfer of debt; the lender is the owner (holder) of the debt between 0 and $T$ and is committed to sell it at $T$ while the borrower commits to repurchase it.

There exist also more complex repos such as three-party repos, involving a basket of securities as collateral and a third-party clearing agent or bank.

### 3.2.3 Other Trades

Other trades, different from buying or selling securities or transfers of debt as described in the two previous sections, are carried out in the money market. We consider here briefly transactions without collateral as well as lending and borrowing securities.

**Lending and Borrowing Without Collateral**

This is the simplest type of lending and borrowing between banks. The lending bank simply puts funds at the disposal of the borrowing bank, by a simple transfer from an account to the other, without any issuing or transfer of securities or any temporary transfer of debt hedging the counterparty's risk. These transactions require the lender to have high confidence in the solvency of the borrower and deny the lender the possibility of refinancing in the market since the debt is not materialized. Consequently, they mainly involve bank borrowers with a top credit rating and debts are usually very short-term (most often overnight). They do make up a non-negligible part of the transactions in the interbank market. They are generally concluded on the phone and directly between the parties (without intermediaries).

**Lending and Borrowing Securities**

This is the case when a security $x$ is lent by an agent L to a borrower B for the period between 0 and $T$. In contrast with all the transactions described in this chapter, *lending and borrowing securities should not be analyzed like lending and borrowing money* (in particular it must not be mistaken for a repo), but rather as a *financial service* provided by L to B: L puts the security $x$ at the disposal of B between 0 and $T$ (B is committed to give back the security at $T$). B does *not* buy the security from L, who remains the owner, and the only cash flow received by L is a *commission fee* of several basis points (typically between 15 and 20 bips) paid by B as compensation for the service provided.[9] Even if B may use the security as he wishes between 0 and $T$ (notably, he can sell it), L retains all the rights and privileges inherent to ownership; in particular, B has to transfer back to L any remuneration (coupon, dividend) paid by security $x$ between 0 and $T$. In addition, L's ownership is guaranteed by B posting collateral (pledging securities) to the benefit of L so that the risk that B does not give back the security at $T$ is covered.

Usually, the lender is a bank or a mutual or hedge fund which hopes to reap commission fees so as to increase the return on its portfolio.

The borrower of the security is motivated by the desire to carry out other transactions allowed by holding it (timely delivery, hedging, carrying out some fiscal arbitrage connected with recovering a tax credit on a dividend, and so on). In particular, borrowing security $x$ allows *shorting* (*short selling*) it. This short position is obtained for B by selling $x$ after L has lent it to him. Let us denote by $V_x(0)$ and $V_x(T)$ the market values of $x$ at 0 and $T$, respectively. The combination of borrowing the security and selling it for cash at 0 brings for B the following cash flows (we assume that no coupon is paid by $x$ from 0 to $T$ and we neglect commission fees):

at 0: $+V_x(0)$: produced by the cash sale of $x$,
at $T$: $-V_x(T)$: buying $x$ to give it back to L.

This is the "reverse" sequence of that produced by the purchase of $x$ at 0 followed by its resale at $T$ (i.e. a long position held from 0 to $T$) and is characteristic of shorting $x$, between 0 and $T$.

---

[9]Plus the transfer by B of any possible remuneration paid by $x$ from 0 to $T$ (see below). The fee rate is in fact very different according to whether the security is rare, thus in demand, or not. The indicated spread of 15 to 20 bips *pro rata temporis* corresponds to a standard situation. It should also be noted that the fee effectively paid by the borrower varies over time since it is computed as the product of the fixed interest rate (*pro rata temporis*) by the variable market value of the security lent.

## 3.3 Participants and Orders of Magnitude of Trades

After a brief description of the main market participants, we will present some data about the main money markets.

### 3.3.1 The Participants

In the money market, some economic agents are structurally lenders, others are always borrowers and yet others are both lenders and borrowers. Some intermediaries facilitate balancing the supply of and demand for funds. The settlement system implements the transactions.

**Structural Lenders**
In most advanced economies, the Central Banks that play a dominant role in regulating interest rates and liquidity are net lenders in the interbank market (see the following section).

The institutional investors, that collect savings and invest them, such as insurance companies and pension or mutual funds, are also structurally investors in the markets, especially in the money market (but also in the bond and stock markets).

Note that although their role is very marginal, individuals can acquire assets in the money market directly, in addition to their holdings of managed money market funds.

**Structural Borrowers**
The main borrowers on the money (and bond) markets are sovereign states. In the money market they issue short-term debt ("Treasury Bills"). In addition, some credit establishments refinance their loans in this market.

**Agents that are both Lenders and Borrowers**
These are:

- Banks and other financial institutions that issue CDs (hence are borrowers) but also invest their cash surpluses in the money market (as lenders). The global banking system (outside the Central Bank) is a net borrower in the interbank market.
- Firms that issue commercial paper or invest their cash surpluses.

**Intermediaries**
Recall from Chap. 1 that two types of financial intermediaries coexist: those who are exclusively brokers and the dealers who also take positions on their own accounts. In

addition, some intermediaries are specialized as dealers in Treasury securities. Among the dealers, we usually find the largest banks in the area.[10]

**The compensation-settlement system for securities**
The operations of safekeeping, clearing, and settlement are taken care of by different custodians, the main one being the Central Security Depositaries (CSD). Schematically the role of a CSD is:

- To calculate the net position in a security for each investor: this is called "clearing."
- To ensure that the buyer pays for[11] and that the seller delivers the securities (transfer of ownership which is mostly dematerialized): this is called "settlement/delivery."
- To produce accounting entries and documentation for the intermediaries with accounts at the CSD.

In some European countries, the role of the central custodian for financial assets (money, bonds, stocks) is assumed by Euroclear. A new platform (TARGET 2-Securities or T2S) offers centralized *delivery-versus-payment* settlement across all European securities markets. CSDs from the different Euro-zone countries are connected to the T2S platform, which increases integration and harmonization in the previously fragmented European system of securities settlement.

The European Central Bank is a main actor in the system and plays a triple role: settlement, counterparty, and execution of monetary policy decisions in the decentralized implementation of the European monetary policy (open market operations, permanent lending, and borrowing facilities).

### 3.3.2 Orders of Magnitude

The largest money market in the world is the USA (about \$4500 billion outstanding). In Europe, the main international market in marketable short-term securities is the European Commercial Paper (ECP) based in London: the equivalent of 860 billion euros in 2019, of which 80% are of maturity less than 3 months. The main component of the ECP market is bank debt: other securities result from securitization operations (asset-backed ECP: ABECP). ECPs are priced in Euros, Dollars, Sterling, or other currencies. For more details and updating, the reader should consult the ICMA website ("International Capital Markets Association" at http://www.icmagroup.org/).

---

[10]For example, in London, the most active on the Euro Commercial Paper market are Citigroup, JP Morgan, Deutsche Bank, Goldman Sachs, Morgan Stanley, HSBC, among others.

[11]Payments are carried out through an intermediary system, such as TARGET in Europe.

The main European national market is the French short-term Titres de Créances Négociables à Court Terme (or NEU CP) market (235 billion euros in 2019, excluding Treasury bills). There are also smaller markets in Belgium, Spain, and Germany (less than 50 billion euros outstanding, for each, excluding Treasury bills).

## 3.4 Role of the Interbank Market and Central Bank Intervention

It is on the interbank market that credit institutions, Central Banks and other financial institutions borrow and lend central bank money. We describe briefly the functions of central bank money and of the interbank market (in Sect. 3.4.1), and then interventions by Central Banks and their role in setting (nominal) interest rates (in Sect. 3.4.2).

### 3.4.1 Central Bank Money and the Interbank Market

#### 3.4.1.1 Central Bank Money, Bank Reserves, and Interbank Settlement Payments

A Central Bank is a bank for banks. This means, that between them, credit institutions settle their debts in the money issued by the Central Bank. Central Bank money is made up of deposits to the Central Bank, essentially held by the banks, in the same way that common currency is essentially made of deposits held by nonbank agents (households, firms, local governments) at the banks. Each credit institution has an account at the Central Bank; interbank settlements are implemented by transfers from account to account at the Central Bank, in a more or less instantaneous manner using powerful payment systems.[12]

The holdings of a bank in its account at the Central Bank are called its "reserves." These, which are assets on the balance sheet of the bank, must not be confused with the accounting "reserves" of a firm (which are a liability). The banking reserves can be considered as "cash" and allow settling debts in the interbank market. Bank accounts at the Central Banks must be positive (assets); more exactly, credit institutions must hold reserves in an amount greater than or equal to a minimum fixed by the Central Bank and called "reserve requirements." Reserve requirements are calculated as a proportion of the demand and time deposits ($<2$ years) and of the securities that represent the bank's debts with a maturity of less than 2 years (notably, certificates of deposit). The calculation is done on a monthly basis for most banks. For example, for Euro-zone banks, the percentage of reserves required is

---

[12] For example, a payment system called TARGET (Trans-European Automated Real-time Gross-settlement Express Transfer) was introduced with the Euro. This system allows nearly immediate settlements, with maximum security and carries out as many national settlements as it does international. It is mainly for interbank transactions but is also used by nonbank agents.

presently fixed at 1% and is revised only infrequently. In the USA, it has been reduced to 0% in March 2020 in connection with the COVID-19 crisis.

Everyday transactions lead some banks to have a surplus of reserves and others to experience a deficit that they must take care of.

Central Bank money, also called the *monetary base*, or sometimes *hot money*, lies at the heart of the mechanisms for money creation and bank lending. Recall that the ratio of the money supply to the monetary base, called the money *multiplier*, depends on a certain number of factors (for instance, the proportion of the money supply held in cash, the percentage of reserve requirements). Denoting by $B$ the monetary base, $M$ the money supply, and $m$ the multiplier, we have: $M = m B$.

As a result of this, the monetary authorities can exercise control over the money supply in two ways:

- By modifying the percentage of the reserve requirement, on which the multiplier $m$ depends negatively.
- By controlling the base $B$, i.e., by adjusting the volume of Central Bank money available to the banking system that makes up the latter's liquidity. This adjustment is implemented in various ways (see infra) but mainly by interventions in the interbank market that are called "refinancing" or "open market operations," which are studied in Sect. 3.4.2.

### 3.4.1.2 A Simple Analysis of the Macroeconomic Effects of Monetary Policy

Roughly speaking, economists agree that when there is spare production capacity, varying the money supply $M$ in the banking system controlled by the Central Bank has a short-term effect on the nominal interest rate, a short- to medium-term impact on the level of real activity (Gross National Product or Net National Income) $y$ and the unemployment rate, and mostly long-term effect on the general price level $P$. The analysis that follows rests on the usually accepted assumptions: the nominal interest rate $r$ is very flexible (it adjusts rapidly to market pressures); the activity level $y$ and unemployment then adjust if the production capacity allows; prices are more rigid (less flexible). The condition for equilibrium between the money supply $M$ (considered as exogenous to the model) and the demand for money then leads to a simple explanation of the effects of monetary policy. The demand for money $m^d(r, y)$ from economic agents (firms, individuals, . . .) is expressed in real terms, that is deflated by inflation, as a decreasing function of the nominal interest rate $r$ (the higher it is, the more the opportunity cost of holding money rather than, e.g., interest bearing notes, goes up) and an increasing function of real income (the larger is the individuals' income, the more they need money to finance their consumption). Let us note that global income $y$ can be identified with total production since the latter is valued on the basis of the income it produces. Expressed in nominal terms, the monetary demand equals $P \, m^d(r, y)$. The condition for equilibrium between the nominal supply of and demand for money is, therefore, the following equation:

## 3.4 Role of the Interbank Market and Central Bank Intervention

$$P\, m^d(r, y) = M. \tag{3.3}$$

Starting from equilibrium, an increase in $M$ creates a surplus of money outstanding which is assumed to be absorbed in three stages:

- By a reduction in $r$ (first stage) because the banks have more liquidity; this lowering of $r$, in turn, increases the demand for money.
- By an increase in production $y$, in response to the demand for goods and services stimulated by the lowering of interest rates (credit is cheaper, . . .), and (often) an accompanying decrease in the unemployment rate.[13]
- Finally, an increase in prices; the reason for this is that agents tend to get rid of their excess money holdings by increasing their demand for consumer goods, and/or that production capacity becomes saturated. This increase in prices can be accompanied by higher nominal interest rates.[14]

We underline the importance of the assumption that production is flexible. If the production capacity is used up, phase 2 of the process described above is not possible and an increase in $M$ results immediately in a temporary lowering of $r$, and then, without any change in the level of activity, in higher prices, possibly in the same proportion as the increase in the money supply that started the process, together with a rise in the nominal interest rate.[15]

In a rather simplistic way, we can say that some of the controversies about economic and monetary policies reflect either difference in diagnoses as to the ability of production to react, in the short term, to an increase in the demand for goods, or differences in implicit horizons: neo-Keynesian economists and most politicians are more focused on the level of activity in the short term, while monetarist, or neoclassical, economists, as well as Central Bankers, often favor long-term equilibrium and the control of inflation.

In the sequel, we adopt a very short-term point of view, since we restrict our analysis to the impact of monetary policy on interest rates (in Sect. 3.4.2).

### 3.4.1.3 The Role of the Interbank Market

It is in the interbank market that banks with surplus reserves lend central bank money to those who momentarily have a deficit. The majority of these transactions are loans without collateral and repos. Furthermore, other transactions between banks, such as

---

[13] The explanations of stages 1 and 2 of this process found in macroeconomic textbooks (adjustments in $r$ and $y$ in the Keynesian model) are based on a graphical presentation of Eq. (3.3) above (the so called *LM* curve) and the market equilibrium condition for goods (the *IS* curve).

[14] A reduction in $M$ would produce, ignoring asymmetries that can be important in some cases but whose analysis is outside the scope of this book, the opposite consequences.

[15] One recovers therefore the framework of the standard neo-classical model with fixed $y$ and variable $P$.

swaps, that are intended to manage interest and exchange risks,[16] not bank liquidity, are also made in the interbank market. In fact, the interbank market plays a *double role: it is the place where the banks manage their cash positions* (manage their liquidity by lending and borrowing central bank money, and manage their interest and exchange risks); it is also *the place where Central Banks intervene* with *open market operations* (OMO), to adjust the banking system liquidity and influence the level of interest rates, as dictated by monetary policy. Such interventions are studied in the next subsection.

## 3.4.2 Central Bank Interventions and Their Influence on Interest Rates

Implementation of monetary policy is part of the mission of Central Banks. In the European Union, monetary policy is conducted by the European System of Central Banks (ESCB) which is made up of the European Central Bank (ECB) located in Frankfurt, and the National Central Banks (NCBs, such as the Bank of Italy, of France, of Germany,...). The Eurosystem is the subset of the ESCB made up of NCBs of the countries in the Euro-zone. The council of governors of the ECB determines monetary policy and the National Central Banks implement it by intervening in the different national financial arenas.

In the USA, the central bank system is called the Federal Reserve System or simply the Fed. It is made up of three large institutions: The Board of Governors, the Federal Open Market Committee, and 12 Federal Reserve Banks with regional responsibility. Roughly, setting monetary policy is done by the first two institutions (similar to the ECB and its board of Governors) and its implementation devolves to the 12 federal banks (like the NCBs in Europe), mainly the Federal Bank of New York.

Conventional Central Bank interventions take two forms: refinancing (funding) banks in the "open market" and providing banks with permanent access to central bank money. Since 2008, "unconventional" monetary operations ("quantitative easing," leading to unprecedented huge purchases by the Central Banks to inject hot money in the liquidity-starved economy) have been implemented in response to various financial crises.

### 3.4.2.1 Open Market Refinancing

These operations are traditionally intended to steer short-term interest rates and control bank liquidity and are implemented *in the interbank market*. They use different techniques of which the most significant is the repo.[17] With such operations, the Central Banks refinance (fund) the banking system, or sometimes

---

[16] See later chapters.

[17] Also used are purchases and sales of securities (negotiable debt securities, debt certificates), unsecured loans, etc.

### 3.4 Role of the Interbank Market and Central Bank Intervention

reduce their refinancing, and thus influence short-term interest rates. In addition to these standard open market operations, "quantitative easing" has massively been undertaken by different Central Banks.

In the sequel, we describe the European Central Banks' operations and provide some hints about those of the Fed.

- *Regular refinancing operations* are done through tender offers, of which the most important take place on a weekly basis, and consist of temporary debt transfers by banks (loans by the Central Bank to the banks), typically for a week or a fortnight, in amounts fixed globally by the Central Bank.[18] These tender offers can take two forms: (1) auctions at fixed rates through which loans are extended to banks that tender at the key lending rate, called the *Main Refinancing Operations (MRO) rate, $r_{MRO}$*; (2) auctions with variable rates in which the winners are the banks who tender *at the best (highest) rates*. In any case, the banks who tender must offer a rate greater than or equal to *the minimum bid rate* set by the European Central Bank. The short-term interest rates on the interbank market are established, in equilibrium, at a level near this intervention rate.

---

**Example 6 (A stylized example of weekly refinancing operation with a variable rate)**

The European Central Bank announces that it will offer 500 million euros. The minimum bid rate is 0.25%. This auction generates funding requests associated with the following rates offered, by banks A, B, C, .... listed in decreasing order of rates offered.

| Bank | Amount requested | Rate offered (%) |
|------|------------------|------------------|
| A | 200 | 0.30 |
| B | 200 | 0.29 |
| C | 300 | 0.27 |
| D | 200 | 0.26 |
| ... | | |

Since the Central Bank is willing to put out 500 million euros and it serves the banks with the best rates, A and B are entirely satisfied, C is partially satisfied (it gets 100 of the 300 requested) and D gets nothing.

According to the auction process adopted for this tender offer at variable rates, two possibilities are open as to the rate(s) payable by the borrowers:

(continued)

---

[18] Other open market interventions, such as longer-term funding (intended to provide liquidity, carried out monthly and with a maturity of 3 months) or "fine-tuning" operations, are also implemented by the ESCB.

> - All borrowers are accommodated at the same rate of 0.27%, i.e., at the marginal rate (a so-called *Dutch* auction).
> - Each pays the rate offered: A then borrows at 0.30%, B at 0.29%, and C at 0.27% (an *American* auction).

During a *fixed rate* auction, banks offer an amount (and not an amount and a rate) in the knowledge that all refinancing is based on the MRO rate (in our example, it could be 0.25%).

*Other refinancing operations.* Refinancing operations have, since 2007–2008, undergone extensive overhauls intended to address different crises: reduction in the MRO and the minimum bid rate, enlarging the list of eligible securities,[19] lengthening the duration of refinancing to Longer-Term Refinancing Operations (LTRO's) which are 3-month liquidity providing operations, and massive unconventional operations of quantitative easing.

Among the nonstandard monetary policy measures recently undertaken, let us mention: 3-years LTRO's; Targeted LTROs providing financing to credit institutions; Asset purchases programs aiming to sustain growth; and in 2020 Pandemic emergency programs (PELTROs and PEPP) designed to lower borrowing costs as response to the COVID-19 emergency.

The same kind of quantitative easing as well as unconventional refinancing programs have been launched by most central banks, in particular by the by the Fed and by the Bank of England.

### 3.4.2.2 Permanent Access to Central Bank Money (Standing Facilities)

Lending facilities (at a cap rate $r^+ > r_{MRO}$ called rate on *marginal lending facility*) and deposit facilities (at a floor rate $r^- < r_{MRO}$, called rate of *deposit facility*) are also available to banks. These are overnight operations made on-demand through which banks with a pressing need for financing can borrow from the Central Bank at its lending rate $r^+$ and those who have surplus cash can place it at the deposit rate $r^-$, for 24 h.

The ECB usually fixes the cap (ceiling) rate $(r^+)$ 20 to 50 basis points above the MRO rate and the floor rate $(r^-)$ at 20 to 50 basis points below. As an example, in September 2019, the intervention rates of the ECB were:

$r_{MRO} = 0\%$; $r^- = -0.50\%$ for deposit facilities; and $r^+ = 0.25\%$ for the rate of marginal lending facility.

---

[19] Only specific securities can be refunded by the Central Bank (they are said to be "eligible"). The ECB and the Fed have substantially lengthened the list of eligible securities to make funding of financially distressed banks easier and to restore trust in sovereign debts issued by some States by making them eligible. For instance, the Fed did it intensively when implementing its Quantitative Easing (QE) policy after the collapse of Lehman Brothers in 2008 and the ensuing financial crisis. This very capacity of Central Banks to alter the list of eligible securities is a powerful instrument of monetary policy.

The short-term rates in the money market (EONIA/$STER, EURIBOR-1 week) thus fluctuate between the cap and floor rates, $r^-$ and $r^+$, and so are *close* to the (in-between) $r_{MRO}$. These different *intervention rates*, or *key driving rates*, and primarily the MRO rate which is involved in main open market refinancing operations, constitute, along with the rate of reserve requirements (which is very infrequently modified) and the list of eligible debts, the principal instruments for monetary policy within the Euro-system.

Analogous techniques and instruments are used by the Central Banks charged with conducting monetary policy in the other large monetary zones (US Federal Reserve System, Bank of England, Bank of Japan, Bank of China).

For example, in the USA, the intervention rate of the Federal Reserve is called the *federal fund target rate* (equivalent to the MRO rate in the Euro-zone); it is fixed by the Federal Open Market Committee. The overnight rate on the interbank market is called *federal fund effective rate* or more simply the *federal fund rate* (equivalent of the EONIA/$STER) and is in general very near to the target rate (*federal fund target rate*). Furthermore, the banks may borrow directly from the Fed at the *discount window* at a rate higher than that of federal funds (this is a mechanism equivalent to lending facilities provided by the ECB).

The cost of funding and the return on lending in the money market are determined from the interbank market rate by adding a margin that takes into account the borrower's creditworthiness (rating). The costs of short-term bank loans charged to the customers are also crucially influenced by the money market rate, as the banks use it as a benchmark to fix their own rates. As a result of these multiple functions, the interbank market is the heart of modern financial systems. *Quantitative easing* by the Fed and the ECB during and after the 2007–2008 financial crisis to prevent economies to stay in a recession is a particularly spectacular illustration.

## 3.5 The Main Monetary Indices

An index or a reference rate represents either the value of a rate prevalent on the market *on a given date* or *its average value during a given period*. In addition, recall that, for the same issuer, the value of a rate depends on the duration (or maturity) of the instrument.

### 3.5.1 Indices Reflecting the Value of a Money-Market Rate on a Given Date

The main indices reflecting the value of a money-market rate on *a given date* are the EONIA progressively replaced by the $STER (pronounced ESTER), the EURIBOR, and the Eurepo® for transactions denominated in Euros. Equivalent indices such as the SONIA, the SFOR, the LIBOR, are computed for transactions denominated in the other major currencies.

**EONIA (Euro Overnight Interest Average), €STR, Federal Fund rate, ...**

The EONIA is an index computed each business day indicating the value of the overnight rate (maturity 24 h) on the interbank market, for transactions denominated in Euros, and not guaranteed by a collateral (unsecured). The EONIA is computed as the average of the rates for unsecured loans, posted on a given day, by a panel of about 60 large banks in the Euro-zone. The overnight rates are computed for various major currencies. As we have seen, in the USA, the main overnight rate is the federal fund rate. Regulators and central banks have announced that, due to possible manipulation, new references to overnight rates will replace the existing overnight indices. In particular, the EONIA is gradually replaced by the Euro short-term rate €STR published by the ECB since October 2019. Like the EONIA, it reflects the unsecured overnight rates but contrary to it, it is grounded on actual transactions, represents borrowing (and not lending) rates and a broad range of financial institutions (and not just banks). The transition period for this substitution ends on January 2022. Analogous transitions are underway in the UK (towards the SONIA), in the USA (toward the SOFR, Secured Overnight Financing Rate), in Switzerland and in Japan.

**EURIBOR (Euro Interbank Offered Rates) and LIBOR (London Interbank Offered Rates)**

The $x$-month EURIBOR is an actual/360 money market yield, computed each business day and indicates the value on this day of the interbank rate for transactions with $x$ months to maturity between prime banks, denominated in Euros and unsecured. Five different EURIBOR rates are computed: 1 week, 1, 3, 6, and 12 months. For example, the EURIBOR-3 months on day $j$ indicates the rate at which Euro-zone banks borrow in Euros, in the interbank market, for 3 months (between $j$ and $j + 3$ months); it is "forward-looking" (or "predetermined"), i.e., known at the start of the period (on $j$). EURIBOR rates are estimated each day and published by the European Money Markets Institute, from the arithmetic mean of the rates offered on the interbank market by a panel of large banks, mostly from the Euro-zone, during the day upon consideration.

Indices equivalent to the EURIBOR, called LIBOR (London Interbank Offered Rate), are computed for transactions denominated in five major currencies (dollar, sterling, yen, Swiss franc, and Euro[20]), seven different maturities, and made up, till the present, the principal reference rates on the international markets. However, due to evidence of manipulation (possible because the estimations are not grounded on actual transactions but on submissions of a panel of banks and on expert opinions) and also because liquidity issues, the LIBOR rates will be replaced by more reliable

---

[20] The Euro-LIBOR reflects, as does the EURIBOR, the rates offered in transactions denominated in euros. In contrast to the EURIBOR, it is computed from a panel containing banks outside the Eurozone (mainly British). The differences in value between the EURIBOR and the Euro-LIBOR are very slight (a few bips).

## 3.5 The Main Monetary Indices

alternative interest indices grounded on the new overnight benchmarks.[21] In fact, most short-term rate indices are on the verge to be either modified or replaced, in compliance with new standards of reliability within a vast international reform of the "Risk-Free Rate" (RFR) indices. As to the EURIBOR, whose reliability has also been disputed for similar reasons, the manner in which it is determined has been modified. It is now anchored, to the extent possible, to actual transactions so as to comply with the required reliability standards; as of June 2020, it is not scheduled to be discontinued.

Let us emphasize that EURIBORs and LIBORs are *forward-looking (or predetermined)* interest rate indices, since, for any interest period, they are *known at the beginning* of that period. For instance, the interests due for a 3-months period $n$ usually paid at the end of $n$ are calculated with the 3-month EURIBOR prevailing at the *beginning* of $n$.

*Eurepo*® are rates equivalent to the EURIBOR (Euros) but refer to repos, thus loans secured with a collateral. The repo rates are also computed for other major currencies. Repo rates are slightly less than the corresponding EURIBORs or other unsecured rates because of the collateral.

Furthermore, the fixed rates of "vanilla" interest rate swaps, in particular OIS swaps, are important interest rate benchmarks; they will be thoroughly studied in Chap. 7.

### 3.5.2 Indices Reflecting the *Average* Value of a Money-Market Rate During a Given Period

Different indices reflect the *average* value (or compound value) of a money-market rate *during* a given period. In contrast with the EURIBORs and LIBORs, they are necessarily *post-determined* (*backward-looking*) since this average is only known at the end of the considered period. Different indices of this type can be computed.

**Compound EURIBORs or LIBORs**

For example, the EURIBOR-3 months capitalized over 4 successive trimesters: $[1 + (n_1/360)\text{Eur}(1)][1 + (n_2/360)\text{Eur}(2)][1 + (n_3/360)\text{Eur}(3)][1 + (n_4/360)\text{Eur}(4)] - 1$ where $\text{Eur}(i)$ denotes the EURIBOR-3 months holding at the start of the trimester $i$ and $n_i$ the number of days in quarter $i$. Such an index is a geometric mean of the four EURIBOR-3 months and reflects the average annual rate obtained by rolling over the four 3-month placements. Indices of this type are disappearing and being replaced by the following compound overnight rates.

---

[21] The substitutes may be grounded on the new overnight rates compounded (for backward-looking indices, see Sect. 3.5.2 below) or on the term structure implicit in OIS swaps or other liquid derivatives (for forward-looking indices).

**Compound EONIA/\$STER, compound SOFR, SONIA, ... (OIS Indices)**

The compound EONIA or \$STER is obtained by capitalizing the overnight rate on business days and using simple interest on Friday, Saturday, Sunday, and holidays, according to what it is possible to obtain by transactions that can actually be carried out.[22] This index is used on European markets, notably as a reference for the variable leg of Overnight Index Swaps (OIS). Indices technically identical but computed from the federal funds rate or from the SOFR are used in the American market and a compound SONIA is used in the UK.

The EONIA/\$STER-Compound between $t$ and $t'$, denoted by I($t, t'$), is computed as follows from the EONIA/\$STER, E($j$), of different days $j$:

- We start by computing $C(t,t')$ which represents the capitalized interest on 1 Euro invested repeatedly overnight from $t$ to $t'$ (capitalized 4 times each week Mon-Tue, Tue-Wed, Wed-Thu and Thu-Fri, and with simple interest over the weekends):

$$1 + C(t,t') = \prod_{j=1}^{o(t,t')} \left[1 + \frac{n_j E(j)}{360}\right]$$

with

$o(t,t')$ = number of business days between $t$ and $t'$

$N_j$ = number of calendar days when the EONIA/\$STER of business day $j$ ($E(j)$) is applied that is 1, or 3 (on Friday), or 2 (on the eve of a holiday), or even, very unusually, 4 if the Friday or the Monday is a holiday.

Let us consider, for example, \$1 invested for 6 days, from the Wednesday of week $n$ (date $t$) and the Tuesday of week $n + 1$ (date $t'$). The capital recovered at $t'$ is:

$$1 + C(t,t') = \left(1 + \frac{1}{360} E(We.n)\right) \left(1 + \frac{1}{360} E(Th.n)\right) \left(1 + \frac{3}{360} E(FR.n)\right)$$
$$\times \left(1 + \frac{1}{360} E(Mo.n + 1)\right)$$

where $E(We.n)$ represents EONIA/\$STER prevailing on the Wednesday of week $n$, etc.

The Overnight-Compound between $t$ and $t'$, annualized, is then:

$$I(t,t') = C(t,t') \frac{360}{Nj(t,t')} \tag{3.4}$$

where $Nj(t,t')$ is the number of calendar days between $t$ and $t'$.

---

[22] For a "roll over" of 24 hour transactions, the first 4 days of the week, and for 3 days on Friday.

## 3.6 Summary

$I(t,t')$ is an annualized rate representing a kind of geometric "average" of the overnight rates between $t$ and $t'$, effectively replicable with a "roll-over" of overnight transactions between $t$ and $t'$.[23]

These different forward- or backward-looking indices are used as references for instruments in the money and bond markets, as we will see in a later chapter.

## 3.6 Summary

- The money market differs from the bond market by the terms of the transactions (short term in the money market), certain practices (e.g., securities trade on an *interest-rate* basis, not a price basis), and by the crucial role played by Central Banks in the interbank market (its main compartment).
- The interest rates used for short-term transactions (under 1 year) are *simple*, in general zero-coupon or *money market and actual/360*, while those used for medium- and long-term securities are compound and actual/actual. Interests may be payed ex post (in arrears) or ex ante (in advance).
- The main short-term securities traded in the money market are *certificates of deposit* (CD) when issued by a bank or a financial institution, *commercial paper* when issued by a nonfinancial corporation and *Treasury bills* when issued by a government.
- A large fraction of the lending/borrowing market is made of *repurchase agreements* (*repos*). A repo is a short-term agreement to sell a security and buy it back (frequently one or two days later) at a slightly higher predetermined price. The implicit *repo rate* is a proxy for the overnight riskless rate. The underlying security is a "collateral" which allows the borrower to pay the lowest interest rate.
- *Lending and borrowing securities*, which differ from lending and borrowing money, is a financial service provided by the lender to the borrower of the securities for a predetermined period at a given cost. It is useful in particular for investors desiring to sell short securities.
- The operations of safekeeping, clearing, and settlement of trades are taken care of by different custodians the main one of which being the *Central Security Depositary*.
- Central Banks exercise control over the *money supply* in two ways: (i) by modifying (very infrequently) the rate of reserve requirements, on which the multiplier depends negatively, (ii) by controlling the *monetary base*, i.e., by adjusting the quantity of Central Bank money available to the banking system through the interbank segment of the money market.
- The main instruments of monetary policy, apart from the rate of reserve requirements, are the Central Bank's different *intervention rates*, primarily the

---

[23] Simple overnight averages are also in use but are less meaningful that compound overnight rates because they are not exactly replicable by a rollover of actual transactions. They may be an acceptable approximation for the compound overnight rates over periods of short duration.

Main Refinancing Rate (MRO) involved in open market operations, the marginal lending rate (for borrowing overnight), the rate of deposit facilities, and the list of *securities eligible* for refinancing in the open market. In addition, massive refinancing operations have been implemented in response to different financial crises.

- A change in *monetary policy* affects interest rates, the banks' liquidity, hence their ability to grant credit, then the investment and consumption decisions by firms and individuals, thereby production growth, employment, and prices.
- An index or a reference rate represents either the value of a rate prevailing on the market *on a given date* (such as the EONIA/$STER, the Federal Fund rate or the SOFR representing overnight rates, or the EURIBORS and LIBORs for different currencies and maturities) or its average value *during a given period* (such as the compound $STER, the SONIA or the compound SOFR).

## Suggestions for Further Reading[24]

Fabozzi, F. (2006). *Fixed income mathematics: Analytical and statistical techniques* (4th ed.). McGraw-Hill Education.
Fabozzi, F. (2016). *Bond markets analysis and strategies* (9th ed.).

---

[24]The money market is frequently treated along with the bond market, and is often explained in basic texts on interest rates. For example:

# The Bond Markets

# 4

It is in the bond market that long-term securities, called bonds, are issued (on the primary market) and traded (on the secondary market). Bonds are negotiable securities that represent a portion of a debt issued by a company or a government. A bond is thus a debt security and gives rise to interest payments (or coupons) and, when it is redeemed, to a reimbursement of the principal. The coupon rate, maturity date, and face value are parts of the *bond indenture* which is the contract binding the bondholders and the issuer.[1]

Like the money market, the bond market is thus an interest rate market. Some differences distinguish the two markets. The first concerns the duration of the securities which are issued: short term for the money market (initial maturity $<1$ year); long term for the bond market (traditionally $>7$ or 10 years). Medium-term securities ("notes") of maturity between 2 and 10 years are, in most countries and in most official statistics, classified as bonds. The second difference between the two markets is in the rules and practices that prevail, notably in how the securities are quoted and traded: in terms of yields on the money market, generally of prices on the bond market.

Bonds can be issued by governments (States), by industrial and commercial firms, and by other legal entities such as municipalities, public establishments, associations, pension funds, and so on. The issuer, other than a government, must satisfy certain requirements as to capital involved, duration, legal announcement and so on, and abide by them.

The bond debt is sometimes backed by a guarantee either in the form of collateral (a mortgage, for instance) or a personal guarantee (a deposit from a third party, e.g., the government). Some contracts contain additional clauses (the *covenants*) which impose restrictions on the issuer with the aim of protecting the holders. These

---

[1] An *indenture* is a formal debt agreement that establishes the terms of a bond issue, while *covenants* are the clauses of such an agreement. Covenants specify the rights of bondholders and the duties of the issuer, such as actions that the issuer is obligated to perform or is prohibited from performing.

© The Author(s), under exclusive license to Springer Nature Switzerland AG 2022
P. Poncet, R. Portait, *Capital Market Finance*, Springer Texts in Business and Economics, https://doi.org/10.1007/978-3-030-84600-8_4

covenants often restrict the conditions under which the issuer may issue future loans or impose the *subordination* of further debts to present ones. Such a subordination covenant means that, in case of bankruptcy, future bonds will not be repaid unless the prior senior debt is fully repaid (a "me-first rule"). Other covenants forbid the borrower from pledging any asset to future lenders, or limit the amounts of dividends paid out to stockholders.

Bondholders are represented by a trust which is charged with maintaining the relationship with the company, with checking that the terms of the bond indenture are observed, and, more generally, with protecting the bondholders' interests.

Banks play an important role in the practical carrying out of transactions (such as network of bank tellers, guarantee of placement) and charge fees for their services. Such fees are one of the components of the cost of issuing bonds.

A standard bond is a special form of long-term security for which the coupon rate (thus the coupons or interests paid out on each maturity date) and the amount to be reimbursed are fixed. A second category of long-term loans is bonds for which the coupon and/or the redemption value are indexed either against market interest rates (bonds with a floating rate) or against the profits of the issuing entity, or against inflation, or some other reference rate. A third category includes bonds with optional clauses such as bonds with bond or equity warrants or convertible bonds.

Furthermore, some loan contracts contain clauses regarding repurchase on the market and/or pre-payment to the benefit of the issuer or the investors.

- A repurchase on the market (buyback) covenant, if present in the contract at issuance (the bond indenture), allows the issuer to avoid reimbursing part of the bonds at the initially planned redemption value by repurchasing them directly on the market. This may be worth doing if the market value of the bonds (which depends on the prevailing market rate) differs from the redemption value. Repurchase thus leads to capital gains (repurchase value < redemption value) or losses (the opposite case) for the issuer. The interest in realizing such profits or losses depends on various considerations, notably fiscal ones. Such a buyback provision is usually limited to a fraction of the capital involved.
- A prepayment covenant can be to the benefit of the borrower (the bond is then termed "callable" from the fact that it includes a call option in the borrower's favor) or of the investor (the bond is called "putable" since it involves a put).[2] Such clauses are generally limited to definite periods. For example, a prepayment option allows, at given instants or for predefined periods of time (called "windows"), either the issuer or the bondholders, according to the terms of the bond indenture, to demand a prepayment of the remaining bonds. Such a prepayment often involves a penalty to be paid by the party requesting it to the party that must abide by it. These covenants, therefore, give the bond an option-like feature which will be examined in later chapters.

---

[2]For the definition of calls and puts, see Chap. 10, Sect. 1.1.

## 4.1 Fixed-Rate Bonds

The chapter is organized as follows. The first section concerns the analysis of standard fixed-rate bonds. The second section succinctly describes different forms of loans with floating rates, indexed rates, and covenants, without undertaking the thorough analysis that will be done in later Chaps. 16 and 17 after the study of floating rate instruments (Chap. 7) and options (Chaps. 10 and 11). The third section deals with operations specific to bond issues. The fourth section briefly presents some international aspects, discusses agents and the functioning of bond markets, and provides the orders of magnitude of the trading handled there.

## 4.1 Fixed-Rate Bonds

We analyze a loan as it is defined when issued (Sect. 4.1.1), study its value at an arbitrary date before maturity (Sect. 4.1.2), and then discuss the way it is quoted on an official market (Sect. 4.1.3). Some bond rate indices are presented in Sect. 4.1.4.

### 4.1.1 Financial Characteristics and Yield to Maturity at the Date of Issue

From a financial point of view, a standard bond debt is characterized, at issue, by the following data: the total amount of the debt; the settlement date (date the investors actually provide payment for the bonds and become owners); the maturity date; the accrual start date (date the first coupon start accruing); the periodicity of coupon payments (annually or semi-annually, infrequently quarterly, or monthly); the nominal interest rate (or face rate, or coupon rate); the nominal value of each bond (face value, par value, or principal) and therefore, given the total amount of the bond issue, the total number of bonds; the repayment price (or value) of the bonds at maturity; and possible covenants concerning repurchase on the market (granted to the issuer) and/or prepayment options (granted to either the issuer or the investors).

In what follows, we assume that the settlement date and the accrual start date coincide (which is most frequently the case) and we call it the issue date.[3]

The gross yield-to-maturity (YTM) of the bond issue is the result of these different characteristics. Let us revisit the main ones to introduce some notation:

- *The coupon* or *face* or *nominal rate,* denoted by $k$ in what follows, and fixed throughout the life of the bonds, is the rate used to calculate the coupon (or interest) due on each bond.

---

[3] Sometimes the accrual start date precedes the settlement date, which benefits the investors as the first coupon payment takes place less than 1 year (or less than 6 months) after the payment for the bonds. This can easily be accounted for in the calculation of the yield to maturity rate (see Chap. 2).

- *The face* or *nominal value*, denoted by $V_n$, is the basis used for calculating the *coupon*: at each due date the coupon for a bond equals the product of the coupon rate by the nominal value:

$$\text{Coupon} = k\, V_n. \qquad (4.1)$$

These days, the nominal values are often \$1000 or 1000 € and those of newly issued Treasury Bonds in Europe (since the introduction of the Euro in 1999) are 1 €. In most Euro-zone countries and in the Euro-bond market, the coupon is paid once a year.[4] In many countries, notably the USA, Britain, and Japan, bonds distribute two coupons each year, these being calculated from an annual face rate $k$, by the formula:

$$\text{Semi-annual coupon} = \frac{k}{2} V_n.$$

In the following, we concentrate on the case of annual coupons, but in various subsections we give a special mention to, and examples of, semi-annual ones. Indeed the formulas and analysis for annual coupons may be applied for semi-annual coupons with the following adjustments: use the half-year as the unit of time, the rate $k/2$ as the face rate and semi-annual rates rather than annual ones.

It is worth noting that there exist zero-coupon (or pure discount) bonds, that is, securities that distribute only one cash flow at their maturity. This one final cash flow can theoretically be decomposed into two parts: repayment of the principal and capitalized interests (see Chap. 6 for the analysis of zero-coupon instruments). Such zero-coupon bonds can result from "stripping" coupon bonds (cf. Sect. 4.2.1).

- *The issue price* (or *initial price*) of the bond, $V_e$ in what follows, is the effective price paid by investors.
- *The repayment price* (or *repayment value*) of the bond, denoted by $V_r$, is the price paid by the issuer to the bondholder at maturity.

It should be noted that the nominal value, the issue price, and the repayment value can differ. In general, but not always, the nominal value and the repayment value coincide. By contrast, the nominal value and the issue price usually differ, the difference being either positive or negative. If the two coincide, the bond is said to be issued "*at par.*"

- The *repayment profile* (also known as *amortization*) is defined by the total number of due dates denoted by $T$, the number of bonds to be settled at each due date and the value of the repayment at that time. If the issue date lies exactly a year before the first due date and the due dates are annual, the total duration of the debt is $T$ years; more generally, the initial lifespan of the debt is the time between the issue date and the last due date (and so is between $T$-1 and $T$ years).

---

[4] Italian bonds distributing semi-annual coupons are an exception to this rule.

## 4.1 Fixed-Rate Bonds

We denote by $n_i$ the number of securities repaid at the $i$th due date. With $N$ denoting the total number of bonds issued, we have: $\sum_{i=1}^{T} ni = N$.

Different amortization schemes are possible: principal repayments can be postponed for one or more years ($n_1 = n_2 = \ldots = n_p = 0$) to give the issuing firm time to generate sufficient cash flows; a constant number of repaid bonds in each period ($n_i = N/T$), in which case the $n_i$ repaid bonds are usually randomly selected; or an increasing number over time in such a way that the total cash payment to bondholders (i.e., the sum of the capital repayment and the interests) stays constant. Today, on all the large markets, *most bonds are of the "bullet" variety*, so all the $N$ bonds are repaid as a block at the maturity date: $n_1 = n_2 = \ldots = n_{T-1} = 0; n_T = N$.

If the loan is not of the "bullet" type,[5] the bonds are repaid during the life of the bond according to a contractually specified payment schedule of due dates $n_1, n_2, \ldots, n_T$.

The yield to maturity (YTM) at issue is, by definition, the Internal Rate of Return (IRR) of the investment for the underwriter who will buy the bonds at issue and retain all of them until maturity. It is also the effective cost of the financing, before taxes and without issuance costs, for the issuer.

For a bullet loan, the computation of the yield to maturity can be done for a single bond since all bonds have identical repayment profiles. If the payment schedule includes reimbursements at intermediate dates, the computation of the yield to maturity has to be done for the whole loan issue.

When the issue price, the nominal value, and the repayment price are equal ($V_e = V_n = V_r$; one says the loan is issued and repaid "*at par*"), the gross yield to maturity equals the nominal interest rate (for an integer number of years).

If the repayment price is fixed above par and the issue value is fixed below par ($V_e < V_n < V_r$), the yield to maturity of the loan exceeds the nominal interest rate.

In a general formulation where the repayment price $V_r^i$ depends on the due date $i$, the cash flow sequence $F_0 \ldots, F_T$, generated by the loan for the collection of its holders (which is, with the opposite sign, the sequence before taxes and issuance costs for the issuing enterprise) can be written:

$$\text{Instant } 0: \quad F_0 = -NV_e$$
$$\text{Instant } 1: \quad F_1 = n_1 V_r^1 + kNV_n$$
$$\ldots$$
$$\text{Instant } i: \quad F_i = n_i V_r^i + k(N - n_1 - \ldots - n_{i-1})V_n$$
$$\ldots$$
$$\text{Instant } T: \quad F_T = n_T V_r^T + k(N - n_1 - \ldots - n_{T-1})V_n = n_T V_r^T + k\, n_T V_n.$$

The gross YTM at issue is the IRR of the sequence $F_0, \ldots, F_T$ above.

---

[5]This is the case for some "Asset Backed Securities" or "Mortgage Backed Securities" (see Chap. 29).

In the simplest and most common case of a bullet loan, the sequence to be used to compute the yield to maturity at issue involves one bond only and is the following:

$$F_0 = -V_e; F_1 = kV_n; F_{T-1} = kV_n; F_T = kV_n + V_r.$$

The yield to maturity at issue is the discount rate $r^*$ such that:

$$V_e = \frac{F_1}{(1+r^*)} + \frac{F_2}{(1+r^*)^2} + \cdots + \frac{F_T}{(1+r^*)^T} \qquad (4.2)$$

---

**Example 1**

Let us consider company X that issues bullet bonds with a term to maturity of 12 years. The face rate is 5%. The issue price is 960 €. The face value and the repayment value are equal to 1000 €.

The cash flow sequence whose IRR is the gross yield to maturity can be written:

$-960, 50, \ldots, 50, 1000{+}50$

(11 times)

, which gives a gross yield to maturity at issue of 5.46% because

$$\frac{50}{1.0546} + \frac{50}{(1.0546)^2} + \cdots + \frac{50}{(1.0546)^{11}} + \frac{1\,050}{(1.0546)^{12}} = 960\ €.$$

Note this rate would be exactly 5% if the bonds were issued at par (that is with an issue price of 1000 €).

---

*The gross yield to maturity at issue* gives the compounded cost of financing, before taxes and without costs, for the issuer. The *net* cost of the loan also depends on the issuance costs and taxes.

To appreciate the influence of issuance costs, we deduct them from the amount initially received ($NV_e$).

Corporate income taxes, as a first approximation, *reduce* the cost of the loan in proportion to the tax rate since paid out interests can be deducted from the taxable amount, which generates tax savings (tax shield). For a tax rate of 33%, for example, a coupon of 100 gives rise to a tax saving of 33; the "net" coupon is therefore only 67 for a for-profit enterprise, as if the "net" face rate had been 0.67 $k$.

Remark that the price at issue is in principle equal to the market price on the issue date and that this latter price depends on the level of bond rates that prevails on the market and the risk of the borrower's signature (its credit rating), according to mechanisms we will explain in the following subsection.

# 4.1 Fixed-Rate Bonds

Note also that Eq. (4.2) given above is only applicable to the case of a first coupon paid exactly 1 year after issue. For a first coupon at an arbitrary time after issuance, the reader is referred to the following subsection (Eq. 4.3).

Finally, let us recall that certain loan contracts have covenants about repurchase on the market and/or prepayment to the benefit of the borrower or the lenders; the preceding analysis does not apply to these cases without extensive modifications.

- The case of bonds with semi-annual coupons is handled in an analogous manner. We have seen examples where cash flows occurred once a year. For a loan that pays semi-annual coupons, the cash flows $F_1$, $F_2$, ..., $F_T$ occur every 6 months and the rate to be used to compute the coupon is $k/2$. So the semi-annual cash flows can be written, for a bullet bond:

$$F_0 = -V_e, F_1 = \frac{k}{2} V_n, \qquad , F_{T-1} = \frac{k}{2} V_n, F_T = \frac{k}{2} V_n + V_r.$$

The YTM of this sequence (the discount rate that makes the present value equal to zero) is, therefore, a semi-annual rate $r^*_s$. Annualizing this semi-annual YTM is done, according to the context, using one of two possible conventions:

- That of compound rates: The corresponding annualized YTM is $r^{*a} = (1 + r^*_s)^2 - 1$ which is sometimes called the *Annual Percentage Yield* (APY) or *effective yield*.
- That of simple rates: $r^{*a\prime} = 2\, r^*_s$, which is called the *Annual Percentage Rate* (APR).

---

**Example 2 (to be contrasted with Example 1)**

Let us consider bullet bonds with a maturity of 12 years with semi-annual coupons. The annual coupon rate is 5%. The issue price is \$960. The nominal value and the repayment value equal \$1000.

The cash flow sequence whose IRR is the semi-annual YTM is written:

$$-960, \underbrace{25, \ldots, 25}_{23 \text{ times}}, 1000 + 25$$, which gives a yield to maturity at issue equal to

2.73% since

$$\frac{25}{1.0273} + \frac{25}{(1.0273)^2} + \ldots + \frac{25}{(1.0273)^{23}} + \frac{1025}{(1.0273)^{24}} = 960$$

This means an annualized YTM, using compound interest (APY), of $(1.0273)^2 - 1 = 5.53\%$, which is slightly larger than in Example 1 since it is preferable to receive a coupon of 25 in mid-year and another coupon of 25 at year's end than a single coupon of 50 at year's end. We notice also that, without compounding (using APR), the annualized rate is $2 \times 2.73\% = 5.46\%$, exactly as in Example 1.

## 4.1.2 The Market Bond Value at an Arbitrary Date; the Influence of Market Rates and of the Issuer's Rating

The yield of an investment in a given bond required by the market at a given time (between two coupon dates) depends on the levels of bond rates prevailing on the market at that time, as well as on the issuer's credit rating in the same market.

### 4.1.2.1 Market Value and Interest Rate

The market value varies with supply and demand: at a point $t$ in the life of a loan, it thus usually differs from the issue value, from the nominal value and from the repayment value. The market value at $t$ essentially depends on the interest rates then prevailing on the bond market and on the credit risk premium then applied to the issuing company.

> **Example 3**
>
> Consider some bonds issued at par, that is, at a face value assumed to be $100. Assume that the repayment value is also equal to $100: the face rate is, therefore, the interest rate then prevailing on the market, for example, 3% (the issuer being considered first class, we assume no credit risk premium). Each bond will entitle the buyer to an annual coupon of $3. At the initial public offering (at $t = 0$), as the issue takes place under current market conditions, the market value of each bond is indeed $100 (the bond is at par):
>
> $$V_0 = \frac{3}{(1.03)} + \frac{3}{(1.03)^2} + \cdots + \frac{3}{(1.03)^N} + \frac{100}{(1.03)^N} = \$100.$$
>
> Assume that shortly after the issue (at $t = 0+$), the rates on the bond market go up by 1%, i.e., 100 basis points (our relevant rate thus becomes 4%). Obviously, investors will not continue to pay $100 for a bond giving a coupon of $3 when they can acquire for the same price bonds offering coupons of $4. The market value of a bond with a coupon of $3 will therefore decrease and settle at a level such that the investment yields exactly 4%, i.e.,
>
> $$V_{0+} = \frac{3}{(1.04)} + \frac{3}{(1.04)^2} + \cdots + \frac{3}{(1.04)^N} + \frac{100}{(1.04)^N} < \$100.$$
>
> The increase in rates has led to a decrease in the market price of the fixed-rate bond.

In fact, the market value, $V$, of a bond equals the discounted value of the cash flow sequence at the due dates $F_1, F_2, \ldots, F_T$ (coupons and repayments) computed on the basis of the discount rate $r$ *equal to the current market interest for bonds with the same risk and the same maturity*.

## 4.1 Fixed-Rate Bonds

When the next cash flow, $F_1$, arrives in a year (at $t = 1$ when we start at $t = 0$), we have

$$V(r_0) = \frac{F_1}{(1+r_0)} + \frac{F_2}{(1+r_0)^2} + \ldots + \frac{F_T}{(1+r_0)^T}. \qquad (4.2')$$

In this formula, the value $r_0$ is given (it is the market rate at date $0^6$) and $V$ is a computed theoretical value.

Furthermore, if $V_0$ denotes the bond's market value at $t = 0$, the bond's yield to maturity at $t = 0$, $r^*_0$, can be calculated from Eq. (4.2):

$$V_0 = \frac{F_1}{(1+r^*_0)} + \frac{F_2}{(1+r^*_0)^2} + \ldots + \frac{F_T}{(1+r^*_0)^T}.$$

Here, it is $V_0$ which is given and $r^*_0$ which is computed.

It is because the yield rate $r^*_0$ cannot, at equilibrium, differ from the rate $r_0$ offered on the market for bonds with comparable maturities and credit risks (therefore $r^*_0 = r_0$, at equilibrium) that the observed value $V_0$ cannot differ, unless there is arbitrage, from the theoretical value $V(r_0)$ given by Eq. (4.2'). For example, if $r^*_0$ exceeded $r_0$, the investors would try to buy the bond under consideration (selling other securities to do so) thus raising its price $V_0$ and lowering its yield $r^*_0$ to the point at which $r^*_0 = r_0$.

In addition, since the present value is a decreasing function of the discount rate, if the rate $r$ (then $r_0$) increases, the bond's value decreases; if in contrast the market rate decreases, the bond's value increases. This sensitivity of market prices to rates when they fluctuate in a random way induces an interest rate risk that we will study more deeply in the next two chapters.

Up to now, we have only considered the case of the next cash flow or coupon happening in exactly 1 year. More generally, let us consider an arbitrary moment $J_c$ days distant from the preceding cash flow. If the date of the preceding coupon is $t = 0$, today's date in fractions of a year is then $\tau = J_c/N_a$ (where $N_a$ is the exact number of days in the year, that is, either 365 or 366 days) and that of the succeeding coupon is $t = 1$, in accordance with the following diagram:

**Diagram 4.1** Next coupon due in less than a year

So the theoretical value of the bond on date $\tau$, denoted $V(r_\tau)$, can be written as a function of the rate $r_\tau$

---

[6]In fact, it is the market rate $r$ plus a "spread." The spread is a positive margin that depends on the issuer's credit rating and increases with the credit risk (see Sect. 4.1.2.2).

$$V(r_\tau) = (1 + r_\tau)^\tau \left[ \frac{F_1}{(1 + r_\tau)} + \frac{F_2}{(1 + r_\tau)^2} + \cdots + \frac{F_T}{(1 + r_\tau)^T} \right] \qquad (4.3)$$

the bracketed term representing the value at date 0 which capitalizes until $\tau$ through the term $(1 + r_\tau)^\tau$.

Besides, for arbitrary $\tau$, the yield to maturity $r^*_\tau$ is given as a function of the observed value $V_\tau$ by

$$V_\tau = (1 + r^*_\tau)^\tau \left[ \frac{F_1}{(1 + r^*_\tau)} + \frac{F_2}{(1 + r^*_\tau)^2} + \cdots + \frac{F_T}{(1 + r^*_\tau)^T} \right]. \qquad (4.3')$$

In contrast to the context of Eq. (4.3), here $V_\tau$ is given and $r^*_\tau$ is derived.

At equilibrium $r^*_\tau = r_\tau$, therefore $V_\tau = V(r_\tau)$.

In the special case of a bullet bond, expressions (4.3) and (4.3') can, respectively, be written

$$V(r_\tau) = V_n(1 + r_\tau)^\tau \left[ \frac{k}{(1 + r_\tau)} + \frac{k}{(1 + r_\tau)^2} + \cdots + \frac{k}{(1 + r_\tau)^{T-1}} + \frac{1+k}{(1 + r_\tau)^T} \right]$$

$$\qquad (4.4)$$

and

$$V_\tau = V_n(1 + r^*_\tau)^\tau \left[ \frac{k}{(1 + r^*_\tau)} + \frac{k}{(1 + r^*_\tau)^2} + \cdots + \frac{k}{(1 + r^*_\tau)^{T-1}} + \frac{1+k}{(1 + r^*_\tau)^T} \right].$$

$$\qquad (4.4')$$

---

**Example 4**

Consider a Treasury bond, with nominal value 1 €, which distributes an annual coupon on the basis of a face interest rate of 5% on June 1 of each year. Suppose we are at March 3 of year $n$, so 89 days from the next coupon. Taking the year $n$ to be a leap year, we are separated in time by $\tau = \frac{366-89}{366} = 0.757$ year from the preceding coupon. The maturity date being June 1 in the year $(n + 10)$, 11 cash flows remain to be distributed. The YTM which corresponds to today's bond market rates is 6%. This bond's value can consequently be reckoned to be:

$$V(6\%) = (1.06)^{0.757} \left[ \frac{0.05}{1.06} + \frac{0.05}{(1.06)^2} + \cdots + \frac{0.05}{(1.06)^{10}} + \frac{1.05}{(1.06)^{11}} \right]$$

$$= 0.963 €.$$

## 4.1 Fixed-Rate Bonds

For a bond paying $N$ semi-annual coupons on the basis of a face rate of $k/2$, formula (4.4) give

$$V(r_\tau^s) = V_n(1 + r_\tau^s)^\tau \left[ \frac{k/2}{(1 + r_\tau^s)} + \frac{k/2}{(1 + r_\tau^s)^2} + \cdots + \frac{k/2}{(1 + r_\tau^s)^{N-1}} + \frac{1 + k/2}{(1 + r_\tau^s)^N} \right].$$

The discount rate $r_\tau^s$ is a semi-annual rate that can be obtained from the annual market rate $r_\tau^a$ with the equations $r_\tau^s = (1 + r_\tau^a)^{0.5} - 1$ or $r_\tau^s = r_\tau^a/2$ according to which definition of the annual rate one uses (APY or APR). The duration $\tau$ represents a fraction of 6 months (number of days /183, for example). The formula allowing us to calculate a semi-annual YTM can be derived by using (Eq. 4.4') analogously, the corresponding annual rate (APY) being $r^{*a} = (1 + r^{*s})^2 - 1$.

### 4.1.2.2 Market Value, Credit Risk, and Rating

The investor must take into account the *credit risk* he or she bears with the issuer as a creditor. This risk has two distinct aspects. First of all, it is the risk linked with *the possibility that the issuer cannot honor its contracts*, so all of (or part of) the capital invested and/or the interests will not be repaid; this is the *risk of default*. Second, it is, for an investor who does not hold the bond until maturity, *the risk that the bond depreciates as a result of an increase in the likelihood of the eventual bankruptcy of the debtor*; in fact it is the *risk of a rating downgrade* of the issuer's credit rating.

Credit risk can be reduced by contract clauses (covenants) under which the issuer submits to some restrictions (notably in regard to future borrowing) or by guarantees (such as a collateral or a third-party bail).

A *rating* allows the investor to assess, from a simple grade, the credit standing of the issuer. Ratings are provided by a specialist firm (a rating agency) on the basis of a financial analysis, an assessment of the management quality, the guarantees offered to the various stakeholders and also statistical models. The agency is independent of the company rated but compensated by it. The principal rating agencies operating in Europe and North America are: Standard and Poors, Moody's, and Fitch.

There is a rating scale for short-term financial instruments and another scale for medium to long-term ones so that each rated company has two types of rating. The first pertains to commercial paper and CDs and the second to bonds and medium-term notes.

Usually, the ratings for long to medium term paper run from AAA (or Aaa), for the best, to C (or D) for the worst. The securities in the four highest categories (AAA, AA, A, and BBB) are termed "investment grade" in contrast to lesser categories termed "high yield" or "junk bonds."

Ratings for short-term securities are specific to each rating agency (such as A-1+, P-1, or F1+, for the best grade given by Standard and Poors, Moody's, and Fitch, respectively.)

Ratings are reviewed at least once a year and can therefore move upward or downward. The yield to maturity required by investors (the market) is obviously a function of the rating provided. For companies wanting to issue securities on financial markets, rating has become an inescapable requirement in the USA, and almost inescapable in Europe.

A rating downgrade usually entails a depreciation of the debt; someone holding the security thus runs a downgrade risk.

The value of the debt depends negatively on the yield to maturity demanded by the market, and the latter increases as the rating drops. The YTM can be split into the market risk-free rate, and a credit spread:

$$\text{Yield to maturity} = \text{market riskless rate} + \text{credit spread}$$

The market yield is the yield that prevails on the bond market for risk-free loans (government bonds) with the same maturity and amortization scheme as the bond under consideration. The credit spread primarily (but not exclusively) depends on the rating.[7] The spread decreases as the issuer's rating increases, and for the same issuer, it increases with the remaining duration of the bond. Generally, it is higher in North America than in Europe. Finally, the credit spread varies over time, increasing in a bear (down) market and decreasing in a bull (up) market.

Now let us look at a bondholder who does not intend to hang on to the bond until maturity: *the primary risk is the possibility of a higher credit spread $s(t)$* (following, e.g., a downgrading of the issuer's rating) which results, other things being equal, in a reduced bond price.

---

**Example 5**

An investor purchases 1000 bonds issued by company O, with a 10-year maturity, face rate of 5.6%, face value 1000 €, at the price of 1,015,148 € corresponding to a yield to maturity of 5.40%. The 10-year Treasury bond yield being 4.80%, the credit spread for O is 60 basis points (5.40% − 4.80% = 0.60%). We suppose that O has just distributed its yearly coupons.

The investor holds these bonds for a year during which bond rates climb by 50 basis points (the Treasury bond yield is thus at 5.30%). Furthermore, as a result of events that influence O's business sector in general, and O's business in particular, O's rating is lowered from A to BB and the credit spread for O also increases by 50 basis points (thus becoming 1.1%). As a result of this, at the end of the year, the investor sells the O bond at a price corresponding to a yield to maturity of 5.30% + 1.1% = 6.40%, which yields 946,521 € or 946.52 € per bond. If the bond rate and the credit spread had remained constant, the yield to maturity of O's bonds would have remained 5.40% and the price at which a single O bond (with a term of 9 years) would have sold would have been 1014 €. The rise in rates and in the risk premium (100 basis points in total) made for a decrease of roughly 68 € in the bond market price, of which

(continued)

---

[7]It also depends on other factors such as market liquidity, as we will see in later chapters.

## 4.1 Fixed-Rate Bonds

about 34 € can be ascribed to the bond rates' increase (interest rate risk) and about 34 € to the increase in the credit spread associated with a lower rating (credit, or downgrade, risk).

We will take up again, in a much more detailed way, the analysis of credit spread and what determines it in Chap. 5 and also in Chaps. 6, 7, and 28 (the last two entirely devoted to credit risk).

### 4.1.3 The Quotation of Bonds

The quotation and trading of bonds are based on the following conventions and definitions.

#### 4.1.3.1 Definitions and Conventions

Bonds are quoted at a percentage of their face values (as if the face value equaled 100 units of the domestic currency) up to two decimal places.

Furthermore, one distinguishes between the "full price" or "dirty price" and the "clean price" (net of accrued interest). The full price is the price that would be paid by the buyer if the face value of the bond were 100. The clean price is the full price reduced by the amount of the "accrued coupon" (relative to a face value of 100). This last accrued interest is computed according to the formula

$$Accrued\ interest = 100\ k\ J_c/N_a \qquad (4.5)$$

where $k$ denotes the face rate, $J_c$ is the number of days since the payment of the last coupon and $N_a$ is the exact number of days in the year (365 or 366).[8]

Defining $\tau = J_c/N_a$ (the fraction of the year which has passed since the payment of the preceding coupon, see Diagram 4.1), Eq. (4.5) is equivalent to

$$Accrued\ interest = 100\ k\ \tau. \qquad (4.5')$$

*The full (dirty) price* at $\tau = J_c/N_a$, denoted by $C_\tau'$, is linked to the value $V$ of the bond by

$$V_\tau = C_\tau'\ V_n/100. \qquad (4.6)$$

---

[8]This is thus an actual/actual convention which is not observed in every country: for example, Germany applies the 30/360 convention to calculate the accrued interest; in the USA the actual/actual convention is used for T-bonds but the 30/360 convention is employed for municipal bonds! Furthermore, for semi-annual coupons $\tau$ denotes a fraction of the half-year.

The full price is therefore the price that would be paid by a buyer if the face value of the bond were 100 (it is in fact $V_n$). For a bullet bond, one can also write

$$C'_\tau = 100(1 + r_\tau)^\tau \left[ \frac{k}{(1 + r_\tau)} + \frac{k}{(1 + r_\tau)^2} + \cdots + \frac{k}{(1 + r_\tau)^{T-1}} + \frac{1 + k}{(1 + r_\tau)^T} \right].$$

$$(4.7)$$

Equation (4.7) can be used either to compute $r_\tau$ from $C'_\tau$ (a yield to maturity calculation) or to calculate $C'_\tau$ from the rate $r_\tau$ (a value calculation).

*The clean price* at $\tau$, denoted by $C_\tau$ is, by definition, the full price reduced by the accrued interest.

$$C_\tau = C_\tau' - 100 \, k \, \tau. \qquad (4.8)$$

Therefore, for a bullet bond

$$C_\tau = 100(1 + r_\tau)^\tau \left[ \frac{k}{(1 + r_\tau)} + \frac{k}{(1 + r_\tau)^2} + \cdots + \frac{k}{(1 + r_\tau)^{T-1}} + \frac{1 + k}{(1 + r_\tau)^T} \right]$$
$$- 100k\tau.$$

$$(4.9)$$

A bond is termed *at par* at $\tau$ if its clean price is equal to its face value: $C_\tau = 100$.

---

**Example 6**

Let us consider a bond X, with face rate of 6.90%, providing an annual coupon on March 8 of each year for 10 years, whose clean price is 112.10 on June 4 of year $n$, and with a face value of 5000 €. On June 4 of year $n$, 88 days have passed since the last coupon was paid out (March 8, year $n$).

The accrued interest is 6.90% × 88/365 = 1.66%.

The full price equals 112.10 + 1.66 = 113.76.

The bond's value (price to be paid on the exchange by a buyer) = 5000 € × 113.76/100 = 5688 €.

Furthermore, its yield to maturity $r^*_\tau$ is such that: $113.76 =$
$(1 + r^*_\tau)^{0.24} \left[ \frac{6.90}{(1 + r^*_\tau)} + \frac{6.90}{(1 + r^*_\tau)^2} + \cdots + \frac{106.90}{(1 + r^*_\tau)^{10}} \right]$, from which $r^*_\tau = 5.28\%$.

---

Recall that for a bond paying semi-annual coupons, formula (4.9) is modified by using a half-year as the period: $T$ is a number of half-years, $r$ is a semi-annual rate, and $\tau$ a fraction of a half-year (number of days /183, for instance).

## 4.1 Fixed-Rate Bonds

### 4.1.3.2 Some Important Properties of Bond Quotes

Certain properties given in the sequel are based on the following result:

**Lemma** For any $r$ and any $T$: $\sum_{i=1}^{T} \frac{r}{(1+r)^i} + \frac{1}{(1+r)^T} \equiv 1.$

Indeed:

$$r \sum_{i=1}^{T} \frac{1}{(1+r)^i} + \frac{1}{(1+r)^T} = r \frac{1-(1+r)^{-T}}{r} + (1+r)^{-T} = 1.$$

**Proposition:**

(a) At the coupon date, the bond is at par if and only if its face rate equals its yield to maturity: For $\tau = 0$ or $1$, $C_\tau = 100$ if and only if $r^*_\tau = k$.

(b) Except on the coupon date, if the face rate equals the yield to maturity, the bond is near (slightly below) par.

**Proof**

Rewrite (4.9) as

$$C(r^*_\tau, k) = 100(1+r^*_\tau)^\tau \left[ \frac{k}{(1+r^*_\tau)} + \frac{k}{(1+r^*_\tau)^2} + \cdots + \frac{k}{(1+r^*_\tau)^{T-1}} + \frac{1+k}{(1+r^*_\tau)^T} \right]$$
$$- 100k\tau.$$

If $r^*_\tau = k \equiv r$ we may write

$$C(r) = 100(1+r)^\tau \left[ \frac{r}{(1+r)} + \frac{r}{(1+r)^2} + \cdots + \frac{r}{(1+r)^{T-1}} + \frac{1+r}{(1+r)^T} \right] - 100r\tau.$$

By the lemma, the term in brackets is equal to 1, therefore

$$C(r) = 100 \left[ (1+r)^\tau - r\tau \right];$$

From this we have

$$C(r) = 100 \text{ for } \tau = 0 \text{ and for } \tau = 1; \text{ in addition, } C(r) < 100 \text{ for } 0 < \tau < 1.$$

The proposition follows since, at the coupon date one can place oneself just after the coupon payment (thus $\tau = 0$; accrued interest $= 0$), or just before the coupon payment (thus $\tau = 1$; accrued interest for 1 year). Furthermore, between two coupon dates, we have $0 < \tau < 1$.

In fact, even for $0 < \tau < 1$, $(1+r)^\tau - r\tau$ is close to $1$[9]: a bond whose yield to maturity equals the coupon rate is thus *approximately at par* between coupon dates. □

---

[9] As the Taylor series expansion of $(1+r)^\tau$ $(= 1 + \tau r + 0.5\tau(\tau-1)r^2 + \ldots)$ with small $r$ shows.

Let us remark that the clean price, $C_\tau$, is different from what would have been, for the same yield to maturity $r^*_\tau$, the full price if the last coupon had just been distributed (which is contrary to an opinion that is widely held); indeed the latter price, a virtual one, can be written

$$X_\tau = 100 \left[ \sum_{i=1}^{T-1} \frac{k}{(1 + r^*_\tau)^i} + \frac{1+k}{(1 + r^*_\tau)^T} \right]$$

Except for $\tau = 0$, this is different from the clean price:

$$C_\tau = 100(1 + r^*_\tau)^\tau \left[ \sum_{i=1}^{T-1} \frac{k}{(1+r^*_\tau)^i} + \frac{1+k}{(1+r^*_\tau)^T} \right] - 100k\tau.$$

*The face rate which would* put *the bond exactly at par is called the "par yield."* From the previous results, we can see that the "par yield" is equal to the yield to maturity on the coupon date but is slightly larger than the latter both before and after the coupon date. The interest of the notion of a "par yield" is essentially conceptual since the difference from the yield to maturity is very little in practice, as Example 7 shows.

**Example 7**
We take up again the preceding example of bond X, with a face rate of 6.90%, distributing a coupon for another 10 years, whose coupon has accrued for 88 days ($88/365 = 0.24$ year) and whose full price equals 113.76; its yield to maturity is 5.28%. Its clean price, 112.10, is computed as follows:

$$C = (1.0528)^{0.24} \left[ \frac{6.90}{(1.0528)} + \frac{6.90}{(1.0528)^2} + \cdots + \frac{106.90}{(1.0528)^{10}} \right] - 1.66$$

$$= 112.10.$$

This clean price differs slightly from what would have been, for a yield to maturity of 5.28%, the full price if the last coupon had just been distributed, namely

$$X = \left[ \frac{6.90}{(1.0528)} + \frac{6.90}{(1.0528)^2} + \cdots + \frac{106.90}{(1.0528)^{10}} \right] = 112.34.$$

In this example the "par yield" is $y_p$ such that $(1.0528)^{0.24} \times$ $100 \left[ \frac{y_p}{(1.0528)} + \frac{y_p}{(1.0528)^2} + \cdots + \frac{1+y_p}{(1.0528)^{10}} \right] - 0.24y_p = 100$; so that $y_p = 5.285\%$, or practically the same value as the yield to maturity.

## 4.2 Floating-Rate Bonds, Indexed Bonds, and Bonds with Covenants

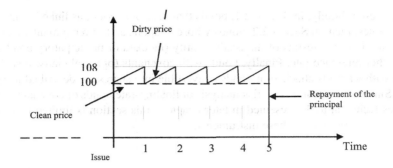

**Fig. 4.1** Evolution of the dirty price of a bond

Let us also note that the clean price process is continuous over time while the full price is discontinuous on the coupon payment dates. For example, if we always have $r^* = k = 8\%$, the "clean" price of bond X in the preceding example remains approximately constant and equal to 100, while the full price varies in a seesaw fashion, taking on the value 108 just before a coupon payment and 100 just after, as shown in Fig. 4.1.

### 4.1.4 Bond Yield References and Bond Indices

The YTMs of Treasury Bonds with different maturities, which prevail on the market on a given date, constitute, for the different financial markets, the main benchmarks for bond rates. They reflect the (risk-free) bond YTM which can be obtained *on a given date* and should not be mixed up with the bond indices which reflect the evolution of the value of a portfolio of bonds. The latter allows computing the yield of a portfolio whose composition is specifically associated with the index over any period of interest. This yield is the result of coupon payments and the variations in value of the bonds caused by variations in market rates. Such representative indices for portfolios of American bonds are computed by Citigroup (Fixed Income Indices LLC), by EFFA (European Federation of Financial Analysts), and JP Morgan (Global Bond Index or GBI) for portfolios of international bonds.

## 4.2 Floating-Rate Bonds, Indexed Bonds, and Bonds with Covenants

Ordinary fixed-rate bonds are just an important part of the bond market. The latter has seen the creation of a huge number of products without predetermined rates, an exhaustive description of which would exceed the bounds of this book. We will limit ourselves here to a very short exposition as the analysis of certain of these products will be undertaken much more deeply in later chapters, with the help of suitable analytic tools.

We present briefly, in Sect. 4.2.1, bonds that distribute coupons linked to market interest rates, then, in Sect. 4.2.2, bonds whose coupons and/or repayment value are related to either the results of the issuing entity (indexed or participatory bonds) or some other reference rate. Finally, bonds with covenants (optional clauses) such as bonds with warrants attached or bonds convertible into shares are described in Sect. 4.2.3. Since the evaluation and risk analysis of floating-rate bonds or covenant bonds requires technical tools presented in later chapters, this section is limited to simple and short descriptions of these instruments.

## 4.2.1 Floating-Rate Bonds and Notes

The coupon of floating-rate bonds and notes varies as a function of the level of rates prevailing on the market. The objective of such an indexing is to limit the influence of rate fluctuations on bond values, in accord with a mechanism thoroughly analyzed in Chap. 7 devoted to floating-rate instruments.

In general, the face rate $i_t$ which is the basis for computing the coupon payment at $t$ results from a formula such as.

$$i_t = \text{reference}(t) + \text{spread}.$$

The spread is fixed at issue and depends on the credit rating of the issuer and on the reference rate; the reference rate depends on the rate conditions that prevail in the market at time $t$-1 (a pre-determined index), or else on those that did prevail, on average, between $t$-1 and $t$ (post-determined index).

As a result, the coupon is "revised" as a function of market conditions. It follows that the capital value of a bond undergoes fluctuations markedly smaller than those undergone by a fixed-rate bond (see Chap. 7). However, it would be an error to consider this market value as entirely unvarying, above all when the reference rate is a long-term rate.

The reference rates used relate either to the money market or to the bond market. Among the monetary reference rates, let us mention the composite EONIA, T-bill rates, and the LIBOR and EURIBOR.

For investors, the choice between a fixed rate and a floating one is based on what they expect of the rates on the one hand, and on the exposure to interest rate risk they are willing to accept on the other.

If the investor predicts an increase in rates, she will prefer variable rate investments; indeed, if she is right, the value of her securities will remain stable and her coupons will increase; in the contrary case of expecting a decrease in rates, she will prefer fixed rates.

Finally, financial assets with floating rates indexed to a monetary reference rate, even if their maturity is far away, behave as far as rates are concerned just like monetary assets and are much less sensitive to fluctuations in market rates than fixed-rate assets with the same maturity (see the following chapter). From this point of view, they are, for many investors, preferable to fixed-rate instruments.

## 4.2.2 Indexed Bonds

In contrast to the floating-rate bonds of the previous subsection, indexing is not here based on an interest rate but rather on an economic variable. Floating rates can be linked to the activities of the issuer (profits, share market price, etc.) or to some other variable such as a market index or a price index.

Different governments have issued several debts indexed either against gold or inflation. In these cases, both capital and coupons are subject to variations as a function of the index. Bonds indexed against inflation deserve a special mention.

The American TIPS ("Treasury Inflation Protected Securities"), British indexed-linked Gilts (see Sect. 4.2.2 infra), and the OATi (OAT indexed to inflation), issued by the French Treasury, are notable examples. Their nominal value (equal to the repayment value) is indexed to the consumer price index (CPI). Both coupons and principal follow the evolution of the CPI. The cash flows produced by a bond indexed in this way are therefore fixed in real terms (in constant \$ or €, that is with a stable purchasing power) and the coupon rate used is thus a real rate and not a nominal one as is the case for other bonds. Besides, this coupon rate is of the same order of magnitude as the bond market rate net of the inflation rate.

## 4.2.3 Bonds with Covenants (Optional Clauses)

We have already mentioned clauses concerning early reimbursement at the behest of the holder ("putable" bonds) or of the issuer ("callable" bonds). These are options that may be exercised at certain time points in the life of the bond. In addition, a large variety of securities giving options to the holder and/or the issuer are commonly written and traded on the bond market. These "hybrid" securities will be subject to detailed analysis in Chap. 17, after Chap. 16 devoted to interest rate options. We restrict the discussion here to a brief description of two of these products.

### 4.2.3.1 Bonds Convertible to Shares

"Convertibles" are bonds that can be converted into shares of the issuing company by their holders if and when it is profitable for them to do so.

- As a bond, each convertible provides the right to a coupon and repayment of the principal.
- Being convertible, either at any instant, or during a period determined in the contract, it gives the holder a possibility of conversion into a contractually specified number of (adjusted) shares.[10]

By investing in this kind of hybrid security, the investor can benefit from the upside potential of the issuing company's share price by exercising the option to

---

[10] The notion of "adjusted share," which is very important in practice, is explained in Chap. 8.

convert. In addition, unlike a shareholder, the bondholder is well protected against a fall in the share price since, if that happens, the convertible behaves like an ordinary bond, at least in so far as the company's credit standing is not seriously damaged.

Because of the embedded option, convertibles offer a smaller coupon than is required for (otherwise comparable) ordinary bonds, which alleviates the financing burden for the issuer.

### 4.2.3.2 Bonds with a Detachable Warrant

There are also bonds to which are *initially* attached warrants. These warrants represent options to buy a given number of the issuing company's shares. They are *detached from the bond just after issuance* so that they can be traded separately from the bonds on the secondary market with a specific quotation.

## 4.3 Issuing and Trading Bonds

We briefly describe the primary and secondary markets in Sect. 4.3.1 and examine in Sect. 4.3.2 the case of long-term government bonds, which are issued according to special procedures ("reopening").

### 4.3.1 Primary and Secondary Markets

In most significant financial exchanges, the primary and secondary markets function according to the following rules:

On the primary market, bonds may be issued to the public by the Treasury, by public or quasi-governmental organizations (regions, local authorities, municipalities, etc.), and by companies (corporate bonds). Non-public entities must exhibit "sufficient" financial health and historical records to have access to the bond market.[11] A public offering (initial or not) is subject to various formalities and legal requirements (prospectuses, ratings, and the like) to ensure that the information provided is fair and complete and investors' interests are protected.

Banks play an important role in the actual handling of corporate bond issues (such as bank windows, investment guarantees, and syndication). The commissions and fees paid to banks make up one major component of issuance costs.

The bond issue can be carried out in one of the two following ways:

- *Syndication,* which is a process whereby an issuer commissions a group of banks to manage the sale of the bonds to the public on its behalf. The syndicate operates either under a *firm commitment* (the underwriting syndicate buys the bonds at the offering price less a margin and then sells them in the market), or under a *best*

---

[11] Such rules vary from country to country.

## 4.3 Issuing and Trading Bonds

*effort agreement* (the banks do not buy the bonds but try to sell them to their customers, acting as simple brokers).

- *Public auction* (or *tender offer*), organized by very large issuers such as the Treasury. The issuer indicates the amounts offered and all the characteristics of the bonds. Banks make competitive bids to acquire them, according to some auction procedures.

In some countries, for some days (or some hours) before the official listing, the bond issue is listed on an interbank market (which non-financial firms and individuals are denied access to) called the *gray market*. How the issue performs on the gray market is a good test of the issuer's financial standing and of the probable market demand.

After issuance, the secondary market provides a greater or lesser degree of liquidity for the bonds. Although, for most of the major financial markets, bonds are listed on an organized market, the majority of trading is carried out in the OTC ("over the counter") market, where prices are often driven by market makers. Trades are carried out by a fast and efficient system of clearing and settlement, the principles of whose functioning were briefly described in Chap. 1, Sect. 1.4.

### 4.3.2 Treasury Bonds and Treasury Notes Issues: Reopening and STRIPS

Bonds issued by the Treasury constitute, in all major countries, a significant part of the bond market and involve special issue procedures. The long-term bonds (in general with an initial term of between 10 and 30 years) are called T-Bonds in the USA, Bunds in Germany, Gilts in Britain, OAT (Obligations Assimilables du Trésor) in France. We have already mentioned that large government debt is issued through a tender offer procedure. In addition, most Treasuries in developed countries issue securities as fungible bonds (with reopening). Furthermore, in many countries sovereign bonds can be stripped, which allows for the issuing of zero-coupon bonds.

#### 4.3.2.1 Fungible Treasury Bonds and Reopening

Most Treasuries in developed countries use a technique that consists in issuing fungible bonds in several installments. The idea is the following: The Treasury issues a "reservoir" loan which is complemented subsequently by issuing new tranches of bonds that are fungible with the preceding ones, i.e., which *have the same coupon rate, repayment, coupon dates, and repayment date*. Their prices at issue obviously depend on the then prevailing market rate. All the bonds, whether created initially or as add-ons are, in fact, identical; they are listed together (i.e., are fungible) and thus are highly liquid and have a huge total market capitalization. This add-on technique is called *"reopening"*. The bonds we study in this section are French OAT or German Bunds. Comparable bonds (gilts, T-Bonds, BTP, JGB, . . .) are described in Sect. 4.4. At the level of analysis that we are carrying out in this section, the main differences between fixed-rate Treasury bonds issued by different governments are the coupon frequency (annual on the Euro-bond market and countries in the Euro-zone, and semi-annual in Anglo-Saxon countries and Japan) and the time conventions for calculating interests (actual/actual in the Euro-zone).

German or French T-Bonds have a face value of 1 €, are repayable at maturity (bullet bonds), with an initial maturity between 7 and 30 years, and are fixed-rate or floating-rate instruments (see Sect. 4.2 above and, for a deeper analysis, Chap. 7).

---

**Example 8**

Let us consider an issue of Treasury Bonds on January 1 of year $n$, to mature on January 1 of $(n + 10)$ (a period of exactly 10 years), providing a 5% coupon every January 1.[12] Assume the 10-year rate for Treasury debt repayable at maturity is 5% and that the bond is issued at par: its yield to maturity is 5% and its issue price, equal to its face value, is 1 ($ or € or any other currency). Suppose that on April 1 of year $n$, 90 days after the initial offering, the Treasury issues another installment of the same bond (maturity date January 1 $(n + 10)$, fixed coupon of 5% every January 1). The bonds issued on 1 January $n$ and those issued on 1 April $n$ are perfectly fungible since they are identical, have the same market value and furthermore are listed as one. Naturally, the issue price of the new installment of Treasury bonds will depend on the bond market rate prevailing on April 1 of year $n$ and will almost surely not be 1. Suppose, for example, that the 10-year bond rate is 4.60% on April 1, $n$. Securities issued on this date will be placed at a price (full or dirty price) of: $(1.046)^{90/365} \left[ \frac{0.05}{1.046} + \frac{0.05}{(1.046)^2} + \cdots + \frac{1.05}{(1.046)^{10}} \right] = 1.0430$.

---

Some of these Treasury bonds can be delivered against futures contracts written on notional long-term bonds, on the Chicago market, EUREX, LIFFE, etc. (see Chaps. 9 and 15).

### 4.3.2.2 Stripping T-Bonds and Creating Zero-Coupon Bonds

Stripping of Treasury bonds into zero-coupon bonds issued on the market are carried out by specialists in Treasury securities in several countries. To explain such operations let us first remark that a bond with fixed payments $F_\theta]_{\theta \, = \, t1, \, \cdots, tN}$ can be considered as a basket of $N$ zero-coupon bonds, the first giving a cash flow of $F_{t1}$ at $t_1$, the second a cash flow of $F_{t2}$ at $t_2, \ldots,$ the $N$th a cash flow of $F_{tN}$ at $t_N$. Thus, as shown in Fig. 4.2 (in black) below, a bond with face value of 1 (unit of the relevant currency) is a basket of $N$ zero-coupon bonds, the first $N$-1, $z_1, \ldots, z_{N-1}$, corresponding to the first $N$-1 coupons $c$ and the $N$th, $z_N$, to the final cash flow of $1 + c$. Bond stripping consists in buying such a bond (for a price $V_0$) and, at the same time, selling separately the $N$ zero-coupon bonds $z_1, \ldots, z_N$ which make it up for a

---

[12]The settlement date of the bond is assumed to be exactly 1 year before payment of the first coupon (the T-bond price is therefore exactly equal to its clean price at issue), which is a special case.

## 4.4 International and Institutional Aspects; the Order of Magnitude of... 141

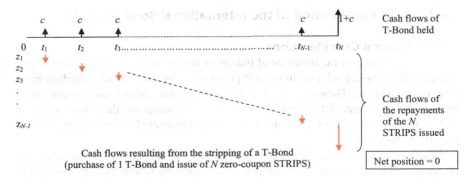

**Fig. 4.2** Stripping a Treasury bond

total, if possible higher, price of $\sum_{i=1}^{N} V_{z_i}$. The series of repayments that these $N$ issues generate (shown in red on Fig. 4.2) exactly offsets the cash flows of the Treasury bond; the position made up in this way leads to null cash flows at $t_1, \ldots, t_N$, and is an arbitrage if initially $\sum_{i=1}^{N} V_{z_i} > V_0$.

Such potential arbitrage opportunities imply that the yields-to-maturity of coupon bonds and the zero-coupon rates must be consistent (their relationships are studied in Chap. 6). The mechanism of stripping Treasury bonds as well as the direct issue of zero-coupon bonds by some countries (notably the US Treasury) explains why there exist long-term zero-coupon bonds on the market. Such zero-coupon bonds are called STRIPS[13] in American terminology and possibly have other names in other countries.

Remark, finally, that in addition to Treasury Bonds which generally have an initial maturity of more than 10 years, sovereign States issue shorter maturity bonds (between 2 and 10 years) which are given the generic name of "Notes." They are called T-Notes in the USA, BOBL in Germany, BTAN in France, etc. They are described in more detail in Sect. 4.4. They can be analyzed like bonds and are often issued using reopening.

## 4.4 International and Institutional Aspects; the Order of Magnitude of the Volume of Transactions

We briefly present the international bond markets in Sect. 4.4.1 and then describe the different national markets and the bonds transacted there in Sect. 4.4.2.

---

[13] This term makes one think of the "undressing" of a bond but in fact stands for "Separate Trading of Registered Interest and Principal of Securities."

## 4.4.1 Brief Presentation of the International Bond Markets

### 4.4.1.1 General Considerations

As much as a result of the behavior of issuers as that of investors who diversify and optimize their trading without limiting it geographically, the financial markets have become worldwide. However, the bond market is not unified, on the one hand because different financial centers have different regulations and their own practices, and on the other because the bonds are mostly denominated in the currency of the country of issue.

Bonds are in particular distinguished according to the issuer's nationality, the currency in which they are denominated and the location where they are issued. On the international markets, we distinguish three main types of bonds:

- "*Domestic bonds*" issued on the market in country X, by a borrower in country X, usually denominated in the currency of country X.
- "*Foreign bonds*" issued on the market in country X, by a borrower from country Y differing from X; they are usually denominated in the currency of X and behave according to the regulations and practices of X. Such foreign bonds are significant in the USA, Britain, Switzerland, Japan, and Germany.[14]
- "*Euro-bonds*" in currency X (not necessarily the Euro) issued by an international syndicate of banks, located in general in markets of several countries Y whose currencies are not X. Such bonds are not subject to the regulations of a particular country although the transactions (issuance, underwriting) carried out in a country are subject to the rules prevailing there.

Note that the boundary between Euro-bonds and foreign bonds is fuzzy and together they form what are called "international bonds." New York is the main primary market for such bonds while London is the leading secondary market.

Prominently among the conditions needed for an international market to prosper features the absence of double taxation. In most of the issuers' countries, a tax is withheld as a fraction of the interest paid to the bondholders. To encourage the issuing of Euro-bonds or foreign bonds and avoid double taxation which could harm the interest in the issuer's country (withholding tax) and in the investor's (income tax), the withholding tax is usually not applied to non-residents.[15]

---

[14]Foreign bonds in dollars, British pounds, Swiss Francs, or Yens are given the exotic names of *Yankee, bulldog, chocolate,* and *samurai* bonds, respectively.

[15]This exemption from the withholding tax is sometimes criticized since it can encourage tax evasion, when the investor possibly omits to declare in his/her own country the interests earned abroad. Bilateral fiscal treaties eliminating some or all double taxation constitute an alternative to exemption from withholding taxes. They allow an investor to subtract the tax paid elsewhere from the tax due locally. This eliminates or limits fiscal evasion, but from the point of view of the investors, it is less simple and less flexible than an exemption from the withholding tax for non-residents.

## 4.4 International and Institutional Aspects; the Order of Magnitude of... 143

### 4.4.1.2 The Euro-Bond Market

Most Euro-bonds are listed in Luxemburg but the real trading is done OTC. The market is kept going by an international network of market makers and dealers grouped as "International Securities Market Association" (ISMA) based in London and Zurich and offset by such centers as Euroclear and Clearstream.

Bonds issued on the Euro-bond market are long-term (bonds) or medium-term (bills or notes). The main ones are:

- Fixed-rate bonds (called "straight bonds"), which make up the majority of issues and pay a fixed annual coupon.
- Floating-rate bonds (floating rate Euro-notes), which pay periodically (often three or two times a year) a coupon tied to a reference rate.
- Convertible bonds, bonds with warrants, perpetuities (consolidated bonds), dual currency bonds, and so on.[16]

The global market in Euro-bonds is made up of about 50% in bonds denominated in dollars and 33% in Euros.

Finally, there exists an international market for bonds issued in *emerging countries* (Latin America, Asia with the exception of Japan, and countries in Eastern Europe). They are usually denominated in dollars and bear some credit risk.

### 4.4.2 The Main National Markets

The American, European, and Japanese markets are the main ones. Out of the more than $100 trillion bonds outstanding in the world, 30% are international bonds. American bonds make up about 40% of the total, European ones more than 25%, and Japanese 15%. Government bonds (sovereign debt) constitute between 40% and 50% of the total except in the case of Japan where they represent 75%.[17]

We now present in more detail the major national markets and the main products that are traded there.

### 4.4.2.1 The American Market

The American bond market is the largest and most liquid in the world. Even though government bonds (sovereign debts) are listed on the NYSE, the majority of the trading is OTC (over the counter) and daily operations are taken care of by market makers (primary dealers). They trade government securities, quasi-governmental

---

[16]Dual currency bonds have coupons and principal denominated in two different currencies.

[17]These are merely orders of magnitude, since the different statistics are not uniform, even as to how they define the securities considered as bonds. The sources used to obtain these orders of magnitude are: the BRI and ECB (for the European fixed-income securities), the European Capital Markets Institute, and the "Securities Industries and Financial Markets Association" (SIFMA) for American data.

bonds, corporate bonds, mortgage-backed securities, and asset-backed securities (see Chap. 29 for definitions and details).

- Government securities are called *T-Bonds* (if their maturity at issue is between 10 and 30 years) or *T-Notes* (if their initial term is between 2 and 10 years). Furthermore, the Treasury issues securities with maturities less than 1 year called *T-Bills* which, in our classification, belong to the money market. T-Bonds and T-Notes share the same technical characteristics: they are "bullet" bonds and have semi-annual coupons.

There are also, as we have seen, bonds indexed against inflation called "Treasury Inflation Protected Securities" or TIPS. The principal which remains due is indexed, which implies that the coupons, which depend on the latter, are also indexed.

Finally, zero-coupon bonds resulting from the stripping of government securities (STRIPS; see Sect. 4.3.2) are also traded.

Government securities are issued through an auction process and most often according to the reopening technique described in Sect. 4.3.2. The secondary market is OTC, open 24 hours and conducted by Treasury bond specialists who allow the general public to access this market and trade with interdealer brokers as intermediaries.

- Quasi-governmental bonds are "municipal bonds" or "agency bonds." "Municipal bonds" are issued by state governments (but not the federal government) and local authorities and often offer fiscal advantages. "Agency bonds" are issued by federal government agencies and quasi-governmental enterprises ("Government-Sponsored Enterprises," particularly those specializing in mortgage-backed credit); they benefit from government guarantees.
- "Corporate" bonds are issued by private enterprises. Among the issuers, "utilities" (those having a concession for the supply of a public service such as post office, telephone, or electricity) are distinguished from industrial and financial enterprises.
- Finally, "Mortgage bonds" are bonds backed by a mortgage guarantee, and their volume is considerable.

### 4.4.2.2 European Bond Markets

The European bond markets have increased dramatically in size since the introduction of the Euro, in part because of the increased liquidity resulting from a single currency. The German, Italian, British, and French bond markets are the largest. With the exception of the German market, the major part is made up of government bonds. Interestingly, the market for European government bonds is larger than its American counterpart.

## 4.4 International and Institutional Aspects; the Order of Magnitude of... 145

(a) *In the Euro-zone*

The bond markets have been somewhat harmonized and the bonds now share numerous features: denomination, bullet profiles, annual coupons calculated on an actual/actual basis, etc. The largest issuers of domestic bonds are Italy, France, and Germany (each between 25% and 30% of the Euro-zone total), with government securities making up about 50% of each market.

In *Germany*, the federal government issues *bunds* (with original maturity between 10 and 30 years), *Bobl* (with an initial maturity of 5.5 years), and *Schatze* (with original maturity of 2 years). They are "bullet" securities, with, in general, a fixed annual rate. Bunds and Bobl are issued using the reopening technique. The Schatze are not quoted at a price, but at an annual rate, and from this point of view belong more to the money market.

The German bond market is characterized by a very large portion of banking debt. These bonds are termed "Pfandbriefe," have an original maturity of between 2 and 10 years and are issued by banks to refinance loans that they have themselves provided to local authorities or to mortgage borrowers. They are thus backed by collateralized debt as a guarantee. Besides, debt certificates called "Schuldscheine," transferable through endorsement, and with initial maturity of 3 to 6 years, represent a large part of the market. The corporate bond segment is traditionally limited, which explains why the large German industrial enterprises make massive use of the Euro-bond market.

In *France*, government bonds represent more than half of is the amount outstanding on the market. They have a maturity between 7 and 30 years at issue, may offer a fixed, floating, or indexed coupon, and may also carry covenants. Treasury bonds (Obligations Assimilables du Trésor or OAT) make up the vast majority of these securities. The French government also issues Medium Term Treasury notes (Bons à Terme Assimilables Négociables or BTAN). The private sector issues different types of long-term bonds as well as Medium Term Notes (Bons à Moyen Terme Négociables or BMTN).

In *Italy*, the primary bond market is dominated by the government which issues CTZ (2-year Certificate del Tesoro Zero Coupon), CCT (Certificate di Credito del Tesoro with a floating rate and a 7-year maturity), and BTP (Buoni del Tesoro Poliennali). BTP, the most significant part of the outstanding bonds, are bullet securities, with a fixed semi-annual coupon, and original maturity between 3 and 30 years, equivalent to American T-notes and T-bonds.

Other active bond markets in the Euro-zone are situated in Spain, where the government issues medium term "Bonos del Estado" and "Obligaciones del Estado" with initial maturities from 10 to 15 years, in Belgium and The Netherlands.

(b) *Outside the Euro-zone*

The significant non-European bond market is the *British market*. Government bonds are called "gilts" and are issued with different maturities (short: less than 7 years; medium: 7–15 years; long: more than 15 years). Just like their American counterparts, the repayment profile is bullet and the coupons are semi-annual. In addition to these classic gilts there are gilts indexed to inflation, with a floating rate and so on. Besides, sterling corporate bonds, whose volume is much smaller than that of the gilts, are nonetheless gaining momentum. How the British market is organized and works are reminiscent of the American market, notably as to the dominant role of OTC trading directed by market makers.

### 4.4.2.3 The Japanese Market

The Japanese bond market, which hardly existed before the start of the 1970s, is presently the second largest national market by volume. Government issues dominate and are mainly made up of JGB (Japanese Government Bonds, or *Gensaki*). JGB are long-term securities (6–20 years at issue, most often 10 years), with a bullet profile, paying a semi-annual coupon, and possibly admissible to reopening. Thus, they are essentially similar to long-term securities issued by the government of the other major countries.

The corporate sector, where bonds are issued by banks, large firms, and utilities, makes up about 25% of the market.

### 4.4.2.4 Other Markets

Finally, let us mention the existence of other bond markets:

- *Canada*, where half of the bonds outstanding is issued by the federal government (GoCs), about a third by the provinces and municipalities (the province of Quebec being among the most active), and the rest by the private sector.
- *Australia*, where the market is dominated by government and quasi-governmental securities (ACGB[18]).
- *China* and *South Korea*, where the business is picking up quickly.

## 4.5   Summary

- Bonds are long-term negotiable securities representing a debt issued by a company (corporate bonds) or a government (sovereign bonds). The coupon (facial, nominal) rate, maturity date, and face (nominal) value all are specified in the *bond indenture* binding the bondholders and the issuer.
- A wide variety of bonds exist: (1) standard bonds whose coupon rate and the amount to be repaid are fixed; (2) bonds whose coupons and/or the repayment value are indexed either on some market interest rate (floating rate) or some other reference (e.g., inflation); (3) bonds with optional components (warrants, convertible or repurchase clauses to the benefit of the issuer or the investors). Besides, bonds are often backed by a guarantee (e.g., collateral, mortgage)

---

[18] Australian Commonwealth Government Bonds.

## 4.5 Summary

and/or contain *covenants* that impose restrictions on the issuer in order to protect the holders ("me-first rules", . . .).

- The bond's nominal value, issue price, and repayment value can differ. If the nominal value and the issue price coincide, the bond is issued "*at par.*" If the nominal value and the repayment value coincide, the bond is redeemed "*at par.*"
- In most Euro-zone countries and in the Euro-bond market, the coupon is annual. It is semi-annual in many countries such as the USA, Great Britain, and Japan.
- Many different *amortization (repayment) schemes* for a bond are possible, but most bonds are of the *bullet* variety, i.e., are repaid as a block at maturity.
- For fixed-coupon bonds, the *yield to maturity* (YTM) at issue is the Internal Rate of Return (IRR) of the investment for the subscriber, i.e., the IRR of the series of coupons and principal repayments up to maturity, the issue price being the initial outflow. A YTM can be computed at any time after the issue, the market price being the outflow. These YTM are equal to the bond market risk-free rate plus a *credit spread*.
- A fixed-coupon bond is valued by discounting its cash flows with a discount rate equal to the market rate for bonds of same risk and maturity.
- Bonds are subject to interest rate risk, *default risk* and to the *risk of a credit rating downgrade* (which implies a market value decrease even in absence of default).
- The full or "*dirty*" quoted market price of a bond is the price that would be paid by the buyer if the face value of the bonds were 100. The "*clean*" price is the full price reduced by the amount of the accrued coupon (relative to a face value of 100). The clean price process is continuous over time while the full price is discontinuous on the coupon payment dates. Cheapness or dearness of a bond is better assessed on the basis of its clean price than on its full price.
- A bond is *at par* if its clean price is equal to 100. A bond whose YTM equals the coupon rate is thus at par exactly on the coupon date and approximately on other dates. The coupon rate which would make the bond exactly at par is called the "par yield."
- The coupon of *floating-rate* bonds varies as a function of the level of rates prevailing on the market, so that their market prices undergo fluctuations markedly smaller than those of fixed-rate bonds.
- The floating rate is the sum of a (varying) *reference rate* plus a *fixed spread* that depends on the issuer's credit rating. The coupon thus is "revised" as a function of market conditions. The reference rates relate either to the money market or to the bond market.
- For *indexed bonds*, indexing is based on an economic variable, either specific to the issuer or an aggregate such as a market price index.
- *Convertible bonds* are hybrids that can be converted into shares of the issuing corporation by their holders if and when it is profitable for them to do so.
- A bond *issue* can be carried out either by *syndication* (a group of banks manages the sale of the bonds to the public) or by *public auction* (or tender offer) among banks.

- *Reopening* by the Treasury is issuing new tranches of Treasury bonds that have the same coupon rate, coupon payment dates, and maturity. The purpose is to increase the liquidity of these bonds in the secondary market.
- *Zero-coupon bonds* can be obtained from coupon bonds by the technique called *stripping* by which the principal repayment and the coupons are disentangled and each and every component is treated as a separate, individual zero-coupon bond.

## Suggested Readings

Bodie, Z., Kane, A., & Marcus, A. (2013). *Investments*, Irwin (10th ed.).
Fabozzi, F. (2006) *Fixed income mathematics: Analytical and statistical techniques* (4th ed.). McGraw Hill Education.
Fabozzi, F. (2016). *Bond markets analysis and strategies* (9th ed.).
Solnik, B., & McLeavey, D. (2013). *Global investments* (6th ed.). Pearson.

# Introduction to the Analysis of Interest Rate and Credit Risks

**5**

In the two preceding chapters devoted to instruments on the money and bond markets, we emphasized several times that a change in market rates leads to a change with the opposite sign in the market prices of securities. Since the rate fluctuations are random this dependency of prices on rates induces another risk termed *interest rate risk*.

We also indicated that the risk of default on the part of a borrower explains and justifies the "credit spread" which is a component of the rate charged by the lender so that on the interest rate market one may write:

Yield to maturity $(t)$ = riskless market rate$(t)$ + credit spread $(t)$.

Finally, we called attention to the fact that credit risk cannot exclusively be understood as the risk of the bankruptcy of the borrower (the default risk in a strict sense), as it also can be the risk of an increase in the credit spread of the borrower (or issuer) whose effect on the price of a security is the same as a market rate increase of the same size.

This chapter is devoted to the study of interest rate and credit risks. It is an introduction as the study of these risks will be taken further throughout the whole of this work. The principal factors explaining the sensitivity to interest rates of fixed-income securities and simple quantitative estimates of interest rate risk are presented in Sect. 5.1. Section 5.2 is devoted to credit risk.

## 5.1 Interest Rate Risk

Two introductory examples, given in Sect. 5.1.1, demonstrate the influence of the duration of the security on its sensitivity to rates. This influence is made precise, and the notions of Variation, Sensitivity, and Duration are defined in Sect. 5.1.2. Formulas for the Variation (or absolute sensitivity), Sensitivity (or Modified Duration), and Duration (or Macaulay Duration) with both proportional rates and continuous rates are derived in Sect. 5.1.3. Some properties of Sensitivity and Variation are

---

© The Author(s), under exclusive license to Springer Nature Switzerland AG 2022
P. Poncet, R. Portait, *Capital Market Finance*, Springer Texts in Business and
Economics, https://doi.org/10.1007/978-3-030-84600-8_5

discussed in Sect. 5.1.4. The interest rate risk affecting a portfolio of assets and liabilities is studied in Sect. 5.1.5. The effect of convexity is explained in Sect. 5.1.6.

## 5.1.1 Introductory Examples: The Influence of the Maturity of a Security on Its Sensitivity to Interest Rates

We will first examine, starting from a numerical example, how the same rate increase affects the price of a long-term security (a) and a short-term security (b).

---

**Example 1**

(a) Let us consider a perpetuity (perpetual annuity) $L$ (thus a long-term security) defined as the right to receive an annual $5 coupon forever, with no reimbursement of principal. If the prevailing very long-term market rate is 5%, the value of the annuity $L$ is $100. Indeed:

- the present value of cash flows of $5 each year forever is equal to $\frac{5}{0.05} = 100$.
- an investment of $100 in L giving $5 yearly does yield a rate of 5%.

Assume now that the long-term rate increases from 5% to 6% (100 basis points): the market value of the annuity $L$ decreases and becomes $\frac{5}{0.06} = 83.33$ (an investment of $83.33 in $L$ brings $5 yearly, i.e., the 6% market rate).

(b) Now consider a (short-term) security $C$ which is due in one year, the date at which it will repay the capital of $100 and a coupon of $5.

If the interest due in one year is 5%, the market value of the security $C$ is $\frac{105}{1.05} = 100$.

If the interest rate climbs to 6%, the value of the security $C$ becomes $\frac{105}{1.06} = 99.06$.

The comparison of the results obtained in the two examples suggests the following remark and conclusion:

*(i)* The *same* rate increase of 1% leads to

- A reduction of 16.67% in the price of the security $L$ ($\frac{100-83.33}{100} = 16.67\%$)
- A reduction of only 0.94% in the value of $C$ ($\frac{100-99.06}{100} = 0.94\%$)

The market value of security $L$ is therefore much more sensitive to rate variations than that of security $C$.

*(ii)* We are tempted to ascribe this difference in sensitivities to the difference in the maturity of the two securities.

---

In fact, the preceding example has a general significance: the market value of long-term securities is more sensitive to interest rate fluctuations than that of short-term securities. Such an assertion is, however, imprecise, as the example below shows with two securities of the same maturity but of different sensitivities.

## 5.1 Interest Rate Risk

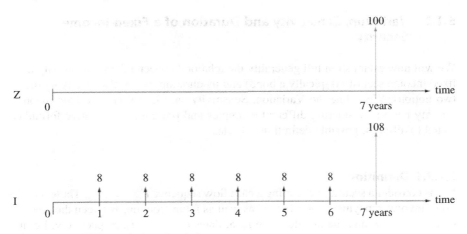

**Diagram 5.1** Different timing of cash flows

> **Example 2**
> (a) First consider a zero-coupon security $Z$ which provides a single payment of $100 over 7 years. For a rate of return of 5% its value is $V_Z(5\%) = \frac{100}{(1.05)^7} = 71.07$.
>
> A rate increase of 1% (100 basis points) lowers its value to $V_Z(6\%) = \frac{100}{(1.06)^7} = 66.51$.
>
> (b) Now consider a *bullet* security $I$, with 7 years to run, which gives an annual coupon of 8% and repays the face value of $100 (plus the last coupon) at the end of 7 years.
>
> For a rate of return of 5% its value is $V_I(5\%) = \sum_{i=1}^{7} \frac{8}{(1.05)^i} + \frac{100}{(1.05)^7} = 117.36$.
>
> A rate increase of 1% lowers its value to $V_I(6\%) = \sum_{i=1}^{7} \frac{8}{(1.06)^i} + \frac{100}{(1.06)^7} = 111.16$.
>
> Remark that a similar increase in rates of 1% leads to:
> – A reduction of 6.42% in the price of security $Z$ ($\frac{71.07-65.51}{71.07} = 6.42\%$)
> – A reduction of only 5.28% in the security $I$ ($\frac{117.36-111.16}{117.36} = 5.28\%$)
>
> The market value of $Z$ is, therefore, more sensitive to rate variations than that of $I$ although the two securities have the same lifetime.
>
> We may understand this result, intuitively, by comparing the positions of the cash flows for $Z$ and $I$ over the course of time (see Diagram 5.1):
>
> The *average position of the cash flows*, plotted over time (the center of gravity of the cash flows), is nearer the present date (0) for $I$ than it is for $Z$; in a sense which we will make more precise in the sequel, $Z$ is "more long-term" than $I$, which explains why it is more sensitive.

## 5.1.2 Variation, Sensitivity and Duration of a Fixed-Income Security

We will now establish in full generality the relation between the rate sensitivity of a fixed income security (typically a bond) and its duration, while also clarifying these two notions. We define the Variation, Sensitivity, and Duration of a fixed-income security before considering different examples and presenting alternative formulas suited to different possible definitions of rates.

### 5.1.2.1 Definitions

Let us consider a security generating a cash flow sequence $F_{t_1}, \ldots, F_{t_n}$. These $n$ cash flows involve repayments of principal as well as future coupons between the present moment ($t = 0$) and the last due date $t_n$, at dates $t_1, t_2, \ldots, t_n$, respectively. Let us rewrite the formula for the market value

$$V(r) = \frac{F_{t_1}}{(1+r)^{t_1}} + \frac{F_{t_2}}{(1+r)^{t_2}} + \ldots + \frac{F_{t_n}}{(1+r)^{t_n}} \equiv \sum_{\theta=t_1}^{t_n} \frac{F_\theta}{(1+r)^\theta}. \tag{5.1}$$

In this formula, $r$ is "the rate of interest" prevailing on the market at present, which equals in equilibrium the yield to maturity of the asset.

An infinitesimal variation $dr$ in the market interest rate triggers a variation $dV$ in the value of the security obtained by taking the derivative of Eq. (5.1):

$$\frac{dV}{dr}(r) = \frac{-t_1 F_{t_1}}{(1+r)^{t_1+1}} + \frac{-t_2 F_{t_2}}{(1+r)^{t_2+1}} + \ldots + \frac{-t_n F_{t_n}}{(1+r)^{t_n+1}}$$

$$\equiv -\frac{1}{(1+r)} \sum_{\theta=t_1}^{t_n} \frac{\theta F_\theta}{(1+r)^\theta}. \tag{5.2}$$

**Definition** *We call $\frac{dV}{dr}$ the bond's "Variation" (or absolute sensitivity to interest rates).*

The "relative" sensitivity is given by the expression $S = \frac{-dV}{V dr}$, which can be (approximately) interpreted as a percentage change of $V$ resulting from an absolute variation of 1% in $r$:

$$S = \frac{-dV}{V dr} = \frac{1}{(1+r)} \sum_{\theta=t_1}^{t_n} \frac{\theta \ F_\theta}{V(1+r)^\theta}. \tag{5.3a}$$

We remark that we can equally well write

# 5.1 Interest Rate Risk

$$S = -\frac{d\ln(V(r))}{dr}.$$ 

(5.3b)

**Definition** $S = \frac{-dV}{Vdr}$ *is called the "sensitivity" of the bond.*
$S$ is also called *"modified Duration"* for reasons that will appear in the sequel.
$S$ thus is a measure of the interest rate risk affecting a fixed-income security.

The variation $\frac{\Delta V}{V}$ caused by a non-infinitesimal variation $\Delta r$ of the interest rate can be estimated using the approximation

$$\frac{\Delta V}{V} = -S\Delta r.$$ 

(5.4)

---

**Example 3**
A bond whose Sensitivity is 6 loses about 6% of its value if there is a rate increase of 1%. Its value increases by about 0.6% if there is a rate decrease of 0.1%.

---

Now let us examine the sum term in Eq. (5.3a) and define

$$D = \sum_{\theta=t_1}^{t_n} \frac{\theta F_\theta}{V(1+r)^\theta} \equiv \frac{F_{t_1}}{V(1+r)^{t_1}}t_1 + \frac{F_{t_2}}{V(1+r)^{t_2}}t_2 + \ldots + \frac{F_{t_n}}{V(1+r)^{t_n}}t_n.$$ 

(5.5)

**Definition** $D$, *given by equation (5.5), is called the "Duration" of the bond.*

The Duration (more precisely the Macaulay Duration) can be understood as the average lifetime of the bond. Indeed, $D$ equals a weighted mean of the periods $t_1$, $t_2$, $\ldots$, $t_n$, corresponding to the different due dates; the weight for a due date $t_i$ is $\frac{F_{t_i}}{V(1+r)^{t_i}}$; this coefficient is equal to the relative contribution of the cash flow $F_{t_i}$ to the bond's value $V$, the due date $t_i$ "weighing" more heavily the larger the discounted cash flow $\frac{F_{t_i}}{(1+r)^{t_i}}$ is (relative to $V$). One sees that the sum of the weights is actually 1.[1]

*The Duration thus corresponds to the mean position (or center of gravity, or barycenter) of the discounted cash flows over time.*

Diagram 5.2 below, based on an analogy with elementary physics, helps understanding Duration as the barycenter of the discounted cash flows.

Let us consider a balance scale (or beam balance) whose beam on the right supports a series of $n$ weights, the $i$th being $\frac{F_{t_i}}{(1+r)^{t_i}}$ Kg positioned at a distance of $t_i$ meters from the fulcrum (knifepoint) at 0.

---

[1] Indeed, since $\sum_{\theta=t_1}^{t_n} \frac{F_\theta}{(1+r)^\theta} = V$, we have $\sum_{\theta=t_1}^{t_n} \frac{F_\theta}{V(1+r)^\theta} = \frac{V}{V} = 1$.

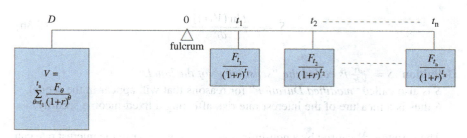

**Diagram 5.2** Macaulay Duration as a barycenter

On the left part of the beam at a distance $D$ (in meters) from the fulcrum, is a weight of $V$ Kg. We note that the sum of the weights on the right ($\sum_{\theta=t_1}^{t_n} \frac{F_\theta}{(1+r)^\theta}$) is equal to the weight $V$ placed on the left. The Duration of the bond corresponds to the distance $D$ that there must be between the weight $V$ and the fulcrum so that the balance is in equilibrium. $D$ is therefore such that:

$$V\,D = \sum_{\theta=t_1}^{t_n} \frac{F_\theta}{(1+r)^\theta}\,\theta.$$

Using both equations (5.3a) and (5.5) shows that the Duration is related to the Sensitivity by the equation

$$D = S(1+r) \qquad (5.6)$$

which is the reason why the sensitivity is often called "modified Duration." The interest rate sensitivity of a bond is a positive function of its "mean lifetime."

### 5.1.2.2 Two Interesting Special Cases: Zero-Coupon Bonds and Perpetual Annuities

As an example and to provide insight as to this concept, let us calculate the Duration of a zero-coupon bond. By definition, such a bond only provides a single payment at $T$ (no remuneration or reimbursement of the principal happens before the maturity date $T$, thus $F_{ti} = 0$ except for $t_i = T$); at $T$ the bond reimburses the principal and pays the capitalized interest, the sum of these two parts corresponding to the cash flow $F_T$.

**Proposition**
*The Duration of a zero-coupon bond equals its lifetime T.*

**Proof**
The bond's Duration, given by equation (5.5), reduces in this case to $D = \frac{TF_T}{V(1+r)^T}$ (since all the payments $F_{ti}$ vanish except the last).
Since, furthermore, $V = \frac{F_T}{(1+r)^T}$, it follows that $D = T$.

## 5.1 Interest Rate Risk

For a zero-coupon bond, the Duration is the lifetime remaining before the last (and only) due date: the center of gravity of the payments is obviously at $T$.

For a loan leading to a distribution of coupons and, *a fortiori*, to reimbursements in intermediate periods, the Duration is less than the remaining lifetime before the final due date.

Let us take up again the examples of bonds $Z$ and $I$ shown in Diagram 5.1 of Example 2.

– Bond $Z$ (zero-coupon) has a Duration of 7 years;

– Bond $I$, with coupons and with a full lifetime of 7 years, has a Duration equal to

$$\frac{1}{117.36}\left[\sum_{t=1}^{6}\frac{8}{(1.05)^t}\,t + \frac{108}{(1.05)^7}7 = 5.74\,years\right]$$

Now let us consider a perpetuity (perpetual annuity) providing yearly payments of $x$ forever.

### Proposition
*The Sensitivity of a perpetual annuity paying $x$ each year equals the inverse of its yield, i.e. $S = \frac{1}{r}$. Its Duration is thus $D = \frac{1+r}{r}$.*

### Proof
$V = \sum_{\theta=1}^{\infty}\frac{x}{(1+r)^\theta} = \frac{x}{r}$ thus $S = -\frac{d\ln v}{dr} = \frac{1}{r}$ and $D = S(1+r) = \frac{1+r}{r}$.

For example, the Duration of a perpetuity whose yield to maturity equals 5% is 21 years (and not infinity).

---

### Example 4
Consider a bond with 10 years to live, a face value of $100, whose nominal rate is 6% while its yield to maturity (equal to the current market rate) is 5%.

Its present value is, therefore, $V = \sum_{t=1}^{9}\frac{6}{(1.05)^t} + \frac{106}{(1.05)^{10}} = 107.72$.

Its Sensitivity is

$$S = \frac{-dV}{V\,dr}$$

$$= \frac{1}{1.05 \times 107.72}\left[\frac{6}{1.05}\times 1 + \frac{6}{(1.05)^2}\times 2 + \cdots + \frac{6}{(1.05)^9}\times 9 + \frac{106}{(1.05)^{10}}\times 10\right]$$

$$= 7.52.$$

Its Duration is $D = 1.05 \times S = 7.89$ years.

(continued)

> The Sensitivity value indicates that an increase in the rates of 1% leads, approximately, to a decrease in the bond's value of 7.52%; in fact, the exact value of the decrease is
>
> $$V(5\%) - V(6\%) = 107.72 - 100 = \$7.72, \text{ or } 7.17\%(7.72/107.72).$$

The foregoing example shows that estimating the variation in the value $\Delta V$ resulting from a variation $\Delta r$ (taken not to be infinitesimal) using the Sensitivity (Eq. (5.4)) is imprecise. This is inherent in differential calculus (which only provides an exact result $dV$ "at the limit" when the rate variation considered is infinitesimal). Indeed the variation $\Delta V$ in the value obtained from the Sensitivity (Eq. (5.4)) becomes more precise as the rate variation $\Delta r$ becomes smaller. In addition, as we will see in Sect. 5.1.6, accounting for the second order (convexity), leads to a much more precise result.

### 5.1.2.3 Practical Computation of the Sensitivity and the Duration

If one wishes to calculate the Sensitivity or the Duration of some bond quickly and does not have an *ad hoc* program to calculate the sensitivity (but does have a calculator or a spreadsheet that can compute present values) rather than applying (3) or (5) it is simpler to compute numerically the relative variation in value resulting from a small variation in rates (0.1%, for example).

> **Example 5**
>
> Let us calculate the Sensitivity and Duration, around par, of a bullet bond with a maturity of 10 years providing an annual coupon of 5%.
>
> We assume that its yield to maturity, $r$, is 5% and that this means that it is presently at par:
>
> $$V(5\%) = \sum_{i=1}^{10} \frac{5}{(1.05)^i} + \frac{100}{(1.05)^{10}} = 100 \; ;$$
>
> For $r = 5.1\%$ we have $V(5.1\ \%) = \sum_{i=1}^{10} \frac{5}{(1.051)^i} + \frac{100}{(1.051)^{10}} = 99.23 \; ;$
>
> $S \approx -\frac{\Delta V}{V \Delta r} = \frac{0.77}{100 \times 0.001} = 7.7;$ whence $D = 7.7 \times 1.05 = 8.09$ years.

### 5.1.3 Alternative Expressions for the Variation, Sensitivity and Duration

The expressions provided in the preceding section were derived from Eq. (5.1) by introducing a discrete yield to maturity. Alternative formulas, suitable for other rate definitions, lead to different expressions for the Variation, Sensitivity and Duration.

## 5.1.3.1 Expressions for *S* and *D* as a Function of Proportional Rates

Let us consider a short-term bond giving a fixed payment $F$ in a fraction of a year $T$ ($T = number\ of\ days/360 < 1$). Calculated with a proportional rate $r$ its value is $V(r) = \frac{F}{1+rT}$. From this we derive:

Its Variation $\equiv \frac{dV}{dr} = -F\frac{T}{(1+rT)^2}$;

Its Sensitivity $\equiv S = -\frac{dV}{Vdr} = \frac{T}{1+rT}$;

Its Duration $\equiv D = T$ since we here have a zero-coupon bond.

If $r$ denotes a proportional rate, the Sensitivity and Duration are linked through the equation $D = S(1 + r\,T)$.

## 5.1.3.2 Expressions for *S* and *D* as Functions of Continuous Rates

### Proposition
*The Sensitivity is the same as the Duration if one uses continuous rates.*

### Proof
First consider a security that generates a sequence of payments $F_{t_1}, \ldots, F_{t_n}$ whose value is obtained by discounting with a continuous rate $r$: $V = \sum_{\theta=t_1}^{t_N} F_\theta e^{-r\theta}$.

$$S = -\frac{dV}{Vdr} = \sum_{\theta=t_1}^{t_N} \frac{F_\theta}{V} e^{-r\theta}\theta = D.$$

Now consider a security that generates a density of payments $\delta F(\theta) = f(\theta)\delta\theta$ in the interval $(\theta, \theta + \delta\theta)$ and starting from the formula (5.1′) using a continuous rate:

$$V(r) = \int_0^T f(\theta)e^{-r\theta}d\theta. \tag{5.1′}$$

Consequently:

$$S \equiv -\frac{dV}{vdr} = \frac{1}{V}\int_0^T \theta f(\theta)\,e^{-r\theta}d\theta. \tag{5.3′}$$

We define, again, the Duration as the center of gravity of the payments realized over time

$$D = \frac{1}{V}\int_0^T \theta f(\theta)\,e^{-r\theta}d\theta; \tag{5.5′}$$

Thus $S = D$.

Expressed with continuous rates, Sensitivity (modified Duration) is equal to Duration (Macaulay Duration). Here, as elsewhere, formulas involving continuous rates are simpler than those with discrete rates.

### 5.1.3.3 A Simple Expression for the Sensitivity as a Function of Zero-Coupon Rates

It is more rigorous to apply a formula involving the vector $\underline{r}$ of zero-coupon rates than to use the yield to maturity of the bond

$$V(\underline{r}) = \sum_{\theta=t_1}^{t_n} \frac{F_\theta}{(1+r_\theta)^\theta}$$

($r_\theta$ denotes the zero-coupon rate with maturity $\theta$, and $\underline{r}$ is the vector $(r_{t1}, \ldots, r_{tm})$).

We consider a constant infinitesimal variation $dr_\theta = dr$ for each rate $r_\theta$, or, in other words, an infinitesimal *parallel* shift of the whole curve.

This special variation of the rates leads to a relative variation (sensitivity) of the bond market value:

$$S = \frac{-dV}{Vdr} = \sum_{\theta=t_1}^{t_n} \frac{\theta F_\theta}{V(1+r_\theta)^{\theta-1}}.$$

This formula, which is analogous to (5.3) although more general, does not allow us however to exhibit Duration defined in a simple way. Nonetheless, using continuous rates leads, as it did before, to an explicit Duration which is equal to the Sensitivity:

$$V = \sum_{\theta=t_1}^{t_N} F_\theta e^{-\theta r_\theta}, \text{which gives} \quad S = -\frac{dV}{Vdr} = \sum_{\theta=t_1}^{t_N} \frac{F_\theta}{V} e^{-\theta r_\theta} \quad \theta = D.$$

This approach makes more explicit the hypothesis that is implicit when one uses the simple Sensitivity or Duration (as they result from definitions (5.3) or (5.5)): all rates are assumed to vary by the same amount, that is, the whole yield curve is subject to a simple translation. This fundamental point will be expanded upon later when we discuss the limitations of these concepts and tools and their extensions.

### 5.1.4 Some Properties of Sensitivity and Duration

We examine successively the influence of rate variations and the passage of time on Sensitivity and Duration.

#### 5.1.4.1 The Influence of Rates on Sensitivity and Duration
**Property 1**
Sensitivity and Duration decrease as rates increase.

Intuitively, it is clear that a raised discount rate decreases the weight of a payment distant in time relative to ones closer and this causes the center of gravity of discounted payments to become closer to the present.

## 5.1 Interest Rate Risk

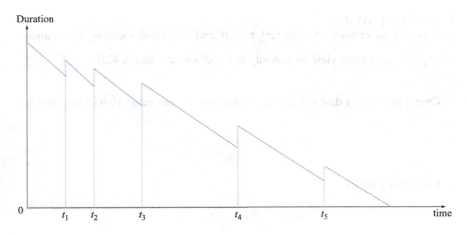

**Diagram 5.3** Evolution of the Duration over time

Furthermore, it is easy to show, by taking the derivative of equations (5.3) or (5.5), that $S$ and $D$ are monotonously decreasing functions of the discount rate $r$.

**Property 2**
In the class of bonds with a bullet repayment for a given maturity, the Sensitivity and Duration decrease with the nominal rate (the coupon rate).

The reason is that the center of gravity of the payments is closer to 0 in time when the coupon rate is higher.

### 5.1.4.2 The Influence of the Passage of Time on S and D
**Property 3**
(i) For a constant return rate and *between* two payment dates, the Duration diminishes with time (in $z$ periods the Duration will have decreased by $z$).
(ii) The distribution of a payment (a coupon or a partial reimbursement of the principal) results in a discrete *increase* in both Sensitivity and Duration. More precisely, the ratio of the two Durations $\frac{D^-}{D^+}$ equals the ratio $\frac{V^+}{V^-}$ where $V^-$ and $D^-$ denote the bond's value and its Duration just before the distribution while $V^+$ and $D^+$ denote the value and Duration just after the distribution.
(iii) As a result, the Duration presents over time a saw tooth pattern as in Diagram 5.3.

The $t_i$ are the dates of distributions (coupons and/or principal); for a constant yield to maturity, the slope of the saw tooth segments equals $-1$.

## Proof of Property 3

(i) First let us consider the instant $t = 0$ and the bond yielding the sequence $\{F_\theta\}_{\theta = t_1,\ldots, t_n}$, whose yield to maturity is $r$ and whose value is $V_0$: $V_0 = \sum_{\theta=t_1}^{t_n} \frac{F_\theta}{(1+r)^\theta}$.

Over $z$ periods, at date $t = z < t_1$, for the same $r$, the value $V_z$ will be given by

$$V_z = (1+r)^z \sum_{\theta=t_1}^{t_n} \frac{F_\theta}{(1+r)^\theta}.$$

From this follows

$$\ln V_z = z\ln(1+r) + \ln V_0.$$

The Sensitivities $S_z$ and $S_0$ (at $z$ and 0, respectively) are thus linked by the equation

$$S_z = -\frac{d\ln V_z}{dr} = -\frac{z}{1+r} - \frac{d\ln V_0}{dr} = -\frac{z}{1+r} + S_0.$$

$D_z$ and $D_0$ therefore satisfy the equation

$$D_z = (1+r)S_z = (1+r)S_0 - z = D_0 - z.$$

(ii) Consider a bond providing a payment (coupon and/or principal) $F$ and denote by $V^-$ and $V^+$ the bond's value just before and just after the distribution, respectively. We can write

$$V^- = V^+ + F; \quad \frac{dV^-}{dr} = \frac{dV^+}{dr}.$$

This implies the equation relating the Sensitivities $S^+$ and $S^-$

$$S^+ = -\frac{1}{V^+}\frac{dV^+}{dr} = -\frac{V^-}{V^+}\frac{1}{V^-}\frac{dV^-}{dr} = \frac{V^-}{V^+}S^-.$$

In the same way

$$\frac{D^-}{D^+} = \frac{(1+r)S^-}{(1+r)S^+} = \frac{S^-}{S^+} = \frac{V^+}{V^-} < 1.$$

(iii) The preceding implies that the passage of time leads to a linear decrease in the Duration (of slope $-1$) except at the distribution dates when the Duration jumps up as in Diagram 5.4. This result is logical: a bond with coupons has an initial duration less than its maturity ($D < T$) and a final duration equal to its maturity ($D' = T' = 0$). Because of this, since the Duration decreases as time does between two distribution

## 5.1 Interest Rate Risk

dates, it must surely increase at those dates. To take up the analogy with a physical balance again, the occurrence of a payment pushes the center of gravity of the remaining payments to the right.

### 5.1.5 The Sensitivity of a Portfolio of Assets and Liabilities or of a Balance Sheet: Sensitivity and Gaps

We are going to consider a portfolio made up of long and short positions, or equivalently from the point of view of the interest rate risk that affects them, of a balance sheet made up of assets and liabilities (debts and equity). The rate risk should be analyzed in different ways and with different instruments according to whether the elements of the balance sheet are valued at market value ("marked-to-market") or on the basis of the remaining principal due. We have already emphasized the difference between these two methods.

Let us denote by 0 the date of the previous coupon (or the time of issue of the bond under examination) and by $t$ the present time ($t > 0$). Recall (see Eq. (2.15) that the principal $C_t$, remaining due at $t$, equals the value of the remaining maturities discounted at the *nominal rate k*, on the date of the preceding coupon (at 0), that is, calling the latter $\{F_q\}_{q = 1, \ldots, n}$ ($0 < t < 1 < \ldots < n$):

$$C_t = \sum_{\theta=1}^{n} \frac{F_\theta}{(1+k)^\theta}.$$

The market value at the same date $t$ results from discounting the maturities at the rate $r(t)$ representing the rate of return of comparable bonds prevailing in the market at $t$:

$$V_t = \sum_{\theta=1}^{n} \frac{F_\theta}{(1+r(t))^{\theta-t}}.$$

The comparison of $C_t$ and $V_t$ leads to the following observations:

- From the issue of the security at 0, all the values $C_t$ for $t \in [0, T]$ are fixed, in particular $C_t$ is independent of the rate $r(t)$ prevailing in the market at $t$, since it only depends on $k$ (usually equal to $r(0)$) and on the history of $F_\theta$.
- Seen from the instant 0, $r(t)$ is stochastic making $V_t$ stochastic (while $C_t$ is known).

The last remark implies that interest rate risk is to be seen differently according to whether one considers the market value $V_t$ or the remaining principal $C_t$.

We will first analyze the interest rate risk affecting a balance sheet or a portfolio of assets and liabilities evaluated in terms of market values before treating the case of a balance sheet whose elements are accounted for in terms of the principal remaining due.

### 5.1.5.1 Interest Rate Risk of a Portfolio of Assets and Liabilities Evaluated at Market Value

We can state the following property concerning the Sensitivity of a portfolio.

**Proposition**
*The Sensitivity $S$ of a portfolio made up of $K$ securities with long or short positions equals the weighted average of the Sensitivities of the component securities, that is $S = \sum_{k=1}^{K} x_k S_k$, where $x_k$ denotes the weight of security $k$ in the portfolio and $S_k$ its Sensitivity.*

**Proof**
Denoting by $V$ the total value of the portfolio and by $V_k$ that of a security $k$ included, we have

$$V = \sum_{k=1}^{K} V_k; x_k = \frac{V_k}{V};$$

$$S = -\frac{1}{V}\frac{dV}{dr} = -\sum_{k=1}^{K}\frac{1}{V}\frac{dV_k}{dr} = -\sum_{k=1}^{K}\frac{V_k}{V}\frac{1}{V_k}\frac{dV_k}{dr} = \sum_{k=1}^{K} x_k S_k.$$

**Remarks**
(*i*) The portfolio can contain both short and/or long positions. If security $k$ is shorted, $V_k$ and $x_k$ are negative but the preceding formulas still hold.
(*ii*) In the preceding, we have assumed the *same variation dr* for the rates of return $\{r_k\}_{k=1,...,K}$ of the different securities. If, in addition, these rates are all the same, that is, $r_k = r$, we have

$$D = (1+r)S = \sum_{k=1}^{K} x_k(1+r)S_k = \sum_{k=1}^{K} x_k D_k.$$

The Duration of the portfolio is therefore the weighted average of the Durations of its components.
(*iii*) The same formulas apply to a balance sheet made up of assets and liabilities evaluated at market value; $V$ is then the financial value of the equity (assets minus debts).

## 5.1 Interest Rate Risk

Nevertheless, in some situations, the value of the assets equals that of the debts: the net value of the portfolio, $V$, is then zero and the sensitivity, just like the weights $x_k$ defined above, are infinite.[2] The previous definitions of the Sensitivity of a portfolio of assets and liabilities are obviously not well suited to such a situation. If the net value $V$ is zero or near zero, assets and liabilities have to be treated separately and the weights defined relative to the asset total $A$ and the liability total $P$ according to whether the components are considered an asset or a liability. The formulas then become

$$\frac{dA}{Adr} = S_A = \sum_{k=1}^{K} x_k S_k \ ; \quad \frac{dP}{Pdr} = S_P = \sum_{k'=1}^{K'} y_{k'} S_{k'} \ ; \quad \frac{dV}{dr} = \frac{dA}{dr} - \frac{dP}{dr}$$

$$= A \ S_A - P \ S_P.$$

If $A = P$, the net value $V$ of the portfolio is not sensitive to the interest rate if and only if the Sensitivity of assets equals the Sensitivity of liabilities ($S_A = S_P$).

### 5.1.5.2 Interest Rate Risk of a Balance Sheet Made Up of Assets and Liabilities Valued in Terms of the Principal Remaining Due

We have remarked that the market value $V_t$ of a security or of a portfolio is stochastic while, if there is no counterparty risk, the principal remaining, $C_t$, is deterministic.

Because of this, interest rate risk is not the same when one works with the market value $V_t$ as it is when one uses the remaining principal $C_t$. In the first case, it affects the capital $V_t$ while in the second, as we see here, it does not affect $C_t$ but it does change *the financial margin*. The latter is defined as the difference between the interests (and possibly fees) produced by the asset considered and the interests (and possibly fees) generated by its financing.

The financial margin, provided at $t$ by an asset with nominal rate $k$ and booked into accounts at $C_t$, indeed depends on its financing and notably on the financing initially contracted (at 0). In this connection, it is useful to contrast the case of a day-to-day financing at a rate $r(t)$ at $t$ and that of a financing wholly put in place at time 0 and such that its remaining principal is always equal to that of the asset $C_t$.

In the first case, the instantaneous financial margin at $t$ is equal to $C_t[k - r(t)]$; an increase $dr$ in the short-term rate brings a reduction of $-C_t dr$ in the margin.

In the second case, the financing at $t$ (in the amount $C_t$) comes entirely from the initial moment at which it was contracted, say at a rate $k'$; the financial margin at $t$ is therefore equal to $C_t [k - k']$ and is completely deterministic.

Actually, the most common situation is intermediate between the first and second cases: the amortization of the assets and the liabilities do not coincide (since there are discrepancies between the durations of assets and their fixed-rate financing). For

---

[2]This is so, notably, for derivative products such as swaps and forward or Futures contracts, which can be considered as portfolios whose initial value is zero by construction: lending at a fixed rate combined with borrowing at a floating rate in the case of a (floating-rate payer) swap, or a long position in the underlying financed by borrowing in the case of a forward contract, for example (cf. Chap. 7).

example, if the repayment of debts is faster than that of the assets, new funds must be raised with the sole purpose of financing the need that arises. If the rates rise, this new financing will be more costly than that which covers the asset originally and the financial margin will be reduced. These differences in the amortization profiles for the assets and the liabilities are therefore the sources of rate mismatches or "Gaps" at different times in the future.

The following introductory example illustrates the computation and the significance of these rate Gaps.

> **Example 6**
> Consider a portfolio P of assets and liabilities (or a balance sheet) which initially is as follows:
>
> | Portfolio P at $t = 0$ | | | |
> |---|---|---|---|
> | Assets | 200 | Liabilities | 160 |
> | | | Equity | 40 |
>
> The assets are made up of two fixed-rate loans with bullet profiles (the principal is to be repaid entirely at maturity). The first loan of $100M will be due in three half-years ($t = 3$) and the second loan, also of $100M, in two half-years ($t = 2$).
>
> The debt of $160M is at a fixed rate and repayable in six months ($t = 1$).
>
> The interest rates on the loans and the debt are, respectively, 6% and 5% semi-annually.
>
> The diagram below shows the evolution with time of the loan amount outstanding (principal still due) of the asset ($A_t$) and the liability ($P_t$) in this portfolio.
>
> (continued)

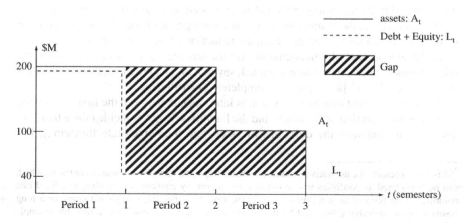

**Diagram 5.4** A balance sheet Gap

> The mismatch (or Gap) in period $t$ (between $t-1$ and $t$), denoted by $G_t$, is by definition
>
> $$G_t = A_t - P_t.$$
>
> In the example, $G_1 = 0$ ; $G_2 = 160$; $G_3 = 60$.

The gap $G_t$ can be interpreted as the amount of new financing (contracted after time 0) needed to finance asset $A_t$.

Because of this, an interest rate increase leads to a *reduction* in the company's financial margin (or the bank's) which will have to pay more for new resources (in the amount $G_t$) needed to finance asset $A_t$ which continues to bring the same interest rate.

More precisely, a rate variation of $\delta\%$ occurring before $t$ gives rise to a variation $\Delta M_t$ in the financial margin over period $t$ equal to

$$\Delta M_t = G_t \times \frac{\delta}{100}$$

(if $G_t$ is positive, it should be read as the risk of a rate increase and, if $G_t$ is negative, the risk of a rate decrease).

---

**Example 7**

Let us take up again the preceding Example 6 of portfolio P.

Assume first that the interest rates remain stable for three half-years so that the new debts are contracted at the starting rate of 5% semi-annually; the semi-annual income statement for P then is as follows:

| Semester | 1 | 2 | 3 |
|---|---|---|---|
| (1) Creditor interests | 6% × 200 = 12 | 6% × 200 = 12 | 6% × 100 = 6 |
| Debtor interests | | | |
| (2) On old debts | 5% × 160 = 8 | 0 | 0 |
| (3) On new debts | 0 | 5% × 160 = 8 | 5% × 60 = 3 |
| Financial margins ((1) − (2) − (3)) | 4 | 4 | 3 |

Now suppose the rate rises by 1% during the first half-year and then remains unchanged so that the half-yearly rate of the new debt is fixed at 6% in periods 2 and 3.

The income statements for P will then be as follows:

| Semester | 1 | 2 | 3 |
|---|---|---|---|
| (1) Creditor interests (unchanged) | 12 | 12 | 6 |
| Debtor interests | | | |
| (2) On old debts | 8 | 0 | 0 |
| (3) On new debts | 0 | 6% × 160 = 9.6 | 6% × 60 = 3.6 |

(continued)

| Semester | 1 | 2 | 3 |
|---|---|---|---|
| Financial margins $((1) - (2) - (3))$ | 4 | 2.4 | 2.4 |

The rate increase of 1% resulted in a decrease in financial margins of
$4 - 2.4 = 1.6 (= 1\% \times G_2)$ in period 2,
$3 - 2.4 = 0.6 (= 1\% \times G_3)$ in period 3.

One can hedge against interest rate risk by making the different gaps vanish, that is by "synchronizing" the amortizations of fixed-income assets and debts (on and off the balance sheet) notably by using non-optional derivative products on interest rates (i.e., swaps, forwards, and futures).

The preceding example can easily be made more general. Consider a balance sheet initially made up of assets, at instant 0, in the total amount of $A(0)$ and liabilities in the amount $P(0) = A(0)$. Denote by $A(t)$ and $P(t)$, respectively, the amounts of *these* assets and *these* liabilities still alive at date $t$ bearing the same rates as at date 0. In the general case, $A(t)$ differs from $P(t)$ since the assets and the liabilities are not amortized simultaneously. The Gap at time $t$ is defined as

$G(t) = A(t) - P(t)$.

From this, at $t$, the financing of asset $A(t)$ resulting from the original balance sheet can be decomposed into $P(t)$ of old financing and $G(t)$ of new financing.

Denote by $k$ the lending rate on the asset, by $k'$ the borrowing rate on the old debts (existing at 0 and still alive at $t$) and by $r(0) + d(t)$ the borrowing rate, paid at $t$, on debts contracted between 0 and $t$ ($d(t) = 0$ if there has been no rate variation between 0 and $t$). The instantaneous margin at $t$ can then be split up as follows:

$Margin(t) = k\,A(t) \quad - \quad k'P(t) - (r(0) + d(t))G(t)$.

Creditor payments      Debtor payments

Consequently, if the Gap $G(t)$ is not zero, rate variations which lead to a $d(t)$ differing from zero impact the financial margin.

The global margin, for this portfolio between 0 and $T$ (the longest horizon for any asset or debt), can be written

$$Margin(0, T) = \int_0^T \left[ k\,A(t) - k'P(t) - (r(0) + d(t))\,G(t) \right] dt.$$

The variation in interest rates between 0 and $T$, producing random differences $\{d(t)\}t_{\in[0,\,T]}$ from a stable situation, thus causes a variation in the margin

$$-\int_0^T d(t)G(t)dt.$$

# 5.1 Interest Rate Risk 167

In conclusion, it should be noted that the gap method is not suitable for the case of assets and liabilities valued at market values, since the risk, both accounting and financial, is then borne on the market value of the principal and no longer on the financial margin; in this connection, the introduction of new accounting norms,[3] which generalize market value (*fair value*) as the basis for accounting for the elements of the balance sheet, reduces the importance of the "Gap" approach in favor of the "market value sensitivity" approach presented in the preceding Sect. 5.1.5.1.

## 5.1.6 A More Accurate Estimate of Interest Rate Risk: Convexity

Sensitivity, as previously defined, only gives acceptable results for small variations in interest rates. Indeed, recall that the variation $\frac{V(r+\Delta r)-V(r)}{V} \equiv \frac{\Delta V}{V}$, caused by a non-infinitesimal rate variation $\Delta r$ can be estimated with the approximate Eq. (5.4) resulting from the expansion of $V(r + \Delta r)$ up to the first order, which we repeat here

$$\frac{\Delta V}{V} = \frac{1}{V}\frac{dV}{dr}\Delta r = -S\Delta r.$$

Such a formula gives results that become less precise as the interest rate variation considered, $\Delta r$, becomes larger.

To obtain a better precision one may calculate $\Delta V$ from an expansion up to order two:

$$V(r + \Delta r) - V(r) = \left(\frac{dV}{dr}\right)\Delta r + \frac{1}{2}\frac{d^2V}{dr^2}(\Delta r)^2 + \underbrace{o(\Delta r^2)}_{\text{negligible}}$$

The term $o(\Delta r^2)$ represents, by convention, a term of higher order than two.[4] Neglecting this term and defining $C \equiv \frac{1}{V}\frac{d^2V}{dr^2}$, the last equation leads to the following approximation formula:

$$\frac{\Delta V}{V} = -S\Delta r + \frac{1}{2}C(\Delta r)^2. \tag{5.7}$$

$C$ is called the convexity of the security. Since the function $V(r)$ is convex,[5] $C$ is positive.

---

[3] IAS (International Accounting Standards) and IFRS (International Financial Reporting Standards).

[4] That is $\lim_{\Delta r \to 0}\frac{o(\Delta r^2)}{\Delta r^2} = 0$.

[5] A function $V(r)$ is convex if the set of points in $\mathbf{R}^2$ $\{v,r \mid v>V(r)\}$ is convex (i.e., the part of the plane above the curve of $V(r)$ is convex, or the concavity of the curve $V(r)$ faces up, *see* Diagram 5.5). If $V(r)$ is twice differentiable, one has the following equivalences:

$V(r)$ convex $\Leftrightarrow \frac{dV}{dr}$ increasing $\Leftrightarrow \frac{d^2V}{dr^2} > 0(C > 0)$.

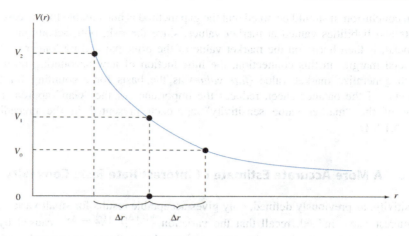

**Diagram 5.5** Market value as a function of interest rate

Thanks to the additional correction involving the convexity C, Eq. (5.7) gives a more precise estimate of the effect of rate variations than the simple sensitivity method (Eq. (5.4)).

For a fixed-income bond whose value is $V(r) = \sum_{\theta=t_1}^{t_n} \frac{F_\theta}{(1+r)^\theta}$, we obtain

$$C = \frac{1}{V}\frac{d^2V}{dr^2} = \frac{1}{V(1+r)^2}\sum_{\theta=t_1}^{t_n} \frac{\theta(\theta+1)F_\theta}{(1+r)^\theta}.$$

Note that for continuous rates we have the simpler formula

$$C = \frac{1}{V}\frac{d^2V}{dr^2} = \frac{1}{V}\sum_{\theta=t_1}^{t_n} e^{-r\theta}F_\theta\theta^2.$$

We can verify that $C$ is positive which implies that the function $V(r)$ really is convex, as Diagram 5.5 shows.

---

**Example 8**
Let us consider a 10-year zero-coupon bond with a face value of $100M and a rate of return $r$ equal to 10%; we have

$$V = \frac{100}{(1.1)^{10}} = 38.554; \quad S = \frac{1}{38.5}\frac{10 \times 100}{(1.1)^{11}} = 9.09; \quad C$$

$$= \frac{1}{38.5}\frac{10 \times 11 \times 100}{(1.1)^{12}} = 90.91$$

---

(continued)

## 5.1 Interest Rate Risk

A variation of $\Delta r$ in the rate has an impact $\Delta V$ on the bond's value which can be estimated using the formula

$$\Delta V = -VS\Delta r + \frac{1}{2}VC(\Delta r)^2 = -350.5(\Delta r) + 45.55 \times 38.55 \times (\Delta r)^2.$$

Assume $r$ decreases from 10% to 9% ($\Delta r = -0.01$): $\Delta V = +\$3.68$.

The exact increase $\Delta V$ in fact equals $+\$3.69$ and the approximation error is therefore $-\$0.01$.

We remark that the answer from the formula using Sensitivity only would have been

$\Delta V = -350.5(\Delta r) = +\$3.5$ (error: $-\$0.19$).

Second-order expansion ($S$, $C$), rather than first-order ($S$ only), thus allows dividing the error by 19!

The convexity of $V(r)$ has the two following consequences:

– Estimating the variation $\Delta V$ from the first-order term ($SV\Delta r$) alone overestimates the absolute value of the effects of a rate increase and underestimates the effects of a rate decrease.

– Convexity of $V(r)$ leads to an asymmetric reaction to the increase or decrease in the rates: for the same absolute rate variation $|\Delta r|$, the change $|\Delta V|$ due to an increase is smaller than the change $\Delta V$ due to a decrease. These results come from Eq. (5.7) and from Diagram 5.5.

### Example 9

Let us consider a perpetual annuity whose return rate is $r$ and whose nominal rate is $k$:

$$V = \frac{k}{r}, S = -\frac{dV}{Vdr} = \frac{1}{r}, C = \frac{1}{V}\frac{d^2V}{dr^2} = \frac{2k}{Vr^3} = \frac{2}{r^2}.$$

For $r = 10\%$ we, therefore, have $S = 10$ and $C = 200$.

From this, a rate increase of 2% brings with it a relative decrease in value, which can be computed using approximation (7), namely

$$-\frac{\Delta V}{V} = S\Delta r - \frac{1}{2}C(\Delta r)^2 = 10 \times 0.02 - 100 \times (0.02)^2 = 16\%.$$

(The result computed using the simple Sensitivity is 20%; the exact result is 16.67%).

A rate decrease of 2% entails an increase in value of approximately

$$\frac{\Delta V}{V} = -S\Delta r + \frac{1}{2}C(\Delta r)^2 = 24\%$$

(the exact result is $+25\%$).

## 5.2 Introduction to Credit Risk

In the previous chapter relating to credit risk (Chapter 4, Sect. 4.1, Sect. 4.1.2.2), we met the *risk of default*, defined as the risk that the principal and/or the interest will, partly or entirely, not be repaid. *We also met the risk of a decrease in the market value of a debt as a result of an increase in the credit spread mainly linked to an increase in the perceived probability of a future bankruptcy (deterioration of credit worthiness) on the debtor's part.* In effect, the spread risk is related to the risk of an increase in the yield to maturity required by investors which may be split into a (risk-free) market rate and a credit spread:

*Yield to maturity = risk-free market rate + credit spread.*

An increase in the credit spread, for a given (risk-free) market rate causes a decrease in the market value of a debt. The credit spread, which depends in an essential way (but not uniquely, as we will see) on the risk of default by the borrower, grows as the issuer's credit rating falls; for a given issuer we observe that it is higher the longer the life of the bond. Finally, it increases in crisis periods and diminishes in good times.

We recall that the main credit rating agencies (Standard and Poor's, Moody's, Fitch, etc.) rate issuers each according to its own scale. We give those of S&P and Moody's as examples (from the best to the worst):

S&P: AAA, AA, A, BBB, BB, B, CCC
Moody's: Aaa, Aa, A, Baa, Ba, B, Caa

Each category is split into three finer categories. For example, in category AA, S&P distinguishes AA +, AA, and AA–. In the corresponding category Aa, Moody's distinguishes Aa1, Aa2, and Aa3.

Summing this up, credit risk encompasses default risk and spread risk; the spread increases with the probability of default as it is perceived by the market and the probability increases if the rating of the issuer deteriorates.

### 5.2.1 Analysis of the Determinants of the Credit Spread

Let us consider an investor who buys a bond with the intention of holding it until maturity.

The risk of default gives a random aspect to the *effective* rate of return on an investment. In fact the rate, which it is conventional to call the yield to maturity, is *computed under the assumption that the issuer will honor all claims* (coupons and principal) both in their amounts and as to their payment dates. This "promised rate" is in fact the maximum rate the investors may hope for and which they obtain under the most favorable circumstances; we denote it by $r^*_{max}$ in this subsection. It is this rate that can be split into a risk-free market rate, denoted $r$, and a credit spread $s$:

## 5.2 Introduction to Credit Risk

$$r^*_{max} = r + s.$$

If there is a risk of default, the rate of return which is effectively realized, denoted by $R$, is indeed random and $E(R) < r^*_{max}$ (E denotes an expectation, a capital letter in italics denotes a random variable). The first reason for the existence of a credit spread is thus to compensate the investor for this reduction in the expected return resulting from the risk of default.

Since $r$ denotes a risk-free rate, in the absence of risk aversion on the part of investors we would have:

$$E(R) = r; \quad s = r*_{max} - E(R).$$

However, the market is characterized by an aversion to risk and because of this demands an *expected return larger than the risk-free market rate*: this is the risk premium in the strict sense, denoted by $\pi$ in what follows. As a result, the credit spread $s$ can be analyzed as the sum of two components:

$s = (r*_{max} - E(R)) + (E(R) - r)$

$s = (r*_{max} - E(R)) + \pi$

The first term in parenthesis is a compensation for the decrease in E(R) and the second term is the risk premium in the *strict sense*.

Furthermore, if the bond's liquidity is not as good as one would wish, a liquidity premium $l$ is added to the yield to maturity $r^*_{max}$ demanded so that the spread is larger and the sum of *three components*:

$s = (r*_{max} - E(R)) + \pi + l,$

i.e. compensation for the decrease in E(R), risk premium and liquidity premium.

**Remarks**

– In our definition, $r$ is the risk-free rate for transactions with the same lifetime as the bond under consideration; as a result, the spread can only be positive. However, in practice, the spread is often computed relative to a rate $r$ which is not completely free of credit risk (a Libor/Euribor rate for example). In this case, the spread can, possibly, be slightly negative, if the transaction is one where the risk of default is virtually zero, notably as a result of a guarantee (a "repo" between two large banks, for example, involving a collateral asset).

– In current practice, the spread $s$ is interpreted, wrongly, as a risk premium in the broadest sense of the term. In fact, it includes compensation for a reduction in $E(R)$ as well as the premiums $\pi$ and $l$ (risk premium in the *strict sense* and liquidity premium).

---

**Example 10**

Let us consider a very liquid bond X whose yield to maturity is 6%, yielding, in principle, a payment of $106 in exactly a year (its market value is thus $100). In fact, with a probability of 1%, the issuer will be bankrupt and the

(continued)

bondholder will receive neither the coupon nor the principal. The bond, therefore, gives randomly a payment of \$106 with a probability of 99% (yield of 6%) and of \$0 with a probability of 1% (yield of $-$ 100%). The expected yield is thus $0.99 \times 6\% - 0.01 \times 100\% = 4.94\%$.

Let us assume that on the market, the one-year rate for bonds free of credit risk is 4.54%. The credit spread of bond X is therefore 1.46% ($= 6\%\text{–}4.54\%$) of which 1.06% can be explained by the decrease in expected yield coming from risk ($6\%\text{–}4.94\%$) and 0.40% by the risk premium in the *strict sense* (the liquidity premium is assumed to be zero here).

An increase $\Delta s$ in the credit spread materializes, other things being equal, by a reduction in the bond market value approximately equal to $\Delta V = - S V \Delta s$.

The increase $\Delta s$ has the same effect as an increase of a similar amount in the market rate.

Also let us remark that for bonds with a large risk of default the amplitudes of the rate variations $\Delta r$ and the spread fluctuations $\Delta s$ are of the same size: in this sense, the interest rate risk and the credit risk can be of the same order of magnitude.

Finally, we draw attention to the fact that since the market conditions evolve over time, the premiums $p$ and $l$ should not be considered as constants but as random variables even for a given default risk. Thus, periods of crisis witness a rise in the market spreads *for a given risk category*. This spread increase is larger for the riskier categories (BB or below).

## 5.2.2 Simplified Modeling of the Credit Spread; the Credit Triangle

The preceding Example 10 is easy to generalize.

Consider a one-period risky bond, worth \$1 at date 0, and promising an interest rate $r^*_{max}$, so a final payment of $1 + r^*_{max}$. In reality, the final payment is random since this bond is affected by the risk of default: with a probability $p > 0$, the debtor will be bankrupt at maturity. Assume meanwhile that upon default the bondholder will recover a fraction $\alpha$ ($0 \leq \alpha < 1$) of the debt ($\alpha$ is called the *recovery rate*). The bond consequently generates a random payment $X$ and a random yield $R$:

$$X = 1 + R = \begin{cases} 1 + r^*_{max} \text{ with probability } 1 - p \text{ (no default)} \\ \alpha(1 + r^*_{max}) \text{ with probability } p \text{ (default)} \end{cases}.$$

From this, we have:

$$\begin{aligned} E(1 + R) &= (1 + r^*_{max})[\alpha p + (1 - p)] \\ &= (1 + r^*_{max})[1 - p(1 - \alpha)] \ (< 1 + r^*_{max}). \end{aligned}$$

Furthermore, with $r$ denoting the risk-free rate, by the definition of the spread we have $r^*_{max} = r + s$.

# 5.2 Introduction to Credit Risk

173

Thus:

$$E(1 + R) = (1 + r + s)[1 - p(1 - \alpha)].$$

To compensate for the risk of default and a possible illiquidity of the bond, the expected yield of the investment must equal $r + p + l$. Therefore, we have:

$$(1 + r + s)[1 - p(1 - \alpha)] = 1 + r + \pi + l.$$

Neglecting terms of order 2 (in $rp$ and $sp$), we obtain the credit triangle:

$$s = \pi + l + p(1 - \alpha), \tag{5.8}$$

risk and liquidity premiums plus a compensation for reduced expectation.

Note that this relationship is often simplified (wrongly neglecting $\pi$ and $l$[6]) to:

$$s \approx p(1 - \alpha). \tag{5.8'}$$

We interpret $(1 - \alpha)$ as the fraction of the principal lost in the case of default ("loss given default") and $p(1 - \alpha)$ as the expected loss of principal ("expected loss") for an exposed principal of \$1. Equation (5.8′) means that the spread should just cover the expected loss.[7]

Equations (5.8) and (5.8′) can be generalized to the case of a multi-period bond with coupons (*see* the Appendix).

---

**Example 11**

Let us consider the (assumed very liquid) bond X from the previous Example 10 which gives, in one year, a random payment equal to \$106 with a probability of 99% (yield of 6%) and to \$0 with a probability of 1%: $p = 0.01$ and $\alpha = 0$. The simple credit triangle (Eq. (5.8′)) gives $s = p = 1\%$. In fact, $s = 1.46\%$ (*see* the preceding example): the simple credit triangle, therefore, leads to underestimating the spread by 0.46% of which 0.40% is explained by the fact that the premium $\pi$ is neglected and 0.06% by the approximate character of the formula.

---

[6]Indeed, the spread $s$ is underestimated if it is calculated using (5.8′), given estimated $\alpha$ and $p$; $p$ is overestimated if computed from (5.8′) given an observed spread $s$ and an estimated $\alpha$.

[7]We will see in Chap. 28, which takes up these matters in much more depth, that equation (5.8′) actually reveals its real meaning when $p$ is interpreted as a risk-neutral (as opposed to the true) probability.

**Example 12**

Let us consider a bond whose yield to maturity is 1.20% in excess of the rate on government bonds with the same lifetime: $s = 1.20\%$. Our estimate for the recovery rate $\alpha$ is 40%.

By equation (5.8′), such a credit spread would be compatible with a probability of default of $p = s/(1 - \alpha) = 1.2\%/0.6 = 2\%$. In fact, the default probability, as can be estimated from the frequencies of defaults by companies with the same rating, is appreciably less than 2%. This is because the risk premium $\pi$ is neglected in this reasoning. These questions will be addressed further and in more depth in Chap. 28.

## 5.3    Summary

- The value $V(r)$ of a fixed-income instrument decreases when the market yield $r$ increases. Thus it bears an *interest rate risk* measured by three closely related quantities: *variation* (or absolute sensitivity) $-dV/dr$, sensitivity (or modified duration) $S = -dV/Vdr$ and (Macaulay) *duration* $D = S(1+r)$, where $r$ denotes a yield to maturity. The larger these measures, the more the instrument's market value is affected by a given change $\Delta r$ in the market interest rates since, up to the first order, $\Delta V = -SV \Delta r$. Sensitivity thus measures, roughly, the percentage decrease in market value induced by a 1% increase in $r$.

- The duration of a fixed-income instrument yielding the stream $\{F_\theta\}$ is $= \sum_{\theta=t_1}^{t_n} \times \frac{\theta F_\theta}{(1+r)^\theta}$. It corresponds to the *average position* (center of gravity) of the instrument's discounted cash-flows over time. For a zero-coupon bond, it is equal to its lifetime. For a coupon bond, it is smaller.

- Alternative expressions for $S$ and $D$ can be written for proportional and continuous rates. Sensitivity and Duration are *equal* if one uses *continuous* discount rates.

- Sensitivity and duration *decrease* as market rates *increase*.

- Between two payment dates, the duration *decreases* exactly with time. On a date of distribution (of a coupon or a fraction of the principal), sensitivity and duration both *increase*.

- Formulas involving discounting with different *zero-coupon rates* rather than a unique yield to maturity are more accurate.

- The sensitivity of a portfolio or a balance sheet is equal to the *weighted average* of the sensitivities of its components. Managing the interest rate risk of a portfolio by a simple control of its global sensitivity rests, implicitly, on assuming *parallel shifts* of the yield curve only.

- The differences in the amortization profiles for the assets and liabilities of a balance sheet are sources of rate mismatches or "*gaps*" at different times in the future.

Appendix 1 175

- Adding to the standard formulas a correction term involving the *convexity* of the bond price ($C = \frac{1}{V} \frac{d^2V}{dr^2}$) results in a more precise estimate of the effect of rate variations on bond values.
- The *credit spread* that is required on a bond for the *credit or default risk* it entails is the sum of a compensation for the decrease in its expected return rate, a risk premium in the strict sense, and (often) a liquidity premium; however, it is usually wrongly interpreted as a risk premium in the broadest sense of the term.
- In the so-called *credit triangle formula*, the first component of the spread above is equal to the product of the probability of default by the "loss given default" (the fraction of the principal lost upon default).

## Appendix 1

### Default Probability, Recovery Rate and Credit Spread

This appendix establishes more generally and more rigorously the credit triangle formula introduced in the body of the text (Sect. 5.2), which connects the credit spread $s$ to the probability $p$ of default and the recovery rate $a$.

Let us consider a bond at par whose face value is \$1 and which distributes a coupon $k\Delta t$ on due dates spaced at constant intervals of length $\Delta t$. $N-1$ coupons are expected before the final payment $N$ which includes the $N$th and last coupon as well as the bond's face value. Since it is at par, its yield to maturity equals $k\Delta t$ and we may write:

$$r*_{max} = k, \text{recalling that } \sum_{i=1}^{N} \frac{k\Delta t}{(1 + k\Delta t)^i} + \frac{1}{(1 + k\Delta t)^N} \equiv 1.$$

Let us assume that this bond is affected by credit risk in such a way that at each due date (payment of a coupon) the issuer has a probability of defaulting of $p\Delta t$.[8] If there is a default, the bondholder will recover a fraction $a$ of the coupon and the principal and nothing more.

The payment in period $i$, for $i = 1, \ldots, N-1$, is therefore

- $k\Delta t$ with probability $(1 - p\Delta t)^i$ (if the issuer does not default during the first $i$ periods).
- $a[k\Delta t + 1]$ with probability $(1 - p\Delta t)^{i-1} p\Delta t$ if the issuer has not defaulted during the first $i - 1$ periods and does default on the $i$th due date.
- 0 with probability $1 - (1 - p\Delta t)^{i-1}$ if the issuer has defaulted during one of the preceding periods.

---

[8] This is the probability conditional on the absence of default in the preceding time periods.

At due date $N$, the possible repayment of the principal, equal to 1, is added in; this 100% reimbursement only occurs if the issuer does not default during the whole lifetime of the bond, therefore with probability $(1 - p\Delta t)^N$.

As a result, the expected payment distributed by this bond in period $i$ can be written

$$(1 - p\Delta t)^i k\Delta t + (1 - p\Delta t)i - 1p\Delta ta[k\Delta t + 1] \quad \text{for} \quad i = 1, \ldots, N - 1$$

$$(1 - p\Delta t)N[k\Delta t + 1] + (1 - p\Delta t)N - 1p\Delta ta[k\Delta t + 1] \quad \text{the } N\text{th period.}$$

The bond's value is equal to the present value of these expected payments, discounted at the risk-free market rate plus a premium (risk and liquidity) i.e. $(r + \pi + l)$. From this we have

$$1 = \sum_{i=1}^{N} \frac{(1 - p\Delta t)^i k\Delta t + (1 - p\Delta t)^{i-1} p\Delta t\alpha[k\Delta t + 1]}{(1 + (r + \pi + l)\Delta t)^i} + \frac{(1 - p\Delta t)^N}{(1 + (r + \pi + l)\Delta t)^N}$$

$$= \sum_{i=1}^{N} \frac{k\Delta t + \dfrac{p\Delta t}{1 - p\Delta t}\alpha(k\Delta t + 1)}{\left(\dfrac{1 + (r + \pi + l)\Delta t}{1 - p\Delta t}\right)^i} + \frac{1}{\left(\dfrac{1 + (r + \pi + l)\Delta t}{1 - p\Delta t}\right)^N}$$

Noting that $\sum_{i=1}^{N} \frac{y}{(1+x)^i} + \frac{1}{(1+x)^N} = 1$ if and only if $x = y$[9], in order that this bond to have a value of 1, it is necessary that

$$k\Delta t + \frac{p\alpha\Delta t}{1 - p\Delta t}(k\Delta t + 1) = \frac{1 + (r + \pi + l)}{1 - p\Delta t}$$

For a small $\Delta t$, we have:

$k\Delta t + p\alpha\Delta t = p\,\Delta t + (r + \pi + l)\Delta t + o(\Delta t)$, where $o(\Delta t)$ is a term of order 2.

Therefore as $\Delta t$ tends to 0, $k$ tends to $p(1-\alpha) + r + \pi + l$. Since the bond is at par, we have: $k = r^*_{max} = p(1-\alpha) + r + \pi + l$, and we recover formula (5.8) for this coupon bond.

## Suggested Reading

Bodie, Z., Kane, A., & Marcus, A. (2013). *Investments* (10th ed.). Irwin.

Fabozzi, F. (2006). *Fixed income mathematics: Analytical and statistical techniques* (4th ed.). McGraw Hill Education.

Fabozzi, F. (2016). *Bond markets analysis and strategies* (9th ed.). Frank J. Fabozzi editor.

Hull, J. (2018). *Options, futures and other derivatives* (10th ed.). Prentice Hall Pearson Education.

---

[9]This can be easily shown by expressing the sum of the first $N$ terms of a geometric series with ratio $1/(1 + x)$. That means financially that a bond is at par if its coupon rate $y$ equals its yield to maturity $x$.

# The Term Structure of Interest Rates

**6**

This chapter is devoted to the analysis of the term structure of interest rates, commonly called the yield curve. Section 6.1 describes the structure of interest rates, defines the yields to maturity of *bullet* bonds as well as zero-coupon rates, indicates how one may construct forward, or futures, transactions starting from cash instruments, and exhibits the different forward rates implicit in the structure of spot rates. Section 6.2 analyzes the economic factors that determine the form of the yield curve, notably its slope. Section 6.3 generalizes the standard indicators of interest rate risk (such as simple duration or variation), which allows assessment of the financial consequences of deformation of the yield curve.

## 6.1 Spot Rates and Forward Rates

On financial markets, the interest rate used for a transaction depends on its maturity. The term structure of spot interest rates (the rate as a function of its maturity) is first briefly introduced. Then we study the yield to maturity of bullet bonds and zero-coupon rates. Finally, we define and calculate the forward rates implicit in this structure.

### 6.1.1 The Yield Curve

#### 6.1.1.1 The Interest Rate as a Function of Its Maturity

We denote by $r_\theta(t)$ the interest rate prevailing at time $t$ on the money or bond market for lending and borrowing transactions (buying or selling bonds), denominated in a given currency, without default risk, and with maturity $\theta$. When there is no ambiguity regarding the date $t$ (it is the present moment) $r_\theta(t)$ is denoted simply by $r_\theta$.

▶ **Definition** *The yield curve at time t is made of the $r_\theta(t)$ for different maturities $\theta$.*

© The Author(s), under exclusive license to Springer Nature Switzerland AG 2022     177
P. Poncet, R. Portait, *Capital Market Finance*, Springer Texts in Business and Economics, https://doi.org/10.1007/978-3-030-84600-8_6

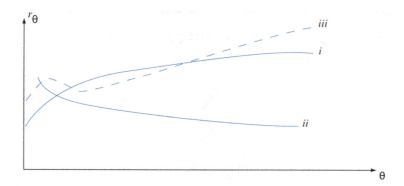

**Fig. 6.1** Three possible shapes of yield curves

The yield curve can be represented graphically as in Fig. 6.1 where the abscissa corresponds to different maturities or terms, and the ordinate axis corresponds to the prevailing market rates. In this figure, amongst a great number of possible and observed yield curves, three commonly encountered ones are shown: (*i*) an ascending curve (also called "normal" since it is the most frequently seen); (*ii*) a descending curve (called "inverted"); (*iii*) a curve with a bump for short maturities and then ascending for medium and long terms.

As we will see in Section 6.2, such forms can be understood, notably as anticipations of the future rates that will prevail in the market.

#### 6.1.1.2 Different Yield Curves for Different Markets and Definitions of Market Rates

This first definition of "the" *spot* yield curve requires three clarifications:

- First, there are for a given instant not one but several yield curves corresponding to the different currencies, and for a given currency to partly segmented financial markets. For example, on the bond market the yield to maturity for bonds of first grade, with a term $\theta$ still remaining until maturity, is not in general exactly the same as the rate for money market instruments with the same term $\theta$. The differences, which are very small, are due to possible differences in regulations (asset restrictions, regulatory ratios, etc.) which possibly distinguish the different markets. In what follows, we will consider a given currency and neglect these other differences. Furthermore, as we have remarked in preceding chapters, for the same maturity the rate depends on the default risk of the issuer. We will examine the effect of credit spreads on yield curves in Subsection 6.2.3 of Sect. 6.2 but here we refer to the yield curve for the money-bond market for transactions assumed free from default risk (for example, government bills and bonds issued by a highly-rated country, etc.).
- Second, the yield curve can be expressed, according to markets and maturities, in terms of either yields to maturity or proportional rates; in each case, there are

# 6.1 Spot Rates and Forward Rates

some details to be worked out since the calculation of interests due depends on the nature of the rate.

– Finally, in computing and analyzing the yield curve, the rates used should be rigorously "pure" rates, that is, relative to zero-coupon bonds. Since such bonds are rare for many maturities, in practice one mainly observes the yields *to maturity* (YTM) of bullet bonds or simply "yields." However, as Sect. 6.1.2 below shows, from the YTM yield curve, one can deduce the implicit curve of zero-coupon rates.

In the remainder of this chapter and most of this book, **the $r_\theta$ will denote zero-coupon rates (compound, proportional, or continuous) while the YTM for a maturity $\theta$ will be denoted by $y_\theta$.**

The definition of zero-coupon rates and the computational method that permits extracting the zero-coupon rates from the YTM curve are examined in the next subsection.

## 6.1.2 Yields to Maturity and Zero-Coupon Rates

### 6.1.2.1 Bullet Bonds and YTM Curves

Recall that a bullet bond provides no repayment of the principal before its maturity but does provide periodic interest payments. Its rate $y_\theta$ depends on its maturity $\theta$ as noted above. The collection of $y_\theta$ for all values of $\theta$ makes up the YTM curve. The latter is difficult to use when one wants to value or discount cash flow sequences with rigor and precision.

Consider the simple case of a bullet bond of term $n$, exempt from default risk, whose next coupon is due in one year, which is not at par and consequently whose coupon rate $k$ differs from the YTM of a bond at par of same maturity $n$. The bullet bond, differing as it does from its equivalent at par bond by the coupon stream, does not provide a yield necessarily equal to that of the latter. Consequently, the formula for valuing \$1 of nominal value, $V = k \sum_{i=1}^{n} \frac{1}{(1+y_n)^i} + \frac{1}{(1+y_n)^n}$, is *ambiguous* and at best only an approximation.[1] This ambiguity (the same rate $y_n$ concerns two different bonds), leads to errors that cannot be neglected, either in theory or in practice.

To overcome this ambiguity on the definition of $y_n$, we will consider the special subset of fixed-rate, non-defaultable bullet bonds which have just paid off their coupon (the next coupon then is due in a year) and *which are at par*. The latter conditions imply that the coupon rate equals the yield to maturity $y_\theta$.[2]

---

[1] The duration of bullet bonds with a maturity $n$ is a decreasing function of $k$, since the center of gravity of the cash flows yet to come gets nearer the current date as the coupon becomes larger.

[2] Thus it is a "par yield" (see Chap. 4, Sects. 1 and 3.2).

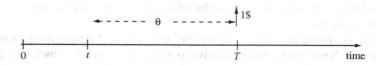

**Fig. 6.2** Time to maturity of a zero-coupon bond

▶ **Definition** *The YTM curve is the set* $\{y_\theta\}_{\theta \in (0,\Theta)}$ *of yields to maturity of bullet bonds at par, in general paying* coupons; $\Theta$ *denotes the longest maturity traded on the market (30 years, for example).*

Generally one considers arbitrary cash flow sequences (not necessarily regularly spaced, possibly involving intermediate repayments of principal, and such that the next payment does not necessarily come at the end of the current period) denoted by $\{F_\theta\}_{\theta \,=\, t_1,\ldots,\, t_n}$ (seen from the present date (0), $t_n$ denotes the date of the last cash flow as well as the remaining duration of the bond). Using $y_{t_n}$ as the discount rate usually is, in this context, badly wrong. It is always more rigorous, and at most times preferable, to employ the zero-coupon rate curve.

### 6.1.2.2 Zero-coupon Bonds and Rate Curves; Discount Factors

By definition, a zero-coupon bond only provides its holder with a single cash flow upon maturity of the bond.

If we are considering a classical lending and borrowing transaction, there is no payment of interest before maturity, the date upon which the entire principal is repaid along with the interest, possibly already capitalized (for maturities longer than one year), as a lump sum payment.

At any time $t$, in the absence of arbitrage, a single rate $r_\theta(t)$ prevails for all risk-free transactions with remaining term $\theta$. To this rate corresponds exactly one discount factor $b_\theta(t) = (1 + r_\theta(t))^{-\theta}$.

Now let us consider the "elementary" zero-coupon bond called B$_T$ paying off $1 at date $T$; an arbitrary zero-coupon bond paying $x$ at $T$ can trivially be treated as $x$ units of B$_T$.

At date $t$, the zero-coupon bond B$_T$ has a maturity $\theta = T - t$, as diagrammed in Fig. 6.2.

The market price of the zero-coupon bond B$_T$ at date $t$ is the same as the discount factor $b_\theta(t)$; it is connected with the zero-coupon rate $r_{T-t}(t)$ for maturity $\theta = T-t$ by the equation $b_\theta(t) = (1 + r_{T-t}(t))^{-(T-t)} = (1 + r_\theta(t))^{-\theta}$.

In the sequel, $T$ or $t$ denote dates while $\theta$ is the term (duration) remaining; the discount factor $b_\theta(t)$ is written as a function of the time $\theta$ remaining.

▶ **Definition** *The set* $\{r_\theta\}_{\theta \in (0,\Theta)}$ *of yields to maturity of zero-coupon bonds, where* $\Theta$ *denotes the longest maturity considered, forms the zero-coupon (ZC) rate curve. To this curve, there corresponds a curve of discount factors* $\{b_\theta\}_{\theta \in (0,\Theta)}$ *as well as the corresponding prices of zero-coupon bonds.*

## 6.1 Spot Rates and Forward Rates

**Remarks**

- In contrast to the YTM curve, the zero-coupon (ZC) rate curve can, and must, be used as a basis for computing the present values of arbitrary cash flow sequences. Indeed, any particular sequence of cash flows $\{F_t\}_{t = t_1,\ldots, t_n}$ can be considered as a collection of zero-coupon bonds ($F_{t_1}$ units of $B_{t_1}$, ..., and $F_{t_n}$ units of $B_{t_n}$); the discounted value (at date 0) of this sequence should then equal that of the collection, so that

$$V(0) = \sum_{\theta=t_1}^{t_n} F_\theta b_\theta = \sum_{\theta=t_1}^{t_n} \frac{F_\theta}{(1+r_\theta)^\theta}.$$

In this equation, which is a theoretically preferable alternative to the discounting formula with a single rate, the discount rate which is applied to $F_\theta$ depends on the maturity $\theta$. Another significant advantage of this approach is that any certain cash flow of arbitrary maturity $\theta$ is discounted at a single rate, specific to that maturity, and this is true whatever the bond with which the cash flow is associated.

- In the markets, almost all bonds with terms less than a year are zero-coupon bonds; on the other hand, most instruments with terms longer than a year have coupons. This is why, for maturities longer than a year, the most observable rates are YTMs. Fortunately, one can derive from a YTM curve $\{y_\theta\}\theta \in (0, \Theta)$ the corresponding ZC rate curve $\{r_\theta\}\theta \in (0, \Theta)$, as we show below.

### 6.1.2.3 Estimating the Zero-coupon Rate Curve from a YTM Curve

Starting from the observed annual YTM curve $(y_n)_{n = 1,\ldots,N}$, we will derive by an iterative process a ZC rate curve $(r_n)_{n = 1,\ldots,N}$ or, equivalently, the curve of discount factors $(b_n)_{n = 1,\ldots,N}$.

First, note that bills and bonds of short remaining maturity[3] are usually zero-coupons. Thus, consider a one-year zero-coupon bond: $r_1 = y_1$ and $b_1 = \frac{1}{1+y_1}$.

Furthermore, if we consider a bullet bond, with a term of 2 years quoted at par (so the coupon rate is the YTM $y_2$) and with a unit nominal value (so the market value is 1) characterized by the sequence $(y_2, 1 + y_2)$, we can write, evaluating it with the discount factors $b_1$ and $b_2$:

$V = 1 = b_1 y_2 + b_2 (1 + y_2)$, which allows us to compute $b_2 = \frac{1-b_1 y_2}{1+y_2}$ and $r_2 = b_2^{-1/2} - 1$.

Similarly, we calculate $b_2, b_3, \ldots b_n, \ldots$, in the order of increasing due dates. The factor $b_n$ (and the corresponding ZC rate $r_n$) is obtained by considering a bullet bond,

---

[3]That is, less than 6 months in the US or UK, and less than I year in the Eurozone.

at par, with $n$ years to run, whose coupon rate is $y_n$, thus generating the cash flow sequence $(y_n, \ldots, y_n, 1 + y_n)$, and by writing its value as

$$1 = y_n \sum_{i=1}^{n-1} b_i + (1 + y_n) b_n$$

and thus

$$b_n = \frac{1 - y_n \sum_{i=1}^{n-1} b_i}{1 + y_n}$$

gives $b_n$ as a function of $y_n$ and the $b_1, b_2, \ldots, b_{n-1}$ determined at the previous steps of the computation. Recall the rate $r_n$ is equal to $b_n^{-1/n} - 1$.

In this way we can, starting from the sequence $y_n$, calculate the corresponding sequences of $b_n$ and $r_n$, step by step from the nearest maturity to the latest.

---

**Example 1**

The 1-year Euribor is 5%, the 2-year YTM is 5.5% and the 3-year YTM is 6%. We will determine the ZC rates compatible with these three rates.

First, the 1-year Euribor must be converted into an exact rate; assuming a year of 365 days, the exact 1-year yield (YTM)[4] is 5% × 365/360 = 5.07%. In addition, since in the Eurozone transactions with terms less than a year are zero-coupons, the 1-year ZC rate is also $r_1 = 5.07\%$.

Now consider a 2-year bond at par; its coupon rate is identical to its YTM (over 2 years) $= y_2 = 5.5\%$. Its value, which is 1, is also equal to the present value of its future cash flows, discounted using the ZC yield curve ($r_1 = 5.07\%$ and $r_2$ is the unknown we wish to find), i.e., $1 = \frac{0,055}{1,0507} + \frac{1,055}{(1+r_2)^2}$. From this, we infer $r_2 = 5.51\%$.

Knowing $r_1$ and $r_2$, we may deduce $r_3$ by expressing it using the value of a 3-year bullet bond at par whose coupon rate is 6%: $1 = 0.06\left(\frac{1}{1,0507} + \frac{1}{(1,0551)^2}\right) + \frac{1,06}{(1+r_3)^3}$, which gives $r_3 = 6.04\%$.

---

Remark first that if the longest available zero-coupon maturity is 6 months and not 1 year (i.e., if bonds pay bi-annual coupons), the same calculations can be implemented with semi-annual instead of annual rates. At the end of the same computing process, the 6-month zero-coupon rate is simply converted into an annual rate.

---

[4]The fact that the Euribor is a proportional rate does not require any conversion here since a proportional rate for one year is also the yield to maturity; conversions are necessary for maturities less than one year.

## 6.1 Spot Rates and Forward Rates 183

Remark also that in the preceding example the slope of the ZC rate curve is slightly steeper than that of the YTM rates, which is a general property in the case of an upward yield curve. For the less common case of an inverted curve, the zero-coupon rates decrease more sharply with maturity than the YTM. This effect of steepening the slope can be intuitively explained by the fact that a zero-coupon bond of maturity $\delta$ is, in terms of Macaulay duration, equivalent to a bullet bond with maturity $\delta' > \delta$: for an increasing yield curve, the zero-coupon rate is thus higher than the YTM for the same maturity. The difference between YTM and ZC rates, which is nil for a flat yield curve, increases with the (absolute value of the) slope of the curve and with the level of interest rates.

The method of deriving discount factors from the YTM curve which has just been described is theoretical and not always practical, or even possible, notably when several bullet bonds of the same maturity are traded at different rates due to different coupon rates or to liquidity or credit-risk differences (which one should be used?), or if certain maturities are missing, or when it is in practice impossible to find bonds with different maturities producing, over their common lifetime, synchronized coupons. One then tries to find a theoretical curve $b_\theta$ of discount factors that best explains the observed prices by using econometric methods or even by explicitly modeling the curve $b_\theta$.

Without getting into details, the problem can be formulated as follows:

- one observes the prices $P_j$ of $m$ bullet bonds $j = 1, \ldots, m$, producing coupons $c^j_\theta$ on the dates $t_{j1}, \ldots, T_j$;
- to a range of discount factors $b_\theta$ correspond $m$ theoretical prices $\widehat{P}_j = \sum_{\theta=t_{j1}}^{T_j} c^j_\theta b_\theta$,
- one seeks the $b_\theta$ which best explains the observed prices, by implementing an appropriate econometric method such as bootstrapping or by minimizing the sum of the squares of the deviations $= \sum_{j=1}^{m} \left( \widehat{P}_j - P_j \right)^2$.[5]

Another difficulty comes from the different risk (premiums or) premia carried by most rates. Traditionally the risk-free yield curve has been extracted from the prices of government bonds and bills of different maturities. The main benchmarks nowadays are constructed with *vanilla swap rates* in particular the fixed rates of the Overnight Index Swaps (OIS) of different maturities. We will explain in Chap. 7 why the fixed rate of a vanilla Interest Rate Swap is « equivalent » and comparable to

---

[5] To deal with the problem of missing maturities $\theta$, an interpolation may be required. To deal with the problem posed by the existence of several bullet bonds of the same maturity traded at different rates, we can represent the curve $b_\theta$ by a function $f(\theta, a_1, \ldots, a_p)$ depending on $p$ parameters $a_1, \ldots, a_p$, with a form for $f$ given a priori (for example, $b_\theta = \exp\text{-}\theta(a_1 + a_2\theta + \ldots + a_p\theta^{p-1})$); on each date, we determine with econometric methods the values of the $p$ parameters $a_i$ which provide the best estimate for the $P_j$ observed (least squares, maximum likelihood, etc.).

**Table 6.1-a** A synthetic forward loan

| Instant | 0 | $n$ | $n + d$ |
|---|---|---|---|
| Lending for $n + d$ periods | $-1$ | $-(1 + r_n(0))^n$ | $(1 + r_{n+d}(0))^{n+d}$ |
| Borrowing for $n$ periods | $+1$ | | |
| Net cash flow (synthetic loan) | 0 | $-(1 + r_n(0))^n$ | $(1 + r_{n+d}(0))^{n+d}$ |

the YTM of a fixed rate security of the same maturity and why the OIS swap rates may be considered as good proxies for the risk-free rates.

### 6.1.3 Forward Interest Rates Implicit in the Spot Rate Curve

#### 6.1.3.1 Equations Involving the YTM

It is essential to understand that there exists *a forward yield curve* implied by the spot yield curve described above.

Indeed, by making simultaneous lending and borrowing transactions with different due dates, an investor can reproduce today (at time 0) an investment with term $d$, starting $n$ periods later (for any $n$ and $d$) and ensure the return rate of this investment. For example, by lending for $n + d$ periods (at the ZC rate $r_{n+d}(0)$) and borrowing for $n$ periods (at the ZC rate $r_n(0)$), an investor can ensure today the rate for a (forward) investment between $n$ and $n + d$.

Table 6.1-a displays the cash flows resulting from this double transaction of lending and borrowing, for a unitary amount repaid at maturity together with the compounded interests).

It is obvious from this table that the double transaction is equivalent to an investment of \$$(1 + r_n(0))^n$ at future date $n$ during $d$ periods leading to a repayment of principal and interest of \$$(1 + r_{n+d}(0))^{n+d}$. Notice that the two cash flows generated by this transaction (at $n$ and $n + d$, respectively) are known at 0, which is an essential condition for its being a forward transaction. This having been made precise, for ease of notation we will in the sequel omit specifying that the rates are those prevailing at the current time 0.

The rate $f_{n,d}$ for such a synthetic forward investment, contracted at 0, but commencing at $n$ and with a duration $d$, is given by the equation $\left(1 + f_{n,d}\right)^d = \frac{(1 + r_{n+d})^{n+d}}{(1 + r_n)^n}$.

$f_{n,d}$ thus represents the forward rate implicit in (or implied by) the current yield curve, for forward lending or borrowing relative to the period $(n, n + d)$. These are called *forward rates*.

We can summarize the foregoing results in the following proposition:

**Proposition**

(i). *Lending for $(n + d)$ periods and simultaneously borrowing for n periods is equivalent to lending forward for d periods, taking place n periods in the future (i.e., relative to the $(n, n + d)$ period);*

(ii). *Borrowing for $(n + d)$ periods and simultaneously lending for n periods is equivalent to borrowing forward for d periods, n periods later;*

# 6.1 Spot Rates and Forward Rates

(iii). *The rate of forward transactions (i) and (ii), denoted by $f_{n,d}$, is fixed today by the current spot rates $r_n$ and $r_{n+d}$ according to the formula holding for zero-coupon rates:*

$$\left(1 + f_{n,d}\right)^d = \frac{\left(1 + r_{n+d}\right)^{n+d}}{\left(1 + r_n\right)^n} = \frac{b_n}{b_{n+d}}. \tag{6.1-a}$$

$f_{n,d}$ is called the (forward) rate for d periods, n periods in the future.

In particular, for $n = d = 1$, formula (6.1-a) reads:

$$1 + f_{1,1} = \frac{\left(1 + r_2\right)^2}{1 + r_1} = \frac{b_1}{b_2}.$$

The relation between a long-term rate (for $p$ periods), the short-term rate (for one period), and the different short-term (one-period) forward rates $f_{i,1}$ writes:

$$(1 + r_p)^p = (1 + r_1)(1 + f_{1,1})(1 + f_{2,1}) \dots (1 + f_{p-1,1}). \tag{6.2}$$

Consequently, the long-term rate $r_p$ can be considered as the (geometric) mean of the short-term *spot* rate $r_1$ and the different short-term *forward rates* $f_{i,1}$.

## 6.1.3.2 Alternative Relationships

The preceding relations involve the compound rates. Alternative relations are preferable when we are dealing with proportional rates or continuous rates.

- When $r_n$ and $r_{n+d}$ are annual proportional rates, $n$ and $n + d$ are fractions of a year (often actual/360) and the synthetic forward rate $f_{n,d}$ resulting from the Table 6.1-b of cash flows (6.1-b) below is such that.

$$1 + d\, f_{n,d} = \frac{1 + (n+d)r_{n+d}}{1 + nr_n} = \frac{b_n}{b_{n+d}}, \quad \text{i.e.,}$$

$$f_{n,d} = \frac{(n+d)r_{n+d} - nr_n}{d(1 + nr_n)} = \frac{b_n - b_{n+d}}{d\, b_{n+d}}. \tag{6.1-b}$$

**Table 6.1-b** Proportional rates

| Instant | 0 | n | n + d |
|---|---|---|---|
| Lending for $n + d$ periods | −1 | | $1 + (n+d)r_{n+1}$ |
| Borrowing for $n$ periods | +1 | $-(1 + n.r_n)$ | |
| Net cash flow (synthetic loan) | 0 | $-(1 + n.r_n)$ | $1 + (n+d)r_{n+d}$ |

186                                6   The Term Structure of Interest Rates

- Second, if $r_n$ and $r_{n+d}$ denote continuous rates, we have:

$$e^{df_{n,d}} = \frac{e^{(n+d)r_{n+d}}}{e^{nr_n}}, \quad \text{i.e.} \quad f_{n,d} = \frac{(n+d)r_{n+d} - nr_n}{d}. \tag{6.1-c}$$

In what follows, we often use continuous rates since they frequently lead to simpler formulas than with discrete rates.

### 6.1.3.3 Numerical Examples

**Example 2**

Assume that on 01-01-$n$ we observe the following zero-coupon curve:

| Maturity | 1 year | 2 years | 3 years | 4 years |
|----------|--------|---------|---------|---------|
| Rate | 5% | 5.6% | 6% | 6.2% |

- The forward rate for transactions with a one-year term, starting 01-01-$n + 1$ (one year in a year) is therefore $f_{1,1} = (1.056)^2/1.05 - 1 = 6.2\%$
- The rate for 1 year in 2 years is $f_{2,1} = (1.06)^3/(1.056)^2 - 1 = 6.8\%$
- The rate for 1 year in three years is $f_{3,1} = (1.062)^4/(1.06)^3 - 1 = 6.8\%$
- One may check that $(1.062)^4 = 1.05 \times (1.062 \times 1.068 \times 1.068)$ (eq. 6.2) (long-term) = (short-term) × (product of forward rates).
- The rate for 3 years in 1 year is $f_{1,3} = [(1.062)^4/1.05]^{1/3} - 1 = 6.6\%$
- ......

**Example 3**

The zero-coupon rates $(r_\theta)$ are found as follows for the 01/04/$n$:

| 3 months | 6 months | 9 months | 12 months | 15 months |
|----------|----------|----------|-----------|-----------|
| 5% | 5.30% | 5.40% | 5.50% | 5.60% |

The corresponding forward rates $f_{3,6}$ (6 months in 3 months) and $f_{9,6}$ (6 months in 9 months) can be written.

$(1 + f_{3,6})^{0.5} = \frac{(1+r_9)^{0.75}}{(1+r_3)^{0.25}}$    or    $(1 + f_{3,6}) = \frac{(1+r_9)^{1.5}}{(1+r_3)^{0.5}}$,    so    $f_{3,6} =$ 5.60%;   $(1 + f_{9,6})^{0.5} = \frac{(1+r_{15})^{1.25}}{(1+r_9)^{0.75}}$, so $f_{9,6} = 5.91\%$.

## 6.2 Factors Determining the Shape of the Curve

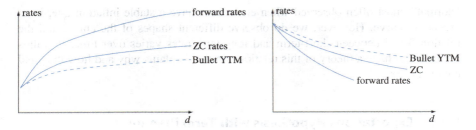

**Fig. 6.3** Normal and inverted interest yield curves

---

**Example 4**
Assume that on 01–01-$n$ we find the following Euribor rate curve

| Maturity | 3 months | 6 months | 1 year |
|---|---|---|---|
| Rate | 4% | 4.5% | 4.8% |

The 3-month in 3 months rate is $4(0.5 \times 4.5\% - 0.25 \times 4\%)/(1 + 0.25 \times 0.04) = 4.95\%$.

The 9-month in 3 months rate is $(4/3)(4.8\% - 0.25 \times 4\%)/(1 + 0.25 \times 0.04) = 5.02\%$.

We are applying formula (6.1-b) here because we have proportional rates (money market, actual/360).

---

We remark that in the previous examples the slope of the forward rate curve is above that of the ZC rates, which itself has a greater slope than the YTM curve; this property is generally true for an upward-sloping curve. In the less frequent case of an inverted curve, forward rates decrease faster with maturity than ZC rates. Figure 6.3 below shows the two cases.

## 6.2 Factors Determining the Shape of the Curve

We recall in Sect. 6.2.1 that the yield curve can exhibit various shapes, before proposing in Sect. 6.2.2 several theoretical explanations for the differences in the observed forms, and analyzing in Sect. 6.2.3 the impact of credit risk and what determines the credit spreads according to maturity.

### 6.2.1 The Curve Shape

As we have already mentioned, depending on the economic circumstances, the yield curve takes on an ascending, flat, or descending shape or exhibits one or several bumps, as in Fig. 6.1 at the beginning of the chapter. The case considered as "normal", most often observed when economies have a stable inflation rate, is an ascending curve. However, we do observe different shapes of the curve, and the relation found between short-term and long-term rates varies over time. It is thus important to have a theory of this relationship that explains why and how the yield curve changes.

## 6.2.2 Expectations Hypothesis with Term Premiums

### 6.2.2.1 The Basic Hypothesis

We will restrict ourselves here to one of the classical models of interest rate dynamics which is both simple and sufficiently general to explain most of the observed interest rate market data.

According to this theory, economic agents compare the *forward* short-term interest rates implicit in the yield curve to their expectations concerning the spot rates which will prevail in the future; they then make transactions intended to take advantage of the differences, if they judge them to be large enough.[6] In equilibrium, *forward* interest rates are equal to corresponding *spot rates* expected in the short term, increased by a term premium which depends on the maturity under consideration, and is generally increasing:

$$f_{n,1}(t) = e_{n,1}(t) + L_n \quad \text{for } n = 1, 2, \ldots \tag{6.3}$$

where $e_{n,1}(t)$ is the expected value at $t$ of the spot rate $r_1(n + t)$ for one period which will prevail $n$ periods forward (the value for $r_1(n + t)$ expected on average by the market at instant $t$), and $L_n$ is the term premium used for transactions with maturity $n$ (assumed independent of $t$).[7] This premium is often considered as compensation for the preference exhibited on average by investors for borrowing for a long time and investing for a short one (term premium).[8]

### 6.2.2.2 Arguments for the Hypothesis

Let us take up again, more carefully, the analysis and transactions that "representative agents"[9] are tempted to undertake when, for example, they expect that a spot rate in a year will be smaller than is the forward rate for "one year in a year" which is implicit in the present-day yield curve, i.e., $e_{1,1} < f_{1,1}$. The speculative transaction consists in constructing a synthetic forward loan guaranteeing a rate $f_{1,1}$ (lending for

---

[6]These are speculative transactions; *see below*.

[7]When all term premiums are assumed to be zero, the hypothesis is called the "Pure Expectations Hypothesis." Since both standard utility theory (which implies risk aversion) and empirical evidence go strongly against this assumption, we analyze directly its generalization, the "Expectations hypothesis with term premiums."

[8]This preference stems from the fact that short-term assets on the balance sheet (similar to monetary assets) are subject to a smaller interest rate risk than longer term ones.

[9]That is, agents whose expectations are the same as those of the market as a whole.

## 6.2 Factors Determining the Shape of the Curve

**Table 6.2** Expectations hypothesis

| Instant | 0 | 1 | 2 |
|---|---|---|---|
| Borrowing for one year | +1 | $-[1 + r_I(0)]$ | |
| Lending for two years | −1 | | $+[1 + r_2(0)]^2$ |
| Borrowing (later on) for one year | | $1 + r_I(0)$ | $-(1 + r_1(0))(1 + r_1(1))$ |
| Net effective cash flow | 0 | 0 | $[1 + r_1(0)][\frac{(1 + r_2(0))^2}{1 + r_1(0)} - (1 + r_1(1))]$ $= [1 + r_1(0)][f_{1,1} - r_1(1)]$ |
| Net expected cash flow | 0 | 0 | $[1 + r_1(0)][f_{1,1} - e_{1,1}] > 0$ |

two years combined with borrowing for one year) which will be financed a year later by borrowing for one year at the rate $e_{1,1}$ that is expected to prevail (but which will be in fact the one-year *spot* rate holding in a year, namely $r_1(1)$; note that this last rate is random as seen from today). The different cash flows involved in such a transaction are tracked in Table 6.2.

This speculative position, consisting today in a long-term asset (2 years) financed by a short-term liability (1 year), is in fact risky since the effective profit will ultimately depend on the future rate $r_1(1)$ that is unknown today. If the profit margin which is hoped for, $f_{1,1} - e_{1,1}$, were greater than the term premium $L_1$ ($f_{1,1} > e_{1,1} + L_1$), agents would carry out such transactions. In that way, they would exert pressure today to raise the short-term rate $r_1(0)$ and lower the long-term rate $r_2(0)$ (i.e. to lower the forward rate $f_{1,1}$) and thereby force the market to return to an equilibrium characterized by the equality $e_{1,1} = f_{1,1} + L_1$.

### 6.2.2.3 The Long-term Rate as the Geometric Mean of Anticipated Short-term Rates Augmented by Premiums

According to the theory just discussed, eq. (6.2), taking into account (6.3), becomes

$$(1 + r_p)^p = (1 + r_1)(1 + e_{1,1} + L_1)(1 + e_{2,1} + L_2) \ldots (1 + e_{p-1,1} + L_{p-1}).$$

$$(6.4)$$

From this, we can see that the long-term rate $r_p$ is a geometric mean of the short-term rate $r_1$ and the expected short-term rates $e_{n,1}$ plus term premiums.

Note that the long-term rate is commonly considered, especially by practitioners, as an average of expected short-term rates. The previous analysis shows that this conception is not accurate enough as it does not consider term premiums.

Figure 6.4 illustrates the role of the premiums $L_n$ and the expectations of rates in the three cases where the market anticipates (a) constant rates, (b) an increase in rates, and (c) a decrease in rates.

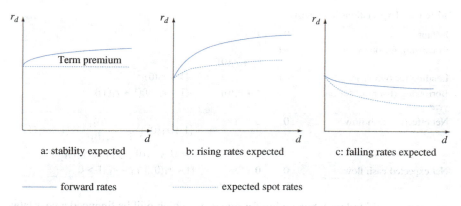

**Fig. 6.4** Effect of expectations and premiums on the spot curve shape

### 6.2.2.4 Implications for the Dynamics of the Yield Curve

One may infer from expectations theory with term premiums the following behavior of the yield (in fact spot) curve:

- When the market expects interest rates to rise (resp., fall), the curve has a tendency to be ascending (resp., descending).
- In periods of stability, the yield curve has a tendency to be slightly increasing (because of the premiums $L_n$).
- When the rates are "abnormally" high, decreases are expected and the curve has a tendency to go down, in spite of the premiums; when the rates are "abnormally" weak, increases are expected and the curve goes up (for two reasons: the effects of expectations and of premiums).
- A rate variation in the short term in general affects the expectations $e_{n,1}$ in a decaying way, and thus also the long-term rate. For example, an increase in the short-term rate results in an increase in the long-term rate of a smaller size.

### 6.2.2.5 The Effect of Expectations and Term Premiums

**Example 5**
Let us assume that on 01-01-2030 we observe the following yield curve for the zero-coupon rates:

|  | 1 year | 2 years | 3 years | 4 years |
|---|---|---|---|---|
| **Spot rate (observed)** | 12% | 13% | 13.6% | 14% |
| **Forward rate (calculated)** | $f_{1,1} = 14.01\%$ | $f_{2,1} = 14.81\%$ |  | $f_{3,1} = 15.21\%$ |

## 6.2 Factors Determining the Shape of the Curve

Assume that over a long period of $T$ years (for example, the last 30 years) one has observed the following mean deviations between the different *forward rates* ($f_{j,1}$) and the *spot rates* over a year ($r_1$):

$$\frac{1}{T}\sum_{t=1}^{T}\left(f_{1,1}(t)-r_1(t)\right)=0.3\ \%\ ;\ \frac{1}{T}\sum_{t=1}^{T}\left(f_{2,1}(t)-r_1(t)\right)$$

$$=0.4\ \%\ ;\ \frac{1}{T}\sum_{t=1}^{T}\left(f_{3,1}(t)-r_1(t)\right)=0.5\ \%\ .$$

If we assume the predictions are not systematically biased,[10] we obtain:

$$\frac{1}{T}\sum_{t=1}^{T}r_1(t)=\frac{1}{T}\sum_{t=1}^{T}e_{j,1}(t)$$

for $j=1,2,3$ and consequently $L_1=0.3\%$; $L_2=0.4\%$ and $L_3=0.5\%$.

One may then deduce that on 01-01-2030 the market predicts three consecutive rate hikes, bringing the predicted one-year rate to, respectively:

$e_{1,1}=14.01\%-0.3\%=13.71\%$ on 01-1-2031
$e_{2,1}=14.81\%-0.4\%=14.41\%$ on 01-1-2032
$e_{3,1}=15.21\%-0.5\%=14.71\%$ on 01-1-2033

If the market had not expected these large increases but stability, from 2030 on, at a level of 12%, the yield curve would have looked as follows at the beginning of 2030:

Rate for 1 year: $=12\%$
Rate for 2 years: $(1.12\times1.123)^{1/2}-1=12.15\%$
Rate for 3 years: $(1.12\times1.123\times1.124)^{1/3}-1=12.23\%$
Rate for 4 years: $(1.12\times1.123\times1.124\times1.125)^{1/4}-1=12.30\%$

### 6.2.3 Influence of the Credit Spread on Yield Curves

Up to now, we have considered both securities and transactions as exempt from credit risk. As we explained in Chap. 5, credit risk means there has to be an addition to the interest rate asked, or a credit spread. In this way, the curve for zero-coupon rates exempt from credit risk (securities issued by AAA governments of the best rank[11]) is below the curve for AA securities, and that in turn is below the curve for A securities, etc.

---

[10]That is they are consistent with what has occurred on average over a long period.

[11]In Europe, it is natural to think of Germany. Indeed, in the Eurozone one observes "sovereign" spreads which may be as much as several tens of basis points between rates with the same maturity

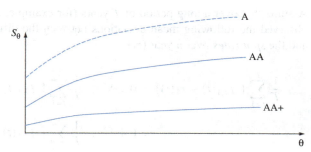

**Fig. 6.5** Spreads for AAA Treasury Bonds plotted against maturity, for various ratings

We will consider in what follows a credit spread $s_\theta$ calculated as the difference between the zero-coupon rate of maturity $\theta$ for an issuer with a given rating, denoted by $r_\theta^*$, and the corresponding rate $r_\theta$ for a AAA government security: $s_\theta = r_\theta^* - r_\theta$. We have already remarked that the "committed rate" $r_\theta^*$ is the maximum rate of return obtained if the issuer does not default. Figure 6.5 shows the typical profile of the ranges $s_\theta$ of these *spreads* according to the quality of signatures: empirically, they rise along with the weakness of the rating (for obvious reasons) and with the length of time to maturity. One also notes that the spreads increase during periods of crisis, the curves being shifted to a higher level as they correspond to lower ratings.

We will often revisit the issue of credit risk and credit spreads. In particular, we will see in Chap. 28 how, from a curve of spreads for a given rating, one may compute the probabilities of default for successive future periods which are implicit in this curve.

## 6.3 Analysis of Interest Rate Risk: Impact of Changes in the Slope and Shape of the Yield Curve

We have seen in the preceding chapter that using the simple variation-duration to estimate and control interest rate risk is based on assuming only parallel shifts of the yield curve. This assumption is contradicted by the empirical evidence. In the first subsection, we explain the limits of the standard tool of Duration-Variation in a context where the slope of the yield curve changes, before defining the more general notions of multi-factor variation-duration suited to the analysis of more complicated and more realistic rate variations than a simple parallel shift of the yield curve.

### 6.3.1 The Risk of a Change in the Slope of the Yield Curve

Let us suppose we want to compare two securities "a" and "b" with different maturities $a$ and $b$ (see Fig. 6.6) whose yield rates are respectively denoted by $y_a$

---

paid by different "virtuous States" in the zone, and several hundreds of basis points when the spread is between Germany and some countries with questionable signatures.

## 6.3 Analysis of Interest Rate Risk: Impact of Changes in the Slope and...

**Fig. 6.6** Flattening of the yield curve

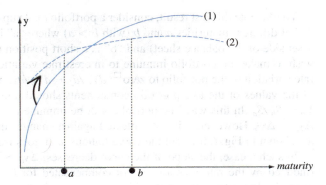

and $y_b$. If the two rates undergo variations $\Delta y_a$ and $\Delta y_b$, the values $V_a$ and $V_b$ of the two securities will show variations $\Delta V_a$ and $\Delta V_b$. Calling $S_a$ and $S_b$ the respective sensitivities of the two securities, the variations in value can be written (as a first-order approximation):

$$\Delta V_a = -V_a S_a \Delta y_a; \Delta V_b = -V_b S_b \Delta y_b, \text{ whence}: \frac{\frac{\Delta V_a}{V_a}}{\frac{\Delta V_b}{V_b}} = \frac{S_a}{S_b}\frac{\Delta y_a}{\Delta y_b}.$$

The assumption $\Delta y_a = \Delta y_b$ underlying the use of the simple rate sensitivity for managing a portfolio leads to two conclusions:

- The relative values vary in proportion to the sensitivities: the rate risk (per Euro or Dollar invested) associated with a bond is therefore proportional to its sensitivity.
- One may compensate for interest rate risk on bond "a" by taking a rate risk on bond "b" with the opposite sign. For example, a portfolio made up of $200 M fixed-rate Treasury Notes with sensitivity 4 as assets and $100 M of debt with sensitivity 8 as liabilities is immune (to a first approximation) to interest rate fluctuations.

In reality, these conclusions should be taken with caution, because the rate variations $\Delta y_a$ and $\Delta y_b$ are in general different since the interest yield curve displays changes in slope: as a result, the relative values do not vary in proportion to the sensitivities (because the factor $\Delta y_a/\Delta y_b$ differs from one and is a random variable) and the preceding conclusions are thus questionable. Notably:

- If the sensitivity of $y_a$ differs from that of $y_b$ the ratio of the sensitivities of "a" and "b" is not equal to $S_a/S_b$.
- Compensating for a rate risk on security "a" by a rate risk on security "b" (with the opposite sign) can only be a rough hedge since it is based solely on the notion of simple sensitivity.

To illustrate this last result, consider a portfolio made up of two securities "a" and "b" of different maturities $a$ and $b$ (with $b > a$) where "a" is a long position (on the asset side of the balance sheet) and "b" is a short position (a liability). Suppose we wish to make the portfolio immune to interest rate variations by setting the global rate variation of the portfolio to zero[12]: $dV_a/dy = dV_b/dy$, or $V_aS_a = V_bS_b$. The ratio of the values of the two portfolio components should thus be chosen so that: $V_a / V_b = S_b/S_a$. In this way, the portfolio will be immune to yield variations such that $\Delta y_a = \Delta y_b$. However, it is not protected against changes in the yield curve such as are shown in Fig. 6.6 where the curve flattens as it goes from (1) to (2).

In such a case, the slope of the curve decreases, $\Delta y_a > \Delta y_b$, and the loss on "a" (induced by the rate increase) is not compensated for by the gain on "b" (in the example illustrated, $\Delta y_b < 0$, which induces an increase in the value of "b" that *increases the loss*).

Let us assume, for example, that the portfolio consists of "a", \$300 M of bonds with sensitivity $= 3$ on the asset side, and "b", \$100 M of debts with sensitivity $= 9$ on the liability side. If $\Delta y_a = \Delta y_b$ the gain on "b" compensates exactly for the loss on "a"; but if, for example, $\Delta y_a = 2\Delta y_b$ the loss on "a" is twice the gain on "b."

More generally, if many securities are involved, a portfolio of assets and liabilities whose total variation is zero is protected against parallel variations in the yield curve but not against changes in its shape (notably changes in the slope).

The tools developed in the following subsection are extensions of simple sensitivity. They allow quantifying the consequences of a non-parallel shift in the yield curve. They provide a more powerful way of managing interest rate risk than simple sensitivity and can serve as a basis for Value-at-Risk methods which will be discussed in Chap. 27.

## 6.3.2 Multifactor Variation and Sensitivity and Models of Yield Curves

There are two ways to develop an instrument capable of accounting for the "multi-dimensional" character of rate risk: the first, using the zero-coupon curve, is explored in Sect. 6.3.2.1 and 6.3.2.2; the second, based on the YTM curve, is examined in Sect. 6.3.2.3.

### 6.3.2.1 Analysis of Non-parallel Variations of the Zero-coupon Curve Using a Model

To keep the formulas simple, we use continuous rates.

If we could use a unique yield $y$, we could write the present value at date 0 of the series of cash-flows paid off by a bond $\{F_\theta\}_{\theta=t_1,...t_n}$ in the form $V(y) = \sum_{\theta=t_1}^{t_N} F_\theta e^{-\theta y}$.

---

[12]The variation in the portfolio value is defined in the same way as the variation of a security value as $dV_p/dy$, where $V_p$ denotes the total value of the portfolio.

## 6.3 Analysis of Interest Rate Risk: Impact of Changes in the Slope and... 195

This would lead to the following variation, sensitivity $S$ (or modified duration) and Macaulay duration $D$:

$$\frac{dV}{dy} = -\sum_{\theta=t_1}^{t_N} \theta F_\theta e^{-\theta y} \; ; \; S = D = -\frac{dV}{Vdy} = \sum_{\theta=t_1}^{t_N} \frac{F_\theta e^{-\theta y}}{V} \theta.$$

In fact, to carry out discounting we will use a whole ZC curve, which is not flat in general, and thus write:

$$V(r_{t_1}, \ldots, r_{t_n}) = \sum_{\theta=t_1}^{t_n} F_\theta e^{-\theta r_\theta}$$

where $r_\theta$ represents the ZC rate with maturity $\theta$.

An arbitrary infinitesimal variation in the ZC curve is described by an *N-tuple* $(dr_{t_1},...,dr_{t_n})$ where, in contrast to the assumption underlying the preceding analyses, the $dr_\theta$ are not all equal to the same differential $d_r$. Such an $n$-dimensional variation creates a price variation for the bond which satisfies

$$dV = -\sum_{\theta=t_1}^{t_n} \theta F_\theta e^{-\theta r_\theta} dr_\theta. \tag{6.6}$$

So, the generalization of the standard analysis leads us to define the $n$ variations $\{\delta_\theta \equiv \frac{\partial V}{\partial r_\theta} = -\theta.F_\theta e^{-\theta r_\theta}\}_{\theta=t_1,...t_n}$ and the corresponding sensitivities $\delta_\theta /V$.

Although this approach has the merit of generality, it is, however, unhandy in practice. The rate risk is measured by an $n$–dimensional vector of sensitivities and in general $n$ is very large; for example, for a complex portfolio involving cash flows for every business day over ten years, $n$ is more than 2500. This is why, to obtain a practical tool, it is necessary to simplify the representation of the ZC curve by using a model.

Such a model expresses the rate $r_\theta(_t)$ observed at time $t$ as a given function $f$ of its maturity $\theta$ and of a limited number $q$ of factors or parameters $h_1, \ldots, h_q$ that vary over time:

$$r_\theta(t) = f(\theta, h_1(t), h_2(t), \ldots, h_q(t)). \tag{6.7}$$

According to the model used, the factors $h$ may, or may not, have a direct financial interpretation but, in all cases, their fluctuations generate the uncertainty that leads to the erratic movements of the ZC curve.

As an example, we introduce a model with two factors $h_1$ and $h_2$ that we will take up again later on. Here, $h_1(t) = r(t)$ represents a short-term rate (for example, the one-month rate) and $h_2(t) = s(t)$ denotes the spread between a long-term rate and a short-term one (for example, the 12-month rate minus the 1-month rate). As a result, a zero-coupon rate of arbitrary maturity $\theta$ is, at any instant $t$, assumed equal to

$$r_\theta(t) = f(\theta, r(t), s(t)),$$

$f$ being a known function or one that is estimated econometrically.

This implies that the set of all ZC curves is determined by the short-term rate and the spread.[13]

Let us go back to the general model (6.7) which leads to

$$dr_\theta = \sum_{j=1}^{q} \frac{\partial f(\theta, \underset{\sim}{h})}{\partial h_j} \, dh_j \quad (where \ \underset{\sim}{h} \equiv (h_1, \ldots, h_q)). \tag{6.8}$$

We will then use both eqs. (6.8) and (6.6) and write

$$dV = -\sum_{\theta=t_1}^{t_n} \theta e^{-\theta r_\theta} F_\theta \sum_{j=1}^{q} \frac{\partial f(\theta, \underset{\sim}{h})}{\partial h_j} dh_j, \quad or$$

$$dV = -\sum_{j=1}^{q} \left( \sum_{\theta=t_1}^{t_n} \theta e^{-\theta r_\theta} F_\theta \frac{\partial f(\theta, \underset{\sim}{h})}{\partial h_j} \right) dh_j. \tag{6.9}$$

The term $\delta_j = \sum_\theta \theta e^{-\theta r_\theta} F_\theta \frac{\partial f(\theta, \underset{\sim}{h})}{\partial h_j}$ represents the variation and $S_j = \frac{\delta_j}{V}$ the sensitivity to the factor (or risk source) $h_j$ ($j = 1, \ldots, q$).

If $q$ is not too large ($q = 2$ to 5), a few sensitivities $S_j$ determine the behavior of $V$.

As an illustration of the general method, let us reconsider the special model with two factors ($q = 2$), $r$ and $s$:

$$r_\theta = f(\theta, r, s) \ ; \quad dr_\theta = \frac{\partial f}{\partial r} dr + \frac{\partial f}{\partial s} ds$$

$r_\theta = r =$ short-term rate; $s =$ spread $=$ long-term rate $-$ short-term rate.

Using (6.6) this gives us

$$dV = -\left[ \sum_{\theta=t_1}^{t_n} \theta F_\theta e^{-\theta r_\theta} \frac{\partial f(\theta, r, s)}{\partial r} \right] dr - \left[ \sum_{\theta=t_1}^{t_n} \theta F_\theta e^{-\theta r_\theta} \frac{\partial f(\theta, r, s)}{\partial s} \right] ds.$$

Let us define

---

[13]Equivalently, one could have written the ZC curve as a function of two rates, the short-term rate $r$ and the long-term rate $l$: $r_\theta = \phi(\theta, r, l)$. The models using $f$ and $\phi$ are equivalent since $f(\theta, r, s) = f(\theta, r; l\text{-}r) = \phi(\theta, r, l)$.

## 6.3 Analysis of Interest Rate Risk: Impact of Changes in the Slope and... 197

$$S_r \equiv \frac{1}{V}\left(\sum_{\theta=t_i}^{t_n}\theta F_\theta e^{-\theta r_\theta}\frac{\partial f}{\partial r}\right) \quad ; \quad S_s \equiv \frac{1}{V}\left(\sum_{\theta=t_i}^{t_n}\theta F_\theta e^{-\theta r_\theta}\frac{\partial f}{\partial s}\right) \tag{6.10}$$

where $S_r$ and $S_s$ are the respective sensitivities with respect to $r$ and $s$.

The function $f$ can be estimated from historical data on the rates, using econometric methods like those shown in the examples in the next subsection 6.3.2.2.

$S_r$ allows us to appreciate the relative change in the portfolio value resulting from a change in the short-term rate $r$, for a *constant spread* $s$ (a change in the level of the ZC curve);

$S_s$ allows us to measure the relative change in the portfolio value resulting from a change in the spread $s$, *for constant* $r$ (a change in the slope of the ZC curve).

The equation $-dV/V = S_r dr + S_s ds$ allows us to evaluate the impact on the portfolio value $V$ of any joint changes in the level $(dr)$ *and* the slope $(ds)$.

Without any additional conceptual complications, we could have used a more general model of the ZC curve and considered several "key rates" instead of two. This type of approach is considered in the analysis of the Value at Risk (VaR) studied in Chap. 27. In this chapter, we present (in Subsect. 6.3.2.2 below) different examples of models with one or two factors.

In general, the $q$ factors and their influences can be determined from an econometric model such as

$$\Delta r_\theta(t) = \sum_{j=1}^{q}\frac{\partial f_\theta}{\partial h_j}\Delta h_j(t) + \varepsilon_\theta(t).$$

$$(Variation/hj) \qquad (residual)$$

Model 3, in subsection 6.3.2.2 below, illustrates such an approach.

The factors come from:

– Factor analysis that finds two or three factors that best explain variations in the interest rates. These are then combinations of rates interpretable as, respectively, the general level of rates, the slope of the ZC curve and its convexity.
– Factors given on a priori grounds whose effects are estimated in some econometric way.
– A theoretical model for the ZC curve (for example, Vasicek's, discussed in Chap. 17) whose parameters must be estimated.

### 6.3.2.2 Examples of Yield Curve Models Applied to Interest Rate Risk Analysis

We will present, as examples and in increasing order of complexity, three models of the yield curve and apply them to analyze the interest rate risk affecting bond portfolios.

## a) Model 1: a zero-coupon yield curve with a single factor

Let us suppose that some variation $r$ of a short-term rate (for example, the 1-month Euribor) is distributed along the ZC curve $r_\theta$ according to the law $\Delta r_\theta = f_\theta \Delta r$, where the $f_\theta$ are coefficients decreasing with $\theta$.

The $f_\theta$ could be, for example, given by the following table:

| $\theta$ | 1 year | 2 years | 3 years | ... | 10 years |
|---|---|---|---|---|---|
| $f_\theta$ | 0.6 | 0.5 | 0.4 | ... | 0.1 |

This means that if $\Delta r = +1\%$ (100 basis points, bips or bps) the 1-year rate goes up by 60 bps, the two-year rate by 50 bps, and so on.

The impact of $\Delta r$ on the value $V$ of a bond $\{F_\theta\}_{\theta = t_1, \dots t_n}$ is measured by

$$\Delta V = \frac{dV}{dr} \Delta r = -\sum_{\theta = t_1}^{t_n} \theta F_\theta e^{-\theta r_\theta} f_\theta \Delta r,$$

while its variation with respect to $r$ writes

$$\frac{dV}{dr} = -\sum_{\theta = t_1}^{t_n} \theta F_\theta e^{-\theta r_\theta} f_\theta.$$

Now consider a position characterized by a series of constant annual cash-flows equal to 100 to be received in the three coming years and assume that initially the ZC curve is flat at a level of 5%. In this example, we use discrete rates and write the initial value of the position as

$$V(r = 5\%) = 100 \times \left(1.05^{-1} + 1.05^{-2} + 1.05^{-3}\right) = 272.3.$$

Assume the short-term rate increases by 0.5% which results in a decaying increase in the long-term rates so that a 1-year rate increases by 0.30%, the 2-year rate by 0.25%, and the 3-year rate by 0.20%, in accordance with the law $f_\theta$ previously postulated.

The new value of the position is

$$V(r = 5.5\%) = 100 \times \left(1.053^{-1} + 1.0525^{-2} + 1.052^{-3}\right) = 271.1.$$

The variation $dV/dr$ computed from this model is equal to 240.[14] We check, using this variation, that $\Delta r = 0.005$ implies $\Delta V = 240 \times \Delta r = 240 \times 0.005 = 1.2 = 272.3 - 271.1$. Remark that, from this change, one may calculate, by simply using multiplication, the impact of any $\Delta r$ without having to calculate the present values of all the cash-flows.

---

[14] This variation can be calculated directly from the mathematical formula giving the variation (for discrete rates) or, more simply, by computing the values for two very close rates, for example: $V(r = 5\%) = 272.32$; $V(r = 5.1\%) = 272.08$; thus, $\Delta V = 0.24$ and $\Delta V/\Delta r = 240 =$ variation.

6.3 Analysis of Interest Rate Risk: Impact of Changes in the Slope and... 199

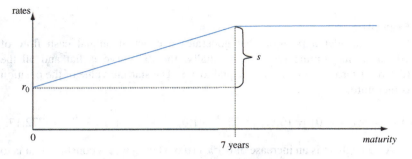

**Fig. 6.7** A simple model of a two-factor yield curve

### b) Model 2: a two-factor ($r_0$ and s) piecewise linear yield curve

Let $r_0$ denote the overnight rate and $s$ the spread between the 7-year rate and the overnight rate.

Suppose the ZC curve is conveniently represented as a function of $r_0$ and $s$ by the piecewise linear curve in Fig. 6.7.

The mathematical model that this figure represents can be written

$$r_\theta = f_\theta(r_0, s) = \begin{cases} r_0 + \dfrac{\theta}{7}s & \text{si } \theta \leq 7 \\ r_0 + s & \text{si } \theta > 7 \end{cases}$$

$$\frac{\partial f_\theta}{\partial r_\theta} = 1 \quad ; \quad \frac{\partial f_\theta}{\partial s} = \begin{cases} \dfrac{\theta}{7} & \text{si } \theta \leq 7 \\ 1 & \text{si } \theta > 7 \end{cases}$$

Since $\frac{\partial V}{\partial r_0} = \left( \sum_{\theta=1}^{n} \theta F_\theta e^{-\theta r_\theta} \frac{\partial f_\theta}{\partial r_0} \right) = \sum_{\theta=1}^{n} \theta F_\theta e^{-\theta r_\theta}$, we have $\frac{\partial V}{\partial r_0} =$ simple variation.

Because of this, the sensitivity for short-term rates is identical to the simple sensitivity.

Furthermore,

$$\frac{\partial V}{\partial s} = -\sum_{\theta=1}^{7} \frac{\theta^2 F_i e^{-\theta r_\theta}}{7} - \sum_{\theta=8}^{N} \theta F_\theta e^{-\theta r_\theta}.$$

This model is a generalization of the standard model. Like the latter it allows us to assess the effect of a parallel shift of the yield curve with modulus δ (by $\delta \frac{\partial V}{\partial r_0}$); in contrast to the simple model of variation, it also allows assessing the effect of a change δs in the spread (slope) given a constant short-term rate (by calculating $\frac{\partial V}{\partial s} ds$), as well as the impact of any combination ($dr_0$, $ds$) by computing $dV = \frac{\partial V}{\partial r_0} dr_0 + \frac{\partial V}{\partial s} ds$.

**Example 6**

Let us consider a position that generates a constant annual cash flow of 100 over the coming ten years. Initially, the ZC curve is flat and all the (discrete) zero-coupon rates are equal to 5%. The starting value of the position is therefore:

$$V(r = 5\%, s = 0) = 100 \times \left(1.05^{-1} + 1.05^{-2} + \ldots + 1.05^{-10}\right) = 772.17.$$

Assume there is an increase of 0.1% (10 bps) in $r_0$ with $s$ constant, that is to say, a parallel shift upwards of the ZC curve. The value of the position will be

$$V(r = 5.1\%, s = 0) = 100 \times \left(1.051^{-1} + 1.051^{-2} + \ldots + 1.051^{-10}\right)$$
$$= 768.44.$$

The change in the level $\frac{\partial V}{\partial r_0}$ can thus be estimated to be $(768.44 - 772.17)/0.001 = -3730$.

Now consider an increase of 0.7% in the spread for a constant $r_0 = 5\%$; the yield curve pivots upwards (an increase in slope) about the short-term rate which remains fixed. The position's value becomes

$$V(r = 5\%, s = 0.7\%) = 100 \times \left(1.051^{-1} + 1.052^{-2} + \ldots + 1.057^{-7}\right.$$
$$\left. + 1.057^{-8} + 1.057^{-9} + 1.057^{-10}\right) = 750.99.$$

The change in the spread $\frac{\partial V}{\partial s}$ can then be estimated to be $(750.99 - 772.17)/0.007 = -3026$.

Every change characterized by the pair $(\Delta r_0, \Delta s)$ has an impact on the value of the position approximately equal to $\Delta V = -3730 \times \Delta r_0 - 3026 \times \Delta s$.

Model 2 has the advantage that it does not require any estimate for the function $f_\theta(r_0, s)$ modeling the ZC curve, and the drawback of being crude; model 3 which follows is more sophisticated.

*c) Model 3: An econometric model with two factors (l, s)*

Here we adopt the long-term rate and the spread as factors:

$l = $ 14-year rate; $s = $ 14-year rate $-$ 1-year rate; 1-year rate $= r_1 = l - s$.

For a maturity $\theta$, we perform a regression over a period of several years of the variations $\Delta r_\theta$ (the dependent variable) over $\Delta l$ *and* $\Delta s$ (the independent variables):

$$\Delta r_\theta(t) = a_\theta\, \Delta l(t) + b_\theta\, \Delta s(t) + \varepsilon_\theta(t).$$

As an example, such a regression providing estimates for $a_\theta$ and $b_\theta$ has been carried out with (old) data for the French and US markets. Results are displayed in Table 6.3.

## 6.3 Analysis of Interest Rate Risk: Impact of Changes in the Slope and...

**Table 6.3** A two-factor yield curve model

| French market (1987/1989) | | | | American market (1930/1979) | |
|---|---|---|---|---|---|
| Maturity $\theta$ | $a_\theta$ | $b_\theta$ | $R^2$ of the regression | $a_\theta$ | $b_\theta$ |
| 1 year | 1 | −1 | 100% | 1 | −1 |
| 2 years | 0.95 | −0.70 | 69% | 0.99 | −0.74 |
| 3.6 years | 1.06 | −0.58 | 78% | 1.02 | −0.54 |
| 5.3 years | 0.99 | −0.37 | 86% | 1.02 | −0.27 |
| 6.5 years | 1 | −0.29 | 85% | 0.95 | −0.18 |
| 7.2 years | 0.98 | −0.26 | 84% | 0.95 | −0.16 |
| 7.9 years | 1.07 | −0.17 | 96% | 0.96 | −0.13 |
| 8.6 years | 0.96 | −0.18 | 89% | 0.97 | −0.11 |
| 13 years | 0.98 | −0.12 | 93% | 1.01 | −0.04 |
| 14 years | 1 | 0 | 100% | 1 | 0 |

The numerical results in Table 6.3 are "well fitted" by the following values[15]:

$$a_\theta = 1; b_\theta = \exp(-k(\theta - 1))$$

with $k = 0.23$ for the French market and $k = 0.28$ in the USA. As a result we have:

$$dr_\theta = d\,1 - \exp(-k(\theta - 1))ds.$$

Let us consider a bond providing the cash flows $F_{T_1}, F_{T_2}, \ldots, F_{T_n}$ and with value $V = \sum_{\theta=T_1}^{T_n} F_\theta e^{-\theta r_\theta}$.

We then have: $dV = - \sum_{\theta=T_1}^{T_n} \theta F_\theta e^{-\theta r_\theta} dr_\theta$, or else:

$$dV = - \sum_{\theta=T_1}^{T_n} \theta F_\theta e^{-\theta r_\theta} dl + \left[ \sum_{\theta=T_1}^{T_n} \theta F_\theta e^{-\theta r_\theta} \exp(-k(\theta - 1)) \right] ds.$$

Consequently, we can explicitly compute the two variations:

– Long-term rate Variation: $\frac{\partial V}{\partial \ell} = - \sum_{\theta=T_1}^{T_n} \theta F_\theta e^{-\theta r_\theta} =$ simple variation;

– Spread Variation: $\frac{\partial V}{\partial s} = \sum_{\theta=T_1}^{T_n} \theta F_\theta e^{-\theta r_\theta} e^{-k(\theta - 1)}$.

These two variations allow one to assess, as a first approximation, the impact $\Delta V$ of any change in the ZC curve defined by $(\Delta l, \Delta s)$.

---

[15] These results are due to S. Schaeffer.

### 6.3.2.3 Analysis of Non-parallel Variations in the YTM Curve

We can here follow another path, sometimes simpler but less rigorous, and work with the yields to maturity instead of ZC rates. The value of a *coupon bond* of remaining time to maturity $x$ free of credit risk can be computed as the value of its cash-flows discounted at the bullet bond yield rate $y_x$ prevailing at date 0. For a nominal value of \$1 this gives

$$V_x(y_x) = k \sum_{\theta=t_1}^{x} e^{-\theta y_x} + e^{-xy_x}. \tag{6.11}$$

Without going back over the difficulties and ambiguities in this approach (notably the uniqueness of $y_x$, which in fact also depends on the coupon rate $k$), let us combine this formula with a model for the yield curve. Let us choose, for example, a two-factor model:

$$y_x = f(x, r, l) \equiv \Psi^x(r, l) \tag{6.12}$$

where $r$ denotes a short-term rate (e.g., the 3-month money market rate) and $l$ a long-term rate (e.g., the 10-year Treasury bond rate).

From this, we see that any infinitesimal change in the yield curve is determined by $(dr, dl)$ according to the equation

$$dy_x = \frac{\partial \psi^x(r, l)}{\partial r} dr + \frac{\partial \psi^x(r, l)}{\partial l} dl,$$

which implies

$$dV_x = \frac{dV_x}{dy_x} \left[ \frac{\partial \Psi^x}{\partial r} dr + \frac{\partial \Psi^x}{\partial l} dl \right]. \tag{6.13}$$

Set $S_x \equiv \frac{1}{V_x} \frac{dV_x}{dy_x} = -$ simple sensitivity of a bond with remaining time to maturity $x$.

Rewrite (6.13) in the form

$$\frac{dV_x}{V_x} = S_x \left[ \frac{\partial \Psi^x}{\partial r} dt + \frac{\partial \Psi^x}{\partial l} dl \right]. \tag{6.14}$$

This allows us to interpret the two sensitivities

$$S_x^r \equiv S_x \frac{\partial \Psi^x}{\partial r} = \text{ sensitivity of the security to } r$$

$=$ relative change in the value $V_x$ induced by a 1% change in $r$, for constant $l$;

# 6.4 Summary

$$S_x^l \equiv S_x \frac{\partial \Psi^x}{\partial l} = \text{ sensitivity to } l$$

= relative change in the value $V_x$ induced by a 1% change in $l$, for constant $r$.

**Remarks**
- Most of the properties and analyses concerning simple sensitivity can be generalized to this case of a two-dimensional sensitivity, notably for a portfolio $z$ defined by a vector of weights $(z_i)_{i = 1,...,N}$:

$$S_z^r = \sum_{i=1}^N z_i S_i^r \; ; \; S_y^l = \sum_{i=1}^N z_i S_i^l$$

where $S_i^r \equiv S_i \frac{\partial f}{\partial r}$ and $S_i^l \equiv S_i \frac{\partial f}{\partial l}$. denote the sensitivities of bond $i$ to $r$ and $l$, respectively.

- Without other major complications, we could have started with a more general model of the yield curve. For example, one can consider three "key rates" instead of two and write

$$y_x = f(x, r, b, l) = \Psi^x(r, b, l)$$

where $b$ is the 4-year bond, for example, which leads to 3 sensitivities relative to short-term, medium-term, and long-term rates, respectively:

$$S_x^r \equiv S_x \frac{\partial \Psi^x}{\partial r} \; ; \; S_x^b \equiv S_x \frac{\partial \Psi^x}{\partial b} \; ; \; S_x^l \equiv S_x \frac{\partial \Psi^x}{\partial l}.$$

---

## 6.4 Summary

- The term structure of interest rates (TSIR), or spot yield curve, at time t is made of the spot interest rates $r_\theta$ prevailing on the money and bond markets for all quoted maturities $\theta$, a given currency, and for non-defaultable instruments.
- The TSIR can assume any shape although it is most often upward sloping.
- It can be constructed from yields to maturity or, preferably, from zero-coupon rates. The zero-coupon rate (ZC) curve is preferable to the YTM (Yield to Maturity) curve for computing the present values of arbitrary cash flow sequences $\{F_\theta\}$: $V = \sum_{\theta=t_1}^{t_n} \frac{F_\theta}{(1+r_\theta)^\theta}$ .
- The non-directly observable ZC rate curve can be derived from the observed YTM curve by an iterative process starting with the shortest maturity up to the longest one.

- Lending for $(n + d)$ periods at a ZC rate $r_{n+d}$ and simultaneously borrowing for $n$ periods at rate $r_n$ is equivalent to lending *forward* for $d$ periods, starting $n$ periods ahead at forward rate $f_{n,d}$ with: $(1 + f_{n,d})^d = (1 + r_{n+d})^{n+d}/(1 + r_n)^n$. Thus, there exists a *forward* rate curve *implicit* in the spot rate curve.
- According to the simple Expectations Hypothesis with term premiums, *forward* interest rates are equal to corresponding spot rates *expected* in the short term, augmented by a *term premium* which depends on the relevant maturity and is generally increasing. Consequently, when the market expects a rise (fall) in interest rates, the TSIR is upward (downward) sloping.
- In practice, due to credit risk *spreads*, there exists for each currency at any one time as *many TSIRs* as there are credit risk *ratings* (AAA, AA, and so on).
- The study of more complicated and realistic rate variations than a simple parallel shift of the yield curve requires the use of *multifactor variation/duration* and of a model of the TSIR. Such a model expresses the rate $r_\theta(t)$ observed at time $t$ as a given function of its maturity $\theta$ and of a limited number of variable factors which are the sources of risk.
- One such model is an *econometric* model with a short term rate $r(t)$ and the slope $s(t)$ of the yield curve as the two factors driving the TSIR: $r_\theta(t) = f(\theta, r(t), s(t))$. It allows assessing, through two different sensitivities, the risks of non-parallel shifts of the yield curve as well as changes on its slope.

## Suggested Readings

**See also the references listed at the end of chapter 17.**
Bodie, Z., Kane, A., & Marcus, A. (2018). *Investments* (11th ed.). Irwin.
Fabozzi, F. (2006). *Fixed income mathematics: Analytical and statistical techniques* (4th ed.). McGraw Hill Education.
Fabozzi, F. (2016). In F. J. Fabozzi (Ed.), *Bond markets analysis and strategies* (9th ed.). MIT Press.
Tuckman, B., & Serrat, A. (2011). *Fixed income securities: Valuation, risk, and risk management* (3rd ed.). Wiley.

# Vanilla Floating Rate Instruments and Swaps

**7**

The two preceding chapters were devoted to the analysis of fixed coupon bonds. If an issue of such bonds is repaid according to a predetermined schedule (with no possibility of prepayment and neglecting the risk of default...) the bonds making up the issue generate at each time a fixed income and so can be analyzed with classical methods.

This chapter is devoted to instruments that provide at times $t = 1, 2, ..., N$ interest payments that are functions of the value $I(t)$ of some reference index. $I(t)$ is supposed to reflect the interest rate conditions prevailing in the market at the instant $t$ or having prevailed for a period preceding $t$. As a result, in general, at any instant in the life of an instrument involving floating rates, the future cash flows are random and classical methods are not adapted to valuing them.

We distinguish two categories of instruments with floating interest rates (*floaters*). The first category is made up of assets like notes and bonds, distributing floating coupons. They are used to transfer liquidity (investment/financing). Their market values fluctuate around their nominal value. According to the rules that govern the contracts and the calculation of coupons, these assets are said to have "forward-looking" (predetermined) rates or "backward-looking" (post-determined) rates. Section 7.1 is devoted to such *floaters*, which are traded on the money and bond markets.

*Interest rate swaps* belong to the second category of floating-rate instruments to which Sect. 7.2 is devoted. An interest rate *swap* can be analyzed as a position with two components of opposite sign (long and short); on the one hand, usually a fixed-rate instrument and, on the other, a floating-rate instrument.[1] In contrast to assets in the first category, the market value of a *swap*, which is initially nil, in general, remains much smaller than its nominal value even though it varies as a function of

---

[1] There are also swaps (called *basis swaps*) that exchange different floating rates, for example, a money market rate for a bond market rate (see Sects. 7.2 and 7.2.4.4).

© The Author(s), under exclusive license to Springer Nature Switzerland AG 2022
P. Poncet, R. Portait, *Capital Market Finance*, Springer Texts in Business and Economics, https://doi.org/10.1007/978-3-030-84600-8_7

interest rates; from this follows that a *swap* is not aimed at transferring liquidity from the investor to the issuer but is rather an instrument for managing interest rate risk.

In this chapter, we study the instruments whose valuation and interest rate risk are the simplest to assess and the most common and accordingly are termed "*plain vanilla*" or "*vanilla.*" They are in fact replicable by a sequence of spot short-term transactions on the money market. The study of vanilla floaters (in Sect. 7.1) will ground our analysis of vanilla swaps (in Sect. 7.2). The analysis of swaps other than vanilla, which requires probabilistic methods, is carried out in Chaps. 16 and 17.

## 7.1 Floating Rate Instruments

After a general discussion of floating-rate instruments (Sect. 7.1.1), we present an analysis of their valuation and the interest-rate risk they bear (Sect. 7.1.2).

### 7.1.1 General Discussion and Notation

Securities with floating interest rates (notes, bills, bonds, etc.) have seen a real boom, in the international market (since the 1970s), in the US market, and in the European markets (since the start of the 1980s). They are called Floating Rate Notes or FRN, or floating rate bonds, and are issued by governments, financial institutions and banks, and nonfinancial corporations. Floating rate loans offered by banks and financial institutions are also widespread. All these products are referred to as floaters. Besides, as we show in Sect. 7.2, the floating leg of an interest swap can be analyzed and valued as a floater.

The purpose of these floaters is to reduce, and if possible eliminate, the interest rate risk that affects market values. The main method consists in adjusting the value of the coupon paid at each period $n$ so that it is a function of the market rate at the period $n$ (and not of the interest rate prevailing at issue). The value of a security that yields a return matching the current market conditions is, in principle, insensitive (or only mildly sensitive) to variations in these conditions.

Different reference indexes (benchmarks) are used and various types of contracts have been created. A distinction should be drawn between forward-looking (or predetermined) and backward-looking (or post-determined) floating rates.

#### 7.1.1.1 Definition and Basic Principles

▶ **Definition** *An asset with a floating interest rate distributes coupons calculated on the basis of a nominal rate revised as a function of the evolving market interest rate:*

*The nominal interest rate and therefore the coupon are functions of a reference index I which in general is related to the money or the bond market. If $C_n$ denotes the $n^{th}$ coupon, then $C_n = f(I_n)$.*

## 7.1 Floating Rate Instruments

It is important to understand why indexing the nominal interest rate on a market rate can reduce the interest rate risk affecting the value of the security. We provide first rough and simple explanations before proceeding to a deeper analysis.

Let us consider a security with a nominal value of 100 and a one-year term, yielding a coupon of 10 in one year.

If the 1-year interest rate is 10% this security's value is $V = 110/1.1 = 100$.

If the interest rate increases to 11% and the coupon remains at 10, the security's price goes down and becomes $V = 110/1.11 = 99.1$. But if, as a result of the increase in the interest rate, *the coupon is adjusted to be* 11, *so that the nominal rate remains identical to the market rate, V is stabilized at* 100 ($V = 111/1.11$ after the rate increase and the corresponding adjustment of the coupon).

More generally, let us write down the formula for the value $V(r, c)$ of a security with nominal value 1 and coupon rate $c$ and whose yield rate $r$ fluctuates according to market conditions:

$$V(r,c) = \sum_{i=1}^{N} \frac{c}{(1+r)^i} + \frac{1}{(1+r)^N}.$$

As we have seen in the preceding chapter, $V(r, c) = 1$ if and only if $r = c$, which means that an asset whose coupon rate is equal to its yield rate is quoted at par.

From this, if the coupon rate is permanently maintained in accordance with the market yield, the security's value is left unaffected by interest rate variations. This is why floating-rate securities have a coupon that fluctuates randomly in a way parallel to market conditions so as to eliminate the interest rate risk affecting their prices.

### 7.1.1.2 Notation

Generally, we consider a security with nominal value 1, duration $N$ periods, and which delivers $N$ floating coupons $C_n$ ($n = 1, ..., N$). The period $n$, by convention, begins at the instant $n-1$ and ends at instant $n$, as in the following diagram 1:

Diagram

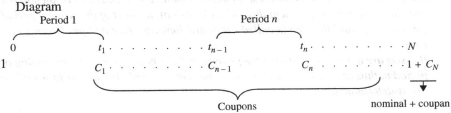

The coupon paid at $n$, for the period $n$, is given by the formula:

$$C_n = (\alpha I_n + m) D_n \quad \text{for } n = 1, \ldots, N \tag{7.1}$$

where:

$I_n$ = an index or money-market reference value (Compound Overnight, EURIBOR, LIBOR, ...);

$\alpha$ = a multiplicative factor (or leverage factor) set at issue (at $t = 0$) and usually equal to 1;

$m$ = an additive spread set at issue, which notably depends on the credit risk for the asset at issue; it does not vary over the life of the asset;

$D_n$ = the duration of the period $n$ as a fraction of a year; in general $D_n$ is computed for a year of 360 days so that $D_n = Nd(t_{n-1}, t_n)/360$ with $Nd(t_{n-1}, t_n)$ = number of days separating dates $t_{n-1}$ and $t_n$.

More generally, if $t$ and $t'$ are two arbitrary dates, we use the notation $D(t, t') = Nd(t, t')/360$.

### 7.1.1.3 Types of Floating-rate Assets and Main Reference Values
(a) *Instruments with forward-looking or backward-looking interest rates*

An important distinction should be retained between two types of floating rates, respectively called in the sequel *forward-looking* (or *predetermined*) interest rates and *backward-looking* (or *post-determined*) interest rates.

In the first case, the coupon is reset each time an interest payment is made, and then it remains unchanged until the next coupon payment date: in this case, the coupon is known (predetermined) at the start of the period.

In the second case, usually, the floater resets more frequently than the payments, as often as daily. The coupon thus will generally reflect an average of the resets since the previous interest payment. For instance, if resets are monthly while payments are made twice a year, each coupon will reflect the average of the six last monthly resets. Obviously, in this case, the coupon is unknown at the start of the corresponding period (it will only be known, at best, at the end of the fifth month of the 6 months coupon period).[2] The previous descriptions are summarized in the following two definitions.

▶ **Definitions**

- *If $C_n$ is reset each time an interest payment is made and then remains unchanged until the next coupon payment date, $C_n$ is known at the start of the coupon period (at time $t_{n-1}$) and the rate is called "forward-looking" rate or "predetermined" rate;*
- *If resets are more frequent than the payments, $C_n$ is not known at the beginning of period $n$, thus at time $t_{n-1}$. In this case, the rate is called "backward-looking" or "post-determined."*

---

[2] Another instance of a post-determined rate occurs even though the number of payment dates is equal to the number of setting dates. It is the case of a floater "in arrears" for which the payment and setting dates coincide, i.e., the coupon paid at the *end* of period $n$, $C_n$, is known only at date $t_n$, when the coupon rate is reset using the (single) current value $I(t_n)$ of the reference rate. This floater "in arrears" is not a vanilla instrument as it is not replicable by actual transactions. Its study is therefore postponed to Chap. 16 where methods involving stochastic calculus are used.

# 7.1 Floating Rate Instruments

Forward-looking (FL hereafter) rate instruments most often refer to an instantaneous index $I$, or at least a quasi-instantaneous one, (for example a LIBOR or a EURIBOR) prevailing at the start of the coupon period (date $t_{n-1}$), and the maturity of the reference matches the length of the coupon period: for instance quarterly interest payments $C_n$ will be indexed on the 3 month-EURIBOR prevailing at $t_{n-1}$, monthly payments will be indexed on the 1 month-LIBOR, and so on.

Backward-looking (BL hereafter) rate instruments refer to an index reflecting the average market conditions during the period $[t_{n-1}, t_n]$ (for example, a compound overnight rate $(OIS(t_{n-1}, t_n)$-index) or a capitalized short-term rate during the period $(t_{n-1}, t_n)$).

These interest rate references as well as the distinction between FL and BL indices have been described in Chap. 3 and will only be briefly recalled in what follows.

(b) *Principal reference benchmarks.*

Different reference values (or indices) are used to compute floating coupons. We list the main ones below.

## – *Benchmarks based on the EURIBOR and the LIBOR*

Recall that the Euribor is the arithmetic mean of the interest rates offered on the interbank market for unsecured transactions by a panel of large banks in the Euro zone during the day. It is established each business day and for each of the 5 maturities mentioned in Chap. 3, Sects. 5 and 2. The Euribor is the reference for different indices among which the principal ones are the 3-month Euribor used to calculate the quarterly coupons or biannual coupons (by capitalizing over 2 successive quarters) and the 6-month Euribor that has also sometimes been used as an FL reference for biannual payments.

On the international market and on the London Exchange, the reference benchmark that has been massively used (until 2020) for floating-rate instruments is the Libor (the equivalent of the Euribor for the Euro, but computed for different currencies).

> **Example 1 Forward-Looking Rate**
> Consider a Floating Rate Note (FRN) of nominal value \$200,000 distributing a coupon every quarter (on the dates 10/3, 10/6, 10/9, and 10/12). A coupon is calculated from the 3-month Libor *prevailing 3 months before*, with an additive spread of 0.60% ($L_3$+ 0.6%); this spread, which remains constant throughout the life of the security, is essentially justified by the credit risk ascribed to the issuer at the time of issue. We reckon 92 days from 10/6 until 10/9.
>
> On 10 September, the coupon distributed is: $C(10/9) = 200,000 \frac{92}{360}$ [Libor $(10/6) + 0.6\%]$.

> **Example 2 Capitalized Euribor**
> A bond of nominal value €300,000 distributes, on 10 March, an annual coupon calculated from the capitalized Euribor 6 months ($E_6 C$).
> On 10 March $n$ (10/3/$n$) we have $1 + E_6 C(10/3/n - 1, 10/3/n) = \left[ 1 + \frac{184}{360} E_6(10/3/n - 1) \right] \left[ 1 + \frac{181}{360} E_6(10/9/n - 1) \right]$.
> (We reckon 184 days from 10 March to 10 September and 181 days from 10 September to 10 March.)
> The coupon distributed on 10 March $n$ is therefore
>
> $$C(n) = 300,000 \left[ 1 + \frac{184}{360} E_6 C(10/3/n - 1) \right] \left[ 1 + \frac{181}{360} E_6 C(10/9/n - 1) \right]$$
> $$- 300,000.$$
>
> We remark that $C(n)$ is the global coupon for an investment for a year consistent with a succession of two half-year investments (a roll-over), with capitalization of interests: this is an investment of €300,000 on 10 March $n-1$ for 6 months with reinvestment of the capital and the first coupon on 10 September $n-1$ for another 6 months.

*– References to overnight rates: EONIA, \$STER, SOFR, ..., OIS-Index*

Recall (from Chap. 3) that different rate indices are calculated from the overnight rate. The main indices are the following:

- The EONIA (Euro Overnight Interest Average) has been the mean rate used by major banks, a given day, for 24-hours interbank transactions denominated in Euros. EONIA is being replaced by the Euro Short Term Rate (€STER pronounced ESTER).
- The daily interbank overnight rate for transactions in Sterlings is reflected by the SONIA (Sterling Overnight Interest Average).
- The Effective Federal Funds Rate (EFFR) and the new Secured Overnight Financing Rate (SOFR) reflect the overnight conditions in the US interbank market.
- A compound overnight index over a given period $(t, t')$ is obtained by capitalizing an overnight rate over this period. For instance, a compound €STER (or Euro-OIS Index) capitalizes the €STER on business days and uses simple interest on Fridays, Saturdays, Sundays, and holidays, as was explained in Chap. 3. In our notation, $1 + C(t, t')$ represents the value at $t'$ of a portfolio worth €1 at $t$ and invested between $t$ and $t'$ at overnight rates (at the daily €STER), with neither withdrawal nor addition of funds, $C(t, t')$ being the capitalized interests. The annualized value of $C(t, t')$ (the OIS index, see Chap. 3) therefore reflects a "geometric average" of the €STER between $t$ and $t'$. These indices are widely used as benchmarks for floating-rate instruments.

## 7.1 Floating Rate Instruments

### Example 3 Backward-Looking (Post-Determined) Rate

A bond distributes on 1 June of each year $n$ a coupon indexed against an OIS-index denoted by $I$, given as a real number (and not as a percentage), and increased by a spread of 50 basis points. Then on 1 June $n$, its coupon is

$$C_n = Nominal \times [C(01/6/n - 1, 01/6/n) + 0.005],$$

$C(01/6/n-1, 01/6/n)$ denoting the capitalized overnight interests during this one-year period. The exact amount of this coupon is only known on June $1/n$ (when the last overnight rate is known, at the end of the coupon period).

### Example 4 Backward-Looking Rate

A bullet bond of nominal €1000, over 7 years, distributes an annual coupon on 1 December, on the basis of 90% of the OIS index at the end of October, increased by an additive spread of 1%:

$$I_n = I(1/11/n - 1, 31/10/n); \alpha = 0.9; m = 1\%;$$

$$C(1/12/n) = [0.9 \times C(31/10/n - 1, 31/10/n) + 0,01] \times 1,000 \text{ €}.$$

Remark that in this example the coupon is not computed from an index related to the 12 months preceding the coupon's date $(01/12/n-1, 01/12/n)$ but is shifted by one month.

– *Other benchmarks*

Other reference rates for floaters are also used, such as a 3-month Treasury bill rate, a prime rate, etc.

The reference index may also be a bond yield such as the market rate on government bonds for a given maturity, or indices calculated from it. In this last (non-vanilla) case, the market price reactions to rate variations can be noticeably different from those when the reference is a short-term money-market rate and are beyond the scope of this chapter.

(c) *Floating rate instruments with a floor or a cap*

Some assets provide a floating coupon whose value, in any event, must be

– either greater than a given *floor*, denoted by $p$ in the sequel, which results from the formula

$$C_n = \text{Max}[p; \alpha I_n + m]$$

– or less than a given *cap P* and given by

$$C_n = \text{Min}[P; \alpha I_n + m]$$

– or between a floor $p$ and a cap $P$:

$$C_n = \text{Max}\ [p; \text{Min}\ (\alpha I_n + m; P)].$$

**Example 5**
– A bond of nominal value €100 pays a BL interest rate equal to the OIS index – 0.3% with a floor of 4%. Therefore:
$C_n = 100\ \text{Max}\ [4\%; C(n, n–1) – 0.3\%]=$
4 € if $C(n, n–1) \leq 4.3\%$
or $100[C(n, n–1) – 0.003]$ if $C(n, n–1) > 4.3\%$.
– A bond of nominal value €100 pays a BL interest rate equal to $C(n, n–1) + 0.1\%$ with a floor of 4.5% and a cap of 10%. Therefore:
$C_n = 4.5$ € if $C(n, n–1) \leq 4.4\%$
$C_n = 100[C(n, n–1) + 0.001]$
   if $4.4\% < C(n, n–1) < 9.9\%$.
$C_n = 10$ € if $C(n, n–1) \geq 9.9\%$.

Such assets can be analyzed as simple instruments with floating interest rates combined with floors bought or caps sold, as we show in Chap. 16.

## 7.1.2 "Replicable" Assets: Valuation and Interest Rate and Spread Risks

In this subsection, we will study the interest rate risk affecting instruments with floating money-market rates. Some of these assets amongst the most traded lead to coupons that can be replicated by a succession of short-term transactions in the money market. In absence of variations in the credit risk of the issuer, they have values always near their nominal values.

Following a general presentation of the problem of valuing floating-rate *replicable* assets (2.1), we analyze the values and risks of FL rate securities (2.2), then of BL rate instruments (2.3), and conclude by examining spread risks (2.4).

## 7.1 Floating Rate Instruments

### 7.1.2.1 Valuation and Risk for Floating-rate Assets: Generalities

(a) *Difficulties in applying classical discounting methods*

In general, a bond is a security that distributes a sequence of $N$ coupons $\{C_n\}_{n=1,...,N}$ and the principal along with the last coupon $C_N$.

If the coupons $C_n$ are all known at the outset, the value of such a security can be found using the standard formula for discounting. However, in the case of floating-rate instruments, $C_n$ is a random variable and the formulas for discounting do not, in themselves, have any meaning since the discounted value of a random variable is itself a random variable that cannot be interpreted as a current price. One could, however, be tempted to interpret $C_n$ as the *mathematical expectation* of the future $n^{th}$ coupon and to use one or several discounting rates incorporating one or several risk premiums.

As such a procedure is not easy to apply (due to the difficulty, or even the arbitrariness, of calculating the expectation of $C_n$ and assessing the risk premium), other methods should be used if possible. The alternative methods explained in this chapter are in fact suitable for the case of *replicable* indexing relative to a money-market benchmark (the *replicable* character is precisely defined below). It turns out that most vanilla instruments are *replicable*. The valuation of more complex instruments, notably those with a bond reference, is theoretically more difficult and based on probability theory and will be handled in Chaps. 16 and 17.

(b) *Interest rate risk and spread risk*

In the sequel, we make the difference between interest rate risk and spread risk. Indeed, the price of a floating-rate instrument on date $t$ can, in general, be expressed as a function $V(r(t), m(t))$ where $r(t)$ is a market interest rate and $m(t)$ the added spread *required by the market* at time $t$ on floating-rate securities similar to the instrument in terms of maturity and credit risk. Even when the interest risk is canceled through the floating rate mechanism, a spread (credit) risk remains in general. Interest rate risk is studied under the assumption of no spread.

### 7.1.2.2 Analysis of Forward-Looking Rate Instruments Depending on a Money-Market Benchmark

After presenting the general method for analysis and precise characterization of *replicable* floaters, we examine the valuation and interest rate risk for FL rate assets based on a *replicable* money-market reference characterized by a multiplicative factor $\alpha$ equal to 1 and a zero additive spread $m$: $C_n = I_n D_n$.

(a) *The general method of analysis: replication by a roll-over of spot operations in the money market*

The method consists in replicating, when possible, the floating-rate instrument by a *roll-over* of short-term lending.

In this subsection, the present instant is denoted by 0 and we consider a floater $x$, with a nominal value of \$1 that will distribute $N$ floating coupons $C_1, C_2, \ldots, C_N$ at $t = t_1, t_2, \ldots, t_N$ as well as the face value of 1 at $t_N$.

The interval $(t_{n-1}, t_n)$ separating two coupon detachments defines a period of $D_n$ year (for example, a quarter if the coupons are every three months, $D_n$ being about 0.25 years). More precisely, in general, $D_n = Nd(t_{n-1}, t_n)/360$ where $Nd(t_{n-1}, t_n)$ stands for the number of days (nights) between $t_{n-1}$ and $t_n$.

A *roll-over* of investments (or loans) is a succession (or a chain) of $N$ loans of \$1 with term one period (of duration $D_n$), renewed from $t_1$ to $t_{N-1}$.

We say that a roll-over, denoted by RO, replicates $x$ if it generates the sequence $C_1, C_2, \ldots, C_N$ at $t_1, t_2, \ldots, t_N$, as well as the nominal capital of \$1 at $t_N$.

The $n^{th}$ investment in the chain (occurring over the period $(t_{n-1}, t_n)$) is clearly carried out according to the market conditions prevailing at the start of the corresponding period (at date $t_{n-1}$) and generates FL interests $J(n)$ paid at the end of the period (at $t_n$). Furthermore, $J(n) = C_n$ if the replication of $x$ is perfect.

The repayment of the principal of the $n^{th}$ loan, immediately reinvested, constitutes the capital lent in the $(n + 1)^{th}$ loan of the chain. In absence of credit risk, since this $n^{th}$ loan of \$1 takes place at $t_{n-1}$ at the interest rate prevailing on that date, $r(t_{n-1})$, RO produces the cash flows given in Table 7.1.

In its conventional acceptation, "plain vanilla" means "standard", "common" or "usual." In the context of floaters, most, if not all, vanilla instruments are replicable and we could make no distinction between these two terms. However, for the purpose of analysis and valuation, it is the replicable character of the vanilla instruments that matters.

▶ **Definition** *An instrument is called replicable (or plain vanilla) if it can be replicated by a roll-over of market transactions: $J(n) = C_n$, for $n = 1, \ldots, N$.*

To make these ideas concrete and the explanation easier, let us consider the special case of an asset with an FL rate, free of credit risk, which distributes a quarterly coupon at dates $t_1, t_2, \ldots, t_N$, based on the FL 3-month Euribor $(E_3)$ prevailing at the beginning of the quarter, which means that $I_n = E_3(t_{n-1}) = 3$-month Euribor at $t_{n-1}$. In addition we assume that $\alpha = 1$ and $m = 0$ and set $D_n = Nd(t_{n-1}, t_n)/360 = $ length of quarter $n$ based on a year of 360 days. This gives, for a nominal of \$1:

$$C_n = D_n E_3(t_{n-1}),$$

and the cash flow series $\underline{A}$, which this asset is a claim on, writes as:

| | $t_1$ | $t_2$ | ......... | $t_n$ | ......... | $t_{N-1}$ | $t_N$ |
|---|---|---|---|---|---|---|---|
| $\underline{A}$: | $D_1$ | $D_2$ | ......... | $D_n$ | ......... | $D_{N-}$ | $1 + D_N E_3(t_{N-}$ |
| | $E_3(0)$ | $E_3(t_1)$ | | $E_3(t_{n-1})$ | | $_1 E_3(t_{N-2})$ | $_1)$ |

## 7.1 Floating Rate Instruments

**Table 7.1** Replication by a roll-over of spot transactions

|  | 0 | $t_1$ | - - - | $t_{n-1}$ | $t_n$ | - - - | $t_{N-1}$ | $t_N$ |
|---|---|---|---|---|---|---|---|---|
| Roll-over |  |  |  |  |  |  |  |  |
| Loan 1 | −1 | $J(1)+1$ |  |  |  |  |  |  |
| Loan 2 |  | −1 | - - - |  |  |  |  |  |
| - - - - |  |  |  |  |  |  |  |  |
| Loan $n$–1 |  |  |  | $J(n-1)+1$ |  |  |  |  |
| Loan $n$ |  |  |  | −1 | $J(n)+1$ |  |  |  |
| - - - - |  |  |  |  | -1 | - - - |  |  |
| Loan $N$–1 |  |  |  |  |  | - - - | $J(N-1)+1$ |  |
| Loan $N$ |  |  |  |  |  |  | −1 | $J(N)+1$ |
| **Net cash flow** | −1 | **J(1)** | **...** | **J(n−1)** | **J(n)** | **...** | **J(N−1)** | **J(N)+1** |

The payment dates $t_1, t_2, ..., t_{N-1}$ are also *reset dates* since the interest rate that will be applied in calculating the next coupon is reset on each date. Remark that the first coupon, $D_1 E_3(0)$, is known on date 0 (today).

It is additionally useful to establish a distinction between the instant preceding the detachment of a coupon and the instant following it; for any date $t_n$, these two instants are respectively denoted by $t_n^-$ and $t_n^+$. The value before detachment and the value after detachment are linked, in absence of arbitrage, by the relation: $V(t_n^-) = V(t_n^+) + C_n$.

The RO that replicates $\underline{A}$ can be implemented in a simple way, with an initial investment of $1: on the coupon date $(t_{n-1})^+$ (reset date), one invests $1 for 3 months and, on $t_n$, collects the coupon $D_n E_3(t_{n-1})$ and reinvests $1 for 3 months, and so on until the date $t_N$ on which one collects the last coupon plus the capital of $1 (see Table 7.2). This RO of investments for 3-months does replicate the cash flow series $\underline{A}$ characteristic of the asset under consideration.

The series $\underline{RO}$ and $\underline{A}$ are identical: one can thus obtain (replicate) $\underline{A}$ from a quarterly investment of $1 at 0 rolled over until $T$.

The foregoing analysis also leads to a criterion that allows us to easily recognize *replicable* FL rate assets.

*Criterion*: A FL rate asset free of credit risk is replicable by a roll-over of the type just described *iff* the following *two* conditions are satisfied:

(i). the coupon $C_n$, paid at $t_n$, must be computed on the basis of the spot interest rate prevailing at time $t_{n-1}$;

(ii). the reference benchmark rate on which the adjustment is based must have a time to maturity that corresponds to the interval $D$ between coupons (3-month Libor if the coupon is quarterly, 6-month Libor if it is semi-annual, etc.); then

**Table 7.2** Replicating a plain vanilla FL rate instrument

|  | 0 | ......... | $t_{n-1}$ | $t_n$ | ......... | $t_N$ |
|---|---|---|---|---|---|---|
| $\underline{RO}$ | $-1$ | | | | | |
| | | | $1 + D_{n-1}\,E_3(t_{n-2})$ | | | |
| | | | $-1$ | $1 + D_n E_3(t_{n-1})$ | | |
| | | | | $-1$ | ......... | |
| | | | | | | $1 + D_N\,E_3(t_{N-1})$ |
| Net $\underline{RO}$ | $-1$ | ......... | $D_{n-1}\,E_3(t_{n-2})$ | $D_n E_3(t_{n-1})$ | ......... | $1 + D_N\,E_3(t_{N-1})$ |

we have, up to a difference of 1, 2, or 3 days, $D_n \approx D$ for any period $n$ (for example, for quarters the $D_i$ range from 89/360 to 92/360 and $D = 0.25$).

We thus write a vanilla (replicable) sequence in the form:

$$
\underline{TR:} \quad
\begin{array}{l|l|l|l|l|l|l}
 & t_1 & t_2 & ......... & t_n & ......... & t_{N-1} & t_N \\
 & D_1 & D_2 & ......... & D_n & ......... & D_{N-1} & 1 + D_N \\
 & r_D(0) & r_D(t_1) & & r_D(t_{n-1}) & & r_D(t_{N-2}) & r_D(t_{N-1})
\end{array}
$$

(b) *Valuation and sensitivity to interest rates of credit-risk free, replicable, FL rate instruments.*

In accordance with the notation already introduced in Chap. 6, let us denote by $b_d(t)$ the value at time $t$ of a zero-coupon bond with time to maturity $d$ and value \$1 at $t + d$; depending on whether the interest rate is proportional or compound we have:

$$
b_d(t) = \frac{1}{1 + d\,r_d(t)}, \quad \text{or} \quad b_d(t) = \frac{1}{(1 + r_d(t))^d}.
$$

$b_d(t)$ therefore represents the discount factor which allows calculating, by simple multiplication, the discounted value at $t$ of the cash flow received at $t + d$. The next proposition gives the value at any time of a *replicable* FL rate asset and quantifies the interest rate risk that affects its value.

## Proposition 1

(i). *Credit-risk free, replicable, FL assets are quoted **at par on the coupon dates**. It means that, for nominal \$1:* $V = 1$ *at* $t_n{}^+$; $V = 1 + C_n (= 1 + current\ coupon)$ *at* $t_n{}^-$ *(where* $t_n{}^+ = $ *the moment just after detachment of* $C_n$ *and* $t_n{}^- = $ *the moment just before);*

(ii). *At an arbitrary date $t$ between the coupon dates* $t_{n-1}$ *and* $t_n$, *a replicable asset generally strays from par. More precisely, at some distance* $d = Nd(t, t_n)/360$ *from its next coupon* $C_n (= D_n\,r_d(t_{n-1}))$ *(see the diagram below), we have:*

## 7.1 Floating Rate Instruments

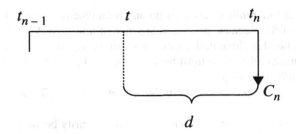

its value is $V_t = \frac{1+C_n}{1+dr_d(t)} = (1 + D_n\, r_d(t_{n-1}))b_d(t)$, where $r_d(t)$ represents the rate for transactions of term d (for example, the d-period Euribor) prevailing at t and $b_d(t)$ is the corresponding discount factor;
its duration is d (its modified duration is that of a zero-coupon bond of duration d).

**Proof of (i)**
We consider the vanilla (replicable) cash flows paid out at times $t_n, \ldots, t_N$:

$$D_n\, r_d(t_{n-1}),\ \ldots\ ,\ D_{N-1}\, r_d(t_{N-2}),\ 1 + D_N\, r_d(t_{N-1}).$$

This is the cash flow sequence generated by the security, viewed from any moment $t$ between $t_{n-1}^+$ and $t_n$.
We give two justifications for part (i) of the proposition.

*First justification of (i)*
On any coupon date (or reset date) $t_{n-1}^+$ one can replicate the sequence $\underline{A}$ by investing \$1 at $t_{n-1}^+$ (just following the detachment of $C_{n-1}$) for a period of duration $D_n$, then cashing in the coupon $C_n$ and reinvesting \$1 for the following period, and so on. This is a roll-over of investments for the periods of length $D_i$ described previously (Table 7.1) but here carried out starting from an arbitrary reset date $t_{n-1}$ (and not only from 0) and now displayed in Table 7.3.
Therefore, this strategy exactly generates $\underline{A}$.

**Table 7.3** Replicating a plain vanilla FL rate instrument starting at any reset date

|    | $t_{n-1}^+$ | $t_n$ | $t_{n+1}$ | $\ldots\ldots$ | $t_N$ |
|---|---|---|---|---|---|
| RO | $-1$ | $1 + D_n\, r_D(t_{n-1})$ |  |  |  |
|    |      | $-1$ | $1 + D_{n+1} r D(t_n)$ |  |  |
|    |      |      | $-1$ |  |  |
|    |      |      |      | $\ldots\ldots$ | $1+ D_N\, r_D(t_{N-1})$ |
|    |      |      |      |  |  |
| $\underline{A}$ | $-1$ | $D_n\, r_D(t_{n-1})$ | $D_{n+1} r D(t_n)$ | $\ldots\ldots$ | $1+ D_N\, r_D(t_{N-1})$ |

One can therefore replicate an asset on any reset date $t_{n-1}^+$, $(n = 1, \ldots, N)$, using an investment of \$1. Because of this, in absence of arbitrage: $V(t_{n-1}^+) = 1$.

Since, just after detachment, the security's value is equal to 1, then just before, in absence of arbitrage, the value must be $1 + Coupon$: $V(t_{n-1}^-) = 1 + C_{n-1}$.

**Second justification of (i)**

Let us consider the end of the life of the security, i.e., $t_N^-$. Then: $V(t_N^-) = 1 + D_N r_D(t_{N-1})$.

As from date $(t_{N-1})^+$, $r_D(t_{N-1})$ is known and the security becomes a fixed-income asset that can be valued by simple discounting: on $t_{N-1}^+$ it is worth $V(t_{N-1}^+) = \frac{1+D_N r_D(t_{N-1})}{1+D_N r_D(t_{N-1})} = 1$. Therefore, the security is at par on the coupon date $t_{N-1}$ and at $t_{N-1}^-$ is worth: $V(t_{N-1}^-) = 1 + C_{N-1} = 1 + D_{N-1} r_D(t_{N-2})$.

But this value becomes certain and known on date $t_{N-2}^+$, the date on which the security is certainly worth $V(t_{N-2}^+) = \frac{1+D_{N-1} r_D(t_{N-2})}{1+D_{N-1} r_D(t_{N-2})} = 1$; it is thus at par on the coupon date $t_{N-2}$.

In this way, by backward induction, starting from the final date $(t_N)$ and moving backward to the initial date $(t = 0)$, we show that the security is at par on all coupon dates $(t_N, t_{N-1}, \ldots, t_n, \ldots, 0)$.

**Proof of (ii)**

Part (ii) of the proposition is a direct consequence of (i). Indeed, consider any date $t$ between $t_{n-1}^+$ and $t_n$, and define $d \equiv Nd(t, t_n)/360$, as in the following diagram:

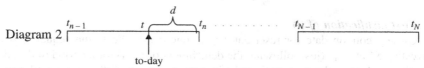

Diagram 2

The next coupon $C_n$ is fixed and one knows from (i) that on $t_N^-$ the security will *with certainty* be valued at $1 + C_n$; this is equivalent to a zero-coupon which is worth, on the date $t$, $V(t) = (1 + C_n) b_d(t) \equiv \frac{1+C_n}{1+d r_d(t)}$ where $r_d(t)$ represents the $d$-period interest rate prevailing on $t$ and $b_d(t)$ is the corresponding discount factor. The duration of this security is therefore $d$ (its modified duration is that of a zero-coupon bond of duration $d$).

> **Example 6 Value and Modified Duration of a *Replicable* FL Floater**
>
> An FL rate, credit-risk free, asset with a total duration of 4 years, distributes a semi-annual coupon referenced to the FL rate 6-month Libor ($L_6$): the coupon $C(s)$ paid on $s$ (at the end of the semester $s$) equals $C(s) = L_6(s-1) \times Nominal \times N_d/360$, where $N_d$ is the number of days in semester $s$.
>
> The maturity dates are fixed as of 1st April and 1st October and the nominal value is \$100 K. The coupons are calculated using a 360-day interest rate; the

(continued)

# 7.1 Floating Rate Instruments

219

October coupon runs 183 days and the April coupon 182 days, and the 6-month Libor on 1 October $n$-1 (the reset date) was 4%.

**Questions:**

(i). What is the market value of this security.

- On the evening of 31st March $n$, just prior to the detachment of the coupon?
- The following day, 1st April $n$, just after detachment?
- on 30th December $n-1$ the date on which the 3-month Libor $L_3 = 5\%$?

(ii). What is the change in the security value as well as the value lost as a result of a hike in the 3-month Libor of 0.2% (20 bips) on 30th December $n-1$?

**Answers:**

(i). Since 1st October $n-1$ one would have known the next coupon $C(n)$ to be paid 1st April $n$:

$C(n) = 100{,}000 \times 0.04 \times 182/360 = \$2022$; therefore:

Value on 30th March $n = 100{,}000 + C(n) = \$102{,}022$.
Value on 1st April $n = \$100{,}000$ (the security is at par).
Value 3 months before, on 30th December $n-1$, the date on which $L_3 = 5\%$:
$V(L_3) = 102{,}022/(1 + L_3 \times D) = 102{,}022/[1 + 0.05 \times 90/360] = \$100{,}762$.

(ii). Let us first write down two formulas applicable to a security with value $V(r)$, giving a unique cash flow $F$ at the end of a period of duration $D$, where $r$ is a proportional interest rate:

$V(r) = F/(1 + rD)$; $dV/dr = -FD/(1 + rD)^2$.
Now let us apply these formulas to the example being considered:
On 30th December $n-1$, the date when $r = L_3 = 5\%$ and $D = 0.25$:
$V(r) = \$100{,}000$ (as computed above);
Variation $= dV/dr = -102{,}022 \times 0.25/(1 + 0.05 \times 0.25)^2 = -\$24{,}880$;

(continued)

So, an increase of the short interest rate of 0.2% leads to a reduction of $V = 0.002 \times 24{,}880 = \$49.80$.

(c) **FL interest rates: an approach using forward rates**

We consider, at time 0, two future times $t$ and $t + D$, with $D = Nd(t, t + D)/360$. Let $D \cdot \tilde{r}_D(t)$ be a random cash flow which will be paid at $t + D$; $\tilde{r}_D(t)$ denotes the proportional spot rate for $D$ periods which will prevail at $t$ (for example, $D$ could be a quarter and $\tilde{r}_D$ the 3-month Euribor); the tilde denotes the random character of $\tilde{r}_D(t)$.

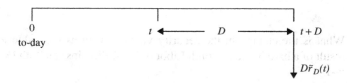

The problem is to value, on date 0, the claim to this random cash flow to be paid on date $t + D$.

Our method depends on the cash flow being replicable. It consists in adding to the existing position made up of the cash flow $D \cdot \tilde{r}_D(t)$ that will be received at $t + D$ the following strategy (decided at 0):

- Borrowing of $1, effective as of the future date $t$ (but arranged at 0) for a quarter, at the interest rate $\tilde{r}_D(t)$ (unknown at 0);
  - Lending $1 at 0 for $t + D$ periods and simultaneously borrowing $1 at 0 for $t$ periods, so as to replicate a $1 forward loan of duration $D$ starting at $t$, at the interest rate $f_{t,D}(0)$ (the forward rate of duration $D$ starting at $t$, as prevailing at 0). The cash flows induced by such a strategy, along with the existing position, are shown in Table 7.4.

This shows one can "exchange" the *uncertain* cash flow $D \cdot \tilde{r}_D(t)$ for a *certain* cash flow $Df_{t,D}(0)$. We can then consider $f_{t,D}(0)$ as the "*certainty equivalent*" of $\tilde{r}_D(t)$.

**Table 7.4** Exchanging a random cash flow for a certain one

|  | $t$ | $t + D$ |
|---|---|---|
| Existing position |  | $D \cdot \tilde{r}_D(t)$ |
| Spot borrowing at $t$ | +1 | $-D \cdot \tilde{r}_D(t) - 1$ |
| Forward loan (negotiated at 0) | −1 | $1 + D \cdot f_{t,D}(0)$ |
| Net cash flow | 0 | $D \cdot f_{t,D}(0)$ |

## 7.1 Floating Rate Instruments

Remark that, contrary to a widely held (mis)conception, $f_{t,D}(0)$ is *not* the interest rate $\tilde{r}_D(t)$ expected at 0 but may be interpreted as the expected value of $\tilde{r}_D(t)$ with a premium added (see Chap. 5).[3]

We can, for transactions carried out on date 0, exchange $\tilde{r}_D(t)$ for $f_{t,D}(0)$ (both available at $t + D$) and vice versa. It is therefore legitimate to value on date 0 the claim to a cash flow $\tilde{r}_D(t)$ available at $t + D$ by discounting the certainty equivalent cash quantity $f_{t,D}(0)$. Denoting by $v(0)$ the value of such a claim, we obtain: $v(0) = b_{t+D}(0)\, f_{t,D}(0)$. Recall that $b_\theta(0)$ is the discount factor, i.e. the value at 0 of a risk-free zero-coupon bond providing \$1 at $\theta$.

When we have to value a series of vanilla cash flows it suffices to discount the certainty equivalent of each cash flow (computed using a forward interest rate) and to sum the values thus obtained. For example, to value on date 0 a sequence of forward-looking cash flows $M_i\, \tilde{r}_D(t_i\text{-}D_i)$, for $t_i = t_1,..., t_N$, available at times $t_i$, we discount the $M_i f_{t_i - D_i, D_i}$ using the discount factors $b_{t_i}(0)$ and add up the results obtained.

The next proposition follows from this analysis:

**Proposition 2**

(i). *Seen as of time 0, $f_{t,D}(0)$ is the certainty equivalent of $\tilde{r}_D(t)$ available at $t + D$; therefore, the value on date 0 of the claim to $\tilde{r}_D(t)$ available at $t + D$ equals $b_{t+D}(0)\, f_{t,D}(0)$.*

(ii). *Consider an FL rate instrument which distributes at times $t_1,..., t_N$, the sequence:*

$$D_1.\, r_{D1}(t_1 - D_1), D_2.\, \tilde{r}_{D2}(t_2 - D_2), ..., D_N, \tilde{r}_{DN}(t_N - D_N)_{+1},$$

*(the last cash flow includes, in addition to the coupon, the principal value of \$1). As of date 0, $t_1\text{-}D_1$ is the date of the last reset (in general this is already past), $t_1\text{-}D_1 \le 0 < t_1 < ... < t_N$, (thus the first cash flow $D_1 r_{D1}(t_1\text{-}D_1)$ is a known quantity on date 0 while the $D_i.\tilde{r}_{Di}(t_i\text{-}D_i)$ are random for $i = 2,..., N$).*

*On date 0, the market value of such a security equals the present value of the future floating payments, computed using forward interest rates starting from the second expected coupon:*

$$V = b_{t_1}(0)\, D_1.r_{D_1}(t_1 - D_1) + \sum_{i=2}^{N} b_{t_i}(0) D_i.\, f_{t_i - D_i, D_i} + b_{t_N}(0). \tag{7.2}$$

Part (*i*) of the proposition results from the preceding analysis, notably from the possibility of exchanging the uncertain cash flow $\tilde{r}_D(t\text{-}D)$ for the certain $f_{t-D,D}(0)$.

Part (*ii*) follows from (*i*). Indeed, in the right-hand side of Eq. (7.2), the first term $(b_{t1}(0)\, D_1.r_{D1}(t_1\text{-}D_1))$ simply represents the present value (at 0) of the first expected

---

[3] We will see in Chap. 16 that the forward interest rate is equal to the expected value of the corresponding spot interest rate if this expectation is calculated using the "forward-neutral" probability measure, *not* the true probability measure.

and known cash flow; the second term $(\sum_{i=2}^{N} b_{t_i}(0)D_i. f_{t_i-D_i,D_i})$ is the present value of the later FL coupons, calculated from their certainty equivalents (with forward rates) according to $(i)$; the last term, $b_{tN}(0)$, is the present value of the nominal \$1 paid at $t_N$.

Notice that Eq. (7.2) is an alternative to Proposition 1 in valuing FL rate instruments. Also remark that Eq. (7.2) leads to a more general valuation method than Proposition 1 since it allows one to value instruments that are not strictly *replicable* by a roll-over because, in Eq. (7.2), $t_i - D_i$ can be different from $t_{i-1}$. For example, a *semi-annual* sequence of 3-*month* Euribor $(E_3(t_i-3\ months)$ paid at $t_i)$ is worth.

$$V = \sum_{i}^{N} b_{t_i}(0) f_{t_i-3month, 3month},$$

even if the consecutive maturity dates $t_i$ *are separated by periods of a half-year* $(D_i = 0.25$ year; $t_i - t_{i-1} = 0.5$ year).

For *replicable* instruments, $(t_i - D_i = t_{i-1})$, Eq. (7.2) simplifies to

$$V = b_{t_1}(0)D_1.r_{D_1}(t_0) + \sum_{i=2}^{N} b_{t_i}(0)D_i. f_{ti-1,Di} + b_{tN}(0) \tag{7.2'}$$

where $t_0 = t_1 - D_1$ denotes the (past) date of the preceding reset.

Employing Eq. (7.2') is an alternative to Proposition 1 for valuing vanilla FL rate floaters, and the two methods are perfectly equivalent as we show in the Appendix.

Two examples (11 and 12) of valuing the floating leg of an interest rate *swap* with Equation (7.2') are presented in the following section.

### 7.1.2.3 Replicable Backward-Looking (Post-Determined) Money-Market Rates

Here we study the value and the interest rate risk for money-market, *replicable*, *backward-looking interest rate assets* characterized by a multiplicative factor $\alpha$ equal to 1 and a zero additive spread $m$: $C_n = I_n D_n$ (*see* subsection 7.1.2.4 for more general cases).

In this case, the coupon $C_n$ as well as the index $I_n$ are unknown on date $n-1$. Generally, the index $I_n$ is a (sort of) geometric mean of interest rates during the coupon period $n$ (time interval $n-1$, $n$). The instrument is said to bear a post-determined or *backward-looking* rate, and $I_n$ and $C_n$ are post-determined because they are unknown at the beginning of coupon period $n$ (in fact, they are often unknown "most" of period $n$).

- As for a forward-looking, a *backward-looking* floater is termed *replicable* if it can be replicated by a *roll-over* of cash transactions on the money market. We can distinguish two families of such BL floaters: those referenced to a capitalized Euribor (or Libor) which can be replicated by a roll-over of transactions with a

## 7.1 Floating Rate Instruments

duration coinciding with the maturity of the reference Euribor (or Libor); and those referenced to an index calculated from an overnight index and replicable by roll-overs of overnight transactions.

The capitalized 3-month Libor over two quarters delivering bi-annual coupons has been a standard reference, in particular for the floating leg of interest rate swaps (see next section) Since the LIBOR references are becoming obsolete while the compound overnight benchmarks make the bulk of the market, we limit our analysis to floating rate instruments referenced to a compound overnight rate such as €STER, SONIA or SOFR (OIS index).

Let us, therefore, consider the case of a floater, referenced to an OIS index. We denote $1 + C(t, t')$ the capital and the interests accrued at time $t'$ by an overnight investment of \$1 at $t$, constantly reinvested between $t$ and $t'$ in a feasible way. Capitalization (compounding) is actually feasible if the transaction straddles two business days (Mon-Tue, Tue-Wed, Wed-Thu, Thu-Fri) and thus effectively implementable; if funds have to be lent for more than 24 hours (e.g., from Friday to Monday, or over a holiday) simple interest is used.

We focus on the value of \$1 of nominal of a credit-risk free floater $x$ referenced to a compound overnight rate $r$.

- Let first consider day $t_0$ just *after a coupon distribution* (or at issue). Since such a floater yields exactly the same cash flows as \$1 invested overnight in the market and appropriately rolled over, its market value is \$1 and it is at par at coupon date $t_0$. Remark that, just *before* the distribution, its value is 1 plus the accrued coupon.
- Let us now generalize the previous result by considering *any business day* $t_i$ situated between two coupon dates $t_0$ and $t_n$. We already know that the value of the floater just before the next coupon will be:

$$1 + C(t_0, t_n) = \prod_{k=1}^{n} [1 + (n_k/360)r(t_{k-1})]$$

where $C(t_0, t_n)$ is the coupon paid on $t_n$,

$n$ = number of business days between $t_0$ and $t_n$ (these days are $t_0, \dots t_k \dots, t_{n-1}$), and,

$n_k$ = number of calendar days when the overnight rate of business day $t_k$ ($r(t_k)$) is applied, that is 1, or 3 (on Friday), or 2 (on the eve of a holiday), or even, very unusually, 4 if Friday or Monday is a holiday.

At date $t_i$ the rates $r(t_{k-1})$ for $k = 0, \dots, i$ taking part in making up $C(t_0, t_n)$ are known while those afterward are unknown. We can thus decompose $1 + C(t_0, t_n)$ into the component known at $t_i$ and a random component:

$$1 + C(t_0, t_n) = \prod_{k=1}^{n} [1 + (n_k/360)r(t_{k-1})]$$

$$= \prod_{k=1}^{i} [1 + (n_k/360)r(t_{k-1})] \prod_{k'=i+1}^{n} [1 + (n_{k'}/360)r(t_{k'-1})]$$

$$\equiv (1 + C(t_0, t_i))(1 + C(t_{i+1}, t_n)).$$

At $t_i$, $C(t_0, t_i)$ is the accrued interest and is *known*, while $(1 + C(t_{i+1}, t_n))$ is a *random*, unknown element. However, the second component can be *replicated* with an investment of \$1 at $t_i$ by a RO of overnight transactions made between $t_i$ and $t_n$. From this, the entire future payoff (or value) $1 + C(t_0, t_n)$ can be replicated by a RO investment, at $t_i$, of $(1 + C(t_0, t_i))$. Therefore, under no arbitrage, the security is worth $1 + C(t_0, t_i)$, which means that it is at par: its market value is always equal to its nominal value plus the accrued interest.

Our results are summed up in the following proposition.

**Proposition 3**
*A backward-looking floater referenced to the compound overnight rate r is replicable and thus always at par, meaning that its market value is equal to its nominal value plus the accrued interest.*

A notable consequence of the preceding proposition is that, in absence of a credit spread variation, such a floater is constantly at par, which means, in this sense, that it can be considered as free of interest rate risk.

### 7.1.2.4 Credit Risk, Spreads, Spread Risk

A spread $m$ is often added to the reference index to calculate the coupon. For example, a FL rate is referenced to the 3-month Libor, $L_3 + m$, with $m = \text{spread} = 0.2\%$ (20 bips). A loan of \$1 then gives rise to coupons $C_n = D_n [L_3(n-1) + 0.2\%]$. Such additive spreads may be applied to both FL and BL rates.

In addition, for some contracts, a multiplicative factor, also called leverage factor, is applied to the index so that the coupon results from the formula $C_n = (\alpha I_n + m) D_n$. When $\alpha > 1$, the floater is termed leveraged (or super floater), and if $\alpha < 1$ the instrument is said under-leveraged.

We study first the reasons for and the consequences of the existence of an additive spread $m \neq 0$, and then we consider instruments with a multiplicative factor $\alpha \neq 1$.

(a) *Existence of an additive spread: reasons and consequences*

We consider here the most frequent case of an additive spread $m \neq 0$ associated with a multiplicative spread $\alpha = 1$. Two reasons, which are related to credit and liquidity risks, respectively, explain the existence of this spread.

## 7.1 Floating Rate Instruments

225

- The first reason, of foremost importance for many issuers, is that they usually pay an interest rate in excess of the interbank rate to compensate for their *credit risk*. We have seen that this credit spread depends on the issuer's rating, the term of the loan, the presence or not of a collateral asset, and the market price of credit risk. Moreover, often rates called risk-free are not completely exempt from credit risk. This is the case, for example, of some important unsecured rates such as the LIBOR or the EURIBOR[4] which are always larger than some secured rates of equivalent maturity such as a repo rate and than the rate of an OIS swap which is virtually free of credit/counterparty risk (see below, Sects. 7.2 and 7.2.3.2 regarding the OIS-LIBOR spread). When the reference rates, like IBORs, are not risk-free, the spread $m$ paid in addition to the reference can thus be considered as a *quality spread* defined as the *additional* premium that a lower quality borrower must pay in the market over a higher rated agent.
- The second reason is related to *liquidity*, and may be magnified if the issuer presents additionally a credit risk. Indeed, from the previous subsections, and especially Propositions 1 and 3, a default-free instrument X, perfectly *replicable* by a roll-over involving, say, an initial investment of $1 as in Table 7.1, would theoretically be worth $1 on a coupon date in the absence of arbitrage opportunity. In fact, even if X and the RO of loans that replicate X in theory generate the same series of cash flows, the market does not assign to the two series exactly the same price. This is because the issuer of X certainly has $1 available during $N$ periods while the agent who repeatedly borrows $1 for a succession of short periods does not enjoy the same guarantee of liquidity. Symmetrically, for an investor X is a less liquid placement of funds than the RO of short-term loans that replicates its cash flows. Moreover, when there is a risk of future credit degradation on the part of the borrower, the lender prefers to roll over short-term loans than to grant a floating rate loan to the same borrower if there is no covenant in the contract allowing an upward revision of the spread, should the borrower's creditworthiness deteriorate. Symmetrically, issuers prefer to lock in their credit spread so as to avoid the risk of an increase. However, this observation does not apply in the (frequent) presence of the previously mentioned covenant authorizing spread revisions. In our previous theoretical analysis, we ignored these liquidity-credit risk issues, which may justify a quality-liquidity spread for the floating rate instrument.

Let us denote by $m(t)$ the spread demanded by the market at time $t$ and by $m$ the spread effectively paid by the considered instrument. As already observed in Chaps. 4 and 5, the required credit spread $m(t)$ is larger if the life of the security is

---

[4]The Euribor–Libor cannot be taken as a benchmark for the riskless rate; recall that, in fact, it is the rate for borrowing without collateral contracted by a banking institution that is not strictly free from risk (even for an AA rating that few banks can claim nowadays). The risk-free rate (defined as that for borrowing by sovereign states with AAA rating) is usually smaller by one or two tens of basis points than the Euribor–Libor. It is also true for a repo rate contracted by the same banking institution due to the presence of some collateral.

longer, increases in periods of crisis to decrease back as market conditions improve and, of course, increases when the creditworthiness of the issuer deteriorates.

If $m(t)$ rises for any of the previously mentioned reasons, without a corresponding adjustment of $m$ the value of the floating-rate security goes down. Let $\Delta m = m - m(t)$ represent the (positive or negative) spread differential that prevents the floater from being at par on coupon dates. The difference $\Delta V$ between the effective market value of the instrument and the value it would have if its spread conformed to current market conditions is therefore equal to the present value of the series $\Delta m_i$ expected on the maturities $i$ still to come: the impact on the price of this differential $\Delta m$ is more pronounced the longer the remaining duration of the security without revision of its spread. In presence of a covenant authorizing periodic spread revisions if the borrower's creditworthiness declines, $\Delta m$ is mostly temporary and the spread risk is small. If there is sizable credit risk, in absence of spread revisions and for long-term securities, the spread risk can be quite large.

---

**Example 7 Spread Risk**

An FL rate security of remaining life one year distributes a quarterly coupon indexed on the FL 3-month Libor ($L_3$) plus a spread of 0.1% (10 bips). Let $C_t$ be the coupon paid at $t$ (end of the quarter):

$$C_t = Nominal\ [L_3(t-1) + 0.1\%] \times \frac{N_d(t-1, t)}{360}$$

The maturities are 1st January, 1st April, 1st July, and 1st October and the nominal value is \$100 M.

On 1st July $n$, the 3-month Libor rate $= 4\%$ and the 6-month Libor $= 4.2\%$.

The durations (in days) of the inter-coupon periods are given in the Table below:

| Reset date $t$: | 01/1/$n$ | 01/4/$n$ | 01/7/$n$ | 01/10/$n$ | 01/1/$n + 1$ |
|---|---|---|---|---|---|
| $Nd(t-1, t)$ | – | 90 | 91 | 92 | 92 |

Let us assume that on 1st July $n$ the spread demanded by the market for this type of instrument increases to 0.2%. There is no covenant allowing for a corresponding adjustment of the security spread. What is the effect of this increase (from 0.1% to 0.2%) on the instrument's market value?

**Answer**

The coupon distributed at $t$ is (in \$M):

$$C_t = 100 \times (L_3(t-1) + 0.1\%) \times \frac{N_d(t-1,t)}{360}$$

$$= 100 \times (L_3(t-1) + 0.2\%) \times \frac{N_d(t-1,t)}{360} - 0.1\% \times 100 \times \frac{N_d(t-1,t)}{360}$$

(continued)

# 7.1 Floating Rate Instruments

227

- The first term in the last equation is the cash flow generated by a floating-rate asset which yields the spread demanded by the market, and which is therefore at par on the coupon date (first October $n$).
- The second term is constant and the corresponding sequence is easily valued by standard discounting. As a result:

$$V(1st\ July\ n) = 100 - \frac{0,1 \times \frac{92}{360}}{(1 + \frac{92}{360} \times 0,04)} - \frac{0,1 \times \frac{92}{360}}{(1 + \frac{184}{360} \times 0,042)} = \$99.95\ M.$$

Consequently, the loss resulting from the increase in the spread required by the market is $0.05 M.

## (b) *The case of a multiplicative factor $\alpha$ differing from unity*

Let the current date (denoted by 0) be a coupon date and consider a $1 nominal instrument X whose $N$ future coupons are computed according to:

$$C_n = (\alpha I_n + m) D_n \quad \text{with} \quad \alpha \neq 1, \text{for } n = 1, \ldots, N.$$

In fact, a $1 asset X whose multiplicative factor $\alpha$ differs from unity can be decomposed into $\alpha$ dollars of an asset Y with unit multiplicative factor and an asset Z generating fixed revenues according to Table 7.5 which makes this decomposition explicit.

It is convenient to define Y, as in Table 7.5, with an additive spread $m$-(0) compatible with that prevailing on the market at current date 0 (in the *replicable case*, it is therefore at par on the coupon date and satisfies Propositions 1 or 3 according to whether it is an FL or a BL rate instrument). Furthermore, asset Z, defined by the explicit cash flows shown in Table 7.5 can be evaluated by standard discounting techniques since it generates certain and known cash flows.

**Table 7.5** Decomposition of an asset X with leverage factor $\alpha \neq 1$

|  | $n = 1, ..., N-1$ | $N$ |
|---|---|---|
| Z (fixed revenues) | $m - \alpha\, m(0)$ | $m - \alpha m(0) + 1 - \alpha$ |
| $\alpha$ $ of Y (asset with $\alpha = 1$) | $\alpha\, [I_n + m\, (0)]$ | $\alpha[I_n + m(0)] + \alpha$ |
| Total $= $ X | $[\alpha I_n + m]$ | $[\alpha I_n + m] + 1$ |

## 7.2 Vanilla Swaps

After a general presentation of swaps and their uses (Sect. 7.2.1), we study standard interest rate swaps, their replication, and evaluation (Sect. 7.2.2) as well as their sensitivity to interest rates (Sect. 7.2.3). A summary listing of the most current types of swaps is given in Sect. 7.2.4.

### 7.2.1 Definitions and Generalities About Swaps

A *swap*, meaning an exchange of two assets, designates various transactions most of which belong to one of two categories called interest rate swaps and currency swaps.[5] Each of these two families includes several members.

This subsection is limited to standard interest rate swaps, while currency swaps and nonstandard swaps are (briefly) presented in subsection 4. Nonstandard swaps will be re-examined more deeply, using probabilistic methods, in Chap. 16.

#### 7.2.1.1 Definition of a Standard Interest Rate Swap

An interest rate swap is an exchange of interests between two parties, denoted respectively by A and B in what follows. In a vanilla fixed-for-floating interest rate swap, the interests correspond to two debts of the same amount, denominated in the same currency, one at a fixed interest rate $k$ and the other at a floating (FL or BL) reference $r(t)$. This reference is a rate (or an average of rates) prevailing at or prior to date $t$.

It is also important to note that the principal of the debts is *not* exchanged, which greatly limits the credit/counterparty risk.

Let us consider a swap of amount $M$ (the *notional* or *principal*), leading to $N$ payments at times $t = t_1, \ldots, t_N$; $D_i$ denotes the duration as a fraction of a year separating $t_{i-1}$ and $t_i$ ($D_i = Nd(t_{i-1}, t_i)/360$).[6] Two cases are distinguished.

- In the first case, A *pays a fixed interest rate and receives a floating rate (see left part of* diagram 3). From A's viewpoint, such a swap is *equivalently and alternatively* called *fixed-rate payer or floating-rate receiver*. The net cash flow for A is algebraically:

$$M \, D_i \, [r(t) - k] \text{ at } t_i = t_1, \ldots, t_N.$$

---

[5]ELS (equity-linked *swaps*) are also traded in a market that is quite large (see Sect. 7.2.4.4).

[6]Or $Nd/365$ in some cases (such as instruments denominated in sterlings).

## 7.2 Vanilla Swaps

Diagram 3: interest rate swaps

Fixed-rate payer                  Fixed-rate receiver

- The second case for A, represented in diagram 3 on the right, is that of a *fixed-rate receiver* or *floating-rate payer swap* (A pays the variable interest rate and receives the fixed rate); the cash flow of the swap is $M D_i [k - r(t)]$ at $t_i = t_1, \ldots, t_N$.

Obviously, A's fixed-rate payer swap is a floating-rate payer swap for B and conversely.

The chosen reference for an interest rate swap may be an OIS index, a Libor, a Euribor, etc. The frequency of the floating interest payments depends on the reference $r(t)$ chosen and is often different from that of fixed interests. For example, in the case of a 3-month Libor reference ($r(t_i) = L3(t_{i-1})$), the swap has quarterly payments of the floating interests and annual or semi-annual payments of the fixed interests), a 1-month Euribor swap has monthly floating-rate interest payments, etc. The length of the period between two payments is called the "*tenor*" (expressed in years, months or days). For an *FL interest rate vanilla* swap, the tenor of the floating leg corresponds to the maturity of the reference rate.

> **Example 8 Overnight Indexed Swap (OIS)**
> A *swap* with notional $M and maturity 10 years, with a semi-annual fixed rate $k$ exchanged twice a year for the capitalized overnight rate (an OIS) generates, at each annual payment date $t$, a cash flow which, for the fixed-rate payer A, is:
>
> $$\$M[C(t-1, t) - k] \text{ for } t = 1, \ldots, 20$$
>
> where $C(t\text{-}1, t)$ is, for instance, the SOFR capitalized on business days and proportional on weekends and holidays during the interval $(t\text{-}1, t)$, e.g. semester $t$. So, this represents the average SOFR in the half-year preceding $t$ (*see previous Section and Chap. 3*). If $C(t\text{-}1,t) > k$, there is a net transfer of funds *from B to A*, and if $C(t,t\text{--}1) < k$, *from A to B*.

### 7.2.1.2 Managing Interest Rate Risk with Swaps

A *k-payer* (*r(t)-receiver*) swap benefits from interest rate hikes, and so can be used to hedge against such moves, and a *k-receiver* (*r(t)-payer*) swap has the reverse quality and can hedge the risk of a fall in interest rates.[7]

---

[7] The reader may remark that the profit-and-loss (P&L) profile of a swap is identical with that of a FRA (forward rate agreement), if the swap has only one payment date. A FRA is therefore analogous to a degenerate (one-period) swap.

> **Example 9 Transforming a Floating rate into a Fixed Rate**
>
> A company expects to borrow, repeatedly, in the money market during the next two years at the 3-month Euribor rate $E_3$. The amount borrowed will be €M and the interest that it expects to pay, a quarter $t$ in the future, will be €M $\times E_3(t-1)/4$ (note that in this example $r(t) = E_3(t-1)$).
>
> To hedge against the risk of an increase in its financing costs caused by a hike in the Euribor rate, the company contracts a *k-payer* swap against the 3-month Euribor of €M which generates quarterly payments of €M$[E_3(t-1) - k]/4$. The financing costs for quarter $t$ are therefore reduced (or increased) by €M$[E_3(t-1) -k]/4$, so that their net amount will be:
>
> $$€M \times E_3(t-1)/4 - M[E_3(t-1) - k]/4 = €M \times k/4.$$
>
> The *swap* transforms floating-rate borrowing into fixed-rate borrowing.

### 7.2.1.3 Benefiting from a Comparative Advantage: The Quality Spread Differential

Two parties with different credit ratings may enter into a swap to benefit from a comparative advantage if the differences in interest rates to which they have access are not the same on the fixed-rate and the floating-rate markets. Let us first give an example of such a situation.

> **Example 10 Benefiting from a Quality spread Differential**
>
> Firm B, rated BBB$^+$, wants to borrow $100 M at a fixed interest rate for 5 years. The conditions offered to B are: Fixed-rate loan over 5 years: 2%; Floating-rate loan over 5 years: compound SOFR + 0.5%.
>
> Firm A, rated AA, can borrow for 5 years at a fixed interest rate of 1% but presently borrows in the money market at the SOFR rate.
>
> To take advantage of the difference (A borrows 1% more cheaply than B at the fixed rate but only 0.5% more cheaply at the floating rate), B will borrow at the floating rate and swap the compound SOFR with A for a fixed rate, for example, 1.25%.
>
> Simultaneously, A will borrow an additional $100 M (more than it initially intended) at a fixed interest rate (and $100 M less at the floating rate).
>
> The result of this transaction for B and for A is as follows:
>
> |   | Borrowing cost | Swap | Net cost of resulting borrowing |
> |---|---|---|---|
> | B | SOFR + 0.5% | −SOFR +1.25% | 1.75% |
> | A | 1% | −1.25% + SOFR | SOFR − 0.25% |

## 7.2 Vanilla Swaps

So this investment allows B to replicate a fixed-rate loan with a savings of 0.25% (= 2% − 1.75%); and A also gains 0.25% on its loan at the SOFR rate.

More generally, the gain of 0.5% (amount of comparative advantage) is shared between the two contracting parties according to their respective bargaining power (assumed equal in this example). Let us remark that this operation implies an additional counterparty risk for A (the risk that B defaults during the 5 years swap life when the swap value is positive for A) which may not be considered tiny, even though the principals are not exchanged, since 5 years is quite a long duration.[8]

This comparative advantage is based on the observed *Quality Spread Differentials* (QSD). The quality spread represents the *additional* premium that a lower quality borrower must pay over a higher rated agent for borrowing over the same maturity. Usually, quality spreads increase with maturity. Moreover, the *quality spread is usually higher for fixed rate than for floating rate financing, for the same maturity.* On the swap market, the usual definition of the Quality Spread Differential is:

*QSD = Fixed-rate quality spread − Floating-rate quality spread for the same maturity.*[9]

In our previous example, the spread differentials between the low rated B and the high rated A is 0.5% for floating and 1% for fixed long-term maturities, hence the QSD is 0.5%. A positive QSD indicates that swapping is in the interest of both parties.

The comparative advantage argument has often been invoked to explain the growth of the swap market. In fact, the mere existence of comparative advantages may seem intriguing: the pressure of massive use of swaps aimed to take advantage of QSDs, which amount to arbitrage opportunities, should prevent their existence. Different answers have been put forward, grounded on various types of market imperfections, hidden covenants or options, risk shifting at the expense of shareholders, bankruptcy costs, agency costs, and more.[10]

A simple explanation for this puzzle may rely on a hidden option usually embedded in the floating-rate debt: As mentioned previously, in the floating-rate market, the lender has often the right to increase the spread in case of deterioration of

---

[8] In general, the swap involves the intermediation of a swap dealer who may bear the counterparty risk. This implies, however, the payment of fees which reduce the comparative advantage incentive.

[9] An alternative definition is used in other contexts: *QSD' = Fixed-rate quality spread for a long maturity − Floating-rate quality spread for a short maturity.*

[10] For alternative explanations of QSDs, see Wall and Pringle (1989).

the borrower's credit worthiness. By contrast, this option does not usually exist in the fixed-rate market. The apparent comparative advantage should then simply reflect the value of this option given to the floating-rate lender which is the more valuable, the smaller is the borrower credit worthiness and the longer is the considered maturity. According to this explanation, and several others, a swap motivated by a QSD *is not a true arbitrage* since an additional risk (an option given) is borne by the already riskier party.

## 7.2.2 Replication and Valuation of an Interest Rate Swap

We first show how a swap can be replicated from simultaneous lending and borrowing before studying how to value a swap.

### 7.2.2.1 Replication of a Swap

We consider a fixed-rate payer swap and then a fixed-rate receiver swap. Recall we denote by $k$ the fixed interest rate and by $r(t)$ the floating (BL or FL) reference rate.

(a) *The fixed-rate payer swap*

Consider a $k$-payer swap, denoted by $S_{r-k}$, yielding $MD_i(r(t) - k)$ during $T$ periods, where $D_i$ denotes the accrual period $t_i$. $S_{r-k}$ can be analyzed and replicated as a *combination of borrowing at the fixed rate $k$ and lending at the floating rate $r(t)$* (both borrowing and lending are bullets, for the same time period $T$ and in the same notional amount $M$). Thus, we can write:

$$k-\text{payer swap} \equiv S_{r-k} = \text{Lending at } r(t) + \text{Borrowing at } k.$$

In an equivalent way, one may consider the $k$-payer swap as replicated by a long position in an asset $A_r$ carrying a floating rate, with a bullet payment, notional value $M$ and maturity $T$, and a short position in a security $A_k$ with fixed rate and the same maturity and notional value. We thus write:

$$S_{r-k} = A_r - A_k. \tag{7.3-a}$$

Seen this way, the notional values of the fixed and floating components are *fictitiously* exchanged at the initial and terminal dates, so that the net cash flow is nil.

Alternatively, one can consider the cash flow series $L_r$ and $L_k$ of the coupons *without any reimbursement of capital,* called, respectively, the variable and fixed "legs" of the swap. The fixed leg $L_k$ gives a non-random cash flow $M k D_i$ on date $t_i$ and the floating leg $L_r$ gives the random cash flow $M D_i r(t_{i-1})$ on date $t_i$ for $i = 1, \ldots,$ $N$ if the benchmark is predetermined (FL) or $M D_i r(t_i)$ if it is post-determined (BL). In contrast with the assets $A_k$ and $A_r$, the legs do not pay out the nominal value \$M at the end of their lives. To summarize, $L_k$ and $L_r$ respectively represent the fixed and floating *legs* of the swap and we can write:

## 7.2 Vanilla Swaps

$$S_{r-k} = L_r - L_k. \tag{7.3-b}$$

The *swap* $S_{r-k}$ is equivalent to a long position in $L_r$ and a short one in $L_k$.

To avoid semantic misunderstandings, in the sequel we will use the terms fixed and floating *components* of the swap to mean $A_k$ and $A_r$, and fixed and variable *legs* to mean $L_k$ and $L_r$. We will use indifferently the terms "floating" and "variable."

### (b) *Fixed-rate k receiver swap*

A *fixed-rate receiver swap* $(S_{k-r})$ is replicated by lending at a fixed rate $k$ and borrowing at a floating rate $r(t)$ or, equivalently, by a position long in fixed-rate security and short in floating-rate security:

$$S_{k-r} = A_k - A_r = \text{fixed component} - \text{floating component} \tag{7.3-c}$$

or, equivalently, without the fictional exchange of principal:

$$S_{k-r} = L_k - L_r = \text{fixed leg} - \text{variable leg}. \tag{7.3-d}$$

### 7.2.2.2 Valuing a Replicable Swap

Generally, since the swap is the algebraic sum of a fixed component (leg) and a floating component (leg), the swap's value is at each time equal to the difference between the two components (legs) values.

In this subsection we analyze the value of a plain vanilla swap, whose floating component is, by definition, plain vanilla. We consider first the date of initiation of the swap (time 0) and then an arbitrary date.

### (a) *Valuation on the initiation date: "pricing" the swap (choosing the fixed rate)*

At the initiation date, the *market value of the swap should be zero* so that the capital outlay for *each* party is zero: $V_0(S) = 0$.

It is the fixed-rate $k$ that is calculated such that the swap initial value is zero, i.e. such that the initial value $V_0(A_k)$ of the fixed component is equal to that of the floating component, $V_0(A_r)$.

Since $A_r$ is at par (vanilla floater), $A_k$ must also be at par. Since $A_k$ is at par, $k$ is a *par yield* (the coupon rate for which security $A_k$ is at par) which should theoretically be equal to the market yield[11] $y_T(0)$ (using the convention $y_\theta(t) = $ yield to maturity of risk-free bullet securities at par, of maturity $\theta$, prevailing in the market at time $t$).

From this, at each moment $t$, the *term structure of fixed rates of plain vanilla swaps* free from credit risk, denoted by $\{k_\theta(t)\}_{\theta \in (0, T\max)}$, *should be identical to the*

---

[11] The *par yield* amounts to the yield to maturity on the coupon date or one period before the coupon date (when the accrued coupon $= 0$), which is almost always the case on the start date of the swap (see Chap. 4 for subtleties about the *par yield* and the yield to maturity).

*term structure* $\{y_\theta(t)\}_{\theta \in (0,Tmax)}$. In this notation $k_\theta(t)$ denotes the fixed rate of the swap of time to maturity $\theta$ for which, on date $t$, the swap market value is zero; or equivalently the interest rate for which the fixed component $A$ *is at par* (the *par yield*). This set of $k_\theta$ for different maturities $\theta$ is called the *swap fixed-rate curve*, the *swap rate curve* or the *swap term curve*. In practice, there are discrepancies between the money market yield curve and that of swap rates. In absence of credit or counterparty risks, small differences can be explained by the differences (as far as liquidity is concerned) between a swap and the portfolio that replicates it (see the previous section). Issues related to credit or counterparty risks enlarge these differences, as explained later on (in Sect. 7.2.3.2).

Following professional jargon, we call *pricing (or quoting) the swap* the process of setting (at $t = 0$) the fixed interest rate $k$ that makes the initial value of the *swap* zero ($k = k_T(0)$) and *valuing the swap* the computation of the (positive or negative) value of the swap at any later date. Professionals often call "long" a fixed-rate payer swap, "buying the swap" meaning taking such a long position.

Market-makers in *swaps* quote the interest rates $k_\theta$ (asks and bids) at which they are prepared to deal a fixed-for-floating plain vanilla swap "at market" or "at par" (at zero value) in different currencies and for different maturities. The bid-ask spread on the $k_\theta$ is very small for vanilla swaps[12] (for small counterparty risks it is usually between 1 and 5 basis points according to the maturity of the swap), which reflects the very good liquidity of plain vanilla swaps. The difference (bid–ask) is of course larger if there is a risk of default by the counterparty (see Sect. 7.2.3.2 below).

Alternatively, swap quotes may be given in terms of a swap spread, which is a differential percentage to be added to or subtracted from the yield of a risk-free instrument of the same maturity (for instance, a Treasury yield). Such a swap spread must not be confused with the bid-ask spread previously described.

Furthermore, one may extract from the swap rates $\{k_\theta(t)\}_{\theta \in (0,Tmax)}$ a curve of zero-coupon rates, exactly as one extracts a curve of zero-coupon rates from the bullet bond yields-to-maturity (see Chap. 6). From the zero-coupon rates so obtained, one derives a curve called the *swap zero-curve*.[13]

The International Swap Dealers Association (ISDA) daily publishes the curve of fixed rates for vanilla swaps which has become a major benchmark. The OIS rate curves become a standard for discounting (quasi) risk-free cash flows for reasons explained in Sect. 7.2.3.2.

### (b) *Valuing a swap at any date*

At any instant $t$ after initiation, the *swap* has, in general, a non-zero value $V_t(S)$ as a result of market rate variations.

---

[12] The fixed rate that market-makers are willing to pay is smaller than the fixed rate they accept to receive.

[13] Or sometimes the LIBOR zero-curve. We will not adopt the latter terminology (which is in any case little used and vanishing) because it is ambiguous.

## 7.2 Vanilla Swaps

Two valuation methods are possible (and allowed under the regulations governing financial institutions): the "decomposition" method and the "offsetting swap" (or "replacement") method. They are in theory equivalent but may give results that (very slightly) differ in practice as a result of rounding and interpolation.

- **Method 1 (decomposition)**

The replication of the swap using a long position associated with a short position in bullet securities, one with a fixed rate and the other with a floating rate, allows us to write

$$V_t(S_{k-r}) = V_t(A_k) - V_t(A_r) \quad ; \quad V_t(S_{r-k}) = V_t(A_r) - V_t(A_k). \tag{7.4}$$

*Under the decomposition method, the two parts $A_k$ and $A_r$ are valued separately.* $V_t(A_k)$ is obtained by simple discounting of the fixed cash flows generated by $A_k$. The discounting is done with zero-coupon rates extracted from the prevailing curve of fixed rates for "at par" swaps $k_\theta(t)$. $A_k$ may differ noticeably from par if interest rates have varied substantially between 0 and $t$.

*On the contrary, on settlement dates $V_t(A_r)$ is at par and, at any other dates, can differ only slightly from par.* Its value is precisely computed by the methods developed in the previous section ($A_r$ satisfies Propositions 1 and 2 for a forward-looking rate and Proposition 3 for a backward-looking compound overnight rate).

More formally, consider a plain vanilla FL rate swap, with a notional value $M = \$1$. Suppose we are at some date (after the initiation) denoted $t = 0$ without loss of generality and placed between two settlement dates $t_0$ and $t_1$ for the floating leg of the swap ($t_0$ is, therefore, a past date). We describe and value the floating component $A_r$ and the fixed $A_k$ for this swap.

- The floating component $A_r$ generates FL cash flows $D_i\, r_D(t_{i-1})$ at instants $t_1, \ldots, t_N$, the first cash flow $D_1\, r_D(t_0)$ being known and the last cash flow including the notional \$1. The FL benchmark rate $r_D(t)$ is a proportional rate such as an IBOR and $D_i = t_i - t_{i-1}$ coincides with the duration $D$ of the benchmark because of its vanilla feature ($D_i = D$, up to 1 day or 2 days discrepancy).

The value of this floating component is then, as stated in Proposition 1, equal to:

$$V_0(A_r) = b_{t1}(0)\, (1 + D_1\, r_D(t_0)), \tag{7.5}$$

or, if one neglects the notional value, i.e., one discounts the series $L_r$ of floating coupons only:

$$V_0(L_r) = b_{t1}(0)\, (1 + D_1\, r_D(t_0)) - b_{tN}(0). \tag{7.5'}$$

The $b_\theta$ are the prices of zero-coupons extracted from the curve $k_\theta(0)$ of fixed rates for at par swaps. (7.5′) results directly from (7.5) by simple subtraction of the present value of the notional \$1.

Alternatively, one can apply Proposition 2 (in fact Eq. (7.2′)) and "replace" the FL vanilla random cash flows $D_i r_D(t_{i-1})$ by their certainty-equivalents $D_i f_{t_{i-1},D_i}$ (where $f_{t_{i-1},D_i}$ is the forward rate of duration $D_i$ in $t_{i-1}$ periods), these certainty-equivalents being valued by simple discounting: $V_0(L_r) = b_{t_1}(0)D_1 r_{D_1}(t_0) +$

$$\sum_{i=2}^{N} b_{t_i}(0)D_i f_{t_{i-1},D_i}.$$

The example examined in Sect. 7.2.2.3 uses, in particular, this last approach (called the "forward rate" method).

– The fixed leg generates at times $t'_1, \ldots, t'_N$ the fixed cash flows $kD'_i$ at $t'_i$ with $D'_i = t'_i - t'_{i-1}$; these cash flows do not necessarily occur on the same dates as the floating cash flows. The value of the fixed *leg* thus writes:

$$V_0(L_k) = k \sum_{i=1}^{N} b_{t_i}(0)D'_i, \tag{7.6}$$

Equivalently, taking into account the notional amount of \$1 at $t'_N$, the value of the fixed *component* writes:

$$V_0(A_k) = k \sum_{i=1}^{N} b_{t_i}(0)D'_i + b_{t_N}(0). \tag{7.6'}$$

Again, the $b_\theta$ are the prices of zero-coupons extracted from the swap rate curve $k_\theta(0)$.

– The value of the s*wap* is a result of the difference in values of the two legs: $V(L_r) - V(L_k)$ (or $V(A_r) - V(A_k)$) for a $k$-payer swap; $V(L_k) - V(L_r)$ (or $V(A_k) - V(A_r)$) for a $k$-receiver swap.

– **Method 2 (offsetting swap)**

Today's date is again denoted by 0 and we consider a fixed-rate-receiver swap initiated on some past date; $T$ was its initial time to maturity (at initiation) and $T'$ its remaining time to maturity as of today ($T' < T$). The fixed-rate prevailing today for swaps with remaining lifetime $T'$ is $k_{T'}(0)$. To simplify the notation, we set $k' = k_{T'}(0)$.

The swap to be valued usually has a fixed interest rate $k \neq k'$ (in principle $k = k_T$ (as of initiation date)).

One may eliminate the randomness affecting the value of the $k$-receiver swap $S_{k-r}$ by contracting an "offsetting swap" at the current market swap rate $k'$, here a $k'$-

## 7.2 Vanilla Swaps

payer swap $S_{r-k'}$ (with no capital outlay). The resulting position $(S_{k-r} + S_{r-k'})$ generates a series of cash flows $k - k',\ldots, k - k'$ (the two floating legs cancel each other). One then discounts this series (with swap zero-coupon rates), the result being the value of the swap.

For valuing a $k$-payer swap, one discounts the series $k' - k,\ldots, k' - k$.

### 7.2.2.3 Examples of Valuing Swaps

Examples 11, 12, and 13 illustrate the different methods of valuing interest rate swaps.

---

**Example 11 (Vanilla Forward-Looking Rate)**

**A.** On 1 January of year $n$ (date 0) the fixed-rate curve for vanilla swaps quoted in pounds Sterling (£) is flat at 4% (whatever their maturities, the fixed interest rates for swaps contracted on this date are all 4%, which implies that the corresponding zero-coupon rates are all equal to 4%).

Let us consider a fixed-rate receiver swap, with a notional value of £100 million, with 4 years to run, and paying the 3-month forward-looking £-Libor rate, contracted at an earlier date, at the fixed rate of 5%. The FL coupons are all paid quarterly (at the end of the quarter, say 31 March, 30 June, 30 September, and 31 December), and the fixed coupon only once a year (at the year's end, 31 December). The 3-month £-Libor on 30 September of the year $n$-1 (the last reset date) was at 4%.

We consider quarters of 0.25 years, without counting the exact number of days.[14] We calculate the value of this swap on 1 January $n$ by three methods, then its value on 30 December $n$-1.

(a) *Market value of this fixed-rate receiver swap on 1 January n (at $0^+$, just after settlement).*

Decomposition method: $V_0^+(S_{k-r}) = V_0^+(A_k) - V_0^+(A_r)$.
Since $A_r$ is at par, $V_0^+(A_r) = 100$.
Since the zero-coupon rates used to discount the cash flows of $A_k$ are all equal to 4%, we obtain:

$$V_0 + (A_k) = 5 \sum_{i=1}^{4} \frac{1}{(1.04)^i} + 100 \frac{1}{(1.04)^4} = 103.63. \qquad \text{It} \qquad \text{follows} \qquad \text{that}$$

$V_0^+(S_{k-r}) = \text{£}3.63 \text{ M}.$

(continued)

---

[14] In fact, for sterling-denominated instruments, the short rates are usually Actual/365 (and not Actual/360 as for Euro- or Dollar-instruments), which makes more accurate the approximation "duration of the quarter = 0.25."

Offsetting swap method: 4-year swaps are traded on 1 January $n$, on the basis of a fixed rate of 4% and since the swap receives a fixed rate of 5%, we have:

$$V_0 + (S_{k-r}) = (5-4) \sum_{i=1}^{4} \frac{1}{(1.04)^i} = £3.63 \text{ M}.$$

Forward rate method (a variant of the decomposition method): since the rate curve is flat, the forward rates are all equal to 4% and we can check that $V_0^+(A_r) =$

$$4 \sum_{i=1}^{4} \frac{1}{(1.04)^i} + 100 \frac{1}{(1.04)^4} = 100 \text{ while } V_0^+(A_k) = 103.63. \text{ The value of the}$$

swap thus is £3.63 M, as with the previous two methods.

(b) *Market value of this fixed-rate receiver swap on 30 December n-1 (at $0^-$, just before settlement).*

It suffices to add to the value at $0^+$ of each component the corresponding coupon (annual for the fixed leg, quarterly for the variable leg):
$V_0^-(A_k) = 103.63 + 5 = 108.63$; $V_0^-(A_r) = 100 + 0.25 \times 4 = 101$; therefore $V_0^-(S_{k-r}) = £7.63$ M.

**B.** One year later, on 1 January of $n + 1$, the swap fixed-rate curve is as follows:

| 1 year | 2 years | 3 years | 4 years |
|--------|---------|---------|---------|
| 6% | 6.5% | 7% | 7.2% |

We compute first the zero-coupon interest rates for 1 year, 2 years, and 3 years implicit in this curve, and then the value of the swap X described under part A (3-month £-Libor against 5%), which now has 3 remaining years to live.

(a) *Calculation of prices ($b_\theta$) and ZC rates ($r_\theta$) from the swap rate curve $k_\theta(0)$* (see Chap. 6 for an exposition of this method):

$$r_1 = k_1 = 6\%, \text{ thus } b_1 = \frac{1}{1.06} = 0.9434.$$

Knowing that a 2-year security with coupon $k_2$ is at par, we get:
$1 = k_2 b_1 + (1 + k_2) b_2 = 0.065 \times 0.9434 + 1.065 \times b_2$, that is, $b_2 = 0.8814$ so that $r_2 = 6.516\%$.

Write down that a 3-year security with coupon $k_3$ is at par:
$1 = k_3 b_1 + k_3 b_2 + (1 + k_3) b_3$.

(continued)

## 7.2 Vanilla Swaps

From this we have $b_3 = \frac{1 - 0.070 \times 0.9434 - 0.070 \times 0.8814}{1.070} = 0.8152$ and $r_3 = 7.048\%$.

(b) *Value of the swap on 1 January n + 1*: Use the numerical values of $b_1$, $b_2$, and $b_3$ found in part a).

By decomposition: $V_0^+(A_r) = 100$; $V_0^+(A_k) = 5 \times (b_1 + b_2) + 105 \times b_3 = 94.72$; $V_0^+(S_{k-r}) = -\pounds 5.28$ M.

By an offsetting swap: $V_0^+(S_{k-r}) = (5 - 7) \times (b_1 + b_2 + b_3) = -\pounds 5.28$ M.

---

**Example 12 (Vanilla Forward-Looking Floating Leg)**

The zero-coupon rates $(r_\theta)$ extracted from the swap rate curve is as follows on 1 April of year $n$:

| 3 months | 6 months | 9 months | 12 months | 15 months |
|---|---|---|---|---|
| 5% | 5.30% | 5.40% | 5.50% | 5.60% |

Consider the date 1 April $n$, and a swap of the 6-month Libor rate for a fixed rate of 6% (fixed-rate receiver) with bi-annual settlements for the floating leg and annual for the fixed leg, contracted on some past date. This swap has 15 months to run. The three settlement dates ahead on the variable leg are 1 July $n$, 1 January $n + 1$ and 1 July $n + 1$ and the annual settlement dates on the fixed leg to come are 1 July $n$ and 1 July $n + 1$; the nominal amount of the swap is \$100 M. The 6-month Libor rate was 4% on 1 January $n$ (the date of the preceding reset). For the sake of simplicity, we do not count the exact number of days and consider that the inter-settlement periods of the variable leg are 0.5 year long, and that quarters are 0.25 years long.

We calculate the *semi-annual* forward interest rates: $f_{3,6}$ (6 months in 3 months) and $f_{9,6}$ (6 months in 9 months):

$$(1 + f_{3,6}) = \frac{(1+r_9)^{0.75}}{(1+r_3)^{0.25}}, \text{ that is, } f_{3,6} = 2.762\%; \quad (1 + f_{9,6}) = \frac{(1+r_{15})^{1.25}}{(1+r_9)^{0.75}}, \text{ that is,}$$

$f_{9,6} = 2.91\%$.

Then we calculate the swap's value using two methods.

Forward rate method:

Value of the fixed leg: $V(L_k) = 6\left(\frac{1}{(1.05)^{0.25}} + \frac{1}{(1.056)^{1.25}}\right) = \$11.89$ M.

Value of the variable leg: $V(L_r) = 2\frac{1}{(1.05)^{0.25}} + \left(\frac{2.762}{(1.054)^{0.75}} + \frac{2.91}{(1.056)^{1.25}}\right) =$ \$7.35 M.

(continued)

In the last calculation, we discount the first known coupon $(0.5 \times 4\% \times 100 = 2)$ and the certainty-equivalents of the following ones (semiannual forward rates $\times$ 100).

Value of the swap $= V(L_k) - V(L_r) = 11.89 - 7.35 = \$4.54$ M.

Decomposition method:

$$V(A_k) = 6\left(\frac{1}{(1.05)^{0.25}} + \frac{1}{(1.056)^{1.25}}\right) + \frac{100}{(1.056)^{1.25}} = \$105.30 \; ; \; V(A_r) = \frac{102}{(1.05)^{0.25}} = \$100.76.$$

Value of the swap $= V(A_k) - V(A_r) = 105.30 - 100.76 = \$4.54$ M.

We recover the value calculated by the preceding method.

---

### Example 13 Overnight-Indexed-Swap

Consider a one-year swap, of notional value $M$, issued on 1 June of year $n$, whose floating leg pays a single coupon, on 1 July of year $n$+1, indexed on an OIS-index $I$. The fixed leg pays also a single coupon of 2% on 1 July of year $n$ +1. The cash flow at maturity on 1 July $n + 1$, for the floating component is:

$$A_{f,n+1} = M \times C(1/7/n, 1/7/n + 1)$$

Recall that $C(1/7/n, 1/7/n + 1)$ stands for the compound interest on a \$1 investment on $1/7/n$, rolled over overnight till the $1/7/n + 1$.

At origination, the swap value was 0 (the fixed rate was set at 2% to obtain a zero value). Assume that today, at the exact middle of the swap life (the $30/12/n$), the fixed rate for OIS swaps of 6 months to maturity is 3% while the value of the compound overnight interests (OIS index) over the second semester of year $n$ is: $C(1/7/n, 30/12/n) = 1.018$. The problem is to value such a swap on date $30/12/n$.

Let us apply first the decomposition method. The future (and random) floating cash flow will be:

$$\begin{aligned} A_{f,n+1} &= M \times C(1/7/n, 1/7/n + 1) \\ &= M \times C(1/7/n, 30/12/n) \, C(30/12/n, 1/7/n + 1) \\ &= M \times 1.018 \times C(30/12/n, 1/7/n + 1). \end{aligned}$$

The term $C(30/12/n, 1/7/n + 1)$ concerning the future 6 month-period is random but can be obtained from an initial \$1 investment, by a roll-over of overnight investments during 6 months, and hence is worth \$1. Consequently, today's value of the floating component is equal to $M \times 1.018$.

The fixed component will yield 6 months from now the certain cash flow $M \times 1.02$, hence its value today is obtained by simple discounting. The

(continued)

## 7.2 Vanilla Swaps

relevant proportional discount rate is the 6-month OIS fixed-rate swap, namely 3% (1.5% over 6 months).

Hence, today's value of the fixed component is equal to $M \times 1.02/1.015$.

From the viewpoint of the fixed-rate payer, this OIS swap is worth: $M \times (1.018 - 1.02/1.015) \# M \times 0.0131$.

Let us apply now the alternative method (offsetting swap) which implies adding today to the considered fixed-rate payer OIS swap a fixed-rate receiver OIS of the same time to maturity and of a notional amount equal to $M \times 1.018$. This new swap having a zero initial value implies, when traded, a zero cash flow. We obtain the following cash flows on date $1/7/n + 1$:

Old fixed-rate payer swap: $M \times 1.018 \times C(30/12/n, 1/7/n + 1) - M \times 1.02$.
New fixed-rate receiver swap: $-M \times 1.018 \times C(30/12/n, 1/7/n + 1) + M \times 1.018 \times 1.015$.
Net cash flow: $0 + M \times (1.018 \times 1.015 - 1.02)$.

The present value of this certain cash flow is the swap value: $M (1.018 - 1.02/1.015) \# M \times 0.0133$, roughly the same result as with the decomposition method.

### 7.2.3 Interest Rate, Counterparty and Credit Risks for an Interest Rate Swap

Fixed-for-floating rate swaps are always affected by a significant interest rate risk (Sect. 7.2.3.1), most often by a small counterparty risk (Sect. 7.2.3.2), and by a credit spread risk when the reference is not a risk-free rate (Sect. 7.2.3.3).

#### 7.2.3.1 Interest Rate Risk: Modified Durations for a Fixed-for-Floating Interest Rate Swap

We will first examine the consequences of a parallel shift and then of a non-parallel variation in the yield curve.

(a) *Simple variation; consequences of a parallel shift of the yield curve*

The equation $V(S_{k-r}) = V(A_k) - V(A_r)$ implies that the variation in the *swap* value is the difference of the variations of its two components:

$$\frac{dV(S_{k-r})}{dr} = \frac{dV(A_k)}{dr} - \frac{dV(A_r)}{dr} = V(A_r) Sens_{Ar} - V(A_k) Sens_{Ak}$$

where $Sens_{Ar}$ is the (small) sensitivity (modified duration) of the floating component and $Sens_{Ak}$ that of the fixed component (which is much larger).

242          7   Vanilla Floating Rate Instruments and Swaps

This last equation allows us to evaluate the consequences of a parallel shift of the yield curve.

---

### Example 14 Valuing and Assessing Interest Rate Risk for a Vanilla Interest Rate Swap

Let us consider a fixed-rate receiver swap, 4 years to maturity on 1 October $n-1$, whose notional value is $100 M and whose variable leg is defined in the preceding example of an FL rate (semi-annual coupon referenced to the *6-month Libor* on 1 April and 1 October) and whose fixed leg gives rise to an *annual* coupon of 5% on $100 M paid on 1 October $n$; 1 October $n + 1$; 1 October $n + 2$ and 1 October $n + 3$.

The reset dates of the variable leg are as follows:

| Date of $k$ reset: | 01 Oct $n-1$ | 01 Apr $n$ | 01 Oct $n$ | 01 Apr $n + 1$ | 01 Oct $n + 1$ |
|---|---|---|---|---|---|
| Rate (flat curve) | 5.5% | 6% | 5.5% | 5.5% | 5% |
| $Nd(k-1, k)$ | – | 182 days | 183 days | 182 days | 183 days |

#### Questions

(i). On 1 October $n + 1$, the yield curve is flat and all swap rates equal 5% (in particular the yield-to-maturity of 4-year Treasury Bills as well as the fixed interest rates for 4-year *plain vanilla* swaps). What is the market value of this swap on 1 October $n + 1$ (after payment of the coupon)?

(ii). On 31 March $n$, the yield curve was flat and all rates were equal to 6%. What was the market value of this swap on the evening of 31 March $n$ (just before payment of the variable coupon) given by the decomposition method?

(iii). On 1 October $n$, all interest rates are 5.5%. Calculate the swap's market value on the evening of 1 October $n$ (after the coupon payment) using two methods (decomposition and offsetting swap).

(iv). On 30 December $n-1$, the 3-month Libor rate as well as all other interest rates are 5%. What is the effect of a uniform rate increase of 0.2% on 30 December $n-1$ on the market value of this swap?

#### Answers

(i). On 1 October $n + 1$, the market value of the swap $= 0$ since both the fixed and the floating components are at par.

(ii). On 31 March $n$, the value of the fixed-rate receiver swap calculated by decomposition is:

(continued)

$$V(S_{k-r}) = V(A_k) - V(A_r).$$

The variable coupon on 1 April $n = C(n) = 5.5\% \times 100,000 \times 182/360 = \$2780.5$ K

$$V(A_r) = 100,000 + C(n) = \$102,780.5 \text{ K}$$

$$V(A_k) = 100\,000(1.06)^{1/2}\left(\sum_{i=1}^{4} \frac{0.05}{(1.06)^i} + \frac{1}{(1.06)^4}\right) = \$99,388.8 \text{ K}$$

Therefore $V(S_{k-r}) = -\$3391.7$ K.

(iii). On 1 October n (after coupons) the swap's value calculated in two ways is:

*Method 1 (decomposition)*

$$V(A_r) = \$100,000 \text{ K}$$

$$V(A_k) = 100\,000\left(\sum_{i=1}^{3} \frac{0.05}{(1.055)^i} + \frac{1}{(1.055)^3}\right) = \$98,651.033 \text{ K}$$

Therefore $V(S_{k-r}) = -\$1348.966$ K.

*Method 2 (offsetting swap)*

$$V(S_{k-r}) = -100\,000\left(\sum_{i=1}^{3} \frac{0.005}{(1.055)^i}\right) = -\$1,348.966 \text{ K}.$$

(iv). Let us start from the following formulas:

$$dV(S_{k-r})/dr = dV(A_k)/dr - dV(A_r)/dr$$

$$A_r = F/(1 + rd), \text{ whence } dV(A_r)/dr = -F\,d/(1 + rd)^2,$$

where $F$ stands for the (known) value of $A_r$ at the next reset (100 + next coupon) and $d$ is the fraction of year remaining to the next reset.

On 30 December $n-1$, $F = 102,780.5$, $d = 0.25$, and all rates are 5%. As a result, a variation of $\Delta r = 0.2\%$ implies, to the first order

(continued)

$$\Delta V(A_r) = -\Delta r \times F \times d/(1+rd)^2$$
$$= -0.002 \times 102,780.5 \times 0.25/(1+0.25 \times 0.05)^2 = -\$50.129 \text{ K}.$$

$V(A_k) = \left( \sum\limits_{i=1}^{4} \frac{F_i}{(1+r)^{i-0.25}} \right)$, whence $dV(A_r)/dr = -\left( \sum\limits_{i=1}^{4} \frac{(i-0.25)F_i}{(1+r)^{i+0.75}} \right)$. To the first order:

$$\Delta V(A_k) = \Delta r \left[dV(A_k)/dr\right] = -0.002 \times 334,845.004 = -\$669.69 \text{ K}.$$

Thus, $\Delta r = 0.2\%$ implies that $\Delta V(S_{k-r}) = -\$669.69$ K+ \$50.129 K $= -\$619.56$ K.

### (b) *Non-parallel shift of the yield curve*

The methods explained in Chap. 5, which allow calculating, from multi-factor variations, the risk from variations in the level, the slope, and more generally the deformation of the yield curve, can be applied to swaps.

Let us consider, for example, a fixed-rate payer swap whose value at any time is

$$V(S_{k-r}) = V(A_k) - V(A_r).$$

The value of the fixed leg is determined by a long-term yield while that of the variable leg depends on a short-term interest rate:

$$V(A_k) = f_1(l); V(A_r) = f_2(r) \text{ where } l = \text{long} - \text{term yield and } r = \text{short} - \text{term rate};$$

$dl$ affects $A_k$ (strongly) while $dr$ affects $A_r$ (weakly):

$$\frac{\partial V(S_{k-r})}{\partial l} = \frac{\partial f_1(l)}{\partial l} < 0 \quad ; \quad \frac{\partial V(S_{k-r})}{\partial r} = \frac{\partial f_2(r)}{\partial r} > 0 \ .$$

From this, a variation $(\Delta l, \ \Delta r)$ of both long and short rates has an impact on the value of the swap given approximately by the formula:

$$\Delta V(S_{k-r}) = \frac{\partial f_1(l)}{\partial l} \Delta l + \frac{\partial f_2(r)}{\partial r} \Delta r.$$

This formula allows us to estimate the impact:

– Of a variation in the long-term yield while the short-term rate remains constant ($\Delta l \neq 0$, $\Delta r = 0$, i.e., a change in the slope of the curve).
– Of a variation in the short-term rate while the long-term remains constant t ($\Delta r \neq 0$, $\Delta l = 0$, i.e., a change in the slope of the curve).

## 7.2 Vanilla Swaps

- Of a parallel shift of the yield curve ($\Delta l = \Delta r$).
- Or of any combination ($\Delta l$, $\Delta r$).

The interest rate risk affecting a fixed-rate receiver swap can be analyzed similarly.

A more precise analysis would involve a model of the zero-coupon rate curve and valuation of $A_k$ and $A_r$ using discounting based on the zero-coupon swap curve.

### 7.2.3.2 Intermediation and Counterparty Risk on Interest Rate Swaps

Suppose that one party of a swap (rated BBB for example) is affected by counterparty risk while the other (rated AA+) is almost free from it. The AA + party who bears the risk of default on the part of BBB is compensated for by a spread: the fixed rate that AA + will accept to give (bid) is smaller, and that which she will demand to receive (ask) will be higher, the greater the counterparty risk she is exposed to. It is important to note that the counterparty risk of a swap does not involve the principal (notional), but *only the market value of the difference in interests*. The spread associated with counterparty risk in swaps then is obviously much smaller than that which would prevail for classic lending and borrowing involving the same counterparties.

In fact, for swaps between prime banks of similar ranking (often around A+ or AA-) this counterparty risk premium on the fixed rate applied to the benefit of the less risky party can be assumed nil for practical purposes or most often borne by a third party, a *financial intermediary*: Actually, the two parties usually do not make the transaction directly but deal through an intermediary. This swap dealer enters into two offsetting swaps with two independent contracts and earns a few basis points partly for taking a counterparty risk on each of these two contracts. Besides, a provision in the swap contract may specify how the transaction is unwound if one of the parties defaults. The most standard consists in presuming the swap is in default as soon as one contracting party defaults on a payment. A possible clearing (called netting) with other positions involving the same counterparty is often provided for in the swap agreement. Netting decreases the already small counterparty risk affecting swaps.

Finally, the swap market, as most OTC derivative markets, has experienced a major evolution since the 2007–2009 financial crisis aiming at eliminating counterparty risks. On pre-crisis times, the two parties of a swap could agree on any clearance or settlement procedure and on the collateral to be posted. In response to the crisis, the OTC derivatives markets, the swap markets, in particular, have been increasingly regulated and the market participants have much less flexibility. First, standard swaps between two financial entities must now be cleared through a central counterparty clearing house (CCP) which takes the counterparty risk and provides clearing and settlement services. These CCP's are appropriately capitalized and regulated to avoid potential systemic risk. Moreover, the CCP reduces its own counterparty risk by netting procedures and by requiring from its counterparties initial margins (guarantee deposits), as well as variation margins (margin calls) in response to price variations.

Second, for nonstandard transactions which may continue to be cleared bilaterally, collateral must be posted and transactions between financial entities are subject to initial and variation margins. Chaps. 9 and 29 will provide more details on this recent evolution toward more regulated and organized OTC markets.

### 7.2.3.3 Credit Spread Risk on the Reference Rate and the LIBOR-OIS Spread

The second type of risk premium often carried by swap rates reflects the *credit risk* borne by the variable leg reference (an IBOR, etc.) due to the fact that this reference is not really riskless. This *risk premium on the variable leg reference implies a risk premium on the swap rate of the fixed leg* (receiving more on the floating rate implies paying more on the fixed rate).

The credit risk premium is smaller the shorter is the duration of the benchmark. It is important to understand why it is riskier to lend for 3 months, without collateral, to a bank rated, say, $AA^-$ (at the 3-month LIBOR), than to lend to several $AA^-$ banks overnight during 3 months. The OIS variable leg for instance can be considered as a series of overnight loans to financial institutions that are rated $A+/AA^-$ *at the start of each loan*. This is the reason why these overnight floating rates are sometimes referred to as "continually-refreshed" $AA^-$ rates.[15] In the case of such a roll-over of continuously refreshed loans to several $AA^-$ banks, the outstanding debt is $AA^-$ *during the whole duration of the roll-over* (one day is not long enough for the current $n^{th}$ debtor to lose their solvency), while a single $AA^-$ debtor can well go bankrupt within a few months or even weeks, especially during a financial turmoil, as it has been experienced during the 2008 crisis. This is the reason why a stream of $AA^-$ overnight loans continuously refreshed during, say, 3 months can arguably be considered quasi riskless, while a 3-month $AA^-$ loan (yielding the 3-month LIBOR), for instance, is definitely not risk-free. A similar explanation was presented in Sects. 7.2.1–7.2.4.4 when discussing the credit spread of a floating rate instrument.

This discussion helps to understand the meaning and the dynamics of the so-called LIBOR-OIS spread defined as the difference between the 3-month LIBOR and the 3-month OIS swap rate. This spread reflects the difference between the credit risk in a 3-month unsecured loan to a bank of "good" credit quality and the credit risk in continually-refreshed one-day unsecured loans to several banks of "good" quality also. The spread prevails because, as just explained, the LIBOR is not risk-free while the compound overnight rate is quasi risk-free. More precisely, the difference between a 3-month LIBOR and a 3-month OIS swap rate reflects the risk of decline of the LIBOR borrower's credit worthiness over three months.

In normal market conditions, the LIBOR-OIS spread is minimal, hovering around 10–15 basis points. However, during the 2008 crisis, it briefly skyrocketed to a

---

[15]Collin-Dufresne and Solnik (2001) were the first to introduce this notion. For an analysis of the Libor-OIS spread, see for instance Hull and White (2003).

record of 364 basis points in October. Since then, it ebbed and flowed, increasing substantially each time banks began to worry about each other's solvency.

The LIBOR-OIS spread is considered a key measure of credit risk within the banking sector and OIS swap rates are preferred to LIBORs as benchmarks for risk-free rates.[16] Moreover, some overnight rates such as the SOFR are relative to *secured* transactions (lending-borrowing overnight in the US Treasury repo market) on which a really risk-free OIS yield curve can be constructed.

A further analysis of default risk affecting swaps will be carried out in the last part of this book, notably in Chaps. 28 (Sects. 1 and 3.3) and 30 (in particular, the Appendix).

### 7.2.4 Summary of the Various Types of Swaps

A wide variety of swaps is traded on the market and the list that follows is not exhaustive. The market trades long-term (> 1 year) as well as short-term swaps (< 1 year), swaps involving a single currency or two currencies, and swaps with one or two variable legs indexed to a large variety of floating rates. We first describe the main types of fixed-for-floating rate swaps, then consider currency swaps and basis swaps, and finally provide some examples of non-standard or second-generation swaps. The description, analysis, and valuation of some non-standard swaps will be taken up again in Chaps. 16 and 17 using probabilistic methods.

#### 7.2.4.1 Fixed-for-Floating Interest Rate Swaps

We distinguish those which are referenced to a short-term, money-market rate from those referenced to a long-term interest rate.

(a) *Swaps referenced to a money-market interest rate*

Let us briefly describe the *plain vanilla* swaps whose valuation and interest rate risk involve the methods developed above (Propositions 1 and 3).

– *Overnight Indexed Swaps (OIS, backward-looking rate)*

The variable leg is referenced to the previously defined OIS index (capitalized EONIA/€STER, SONIA, SOFR, or Federal-funds rate). In theory, as we have seen, the backward-looking variable leg can be replicated by a roll-over of overnight transactions and the OIS-index belongs to the BL rate family.

Given its flexibility, such a reference allows covering any period, any starting date, and any duration, even a very short one. This is one of the reasons why OIS swaps often have relatively short lives (less than a year, even three months or less).

---

[16]This is not entirely independent of the Libor rigging scandal uncovered by the *Financial Times* on July 27, 2012.

However, OIS swaps as long as 5–10 years are also common. Swaps of one year or less deliver generally a single payment at maturity, while longer-term OIS swaps have often quarterly settlements.

OIS swaps are massively traded plain vanilla BL rate instruments and the OIS-rate curves (the $k_\theta$) are now the benchmarks for risk-free interest rates for many maturities and currencies, as explained in Sect. 7.2.3.2 above.

*– Euribor or Libor rates*

The variable leg is indexed to some Euribor $E_x$ or Libor $L_x$. It could be an FL rate or, less frequently, a BL rate (capitalized $E_xC$ or $L_xC$). In both cases, these swaps are replicable.

LIBOR referenced swaps will progressively disappear, being replaced by OIS swaps and perhaps by some more reliable new forward-looking references such as the fixed rate $k_\theta(t)$ of a plain vanilla swap with a fixed maturity $\theta$.

(b) *Swaps referenced to a long-term interest rate; constant maturity swaps*

In the case of *constant maturity swaps* (CMS) the floating leg gives rise to payments of coupons indexed against a government bond rate of given maturity or, more frequently, the fixed interest rate $k_\theta(t)$ of a plain vanilla swap with a fixed maturity $\theta$ that goes beyond the swap's reset period, like for instance the 5-year swap rate. These swaps are not plain vanilla because the variable leg cannot be replicated. Their analysis requires methods discussed in Chaps. 16 and 17.

CMS can be used to speculate on, or hedge against, a change in the slope of the yield curve. For instance, when a firm anticipates that the yield curve will become steeper, it may enter into a CMS paying a short-term compound overnight rate and receiving the five-year swap rate.

### 7.2.4.2 Currency Swaps

A currency swap can be defined as lending denominated in a given currency associated with borrowing for the same duration but denominated in a different currency, one being at a fixed rate and the other at a floating rate (*currency-interest swap*), or both at fixed rates (*currency swap*).

In both cases there is then:

- An exchange of interests denominated in the currencies involved, the variable leg having most often been referenced to a Libor;
- A transfer of principals. Initially, the two amounts exchanged have the same value since the parties use the spot exchange rate between the two currencies prevailing at that moment (for this reason, usually the transfer does not physically take place). At maturity, the principals are again exchanged, but in a reverse sense as for the reimbursement of any loan. There the transfer *almost always* takes place since, as the spot exchange rate has changed, the nominal values of the two legs differ. This implies that one party has a currency conversion gain and the other a

## 7.2 Vanilla Swaps

loss. Seen from its origination, such a swap involves, in addition to an interest rate risk, a (minor) exchange rate risk on the leg with interests denominated in foreign currency and a (major) exchange rate risk on the principal. The latter risk obviously increases with the volatility of the exchange rate and the duration of the swap.[17]

The market value of a swap is nil initially. In the case of the exchange of a fixed interest rate for a floating one (*currency-interest swap*), the variable leg is most often plain vanilla and satisfies Proposition 1 or 3, according to whether it involves an FL or a BL rate. The fixed leg is valued by usual discounting using the yield curve for the relevant currency.

In the case of an exchange of a fixed interest rate for another fixed one (*currency swap*), each of the two fixed rates is based on the rate that prevails for the relevant currency. The two fixed legs can be valued at any date in their respective currency by discounting using the proper yield curve.

---

**Example 15 Currency Swap (Currency-Interest Swap)**

An exchange of €90 M for $100 M (an exchange rate of 0.90 € for $1 is assumed to hold at initiation, $t = 0$), for 12 months, as well as the interests accruing to these two principals calculated, respectively, with a fixed rate of 2% semannually for euros and the 6-month FL $-Libor for dollars, paid semi-annually. Denote by 0, 1, and 2, respectively, the initial instant, the first settlement date, and the second (and last) settlement date (separated by a half-year).

At instant 0, the 6-month $-Libor rate is 2.2% semi-annually. At instants 1 and 2, the interest rate and the exchange rate are as shown in the table below:

|  | Instant 0 | Instant 1 | Instant 2 |
|---|---|---|---|
| 6-month $-LIBOR semi-annual | 2.2% | 2.4% | 2.5% |
| Exchange rate: € against $ | 0.90 | 0.92 | 0.94 |

The cash flows generated by this swap thus are as follows:

| 0 | 1 | 2 | |
|---|---|---|---|
| Principal | Interests | Interests | Principal |
| − 90 M€ | 1.8 M€ | 1.8 M€ | + 90 M€ |
| + 100 M $ | −100 × 2.2% = − 2.2 M$ | −100 × 2.4% = − 2.4 M$ | − 100 M$ (= −94 M €) |

---

[17] The existence of this exchange rate risk on the principal is the reason why, in some cases, the principals are not exchanged at all and thus are purely notional.

> Remark that the increase in the \$–Libor as well as the fall of the euro against the dollar both lead to a loss for the holder of this swap.

### 7.2.4.3 Basis Swaps (Floating-for-Floating)

– A *basis swap* is an interest rate swap or a currency swap involving two distinct floating interest rates.

– A *basis swap* of interest rates on a *single* currency will consist, for example, in the exchange of the 3-month Libor rate for the 6-month Libor rate. In other cases, a long-term floating rate is swapped for a short-term BL or FL floating rate. For example, a bond rate or a *swap* rate $k_\theta(t)$ (for a fixed $\theta$) is exchanged for the 3-month Euribor$(t)$ + *spread*. The *spread is the subject of negotiation*. In the first example ($L_{3months}$ for $L_{6months}$) the two legs are plain vanilla and the equilibrium spread is theoretically nil and in practice small.[18] In the second case (long rate for $E_{3months}$) the leg referenced to the long rate is not vanilla and the spread can in theory and in practice be substantial.

These interest rate basis swaps are also termed *yield curve swaps*.

– A *currency basis swap* allows, for example, exchanging a Euribor-based index for a dollar-Libor-based index. As in the case of currency swaps, there is most often a transfer of capital at the end of the contract.

We leave as an exercise to the reader the proof that a fixed-to-floating rate currency swap can be constructed from an interest rate swap (fixed rate – floating rate) and a currency basis swap.

### 7.2.4.4 Nonstandard Swaps

There exists a large variety of more or less sophisticated swaps. We limit the list here to several typical examples as we resume the analysis in Chap. 16, Section 2.

– Forward swaps: their fixed interest rate is determined on the initiation date 0 but they begin on a later date $t$. If the variable leg is plain vanilla, the methods and results valid for plain vanilla instruments also apply (they have zero value at 0, the variable leg is at par on a coupon date, and so on). Forward swaps are studied in Chap. 16.

– Amortization (step-down) swaps: the principal $M$ (or nominal or notional value of the swap) which is used as the basis for computing the floating and fixed interest

---

[18] In absence of a difference in the two counterparty risks (affecting the two legs), the spread should be nil.

## 7.3 Summary

rates is amortized according to a predetermined schedule. If the variable leg is plain vanilla (i.e., if it satisfies the criteria given in Section 7.1, even if the capital is amortized), it is subject to methods previously explained (notably they have zero value at 0, and the variable leg, appropriately defined, is at par on a coupon date). These swaps will also be analyzed in Chap. 16.

- Libor in arrears swaps: the coupon on the FL leg is paid on the basis of a Libor (or Euribor) rate in arrears, at the end of a period, rather than in advance: also the cash flow transferred at $t_i$ is $D_i\, r(t_i)$ (and not $D_i\, r(t_{i-1})$ as in the vanilla case). We show in Chap. 16 how an adjustment of the forward interest rate allows valuing such swaps.
- Zero-coupon swaps: one of the two legs (generally the fixed leg) gives rise to a single payment; the valuation method is the same as in the vanilla case if the variable leg can be replicated.
- Step up or down swaps: the fixed interest rate is not constant but rises (step up) or falls (step down) in a predetermined manner. If the variable leg can be replicated, these swaps are valued as vanilla swaps.
- Index and equity swaps (or *equity-linked swaps,* ELS).
- In an index swap, two agents agree to pay (receive) on fixed dates and during a given period the interest on a loan and to receive (pay) the return on a stock market index. This return is paid (received) if it is negative. The main use of the index swap is to replicate an index fund.
- More generally, ELS (*equity-linked swaps*) involve, in at least one of their two legs, the total return of a stock, a basket of stocks, or a stock index.
- *Quanto swaps* (or differential swaps, "*diff swaps*" for short): the payments of the two legs are denominated in the same currency but the two benchmark interest rates (from which are calculated the interests paid by the two legs) are relative to two different currencies. An example would be the exchange of the 3-month Euribor for the 6-month £-Libor, with a notional amount in euros, or in sterling, in dollars or yens.

---

## 7.3  Summary

- The term "floating" refers to both forward-looking and backward-looking rates.
- In the case of *forward-looking* rates the coupon is reset each time an interest payment is made, and then remains unchanged until the next coupon payment date: in this case, the coupon is known at the start of the period and the coupon is *predetermined*.
- In the case of *backward-looking* rates, the floater resets more frequently than the payments, as often as daily: the coupon is unknown at the start of the period and thus is *post-determined*.
- Many references are used to compute floating coupons, such as Libor, Euribor, or compound overnight rates such as €STER, SOFR, or SONIA (OIS indices).
- The analysis and valuation of floating-rate assets consist in replicating, when possible, the floating-rate instrument by a *roll-over* of short-term lending.

- In absence of credit risk, replicable, floating-rate assets are quoted *at par on the coupon date.* Between the coupon dates, a replicable asset may somewhat stray from par but OIS indices remain at par.
- In presence of credit risk, when the margin does not match the current credit risk, the floating rate asset strays from par.
- The *certainty equivalent* of a replicable vanilla stream of predetermined (forward-looking) coupons is a stream of coupons computed with the appropriate current forward rates.
- A *fixed-rate receiver* (or *floating-rate payer*) *swap* pays the floating interest rate and receives the fixed rate. A *fixed-rate payer* (or *floating-rate receiver*) *swap* pays the fixed interest rate and receives the floating rate.
- A *fixed-rate payer* swap benefits from interest rate hikes, while a *fixed-rate receiver* swap has the reverse quality. Both instruments can thus be used to manage interest rate risk. A comparative advantage due to a quality spread differential (QSD) may also allow to lower the financing costs of the two parties.
- A *fixed-rate receiver* swap is replicated by lending at a fixed interest rate loan coupled with borrowing at a floating rate, these two positions being called the two legs of the swap. Symmetrically, *a fixed-rate payer* swap is replicated by borrowing at a fixed interest rate with lending at a floating rate.
- In general, the value of a swap is equal to the difference of the value of its two legs.
- At the instant the swap is initiated, its *market value should be zero* so that the capital outlay for each party is zero. Then, at initiation, the floating leg being at par (vanilla case), the fixed leg should also be at par.
- Consequently, the fixed rate should be equal to the yield to maturity of risk-free bullet securities (with the same maturity as the swap) prevailing in the market at the initiation time.
- In practice, discrepancies, due to credit risk and liquidity, may prevail between the *term structure of fixed rates of vanilla swaps and the term structure* of bullet securities. The OIS fixed swap rates term structure is a relevant benchmark for risk-free rates.
- At any time after its initiation, a swap has, in general, a *non-zero value* as a result of variation in market interest rates to which mainly the fixed leg is sensitive.
- A *currency swap* can be defined as a loan associated with borrowing for the same duration but denominated in a different currency, one being at a fixed rate and the other at a floating rate (*currency-interest swap*), or both at fixed rates (*currency swap*).

## Appendix

### Proof of the Equivalence Between Eq. (7.2′) and Proposition 1

Let us start with eq. (7.2'): $V = b_{t_1}(0)D_1 r_{D_1}(t_0) + \sum_{i=2}^{N} b_{t_i}(0)D_i f_{t_{i-1},D_i} + b_{tN}(0).$

where $t_0 = t_1 - D_1$ denotes the past date of the preceding reset and, more generally, $t_{i-1} = t_i - D_i$ .

First notice that $D_i f_{t_{i-1}, D_i} = \frac{b_{t_i - D_i}}{b_{t_i}} - 1 = \frac{b_{t_{i-1}}}{b_{t_i}} - 1 = \frac{b_{t_{i-1}} - b_{t_i}}{b_{t_i}}$ . Replacing in eq. (7.2') $D_i f_{ti-1, Di}$ by $\frac{b_{t_{i-1}}(0) - b_{t_i}(0)}{b_{t_i}(0)}$, we obtain:

$$V = b_{t_1}(0) D_1 r_{D1}(t_0) + \sum_{i=2}^{N} b_{t_i}(0) \frac{b_{t_{i-1}}(0) - b_{t_i}(0)}{b_{t_i}(0)} + b_{t_N}(0)$$

$$= b_{t_1}(0) D_1 r_{D_1}(t_0) + \sum_{i=2}^{N} [b_{t_{i-1}}(0) - b_{t_i}(0)] + b_{t_N}(0) \; ;$$

$$= b_{t_1}(0) D_1 r_{D1}(t_0) + b_{t_1}(0),$$

whence $V = b_{t1}(0) (1 + C_1)$, where $C_1 = D_1 r_{D1}(t_0)$ is the next expected coupon (a known quantity), in agreement with Proposition 1-(ii): The two methods, therefore, lead to the same valuation.

## Suggested Reading

### Books

Brown, K. C., & Smith, D. J. (2005). *Interest rates and currency swaps: A tutorial.* The Research Foundation of AIMR.
Corb, H. (2012). *Interest rate swaps and other derivatives.* Columbia University Press.
Hull, J. (2018). *Options, futures and other derivatives* (10th ed.). Prentice Hall Pearson Education.
Kolb, R. (2007). *Futures, options and swaps* (5th ed.). Blackwell.

### Articles

Artzner, P., & Delbaen, F. (1990). 'Finem Lauda' or the risks in swaps. *Insurance: Mathematics and Economics, 9*, 295–303.
Collin-Dufresne, P., & Solnik, B. (2001). On the term structure of default Premia in the swap and Libor markets. *The Journal of Finance, 56*(3), 1095–1115.
Hull, J., & White, A. (2013). LIBOR vs. OIS: The derivatives discounting dilemma. *Journal of Investment Management, 11*(3), 14–27.
Hull, J., & White, A. (2016). Multi-curve modeling using trees. In *Innovations in derivative markets* (Proceedings in mathematics and statistics) (Vol. 165, pp. 171–189). Springer.
Litzenberger, R. H. (1992). *Swaps* plain and fanciful. *Journal of Finance, 47*(3), 831–850.
Wall, L. D., & Pringle, J. J. (1989). Alternative explanations of interest swaps: A theoretical and empirical analysis. *Financial Management, 18*(2), 59–73.

# Stocks, Stock Markets, and Stock Indices 8

Stocks and shares are the parts of equity capital, or more simply equity, which materialize property rights on the firm's assets and its future production and lie at the heart of the capitalist system. This is in sharp contrast with bonds or notes, which materialize debts. Shares are issued upon the creation of a company or corporation and, later on, whenever new investments require an increase in equity. The sum of all ordinary shares makes up the firm's common stock. These issues are carried out on a financial market (i.e., a stock exchange[1]) or over the counter (e.g., when they are reserved for new partners). Corporations issue shares on the *primary (stock) market* (new securities thus are created). The *secondary market* ensures the liquidity of existing shares. The market value of shares quoted on the stock market fluctuates randomly, according to supply and demand. The market returns of a whole industry or an official stock exchange are estimated using stock market indices.

This chapter is organized as follows. Section 8.1 is devoted to shares (general notions, and analysis and valuation of stocks). Section 8.2 describes the statistical characteristics of future prices and the mathematical representation of their dynamics as stochastic processes. Section 8.3 presents stock markets, their organization, and their functioning. Section 8.4 analyzes stock market indices.

## 8.1 Stocks

This section presents some fundamental notions concerning capital funds, equity, and shares. If some of these notions are related to corporate finance, they would be neglected at the cost of insufficient comprehension of the accounting, economic and financial logics underlying capital transactions. After presenting some fundamental

---

[1] The use here of the English word stock, in its old meaning of stick as in German, dates back to the days when transactions and debts were recorded on wooden sticks that were broken apart so that only the counterparties, each holding one end, could put them together seamlessly when the transaction was completed. It was an old method of providing authentication.

© The Author(s), under exclusive license to Springer Nature Switzerland AG 2022     255
P. Poncet, R. Portait, *Capital Market Finance*, Springer Texts in Business and Economics, https://doi.org/10.1007/978-3-030-84600-8_8

notions regarding the role of equity and capital transactions in Sect. 8.1.1, we analyze shares in Sect. 8.1.2, the data relevant for computing stock returns in Sect. 8.1.3, and introduce briefly, in Sect. 8.1.4, some simple methods presently in use for valuing shares or common stock.

## 8.1.1 Basic Notions: Equity, Stock Market Capitalization, and Share Issuing

### 8.1.1.1 General Considerations

According to the legal status of the issuing organization, portions of the equity capital are called partnership shares or company shares or stocks. The firm's equity is also called its own capital. The share or stock confers upon its owner the quality of an *associate* and generally gives a right to a fraction of i) the dividends (taken from the profits), ii) the *reserves* (earlier undistributed profits) possibly distributed by the company, and iii) the firm's *net assets in case of liquidation*.[2]

The share or stock also gives the holder the right to oversee the firm's management by way of the *right to vote* in the shareholders' yearly meetings. The task of managing the company outside these meetings is delegated to the board of directors. This board itself is controlled by the board of overseers. Its president, who sometimes serves also as CEO (chief executive officer), is thus mandated by the shareholders to manage the firm in their best interests.

It is essential to note the difference between debt securities, studied in preceding chapters, and equity capital (shares, stocks), which is this chapter's concern. In contrast to debt securities, equity shares are property rights. For shareholders, this essential quality of being owners leads to rights over the firm's management decisions but, on the other hand, leads to a lesser priority in the awarding of available cash flows generated by the firm's activities. Indeed, payment of interests and reimbursement of debt principal(s) at maturity takes priority over paying dividends and (eventually) repayment of equity. This priority rule of debts over equity implies that the latter absorbs most risks and as a result provides a guarantee to debt holders.

The following simplified example emphasizes the priority rule.

---

**Example 1**

Let us consider a firm that does not invest any more in plant, equipment, or working capital. It neither issues nor reimburses any debt during the period under consideration. Its activity generates annual revenues $R$, and nonfinancial expenses $C$ (mainly salaries, purchase of materials, energy consumption, administrative, and marketing costs). The difference $R - C$ is called *EBITDA* ("Earnings Before Interests, Taxes, Depreciation, and Amortization").[3]

(continued)

---

[2] That is, the product of liquidation of the firm's assets, after repayment of all debts.

[3] We assume a constant working capital.

## 8.1 Stocks

Random variables are written in boldface. The annual depreciation and amortization is an amount $DA$. The company has financed its assets of total value $K$ by issuing a bullet bond of amount $D$ bearing an interest rate $i$ and by issuing an amount $(K - D)$ of equity. Therefore, the cash flow before interests and taxes is equal to **EBITDA**.

The gross operating profit (before taxes) is, by definition, **EBITDA** net of depreciation, amortization, and financing costs: **Profit before tax = EBITDA − DA − iD**.

The distribution of available cash flow, here equal to **EBITDA,** is subject to the following rules:

- *The cash flow before interests is to be assigned as a priority to debt servicing (payment of interest and reimbursement of maturing debts, the latter being here assumed nil).*
- *It is only after the financing costs iD and the depreciation/amortization DA have been deducted, and if the profit before taxes is positive, that, after paying out taxes, the shareholders can receive dividends. The shareholders thus are clearly at the end of the chain of priorities.*

As a result, in this example, three cases may occur:

1st case: **EBITDA** $<$ $iD$. The cash flow is not sufficient to pay the financing costs and the company is theoretically unable to meet its payment obligations (which could lead to its bankruptcy and liquidation).[4]

2nd case: $iD \leq$ **EBITDA** $\leq DA + iD$ and the gross operating profit is negative; therefore the firm cannot distribute dividends, but can pay its financial costs and thus does not default.

3rd case: **EBITDA** $> DA + iD$ and the company makes a profit. As a result, on the pre-tax income (**EBITDA** $- DA - iD$), it pays the corporate taxes and may (but does not have to) pay from the balance dividends to its shareholders.

Note that the probability of default by the firm decreases as the proportion of its equity grows: Prob (**EBITDA** $< iD$) is an increasing function of $D$ and so a decreasing function of own capital.

Finally, it is intuitively obvious that since debt expenses (financing costs and reimbursement of the principal) have priority over the remuneration of equity, the randomness that affects the return on equity increases with the level of indebtedness. In fact, the financial *leverage* effect (see Chap. 2) is a result of the fact that debt expenses are fixed and have priority.

---

[4] We have simplified this example for the sake of exposition: in particular, we assume that the firm cannot issue new debts or equity (which would allow it to avoid default, at least temporarily). Furthermore, a distressed firm may benefit from legal processes that shield it from bankruptcy (such as rescheduling or partial write-down of its debt).

### 8.1.1.2 Some Definitions About Equity and Market Capitalization (Total and Floating)

When the company is quoted on a stock market, a share has, at any instant $t$, a market value which we denote by $S(t)$. This market value must be distinguished from the nominal value of the share. The latter value, which is generally a *fixed* quantity such as $10, is the basis for calculating the company's share capital, which is the product of the nominal value by the number of shares outstanding. This notion is essentially relevant for accounting and legal purposes, but is largely irrelevant to finance. The equity book value (accounting shareholders' equity) is the sum of share capital and retained earnings (i.e., accumulated earlier undistributed profits). *This accounting (book) value in general differs from the market value*, which is also called *the market capitalization* (product of the market price of a share by the total number of shares).

We will have to distinguish in the sequel the (total) market capitalization from the *floating market capitalization* (or simply *floating capitalization*) made up of just the shares held by the public. The floating capitalization is equal to the total capitalization reduced by the fraction that is not actually tradable on the market because it is held back by the so-called "core shareholders" who control the company. These shares are owned by the founders of the company, or the government, or stable blocks of shareholders who invested in the firm for strategic, long-term reasons. Floating capitalization is more relevant than total market capitalization for estimating the liquidity of a stock. It is the quantity used, as we will see in Section 8.3, in the computation of the weights in most market price indices. In the rest of this chapter, however, without explicit mention to the contrary the term market capitalization will mean the total value of all the shares outstanding.

Market capitalization thus represents the weight of a company in the financial market. *Blue Chips* are corporations with large capitalizations, financial solidity, established reputation, and large profits.[5]

### 8.1.1.3 Different Forms of Issue: Partnership Shares and Stocks

With the approval of the existing shareholders, a company can decide to issue new shares.

Note first that shares can be paid for by one (or several) of the following means:

- Cash.
- Nonmonetary assets, for example, real estate or loan to a third party.
- Conversion of debt(s) owed by the company (the debt holders swap their debt(s) for shares).

---

[5]This term is inherited from the blue color of the most valuable chips in a casino.

# 8.1 Stocks 259

Moreover, creating new shares may also result from simply distributing free shares to existing shareholders (with or without incorporation of retained earnings into the share capital[6]).

Generally, issuing (floating) new equity capital is subject to constraining rules (such as calling of a shareholder meeting and legal publicity). Issuing new shares is done either on the market, by offering them to the public (flotation), or over the counter (OTC) to private investors. Public flotation is subject to prior appropriate information and publicity and is supervised by the relevant market authorities (e.g., the Security Exchange Commission (SEC) in the USA, Autorité des Marchés Financiers (AMF) in France, Financial Conduct Authority (FCA) in the UK, and BaFin in Germany).

Changes in equity capital are thus constrained by a high degree of legal formality and prove in practice to be complicated. Banks, however, are familiar with such transactions and can help. Against compensating fees, they can undertake part of the process and some operational steps (creating the necessary documentation and calling meetings, providing trading facilities and commercial services for placing shares, granting guarantees of credit worthiness, or even securing a fixed issuing price, etc.).

Also, note that there is a large variety of possible legal arrangements concerning ownership of shares, which vary between two extremes:

- The shares issued by a public limited liability company are anonymous (joint stock company) and shareholders have a financial responsibility limited to their investment. The value of such shares can therefore never be negative. In addition, such shares are generally freely negotiable as each shareholder is unconcerned by the identity of others.
- When the shares are issued by a partnership, limited liability does not exist and partners guarantee personally and entirely the firm's debts with their own wealth (implying that the shares may have a negative value) and usually cannot dispose of their shares without the other partners' consent.

According to the legal nature of the issuing company, shares thus sharply differ, notably in terms of risk and liquidity. In the sequel, we only consider shares that cannot have a negative value (due to limited liability) and are freely negotiable.

Finally, remark that a company may issue:

---

[6]If retained earnings are incorporated into the capital, the accounting counterpart of the resulting increase in equity is a decrease in retained earnings. If not, the accounting counterpart is a proportional reduction in the nominal value of each share (total share capital and retained earnings both remain constant). Therefore, the cost-free share distribution is in this case a mere *stock split* (old shares are "divided up" as are their market value and nominal value). It is important to note that, in both cases, the asset side of the balance sheet is unaffected, no new resources being brought to the firm.

- Common stock or preferred stock, the latter usually carrying a priority over common stock, especially regarding the dividends and the claim to the firm's assets in case of liquidation, but having less voting rights.
- Stock without voting rights, with single voting rights or double voting rights.

### 8.1.1.4 Listing and Initial Public Offering (IPO)

Shares may or may not be listed on a stock market. Shares of a company are listed when being issued on the stock market for the first time. This generally occurs when the company reaches a certain size, so the public offering is to complete or replace the funds provided by the founding stockholders. Floating new shares on the market allows access to new sources of financing, and makes the holdings of the original shareholders liquid.

However, public listing involves some issuing costs initially and other expenses afterward, because of the regulation that requires the transparency appropriate to public offerings. Notably, every company listed must provide periodic information to investors, warnings of financial transactions, and any information that might have an impact on the stock market price.

The firm may entrust all or part of the shares issued on the market to one or more financial institutions (banks or investment houses). They are in charge of placing the shares with different groups of investors in the days preceding the eventual listing.

### 8.1.1.5 Reduction of Equity Capital and Share Repurchase

Just as a company may decide to increase its equity capital, it can also decide to reduce it to remunerate the shareholders by giving them back part of their investment.[7] Most often the company is mature and has hoarded cash (the reduction of equity on the liability side of the balance sheet has as a counterpart a decrease in cash on the asset side). The reduction of equity capital can be carried out by distributing liquid assets to the shareholders or by repurchasing shares on the stock (or OTC) market. In the first case, the process is tantamount to a distribution of dividends to all shareholders. In the second, each shareholder is free to participate in the repurchase program or not. *Repurchased shares are cancelled*: the relative power of the remaining shareholders thus increases as equity is reduced.

A listed company may also repurchase on the market, as any investor, or through a public offering, its own shares and *keep* them. If so, its equity capital is *not* reduced. The company, therefore, retains the possibility of later reselling these shares if necessary for future financing. Although shares held in this way by their own issuer are not cancelled (and the capital not reduced), they nonetheless do not have voting rights nor receive dividends.

---

[7] Other motivations are to allow the withdrawal of minority shareholder(s) or to absorb losses.

# 8.1.2 Analysis of Stock Issues, Dilution, and Subscription Rights

We consider a company quoted on the stock market that issues new shares. For the sake of analysis, the process is assumed to be carried out instantaneously and we consider the circumstances surrounding the date of issue $\tau$ ($\tau^-$ is just before, $\tau^+$ is just after).

### 8.1.2.1 Impact of the Issue on Share Value and Market Capitalization

The market value $S(t)$ of one share and the market capitalization $C(t)$ of the company vary over time. Notably, issuing equity changes the firm's market value. We consider a share issue occurring on date $\tau$ and denote by $S^-$ (short for $S(\tau^-)$) the market value of a share just before the issue and by $S^+$ its value just after.

The firm chooses the price at which it will sell the new shares, or *issue price*, denoted by $S_e$. As we will see later, the issue price is usually fixed at a level close to but lower than the current market price, except for some issues with rights offering for which the discount may be substantial. The issue price, which results from supply and demand, has no reason to equal the face (nominal) value; most often the market and issue prices are larger than the nominal value, as a result mainly of the worth of both retained earnings and opportunities for future growth (the positive net present value of future investments).

The difference between the issue price $S_i$ and the face value is called the *issue premium*.

Furthermore, $S_i$ should be chosen below the current market price because no investor would consider paying, to subscribe to a new share, a higher price than he could pay to acquire the same share on the secondary market.

In summary, we use the notation:

- $S^-$ the market value of a share just before issue.
- $S^+$ the market value just after issue.
- $S_n$ its face or nominal value.
- $N$ the initial number of shares, that is, just before the new issue.
- $n$ the number of new shares issued, so $N + n$ is the total number of shares after issue,
- $S_i$ the issue price of the new shares; note that in general one observes: $S_n < S_i < S^+ < S^-$.

The market capitalizations, just before and after the issue, respectively write: $C^- = NS^-$ and $C^+ = (N + n) S^+$. Theoretically, they must satisfy

$$C^+ = C^- + nS_i \tag{8.1-a}$$

and

$$S^+ = \frac{NS^- + nS_i}{N + n} \tag{8.1-b}$$

Equation (8.1-a) simply expresses the fact that the theoretical value of the company increases by the amount contributed by new shareholders ($nS_i$); eq. (8.1-b) results from (8.1-a) and the fact that $S^+ = \frac{C^+}{N+n}$ and $C^- = NS^-$.

> **Example 2**
> Consider a company whose capital is, on 2 January, made up of $N = 40{,}000$ shares, whose unit face value is $S_n = \$100$, and market value is $S^- = \$300$. On 3 January, the company issues $n = 10{,}000$ shares at a price $S_i = \$240$.
>
> The share market value $S^+$ immediately after the issue settles at a level \$12 below $S^-$:
>
> $$S^+ = \frac{(40{,}000 \times 300) + (10{,}000 \times 240)}{50{,}000} = \$288.$$

### 8.1.2.2 Protection of Former Shareholders and Subscription Rights

We have noted that an equity issue implies a decrease in share price, called *dilution*. This fall in value, explained by the fact that the new shares (which have the same ownership rights as existing ones) are issued at a lower price than the current market price, is a concern. Actually, in the absence of compensation, the existing shareholders would experience a financial loss and thus oppose possibly desirable (from the firm's viewpoint) equity issues. This is why, by law in some countries and through rights offerings in others, the existing shareholders may benefit from a preferential subscription right to new shares in proportion to the number of shares they hold: an old share gives the right to subscribe to $\frac{n}{N}$ new ones at the price $S_i$ (instead of $S^+$). This right can take the form of a *negotiable coupon*, called *subscription right* or *warrant*, associated with each old share. We denote its value by $D$.

**Proposition 1** *The value $D$ of the subscription right exactly compensates for the fall in the share market price, i.e.,*

$$D = S^- - S^+ \tag{8.2-a}$$

or

$$D = \frac{n(S^+ - S_i)}{N}, \text{ or} \tag{8.2-b}$$

# 8.1 Stocks 263

$$D = \frac{n(S^- - S_i)}{N+n}.$$  (8.2-c)

***Proof*** We can find different ways of expressing $D$ by remarking that an old share provides a right to subscribe to $\frac{n}{N}$ new share at the price $S_i$ instead of $S^+$, which means that $D = \frac{n}{N}(S^+ - S_i)$, which is eq. (8.2-b).

Replacing $S^+$ by $\frac{NS^- + nS_i}{N+n}$ (eq. (8.1-b)) in (8.2-b), we obtain (8.2-c).

(8.1-b) also implies that $S_i = S^+ + \frac{n}{N}(S^+ - S^-)$ which by substitution into (8.2-b) gives (8.2-a).

> **Example 3**
> Let us take up the previous example again: $D = (10{,}000/40{,}000)\,(288{-}240)$, i.e., a value of \$12 per share, which exactly cancels out the loss of \$12 on the share value.

A special case occurs when the increase in equity capital is done by incorporating retained earnings. This is a pure accounting ("cosmetic") operation, as no new resources are brought to the firm, which leads to a *distribution of free shares*. The value $D$ of the right to receive the free new shares is given by the preceding formulas, noting that the value of the financial contribution to the firm and the issue price are both zero ($S_i = 0$).

Thus, we obtain

$$C_b{}^+ = C_b{}^-; \quad S^+ = \frac{N}{N+n}S^- \; ; \quad D = \frac{n}{N+n}S^- = \frac{n}{N}S^+.$$  (8.3)

Another special case, close to the previous one and rather frequent, is that of stock splits. Here, there is not even an increase of equity capital by incorporation of retained earnings, but a simple increase in the number of shares (by a specific multiple m) and a proportional decrease in the nominal value and the market value per share (which are multiplied by 1/m), so that the firm's nominal equity and market capitalization remain constant. This apparently useless decision is, in general, motivated by the desire that the price of one share does not become too high and so remains accessible to small portfolios or investors, which may influence favorably the liquidity of the stock.

## 8.1.3 Market Performance of a Share and Adjusted Share Price

In absence of subscription rights, distribution of free shares, or stock split, the evolution of the market share price during a given period reflects its true performance during the same period (ignoring dividends, however). Such is not the case if, in the period considered, one or more stock issues have taken place that have reduced the share's market price, all else being equal. These mechanical drops in price do not affect the real market performance since the subscription or attribution rights exactly balance their effect. A careless assessment of the evolution of the share's market price can therefore lead to underestimating its real performance.

To eliminate this error, we evaluate its performance based on the evolution of the self-financing portfolio of a fictional shareholder who upon each equity issue neither invests in nor withdraws funds from the company. When an equity issue occurs, such a shareholder trades her right $D$ to acquire new shares (at the price $S^+$), or collects the free shares, and the number of her shares increases by the proportion $\frac{D}{S^+}$; if she had one share before the issue, she has $1 + \frac{D}{S^+}$ new shares after it. The coefficient $1 + \frac{D}{S^+}$ is the *adjustment coefficient,* denoted by $\alpha$. It is important to remark that the wealth of such a shareholder is unaffected on the day of issue (it shows no discontinuity): immediately before the issue, the portfolio is worth $S^-$ and just after $\left(1 + \frac{D}{S^+}\right)S^+ = S^+ + D = S^-$.

Note that the adjustment coefficient is also equal to $\alpha = \frac{S^-}{S^+}$ in the general case and to $\frac{N+n}{N}$ in the special case of free additional shares. In effect, we have $\alpha = 1 + \frac{D}{S^+} = 1 + \frac{S^- - S^+}{S^+} = \frac{S^-}{S^+}$, in the general case and $\alpha = 1 + \frac{D}{S^+} = 1 + \frac{n}{N}\frac{S^+}{S^+} = 1 + \frac{n}{N} = \frac{N+n}{N}$ when free shares are distributed.

This systematic and rigorous method allowing us to assess market performance during a period in which the company has carried out equity operations depends on the notion of *adjusted share.* Fix, for example, a reference moment $t = 0$.

▶ **Definition** *At any instant t, the adjusted share price is defined as* $S^a(t) = S(t) \times \prod_{j=1}^{m} \alpha_j$, *where:*

*$S(t)$ is the current market value of the share.*
*$m$ is the number of equity operations carried out between 0 and t,*
*$\alpha_j$ is the adjustment coefficient for the $j^{th}$ equity operation.*

*The dynamics that count is that of the price of the adjusted share.* Note that in particular the adjusted share's price suffers no discontinuity on the day of an equity issue in contrast to the current share price.

---

**Example 4**

Let us pursue the previous example of a company whose capital consists, on 2 January $n$, of $40,000$ shares quoted at \$300 and issuing on (the morning of) 3 January 10,000 new shares at \$240. The value of the subscription right is \$12 and the new share price should be, in principle, \$288 on (the evening of) 3 January, just after the issue.

A shareholder who has *one* share on 1 January and who carries out a cash-neutral transaction, sells his subscription rights for \$12 and buys with that cash $\frac{12}{288}$ new shares. Thus after the transaction (4 January) he has $1 + \frac{12}{288} = 1.04$ shares.

(continued)

Remark that the number of shares he holds has increased by the fraction $\frac{12}{288} = \frac{D}{S}$ and also that $(1 + 12/288) = \frac{300}{288} = \frac{S^-}{S^+}$.

Suppose now that, on 20 June of year $n$, the company distributes one free share for 5 old shares; on 21 June our fictional shareholder with then have $(1 + 12/288)(1 + 1/5) = 1.25$ shares.

Suppose further that the market price of the share on 1 July is \$280, i.e. \$20 less than at the beginning of January. It would be wrong to conclude that the market performance of the share is negative. In fact, the value of a portfolio self-financed by the shareholder who carried out cash-neutral transactions from the start of January to the end of June has increased by \$50 per share. Indeed, for each initial share costing \$300 at subscription, he owns on 1 July 1.25 shares worth $1.25 \times 280 = \$350$: his market holding has grown by $50/300 = 16.67\%$ in six months, ignoring the possible dividends that may have been received during this period.

Thus, the evolution that really matters is that of the price of the *adjusted share*. Taking as our basis 1 share on 1 January $n$, the adjusted share is equivalent to

- 1 current share at the beginning of January $n$
- 1.04 current shares between 3 January and 20 June
- 1.25 current shares from 20 June until the next equity issue.

The following Table shows the real market performance of the share during the first six months of the year:

| Date | 01 January $n$ | 01 July $n$ |
|---|---|---|
| **Price of the adjusted share (basis 01 January n)** | 300 | 350 |

The reference moment chosen (the date upon which the portfolio contains one share) can be the instant 0 (start of the period, as in the example) or the final instant (end of the period). Reconsidering the example above with the basis fixed as 1 July $n$, the adjusted share price equals: \$280 on 1 July $n$; and $300/1.25 = \$240$ on 1 January $n$.

If dividends are paid on dates $t_1, \ldots, t_m$ in the period $(0, t)$ under consideration, one must include them in the calculation of the total return over this period. Two different methods, yielding slightly different results, can be implemented to this aim:

(i). The total return can be defined as the IRR (Internal Rate of Return) of the cash flow series:

| Date | 0 | $t_1$ | ......... | $t_m$ | $t$ |
|---|---|---|---|---|---|
| Cash flow | $-S^a(0)$ | $Div^a(t_1)$ | | $Div^a(t_m)$ | $S^a(t)$ |

$Div^a$ is the dividend per *adjusted share*, i.e., the dividend that would be received by a portfolio comprising the total number of *current* shares, which has increased by all cash-neutral transactions resulting from new capital issues.

(ii). The adjustments can encompass dividend payments. In this case, the fictional self-financing investor acquires new shares with the dividend proceeds (at the price $S^+$ ex-dividend). An adjustment coefficient $\alpha$ is then calculated at each dividend distribution, which is conceptually identical to the one previously explained and obeying to the same formulas ($\alpha = 1 + D/S^+ = S^-/S^+$, $D$ denoting here a dividend, and $S^a(t) = S(t) \times \prod_{j=1}^{m} \alpha_j$). The evolution of the (fully) price adjusted share $S^a(t)$ can thus be established and its performance can be measured without bias. The annualized return between 0 and t is simply equal to $(S^a(t)/S^a(0))^{1/t} - 1$.

## 8.1.4 Introduction to the Valuation of Firms and Shares; Interpretation and Use of the PER

Since the market price of a share reflects the evaluation of a firm by the market, we briefly present several valuation methods. A deeper and more thorough exposition would involve concepts from corporate finance that are beyond the scope of this book. In addition, this subsection is restricted to an elementary presentation of the valuation methods that will be most relevant and accessible to most readers. We cover in turn *static* then *dynamic* methods, the *PER* (Price Earnings Ratio), and *mixed* methods. Since most of the approaches depend on discounting future cash flows, we provide a brief analysis of the appropriate discount rate.

### 8.1.4.1 Valuation Using Static or Asset-Based Methods

First, recall that the value of equity capital equals, by definition, the difference between the value of the assets and that of all debts. These values being subject to various assessments, different valuations of the equity are possible.

The most simplistic valuation method for shares is to divide the *book value* of the equity on the company's *balance sheet* by the number of shares; one obtains the *share book value*.

An improved variant of this method consists in, after a critical examination of each item on the balance sheet, adjusting its value, notably by using a *fair value* basis. The *fair value* is what would result from an agreed revaluation either at the market value (this can be found for the asset or liability if quoted on a market, which is the case for numerous financial products), or at a price which would likely be agreed on in an over-the-counter transaction (when the asset or liability is not quoted). Book valuation on a *fair value* basis is in general use in America and Europe for larger corporations.

# 8.1 Stocks

By revaluing the different assets and liabilities, one obtains a value, called *intrinsic value*, of the equity that is more satisfying than the book value.

One can also estimate the *liquidation value* of the company by estimating the selling price of the assets under conditions of greater or lesser urgency. It is therefore necessary to deduct the costs of liquidation (debt refunding, liquidation costs, compensations, etc.). The ratio of the net liquidation value to the number of shares is the *share liquidation value*.

Static and asset-based methods have the major drawback of ignoring that the value of a company, except upon liquidation, is essentially linked to its capacity to generate future profits and distribute dividends. For example, among several firms whose book values may be equal, the company with the best results and strongest growth is obviously the most valuable.

It is therefore essential to value a company on the basis of its present and potential profitability. This valuation is the goal of dynamic methods.

### 8.1.4.2 Dynamic Methods

These methods depend on the discounting of the future cash flows that the firm will generate. According to whether one computes the value of the equity or the total value (debts included), the relevant future cash flows are different. In the first case, they are those collected by shareholders (future dividends net of possible equity issues or, in other words, dividends per adjusted share). In the second, they are those received by shareholders and creditors (therefore including debt interests and repayments net of possible debt issues). We denote them by $Div(t)$ and $F(t)$, respectively. More precisely, the current date is date 0, $Div(t)$ and $F(t)$ represent random future cash flows, $E_0(X)$ is the expectation of a random variable $X$ as of date 0, $Div_t \equiv E_0(Div(t))$ and $F_t \equiv E_0(F(t))$. The reader should carefully distinguish in what follows $F(t)$ from $F_t$ and $Div(t)$ from $Div_t$, i.e., a random variable from its expectation.

By considering a series of future cash flows occurring on dates $t = 1, \ldots, T$, the total value $V$ (shares and debts) and the total value $C$ of the shares (market capitalization) write, respectively:

$$V(0) = \sum_{t=1}^{T} \frac{F_t}{(1+r)^t} + \frac{V_T}{(1+r)^T} \tag{8.4}$$

$$C(0) = \sum_{t=1}^{T} \frac{Div_t}{(1+r_e)^t} + \frac{C_T}{(1+r_e)^T}, \tag{8.5}$$

where $r$ is the average cost of resources or weighted average cost of capital (*wacc*, see Chap. 2 and Sect. 8.1.4.5 below), $r_e$ is the return required by the shareholders or cost of equity, $V_T \equiv E_0(V(T))$ and $C_T \equiv E_0(C(T))$.

To use formula (8.4), we must forecast future cash flows $F$. Discussing the available forecasting methods is beyond the scope of this book.[8] In the sequel, we use equation (8.5) in applications based on very simple assumptions regarding expected dividends.

The calculation can be done, interchangeably, either with the market capitalization (equal to the present value of all the future dividends distributed) or with the share value (equal to the present value of the dividends *per share*). In the second case, we write

$$S(0) = \sum_{t=1}^{T} \frac{div_t}{(1+r_e)^t} + \frac{S_T}{(1+r_e)^T} \tag{8.6}$$

where $S(t)$ stands for the (adjusted) value of a share on date $t$, $S_T \equiv E_0(S(T))$ is the expected value of $S(T)$ and $div_t \equiv E_0(div(t))$ is the dividend per (adjusted) share anticipated at $t$.

Often $S_T$ is estimated by assuming that from $T$ onward dividends will stay constant, which implies that $S_T = \frac{div_{T+1}}{r_e} = (1+g)^T \frac{div_1}{r_e}$. Then note that $\frac{S_T}{(1+r_e)^T} = \left(\frac{1+g}{1+r_e}\right)^T \frac{div_1}{r_e}$ tends to 0 as the horizon $T$ increases, if and only if $g < r_e$.

Note also that eq. (8.6) is compatible with an alternative formula according to which the value $S(t)$ of the share on date $t$ (just after distribution of $div(t)$) is equal to the present value of the dividend $div_{t+1}$ and the share price anticipated (at $t$) for the date $t + 1$:

$$S(t) = \frac{E_t(div(t+1)) + E_t(S(t+1))}{1+r_e}. \tag{8.7}$$

*where* $E_t(.)$ denotes a conditional expectation on date $t$, a date on which the values of the shares and dividends on $t + 1$ are random.[9] We assume that $r_e$ is constant.

Writing eq. (8.7) for $t = 0$, we have

(i). $S(0) = \frac{E_0(div(1))}{1+r_e} + \frac{E_0(S(1))}{1+r_e}$.

When $t = 1$, we have

(ii). $S(1) = \frac{E_1(div(2))}{1+r_e} + \frac{E_1(S(2))}{1+r_e}$.

Substituting (*ii*) into (*i*) yields

---

[8] The cash flow $F_t$ to be split between shareholders and debt holders is called the *free cash flow*. It is equal to the EBITDA net of the investments in physical plant and operating items, and net of taxes (these being calculated assuming no debts).

[9] Recall that in this notation $div(t)$ is random and $div_t = E_0(div(t))$.

## 8.1 Stocks

$$S(0) = \frac{E_0(div(1))}{1+r_e} + \frac{1}{1+r_e} E_0 \left( \frac{E_1(div(2))}{1+r_e} + \frac{E_1(S(2))}{1+r_e} \right).$$

Using a well-known property of conditional expectations: $E_0[E_1(X(t)] = E_0(X(t)) \equiv X_t$, we obtain

$$S(0) = \frac{E_0(div(1))}{1+r_e} + \frac{1}{1+r_e} \left( \frac{E_0(div(2))}{1+r_e} + \frac{E_0(S(2))}{1+r_e} \right)$$

$$\equiv \frac{div_1}{1+r_e} + \frac{div_2}{(1+r_e)^2} + \frac{S_2}{(1+r_e)^2}.$$

Repeating this process $T$ times yields (8.6).

Equation (8.6) is often used under restrictive conditions. A simple and popular assumption is that the expected dividends increase indefinitely at a constant rate, denoted by g (growth). In this case:

$div_t = div_{t-1}(1+g) = \ldots = div_1(1+g)^{t-1}$ and eq. (8.6) rewrites[10]

$$S(0) = \sum_{t=1}^{T} \frac{(1+g)^{t-1} div_1}{(1+r_e)^t} + \frac{S_T}{(1+r_e)^T}$$

which simplifies, as $T$ goes to infinity with the condition $g < r_e$, to

$$Gordon - Shapiro\ formula : S(0) = \frac{div_1}{r_e - g}. \tag{8.8}$$

**Remarks**
- Equation (8.8), which is much used by practitioners, is known as the Gordon-Shapiro (or simply Gordon) formula.[11] Although simple and practical, it depends nonetheless on the choice of a growth rate g assumed to be constant forever.
- For (8.8) to hold, it is necessary that $g < r_e$. If not, $S(0)$ is infinite, which reflects the inconsistency of assuming a growth rate constant and larger than the total return required by the market.

### 8.1.4.3 The PER Method

This method is based on the PER (*Price Earnings Ratio*) which is seldom called also the "earnings capitalization coefficient" or "multiplier." Valuation can equivalently be performed using the firm's equity market value and total earnings or using the adjusted share price and the adjusted earnings per share. In the second case, we write

---

[10] $div_1$ can itself be thought of as equal to $div_0$, which was just paid out, multiplied by $(1 + g)$.

[11] This formula holds immediately after distribution of the dividend $div_0$ (the ex-coupon price). If the dividend $div_0$ is going to be paid out, then one has: $S(0) = div_1/(r_a\text{-}g) + div_0$ (price cum dividend).

$$PER(x) = \frac{(observed)\ S(x)}{Earnings\ per\ share\ (x)}. \tag{8.9-a}$$

When the PER is used for valuation purposes, $S(x)$ is the unknown and PER(x) is given: the value of share $x$ is estimated by multiplying the earnings of $x$ by the PER observed for listed companies that are comparable to $x$ as far as activity, risk, and growth are concerned. Therefore, we have

$$\begin{aligned}(theoretical)\ S(x) = {}&PER(comparable\ companies)\\ &\times\ earnings\ per\ share\ of\ x\end{aligned} \tag{8.9-b}$$

In practice, there is for each industry (or business sector) an empirical estimate of the PER, based on the market values of listed companies and the dividends they provide. When applied to firm $x$, this PER is modified to account for its specific risk and growth potential. As explained more precisely in the next subsection, the PER decreases with the interest rate and the risk affecting the share price and increases with the expected growth of the share price.

As intuition suggests, the PER is linked to the expectation of discounted dividends. Indeed, (8.6) implies

$$PER = \frac{\sum_{t=1}^{T} \frac{div_t}{(1+r_e)^t} + \frac{S_T}{(1+r_e)^T}}{Earnings\ per\ share}. \tag{8.10}$$

First, let us consider the very special case of entirely distributed earnings, which the market expects to remain constant in the future. From this, $(1/PER)$ equals the "pseudo-return rate": $(Earnings\ per\ share\ /\ Share\ value)$. The $PER$ can thus be interpreted as a capitalization multiple or a kind of inverse of the required rate of return.

In the less restrictive case of a dividend that is a constant fraction $\lambda$ of the earnings per share assumed to grow at the constant rate $g$, the Gordon-Shapiro formula leads to:

$$S = \frac{Div\ per\ share}{r_e - g} = \frac{\lambda\ (Earnings\ per\ share)}{r_e - g}, \text{whence}:$$

$$PER = \frac{S}{Earnings\ per\ share} = \frac{\lambda}{r_e - g}. \tag{8.11}$$

Consequently, the $PER$ increases with $g$ and decreases with $r_e$. Since $r_e$ rises with the interest rate and with the risk affecting the future dividend stream, the PER increases with:

- Optimistic expectations of the firm's growth ($g$).
- A low risk-free interest rate prevailing in the market.
- Smaller perceived risk affecting expected cash flows.

## 8.1 Stocks

We remark that it would be wrong to conclude, from incorrectly interpreting equation (8.11), that the PER and share value increase with the distribution rate $\lambda$. Indeed, there is an *inverse* relationship between $g$ and $\lambda$: the smaller the part $\lambda$ of the earnings paid out (the larger the retained earnings), the higher the firm's future growth. Therefore, equation (8.11) can rewrite more explicitly as: $PER = \frac{\lambda}{r_e - g(\lambda)}$. Therefore, no conclusion can be drawn *from this model* as to the influence of the distribution rate $\lambda$ on share values.[12]

---

**Example 5**

Company Z is not listed yet on the market and wants to estimate the value of its shares with a view to issuing them on the market. Its earnings are \$50 M, of which \$25 M will be distributed as dividends. Company Z's management and its banking advisors think that future earnings and dividends will continue to increase at *the current* yearly rate of 4%. They also think that the financial market will have the same expectation. They estimate at 9% the return currently required by the market for investing in companies with similar risks in the same business sector. This required return is estimated by adding to the long-term risk-free rate, currently 4%, a risk premium of 5%.

Applying the Gordon-Shapiro formula (8.8), the theoretical value of the share is equal to

$$S = \$25/(0.09 - 0.04) = \$500.$$

The *PER* to be applied to the share Z, taking into account its potential growth, its risk, and the state of the economy at large, is therefore \$500/\$50 = 10.

---

### 8.1.4.4 Mixed Methods

These are based on both a static estimate of the firm's equity (using one of the methods described in Sect. 8.1.4.1) and expected future earnings. For example, one can use a *weighted average*

$$C = \alpha \, BE + (1 - \alpha) \, PV(Div) \tag{8.12}$$

where $BE$ denotes the book value of equity as of the last issued balance sheet, $PV(Div)$ the present value of future dividends, and $\alpha$ a weighting coefficient between 0 and 1.

---

[12] The influence of dividend distribution on share prices is a classical and complicated question that involves corporate finance, and is outside this book's scope. Let us just state one of the famous theorems of Modigliani and Miller (1958, 1963): in *perfect* markets, for a given investment policy and a given capital structure, dividend policy is *irrelevant*. It has *no* influence whatsoever on the firm's value (or the share price), since the gain (loss) occurring when the dividend increases (decreases) is perfectly offset by a loss (gain) on the share market price.

### 8.1.4.5 The Choice of the Discount Rate

The discount rate $r$ to be used in formulas containing cash flows $F$ (equation (8.4)), which involves all the financial stakeholders (i.e., creditors and shareholders) has to reflect the cost of financial resources (debts and equity capital) used by the firm. The most frequent method is the WACC (*weighted average cost of capital*), defined by the formula

$$r = WACC = (1 - l) \, r_e + l \, (1 - \tau) \, r_d \qquad (8.13)$$

where $l$ *(leverage)* is the fraction of the debt and $(1-l)$ the fraction of equity ($0 \leq l \leq 1$), $\tau$ is the firm's profit tax rate (25%, for example), $r_e$ is the cost of equity (or the minimum return rate required by the shareholders) and $r_d$ the average cost of debt before taxes.

Most of the equations derived in this section involve the present value of the dividends: the relevant discount rate is therefore $r_e$, the cost of equity or the expected return required by shareholders. It increases with the risk perceived by them. Generally, one may write

$$r_e = r_f + \text{risk premium,}$$

where, here, $r_f$ is the risk-free rate. We show in Chap. 22 (devoted to the CAPM) the following relationship:

$$r_e = r_f + \beta \, (E(R_m) - r_f),$$

where $E(R_m)$ is the expected return on the whole market and $\beta$ is a coefficient reflecting how sensitive the company's stock return is to variations in the market return. The share's $\beta$, and so $r_e$, increase with the risk affecting the company's business. They also increase with the company's debt since the risk for shareholders, for a given level of business activity, rises with the firm's indebtedness (higher leverage).[13] We examine these various issues more thoroughly in Chaps. 21 and 22 devoted to portfolio theory.

---

[13] In spite of the fact that $r_e$ *rises* with $l$, an increase in leverage *reduces*, according to standard financial theory, the cost of capital (wacc) $r$, because $r_d$ is less than $r_e$ and therefore $(1-\tau)r_d$ is much less than $r_e$. This well-known result is one of the theorems due to Modigliani and Miller in the case of a positive corporate income tax $\tau$. However, this result does not hold when personal income taxes, bankruptcy costs, agency costs, or other market imperfections are introduced in the analysis.

## 8.2 Return Probability Distributions and the Evolution of Stock Market Prices

In this section, we analyze the price of a stock issued by a limited responsibility company from both statistical and probabilistic points of view. This price evolves as a random process whose value can never be negative. In Sect. 8.2.1 we consider a given *future moment T*, a date on which the share price is a *random variable* whose probability distribution we will examine first theoretically and then empirically (static analysis). More generally, this share's price follows a *stochastic process* from time 0 to time $T$ which we study first mathematically then empirically in Sect. 8.2.2 (dynamic analysis). This last subsection depends on notions involving probability theory, stochastic processes, and stochastic calculus methods that we discuss in Chap. 18.

Note that in this Section we use "*stock*" (price or return) instead of "*share*," when there is no ambiguity, to conform to standard terminology.

### 8.2.1 Stock Price on a Future Date, Stock Return, and Its Probability Distribution: Static Analysis

The definition and computation of the logarithmic return are recalled in Sect. 8.2.1.1 and the probability distribution of future prices up to a given horizon $T$ is examined in Sect. 8.2.1.2.

#### 8.2.1.1 A Refresher on Return and Log-Return Calculations

Let us start at time 0 and consider some future date $T$. Consider a stock whose price is $S(0)$ at 0 (a known quantity) and $S(T)$ on date $T$.

- First assume $S(T)$ is not random (it is certain or deterministic) and does not distribute any dividend between 0 and $T$ (e.g., this could be a risk-free zero-coupon bond maturing on date $T$).

In what follows, for brevity at the cost of being slightly inaccurate, we use "return" for "rate of return." According to the definitions in Chap. 2:

- The non-annualized period return $R'_{0,T}$ over $(0,T)$ is such that $1 + R'_{0,T} = \frac{S(T)}{S(0)}$.
- The annualized period return $R''_{0,T}$ over $(0,T)$ is such that $\left(1 + R''_{0,T}\right)^T = \frac{S(T)}{S(0)}$.
- The non-annualized continuous return $R_{0,T}$ is such that $e^{R_{0,T}} = \frac{S(T)}{S(0)}$, i.e.,

$$R_{0,T} = \frac{\ln(S(T))}{\ln(S(0))} = \ln(S(T)) - \ln(S(0)).$$

$R_{0,T}$ is called the logarithmic return or log return.

- The annualized log-return $R$ is such that $e^{RT} = \frac{S(T)}{S(0)}$, i.e.,

$$R = \frac{1}{T}[\ln(S(T)) - \ln(S(0))]. \tag{8.14}$$

In the sequel, we will often use the annualized continuous or log-return which can be seen as the continuous rate $R$ that, applied to $S(t)$, generates, in the interval $(t, t + dt)$, $dS = RS(t)dt$, which implies $S(T) = S(0)\, e^{RT}$.

- Now assume that $S(T)$ is random, which is always the case for the future stock price; the previous definitions still apply, notably that of the log-return $R$ given by (8.14) which is in this context an equation involving random variables.
- Assume that a dividend $Div(T)$ is paid out on date $T$. *To compute the return over* (0, $T$), including dividend and capital gains, we consider time $T^-$ (just before distribution) or $T^+$ (just after).

The first calculation writes $R = \frac{1}{T}[\ln(S(T^-)) - \ln(S(0))]$.

The second writes $R = \frac{1}{T}[\ln(S(T^+) + Div) - \ln(S(0))]$, and gives the same result $R$ since $S(T^+) = S(T^-) - Div$ (the distribution causes the price to fall by exactly the amount of the dividend under no arbitrage). Note, again, that the distribution does not entail any value discontinuity.

In the sequel, the term *return* will be used generically, whether the dividends are included or not, according to the context.

Adopting a unit period less than a year and periodic returns (e.g., monthly), we decompose the total period (0, $T$) into successive unit periods (0, 1), (1, 2), ..., ($t$–1, $t$), ..., ($T$–1, $T$) and write $R_t = \ln(S(t)) - \ln(S(t-1))$, for $t = 1, \ldots, T$. We then have

$$R_{0,T} = \ln(S(T)) - \ln(S(0)) = \sum_{t=1}^{T} R_t \tag{8.15}$$

which implies that the global log-return over (0,$T$) is equal to the sum of successive log-returns.

It is instructive to compare this simple composition law (8.15) for log-returns with that (less simple) for arithmetic returns: $1 + R'_{0,T} = \prod_{t=1}^{T}(1 + R'_t)$.

Remark finally that, while the composition over successive periods is simpler for log-returns than for arithmetic returns, the contrary prevails when we aggregate returns within one period. The arithmetic return of a *portfolio* is simply the weighted mean of the returns of its components while the log-return of the same portfolio is not a simple function of the log-returns of its components, which is an obvious disadvantage.

## 8.2 Return Probability Distributions and the Evolution of Stock Market Prices 275

### 8.2.1.2 Probability Distributions of Future Stock Prices and Returns

Denote by 0 today's date and consider some future date, arbitrary but fixed, denoted 1 without loss of generality. The price $S(1)$ and so the arithmetic return $R' = \frac{S(1)+Div-S(0)}{S(0)}$ and the log-return $R = [\ln(S(1) + Div) - \ln(S(0))]$ are random variables which can be characterized by their probability distributions.

For simplicity, we sometimes assume the price $S(1)$, and therefore the arithmetic return $R'$ (if the dividend is deterministic), to follow normal laws. This is notably the assumption that underlies classical portfolio theory (due to Markowitz; see Chap. 21). It is obvious that such an assumption cannot hold for the stock price of a limited liability company (which cannot be negative) since a normally distributed random variable always has a positive probability of taking on negative values.

An alternative hypothesis, a priori more satisfactory, consists in assuming that $S(1)$ is log-normal, i.e., has a *normal log-return $R$*.[14] Since a log-normal variable does not take on negative values, such a representation is compatible with limited liability. In addition, and crucially, log-normality is consistent with the weak form of efficiency that characterizes, as a first approximation, the stock market. Recall that efficiency implies it is impossible to use direct knowledge of past prices to obtain "abnormal" returns. However, an autocorrelation in the price or log-return variations could be exploited to obtain such profits. As a result, it is natural to associate weak efficiency with serial independence of log-returns.[15] And serial independence does imply that log-returns are normally distributed, and thus that prices are log-normal. In effect, by reusing (8.15) and dividing the period (0, 1) into an arbitrarily large number of subintervals of length $1/T$, we can write $R \equiv R_{0,1} = \sum_{t=1}^{T} R_t$. Consequently, log-return $R$ can be written as the sum of an arbitrarily large number of log-returns $R_t \equiv \ln S_t - \ln S_{t-1}$. When these are independent, $R_{0,1}$ is a normal random variable by the central limit theorem.

For these reasons, *the log-normality of stock prices has become a standard assumption.*

However, the *empirical evidence only partially vindicates* this assumption.

Figure 8.1 shows the empirical distribution typically observed for the log-return of a stock price (blue and solid line) as well as the theoretical normal (Gaussian) distribution (dashed line) with the same mean and standard deviation.

A comparison of these two distributions exhibits three differences.

– The first is that the Gaussian density is symmetric about its mean while the empirical density shows asymmetry. It is measured by a coefficient of asymmetry or *skewness* associated with the third moment of the distribution (see the Appendix to the chapter).

---

[14]Recall that if the log of a random variable obeys a normal law, this variable is log-normal.

[15]Independence is a stronger hypothesis than absence of correlation, but these two properties are equivalent in the (unique) case of normal variables. Actually, efficiency does not require serial independence of returns, but only their lack of serial correlation.

**Fig. 8.1** Empirical distribution of log-returns

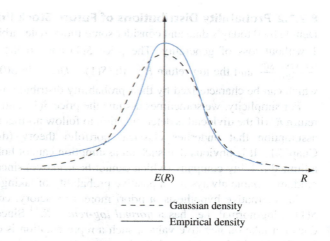

The coefficient is zero for a symmetric distribution (such as the normal distribution) and negative in the case of log-returns, which means that the probability of small values (compared to the mean) is larger than that of large values (see Fig. 8.1).

- The second difference comes from the fact that the empirical distribution of log-returns contains a larger proportion of values much smaller than the mean than the normal distribution (the $R$ distribution has a fat left tail). This is also true, to a lesser extent, of the values much larger than the mean. The existence of fat tails (left and/or right) is revealed by the kurtosis, associated with the fourth moment of the distribution of $R$ (see the Appendix). A distribution whose kurtosis is larger than that of the normal distribution is *leptokurtic*. It is characterized by a higher probability that extreme values (very low and/or very high) occur than the Gaussian distribution. This is the case for stock log-returns. This property, associated with the asymmetry to the left noted before, implies *a left tail fatter than that of the normal distribution* (it is also fat on the right but less noticeably).
- The third difference is that the maximum of the log-returns density is larger than the maximum of the Gaussian density with the same standard deviation, and that the histogram is narrower about this maximum (see Fig. 8.1).[16]

## 8.2.2 Modeling a Stock Price Evolution with a Stochastic Process: Dynamic Analysis

A stock price $S(t)$ evolves randomly, i.e., is a stochastic process. This subsection is devoted to the mathematical representation of this evolution in the interval $(0, T)$. While previously we considered the random variable $S(1)$ relative to the instant

---

[16] Such a profile is said to have a *thin waist* and is characteristic of leptokurtosis, "leptos" meaning "sharp" in Greek.

## 8.2 Return Probability Distributions and the Evolution of Stock Market Prices

1 (arbitrary but fixed), here we are interested in how the price $S(t)$ changes from its initial value $S(0)$ to its final value $S(T)$ (a film, in contrast with a photo). Naturally, the value at an arbitrary moment is a point in the price's motion and the dynamic representation must be consistent with the results from static analysis: for every date $t$ the model representing the process should lead to a probability distribution for the random variable $S(t)$ consistent with the preceding analysis and empirical observations.

The process $S(t)$ can be written in discrete or continuous time. In discrete time, we consider only special instants such as $S(0)$, $S(1)$, ..., $S(T)$. In continuous time $S(t)$ is observed for all values of $t \in (0, T)$. Most often, continuous time models assume that the stock market is always open for trade and that transactions take place continuously at no cost.

The simplest model consists in supposing that price increments $\Delta S \equiv S(t + \Delta t) - S(t)$, over successive time intervals $(t, t + \Delta t)$, are independent. Therefore, the price follows a random walk. This hypothesis implies normality of the price by the central limit theorem. In continuous time, the price process can obey a Brownian motion. We have emphasized that this assumption of normality is unworkable when applied to prices of securities that cannot take on negative values. A model leading to log-normal prices is to be preferred. Such a log-normal model of prices can be based on independence of the increments $\Delta \ln S \equiv \ln S(t + \Delta t) - \ln S(t)$ of logarithms of the prices. Indeed, such a serial independence of log-returns leads, by the central limit theorem, to log-normality of prices, and thus to normality of log-returns. Despite its empirical disadvantages examined in the previous subsection, it is this log-normal model that is most often assumed and we start with it.

Understanding the analysis that follows requires familiarity with introductory stochastic calculus, which the reader will find in Chap. 18.

### 8.2.2.1 Representation of Price Evolution Using a Geometric Brownian Motion

- The simplest log-normal model is in discrete time and assumes that the variations $\Delta \ln S$ of the logarithms of prices are independent and identically distributed when the increments are calculated on disjoint time intervals of the same length. Assuming the expectation $m$ and the variance $\sigma^2$ of the changes in unit time to be finite, and writing $\ln S(t) - \ln S(0)$ as a sum of independent increments, we have:

$$\ln S(t) - \ln S(0) \equiv \sum_{s=1}^{t} (\ln S(s\text{-}1)), \text{ which gives.}$$

$$E[\ln S(t) - \ln S(0)] = \sum_{s=1}^{t} m = m\ t\ ; \quad \text{variance } [\ln S(t) - \ln S(0)] = \sum_{s=1}^{t} \sigma^2 = \sigma^2 t.$$

The expectation and variance of price logarithms thus increase linearly with time. One can also conclude that $\ln S(t + \Delta t) - \ln S(t)$ is distributed according to $N(m\Delta t, \sigma^2 \Delta t)$, where $N()$ denotes a normal distribution, and so write

$$\ln S(t + \Delta t) - \ln S(t) \equiv \Delta \ln S = m\Delta t + \sigma \Delta W, \tag{8.16}$$

with $\Delta W$ distributed according to $N(0, \Delta t)$.

- In continuous time, the equation corresponding to (8.16) is the stochastic differential equation (SDE) governing, in the interval $(t, t + dt)$, the *arithmetic* Brownian motion of $\ln S(t)$ (it is *geometric* for $S(t)$):

$$d\ln S(t) = mdt + \sigma dW(t), \tag{8.16-a}$$

where $m$ and $\sigma$ are two constants and $W(t)$ is a standard Brownian motion or Wiener process (see Chap. 18).[17]

We recall that:

(i). The solution to this SDE is: $\ln S(t) = \ln S(0) + m\,t + \sigma W(t)$, or $S(t) = S(0) e^{mt + \sigma W(t)}$.

(ii). Applying Itô's Lemma to $\ln S$, we get[18]: $\frac{dS}{S} = \left(m + \frac{\sigma^2}{2}\right) dt + \sigma dW(t)$;

or defining $\mu \equiv m + 0.5\sigma^2$:

$$\frac{dS}{S} = \mu dt + \sigma dW(t) \quad \Longleftrightarrow \tag{8.17-a}$$

$$S(t) = S(0) e^{(\mu - 0.5\sigma^2)t + \sigma W(t)}. \tag{8.17-b}$$

(iii). Equations (17) are relative to a *geometric Brownian motion* with the following properties: $S(t)$ is log-normal (*see* 8.17-b); and the increments $\Delta \ln S$ are independent (for disjoint intervals) and identically distributed according to the law $N((\mu - 0.5\sigma^2) \Delta t, \sigma^2 \Delta t)$ (when computed for intervals of the same length $\Delta t$).

**Interpretation of Eq. (8.17-a) and Remarks**

(i). Equation (17) only holds on a time interval during which the stock does not pay out a dividend since, otherwise, the price falls by the amount distributed on the date of payment. In absence of dividends, the relative increment $dS/S$ is the stock return.

(ii). The equation of a geometric Brownian motion in the form (8.17-a) is not defined for $S = 0$. In fact, when $S(t)$ is near 0, it is preferable to write:

---

[17]The standard Brownian motion (or Wiener process) is characterized by independent Gaussian increments, with zero mean and variance equal to the length of the interval on which it is calculated.

[18]$d \ln S = \frac{dS}{S} - \frac{(dS)^2}{2S^2} = \frac{dS}{S} - \frac{(d \ln S)^2}{2}$, which implies: $\frac{dS}{S} = d \ln S + \frac{\sigma^2}{2} dt = \left(m + \frac{\sigma^2}{2}\right) dt + \sigma \, dW(t)$.

## 8.2 Return Probability Distributions and the Evolution of Stock Market Prices

$$dS = \mu S dt + \sigma S dW.$$

Therefore, $dS = 0$ when $S = 0$, which means that the motion stops with certainty and the price remains zero ever after: the process has an absorbing barrier at 0. The price process for a stock with limited liability has this property. In effect, to prevent arbitrage, a security whose value can never be negative cannot attain the value zero as long as there is the slightest probability that it has a positive value in the future. If it were zero, an arbitrage would consist in acquiring that stock for nothing in the hope of a price increase.

(iii). Equation (8.17-a) expresses the return $dS/S$ over the interval $(t, t + dt)$ as a sum of its expectation and a random part, as indicated below:

$$dS/S = \mu dt + \sigma dW$$
$$[E(dS/S)][random\ term\ (E(\sigma dW) = 0;\, \mathrm{var}(\sigma dW) = \sigma^2 dt)]$$

- The term $\mu dt$ is non-random and equals the expectation of the return $E(dS/S)$;

$\mu = \frac{1}{dt} E\left(\frac{dS}{S}\right)$, therefore, is the expectation of the instantaneous return, i.e., the *drift* of the process.
- The term $\sigma dW$ is random, with zero mean and variance $\sigma^2 dt$ (since $E(dW) = 0$ and $var(dW) = dt$). From this, $\sigma^2 = \frac{1}{dt} var\left(\frac{dS}{S}\right)$ gives the instantaneous variance of the return and $\sigma$ is its square root,[19] *called the volatility of the stock price* (i.e., *the diffusion parameter*).

(iv). The volatility is also the square root of the instantaneous variance of the log-return since $\sigma^2 = \frac{1}{dt} var[\ln(S(t + dt)) - \ln(S(t))]$, from Equation (8.16).

In addition, for a geometric Brownian motion $S$, we have: $var\,[\ln\,(S(t + \Delta t)) - \ln (S(t))] = \sigma^2 \Delta t$, for any period $\Delta t$ over which the log-return is calculated.

(v). The geometric Brownian motion characterized by Equation (17) is a very special case of an Itô process that writes

---

[19] It is not rigorous to assert that $\sigma$ is the instantaneous standard deviation of the return $dS/S$ because $\frac{1}{dt}\sigma\left(\frac{dS}{S}\right)$ is undefined (from the fact that $\sigma\left(\frac{\Delta S}{S}\right)$ is of order $\sqrt{\Delta t}$).

$$\frac{dS}{S} = \mu(t)dt + \sigma(t)dW \tag{8.18}$$

In equation (8.18), the instantaneous mean $\mu(t)$ and the volatility $\sigma(t)$ are not constant but vary, in general stochastically (e.g., they may depend on the level of the price $S(t)$). In fact, how relevant is the hypothesis of a geometric Brownian movement whose parameters $\mu$ and $\sigma$ are constant should be assessed from both theoretical and empirical points of view.

### 8.2.2.2 Mean Return: Interest Rate and Risk Premium

It is important to remark first that the instantaneous mean return is the sum of the risk-free rate here denoted by $r(t)$ and a risk premium. The premium is required by investors allergic to risk so that the market is in equilibrium. From this, we write:

$$\mu(t) = r(t) + \theta(t)$$
$$[mean\ return]\ [riskless\ rate] \qquad [risk\ premium] \tag{8.19}$$

Except if one is willing to assume, heroically, that variations in $r(t)$ and $\theta(t)$ constantly and exactly cancel each other, $\mu(t)$ can only be constant under the double assumption of constant interest rate $r$ and constant premium $\theta$.

One often supposes, for the sake of simplicity, a constant interest rate when its random property is not essential to the problem at hand. Such is notably the case in the standard binomial and Black-Scholes option models studied in Chaps. 10 and 11.

Actually, interest rates vary randomly all the time and it is important to understand the effect their variation has on the price of a financial asset and its dynamics. For example, an increase in the risk-free rate $r(t_0)$ at time $t_0$ translates into: ($i$) an instantaneous *reduction* in the stock price $S(t_0)$ (since it is equal to the present value of future dividends and the discount rate has gone up), and ($ii$) an *increase* in the anticipated growth rate $\mu$ of the stock price, by virtue of equation (8.19). Figure 8.2 displays this dual effect.

A decrease in interest rates leads to the opposite effects: an immediate surge in the price, followed by a subsequent less favorable trend.

The second component of the growth rate $\mu(t)$ is the risk premium $\theta(t)$. It is proportional to the risk intensity affecting the stock upon consideration, the proportionality coefficient being the market risk premium:

$$\theta(t) = Market\ risk\ premium \times risk\ affecting\ the\ stock$$

The market risk premium depends essentially on the risk aversion of the "average" investor. The risk affecting the stock is estimated using various theoretical models (Capital Asset Pricing Model, Arbitrage Pricing Theory, ...) For instance, according to the CAPM, the risk affecting a security may be measured by its beta (see Chap. 22). Denoting by $R_M$ the return on a portfolio representative of the market

## 8.2 Return Probability Distributions and the Evolution of Stock Market Prices

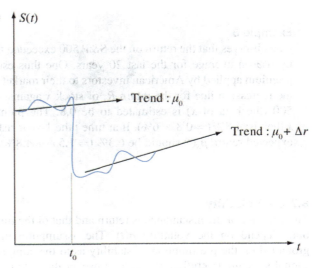

**Fig. 8.2** Effect of an increase $\Delta r$ in the interest rate on a security price $S(t)$

as a whole, whose beta is by construction equal to one, the market risk premium is equal to $E(R_M) - r$ and the stock risk premium to $(E(R_M) - r) \times \beta$.

Assuming a constant risk premium is based on the assumption of both a constant risk on the stock (no change in the firm's business and its financial leverage) and a constant market risk premium (no change in the average investor's risk aversion). The empirical evidence contradicts less obviously the assumption of a constant market risk premium than that of a constant interest rate, at least in the short and medium terms and excluding crises.

Furthermore, empirical estimates of the average investor's risk aversion lead to risk premiums much lower than those we observe on most stock markets (whose actual order of magnitude is around 6% per year). Academics refer to this stylized fact as the *equity premium puzzle*.

A standard and simple method for estimating the risk premium to be applied to a particular stock consists in first assessing the market risk premium $(E(R_M) - r)$. This premium is not directly observable since the expected market return $(E(R_M))$ is unobservable. It is estimated by assuming it is constant and equal to the mean of the excess return $R_M(t) - r(t)$ observed over a long period. $R_M(t)$ denotes the return of some global market index (Dow Jones, S&P 500, FTSE, Eurostoxx, Dax, CAC40, Nikkei, Shanghai composite, etc.) and $r(t)$ is a risk-free rate for a maturity of typically one year. Depending on the (long) period used for the estimation, one gets a market premium lying between 5% and 7%.

To estimate the premium $E(R_x) - r$ applicable to a specific stock $x$, one multiplies the market premium by the stock's estimated market beta. This last estimate is generally obtained by a linear regression of $R_x$ on $R_M$. We revisit these issues in much more detail in Chaps. 22 and 23.

**Example 6**

One observes that the return on the S&P 500 exceeded the 1-year risk-free rate by 6% on average for the last 20 years. One thus estimates at 6% the risk premium applied by American investors to their market portfolio. The slope of the regression line for the return $R_x$ of stock $x$ against the return on the S&P 500 (the beta of $x$) is estimated to be 0.8. The premium to be applied to $x$ then is: 4.8% ($= 0.8 \times 6\%$). If at time $t$ the 1-year risk-free rate is 1.5%, the expected return $\mu_x(t)$ should be 6.3% ($= 1.5\% + 4.8\%$).

### 8.2.2.3 Volatility

The variance of the instantaneous return and that of the future price at some horizon both depend on the volatility $\sigma(t)$. The assumption of a constant volatility is grounded on the presumption of stability as to the company's activity (assets) and capital structure (Debt/Equity). Note however that a fall in the stock price entails a larger ratio *market value of debts / market value of equity*. The induced increase in the firm's financial leverage makes the volatility $\sigma(t)$ rise. This dependency makes the assumption of constant volatility doubtful. It is in particular questioned by empirical studies that detect a negative correlation between $S(t)$ and $\sigma(S(t))$.

Volatility estimates can be obtained by using a simple model of price variations of the geometric Brownian motion type (with constant volatility) or on more complicated models with variable volatility, whether stochastic or not. In all cases, an history of stock price observations at past dates $0, 1, \ldots, N$, allows calculating $N$ log-returns over periods of constant length (days, weeks, months):

$$R_1 \equiv \ln(S(1)) - \ln(S(0)), \ldots, R_t \equiv \ln(S(t)) - \ln(S(t-1)), \ldots, R_N$$
$$\equiv \ln(S(N)) - \ln(S(N-1)).$$

- Simple models with constant volatility estimate this volatility using standard estimators:

$$\widehat{\sigma}_p = \sqrt{\frac{1}{N-1} \sum_{i=1}^{N} \left(R_i - \widehat{\mu}_p\right)^2} \text{ with } \widehat{\mu}_p = \frac{1}{N} \sum_{i=1}^{N} R_i.$$

$\widehat{\sigma}_p$ is an estimate for the periodic volatility (weekly if returns are weekly, daily if they are daily, and so on). To find the annual volatility $\widehat{\sigma}$, the periodic volatility $\widehat{\sigma}_p$ is multiplied by $\sqrt{p}$, where $p$ denotes the number of periods in a year (e.g., 52 if log-returns $R_i$ are weekly).

## 8.2 Return Probability Distributions and the Evolution of Stock Market Prices

– Some more realistic models assume a variable volatility.

The simplest of them just uses a larger weight for more recent observations ($R_1$, $R_2$,..) than for older ones ($R_N$, $R_{N-1}$, . . .). It applies exponentially *decreasing* weights, which leads to the estimator:

$$\widehat{\sigma}_p = \sqrt{\frac{1-\lambda}{1-\lambda^N} \sum_{i=1}^{N} \lambda^{i-1} \left(R_i - \widehat{\mu}_p\right)^2}$$

where $\lambda$ is a parameter $<1$ and the sum of the weights $\lambda^{i-1}(1-\lambda)/(1-\lambda^N)$ is one. The most recent observations ($i = 1, 2, \ldots$) are more heavily weighted (and the older ones correspondingly less) as $\lambda$ becomes smaller.

Other more complicated representations attempt to accommodate the heteroskedasticity of log-returns[20] as well as certain observed properties of price dynamics. Among these, the most significant is the fatness of the tails in the empirical distribution (leptokurtosis) of log-returns, and the succession of periods (the number of which is random) with high empirical volatility and with low volatility.[21] These properties of the observed price evolution, which are incompatible with the geometric Brownian motion, have motivated the development and use of the ARCH process (AutoRegressive Conditional Heteroskedasticity) and the GARCH process (Generalized ARCH). The Appendix to this chapter briefly describes these processes, as well as some of their extensions (Exponential GARCH, Integrated GARCH, Fractionally Integrated GARCH, etc.).

Finally remark that any model of the ARCH type, which intends to describe the leptokurtic property and heteroskedasticity of empirical distributions, keeps using Gaussian distributions (see the Appendix). The advantage is that the Gaussian framework is by far the easiest to manipulate. Among its drawbacks, let us mention that the undesirable leptokurticity does not usually disappear completely after ARCH treatment (which explains the inflation in the number of models of GARCH type) and that the possible effect of *long memory* on the volatility (observed in some markets) is not accounted for. An alternative model initiated by B. Mandelbrot in the 1960s and refined since then, consists in abandoning the Gaussian framework for the more general one of $\alpha$-stable laws (also called Pareto-Lévy, of which the normal law is only a special case). Although their study is beyond the scope of this book, we point out that these distributions can have an infinite variance, lead to processes with discontinuous paths, and seem well suited to modeling phenomena more turbulent than the Brownian motion.

---

[20] The term reflects the fact that the variance of log-returns varies over time.

[21] This phenomenon, called *volatility clustering* (persistence of the volatility), is that large (small) variations in the absolute value of the price are usually followed by large (small) variations in absolute value (of whichever sign).

## 8.3 Placing and Executing Orders and the Functioning of Stock Markets

This section, unusually for this book, has a more operational than analytic stand. The potential investor should really know the basic principles of the organization and functioning of stock markets and the main issues at stake, as well as the crucial role of the fund management industry. We present in succession the placing of market orders (Sect. 8.3.1), the system of clearing and settlement (Sect. 8.3.2), delegated fund management (Sect. 8.3.3), and some key statistics about the world's main stock markets.

### 8.3.1 Types of Orders

Quoted stock prices on the stock market result from the tension between buy and sell orders, collected in an order book, leading, as much as possible, to Walrasian equilibrium prices. We refer the reader to Chap. 1 (Sect, 3.4) for an exposition of this mechanism for setting prices by crossing the bids and asks in the order book (fixing or continuous quotation) and a numerical example.

Actually, the mechanism for setting prices and assigning shares described in Chap. 1 has to be refined since orders in the book are not entirely of the same nature because of the different motivations and expectations of the investors. We describe briefly below the rules that apply to all orders and then the properties of the main market orders and their priorities in the order book.

To submit a stock market order, it is necessary to be the owner of an account opened at a registered intermediary, who is either a broker or a dealer.[22]

A market order is made up of *general* clauses (buy or sell, types and numbers of shares to trade) and *specific* clauses concerning how long the order is to stand and the price conditions under which the order is to be executed (see below).

Order execution is subject to priority rules, first by price and then chronologically:

- The buy order with the largest limit (greatest maximum price) is executed before all buy orders with smaller limits; symmetrically, the sell order with the smallest limit (least minimum price) is dealt with before all other sell orders with higher limits.
- Orders of the same type and with the same limits are executed in the exact sequence in which they arrive ("first come, first served").

---

[22]Brokers simply collect orders from their customers while dealers, who tend to be more specialized, also purchase and sell securities or other assets for their own account. Both are compensated by commission fees paid by their customers.

## 8.3 Placing and Executing Orders and the Functioning of Stock Markets

The *validity time* of a market order, whatever it may be, obeys the following conventions. The order may be specified by the investor as good for a *day* (if it is not executed during the stock market's business hours, it is withdrawn), for a *definite duration* ("valid until ...") not ever exceeding the end of a calendar month, or *revocable* (valid to the end of the calendar month, unless cancelled by the client). By default, any order is revocable.

We now describe details of the main types of stock market orders mentioning their respective advantages and disadvantages. There exist other special types of orders but they are more rarely used. The interested readers should contact their broker.

### 8.3.1.1 Limit Orders

This is the standard order. The investor fixes a *maximum price at which to buy* or a *minimum price at which to* sell. When the market opens, *all* buy orders with high limits and *all* sell orders with low limits are executed. Orders with limits equal to the opening price are (partially or totally) executed or not according to the available supply and the "first come, first served" rule. During the stock market session, an order is, partially or entirely, executed or not according to how the price evolves and whether there is an adequate counterparty. A limit order is popular because it *allows control of the execution price and protects against local price variations*.

### 8.3.1.2 Market Orders

A market order is an order to buy or sell a security immediately, without a price limit. This type of order then is guaranteed to be executed, but its execution price is not guaranteed. It generally will execute at or near the current bid (for a sell order) or ask (for a buy order) price. However, the last traded price is not necessarily the price at which it will execute, as this price depends on bids and asks present in the order book. A market order allows the investor to obtain the best possible price at the time it is sent to the market, but this price is unknown and depends on the stock's liquidity.

### 8.3.1.3 Stop-Loss Orders

Such an order (also called a stop order) is an order to buy or sell a stock once its market price hits the value specified by the investor, known as the stop price. The stop-loss order becomes a market order when the stop price is reached. Such an order is generally intended at *protecting an existing market position against a possible market trend reversal* (hence its name of "stop-loss"), or is a bet on the *continuation of a trend* (called *momentum*).

### 8.3.1.4 Futures

Futures transactions for stocks are sometimes possible, market authorities permitting. Against a commission, an investor may postpone the settlement for and delivery of certain stocks traded on the official market. As a result, the buyer benefits from the financial leverage characterizing a forward or futures contract since the amount of cash necessary for the transaction is only a fraction of the spot price.

## 8.3.2 The Clearing and Settlement System

When a security is traded, the change in ownership implies a complex clearing and settlement system whose effectiveness is of considerable practical significance.

The *clearing and settlement* system is the organization that makes possible transfers of property and payments. The buyer pays the price to the seller who simultaneously delivers the shares to the buyer. This means there have to be two circuits involving different intermediaries: one relative to securities and the other to payments.

### 8.3.2.1 Transfer of Securities

To facilitate the transfer of property, the holding of securities must be registered with a *local depository* (*custodian* or *custody service*) whose role is to maintain an up-to-date portfolio of its clients' shares and to liaise with the *Central Securities Depository*. All the custodians holding securities on behalf of their clients must in effect send instructions to the *Central Depository* about the purchase or sale of these securities. The central depository (e.g., Fedwire in the USA and the international EUROCLEAR group for Europe) plays an essential role since it is necessary to know at any instant how many securities are in circulation and how they are divided up among the various local depositories. For the same reason, all transactions (including payment of dividends and coupons) pass through the channel of the Central Depository.

Incidentally, let us remark that some large banks offer to other intermediaries in the market a service called *global custody*, which consists in managing on behalf of the intermediaries all the transactions (in securities and cash) connected with portfolios and, possibly including the valuation of portfolios and calculation of their returns. In this way, the client (a market intermediary) delegates to the *global custodian* everything not directly connected with decisions about investment or divestiture and market trading. The *global custodian* does not trade and does not manage portfolios.

Figure 8.3 illustrates the most general situation (with a global custodian): the issuer of the order, an institutional investor, for example, negotiates with a counter-party[23] the purchase or sale of securities with a broker as intermediary (step 1). After the agreement of the two parties (step 2), the order issuer sends instructions to his/her global custodian concerning the payment and delivery to be made (step 3). This custodian contacts the custodian of the local depository in contact with the central depository for the concerned security. The counterparty of the order issuer proceeds in the same way on his/her side, and the two separate sets of instructions for payment and delivery are reconciled at the Central Depository (step 4). The latter registers the transaction in the name of the local depository, which itself registers it in the name of the global custodian, who in turn registers it in the name of its client who issued the

---

[23]This counterparty obviously proceeds in the same way as the issuer of the order, but this is ignored in Fig. 8.3 to avoid unnecessary complication.

## 8.3 Placing and Executing Orders and the Functioning of Stock Markets

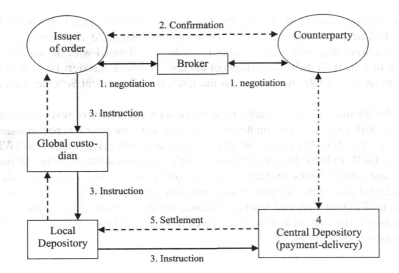

**Fig. 8.3** Complete scheme of clearing and settlement

order. If there are no difficulties (such as instructions that are partly incorrect), the transaction is settled (step 5).

**Remark**
Securities were dematerialized in the 1980s in most countries. Therefore, they are just recorded in accounts held by authorized financial intermediaries, who must, as we have seen, open an account with a Central Depository.

### 8.3.2.2 Transfer of Cash: The Payment System
In addition to the transfer of property of the securities, the clearing-settlement system manages the transfer of money through the banking system. What follows is a sketched discussion of the organization of *payments between banks* (in the final analysis, it is in bank accounts that debits and credits take place).

The interbank payment system being centralized, all money movements pass through a single institution, in principle the Central Bank of the relevant country, which is the compulsory channel for any exchange of funds between two banks in the same country. The same principles drive analogous systems in most countries. Two types of systems are generally used. They do not produce the same amount of risk for the involved financial market(s):

- Real-Time Gross Settlement (RTGS): Each instruction is instantly carried out (in real time) and irrevocably for the bank account that issues it, without waiting for the end of the business day. The use of gross settlement is usually restricted to large amounts. Target2 is the system used in the Eurozone (plus 5 other European countries). Fedwire also operates on this basis.

– The Net Settlement system: over the day, the cash amounts are simply credited to, or debited from, the account of the instructing bank. At a predefined hour, transactions stop, and each member is credited or debited with the *net* amount resulting from all their transactions of the day. The bank transfers become at that moment irrevocable. An example is the UK's BACS payment Schemes Limited.

At the international level, banks need quick, safe, and effective ways of communicating with their foreign counterparts (outside their monetary zone) to execute fund transfers. For this reason in 1973 was created the private network SWIFT (Society for Worldwide Interbank Financial Telecommunication) by a consortium of banks and central banks (initially European ones) whose goal was to provide an international electronic financial communication network connecting the principal agents in different financial markets (shares, bonds, interest and exchange rates, derivatives). The system is nowadays decentralized, and banks directly communicate between themselves.

### 8.3.3 Investment Management

The proportion of financial assets owned by individuals, notably stocks, managed by professionals keeps growing and replacing direct individual management. There are many reasons for this, such as the reduction in transaction costs for large market orders and the accompanying increase for small ones, the creation of corporate pension and savings funds, the increased complexity of savings products, the explosion in derivative instruments, the reduction in fiscal and regulatory diversity in (mainly) the Euro-zone, which makes international diversification more attractive, and the increasing use by professional managers of sophisticated tools beyond the grasp of most individual investors. Since the issues associated with investment management are considerable, we provide below some explanations.

#### 8.3.3.1 General Principles

Investment management (or asset management[24]) is done either on a "proprietary" basis (e.g., by banks or insurance companies who are investing their own assets) or on behalf of a "third party" (client). In the second case, it may be either discretionary (transactions are chosen freely by the managers, not their clients) or non-discretionary (the clients fully decide). With discretionary management, investors delegate the management of their portfolio to an asset manager, often an Asset Management Company (AMC hereafter), usually under an investment mandate against commissions and fees. This is why management for a third party is called delegated management.

We present in subsection (8.3.3.2) the principles governing such a mandate, which is the legal foundation for management on a third party's behalf.

---

[24]We use the two expressions indifferently.

## 8.3 Placing and Executing Orders and the Functioning of Stock Markets

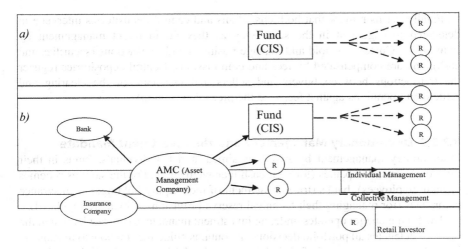

**Fig. 8.4** Asset Management on behalf of a third party

Asset management may be delegated by individual or institutional investors, and maybe collective. In the last case, it takes place through investment funds, also called Collective Investment Schemes (CIS), or more simply funds, which pool the resources from different individual investors and invest them on a unique portfolio. Investors do not own the securities which the fund invests in, but own shares of the fund itself. The equity capital of "open-end funds" is variable (funds ebb and flow from the clients; see Sect. 8.3.3.3 below) and that of "closed-end funds" is fixed.

In discretionary management, the "principal" (investor) and the "agent" (manager) sign a contract (mandate) defining the rights and obligations of each of the parties. In collective management, the investor's subscription to the mutual fund's shares constitutes acceptance of the mandate. Moreover, the mutual fund can itself delegate its management to an AMC that manages the portfolio in conformity with the rules written in the fund's information brochure. Let us note that AMCs can undertake both discretionary management (for individuals) and collective or institutional management. Figure 8.4 presents the organization of the asset management industry in a simplified form.

AMCs can also be members of a banking or financial group. Actually, most banks and insurance companies of large and medium size have created subsidiaries specialized in asset management. These subsidiaries' business is to manage the assets of either the mother corporation or their clients (discretionary and/or collective management).

AMCs must raise a minimum level of own equity. They are subject to regulations governing certification, ethical conduct, and supervision from the relevant supervisory authority (e.g., the Securities and Exchange Commission in the USA, the Financial Service Authority in the UK, the Autorité des Marchés Financiers in France, the BaFin in Germany, etc.).

Finally, let us remark that both custodians and central depositories intervene in delegated management in the same way as they do in direct management by individuals. Custodians hold and manage positions and transactions (securities and cash), and are compensated by fees and commissions. Central depositories register the transactions between buyers and sellers and take care of the clearing and settlement operations against fees (see the preceding section).

### 8.3.3.2 Discretionary Management and the Investment Mandate

Discretionary management by AMCs is aimed at rich individuals, funds in their different forms, companies (for their cash management, and/or the savings schemes of their employees), banks (for management of their equity holdings), or insurance companies (for managing their technical reserve provisions, which are regulated).

The fund manager operates under an investment mandate, a contract by which the investor delegates all portfolio decisions. It contains guidelines on how to manage an individual amount or a pool of capital, based on the risk/return tradeoff chosen by the investor(s). It mentions in particular: the capacity of the investor; the objectives of the portfolio, with possibly the adopted benchmark(s); the manager's compensation; the financial instruments to invest in, and possibly a specific authorization to participate in derivative markets; how reporting is done; and how performance is measured (possibly by reference to a benchmark).

### 8.3.3.3 Collective Management and the Workings of Funds

Collective Investment Schemes (Investment funds or Investment Companies in the USA) usually function as open-end funds with variable equity capital.

(i). *How open-end funds with variable equity capital work.*

Closed-end funds have an equity capital fixed initially, which remains constant later on; however, after their issue, the shares of a closed-end fund may be bought and sold on the secondary market. By contrast, open-end CIS continually issue shares or units of account against the funds they collect, hence their equity capital varies constantly. The collected funds then are invested in stocks, bonds, or monetary assets. Mutual funds are the most representative (but not the only) members of the open-end CIS family.

If the open-end fund has no debt, the value of the shares or units making up the funds' equity capital (liabilities) is equal at any moment to that of the securities in its portfolio (assets) through a specific financial mechanism explained below. Consequently, the financial performance of a *single* share (or unit) of the mutual fund follows that of the overall portfolio and the *total* value of the mutual fund depends on the portfolio's market value, called *Net Asset Value* (NAV) or *Liquidation Value*. This *total* value fluctuates with withdrawals and subscriptions by investors on the one hand, and with the evolution of market prices on the other. In contrast, the value of a *single* particular share or unit depends only on market price fluctuations.

## 8.3 Placing and Executing Orders and the Functioning of Stock Markets 291

- More generally, in addition to equity, the liabilities of a CIS may include debts. Besides, different operating costs affect negatively the return for investors, which as a result is slightly less than the return on the portfolio of assets and liabilities of the CIS.[25] We adopt the following notation:
- $n(t)$: The number of units (or shares) making up the open-end fund equity capital.
- $NAV(t)$: The net asset value of the fund = the value of the securities in the portfolio (assets) minus the value of the debts = the total liquidation value of the fund.
- $v(t)$: The NAV per share, which is equal to total NAV divided by number of shares: $v(t) = \frac{NAV(t)}{n(t)}$.

Remark that the portfolio of a CIS will contain, as the case may be, shares, bonds, monetary assets, and, under certain conditions, debt on the liability side, short positions in different instruments and derivative products.

The NAV per share is computed and published periodically. "Large" mutual funds are required to calculate their liquidation value and publish it daily.

The NAV per share is the basis for the calculation of the subscription price $v_s(t)$ paid by a buyer as well as for the price of redemption (buy-back) or transfer of a unit $v_r(t)$.

The subscription price is slightly higher than the NAV due to the cost of joining the fund. This entry cost is calculated using a rate $c_s$ so that $v_s(t) = (1 + c_s)v(t)$.

The redemption price is smaller than the NAV because of the fees paid for leaving the fund; it is calculated with a rate $c_r$ and thus is equal to $v_r(t) = (1 - c_r)v(t)$.

The entry and exit fees as well as the management fees (calculated as a percentage of the NAV) allow paying the various costs that result from the management activity itself (brokerage fees, deposit services, etc.) and compensating the fund manager. At any time t, the difference $(v_s(t) - v_r(t) = (c_s + c_r)v(t))$ can be interpreted as a bid-ask spread.

The open-end fund sells its own shares on demand and uses the proceeds to increase its portfolio. Conversely, it buys back its own shares by selling, if necessary, part of its assets. It creates thus a (theoretically) infinitely elastic supply/demand for its own shares at the NAV price (with a small bid-ask spread). The possibility of selling or buying shares on the basis of the NAV, from the fund itself, at least once a day, "pegs" the market price of the open-end fund share at its NAV value.

We just described how standard mutual funds work. An ETF (Exchange-Traded Fund) can be considered as an open-end variant of the standard mutual fund with the additional feature of shares being traded continuously, like stocks. An arbitrage mechanism, involving institutional investors, minimizes the potential deviation between the current market price of an ETF share and its NAV.

In contrast, the market value of a closed-end fund depends on supply and demand on the secondary market and may be different from its NAV.

---

[25] But not necessarily smaller than the return on the same investment undertaken directly by the investor, because of the transaction costs that, under all circumstances, are to be paid.

## (ii). *The benefits of collective management.*

Collective management has a number of advantages. The first is that it does allow less well-off investors to diversify their portfolios with a limited expenditure. The second is the reduction in transaction costs resulting from buying and selling units or shares of a mutual fund as opposed to buying and selling each element of the fund's portfolio. A final advantage, although often much less obvious than the preceding ones, is the relative skill of the professional manager. The possibility that the manager outperforms a non-professional who follows a passive strategy is questionable, especially if the extra fees are accounted for.

## (iii). *Different types of funds.*

There is a wide variety of Investment Funds. One can distinguish several categories based on the legal rules under which they operate as well as on the composition of their portfolios or their risk-return objective. Mutual funds, hedge funds, and most ETFs are open-end CISs. Mutual funds exist in most developed countries and operate under slightly different but similar rules under different names: mutual funds in the USA, SICAV[26] in most countries in Europe, OEIC in the UK,[27] etc. In the USA, Unit Investment Trusts (UIT) may offer a fixed (unmanaged) portfolio of securities and have a predetermined termination date, and Collective Investment Trusts (CIT) are often a type of tax exempt-low fees pooled investment vehicle available in the context of a retirement plan. There are also differences in the conditions under which CIS are marketed. For instance, some funds, such as mutual funds and ETFs, are sold to the public and some others are only sold privately, such as hedge funds. In Europe, some funds called "coordinated UCITS" (Undertakings for Collective Investments in Transferable Securities) hold a "European passport" whereby they are authorized to market their products throughout the European economic union.[28]

CIS also differ by their portfolio composition: stocks; bonds; monetary assets; with guaranteed capital; diversified; speculative (using volatile derivatives, such as options or structured products). This composition depends on the investment target and on the risk-return tradeoff accepted by the investor(s). Some CIS, either Indexed funds (which are mutual funds) or ETFs, aim at replicating a given Index while others deviate more or less from a benchmark (see Chap. 25 for a thorough analysis of benchmark-based strategies). Some Investment funds may invest in specific geographical or industrial sectors.

---

[26] French acronym for Société d'Investissement à Capital Variable. They are present under this name in Spain, Italy, Belgium, France, Switzerland, Czech Republic, Luxembourg,and Malta.

[27] Open-Ended Investment Company.

[28] Those UCITS with a "European passport" have to obey certain regulations as to their governance structure, activities, portfolio composition, and information made publicly available. As of mid-2020, there is concern and uncertainty on whether or not UK-domiciled funds will keep their passport.

## 8.3 Placing and Executing Orders and the Functioning of Stock Markets    293

**Table 8.1** The seven main stock markets in December 2018 (Capitalizations in trillions of $)

| NYSE (New York) | Nasdaq | Tokyo | London | Shanghai | Hong Kong | Euronext |
|---|---|---|---|---|---|---|
| 22.92 | 10.86 | 5.68 | 4.59 | 4.03 | 3.94 | 3.93 |

(iv). *Size of the asset management industry.*

The size of this industry is huge. Worldwide, at the end of 2019, the assets under management (stocks being only one of the components) amounted to $90 trillion, with $42 trillion in the USA, $22 trillion in Europe (about 76% were UCITS), and $18 trillion in Asia.[29]

### 8.3.4 The Main Stock Markets

The world market capitalization is about $94 trillion as of December 2019 (hence before the COVID-19 crisis), i.e., the same order of magnitude as the total of Gross National Products. It is made up of roughly 45% in North America, 31% in Asia-Pacific, and 23% in Europe and Middle East. Furthermore, although the GNPs of Europe and North America are comparable, the American market capitalization is about 50% larger than Europe's, which is indicative of a relatively large degree of "marketization" of American finance (the total amount of stocks, bonds, debts, and bank assets being comparable for the two continents). An analogous phenomenon is observed in Europe, where UK's market capitalization, relative to its GNP, is twice as large as in the other European countries.

Table 8.1 shows the capital handled in shares on the main stock markets of the world, at the end of December 2018 (source: World Federation of Exchanges).

According to this table, the most important stock markets are the *NYSE, the Nasdaq,* and the *Tokyo Stock Exchange.*

*Euronext* was born in 2000 out of the fusion of the stock and derivatives markets in Amsterdam, Brussels, and Paris. In 2002, it merged with the similar Portuguese (Lisbon) market and bought the LIFFE (the derivatives market based in London).In April 2007, the New York Stock Exchange (NYSE) and Euronext merged to form the NYSE-Euronext as a holding company divided into two entities: NYSE-New York and NYSE-Paris. The transcontinental group NYSE-Euronext was then acquired (for $11 billion) in November 2013 by the American exchange operator ICE (InterContinental Exchange). In 2016, however, NYSE and Euronext became completely separate entities.

The *London Stock Exchange* (LSE), which is more than 300 years old, remains one of the principal stock and bond markets in Europe and in the world, both in the volume of equity and bond issues and the volume of transactions on the secondary market. It provides the same range of services as Euronext (more than 300 financial

---

[29] Source: The BCG report on asset management (2020).

brokers over the world are members). As to derivative products, it is specialized in warrants (transferable securities analogous to long-term options) written on individual securities or stock indices.

The other main European stock exchanges are located in Frankfurt (*Deutsche Börse*), Milan, Zurich, and Madrid.

In the USA, the main stock exchanges are the *New York Stock Exchange (NYSE-New York* [www.nyse.com], the *NASDAQ* [www.nasdaq.com] and, far smaller, the *American Stock Exchange (*called *AMEX)* [www.amex.com].

Created in 1792, the *NYSE-New York* is by far the largest market in the world: on average 1.5 billion shares are traded daily, with a value of over 50 billion dollars. As far as the market value of domestic stocks alone is concerned, out of a world total at the end of 2018 of about \$78 trillion, the NYSE made up roughly \$23 trillion (first rank). The *NASDAQ* is an electronic market specializing in new technologies and communications, and its market capitalization on the same date ran to \$10.9 trillion, making it the second-largest market in the world. Today, ICE is the owner of 5 stock market locations in the US: NYSE, AMEX, ARCA (Archipelago Exchange), EDGE, and LIFFE U.S., the latter purchased from the London futures market and specializing in options and futures.

In the rest of the world the principal exchanges are: the *Tokyo Stock Exchange* (third in the rankings) [www.tse.org.jp/english], the *Toronto Stock Exchange* [www.tse.com], the *Sao Paolo Stock Exchange* [www.bovespa.com], the Mexican *Bolsa de Mexico* [www.bmv.com], the *Australian Stock Exchange* [www.asx.com.au], the *Singapore Stock Exchange* [www.sgx.com], the *Hong Kong Stock Exchange* [www.sehk.com.hk], and the *Shanghai Stock Exchange* [www.sse.com.cn].

Finally, we observe that, in the last fifteen years or so, financial transactions have undergone significant changes linked, mostly, with the development of electronic networks. In this connection, note the creation of numerous MTF (Multilateral Trading Facilities) mainly facilitating the connections between various networks. Many traditional exchanges have set up their own MTF.[30] In 2018, MTFs carried more than 33% of all stock transactions on secondary markets in Europe.

## 8.4    Stock Market Indices

A market index represents the value of a portfolio at any instant and its evolution allows estimating the return on the portfolio between any two dates. In this chapter, we examine indices that represent the value of a stock portfolio.

An index reflects the value of a portfolio of $N$ stocks; it is thus characterized by these $N$ securities (its basis, or coverage) as well as by the way it is calculated (its weighting). The coverage and weights at a reference date (the start date of the index) allow computing the index value at any time.

The role of an index is threefold:

---

[30] Such as Smartpool (NYSE) and Turquoise (LSE).

## 8.4 Stock Market Indices

- To represent the evolution of a collection of market values; in this respect it allows measuring the performance of a market, and by comparing it with another index, to compare the performance of two different markets.
- To serve as a benchmark in estimating the performance of investors or managers of stock portfolios, or as a management tool for the latter (*see* index management in Chap. 24).
- To constitute an underlying asset for derivative products (forward contracts, *swaps,* or options).

We now consider the composition of, and different ways of calculating, an index (Sect. 8.4.1) and then briefly describe the world's main market indices (Sect. 8.4.2).

### 8.4.1 Composition and Calculation

#### 8.4.1.1 The Composition of an Index

A stock index is but a portfolio. In the following, we simply identify the index with the portfolio it represents. The choice of the $N$ stocks making up the portfolio depends in the first place on the reason for creating the index, which is, in general, to represent a market, a market sector, or a subpopulation of stocks, for example:

- A geographical sector such as a country (the USA, UK, Japan, France, Germany, etc.), a group of countries, an economic or monetary zone (Europe, Asia, etc.), or even the world.
- An economic sector (financial institutions, industrial or technological firms, etc.).
- Companies with large, moderate, or small capitalizations.

The index's $N$ stocks often correspond to the largest and most liquid market capitalizations (in terms of free float or by transaction volume) in the represented sector. Since the relative size of market capitalizations varies over time, it is necessary to revise the composition of an index periodically by replacing a stock having lost too much value with one of increased value.

The size of the index (the value of $N$) results from a tradeoff between conflicting requirements:

- The desire of having a simple index, which is liquid and easily reproducible, which suggests using a small number $N$ of securities.
- The desire of having a comprehensive and representative sample, which suggests choosing a larger $N$.

Once the $N$ stocks are selected, like any portfolio the index can be defined in two ways:

- By the *weights* given each of the $N$ stocks, with the weights summing to one.
- By the *numbers $n_i$* of shares of each stock $i$ ($i = 1, \ldots, N$).

In effect, an index can be weighted according to market capitalizations or in different ways. Furthermore, an index can take account of price changes only (to reflect capital gain or loss only) or can integrate the dividends into its calculation to reflect total return.

### 8.4.1.2 Weighting by Market Capitalizations (Total or Floating)

Indices weighted by market capitalizations are defined as portfolios with weightings proportional to the component stocks' market capitalizations. As a result, the weight of each stock reflects its relative importance in the sector that the index represents.

This way of computing weights is the most natural, and, on a priori *grounds,* the most satisfactory. Let us note, however, that two variants of weighting by market capitalizations exist: the weights may be calculated from the total market capitalization or from just the floating part of the capitalization (i.e., the part not held by core shareholders who in principle keep their shares for long-term strategic reasons). The principles and the formulas used are the same in both cases (just the meaning of the capitalization variable is changed).

Despite its apparent simplicity, developing an index weighted by market capitalizations does raise some technical problems, which we explain briefly.

- First, consider a period $(s, t)$ *during which the companies involved in the index issue no new* shares *and no revision of the composition of the index is undertaken.* The index portfolio is simply *obtained by a static strategy* ("buy and hold") involving no transaction between $s$ and $t$; therefore, for each component $i$, the index's portfolio contains a *constant number of shares* equal to the initial $n_{is}$. Remark that, even if the numbers of shares remain constant, the *weights vary*; indeed, the weights of stocks with a relatively strong price growth increase to the detriment of those of stocks whose prices fall (or rise less than the average).

The index's value $I$ thus evolves, over the period $(s, t)$, according to the following equation:

$$I(t) = I(s) \frac{\sum_i n_{is} S_i(t)}{\sum_i n_{is} S_i(s)}. \tag{8.20}$$

Here, $n_{is} = n_{it}$ is the (constant) number of shares of $i$, at $s$, and between $s$ and $t$; only shares in the free float are counted if weighting is in terms of floating market capitalizations.

$S_i(t)$ = price of share $i$ at $t$; $n_i S_i$ = market capitalization (total or floating) of $i$.

Equation (8.20), written for the interval $(t, t + 1)$ of length one period (period $t$), becomes

## 8.4 Stock Market Indices

$$\frac{I(t+1)}{I(t)} = 1 + \sum_i x_{it} R_{it} = 1 + R_{It}$$

where $x_{it} = \frac{n_{it} S_i(t)}{\sum_{j=1}^{N} n_{jt} S_j(t)}$ is the weight of $i$ in the index, $R_{it}$ represents the return of share $i$ during the period $t$ and $R_{It}$ that of the index. Recall that this return is only made up of the change in capital value (it does not include dividends). This last relation simply reflects the fact that the (arithmetic) return on a portfolio is the weighted average of the returns of its components.

- Next, consider the still simple case of distribution of free shares $i$, or stock split, implying a constant market capitalization but a proportional decrease in the price per share $i$. It suffices in this case to increase the number $n_{it}$ of shares involved in the computation of the index in proportion with the increase of the number of shares. By doing so, the split does not alter the weight for $i$ in the index and a discontinuity at the split date is avoided. Such a discontinuity would fallaciously reflect a negative return.

- Now consider the more complex general case where, during the period $(s, t)$, companies in the index issue new non-free shares to increase their equity capital. As a result, the numbers $n_i$ should not stay constant. First, capitalizations increase (theoretically by the amount of the new shareholders' contribution) and second, the share price is mechanically lower if the new shares are issued below the market price, with rights offering (dilution). These two effects should be appropriately reflected in the index value, which must express the new capitalizations (implying a new array of weights) while avoiding any discontinuity in its evolution. The second condition requires an adjustment to enforce the continuity of the index, in the absence of which wrong returns would be computed.[31] Therefore, if there is an equity issue between s and t, the computation of the index must be revised so as to preserve its continuity and reflect the new market capitalizations, according to the equation:

$$I(t) = I(s) \frac{\sum_{i=1}^{N} n_{it} S_i(t)}{\alpha \sum_{i=1}^{N} n_{is} S_i(s)} \tag{8.21}$$

where $\alpha$ is an adjustment factor and the $n_i$ are changed to reflect the new weights. The adjustment factor ensures that, at the revision date $u$, the index is continuous ($I(u^-) = I(u^+)$).

---

[31] As seen above (Sects. 8.1 and 8.1.3), only the adjusted price variation constitutes a genuine return, and only the adjusted share price shows no discontinuity as a result of an equity issue.

- A problem analogous to that we have just examined arises, and has a similar solution, if the index's composition is revised (i.e., when some stocks leave the index and others enter).

Suppose, for example, that on date $u$ $(s < u < t)$ a new stock 1 replaces the old stock 1. In equation (8.21), which still holds, $n_1$ and $S_1$ then do not refer to the same asset on dates $s$ or $t$. In addition, if there is no adjustment ($\alpha = 1$), the index values $I(u^-)$ and $I(u^+)$ just before and just after the replacement will be different, which is not desirable. Distinguishing for clarity the old security as 1 and the new as $1'$, we define

$$\alpha = \frac{n_{1'u}S_{1'}(u) + \sum_{i=2}^{N} n_{iu^+}S_i(u)}{n_{1u}S_1(u) + \sum_{i=2}^{N} n_{iu^-}S_i(u)},$$

and then apply (8.21), which means that for any $s < u$ and $t > u$

$$I(t) = \frac{1}{\alpha}I(s)\frac{n_{1't}S_{1'}(t) + \sum_{i=2}^{N} n_{it}S_i(t)}{n_{1u}S_1(u) + \sum_{i=2}^{N} n_{iu}S_i(u)}$$

and which implies that $I(u^-) = I(u^+)$ (i.e., continuity of the index is ensured).

Under the assumption that $m$ operations requiring adjustments (equity issues, index composition changes) take place from $s$ to $t$, we write

$$I(t) = \frac{1}{\alpha_{s,t}}I(s)\frac{\sum_{i=1}^{N} n_{it}S_i(t)}{\sum_{i=1}^{N} n_{is}S_i(s)}$$

with $\alpha_{s,t} = \prod_{j=1}^{m} \alpha_j$.

where $\alpha_j$ denotes the adjustment factor for the $j^{th}$ operation.

### 8.4.1.3 Other Weightings and Ways of Calculating the Index

Although indices derived from weightings by capitalizations (total or floating) are the most typical and most common, other weightings are adopted because of their

## 8.4 Stock Market Indices

computational simplicity and, in some cases, because of the ease with which the index can be duplicated by a real portfolio.[32] We can mention:

- Price-weighted indices, which reflect the evolution of a portfolio containing *precisely one share* of each company included in the index. Such an index evolves between $t–1$ and $t$ according to the equation

$$I(t) = I(t-1) \frac{\sum_{i=1}^{n} S_i(t)}{\alpha_t \sum_{i=1}^{n} S_i(t-1)}.$$

Thus, the index contains one share of each of its components, whatever their market capitalizations (the number of shares issued by each company is, oddly, irrelevant!).

The Dow Jones Industrial Average (or simply Dow Jones) and the Japanese Nikkei 225 are the two most striking examples of price-weighted indices.

Two main arguments are in favor of the apparently economically absurd principle of price weighting. On the one hand, building a portfolio duplicating the index (a static strategy with an equal number of shares of each stock) is very easy. On the other, the correlation observed empirically between price-weighted indices and the corresponding index weighted by market capitalizations is often (surprisingly) high.[33]

- Return-based indices (which account for dividends). Indices as described above, which are the most common, only reflect the evolution of market prices and thus are *price-based* indices. Since they do not include dividends, their return reflects capital gains only. This is why other indices (or other versions of the previously defined indices) include dividends. Dividends are thus deemed reinvested in the index's portfolio which is thus self-financing (there is no cash inflow or outflow). Reinvestment of dividends implies a rebalancing of the portfolio reflected by the index that maintains its continuity at *distribution dates*, namely

$$I(t^-) = \sum_{i} n_{it-} S_i(t^-) = \sum_{i} n_{it-} \left( S_i(t^+) + Div_i(t) \right) = I(t^+) = \sum_{i} n_{it+} S_i(t^+).$$

These indices are *return-based* indices. They most often use market capitalization weights, as in the preceding equation. There are, however, *price-weighted return-*

---

[32] This point is particularly important when the index is used as a benchmark for passively managed funds (*see* Chap. 24).

[33] For example, the correlation coefficient between the Dow Jones and the S&P 500 is about 0.97.

**Table 8.2** National European indices

| Name | Location | Number of stocks |
|---|---|---|
| DAX | Frankfurt | 30 |
| FTSE (Footsie) | London | 100 |
| CAC 40 | Paris | 40 |
| SMI | Zurich | 25 |
| MIB | Milan | 30 |
| IBEX | Madrid | 35 |
| AEX | Amsterdam | 25 |

*based indices.* The returns then are computed simply as arithmetic or geometric (for the Value Line index) averages of individual returns.

As we will see in the following section, most of the significant indices are computed using market capitalizations (Standard & Poor's, Stoxx, FTSE, Nikkei 300, CAC 40, Dax, etc.), but one does find some notable exceptions: Dow Jones indices are price-weighted; Value Line is a geometric mean of returns.

### 8.4.2 The Main Indices

The main stock exchanges and some index providers publish a battery of market indices that represent the market, an economic sector, or a specific collection of stocks. We only mention the main ones.

#### 8.4.2.1 North American Indices
The main US indices are the following:

- The Standard and Poor's Indices (weighted by capitalizations), of which the most important is the S&P 500, which includes 500 companies chosen on the basis of size and liquidity and accounts for more than 70% of the US market capitalization.
- The Dow Jones Indices of which the best known, the DJ Industrial Average (DJIA), is computed as the average of 30 prices of American Blue Chips.
- The NASDAQ 100 and NASDAQ composite indices, covering technology companies.
- Indices for the NYSE, the AMEX, the Wilshire 5000, Value Line, and the MMI (Major Market Index) on which futures and options contracts are traded.

In Canada, the S&P/TSX 60 covers 60 large Companies listed in the Toronto Stock Exchange and the S&P/TSX. Composite covers about 250 companies representing roughly 70% of the Toronto Stock Exchange.

#### 8.4.2.2 European Indices
We distinguish national indices (relative to different countries) from global indices involving the whole of Europe or a number of European countries.

## 8.4 Stock Market Indices

- Table 8.2 displays the main European national *Blue Chips*.
- The leading European global indices are designed by Stoxx, an index provider owned by Deutsche Börse Group, and are either restricted to the Euro-zone (Euro Stoxx index) or covering all Europe (the Stoxx Europe Indices or simply Stoxx). They are all floating capitalization-weighted. Price and return indices are available. Different extensive, restricted, and sector indices are also available, among which we mention:
- The Stoxx Europe 600, which contains 600 stocks of large mid and small capitalizations of 17 European countries (not restricted to the Euro-zone), representing approximately 90% of the European stock free-float capitalization.
- The Stoxx Europe 50 (for all Europe) and the Euro Stoxx 50 (limited to the Eurozone), which contain 50 European Blue Chips.

### 8.4.2.3 Main Asian Indices

- There are various Asian national indices, of which:
  - In Japan, the main stock market index is the Nikkei 225. It includes 225 Japanese Blue Chips and is computed, like the DJIA, as the sum of 225 uniformly weighted prices.
  - In China, the main stock market indices are the Shanghai Stock Exchange (SSE) composite, which tracks all stocks traded in the Shanghai Stock Exchange, the Hang Seng covering the largest companies trading in the Hong Kong Stock Exchange, the Shanghai Shenzhen CSI 300 and the SZSE composite.
  - In India, the main stock market indices are the National Stock Exchange (NSE) Nifty50 and the Bombay Stock Exchange (BSE) Sensex.
  - In Korea, the KOSPI (Korea composite Stock Price Index) covers the common stocks traded in the Korean Stock Exchange.
- Besides, the S&P Asia 50 is an Asian global index (part of the S&P Global 1200) that includes companies from Hong Kong, South Korea, Singapore, and Taiwan.

### 8.4.2.4 Main Worldwide Global Indices

- The MSCI (Morgan Stanley Composite Index) indices are capitalization-weighted and include dividends; they are calculated as global indices (World, Pacific, North American, European, etc.) or national indices (for all major countries).
  The MSCI World includes 1644 stocks from 23 developed countries in the world. It does not include stocks from emerging countries but the MSCI All Country World Index (ACWI) covers both developed and emerging countries. It is computed with and without dividends, and in different currencies.
- Among the other most popular world-global stock market indexes, we can mention: The S&P Global 100 Index (100 very big multinational capitalizations), the S&P Global 1200 Index (composed of 7 regional indices, it includes 1200 companies, among which those of S&P Global 100, covers 10 sectors, 31 countries, and approximately 70% of the world capitalization), the Dow Jones Global Titans 50 and the FTSE All-World Index.

## 8.5 Summary

- Stocks and shares are the parts of the *equity* issued by a firm and materialize property rights on the firm in contrast to debt securities. They can be issued and traded on the stock market.
- They give a *claim* to the dividends and to the firm's net assets in case of liquidation. They also give the right to oversee the firm's management by way of the right to vote in the shareholders' meetings.
- Payment of interests and reimbursement of debt principal(s) have *priority* over payment of dividends and liquidated assets. This priority of debts over equity implies that the latter absorbs most risks and provides a guarantee to debtholders.
- The *market* value of a share is to be distinguished from its *nominal* value (which is an accounting, not a financial, notion).
- There is a wide variety of possible legal arrangements implying, in particular, varying *degrees of responsibility* of the shareholders. The shares of a limited liability company are anonymous, and the shareholders have a responsibility *limited* to their investments; the value of such shares thus cannot be negative.
- When the firm issues new shares, current shareholders have a *preferential subscription* right to new shares.
- The market value of a share is theoretically equal to the discounted value of all future dividends. The discount rate $r_e$ should be the *total return required* by the shareholders that includes a risk premium. A simple model grounded on an assumption of a constant and infinite dividends growth at rate g yields the Gordon-Shapiro formula which amounts to discount, up to infinity, the current dividend at a discount rate $(r_e - g)$.
- By definition, the *PER* (Price Earnings Ratio) is the ratio of the observed share price to the earning per share. When the PER is used for valuation purposes, the share value is estimated by multiplying the earnings per share by the PER observed for companies comparable in terms of activity, risk, and growth.
- The arithmetic return of a share over the period (0, 1) is $R = [S(1) - S(0)] /S(0)$, where $S$ denotes the stock price; in presence of a dividend $(Div)$, one has: $[S(1) - S(0) + Div] /S(0)$.
- The *log-return* over the same period (0, 1) is $\ln S(1) - \ln S(0)$.
- Over several periods, log-returns are *additive*.
- A standard way to take risk into account consists in assuming, as of time 0, S (1) log-normal, implying a *normal log-return*. Since a log-normal variable does not take on negative values, such a representation is compatible with limited liability.
- However, the empirical evidence *only partially* confirms the assumption of log-return normality, as the observed log-return distributions show skewness and a left tail thicker than that of the normal distribution characterizing a higher probability of negative extreme values.
- The log-normality assumption leads to a representation of the continuous time returns by geometric Brownian motions: $dS/S = \mu dt + \sigma dW$, where $\mu$ is the instantaneous mean return (drift) and $\sigma$ the volatility (diffusion parameter).

## 8.5 Summary

- $\mu$ is the sum of the risk-free interest rate and a risk premium. The *premium* is required by risk-averse investors and is set at a level compatible with market equilibrium.
- The simplest *estimation* of the volatility $\sigma$ relies on a sample of returns over past time intervals of equal duration and on a discretized version of the geometric Brownian motion assumed for these returns.
- Stock market orders are *executed* through a broker or a dealer. They contain *general* clauses (buy or sell, types and numbers of shares to trade) and *specific* clauses (duration of the order, price conditions). Specific clauses are numerous ("limit" orders, "at any price," "stop loss," market orders, etc.); for instance, in contrast with an order "at any price," a limit order fixes a maximum price at which to buy or a minimum price at which to sell.
- The *clearing and settlement* system ensures the transfer of property as well as the payment implied by a stock transaction. This process involves different intermediaries.
- *Asset management* is done either by banks or by insurance companies investing on their own account or on behalf of a "third party." In the second case, investors *delegate* the management to an asset management firm in the form of a mandate against commissions. Delegated asset management may be for individual investors, for institutional investors or collective through an investment fund.
- *Mutual funds* usually function as open-end funds with variable equity capital. The total value of the shares making up the fund's equity capital is equal at any time to the net asset value of the fund's portfolio.
- The North American stock markets represent almost 50% of the world capitalization. The main stock markets are the New York Stock Exchange (NYSE) and the Nasdaq in the USA, the Tokyo, Shanghai and Hong Kong markets in Asia, and the Euronext and the London Stock Exchange in Europe.
- A stock *index* may be considered as a portfolio of stocks representing a market, a market sector, or a subpopulation of stocks, for example, a geographical sector such as a country, an economic sector (financial, industrial or technological firms, etc.), or companies with large, moderate, or small capitalizations.
- The index evolution allows estimating the return between any two dates on the portfolio represented by the index. It can be used as a *benchmark* for portfolio managers, or as an underlying asset for derivatives.
- As any portfolio, an index is defined by its composition (N stocks and N weights). Indices weighted by market capitalizations are the most common but other weighting schemes also exist.
- The main US indices are the Dow Jones Industrial Average, the NASDAQ Composite, and the S&P 500 Index. The main global European indices are the Stoxx indices, in particular the Euro Stoxx 50. Important national indices are the FTSE (England), the Dax (Germany), the CAC40 (France), the Nikkei 225 (Japan), and the Shanghai and Shenzhen indices (China).

# Appendix 1

## Skewness and Kurtosis of Log-Returns

We present successively a measure of asymmetry (*skewness*), a measure of flattening (*kurtosis*), and a measure of the two properties combined.

### (a) *Asymmetry or skewness*

The asymmetry of the probability density of a random variable $R$ can be estimated using its *skewness Sk,* defined as the third moment of the distribution of R standardized with respect to its mean and standard deviation:

$$Sk = E\left(\frac{R - E(R)}{\sigma(R)}\right)^3 = \frac{E[R - E(R)]^3}{\sigma(R)^3} .$$

This coefficient vanishes for a symmetric distribution (such as the normal distribution) and is negative if the density of very small values (in comparison to the mean) exceeds that of very large values (as is the case for stock returns shown in Fig. 8.1).

The empirical estimator for $Sk$ is computed from $n$ observations by the formula

$$\widehat{Sk} = \frac{\frac{1}{n-1} \sum\limits_{t=1}^{n} (R_t - \overline{R})^3}{\widehat{\sigma}(R)^3},$$

where $\overline{R}$ and $\widehat{\sigma}$ denote the empirical mean and standard deviation.

The empirical skewness of log-returns for stocks is mostly negative, which means that the frequency of very negative returns is greater than that of very positive returns (asymmetry to the left).

### (b) *Kurtosis*

The probability density of extreme values (very much smaller or very much larger than the mean) can be either higher or lower than that for the normal distribution. The existence of fat tails (to the left or to the right, or both), or thin tails, is estimated by the *kurtosis K* defined as the fourth moment of the distribution of the variable $R$ (standardized with respect to its mean and standard variation):

$$K = E\left(\frac{R - E(R)}{\sigma(R)}\right)^4 = \frac{E[R - E(R)]^4}{\sigma(R)^4}.$$

Standard normal distributions have a kurtosis of 3. A distribution with kurtosis larger than 3, called leptokurtic, is characterized by a density of extreme values (very

Appendix 2                                                                                      305

large or very small) greater than that of a normal variable with the same expectation
and standard deviation and a smaller proportion of near-average values; conversely,
a kurtosis smaller than 3 (a platykurtic distribution) means a small proportion of
extreme values.

The kurtosis $K$ is estimated from a sample of $n$ observations using

$$\widehat{K} = \frac{\frac{1}{n-1} \sum_{t=1}^{n} \left(R_t - \overline{R}\right)^4}{\widehat{\sigma}(R)^4}.$$

The empirical distributions of stock log-returns are usually leptokurtic.

### (c) *The combination of a left asymmetry and leptokurtosis*

The empirical distribution of stock returns and stock indices are asymmetric to the
left ($\widehat{S}_k < 0$) and leptokurtic ($\widehat{K} > 3$).[34] These two properties imply *a left distribution
tail that is fatter than that of the normal distribution* with the same expectation and
standard deviation, as in Fig. 8.1.

Finally, the *Jarque-Bera statistic* combines skewness and kurtosis and writes:

$$JB = \frac{n-2}{6} \left(\widehat{Sk}^2 + \frac{1}{4}\left(\widehat{K} - 3\right)^2\right).$$

This statistic means we can test the hypothesis of normality for the law of $R$ by
using the fact that if normality holds, then $JB$ follows a $\chi^2$ distribution with 2 degrees
of freedom (using $Sk$ and $K$).

---

# Appendix 2

## Modeling Volatility with ARCH and GARCH

The empirical distributions of log-returns are characterized not just by the fatness of
their tails (leptokurtosis) but also by the time variation of their volatility
(heteroskedasticity). Robert Engle (1982) was the first to suggest modeling price
variations using variances that change with time, in a way compatible with the
existence of a market with successive calm and turbulent periods (*volatility clustering, or volatility persistence*) and the phenomenon of *leptokurtosis*. Assume that
stock log-returns are of the form

---

[34] For example, during the period 1985–2005, the empirical skewness and kurtosis of the Dow Jones
Industrial Average (DJIA) were respectively equal to − 0.9 and 7.5.

$$R_t = \mu + \varepsilon_t$$

where $\mu$ is a possibly vanishing constant with[35]

$$\varepsilon_t \sim N(0, \sigma_t) \qquad (8.\text{A-1})$$

and $N$ denotes the normal law. Because of this, the Gaussian framework is preserved. Engle then assumes that the variance $\sigma_t{}^2$ is a linear combination of the squares of the $p$ most recent perturbations observed, as in the equation

$$\sigma_t^2 = \omega + \sum_{i=1}^{p} \alpha_i \varepsilon_{t-i}^2 \qquad (8.\text{A-2})$$

with $\omega$ a positive constant and the $\alpha_i$ nonnegative constants, for every $i$, to ensure the positivity of the variance.

Equations (8.A-1) and (8.A-2) define a process called $ARCH(p)$, for "AutoRegressive Conditional Heteroskedasticity" of order $p$. The number $p$ which counts significant past perturbations and the values of the parameters should be estimated with the usual econometric procedures (the standard method used is Maximum Likelihood).

The *conditional* moments (conditioned on the information $I_{t-1}$ available at time $t-1$, namely the past values in the time series of log-returns) of $\varepsilon_t$ are given, respectively, by

$$E(\varepsilon_t | I_{t-1}) = 0 \quad \text{and} \quad \text{Var}(\varepsilon_t | I_{t-1}) = \sigma_t{}^2 = \omega + \sum_{i=1}^{p} \alpha_i \varepsilon_{t-i}^2$$

thus the conditional variance varying through time.

These conditional moments are the ones that are relevant and make the model interesting.[36] Let us take the example of an $ARCH(1)$. The conditional variance of $\varepsilon_t$ knowing $\varepsilon_{t-1}$ depends on $\varepsilon_{t-1}$. If this last perturbation is large (small), $\text{Var}(\varepsilon_t | I_{t-1})$ will be large (small). Thus the marginal (empirical) law of $\varepsilon_t$ is not Gaussian and exhibits leptokurtosis.

---

[35] There is yet a more general class of models (e.g., ARCH-in-Mean) in which the expectation $\mu_t$ is not constant but itself depends on volatility (the present volatility, and possibly the past).

[36] The $ARCH(p)$ process defined by equations (A-1) and (A-2) is stationary (its distribution laws remain unchanged by translation in time) if the roots of equation $1 - \sum_{i=1}^{p} \alpha_i x^i = 0$ are outside the unit circle (for example, with $\omega > 0$ and $\alpha_i \geq 0$, $\forall i$, $\sum_{i=1}^{p} \alpha_i < 1$). The stationarity of a process is a crucial property for parameter estimation and forecasting.

# Appendix 2                                                                                                 307

This model was extended by Bollerslev (1986) to Generalized ARCH, or $GARCH(p, q)$. The variance $\sigma_t^2$ is a linear combination of the squares of the $p$ last observed perturbations as well as of the $q$ most recently observed variances[37]:

$$\sigma_t^2 = \omega + \sum_{i=1}^{p} \alpha_i \varepsilon_{t-i}^2 + \sum_{j=1}^{q} \beta_j \sigma_{t-j}^2 \qquad (8.A\text{-}3)$$

with the following conditions: $\omega > 0$, $\alpha_i \geq 0$, $\forall i$, $\beta_j \geq 0$, $\forall j$, and $\sum_{i=1}^{p} \alpha_i + \sum_{j=1}^{q} \beta_j < 1$.

Remark that, most often, a simple model of type $GARCH(1,1)$ is enough to explain the data.[38]

In models of the $ARCH$ type, it is the shocks $\varepsilon_t$ that modify the variance, the change depending on the shock amplitudes and not on their signs. However, it seems that for many markets negative shocks have a more pronounced effect on volatility than positive ones. To account for this phenomenon, different extensions of $GARCH$ models have been proposed. For example, the $EGARCH(1,1)$ or exponential $GARCH$ model due to Nelson (1991) is of the form:

$$\ln \sigma_t^2 = \omega + \beta \ln \sigma_{t-1}^2 + \alpha \left| \frac{\varepsilon_{t-1}}{\sigma_{t-1}} \right| + \gamma \frac{\varepsilon_{t-1}}{\sigma_{t-1}} \ .$$

Since the left-hand side of this equality is the logarithm of the conditional variance, conditional variances cannot be negative, even in the absence of other restrictions on the parameters. In addition, the model's asymmetry is reflected in the coefficient $\alpha$, the impacts not being symmetric if it is non-zero.

A second model often tried is the $TARCH$ (*Threshold ARCH*) model whose equation is.

$$\sigma_t^2 = \omega + \alpha \varepsilon_{t-1}^2 + \gamma \varepsilon_{t-1}^2 d_{t-1}$$

with $d_t = 1$ if $\varepsilon_t < 0$ and $d_t = 0$ otherwise, and $\alpha$ and $\gamma > 0$. Good ($\varepsilon_t > 0$) and bad ($\varepsilon_t < 0$) news thus do not have the same impact on the conditional variance (respectively $\alpha$ and $\alpha + \gamma$).

---

[37] Remark the analogy with $ARMA(p, q)$ models used for the behavior of the expectation (of log-returns, for example).

[38] From a theoretical point of view, such modeling is important because a $GARCH(1,1)$ process converges, as the time step tends to zero, to a continuous-time process with stochastic volatility and continuous realizations (Nelson (1990)).

## Suggestions for Further Reading

### Book Chapters

Bodie, Z., Kane, A., & Marcus, A. (2014). *Investments* (10th ed.). Irwin.
*Campbell, J., Lo, A., & MacKinley, C. A. (1997). *The econometrics of financial markets*. Princeton University Press *.

### Articles

Asness, C. S. (2000). Stocks versus bonds: Explaining the equity risk premium. *Financial Analysts Journal, March-April*, 96–113.
Christie, A. A. (1982). The stochastic behavior of common stock variances: Value, leverage and interest rate effects. *Journal of Financial Economics, 10*, 407–432.
Fama, E. (1965). The behavior of stock market prices. *Journal of Business, 38*, 34–105.
Fama, E. (1970). Efficient capital markets: A review of theory and empirical work. *Journal of Finance, 25*, 383–317.
Kon, S. J. (1984). Models of stock returns: A comparison. *Journal of Finance, 39*, 147–165.
Stoll, H., & Whaley, R. (1990). The dynamics of stock index and stock index futures returns. *Journal of Financial and Quantitative Analysis, 25*, 441–468.

### For an Online Comparative Description of Investment Funds from Different Countries

ILP Abogados, Collective investment schemes (CIS) in Spain, France, the US, the UK and India, https://www.ilpabogados.com/en/collective-investment-schemes-cis-in-spain-france-the-us-the-uk-and-india/, 2019.

### For an Online Description and Analysis of the Asset Management Industry

Boston Consulting Group BCG. Global-Asset-Management, 2020.

# Part II
# Futures and Options

This second part, made up of Chaps. 9–20, is devoted to derivative products (options and futures) and to the theoretical framework and the mathematical methods necessary for their analysis and understanding. It presents the main models, stochastic calculation tools, and probabilistic theories on which modern methods of valuing contingent assets and assessing financial risks are based. This part, much more technical than the previous one, is aimed at graduate market finance students as well as market professionals. The reading of some materials requires the mathematical background (analysis, differential and integral calculus, and probability theory) that can be acquired during the last 2 years of undergraduate studies with a quantitative orientation. The most technical sections are marked with an * (or, very rarely, ** for the most difficult). Readers with this minimum mathematical background will find in Chaps. 18–20 all the additions concerning stochastic calculus and probabilistic theories, necessary and sufficient for a thorough understanding of modern market finance.

Chapter 9 is devoted to forward and futures contracts.

Chapter 10 is devoted to options and the binomial (Cox-Ross-Rubinstein) valuation model.

Chapter 11 analyzes in detail the Black and Scholes model and its extensions (dividends, stochastic interest rate, variable volatility, etc.). We warn the reader unfamiliar with stochastic calculus that Chap. 18 (dealing with Itô calculus) is a must read before continuing to study this part.

Chapter 12 describes the implementation of static and dynamic strategies for option portfolios (Greek parameters, implicit volatilities, smile and skew effects, etc.).

Chapter 13 deals with most aspects of American options.

Chapter 14 presents the main exotic options and provides for their valuation formulas.

Chapter 15 analyzes interest rate futures, with an emphasis on notional security contracts and the new STIR contracts.

Chapter 16 discusses the valuation of fixed-income products, non-vanilla interest rate products, and hybrid securities (such as convertible bonds).

310                                               Part II  Futures and Options

Chapter 17 presents stochastic models of the term structure of interest rates grounded on the dynamics of either the spot rate or the instantaneous forward rate.

Chapters 18–20 are devoted to mathematical tools and the foundations of financial market theory. Chapter 18 presents the elements of stochastic calculus constituting the minimum technical background necessary for a rigorous study of options. The foundations of financial market theory, as well as some probabilistic tools (change in probability measure, change in numeraire, etc.) used in the advanced developments of the book, are studied in Chaps. 19 and 20.

# Futures and Forwards

# 9

A forward or a futures contract fixes today the terms of a transaction to be carried out on a future date. They have been known and employed since antiquity. However, such transactions have significantly increased in number over the last several decades, first in agricultural commodity markets and then in financial markets.

We can distinguish different markets according to the underlying assets that are traded, their geographic location and their organization.

– As far as the underlying asset is concerned, we distinguish physical assets from financial assets.

In the first case, it can be a commodity (such as wheat, corn, soybean, sugar, or petroleum), a metal (gold or silver, for example), or any other merchandise whose quality is well specified. As to financial assets, we find mainly fixed income securities (short-, medium- or long-term), stocks, stock market indices, and currencies.

– Significant markets are located in Chicago (CME Group), New York, Philadelphia, London, Frankfurt, and Tokyo, to mention only the most important ones. Note that nationals from one country can trade in all foreign markets, thus making the latter more integrated.
– The markets also differ in their organizational structures (rules, procedures, types of traders, and so on). Some of them are official, organized markets, and others are over-the-counter (OTC).

In organized exchanges (such as the CME Group, EUREX, and ICE Futures Europe), contracts are standardized in their amounts, maturities, and the quality of the underlying asset. They also have exchange *clearing houses* to serve as intermediaries and as a result practically eliminate counterparty risk. They are subject to regulations and well-defined procedures, and trade futures (not forward)

---

© The Author(s), under exclusive license to Springer Nature Switzerland AG 2022
P. Poncet, R. Portait, *Capital Market Finance*, Springer Texts in Business and
Economics, https://doi.org/10.1007/978-3-030-84600-8_9

contracts. This implies that the contracts are subject to daily margin calls (see Sects. 9.1 and 9.1.2, for the definition and precise description of futures).

On a pure over-the-counter (OTC) market, the parties agree to the terms of a contract at their own convenience. These contracts are usually of the forward type (see Sects. 9.1 and 9.1.1).

Recall from Chap. 1 that the advantage of OTC transactions is their flexibility (the contract, created to the parties' best interests, can involve any sort of underlying, maturity, or amount), but that exchange-traded markets bring more liquidity and security (by eliminating counterparty risk). This last advantage of those markets induced regulators and market participants to promote an increasing level of organization in OTC markets to such an extent that the traditional distinction between exchange market and OTC becomes less and less relevant on practical grounds.

Section 9.1 provides a general presentation of futures and forwards. Section 9.2 examines the relationship between the futures price and the spot price. Section 9.3 considers hedging, and Sect. 9.4 presents the main contracts according to their underlying assets.

## 9.1 General Analysis of Forward and Futures Contracts

In this chapter, except when the contrary is made explicit, 0 will denote the date of a contract's issuance, $T$ its expiration date (maturity), and $t$ a future date before expiration ($0 \leq t \leq T$).

### 9.1.1 Definition of a Forward Contract: Terminology and Notation

#### 9.1.1.1 General Definitions

Although the following descriptions and definitions, strictly speaking, only apply to forward contracts, most of them are valid for futures as well. The difference between both types of contracts will be clarified in Subsection 9.1.2.

A forward (contract) signed at 0 is for the buyer the *right* and the *obligation* to buy upon maturity $T$ a definite object, called the underlying asset, *at a price fixed on date* 0; for the seller, this contract brings the right and the obligation, symmetrically to those of the buyer, to sell the underlying at that time for the agreed price.

The underlying of a contract can be a physical asset (commodity), a financial asset (a loan, a share, a negotiable debt), or a market index. By what one may consider an abuse of language, one says that the buyer purchases the contract, and the seller sells it. One also says that the buyer has "a long position" on the underlying or on the contract, and the seller a short position, or has "shorted" it.

A forward (and futures) contract must define precisely:

- The underlying asset, that is, its "quality," especially in matter of commodities.
- The size of the contract, that is, the number of units of the underlying asset.

## 9.1 General Analysis of Forward and Futures Contracts 313

- The delivery location, which is particularly important when the underlying is a physical asset.
- The delivery date, which we assume to coincide with the contract expiration date[1]; note that the contract is often specified with its delivery month (e.g., Eurobond-September).

Finally, remark that there is no monetary transfer between buyer and seller at the origination of the contract.

### 9.1.1.2 Notation

$F_0^T$, or simply $F_0$ if there is no possible ambiguity as to the expiration date, denotes the price for the underlying stipulated at 0 for delivery and payment at time $T$; $F_0$ is called the forward (or futures[2]) price of the underlying or also the contract price at 0. More generally, $F_t$ will denote the price at time $t$, for $t \in [0, T]$.

$S_t$ is the spot price of the underlying on some date $t$, i.e., the price set at $t$ for immediate payment and delivery.

Generally, $F_t$ *differs from* $S_t$ and *we define the basis as the difference between the futures price and the* spot *price*: $B_t \equiv F_t - S_t$. A more explicit notation is $B_t^T$ ($B_t^T \equiv F_t^T - S_t$ with $t \leq T$).

At the specific time $t = T$, the forward (or futures) price is the same as the spot price, under no arbitrage (with $F_T^T$, or simply $F_T$, denoting the prevailing price at $T$ for delivery and payment at $T$, thus the spot price), so that $S_T = F_T$ and therefore $B_T = 0$.

Generally, $B_t$ tends to 0 as $t$ goes to $T$ (a phenomenon called *convergence* of the basis to zero).

Remark that it is important to avoid confusion between the *contract* (the right and obligation to buy the underlying) and the underlying *asset* itself; in particular, we must distinguish the *value V* of the contract and the price $F$ which is written in the contract, and confusingly called the contract *price*. The *value* of the forward (and futures) contract is zero at origination (since the contract can be bought or sold without any transfer of funds[3]). As time progresses, the value of the forward contract can be positive, negative, or zero, most often much less in absolute value than the *price F*, as we will see in the forthcoming Subsection 9.1.4. On the contrary, as we will also see in Subsection 9.1.4, the value of a futures contract is (almost) constantly equal to zero.

---

[1] In practice, the delivery date is several days after the contract's maturity date. The latter is defined as the last date when the contract may be traded (after the maturity date one may neither buy nor sell the contract). In the following analyses, these two dates will be treated as the same for simplicity.

[2] For the sake of simplicity, we do not distinguish in our notations here the forward and the futures prices (both denoted by $F$), which, in general, are in fact slightly different (see Proposition 1).

[3] However, as we will see in the sequel, on organized markets a guarantee (*deposit*) is required from both the buyer and the seller.

314                                                                                              9 Futures and Forwards

### 9.1.1.3 Gains for the Buyer and the Seller

Buying a forward or a futures contract and holding it between 0 and $T$ *leads to a gain or a loss*, also termed a *margin*, for which the formula is $F_T - F_0$. Symmetrically, the sale of a contract leads to a margin equal to $F_0 - F_T$. Indeed, buying at $T$ (resp. selling) the underlying for a price $F_0$, while its value is $S_T = F_T$, implies a gain or loss of $F_T - F_0$ (resp. $F_0 - F_T$).

## 9.1.2  Futures Contracts: Comparison of Futures and Forward Contracts

Two methods, respectively known as forward and futures, are conceivable for settling transactions and paying margins.

### 9.1.2.1 Forwards and Futures

- A pure forward contract is what has just been described. There is neither physical nor financial exchange, before the due date $T$. At $T$, the margin $F_T - F_0$ can be settled in one of the following two ways:
- By *physical delivery* of the underlying, worth $S_T = F_T$, against the payment of $F_0$ (gain or loss for the buyer $= F_T - F_0$).
- By transfer, from the seller to the buyer, of cash in the amount of $F_T - F_0$ if $F_T > F_0$, or of $F_0 - F_T$ from the buyer to the seller if $F_0 > F_T$. In the latter case, called *cash settlement*, no physical delivery of the underlying actually occurs. It is important to note that cash settlement assumes that at maturity the price $F_T$, which is also equal to the underlying asset's spot price $S_T$ can be *observed or calculated* unambiguously and without question (and without any chance of rigging).

Physical delivery and cash settlement are obviously equivalent from a theoretical point of view, since the payoff is the same in the two cases: the buyer gains $F_T - F_0$ if $F_T > F_0$ (and loses $F_0 - F_T$ in the opposite case), the loss (or profit) of the seller being symmetric.

Usually, over-the-counter transactions are of the forward type.[4]

- "Futures" contracts (always written in the plural to avoid confusion with the adjective "future") differ from the preceding in that the global margin $F_T - F_0$ is paid progressively between 0 and $T$ and not in a lump sum at expiration $T$. More precisely, the buyer is credited or debited on each day $t$ with the daily margin equal to $F_t - F_{t-1}$ according to the latter's sign, while the seller is debited or credited on day $t$ the symmetric cash flow $F_{t-1} - F_t$. Furthermore, it is important to

---

[4]Note, however, that the parties tend increasingly to proceed to margin calls, as in the case of futures (see Sect. 9.1.2.2), to decrease the counterparty risk for each of them. The frequency of these margin call payments is determined over-the-counter at the origination of the forward contract. The financial crisis of 2007–2008 has amplified this tendency.

## 9.1 General Analysis of Forward and Futures Contracts

**Table 9.1** Comparison of (pure) forward and futures contracts

| Day | 0 | 1 | .......... | $t$ | .......... | $T$ | Total |
|---|---|---|---|---|---|---|---|
| Forward | 0 | 0 | .......... | 0 | .......... | $F_T - F_0$ | $F_T - F_0$ |
| Futures | 0 | $F_1 - F_0$ | .......... | $F_t - F_{t-1}$ | .......... | $F_T - F_{T-1}$ | $F_T - F_0$ |

note (in order to understand the discussion below about counterparty risk) that in an organized market (which is usually the case for futures) the buyer and seller "do not know each other": both the buyer and seller have the clearinghouse as their counterpart, so that neither bears a counterparty risk (assuming the clearinghouse never defaults on its obligations). Daily payments between sellers and buyers are made exclusively through this clearinghouse as intermediary.

Like forward contracts, futures not settled before their expiration $T$ can, according to the contract specificity, lead to the actual *delivery* of the underlying asset or not (*cash settlement*). The same conditions apply to both futures and forwards, as to the observable and undisputable nature of the closing price $F_T$.

Futures are always traded in exchange markets. As for OTC markets, an increasing degree of regulation and organization has been introduced since the subprimes crisis, aiming to minimize counterparty risks. Standard OTC derivatives between financial entities are now usually cleared through central clearing counterparties (CCPs), sort of exchange clearinghouses requiring an initial deposit and implementing margin calls. Nonstandard OTC operations, however, are still cleared bilaterally but collateral must be posted and margin calls implemented. While the frontier between OTC-forward and Exchange traded futures has been blurred by these important market reforms, the distinction between the two procedures remains conceptually and theoretically relevant. In the sequel of this chapter, "forward" systematically refers to "pure forward" (no clearing, no margins) without further precision, but details on the subject will be provided in later chapters.

### 9.1.2.2 Comparison of (Pure) Forward and Futures Contracts

To compare the two contracts, Table 9.1 displays the cash flows generated by a long position on either a forward or a futures contract.

This table entails the following remarks:

- For a forward transaction, the investor does not receive (resp. does not pay) any gain (resp. loss) on dates $t < T$. The global margin $F_T - F_0$ will be cashed in (or out) at expiration $T$. By contrast, with futures, the global margin $F_T - F_0$ gives rise to successive positive or negative *margin calls*.
- In both cases initiating a contract does not, in theory, require any initial investment of funds.
- Futures, like forwards not settled before their maturity $T$, can lead, as the contract may happen to require, to a physical delivery of the underlying or not. If they want delivery, the buyers pay the price $F_T$ at $T$ in the case of futures while they pay $F_0$ with forwards. If they choose not to be delivered the underlying at

maturity (as in Table 9.1), they cash in or out the final margin $F_T - F_{T-1}$. By contrast, the global margin $F_T - F_0$ is paid or received at date $T$ in the forward case.

- In practice, transactions on the futures market differ from the model described above in two ways:
  - The contract requires an initial guarantee (a *deposit*) from both the buyer and the seller, and therefore may involve an initial disbursement. However, such a deposit will be small compared to the amount of the transaction[5] and in addition might earn interest (notably when it takes the form of Treasury Bills), which does justify neglecting such a feature.
  - The last trades, and consequently the fixing of the closing price $F_T$, take place a few days before delivery and final payment to allow time for preparation of paperwork that may be required for delivery. This small time gap is ignored in what follows.
- The system of successive margin calls makes for a greater guarantee of a futures transaction against the risk of possible default by the losing party than that for a forward transaction where the loss (which may have accumulated to reach a high level) is paid all at once on date $T$. This is because if the losing party cannot pay their (daily) margin call, the clearinghouse forces them out of the market and winds up their position by paying the loss using the initial deposit. Since the latter is always larger than the maximum daily variation in the contract price the official market authorizes, the clearinghouse in principle cannot lose money. In addition, an organized market is a legal entity with its own equity capital that allows it to underwrite unpredictable losses resulting from participants' cancelled positions. This double safety net (daily margin calls combined with an initial deposit and the presence of the clearing house's equity capital) strengthens the safety of futures contracts to such an extent that in practice they can be considered as free from counterparty risk.
- Finally, the difference in "timing" of the margin calls has an influence on their value; calculated, for example, at $T$, the capitalized value of the margins is $F_T - F_0$ in the forward case, but $\sum_{t=1}^{T} e^{(T-t)r_{T-t}(t)}[F_t - F_{t-1}]$ for futures, where $r_{T-t}(t)$ denotes the (continuous) rate prevailing at time $t$ for transactions with duration $T-t$.

Capitalization of the margins associated with the earnings of the investment (or the cost of financing) of the intermediate margins $F_t - F_{t-1}$ in the interval $[t, T]$ is, from an actuarial point of view, the difference between forward and futures contracts. This difference is often ignored in financial analyses, as a first approximation, since treating futures as if it was a forward does provide useful simplifications

---

[5]Typically, it ranges from 1% to 2%, and is always less than or equal to 5%.

## 9.1 General Analysis of Forward and Futures Contracts

**Table 9.2** Unwinding a position at $t < T$

|  | $t$ | $t+1$ | ------------- | $T-1$ | $T$ |
|---|---|---|---|---|---|
| Futures |  |  |  |  |  |
| Contract bought at 0 | $F_t - F_{t-1}$ | $F_{t+1} - F_t$ | ------------- | $F_{T-1} - F_{T-2}$ | $F_T - F_{T-1}$ |
| Contract sold at $t$ (eve) |  | $F_t - F_{t+1}$ | ------------- | $F_{T-2} - F_{T-1}$ | $F_{T-1} - F_T$ |
| Net cash flow: |  |  |  |  |  |
|  | $F_t - F_{t-1}$ | 0 |  | 0 | 0 |
| Forward |  |  |  |  |  |
| Contract bought at 0 | 0 | 0 | ------------- | 0 | $F_T - F_0$ |
| Contract sold at $t$ | 0 | 0 | ------------- | 0 | $F_t - F_T$ |
| Net cash flow: |  |  |  |  |  |
|  | 0 | 0 |  | 0 | $F_t - F_0$ |

and the discrepancy is small in practice.[6] Yet, to be rigorous, the forward price and the futures price of the same asset with the same expiration should be distinguished. Such attention to rigor is needed and the forward-futures distinction is not to be ignored when we start to consider more theoretical questions (especially in option valuation). We, therefore, establish the following proposition, but prove it and discuss it further in the Appendix.

**Proposition 1**

*The futures and forward prices of the same underlying and for the same expiration coincide in theory in a world where interest rates evolve deterministically but (slightly) differ, by a convexity adjustment term, in the real world where interest rates are stochastic.*

### 9.1.3 Unwinding a Position Before Expiration

So far, a contract has been assumed to be held until its expiration date $T$. In order to explain the mechanics and consequences of winding up contracts before maturity, consider at a date $t$ a long position acquired on date 0 in the past ($0 < t < T$).

It is possible, on the eve of day $t$, to sell at a price $F_t$ a contract expiring at $T$, in this way adding to the initial position a new position with the opposite sign (such a transaction leads to no cash flow at $t$). The cash flow stream generated by the existing long position coupled with the new, short position is illustrated, for forward and futures transactions, in Table 9.2.

Table 9.2 calls for the following remarks:

---

[6] There are discrepancies between forward and futures prices, which can be ascribed to differences in counterparty risk (which is virtually inexistent for futures), in liquidity (often much higher for futures), in transaction costs (often smaller for futures), in taxes, etc.

318                                                                    9   Futures and Forwards

- With futures, the existing long position is cancelled by the sale of a contract at $t$ since all the net cash flows are zero from that date on.
  Since the sale does not require putting up any upfront money and futures are standardized, *a futures position can be cancelled (unwound) at any time,*[7] *simply and at no cost.* In fact, the agent has two possibilities: unwind his position before expiration; or wait for maturity and deliver (or take delivery of) the underlying asset if the contract requires delivery and not cash payment. In practice, the vast majority of futures transactions on the financial markets are unwound before expiration.
- For forward transactions, the contract sale at $t$ does not cancel the net cash flow to come (on date $T$) but does fix its value at $F_t - F_0$ *which is known today (at t)*. In practice, unwinding a forward contract before its maturity is often much more problematic than for a futures contract, which gives the latter a significant liquidity advantage.
- The total net gain (*not* taking into account discounting of the margins) is the same in both cases, namely $(F_t - F_0)$, since, for futures, previous to the cash flow $(F_t - F_{t-1})$ the buyer has received or disbursed the margins $\sum_{j=1}^{t-1} \left( F_j - F_{j-1} \right) = F_{t-1} - F_0$.

## 9.1.4   The Value of Forward and Futures Contracts

We have already alerted the reader to the possible confusion between the contract (which, involving as it does rights and obligations, is itself an asset) and the underlying asset on which the contract is written (which may be a physical or a financial asset). This warning is equally valid for the possible confusion between the price of the underlying and especially its forward price $F$, as stipulated in the contract, and the value $V$ of the contract that gives the right and obligation to pay $F$ to obtain the underlying at time $T$.

The confusion is aggravated by the terminology, doubtless unfortunate, according to which the price $F$ *stipulated* in the contract is called the contract *price* (and not, for instance, the contracted price).

Table 9.2 presented in the preceding subsection allows us to formulate the following proposition about the value $V$ of a forward or futures contract.

### Proposition 2
  (i). *The value of a futures contract is always zero.*
 (ii). *The value $V^0(t)$ at t of a forward with maturity T, issued on date 0 at a price $F_0$, is the discounted value of the variation of the price between 0 and t:*

---

[7] On organized markets, where most futures transactions are handled, this unwinding amounts to a pure and simple cancellation of the agent's position by the clearinghouse.

## 9.2 Cash-and-carry and the Relation Between Spot and Forward Prices

$$V^0(t) = (F_t - F_0)e^{-(T-t)r_{T-t}^{(t)}} \tag{9.1}$$

*where $r_{T-t}(t)$ denotes the continuous zero-coupon interest rate for a duration of $(T-t)$ prevailing at t.*

**Proof**

Both for a forward and a futures contract, this proposition's justification depends on the possibility of unwinding as described in Subsection 9.1.3.

(i). Any futures contract can be cancelled at any time without cost, and so its value must be zero.[8]

One may, at any instant $t$, by selling a forward at the price $F_t$ which requires no upfront payment, fix at $F_t - F_0$ the value of the cash flow *at* $T$ induced by the existing position; the value at time $t$ of a contract is thus the discounted cash flow at time $t$, which is represented by equation (9.1).

---

## 9.2 Cash-and-carry and the Relation Between Spot and Forward Prices

Two theories coexist as to the relation between the forward price and the spot price. The first, as analyzed in Sect. 9.2.1, amounts to a parity relationship called *cash-and-carry* which should obtain in the absence of arbitrage opportunities (AAO); it can only be applied rigorously to a forward contract.[9] The second, succinctly described in Sect. 9.2.2, links the forward price to the expected spot price.

The cash-and-carry relationship applies to contracts written on stock market indices, interest rates, and exchange rates, while the relation between a forward price and the expected spot price is useful for analyzing contracts involving commodities.

### 9.2.1 Arbitrage, Cash-and-Carry, and Spot-Forward Parity

Recall that a pure arbitrage is a transaction free of all risk of a negative cash flow, but involving the possibility of a positive cash flow with non-zero probability. Also remember that opportunities for arbitrage may occur any time that an asset may be replicated by some combination of other assets.

---

[8] Rigorously speaking, this result holds exactly only in the evening after the margin call, or for continuous time under the hypothesis of continuous margin calls.

[9] This only applies to futures in a world with deterministic rates.

**Table 9.3** Cash-and-carry arbitrage

| | 0 | $T$ |
|---|---|---|
| Purchase of the underlying, capitalized earning and delivery | $- S_0$ | $+R$ |
| Forward sale and cash receipt upon delivery of the underlying at $T$ | 0 | $+ F_0$ |
| Borrowing | $+ S_0$ | $- (S_0 + I)$ |
| Net cash flow | 0 | $F_0 - S_0 - (I - R)$ |

In the context of forward markets, possible arbitrage opportunities mostly involve one or more forward contracts, some spot transactions on the underlying, and lending or borrowing.

The classic case is a *cash-and-carry* arbitrage. The condition that an opportunity for *cash-and-carry* arbitrage be absent is called *spot-forward parity*. The precise form of this relationship depends on the context and the specific details of the contract.

### 9.2.1.1 Cash-and-carry Arbitrage and Spot-Forward Parity: Fundamental Formulation

Cash-and-carry arbitrage results from buying the underlying asset against cash (at spot price $S_0$) and holding (carrying) it from 0 to $T$, this purchase being financed by borrowing (the sum $S_0$). This is combined with the forward sale of the underlying at the forward price $F_0$. The operation ends at $T$ with the delivery of the underlying asset for $F_0$ in cash. Call $I$ the interest paid on the loan. If holding the support entails receiving a cash-flow (such as a coupon or a dividend) between 0 and $T$, denoting by $R$ the value *capitalized at $T$* of this earning, the transaction is profitable (and so constitutes an arbitrage) if $F_0 - S_0 > I - R$, as in Table 9.3.

Since there is no cash payment required at time 0 and all the cash flows at time $T$ are known on date 0, under AAO the following inequality must hold: $F_0 - S_0 \leq I - R$.

If $F_0 - S_0 > I - R$, the arbitrage consists in taking an opposite position (sale of the underlying, lending, and purchase of the contract); such a transaction is termed a *reverse* cash-and-carry. Sometimes short positions on the underlying are difficult, or even impossible, to achieve. Reverse cash-and-carry is then only an option for those who already own the underlying. The latter, by selling the underlying for cash and buying the forward contracts, exert upward pressure on $F$ and a downward one on $S_0$ until the equality $F_0 - S_0 = I - R$ holds.[10]

From this we see that only the following *equality* is compatible with AAO:

$$F_0 - S_0 = I - R. \tag{9.2}$$

---

[10]However, these transactions are impossible for some commodities, as we will see in Sect. 9.4 and 9.4.1.

## 9.2 Cash-and-carry and the Relation Between Spot and Forward Prices

The left-hand side of equation (9.2) is the basis; the right-hand side is the cost $I$ of financing the underlying from 0 to $T$ net of the earnings $R$ produced by the latter in the same period $(0, T)$. This difference $I - R$ is called the *net cost of carry*, or simply the cost of carry. *Equation (9.2)* thus means that, *under no arbitrage, the basis is equal to the cost of carry; this is the spot-forward parity.*

According to whether the *cost of carry* is positive or negative, the basis is therefore positive or negative. In the first case (basis $> 0$ so $F > S$), one says that the forward price is at a *premium* compared to the spot price; in the other case $(F < S)$, the forward price is *at a discount* relative to the spot price.

Like any AAO condition, equation 9.2 expresses the redundancy of assets involved in the arbitrage that can be carried out when it is not satisfied. In that case the three securities, the underlying, the forward contract, and the zero-coupon debt (borrowing-lending) with maturity $T$, are here redundant. For example, it is possible to replicate on the date 0 a forward contract with maturity $T$ (leading to a payoff $F_T - F_0$) with a portfolio containing the underlying and a loan of maturity $T$ (which yields at $T$ the amount $S_T + R - S_0 - I = F_T - F_0$). It is also possible to replicate a loan by buying the underlying and selling it forward.

### 9.2.1.2 Alternative Formulations of the Spot-Forward Parity

How to write the interest $I$ and the yield $R$ (capitalized at $T$) depends on the context. For example, according to whether the rate $r$ being used for calculating the interest $I$ is proportional, discrete, or continuous, we will write:

$$I = S_0 \, r \, T; \quad I = S_0 \left[ (1 + r)^T - 1 \right] \quad \text{or} \quad I = S_0 \left[ e^{rT} - 1 \right].$$

In the same way, the yield $R$ on the underlying can take various forms. For instance, it may be a discrete sum (dividend, coupon) $D$ paid out on some date between 0 and $T$, or a continuous flow between 0 and $T$.

Alternative ways of writing the spot-forward parity correspond to these different cases.

(a) A first way involves the price $B_T(t)$ of a zero-coupon bond yielding \$1 at $T$.

Recall that $B_T(0) = \frac{1}{(1+r_T(0))^T}$ at $t = 0$ and that $B_T(t) = \frac{1}{(1+r_{T-t}(t))^{T-t}}$ at arbitrary $t \, (\leq T)$.

According to whether the underlying does not distribute anything between $t$ and $T$ or distributes an amount $D$ at $T' < T$, the spot-forward parity is written respectively as:

$$S_t = F_t \, B_T(t) \quad \text{or} \tag{9.3-a}$$

$$S_t = F_t \, B_T(t) + D \, B_{T'}(t). \tag{9.3-b}$$

Let us first prove (9.3-b) by considering the synthetic loan made up of a purchase at $t$ of the underlying that distributes $D$ at $T' < T$, combined with the sale at $t$ of a contract expiring at $T$ written on the same underlying; the latter is held until $T$, the date on which it is delivered, honoring the forward contract, against a payment of $F_t$. Such a transaction yields the cash flow series $(- S_t, D, F_t)$ whose discounted value on $t$ is equal to $- S_t + D\, B_{T'}(t) + F_t\, B_T(t)$ which should be zero. This implies (9.3-b); (9.3-a) is a special case of (9.3-b) with $D = 0$.

Remark that, absent any cash flow before maturity $T$, (9.3-a) implies $F_t = S_t\,/B_T(t)$, which means we may interpret the forward price as a price expressed in terms of a zero-coupon bond price (equivalently, as a price using the zero-coupon price as the *numéraire*). This interpretation will prove very useful for the theoretical developments in later chapters.[11]

The main results of the preceding analyses are summarized in the following proposition:

**Proposition 3 Spot-Forward Parity**
- *A forward contract with maturity $T$ can be replicated by buying the underlying asset and borrowing for period $T$.*
- *Under AAO, the basis equals the cost of carry (cost of holding the underlying asset from 0 to $T$ net of the possible yield on the asset over the same period).*
- *The spot-forward parity can be written $S_t = F_t\, B_T(t)$ or $S_t = F_t\, B_T(t) + D\, B_{T'}(t)$ according to whether the underlying asset distributes nothing before $T$ or distributes $D$ on date $T' < T$.*

(b) A second way of expressing the spot-forward parity can be used when earnings are in the form of continuous payments calculated from a *yield rate* denoted by $r^*$. More precisely, we assume that the earnings are proportional to the underlying's price $S_t$ and equal to $r^* S_t\, \delta t$ in the interval $(t, t + \delta t)$. The parity then can be written in an alternative form to (2) or (3) and is deduced from a slightly different and more complicated arbitrage than that previously described. Indeed, in this context, a cash-and-carry involves on the date 0 initially acquiring the underlying and between 0 and $T$ the immediate reinvestment in the same underlying of any dividend or coupon. Such a strategy of capitalization (which is akin to that of a mutual fund composed of a single asset and continuously reinvesting in that asset) leads to *an exponential increase with rate $r^*$ in the number of* shares *of the underlying present in the portfolio.*[12]

More precisely, the cash-and-carry arbitrage implies the following decisions on date 0:

---

[11]One can also remark that (9.3-b) implies $F_t = S_t\,/B_T(t) + D\, B_{T'}(t)\,/B_T(t) = S_t\,/B_T(t) + D(1 + f_{T\text{-}T'}(t))^{T\text{-}T'}$ where $f_{T\text{-}T'}(t)$ is the forward rate in force at $t$ for the future period $(T', T)$; $D(1 + f_{T\text{-}T'}(t))^{T\text{-}T'}$ is thus the yield $R$ capitalized at $T$.

[12]If $n(t)$ denotes the number of shares in the portfolio at instant $t$, the dividends received and immediately reinvested in the interval $(t, t + dt)$ amount to $nr^* S_t\, dt$ which allows acquiring $dn = r^* n dt$ new shares so that $dn/n = r^* dt$, and $n(t) = n(0)\, e^{r^* t}$.

## 9.2 Cash-and-carry and the Relation Between Spot and Forward Prices

**Table 9.4** Cash-and-carry arbitrage

| | 0 | T |
|---|---|---|
| Purchase of $e^{-r^*T}$ unit of the underlying | $-S_0 e^{-r^*T}$ | Delivery of |
| Immediate reinvestment of coupons between 0 and T: *1 unit* | | the underlying |
| *of underlying delivered at T* | | |
| Forward sale on date 0 | 0 | $+F_0$ |
| Payment upon delivery of the underlying at T | | |
| Borrowing at rate r | $+S_0 e^{-r^*T}$ | $-S_0 e^{(r-r^*)T}$ |
| Net cash flow | 0 | $F_0 - S_0 e^{(r-r^*)T}$ |

- Buying $e^{-r^*T}$ unit ($<1$) of the underlying.
- Borrowing an amount $S_0 e^{-r^*T}$ at the interest rate $r$.
- Selling a forward contract with maturity $T$ at the price $F_0$.

These transactions taken together lead to zero initial cash flow.

Between 0 and $T$, the strategy requires the immediate reinvestment of any cash inflows, which generates an exponential increase in the number of units in the portfolio, so that *one unit* of the underlying is held on the final date $T$.

At $T$, the underlying is exchanged for the payment of the price $F_0$ and the loan is repaid, which implies a disbursement equal to $S_0 e^{(r-r^*)T}$.

These transactions, together with the cash flows to which they give rise, are described in Table 9.4.

The preceding results can be summarized as a proposition:

**Proposition 4 (Spot-Forward Parity)**
*For an underlying which continuously distributes earnings at a rate $r^*$, the following parity relation holds in the absence of arbitrage opportunities:*

$$F_0 = S_0 e^{(r-r^*)T}. \tag{9.4}$$

*In this context, the cost of carry is $S_0 (e^{(r-r^*)T} - 1)$.*

(c) Different costs and difficulties bound up with carrying out arbitrage strategies, notably with reverse cash-and-carry, explain why the parity *does not hold exactly*, in most cases:

- Both spot and forward transactions are saddled with transaction costs.
- The lending and borrowing interest rates are not identical.
- The short position needed in the reverse cash-and-carry may require borrowing the underlying if the arbitrageur does not hold the latter; but borrowing the underlying may prove costly or even impossible. In the same vein, short positions may be restricted by regulations.
- The delivery may be more or less costly, especially for raw materials and some other commodities, which explains why the forward price can differ according to the delivery location.

For all these reasons the parity relation is written, in its most general form, as a double inequality expressing that, in the absence of arbitrage, the basis $F - S$ belongs in an interval of variable size:

$$I - R - e_1 \leq F - S \leq I - R + e_2.$$

Here, $e_1$ and $e_2$ denote two positive numbers, were most often $e_2 > e_1$ as a result of the difficulties inherent in a reverse cash-and-carry.

Finally, for contracts written on commodities, a cash-and-carry transaction consists in buying the commodity and holding it until the contract matures; the carrying cost can therefore include warehousing expenses and depreciation which add on to the financing cost $I$ (or can be considered as a *negative* earning $R$). We will see that this parity only holds for contracts written on storable commodities and that the reverse cash-and-carry can be an issue even for these contracts.

We will subsequently use alternative formulations of cost of carry and spot-forward parity which are adapted to the specific contracts being considered.

### 9.2.2 Forward Prices, Expected Spot Prices, and Risk Premiums

In the numerous cases of cash-and-carry activity that only concern *observable* variables (e.g., $S_0$, $I$, $R$, and $T$ in equation (9.2)), the value of $F_0$ can be specified as a function of these variables. For situations where cash-and-carry does not apply well (products that cannot be stored, uncertainty or ambiguity over $R$, etc.), another approach to evaluating $F_0$ is needed. As we have already seen, the profit realized on a forward contract expiring on date $T$ and bought on date 0 is equal to $S_T - F_0$. From this, we see that the equilibrium price $F_0$ depends on the market expectations for $S_T$, namely the value at expiry of the spot price of the underlying.[13] *If there is no risk premium required by the market, then one should have at equilibrium $F_0 = E(S_T)$,* where $E(S_T)$ denotes the expectation for the spot price on $T$ as judged by the market on date 0.

As a result, only the risk premium required by the market can explain a difference between the forward price and the expected spot price. Let us adopt the buyer's point of view and denote by $p$ the risk premium defined by

$$p = E(S_T) - F_0.$$

According to whether the premium $p$ is positive or negative, the buyer's expected gain will be positive or negative (the seller's expectation has the opposite sign). The situation which is considered normal is that with a positive premium. This situation, called *normal backwardation,* means *the forward price is less than the expected* spot *price* and implies a positive expected gain for the buyer. The idea that this is

---

[13]This assertion, although less useful in the (numerous) cases where the cash-and-carry relation holds, remains true since the equilibrium price $S_0$ does itself depend on market expectations.

"normal," as above all advocated by the English economists Keynes and Hicks, goes back to contracts on commodities. It assumes that most sellers of contracts are producers or brokers who have or will have inventories of the underlying commodity on date $T$ and want to hedge the risk of a fall in prices, even if it requires paying a premium. It also assumes that a good number of buyers are speculators who accept taking on the risk if they can charge a positive premium $p$.

The opposite situation prevails when the forward price exceeds the expected spot price and is called *contango*.

Another approach is based on portfolio theory (see Chap. 22) and links the sign of the premium $p$ to the sign of the beta of $S_t$. According to this theory, the expectation of the profit $S_T - F_0$ resulting from buying a contract should be positive if $S_T$ is positively correlated with the market, but negative in the opposite case since then the contract can be used to reduce risk.

Empirical estimates rather favor the hypothesis of a positive premium $p$ (*normal backwardation*), above all for commodities contracts (for which the premium would be of the order of 5%).

We warn the reader of a widely used but questionable terminology, according to which *normal backwardation* means that the spot price is higher than the forward price (negative basis) and *contango* that the basis is positive: it is then the current spot price (and not the expectation of its value at $T$, which is today unobservable) that is compared to the forward price. This phrasing that can be a source of unfortunate confusion will not be employed in this book.

Finally, in theoretical analyses that will be developed in more advanced chapters of this book, expectations will be expressed under probability measures other than the "true" or "historical" or "physical" probability $P$. Risk-neutral expectations, for instance, lead to zero risk premiums. It will then be shown that the *futures price* is equal to the risk-neutral expected spot price: $F_0 = E^Q(S_T)$, $Q$ standing for the "risk-neutral probability," an extremely useful theoretical device different from $P$. In the same vein, another useful theoretical construction will be the "forward-neutral" probability $Q_T$ under which the *forward price* is equal to the expected spot price.

## 9.3 Maximum and Optimal Hedging with Forward and Futures Contracts

Hedging is carried out by investors who hold assets or portfolios which they cannot (or do not wish to) trade, at least temporarily.[14] We will suppose here that these investors hedge the undesirable risks associated with this investment with forward or futures contracts.

---

[14] Such a situation does occur for agricultural products awaiting harvest, and in some other cases because of constraints resulting from regulations, customers or senior management, or of transaction or information costs.

In fact, two sorts of trades deserve to be distinguished: on the one hand, *perfect* or *maximum* hedging; on the other, *optimal* hedging. Maximum (resp., perfect) hedging minimizes (resp., cancels) the risk of a portfolio up to a given horizon $H$, while optimal hedging is only partial since it seeks the best tradeoff between risk and expected return from the investor's viewpoint. Optimal hedging depends on the investor's preferences (risk aversion) and always contains some degree of speculation.

In this subsection, we will treat futures as if they were forward contracts for the sake of simplicity (i.e., we will not take into account the costs or benefits of margin calls associated with futures).

Also for simplicity, we suppose the agent has mean-variance preferences. In the pure hedging case, studied in Sect. 9.3.1, the agent thus minimizes the variance of her position's value on date $H$, while in a more general optimization, examined in Sect. 9.3.2, she looks for the best tradeoff between the variance and the expectation of her position's value on date $H$.

### 9.3.1 Perfect or Maximum Hedging

A perfect or maximum hedge is intended to reduce as much as possible the risk that affects, at a future time $H$ (the horizon), a given pre-existing position. It is realized by taking an additional, random position negatively correlated with the existing risk and wound up at $H$.

A hedge allowing for the total elimination of risk is called perfect. However, it is in general only imperfect, leaving a residual risk, even if it is maximal. In considering a hedge using a forward contract, the latter should have an underlying "as near as possible" to the existing asset. In the special case where the underlying is the asset to hedge itself, the hedge is perfect: it is enough to sell the existing asset forward with maturity $H$ to ensure the price $F_0$ on that date.

#### 9.3.1.1 A Model of Maximum Hedging

The following model expresses, in the simplest way possible, the problem of an existing asset and a forward contract whose prices are not perfectly correlated.

Let $t = 0$ be the present date, and consider an agent, whose horizon is $H$, who has to hold a position composed of $m$ units of an asset whose unit spot price on date $t$ is $C_t$. The agent is concerned by the risk that affects, on date $H$, the value $W_H = mC_H$ of their existing position, and decides, to hedge this risk, to trade at time 0 $x$ forward contracts ($x > 0$ indicates a long position; $x < 0$ a short position). Usually, there is no forward contract whose maturity $T$ coincides with the hedge's horizon $H$. In what follows, we assume that the maturity of the contract used for hedging is $T > H$ (we will later consider the case where such a contract does not exist, the only maturities traded being shorter than $H$). We continue to use the notation $F_t^T$, or simply $F_t$, for the price of a contract with maturity $T$ prevailing at time $t$.

The algebraic value of the profit generated by this position (long or short) made up of $x$ forward contracts is thus $x(F_H - F_0)$.

## 9.3 Maximum and Optimal Hedging with Forward and Futures Contracts

The value $V_H$ on date $H$ of the hedged position can as a result be written

$$V_H = m\tilde{C}_H + x(\tilde{F}_H - F_0). \tag{9.5}$$

In equation (9.5), the random terms are signaled by a tilde; the reader should remember this distinction which will not be repeated in the sequel, so as not to overburden the notation. Also for simplification, the various variances and covariances are denoted as follows:

$$\sigma_V{}^2 \equiv \text{var}(V); \sigma_C{}^2 \equiv \text{var}(C_H); \sigma_W{}^2 \equiv \text{var}(W_H); \sigma_F{}^2 \equiv \text{var}(F_H); \sigma_{CF}$$
$$\equiv \text{cov}(C_H, F_H).$$

Note that $\sigma_V^2(x) \equiv var(mC_H + x(F_H - F_0))$ means that the variance of the position value $V_H$ is computed for a hedge $x$.

**Proposition 5**
*Under the preceding assumptions and notations:*

(i). *An agent who wants to minimize the risk as measured by $\sigma_V{}^2$ will choose $x$ such that:*

$$x = x* = -m\frac{\sigma_{CF}}{\sigma_F^2}. \tag{9.6}$$

(ii). *The maximum relative reduction of the variance $\frac{\sigma_W^2 - \sigma_V^2(x*)}{\sigma_W^2}$ which allows a maximum hedging $x*$ is equal to the square of the correlation coefficient between $C_H$ and $F_H$ (the determination coefficient, denoted by $R^2$).*

(iii). *The necessary and sufficient condition for the variance of the hedged position $\sigma_V^2(x*)$ to be nullified by maximum hedging $x*$ is that $C_H$ and $F_H$ be perfectly correlated.*

**Proof**

(i). The condition required by an agent who is trying to maximize hedging is

$$\underset{x}{\text{Min}} \left[\sigma_V^2\right] = \underset{x}{\text{Min}}\{\text{var}\,(mC_H + x(F_H - F_0))\}$$

$$= \underset{x}{\text{Min}}\{m^2\sigma_C^2 + x^2\sigma_F^2 + 2mx\sigma_{CF}\}.$$

The minimum occurs if $\frac{d}{dx}(m^2\sigma_C^2 + x^2\sigma_F^2 + 2mx\sigma_{CF}) = 0$, therefore for $x = x^* = -m\frac{\sigma_{CF}}{\sigma_F^2}$

(ii). Let us denote $\sigma_V^2(x^*) = var\ \{(mC_H + x^*F_H)\}$ (the minimum variance attainable):

$$\sigma_{V^2}(x*) = m^2\sigma_{C^2} + (x*)^2\sigma_{F^2} + 2mx*\sigma_{CF}$$

$$m^2\sigma_C^2 + m^2\frac{\sigma_{CF}^2}{\sigma_F^2} - 2m^2\frac{\sigma_{CF}^2}{\sigma_F^2} = m^2\left[\sigma_C^2 - \frac{\sigma_{CF}^2}{\sigma_F^2}\right]$$

Since $W = mC$, it follows that $\sigma_W^2 = m^2\sigma_C^2$, which gives

$$\frac{\sigma_W^2 - \sigma_V^2(x*)}{\sigma_W^2} = \frac{\sigma_{CF}^2}{\sigma_F^2\sigma_C^2} = R^2.$$

From (ii) we have $\sigma_V^2(x*) = 0 \quad \Leftrightarrow \quad R^2 = 1 \quad \Leftrightarrow C_H$ and $F_H$ are perfectly correlated, which ends the proof.

## Remarks
- This set of assumptions is rather restrictive; in particular, once chosen, the hedge $x$ cannot be adjusted between the dates 0 and $H$. More complex models, which this book does not cover, provide dynamic strategies of adjustment of hedging positions between 0 and $H$.[15]
- In general, $C$ and $F$ are positively correlated, therefore $\sigma_{CF} > 0$ and $x^* < 0$: to an existing long (short) position corresponds a short (long) hedge position.
- The ratio $\frac{\sigma_W^2-\sigma_V^2(x*)}{\sigma_W^2} = R^2$ is called the *hedge effectiveness*: when $R^2 = 0$, the hedge does not allow reducing the variance and its effectiveness is nil; if $R^2 = 1$, the variance vanishes and the hedge is perfect. As a general rule, a hedge allows reducing variance but not its cancellation: it is thus imperfect and some residual risk remains which we analyze in the next subsection.
- To normalize the size of the hedge one uses the *hedge ratio h*. This is defined as the ratio $-x/m$. Equation (9.6) implies that the maximal hedge ratio $h^*$ equals

$$h* = \frac{\sigma_{CF}}{\sigma_F^2} = \rho_{CF}\frac{\sigma_C}{\sigma_F},$$

where $\rho_{CF}$ denotes the correlation coefficient between the price variations for $C$ and $F$. This ratio will only *exceptionally* be equal to one.[16]

---

[15] See, for example, the book by A. Lioui and P. Poncet: *Dynamic Asset Allocation with Forwards and Futures*, Springer, 2005.

[16] It would indeed be necessary that $\rho_{CF}\sigma_C$ equals $\sigma_F$, which would happen for example if the prices of $C$ and $F$ were perfectly correlated and had identical variances: if, in particular, the contract's underlying is the same as the asset to be hedged, the hedge is perfect and it suffices to sell forward the existing asset with maturity $H$ to be sure to receive on that date the price $F_0$.

## 9.3 Maximum and Optimal Hedging with Forward and Futures Contracts

- It is important to note that the maximal hedge ratio $h*$ has a precise statistical significance which makes its use and interpretation easy. By running a linear regression of the spot price variations on the forward price changes we get

$$\Delta C_t = a + \beta \, \Delta F_t + e_t \quad \text{for } t = 0, \ \ldots, H;$$

The regression coefficient $\beta$ is, in fact, none other than $\sigma_{CF}/\sigma^2_F$, which is $h*$.

---

**Example 1**

Simple hedging and hedging by regression of price variations.

(i). An investor holds a well-diversified portfolio of French stocks which, at a first glance, replicates the French stock market index (CAC40) and presently is worth, on 1 January, 16 million Euros. Her investment horizon is the end of December and she decides to hedge by using a contract on the CAC40 expiring in December. The current value of the CAC40 index is 4000 and the multiplier for the contract is 10 Euros, so the index is "worth" 40,000 Euros. The December contract costs 4040 and thus is "worth" 40,400 Euros. The risk-free interest rate is 3.50%.

The hedge ratio $h*$ is about 1 to a first approximation, so the investor would sell $x* = -m = -(16 \times 10^6)/(4000 \times 10) = -\mathbf{400}$ contracts, at a price of 4040.

At the end of December, the value of the position would be, with certainty, $4040 \times 10 \times 400 = 16.16$ million Euros, plus any dividends paid out by the stocks in the portfolio (assumed risk-free for simplicity), i.e., 2.5% of the value of the portfolio (thus $16 \times 2.5\% = 0.4$ million). The total would then increase to $16.16 + 0.4 = 16.56$ million, yielding a risk-free annual rate of $(16.56 - 16)/16 = 3.50\%$.

The latter is the risk-free interest rate, which shows that the initial price of the December contract (4040) was compatible with the absence of arbitrage opportunities and with the cash-and-carry relationship.

(ii). In fact, the investor knows that her portfolio only approximately tracks the CAC40 and has last year run, using weekly data, the linear regression of the value of her portfolio, excluding dividends, on the price of the December contract. The results are as follows:

$$\beta = 1.01; \quad R^2 = 0.970.$$

(continued)

The ratio $h*$ is therefore not equal to 1, but to 1.01 and the optimal number of contracts to be sold is not $-400$ but $-400 \times 1.01 = -404$. The effectiveness of this hedge, which is not perfect, is 97%.

Consider two possible cases for the CAC40 value and the investor's position at the end of December, assuming in both cases that the portfolio's actual beta remains 1.01.

– Case 1: The index, and the December contract, are worth 4400, and so the contract price has increased by $360/4040 = 8.91\%$. The portfolio, excluding dividends, therefore is worth $16 \times 1.0891 \times 1.01 = 17.6$ millions. Including 0.4 million in dividends, it is consequently worth 18 millions. The loss on the 404 contracts that were sold amounts to $404 \times (4400 - 4040) \times 10€ = 1.4544$ million.

The hedged position consequently has the value 16,545,600 Euros, which makes for an annual yield of 3.41%, very close, but not equal, to the preceding result (3.50%). Let us note that, without a forward hedge, the yield would have been 12.50% since the market went up sharply.

– Case 2: The CAC40 index, and the December contract, are worth 3500 at maturity, so the contract has lost $-540/4040 = -13.366\%$. The portfolio, excluding dividends, has thus lost $-13.366\% \times 1.01 = -13.50\%$. It is, therefore, worth $16 \times (1 - 0.135) = 13.84$ millions, or, with dividends included, 14.24 millions. The profit on the 404 contracts that were sold amounts to $404 \times (4040 - 3500) \times 10€ = 2.1816$ millions.

The hedged position is consequently worth 16,421,600 Euros, that is, in spite of the large drop in the market ($-12.5\%$ for the CAC40 without dividends), an annual yield of 2.64%, near, though not identical to, the preceding results (3.50% and 3.41%). Remark that, without forward hedging, the yield would have been a disastrous $-11.00\%$ due to the big drop in the market.

### 9.3.1.2 Basis and Correlation Risks

Let us take up again Equation (9.5) making the notation for the maturity $T$ forward prices more explicit:

$$V_H = mC_H + x\left(F_H{}^T - F_0{}^T\right).$$

Denote by $S_t$ the cash price of the forward contract's underlying on date $t$. This underlying is in general different from the security to be hedged so that $S_t \neq C_t$. We rewrite the equation above as

## 9.3 Maximum and Optimal Hedging with Forward and Futures Contracts 331

$$V_H = mC_H + x\left(F_H{}^T - S_H + S_H - S_0 + S_0 - F_0{}^T\right).$$

Denoting by $B_t{}^T = F_t{}^T - S_t$ the basis on date $t$ of the contract, the last equation becomes

$$V_H = mC_H + x\left(B_H{}^T + S_H - S_0 - B_0{}^T\right). \qquad (9.7)$$

– First let us assume that the agent hedges with a contract whose expiration date coincides with the hedge's horizon, which implies

$$T = H, \text{thus } B_H{}^T = B_T{}^T = 0, \text{from which } V_H = mC_H + x\left(S_H - S_0 - B_0{}^T\right).$$

As we saw in Proposition 5, the variance of $V_H$ can be cancelled if and only if $C_H$ and $S_H$ are perfectly correlated; for an imperfect correlation, any hedge, even a maximal one, leaves a random residual which is the *correlation risk*, strictly speaking.

– Let us now assume $T > H$, which implies that $B_H{}^T$ differs from 0 and is stochastic. This random variable $B_H{}^T$, which is present in equation (9.7) for $V_H$, is generally not perfectly correlated with $C_H$[17]; it constitutes an additional risk and implies that $S_H + B_H{}^T = F_H{}^T$ cannot be perfectly correlated with $C_H$, even if $S_H$ is. This extra risk is called the *basis risk*.

Let us remark that often the correlation risk (imperfect correlation of $S_H$ and $C_H$) and the basis risk (randomness of $B_H{}^T$) are not distinguished, in contrast to the preceding analysis. They are therefore merged in a global risk called (somewhat ambiguously) the "basis risk" associated with the imperfect correlation of $F_H{}^T$ and $C_H$ and which is in fact, as we have just seen, usually only partly attributable to the basis $B_H{}^T$.

---

### Example 2

Let us return to the preceding example but assume that, on the one hand, the December contract expires not on Monday the 31st (still a business day) but on Friday the 28th and, on the other, that the actual beta as of the end of December is not 1.01 but 1.02. Let us further assume the contract is worth 4400 (as in

(continued)

---

[17] $B_H$, equal to the carrying cost for the period $(H, T)$, therefore depends on the short-term rate $r_{T-H}(H)$ which is usually imperfectly correlated (or even uncorrelated) with the price $C_H$ of the asset to be hedged.

> Example 1, Case *ii*-1) on December 28. The loss on the 404 contracts is therefore 1.4544 million. On that date, the portfolio is worth $16 \times 1.0891 \times 1.02 = 17.774112$ millions. Assume, however, that between December 28 and December 31 both the index and the portfolio go up by 0.427%. Apart from dividends, the portfolio is thus worth 17.85 million, to which 0.4 million in dividends should be added. The hedged portfolio thus is valued at $17.85 + 0.4 - 1.4544 = 16.7956$ million, and the annual yield becomes 4.97%, in contrast to the 3.41% achieved in the previous example (Case *ii*-1).

### 9.3.1.3 Hedging by Rolling Over Forward Contracts

In some cases, the horizon $H$ of the hedge period falls *after* the maturity $T$ of the longest available contract. The hedge then consists in arranging a succession (rollover) of $n$ forward transactions with the periods $(T_0, T_1)$, $(T_1, T_2)$, ..., $(T_{n-1}, T_n)$ where $T_0 = 0$ and $T_n = T$. To start with, let us assume that each of these forward transactions lasts until its maturity: $T_i$ denotes the maturity date of the $i^{th}$ contract in the sequence, the date on which the following contract with expiration date $T_{i+1}$ takes over. The cumulative effect of the margins produced by such a rollover for a purchased contract (intended to hedge a short position) is therefore equal to.

$$\sum_{i=1}^{n} \left( F_{T_i}^{T_i} - F_{T_{i-1}}^{T_i} \right) = \left[ S_{T_1} - F_0^{T_1} + S_{T_2} - F_{T_1}^{T_2} + \ldots + S_{T_{n-1}} - F_{T_{n-2}}^{T_{n-1}} + S_T - F_{T_{n-1}}^{T} \right]$$

$$= -F_0^{T_1} - \left( F_{T_1}^{T_2} - S_{T1} \right) - \ldots - \left( F_{T_{n-1}}^{T} - S_{Tn-1} \right) + S_T,$$

so that

$$\sum_{i=1}^{n} \left( F_{T_i}^{T_i} - F_{T_i}^{T_i} \right) = S_T - F_0^{T_1} - \sum_{i=2}^{n} B_{T_{i-1}}^{T_i}$$

where $B_{T_{i-1}}^{T_i}$ denotes the basis prevailing at $T_{i-1}$ for an expiry at $T_i$.

We can infer from the last equation the following features:

- If the price $S_T$ of the contract is correlated with the market price $C_T$ of the security to be hedged, such a rollover can be very useful.
- However, this induces $(n-1)$ basis risks, or additional sources of uncertainty $B_{T_{i-1}}^{T_i}$

  that limit its efficiency.

Note that the agent is not required to hold on to each of his forward positions making up this rollover until it expires: indeed, he can unwind the $i^{th}$ transaction

## 9.3 Maximum and Optimal Hedging with Forward and Futures Contracts 333

concerning a contract with expiry $T_i$ at some instant $t_i$ before $T_i$ while setting up at the same time a new contract with expiry $T_{i+1}$. The reader will check that such a strategy gives rise to a global margin given by the following equation:

$$\sum_{i=1}^{n} \left( F_{t_i}^{T_i} - F_{t_{i-1}}^{T_i} \right) = -F_0^{T_1} - \left( F_{t_1}^{T_2} - F_{t_1}^{T_1} \right) - \ldots - \left( F_{t_{n-1}}^{T_n} - F_{t_{n-1}}^{T_{n-1}} \right) + F_H^{T_n}.$$

### 9.3.2 Optimal Hedging and Speculation

Recall that a speculative trade is a risky transaction motivated by the expectation of a gain. This expectation, along with its associated risk assessment, result from the investor's forecasts.

Speculation, thus defined, is in fact a component of rational portfolio selection, as shown by the following simple model, set in the framework of the hedging model exposed in Subsection 9.1.1. As in the latter, the agent is interested in the value $V_H$ of his portfolio on date $H$. $V_H$ is derived from the initial position then worth $mC_H$ plus the profit/loss resulting from the forward position, such that $V_H = mC_H + x (F_H - F_0)$. However, contrary to the assumption underlying the preceding model, we assume here that investors are not exclusively concerned with risk but are *equally* worried about the expected value of their holdings at $H$. We use the notation.

$$E_V(x) \equiv E(V_H); \quad \sigma V^2(x) \equiv \mathrm{var}(V_H).$$

Then the optimization problem writes[18]:

$$\underset{x}{\mathrm{Max}}\{f(E_V(x), \sigma V^2(x))\}.$$

The function $f$ represents the preferences of the investor who is supposed to desire as large an expected portfolio value $E_V$ as possible with as small a variance $\sigma_V^2$ as possible. From this, assuming that the function $f$ has first derivatives, we impose

$$f_1 \equiv \frac{\partial f}{\partial E_V}\left(E_V, \sigma_V^2\right) \geq 0 \; ; \quad f_2 \equiv \frac{\partial f}{\partial \sigma_V^2}\left(E_V, \sigma_V^2\right) \leq 0.$$

Furthermore, $E_V = m\, E\,(C_H) + x\, E\,(F_H - F_0)$, from which follows

$$\frac{\partial E_V}{\partial x} = E_{\Delta F} \text{ where } E_{\Delta F} \equiv E(F_H - F_0).$$

Also, $\sigma_V^2 = m^2 \sigma_C^2 + x^2 \sigma_F^2 + 2mx\sigma_{CF},$

---

[18] The agent uses the mean/variance criterion whose foundations are discussed in Chap. 21.

so that: $\frac{\partial \sigma_V^2}{\partial x} = 2[x\sigma_F^2 + m\sigma_{CF}]$.

As a result, the condition that the optimal position $x^{**}$ must satisfy can be written

$$f_1 \frac{\partial E_V}{\partial x} + f_2 \frac{\partial \sigma_V^2}{\partial x} = f_1 E_{\Delta F} + f_2(2x^{**}\sigma_F^2 + 2m\sigma_{CF}) = 0,$$

which requires the optimal choice of the forward position given in Equation (9.8):

$$x = x^{**} = \underbrace{-m\frac{\sigma_{CF}}{\sigma_F^2}}_{hedging} \underbrace{-\frac{1}{2}\frac{f_1}{f_2\sigma_F^2}E_{\Delta F}}_{speculation} \tag{9.8}$$

This optimal position thus has two components:

- The first, $-m\frac{\sigma_{CF}}{\sigma_F^2} = x^*$, is the one that leads to minimum variance (see Equation (9.6), of which (9.8) is a generalization); strictly speaking, it represents the hedge.
- The second, $-\frac{1}{2}\frac{f_1}{f_2}\frac{E_{\Delta F}}{\sigma_F^2}$, can be interpreted as a speculative component, which is implemented by taking a position that *increases the risk* and is motivated by the expected variation $\Delta F$.

In this connection, let us first note that $-\frac{1}{2}\frac{f_1}{f_2}\frac{E_{\Delta F}}{\sigma_F^2}$ increases with the forward price's anticipated rise $E_{\Delta F}$, since $\frac{f_1}{f_2} < 0$, and decreases as the risk $\sigma_F^2$ increases.

Next remark that the speculative component increases with the ratio $-\frac{f_1}{f_2}$, expressing the marginal importance given to the expectation relative to that given to the variance, a ratio that may be viewed as a measure of *risk tolerance*.

The second component was absent in (9.6), since the risk tolerance $-\frac{f_1}{f_2}$ was implicitly zero for an investor seeking minimal risk.

The main merit of the model just described is that it highlights the factors occurring in speculative investment as well as the duality in the rational agent's optimal position, which is *a combination* of hedging and speculating.

## 9.4 The Main Forward and Futures Contracts

We will distinguish the main forward and futures contracts according to their underlying asset. We will not take on in this chapter some contracts involving interest rates (notional bonds, Libor, Euribor, OIS rates, etc.) which are explained in Chap. 15. Furthermore, for simplicity futures contracts will be analyzed as if they were forwards. Subsection 9.4.1 refers to contracts on commodities, Sect.9.4.2 to currency contracts, and Sect. 9.4.3 to contracts on financial instruments and market indices.

## 9.4 The Main Forward and Futures Contracts

### 9.4.1 Contracts on Commodities

#### 9.4.1.1 Brief Summary of Contracts and Markets

Futures on commodities, traded on organized markets, allow producers and dealers to hedge against the risk of price changes. Speculators who bet on the direction of this variation and, more generally, investors that hold portfolios of futures contracts, ensure the market's liquidity. In addition, futures on commodities may allow better portfolio diversification, and thus improve the risk/return tradeoff.

While contracts on financial securities are almost always settled prior to expiration, contracts on commodities more frequently result in physical delivery.

The main *commodities* underlying futures contracts are:

- Metals: Gold, silver, copper, aluminum, platinum, nickel, tin, zinc, and palladium.
- Raw materials connected with energy: Gas, gasoline (petroleum products), propane, and coal as well as electricity.
- Agricultural and food materials and products: Beef, pork, milk, wheat, corn, soy, sugar, cocoa, coffee, orange juice, wood, and fertilizer.

The main markets for futures contracts on commodities are in Chicago (CME Group), New York (New York Mercantile Exchange, NYMEX which is actually part of the CME group), and London (London Metal Exchange and ICE).

Like other financial instruments, commodity futures are traded for arbitrage, speculation, or hedging motives.

> **Example 3**
> As an example of this last case, let us consider a metalworking company whose production process uses aluminum. On average, it uses the equivalent of 220,000 pounds of aluminum each month and would like to fix today the price it will have to pay for this metal a month from now, fearing a rise in the international market price. On the NYMEX, aluminum contracts are for 44,000 pounds. The current futures price is \$0.805 per pound. The company will buy 5 contracts (220,000/44,000) and will get these 220,000 pounds delivered in one month by paying \$177,100, whatever the spot price of aluminum on that date.

#### 9.4.1.2 Relation Between Forward and Spot Prices, Warehousing Cost, and Convenience Yield

For commodities contracts, the derivation of the spot-forward parity depends essentially on whether the underlying asset can be stored or not.

– When the contract's underlying is storable, spot-forward parity results from the cash-and-carry relation derived under AAO. As already mentioned, a cash-and-carry trade involves buying and storing the underlying of the contract. Warehousing is in general costly. According to whether its cost between 0 and $T$ for a unit of the commodity is equal to some amount $J$ independent of its price, or is proportional to its value and calculated on the basis of a continuous rate $j$, and depending on the way interests are computed, we can write spot-forward parity at date 0 in different ways, for example:

$$F_0 = S_0 + I + J \tag{9.9-a}$$

$$F_0 = (S_0 + J)\, e^{rT} \tag{9.9-b}$$

$$F_0 = S_0\, e^{(r+j)T}. \tag{9.9-c}$$

These equations express the equality of the basis and the cost of carry. The latter is made up of two parts: the cost $(I, r)$ of financing the purchase of the underlying over the period $(0, T)$ and the storage cost $(J, j)$.

---

**Example 4**

In New York on March 18, silver (in US cents per ounce) is quoted at 733.50 spot. A December contract can be traded for 753.80. Applying formula (9.9-c), taking $T = 0.75$, gives $(r + j) = \ln(F_0/S_0)/T = 3.64\%$. Knowing that the 9-month US rate is 3.47%, we deduce that the storage cost of silver amounts on that day to 0.17% annually.

---

These equations assume that physically holding the underlying does not entail any earning or advantage. However, in some situations, possession of a commodity is considered an advantage for the holder. This is notably true for industrial enterprises that use the material in their production processes or for dealers who hold inventories so they can make immediate deliveries. Actually possessing the commodity covers them against the risk of shortages which could lead to supply disruption and have nasty consequences for their production or sales. In this case, physically holding the commodity is an advantage that should be valued and reduce the cost of carry. Although difficult to estimate since it is very volatile, this advantage is often quantified as a *convenience yield* (return rate on the physical commodity) which we will denote by $c$. This means that the benefit resulting from holding a commodity with price $S_t$ during the interval $(t, t + \delta t)$ is reckoned equivalent to a financial return of $cS_t\, \delta t$. Such a convenience yield, $c$, which plays the same role as the $r^*$ in Equation (9.4), lowers the cost of carry, as if the interest rate were $r - c$ and not $r$ (or as though the storage cost were to be calculated using a rate $j - c$). From this we see that spot-forward parity should not be expressed in one of the forms (9.9-a), (9.9-b), or (9.9-c) but rather in the following more general relation

## 9.4 The Main Forward and Futures Contracts

$$F_0 = S_0 \, e^{(r-c+j)T}. \tag{9.9-d}$$

Remember that the convenience yield is linked to the liquidity risk that influences the commodity market and rises as the risk does. The convenience yield is thus small if manufacturers and dealers have swollen inventories and rise when their inventories are lower. Partly because the available inventories of commodities fluctuate, the convenience yield is not a fixed quantity; it varies in a random way, which complicates using it and limits its interest.[19]

---

**Example 5**

On the London Metal Exchange, on March 18, nickel (in \$/ton) has a spot price of 16,232. A contract with a maturity of 3 months (exactly) is quoted at 15,888. The 3-month interest rate on the international dollar (quoted in London) is 3.03% and storage cost is estimated at 0.34% annually. The convenience yield for nickel implicit in the contract's rate is such that $(r - c + j) = (3.37 - c) \% = \ln(F_0 / S_0)/T = \ln(15{,}888/16{,}232)/(1/4) = -8.57\%$, which implies $c = 11.94\%$ on an annual basis, which is very large.

---

Note that reverse cash-and-carry transactions, which involve selling the underlying and buying a forward contract, can only be entertained for physical commodities that can be held as investments (gold or silver, for example). Indeed, investors holding such assets in a portfolio have an interest in selling them on the spot market and buying them forward when $F_0 < S_0 \, e^{(r-c+j)T}$; in this way they exert a pressure on prices which tend to correct the imbalance. Meanwhile, some commodities are only stored by dealers or manufacturers with the goal of ensuring the continuity of their sales and manufacturing which would be damaged by delivery delays or disruption in the production chain.[20] Products stocked with this intention have to be physically available and those who hold them are reluctant to dispose of them to carry out a reverse cash-and-carry. For such products, the basis may be less than the cost of carry because cash-and-carry arbitrages can only guarantee the inequality $F_0 \leq S_0 \, e^{(r-c+j)T}$.

– For underlying assets that cannot be stored, such as electricity, cash-and-carry arbitrages (and *a fortiori* reverse ones) cannot be carried out and equations (9) no longer hold. The relationship between spot and forward prices can no longer be

---

[19] The attentive reader will notice that here we have in fact a variable used to adjust the cost of carry so as to make it equal to the observed basis, and that we are trying to rationalize it *ex post* in an ad hoc way. There are, however, models that attempt, with more or less success, to forecast it ex ante on the basis of the behavior of supply of and demand for the commodity. These models are outside the scope of this book.

[20] Those are inventories essentially held for transaction and/or precautionary motives. In fact, there is often also a component, even if a small one, of the inventory which can be attributed to the expectation of an increase in the price of the asset held.

derived from arbitrage considerations but rather from the economic equilibrium between supply and demand on the commodity market over several periods. This equilibrium is especially dependent on the current state of the market for the commodity and expectations for it at the contract's maturity (see Sect. 9.2 and 9.2.2 in this chapter which provides an elementary analysis of the relation between the forward price and the expectation of the future spot price). A full economic analysis of the behavior of the equilibrium for a commodity market over time is outside the scope of this book. Just note one common empirical characteristic of numerous commodity markets: their quoted prices exhibit reversion to the mean. Because of this, if the price is, at a given moment, higher (resp., lower) than its long-term mean, the price variations observed will have a tendency to be negative (resp., positive), which brings the price back to its mean.

## 9.4.2 Contracts on Currencies (Foreign Exchanges)

### 9.4.2.1 Brief Summary of Contracts and Markets

The buyer (the seller) of a currency contract undertakes to buy (to sell) on a certain date $T$, a certain quantity of currency at a price fixed today in some other currency.

OTC forward contracts still make up the majority of such contracts but the proportion of futures is increasing.

– Forward currency contracts, also called FX or Forex contracts, are traded on the foreign exchange market, with no organized secondary market; they are not standardized, although the maturities that are mostly traded are 1, 2, 3, 6, or 12 months. Trades are carried out by telephone, with bid and ask spreads. Most of these OTC transactions are carried through to their maturities and settled by actual delivery of the underlying currency.

A forward exchange rate is often quoted as the spread with the spot rate, i.e., a premium (if $F - S > 0$) or a discount ($F - S < 0$).

For example, if the spot exchange rate \$/€ is at 1.120 and the forward rate in 3 months is 1.125, a dealer on the exchange market will announce a premium of 0.005 \$ per euro. This spread is then added to the spot exchange rate to get the forward rate. Conversely, a discount is subtracted from the spot exchange rate to get the forward rate.

The premium or discount is often expressed as a percentage of the spot rate ($\frac{|F-S|}{S}$) and sometimes the difference $F - S$ is termed by the dealers the *swap* rate.

– Currency futures are traded on organized markets, with a clearinghouse, standardized contracts (as to amounts and maturities), and margin calls each business day. The bulk of currency futures trades are carried out on the Chicago Mercantile Exchange which handles the major currencies (Euro, Sterling, Yen, Canadian dollar, Swiss franc, Mexican peso, etc.) quoted against the US dollar.

## 9.4 The Main Forward and Futures Contracts

**Table 9.5** Cash-and-carry arbitrage on currency

|  | 0 | $T$ |
|---|---|---|
| Purchase of $e^{-r*T}$ unit of currency X; Immediate placement of this from 0 to $T$ giving 1unit of $X$ on date $T$ | $-S_0 e^{-r*T}$ | *Delivery of a unit of X* |
| Forward sale at 0 and payment received upon delivery at $T$ of 1 unit of $X$ | 0 | $+F_0$ |
| Borrowing (in €) then repayment at $T$ | $+S_0 e^{-r*T}$ | $-S_0 e^{(r-r*)T}$ |
| Net cash flow | 0 | $F_0 - S_0 e^{(r-r*)T}$ |

The market is very liquid and allows even transactions in small amounts. Quoting is done by open outcry, with no bid-ask spreads.[21] In contrast with forward transactions on the interbank market, futures are usually settled before their maturity and then do not lead to delivery of the underlying currency, with occasional exceptions.

Foreign exchange forward/futures contracts can be used to hedge against exchange rate risk or to speculate on a currency going up (with a forward buy) or down (with a forward sale) with respect to another one. The great liquidity of the exchange markets for the main currencies implies it is easy to open and close positions, even large ones, and allows for considerable leveraging.

### 9.4.2.2 Analysis and Valuation

In the following, we consider, on date 0, a forward contract for currency $X$ *(against the euro)* and with expiration $T$. The contract is written on one unit of currency $X$ and currently quotes $F_0$ euros. The currency $X$ is quoted today at $S_0$ euros, on the spot market ("direct quotation"). We denote by $r$ the currently prevailing interest rate on transactions in euros with maturity $T$ and by $r*$ that for currency $X$.

Under no arbitrage (cash-and-carry or reverse), the spot-forward parity should hold; this can be written, according to whether the rates $r$ and $r*$ are simple, compound discrete or continuous:

$$F_0 = S_0(1 + (r - r*)T); \quad F_0 = S_0 \frac{(1+r)^T}{(1+r*)^T}; \quad F_0 = S_0 e^{(r-r*)}. \tag{9.10}$$

Table 9.5, which is very similar to Table 9.4, shows the cash-and-carry arbitrage that leads to an arbitrage profit if $F_0 > S_0 e^{(r-r*)T}$; all cash flows are in Euros.

An arbitrage profit appears when $F_0 > S_0 e^{(r-r*)T}$. In case $F_0 < S_0 e^{(r-r*)T}$, the arbitrage consists in taking the opposite position (a reverse cash-and-carry) involving, on date 0, the sale of $e^{-r*T}$ units of currency $X$, a loan of $S_0 e^{-r*T}$ euros and the purchase of a forward contract.

---

[21] Brokers are compensated by fees.

340 9 Futures and Forwards

A *premium* ($F > S$) prevails when the domestic rate $r$ (here, on the euro) exceeds the foreign rate $r^*$ (here, on currency $X$) and a *discount* when the opposite holds ($r^* > r$).

> **Example 6**
> Let us assume that on the international market for euros, the 6-month rates for the dollar and the euro are, respectively, 3.29% and 2.19% (discrete compound interest). The spot exchange rate for the euro is 1.3315 dollars. Its forward exchange rate (6 months) is thus, applying the middle formula (9.10) (*reversing the roles of r and r\* since here we purchase dollars, not euros (this is an "indirect quotation")*): $1.3315 \times (1.0329)^{0.5} / (1.0219)^{0.5} = 1.3386$, so we have a premium (on the dollar) of 0.0071, or a discount (on the euro).

### 9.4.3 Forward and Futures Contracts on Financial Securities (Stocks, Bonds, Negotiable Debt Securities), FRA, and Contracts on Market Indices

We analyze now forward contracts whose underlying is a financial instrument (stock, negotiable debt security, and so on) or a market index. We will not consider here the case of futures on notional bonds or STIR contracts on EURIBOR/LIBOR or overnight averages, which are the subject of Chap. 15. Furthermore, we will not distinguish futures and forward contracts and will treat the former as if they were forwards.

#### 9.4.3.1 Brief Summary of Contracts and Markets

One may distinguish the contracts by three types of underlying: fixed-income (debt) instruments, market indices, and stocks.

- The first futures contracts on interest rates were issued in the mid-1970s by the two large markets in Chicago, which merged in 2007 to become the CME Group.[22] Today futures on numerous underlying assets are traded there, amongst which the most significant are those in notional US Treasury bonds and Eurodollar-3 month bonds.

---

[22] The CME Group was born as a result of the takeover of the CBOT (Chicago Board of Trade) by the CME (Chicago Mercantile Exchange). It combines four large markets: the CME, the CBOT, the NYMEX and the COMEX and offers a large variety of futures contracts and options. It is the largest market of its kind in the world.

## 9.4 The Main Forward and Futures Contracts

After different restructurings mentioned in preceding chapters, the two largest forward rate markets in Europe are currently EUREX (Frankfurt) and ICE Futures Europe.[23] The main futures interest rate contracts will be studied in Chap. 15.

There are also similar futures traded in Sidney, Singapore, Tokyo, Montreal, and Rio de Janeiro.

Furthermore, trading of other fixed-income instruments (bonds, bills, or negotiable debt securities with different maturities) can also take place, and FRA (*forward rate agreements*, see Subsect. 9.4.3.4) can be also traded on the main OTC markets.

- Contracts on indices are defined for a stock market index which is dimensionless (a pure number) but, when multiplied by a number of monetary units determined by the market authorities or the sponsors of the index, then represents the value of a well-defined portfolio, in accordance with what we have explained in Chap. 8 where the main indices were presented. The contract's size is, in the same way, a multiple of the index. The most active is the North American (futures) contract traded by the CME Group, and written on the S&P 500 with a standard multiple of 250 dollars. In Europe the main markets are ICE and EUREX. In those markets some very active contracts are traded, such as the CAC 40 in France (the multiple is €10) or the FTSE 100 in the UK (the multiple is £10). The large stock markets in the Asia-Pacific area (Sydney, Toronto, Tokyo, Singapore) and Latin America also offer contracts on indexes such as the Nikkei 225.
- Forward contracts on stocks are also traded OTC.[24] By postponing the payment, the buyers benefit from the leverage effect associated with forward instruments since the deposits required by their brokers is only a fraction of what they would need to pay in cash on the spot market.[25]

### 9.4.3.2 Analysis and Valuation of a Contract on Fixed-Income Instruments or Stocks

Denote today's date by $t$, and consider a forward contract *signed at some instant* $0$ *in the past* $(0 \leq t)$ which requires the seller to deliver at $T$ some financial instrument (stock, bond, negotiable debt). In exchange, the buyer will pay the price $F_0$ agreed upon at date $0$.

This forward contract can easily be valued by applying the methods previously described.

---

[23] The NYSE-EURONEXT-LIFFE has been renamed ICE Futures Europe after several mergers and acquisitions that left it under the ownership of the ICE (Intercontinental Exchange).

[24] An interesting case is the official French system SRD (Service de Règlement Différé) created in 2000 where traders can take forward positions on some listed French stocks obeying minimum requirements.

[25] The deposit requirements are as follows: (i) 20% in cash, Treasury Bonds or monetary mutual funds; (ii) 25% in listed bonds, negotiable debt securities or bond mutual funds; and (iii) 40% in listed stocks or equity mutual funds.

| | | | |
|---|---|---|---|
| 0 | $t$ | $s$ | $T$ |
| (initiation) | (to-day) | (coupon $D$) | (contract maturity) |

Let us call $R$ the value of the revenue expected by the holder of the security between $t$ and $T$, *capitalized on date $T$*. This is the value of the coupon(s) or dividend (s) distributed by the security and capitalized on date $T$ at the appropriate forward rate. We assume its (their) amount(s) is (are) known on current date $t$. For example, if only one dividend distribution $D$ occurs between $t$ and $T$, on date $s$ ($t \leq s \leq T$), denoting by $f_{s,T-s}(t)$ the forward rate for transactions lasting $T-s$ periods and starting at $s$ which is implicit in the spot yield curve on date $t$ (today), we have

$$R = De^{(T-s)f_{s,T-s}^{(t)}}.$$

The spot forward parity equation, expressing the equality between the basis and the cost of carry, then writes:

$$F_t - S_t = I - R = S_t \left( e^{(T-t)r_{T-t}^{(t)}} - 1 \right) - De^{(T-s)f_{s,T-s}^{(t)}},$$

or

$$F_t + De^{(T-s)f_{s,T-s}^{(t)}} = S_t \, e^{(T-t)r_{T-t}^{(t)}}.$$

This equation is written for continuous rates; with simple discrete rates it becomes

$$F_t + D[1 + (T - s) f_{s,T-s}(t)] = S_t [1 + (T - t)r_{T-t}(t)].$$

Furthermore, on date $t$, a forward contract, *with a prior start date $0$ and an initial price $F_0$*, has a value $V_0(t)$ given by:

$$V_0(t) = (F_t - F_0) \, e^{-(T-t)r_{T-t}^{(t)}}. \tag{9.11}$$

By contrast, recall that a futures contract has zero value (each evening after the margin call).

---

**Example 7**

Shares in Accor are quoted spot at €37.56. In 3 months, they will distribute a dividend of €1.50. The proportional 3-month and 9-month rates are, respectively 2.15% and 2.35%. The 6-month forward rate in 3 months is therefore (working with months and not with days, since the difference is immaterial here):

(continued)

## 9.4 The Main Forward and Futures Contracts

$$1 + (6/12)f_{3.6} = (1 + (9/12) \times 2.35\%)/(1 + (3/12) \times 2.15\%)$$
$$= 1.012185.$$

An investor who wants to make a forward purchase of the share for delivery in 9 months will consequently pay a forward price of

$$F_t = S_t \left[1 + (T - t)r_{T-t}(t)\right] - D[1 + (T - s)f_{s,T-s}(t)]$$
$$= 37.56 \left[1 + (9/12) \times 2.35\%\right] - 1.50 \left[1.012185\right] = €36.70.$$

### 9.4.3.3 Analysis of Forward Contracts on Fixed-Income Securities

A forward contract on a fixed-income security can be valued in two different but equivalent ways:

– By applying the methods of the preceding Sect. 9.4.3.2, which are valid for different contracts on financial instruments.
– By considering the forward contract itself as a fixed-income security and applying the standard methods developed in preceding chapters.

Here we will follow this second way and find formulas equivalent to those obtained in Sect. 9.4.3.2, though sometimes written in a different form.

Denote by $t$ today's date, and consider a forward contract expiring on date $T$ that was signed on date 0 in the past for a price $F_0$. The contract obliges the seller to deliver on date $T$ the security that pays out the stream of cash flows $\{E_\theta\}_{\theta = t_1,\ldots,\, t_n}$ with $T \leq t_1$. Thus, the contract can be seen by the buyer as the series of cash flows:

$$
\begin{array}{ccccc}
0 & t & T & t_1, \ldots\ldots, t_n \\
& & -F_0, & E_{t_1}, \ldots\ldots, E_{t_n}
\end{array}
$$

First remark that, seen from the moment $t$, this payment schedule, set up previously at time 0, today has a value $V_0(t)$ given by the equation

$$V_0(t) = -F_0 e^{-r_{T-t}(T-t)} + \sum_{\theta=t_1}^{t_n} E_\theta e^{-r_{\theta-t}(\theta-t)} \tag{9.12}$$

where $r_\theta$ is the zero-coupon rate of duration $\theta$ prevailing today (at $t$).

Since the forward contract started today (at $t$) has zero value $V_t(t)$ (the price of the contract $F_t$ is such that the transaction's Net Present Value is zero for both buyer and seller since there is no payment from one to the other), we have:

$$-F_t e^{-r_{T-t}(T-t)} + \sum_{\theta=t_1}^{t_n} E_\theta e^{-r_{\theta-t}(\theta-t)} = 0. \tag{9.13}$$

Equations (9.12) and (9.13) therefore give $V_0(t) = (F_t - F_0)e^{-r_{T-t}(T-t)}$, which is just equation (9.11).

This approach allows us also to express the spot-forward relation in a form specific to forward contracts on interest rates. Indeed, assume that the underlying security distributes, between $t$ and $T$, cash flows denoted $\{E_s\}_{s=s_1,\dots,s_m}$, with $t < s_1 < \dots < s_m < T$, so that the complete schedule of cash payments distributed can be written.

$$
\begin{array}{c|c|c c c|c|c c c}
0 & t & s_1, \dots, s_m, & T & t_1, \dots, t_n \\
 & & E_{s_1}, \dots, E_{s_m}, & & E_{t_1}, \dots, E_{t_n}
\end{array}
$$

From this we see that $S_t = \sum_{s=s_1}^{s_m} E_s e^{-r_{s-t}(s-t)} + \sum_{\theta=t_1}^{t_n} E_\theta e^{-r_{\theta-t}(\theta-t)}$, which combined with equation (9.13) yields the spot-forward parity:

$$S_t = F_t e^{-r_{T-t}(T-t)} - \sum_{s=s_1}^{s_m} E_s e^{-r_{s-t}(s-t)}.$$

This equation is a generalization of (9.3-b).

### 9.4.3.4 Forward Rate Agreement (FRA)
– We start with some preliminary remarks:

We consider three successive dates $0, T, T + D$ where $T$ and $D$ are fractions of a year calculated as the ratios of a number of days divided, for instance, by $360$,[26] and $0$ is today's date. Also consider a forward rate $f$ (proportional, with the same convention actual/360 as D) and an amount $M$ both *determined on date* 0, as well as three transactions *initiated on date* 0 and generating three cash flow series $\underline{F}_1, \underline{F}_2, \underline{F}_3$, during the period $(T, T + D)$ of length $D$:

---

[26]When the rates are set up on an actual or 365 basis, these durations are calculated on an actual/actual or actual/365 basis.

## 9.4 The Main Forward and Futures Contracts

$$
\begin{array}{llll}
dates: & 0 & T & T+D \\
\underline{F}_1: & 0 & M & -M(1+fD) \\
\underline{F}_2: & 0 & 0 & M(r_D(T)-f)D \\
\underline{F}_3: & 0 & M\dfrac{(r_D(T)-f)D}{1+Dr_D(T)} & 0
\end{array}
$$

It is important to note that $r_D(T)$ represents the spot rate that will prevail on date $T$ for transactions of duration $D$ and as such is an unknown quantity today.

First note that the transaction $\underline{F}_1$ is borrowing forward (contracted on date 0 and implemented between $T$ and $T+D$ on the basis of a forward rate $f$), that $\underline{F}_2$ is a plain vanilla borrowing *swap* (with the fixed rate $f$) made up of a single payment[27] and that an FRA takes usually the form of $\underline{F}_3$.

Also note that these three transactions are *equivalent* since, starting from the cash flows generated by one of them, one can obtain those of the other two. For example, if one has borrowed forward (which generates $\underline{F}_1$), it is enough to lend the amount $M$ on the spot market on the date $T$ for a duration $D$ and on the basis of a (future) spot rate $r_D(T)$, to end up with $\underline{F}_2$. One can also easily transform $\underline{F}_2$ into $\underline{F}_3$ by borrowing the amount $M\frac{(r_D(T)-f)D}{1+Dr_D(T)}$ on date $T$ for a duration $D$.

Each of these three investments thus allows guaranteeing today (on date 0) the interest rate of a loan or an investment of an amount $M$ to be realized in the future (between $T$ and $T+D$).

– Like $\underline{F}_1$, $\underline{F}_2$, or $\underline{F}_3$, and by definition, an FRA is a transaction allowing one to guarantee today (on date 0) the interest rate of a loan (by buying an FRA) or of an investment (by selling an FRA) of a sum $M$ to be made in the future (between the dates $T$ and $T+D$). In fact, the FRA is generally of the form $\underline{F}_3$, thus the winning party will receive on date $T$ the difference in interests[28]:

$$
\pm M\frac{(r_D(T)-f)\,D}{1+Dr_D(T)}.
$$

$M\frac{(r_D(T)-f)D}{1+Dr_D(T)}$ is the cash flow received (if positive) or paid (if negative) by the buyer who profits (loses) from an increase (decrease) in rates. Buying an FRA thus permits *fixing* the cost of a future borrowing. Selling an FRA permits *fixing* the interest earned on some future lending (the seller receives the cash flow $M\frac{(r_D(T)-f)D}{1+Dr_D(T)}$).

---

[27] Initiated on date 0 and not on $T$; this is therefore a *forward swap* (see Chap. 7, Sect. 2 and 4.3.5).

[28] In certain markets (outside the Euro Zone), the rate $r_D(T)$ used could be a discounted one and not post-counted as here, and the same for the forward rate. The cash flow would then be $\pm M(r_D(T)-f)D$.

The reader will remember that:

- If 0 represents the start date of the FRA, under no arbitrage, we must have: $f = f_{T,D}(0)$ (the forward for $D$ periods in $T$ periods prevailing at the start of the FRA), since the market rate $f$ on forward loans of type $\underline{F}_1$ is $f_{T,D}(0)$.
- A vanilla *swap* can be considered as a portfolio of successive FRAs, and an FRA can be considered as a *swap* involving a single payment (see the stream $\underline{F}_2$).

---

**Example 8**

Company X wants to borrow 10 M\$ for 3 months in 6 months.

It would like to protect itself against the risk of an increase in the \$ interest rate and to fix the cost of the loan. It buys an FRA at the guaranteed rate of 6% for a sum of 10 M\$.

If the \$ interest rate reaches 8% in 6 months, the counterparty will pay to X:

$10 \times (0.08 - 0.06) \frac{91}{360 + 91 \times 0.08} = 0.049309$ M\$,

compensating for the rise in the rate (2% $\times$ 10 M\$ over 3 months). Indeed, note that 0.049309 available in 6 months is the same as $2\% \times 10 \times \frac{91}{360}$ available in 9 months.

We also remark that, in an arbitrage-free market, 6% equals the forward rate for 3 months in 6 months which is implicit in the spot yield curve.

---

**Remarks**
- The last remark in the example above is obviously completely general. The rate guaranteed by the FRA between dates $T$ and $T + D$ is equal, under no arbitrage, to the forward rate $f_{T,D}(0)$ implicit in the spot yield curve.
- The counterparty risk affecting an FRA involves only the difference in interest rates and *not* the principal $M$ (in contrast with a forward position replicated by two spot transactions, which also involves a counterparty risk on the principal).
- Short Term Interest Rate futures (STIRS) are the "futures equivalent" of FRAs (which are forward), as explained in Chap. 15.
- FRAs are over-the-counter transactions carried out in the main money markets and are very liquid if they are defined using the 3- or 6-month rates for a major currency ($D = 3$ months or 6 months and generally $T$ is less than a year).

### 9.4.3.5 Forward Contracts on a Market Index

For a contract on an index $I$ the payoff is the difference $I_T - F_0$ (times the multiple $m$) where $I_T = F_T$ denotes the index value on $T$ and $F_0$ is the price agreed on the start date 0. Contracts on indices are governed by the same principles as other forward contracts even if the index itself is not materialized. In fact, $I_t$ (here implicitly

## 9.4 The Main Forward and Futures Contracts

multiplied by a monetary unit, such as 250$) at any moment represents the value of *a portfolio which must be considered as the asset underlying* the contract.

Because of the intangible character of the index, and practical complications resulting from delivery of the securities comprising the underlying portfolio, these contracts *are usually not settled by actual delivery but by cash.*

Cash-and-carry arbitrages (and their reverses), however, involve transactions in relevant securities; they consist in (*i*) taking a long position on the underlying portfolio (the one whose value the index reflects) by buying shares of its components in proportion to their weights in the index, (*ii*) borrowing to finance these purchases, (*iii*) selling a forward contract, (*iv*) cashing in the dividends on the portfolio held then (*v*) winding up all the positions at maturity. It is often handy to calculate the dividends using a continuous yield rate $r*$ since the distributions from the securities making up the underlying portfolio can be numerous and happen frequently.[29] Since $I_t$ equals the value on date $t$ of the underlying portfolio that replicates the index, the spot-forward parity equation can be written, with the usual notation

$$F_t = I_t \, e^{(r_{T-t}^{(t)} - r*)(T-t)}.$$

Furthermore, the value of a futures contract is always zero and the value at $t$ of a forward contract, *initiated on date* 0 at price $F_0$, is

$$V_0(t) = (F_t - F_0) \, e^{-(T-t) \, r_{T-t}^{(t)}}.$$

Actually, as we have seen, contracts on indices are defined as multiples of the reference index. The multiple $m$ is therefore a scale factor that determines the contract's size. The margin provided by a long position in a contract with multiple $m$ opened on date 0 with initial price $F_0$ and wound up on date $t$ at price $F_t$ thus will equal $m \, (F_t - F_0)$ monetary units. This global amount is in fact broken up in daily margins $m \, (F_t - F_{t-1})$ in the case of futures.

---

**Example 9**

The FTSE Eurofirst 100 quotes at the end of March 3579. The British 3-month interest rate is 4.9%. The dividend yield up to the end of June is 2.1%. The June contract should then be quoted $3.579 \, e^{(0.049-0.021)(3/12)} = 3.604$.

---

Contracts on market indices can be used to hedge a stock portfolio against market fluctuations. An important parameter in placing a hedge will be the sensitivity (beta; see Chap. 22, Sects. 9.1 and 9.1.3) of the portfolio relative to the index that is the contract's underlying.

---

[29] The rate $r*$ depends on the period (0, $T$): indeed dividend distributions are often concentrated at certain times of the year (May and June, for example).

## 9.5 Summary

- A forward or a futures (contract) is for the buyer the *right* and the *obligation* to buy upon maturity a definite object, called the underlying (asset), *at a fixed price*. For the seller, this contract brings the symmetrical right and obligation to sell. The underlying can be a commodity, a financial asset, or a market index.
- Generally, at any time before the contract maturity, the *forward or futures price differs* from the spot *price* and the difference between the two is called the *basis*. At maturity, the two prices *coincide* under no arbitrage and the basis thus is *zero*.
- Buying a forward or a futures contract at time 0 at a price $F_0$ and holding it up to maturity $T$ leads to a gain or a loss, also called *margin*, equal to $F_T - F_0$. By symmetry, the sale of a contract leads to the margin $F_0 - F_T$. The buyer thus benefits from a price increase ($F_T > F_0$) and the seller from a price decrease ($F_0 > F_T$).
- The margin $F_T - F_0$ of a forward contract can be settled either by *physical delivery* of the underlying, against the payment of $F_0$ or by cash settlement, i.e., a transfer, from the loser to the winner, of a cash amount equal to $+/-(F_T - F_0)$.
- Futures contracts differ from forwards in that the global margin $F_T - F_0$ is paid progressively (usually on a daily basis) between 0 and $T$ and not in a lump sum at expiration $T$.
- The prices of forwards and futures written on the same underlying and with the same maturity do, in general, differ slightly (except at maturity). However, under *deterministic interest rates*, the futures price equals the forward price.
- *Unwinding a position*, on any date before expiration, amounts to adding to the initial position a new position with the opposite sign. Such a transaction is always done at zero cost for a futures. This implies that the market *value* (as opposed to its price) of a *futures* is always nil. This is not true for a *forward*, whose *value* may be positive or negative.
- Cash-and-carry arbitrage consists in buying spot the underlying asset and holding (carrying) it until maturity $T$, this purchase being financed by borrowing. This position is combined with the forward sale of the underlying. Cash-and-carry and reverse cash-and-carry operations lead, under no arbitrage, to the *spot-forward parity*: $F_t - S_t = cost\ of\ carry$ (Interest on the borrowing minus Return on holding the underlying). Different formulas for the spot-forward parity exist which are adapted to each specific underlying.
- A *risk premium* required by the market implies that the forward price differs (slightly) from the future spot price expected at maturity.
- *Hedging* reduces the risk that affects, at a future time, an existing position. It is realized by taking an additional position negatively correlated with the existing risky position.
- Hedging can usually be only *imperfect*, since it leaves a residual risk, except when the hedging instrument is perfectly correlated with the existing asset.
- When hedging with a forward, the underlying should be "as close as possible" to the existing asset and the maturity of the contract close to this horizon (to avoid basis risk).

Appendix 349

- The main *commodities* underlying futures contracts are metals, energy, and food materials and the main organized markets for commodity futures are in Chicago (Chicago Mercantile Exchange or CME), New York, London, and the international trading platform ICE.
- The buyer (the seller) of a *currency* contract is committed to buying (to sell) at a certain horizon, a certain quantity of currency at a price fixed today in some other currency.
- Forward currency contracts, also called FX or Forex contracts, are often quoted as the spread with the spot rate: at a premium (if $F - S > 0$) or at a discount ($F - S < 0$).
- Currency futures are traded on organized markets, the main one being the CME, which handles the major currencies.
- Forward contracts on a *financial instrument* are written on negotiable debt securities, market indexes, or stocks.
- Futures on numerous financial assets are traded in the CME in America and in EUREX (Frankfurt) and ICE Futures Europe (they are studied in Chap. 15).
- The two largest futures rate markets in Europe are EUREX (Frankfurt) and ICE.
- A forward contract on a *fixed-income security can be priced* in two equivalent ways:
- By valuing the underlying spot and then applying the spot-forward parity which yields the forward price.
- By considering the forward itself as a fixed-income security and computing the present value of its cash flows.
- The payoff of a *contract on an index $I$,* started at time zero, is the difference $I_T - F_0$ (times the multiple $m$) where $I_T = F_T$ denotes the index value at maturity. Cash-and-carry types of arbitrage involve transactions in the assets composing the underlying portfolio (reflected in the index) in proportion to their weights.

## Appendix

### The Relationship Between Forward and Futures Prices

The prices of forward and futures contracts written on the same underlying and with the same maturity $T$ do, in general, differ slightly. However:

**Proposition 1**
*Under AAO and with deterministic interest rates, the futures price equals the forward price.*

**Proof**
Let us use the notation:

$S_t$: the spot price on $t$ of an asset $x$ .
$F_t$: the price at $t$ of a *futures* contract with maturity $T$ written on $x$ ($F_T = S_T$) .

$\Phi_t$: the price at $t$ of a *forward* contract on $x$ with the same maturity ($\Phi_T = S_T$).
$r_{T-t}(t)$: the continuous rate prevailing at the instant $t$ for transactions with duration $T-t$ and assumed to be deterministic.
Consider the period $(0, T)$, just $T$ days long, the $j^{th}$ day being bracketed by the instants $(j-1, j)$ for $j = 1, \ldots, T$.

First, let us consider strategy A consisting in holding during day $j$ (from $j-1$ to $j$) $n(j-1)$ futures contracts with maturity $T$, with

$$n(j - 1) = e^{-(T-j)r_{T-j}^{(j)}}.$$

This strategy provides each day $j$ a margin $= n(j-1)[F_j - F_{j-1}]$ which one loans or borrows between the dates $j$ and $T$. The sum of these margins capitalized at $T$ can be written

$$\sum_{j=1}^{T} e^{(T-j)r_{T-j}(j)} n(j-1)[F_j - F_{j-1}] =$$

$$\sum_{j=1}^{T} [F_j - F_{j-1}] = F_T - F_0 = S_T - F_0.$$

**Remark**
- Strategy A only involves a single cash flow (random, taking place at $T$ and equal to $S_T - F_0$) .
- Strategy A is *only possible if rates are deterministic* since in the contrary case $n(j-1)$ cannot be determined at time $j-1$ as a function of the rate $r_{T-j}(j)$ which will only become known at time $j$.

Now, let us consider strategy B consisting of buying on date 0 a forward contract at the price $\Phi_0$ and held to maturity; B only leads to a single (random) cash flow on date $T$, which writes:

$$\Phi_T - \Phi_0 = S_T - \Phi_0.$$

If $\Phi_0 > F_0$ the strategy "A–B" consisting in adopting strategy A and selling a forward contract would generate a single, *certain*, cash flow on date $T$ equal to $S_T - F_0 - [S_T - \Phi_0] = \Phi_0 - F_0 > 0$.
This would therefore constitute an arbitrage. Strategy "B–A" would be an arbitrage if $F_0 > \Phi_0$. Consequently, at any instant denoted by 0, under AAO the equality $F_0 = \Phi_0$ must hold.

# Appendix

351

## Proposition 2

*When interest rates are stochastic, the futures price exceeds the forward price if the interest rates are positively correlated with the underlying asset price, and falls below in the reverse case.*

This result has an intuitive explanation. If rate and price are positively correlated, for a price hike the positive margin received by the holder of the futures contract will be invested at a rate presumably increasing. In the reverse case the negative margin will be borrowed at a falling rate. In this way, for a buyer of futures, the positive effect of an appreciation is amplified and the negative effect of a depreciation is damped: the existence of margin calls benefits the buyer, which explains why the futures price is (slightly) higher than the forward price if the interest rate and the underlying asset price are positively correlated. The opposite is true if rate and price are negatively correlated.

*This intuitive result can be proved with a continuous-time stochastic model. Consider the risk-neutral probability $Q$ (different from the true probability $P$) under which the futures price $F_t$ is a martingale.[30] Since $F_T = S_T$, we obtain

$$F_t = E_t^Q [S_T] \qquad (9.\text{A-}1)$$

where $E_t^Q$ is the conditional expectation under $Q$ conditioned on the information available at $t$.

Applying Itô's Lemma to the forward price, that is to the ratio $\Phi_t(S, B) = \frac{S_t}{B_T(t)}$, where $B_T(t)$ denotes the price of the maturity-$T$ zero-coupon bond at $t$, we have

$$\frac{d\Phi_t}{\Phi_t} = \frac{dS_t}{S_t} - \frac{dB_T(t)}{B_T(t)} - cov\left(\frac{dS_t}{S_t}, \frac{dB_T(t)}{B_T(t)}\right) + var\left(\frac{dB_T(t)}{B_T(t)}\right)$$

$$\frac{dS_t}{S_t} - \frac{dB_T(t)}{B_T(t)} - cov\left(\frac{d\Phi_t}{\Phi_t}, \frac{dB_T(t)}{B_T(t)}\right).$$

And since, under $Q$, the derivatives of $\delta S/S$ and $\delta B/B$ are both equal to the risk-free rate $r(t)$, the derivative of $\frac{d\Phi_t}{\Phi_t}$ equals $- cov\left(\frac{d\Phi_t}{\Phi_t}, \frac{dB_T(t)}{B_T(t)}\right)$ which is denoted by $- \sigma_{\Phi B}(t)\delta t$. Using the notation $\sigma_\Phi$ for the volatility of the forward price and $W$ for a Brownian process under $Q$, we can then write

$$\frac{d\Phi_t}{\Phi_t} = -\sigma_{\Phi B}(t)dt + \sigma_\Phi(t)dW,$$

---

[30] This result is proved in Chap. 19; it is a simple consequence of the fact that taking a position in a futures contract which requires no disbursement of funds on date $t$ leads to a margin $dF$ on $t + dt$: as a result, $E_t^Q dF = 0$, otherwise the expected return would be $+/- \infty$, which is incompatible with the risk-neutrality that is supposed to hold under Q. $F_t$ is therefore a $Q$-martingale.

which implies that $\Phi_t e^{\int_0^t \sigma_{\Phi B}(u)du}$ is a martingale with respect to $Q$; from this follows:

$$\Phi_t e^{\int_0^t \sigma_{\Phi B}(u)du} = E_t Q \left[ \Phi_T e^{\int_0^T \sigma_{\Phi B}(u)du} \right], \text{ or else}$$

$$\Phi_t = E_t^Q \left[ S_T e^{\int_t^T \sigma_{\Phi B}(u)du} \right]. \tag{9.A-2}$$

Assuming that $\sigma_{\Phi B}(t)dt \equiv \text{cov}\left(\frac{d\Phi_t}{\Phi_t}, \frac{dB_T(t)}{B_T(t)}\right)$ is deterministic, (9.A-2) simplifies to:

$$\Phi_t = e^{\int_t^T \sigma_{\Phi B}(u)du} E_t^Q[S_T],$$ which, by (9.A-1), implies the following relation between forward and futures prices:

$$\Phi_t = e^{\int_t^T \sigma_{\Phi B}(u)du} F_t. \tag{9.A-3}$$

(9.A-3) shows that the futures price exceeds the forward price if and only if $\int_t^T \sigma_{\Phi B}(u)du < 0$, i.e., if and only if the covariance between the relative variations of the forward price and the zero-coupon *price*, which will hold on average from now until the end of the contract, is negative, therefore if and only if the covariance between the variations of the forward price and those of the zero-coupon *rate* is positive.

---

## Suggestions for Further Reading

### Books

Hull, J. (2018). *Options, futures and other derivatives* (10th ed.). Prentice Hall Pearson Education.
Kolb, R. (2007). *Futures, options and swaps* (5th ed.). Blackwell.
*Lioui, A., & Poncet, P. (2005). *Dynamic asset allocation with forwards and futures*. Springer.

### Articles

Black, F. (1976). The pricing of commodity contracts. *Journal of Financial Economics, 3*, 167–179.
Cox, J., Ingersoll, J., & Ross, S. (1981). The relation between forward and futures prices. *Journal of Financial Economics, 9*, 321–346.
Jarrow, R. A., & Oldfield, G. S. (1981). Forward contracts and futures contracts. *Journal of Financial Economics, 9*, 373–382.
*Richard, S., & Sundaresan, M. (1981). A continuous time model of forward and futures prices in a multi-good economy. *Journal of Financial Economics, 9*, 347–372.

# Options (I): General Description, Parity Relations, Basic Concepts, and Valuation Using the Binomial Model

# 10

Option theory, which was developed at the start of the 1970s by Black, Scholes, and Merton, constituted a major advance in economic and financial theory. The applications of this theory extend well beyond its use for options. Not only do numerous financial products have option components (convertible bonds, caps and floors, hybrid products, ..., and even the bonds and shares issued by limited companies where there is a risk of bankruptcy[1]) but many decisions have an aspect that can only be understood in terms of options (investments,[2] analysis of credit risk, etc.). Option theory provides tools that not only allow to price optional components but also to manage portfolios of assets and liabilities that may include them. By greatly improving our understanding of financial mechanisms and risk management, option theory has significantly contributed to the increase in activity on financial markets.

This theory is made up of different models based on more or less restrictive hypotheses. These can be split into two categories, according to whether they model time as discrete or continuous.

The current chapter, made up of four sections and three appendices, is devoted to the theory of options in discrete time. The first section gives a general description and presents the basic definitions with an emphasis on the call-put parity, from which follow a number of important properties. The second section introduces the simplest valuation model, with one period and two states, which can easily be extended to the discrete multi-period model which is due to Cox et al. (1979) and is explained in the third section. The fourth section explains how to choose the model parameters and how the solution converges toward that of the continuous time model due to Black

---

[1]The possibility of filing for bankruptcy gives equity an optional aspect; this point of view is productive for analyzing corporate finance and credit risk, as we will see in Chaps. 28, 29, and 30.

[2]Most investments allow future choices depending on the situation that will prevail; thus they include optional components which contribute to their net present value and which are termed "real options."

---

© The Author(s), under exclusive license to Springer Nature Switzerland AG 2022

P. Poncet, R. Portait, *Capital Market Finance*, Springer Texts in Business and Economics, https://doi.org/10.1007/978-3-030-84600-8_10

353

and Scholes (1973). The latter model, and some of its extensions, will be handled in Chap. 11.

## 10.1 Basic Concepts, Call-Put Parity, and Other Restrictions from No Arbitrage

After some general definitions and an initial description of the value of an option as a function of the price of its underlying asset (Sect. 10.1), the standard call-put parity relation is given in Sect. 10.2, other parity relations are offered in Sect. 10.3, while different restrictions and relations which hold when there is no arbitrage opportunity are established in Sect. 10.4.

### 10.1.1 Definitions, Value at Maturity, Intrinsic Value, and Time Value

In the widest sense of the term, an option is an asset for which the buyer pays the seller a sum of money (often called the option's *premium*) on the initial date and receives, as a counterpart, on a future date, a positive or zero cash flow (called the option's *payoff*) whose amount depends on the evolution of the underlying asset price. Resulting from this very broad definition there is a vast category of assets called *contingent claims*; in this chapter we introduce just the standard (or vanilla) options which we now define more precisely.

There are two types of options: options to buy (calls) and options to sell (puts). A call option confers the right (but does not impose the obligation) to buy the underlying asset (*underlying* for short) at a price $K$ called the *strike* (or exercise price); $K$ is fixed by the contract when the option is issued (date 0).

A put option confers the right (but does not impose the obligation) to sell the underlying asset at the exercise price $K$.

One says an option is *European* when *it can only be exercised at the end of the contract*. This date is called the *maturity, or expiration date, or expiry*; it will be denoted by $T$. The option is said to be *American* if its holder has the right to exercise it at *any* time during the interval $(0, T)$. The option can only be exercised once during this period. Most options listed on organized markets are American options. In this chapter, we will concentrate on European options as they are easier to value than American options; the latter case will just be touched on in Sect. 13.1.1 and then treated in more depth in Chap. 13.[3]

---

[3] We also will not be treating here the options called exotic, which are studied in Chap. 14, some of which have a payoff that depends on the path of the underlying's price between 0 and $T$ and not just on the value on date $T$ (*path dependent* options such as American, Bermudian, on an average, barrier and lookback options).

## 10.1 Basic Concepts, Call-Put Parity, and Other Restrictions from No Arbitrage

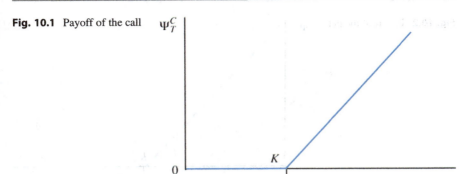

**Fig. 10.1** Payoff of the call

Denote by $S_t$ the underlying's market price on date $t \geq 0$; to make things more concrete we will sometimes identify the underlying with a stock, but that is not necessarily the case.

Someone who holds a European call thus has the right to buy on a date $T$, for a price $K$, the underlying asset which is worth $S_T$. So he will exercise that right if and only if $S_T > K$, in which case his position will be worth $S_T - K$; if $S_T \leq K$ the option will be foregone and the position will be worth nothing. The owner of a European put will exercise her option to sell if and only if $S_T < K$. On date $T$, the value of the option is therefore

$$\Psi_T^c = \max(S_T - K; 0) = C(S_T) \text{ for a call;}$$

$$\Psi_T^p = \max(K - S_T; 0) = P(S_T) \text{ for a put.}$$

Exercising the option on $T$ can also take the form of a payment by the seller (also called the issuer) to the benefit of the option holder, with a cash flow equal to the final price of the option $Y_T$; if the option is not exercised then there is no delivery of the underlying. In every case, the value of the final cash flow $Y_T$ is called the option's *payoff*.

Since the option is a zero-sum game between the buyer and the seller (or issuer) of the option, the (short) position of the latter will have a negative or null value:

$$-\Psi_T^c = -C(S_T) \text{ for a call;}$$

$$-\Psi_T^p = -P(S_T) \text{ for a put.}$$

It is usual to graph the final value or payoff of an option as a function of the final value of the underlying asset. The two figures below illustrate the payoffs of a call and a put, respectively $\Psi_T^c$ and $\Psi_T^p$, as a function of $S_T$. Note these figures represent the option value only at maturity date $T$.

The main issue in this chapter is to price the option premium on date 0 or, more generally, at some intermediate date $t < T$. So what is needed is graphs not for the maturity date $T$ as in Figs. 10.1 and 10.2, but on some earlier date $t < T$. Figs. 10.3

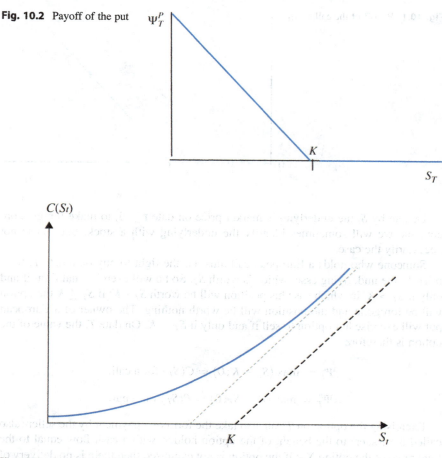

**Fig. 10.2** Payoff of the put

**Fig. 10.3** Approximate form of a call's premium before expiry as a function of the underlying's value (see Fig. 10.5 for a more precise graph)

and 10.4 give an *approximate* general idea of these new graphs. A more exact and precise representation, which shows the exact asymptote of the curve of $C(S)$ as $S$ tends to infinity as well as the behavior of $P(S)$ as $S$ tends to 0, will be given later and illustrated in Figs. 10.5 and 10.6.

The solid blue curves represent the market price of the option on an intermediate date $t$.[4] The dashed lines (starting from K) reproduce the preceding Figs. 10.1 and 10.2 and represent the option's *"intrinsic value."* The latter, denoted by *IV*, is defined as the value the option would have if it had reached maturity, i.e., on date $t$:

---

[4] As one will see in Fig. 10.6, the graph in Fig. 10.4 is relevant to the American put only.

## 10.1 Basic Concepts, Call-Put Parity, and Other Restrictions from No Arbitrage

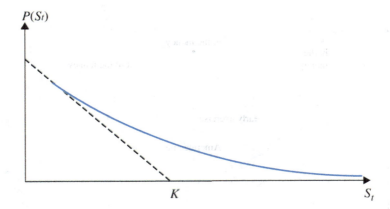

**Fig. 10.4** Approximate form of a put's premium before expiry as a function of the underlying's value (see Fig. 10.6 for a more precise graph)

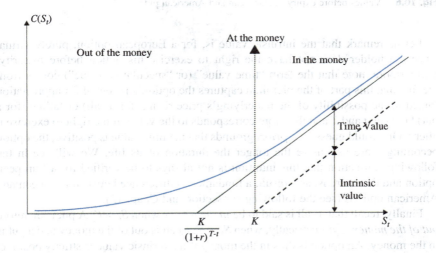

**Fig. 10.5** Premium of the call before expiry as a function of the underlying's price

$$IV_c(t) = \max(S_t - K; 0) \text{ for a call};$$
$$IV_p(t) = \max(K - S_t; 0) \text{ for a put}.$$

It is useful to split an option's premium in two parts: its intrinsic value (IV) and its *time value* (TV). Thus, by definition

$$O_t = IV(t) + TV(t),$$

where $O_t$ is the premium (value) of the option on date $t$.

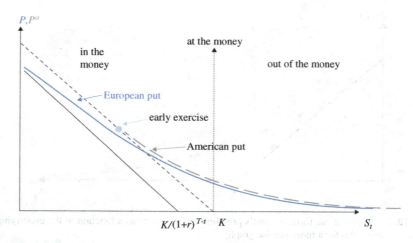

**Fig. 10.6** Values before expiry of European and American puts

Let us remark that the intrinsic value is, for a European option, purely virtual since the holder does not have the right to exercise his option before maturity. Furthermore, note that the term "time value" (or "speculative value") comes from the fact that this part of the premium captures the option's potential for appreciation linked to the possibility of the underlying's price rising (for a call) or falling (for a put) between $t$ and $T$ and therefore corresponds to the value of the right to exercise it after $t$. One could guess on *a priori* grounds that the time value is positive, the option becoming more expensive the longer the duration of its life. We will see in the following subsection that this intuition is not always to be verified for a European option and that there is, as a result, a fundamental difference between European and American options (see the following subsection and Chap. 13).

Finally, recall that a call is said to be *in the money* when $S_t > K$. A put is therefore *out of the money*. Symmetrically, when $S_t < K$, a call is out of the money and a put is in the money. An option is thus in the money if its intrinsic value is strictly positive. In addition, calls and puts are said to be *at the money* when $S_t = K$.

## 10.1.2 The Standard Call-Put Parity

Consider a European call and put, written on the same underlying asset whose value is $S$, with the same expiration date $T$ and the same strike price $K$. We call the "standard case" that of options written for spot *securities that do not distribute either dividends or coupons from now until expiry T*. Also consider the securities lending market, whose underlying is the collection of all zero-coupon securities. We use $r$ (short for $r_{T-t}(t)$) for the discrete yield to maturity of the zero-coupon security with maturity $T$ (so duration $T-t$) which holds on date $t$.

## 10.1 Basic Concepts, Call-Put Parity, and Other Restrictions from No Arbitrage 359

Now consider two portfolios A and B: A is made up of the underlying and the put while B is composed of a zero-coupon paying $K$ on date $T$ and the call. We will show that the values of these two portfolios are identical. On date $T$, the value $V_A(T)$ of portfolio A equals the sum of $S_T$ (the stock) and max $(K - S_T, 0)$ (the put). We obtain

$$V_A(T) = S_T + \max (K - S_T, 0) = \max (S_T, K).$$

Portfolio B has the value of the sum of $K$ (for the zero-coupon) and max $(S_T - K, 0)$ (the call), that is max$(S_T, K)$.

The two portfolios, therefore, have the same value on the maturity date $T$. Since the two portfolios do not distribute any cash in the meanwhile (the stock is supposed not to pay out dividends between the dates 0 and $T$), the assumed absence of arbitrage opportunity (AAO) implies that the two portfolios' values are the same for all $t \leq T$. Indeed, if that were not the case, it would be enough to buy the cheapest portfolio and to sell the costliest to reap an arbitrage profit. Therefore, we obtain the following equality on date $t$:

$$P_t + S_t = \frac{K}{(1 + r)^{T-t}} + C_t,$$

where $P_t$ is the value of the put, $C_t$ that of the call, $S_t$ the stock's, and $r$ represents here the zero-coupon rate on date $t$ for transactions of duration $T–t$; $\frac{K}{(1+r)^{T-t}}$ is thus the value at $t$ of the zero-coupon paying $K$ on date $T$. This equation leads to the following proposition.

**Proposition 1 Call-put parity for a spot underlying asset without dividends**
*Under the no-arbitrage rule, the premiums for the call and the put with the same characteristics, the underlying's spot price, and the discounted strike are related by the equation.*

$$C_t - P_t = S_t - \frac{K}{(1 + r)^{T-t}} \tag{10.1a}$$

*or else, using a continuous interest rate,*

$$C_t - P_t = S_t - K\, e^{-r(T-t)}. \tag{10.1b}$$

Note that the parity equation holds when there is no arbitrage (AAO) because the four assets (the underlying, the call, and the put with same strike price and maturity $T$, and the zero-coupon with maturity $T$) are redundant. A portfolio made up of any three of them replicates the fourth: for example, the call is equivalent to a long position in the corresponding put and in the underlying plus a short position (borrowing) on a zero-coupon paying $K$ on date $T$ (that is why $C = P + S - \frac{K}{(1+r)^{T-t}}$). The zero-coupon is replicated by a long position in the put and in the

360          10   Options (I): General Description, Parity Relations, Basic Concepts,...

underlying and a short position in the call (therefore $\frac{K}{(1+r)^{T-t}} = P + S - C$. And so on.

We can deduce as a corollary of Proposition 1 the lower bounds that must hold under AAO for the premiums of European calls and puts.

**Corollary**

*The value on date $t$ of a European call with maturity $T$ and strike $K$ satisfies the following inequalities:*

$$C_t \geq 0,$$

$$C_t \geq S_t - \frac{K}{(1+r)^{T-t}}. \tag{10.2}$$

*For a European put we have:*

$$P_t \geq 0,$$

$$P_t \geq \frac{K}{(1+r)^{T-t}} - S_t. \tag{10.3}$$

**Proof**

Under AAO, $C_t$, and $P_t$ are positive since the payoffs on the two options are certain to be positive or zero. The other inequalities are derived by applying the parity equation. Indeed, since.

$$C_t = S_t - \frac{K}{(1+r)^{T-t}} + P_t.$$

and $P_t$ is positive, inequality (10.2) follows. Since the parity equation can also be written.

$$\frac{K}{(1+r)^{T-t}} - S_t + C_t = P_t,$$

the requirement $C_t \geq 0$ implies inequality (10.3).

Now we can formulate several important remarks about the most common case when the underlying asset has a value that can never be negative (for instance, a share of limited company or a bond). If at some given date $t$ the value $S_t$ is zero, AAO implies that future prices of the underlying asset are certainly zero. Indeed, let us assume $S_t = 0$, and that there is a future date $s \geq t$ on which Proba $[S_s > 0] > 0$. Knowing that $S$ can never take on a negative price, buying, on date $t$, the underlying (with price zero) followed by the sale on date $s$ (for a price that is certainly non-negative but possibly positive) is an arbitrage opportunity. Assuming AAO, therefore, $S_s = 0$ with certainty for any date $s \geq t$. We say that the value 0 is an

## 10.1 Basic Concepts, Call-Put Parity, and Other Restrictions from No Arbitrage

absorbing value for the underlying. In particular, if $S_t = 0$, then necessarily $S_T = 0$ and the call's value is nil because its payoff is then certain and equal to 0:

$$S_t = 0 \text{ implies } C_t = 0.$$

Furthermore, if $S_t = 0$ the payoff of the put on date $T$ is with certainty equal to $K$, whence

$$S_t = 0 \text{ implies } P_t = \frac{K}{(1+r)^{T-t}}.$$

We will also admit that $P(S_t)$ tends to 0 as $S_t$ tends to infinity[5]: $\lim_{S_t \to \infty} P(S_t) = 0$.

We can now reconsider Figs. 10.3 and 10.4 with greater precision and draw Figs. 10.5 and 10.6.

We note that, in Fig. 10.5, the curve for $C(S)$ tends asymptotically, as $S$ tends to infinity, to the half-line with slope $45°$ starting from the point $(\frac{K}{(1+r)^{T-t}}, 0)$ (since $C(S)$ $- (S - \frac{K}{(1+r)^{T-t}}) = P(S)$ tends to 0 as $S$ tends to infinity). We also note that the time value of a European call is always strictly positive. Indeed, for any positive $r$, inequality (10.2) implies:

$$C_t \geq S_t - \frac{K}{(1+r)^{T-t}} \geq S_t - K.$$

In contrast, this property does not always hold for European puts: the curve for the put's premium starts at the point $\frac{K}{(1+r)^{T-t}}$, which is strictly below $K$ on the coordinate axis, and crosses the intrinsic value for a strictly positive value of $S_t$. This implies that the time value of a European put is negative if the latter is deep in the money.

This remark will allow highlighting important properties of American options, some of which are discussed in Sect. 10.4 below.

### 10.1.3 Other Parity Relations

Outside the standard case, either because the underlying asset pays out some dividend during the remaining life of the option or because it is not a spot asset but a forward or a futures contract, the call-put parity needs to be amended. We first examine the case of European options on spot securities that pay out dividends or coupons before handling the case of European options on forward contracts.

---

[5]Intuitively, the conditional probability of exercising the option $\mathrm{Proba}(S_T < K \mid S_t) \to 0$ as $S_t \to \infty$, so the expectation of the payoff $\mathrm{E}(Y^P{}_T) \to 0$ (since $\mathrm{E}(Y^P{}_T) < K\,\mathrm{Proba}(S_T < K \mid S_t)$).

### 10.1.3.1 Call-Put Parity for European Options Written on an Underlying Spot Asset Paying Discrete Dividends

The prospect of a dividend or coupon distribution before an option's maturity (which will be reflected by a fall in the price of the underlying) reduces the call's price, adds to the put's price and affects call-put parity. The following proposition gives this relation in a context where the underlying asset does distribute dividends or coupons.

### Proposition 2

*Consider a European call and a European put, with the same strike $K$ and exercise date $T$, written on the same underlying spot asset which distributes dividends on date $\tau$, between $t$ and $T$, whose present value at current date $t$ is denoted $D*$. Under AAO the following put-call parity holds at time $t$ ($\leq T$):*

$$C_t - P_t = (S_t - D*) - \frac{K}{(1+r)^{T-t}}. \tag{10.4}$$

### Proof

Assume that at some arbitrary instant $t$, Eq. (10.4) is not satisfied; for example, that $C_t - P_t - (S_t - D*) + \frac{K}{(1+r)^{T-t}} > 0$; first consider a single dividend $D$ occurring on date t. The arbitrage (which is called a *conversion* by professionals) consists of two parts:

–Selling the call at the initial instant $t$, buying the put and the underlying, and borrowing \$ $\frac{K}{(1+r)^{T-t}}$ for the period $T$–$t$ and $D*$ for the period t- $t$; so the positive amount $C_t - P_t - S_t + D* + \frac{K}{(1+r)^{T-t}}$ *is cashed in at initial date t.*

–Repaying the loan which is due on t with the dividend which is paid out on the same date (the net cash flow is thus zero).

Upon expiry at $T$, the full net value of the position is zero, in the two cases that can occur, as shown in the table below ($S_T < K$; $S_T \geq K$):

| | $S_T < K$ | $S_T \geq K$ |
|---|---|---|
| Long Put | $K - S_T$ | 0 |
| Underlying | $S_T$ | $S_T$ |
| Short Call | 0 | $-(S_T - K)$ |
| Loan due | $-K$ | $-K$ |
| Net value of the position | 0 | 0 |

This transaction, which produces a positive cash flow upon initiation and does not entail any negative cash flow later on, is an arbitrage.

If market prices are such that $C_t - P_t - (S_t - D*) + \frac{K}{(1+r)^{T-t}} < 0$, the arbitrageur takes the opposite position (called an *inversion* or *inverse conversion*) with again an arbitrage profit.

This reasoning can easily be extended to several dividends. Indeed, if several coupons $D_1, D_2, \ldots, D_N$ are distributed by the underlying between $t$ and $T$ on the dates $t_1, t_2, \ldots, t_N$, the strategy is, on date $t$, to purchase the underlying $S$ and to contract a loan in the amount of $D*$; the latter is repaid on $t_1, t_2, \ldots, t_N$ according to a

## 10.1 Basic Concepts, Call-Put Parity, and Other Restrictions from No Arbitrage 363

payment schedule $D_1, D_2, \ldots, D_N$, synchronized with the dividends: the investment of $S_t - D^*$ yields $S_T$ on $T$, with no net cash flow between $t$ and $T$ to alter the self-financing nature of the strategy, and so the previous argument applies.

### 10.1.3.2 Call-Put Parity for European Options on Forward Contracts

This subsection is about European options on a forward contract. Recall that for the buyer this means the right and the duty to purchase upon expiry date $T'$ a fixed good (the underlying) at a price fixed at the start, and for the seller the right and duty, symmetrical to those of the buyer, to sell the underlying asset at the agreed price. We will consider a forward contract with expiry date $T'$, whose price quoted on date $t$ will be denoted by $F_t$[6] and European options expiring on $T \leq T'$. Without entering into details about how to exercise the options, a European option on a forward contract can simply be defined by its payoff, paid on $T$ by the seller to the option holder: $\max(F_T - K, 0)$ for a call and $\max(K - F_T, 0)$ for a put.

### Proposition 3

*Consider a European call and put, with the same strikes and exercise dates $K$ and $T$, written on the same forward contract with expiry $T' \geq T$, whose quoted price is $F_t$ on date t.*

*In the absence of arbitrage, the following parity equation holds at t:*

$$C_t - P_t = \frac{F_t - K}{(1+r)^{T-t}}. \tag{10.5}$$

### Proof

Let us assume that at some time $t$, Eq. (10.5) does not hold; for example, that $C_t - P_t - \frac{F_t-K}{(1+r)^{T-t}} > 0$.

The arbitrage consists of selling the call, buying the put and a forward contract (worth 0) and lending $\frac{F_t-K}{(1+r)^{T-t}}$ for a period $T-t$ (or borrowing if $F_t - K < 0$); this yields the positive cash amount $C_t - P_t - \frac{F_t-K}{(1+r)^{T-t}}$.

On date $T$, the overall net value of the position is zero in both cases that can occur $(F_T < K; F_T \geq K)$, as the table below shows:

|  | $F_T < K$ | $F_T \geq K$ |
|---|---|---|
| Put bought | $K - F_T$ | 0 |
| Contract bought | $F_T - F_t$ | $F_T - F_t$ |
| Call sold | 0 | $-(F_T - K)$ |
| Loan repaid | $F_t - K$ | $F_t - K$ |
| Net value of the position | 0 | 0 |

---

[6]To simplify the notation, since to be completely explicit one would have to write: $F_{T'}(t)$.

364            10 Options (I): General Description, Parity Relations, Basic Concepts,...

This transaction is an arbitrage, since it returns a positive cash flow at $t$ and this is not balanced by any subsequent negative cash flow.

If $C_t - P_t - \frac{F_t - K}{(1+r)^{T-t}} < 0$, the arbitrage consists in taking the opposite position.

**Remarks**
– This call-put equation holds for options on forward contracts. Rigorously speaking, it does not hold for options on futures contracts unless the rates are deterministic; indeed, in that case, forward and futures prices coincide, while they differ when rates can vary randomly (see Chap. 9).
– The reason for the price $F_t$, like $K$, to be discounted, while $S_t$ is not in Eq. (10.1), is that the contract's buyer does not pay out $F_t$ today: the present value of $F_t$ is $\frac{F_t}{(1+r)^{T-t}}$.
– Contrary to the case of European options on spot underlying assets not delivering dividends, the European call and put written on a forward are worth the same if they are at the money ($F_t = K \Rightarrow C_t = P_t$).

## 10.1.4 Other Arbitrage Restrictions

The comments above, in particular, the call-put parity relations (each holding under appropriate conditions) are based on simple AAO considerations and use no complicated mathematics. Other relations, also simply justified, hold as well under AAO. We will continue to consider options written on spot assets but, in contrast to the preceding, we will not limit ourselves to European options.

$C$ and $P$ will always denote the values of a European call and put while $C^a$ and $P^a$ denote their American counterparts.[7] We will view $C$, $C^a$, $P$ and $P^a$ as functions of $S$, $K$, and $(T - t)$ while sometimes making one of the arguments explicit. Furthermore, when both European and American premiums appear in the same equation they will obviously refer to options written on the same underlying, with the same exercise date and strike.

### 10.1.4.1 Some Simple Relations Satisfied by European and American Options

Most of the following relations are grounded on call-put parity, AAO, and/or simple logic and the reader is invited, as an exercise, to demonstrate them without referring to the corresponding justifications.

*In the standard case (an underlying spot asset), under no arbitrage, options values satisfy the following inequalities:*

(i) $S \geq C^a \geq Max\ (S - K,\ 0)$

---

[7]We will denote by $C$, $C^a$, $P$, $P^a$, and $S$ the values of options and their underlying on an arbitrary date $t$, except in the proof in Sect. 10.1.4.2 where we will be more specific and use $C_t$, $C^a_t$, $P_t$, $P^a_t$, and $S_t$.

## 10.1 Basic Concepts, Call-Put Parity, and Other Restrictions from No Arbitrage 365

$K \geq P^a \geq Max\ (K - S, 0)$

(ii) $C^a(T_2-t) \geq C^a\ (T_1-t)\ if\ T_2 > T_1$

$P^a(T_2-t) \geq P^a(T_1-t)\ if\ T_2 > T_1$

(iii) $C^a = S\ if\ K = 0$

$P^a = 0\ if\ K = 0$

(iv) $C^a \geq C \geq S - \frac{K}{(1+r)^{T-t}}$ *(the second inequality only holds if there is no dividend distribution by the underlying between t and T).*

**Justifications**

*(i)* – If $C^a < S - K$, one purchases the call and exercises it immediately, which yields an instant arbitrage gain.

– If $C^a > S$, one buys the underlying and sells the call, which leaves a positive net cash flow; if the option is exercised, one gets $K > 0$ in exchange for the asset; if not, one holds in the portfolio an asset with value $S \geq 0$.

– The same argument holds for $P^a > Max\ (K - S, 0)$.

– If $P^a > K$, one sells the put and places the cash amount ($> K$) at (assumed positive) interest on the money market. The transaction is a win even if the put is exercised by the counterparty (one pays $K$ with part of the placement and still gets $S$).

*(ii)* An American option with a far-off expiration date $T_2$ confers the same rights as its equivalent with a closer expiration $T_1$ *plus* exercise rights between $T_1$ and $T_2$: the value of an American premium thus cannot decrease with the duration of the option.

*(iii)* If $K = 0$ and $S > C^a$, one sells the underlying to buy the call to exercise it immediately which cancels the position in the underlying. If $K = 0$ and $S < C^a$, one sells the call to purchase the underlying; a possible exercise of the call by the counterparty causes delivery of the underlying.

*(iv)* $C^a \geq C$ since an American call provides more rights than its European counterpart.

$C \geq S - \frac{K}{(1+r)^{T-t}}$ by call-put parity (with no dividend distribution) and since $P \geq 0$.

### 10.1.4.2 Irrelevance of the Early Exercise Right for an American Call Written on a Spot, Non-dividend Paying Asset

The following proposition is very important since valuation models sometimes assume the options to be European, while they are American. For an American call written on a spot asset with no distributions, these models are strictly exact (this is the *only* case, but it is frequent). In addition, in many analyses regarding options in various contexts, the irrelevance of the possibility of early exercising the call on a spot, non-paying dividend asset is invoked.

**Proposition 4**

*When an American call expiring on T is written on a spot asset that does not distribute any dividend or coupon before T, it should never be early exercised; therefore $C^a = C$.*

366    10   Options (I): General Description, Parity Relations, Basic Concepts,...

*The previous result does not hold for American puts since, in some situations, the (deep-in-the-money) American put should be early exercised, hence $P^a > P$.*

## Proof

For any $r > 0$ and $t < T$, by (iv) and call-put parity, *the following inequalities hold* at any date $t < T$:

$$C^a_t \geq C_t = P_t + S_t - \frac{K}{(1+r)^{T-t}} > P_t + S_t - K > S_t - K, \text{so that} \quad C^a_t > S_t - K.$$

The principle is simple: since, at $t < T$, a call value is always strictly larger than its payoff (as we noted for Fig. 10.5), it should never be exercised before its maturity $T$. Consequently, the early exercise right embedded in an American call is worthless here and $C^a = C$ on any date $t$.

We also remarked (in the commentary on Fig. 10.6) that a European put which is deep-in-the-money has a negative time value ($C$ is then very near 0 and $P \approx \frac{K}{(1+r)^{T-t}}$ $S < K - S$). By contrast, an in-the-money American put is always worth $P^a \geq K - S$ since it may be exercised at any moment. As a result, we have $P^a > P$. The difference, which is the expected value of exercising the American put early, increases the more in-the-money the put is.


## Comments and Clarifications

– First, it is worth noting that the fact that a call should not be exercised early *does not imply* that it should be held by the investor until expiration. It means that an investor willing to get rid of the call for any reason (need of cash, expectation of a decrease in the underlying price, etc.) should *sell* the call at the market price which, in AAO, is *strictly* larger than the intrinsic value. Indeed, if the call market price were smaller ($C_t \leq S_t - K$), the call-put-parity (which is an AAO condition) would be violated and arbitrageurs would buy the call, finance their purchase by short selling the underlying asset and simultaneously lend with no risk the sum $K/(1 + r)^{T-t}$, thus cashing in a *strictly positive* flow *larger than or equal to* $K(1-1/(1 + r)^{T-t})$ while being sure the future cash flow (on date $T$) is zero.

– The results of proposition 4 should also be related to the time value of the options. In the case of a call struck on a non-dividend paying asset $C_t = P_t + S_t - K/(1 + r)^{T-t}$, hence its time (or speculative) value $TV^c_t = C_t - (S_t - K)$ is equal to $P_t + K(1-1/(1 + r)^{T-t})$.

The time value of this call thus is made of *two components*: (i) $P_t$, that may be interpreted as an insurance component (the put insures that, in case of an adverse evolution of the underlying asset price, the call will not take on negative values), and (ii) $K(1-1/(1 + r)^{T-t})$, which is the benefit of *paying* the strike $K$ at maturity $T$ instead of now (at $t$), (sometimes referred to as the time value of $K$). For a call, these two components are positive, so that $TV^c_t > 0 \Leftrightarrow C_t > (S_t - K)$.

– In the case of the put, its time value writes $TV^P_t = C_t - K(1-1/(1 + r)^{T-t})$. The insurance component $C_t$ is positive but the time value of the strike, $- K(1-$

## 10.2 A Pricing Model for One Period and Two States of the World

**Table 10.1** Convexity of option values relative to the strike price

| | $S_T < K_1$ | $K_1 \leq S_T < K_2$ | $K_2 \leq S_T < K_3$ | $K_3 \leq S_T$ |
|---|---|---|---|---|
| Purchase of $\lambda C_1$ | 0 | $\lambda(S_T - K_1)$ | $\lambda(S_T - K_1)$ | $\lambda(S_T - K_1)$ |
| Sale of $C_2$ | 0 | 0 | $-(S_T - K_2)$ | $-(S_T - K_2)$ |
| Purchase of $(1-\lambda)C_3$ | 0 | 0 | 0 | $(1-\lambda)(S_T - K_3)$ |
| Total value | 0 | $\geq 0$ | $-(1-\lambda)S_T - \lambda K_1 + K_2$ $= -(1-\lambda)S_T + (1-\lambda)K_3$ $= (1-\lambda)(K_3 - S_T) > 0$ | 0 |

$1/(1 + r)^{T-t}$), is negative since the holder of the put *receives* the strike (thus the sooner the better). When $S_t$ is small, $C_t$ is very small (it tends to $0^+$) and may be outweighed by the negative time value of $K$, making the time value of a European put negative.

– The conditions of early exercise of American options as well as the value of the early exercise right are more thoroughly studied in Chap. 13 devoted to American options.

### 10.1.4.3 Convexity of Option Prices with Respect to the Strike

Let us prove a final property:

$C(K)$, $C^a(K)$, $P(K)$, and $P^a(K)$ *are convex functions of* $K$.

**Proof**

Let us show $C(K)$ is convex, i.e.

$\forall K_1$ and $K_3$ ($K_3 > K_1$) and $\forall \lambda \in ]0,1[$: $C(\lambda K_1 + (1 - \lambda) K_3) < \lambda C(K_1) + (1 - \lambda) C(K_3)$.

Define $K_2 = \lambda K_1 + (1 - \lambda) K_3$. Then $K_1 < K_2 < K_3$.

Assume that, on the market, we observe the opposite inequality (> instead of <). Selling the call $C_2$ (overpriced) with strike $K_2$ and worth $C(K_2)$ and buying $\lambda$ calls $C_1$ with value $C(K_1)$ and $(1 - \lambda)$ calls $C_3$ with value $C(K_3)$ will yield a *strictly positive* cash flow. Upon maturity $T$ the portfolio value, as a function of the underlying's price $S_T$, will be such as described in Table 10.1.

Since the initial cash flow is positive and the final value is non-negative, the assumed inequality implies an arbitrage, ruled out under AAO. The proof is similar for $C^a(K)$, $P(K)$, and $P^a(K)$.

---

## 10.2 A Pricing Model for One Period and Two States of the World

This section presents a first model of pricing, a particular case of the binomial model due to Cox et al. (1979) treated in Sect. 10.1.3. It is a simple toy model, with only one period (between dates 0 and 1) and a two-state world for the underlying asset with price $S$. Although simple, this model allows us to introduce *all* the fundamental concepts used in option pricing.

**Fig. 10.7** One-period model

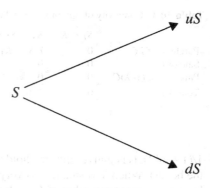

### 10.2.1 Two Markets, Two States

Let us consider two basic markets; that for a risky asset, which we will call a stock (and which will be the underlying for the option introduced later) and that for riskless lending and borrowing. The value on date 0 of the share of stock is known and equal to $S$. On date 1, at the end of the period of uncertainty, the share can take one of two prices denoted $uS$ and $dS$ such that $u > d$ ($u$ for "up", $d$ for "down").[8] The multiplicative parameters $u$ and $d$ can be thought of as the raw returns (1 + the rate) on the share in the two economic situations (or states of the world, "up" and "down"). Thus we have the following situation for the share price (Fig. 10.7):

Now let us consider lending and borrowing transactions, in the form of buying or selling a riskless asset, which returns, for $1 invested on date 0, $ $(1 + r)$ for sure on date 1.

First note that the no-arbitrage rule (AAO) imposes the following condition:

$$d < 1 + r < u.$$

Indeed, first assume $d > 1 + r$: the portfolio made up of a single share, whose purchase on date 0 is financed by a loan of amount $S$, requires zero investment on date 0 although its value on date 1 is positive for any state of the world:

$$dS - (1+r)S > 0 \text{ for state } d,$$

$$uS - (1+r)S > 0 \text{ for state } u.$$

Such a portfolio is therefore an arbitrage.

If $u < 1 + r$, it is enough to sell the share and invest the receipts from the sale at the rate $r$ to achieve a profit that is always positive on date 1, which also constitutes an arbitrage.

---

[8] In fact, case $d$ does not necessarily correspond to a price drop since the only condition required of $d$ is $d < (1 + r)$ (see later on). However, since we do observe ups and downs in stock prices, we assume $d < 1$.

## 10.2 A Pricing Model for One Period and Two States of the World 369

Now let us consider an option on the share with price $S$ described above, for example, a call with strike $K$ and maturity 1. For the scenario in which the share price rises (the event $S_1 = uS$), the option payoff equals $C_1^u = \max(uS - K, 0)$.

For a falling price the payoff equals $C_1^d = \max(dS - K, 0)$.

We will show in the next subsection that such a call's payoff can be obtained with a portfolio involving only the two base securities (the share and the risk-free asset).

### 10.2.2 Hedging Strategy and Option Value in the Absence of Arbitrage

We define a *hedging strategy* (or *replication strategy*) as taking a (uniquely defined) position on $t = 0$ that allows the seller to replicate the option's payoff on date 1, and thus honor the contract. We now show how to make up such a hedge portfolio.

On date 0, the seller of a call can invest in the two base assets, the share and the risk-free asset. Denote by a the quantity of shares bought (a > 0) or sold (a < 0) on date 0, and b the amount placed in the risk-free asset (b positive means lending, b negative borrowing) on date 0. We assume it is possible to buy and sell with no limits both the share and the risk-free asset, which in addition are perfectly divisible. The collection of *admissible* portfolios (ones that could in practice be realized) is thus represented by the pair of real numbers (a, b). The initial value on date 0 of the portfolio (a, b) is given by

$$V_0 = \alpha S + \beta.$$

On date 1, this portfolio's value depends on how the share and the risk-free asset prices have evolved. If the share went up then

$$V_1^u = \alpha uS + \beta(1 + r),$$

and in the opposite case

$$V_1^d = \alpha dS + \beta(1 + r).$$

We can now choose (a, b) so that the portfolio's value on date 1 is equal to the value of the option in both states of the world. It is sufficient to solve the system of two equations in the two unknowns a and b

$$
\begin{aligned}
\alpha uS + \beta(1 + r) &= C_1^u \quad (= Max(uS - K, 0)) \\
\alpha dS + \beta(1 + r) &= C_1^d \quad (= Max(dS - K, 0))
\end{aligned}
\tag{10.6}
$$

whose solution $(\alpha^*, \beta^*)$ is the hedge portfolio. Indeed, since such a portfolio has on date 1 for either state $u$ or $d$ the same value as the option, it is enough for the seller to buy it on date 0 to be certain to honor (cover) the contract; indeed, he will sell the

370      10 Options (I): General Description, Parity Relations, Basic Concepts,...

hedge portfolio on date 1 and settle the option payoff with what he receives from the sale. The solution to (10.6) gives

$$\alpha^* = \frac{C_1^u - C_1^d}{uS - dS}$$
$$\beta^* = \frac{C_1^u - \alpha^* uS}{1 + r} \tag{10.7}$$

In practice, what is important is the value of a*, called the *hedge ratio*, which we note can formally be written as $\alpha^* = \frac{\Delta C}{\Delta S}$, where the operator $\Delta$ denotes the value difference between the two states *up* and *down*.

The needed investment $V_0$ or the initial value of the hedge portfolio $(\alpha^*, \beta^*)$ is therefore

$$V_0 = \alpha^* S + \beta^* = \frac{C_1^u - C_1^d}{u - d} + \frac{C_1^u - \frac{C_1^u - C_1^d}{u - d} u}{1 + r},$$

which can be rewritten as

$$V_0 = \frac{1}{1 + r} \left( C_1^u \frac{(1 + r) - d}{u - d} + C_1^d \frac{u - (1 + r)}{u - d} \right).$$

This equation leads to a crucial interpretation which will be expanded upon in Sect. 10.3.

Since the hedge portfolio has a value on date 1 equal to that of the option $(C_1)$ in each state, under AAO it should have the same value as the option on date 0: $V_0 = C_0$.[9] Let us remark that the reasoning above also can be applied to a put (in that case, replace $C$ by $P$ in all formulas). As a result, we can assert that under AAO *the premium of an option on date 0 equals the initial value of a hedge portfolio* and can be written

$$C_0 = \frac{1}{1 + r} \left( C_1^u \frac{(1 + r) - d}{u - d} + C_1^d \frac{u - (1 + r)}{u - d} \right), \text{ for a call} \tag{10.8a}$$

$$P_0 = \frac{1}{1 + r} \left( P_1^u \frac{(1 + r) - d}{u - d} + P_1^d \frac{u - (1 + r)}{u - d} \right), \text{ for a put.} \tag{10.8b}$$

The hedge portfolio *replicates* or *synthesizes* the option. The option is just being cloned, and the clone (the replicating portfolio) is called a synthetic option.

---

[9] If the hedge portfolio's value $V_0$ is less than $C_0$, it is sufficient, on date 0, to build this portfolio and sell the option (cashing in $C_0 - V_0$) and, on date 1, to settle the option's payoff with what results from the sale of the hedge portfolio; this amounts to an arbitrage. In the opposite case $(V_0 > C_0)$, taking the reverse position (buying the call and selling the portfolio) also is an arbitrage.

## 10.2 A Pricing Model for One Period and Two States of the World

Also notice that it is unnecessary to introduce *probabilities* to get the price of an option! Although they play no role here, in what follows we sometimes consider the probability $p$ of the event *up* and the probability $(1 - p)$ of the event *down*. Many agents concentrate their efforts on estimating these probabilities, for example, using historical returns. Indeed, the expectation of the share price growth rate $m_S$ (which, in the absence of dividends, amounts to its expected return rate) does depend on $p$:

$$m_S = p\, u + (1 - p)d - 1.$$

But we have just shown that the premium does not depend on this statistical probability $p$ (also called "real," "historical," "physical," or "true") of an increase in the price of the underlying or its expected return $m_S$, which may at first seem counterintuitive.

---

**Example 1**

Let us consider a very simple example. Suppose we have an umbrella valued at $S = \$100$, today on date 0. Tomorrow (date 1), the umbrella's price goes up by 10% if it rains (state *up*), and falls by 10% if it does not rain (state *down*): $uS = 110$ and $dS = 90$. Furthermore, assume the rate for lending and borrowing is zero (current accounts yield no interest). What is the price of a call on the umbrella, with a strike price of $100, and maturity 1? We know (information that is in fact useless by our remark about the statistical probability) that the weather service predicts (correctly) that the real probability of rain is $p = 90\%$.

*Solution:* Consider the hedge portfolio $(\alpha^*, \beta^*)$. We obtain from Eqs. (10.7)

$$\alpha^* = \frac{\Delta C}{\Delta S} = \frac{10 - 0}{110 - 90} = 0.5.$$
$$\beta^* = \frac{10 - 0.5 \times 110}{1 + 0} = -45.$$

The option seller can therefore replicate the option's cash flow by acquiring 0.5 of an umbrella and borrowing $45. Making use of Proposition 4, the premium on date 0 equals

$$C_0 = V_0 = 0.5 \times 100 - 45 = 5\$.$$

Selling the option for $5 allows the seller to buy the hedge portfolio $(\alpha^*, \beta^*)$.

*Remark:* An uninformed reader could be tempted to value the option premium using expectations involving "statistical probabilities": if it were to rain, the seller would have to pay $10. This event's probability is 90%. In the contrary case, she has nothing to pay. The mean cash outflow is therefore 0.9 times $10 = \$9. And therefore the estimated premium is $9. Such a (wrong) valuation thus leads to greatly overestimating the value of the call, which as we have shown is only worth $5.

372 10 Options (I): General Description, Parity Relations, Basic Concepts,...

This example illustrates the fact that the premium *does not depend on the statistical probability* of the event "it is raining" occurring. It is, however, possible to use probability calculus in interpreting the option premium. This interpretation is contained in the next subsection.

### 10.2.3 The "Risk-Neutral" Probability

Consider Eq. (10.8) again giving the price of an option. In this equation the option's payoff $C_1^u$ for a rising market is weighted by a multiplicative factor that we denote by $q$:

$$q = \frac{(1+r) - d}{u - d}.$$ (10.9)

In addition, in the same Eq. (10.8), the multiplier for $C_1^d$ is $1 - q$:

$$1 - q = \frac{u - (1 + r)}{u - d}.$$ (10.10)

Recall that assuming AAO implies $d < 1 + r < u$. These inequalities imply in particular that both $q$ and $(1-q)$ lie between 0 and 1. In addition, their sum is equal to 1. Thus, they can be interpreted as probabilities.

Therefore, one may consider the pair $(q, 1 - q)$ as a probability distribution for the events *up* and *down*. With such probabilities, the expected payoffs of a call and a put can be written, respectively, as.

$$Eq[C1] = q \ C_1^u + (1 - q)C_1^d; \quad Eq[P1] = q \ P_1^u + (1 - q)P_1^d.$$ (10.11)

Equations (10.8) and (10.11) allow us to formulate the following proposition:

**Proposition 5**
*The value of an option can be expressed as the discounted expectation of its payoff; this expectation is calculated using the probability distribution termed "risk-neutral" $(q, 1 - q)$, defined by Eqs. (10.9) and (10.10), and discounting the price at the risk-free rate r. As a result one writes.*

$$C_0 = \frac{1}{1 + r} \ E^q[C_1]; \quad P_0 = \frac{1}{1 + r} \ E^q[P_1].$$ (10.12)

*The following proposition provides an explanation for the phrase "risk-neutral" used to describe the probability $(q, 1 - q)$, a description that can mislead the uninformed.*

**Proposition 6**
*Under the risk-neutral probability all assets (stock, option, risk-free asset) have the same expected return, equal to the risk-free rate.*

## 10.2 A Pricing Model for One Period and Two States of the World

Proposition 6 means that no risk premium is required on risky assets when the expectation is calculated with the probability $q$, from which the term "risk-neutral" has been forged.

**Proof**

Let us consider a stock with price $S$, whose rate of return is $R$ and compute, using the probability $q$, the expectation of this rate. We have remarked that $u$ and $d$ represent, respectively, 1 plus the return rate in states $u$ and $d$: $R$ can therefore take on two values, $R_u = u - 1$ and $R_d = d - 1$.

Thus we obtain the expectation:

$E^q[R] = q(u - 1) + (1 - q)(d - 1) = q u + (1 - q) d - 1$; or by Eqs. (10.9) and (10.10),

$$E^q[R] = \frac{(1 + r) - d}{u - d} u + \frac{u - (1 + r)}{u - d} d - 1$$

$= r$, after simplification.

Now consider an option with price $C$; multiplying the two sides of Eq. (10.12) by $1 + r$ we get

$$E^q[C_1] = C_0(1 + r).$$

We deduce that the expectation under $q$ of the call's return is equal to $r$.

The same result holds for the put.

To illustrate this last proposition, re-examine Example 1 about the market in umbrellas, and let us compute the yield on the risky asset using $q$ ($q = 0.5$ by applying Eq. (10.9)):

$$E^q\left[\frac{S_1 - S_0}{S_0}\right] = 0.5 \times 10 \ \% - 0.5 \times 10 \ \% = r = 0 \ \%.$$

One recovers the result stated in Proposition 6: the mean return using $q$ equals the risk-free rate $r = 0\%$. Note that the same computation, using a probability $p = 90\%$, gives

$$E^p\left[\frac{S_1 - S_0}{S_0}\right] = 0.9 \times 10 \ \% - 0.1 \times 10 \ \% = 8 \ \%.$$

Let us emphasize that the risk-neutral probability $q$, and then the option pricing formula, **do not depend on the "true" probability $p$**, which may seem counterintuitive, and that the **"true" expected return** computed with $p$ is strictly **greater than** $r$. The difference is called the *excess return* (or *risk premium*) and is interpreted as the required compensation for the randomness of the underlying's return.

Finally, let us note that this result holds generically for any portfolio of assets and/or derivatives whose payoff at date 1 writes $\psi(S_1)$. Its value at date 0 is equal to

374          10  Options (I): General Description, Parity Relations, Basic Concepts,...

the discounted value of its risk-neutral expectation: $E^q[\psi(S_1)]/(1 + r)$. This can be proved, in the same way as above, by building a hedge portfolio involving the underlying assets, using a hedge ratio equal to $(\psi(uS_0)-\psi(dS_0))/((u-d)S_0)$.

The following subsection introduces the notion of market price of risk on which the excess return depends.

### 10.2.4  The Risk Premium and the Market Price of Risk

We consider the return on assets (stocks, options, risk-free shares) described by a historical or statistical probability distribution $p$. Although the risk-free asset's return is certainly equal to $r$, one should expect a better return for risky assets such as stocks or options. Let us first consider a stock. We recall that $m_S$ is the expectation of its return and denote its variance by $\eta_S$ and its standard deviation by $\sigma_S = \sqrt{v_S}$. By definition:

$$
\begin{aligned}
m_S &= E^p\left[\frac{S_1 - S_0}{S_0}\right] \\
&= pu + (1 - p)d - 1
\end{aligned}
\tag{10.13}
$$

and

$$
\begin{aligned}
v_S &= E^p\left[(R - m_S)^2\right] \\
&= p(u - 1 - m_S)^2 + (1 - p)(d - 1 - m_S)^2 \\
&= p(1 - p)(u - d)^2
\end{aligned}
$$

So the standard deviation is equal to

$$
\sigma_S = (u - d)(p(1 - p))^{\frac{1}{2}}.
\tag{10.14}
$$

The excess return $(m_S - r)$ can be interpreted as the *risk premium*. This premium is positive if and only if the probability $p$ is strictly larger than the risk-neutral probability $q$. Indeed, we easily deduce from the definition of the expected return (see Eq. 10.13) that

$$
E^p\left[\frac{S_1 - S_0}{S_0}\right] > E^q\left[\frac{S_1 - S_0}{S_0}\right] \Leftrightarrow p > q.
$$

Since the risk affecting $S$ is characterized by the standard deviation $\sigma_S$, it is natural to think this premium is proportional to the risk and to write "risk premium = $\lambda \times$ risk", or

## 10.2 A Pricing Model for One Period and Two States of the World

$$m_S - r = \lambda \sigma_S$$

where $\lambda$ is thought of as the market price of "one unit of risk"; from which comes the following definition:

**Definition** *The market price of risk is defined as the ratio of the risk premium (or excess return) over the standard deviation of the return. Computed for a stock S, we obtain*

$$\lambda = \frac{m_S - r}{\sigma_S}. \tag{10.15}$$

We can then establish the following proposition, which expresses the fact that the market price of the risk resulting from the randomness of $\Delta S$ is the same no matter what generates this risk.

**Proposition 7**
*Under AAO, an option's risk premium is computed using the (same) market price of risk $\lambda$ (given by Eq. 10.15). More precisely,*

$$m_C - r = \lambda \sigma_C \quad \text{for a call,} \tag{10.16a}$$

$$m_P - r = \lambda \sigma_P \quad \text{for a call,} \tag{10.16b}$$

**Proof**
First, let us consider a call whose rate of return has expectation $m_C$ and standard deviation $\sigma_C$ given by

$$m_C = \frac{pC_1^u + (1-p)C_1^d}{C_0} - 1,$$

$$\sigma_c^2 = p\left(\frac{C_1^u - C_0}{C_0} - m_S\right)^2 + (1-p)\left(\frac{C_1^d - C_0}{C_0} - m_S\right)^2$$

so that

$$\sigma_C = \frac{C_1^u - C_1^d}{C_0}(p(1-p))^{\frac{1}{2}}. \tag{10.17}$$

Proposition 7, to be proved, states that

$$\lambda \equiv \frac{m_S - r}{\sigma_S} = \frac{m_C - r}{\sigma_C}.$$

As a first step, using Eqs. (10.14) and (10.17), we have

$$\frac{\sigma_C}{\sigma_S} = \frac{C_1^u - C_1^d}{u - d} \frac{1}{C_0} = \frac{\Delta C}{\Delta S} \frac{S_0}{C_0}.$$

Since the ratio $\frac{\Delta C}{\Delta S}$ is the hedge ratio $\alpha^*$ calculated in the preceding Sect. 10.2.2, we get

$$\frac{\sigma_C}{\sigma_S} = \frac{\alpha^* S_0}{C_0}. \tag{10.18}$$

The second step consists in comparing the excess returns $m_S - r$ and $m_C - r$.

We showed in Sect. 10.2 that the option's value equals the hedging portfolio's value on dates 0 and 1: $C_0 = \alpha^* S_0 + \beta^*$; $C_1 = \alpha^* S_1 + \beta^*(1 + r)$. Therefore, we have

$$m_C = \frac{E^p[\alpha^* S_1 + \beta^*(1 + r)] - (\alpha^* S_0 + \beta^*)}{C_0} = \frac{\alpha^*(E^p[S_1] - S_0) + \beta^* r}{C_0}.$$

Replacing $\beta^*$ by $C_0 - \alpha^* S_0$, we find

$$\begin{aligned} m_C &= \frac{\alpha^*(E^p[S_1] - S_0) + (C_0 - \alpha^* S_0)r}{C_0} \\ &= \frac{\alpha^*(E^p[S_1] - S_0 - rS_0) + C_0 r}{C_0} \\ &= \frac{\alpha^* S_0(m_S - r)}{C_0} + r. \end{aligned}$$

From this we deduce $m_C - r = \frac{\alpha^* S_0}{C_0}(m_S - r)$.

And using Eq. (10.18), $m_C - r = \frac{\sigma_C}{\sigma_S}(m_S - r) = \lambda \sigma_C$, which is Eq. (10.16a).

Equation (10.16b), which is used for the put, can be derived similarly, starting from the portfolio ($\alpha^{**}, \beta^{**}$) duplicating the put where $\alpha^{**} = \frac{\Delta P}{\Delta S} < 0$ and $\beta^{**} = P_0 - \alpha^{**} S_0$.

Remark that (see Eq. 10.16b) the put's risk premium is negative. The put in effect is a *risk reducer* since $\Delta P$ and $\Delta S$ have opposite signs: the underlying's price risk can be eliminated by the *purchase of $\alpha^{**}$* puts.

One can make the relation between the probabilities $p$ and $q$ more explicit by involving the market price of risk. Considering its definition $\lambda = \frac{m_S - r}{\sigma_S}$ and Eqs. (10.13) and (10.14), one proves that

$$\lambda = \frac{p - q}{(p(1 - p))^{\frac{1}{2}}}.$$

We see, in this very simple model, that the price of risk is proportional to the difference between the historical probability and the risk-neutral probability. One often says that the latter is equal to the historical probability *adjusted for risk*. Indeed, we have

$$q = p - \lambda(p(1-p))^{\frac{1}{2}}$$

where the term $-\lambda(p(1-p))^{\frac{1}{2}}$ can be seen as an adjustment factor (or deflator) proportional to the price of risk. These important properties will be preserved in the different models (in both discrete and continuous time) studied later. Recall above all that the *risk-neutral* probability is a probability constructed mathematically outside a statistical framework, so is not the same notion as a *real* or *historical* probability.

## 10.3 The Multi-period Binomial Model

In this section, we examine the multi-period model due to Cox et al. (1979) and called the CRR or *binomial* model; it is a generalization of the single-period two-state model discussed in Sect. 10.1.2. It is a simple enough extension since it is obtained by iterating the one-period model. The framework is given in Sect. 10.1, the notions of stochastic processes and of martingales are introduced in Sect. 10.2 and the valuation of European options is treated in Sect. 10.3. The least easy part, explored in Sect. 10.1.4, is studying the convergence of the binomial model, correctly adjusted, to the famous Black-Scholes model (1973) discussed in the next chapter. It will, in particular, be shown that the values of European calls and puts computed from the binomial model converge to those given by the Black-Scholes formula.

### 10.3.1 The Model Framework and the Dynamics of the Underlying's Price

Let us take up again as a first step the one-period model where the risky asset (the stock) is worth $S_0$ at the start of the period and can take on only two values $uS_0$ or $dS_0$ at the end of the period, where the parameters $u$ (up) and $d$ (down) are multiplicative factors. The multi-period model is built by iterating the one-period model and involves the dates $0, 1, \ldots, N$ numbered in chronological order. During the second period $(1, 2)$, for example, the stock value's variation is governed by a one-period model with the *same rates for a rise or fall in price $u$ and $d$* as in the first period, and with initial condition the value $S_1$ attained at the end of the first period: $S_2 = uS_1$ or $S_2 = dS_1$. This same operation is repeated for $N$ periods so that $S_i = uS_{i-1}$ or $S_i = dS_{i-1}$ for $i = 1, \ldots, N$. Diagram 10.1 shows the set of possible stock values on each date $0, 1, \ldots, N$. Such a diagram is called a *tree* because it is constructed by branching out from the initial point (or *root*) $S_0$. We call a *node* any value possibly taken on by the stock on the dates $1, \ldots, N$.

An important property of this model should be noted: the branches of the tree *recombine*. This means that each upper branch of the tree is connected to the lower branch because a rise followed by a fall leads to the same point as a fall followed by a rise. Consider, for example, the date 2. If the stock price first fell, then rose, its value

**Diagram 10.1** The tree of possible values for the underlying security

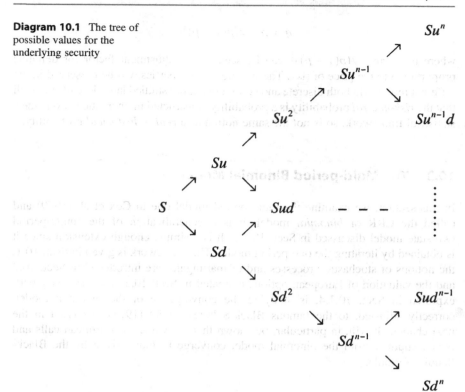

is $S_0 du$; the same if it first rose then fell. This is because the multiplication is commutative $((S_0 d)u = (S_0 u)d)$. This property is important because it limits the number of possible states of the world: on date $N$, the stock can only take on $N + 1$ possible values $S_0 d^N$, $S_0 u d^{N-1}$, ..., $S_0 u^N$. Indeed, only the number of rises and falls counts since the order in which these rises or falls happen does not change the stock price. Such is not the case for a tree which does not recombine. By contrast, a tree is termed "exponential" if each branch remains separate from its neighbors. In that case, on date $N$ the number of different possible values is $2^N$ (instead of $N + 1$).

In what follows we consider large values for $N$. A computational algorithm for a recombining tree grows with the number of nodes as $N^2$ (the total number of different states (nodes) between 0 and $N$ is equal to $\frac{N(N+1)}{2}$), in contrast to an algorithm involving a non-recombining tree whose growth is exponential.

Another difference worth noting between recombining and exponential trees concerns the structure of the information revealed by the stock price. Consider some date $i$ between 1 and $N$. With an exponential tree, the reader will see that knowing the value of the stock price means one knows the price history since date 0. In fact, the branch of the tree on which one finds oneself permits climbing back (down) to the root, with no possibility of losing the way, since all branches are

## 10.3 The Multi-period Binomial Model

379

separate. By contrast, in a recombining tree, it is not possible to find such a reverse path since each node has two predecessors.

Finally, note that the value of a risk-free asset evolves as follows ($r$ denoting an actuarial period rate):

$$1 \to 1 + r \to (1 + r)^2 \to \ldots \to (1 + r)^N.$$

### 10.3.2 Risk-Neutral Probability and Martingale Processes

Applying the analysis in the preceding section for the one-period model, we endow the binomial tree with the risk-neutral probability $(q, 1 - q)$.

Thus consider that at each period the probabilities of a rise or a fall, respectively, are

$$q = \frac{(1 + r) - d}{u - d}; \quad 1 - q = \frac{u - (1 + r)}{u - d}.$$

Note the pair $(q, 1 - q)$ is independent of the value of $S$. It *uniformly* affects each one-period branches making up the tree. Furthermore, we assume the rises and falls are *independent* from one period to another; there is no effect carrying through over time. Consequently, the values of $S_{i + 1}/S_i$, and therefore the successive returns, are independently and identically distributed.

We now show that, with this probabilistic model of the stock's price, the values discounted by the interest rate $r$ of the two assets present in the market, i.e., the stock and the risk-free asset, are governed by *martingale* processes. This property will later be broadened to all contingent assets.

**Definitions** (a) *One calls a process "adapted to S", and denotes it by* $(M_i)_{i = 0, \ldots N}$, *any family of random variables the* $i^{th}$ *of which, denoted* $M_i$, *has a value known as soon as the* $i^{th}$ *value* $S_i$ *of the stock's price is revealed.*

(b) *A process* $(M_i)_{i = 0, \ldots N}$ *adapted to S is a "martingale" if, on any date i, the conditional expectation of* $M_{i + 1}$ *knowing* $S_i$ *is* $M_i$, *that is*

$$E[M_{i+1} \,|\, S_i] = M_i \text{ for } i = 0, \ldots, N. \tag{10.19a}$$

Equation (10.19a) means that, if one considers date $i$ (a generic "today") and one knows the current values of the stock and the adapted process ($S_i$ and $M_i$), the expectation of the future value $M_{i + 1}$ ("tomorrow") is equal to the initial value ("today") $M_i$.

We will show by recurrence that property Eq. (10.19a) is equivalent to

$$E[M_j|S_i] = M_i \text{ for all } i \text{ and all } j = i+1, \ldots, N. \tag{10.19b}$$

It is obvious that Eq. (10.19b) implies Eq. (10.19a) (it suffices to set $j = i + 1$).

Now assume Eq. (10.19a) is true for all $i$ ($M_i$ is, therefore, a martingale). The theorem on iterated expectations entails

$$E[M_j|S_i] = E[E[M_j|S_{j-1}]|S_i].$$

But, from (19-a):$E[M_j|S_{j-1}] = M_{j-1}$. Whence.

$$E[M_j|S_i] = E[M_{j-1}|S_i].$$

Therefore, it is sufficient to repeat the same argument upping the time to $j-2$ to show that$E[M_j|S_i] = E[M_{j-1}|S_i] = E[M_{j-2}|S_i]$;
And so on for the $j-3, j-4, \ldots$, up to $j-k = i + 1$:

$$E[M_j|S_i] = E[M_{j-1}|S_i] = \ldots = E[M_{i+1}|S_i].$$

Since $E[M_{i+1}|S_i] = M_i$ (by 10.19a) we can thus conclude that $E[M_j|S_i] = M_i$.

Therefore, we see that the martingale could as well have been defined using either Eqs. (10.19a) or (10.19b), since the two equations are equivalent.

We also remark that a process that is a martingale for one probability distribution is not necessarily one for a different distribution, since the expectations for the different distributions are different. Especially, in the context of the binomial model, a process that is a martingale when expectations are computed using the probability $q$, is not one under the probability $p$.

### Proposition 8
*The process describing the discounted value of a stock using the risk-free rate r, that is $M_i = \frac{S_i}{(1+r)^i}$ for $i = 0, \ldots, N$, is a martingale under the probability q. The same property holds for the discounted value of a risk-free asset (namely the constant 1).*

### Proof
We start by checking the proposition for a risk-free asset whose discounted value is $\frac{(1+r)^i}{(1+r)^i} = 1$ and therefore follows a martingale (the expectation of a certain variable always equal to 1 is obviously equal to 1).

Now consider the process $\frac{S_i}{(1+r)^i}$ for the discounted stock price and let us show it satisfies Eq. (10.19a) which, in our context, can be written

$$E^q\left[\frac{S_{i+1}}{(1+r)^{i+1}}\,|\,S_i\right] = \frac{S_i}{(1+r)^i}.$$

## 10.3 The Multi-period Binomial Model

Recall that the evolution of $S$ from $i$ to $i + 1$ follows the one-period two-state model studied in the preceding section, and that we have shown that

$$E^q\left[\frac{S_{i+1}}{(1+r)} \mid S_i\right] = S_i.$$

Multiplying each term by $\frac{1}{(1+r)^i}$, we find

$$E^q\left[\frac{S_{i+1}}{(1+r)^{i+1}} \mid S_i\right] = \frac{S_i}{(1+r)^i}.$$

Thus $\left(\frac{S_i}{(1+r)^i}\right)_{i=0,\dots N}$ does satisfy Eq. (10.19a) and thus is a martingale.

We now have at our disposal the main analytic tools for valuing options.

### 10.3.3 Valuation of an Option Using the Cox-Ross-Rubinstein Binomial Model

Valuation of an option in the multi-period framework obeys the same principles as the one-period model. It relies on a recursive backward application of the one-period model, a method that can be applied to a wide variety of situations. For the binomial model, a closed-form solution obtains.

#### 10.3.3.1 Recursive Backward Application of the One-Period Model

We first describe the dynamics of an option value (Diagram 10.2) with a recombining tree that corresponds to the tree describing the evolution of the under-lying asset price (Diagram 10.1).

We use the following notation:

–The tree node at date $i$ and corresponding to $j$ price increases in any order between 0 and $i$ ($j \leq i$) will be denoted by $(i,j)$.

– For each node $(i,j)$ of the tree, the option's value is written $C_{i,j}$.

– $Y_{N,j}$ is the final value of the option when the state of the world is $j$, that is, at node $(N, j)$.

On date $N$ and for state $j$ (at node $(N, j)$), we have.

– $S_{N, j} = S_0\, u^j\, d^{N-j}$;

– $Y_{N, j} = C_{N, j} = \text{Max}\,(S_0 u^j\, d^{N-j} - K, 0)$ for a call;

– $Y_{N, j} = P_{N, j} = \text{Max}\,(K - S_0 u^j\, d^{N-j}, 0)$ for a put.

In what follows, we refer to a *call* for ease of exposition, but the method is exactly the same for a put. The evolution of the call value over the successive unit periods $(0, 1), (1, 2), \dots, (i, i + 1), \dots, (N-1, N)$, is represented in Diagram 10.2.

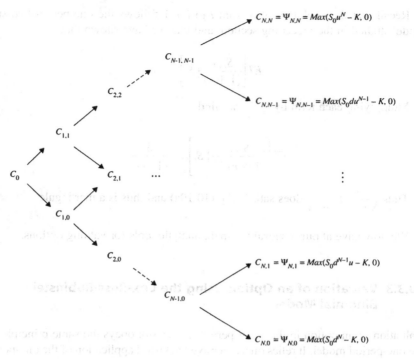

**Diagram 10.2** The tree of possible values for a Call

From each node $(i, j)$ the call, whose value is $C_{ij}$ ($j$ "ups" up to date $i$) can take on, at date $i + 1$, two values, $C_{i+1, j+1}$ or $C_{i+1, j}$, according to whether the stock price rises or falls during the period $(i, i + 1)$. Remark that the branches of the tree recombine, as in Diagram 10.1.

The option valuation method uses a recursive application of the one-period model. It begins with the last period $(N{-}1, N)$, as the $N + 1$ values of the option are known [$Max(S_0 d^{N-i} u^i - K, 0)$] at maturity $N$, and runs backward in time to date 0. The method is based on the risk-neutral probability which allows one to compute the option price as a discounted expectation just as in the one-period model. To save space, we do not make explicit the *multi-period replication strategy* that is *implicit* when one applies recursively the one-period model. This strategy, which replicates at each date $i$ the next (random) value $C_{i+1}$, still is *self-financing*.[10] In this

---

[10] A dynamic strategy or portfolio is called self-financing if the investor does not disburse or withdraw funds on the intermediate dates $i = 1, \ldots, N{-}1$. More precisely, on each date $i = 1, \ldots, N{-}1$, the change in the portfolio's composition leads to neither injection nor withdrawal of funds, the purchase of some assets being financed by the sale of others. Because an option is self-financing (there are no intermediate payments), the portfolio strategy that replicates it necessarily also is.

## 10.3 The Multi-period Binomial Model

multi-period framework, we show generically that the discounted price process for the option is a martingale.

**Proposition 9**
*Let $C_i$ be the price on date $i$ of an option with maturity $N$. Consider the process for the present value of the price:* $\left[\frac{C_i}{(1+r)^i}\right]$ $i = 0, \dots, N$. *Using the risk-neutral $q$, this process is a martingale, from which we have*

$$\frac{C_i}{(1+r)^i} = E^q\left[\frac{C_j}{(1+r)^j} \Big| S_i\right] \quad \text{for all } i \text{ and } j > i. \tag{10.21}$$

*In particular we have*

$$C_0 = E^q\left[\frac{C_N}{(1+r)^N}\right] \tag{10.22a}$$

$$C_i = E^q\left[\frac{C_{i+1}}{1+r} \Big| S_i\right]. \tag{10.22b}$$

*Formulas (10.21) and (10.22) hold equally for a put (just replace $C$ with $P$).*

**Proof**
We have already shown this property for the one-period two-state model. Now consider the situation on date $N-1$: since we know the actual value of the stock $S_{N-1}$ we have

$$C_{N-1} = E^q\left[\frac{C_N}{1+r} \Big| S_{N-1}\right].$$

Now consider date $N-2$. Knowing the value of the stock on date $N-2$, the model allowing us to go to date $N-1$ is still a one-period two-state model and is valid for valuing on date $N-2$ a right on $C_{N-1}$. So we have

$$C_{N-2} = E^q\left[\frac{C_{N-1}}{1+r} \Big| S_{N-2}\right].$$

The same reasoning can be used for any date $i = 0, \dots, N-1$; therefore

$$C_i = E^q\left[\frac{C_{i+1}}{1+r} \Big| S_i\right].$$

This result implies that the discounted price is a martingale and that (see the equivalence of (10.19a) and (10.19b)) for all $j > i$

$$\frac{C_i}{(1+r)^i} = E^q\left[\frac{C_j}{(1+r)^j}|S_i\right].$$

Equations (10.21) and (10.22) thus obtain.

*Proposition 9 thus gives us a way of computing step by step, in the tree, the premium as a discounted expectation of the value on the following date (Eq. (10.22b)).*

Therefore, the tree of premium values can be entirely filled out by proceeding backward from a start at $i = N-1$.

### Example 2

Consider a tree for two periods, whose parameters describing stock price increases and decreases are $u = 1.05$ and $d = 0.95$. Furthermore, assume the risk-free rate for one period is 1% and that the starting value of the stock is $100. Modeling the evolution then gives the following values for the stock on dates 0, 1, 2:

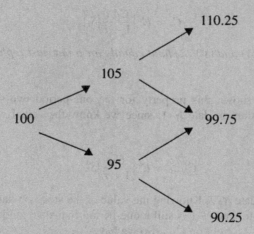

Let us examine a call with maturity 2 and strike 95.

The tree for premiums $C_{i,j}$ is shown on the graph below (the values on date 2 are the payoffs of the call for the three possible states *uu*, *ud*, *dd*).

(continued)

## 10.3 The Multi-period Binomial Model

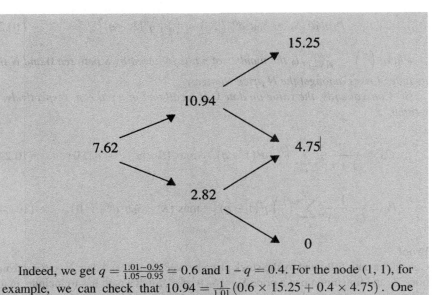

Indeed, we get $q = \frac{1.01-0.95}{1.05-0.95} = 0.6$ and $1 - q = 0.4$. For the node $(1, 1)$, for example, we can check that $10.94 = \frac{1}{1.01}(0.6 \times 15.25 + 0.4 \times 4.75)$. One proceeds in the same fashion for the other nodes.

To sum up, the preceding analysis is based on the very simple recurrence $C_i = E^q\left[\frac{C_{i+1}}{1+r}\big|S_i\right]$, or, for node $(i, j)$: $C_{i,j} = \frac{q\,C_{i+1,j+1}+(1-q)\,C_{i+1,j}}{1+r}$. Starting from $i = N-1$ and working backward step by step up to $i = 0$, we can find all the $C_{i,j}$ to end up with the price $C_0$ we want. By applying this recursive method we avoid the direct computation of $C_0 = \frac{E^q[C_N]}{(1+r)^N}$, which is a major advantage in more complicated situations. However, in the standard binomial model, a closed-form formula can be obtained for this expression by a combinatorial argument that we now explain.

### 10.3.3.2 A Closed-Form Solution for the Premiums of Calls and Puts

The algorithmic (recursive) method involving trees and the martingale approach previously explained apply to a wide variety of situations. For the binomial model, the martingale approach yields a closed-form solution since the law of probability of the $(N + 1)$ final states is the well-known binomial distribution.

Consider the start date 0 and the $N + 1$ states of the world at the horizon date $N$. The following proposition gives the risk-neutral probability of each state and the resulting option values.

**Proposition 10**
*(i) The risk-neutral probability of the event $\{S_N = S_0\,u^j\,d^{N-j}\}$ (the underlying asset price has increased $j$ times in any order), given $S_0$, is equal to.*

$$Proba^q \left(S_N = S_0 u^j d^{N-j} | S_0 \right) = \binom{N}{j} q^j (1-q)^{N-j} \qquad (10.23)$$

where $\binom{N}{j} = \frac{N!}{j!(N-j)!}$ is the number of paths followed by $S$ between 0 and $N$ that includes $j$ rises amongst the $N$ price changes.

(ii) Consequently, the value on date 0 of a call and of a put can, respectively, be written

$$C_0 = \frac{1}{(1+r)^N} \sum_{j=0}^{N} \binom{N}{j} q^j (1-q)^{N-j} \max \left(S_0 u^j d^{N-j} - K; \ 0\right) \qquad (10.24a)$$

$$P_0 = \frac{1}{(1+r)^N} \sum_{j=0}^{N} \binom{N}{j} q^j (1-q)^{N-j} \max \left(K - S_0 u^j d^{N-j}; \ 0\right). \qquad (10.24b)$$

**Proof**

Consider a path followed by the stock price between date 0 and date $N$ starting from a price $S_0$ and finishing up at a price $S_N = S_0 \, u^j \, d^{N-j}$. Such a path has to include, in any order, $j$ increases (and so $N-j$ decreases). Counting them, there are therefore $\binom{N}{j} = \frac{N!}{j!(N-j)!}$ possible paths ending in $S_0 \, u^j \, d^{N-j}$. The probability of each one of these paths is $q^j (1-q)^{N-j}$ (the probability of an increase in $q$, so that of a decrease is $(1-q)$, and the transitions are independent of each other). Adding up the probabilities of all $\binom{N}{j}$ possible paths, we get Eq. (10.23).

By Proposition 9 (Eq. 10.22), we can write the value on date 0 of a call as an expectation and so derive Eq. (10.24a).

$$C_0 = \frac{1}{(1+r)^N} E_q \left[ max \left(S_N - K; 0\right) \right] = \frac{1}{(1+r)^N} \sum_{j=0}^{N} \binom{N}{j} q^j (1-q)^{N-j}$$

$$max \left(S_0 u^j d^{N-j} - K; \ 0\right).$$

For a put, the equation $P_0 = E^q \left[ \frac{P_N}{(1+r)^N} \right] Z$ leads to Eq. (10.24b), using exactly the same reasoning.

## 10.4 Calibration of the Binomial Model and Convergence to the Black-Scholes Formula

Having interpreted option premiums as expressed in Eqs. (10.24) using the probabilities of exercise, we explain how to choose the parameters $q$, $u$, and $d$ and show that the premiums computed using the binomial model, correctly calibrated, converge to those derived from the Black-Scholes formula.

### 10.4.1 An Interpretation of Premiums in Terms of Probabilities of Exercise

Denote by $h$ the smallest number of price increases needed so that the call expires in-the-money, that is, the smallest integer $j$ such that $S_0\, u^j\, d^{N-j} > K$. With that notation we have

$$\max\left(S_0\, u^j\, d^{N-j} - K, 0\right) = 0 \text{ for } j < h; \ \max\left(S_0\, u^j\, d^{N-j} - K, 0\right)$$
$$= S_0\, u^j\, d^{N-j} - K \text{ for } j \geq h.$$

Then $h$ is such that $S_0\, u^h\, d^{N-h} > K$ and $S_0\, u^{h-1} d^{N-h+1} \leq K$, which implies that

$$h = \text{integer part of} \left[1 + \frac{\ln\left(K/S_0 d^N\right)}{\ln\left(u/d\right)}\right].$$

So in this way, we may truncate the sum in (10.24a) to get rid of the cumbersome function max(.) and write

$$C_0 = \frac{1}{(1+r)^N} \sum_{j=h}^{N} \binom{N}{j} q^j (1-q)^{N-j} (S_0 u^j d^{N-j} - K).$$

Separating the terms in $S_0$ from the terms in $K$, we obtain the following two sums:

$$C_0 = \frac{1}{(1+r)^N} \sum_{j=h}^{N} \binom{N}{j} q^j (1-q)^{N-j} S_0 u^j d^{N-j} - \frac{K}{(1+r)^N}$$
$$\times \sum_{j=h}^{N} \binom{N}{j} q^j (1-q)^{N-j} \tag{10.25}$$

The second sum involving $\frac{K}{(1+r)^N}$ has an easy interpretation. Namely, from the definitions of the probability density (see Proposition 10a) and of $h$:

$$\sum_{j=h}^{N} \binom{N}{j} q^j (1-q)^{N-j} = \text{Proba}^q (S_N \geq K),$$

that is, this term is the risk-neutral probability that the stock price exceeds the strike on the expiration date.

Now let us consider the first sum in (10.25):

$$\frac{1}{(1+r)^N} \sum_{j=h}^{N} \binom{N}{j} q^j (1-q)^{N-j} S_0 u^j d^{N-j}$$

$$= \sum_{j=h}^{N} \binom{N}{j} \left(\frac{qu}{1+r}\right)^j \left(\frac{(1-q)d}{1+r}\right)^{N-j} S_0 \tag{10.26}$$

This expression can also be interpreted as a "probability" for the event $S_N \geq K$. Indeed, we set

$$q' = \frac{qu}{1+r}; \quad 1-q' = \frac{(1-q)d}{1+r}.$$

Let us interpret $q'$ as the probability of an upward move and rewrite (10.26) as

$$\sum_{j=h}^{N} \binom{N}{j} (q'^j)(1-q')^{N-j} S_0.$$

Then this sum can be seen as

$$S_0 \, Prob^{q'}(S_N > K).$$

Therefore, we are led to the following proposition, which is the basis for our proof of convergence to the Black-Scholes formula (see Appendix 2).

**Proposition 11**
*The value of a call with maturity N and strike K on date 0 is the difference of two terms: one is proportional to the current value $S_0$ of the underlying, the other proportional to the discounted value of the strike K, and both are proportional to a probability of exercising the option at maturity:*

$$C_0 = S_0 \, Prob^{q'}(S_N > K) - \frac{K}{(1+r)^N} Prob^q(S_N > K). \tag{10.27}$$

*By the call-put parity, one obtains for the* put's *value:*

$$P_0 = \frac{K}{(1+r)^N} Prob^q(S_N \leq K) - S_0 Prob^{q'}(S_N \leq K). \tag{10.28}$$

**Proof**
We have already shown Eq. (10.27). Equation (10.28) can be obtained by repeating the reasoning in the call case but replacing $\max(S_N - K, 0)$ with $\max(K - S_N, 0)$ and truncating the sums over $j$ at the value h. Another proof uses the call-put parity equation:
$C_0 - P_0 = S_0 - \frac{K}{(1+r)^N}$.

10.4 Calibration of the Binomial Model and Convergence to the Black-Scholes Formula 389

Therefore, using expression (10.27), we have:

$$
\begin{aligned}
P_0 &= C_0 - S_0 + \frac{K}{(1+r)^N} \\
&= \frac{K}{(1+r)^N}(1 - \text{Proba}^q(S_N > K)) - S_0(1 - \text{Proba}^{q'}(S_N > K)) \\
&= \frac{K}{(1+r)^N}\text{Proba}^q(S_N \leq K) - S_0\text{Proba}^{q'}(S_N \leq K),
\end{aligned}
$$

which ends the proof.

## 10.4.2 Calibration and Convergence

Application of the binomial model depends on the choice of certain parameters, notably $u$ and $d$. This choice, called calibration, must be done so as to reflect the characteristics of the price process $S(t)$ which the binomial model is supposed to represent. This calibration is done by considering a continuous time model, which particularly involves the notion of volatility and is explained in the next Chap. 11. If the binomial model is appropriately calibrated, its results tend asymptotically to those of the Black-Scholes model. The reader who is not familiar with the Black-Scholes model can skip this subsection and take it up again after having gone over Chap. 11.

### 10.4.2.1 Calibration of the Binomial Model

The pricing formulas obtained from the discrete-time form of Eqs. 10.24 tend asymptotically to their analogues in continuous time (Black-Scholes model). Intuitively, to get convergence, we need to let $N$ tend to infinity on a time interval $(0, T)$ of finite duration. The parameters $u$, $d$, and $r$ then represent the returns on ever smaller time intervals. Therefore, we will consider the binomial model as a discretization of a continuous model.

($i$) Time discretization: Let $T$ be the maturity of the option we wish to price. One period of the binomial model with $N$ periods thus is $T/N$ long.

($ii$) Interest rate discretization: $r_N$, which is indexed by the number $N$ of periods in the interval $[0, T]$, gives the discount rate for each period. Here we let $\rho$ be the continuous zero-coupon rate prevailing during the interval $[0, T]$. So we have $(1 + r_N)^N = e_r{}^T$; or, in another form

$$
1 + r_N = e^{\rho T/N}. \tag{10.29}
$$

Furthermore, the choice of parameters $u$ and $d$ also depends on the length of the period in question, and for a duration $T/N$ we denote them by $u_N$ and $d_N$.

To $u_N$ and $d_N$ corresponds a risk-neutral probability $q_N$ for the state $u_N$ given by:

$$q_N = \frac{1 + r_N - d_N}{u_N - d_N} = \frac{e^{\rho T/N} - d_N}{u_N - d_N}. \tag{10.30}$$

To make the model parameter choice completely determined we have still to fix $u_N$ and $d_N$.

The choice of these two parameters essentially depends on the volatility of the underlying asset. This volatility, traditionally denoted by $\sigma$, is the key parameter in pricing formulas of the Black-Scholes-Merton type (see Chaps. 11 and 12). For the moment, it is enough to know that $\sigma^2$ is defined as the variance of the change in the logarithm of the underlying's price over a unit of time (the time unit usually chosen is a year):

$$\sigma^2 = \mathrm{var}(\ln S_{t+1} - \ln S_t)$$

Additionally, we assume the variance of the change in $\ln S_t$ is proportional to the length of the interval during which it varies (here $T/N$). Therefore:

$$\mathrm{var}(\ln S_{i+1} - \ln S_i) = \sigma^2 T/N.$$

Intuitively, the more $u_N$ and $d_N$ differ, the higher the volatility that results. These two parameters should be chosen to achieve the desired volatility (which is assumed to hold for the underlying's price).

If one only requires that the two parameters $u_N$ and $d_N$ satisfy a single condition concerning the volatility, then an infinite number of calibrations are possible. Here we single out two possible calibrations.

Calibration 1 (Cox, Ross, Rubinstein)

$$: \begin{cases} u_N &= e^{\sigma\sqrt{\frac{T}{N}}} \; ; \quad d_N = \dfrac{1}{u_N} = e^{-\sigma\sqrt{\frac{T}{N}}} \\[2ex] q_N &= \dfrac{e^{\rho\frac{T}{N}} - e^{-\sigma\sqrt{\frac{T}{N}}}}{e^{\sigma\sqrt{\frac{T}{N}}} - e^{-\sigma\sqrt{\frac{T}{N}}}} \end{cases} \tag{10.31}$$

Calibration 2 (Jarrow and Rudd) :

$$\begin{cases} u_N = e^{(\rho-\frac{1}{2}\sigma^2)\frac{T}{N}+\sigma\sqrt{\frac{T}{N}}} \\[1ex] d_N = e^{(\rho-\frac{1}{2}\sigma^2)\frac{T}{N}-\sigma\sqrt{\frac{T}{N}}} \\[1ex] q_N = 0.5 \end{cases} \tag{10.32}$$

Calibration 1 is frequently used in practice. Calibration 2 is theoretically better and justified in Appendix 1, where we use it to show the convergence of the binomial model to the Black-Scholes model.

## 10.4.2.2 Convergence of the Binomial Model Results to Those of Black and Scholes

Denote by $C_0^{(N)}$ and $P_0^{(N)}$ the prices of a call and a put with maturity $T$ and strike $K$ priced using the binomial tree calibrated in this way. The next proposition establishes the convergence of $C_0^{(N)}$ and $P_0^{(N)}$ to the continuous time prices obtained by Black and Scholes.

**Proposition 12**

*Assume the binomial model has been calibrated using Eqs. (10.32) (Calibration 2). If $N \to \infty$, the price $C_0^{(N)}$ has as a limit value that is given by the Black-Scholes formula, namely.*

$$C_0^{(N)} \to S_0 N(d_1) - Ke^{-\rho T} N(d_2)$$

*where $N(.)$ stands for the distribution function of the centered reduced Normal law and with*

$$d = \frac{\ln \frac{S_0}{K} + \left(\rho + \frac{1}{2}\sigma^2\right)T}{\sigma\sqrt{T}}; \quad d_2 = \frac{\ln \frac{S_0}{K} + \left(\rho - \frac{1}{2}\sigma^2\right)T}{\sigma\sqrt{T}}.$$

*The put price $P_0^{(N)}$ has the limit $P_0^{(N)} \to Ke^{-\rho T}N(-d_2) - S_0 N(-d_1)$.*

The proof of this proposition is given in Appendix 2.

In practice, in most situations, results from the binomial model are reasonably close to Black-Scholes prices for values of $N$ (number of periods) above 30.

This convergence result finishes our exposition of the discrete-time binomial model and marks a transition to the next chapter concerning the continuous model; the latter, in its standard form, is indeed nothing else than the limit of the discrete model as the time steps become infinitesimal.

---

## 10.5 Summary

- A call option confers to its holder the right (but not the obligation) to buy the underlying asset at the *strike* price $K$ (or exercise price); $K$ is fixed by the contract when the option is issued (date 0).
- Similarly, a put option confers the right (but not the obligation) to sell the underlying asset at the exercise price $K$.
- An option is *European* when *it can only be exercised at the end of the contract.* This date $T$ is called the *maturity*, or *expiration date*. The option is *American* if its holder has the right to exercise it at *any* time in the interval $(0, T)$.
- With $S_t$ denoting the underlying's price on date $t$, at maturity $T$ the value (payoff) of the option is $C(S_T) = \text{Max}(S_T - K, 0)$ for a call, and $P(S_T) = \text{Max}(K - S_T, 0)$ for a put. This value thus is piecewise linear. On any date $t < T$, this value is a convex curve.

- Under the no-arbitrage rule, the call-put parity assigns a *known, model-free* value for the difference between the call and the put prices, when the two options are written on the same underlying, with the same strike $K$ and maturity $T$. For an underlying spot asset paying no dividend up to the options' maturity, this value is the difference between the current underlying's spot price and the discounted value (using the riskless rate and from date $T$) of the strike.
- This call-put parity is a no-arbitrage condition implied by the *redundancy* of the *four assets* (Stock, Call, Put, riskless asset) as any three of them can replicate the fourth.
- The value of an option can be decomposed into its intrinsic value (the difference between the spot price and the strike, for a call, the opposite for a put) and its time or speculative value.
- The time value is always positive for a call on a spot asset paying no dividend up to the options' maturity and for an American Put. Consequently, an American call, in this case, should never be optimally exercised before expiration, so that its value is equal to that of its European counterpart. In contrast, an American put may optimally be exercised early, so that its value is larger than that of its European counterpart.
- Variants of the Call-Put parity can be obtained for different underlying assets or contracts.
- The *single-period two-state pricing model* (the binomial model due to Cox-Ross-Rubinstein) involves three assets: *the underlying,* which is a spot asset, say a *stock,* the *risk-free* asset, and a *call* (or a put) with strike $K$ and maturity date 1.
- These 3 *assets are redundant* in perfect markets and under the no-arbitrage rule. In particular, an adequately chosen portfolio at date 0, containing stocks and risk-free assets, replicates the payoff of the call at date 1 in both states. Equivalently, an adequately chosen portfolio at date 0, containing stocks and the call, replicates the certain payoff of the riskless asset at date 1.
- Each of the two approaches leads to the value of the call computed as the expectation of its discounted payoff, the expectation involving the probability $q$, which is different from the "true" probability $p$ which is in fact irrelevant.
- Moreover, $q$ can be interpreted as a *risk-neutral probability* under which all assets (stock, option, risk-free asset) have the same expected return, equal to the risk-free rate as no risk premium is required on risky assets.
- *The (multi-period) binomial model* is obtained by iterating the one-period two-state model over $N$ successive periods, the multiplicative coefficients $u$ and $d$ remaining constant and the price ups and downs being assumed independent from one period to another. This procedure yields a *recombining tree.*
- One then applies, *recursively, the single-period two-state model, from period N to period 1* to compute all the possible values of the call, at all dates and in all states. The computation involves the *risk-neutral probability q.*
- The discounted underlying price and the discounted call price are *martingales under the* risk-neutral probability $q$. More generally, the discounted value of any self-financing spot asset or portfolio is also a $q$-martingale.

Appendix 1 393

- This implies that any option *price is equal to its expected discounted payoff,* the expectation being computed under the risk-neutral probability.
- The algorithmic method involving trees as well as the martingale approach can be applied to a wide variety of contexts. In the case of the binomial model, the martingale approach yields a closed-form solution that has the same mathematical structure as the (continuous time) Black-Scholes formula (see next chapter).
- This is why, when the binomial model is appropriately calibrated, its results *tend asymptotically to those of the Black-Scholes* model, as the number $N$ of periods increases. An appropriate calibration implies the choice of the relevant value of $r$ and of the couple $u$ and $d$ yielding the right volatility of $S$. At least two satisfactory alternative calibrations are used in practice.

## Appendix 1

### Calibration of the Binomial Model

We start by computing the first three moments of a Bernoulli variable $X$ that can take on two values $u$ and $d$ with respective probabilities $q$ and $1 - q$.

$$
\begin{aligned}
E(X) &= qu + (1 - q)d \\
var(X) &= E(X^2) - [E(X)]^2 \\
&= qu^2 + (1 - q)d^2 - q^2u^2 - (1 - q)^2d^2 + 2q(1 - q)ud \\
&= u^2q(1 - q) + d^2(1 - q)q + 2q(1 - q)ud,
\end{aligned}
\tag{10.33}
$$

so

$$
var(X) = q(1 - q)(u - d)^2
$$

$$
\begin{aligned}
E[X - E(X)]^3 &= q[u - qu - (1 - q)d]^3 + (1 - q)[d - qu - (1 - q)d]^3 \\
&= q(1 - q)^3(u - d)^3 + (1 - q)q^3(d - u)^3 \\
&= (u - d)^3[q(1 - q)((1 - q)^2 - q^2)],
\end{aligned}
\tag{10.34}
$$

and it follows that

$$
E[X - E(X)]^3 = (u - d)^3 q(1 - q)(1 - 2q).
\tag{10.35}
$$

We justify Calibration 2 (Jarrow and Rudd) given in Eqs. (10.32).

– On the one hand, in the Black-Scholes model, the logarithm of the stock price moves according to a Brownian process with independent increments of variance $= \sigma^2$ and mean $= \rho - \frac{\sigma^2}{2}$ per unit of time (i.e., $\sigma^2 \, T/N$ and $(\rho - \frac{\sigma^2}{2})T/N$ for a period of length $T/N$). Furthermore, the third moment of each increment of $\ln S$ vanishes since it is a normal random variable (and hence has a symmetric distribution). Letting

$\Delta_i \ln S = \ln(S_{i+1}) - \ln(S_i)$ (for the increment in an interval of duration $T/N$), we can therefore find binomial model parameters such that $\Delta_i \ln S$ has the mean $\left(\rho - \frac{1}{2}\sigma^2\right)\frac{T}{N}$ and the variance $\sigma^2 \frac{T}{N}$ (and, possibly, a vanishing third moment).

– The discretized binomial model means that $\Delta_i \ln S = \ln(S_{i+1}) - \ln(S_i) = \ln(S_{i+1}/S_i)$ is a binomial variable $X$ taking on the two values $\ln u_N$ and $\ln d_N$ with respective probabilities $q_N$ and $1 - q_N$. The first two moments of $\Delta_i \ln S$ are thus given by Eqs. (10.33) and (10.34) (when we replace $u$ and $d$ by $\ln u_N$ and $\ln d_N$, and $q$ by $q_N$). Equate the first two moments of the binomial process with those of the Black-Scholes process:

$$ q_N \ln u_N + (1 - q_N) \ln d_N = \left(\rho - \frac{1}{2}\sigma^2\right)\frac{T}{N} \tag{10.36} $$

$$ q_N (1 - q_N)(\ln u_N - \ln d_N)^2 = \sigma^2 \frac{T}{N}. \tag{10.37} $$

An infinity of triples $q_N$, $\ln u_N$, $\ln d_N$ can satisfy these two equations.

If we impose the additional condition that the third moment is zero, the three variables sought will have to obey a third equation:

$$ (\ln u_N - \ln d_N)^3 q_N (1 - q_N) (1 - 2q_N) = 0. $$

This last equation requires the choice $q_N = \frac{1}{2}$.

From this, $\ln u_N$ and $\ln d_N$ are fixed by (10.36) and (10.37) with $q_N = 0.5$, which gives

$$ \ln u_N = \left(\rho - \frac{1}{2}\sigma^2\right)\frac{T}{N} + \sigma\sqrt{\frac{T}{N}} \quad \text{and} \quad \ln d_N = \left(\rho - \frac{1}{2}\sigma^2\right)\frac{T}{N} - \sigma\sqrt{\frac{T}{N}}. $$

(Without necessarily solving these equations, the reader could just check they are satisfied for these two values). The triple $q_N$, $\ln u_N$, $\ln d_N$ is indeed that of Calibration 2 (Eqs. (10.32)).

Let us note that $\ln(S_N) - \ln(S_0) = \sum_{i=1}^{N} [\ln(S_i) - \ln(S_{i-1})]$ is a sum of independent random variables that, by the central limit theorem, converges in law as $N$ goes to infinity to a normal random variable with variance $\sigma^2 T$ and mean $\left(\rho - \frac{1}{2}\sigma^2\right)T$. These are precisely the properties characteristic of the continuous Black-Scholes model (see the next chapter).

To finish, let us remark that for the values of $u_N$ and $d_N$ in Calibration 1 (Eqs. (10.31)), the variance of $\Delta_i \ln S$ only equals $\sigma^2 \frac{T}{N}$ for $q_N = 0.5$, which is not the exact value one has from Eqs. (10.31). However, the reader can check that the desired variance is attained asymptotically (since $\lim_{N \to \infty} q_N = 0.5$).

*Appendix 2                                                                                              395

## *Appendix 2

### Proof of Proposition 12
(convergence of the Cox-Ross-Rubinstein formula to the Black-Scholes formula)

We assume that the binomial model is calibrated using Eqs. (10.32) (Calibration 2 justified in Appendix 1), start from the results of Proposition 12, and show that

$\text{Proba}^{q_N}(S_N \geq K) \to N(d_2)$ and $\text{Proba}^{q'_N}(S_N \geq K) \to N(d_1)$ if $N \to \infty$, with $d_1 = \frac{\ln(S_0/K)+(\rho+\sigma^2/2)T}{\sigma\sqrt{T}}$ and $d_2 = \frac{\ln(S_0/K)+(\rho-\sigma^2/2)T}{\sigma\sqrt{T}}$.

First, consider the value of $S_N$ as a function of the discrete random variable $j$, the number of increases during the period $[0, T]$; since $S_N = S_0 \, u_N^j \, d_N^{N-j}$ and $u_N$ and $d_N$ are given by Eq. (10.32), we have

$$S_N = S_0 e^{\left(\rho-\frac{1}{2}\sigma^2\right)T} e^{2j\sigma\sqrt{\frac{T}{N}}-N\sigma\sqrt{\frac{T}{N}}}.$$

We deduce from this that the event $S_N \geq K$ translates for $j$ as

$$j \geq \frac{\ln(K/S_0) - \left(\rho - \frac{1}{2}\sigma^2\right)T}{\sigma\sqrt{T}} \frac{\sqrt{N}}{2} + \frac{N}{2}. \tag{10.38}$$

Now let us examine the properties of the random variable $j$ which is a binomial variable since it is the sum of $N$ independent and identically distributed Bernoulli random variables. Indeed, define $X_i$ for $i = 1, \ldots, N$ by: $X_i = 1$ if there is an increase between $i$ and $i + 1$, and $X_i = 0$ otherwise.

Then $\sum_{i=1}^{N} X_i$ counts the number of increases occurring during the $N$ periods between dates $0$ and $T$, and we have

$$j = \sum_{i=1}^{N} X_i.$$

The Central Limit Theorem applied to $\frac{1}{N}\sum_{i=1}^{N} X_i$ gives:

$$\frac{\frac{1}{N}\sum_{i=1}^{N} X_i - \frac{1}{N}E\left(\sum_{i=1}^{N} X_i\right)}{\frac{1}{N}\sigma\left(\sum_{i=1}^{N} X_i\right)} = \frac{\sum_{i=1}^{N} X_i - NE(X_i)}{\sqrt{N}\sigma(X_i)} \to N(0,1).$$

In addition, $E(X_i) = q_N$ and $\sigma^2(X_i) = q_N(1 - q_N)$. We then have

$$\frac{j - Nq_N}{\sqrt{Nq_N(1 - q_N)}} \to N(0,1).$$

We will make $\frac{j-Nq_N}{\sqrt{Nq_N(1-q_N)}}$ appear in Eq. (10.38):

$$\frac{j - Nq_N}{\sqrt{Nq_N(1 - q_N)}} \geq \left(\frac{\ln \frac{K}{S_0} - (\rho - \frac{1}{2}\sigma^2)T}{2\sigma\sqrt{T}}\sqrt{N} + \frac{N}{2} - Nq_N\right) \frac{1}{\sqrt{Nq_N(1 - q_N)}}.$$

$$(10.39)$$

Let us use the calibration $q_N = 1/2$ (Calibration 2, Eqs. (10.32)). We then obtain $E(X_i) = q_N = 1/2$ and $\sigma^2(X_i) = q_N(1 - q_N) = 1/4$. From this we have

$$\frac{\sum_{i=1}^{N} X_i - NE(X_i)}{\sqrt{N}\sigma(X_i)} = \frac{j - N/2}{\sqrt{N}/2} \rightarrow N(0, 1).$$

Equation (10.39) becomes

$$\frac{j - N/2}{\sqrt{N}/2} \geq \left(\left(\frac{\ln \frac{K}{S_0} - (\rho - \frac{1}{2}\sigma^2)T}{\sigma\sqrt{T}}\frac{\sqrt{N}}{2} + \frac{N}{2}\right) - \frac{N}{2}\right)\frac{1}{\sqrt{N}/2}$$

$$= \left(\frac{\ln \frac{K}{S_0} - (\rho - \frac{1}{2}\sigma^2)T}{\sigma\sqrt{T}}\right) \equiv -d_2. \qquad (10.40)$$

Thus we see that

$$\mathrm{Proba}^{q_N}(S_N \geq K) = \mathrm{Proba}^{q_N}\left(\frac{j - N/2}{\sqrt{N}/2} \geq -d_2\right) = \mathrm{Proba}(U \geq -d_2)$$

where $U$ is a reduced centered normal law under the probability being considered. $N(u)$ is the standard notation for the distribution function of such a law:

$$N(u) = Prob(U \leq u).$$

By the law's symmetry we have

$$Prob(U \leq d_2) = Prob(U \geq -d_2) = N(d_2).$$

That finishes the proof of convergence for the second term of the call $C_0^{(N)}$. In fact, we have shown

$$\mathrm{Proba}^{q_N}(S_N \geq K) \rightarrow N(d_2).$$

To prove that $\mathrm{Proba}^{q'_N}(S_N \geq K) \rightarrow N(d_1)$, it is sufficient to repeat the reasoning while assigning $q'_N$ to the event $X_i = 1$ with $q'_N = \frac{q_N u_N}{e^{\rho T/N}}$. Using Eq. (10.32) and a Taylor's series expansion we obtain

$$q'_N = \frac{1}{2}\left(1 + \sigma\sqrt{\frac{T}{N}} + o\left(\sqrt{\frac{T}{N}}\right)\right); q'_N(1 - q'_N) = \frac{1}{4}\left(1 + o\left(\sqrt{\frac{T}{N}}\right)\right).$$

So we have again inequality Eq. (10.39) with $q'_N$ instead of $q_N$, by rewriting the second member as:

$$\left(\left(\frac{\ln\frac{K}{S_0} - \left(\rho - \frac{1}{2}\sigma^2\right)T}{\sigma\sqrt{T}}\sqrt{N} + \frac{N}{2}\right) - \frac{N}{2}\left(1 + \sigma\sqrt{\frac{T}{N}} + o\left(\sqrt{\frac{T}{N}}\right)\right)\right)\frac{2}{\sqrt{N}\sqrt{\left(1 + o\left(\sqrt{\frac{T}{N}}\right)\right)}}$$

$$= \left(\left(-d_2\frac{\sqrt{N}}{2}\right) - \frac{N}{2}\left(\sigma\sqrt{\frac{T}{N}} + o\left(\sqrt{\frac{T}{N}}\right)\right)\right)\frac{2}{\sqrt{N}\sqrt{\left(1 + o\left(\sqrt{\frac{T}{N}}\right)\right)}}$$

$$= -d_2 - \sigma\sqrt{T} + \sqrt{N}o\left(\sqrt{\frac{T}{N}}\right)$$

This term goes to $-d_2 - \sigma\sqrt{T} = -d_1$ as $N$ tends to infinity, which proves the result.

The proof of convergence for a put can be inferred from this by using the call-put parity, or by replacing the event $S_N \geq K$ by the event $S_N \leq K$.

## Suggestions for Further Reading

### Books

Cox, J., & Rubinstein, M. (1985). *Options markets*. Prentice Hall.
Hull, J. (2018). *Options, futures and other derivatives* (10th ed.). Prentice Hall Pearson Education.
Kolb, R. (2007). *Futures, options and swaps* (5th ed.). Blackwell.
McMillan, L. G. (1992). *Options as a strategic investment*. New York Institute of Finance.

### Articles

Cox, J., Ross, S., & Rubinstein, M. (1979). Option pricing, a simplified approach. *Journal of Financial Economics, 7*, 229–264.
Jarrow, R., & Rudd, A. (1982). Approximate option valuation for arbitrary stochastic processes. *Journal of Financial Economics, 10*, 347–369.
Stoll, H. (1969). The relationship between put and call option prices. *Journal of Finance, 31*, 319–332.

# Options (II): Continuous-Time Models, Black–Scholes and Extensions

# 11

This chapter presents the famous continuous-time model due to Black and Scholes (1973) as well as some of its extensions, such as those due to Merton, Black, Garman and Kholhagen, or Margrabe. These different models, derived from that pioneered by Black and Scholes, are used in every financial market in the world to value options with differing underlying assets and to manage option positions. They are the most obvious example, if not the only one, of a theoretical development having had a decisive influence on the real economic world and earned Myron Scholes and Robert Merton the Nobel Prize in Economics 1997.[1]

The different formulas for valuing the premiums of European calls and puts are justified by using Itô calculus. This chapter, therefore, requires a basic knowledge of stochastic calculus and of Brownian motion. We refer the reader to Chaps. 18 and 19 for an exposition of these mathematical tools.

The first section is devoted to the standard Black–Scholes model. It is adapted to European options written on spot assets that do not distribute dividends or coupons before the option expires. The second section provides several extensions to the standard model, suitable for different types of underlying asset, amongst which the more general (Black–Scholes–Merton or BSM) model with deterministic volatility and the Heston model with stochastic volatility.

## 11.1 The Standard Black-Scholes Model

The presentation follows that used for the binomial model in the preceding chapter. The reader is invited to observe the correspondence between the discrete and continuous cases at each stage of the argument. Being aware of the analogies will usefully guide their intuition. To avoid irritating repetition the standard Black-Scholes model will be denoted by BS.

---

[1] Unfortunately, Fisher Black died (in 1995) before the prize was awarded.

© The Author(s), under exclusive license to Springer Nature Switzerland AG 2022
P. Poncet, R. Portait, *Capital Market Finance*, Springer Texts in Business and
Economics, https://doi.org/10.1007/978-3-030-84600-8_11

399

## 11.1.1 The Analytical Framework and BS Model's Assumptions

We will continue to consider a market that is always open, free of transaction costs, and made up of two basic securities: a risky asset, which we will call the stock,[2] and a risk-free asset (a bond) which can be bought (lending) or sold short (borrowing) at a constant continuous interest rate denoted by $r$. We will use $S(t)$ or simply $S_t$ for the stock's market price on date $t$.

The stock distributes no dividend (at least up to the option maturity) and its initial value $S_0$ is known on date 0. The random evolution of the price $S_t$, starting from the initial condition $S_0$, is governed by a geometric Brownian motion:

$$\frac{dS_t}{S_t} = \mu dt + \sigma dW_t. \tag{11.1}$$

The parameters $\mu$ and $\sigma$, which are, respectively, the drift and the volatility of the stock price growth rate are assumed to be constant in the standard version of the model.[3] $(W_t)_{t \geq 0}$ is a standard Brownian motion. The model given by (11.1) is based on the physical (or true, or historical) probability $P$. The drift $\mu$ thus represents the mean stock return insofar as it can be estimated from market data.

We denote by $\beta_t$ the value of the risk-free asset on date $t$, that is, the value of $1 invested on date 0 and capitalized at the rate $r$, given by the differential equation

$$\frac{d\beta_t}{\beta_t} = rdt. \tag{11.2}$$

Since in BS $r$ is assumed constant,[4] the yield curve is flat (all rates are equal to $r$, whatever their maturity).[5]

Remark that Eqs. (11.1) and (11.2) have known solutions. The first may be deduced from Itô calculus (see Chap. 18) and can be written

---

[2]The Black–Scholes formula, first conceived to handle the problem of pricing stock options, can easily be extended to other underlying assets (exchange rates, raw materials and other commodities, interest rates) as we will see in the rest of the chapter.

[3]In fact, assuming the constancy of m is not required to end up with the Black–Scholes equation (it is enough that $r$ be constant) as we will see that this (drift) parameter is not used in the option valuation process (see the binomial model too).

[4]Equation (11.2) is an ordinary differential equation since $r$ is constant. If, however, the instantaneous rate were supposed to obey a stochastic process, the form of the equation would remain unchanged, but it would be a stochastic differential equation.

[5]In the opposite case, the same investment of $ 1 would yield two different and certain amounts at $T$ according to whether it was placed at the overnight capitalization rate $r$ or invested in a zero-coupon bond (that is, $e^{rT}$ in the first case and $e^{r_T T}$ in the second), which would clearly constitute an arbitrage opportunity.

# 11.1 The Standard Black-Scholes Model 401

$$S_T = S_0 \, e^{(\mu - \sigma^2/2)T + \sigma WT},$$  (11.3)

and the second, more simply as $\beta_T = e^{rT}$, since $\beta_0 = 1$.

In addition to the two basic assets, we will consider continuously tradable European options written on the risky asset, whose values we will determine with a no-arbitrage argument.

To sum up, the standard BS model is based on the following assumptions:

- The market is always open and free of transaction costs.
- The underlying is a spot asset, distributes no dividend, and its price is governed by the process with constant volatility given by (11.1).
- The risk-free interest rate is constant.
- Options considered are of European type.

## 11.1.2 Self-Financing Dynamic Strategies

As in the discrete case, we will solve the problem of pricing the premium for an option by seeking a self-financing dynamic strategy that replicates the option's payoff on the expiration date $T$. To start with, we define the notion of a self-financing strategy and we deduce the stochastic differential equation that describes the evolution of its value.

Let us consider $\pi_t$, the value on date $t$ of a portfolio invested in a stock and a risk-free bond. Denote by $\delta_t$ the number of stocks held in the portfolio. The amount invested in the risk-free asset is thus equal to

$$\pi_t - \delta_t \, S_t.$$  (11.4)

We deduce that the evolution of the portfolio value $\pi_t$ between two infinitesimally near dates $t$ and $t+dt$ is described, *with no change in the composition of the portfolio*, by the following differential equation:

$$d\pi_t = (\pi_t - \delta_t \, S_t)rdt + \delta_t \, dS_t$$  (11.5)

where $rdt$ is the yield from the risk-free asset, and $dS_t$ the capital gain from the stock.

In fact, Eq. (11.5) only takes account of the **capital gain or loss** realized between $t$ and $t+dt$ and does **not** include possible cash flows (deposits or withdrawals) induced by changes in the portfolio composition. Indeed, the investor can actually choose to increase (or decrease) the stock quantity $\delta_t$ by a purchase (or a sale). The quantity $\delta_{t+dt}$ thus differs from $\delta_t$ after such a portfolio rebalancing. In a *self-financing* strategy, this rebalancing does not affect the portfolio's value: disbursements as a result of buying stock, or receipts from selling stock affect the amount placed in the risk-free asset which they decrease or increase, respectively. As

# 402  11 Options (II): Continuous-Time Models, Black–Scholes and Extensions

a result, the differential Eq. (11.5) does describe the evolution of the value $\pi_t$, even for a portfolio whose composition varies.[6]

We will adopt Eq. (11.5) as the stochastic differential equation of a *self-financing* dynamic strategy. It must be supplemented by the initial condition $\pi_0$, i.e. the value of the original investment on date 0.

Note that giving the process $(\delta_t)_{t\geq 0}$ and the initial condition $\pi_0$ suffices to completely determine the parameters of the stochastic differential Eq. (11.5). Thus, to describe a self-financing strategy it is sufficient to give the initial investment and the process governing how much stock should be held on each date $t$. The amount to be invested in the risk-free asset then follows by subtraction [see Eq. (11.4)].

## 11.1.3 Pricing Using a Partial Differential Equation and the Black–Scholes Formula

### 11.1.3.1 The Fundamental Idea

The idea underpinning BS theory can be summed up as follows: look among the self-financing strategies for those that replicate the call and whose values can be given as a deterministic function[7] of $S_t$ and $t$, which will be denoted by p$(t, S_t)$. Indeed, as we will show, there exists a self-financing portfolio made up of the underlying asset and the risk-free asset whose value on date $T$ is equal to the payoff $\Psi_T$ of the option; in the absence of arbitrage, at any instant, the option's value synthesized in this way is equal to that of the replicating portfolio.

### 11.1.3.2 The Partial Differential Equation for Pricing

So we seek a process $(\pi_t)_{t\geq 0}$ giving the value of a self-financing portfolio with value on expiration date $T$ the same as the payoff to be replicated. We choose *a priori* a sufficiently regular function $\pi(.,.)$ to which we can apply Itô's Lemma (it should be once continuously differentiable in the first variable and twice in the second):

---

[6]More precisely, for any portfolio (whether self-financing or not), Itô's Lemma implies

$$d\pi_t = d(n_t\,\beta_t + \delta_t\,S_t) = n_t\,d\beta_t + \beta_t\,dn_t + \delta_t\,dS_t + S_t\,d\delta_t + dS_t\,d\delta_t$$
$$= [n_t\,d\beta_t + \delta_t\,dS_t] + [\beta_t\,dn_t + (S_t + dS_t)\,d\delta_t].$$

Here $n_t$ denotes the number of risk-free assets in the portfolio on date $t$ and $n_t\,\beta_t = (\pi_t - \delta_t\,S_t)$; the first bracketed term gives the capital gain (equal to $(\pi_t - \delta_t\,S_t\,)rdt + \delta_t\,dS_t)$ while the second term equals the algebraic cash flow (net value of the securities bought and sold) on date $t{+}dt$. It is the second term in brackets which is zero for a self-financing strategy, for which $d\pi_t = n_t\,d\beta_t + \delta_t\,dS_t = (\pi_t - \delta_t\,S_t\,)rdt + \delta_t\,dS_t$.

[7]This restriction, justified in the course of the argument, is logical considering the fact that knowledge of the stock's value on date $t$ should be sufficient to establish the option price.

## 11.1 The Standard Black-Scholes Model

$$d\pi(t, S_t) = \frac{\partial \pi}{\partial t}(t, S_t)dt + \frac{\partial \pi}{\partial S}(t, S_t)dS_t + \frac{1}{2}\frac{\partial^2 \pi}{\partial S^2}(t, S_t)dS_t^2.$$

From Eq. (11.1), we deduce $dS_t^2 = \sigma^2 S_t^2 dt$ and, by substituting that in the preceding equation, we obtain

$$d\pi(t, S_t) = \left(\frac{\partial \pi}{\partial t}(t, S_t) + \frac{1}{2}\frac{\partial^2 \pi}{\partial S^2}(t, S_t)\sigma^2 S_t^2\right)dt + \frac{\partial \pi}{\partial S}(t, S_t)dS_t. \tag{11.6}$$

Any sufficiently regular function of $t$ and $S$ satisfies Eq. (11.6) but does not necessarily constitute the value of a self-financing portfolio; in order that $\pi_t(t, S_t)$ represents at any time $t$ the value of such a portfolio, the variations $d\pi(t, S_t)$ must also satisfy Eq. (11.5), so that both (11.5) and (11.6) hold. By singling out the terms in $dS_t$ and in $dt$ in (11.5) and (11.6), one obtains, respectively:

$$\delta_t = \frac{\partial \pi}{\partial S}(t, S_t).$$

and

$$(\pi(t, S_t) - \delta_t S_t)\, r = \frac{\partial \pi}{\partial t}(t, S_t) + \frac{1}{2}\frac{\partial^2 \pi}{\partial S^2}(t, S_t)\sigma^2 S_t^2.$$

Replacing $\delta_t$ by $\frac{\partial \pi}{\partial S}(t, S_t)$ in this last equation and rearranging terms we get

$$\frac{\partial \pi}{\partial t}(t, S_t) + \frac{\partial \pi}{\partial S}(t, S_t)rS_t + \frac{1}{2}\frac{\partial^2 \pi}{\partial S^2}(t, S_t)\sigma^2 S_t^2 = r\pi(t, S_t). \tag{11.7}$$

This equation is known as the Black–Scholes partial differential equation (PDE), or the *pricing* PDE (sometimes valuation PDE). It *applies to any self-financing portfolio*, in particular to a call or a put.

Finally, for this self-financing portfolio to synthesize the option, its value at maturity has to be, with certainty, equal to the payoff of the option: $\pi(T, S_T) = \Psi_T(T, S_T)$.

The latter equation constitutes the boundary condition that completes the PDE for BS.

It, therefore, suffices that, under AAO, the replicating portfolio's value is, at any instant, equal to that of the asset which is replicated, that is,

$\pi(t, S_t) = C(t, S_t)$ for any $t$ and $S_t$, if $\Psi_T$ is the payoff of a call;
$\pi(t, S_t) = P(t, S_t)$, if $\Psi_T$ is the payoff of a put.

These different equations are the main results of continuous-time pricing theory; given their importance we put them together in the following proposition:

## Proposition 1

*Consider a European call with payoff $\Psi_T(T, S_T) = max(S_T - K, 0)$. The premium on date t for such an option is the solution to the pricing PDE*

$$\frac{\partial C}{\partial t}(t, S_t) + \frac{\partial C}{\partial S}(t, S_t)rS_t + \frac{1}{2}\frac{\partial^2 C}{\partial S^2}(t, S_t)\sigma^2 S_t^2 = rC(t, S_t), \tag{11.7'}$$

*with the (final) boundary condition*

$$C(T, S_T) = \Psi_T(T, S_T). \tag{11.8}$$

*The same equations hold for a* put; *it is sufficient to replace C with P in (11.7) and (11.8) and to set $\Psi_T(T, S_T) = max(K - S_T, 0)$.*

*The self-financing dynamic strategy replicating the final value of the option on date T is defined by the process $(\delta_t)_{t \geq 0}$ specifying, for any t, the quantity of the stock that should be held and by the required initial investment $\pi_0$ such that*

$$\delta_t = \frac{\partial C}{\partial S}(t, S_t) \tag{11.9}$$

*and*

$$\pi_0 = C(0, S_0).$$

## Remarks

- The PDE (11.7) (or (11.7')) is fully specified when one adds the final boundary condition (11.8). From now on we will call it PDE (11.7–11.8).
- Equation (11.9) specifies, for each date $t$, the number of stock units to include in a portfolio so that it exactly replicates the option's cash flow; it is thus the response, as we will see later, to a very important question in practice. This self-financing hedging strategy is fully described as soon as the initial amount of investment, which we denote $\pi_0 = C(0, S_0)$, is calculated. Calculating $\pi_0$, like calculating $\delta_t = \frac{\partial C}{\partial S}(t, S_t)$, requires solving the PDE (11.7–11.8).
- The pricing Eq. (11.7) does not involve the stock's drift parameter $\mu$. Such an independence of the option value regarding $\mu$ is equivalent to the independence, in the discrete case, of the premium from the physical probability $p$. This is a remarkable result to which we will return later, notably in Sect. 11.1.4 where our analysis is based on a probabilistic interpretation.
- The pricing Eq. (11.7) only depends on the self-financing condition and so holds for any self-financing portfolio made up of an underlying and a risk-free asset, or for any security that does not distribute cash payouts and can be replicated by such a portfolio. In this mathematical model, these different self-financing securities and portfolios are only distinguished by their final boundary conditions: $\pi(T, S_T) = max(S_T - K, 0)$ for a call; $\pi(T, S_T) = max(K - S_T, 0)$ for a put, or $\pi(T, S_T) =$ some other payoff, for a different type of security.

## 11.1 The Standard Black-Scholes Model

- AAO is the result of the *local* redundancy of the three securities: option, underlying asset, risk-free asset. This redundancy means that any one of the three securities can be synthesized in the infinitesimal interval $(t, t+dt)$ by an appropriate combination of the two others. Up to needing to change the combination continuously, and since the interval $(0, T)$ can be decomposed into a sum of such infinitesimal intervals, replication can be achieved over the whole period $(0, T)$.
- The argument given is based on the replication of the option by a self-financing dynamic combination of the underlying asset and the risk-free asset. We could have just as well synthesized the risk-free asset using a self-financing portfolio made up of the option and the underlying asset (this was in fact the approach in the 1973 original article by Black and Scholes). The argument would thus have consisted in expressing that under AAO the replicating portfolio's yield equals $r$, which leads rather directly to the PDE (11.7).[8]

### 11.1.3.3 The Black–Scholes Pricing Formula (1973)

It remains to write down a solution to the PDE (11.7–11.8). In the special case where the payoff is that of a standard call ($\Psi_T = max\,(S_T - K, 0)$) or a standard put ($\Psi_T = max\,(K - S_T, 0)$), there is an explicit solution,[9] known as the Black–Scholes formula, given in the following proposition.

**Proposition 2 The Black-Scholes Formula** *The value $C(t, S_t)$ of a European call, as a solution to the PDE (11.7–11.8), is given by*

$$C(t, S_t) = S_t\, N(d_1) - K\, e^{-r(T-t)} N(d_2) \tag{11.10}$$

*with*

$$d_1 = \frac{\ln \frac{S_t}{K} + \left(r + \frac{1}{2}\sigma^2\right)(T - t)}{\sigma\sqrt{(T - t)}}; \quad d_2 = d_1 - \sigma\sqrt{(T - t)} \tag{11.11}$$

*where $N(u)$ denotes the distribution function of the centered reduced normal law:* $N(u) = \int_{-\infty}^{u} \frac{1}{\sqrt{2\pi}} e^{-\frac{x^2}{2}} dx$.

*The value $P(t, S_t)$ of a European put is given by*

$$P(t, S_t) = K\, e^{-r(T-t)} N(-d_2) - S_t\, N(-d_1). \tag{11.12}$$

The proof of these formulas, as given in the original article by Fisher Black and Myron Scholes, uses very complicated and unintuitive changes of variables that allow simplifying the PDE (11.7–11.8) and transforming it into the heat propagation equation whose solution is known from physics. More simply, it is possible to

---

[8] Replicating the risk-free asset instead of the option is simpler but less instructive for beginners.
[9] Regrettably, there often does not exist an explicit solution for a non-standard payoff; it then becomes necessary to employ a numerical method to approximate the solution.

406     11   Options (II): Continuous-Time Models, Black–Scholes and Extensions

verify directly that the functions $C(t, S_t)$ and $P(t, S_t)$ as given in Proposition 2 satisfy the PDE (11.7–11.8). In any case, this verification is a long calculation, and it is more elegant and simpler to use probability theory. We refer the reader to Sect. 11.1.4.3 below for a proof of Proposition 2 based on a simple computation of expectations.

We just note here that the put can be valued as a function of the premium for the call using the call-put parity:

$$P(t, S_t) = C(t, S_t) - S_t + K\,e^{-r(T-t)}.$$

The reader can, therefore, without real loss, lighten their mnemonic load and just remember Eqs. (11.10) and (11.11) which allow calculating $C(t, S_t)$.[10]

---

**Example 1**

We will compute the premiums for a European call and put, written on a cash underlying asset worth \$500, with a strike \$520 and 90-day expiration. The annual (discrete) interest rate is 5% for 3-month transactions and the volatility is estimated as 5.547% from weekly data. We will count a year as 52 weeks and use the fact that the volatility increases as the square root of the length of the period over which it is defined.

The computation gives: $T = 90/365 = 0.2465753$; $r = \ln(1.05) = 0.0488$;
$s = 0.05547\sqrt{52} = 0.4$;
$d_1 = -0.03757945$; $d_2 = -0.236205$; $N(d_1) = 0.48501$; $N(d_2) = 0.40664$.
$C = 500\,N(d_1) - 520\,e^{-rT}\,N(d_2) = 33.583$ and
$P = C - S + K\,e^{-rT} = 47.365$.

---

## 11.1.4 Probabilistic Interpretation

### 11.1.4.1 The Fundamental Idea

Since the appearance of the seminal article of BS in May 1973, the academic and professional worlds have been shaken by one, at first sight very counter-intuitive, aspect of the results, which seemed at odds with all previous known formulas; the expected growth rate $\mu$ of the underlying asset's price (the expectation of its return, since it does not provide dividends) *does not appear* in formulas (11.7–11.11) for pricing the premium. Indeed, the only return parameter on which the premium

---

[10]Equation (11.12) can be inferred from (11.10) and the put-call parity equation:

$$P = SN(d_1) - Ke^{-r(T-t)}\,N(d_2) - S + K\,e^{-r(T-t)} = K\,e^{-r(T-t)}(1 - N(d_2)) - S(1 - N(d_1))$$
$$= K\,e^{-r(T-t)}\,N(-d_2) - S\,N(-d_1).$$

## 11.1 The Standard Black-Scholes Model 407

depends is the risk-free rate $r$. So there arose the idea[11] that the Black–Scholes formula, which connects the "real-world" value of an option with that of its underlying asset and some other parameters $(K, \sigma, r, T)$, must also hold in a virtual risk-neutral universe, in which the risk premiums are consequently all zero and *the expected returns are all equal to the risk-free rate*. In such a universe, security prices can be computed simply using expected payoffs discounted at the risk-free rate. Since computations are simpler in a risk-free world than in the real world, and pricing formulas are the same in both universes, the idea arose to carry out pricing computations *as if* the world were risk-neutral. One should not lose sight of the fact that the probability measure for future events in the risk-free world, denoted by $Q$ in what follows, is different from the probability measure $P$ which describes the real world. From this point of view there is a strict parallelism between the continuous model that is our present subject and the discrete model studied in the preceding chapter, for which we emphasized the difference between the risk-neutral probability $q$ and the historical (or physical) probability $p$.

### 11.1.4.2 Price Dynamics in the Risk-Neutral Universe and the Value of an Option as an Expectation

Consider the risk-neutral world which is governed by the probability measure $Q$, in which the expectations of instantaneous returns for all assets (risky or not) are equal to the risk-free rate $r$. The price dynamics of risky assets thus differs from the real dynamics. In particular, the price of the underlying asset is no longer governed by (11.1) but by the stochastic differential equation

$$\frac{dS_t}{S_t} = rdt + \sigma dW_t^Q \tag{11.13}$$

*where* $(W_t^Q)_{t \geq 0}$ is standard Brownian motion with respect to $Q$.

Equation (11.13) *simply expresses the fact that* the $Q$-expectation of the underlying asset's return is the risk-free rate $r$, therefore that $S_t$ is a geometric Brownian motion whose drift term equals $r$. As we will see, the PDE (11.7) expresses just that. However, first, let us recall the definition of a martingale (the discrete version of which we met in the preceding chapter) and state a result that will be of subsequent use.

**Definition** *A stochastic process* $(Y_t)_{t \geq 0}$ *is a martingale if for any t and t' with* $t' \geq t$:

$$E[Y_{t'} \mid Y_t] = Y_t.$$

We will also use the results of the lemma below.

---

[11] This brilliant idea is especially due to Steve Ross.

# 11  Options (II): Continuous-Time Models, Black–Scholes and Extensions

**Lemma**

(i) *A process $(Y_t)_{t\geq 0}$ whose evolution is governed by $dY_t = a(t)dt + b(t)dW_t$ is a martingale if and only if $a(t) = 0$.*

(ii) *If a process $(X_t)_{t\geq 0}$ is governed by $\frac{dX_t}{X_t} = rdt + \sigma(t)dW_t$, the process $(Y_t)_{t\geq 0}$ defined by $Y_t = X_t\, e^{-rt}$ is a martingale.*

**Proof**

(i) $dY_t = b(t)\, dW_t$ is equivalent to $Y(t') = Y(t) + \int_t^{t'} b(u)dW(u)$ for $0 \leq t \leq t'$, whence

$E[Y(t') \mid Y(t)] = Y(t)$ for $t' \geq t$ (since each and every $dW(u)$ in the integral has expectation zero), which amounts to the actual definition of a martingale. Conversely, if $Y(t)$ is a martingale: $E[Y(t) + dY \mid Y(t)] = Y(t)$, from which $E[dY \mid Y(t)] = 0 = a(t)dt$. Therefore, $a(t) = 0$.

(ii) Itô's Lemma applied to $Y_t = X_t\, e^{-rt}$ gives $dY_t = e^{-rt}\, dX_t - r\, X_t\, e^{-rt}\, dt$.

Dividing the first term in the last equation by $Y_t$ and the second by $X_t\, e^{-rt}$, we obtain: $\frac{dY_t}{Y_t} = \frac{dX_t}{X_t} - rdt$.

Since $\frac{dX_t}{X_t} = rdt + \sigma(t)dW_t$, we have

$\frac{dY_t}{Y_t} = \sigma(t)dW_t$, or $dY_t = Y_t\sigma(t)dW_t$,

and so, by (i), $Y_t$ is indeed a martingale.

Now we are in a position to show that the premium for an option, that is, the solution to (11.7–11.8), can be written as an expectation, which is analogous to the case of the discrete model.

**Proposition 3**

*The solution to the PDE (11.7–11.8) can be written*

$$C(t, S_t) = E^Q\!\left[\Psi_T\, e^{-r(T-t)} \mid S_t\right] \tag{11.14}$$

*where $E^Q$ denotes the expectation with respect to $Q$ (i.e., in the risk-neutral universe), conditional on the known value $S_t$.*

*Using the probability measure $Q$, and whatever the form of the final payoff $\Psi_T$, the process describing the price discounted at the risk-free rate given by $Y_t = C(t, S_t)\, e^{-rt}$ is a $Q$-martingale.*

**Proof**

Use $Q$ and view (11.13) as defining the evolution of $S_t$. First let us show that $Y_t = C(t, S_t)\, e^{-rt}$ is a $Q$-martingale.

Itô differentiation applied to $C(t, S_t)$ gives

$$dC_t = \left[\frac{\partial C}{\partial t}(t, S_t) + \frac{\partial C}{\partial S}(t, S_t)rS_t + \frac{1}{2}\frac{\partial^2 C}{\partial S^2}(t, S_t)\sigma^2 S_t^2\right]dt + \sigma S_t\frac{\partial C}{\partial S}(t, S_t)dW_t.$$

## 11.1 The Standard Black-Scholes Model 409

From PDE (11.7′), the bracketed term equals $rC_t$ ; as a result, $\frac{dC_t}{C_t} = rdt + \frac{\sigma S_t}{C_t} \frac{\partial C}{\partial S}(t, S_t)dW_t$.

$Y_t = C(t, S_t) e^{-rt}$ is, therefore, a martingale, by lemma (*ii*).

We recognize one of the preceding statements according to which the BS PDE expresses the equality of the expectation with respect to $Q$ of the option return and the risk-free rate $r$.

Equation (11.14) is a simple consequence of the martingale property of $Y_t$ which entails

$Y_t = E^Q[Y_T \mid S_t]$.

Since $Y_t = C(t, S_t) e^{-rt}$, we find

$C(t, S_t) = E^Q[C(T, S_T) e^{-r(T-t)} \mid S_t] = E^Q[\Psi_T e^{-r(T-t)} \mid S_t]$.

Remark that the proof showing $C(t, S_t) e^{-rt}$ is a $Q$-martingale is based on the single Eq. (11.7) which *only* expresses the self-financing nature of the replication. As a result, the same holds for an arbitrary self-financing portfolio and we can state the following proposition:

**Proposition 4** *The value $\pi(t, S)$ of* **any self-financing portfolio** *combining an underlying asset, a risk-free asset, and derivatives written on the underlying is governed by the PDE* (11.7) *and, consequently, with respect to* **the risk-neutral** *probability measure Q:*

- *The expected growth rate of $\pi(t, S_t)$ is equal to r.*
- *The discounted value $\pi(t, S_t) e^{-rt}$ is a Q-martingale.*

### 11.1.4.3 Proof of Proposition 2 (Black–Scholes Formula) by Integration

We will now prove the BS [Proposition 2, Eqs. (11.10) and (11.11)], by calculating the expectation of the payoff discounted at the risk-free rate.

Recall that the integral solution of Eq. (11.13) giving the evolution for the stock price, using the probability measure $Q$, between the dates $t$ and $T$ is

$$S_T = S_t e^{\left(r - \frac{\sigma^2}{2}\right)(T-t) + \sigma\left(W_T^Q - W_t^Q\right)}. \tag{11.15a}$$

We know that the increment of the Brownian motion $W_T^Q - W_t^Q$ obeys the centered normal law with variance $T - t$. Denote by $U$ a reduced centered normal random variable with probability density $\varphi(u)du = \frac{1}{\sqrt{2\pi}} e^{-\frac{u^2}{2}} du$ and write $W_T^Q - W_t^Q = \sqrt{T-t}\,U$; from this comes

$$S_T = S_t e^{\left(r - \frac{\sigma^2}{2}\right)(T-t) + \sigma\sqrt{T-t}\,U}. \tag{11.15b}$$

First let us consider the case of a call with maturity $T$, strike $K$ and payoff $\Psi_T = \max(S_T - K, 0)$. Using Proposition 3 (Eq. 11.14), we have

$$C(t, S_t) = E^Q \left[ \max (S_T - K, 0) e^{-r(T-t)} \mid S_t \right].$$

Using (11.15b), we obtain

$$C_t = E^Q \left[ \max \left( S_t e^{\left(r - \frac{\sigma^2}{2}\right)(T-t) + \sigma \sqrt{T-t} U} - K, 0 \right) e^{-r(T-t)} \right] \tag{11.16}$$

and

$$C_t = \int_{-\infty}^{\infty} \max \left( S_t e^{\left(r - \frac{\sigma^2}{2}\right)(T-t) + \sigma \sqrt{T-t}\, u} - K, 0 \right) e^{-r(T-t)} \frac{1}{\sqrt{2\pi}} e^{-\frac{u^2}{2}} du.$$

Note that the maximum occurring in this expression vanishes when

$$S_t e^{\left(r - \frac{\sigma^2}{2}\right)(T-t) + \sigma \sqrt{T-t}\, u} - K < 0,$$

that is, when on date $T$, the option is out of the money and expires without value. This inequality holds if and only if

$$\ln(S_t) + (r - \sigma^2/2)(T - t) + \sigma\sqrt{T - t}\, u < \ln(K), \text{i.e. } u < -d_2, \quad \text{where}$$

$$d_2 = \frac{\ln \frac{S_t}{K} + \left(r - \frac{\sigma^2}{2}\right)(T - t)}{\sigma \sqrt{T - t}}.$$

The limits of the integration over $u$ can thus be moved from $]-\infty, \infty[$ to $[-d_2, \infty[$.

Thus we need to compute

$$C_t = \int_{-d_2}^{\infty} \left( S_t e^{\left(r - \frac{\sigma^2}{2}\right)(T-t) + \sigma \sqrt{T-t}\, u} - K \right) \frac{1}{\sqrt{2\pi}} e^{-r(T-t)} e^{-\frac{u^2}{2}} du,$$

which can be split into two terms

$$C_t = \int_{-d_2}^{\infty} S_t e^{\left(r - \frac{\sigma^2}{2}\right)(T-t) + \sigma \sqrt{T-t}\, u} \frac{1}{\sqrt{2\pi}} e^{-\frac{u^2}{2}} e^{-r(T-t)} du - K \int_{-d_2}^{\infty} e^{-r(T-t)} \frac{1}{\sqrt{2\pi}} e^{-\frac{u^2}{2}} du.$$

Set $I_1 = \int_{-d_2}^{\infty} S_t e^{\left(r - \frac{\sigma^2}{2}\right)(T-t) + \sigma \sqrt{T-t}\, u} \frac{1}{\sqrt{2\pi}} e^{-\frac{u^2}{2}} e^{-r(T-t)} du;\ I_2 = K \int_{-d_2}^{\infty} e^{-r(T-t)} \frac{1}{\sqrt{2\pi}} e^{-\frac{u^2}{2}} du.$

First handle the second term, $I_2$, which is simpler than $I_1$.

We can bring the term $e^{-r(T-t)}$, which does not vary, out from inside the integral and write

## 11.1 The Standard Black-Scholes Model

$$I_2 = Ke^{-r(T-t)} \int_{-d_2}^{\infty} \frac{1}{\sqrt{2\pi}} e^{-\frac{u^2}{2}} du.$$

Noting that the probability density $\varphi(u) = \frac{1}{\sqrt{2\pi}} e^{-\frac{u^2}{2}}$ is a symmetric function, we obtain by changing the integral's limits

$$I_2 = Ke^{-r(T-t)} \int_{-\infty}^{d_2} \frac{1}{\sqrt{2\pi}} e^{-\frac{u^2}{2}} du.$$

This integral is the distribution of the reduced centered normal law evaluated at $d_2$, which we will denote by $N(d_2)$. Thus, we have

$$I_2 = Ke^{-r(T-t)} N(d_2).$$

Now let us examine the first term, $I_1$, in the expression for $C_t$:

$$I_1 = \int_{-d_2}^{\infty} S_t e^{\left(r - \frac{\sigma^2}{2}\right)(T-t) + \sigma\sqrt{T-t}u} \frac{1}{\sqrt{2\pi}} e^{-\frac{u^2}{2}} e^{-r(T-t)} du.$$

Noticing that the terms in $e^{r(T-t)}$ and $e^{-r(T-t)}$ cancel out and that $S_t$ can be brought out from inside the integral, and then regrouping the exponentials, we get

$$I_1 = S_t \int_{-d_2}^{\infty} \frac{1}{\sqrt{2\pi}} e^{-\frac{u^2 - 2\sigma\sqrt{T-t}\, u + \sigma^2(T-t)}{2}} du.$$

There is a perfect square in the exponent: $\left(u - \sigma\sqrt{T-t}\right)^2$. By changing the variable to $v = u - \sigma\sqrt{T-t}$, we have

$$I_1 = S_t \int_{-d_2 - \sigma\sqrt{T-t}}^{\infty} \frac{1}{\sqrt{2\pi}} e^{-\frac{v^2}{2}} dv.$$

Using the same symmetry argument as before (adjusting the integration limits) we have therefore

$$I_1 = S_t \int_{-\infty}^{d_1} \frac{1}{\sqrt{2\pi}} e^{-\frac{v^2}{2}} dv,$$

where $d_1 = d_2 + \sigma\sqrt{T-t}$. This expression contains the distribution function of a reduced centered normal law at $d_1$ denoted $N(d_1)$. The first term $I_1$ in the call premium therefore equals

$$I_1 = S_t N(d_1).$$

Since, $C_t = I_1 - I_2$, we have obtained the BS formula by direct computation of the option's discounted expected payoff.

412      11 Options (II): Continuous-Time Models, Black–Scholes and Extensions

The BS formula for a put can be justified by a similar computation, or more simply deduced from the call's formula and call-put parity.

In the sequel, we will use the *BS pricing function* defined by

$$F(t,x) = e^{-r(T-t)} E^Q \left[ \max \left( x e^{\left(r - \frac{\sigma^2}{2}\right)(T-t) + \sigma \sqrt{T-t} U} - K, 0 \right) \right] \qquad (11.17)$$

where $U$ is a reduced centered Gaussian variable under the probability measure $Q$.
We have just shown

$$F(t,x) = x N(d_1) - K e^{-r(T-t)} N(d_2), \qquad (11.18)$$

where

$$d_1 = \frac{\ln \frac{x}{K} + \left(r + \frac{1}{2}\sigma^2\right)(T-t)}{\sigma \sqrt{(T-t)}}; \quad d_2 = d_1 - s\sqrt{(T-t)}.$$

The BS formula leads to a call premium that one may write as $C(t, S_t) = F(t, S_t)$.

Extensions of this standard model that we now give lead to pricing formulas that can be written very easily in terms of the pricing function $F(t, x)$.

## 11.2    Extensions of the Black–Scholes Formula

This section offers several extensions of the BS formula for underlying assets other than dividend-free stocks. Section 11.2.1 considers an underlying that does provide a payment stream (dividends, coupons, etc.) which we will assume is either continuous or discrete. In Sects. 11.2.2, 11.2.3, and 11.2.4 we deal with some options written on other underlying assets: commodities, exchange rates, and futures contracts. We show, in Sect. 11.2.5, how the BS formula can be adapted to the case of a variable (but still deterministic) volatility. We proceed, in Sect. 11.2.6, to pricing options when interest rates are stochastic, and present the general Black–Scholes–Merton (BSM) model. The particular case of options on interest rates is, however, deferred to Chaps. 16 and 17. In Sect. 11.2.7, we analyze exchange options originally studied by Margrabe (1978). We finally present, in Sect. 11.2.8, the problem posed by a stochastic volatility, focusing on Heston's model (1993).

### 11.2.1   Underlying Assets That Pay Out (Dividends, Coupons, etc.)

Most assets underlying options provide some payment, which translates into an instantaneous reduction in their market price at the moment the payment is made. Recall that options are never protected against such payments. As a result, all else being equal, the value of a call goes down and the value of a put goes up with the amount paid out before expiry on date $T$.

## 11.2 Extensions of the Black–Scholes Formula

In the following, we will refer for convenience to the underlying asset as a stock, but all the results exactly hold for other underlying spot assets. The payment by the underlying asset will be called either a coupon or a dividend. The first step is to specify a model for this payout. First, we analyze the classic model in which the stock pays a continuous dividend; this allows for an explicit formula for European option premiums and constitutes a paradigm from which pricing formulas are derived for other underlying asset types considered later on (commodities, exchange rates, futures contracts). We will subsequently examine the case of a discrete dividend paid on a specific date.

### 11.2.1.1 Model with Continuous Dividends

The process $(S_t)_{t\geq 0}$ describing the underlying stock is assumed to satisfy the stochastic differential equation of BS type (11.1) which we recall is

$$\frac{dS_t}{S_t} = \mu dt + \sigma dW_t,$$

where $\mu$ is a constant drift parameter, $\sigma$ is a volatility parameter (also constant) and $(W_t)_{t\geq 0}$ is a standard Brownian motion with respect to $Q$ (we use immediately the risk-neutral probability measure).

The first question which we need to answer is: under the risk-neutral measure $Q$, what should the drift $\mu$ of the process $(S_t)_{t\geq 0}$ be? We know that if the stock did not pay a dividend, $\mu$ would equal $r$, the risk-free rate. This must be changed in presence of a dividend.

In the continuous-dividend model, time is chopped into infinitesimal pieces $dt$. At the end of the period $[t, t+dt]$ the stock pays whoever holds it a coupon for cash in the amount of $c\, S_t\, dt$, where $c$ is the *dividend yield*, which this model assumes is constant.

Notice that the *amount* of dividend paid on date $t+dt$, for the period $[t, t+dt]$:

- Is proportional to the stock value $S_t$ at the start of the period;
- Is proportional to the length $dt$ of the period;
- Is known at the start of the period (at $t$) but is unknown on date 0 since it depends on the value $S_t$ which itself is not known on date $0$[12].

Notice also that the total return on a stock during the period $[t, t+dt]$, which results from the capital gain $dS_t$ *plus* the dividend $c\, S_t\, dt$, can be written

$$dR = \frac{dS_t + cS_t dt}{S_t} = \frac{dS_t}{S_t} + cdt.$$

From this, using (11.1), we derive $dR = (\mu + c)\, dt + \sigma\, dW_t$.

---

[12] It is, however, important to notice that the dividend received on date $t+dt$ contains *no innovation* (it is certain on date $t$). This is the reason why the dividend affects the process's drift and not the Brownian part of the stochastic differential equation (11.19a).

414        11  Options (II): Continuous-Time Models, Black–Scholes and Extensions

This result, which is more or less obvious, means that the return rate $dR$ on a stock is the sum of the capital gain rate ($dS/S$) and the dividend yield $c$.

Since in the risk-neutral universe the expectation of the return rate on this investment equals the risk-free rate, we have $\mu + c = r$.

It follows that the value $S_t$ of a stock paying a continuous coupon with yield $c$ displays the following dynamics under the risk-neutral probability $Q$:

$$\frac{dS_t}{S_t} = (r - c)dt + \sigma dW_t, \tag{11.19a}$$

which implies

$$S_T = S_t e^{(r-c-\frac{\sigma^2}{2})(T-t)+\sigma(W_T^Q - W_t^Q)}, \text{ or else,}$$
$$S_T = S_t \ e^{(r-c-\frac{\sigma^2}{2})(T-t)+\sigma\sqrt{T-t}U} \ . \tag{11.19b}$$

These equations mean that, in the risk-neutral universe, *the rate of growth of the underlying asset price equals the risk-free rate reduced by the dividend yield c.*

One should not be surprised that paying a dividend reduces the growth of the stock's market value (which changes, under the measure $Q$, from $r$ to $r - c$). We have already seen, several times, that paying out an amount $x$ leads to a reduction of $x$ in the value of the security but that the total return of the investment is not affected by the dividend[13] since the portfolio adds up the increase in capital value ($dS$ in the continuous time model) *and* the dividend ($c \ S_t \ dt$) which the investor can use to reinvest in the stock, in the riskless asset or in any other instrument of their choice.

The stock that pays out (denoted by $s$ in what follows) does create one difficulty in the analysis in that it is not a self-financing investment. That is why it is useful, at this juncture, to introduce a self-financing portfolio made up exclusively of the stock $s$.

So let us consider a portfolio $\phi$ initially made up of $n_0$ *units of* stock $s$ with price $S_0$ and whose initial value is $n_0 S_0$. At later dates, to the $n_0$ initial units of stock are added the continuous dividends cashed in. More precisely, assume that each dividend payment is immediately reinvested fully in stock $s$. This reinvestment has two consequences: the portfolio $\phi$ is self-financing; and the number of units of $s$ in the portfolio $\phi$ increases continuously. Thus we can interpret $\phi$ as a capital fund invested only in stock $s$.

More formally, call $n_t$ the number of $s$ shares in the portfolio $\phi$, on date $t$; since the amount of dividend paid, $n_t \ c \ S_t dt$, is invested immediately in shares, the number of shares of $s$ held in $\phi$ increases by $dn_t$ with $dn_t = \frac{n_t c S_t dt}{S_t} = n_t cdt$; that is, $\frac{dn_t}{n_t} = cdt$, therefore $n_t = n_0 e^{ct}$.

---

[13] Readers familiar with the theory of corporate finance will recognize one of the theorems due to Modigliani and Miller about the irrelevance of dividend policy as to the firm's market value, in perfect markets.

## 11.2 Extensions of the Black–Scholes Formula

The number of $s$ shares contained in the portfolio $\phi$ thus increases exponentially at the rate $c$ and, on any date $t$, the fund's value thus is $V_\phi(t) = n_0 e^{ct} S_t$.

Now consider on date $t$ an investment of \$ $S_t e^{-c(T-t)}$ in a portion of the fund $\phi$, which buys $e^{-c(T-t)}$ shares of $s$; this investment provides on date $T$ one unit of stock $s$ worth $S_T$: *the self-financing strategy $\phi$ thus permits replicating the cash flow $S_T$ at $T$ using an investment of* \$ $S_t e^{-c(T-t)}$ *on date t.*

Taking into account this self-financing strategy $\phi$ will now allow us to establish in a simple way the call-put parity equation, as well as the pricing formulas for options written on spot assets distributing a continuous dividend.

### Proposition 5
*Consider a European call and put, with strike price $K$, expiration date $T$, written on a spot asset with price $S$ which pays a continuous dividend at the constant rate $c$; $r$ is the continuous interest rate prevailing at $t$ for a transaction of duration $T - t$.*

*Under AAO, the call-put parity equation can be written*

$$C_t - P_t = e^{-c(T-t)} S_t - K e^{-r(T-t)}. \tag{11.20}$$

### Proof
Suppose, that at some instant $t$, Eq. (11.20) is not satisfied; for example, $C_t - P_t - e^{-c(T-t)} S_t + K e^{-r(T-t)} > 0$. An arbitrage exploiting this anomaly requires transactions at the initial instant $t$ as well as throughout the period $(t, T)$.

- Initially, the arbitrageur sells the call, buys the put and $e^{-c(T-t)}$ shares, and borrows on the money market \$ $K e^{-r(T-t)}$ for a period $T - t$; this yields the positive cash amount $C_t - P_t - e^{-c(T-t)} S_t + K e^{r(T-t)}$.
- Between $t$ and $T$, the arbitrageur follows the strategy $\phi$ defined above that consists in using the dividends paid to acquire new shares; as we showed, the number of shares in the portfolio increases exponentially at a rate $c$ and the portfolio contains *only the stock on the final date $T$*. We emphasize that the arbitrageur thus neither receives nor disburses any cash flow between 0 and $T$.
- On the expiry of the options, there can be two situations: $S_T < K$ or $S_T \geq K$. They are shown in the two right columns of the Table below displaying the value on date $T$ of each component of the position as well as its net value, which is zero in both situations and thus certain.

| | $S_T < K$ | $S_T \geq K$ |
|---|---|---|
| Put purchased | $K - S_T$ | 0 |
| $e^{-c(T-t)}$ shares bought | $S_T$ | $S_T$ |
| Call sold | 0 | $-(S_T - K)$ |
| Loan refund (money market) | $-K$ | $-K$ |
| Net value of the position | 0 | 0 |

This trade, which yields a positive cash flow initially, and implies no negative cash flow as it unfolds, is therefore an arbitrage.

Under the contrary assumption where $P_t - C_t + e^{-c(T-t)}S_t - K e^{-r(T-t)} > 0$, the arbitrage consists in an initial sale of the put, combined with a purchase of the call, selling $e^{-c(T-t)}$ shares short and lending on the money market, followed by continuous short selling of shares to finance the dividends due on the short position.

Remark that this relationship, like all other forms of call-put parity (see the preceding chapter), continues to hold in a context of stochastic interest rates and that the strike price should be discounted at the zero-coupon rate corresponding to its maturity ($T - t$ on date $t$).

We can now price European options written on a spot asset that provides a continuous payment. Here the rates $r$ and $c$, as well as the volatility $\sigma$, will be assumed constant. The results in Proposition 6 below are an extension of the BS model, often called the Merton model from the name of its author (the model was published in June 1973, just after BS in May).

### Proposition 6 The Merton Model

*Consider a European call and put, with strike price K, expiration T, written on a stock quoted at a spot price S, paying a continuous dividend at the constant rate c; in addition, r (the risk-free rate) is assumed to be constant, as is the volatility $\sigma$ of S.*

*The call's value is given by*

$$C(t, S_t) = S_t\, e^{-c(T-t)}N(d_1) - K\, e^{-r(T-t)}N(d_2). \tag{11.21}$$

*The put's value is given by*

$$P(t, S_t) = K\, e^{-r(T-t)}N(-d_2) - S_t\, e^{-c(T-t)}N(-d_1), \tag{11.22}$$

*where N(u) denotes the distribution function of a reduced centered normal law and*

$$d_1 = \frac{\ln\frac{S_t}{K} + (r - c + \frac{1}{2}\sigma^2)(T - t)}{\sigma\sqrt{(T - t)}}; \quad d_2 = d_1 - \sigma\sqrt{(T - t)}. \tag{11.23}$$

*We give two very simple proofs of Proposition 6. The first is mathematical and based on the Black–Scholes pricing function F(.), defined in the preceding section (Eq. (11.17) and (11.18)). The second is more financial in character and involves the self-financing strategy $\phi$ defined earlier.*

### First Proof

Start by considering the call written on an asset whose value is governed by Eq. (20). Its price obtained from the risk-neutral expectation of the discounted final payoff can be written

## 11.2 Extensions of the Black–Scholes Formula

$$C(t, S_t) = E^Q \left[ \max \left( S_T - K, 0 \right) e^{-r(T-t)} \right]$$

$$= E^Q \left[ \max \left( S_t e^{\left(r-c-\frac{\sigma^2}{2}\right)(T-t)+\sigma\sqrt{T-t}U} - K, 0 \right) e^{-r(T-t)} \right]$$

$$= E^Q \left[ \max \left( \left( S_t e^{-c(T-t)} \right) e^{\left(r-\frac{\sigma^2}{2}\right)(T-t)+\sigma\sqrt{T-t}U} - K, 0 \right) e^{-r(T-t)} \right].$$

Recall Eq. (11.17) that defines the pricing function $F$:

$$F(t, x) = E^Q \left[ \max \left( x e^{\left(r-\frac{\sigma^2}{2}\right)(T-t)+\sigma\sqrt{T-t}U} - K, 0 \right) e^{-r(T-t)} \right].$$

Comparison of the two equations immediately gives

$$C(t, S_t) = F\left( t, S_t e^{-c(T-t)} \right).$$

From this, Eq. (11.18) implies (11.21) and (11.23).

As for the put, its pricing formula can be obtained in the same way, or more simply, starting from the formula for the call and using the call-put parity (11.20).

One can remark that Merton's formulas (11.21–11.23) can be derived from those of Black–Scholes by replacing $S_t$ by $S_t e^{-c(T-t)}$. This result is intuitively clear: the distribution paid out decreases the value of the stock, exponentially, at the rate $c$. So, other things being equal, it is equivalent to have on date $t$ an option on an asset that does not pay out and is worth $S_t e^{-c(T-t)}$ or an option on an asset worth $S_t$ with continuous distribution rate $c$. This intuition, vindicated by a very simple argument based on the strategy $\phi$ we defined earlier, leads to the second proof of Proposition 6.

**Second Proof**

Let us consider the previously defined self-financing $\phi$. We have seen that a share of this fund whose value is $S_t e^{-c(T-t)}$ on date $t$ will be worth $S_T$ on $T$. A call with strike price $K$ and expiration date $T$, written on such a portion of the fund $\phi$, generates a payoff on date $T$ equal to Max $(S_T - K, 0)$ and is therefore strictly equivalent to a call written on the stock $s$ itself. Above all, this call on $\phi$ can be priced using the standard BS formula (since $\phi$ does not pay out), if one introduces $S_t e^{-c(T-t)}$ in place of $S_t$ as the underlying price, which justifies the Merton model, and finishes the proof.

The main point is that *all these formulas,* which can be applied when there is a continuous distribution, are derived from BS by *everywhere* replacing $S_t$ by $S_t e^{-c(T-t)}$. This substitution should be made in the call price (including the formulas for $d_1$ and $d_2$), the put price and the call-put parity. So the reader can without any loss just remember the ordinary parity equation, the BS formula and the rule for adjusting for a dividend distribution.

# 11 Options (II): Continuous-Time Models, Black–Scholes and Extensions

> **Example 2**
>
> (a) Use the same data as in Example 1 (Sects. 1–3.3), but with the underlying asset distributing a continuous dividend at the rate $c = 3\%$. The call and put are then worth, respectively, $ 31.823 and $ 49.29.
>
> (b) Consider a call that is at-the-money, with 4 months to run, written on a stock index with a value of $500, whose volatility is 20%. The index value does not include the dividends distributed by the stocks of which it is composed. These are estimated to be, for the next 4 months, 0.2%, 0.3%, 0.3%, and 0.2% of the index value. With an average payout rate from the underlying of 3% (the annualized mean of the four expected monthly payouts) and a continuous interest rate of 5%, the value of the call is $ 24.40.

To end this subsection, we note that the adjustment of the standard formula required by a dividend distribution can also be interpreted in terms of the present value of the dividends expected from time $t$ to $T$. Indeed, let us compare buying an asset $s$ at a price of $ $S_t$ on date $t$ with an investment of $ $S_t e^{-c(T-t)}$ in the fund $\phi$. In both cases, the portfolio's final worth at $T$ is $ $S_T$. Except for the initial expense, the only difference between these two investments is that $\phi$ distributes nothing and $s$ distributes dividends between $t$ and $T$: the difference in the initial value of $S_t - S_t e^{-c(T-t)}$ thus corresponds to the present value (at $t$) of the dividends distributed by $s$ between $t$ and $T$.[14] *The adjustment consisting in substituting $S_t e^{-c(T-t)}$ for $S_t$ simply amounts to deducing from the quoted price $S_t$ of the underlying the value at $t$ of the dividends that can be expected during the life of the option.* This adjustment also appears, and more simply, in the discrete dividend model, as we now show.

## 11.2.1.2 Model with a Discrete Dividend

The continuous dividend model is useful in different situations (options on indices, on exchange rates, on very long options, also called warrants) where many dividends are expected before the option expires and can, more or less conveniently, be expressed by a continuous stream of cash flows. Still, stocks usually pay an annual dividend (sometimes quarterly in the USA) whose amount is known some months before the date of payment. So investors are faced with a small number of discrete dividends rather than with a continuous stream. Let $s$ denote the option's underlying asset with price $S$, $t$ the current date (today) of the pricing, $D$ the value of the

---

[14]This result, obtained from a financial analysis, can also be derived by calculation: the present value of the dividend cash flows $c\,S_x\,dx]_{x\in(t,\,T)}$, computed under the measure $Q$, is equal to $D* = E^Q\left[\int_t^T e^{-r(x-t)}cS_x dx/S_t\right]$; and since $e^{-(r-c)x}S_x$ is a $Q$-martingale, we have $E^Q[S_x e^{-(r-c)(x-t)}/S_t] = S_t$, from which we see $D* = c\,S_t\int_t^T e^{-c(x-t)}dx_t = S_t(1 - e^{-c(T-t)})$.

## 11.2 Extensions of the Black–Scholes Formula

dividend(s) to be paid between dates $t$ and $T$ that we assume is known in advance,[15] and $D*$ the value of this (these) dividend(s), discounted on date $t$. In the same spirit as the preceding subsection, we define a portfolio $\phi$, self-financing between $t$ and $T$, whose value on date $T$ will be with certainty $S_T$.

For simplicity, we will first consider the case of a single dividend $D$ expected on date $\tau \in (t,T)$.

The strategy $\phi$ consists in buying on date $t$ the stock $s$ at the price $S_t$ and borrowing $D*$ until $\tau$, thus for a period $\tau - t$; the stock is held until $T$, and on date $t$ the dividend it pays out allows repaying the loan. More precisely,

- On date $t$, the investment required, thus the value of the portfolio $\phi$, is $V_f(t) = S_t - D*$.
- On date $\tau$, the repayment of the loan in an amount $D$ is ensured by the dividend of the same size paid out at the same instant by the stock $s$ held in the portfolio; no cash flow is either received or disbursed on this date, and the strategy thus remains self-financing.
- On date $T$, the stock $s$ held is worth $S_T$, thus $V_f(T) = S_T$.

In the case when several coupons $D_1, D_2, \ldots, D_N$ are distributed by the underlying between $t$ and $T$ on dates $t_1, t_2,\ldots, t_N$, the strategy $\phi$ consists, on date $t$, in buying the underlying asset $s$ and contracting a loan for $D*$; the loan is repaid on $t_1, t_2, \ldots, t_N$ according to the schedule $D_1, D_2, \ldots, D_N$, synchronized with the dividends: the preceding argument can therefore be applied and the investment of $S_t - D*$ does give $S_T$ on $T$, with no other transfer of cash between $t$ and $T$, and does not change the self-financing character of the strategy.

It is also the possibility of implementing such a strategy that is behind the justification given in the previous chapter of the call-put parity with dividends and amounting to the condition of no arbitrage.

We rewrite this parity equation for further reference:

$$C_t - P_t = (S_t - D*) - K\, e^{-r(T-t)}. \tag{11.24}$$

Adapting the BS model to this situation with discrete coupons is the content of the following proposition:

### Proposition 7 Discrete Dividends
*Consider a European call and put, with identical strike prices $K$ and expiration dates $T$, written on the same spot asset paying dividends between $t$ and $T$ whose present value on date $t$ is denoted by $D*$. For constant risk-free rate $r$ and volatility*

---

[15] When these dividends are not known in advance, it is necessary to define their expected discounted risk-neutral values. Note that this information can be deduced by comparing the spot price to the forward price if the latter is quoted on the market. Cash-and-carry does indeed involve the risk neutral expectation of the dividend expected between the date of quotation and the expiry of the forward contract.

# 11 Options (II): Continuous-Time Models, Black–Scholes and Extensions

$\sigma$, the option prices are given by the Black–Scholes formula (Proposition 2) in which $S_t$ is replaced by $S_t - D^*$:

$$C(t, S_t) = (S_t - D^*) N(d_1) - K e^{-r(T-t)} N(d_2), \tag{11.25}$$

$$P(t, S_t) = K e^{-r(T-t)} N(-d_2) - (S_t - D^*) N(-d_1), \tag{11.26}$$

where

$$d_1 = \frac{\ln \frac{S_t - D^*}{K} + \left(r + \frac{1}{2}\sigma^2\right)(T-t)}{\sigma\sqrt{(T-t)}}; \quad d_2 = d_1 - \sigma\sqrt{(T-t)}. \tag{11.27}$$

**Proof**

Consider a European option, with strike price $K$ and expiry $T$, written on a share of a fund $\phi$ worth $(S_t - D^*)$ at $t$ and $S_T$ at $T$. Its value is given by the BS model, which is applicable since $\phi$ distributes nothing implying that it can be described by (11.25)–(11.27). But the payoff of such an option is identical to that of a European option of the same sort and same strike and expiration date written on $s$. The two options thus have the same value at any time, which shows that an option on $s$ is also priced by (11.25)–(11.27), just as we asserted.

Another argument, which leads to the results of Proposition 7, is more specific on the process followed by the underlying price $S(t)$ and rules out possible inconsistencies. Assume for the moment that a single certain dividend $D$ is distributed on date $\tau$, $t < \tau < T$. It would not be consistent to assume that $S_t$ follows a GBM between $t$ and $\tau^-$ (the eve of the distribution date) because $S(\tau^-)$ would then be lognormal, implying that the event $S(\tau^-) < D$ has some (small but) positive probability, which is incompatible with the assumption that $D$ is certain. The solution to this theoretical difficulty consists in distinguishing two components in the underlying asset price $S(t)$: the first, risk-free, denoted $D^*(t)$, represents the value of dividends expected on $\tau < T$ discounted on $t < \tau$; the second, denoted $S'(t)$, represents the remainder of the stock value that is assumed to obey a GBM with constant volatility $\sigma$. The stock value then is $S(t) = S'(t) + D^*(t)$. From date t onward, $D^* = 0$ and $S = S'$. To interpret such a disentangling, one may imagine that the company buys a risk-free zero-coupon providing $D$ on $\tau$ which is used to pay the dividend and that it invests the rest of its assets in the usual risky activities. The risky component $S'(t)$ follows a GBM and $S(t)$ is always $> D^*(t)$. In presence of several dividends to be distributed before $T$, the previous argument can be easily extended by imagining that the company buys several zero-coupon bonds (one per dividend) each of them being assigned to the payment of the corresponding dividend. The European option's value on date 0, if it expires on any date $T > \tau$, can be computed using only the component $S'(0)$. An analogous argument was developed in Chap. 10 when, to implement the binomial model, we presented a solution for modeling the price evolution of a stock paying a discrete dividend.

## 11.2 Extensions of the Black–Scholes Formula

The reader will also notice a close similarity between this model for discrete dividends with that for continuous dividends. The corresponding formulas are obtained from the standard equations by replacing $S_t$ with $S_t e^{-ct}$ or with $S_t - D^*$, respectively: in both cases, one has to subtract from the stock price $S_t$ the present value of the expected future dividends ($S_t (1 - e^{-ct})$ or $D^*$).

Yet again, let us remark that the validity of the parity given in (11.24) is not subject to any assumption about the evolution of interest rates; note simply that for a non-constant rate the yield curvecost is usually not flat, and so the dividends as well as the strike price have to be discounted using the rate corresponding to the option maturity.

*On the contrary, the pricing model* (11.25–11.27) assumes constant rates, i.e., a flat rate curve, in the standard version we have just explained (cf. Sect. 11.2.6 below for an extension to stochastic rates).

> **Example 3**
> Continue with the same data as in Example 1 (Sect. 11.1.3.3), with an additional payment of a dividend of $14 in 53 days (discounted value = $13.90). The reader will check that the call is not worth more than $27.23 but the put, in contrast, is worth $54.91.

### 11.2.2 Options on Commodities

The case of options on commodities can be easily treated using the Merton model from the preceding subsection. We introduced in Chap. 9 the notion of *convenience yield* denoted by $c$, which expresses at the same time the possible advantage of physically being in possession of the commodity, the cost of storing it and the liquidity of its market (available inventories that may be expected in the future, etc.). The *convenience yield* works, in a technical sense, exactly like a dividend that reduces (if it is positive) the expectation of price growth for the commodity on the spot market (for this reason we have kept the same notation $c$ to denote the two concepts). The dynamics of the price $S$ of a commodity is therefore generally given in the risk-neutral BS universe as the solution to a stochastic differential equation of type (20) namely

$$\frac{dS_t}{S_t} = (r - c)dt + \sigma dW_t,$$

where the risk-free rate $r$, the convenience yield $c$, and the volatility $\sigma$ are all assumed to be constant and $(W_t)_{t \geq 0}$ is a standard Brownian motion under the risk-neutral probability measure $Q$. This model of the law of motion allows us to conclude that Propositions 5 and 6 hold for the case of an option on commodities, interpreting $c$ as the convenience yield.

In any case, note that most options on commodities are written on futures. The result we have just asserted only holds for options on the spot (cash) assets. In practice, an agent will often use Black's formula given in Sect. 11.2.4 below (it will then not take account of the convenience yield).

### 11.2.3 Options on Exchange Rates

Options on exchange rates can also be priced using Merton's formula (Proposition 6). To prove this result, we need to write down the risk-neutral evolution of the exchange rates and to show that this evolution is similar to that of a stock that pays out a continuous dividend. This argument goes back to Garman and Kohlhagen (1983), which explains the name usually given to the pricing formula for exchange rate options, although it is formally identical to Merton's. Consider a spot exchange rate (for cash) whose value on date $t$ we denote by $S_t$. An exchange rate gives the value of a unit of foreign currency in local money (e.g., $S_t$ is the value of $\$ 1$ in Euros). Denote by $f$ the foreign economy and by $d$ the domestic one. Assume furthermore that each economy has a borrowing and lending market on which the corresponding rates, that are assumed constant, are, respectively, $r_f$ and $r_d$. We continue to work in the BS framework to describe the stochastic evolution of $S_t$. With the risk-neutral measure $Q$, the process $(S_t)_{t \geq 0}$ follows a Brownian motion

$$\frac{dS_t}{S_t} = \mu dt + \sigma dW_t. \tag{11.28}$$

The first question is to determine the drift parameter $\mu$ for the exchange rate.

Using the no-arbitrage rule as was done for a dividend paying stock, we build a self-financing portfolio $\phi$ whose value in domestic currency depends explicitly on the exchange rate. We then write that the risk-neutral drift of the portfolio equals $r_d$ in the domestic economy. Consider the following self-financing portfolio $\phi$: on date 0, a unit of domestic currency can be exchanged for a nominal $\frac{1}{S_0}$ in foreign funds. The foreign currency thus obtained is invested at the foreign risk-free rate $r_f$ with reinvestment (capitalization) of the interests.

The value of the portfolio $\phi$ outstanding in foreign currency on date $t$ is therefore $\frac{1}{S_0} e^{r_f t}$. In domestic currency, it is therefore

$$\pi_t = \frac{S_t}{S_0} e^{r_f t}. \tag{11.29}$$

It remains to deduce from Eq. (11.29) the risk-neutral evolution of $\pi_t$. Using Itô's lemma we have

## 11.2 Extensions of the Black–Scholes Formula

$$d\pi_t = \frac{1}{S_0}e^{r_f t}dS_t + r_f \frac{S_t}{S_0}e^{r_f t}dt.$$

Dividing each term in this equation by (11.29) we obtain

$$\frac{d\pi_t}{\pi_t} = \frac{dS_t}{S_t} + r_f dt.$$

Replacing $dS_t$ by its expression in (11.28), we get

$$\frac{d\pi_t}{\pi_t} = \left(\mu + r_f\right)dt + \sigma dW_t.$$

Now using the property that any self-financing portfolio in currency $d$ has a risk-neutral drift of $r_d$ we deduce that

$$\mu + r_f = r_d.$$

This is expressed in the following proposition:

**Proposition 8** *In the risk-neutral universe where the probability measure $Q$ prevails, the exchange rate $S$ (giving the price of currency f in terms of currency d) is governed by the following stochastic differential equation:*

$$\frac{dS_t}{S_t} = (r_d - r_f)dt + \sigma dW_t \qquad (11.30)$$

*with the risk-neutral drift of the exchange rate equal to the difference between the domestic and the foreign rates.*

Intuitively, this result can be explained very simply: an investment in foreign currency yields, in terms of domestic currency, $r_f$ plus the growth in the exchange rate. In the risk-neutral universe, this return (which is stochastic because of the exchange rate) must have an expectation of $r_d$ ; since the expected exchange rate variation equals $\mu$, we necessarily have $r_f + \mu = r_d$, under $Q$.

Notice that the foreign rate works for the exchange rate like the (known) dividend rate does for a stock. Now we can apply the results of Proposition 6, just replacing the parameters $r$ by $r_d$ and $c$ by $r_f$, to obtain the Garman-Kohlhagen result.

### Proposition 9 The Garman-Kohlhagen Model

*Under this subsection's assumptions and notations, the value of a call on the exchange rate $S$ is given by*

$$C(t, S_t) = S_t\, e^{-r_f(T-t)}N(d_1) - K\, e^{-r_d(T-t)}N(d_2), \qquad (11.31)$$

*and that of a* put *by*

$$P(t, S_t) = K e^{-r_d(T-t)} N(-d_2) - S_t e^{-r_f(T-t)} N(-d_1) \qquad (11.32)$$

$$\text{where} \quad d_1 = \frac{\ln\frac{S_t}{K} + (r_d - r_f + \frac{1}{2}\sigma^2)(T-t)}{\sigma\sqrt{T-t}}; \quad d_2 = d_1 - \sigma\sqrt{(T-t)}. \quad (11.33)$$

*In addition, the call-put parity writes*

$$C_t - P_t = S_t e^{-r_f(T-t)} - K e^{-r_d(T-t)} \qquad (11.34)$$

The formulas above are regularly used in foreign exchange markets.

### 11.2.4 Options on Futures and Forwards

Options on futures, and occasionally on forwards, are very frequently used in organized markets. This is especially true for commodities, for which there is practically no market for options written on spot prices (for obvious reasons concerning delivery time delays).

We consider now, without specifying the nature of the underlying asset, a futures contract with maturity $T'$ whose price quoted on date $t$ is denoted $F_{T'}(t)$, or simply $F(t)$.

We first state and then prove a proposition about futures prices $F(t)$. The first proof is rigorous, the second is more intuitive.

**Proposition 10**
*In the risk-neutral universe where the probability Q prevails, the quoted price F(t) of a **futures** contract is a martingale; assuming a constant volatility, we write*

$$\frac{dF(t)}{F(t)} = \sigma dWt \qquad (11.35)$$

*where $(W_t)_{t \geq 0}$ is a standard Brownian motion under Q.*

**First Proof**
Consider a self-financing portfolio $\phi$ containing at any time $t$ some risk-free asset with value $\pi_t$ and a futures contract with maturity $T'$ whose price quoted on date $t$ is $F(t)$. Recall the market value of this contract is a constant zero since the margin calls $dF(t) = F(t+dt) - F(t)$ are assumed to be paid continuously. Positive margins generated by the contract are immediately used to acquire the risk-free asset (lending) and negative margins are covered by selling the risk-free asset (borrowing) in such a way as to retain the self-financing character of portfolio $\phi$. Now $\pi_t$ gives the value at $t$ of portfolio $\phi$ as well as its component invested in the risk-free asset (since the value of a futures contract is zero). As a result, the evolution of $\pi_t$, from $t$ to $t+dt$, can be written

## 11.2 Extensions of the Black–Scholes Formula

$$d\pi_t = r_t\,\pi_t\,dt + dF(t). \tag{11.36}$$

The first component, made up of the interest generated by the risk-free asset, can be calculated using the rate $r_t$ which we *do not have to assume constant, or even deterministic*. The second term gives the margin on the futures contract.

In the risk-neutral universe, the futures price is assumed to be governed by the stochastic differential equation

$$dF(t) = F(t)(\mu_t\,dt + \sigma\,dW_t). \tag{11.37}$$

Proposition 10 [Eq. (11.35)], which we must prove, asserts that $m_t = 0$. From (11.36) and (11.37), we obtain

$$d\pi_t = (\,r_t\,\pi_t + F(t)\mu_t\,)\,dt + \sigma\,F(t)dW_t. \tag{11.38}$$

Recall that $\phi$, whose value is given by $\pi$, is self-financing: its drift under $Q$, therefore, equals $r_t$: $d\pi_t = r_t\,\pi_t\,dt + \sigma'_t\,\pi_t\,dW_t$. Comparison of this last equation with (11.38) then allows us to conclude that $m_t = 0$, which ends the proof of the Proposition.

### Second Proof

Consider the purchase on date $t$ of a futures contract with expiry $T'$ offered at $F(t)$; the transaction produces a margin on $t + dt$ of $dF(t) = F(t+dt) - F(t)$. Since there is no disbursement of funds at $t$, in the risk-neutral universe one should have $E[dF(t)] = 0$, lest the expectation of the yield is $(\pm)$ infinite. By integration, for any $t' > t$, we have:

$$E[F(t')] - F(t) = E\int_t^{t'} dF(u) = \int_t^{t'} E[dF(u)] = 0,$$

which proves the martingale character of $F(t)$. As in the previous proof, the rate $r_t$ might be stochastic.

Thus, a futures price follows a martingale process under the risk-neutral measure. This result holds for a forward price if it is equal to the futures price, i.e. in a world with deterministic rates, but does not when the two prices differ, i.e., when rates are stochastic (see Chap. 9).

One should also remember that the comparison of the evolutions of the futures and spot prices (Eqs. (20) and (11.35)) allows one to consider that a futures contract *behaves like a* spot *asset with continuous dividend distribution* at the rate $c = r$, which cancel the risk-neutral drift.

Now consider a European call (put) on a future, with maturity $T \leq T'$, and a final payoff $\max(F(t) - K, 0)$ ($\max(K - F(t), 0)$ for the put). It is sufficient to apply Propositions 5 and 6, while replacing the parameter $c$ by $r$, to obtain the call-put parity and the Black pricing formula.

# 11 Options (II): Continuous-Time Models, Black–Scholes and Extensions

Recall that the call-put parity presented in the preceding chapter and holding for options on forward contracts writes

$$C_t - P_t = e^{-r(T-t)} [F(t) - K] \qquad (11.39)$$

where $r$ is the interest rate in force at $t$ for transactions with duration $T - t$.

This parity equation applies to forward contracts and does not rely on the assumption of constant interest rates; for futures, it holds only for *non-stochastic* rates (in which case the forward and futures prices are identical).

The standard Black formula given in the proposition which follows does assume constant rates and so applies to both forwards and futures.

**Proposition 11 The Black Model**
*Consider a European call and put, with strike price K, expiration date T, written on a forward or futures contract expiring on T' ($\geq T$) whose price quoted on date t is F (t). The risk-free rate r and the volatility $\sigma$ of the forward or futures price are assumed constant. The values of the call and put are then given respectively, by*

$$C(t, F(t)) = e^{-r(T-t)} [F(t) N(d_1) - K N(d_2)], \qquad (11.40)$$

$$P(t, F(t)) = e^{-r(T-t)} [K N(-d_2) - F(t)N(-d_1)] \qquad (11.41)$$

$$\text{where} \quad d_1 = \frac{\ln \frac{F(t)}{K} + \frac{1}{2}\sigma^2(T - t)}{\sigma\sqrt{T - t}}; \quad d_2 = d_1 - \sigma\sqrt{(T - t)} \qquad (11.42)$$

Note that the two terms for the premiums of the call and the put are discounted at the risk-free rate $r$. Indeed, the rate $r$ plays the role of a dividend paid by a forward/futures contract. One must therefore discount the futures price $F(t)$ in the first term of the call's premium (similarly the second term in the put's premium). We further notice that $r$ has vanished in the definition (11.42) for the quantities $d_1$ and $d_2$. Indeed, the probability density of the futures price on date $T$ (see the integral solution to (11.35)) depends neither on the interest rate nor on the dividend rate paid by the underlying asset. Finally, observe that the value of the option on a futures contract does not depend on whether or not the underlying asset provides a dividend, in contrast to an option on the asset itself, which does. The possible impact of a dividend payment is in fact completely covered by the spot-forward (cash-and-carry) relationship.

It is important to emphasize that:

- In a world with *constant* rates (or more generally, *deterministic* ones[16]), all the preceding results continue to hold for both forwards and futures.

---

[16]If they are deterministic, it is sufficient to replace the term $e^{-r(T-t)}$ with $\exp - \int_t^T r(s)ds$.

## 11.2 Extensions of the Black–Scholes Formula

- In a world with *stochastic* rates, the forward price *is not* a martingale under $Q$, and the evolution (11.35) is not that of a forward price. Moreover, the Black model (like the BS model) must be adapted to stochastic rates. This adjustment will be made in Sect. 11.2.7 below.

---

**Example 4**

Consider a call and a put written on the notional September Eurobund contract $n$, with strike \$101. The September contract is offered at \$100.80, its expiry is in 4 months (122 days) and its volatility is estimated at 6%. The (discrete) 4-month money rate is 5.

Applying the Black formula, we have $C = 1.278$ and $P = 1.474$ and one may check the call–put parity:

$$C - P = (F - K)\, e^{-r(T-t)} = -0.196.$$

---

### 11.2.5 Variable But Deterministic Volatility

The BS formula was derived under the assumption of constant volatility for the underlying asset. Fortunately, it continues to hold, as we now show, for a variable but still deterministic volatility (we assume the evolution of the volatility between $t$ and $T$ is known on date $t$). The reason this works is that the Gaussian character of the log return is preserved. Stochastic volatility poses much more difficulty as we will see in Sect. 11.2.8.

Let us call $s_t$ the stock's volatility on date $t$. For clarity of exposition, and without any real loss in generality, we assume that the stock does not pay any dividend. If this is not the case it is sufficient to replace the risk-neutral drift of $S$ by $(r - c)$. Therefore, we assume that the risk-neutral evolution of $S$ is given by

$$\frac{dS_t}{S_t} = rdt + \sigma_t dW_t. \tag{11.43}$$

The only change from the stochastic differential Eq. (11.13) for $S_t$ in the standard model is that the parameter $s_t$ is not a constant but a deterministic function of time.

We saw above that the premium for a European option is computed as the expectation of its discounted future payoff. We are going to study the probability density of the stock price $S_T$, and compare it to the density we obtained in the constant volatility case. Applying Itô's formula to the process $(Y_t)_{t \geq 0}$ defined by $Y_t = \ln(S_t)$ gives

$$dY_t = \frac{1}{S_t} dS_t - \frac{1}{2} \frac{1}{S_t^2} \sigma_t^2 S_t^2 dt.$$

Upon replacing $dS_t$ with the expression for it from (11.43), we have

$$dY_t = rdt + \sigma_t dW_t - \frac{1}{2}\sigma_t^2 dt$$
$$= \left(r - \frac{1}{2}\sigma_t^2\right)dt + \sigma_t dW_t$$

By integrating this stochastic differential equation from $t$ to $T$, we obtain

$$Y_T - Y_t = \ln\frac{S_T}{S_t} = \int_t^T \left(r - \frac{1}{2}\sigma_s^2\right)ds + \int_t^T \sigma_s dW_s$$
$$= r(T-t) - \frac{1}{2}\int_t^T \sigma_s^2 ds + \int_t^T \sigma_s dW_s.$$

This last expression involves two integrals, $\int_t^T \sigma_s dW_s$ and $\int_t^T \sigma_s^2 ds$. We know (see Chap. 18) that the stochastic integral $\int_t^T \sigma_s dW_s$, if $\sigma_s$ is a deterministic function of $s$, is a centered Gaussian with variance $\int_t^T \sigma_s^2 ds$, just as $\int_t^T \sigma dW_s = \sigma(W_T - W_t)$ was for BS a centered Gaussian of variance $\sigma^2(T-t)$. We thus have obtained the following:

**Proposition 12**
*If the volatility of $S_t$ is a deterministic function $s_t$ of time, the logarithm of the ratio $S_T/S_t$ follows a Gaussian distribution:*

- *with variance $\nu_S = \int_t^T \sigma_s^2 ds$*
- *and mean $r(T-t) - \frac{1}{2}\int_t^T \sigma_s^2 ds$.*

*Denote $\bar{\sigma}^2$ the mean instantaneous variance of the underlying asset between $t$ and $T$ as defined by*

$$\bar{\sigma}^2 = \frac{1}{T-t}\int_t^T \sigma_s^2 ds = \frac{\nu_s}{T-t} \tag{11.44}$$

*so we may write*

$$S_T = S_t e^{\left(r - \frac{\bar{\sigma}^2}{2}\right)(T-t) + \bar{\sigma}\sqrt{T-t}U} \tag{11.45}$$

*where $U$ is a reduced centered Gaussian.*

This proposition tells us that the only difference between the case of a deterministic variable volatility and a constant volatility $\sigma$ is that the *mean volatility*[17] $\bar{\sigma}$ replaces $\sigma$ in the expression for $S_T$ (compare in particular (11.15b) with (11.45)). All

---

[17] To be exact, $\bar{\sigma}$ is the square root of the mean instantaneous variance and not the mean volatility; to call it the mean volatility is an abuse of language.

## 11.2 Extensions of the Black–Scholes Formula

the valuation formulas for options whose payoffs depend on $S_T$ can thus be derived from the formulas for a constant volatility by simply replacing $\sigma$ by $\bar{\sigma}$. These assertions are summed up in the following proposition:

**Proposition 13**
*If the volatility of the underlying asset is not constant but can be considered deterministic, and $\bar{\sigma}$ is the underlying asset's "mean volatility" between $t$ and $T$ as defined in Eq. (11.44), the Black–Scholes models developed in the previous subsection (and described by Propositions 2, 6, 7, 9, and 11) retain their validity if one simply replaces $\sigma$ by $\bar{\sigma}$ in all formulas.*

This result is very intuitive: it is the underlying asset's mean volatility $\bar{\sigma}$ between $t$ and $T$ which determines the variance $v_S = (T - t)\bar{\sigma}^2$ of $\ln(S_T)$, so it is not surprising that it determines the price of a European option that expires on $T$. These results show that the volatility parameter $\sigma$ in BS formulas can rather be considered as quantifying the variance of the logarithm of the price $S_T$ than as a constant parameter that would fix the model for the stock price dynamics. Also remark that the assumption essential to ending up with a BS-like formula is the Gaussian property of $\ln S_T$, and that the significant indicator is the variance of this Gaussian distribution. Notice in any case that the Gaussian property of the price's logarithm only holds when the function $\sigma_t$ depends deterministically on time. Any other model of volatility (e.g., diffusion models with stochastic volatility) departs from the BS formula.

### 11.2.6 Stochastic Interest Rates: The Black–Scholes–Merton (BSM) Model

In this section, we study the influence of the interest rate being stochastic on the BS formula. Just as it is unlikely that the "true" volatility parameter for a stock remains constant over time, the risk-free rate is not stable throughout the life of an option. In fact, the money market obeys the laws of supply and demand and the short-term interest rate evolves continuously and randomly. We now show that the stochastic nature of the interest rate does *not* invalidate the Black–Scholes model. This is a very significant result that demonstrates the robustness and adaptability of the model and indicates how to use it in an appropriate way (especially as far as the rates and volatility chosen are concerned). There are formulas very similar to those of the Black–Scholes model, which make the so-called Black–Scholes–Merton (BSM) model the most general one treated in this chapter. However, one strong assumption still weakens the BSM model: the distribution of $\ln(S_T)$ has to remain Gaussian.

As we have already noted, as long as the interest rate is assumed constant, there is no need to distinguish rates for different maturities. Indeed, a short-term loan at rate $r$ with continuous reinvestment of both capital and interest which provides with certainty $e^{rT}$ in $T$ periods is equivalent to a loan of duration $T$ at rate $r_T$ which gives $e^{r_T}$: under AAO, $r_T = r$ for any maturity $T$ and the yield curve is flat. Such is not the case when the rates are stochastic (since the reinvestments of short-term loans have to be made at rates that are random as seen from the start) and so one has to distinguish the different rates $r_T$. We denote by $B_T(t)$ the price on date $t$ of a zero-coupon bond providing \$ 1 on date $T$ and which can be written, in terms of the rate $r_{T-t}(t)$ for maturity $(T-t)$,

$$B_T(t) = e^{-r_{T-t}(t)\,(T-t)}. \tag{11.46}$$

We also consider the forward price (of the underlying asset) with expiration date $T$ (identical to the expiration of the option being valued). This forward price is denoted by $\Phi_T(t)$ to distinguish it from the futures price $F_T(t)$, since they differ under stochastic rates. If there is no cash distribution by the underlying, whose spot price is written $S_t$, the spot-forward equation allows us to write

$$\Phi_T(t) = \frac{S_t}{B_T(t)}. \tag{11.47}$$

Recall that $F_T(T) = S_T$ (the spot and forward prices coincide at expiration), and that $B_T(T) = 1$.

If we take the volatility $s_t$ of the forward price to be deterministic, we have

$$\frac{d\Phi_T(t)}{\Phi_T(t)} = m_t dt + \sigma_t dWt \tag{11.48}$$

(the drift term $m_t$ obviously depending on the probability under which the dynamics is written).

Finally, use $\bar{\sigma}$ for the mean over $(t, T)$ of the price volatility (either spot or forward[18]) of the underlying stock defined by $\bar{\sigma}^2 = \frac{1}{T-t}\int_t^T \sigma_x^2 dx$, or again

$$\bar{\sigma}^2 = \frac{1}{T-t}[\sigma(\ln S_T)]^2 = \frac{1}{T-t}[\sigma(\ln \Phi_T(T))]^2. \tag{11.49}$$

We are now in a position to present the Black–Scholes–Merton, or BSM, model which is also called the "Gaussian model." This model is very important for both its theoretical foundations and its practical applications. A long proposition and numerous comments will be devoted to it. The BSM model will be taken up again in Chaps. 16 and 17 where we analyze interest rate models and the valuation of interest rate derivatives.

**Proposition 14 The BSM Model**
*Consider a European call and put with the same strike $K$ and expiration date $T$, written on the same spot asset priced at $S_t$. Assume that the option's price at expiry $T$*

---

[18]The instantaneous variances of the forward price $\Phi_T(t) = S(t)/B_T(t)$ and the spot price $S(t)$ differ $(\sigma_\Phi^2(t) = \sigma_S^2(t) + \sigma_B^2(t) - 2\sigma_{SB}(t)$ ), but their average variances between $t$ and $T$ are equal (cf. (11.49)), since $\Phi_T(T) = S(T)$.

## 11.2 Extensions of the Black–Scholes Formula

*is log-normal, and set $\bar{\sigma} = \frac{\sigma(\ln(S_T))}{\sqrt{T-t}}$. So defined, $\bar{\sigma}$ can be interpreted as the average volatility of the logarithm of the price $S_t$ between 0 and T.*

(i) *For a non-paying-out underlying asset, the call and* put premiums *are given by the standard Black–Scholes formulas in which the average volatility of the logarithm of the price, $\bar{\sigma}$, is substituted for the instantaneous volatility, and the zero-coupon rate $r_{T-t}(t)$ with maturity T is substituted for the short-term rate r. Thus, we write*

$$C_t = S_t\, N(d_1) - K\, e^{-r_{T-t}^{(t)\,(T-t)}} N(d_2) \tag{11.50a}$$

*and*

$$P_t = K\, e^{-r_{T-t}^{(t)\,(T-t)}} N(-d_2) - S_t\, N(-d_1) \tag{11.50b}$$

*where*

$$d_1 = \frac{\ln\frac{S_t}{K} + \left(r_{T-t}(t) + \frac{1}{2}\bar{\sigma}^2\right)(T-t)}{\bar{\sigma}\sqrt{T-t}}; \quad d_2 = d_1 - \bar{\sigma}\sqrt{(T-t)}$$

(ii) *For a dividend paying underlying asset, the above formulas remain valid if one replaces $S_t$ by ($S_t$ –* present value *of the discrete* dividends *expected between t and T), or by $S_t\, e^{-c(T-t)}$ for a continuous payout at rate c.*

(iii) *In both cases, one may apply a generalized Black formula (Eqs. 11.40–11.42 written with the rate $r_{T-t}$ instead of r, $\Phi$ instead of F, and $\bar{\sigma}$ instead of $\sigma$):*

$$\begin{aligned} C_t &= e^{-r_{T-t}^{(t)\,(T-t)}} \left[\Phi_T(t)\, N(d_1) - K\, N(d_2)\right]; \\ P_t &= e^{-r_{T-t}^{(t)\,(T-t)}} \left[K\, N(-d_2) - \Phi_T(t)\, N(-d_1)\right] \end{aligned} \tag{11.51}$$

$$\text{with} \quad d_1 = \frac{\ln\frac{\Phi_T(t)}{K} + \frac{1}{2}\bar{\sigma}^2(T-t)}{\bar{\sigma}\sqrt{(T-t)}}; \quad d_2 = d_1 - \bar{\sigma}\sqrt{(T-t)}.$$

### *Proof

*The argument uses a change in numeraire (unit of account) the principle of which is described in this chapter's Appendix 2, and more thoroughly in Chap. 19. It is recommended that the reader who skips this proof pays nonetheless close attention to the remarks that follow it.*

We start from a known result (the generating function for a Gaussian law); if $X$ is a normal random variable, we have

$E(e^X) = e^{0.5(\sigma(X))^2 + E(X)}$, from which follows

# 11 Options (II): Continuous-Time Models, Black–Scholes and Extensions

$$e^{E(X)} = E(e^X)e^{-0.5(\sigma(X))^2} \qquad (11.52)$$

As a result, for any log-normal $Y$

$$Y = E(Y)e^{-0.5(\sigma(\ln(Y)))^2 + \sigma(\ln(Y))U} \quad \text{where } U \text{ is } N(0,1). \qquad (11.53)$$

Indeed, setting $X = \ln(Y)$, (11.52) implies $e^{E(\ln(Y))} = E(Y)e^{-0.5(\sigma(\ln(Y)))^2}$, and so $Y \equiv e^{\ln(Y)} = e^{E(\ln(Y)) + \sigma(\ln(Y))U} = E(Y)e^{-0.5(\sigma(\ln(Y)))^2}e^{\sigma(\ln(Y))U}$, which is (11.53).

(i) A spot underlying asset not paying any dividend between $t$ and the option expiry date $T$.

We use a result of the proposition in Appendix 2 generalizing Proposition 4: there exists a probability distribution for which the value of a self-financing asset denominated in the numeraire $B_T(t)$ (so its forward price) is a martingale. This probability, termed *forward-neutral*, is denoted by $Q_T$ in the following. We can thus write

$$\frac{S_t}{B_T(t)} = \Phi_T(t)$$

$$= E_{QT}\left[\frac{S_T}{B_T(T)}\right] \quad \left(\frac{S_t}{B_T(t)} \text{ is a } Q_T\text{-martingale since } S \text{ does not pay out}\right),$$

from which

$$\frac{S_t}{B_T(t)} = E_{QT}[S_T] \quad (\text{since } B_T(T) = 1). \qquad (11.54)$$

Applying the same principle to the call priced at $C_t$ with expiry $T$, we have

$$\frac{C_t}{B_T(t)} = E_{QT}\left[\frac{C_T}{B_T(T)}\right] = E_{QT}[\max(S_T - K, 0)]. \qquad (11.55)$$

In addition, by virtue of Eq. (11.53) written for the measure $Q_T$ (with $Y = S_T$), we have

$$S_T = E_{QT}(S_T)e^{-0.5(\sigma(\ln(S_T)))^2 + \sigma\ln(S_T)U} = E_{QT}(S_T)e^{-\frac{\bar{\sigma}^2}{2}(T-t) + \bar{\sigma}\sqrt{T-t}\,U}$$

$$\text{with } \bar{\sigma}^2 = \frac{(\sigma(\ln(S_T)))^2}{T-t}.$$

Furthermore, since by (11.54) $\frac{S_t}{B_T(t)} = E_{QT}(S_T)$, we get $S_T = \frac{S_t}{B_T(t)}e^{-\frac{\bar{\sigma}^2}{2}(T-t) + \bar{\sigma}\sqrt{T-t}\,U}$.

## 11.2 Extensions of the Black–Scholes Formula

As a result, (11.55) implies

$$\frac{C_t}{B_T(t)} = E_{Q_T}[\max(S_T - K, 0)]$$

$$= E_{Q_T}\left[\max\left(\frac{S_t}{B_T(t)}e^{-\frac{\bar{\sigma}^2}{2}(T-t)+\bar{\sigma}\sqrt{T-t}U} - K, 0\right)\right]; \quad \text{or}$$

$$C_t = e^{-r}{}_{T-t}{}^{(t)}{}^{(T-t)}E_{Q_T}\left[\max\left(S_t e^{\left(r_{T-t}(t)-\frac{\bar{\sigma}^2}{2}\right)(T-t)+\bar{\sigma}\sqrt{T-t}\,U} - K, 0\right)\right]. \quad (11.56)$$

This result shows that the BS formula (51) remains valid in a context with stochastic rates and a log-normal underlying price, if one uses a zero-coupon of maturity $T - t$ and the average volatility of the price over $(t, T)$ as arguments in the pricing function (compare (11.56) to (11.17)).

(ii) A spot asset with dividend(s) between $t$ and $T$. The arguments developed in the preceding paragraphs for the case with dividend(s) can be applied here, *mutatis mutandis*.

(iii) Black's formula. It is sufficient to replace $\frac{S_t}{B_T(t)} = e^r{}_{T-t}{}^{(t)(T-t)}S_t$ *with the forward price $\Phi_T(t)$ in Eq. (51) to end up with the formula (11.51), which is Black's (Eqs. 11.40–11.42 written with $r_{T-t}$, $\Phi$ and $\bar{\sigma}$ in place of r, F and σ).*

**Important Remarks**

- First of all, it is *necessary*, when one uses the BS formula, to choose as the discount rate the rate for a zero-coupon with maturity $T - t$ (rather than the overnight rate (EONIA) quoted on the money market).
- In fact, the BS formula is unchanged by introducing stochastic rates if one retains the Gaussian nature of the model. The *critical assumption* is the normality of the distribution of the underlying price's logarithm, ln $S_T$. The parameter $\bar{\sigma}^2$ in the BSM formula is related to the variance of $\ln S_T$, $[\sigma(\ln S_T)]^2$, by $[\sigma(\ln S_T)]^2 = (T - t)\bar{\sigma}^2$.
- As a result, the volatility parameter $\bar{\sigma}$ in the BSM formulas should be thought of as quantifying the logarithm of the price at expiry $S_T$ *rather than as a fixed parameter that specifies the stock's volatility in the period between the option's dates of* valuation $t$ and expiry $T$. These remarks naturally lead us to devote the next chapter to deploying the BS formula, for which the choice of the volatility $\bar{\sigma}$ is the cornerstone.
- As indicated in Proposition 14 (*iii*), Eqs. (51) and (11.51) can be written, in an alternative but equivalent way, as a function of the forward price and so can be treated like Black's model.

These same equations hold both with and without pay-out from the underlying asset (the effect of the dividend being incorporated in the forward price).

Finally, let us point out that there exist several versions of the BSM model with stochastic rates which are suited to the valuation of different derivatives on interest rates (BSM-rate, BSM-price, etc.). These versions are described in Chap. 16 which deals with the modeling of interest rates and the valuation of instruments based on them. There we again take up the analysis of the BSM model with new clarifications and numerous applications.

## 11.2.7 Exchange Options (Margrabe)

A type of option that has become popular, and that potentially can involve all types of underlying assets, is the option of exchanging one risky asset (with price $S$) for another (with price $X$). The original valuation of such an option was published, using very heavy technical machinery, by W. Margrabe (1978). The payoff of such a European option, with expiry $T$, writes

$max\ (S_T - X_T, 0)$

where the initial values of both stocks (or assets) are usually normalized so that $S_0 = X_0$. Furthermore, supposing that neither stock pays out a dividend, under the risk-neutral measure $Q$ the price evolution of the two stocks is assumed to be described by

$$\frac{dS_t}{S_t} = r_t dt + \sigma_S(t) dW_S(t) \quad \text{and} \quad \frac{dX_t}{X_t} = r_t dt + \sigma_X(t) dW_X(t)$$

where $r_t$ is the risk-free rate, *which can be stochastic*, the individual volatilities are allowed to be variable and the two Brownian processes $W_S$ and $W_X$ are in general correlated.

The value of such an option is given in the following proposition.

**Proposition 15 Margrabe's Formula**
*Let $S$ and $X$ be two risky assets that do not distribute any dividend from the valuation date to the option maturity.*

(i) *When the volatility $\sigma(t)$ of the ratio $S_t/X_t$ is deterministic and its average value between $t$ and $T$ is $v$, the value of the European option allowing an exchange of $S$ for $X$ is given by a BS-type formula:*

$$C_t = S_t N(d_1) - X_t N(d_2) \tag{11.57}$$

*where*

## 11.2  Extensions of the Black–Scholes Formula

$$d_1 = \frac{\ln \frac{S_t}{X_t} + \frac{1}{2}v^2(T-t)}{v\sqrt{(T-t)}}; \quad d_2 = d_1 - v\sqrt{(T-t)} \qquad (11.58)$$

and

$$v^2 = \frac{1}{T-t}\int_t^T \sigma^2(u)du \quad \text{with} \quad \sigma^2(u) = \sigma_S^2(u) + \sigma_X^2(u) - 2\sigma_{SX}(u)$$

(ii) *If there is some dividend from one or other of the assets, it suffices to replace, in Eqs. (11.57) and (11.58), the current price by the price net of dividends expected between t and T discounted to date t.*

### *Proof

The proof proceeds by a change in numeraire (the reader who wishes to skip this should go directly to the remarks following the proof): we choose $X$ as the numeraire, which is possible since this asset is assumed not to pay out dividends, and consider the probability measure that makes asset prices denominated in this numeraire martingales (see Appendix 2).

- Proof of (i).

The option *payoff* can be written, factorizing out $X_T$, as $C_T = X_T \max\left(\frac{S_T}{X_T} - 1, 0\right)$

Dividing the two sides by $X_T$, we have

$$\frac{C_T}{X_T} = \max\left(\frac{S_T}{X_T} - 1, 0\right).$$

Consider the probability $Q_X$ which makes the price of any asset price denominated in the numeraire $X$ a *martingale*. Under $Q_X$, $Y_t = S_t/X_t$ thus is a martingale and we can write

$$dY_t = Y_t\sigma(t)dW_Y \Leftrightarrow Y_T = Y_t\exp\left(-\frac{1}{2}\int_t^T \sigma^2(u)du + \int_t^T \sigma(u)dW_Y\right)$$

where $W_Y$ is a Brownian motion with respect to $Q_X$, and $\sigma^2(u) = \sigma_S^2(u) + \sigma_X^2(u) - 2\sigma_{SX}(u)$ represents the instantaneous squared volatility of $Y$. This can be rewritten as

$$Y_T = Y_t\exp\left(-\frac{1}{2}(T-t)v^2 + v\sqrt{T-t}\ U\right),$$

where $U$ is $N(0,1)$ and $v$ is the "average volatility" of $Y$ from 0 to $T$ ($v^2 = \frac{1}{T-t}\int_t^T \sigma^2(u)du$).

*Since $C_t/X_t$ is also a $Q_X$-martingale, we have*

$$\frac{C_t}{X_t} = E^{Q_X}\left[\frac{C_T}{X_T}\right] = E^{Q_X}\left[\max\left(\frac{S_T}{X_T} - 1,\ 0\right)\right] = E^{Q_X}\left[\max\left(Y_T - 1, 0\right)\right]$$

$$= E^{Q_X}\left[\max\left(Y_t \exp\left(-\frac{1}{2}(T-t)\nu^2 + \nu\sqrt{T-t}\ U\right) - 1, 0\right)\right].$$

The Black-Scholes pricing formula, therefore, gives (11.57) and (11.58).

- Proof of (*ii*). If either one of the risky assets pays out dividends, it cannot be directly chosen as the numeraire since it is not self-financing. But, using a previously discussed method, it is sufficient to set up the self-financing fund made up of this asset, whose value on date $T$ is $S_T$ (or $X_T$) and whose value at $t$ is $S_t$ (or $X_t$) minus the present value of the dividends expected between $t$ and $T$. This self-financing fund is a suitable numeraire. The preceding argument thus leads to (*ii*).

**Remarks**
- The risk-free rate can be *stochastic* since it is the evolution of the ratio $S_t/X_t$ which matters, and $r_t$ cancels from the numerator and denominator of this ratio.
- If the volatility of the ratio is deterministic, one recovers a formula of BS type: the present price $X_t$ of the asset then replaces the strike price $K$ and the volatility of $S_t/X_t$ replaces that of $S_t$. Indeed, *the exchange option is a generalization of the standard (vanilla) option*, and *Margrabe's formula generalizes BS*; indeed, the vanilla option can be reinterpreted as an option for the exchange of the risky stock ($S_T$) for a sure cash amount ($K$), the latter being the risk-free special case of a risky asset.
- *It is possible to generalize formula (11.57) to the $n$-asset case (which we leave to the reader). Assume the option payoff can be written

$$\max\left(0, \sum_{k=1}^{n} a_k X_k(T)\right)$$

where the $a_k$ are non-zero real numbers and the $X_k$ represent asset prices, whether risky or not. For example, if $n = 2$, $a_1 = 1$, and $a_2 = -1$, we recover the exchange option, and if, in addition, the second asset is worth $K$ with certainty on the expiration date, then we have the standard call. Therefore, the general solution to (11.57) becomes

$$\sum_{k=1}^{n} a_k X_k(t) Q_{X_k}(A) \tag{11.59}$$

where $A$ is the event "the option expires in the money" and $Q_{Xk}$ is the probability measure associated with the numeraire $X_k$.

## 11.2.8 Stochastic Volatility (*)

After a short discussion of why a stochastic volatility model is generally needed in practice, we present Heston's model (1993) and briefly mention an alternative for fixed-income underlying assets. In contrast to the case of variable but deterministic volatility studied in Sect. 11.2.5, the Gaussian property is not maintained if the volatility is stochastic, which makes the subsection much more difficult than the previous ones.[19] Furthermore, we will not detail the proofs.

### 11.2.8.1 Justification for the Model

We saw in Sect. 11.2.5 that the BS formula remains valid for a variable but deterministic volatility (assuming the entire evolution of the volatility between $t$ and $T$ to be known on date t) with the essential point being that log-normality of the underlying's price is maintained.

In actual markets, this assumption is generally not valid. First, recall that the empirical distributions of log-returns do not conform to a Gaussian model: they exhibit thick tails (the leptokurtosis phenomenon according to which there are too many observations around the mean and, above all, too many extreme events observed in both tails) and are, in addition, often asymmetric, the probability of observing severe negative returns being greater than that of very high returns. Also, notice that the time evolution of the volatility often seems to be erratic.

We will examine here the first aspect (thick tails) which unfortunately calls into question the very convenient Gaussian framework.[20] The simplest way of taking this into account is, for a given underlying and a *fixed maturity T*, to calculate the volatility of the underlying implicit in the market price of the option. The idea is very simple and will be explored further in the next chapter. Instead of using the BS formula to find the price of an option from (known values of $S$, $K$, $r$, $T$ and) an estimated volatility $\sigma$, we "reverse" the formula to find, based on the option's market price (and $S$, $K$, $r$, and $T$), the corresponding volatility $\sigma_i$ anticipated by the market, which we then call the implied volatility (implied by the option's market price).

According to BS theory, for a given maturity, the implicit volatility (of the underlying) should be the same for all quoted strikes. In practice, the curve representing the volatility as a function of the strike is far from being flat and has a slope and convexity, which differ according to the type of underlying asset. This curve is generically called a *smile*.[21] This smile, or skew, is caused by the fact that the market "knows" that the probability of a crash event (a big loss, or, less

---

[19] We strongly encourage the reader to read, before studying this section, the next Chap. 12 (option strategies, Sect. 12.2) and parts of Chaps. 19 and 20 (theoretical framework, especially the notion of an incomplete market, and the related mathematical tools).

[20] In this sense, this section cannot be considered an extension of the standard BSM model since, for the first time, we leave the Gaussian framework.

[21] Or *skew*, if the smile is not symmetric. In the following chapter, we find a detailed discussion of this phenomenon as well as of that called the "term structure of volatilities" which is observed by fixing the strike $K$ and varying the expiry date $T$ of the options (written on the same underlying).

438    11 Options (II): Continuous-Time Models, Black–Scholes and Extensions

frequently, a huge gain) is much larger than would follow from a normal distribution. The market thus corrects the model by using a different volatility for in- and out-of-the-money options than that used for at-the-money options. This systematic observation *is therefore not compatible* with the BS model, as it suggests the price distribution for the underlying is not log-normal.

Amongst numerous attempts to explain the smile, aside from the inevitable imperfections of the market,[22] one finds the introduction of jumps in the underlying's price motion, the use of mixtures of Gaussians, or of Lévy processes, local volatility models,[23] and stochastic volatility models which are the subject of this section.

Models with stochastic volatility assume generally that the volatility of the underlying asset (or rather its square, the variance) on the one hand follows a stochastic mean-reverting process,[24] and on the other hand, is correlated with the relative variations of the underlying's price. Formally, this can be written

$$dS_t = S_t\left(\mu_t dt + \sqrt{V} dW_t^1\right) \tag{11.60}$$

$$dV_t = k(\theta - V_t)dt + \sigma V_t^\alpha dW_t^2 \tag{11.61}$$

where $V$ is the variance, the correlation $\rho$ between the two Brownian processes is assumed constant, $\theta$ is the long-term mean of the variance, $k$ the speed of reversion to the mean, and $\sigma$ the volatility of the variance process, known under the (imprecise but suggestive) name "vol of vol," or more fully as "volatility of the volatility." The models are distinguished essentially by the value they assign to a coefficient.

Hull and White (1987) studied the general case for an arbitrary positive $\alpha$, and gave a closed-form but approximate solution to valuing an option for the very special case in which not only is $\alpha$ equal to 1, but both $k$ and r vanish. The variance process is then none other than a geometric Brownian motion without drift and uncorrelated with the underlying's price movements. Since the model is not very realistic and Heston's model (1993) fits better with empirical observations, we will explain only the latter which is in favor of professionals.

### **11.2.8.2 The Heston Model (1993)**

The Heston model corresponds to $\alpha = 0.5$ with arbitrary $k$, $\theta$ and $\rho$, which implies it is very flexible. We, therefore, have the system of equations

---

[22] Such imperfections involve, for example, the price spreads quoted by market-makers, the existence of minimum increments for quotes (ticks), the sometimes imperfect synchronization of prices recorded for options and their underlyings, etc. For details, see Hentschel (2003).

[23] See, for example, Dupire (1994) and Gatheral (2006).

[24] This assumption is the most plausible; a very "nervous" market will become calmer, and a "quiet" market wakes up more or less abruptly some day. For more details, see the next chapter.

## 11.2 Extensions of the Black–Scholes Formula

$$dS_t = S_t\left(\mu_t dt + \sqrt{V_t}dW_t^1\right) \tag{11.62}$$

$$dV_t = k(\theta - V_t)dt + \sigma\sqrt{V_t}dW_t^2.$$

Before formulating the result, it is important to pay attention to the fact that in stochastic volatility models, the market is *incomplete* (see Chap. 19, Sect. 19.1). The reason for this is that there are two sources of risk (modeled by Brownian motions) and only the underlying asset (and the risk-free asset) is available to replicate the option. Even if the market price of the risk associated with the underlying vanishes, as in the BS model, the market price of the volatility risk remains in the valuation equation. If there existed a tradable asset perfectly correlated with the variance of the underlying's price, the market would then become complete and the market price of the volatility risk would not need to be specified in order to price the option.

It is possible to show by constructing an arbitrage portfolio from two different options written on the underlying that the Black–Scholes PDE (11.7), if the volatility (squared) satisfies (11.62), can be generalized for an option with value $C$ to

$$\frac{\partial C}{\partial t} + \frac{\partial C}{\partial S}rS + \frac{1}{2}\frac{\partial^2 C}{\partial S^2}VS^2 + \rho\sigma VS\frac{\partial^2 C}{\partial V\partial S} + \frac{1}{2}\frac{\partial^2 C}{\partial V^2}\sigma^2 V - rC$$

$$= (k(V - \theta) - \phi)\frac{\partial C}{\partial V} \tag{11.63}$$

where $\phi$ is the market price of the volatility risk (MPVR) which is unknown and has to be estimated.

Amongst the plausible candidates for a function giving the MPVR, Heston chose one that on the one hand allows for a closed-form solution, and on the other is compatible with financial theory (see Chaps. 19–22) in that this price is proportional to the variance $V$ of the underlying's return.[25]

Thus choosing the MPVR as Heston does, $\phi = \lambda V$, and defining $k' \equiv k - \lambda$ and $k'\theta' \equiv k\theta$, we have

$$\frac{\partial C}{\partial t} + \frac{\partial C}{\partial S}rS + \frac{1}{2}\frac{\partial^2 C}{\partial S^2}VS^2 + \rho\sigma VS\frac{\partial^2 C}{\partial V\partial S} + \frac{1}{2}\frac{\partial^2 C}{\partial V^2}\sigma^2 V - rC$$

$$= (k'(V - \theta'))\frac{\partial C}{\partial V}. \tag{11.64}$$

To simplify this expression, and in particular to remove the interest rate, it is convenient to use the forward price $F_{t,T} = S_t e^{r(T-t)}$ and to define $x_t = \ln(F_{t,T}/K)$ and $t = (T - t)$. Applying Itô's Lemma to $F_{t,T}$ and looking at the option's forward

---

[25] In theory we have to solve a model of general equilibrium built from investors' optimization programs based on their utility functions. In fact, it is the only way to value primitive assets, which cannot be valued from the no arbitrage rule alone, since they are primitives and not derivatives.

440      11 Options (II): Continuous-Time Models, Black–Scholes and Extensions

value (projected on date $T$) $D$, $D = C_t e^{r(T - t)}$, to get rid of $r$,[26] Eq. (11.64) becomes

$$-\frac{\partial D}{\partial t} - \frac{1}{2}\frac{\partial D}{\partial x}V + \frac{1}{2}\frac{\partial^2 D}{\partial x^2}V + \rho\sigma V\frac{\partial^2 D}{\partial V\partial x} + \frac{1}{2}\frac{\partial^2 D}{\partial V^2}\sigma^2 V$$
$$= (k'(V - \theta'))\frac{\partial D}{\partial V}. \qquad (11.65)$$

Duffie et al. (2000) showed that the solution to this partial differential equation has the form

$$D(x, V, \tau) = K[e^x P_1(x, V, \tau) - P_2(x, V, \tau)] \qquad (11.66)$$

where $P_1$ and $P_2$ are probabilities. From the definition of $x$ as $\ln(F/K)$, we recover *exactly* the same structure as in the BSM solution (what other structure as simple as this could one imagine for the price of an option?).

What comes next is technically demanding, and especially so is the use of the complex Fourier transform, which is needed if one is to escape the Gaussian framework, since one can no longer compute the values of the two integrals (one for $x$ and the other for $K$) needed to get an option price using the normal law's density function.[27] We give directly the final result as a proposition.[28]

**Proposition 16 Heston's Model**
*The value at $t = 0$ of a European call written on a* spot *asset paying no dividends and whose volatility is stochastic, is given by*

$$C_0 = S_0 P_1(x, V, \tau) - Ke^{-rT}P_2(x, V, \tau) \qquad (11.67)$$

*where* $P_j(x, V, \tau) = \frac{1}{2} + \frac{1}{\pi}\int_0^\infty \text{Re}\left\{\frac{\exp\left(A_j(\kappa, \tau) + B_j(\kappa, \tau)V + i\kappa x\right)}{i\kappa}\right\}d\kappa$    for $j = 1, 2$,

*Re(f) is the real part of the function f, and i is the complex number such that $i^2 = -1$,*

---

[26] To get some insight into this simplification, look at the Black formula (11.40) or (11.41) and multiply it by $e^{r(T - t)}$. There is then no longer any interest rate on the right-hand side of the resulting equation.

[27] The Fourier transform of a function $f(x, V, t)$ is $g(\kappa, V, \tau) = \int_{-\infty}^{\infty} e^{-i\kappa x} f(x, V, \tau)dx$, where $i$ is such that $i^2 = -1$.

[28] Heston's proof is elliptic. The interested reader will find a fuller proof, and a discussion of the model, in the book by Gatheral (2006).

## 11.2 Extensions of the Black–Scholes Formula

$$A_j(\kappa, \tau) = r\kappa i\tau + \frac{a}{\sigma^2}\left\{(b_j - \rho\sigma\kappa i)\tau - 2\ln\left[\frac{1 - g_j\exp(d_j\tau)}{1 - g_j}\right]\right\}$$

$$B_j(\kappa, \tau) = \frac{b_j - \rho\sigma\kappa i + d_j}{\sigma^2}\left[\frac{1 - \exp(d_j\tau)}{1 - g_j\exp(d_j\tau)}\right]$$

$$g_j = \frac{b_j - \rho\sigma\kappa i + d_j}{b_j - \rho\sigma\kappa i - d_j}, d_j = \sqrt{(\rho\sigma\kappa i - b_j)^2 - \sigma^2(2u_j\kappa i - \kappa^2)}$$

and $u_1 = 1/2$, $u_2 = -1/2$, $a = k\theta$, $b_1 = k + \lambda - \rho\sigma$ and $b_2 = k + \lambda$.

### Remarks

- The model's inputs, which must be obtained by calibration, are
  - $\sigma$, the vol of vol
  - $\rho$, the correlation between the two Brownian motions
  - $b_2 (= k + \lambda)$, the speed $k$ of mean-reversion and the proportionality coefficient $\lambda$ of the MPVR not needing to be separately estimated
  - $\sqrt{V_0}$, the initial volatility of the underlying's price, and
  - $a (= k\theta)$, the product of the speed of mean-reversion by the long-term value of the variance of the underlying's price.
- The two integrations necessitate resorting to standard numerical methods and can be very quickly done, *once* the parameters have been found by calibration from vanilla options quoted on the market.
- The put price is obtained from the standard call-put parity.
- If the underlying asset distributes a coupon (dividend) or is an exchange rate, one needs to replace $S_0$ by $S_0 e^{-cT}$ and $r$ by $(r - c)$ in the expressions for $A_j(.)$, where $c$ is the coupon rate or foreign risk-free rate.
- If the underlying asset is a forward contract, one must replace $S_0$ by $F_{0,T}e^{-(r-c)T}$.
- It is possible to show that the form of the smile depends crucially on the correlation parameter $\rho$. For example, since the smile is decreasing for stocks and stock indices (a skew), it is necessary to introduce a negative $\rho$ in the model to obtain this profile (one needs $\rho$ to be zero to have a symmetric smile). There is in fact a negative correlation on these markets between the "vol" and the "vol of vol" so that the smile is generally a decreasing skew.
- One of the model's advantages is that its derivatives with respect to $S$ and $V$ (which are "Greek" parameters; see the next chapter) can easily be found since the functions $A(.)$ and $B(.)$ are independent of $S$ and $V$.

In spite of its merits and being more realistic than the BSM model, this model has two main limitations:

- It is above all useful for rather short-lived options (admittedly the most common sort), typically with a maturity of a year or less, as by the central limit theorem the log-return processes tend in the long term to become Gaussian again, which noticeably flattens the smile for long options, contrary to observations.[29]

- For short enough options, one may obtain an excellent "fit" with the quoted market prices because of the large number of parameters to be calibrated (by definition, the more parameters (correctly) calibrated a model has, the easier it is to get a better fit); however, to find the right calibration is rather difficult and tedious.

### 11.2.8.3 An Alternative Model

An alternative model to Heston's is SABR (an acronym for Stochastic, Alpha, Beta, and Rho), due to Hagan et al. (2002), which allows obtaining *volatility skews* whose form and dynamics are realistic for interest rate instruments.

The authors postulate the following for the forward price $F$ and its volatility:

$$dF_t = \alpha_t F_t^\beta dW_t^1$$

and

$$d\alpha_t = \nu \alpha_t dW_t^2$$

with a correlation $\rho$ between the two Brownian motions.

This represents the simplest stochastic volatility model which is homogeneous in $F$ and $\alpha$. The significant parameters are $\alpha$, $\beta$ and $\rho$, from which the model's name is derived. Notice that the choice $\beta = 0$ would correspond to the Gaussian case and $\beta = 1$ to the log-normal model.

SABR leads to closed-form solutions for option prices, and for their implied volatilities as functions of price $F$ and strike $K$.

In contrast to the Black model for which the smile is always flat (though its level can vary), and to local volatility models which predict that when the price of the underlying drops (resp. rises) the smile is translated up (resp. down), one observes the opposite movement on the market for interest rate instruments. SABR is able to reproduce a level (which is easy) and a dynamics (which is difficult) for the skew which fit rather well empirical observations.

The appendices of this chapter further develop the material used in Sect. 11.1.4 above. The reader unfamiliar with probability theory may skip them. A more systematic and complete exposition of these questions is given in Chap. 19.

---

[29] One does not have this problem with models based on Lévy processes (for example, the "Finite Moment Log-Stable" type) since, as such processes do not have a finite second moment, the central limit theorem does not apply.

## 11.3 Summary

- The standard Black–Scholes (BS) model is adapted for pricing European options written on *spot assets not distributing* dividends or coupons before the option expiration. The evolution of the underlying price $S_t$ obeys a geometric Brownian motion, with a constant *volatility* $\sigma$, and an arbitrary drift (expected growth rate) $\mu$. The risk-free rate $r$ is also assumed constant.
- The BS model is grounded on the theoretical possibility of designing a *self-financing dynamic strategy* that replicates the option's payoff on the expiration date $T$. The replicating portfolio is a continuously rebalanced combination of the underlying and the risk-free asset.
- In absence of arbitrage, at any instant, the option's value synthesized in this way should be equal to that of the replicating portfolio. This implies a no-arbitrage condition that takes the form of a partial differential equation whose solution is the BS pricing formula.
- The pricing function BS gives the option value $O_t$ as a function of $S_t$, $r$, $\sigma$, $K$, and $(T\text{-}t)$: $O_t = BS(S_t, r, \sigma, K, T\text{-}t)$
- Counter-intuitively, the expected growth rate (drift) $\mu$ of the underlying's price is absent from the BS formula.
- This suggested that the BS formula must also hold in a virtual *risk-neutral universe*, in which *the expected returns are all equal to the risk-free rate*, such that all asset prices can be simply computed as *expected payoffs (or future values) discounted at the risk-free rate*.
- The probability $Q$ for future events in the risk-free world differs from the probability prevailing in the real world. There is a strict parallelism with the results obtained in the discrete time model (*risk-neutral probability q* vs historical (or true) probability $p$).
- In the risk-free world the discounted values of all self-financing assets and portfolio, risky or not, follow martingales: current values are $Q$-expectations of future discounted values. The martingale approach leads to powerful tools for pricing contingent claims.
- BS formula has been extended for pricing options on underlying assets other than dividend-free stocks.
- When the underlying asset pays, before $T$, discrete dividends whose present value is $D^*$, the option premium is given by the BS formula where $S_t$ is replaced by $S_t$ - $D^*$: $O_t = BS(S_t - D^*, r, \sigma, K, T\text{-}t)$.
- For underlying spot assets paying a dividend at a continuous rate $c$, pricing formulas are derived from those of Black–Scholes by replacing $S_t$ by $S_t e^{-c(T-t)}$: $O_t = BS(S_t e^{-c(T-t)}, r, \sigma, K, T\text{-}t)$ (Black–Scholes–Merton model).
- The Black–Scholes–Merton (BSM) model applies to options on other types of underlying assets such as commodities ($c$ then is the *convenience yield* representing the possible advantage of disposing of the commodity physically net of the storing cost) and currencies ($c$ then is the foreign interest rate, and $r$ the domestic rate).

- In the risk-neutral universe, the quoted price $F(t)$ of a *futures* contract is a martingale and behaves like a spot asset with continuous dividend distribution at the rate $c = r$. This implies that an option on futures can be priced by the BSM formula where $c$ is replaced by $r$: $O_t = BS(S_t e^{-r(T-t)}, r, \sigma, K, T\text{-}t)$ (Black model).
- If the volatility $\sigma(t)$ of the underlying asset is not constant but still is assumed deterministic, BSM models retain their validity if one simply replaces in all formulas $\sigma$ by the *square root of the average variance* expected between $t$ and $T$ (often called *average volatility* by an abuse of language).
- BSM formulas remain valid with *stochastic rates and a log-normal underlying price*, if one uses a zero-coupon of maturity $(T - t)$ and the average volatility over $(t, T)$ as arguments in the BS formula. Known future dividends must be discounted with the relevant zero-coupon rates (Gaussian model).
- An option of *exchanging* one risky asset (with price $S$) for another (with price $X$) at date $T$ yields a payoff $max\ (S_T - X_T, 0)$. It can be priced by a BSM-type formula, where $X$ corresponds to $K$ and the volatility of $S_t/X_t$ to $\sigma$ (Margrabe model).
- If, more realistically, the *volatility is considered stochastic* the Gaussian property is not maintained and more complex models like Heston's or SABR must be used.

## Appendix 1

## Historical and Risk-Neutral Probabilities and Changes in Probability

Here we provide details on the definition of the risk-neutral measure $Q$ using the Radon-Nikodym derivative $\frac{dQ}{dP}$ and Girsanov's Theorem (see Chap. 19 for more rigor and details).

We introduced in Sects. 11.1.1–11.1.4 the new probability measure that we denoted by $Q$. With this probability measure, the stock price dynamics differs from that of the reference model (11.1) to be described by (11.13): the expected growth rate of the price (that is, the expected return), $\mu$, is replaced by the risk-free rate $r$. Thus the stock, under the measure $Q$, does not compensate for its risk (i.e., its volatility $\sigma$) with an expected risk premium (or excess return). We have in addition remarked that, under $Q$, the expected growth rate of the option value is also equal to the risk-free rate $r$. Indeed, generally, under $Q$, the expected return rate of *any* risky asset equals $r$. This means that, *in the universe where the probability measure $Q$ prevails, all risk premiums vanish*. This probability measure is for that reason called the risk-neutral measure. One can make the relation between the probability measures $P$ and $Q$ clearer by introducing the market price of risk, which we denoted by $\lambda$ in the discrete case, that increases the expected return rate on the stock above the risk-free rate $r$, in proportion to its volatility. Therefore, we use the notation

# Appendix 1 445

$$\lambda = \frac{\mu - r}{\sigma}. \tag{11.68}$$

## Proposition

*The risk-neutral measure $Q$ introduced above is related to the physical (historical) measure $P$ by the Radon–Nikodym derivative*

$$\frac{dQ}{dP} = e^{-\lambda W_t - \frac{\lambda^2}{2}t}. \tag{11.69}$$

## Proof

This proposition is a direct application of Girsanov's Theorem (see Chaps. 18 and 19). From the definition of $\lambda$ we deduce

$$\mu = r + \lambda\sigma.$$

Returning to expression (11.1), we obtain under $P$:

$$\frac{dS_t}{S_t} = \mu dt + \sigma dW_t = rdt + \sigma(dW_t + \lambda dt).$$

Let us check that this stochastic differential equation is equivalent to Eq. (11.13) describing $S$ under $Q$. Indeed, by an application of Girsanov's theorem, we know that the process $\left(W_t^Q\right)_{t \geq 0}$ defined by $dW_t^Q = dW_t + \lambda dt$ is a standard Brownian motion for the measure $Q$ as defined in (11.69). Thus the preceding stochastic differential equation becomes under $Q$:

$$\frac{dS_t}{S_t} = rdt + \sigma dW_t^Q.$$

Proposition 4 in the body of the text expresses the fact that the value of any self-financing portfolio discounted at rate $r$ has the same $Q$-martingale property. In particular, this is so for the self-financing portfolio replicating the option, and thus for the option itself. Just as in the discrete case the pricing of options depends on the risk-neutral probability of an "up" denoted by $q$, in the continuous case calculating the option premium can be interpreted as taking the expectation of the discounted future payoff with respect to the risk-neutral probability measure $Q$. This probability measure plays an essential role in option theory. It is also called the "pricing probability." We show in the following Appendix 2 that this property can be broadened to other pricing probabilities, associated with other ways of discounting (see also Chap. 19).

# Appendix 2

## Changing the Probability Measure and the Numeraire

Here we present the main results of the "numeraire theory," which will allow considerable simplification in the computation of option premiums. Chapter 19 presents this theory more fully.

### Definition and Examples

**Definition** *A numeraire is a self-financing portfolio whose value is strictly positive, surely and constantly. Any numeraire can be chosen as the unit for measuring values.*

A natural numeraire is the domestic currency capitalized at the risk-free rate, i.e. the risk-free asset denoted by $\beta$. If $\pi_t$ denotes the monetary value of a self-financing portfolio or asset, its value denominated in the numeraire $\beta$ can be written, assuming a constant rate $r$,

$$\frac{\pi_t}{\beta_t} = e^{-rt}\pi_t.$$

As a result, values denominated in the numeraire $\beta$ represent discounted values, and when the numeraire $\beta$·is selected, the value of any self-financing portfolio or asset denominated in this numeraire is a $Q$-martingale.

Among other numeraires that could be chosen, we find: any zero-coupon bond, a stock (provided that it does not pay out dividends), a stock index (with the same restriction), or a foreign currency.

### Existence of a Martingale Measure for Each Numeraire

The next proposition gives the correspondence between a numeraire $N$ and a probability measure $Q_N$ that makes asset prices martingales if they are denominated in the numeraire $N$.

### Proposition

*To each numeraire $N$ with value $N_t$ on date $t$ is associated a martingale probability measure that we denote by $Q_N$ such that the value of any self-financing portfolio or asset denominated in the numeraire $N$ is a martingale with respect to $Q_N$. More formally, to each numeraire $N$ corresponds a probability measure $Q_N$ such that $\frac{\pi_t}{N_t}$ is a $Q_N$-martingale for any $\pi_t$ representing the monetary value of a self-financing asset.*[30]

*$Q_N$ is related to the risk-neutral measure $Q$ by the Radon-Nikodym derivative*

---

[30] In fact, the existence of $Q_N$ requires the absence of arbitrage opportunities, and its uniqueness requires that the market is complete.

# Appendix 2

$$\frac{dQ_N}{dQ} = \frac{N_T}{N_t} \frac{\beta_t}{\beta_T}. \tag{11.70}$$

**Proof**

The risk-neutral probability measure $Q$ is in a natural way associated to the numeraire $\beta$. We can now identify for another numeraire $N$ a probability measure that makes the values $\frac{\pi_t}{N_t}$, recorded in units of the numeraire $N$, martingales. We want to define a probability measure $Q_N$ such that for every $T > t \geq 0$ and for any $(\pi_t)_{t \geq 0}$ giving the value of a self-financing portfolio, we have

$$\frac{\pi_t}{N_t} = E^{Q_N}\left[\frac{\pi_T}{N_T} | S_t\right]. \tag{11.71}$$

Since $\pi_t$ does represent the value of a self-financing portfolio we know that

$$\frac{\pi_t}{\beta_t} = E^Q\left[\frac{\pi_T}{\beta_T} | S_t\right].$$

Multiplying each term of this last equation by $\frac{\beta_t}{N_t}$, we obtain

$$\frac{\pi_t}{N_t} = E^Q\left[\frac{\pi_T}{\beta_T} \frac{\beta_t}{N_t} | S_t\right].$$

Defining $\frac{dQ_N}{dQ} = \frac{\beta_t}{\beta_T} \frac{N_T}{N_t}$, this becomes

$$\frac{\pi_t}{N_t} = E^Q\left[\frac{\pi_T}{N_T} \frac{dQ_N}{dQ} | S_t\right].$$

Noticing that $E^Q\left[\frac{\pi_T}{N_T} \frac{dQ_N}{dQ} | S_t\right] = E^{Q_N}\left[\frac{\pi_T}{N_T} | S_t\right]$, by the definition of a change of the probability measure from $Q$ to $Q_N$, we check that Eq. (11.71) holds. Thus we obtain simultaneously that $\left(\frac{\pi_t}{\beta_t}\right)_{t \geq 0}$ is a martingale under $Q$ and $\left(\frac{\pi_t}{N_t}\right)_{t \geq 0}$ is a martingale under $Q_N$, if $Q_N$ is defined by $\frac{dQ_N}{dQ} = \frac{\beta_t}{\beta_T} \frac{N_T}{N_t}$. The proof will be complete if we verify that the random variable $\frac{\beta_t}{\beta_T} \frac{N_T}{N_t}$, chosen to be the Radon-Nikodym derivative, is positive and has expectation 1 under $Q$. Recall that from the definition of a numeraire $N_t$ is, on the one hand, a self-financing portfolio, and, on the other, positive (which ensures the positivity of $\frac{\beta_t}{\beta_T} \frac{N_T}{N_t}$). Furthermore, since it is self-financing it has the property shown in Proposition 4 that $\left(\frac{N_t}{\beta_t}\right)_{t \geq 0}$ is a martingale under $Q$. Therefore, we have for all $T > t \geq 0$:

$$\frac{N_t}{\beta_t} = E^Q\left[\frac{N_T}{\beta_T}|S_t\right].$$

So multiplying each term by $\frac{\beta_t}{N_t}$ we have

$$E^Q\left[\frac{N_T}{\beta_T}\frac{\beta_t}{N_t}|S_t\right] = 1,$$

which ends the proof.

It is sometimes convenient to use probability measures different from $P$ and from $Q$ to simplify calculations. It will always be interesting to examine the numeraires associated with these new measures, for they have a financial interpretation that clarifies the computations and guides the intuition.

### Application to the Numeraire S
Let us choose as numeraire a stock $S$, priced at $S_t$, which does not distribute dividends from 0 to $T$.

The probability measure $Q_S$ is defined, starting from $Q$, by its Radon–Nikodym derivative $\frac{dQ_S}{dQ} = \frac{\beta_t}{\beta_T}\frac{S_T}{S_t}$.

We know that

$$\frac{\beta_t}{\beta_T} = e^{-r(T-t)}$$

and from Eq. (11.15a)

$$\frac{S_T}{S_t} = e^{\left(r-\frac{1}{2}\sigma^2\right)(T-t)+\sigma\left(W_T^Q-W_t^Q\right)}. \tag{11.72}$$

As a result, we have

$$\frac{dQ_S}{dQ} = \frac{\beta_t}{\beta_T}\frac{S_T}{S_t} = e^{-\frac{1}{2}\sigma^2(T-t)+\sigma\left(W_T^Q-W_t^Q\right)}. \tag{11.73}$$

This result helps us interpret below the BS formula.

# Appendix 3

## Alternative Interpretations of the Black–Scholes Formula

In this appendix we will again take up the proof of the BS formula given in Sect. 1, §4, by selecting at each step in the calculation well-chosen probability measures that allow us to interpret the coefficients $N(d_1)$ and $N(d_2)$ in the BS formula in terms of probabilities for the exercise of the option.

Therefore, let us return to the computation of the expectation of the discounted cash flow of a call whose payoff is $\max(S_T - K, 0)$ and whose value on date $t$ is $E^Q[\max(S_T - K, 0)\, e^{-r(T-t)} \mid S_t]$. We saw that this expectation can be split into a sum of two terms, the first of which is proportional to $S_t$ and the second to $Ke^{-r(T-t)}$. We can recover this decomposition, without using the expression written with integrals, by introducing the indicator function of the set of $S_T$ values for which the option will be exercised at $T$. We use the notation

$$\mathbf{1}_{S_T \geq K} = 1 \quad \text{if } S_T \geq K$$

$$= 0 \quad \text{otherwise.}$$

We use the following property that holds for any probability measure $Q_x$:

$$E^{Q_x}[\mathbf{1}_{S_T \geq K}] = \int_{S_T > K} dQ_x = Q_x(S_T \geq K), \tag{11.74}$$

that is the probability of exercising the option under $Q_x$.

Now we can write the maximum appearing in the call's payoff as the sum of two terms:

$$\max\ (S_T - K, 0) = S_T\, \mathbf{1}_{S_T \geq K} - K\, \mathbf{1}_{S_T \geq K},$$

since if the maximum is zero then so is the indicator function. Applying Proposition 3, one has therefore to compute the sum of two expectations

$$C(t, S_t) = E^Q[S_T\, e^{-r(T-t)}\, \mathbf{1}_{S_T \geq K} \mid S_t] - E^Q[Ke^{-r(T-t)}\, \mathbf{1}_{S_T \geq K} \mid S_t].$$

We see here the usual two terms which can be considered separately and about which we know that the first equals $S_t\, N(d_1)$ and the second $Ke^{-r(T-t)}\, N(d_2)$. First, examine the easier second term: since $Ke^{-r(T-t)}$ is deterministic, it can be pulled out of the expectation to give

$$Ke^{-r(T-t)}\, E^Q[\mathbf{1}_{S_T \geq K} \mid S_t].$$

Using property (11.74) we see that

$E^Q[\mathbf{1}_{S_T \geq K} \mid S_t] = Q(S_T \geq K \mid S_t)$, that is, the probability of exercising the option under $Q$.

Therefore, for the second term, we obtain

$$E^Q\left[Ke^{-r(T-t)}\mathbf{1}_{S_T \geq K} \mid S_t\right] = Ke^{-r(T-t)}Q(S_T \geq K \mid S_t).$$

It is now enough to recall that $Q(S_T \geq K \mid S_t) = N(d_2)$ to recover an interesting result we have already mentioned: *the term $N(d_2)$ which multiplies $Ke^{-r(T-t)}$ in the Black-Scholes formula is the risk-neutral probability of exercising the call.*

Now let us consider the first term: $E^Q[S_T e^{-r(T-t)}\mathbf{1}_{S_T \geq K} \mid S_t]$.

Replacing $S_T$ by the solution given in Eq. (11.72) and simplifying, we obtain

$$E^Q\left[S_T e^{-r(T-t)}\mathbf{1}_{S_T \geq K}\mid S_t\right] = S_t E^Q\left[e^{-\frac{1}{2}\sigma^2(T-t)+\sigma\left(W_T^Q-W_t^Q\right)}\mathbf{1}_{S_T \geq K}\mid S_t\right].$$

Here we can usefully choose $S$ as our numeraire and bring in the probability measure $Q_S$, that makes martingales all prices denominated in $S$ and is defined by its Radon–Nikodym derivative given in Eq. (11.73) (Appendix 2), which we reproduce:

$$\frac{dQ_S}{dQ} = e^{-\frac{1}{2}\sigma^2(T-t)+\sigma\left(W_T^Q-W_t^Q\right)}.$$

Then we see that the expectation simplifies as follows:

$$S_t E^Q\left[e^{-\frac{1}{2}\sigma^2(T-t)+\sigma\left(W_T^Q-W_t^Q\right)}\mathbf{1}_{S_T \geq K}\mid S_t\right] = S_t E^Q\left[\frac{dQ_S}{dQ}\mathbf{1}_{S_T \geq K}\mid S_t\right],$$
$$= S_t E^Q{}_S\left[\mathbf{1}_{S_T \geq K}\mid S_t\right],$$
$$= S_t Q_S(S_T \geq K\mid S_t) \text{ (by (75)).}$$

Thus, we have $N(d_1) = Q_S(S_T \geq K \mid S_t)$, that is the *probability of exercising the option under the martingale measure associated with the numeraire $S$.*

Consequently, we can write the call's value as a function of two probabilities for the event $S_T \geq K$ ("the call expires in the money"):

$$C_t = S_t\, Q_S(S_T \geq K) - Ke^{-r(T-t)}\, Q(S_T \geq K).$$

Besides, result (11.73) allows us to understand why the difference between $d_1$ and $d_2$ is equal to $\sigma\sqrt{T-t}$ (recall from Sect. 11.1.4 that to compute the integral giving $N(d_1)$ we made the change of variable $v = u - \sigma\sqrt{T-t}$).

Indeed, let us rewrite, starting from the definition of $d_1$, the term $d_2 = d_1 - s\sqrt{(T-t)}$ in Eq. (11.11) as

$$d_2 = \frac{\ln \frac{S_t}{K} + \left(r - \frac{1}{2}\sigma^2\right)(T - t)}{\sigma\sqrt{(T - t)}}.$$

By (11.73) and Girsanov's theorem, the Brownian motion under $Q_S$ is related to the Brownian motion under $Q$ by

$$dW_t^{Q_S} = dW_t^{Q} - \sigma dt.$$

Process (11.13) under $Q$ thus can be rewritten under $Q_S$:

$$\frac{dS_t}{S_t} = \left(r + \sigma^2\right)dt + \sigma dW_t^{Q_S}$$

with the extra term $+\sigma^2$ in the drift; consequently, the $-\frac{1}{2}\sigma^2$ present in $d_2$ (under $Q$) becomes $-\frac{1}{2}\sigma^2 + \sigma^2 = +\frac{1}{2}\sigma^2$ present in $d_1$ (under $Q_S$).

## Suggested Reading

### Books

Baxter, M., & Rennie, A. (1999). *Financial calculus*. Cambridge University Press.
Cox, J., & Rubinstein, M. (1985). *Options markets*. Prentice Hall.
**Gatheral, J. (2006). *The volatility surface: A practitioner's guide*. Wiley Finance.
Hull, J. (2018). *Options, futures and other derivatives* (10th ed.). Prentice Hall Pearson Education.
**Lewis, A. L. (2000). *Option valuation under stochastic volatility*. Finance Press.

### Articles

Amin, K., & Jarrow, R. (1991). Pricing foreign currencies options under stochastic interest rates. *Journal of International Money and Finance, 10*, 310–229.
Black, F. (1976). The pricing of commodity contracts. *Journal of Financial Economics, 3*, 167–179.
Black, F., & Scholes, M. (1973). The pricing of options and corporate liabilities. *Journal of Political Economy, 81*, 637–659, (seminal article).
Black, F., & Scholes, M. (1972). The valuation of option contracts and a test of market efficiency. *Journal of Finance, 27*, 399–418.
Cox, J., & Ross, S. (1976). The valuation of options for alternative stochastic processes. *Journal of Financial Economics, 3*, 145–166.
Duffie, D., Pan, J., & Singleton, K. (2000). Transform analysis and asset pricing for affine jump diffusions. *Econometrica, 68*, 1343–1376.
Dupire, B. (1994). Pricing with a smile. *RISK, 7*, 18–20.
Garman, M. B., & Kohlhagen, W. (1983). Foreign currency option values. *Journal of International Money and Finance, 2*, 231–237.
Geman, H., El Karoui, N., & Rochet, J. C. (1995). Changes of numeraire, changes of probability measure and option pricing. *Journal of Applied Probability, 32*, 443–458.
Gesser, V., & Poncet, P. (1997). Volatility patterns: theory and some evidence from the dollar-mark option market. *The Journal of Derivatives, 5*(2), 46–61.

Hagan, P., Kumar, D., Lesniewski, A., & Woodward, D. (2002). Managing smile risk. *Wilmott Magazine*, July, pp. 84–108.

Hentschel, L. (2003). Errors in implied volatility estimation. *Journal of Financial and Quantitative Analysis, 28*(4), 779–810.

Heston, S. (1993). A closed form solution for options with stochastic volatility with application to bonds and currency options. *Review of Financial Studies, 6*, 327–343.

Hull, J., & White, A. (1987). The pricing of options on assets with stochastic volatilities. *Journal of Finance, 42*, 281–300.

Jamshidian, F. (1993). Option and futures evaluation with deterministic volatilities. *Mathematical Finance, 3*(2), 149–159.

Margrabe, W. (1978). The value of an option to exchange one asset for another. *Journal of Finance, 33*, 177–185.

Merton, R. (1973). The theory of rational option pricing. *Bell Journal of Economics and Management Science, 4*, 141–183. (seminal article).

Miltersen, K., & Schwartz, E. (1998). Pricing options on commodity futures with stochastic term structures of convenience yields and interest rates. *Journal of Financial and Quantitative Analysis*, 33–59.

Ramaswamy, K., & Sundaresan, S. M. (1985). The valuation of options on futures contracts. *Journal of Finance, 40*, 1319–1340.

Shu, J., & Zhang, J. (2004). Pricing S&P 500 index options under stochastic volatility with the indirect inference method. *Journal of Derivatives Accounting, 1*, 1–16.

Smith, C. W. (1976). Option pricing: a review. *Journal of Financial Economics, 3*, 3–54.

Stoll, H., & Whaley, R. (1986). New option instruments: arbitrageable linkages and valuation. *Advances in Futures and Options Research, 1A*, 25–62.

# Option Portfolio Strategies: Tools and Methods

# 12

A portfolio containing options is exposed to specific risks that need monitoring. Models for option pricing, the most standard of which have been discussed in the preceding chapter, are the foundations of this management. Nonetheless, their implementation does pose a number of difficult problems, mainly connected with the restrictive assumptions on which these models are based. Practitioners have developed additional tools which mitigate to greater or lesser extents some of the models' deficiencies, facilitate their application and underlying strategies for managing portfolios and risks. These tools and methods were built from an engineering rather than a strictly mathematical viewpoint, i.e., without avoiding technical sophistication but accepting compromises with the theory to make them practical. It is in this spirit that we wrote this chapter.

The main monitoring tools derived from pricing formulas are those based on the notion of implied volatility and on the sensitivities of the value of an option to the different model parameters. From a portfolio management perspective, it is useful to distinguish static from dynamic strategies. A static strategy consists in building an initial portfolio which is then left unchanged until it is sold out. Dynamic strategies involve more or less frequent rebalancing of the portfolios over time. It is above all the latter strategies which are based on tools derived from the models mentioned above.

Section 12.1 presents the main static strategies. Section 12.2 examines the important notions of historical volatility, implied volatility, *smile* (or *skew*) and volatility surface. Section 12.3 analyzes the sensitivities of options to the different variables that influence their values (the so-called "Greek" parameters). Section 12.4 offers strategies involving revisions and using Greek parameters as indicators for monitoring, in particular those strategies that are called *delta-neutral*. Some more complicated developments, or those with a more mathematical character, are presented in the appendices to this chapter.

---

© The Author(s), under exclusive license to Springer Nature Switzerland AG 2022
P. Poncet, R. Portait, *Capital Market Finance*, Springer Texts in Business and Economics, https://doi.org/10.1007/978-3-030-84600-8_12

## 12.1 Basic Static Strategies

Static strategies are those which do not require any rebalancing until they are unfolded. We consider here portfolios combining calls and puts written on the same underlying asset, which can also be present in the portfolio. The options have the same expiry $T$ (unless otherwise mentioned, as at the end of Sect. 12.1.2) but their strikes may be different.

We explain briefly the general method of analysis (Sect. 12.1.1). Then we present basic strategies most frequently adopted (Sect. 12.1.2). The general principle of replicating the payoff of a contingent asset with a static portfolio of options is explained in Sect. 12.1.3.

### 12.1.1 The General P&L Profile at Maturity

Consider a position made up of $z$ underlying assets priced at $S(t)$, $\{x_i\}_{i=1\,\ldots\,,n}$ calls on the underlying with strikes, respectively, equal to $\{E_i\}_{i=1\,\ldots\,,n}$ and $\{y_j\}_{j=1\,\ldots\,,m}$ puts with strikes $\{K_j\}_{j=1,\,\ldots\,,m}$. We use the following convention: if $z$, $x_i$, $y_j > 0$, the securities are purchased and if $z$, $x_i$, $y_j < 0$, they are sold short. The strikes of the options are ranked in increasing order.

The P&L (profit and loss) upon expiration on date $T$ is, by definition, the position's final value (on $t = T$), net of the initial investment needed to set it up on date $t = 0$ ($G = V(T) - V(0)$),[1] that is

$$G(S(T)) = \sum_{i=1}^{n} x_i \left([S(T) - E_i]^+ - C_i\right) + \sum_{j=1}^{m} y_j \left([K_j - S(T)]^+ - P_j\right) + z(S(T) - S(0))$$

$$= \sum_{i=1}^{n} x_i \left([S(T) - E_i]^+\right) + \sum_{j=1}^{m} y_j [K_j - S(T)]^+ + zS(T) - \Pi - zS(0)$$

where $\Pi \equiv -\Sigma x_i C_i - \Sigma y_j P_j$ is the total of the premiums paid and received on $t = 0$ ($\Pi$ positive or negative).

Denote, respectively, by $x(S)$ and $y(S)$ the net number of calls and puts *in-the-money* held (or sold) at expiration when $S(T) = S$.

**Proposition 1**
*The P&L profile's slope $G(S)$ at expiration is equal to $x(S) - y(S) + z$.[2]*

*This slope varies from $(z - \sum_{j=1}^{m} y_j)$ to $(z + \sum_{i=1}^{n} x_i)$ as $S$ moves from $0$ to infinity.*

---

[1] In an alternative, more precise definition the initial investment is capitalized at $T$: $G = V(T) - V(0) e^{rT}$.

[2] Except at the points $E_i$ and $K_j$ since the functions Max$(S-E_i, 0)$ and Max $(K_j-S, 0)$ are not differentiable there.

## 12.1 Basic Static Strategies

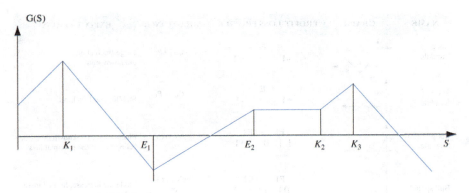

**Fig. 12.1** P&L profile G(S) at maturity

$G(S)$ is a piecewise linear affine function (with changes in the slope at each strike $E_i$ or $K_j$) like the one illustrated in Fig. 12.1, with the lengths of the pieces depending on the gaps between the different strikes $E_i$ and $K_j$.

**Proof**
Each underlying contributes +1 to the slope of $G(S)$. Each *in-the-money* call contributes +1 to the slope of $G(S)$ if it is purchased, and −1 if it is sold, and each *in-the-money* put contributes −1 if purchased and +1 if sold. Options that are *out-of-the-money* contribute zero to the slope. The slope of $G(S)$ thus equals $x(S) - y(S) + z$.

If $S = 0$ all the calls are out-of-the-money ($x(S) = 0$) and all the puts are in-the-money ($y(S) = \sum_{j=1}^{m} y_j$). If, starting from $S = 0$, we consider the increasing values of $S$, the puts successively become out-of-the-money while the calls become in-the-money. As a result, $G(S)$ has a break in its slope for each value of $S$ equal to one of the strikes $E_i$ or $K_j$, and the profile of $G(S)$ is indeed "a rocket trajectory" (continuous and piecewise linear) as in Fig. 12.1. If $S$ is greater than the largest of the strikes ($E_i$ and $K_j$), all the puts are out-of-the-money and all the calls are in-the-money ($y(S) = 0$ and $x(S) = \sum_{i=1}^{n} x_i$).

### 12.1.2 The Main Static Strategies

In Fig. 12.2, which gives the principal elementary strategies, a long call is represented by [0, 1] (which means that the slope of $C(S)$ is equal to 0 for $S$ below the strike and equal to 1 for $S$ above the strike) and a long put by [−1, 0]. The slope profile of a portfolio is obtained by summing the slopes of the different components. Purchasing a straddle, for example, is done by buying a call and a put with the same strike $E$ (the position will be denoted $(+ C_E, + P_E)$), which leads to the slope profile

## 12 Option Portfolio Strategies: Tools and Methods

| NAME | GRAPH | PROFIT-LOSS PROFILE | SOLUTION(S) | EXPECTATIONS |
|---|---|---|---|---|
| Long Straddle | | $\begin{array}{c} E \\ -1 \quad \mid \quad +1 \end{array}$ | $+C_E, +P_E$ | Large fluctuation, unknown sign |
| Short Straddle | | $\begin{array}{c} E \\ +1 \quad \mid \quad -1 \end{array}$ | $-C_E, -P_E$ | Stability, no loss limit |
| Long Strangle | | $\begin{array}{c} E1 \qquad E2 \\ -1 \mid 0 \mid +1 \end{array}$ | $+P_{E1}, +C_{E2}$ | Fluctuation in unknown direction "clipping" of the maximal loss |
| Bull spread | | $\begin{array}{c} E1 \qquad E2 \\ 0 \mid +1 \mid 0 \end{array}$ | $+C_{E1}, -C_{E2}$ or $+P_{E1}, -P_{E2}$ | Moderate increase, limited losses if decrease |
| Bear spread | | $\begin{array}{c} E1 \qquad E2 \\ 0 \mid -1 \mid 0 \end{array}$ | $-C_{E1}, +C_{E2}$ or $-P_{E1}, +P_{E2}$ | Moderate decrease, limited losses if increase |
| Long Butterfly | | $\begin{array}{c} E1 \qquad E2 \qquad E3 \\ 0 \mid +1 \mid -1 \mid 0 \end{array}$ | $+C_{E1}, +C_{E3}, -2C_{E2}$ or $+P_{E1}, +P_{E3}, -2P_{E2}$ | Stability, but possible fluctuation, with unknown sign, and limited losses |
| Short Condor | | $\begin{array}{c} E1 \quad E2 \quad E3 \quad E4 \\ 0 \mid -1 \mid 0 \mid +1 \mid 0 \end{array}$ | $-C_{E1}, -C_{E4}, +C_{E2}, +C_{E3}$ or $-P_{E1}, -P_{E4}, +P_{E2}, +P_{E3}$ | Large fluctuation in unknown direction, small gains and losses |

**Fig. 12.2** Summary table of the main static strategies

$[-1, +1]$. A bull spread, for example, can be built by buying a call with strike $E_1$ and selling a put with strike $E_2$ ($+C_{E1}, -C_{E2}$) when $E_2 > E_1$, and also leads to the profile $[0, +1, 0]$, as the more detailed table below shows.

| | $E_1$ | | $E_2$ |
|---|---|---|---|
| $+C_{E_1}$ | 0 | +1 | +1 |
| $-C_{E_2}$ | 0 | 0 | −1 |
| Result | 0 | +1 | 0 |

Note that the same profile can also be obtained using puts ($P_{E1}, -P_{E2}$).

We note that the P&L profiles associated with specific elementary strategies, as presented in the table above, show only the slopes $+1$, $-1$ or $0$ according to the value of $S$. As we have just seen in the exposition of the general method of analysis, that has not to be the case. For example, a *call ratio backspread* is the sale of a call $E_1$ and the purchase of $x$ ($>1$) calls $E_2$ ($>E_1$), which gives a profile $[0, -1, +(x-1)]$. A *put*

## 12.2 Historical and Implied Volatilities, Smile, Skew and Term Structure

*ratio spread* is made up of the purchase of a put $E_2$ and the sale of $x$ ($>1$) puts $E_1(<E_2)$, leading to a profile at the expiration of $[(x-1), -1, 0]$. The number of such possible strategies is obviously infinite.

Furthermore, some strategies use options with *different* maturities $T$. Typically, these are spread strategies (bullish or bearish), which involve, like those in the summary table (sometimes called *vertical* spreads, or simply spreads), a call (or put) purchased and a call (or put) sold. The spread is said to be *horizontal* or *calendar* if the two options have the same strike and $T_1 \neq T_2$. The spread is said to be *diagonal* if the two options have different strikes and expirations. In both cases, it is convenient to set up the payoff diagram at maturity (say $T_1$) for the shorter option. The graph of the position's value as a function of the underlying's price is then no longer piecewise linear, contrary to the case in the summary table, because whichever of the options lives on has a positive time value and so its value is a convex function of the underlying's price which is not piecewise linear.

### 12.1.3 Replication of an Arbitrary Payoff by a Static Option Portfolio (*)

We have just shown how one may construct a static portfolio which generates an arbitrary piecewise linear payoff such as that in Fig. 12.1 or those described in the summary table. Conversely, any piecewise linear payoff can be obtained by a combination of a finite number of options. More generally, one can show that an arbitrary payoff $\Psi(S_T)$, given *a priori*, where $\Psi$ is a sufficiently regular (not necessarily piecewise linear) function can theoretically be obtained using a static portfolio of standard European options with the same expiration $T$ and different strikes. This portfolio involves (except for piecewise linear payoffs) an infinite number of options, and, as a result, its construction is more theoretical than practical since it requires a continuum of price quotations. More precisely, the static duplication of this payoff $\Psi(x)$ is accomplished by a position made up of $\frac{\partial^2 \Psi}{\partial K^2}(x)dx$ options (call or put) with strike $x$ ($x$ taking on all the values for which $\Psi(x) \neq 0$). This result, which will be used in the analysis of exotic options, is proved and explored in Appendix 3.

### 12.2 Historical and Implied Volatilities, Smile, Skew and Term Structure

*Among the five variables and parameters $S$, $K$, $T$, $r$, and $\sigma$ entering in formulas of* Black-Scholes type, only the volatility $\sigma$ is not directly observable. In addition, the volatility is not constant, contrary to the standard assumption. It is thus in the choice of the value to be given this central parameter that the problems in implementing the models lie. Numerous techniques have been worked out to underlying this choice. In this section, we explain the standard method for estimating the volatility using the underlying asset price history, the notion of implied volatility and its use, as well as the definitions of *smile, skew* and *volatility surface*.

## 12.2.1 Historical Volatility

In the Black-Scholes model (BS from now on) and the models derived from it, the volatility $\sigma$ is the instantaneous and annualized standard deviation of the returns on the underlying. This volatility is primarily due to the arrival of new relevant information which produces price variations (since, by definition, unexpected). Such information may concern the issuing company, general market conditions, or the business sector. In addition, some buy and sell orders cannot be attributed to such news, but result randomly from the liquidity needs of agents (then called *liquidity traders*).

The volatility parameter s is not observable, but an empirical estimate for it is easy to find if returns, and thus also log returns,[3] are independent and identically distributed, which implies that the volatility is constant. It is then sufficient to take $n$ ($\geq 30$) past log-returns $R_i = \ln S_{i+1} - \ln S_i$, calculated on intervals $i$ of constant length (days, weeks, months) called "periods," and to calculate their empirical mean

$$\widehat{\mu}_p = \frac{1}{n} \sum_{i=1}^{n} R_i,$$

and empirical standard deviation

$$\widehat{\sigma} = p \sqrt{\frac{1}{n-1} \sum_{i=1}^{n} \left(R_i - \widehat{\mu}_p\right)^2}, \tag{12.1}$$

$\left(\widehat{\sigma}_p\right)^2$ is an unbiased estimate of the square of the volatility for the period (weekly if returns are weekly, daily if they are daily, and so on). Then, to obtain the annual volatility denoted by $\widehat{\sigma}$, which is generally the parameter present in pricing formulas,[4] we multiply the period volatility $\widehat{\sigma}_p$ by $\sqrt{p}$, where $p$ denotes the number of periods contained in a year; for example, $\sqrt{52}$ if observations are weekly, or $\sqrt{256} = 16$ (to have a round number) if they are daily (since there are on average 252 days in the business year[5]). The estimate thus obtained is called the (annual) *historical* or *physical* volatility. Notice that this method of annualizing the volatility is not really justified unless successive log-returns are independently distributed, a restrictive assumption one can make unless the market is illiquid and inefficient (*see* Chap. 1).

---

[3] Recall that, in the BS model, between two dates $i$ and $i + 1$ separated by a period of duration $\Delta t$ ($\Delta t = 1/p$ in what follows) we have $\ln S(i+1) - \ln S(i) = \left(\mu - \sigma^2/2\right)\Delta t + \sigma\sqrt{\Delta t}U$, where $U$ follows a reduced centered normal law. We use the notation $\mu_p \equiv (\mu - \sigma^2/2)\Delta t$ and $\sigma_p \equiv \sigma\sqrt{\Delta t} = \sigma\frac{1}{\sqrt{p}}$.

[4] It is usual to express $r$ and s as annual rates and $T$ in fractions of a year. Any other choice of period is legitimate, if the dimensions of $r$, $\sigma$ are $T$ mutually coherent. $\sigma$ is multiplied by $\sqrt{p}$, since it is the variance $\sigma^2$ which is proportional to time.

[5] We calculate "as if" no new information turns up on weekends or holidays.

## 12.2.2 The Implied Volatility

As the volatility s of the underlying is not observable, it is necessary to estimate it empirically, the simplest case being when returns are independent and identically distributed. While the assumption of serial non-correlation of returns is acceptable, in contrast, the assumption that distributions are identical is very restrictive. Volatility, in particular, is not constant since the markets are characterized by periods (of random duration) of large, medium, and small fluctuations. *Under such conditions, we have seen that the volatility which determines the option's price is the expected mean volatility between the dates of pricing and the expiration date of the option* (see in Sect. 11.2.5, the model with variable but deterministic volatility) since it is the volatility which determines the standard deviation of $S_T$ on which the option's price ultimately depends.[6] Unfortunately, estimating the future volatility of the underlying asset is a difficult and controversial exercise.

However, there does exist an acceptable solution when the market explicitly quotes option prices with underlying assets that are identical or similar. In that case, the BS formula can be "inverted" to yield the volatility $\sigma$ implied by the market price ($C$ or $P$) of recently quoted options.

Let us consider, for example, a call: to each price $C$ for this call corresponds, using the relevant pricing formula (e.g., the BS formula), a single volatility $\sigma$:[7]

$$C = BS(\sigma; \underline{x}) \text{ implies } \sigma = BS^{-1}(C; \underline{x}),$$

where $\underline{x}$ denotes the vector of remaining parameters and variables which are observable.

Since, for a given $\underline{x}$, the standard models give us a continuous bijection $BS(.)$ between $C$ (or $P$) and $\sigma$, the numerical computation of $\sigma = BS^{-1}(C; \underline{x})$ poses no problems and iterative methods (such as Newton–Raphson) converge rapidly.

We thus make use of this implied volatility $\sigma$ for valuation of the option under consideration, which is different from but written on the same underlying as the one for which we have extracted the implied volatility.

It is important to emphasize the fact that *a standard option's price (call or put) increases with the implied volatility*. A small volatility, therefore, corresponds to a small price, and a high volatility to a high price. For professional traders, what matters for an option is the level of its implied volatility, not its monetary value (price). It is indeed meaningless to say that an option is expensive or cheap according to its price: all else being equal, an in-the-money option is obviously more expensive than an out-of-the-money option, so that we should ignore the intrinsic value of the option and focus on its time or speculative value. Thus an option is dear or not according to the volatility implicit in its price: this volatility is to be compared with

---

[6]More exactly, as the more general model with stochastic interest rates shows, it is the average (between $t$ and $T$) of the volatility of the underlying's forward price which determines the standard deviation of the returns on $S_T$ and therefore the option price.

[7]Since the function $F$ is continuous and monotonously increasing in $\sigma$.

the average volatility observed or with the implied volatilities of options written on the same underlying with different strikes and/or expiration dates.[8] This notion is crucial, as we will see later in studying the monitoring of option positions.

It is also necessary to understand that the use of implied volatility provides a single framework, *in principle* applicable to all options written on the same underlying and with the same duration; this is a significant advantage when one must quickly estimate how expensive an option is. Indeed, the price of an option not only depends on the volatility of its underlying but also on the duration of the option and on the ratio $S/K$ between the underlying's current price and the strike (the *moneyness* of the option). The volatility is a parameter intended to characterize the underlying asset and should be the same for all options of the same maturity written on the same underlying, irrespective of their strike. The volatility is not necessarily independent of the option duration, since the average volatility expected by the market from the present to an expiration date can differ for a different expiration date, from which arises the interest in the generalization of the BS model to a variable volatility (see Sect. 11.2.5). Suppose, for example, that one wishes to estimate the price of an option on a stock market index, and that one has observed that the volatility of the index (historical and implied) is about 20%; on *a priori grounds*, it is unnecessary to consider the strike to conclude that the option price is too high if it carries an implied volatility of 30%.

In mitigation of that last remark, note that the standard models of BS type, which provide the pricing function $F$ used to calculate premiums and implied volatilities, are all based on strong simplifying assumptions (especially that of a constant or deterministic volatility). It is thus not surprising to observe that, contrary to these models' assumptions, options on the same underlying and of the same duration are *not* in fact traded at the same implied volatility whatever their strikes. These questions will be considered in more detail in the following subsection devoted to the *smile or skew* and the term structure of volatilities.

To sum up, volatility is used to represent the price of an option. Indeed, all the parameters and variables which influence an option's value are exogenous for they are unambiguously known or observable in the market.[9] An option is consequently more or less expensive according to the level of volatility at which it is traded. In practice, traders do think in terms of volatility, and computer programs deduce from it the option value in monetary terms (the premium).

---

[8] This type of (uniquely correct) reasoning is pushed to its logical limit on the official market for exchange rate options in Philadelphia (USA). There options are quoted not in dollars or any other currency, but directly in volatilities. Once an option has been traded on the basis of a volatility agreed by the two parties, the market's back-office calculates the price, for example in dollars, using the Garman–Kohlhagen formula (see Chap. 11, Proposition 9).

[9] We are ignoring here the problem of dividends, the estimation of which may be difficult if the option is long-lived and its underlying asset pays several, usually random, dividends (which is the case for numerous warrants written on stocks).

## 12.2 Historical and Implied Volatilities, Smile, Skew and Term Structure

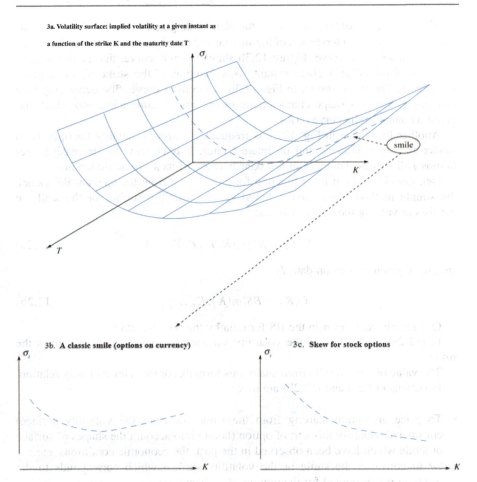

**Fig. 12.3** Volatility surface and smile

### 12.2.3 Smile, Skew, Term Structure, and Volatility Surface

Although the volatility is a characteristic of the underlying asset, we observe that the implied volatilities of options written on the same asset depend on their strikes and expiration dates.

#### 12.2.3.1 Definitions and Use of the Volatility Surface and the Smile or Skew

The curve giving the implied volatility as a function of the strike is called the *smile*, or *skew*, depending on its shape, and that giving the volatility as a function of the duration of the option is called the *term structure*. More generally, the implied volatility is a function of two variables $\sigma_i(K, T)$ which can be graphed in three dimensions to generate a *volatility surface* such as in Fig. 12.3a.

The intersection of the surface and the plane orthogonal to the $T$-axis is the smile (or skew) curve; the intersection of the surface and the plane orthogonal to the $K$-axis is the term structure curve. Figure 12.3b shows a smile curve, that is, the implied volatility observed at a given instant, as a function of the strike $K$, for a given maturity $T$. The smile shown in Fig. 12.3b is a convex curve, first decreasing then increasing: this is a shape characteristic of options on currencies, and which has given its name to the curve $\sigma_i(K)$.

Another form seen in Fig. 12.3c is frequently observed, notably for options on stocks and stock indices, and on interest rates. It is convex but decreasing over (almost) all strikes. Traders call it a *skew* because of its asymmetric feature.

Denoting by $F$ the pricing function for a call taken from the relevant BS model, the simple method often used is based on the volatility surface or the smile; it consists in valuing the call as follows:

$$C(K,T) = BS(\sigma_i(K,T),K,T,\ldots) \tag{12.2a}$$

and, for a given expiration date $T$,

$$C(K) = BS(\sigma_i(K),K,\ldots). \tag{12.2b}$$

One simply replaces $\sigma$ in the BS formula by the function $BS^{-1}$.

In (12.2a), $\sigma_i(K,T)$ gives the volatility surface and in (12.2b), $\sigma_i(K)$ gives the smile.

The value of a put results from analogous formulas or the relevant parity relation. Equations (12.2a) and (12.2b) are used:

- To price an option starting from the smile (or from the volatility surface) considered usual for this sort of option (taking into account the shapes of surface or smile which have been observed in the past, the economic conditions, etc.),
- Or to construct the smile or the volatility surface which corresponds to the different prices quoted for the options at a given time.

### 12.2.3.2 Explanations for the Existence of the Volatility Term Structure and the Smile; The Method's Coherence

It is not surprising that the implied volatility depends on $T$ and $K$, and more sophisticated models than BS exhibit such dependence.

For example, the model with a variable but deterministic volatility presented in the preceding chapter shows that the parameter s introduced in the BS formula is the mean of the volatility over the period $(t, T)$. As a result, if one does not expect that the volatility will be constant, the relevant volatility depends on the option expiration date. The occurrence of a term structure for volatilities is consequently *perfectly compatible* with the BS model.

In contrast, the smile (or skew) corresponding to a given maturity $T$ is *not compatible* with the BS model but can be explained by the fact that the underlying's price distribution $S_T$ is not log-normal.

## 12.3 Option Sensitivities (Greek Parameters) 463

Let us consider, for example, the value $C(K)$ of a call as a function of the strike which obeys a smile described in Eq. (12.2b). We show (*see* Appendix 2) that under the risk-neutral probability $Q$, the probability density of the underlying's final value $S_T$,

$$\phi_T(K) = \frac{Q(S_T \in [K, K + dK])}{dK},$$

is equal to the capitalized value of the second derivative of the option's value (for calls or puts) with respect to the strike

$$\phi_T(K) = e^{rT} \frac{\partial^2 C}{\partial K^2}(K).$$

Therefore, to a price $C(K)$ corresponds a probability density $\phi_T(K)$. Furthermore, if the pricing function is given, for example, that of BS, to a price range for $C(K)$ corresponds a smile curve $\sigma_i(K)$ according to Eq. (12.2b). Thus, *to a smile curve corresponds a probability density $\phi_T(K)$ for $S_T$.* This two-fold correspondence writes:

$$C(K) = BS(\sigma_i(K), K, \ldots), \tag{12.2b}$$

$$\phi_T(K) = e^{rT} \frac{\partial^2 BS(\sigma_i(K), K)}{\partial K^2}. \tag{12.3}$$

Therefore, to a particular form of the smile corresponds a particular density for $S_T$ given by (12.3), which we may call *the density implied in the* smile. For example, for $\sigma_i(K) = \sigma$ independent of $K$, the BS model assigns to $S_T$ a log-normal distribution under the risk-neutral probability. *On the contrary*, the distribution for $S_T$ implied by a *non-horizontal* smile $\sigma_i(K)$ is not log-normal, contradicting the main assumption underlying the BS model. In particular, we can show that the probability density for $\ln(S_T)$, implied by a *skew as in* Fig. 12.3c, is not Gaussian: it is characterized by an asymmetry assigning larger probabilities to values much below the mean (that is, a thick tail to the left). These issues are developed and made more precise in Appendix 2.

## 12.3 Option Sensitivities (Greek Parameters)

To use options optimally and to efficiently manage a portfolio containing them, it is necessary to quantify the impact of changes in the variables or parameters that influence option values. This section is devoted to examining different partial derivatives which measure this impact. BS-like models allow an explicit calculation of all such derivatives.

## 12.3.1 The Delta ($\delta$)

*By definition, the delta is the first partial derivative of the premium with respect to S;* it thus reflects the sensitivity of the option's value to the underlying's market price. If we have a pricing formula, such a partial derivative is easy to calculate. In the BS model we have

$$\delta_C \equiv \frac{\partial C}{\partial S} = N(d_1), \quad \text{therefore} \ 0 < \delta_C < 1 \tag{12.4a}$$

$$\delta_P \equiv \frac{\partial P}{\partial S} = -1 + N(d_1), \quad \text{therefore} \ -1 < \delta_P < 0 \tag{12.4b}$$

The proofs of these equations can be found in Appendix 1 along with the delta values resulting from other pricing models.

Note that by call-put parity $(C - P = S - Ke^{-}rT)$, we have $\delta_C - \delta_P = \delta_S = 1$ for an option on a spot asset that does not pay a dividend.

It follows from its mathematical definition that delta is the slope of the tangent to the curve representing the option price as a function of the underlying's price, and that, to the first order, for a small variation of the underlying price one has

*Variation in the option value* $= \delta_{option} \times$ *Variation in the underlying price.*

Finally, it results from option pricing theory (see the preceding chapter and the proof of the BS formula) that the (risk-neutral) probability that the option expires in-the-money is $\text{Prob}_Q (S(T) > K) = N(d_2)$. But $d_2$ is close to $d_1$ at least for options whose expiration is not too far away or for those whose underlying's volatility is not too high $(d_2 = d_1 - \sigma\sqrt{T - t})$. The delta of a call is therefore approximately equal to the (risk-neutral) exercise probability for short-term options or those written on assets with small volatility.[10]

Figure 12.4 shows delta as a function of the underlying's price for a given strike.

The shape of each curve can be deduced from the graph of the premium as a function of $S$ whose tangent it represents; exact values can be obtained directly using Eq. (12.4).

It is important to remember several fundamental points:

- A call has a positive delta between 0 and 1.
- A put has a negative delta between 0 and –1.
- The underlying asset has a delta of 1; a risk-free asset has a delta of zero.
- The absolute value of the option's delta is approximately equal to the exercise probability.

---

[10] As shown in one of the appendices of the preceding chapter, $N(d_1)$ is exactly the exercise probability for the option under the measure $Q_S$ associated to a special numeraire, namely the underlying itself: $\text{Prob}_{Q_S}(S(T) > K)$.

## 12.3 Option Sensitivities (Greek Parameters)

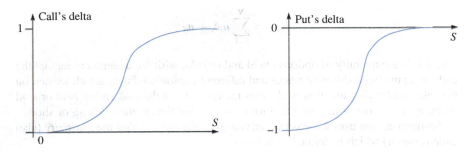

**Fig. 12.4** Deltas

- An option which is at-the-money often has the absolute value of its delta *near* 0.5[11] (intuitively, one thinks that it has approximately one chance in two of expiring in-the-money).
- An option deep in-the-money has a delta of absolute value near to 1 (it is highly probable that it will be exercised).
- An option which is deep out-of-the-money has a delta that is nearly zero (there is practically no chance of its being exercised).

The interest in the delta becomes obvious when we have to consider building a portfolio that is locally insensitive to a (small) variation in the price of the underlying: a portfolio made up of a long call and $-N(d_1)$ short underlyings has a global delta of zero, just like one composed of a single long put plus $(1 - N(d_1))$ long underlyings. More generally, a position in underlyings, calls and puts whose global delta vanishes is called *delta-neutral*. This is, for example, the position of market makers who usually refuse to make a bet as to the direction of the underlying's price variation, except possibly for very short periods (intraday).

More generally, it is possible by modifying the portfolio composition to engineer any level of sensitivity to the underlying's price variation (that is, to achieve an arbitrary degree of hedging) since the portfolio's delta is the weighted sum of its constituents' deltas:

---

[11] If an option is in-the-money, $d_1 = \frac{r+\sigma_2/2}{\sigma}\sqrt{T}$ is near 0 for most of the current values of the parameters. More precisely: $N(0) = 0.5$, therefore $N(d_1) \tilde{N}(0) + d_1 N'(0) = 0.5 + \frac{r+\sigma_2/2}{2\pi\sigma}\sqrt{T}$. The delta of a call in-the-money is therefore a little larger than 0.5, and that of a put a little larger than $-0.5$.

$N(d_1)$ is even closer to this exercise probability for options called *at-the-money-forward* for which $S(0)$ is not equal to the strike $K$ but to the discounted strike $Ke^{-rT}$ (recall the call-put parity). Indeed, the rate $r$ present in the preceding expressions then vanishes.

$$\delta_{\text{port}} = \sum_{i=1}^{N} n_i \delta_i + n_S$$

where $n_i$ is the quantity of options $i$ held and/or sold, with the understanding that the calls and puts (with different prices and different expiration dates) are all written on the same underlying and that $n_S$ denotes the quantity of the underlying held or sold short; $n_i$ is positive or negative according to whether the position is long or short.

Sometimes one uses a measure derived from the delta called the *elasticity* (also called *omega*) which is defined as follows

$$\Omega_C \equiv \frac{dC/C}{dS/S} = \delta_C \frac{S}{C} \quad (\geq 1); \Omega_P \equiv \frac{dP/P}{dS/S} = \delta_P \frac{S}{P} \quad (\leq 0).$$

The elasticity allows us to think in terms of percentage variation and is very useful in calculating the risk premium, the volatility and the beta of an option as a function of the risk premium, the volatility and the beta of its underlying asset (see Chap. 22 for details on the beta, and Sect. 12.3.7 for the use of this elasticity).

---

**Example 1**

Consider a European call and put, written on a spot asset worth \$500, with a strike \$520 and expiration in 90 days. The (discrete) 3-month interest rate is 9.5% and the underlying's volatility is estimated to be 5.547% using weekly data.

Computations give: $(T-t) = 90/365 = 0.2465753$; $r = \ln (1.095) = 0.09075$;

$$\sigma = 0.05547 \times \sqrt{52} = 0.4; \quad Ke^{-r(T-t)} = 508.493.$$

$$d_1 = [\ln(S/K) + (r + \sigma^2/2)(T - t)]/\sigma\sqrt{T - t} = 0.0145127; \quad d_2$$

$$= d_1 - \sigma\sqrt{T - t} = -0.1841101.$$

$$N(d_1) = 0.50579056; \quad N(-d_2) = 0.5703644; \quad N(d_2) = 0.42696356.$$

$$C = 500 \, N(d_1) - 508.493 \, N(d_2) = 35.787 \approx 35.79.$$

$$P = C - S + Ke^{-r(T-t)} = 35.787 - 500 + 508.493 = 44.28.$$

The delta for the call equals $N(d_1) = 0.506$, and for the put $-0.494$, the elasticity of the call is 7.069 and of the put $-5.578$: the call is 7 times more sensitive and the put 5.6 times more sensitive than their common underlying asset.

## 12.3 Option Sensitivities (Greek Parameters)

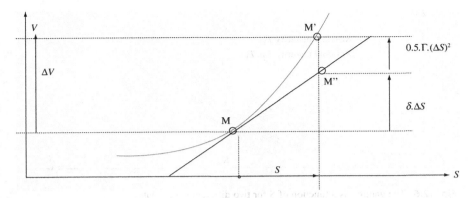

**Fig. 12.5** Under the influence of $\Delta S$, the point M moves to M'. If the convexity vanished ($\Gamma = 0$), it would move to M''

### 12.3.2 The Gamma ($\Gamma$)

The gamma is the only (standard) Greek parameter that is not the first partial derivative of the value of an option. *By definition, the gamma is the derivative of the delta, or the second derivative of the option's value with respect to the underlying's price.* For all options priced using the BS model we obtain (see Appendix 1 for the proof):

$$\Gamma_C \equiv \frac{\partial^2 C}{\partial S^2} = \frac{\partial \delta_C}{\partial S} = \frac{N'(d_1)}{S\sigma\sqrt{T-t}} \quad (12.5a)$$

$$\Gamma_P \equiv \frac{\partial^2 P}{\partial S^2} = \frac{\partial \delta_P}{\partial S} = \frac{N'(d_1)}{S\sigma\sqrt{T-t}} = \Gamma_C \quad (12.5b)$$

where $N'(z) = \frac{1}{\sqrt{2\pi}} e^{-z^2/2}$ is the probability density of a reduced centered Gaussian, and $T - t$ is the remaining time to live for the option.

Notice that the gamma of the underlying is zero and that by call-put parity (taking its derivative twice), we have $\Gamma_C = \Gamma_P$, regardless of the pricing model employed.

Furthermore, like the delta, the gamma is additive so that $\Gamma_{port} = \sum_{i=1}^{N} n_i \Gamma_i$.

The gamma of an option portfolio plays an important role and largely determines its management. A portfolio whose value is a convex function of the underlying's price *S has a positive gamma.* Conversely, the value of a position with a negative gamma is a concave function of S.

The gamma can be viewed as the sensitivity of delta to variations in the underlying price; it is related to the curvature of the curve representing the option price as a function of S (Fig. 12.5).

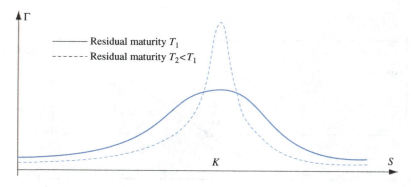

**Fig. 12.6** The gamma as a function of S for two different expiry dates

> **Example 2**
> In the framework of Example 1, the call's gamma (and the put's) is 0.004 (since $N'(d_1) = 0.3989$): for an increase of one dollar in the underlying price, the call's delta (or the put's) increases (decreases in absolute value) by 0.004: $\delta_C = 0.510$ and $\delta_P = -0.490$.

In agreement with Fig. 12.6, the curve for the gamma as a function of the underlying's price has a bell shape with a maximum of $S$ close to the strike.

We note that the gamma of an option that is roughly at-the-money increases as it approaches expiration, while it decreases for options a long way in and out of the money. This result can be interpreted graphically: as the expiration date approaches, the option price curve approaches its intrinsic value until it shows a discontinuity at the strike. Near to the expiration date, the curvature is thus infinite at the discontinuity and vanishes elsewhere. The gamma is then effectively zero everywhere except at-the-money where its curvature is infinite (a Dirac delta function); traders, therefore, call it a "Dirac gamma." As we will see, this situation is uncomfortable and difficult to manage because small variations in the underlying price about the strike mean that the hedge required for the seller of the option changes continually from ($\pm$) 1 underlying (when the option is in-the-money) to 0 underlying (when it is not).

### 12.3.3 The Vega ($v$)

Contrary to one of the assumptions underpinning the basic models, the underlying's volatility $\sigma$ is, in practice, not constant or even deterministic. In devising their strategies, traders must take account of this additional source of risk associated with a future random volatility. This is why it is useful to compute the sensitivity of option values to a change in $\sigma$.

## 12.3 Option Sensitivities (Greek Parameters) 469

*By definition, the vega is the partial derivative of the underlying's premium with respect to $\sigma$.*[12]

Using BS formula, ignoring the fact that the model has been derived on the assumption that $\sigma$ is constant, yields:

$$v_C \equiv \frac{\partial C}{\partial \sigma} = \sqrt{T-t}\, S\, N'(d_1) \qquad (12.6a)$$

$$vP \equiv \frac{\partial P}{\partial \sigma} = \sqrt{T-t}\, SN'(d_1) = vC. \qquad (12.6b)$$

The formulas' proofs are given in Appendix 1 (which also provides vega values for other pricing models). Remark that due to the call-put parity (or the first derivative of it), we have $v_C = v_P$, no matter which model is employed.

It could seem fallacious to calculate the vega using the BS model which assumes a constant volatility. In fact, it is simply necessary to interpret it as the variation in an option's value when the market revises its expectations and expects from now up to the expiration date an average volatility of ($\sigma + d\sigma$) and no longer $\sigma$, which does not contradict the spirit (and the mathematics[13]) of the model.

Looking at Eqs. (12.5) and (12.6), we also see that

$$v = \sigma(T-t)S^2\Gamma.$$

*From this, for given S, $\sigma$ and (T–t), the vega and the gamma are proportional.* However, ordinarily the first decreases with time while the second increases.

The reader should recall the following properties:

- The vega for the underlying is zero.
- The vega of a standard option (call or put) is positive.
- The vega of an option increases with its duration (like the square root of time), and therefore decreases with the passage of time (*see* Fig. 12.7b).
- The vega of an option is maximum when it is close to being in-the-money (*see* Fig. 12.7a).
- The vega is additive (as are all the Greeks), so that the vega of a portfolio is the weighted sum of the vegas of its components.

In practice, it is essential to know this sensitivity; other things aside, a long position (positive vega) is a winner (loser) when the implied volatility increases (decreases); the contrary being true for a short position (negative vega). A hedge is thus all the better, other things being equal, if the portfolio vega is near zero.

---

[12] Vega (the name of a star) is not a Greek letter, in contrast with the names of other common partial derivatives. At first it was called lambda ($\lambda$) or zeta ($\zeta$), but the use of vega came into vogue with professionals. One continues for convenience to use generically the phrasing "Greek parameters" (even for cross second derivatives involving the volatility, such as the *vanna* or the *voma*).

[13] In contrast with the other Greeks, which are sensitivities to the model's *variables*, the vega is a sensitivity to the model's *parameter*.

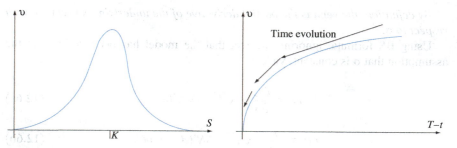

**Fig. 12.7** The vega

> **Example 3**
> In the same setting as the preceding examples, the vegas of the call and the put equal 0.990: the option value consequently goes up by $0.99 if the underlying's volatility increases by 1 % (from 40 % to 41 %).

### 12.3.4 The Theta ($\theta$)

Another cause of change in the value of an option is the passage of time. We have seen that, with some exceptions (notably European puts that are *deep in-the-money*), the value of an option increases with its maturity, and thus its value decreases as time goes by, other things being unchanged.

*By definition, the theta is the partial derivative of the premium with respect to time.*

With the BS model, we obtain

$$\theta_C \equiv \frac{\partial C}{\partial t} \equiv -\frac{\partial C}{\partial T} = -(\frac{S\sigma}{2\sqrt{T-t}}N'(d_1) + rKe^{-r(T-t)}N(d_2)) < 0 \quad (12.7a)$$

$$\theta_P \equiv \frac{\partial P}{\partial t} \equiv -\frac{\partial P}{\partial T} = -\frac{S\sigma}{2\sqrt{T-t}}N'(d_1) + rKe^{-r(T-t)}(1-N(d_2)) \lessgtr 0 . \quad (12.7b)$$

The proof is given in Appendix 1 (the thetas resulting from other models are also provided).

Remark that because of the call-put parity (or its derivative), we have $\theta_P = \theta_C + rKe^{-r(T-t)}$, which shows that $\theta_P > \theta_C$ and explains why European puts which are *deep in-the-money* can have a positive theta.

As for the other Greek parameters, it is important to recall the following basic properties:

## 12.3 Option Sensitivities (Greek Parameters)

- Options in general have a negative theta; a long position in options (usually) has a negative theta, a short position a positive theta.
- The underlying has no theta.
- An option's theta is maximum near at-the-money.
- For options around at-the-money, theta increases (in absolute value) as the option nears its expiration.
- As the theta is additive, a portfolio's theta is the weighted sum of the thetas of its components.

> **Example 4**
> In the context of the preceding examples, the theta of the call equals –0.274 and that of the put –0.148, so the options lose, approximately, 27 and 15 cents a day, respectively, all else being equal.

### 12.3.5 The Rho ($\rho$)

The rho measures the influence of a variation in the interest rate $r$ on the option's value.

*By definition, the rho is the partial derivative of the premium with respect to the interest rate.*

As indicated in the preceding chapter, the call value goes up and the put value goes down as the interest rate increases. With the BS model we obtain:[14]

$$\rho_C \equiv \frac{\partial C}{\partial r} = (T - t)Ke^{-r(T-t)}N(d_2) > 0, \tag{12.8a}$$

$$\rho_P \equiv \frac{\partial P}{\partial r} = (N(d_2) - 1)(T - t)Ke^{-r(T-t)} < 0. \tag{12.8b}$$

The proofs for these equations can be found in Appendix 1.

The sensitivity to interest rates that rho expresses can usefully be interpreted starting from the portfolio which synthesizes the option: recall that the replicating portfolio is made up of a zero-coupon bond whose maturity coincides with the option's (a short position for a call and a long one for a put) and a quantity of the underlying equal to delta (a long position for a call, and short for a put). The risk reflected in the rho is that of a variation in the price of a zero-coupon bond; from this viewpoint, it is "symmetric" to the risk of variation in the underlying's price linked to the delta. In any case, this sensitivity is often ignored since its quantitative significance is usually not of the same order as that of delta because the zero-

---

[14] It is legitimate to apply the BS model in a context with variable rates, even stochastic rates; as we have seen in the preceding chapter, the relevant rate is the rate for maturity $T$ and the relevant volatility is that of the forward price.

coupon volatility is most often much less than the underlying's, especially for options with short maturities. In addition, stockbrokers often delegate the management of interest rate risk to specialists, instructing them regularly to buy (or sell) on their behalf the zero-coupon bonds they need to zero out the rho sensitivity.

> **Example 5**
> In the setting of preceding examples, the call's rho is 0.535 and the put's is − 0.718: approximately, the call makes 54 cents and the put loses 72 cents if the interest rate goes up from 9.5 % to 10.5 %.

The rate risk associated with an option on interest rates, i.e. one whose underlying is a fixed-income asset or a futures/forward contract on such an asset, deserves a special attention. In this case, the underlying's price $S(t,r)$ is determined by the interest rate $r$ and it is useful to write the value $O$ of the option as $O(t, S(t,r))$, where $O(.)$ is either a call or a put. As a result, we have:

$$\frac{dO}{dr} = \frac{\partial O}{\partial r} + \frac{\partial O}{\partial S}\frac{\partial S}{\partial r} = \rho + \delta \times Variation(S).$$

The option's interest rate sensitivity (a total derivative, not a partial one) can be analyzed as the *sum of two components*: the direct variation equal to the rho (also equal to the direct variation of the zero-coupon part of the replicating portfolio), and the indirect variation (much more significant for fixed-income underlyings) which equals the product of the option's delta by the underlying's variation.

### 12.3.6 Sensitivity to the Dividend Rate

The sensitivities for standard options are derived in Appendix 1 using Merton's model where the underlying pays out a dividend (or coupon) at a continuous rate. It is then useful to compute the sensitivity to the dividend rate $c$ *(which, curiously, was given no (Greek) name)*.

It is easy to derive, using the results in Appendix 1,

$$\frac{\partial C}{\partial c} = -(T - t)Se^{-c(T-t)}N(d_1) < 0,$$

$$\frac{\partial P}{\partial c} = (T - t)Se^{-c(T-t)}(1 - N(d_1)) > 0.$$

The dividend risk (linked to the uncertainty about $c$), is often ignored in theoretical presentations although it can be significant in practice. It may therefore be important to hedge this risk in a specific way, if the market allows it, by using suitable tools. These are futures (futures prices do not carry the underlying's dividend risk) and dividend *swaps* that allow fixing the expected dividend streams.

## 12.3 Option Sensitivities (Greek Parameters) 473

### 12.3.7 Elasticity and Risk-Expected Return Tradeoff

Recall from Sect. 12.2.3.1 that the option elasticity (omega), denoted by $\Omega_C$ *(call) or* $\Omega_P$ *(put), measures the sensitivity of the relative variation in the option value to that of its underlying asset:*

$$\Omega_C \equiv \frac{\partial C/C}{\partial S/S} = \frac{S}{C}\delta_C, \text{ and } \Omega_P \equiv \frac{\partial P/P}{\partial S/S} = \frac{S}{P}\delta_P.$$

*It is easy and important to see that* $\Omega_C$ *is larger than 1, as* $S.\delta_C = S.N(d_1)$ and $C = S.N(d_1) - Ke^{-r(T-t)}.N(d_2)$. Also, $\Omega_P$ is obviously negative (because of $\delta_P$) and, in general, smaller than -1 (except in the very rare case where the underlying's price is smaller than half the strike).

In addition, omega increases with the strike, for a given underlying's price (and decreases with the latter, for a given strike). Indeed, the call becomes more out-of-the-money and its loss of value is larger than the decrease in its delta ($S/C$ increases more than $\delta_C$ decreases). The put becomes more in-the-money and its gain of value is larger than the increase in the absolute value of its delta, so that $|\Omega_P|$ *decreases and* $\Omega_P$ *increases (towards zero).*

*The correct assessment of the risk-expected return tradeoff of an option requires the knowledge of its elasticity. Within the standard* capital *asset pricing model (CAPM, presented in Chap. 22), the expected return rate in excess of the riskless rate (r) on any risky asset must in* equilibrium *be equal to (Eq. (22.5) of Chap. 22):*

$$(*) \ (\mu_i - r) = \beta_i(\mu_M - r)$$

where $\mu_i$ and $\mu_M$, respectively, denote the expected return rates on asset $i$ and on the whole financial market $M$, and $\beta_i$, the beta of asset $i$, denotes its risk, measured by the covariance of its return with the market return divided by the variance of the market return ($\sigma_{i,M}/\sigma_M^2$).

The call elasticity can be rewritten as:

$$\Omega_C \equiv \frac{\partial C/C}{\partial S/S} = \frac{\partial C/C}{\partial M/M} \cdot \frac{\partial M/M}{\partial S/S} = \beta_C \sigma_M^2 \frac{1}{\beta_S \sigma_M^2} = \frac{\beta_C}{\beta_S}.$$

From which we have at once:

$$\beta_C = \Omega_C \beta_S,$$

and, applying equation (*) to $i = S$ then to $i = C$:

$$(\mu_C - r) = \Omega_C(\mu_S - r).$$

Thus, the risk premium attached to a call is larger than that associated with its underlying by the multiple $\Omega_C$, as its risk *(measured by its beta) is higher by the same multiple. The more the call is out-of-the-money, the riskier it is and the larger*

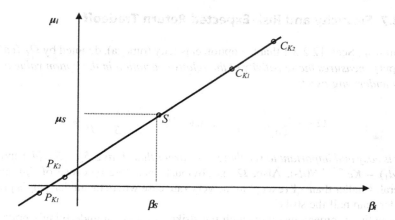

**Fig. 12.8** The risk-expected return tradeoff

is its expected excess return. It is noteworthy that, even though the underlying's beta was supposed constant through time, the call beta can never be constant since $\Omega_C$ changes constantly according to the variations in the underlying's and option's values.

The same two relationships hold for a put. However, since $\Omega_P$ is negative, the risk premium of the put is negative (assuming that that of the underlying is positive, which is generally the case)! The put has a large volatility and yet its expected return is smaller than that of the riskless asset, and even most of the time negative, since its beta is negative: when associated with a long position on its underlying, the global risk of the resulting portfolio is reduced, since its price co-varies negatively with the underlying's price. In other words, the put is a strong risk diversifier, which makes it desirable and justifies its low expected return.

These results are reflected in Fig. 12.8, which displays the linear relationship between risk premium and beta for a typical underlying, two calls and two puts of the same maturity but different strikes $K_1$ and $K_2$. Without loss of generality, we have assumed $K_1 < K_2$.

Finally, let us note that the call and put volatilities are also linked to the volatility of their underlying by the (absolute value of) their elasticity:

$$\Omega_C \equiv \frac{\partial C/C}{\partial S/S} = \frac{\sigma_C}{\sigma_S} \quad \Rightarrow \sigma_C = \Omega_C \sigma_S.$$

Similarly, we obtain for the put: $\sigma_P = |\Omega_P|\sigma_S$.

## 12.4 Dynamic Management of an Option Portfolio Using Greek Parameters

The sensitivities defined in the previous section allow us to describe and control the variations in portfolio value over short time intervals. Because of this, they are indispensable tools for managing the risk inherent in portfolios including options, and for carrying out dynamic strategies for such portfolios.

### 12.4.1 Variation in the Value of a Position in the Short Term and General Considerations

We explain here, using the Greek parameters, the variation in value $V(S, \sigma, r, t)$ *of a portfolio containing a stock (or any asset) with price S* as well as different options written on it. The variation is computed on a small, but not infinitesimal, time interval $(t, t+\Delta t)$ and depends on the variations $\Delta S$ and $\Delta \sigma$ during this period.

A simple expansion of $V$ retaining only the terms most important in practice[15] writes

$$\Delta V \approx \delta \cdot \Delta S + \frac{1}{2} \cdot \Gamma \cdot (\Delta S)^2 + \theta \cdot \Delta t + \upsilon \cdot \Delta \sigma + \rho . \Delta r. \tag{12.9}$$

The parameters $\delta, \Gamma, \theta, \upsilon$ and $\rho$ are therefore tools for managing the three risk sources associated with variations in $S$, $\sigma$ and $r$, and the passage of time.[16]

Let us remark that to make the portfolio insensitive to variations in the underlying, it is necessary to cancel its delta. This hedge, however, can only be local and thus imperfect because of the gamma term.

In addition, to immunize the position against volatility risk, it is necessary to make its vega vanish. Since an option vega is positive and that of the underlying is zero, a portfolio with a zero vega must be both long and short options. The same applies to a portfolio to have zero gamma.

*In the sequel, we ignore the risk associated with the parameter* $\rho$.[17]

---

[15] We neglect the influence of terms "of order 2" (in fact, they are of order 1) except for $(\Delta S)^2$; the latter does have a homogeneous term in $s^2 S^2 \Delta t$ (indeed $\Delta W^2 = \Delta t$). Other terms in the series expansion of order 2, also homogeneous in $\Delta t$ $((\Delta s)^2$ and $(\Delta s \, \Delta S))$, are usually considered of lesser importance in practice.

[16] We do not take into account the variation in the possibly existing dividend rate.

[17] More precisely, we assume this risk is independently monitored by taking positions in zero-coupons matching the options' maturities.

476          12 Option Portfolio Strategies: Tools and Methods

## 12.4.2 Delta-Neutral Management

### 12.4.2.1 Preliminaries

Consider the BS universe where s is constant. We analyze here, using Greek parameters, the variation in the value $V(S, t)$ of a self-financing portfolio containing a stock with price $S$, different options written on this stock and a risk-free asset. According to the theory presented in the preceding chapter, such a portfolio satisfies BS's SDE:

$$\frac{E^Q(dV)}{dt} = \frac{\partial V}{\partial t} + \frac{\partial V}{\partial S} rS + \frac{1}{2} \frac{\partial^2 V}{\partial S^2} S^2 \sigma^2 = rV.$$

This SDE shows that the risk-neutral expectation of the portfolio's return equals the risk-free interest rate: $\frac{E^Q(dV)}{Vdt} = r$.

Another interpretation, which we find just as useful, is to rewrite the BS equation as

$$E^Q(dV) - rVdt = 0.$$

The first term ($E^Q(dV)$) is the expectation of the profit (or loss) in the interval ($t, t +dt$), due to capital gains (or losses) only since the portfolio is self-financing, while the second term ($rVdt$) is the cost of financing, or carrying, this portfolio whose value is $V$ over the same infinitesimal time interval.[18] The SDE, therefore, means that *the risk-neutral expectation of the P&L (Profit and Loss), net of financing costs is zero.*

Making use of the Greek parameters, BS's SDE can also be written

$$\theta + \delta rS + \frac{1}{2}\Gamma S^2 \sigma^2 = rV. \tag{12.10}$$

BS's SDE, therefore, leads directly to Eq. (12.10) relating the three parameters $\theta$, $\delta$, and $\Gamma$.

A portfolio is called *delta-neutral* if its delta vanishes; in that case, which is fundamental in practice, (12.10) becomes

$$\theta + \frac{1}{2}\Gamma S^2 \sigma^2 = rV. \tag{12.11}$$

If, in addition, the delta-neutral portfolio has a zero value $V$, borrowing and short positions financing long positions (which is often the case for managed option

---

[18] As a rule, the initial value of a self-financing portfolio, $V(0)$, is not zero and so brings with it a positive (negative) holding cost (gain) linked to the risk-free rate $r$ over the agent's investment horizon.

## 12.4 Dynamic Management of an Option Portfolio Using Greek Parameters 477

portfolios in trading rooms), or if one takes account of the carrying cost (or gain) $rV$ for $V$ positive (negative) in calculating $\theta$,[19] Eq. (12.11) becomes

$$\theta + \frac{1}{2}\Gamma S^2 \sigma^2 = 0. \tag{12.12}$$

Therefore, the $\theta$ and the $\Gamma$ of a delta-neutral, zero-value portfolio must have opposite signs. Roughly, a portfolio long in options is $\Gamma$-positive and $\theta$-negative while the reverse situation prevails for a portfolio short in options.

While a positive theta is an advantage (since the portfolio appreciates as time goes on), further on we will see that a positive gamma is also desirable: as a result, delta-neutral management involves a *trade-off* between $\theta$ and $\Gamma$ (*see* Sect. 12.4.2.3 below).

In practice, a portfolio obeys a delta-neutral strategy if its composition is frequently (but, obviously, not continuously) adjusted so that its delta never is "too" far from zero. In order that the portfolio's delta be zero at the instant $t$, components must be selected so that $\sum_{i=1}^{N} n_i(t)\delta_i(t, S) + n_S(t) = 0$. But, from that instant onward, the delta will no longer stay at zero (simply because the passage of time and variation in the price of the underlying change the options' deltas): to avoid excursions beyond a given tolerance threshold, the portfolio's composition must periodically be revised. Here lies the prime importance of the gamma.

### 12.4.2.2 Impact of the Underlying's Price Variation on a Delta-Neutral Position According to the Sign of Gamma

We analyze the behavior of a $\delta$-neutral portfolio when there is a non-infinitesimal movement $\Delta S$ in the price of the underlying. To start with, we ignore the effect of time's passing and assume $\sigma$ and $r$ are constants as in the BS world. The analysis makes use of Fig. 12.9. At time $t = 0$, the price of the underlying is $S_0$ ($\$100.50$ in the example shown in Fig. 12.9) and the position is assumed to be delta-neutral; a short time later (e.g., the following day) the underlying is worth $S_0 + \Delta S$ and $\delta$ as well as the portfolio value have changed.

This assertion leads to the following *interpretation* of the gamma which is the most useful in practice: $\Gamma \Delta S$ is the *number of underlying* shares needed to buy or sell so that the portfolio recovers delta-neutrality in the wake of a variation $\Delta S$ in the price of the underlying.[20] Indeed, such a change implies $\Delta \delta \approx \Gamma \Delta S$; since the underlying's gamma is zero and its delta is equal to one, the variation in the portfolio's delta must be offset by a variation in the opposite direction in the number of underlying shares equal to $\Gamma \Delta S$.

---

[19] What is relevant is the *net P&L*, as the carrying cost (gain) $rV$ reduces (increases) $\theta$ so that (12.12) holds.

[20] The proposition does in fact hold more generally than it may seem and is also applicable to positions that are not delta-neutral but may have any (target) delta. The gamma is the number of shares it is required to buy or sell to recover the desired, target delta.

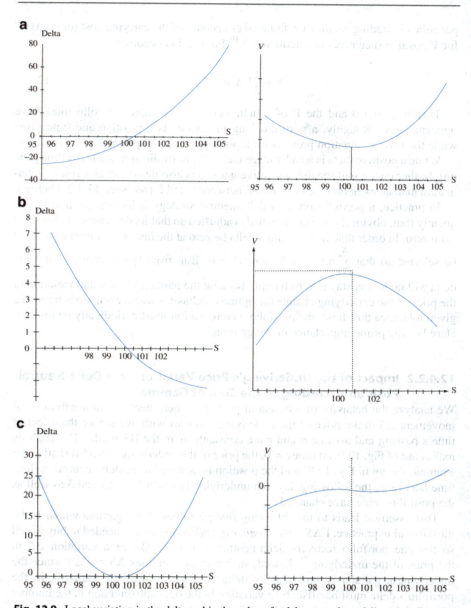

**Fig. 12.9** Local variation in the delta and in the value of a delta-neutral portfolio. (**a**) A δ-neutral portfolio (at 100.5); Γ > 0. (**b**) A δ-neutral portfolio; Γ < 0. (**c**) A δ-neutral portfolio; Γ neutral

## 12.4 Dynamic Management of an Option Portfolio Using Greek Parameters 479

> **Example 6**
> Consider a portfolio made up of 1 call purchased and 0.51 underlying sold. Since the call's delta is 0.51, the portfolio is delta-neutral ($1 \times 0.51 - 0.51 \times 1 = 0$). Assume the call's gamma is 0.06 and that the underlying's price goes up from 100 to 101. The P&L for the position is indeed zero since the call rises in value by its delta (+0.51) but the short position in the underlying loses $1 \times (-0.51)$. However, the global delta of the position is not zero anymore but has become $0.57 - 0.51 = 0.06$, the call's gamma. Therefore, it is necessary to sell 0.06 underlying to restore the delta-neutral position. An analogous argument handles the case where the underlying price falls from 100 to 99. One needs to buy 0.06 underlying. It is important to note that these trades in the underlying *only modify the delta*, and not the other Greek parameters.

In agreement with what can be seen in Fig. 12.9, in which initially ($t = 0$) and for $S = S_0$ the portfolio's gamma is initially either positive, negative, or zero, the variation $\Delta S$ will have a different effect on the portfolio's delta and value ($S_0 = 100.5$ in the example graphed in Fig. 12.8).

- If $\Gamma(S_0) = \frac{\partial^2 V}{\partial S^2}(S_0) > 0$, the portfolio's value is a convex function of $S$ in the neighborhood of $S_0$ and it will have positive variation whatever the sign of $\Delta S$ (*see* Fig. 12.9a).
- If $\Gamma(S_0) < 0$, the portfolio's value is a concave function of $S$ and the variation will be negative whatever the sign of $\Delta S$ (*see* Fig. 12.9b).
- If $\Gamma(S_0) = 0$, $V(S)$ will have an inflexion point at $S_0$ and the impact of $\Delta S$ will be very small (*see* Fig. 12.9c).

In fact, the local variation in a delta-neutral position's value caused by the underlying's price variation $\Delta S$ writes:

$$\Delta V = \frac{1}{2}\Gamma(\Delta S)^2.$$

$V(S)$ is therefore locally a parabola whose minimum is the spot price $S_0$ for which the delta vanishes. A position with a positive gamma, therefore, benefits from any change $\Delta S$ (positive or negative) in the underlying's price while the same movements produce losses in the value of a portfolio with negative gamma.

The convexity characterizing a $\Gamma$-positive position thus is a great advantage, popular with traders.

As one may conclude from Example 6, the position is all the more profitable because the number of times it is necessary to adjust the holdings of the underlying is large (the volatility about the drift is high), because the underlying is sold when it is expensive (101) and bought when it is cheap (99) and/or because the variation in the underlying price is pronounced (e.g., 102 and 98). Since buying an option (to be

$\Gamma$-positive) amounts to initially buying volatility, an investor benefits from the effective average volatility until the option maturity being *as large as possible* so that the gains resulting from trading the underlying offset (and more) the premium paid initially. That is what *gamma monitoring* is about. If, on the contrary, the effective average volatility is less than what was purchased, managing the gamma will not be enough to generate a positive net P&L, and money will be lost.

Symmetrically, it is in the interest of a seller of options with a $\Gamma$-negative position that the effective volatility be *as small as possible* (after selling) since trading the underlying at a loss (buying at 101 and selling at 99) will be needed less frequently and each loss will be smaller.

Considered from this point of view, trading options whose gamma is to be monitored can be analyzed as a bet on the difference between the volatility bought or sold initially (the *implicit* volatility) and the future average volatility that will occur (the *realized* volatility).

However, as we already remarked in Sect. 12.4.2.1 and we show in the following subsection, it is worth understanding that the advantage (disadvantage) of a positive (negative) gamma has an opposite counterpart in theta.

Finally, we must emphasize that this analysis only makes sense locally. The position can only be described by a parabola near the underlying price $S_0$ for which a $\delta$-neutral hedge was devised. On a larger scale, the curve is not a parabola and, generally, it is not even symmetrical.

### 12.4.2.3 Variation in the Value of a Delta-Neutral Position According to the Signs of $\Gamma$ and $\theta$

For pedagogic clarity, we restricted ourselves in the preceding subsection to studying the effects of underlying price variations $\Delta S$ assumed to be instantaneous (jumps). In fact, such a change $\Delta S$ usually occurs in the time interval $(t, t+\Delta t)$. Therefore, we now analyze the results for a delta-neutral position *taking into account the passage of time*. We have seen that for a delta-neutral option position, theta and gamma have opposite signs.[21] In this way, other things being equal, a position with positive gamma loses value over time while a position with negative gamma appreciates. The advantage provided by the convexity of a positive-gamma position is offset by a corresponding negative theta. The trader must therefore strike a balance between $\theta$ and $\Gamma$ (*gamma/theta tradeoff*).

More precisely, still assuming a constant volatility, over an interval of length $\Delta t$ the change in the value of a delta-neutral portfolio (has a parabolic shape with respect to $\Delta S$ and) can be written

$$\Delta V \approx \frac{1}{2}\Gamma(\Delta S)^2 + \theta\Delta t.$$

---

[21] That is always true for a position with zero value and, if not, it is still true if one takes the portfolio's financing cost into account.

## 12.4 Dynamic Management of an Option Portfolio Using Greek Parameters

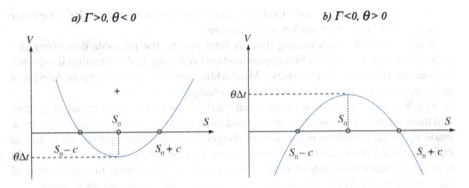

**Fig. 12.10** Variation in a position's value over an interval $\Delta t$. (a) $\Gamma > 0, \theta < 0$, (b) $\Gamma < 0, \theta > 0$

A long option position (usually) has a positive gamma and a negative theta. The parabola points downward and the ordinate of its extremum is $\theta \Delta t$ (negative) (see Fig. 12.10a).

Conversely, a negative-gamma position is locally represented by a parabola that points upward and whose extremum is positive (see Fig. 12.10b).

Therefore, for a given time interval $\Delta t$, there is a change $\Delta S = c$ in the underlying's price that does allow an exact compensation for the passage of time.

For any change in the underlying price with absolute value $|\Delta S| > c$, a position with $\Gamma > 0, \theta < 0$ wins and a position with $\Gamma < 0, \theta > 0$ loses, and the opposite results are obtained when $|\Delta S| < c$.

Setting $\Delta V \approx \frac{1}{2}\Gamma c^2 + \theta \Delta t = 0$, we conclude that the critical variation value $c$, called the break-even point, is

$$c \approx \sqrt{-2\frac{\theta \Delta t}{\Gamma}}.$$

Then using Eq. (12) which relates the parameters $\Gamma$ and $\theta$, we obtain

$$c \approx \sigma S \sqrt{\Delta t}.$$

This magnitude is (to the first order) a standard deviation of the underlying's variation distribution, $\sigma(\Delta S)$.

Thus, if the *realized* volatility turns out to equal s, the break-even point is attained or exceeded in about 32% of cases.

The delta-neutral strategy is therefore no panacea. We have already remarked before that the risk-neutral expectation of $\Delta V$ either vanishes (Eq. (12.12), which holds if $V = 0$) or equals the portfolio financing cost over the period under consideration $(t, t+\Delta t)$ (Eq. (12.11)). Therefore we remark unsurprisingly, since the market is supposed to be arbitrage-free, that *in terms of the risk-neutral expectation the effects of $\sigma$ and $\Gamma$ exactly cancel each other*. As a result, implementing a winning delta-neutral strategy over the long term depends on a trader's *superior* forecasting ability, or on market *imperfections*.

It is also worth remembering that, as time passes, the parabola describing $\Delta V$ (Fig. 12.10a, b) moves (to a first approximation) vertically in the direction in which it is pointing (since $|\theta \, \Delta t|$ increases). Meanwhile, one has also to keep in mind that gamma, and thus the parabola's curvature, changes over time.[22]

Finally, we also note that options that "weigh" most heavily in terms of gamma and theta are options around the money and close to expiry. In this respect, a position made up of such options is potentially dangerous since these options are nearly at their intrinsic values and their gamma is very high and very sensitive. This fact, in practice, means that sellers of options who practice delta-neutral management tend most often to get rid of options that are close to the money before they expire.

#### 12.4.2.4 Taking into Account Variations in Volatility

So as to adjust to market conditions, options traders regularly adjust the volatilities used in pricing the options they handle and revalue those they have in their portfolio. This reevaluation leads to profits and losses linked to their vegas.

Roughly speaking, active management often consists in adopting option positions with a certain level of volatility and conducting delta-neutral management with the goal of selling (buying back) them at a higher (lower) volatility level, thus generating vega-related gains. The idea is to build a position adapted to the prediction regarding the volatility's future evolution (one chooses a positive vega if an increase in volatility is forecast and negative in the reverse case) and to the short-term evolution of the underlying (the gamma/theta trade-off).[23] We have seen that this also holds for the gamma (vega and gamma are positive for a long option position, negative for a short one), which is often a source of confusion to uninformed agents). What distinguishes these two Greek parameters is that the profit from gamma requires (more or less) continuous rebalancing while the gain from the vega is instantaneous and cannot be managed in the strict sense of the word.

#### 12.4.2.5 Dynamic Pseudo-Arbitrages

The least random sort of position would be delta-, gamma-, and vega-neutral. Some strategies of *pseudo-arbitrage* (i.e., leaving a theoretically small residual risk) can be developed on such a basis. We are looking for a position $x$, which is $\delta$-neutral, and as $\upsilon$-neutral and $\Gamma$-neutral as possible, whose expected return $i$ is different from the interest rate $r$: if $r < i$, one takes a long position on $x$ financed by borrowing at the rate $r$, and if $r > i$, one takes the reverse position. Such pseudo-arbitrages can involve an option and its underlying, or several options and their common underlying.

---

[22] If the option is near the money, its gamma increases and the parabola becomes sharper. Conversely, the gamma of an option far from the money decreases and the parabola becomes flatter.

[23] It is, in any case, important to note that, for market makers, overturning positions is constrained by their statutory obligation to serve the market. Recall that the role of market makers is to ensure the market's liquidity: within the limits of the *bid-ask* spreads that they post, they must buy or sell a minimum quantity of options to any market participant who requests it.

## 12.4 Dynamic Management of an Option Portfolio Using Greek Parameters 483

The principle behind option/stock pseudo-arbitrage is simple: one estimates the theoretical value of an option with a BS model and compares the resulting value with the effective value observed on the market; then one builds a δ-neutral portfolio by combining stock and option. This type of strategy depends on an estimate for the future volatility deemed more accurate than the volatility implied by the option market price, and suffers, beyond the risk inherent in any estimate, from the imperfections of the valuation model used. This drawback is partially or wholly avoided with an option/option or option/option + stock pseudo-arbitrage. The principle consists, in its crudest form ignoring the smile phenomenon we will mention later on, in identifying two options O1 and O2 with different implicit volatilities (but same maturity). If, for example, the volatility of O1 is larger than that of O2, the latter is undervalued, not in an absolute sense, but *relative to* O1. Thus one sells O1 and buys O2 so as to make a δ-neutral portfolio.[24] This avoids estimating the relevant volatility. Furthermore, the errors of a model which would be biased to either over-value or under-value option prices partially compensate one another (since the position is both long and short options).

A more sophisticated version of the method consists in comparing the implied volatilities of the two options relative to a smile curve or even to a volatility surface. To consider O1 over-valued relative to O2, it is necessary that the volatility surface (or smile curve) considered as normal divides the two implicit volatilities, that of O1 being above and that of O2 below the surface (or curve). Typically this sort of pseudo-arbitrage involves several options (calls and puts with different strikes and expiry dates) as well as the underlying asset; with several options, it is possible to simultaneously control for several Greek parameters.[25]

For example, market participants can benefit from the smile if they judge it to be too pronounced. To achieve that, they sell options that are out-of-the-money and buy options more nearly in-the-money. That is the situation illustrated in Fig. 12.11: the different options are represented by little circles, the observed smile curve (which best fits these different options' implied volatilities) is the continuous line, while the normal smile curve is the dashed line. A position like this usually involves a positive gamma and a negative theta since nearly in-the-money options (to be bought) have a greater convexity and a more negative theta than those far-from-the-money (to be sold).

Such a position, although gamma-positive, is in any case not without risks if it is not closely monitored. For example, if the underlying's price goes up, the gamma of the calls in-the-money decreases while that of calls initially out-of-the-money increases as they become near-the-money. The position as a whole then can acquire a negative gamma, which can imply significant risk. Thus, there does exist, in fact, a

---

[24] Precisely, one sells a number $\delta_2$ of options O1 and buys a number $\delta_1$ of options O2 so that $-\delta_2\delta_1 + \delta_1\delta_2 = 0$.

[25] To control for the delta, there has to be at least one option (in addition to the underlying); for every additional Greek to be controlled, there has to be at least one additional option.

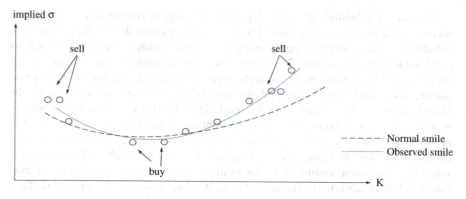

**Fig. 12.11** Options/options pseudo-arbitrages involving the smile

directional risk that can materialize if the underlying's price changes rapidly and/or rebalancing is not fast enough.

### 12.4.2.6 Obtaining Greek Parameters of Any Sign

We have seen that an option purchased has its gamma and vega positive and theta (generally) negative, while the opposite holds if the option is sold. In practice, a trader may wish to hold, for example, a portfolio for which the gamma and vega have opposite signs. Although it seems clear that this cannot be attained with a single option, it is easy to obtain with a basket of options with different characteristics (strikes and expiration dates). The basket can be very complicated (generally a whole *book* of options) but we can show with a simple example that holding only two options can be enough to produce a portfolio whose gamma and vega are of opposite signs.

---

**Example 7**

Suppose the stock ABC is presently quoted at 100. The volatility the market forecasts is 40%. The yield curve is flat at 4.1% (discrete rates). Assume that a trader expects in the short term (a month) that ABC's price will undergo an abrupt change (of unknown sign), as the market is nervous about ABC's uncertain investments. He also expects that the volatility will then reverse to its previous level after 6 months time, the market calming down due to the resolution of uncertainty. The trader consequently wants to build a delta-neutral position (he is unsure of the direction in which ABC's price will change), which is gamma-positive (to profit from big short-term movements) and vega-negative (to benefit from the eventual reduction in volatility).

He can easily attain this goal with a calendar spread consisting of buying a call at-the money-forward ($K = 99$) and short-lived (31 days) and selling a call of the same strike but with longer maturity (183 days). Indeed, we know that

(continued)

the gamma rises, all things being equal, as expiration nears and the option is nearly at-the money-forward. To avoid problems with rounding we consider 1000 calls bought and sold.

Taking market data into account, results from BS lead to the following numbers (for one option):

|       | Shorter call bought | Longer call sold | Net position |
|-------|---------------------|------------------|--------------|
| Value | 5.32                | 12.65            | –7.33        |
| Delta | 0.569               | 0.599            | –0.029       |
| Gamma | 0.034               | 0.014            | 0.020        |
| Vega  | 0.115               | 0.274            | –0.159       |
| Theta | –0.080              | –0.036           | –0.044       |

The trader thus received $7330 for the options, the position has a positive gamma of 20, a negative vega of 159, and is delta-neutral if 29 units of the underlying are purchased (the net cash inflow is, therefore, $4430). The global theta is negative (–44) since for options that are nearly at-the money-forward the absolute value of theta increases as the option's live becomes shorter.

## 12.4.3 A Tool for Risk Management: The P&L Matrix

To measure and control risks from price variations in the underlying or from volatility changes, one can use a P&L matrix (also called risk matrix). This is a double-entry table whose entries are the position's changes in value for a range of variations in $S$ centered on the present price (as abscissa) and a range of variation in volatility (as ordinates) for a given short expiration. The most negative value in the table represents the largest possible loss (with a small but not negligible probability). This maximum loss must at the end of each day's trading be less than a threshold fixed by the risk control department of the financial institution or company concerned. This risk matrix is an important tool in managing a portfolio whether at a global or an individual level. In particular, the trader can see the impact of any operation on the matrix and adjust the portfolio so as to respect the limits that have been set by the hierarchy.

Let us remark once again that options with short maturity and not too far from the money have large convexity and therefore a particularly significant influence on the P&L recorded in the matrix. So as not to go outside the limits fixed by their desk, traders can buy such options, at a significant cost in terms of theta.

> **Example 8 (Simplified)**
>
> Below we give a risk matrix centered on an underlying price of $37.80 and a volatility of 35% (the situation today). For these values of $S$ and $\sigma$ (entry cell in bold font), the P&L = **0** (the left number) and delta equals **0.14** (the number on the right). Each cell gives the P&L and the delta of the position corresponding to different values of $S$ and $\sigma$. If, for example, tomorrow the underlying's price rises to $40.40 while the volatility remains constant at 35%, the position's value will rise by $0.4K and its delta will be 0.20. Note that this P&L of +0.4 resulting from the price rise is approximately $0.14 \times (40.4 - 37.8)$ ($\Delta V = \delta\, \Delta S$).
>
> $S$ = price of 1 share in \$; Position with 1000 shares; P&L in \$K
>
> | $S$ | $\sigma$ | 27% | 31% | 35% | 39% |
> |------|------|------|------|------|------|
> | 29.8 | | − 4.1; −1.85 | −1.9; −1.51 | 2.8; −0.96 | 7.9; −0.40 |
> | 32.5 | | −7.9; −1.48 | −4.8; −0.71 | 0.7; −0.27 | 7.0; 0.15 |
> | 35.2 | | −11.1; −1.0 | −6.3; −0.38 | −0.4; 0.09 | 5.7; 0.61 |
> | 37.8 | | −13.8; −0.88 | −7.3; −0.34 | **0; 0.14** | 7.4; 0.67 |
> | 40.4 | | −16.7; −0.33 | −8.4; −0.19 | 0.4; 0.20 | 9.5; 0.79 |
> | 43.1 | | −16.1; 0.38 | −7.5; 0.51 | 1.9; 0.77 | 12.3; 1.02 |
> | 45.8 | | −14.0; 1.34 | −5.3; 1.62 | 4.9; 1.78 | 16.0; 1.84 |

Usually, the P&L matrix has many more rows and columns than in this sketched example. Most of the time it is required that the position leads to a matrix whose entries are all above a certain threshold called the *"disaster limit"* corresponding to the maximum loss possible in a "catastrophic" variation in the market price (in the example $16.7K, 5th row, 2nd column).

## 12.5  Summary

- *Static (or buy and hold) strategies* do not require any rebalancing between their initiation on date 0 until they are unfolded on date $T$.
- The *P&L (profit and loss)* of a static strategy is the portfolio's final value on $T$, net of the initial investment needed to implement it on date 0, i.e., $V(T) - V(0)$.
- An elementary static strategy involves an underlying asset and different options written on it and expiring on the same date $T$. Its P&L is a *piecewise linear function* of the final asset value, the slope of which changes at each different strike of the options.
- *Typical elementary static* strategies are spreads, straddles, strangles, butterflies and condors.

## 12.5 Summary

- The volatility $\sigma$ of the underlying is the only not directly observable argument of Black-Scholes type (BS) formulas. It has thus to be estimated (*historical volatility*) or inferred from option prices (*implicit volatility*).
- The historical volatility of the underlying is an empirical estimation obtained from previously observed returns assumed to be i.i.d., which implies a constant volatility. In fact, the volatility is *time-varying*, and the historical volatility only reflects the *past*, while the $\sigma$ relevant for option pricing is an average of *expected future* volatilities, impossible to estimate statistically.
- This issue explains the need for the notion of *implied volatility*. Inverting the relevant BS formula yields the unique volatility $\sigma_i$ implied by an *observed* option price (O): for a given O, there is only one $\sigma$ such that $O = BS(\sigma; \underline{x})$ and one writes: $\sigma_i = BS^{-1}(O_{observed}; \underline{x})$.
- Since an option price increases with its $\sigma_i$, an option is *dear or cheap*, not in monetary terms (which strongly depends on its moneyness), but according to its $\sigma_i$ compared to the volatility implicit in other option prices: $\sigma_i$ is the useful benchmark for dearness since it should be *common* to all options written on the *same* underlying.
- However, we observe that the implied volatilities of options written on the same asset depend on both their strikes and expiration dates. This « anomaly » stems from the *over-simplification* of the BS model whose assumptions do not conform to reality (such as normality of returns, constant volatility).
- The curve giving observed $\sigma_i$ as a function of the strike is called the *smile*, or *skew*, depending on its shape, and the curve giving observed $\sigma_i$ as a function of the maturity is called the *volatility term structure*. The function $\sigma_i(K, T)$ graphed in three dimensions is the *volatility surface*.
- In methods more sophisticated than that assuming a unique volatility, the $\sigma_i$ relevant for pricing purposes is drawn from a smile, a term structure, or a volatility surface *empirically constructed*.
- Partial derivatives of the BS-like pricing functions, dubbed the "*Greeks*," quantify the sensitivity of an option price O to changes in S, $\sigma$, t or r.
- By definition, $Delta_O \equiv \frac{\partial O}{\partial S}$. Exact values can be obtained from BS-type formulas. It is also the *hedge ratio*. A call has a positive delta increasing (with S) from 0 to 1. A put has a negative delta increasing from $-1$ to 0.
- To the first order, for a small variation $\Delta S$, we have: *Variation of O = Delta$_o$×* $\Delta S$.
- To the second order, we have: *Variation of O = Delta$_o$× $\Delta S$ + Gamma$_o$ × $(\Delta S)^2$*, with $Gamma_o \equiv \frac{\partial^2 O}{\partial S^2}$, which is *positive*. Gamma is the only standard "Greek" which is a second derivative.
- $Vega_O \equiv \frac{\partial O}{\partial \sigma}$ is *positive*. For small variations of volatility $\sigma$, we have: *Variation of O =Vega$_O$× $\Delta\sigma$*.
- $Theta_O \equiv \frac{\partial O}{\partial t} \equiv -\frac{\partial O}{\partial T}$ is (in general) *negative*. For a small variation of time: *Variation of O = Theta$_O$ × $\Delta t$.*

- $Rho_O \equiv \frac{\partial O}{\partial r}$ is *positive* for a *call* and (in general) *negative* for a put. For small variations of the interest rate: *Variation of $O = Rho_O \times \Delta r$*.
- Partial derivatives being additive, the delta (gamma, vega, theta, rho) of a *portfolio* is the *sum of the deltas* (gammas, vegas, thetas, rhos) of its components. The Greeks allow describing and controlling the variations in the portfolio value over short time intervals and thus are crucial management tools.
- A portfolio obeys a *delta-neutral strategy* if it is continuously rebalanced such that its delta is always zero. In the arbitrage-free world of BS, such a portfolio must yield a sure return equal to $r$.
- In the real world, the portfolio is frequently (and not continuously) rebalanced in order to keep its delta « close » to zero. In addition, the real world differs from the ideal world of BS. Therefore a delta-neutral portfolio is *not totally insensitive to a non-infinitesimal change $\Delta S$*.
- A delta-neutral portfolio benefits (loses money) from any change $\Delta S$ if its gamma is positive (negative). The advantage of a gamma-positive position is balanced by a drawback: its theta is negative, implying depreciation over time (*gamma/theta tradeoff*). In addition, it is vega-positive, hence bears the risk of a decrease in the underlying's volatility.
- *Pseudo-arbitrages* (i.e. those leaving a *residual risk*) may involve delta-neutral and vega-neutral positions containing *presumably mispriced* options. For instance, short positions on options considered over-valued because their $\sigma_i$'s lie above the smile are coupled with long positions on other options on the same underlying whose $\sigma_i$'s lie below the smile.
- A *P&L matrix* is a useful risk management tool consisting in a double-entry table whose entries are the portfolio's changes in value (P&L) for a range of variations in $S$ and in volatility.

---

## Appendix 1

### Computing Partial Derivatives (Greeks)

To simplify notation, in this appendix we always consider that the current time is $t = 0$.

### The Black–Scholes Model
**Preliminary Result**

Let us start with a result that simplifies calculating Greek parameters. Since the Black–Scholes model writes

$$C = S\, N(d_1) - K e^{-rT} N(d_2)$$

with

# Appendix 1

489

$$d_1 = (\ln(S/K) + (r + \sigma^2/2)T)/\sigma\sqrt{T} \quad \text{and} \quad d_2 = d_1 - \sigma\sqrt{T},$$

we have

(i) $SN'(d_1) = Ke^{-rT}N'(d_2)$

where $N'(z) \equiv \frac{1}{\sqrt{2\pi}}e^{-z^2/2}$ is the density of the reduced centered normal distribution.

**Proof**

From the definition of $d_1$:

$$\ln(S/K) + (r + \sigma^2/2)T = d_1\sigma\sqrt{T}$$

$$\Rightarrow \ln S - \ln K + rT = d_1\sigma\sqrt{T} - \frac{\sigma^2 T}{2} = \frac{1}{2}\left(d_1^2 - \left(d_1 - \sigma\sqrt{T}\right)^2\right)$$

$$\Rightarrow \ln S + \ln\left(\frac{1}{\sqrt{2\pi}}\right) - \frac{d_1^2}{2} = \ln K - rT + \ln\left(\frac{1}{\sqrt{2\pi}}\right) - \frac{d_2^2}{2}$$

which proves (i).

## (a) *Delta*

$$\delta_C = \frac{\partial C}{\partial S} = \frac{\partial}{\partial S}\left(SN(d_1) - Ke^{-rT}N(d_2)\right)$$

$$= N(d_1) + \frac{SN'(d_1)}{S\sigma\sqrt{T}} - \frac{Ke^{-rT}N'(d_2)}{S\sigma\sqrt{T}} = N(d_1) \quad \text{from (i)}$$

Moreover, from call-put parity we have $P = C - S + Ke^{-rT}$, from which

$$\delta_P = \frac{\partial P}{\partial S} = \frac{\partial C}{\partial S} - 1 = N(d_1) - 1.$$

## (b) *Gamma*

$$\Gamma_C = \frac{\partial \delta_C}{\partial S} = \frac{\partial N(d_1)}{\partial S} = \frac{N'(d_1)}{S\sigma\sqrt{T}}$$

since $N(d_1) = N((\ln(S/K) + (r + \sigma^2/2)T)/\sigma\sqrt{T})$.
$\Gamma_P = \Gamma_C$ from call-put parity (its second derivative).

## (c) *Vega*

$$v_C = \frac{\partial C}{\partial \sigma} = SN'(d_1)\frac{\partial d_1}{\partial \sigma} - Ke^{-rT}N'(d_2)\frac{\partial d_2}{\partial \sigma},$$

from which

$$v_C = SN'(d_1)(\sqrt{T} - d_1/\sigma) - Ke^{rT}N'(d_2)\left(-\sqrt{T} - d_2/\sigma\right)$$

since $d_1 = (\ln(S/K) + r\,T)/\sigma\sqrt{T} + \sigma\sqrt{T}/2$ and

$$\frac{\partial d_1}{\partial \sigma} = \frac{-(\ln(S/K) + rT).\sqrt{T}}{\sigma^2 T} + \frac{\sqrt{T}}{2} = -\frac{d_1}{\sigma} + \sqrt{T}$$

and

$$\frac{\partial d_2}{\partial \sigma} = \frac{-(\ln(S/K) + rT).\sqrt{T}}{\sigma^2 T} - \frac{\sqrt{T}}{2} = -\frac{d_2}{\sigma} - \sqrt{T}$$

Consequently, and since $\frac{d_2}{\sigma} = \frac{d_1}{\sigma} - \sqrt{T}$, we have

$$v_c = SN'(d_1)\sqrt{T} - \frac{d_1}{\sigma}\left(SN'(d_1) - Ke^{rT}N'(d_2)\right)$$

$$= S\sqrt{T}N'(d_1)$$

from (*i*). Furthermore, $v_P = \frac{\partial P}{\partial \sigma} = v_C$, by call-put parity.

## (d) *Theta*

The direct proof gives us

$$\theta_C = \frac{\partial C}{\partial t} = -\frac{\partial C}{\partial T} = -SN'(d_1)\left[-\frac{1}{2}\frac{\ln(S/K)}{\sigma}T^{-3/2} + \frac{1}{2}\left(\frac{r}{\sigma} + \frac{\sigma}{2}\right)T^{-1/2}\right]$$

$$+ \left[-rKe^{-rT}N(d_2) + Ke^{-rT}N'(d_2)\left[-\frac{1}{2}\frac{\ln(S/K)}{\sigma}T^{-3/2} + \frac{1}{2}\left(\frac{r}{\sigma} - \frac{\sigma}{2}\right)T^{-1/2}\right]\right]$$

$$= -SN'(d_1)\frac{\sigma}{2}T^{-1/2} - rKe^{-rT}N(d_2)$$

$$- \left[SN'(d_1) - Ke^{-rT}N'(d_2)\right]\left[-\frac{1}{2}\frac{\ln(S/K)}{\sigma}T^{-3/2} + \frac{1}{2}\left(\frac{r}{\sigma} - \frac{\sigma}{2}\right)T^{-1/2}\right]$$

# Appendix 1

$$= -\frac{S\sigma}{2\sqrt{T}}N'(d_1) - rKe^{-rT}N(d_2) \text{ by } (i).$$

One can, alternatively, recall that the option price satisfies the Black-Scholes partial differential equation in its form (12.10). Hence we deduce the following equation relating sensitivities

$$\theta + \delta thickmathspacer S + \frac{1}{2}\Gamma thickmathspace S^2\sigma^2 = rC$$

and therefore

$$\theta_C = \frac{\partial C}{\partial t} = r(C - dS) - 0.5\Gamma\sigma^2 S^2.$$

Replacing the gamma and delta by the expressions given above, we recover the expression for theta.

Furthermore, again from call-put parity,

$$\theta_P = \frac{\partial P}{\partial t} = -\frac{\partial P}{\partial T} = -\frac{\partial}{\partial T}\left(C - S + Ke^{-rT}\right)$$

$$= -\frac{\partial C}{\partial T} + rKe^{-rT} = -\frac{S\sigma}{2\sqrt{T}}N'(d_1) + rKe^{-rT}(1 - N(d_2)).$$

### (e) *Rho*

$$\rho_C = \frac{\partial C}{\partial r} = SN'(d_1)\frac{\sqrt{T}}{\sigma} - \left[Ke^{-rT}N'(d_2)\frac{\sqrt{T}}{\sigma} - T.Ke^{-rT}N(d_2)\right]$$

$$= TKe^{-rT}N(d_2)(> 0) \text{ from } (i).$$

The rho of a put can then be obtained from call-put parity:

$$\rho_P = \frac{\partial P}{\partial r} = \frac{\partial}{\partial r}\left(C - S + Ke^{-rT}\right) = TKe^{-rT}N(d_2) - TKe^{-rT}$$

$$= (N(d_2) - 1)TKe^{-rT} \ (< 0).$$

## Other Models

We start from the Black-Scholes formula:

$$C = BS(S) \equiv S\,N(d_1(S)) - Ke^{-rT}\,N(d_2(S))$$

492    12 Option Portfolio Strategies: Tools and Methods

with $d_1(S) = \left(\ln(S/K) + (r + \sigma^2/2)T\right)/\sigma\sqrt{T}$ and $d_2(S) = d_1(S) - \sigma\sqrt{T}$; moreover, we know $BS' = N(d_1)$.

### (a) *Black's model (option on a forward contract quoted F)*

The model gives the premium as a function of $F$: $C = BS\left(Fe^{-rT}\right)$ where the Black-Scholes $BS(.)$ is defined above and $d_1(F) = \left(\ln(F/K) + \sigma^2 T/2\right)/\sigma\sqrt{T}$. From this, using the Greek parameters of the Black-Scholes model, we obtain almost immediately

$\delta_C = \frac{\partial C}{\partial F} = \frac{\partial}{\partial F}BS\left(Fe^{-rT}\right) = e^{-rT}BS'\left(Fe^{-rT}\right)$, from which: $\delta_C = e^{-rT}N(d_1)$.

$\Gamma_C = \frac{\partial^2 BS(Fe^{-rT})}{\partial F^2} = e^{-2rT}BS''\left(Fe^{-rT}\right) = e^{-2rT}\frac{N'(d_1)}{Fe^{-rT}\sigma\sqrt{T}}$, from which

$\Gamma_C = e^{-rT}\frac{N'(d_1)}{F\sigma\sqrt{T}}$.

$$v_C = \frac{\partial C}{\partial \sigma} = Fe^{-rT}\sqrt{T}N'(d_1),$$

and

$$\theta_C = -\frac{\partial C}{\partial T} = -\frac{dBS(T, Fe^{-rT})}{dT} = -\frac{\partial BS}{\partial T} + Fre^{-rT}BS', \quad \text{whence}:$$

$$\theta_C = -\frac{Fe^{-rT}\sigma}{2\sqrt{T}}N'(d_1) - rKe^{-rT}N(d_2) + Fre^{-rT}N(d_1).$$

The Greek parameters related to a put can be obtained from those for the analogous call using the parity relationship written as $P = C - e^{-rT}(F - K)$, whence:

$$\delta_P = \delta_C - e^{-rT}; \quad \Gamma_P = \Gamma_C; \quad v_P = v_C; \quad \theta_P = \theta_C - Fre^{-rT}.$$

### (b) *Merton's Model (When the Underlying Distributes a Continuous Dividend c)*

This model writes: $C = BS(Se^{-cT})$; $d_1 = (\ln(S/K) + (r - c + \sigma^2/2)T)/\sigma\sqrt{T}$ and $d_2 = d_1 - \sigma\sqrt{T}$.
From this, we have:

$$\delta_C = e^{-cT}BS'\left(Se^{-cT}\right) = e^{-cT}N(d_1); \quad \Gamma_C = e^{-2cT}BS''\left(Se^{-cT}\right) = e^{-2cT}\frac{N'(d_1)}{Se^{-cT}\sigma\sqrt{T}}$$

$$= e^{-cT}\frac{N'(d_1)}{S\sigma\sqrt{T}};$$

# Appendix 2

$$v_C = \frac{\partial BS(Se^{-cT})}{\partial \sigma} = Se^{-cT}\sqrt{T}N'(d_1); \quad \theta_C$$

$$= -\frac{Se^{-cT}\sigma}{2\sqrt{T}}N'(d_1) - rK e^{-rT}N(d_2) + cSe^{-cT}N(d_1).$$

The Greek parameters for a put can be obtained from those for a call using the parity relation written as $P = C - (Se^{-cT} - Ke^{-rT})$.

---

# Appendix 2

## Option Prices and the Underlying Price Probability Distribution

The following proposition highlights the relationship between the prices of standard options and the probability distribution for the future value $S_T$ of the underlying.

### Proposition
*The probability density under $Q$ of the final value $S_T$ of the underlying, $\phi_T(k) = \frac{Q(S_T \in [k, k+dk])}{dk}$, is equal to the capitalized value of the second derivative of the option (a call or a put) with respect to the strike denoted by $k$:*

$$\phi_T(k) = e^{rT}\frac{\partial^2 C}{\partial k^2}(k) = e^{rT}\frac{\partial^2 P}{\partial k^2}(k). \tag{12.13}$$

### Proof
This result follows directly from the definition of the premium for calls or puts as the risk-neutral expectation of the discounted final payoff. As an example, take a call with strike $k$ whose value at a distance $T$ from its expiration date is

$$C(k) = E^Q[\text{Max}(S_T - k, \ 0)e^{-rT}] = \int_k^\infty (x - k)e^{-rT}\phi_T(x)dx.$$

Taking the derivative of the right-hand side with respect to k, we have

$$\frac{\partial C}{\partial k}(k) = \int_k^\infty e^{-rT}\phi_T(x)dx.$$

Taking the derivative a second time, we obtain

$$\frac{\partial^2 C}{\partial k^2}(k) = e^{-rT}\phi_T(k).$$

An analogous proof can be used for the put.

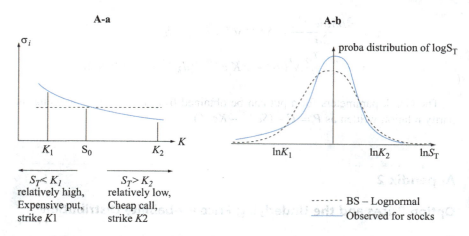

**Fig. 12.12** Relation between the shape of the smile and that of the distribution of $S_T$

This proposition shows that knowing the prices of vanilla options in theory allows us to deduce the distribution for the underlying's future value implicitly used by the market for valuing options. This would be perfect knowledge in the purely theoretical case of a continuum of prices. In practice, of course, price discontinuity makes it imperfect and not so easy to use. In any event, it is theoretically possible to compare the distribution for $S_T$ which is implicit in option market prices with the theoretical distributions used in valuation models. *In the special case of BS, the underlying's distribution is assumed to be log-normal.* In practice, the distribution deduced from vanilla option prices using Eq. (12.13) (*T*-density) seems to differ from the log-normal model. Empirical studies clearly show that, for stocks as underlyings, the left tail of the distribution for small values of $S_T$ is noticeably thicker than what follows from the Gaussian model. And *this form of distribution for $S_T$ can explain why the smile is downward sloping* (a *skew*, as in Fig. 12.12a), which is precisely the usual case for stock options. This connection between the shape of the distribution for $S_T$ and the smile or skew can be seen from Fig. 12.12a, b. Suppose the smile is decreasing like the solid line in Fig. 12.12a; also in Fig. 12.12a the horizontal smile assumed in the BS model is shown as a dotted line which serves as a reference. Assume further that the distribution of $\ln(S_T)$ is asymmetrical, with a thick left tail, as drawn with a solid line in A-b ; the normal distribution (as hypothesized in BS) is also shown in the same figure, drawn dotted as a reference. The solid (dotted) lines in A-a thus correspond to the solid (dotted) lines in A-b.

On the abscissa axis $K$ in Fig. 12.12a we find the underlying's value $S_0$ on date 0 (today). For an option with a low strike, such that $K_1 < S_0$, options are expensive in comparison to a standard BS price; the opposite is true for an option with a high strike, such that $K_2 > S_0$, which is relatively cheap.

Now let us consider *a put deep out-of-the-money* with strike $K_1$; it is logical that it should be relatively costly (*see* A-a) since, as A-b shows, the probability of exercise ($S_T < K_1$) and the expected gain upon exercise are larger than for a log-normal distribution. It also is reasonable that a *call deep out-of-the-money* $K_2$ should be

## Appendix 2

cheap since the right tail of the distribution for $S_T$ is thinner than in the log-normal case. Finally, let us remark that, from the call-put parity, a relatively expensive put implies a relatively expensive call, and conversely, since $C - P = S - Ke^{-rT}$, independently of any assumption regarding the actual distribution of $S_T$.

The relative dearness of stock options with low strikes and the relative cheapness of options with high strikes (tantamount to a decreasing implied volatility curve) are therefore compatible with empirical results on the distributions of stock yields.

Similarly, a U-shaped smile curve ($s_i (K)$ decreasing then increasing again) can be explained by an empirical distribution for $S_T$ with thick tails on both left and right.

This result shows that some assumptions of the Black–Scholes (BS) model are not realistic and, because of this, the model yields biased results. When applying BS formula, taking the smile into account corrects for this bias only imperfectly. The fundamental assumption that should be questioned is the constancy of the volatility parameter (which implies the log-normality of the underlying's distribution).

An approach that is theoretically more rigorous and more general is to look for a model of which BS would be a special case, and which would be compatible with the whole volatility surface observed at a given instant (and not only with the smile curve relative to a specific expiration date $T$). Mathematically one has to find a model for the evolution of the underlying price which provides, for any future value of the underlying, a probability distribution equal to the distribution implied by the surface.

Several answers appear satisfactory and are used in practice. The first is to work with a model for the volatility $s(t,S)$ that depends entirely on the time remaining until the option's expiration date and the current price of the underlying. Such models can also explain the smile. It is, for example, in line with the empirical evidence and the theory that a stock's volatility is a *decreasing* function $s(S)$ of its price $S$. At the theoretical level, a drop in the stock price leads to an increase in the ratio (debt/market capitalization) (financial leverage); this increase in leverage implies an increase in the risk affecting the equity, and so an increased stock volatility. The fact that the stock volatility is higher for small than for large values of S explains the thick left tail of the distribution for $S_T$ (*see* Fig. 12.12b) and *the volatility skew* (*see* Fig. 12.12a).

A complementary explanation for this dual phenomenon of volatilities asymmetric to the right and returns asymmetric to the left is the *feedback effect*. Following a large positive jump in the price, the stock volatility increases which exerts pressure to reduce its return. Indeed, since the *expected* (*ex ante*) return must increase to offset the increased risk, the price must fall and with it the *realized* (*ex post*) return. This feedback phenomenon thus *mitigates* the amplitude of *positive* returns. In the same way, a large negative jump in the price causes an increase in volatility which forces the return down for the same reason as above. Thus the feedback effect *accentuates* the amplitude of *negative* returns. Consequently, this dissymmetric feedback effect contributes to the left asymmetry in the return distribution and the right asymmetry in the volatility distribution.

The interest of these models which express the volatility as a function $s(t,S)$ lies in their relative simplicity; in particular, they can be implemented using Cox-Ross-

Rubinstein recombining trees. An alternative solution consists of models with stochastic volatility, in which the variable $s_t$ is itself the solution to a stochastic differential equation. Such models present the advantage to take into account the *specific risk of a change in the volatility.*[26]

# Appendix 3

## Replication of an Arbitrary Payoff with a Static Option Portfolio

Let us consider European calls and puts with the same underlying and the same expiration date $T$. Let us denote by $F_{0,T}$ the forward price[27] of the underlying on date 0 for an expiration date $T$. An option with strike $k$ is said to be *at-the-money-forward* when $F_{0,T} = k$. Subsequently, $C(k)$ and $P(k)$ respectively denote the premium at 0 of a call and a put with expiration date $T$ and strike $k$. The following proposition describes the replication of an arbitrary payoff $\Psi(S_T)$ by a portfolio of calls and puts out-of-the-money-forward.

### Proposition

*Let us consider a security $\Psi$ which generates a single payoff on date $T$ defined by a sufficiently regular function $\Psi(S_T)$.*

- *This security is worth on date 0:*

$$V_0 = \Psi(\Phi_{0,T})e^{-rT} + \int_0^{\Phi_{0,T}} \frac{\partial^2 \Psi}{\partial k^2}(k)P(k)dk + \int_{\Phi_{0,T}}^{\infty} \frac{\partial^2 \Psi}{\partial k^2}(k)C(k)dk. \quad (12.14)$$

- *The static duplication of the security Y is guaranteed by a position with $\frac{\partial^2 \Psi}{\partial k^2}(k)dk$ options (calls or puts) out-of-the-money and with strike $k$.*

One says the vanilla calls and puts *span* the set of regular payoffs, for a given expiration date.

### Proof

The proposition can be deduced from the following result: For any $x > 0$,

---

[26]That is, the risk which is not generated by the variations in $S$ and whose market price is not necessarily zero. See, for instance, Heston's model (last Section of Chap. 11).

[27]Recall that for an asset not paying dividends: $F_{0,T} = S_0 e^{rT}$. Furthermore, we have in all cases $F_{T,T} = S_T$.

# Appendix 3    497

$$\Psi(x) = \Psi(\Phi_{0,T}) + \frac{\partial \Psi}{\partial k}(\Phi_{0,T})(x - \Phi_{0,T}) + \int_{\Phi_{0,T}}^{\infty} \frac{\partial^2 \Psi}{\partial k^2}(k)(x - k)^+ dk + \int_0^{\Phi_{0,T}} \frac{\partial^2 \Psi}{\partial k^2}$$

$$\times (k)(k - x)^+ dk.$$

This equation can be justified with a bit of functional analysis. By definition of the first derivative, we have

$$\Psi(x) - \Psi(\Phi_{0,T}) = \int_{\Phi_{0,T}}^x \frac{\partial \Psi}{\partial k}(k)dk.$$

Integrating by parts (consider $u = x - k$ and $v = \frac{\partial \Psi}{\partial k}(k)$), we find

$$\Psi(x) - \Psi(\Phi_{0,T}) = -\left[(x - k)\frac{\partial \Psi}{\partial k}(k)\right]_{\Phi_{0,T}}^x + \int_{\Phi_{0,T}}^x (x - k)\frac{\partial^2 \Psi}{\partial k^2}(k)dk.$$

But $x - k = (x - k)^+ - (k - x)^+$. Hence

$$\Psi(x) - \Psi(\Phi_{0,T}) = (x - \Phi_{0,T})\frac{\partial \Psi}{\partial k}(\Phi_{0,T}) + \int_{\Phi_{0,T}}^x (x - k)^+ \frac{\partial^2 \Psi}{\partial k^2}(k)dk$$

$$- \int_{\Phi_{0,T}}^x (k - x)^+ \frac{\partial^2 \Psi}{\partial k^2}(k)dk.$$

Remark that for any $x > 0$, one may replace the integration limit $x$ in the first integral by infinity (since, if $k > x$, $(x - k)^+$ vanishes), and the limit $x$ in the second integral by 0 (since, if $k < x$, $(k - x)^+$ vanishes); as a result we get

$$\Psi(x) - \Psi(\Phi_{0,T}) = (x - \Phi_{0,T})\frac{\partial \Psi}{\partial k}(\Phi_{0,T}) + \int_{\Phi_{0,T}}^{\infty} (x - k)^+ \frac{\partial^2 \Psi}{\partial k^2}(k)dk$$

$$- \int_{\Phi_{0,T}}^0 (k - x)^+ \frac{\partial^2 \Psi}{\partial k^2}(k)dk.$$

Rewriting this, we have

$$\Psi(x) = \Psi(\Phi_{0,T}) + (x - \Phi_{0,T})\frac{\partial \Psi}{\partial k}(\Phi_{0,T}) + \int_{\Phi_{0,T}}^{\infty}(x - k)^+ \frac{\partial^2 \Psi}{\partial k^2}(k)dk$$

$$+ \int_0^{\Phi_{0,T}}(k - x)^+ \frac{\partial^2 \Psi}{\partial k^2}(k)dk$$

This last equation shows that the replication of the payoff $Y(S_T)$ is ensured by a static position containing $\frac{\partial^2 \Psi}{\partial k^2}(k)dk$ options (call or put) out-of-the-money with strike $k$.[28]

Expressing the discounted risk-neutral expectation of each member of the last equation for $x = S_T$ and noting, on the one hand, that $E^Q[(S_T - \Phi_{0,T})e^{-rT}] = 0$, and, on the other, that $E^Q[(S_T - k)^+ e^{-rT}] = C(k)$ and $E^Q[(k - S_T)^+ e^{-rT}] = P(k)$, we end up with Eq. (12.14).

The proposition proves that any payoff can be replicated by a static strategy based on vanilla call and put options. In a very intuitive sense, it is the gamma (second derivative of the payoff) on the expiration date that determines the quantity of standard options to hold in the replicating portfolio.

## Suggestions for Further Reading

### Books

Baxter, M., & Rennie, A. (1999). *Financial Calculus*. Press.
*Bergomi, L. (2016). *Stochastic volatility modeling*. Chapman & Hall.
*Fusai, G., & Roncoroni, A. (2008). *Implementing models in quantitative finance: Methods and cases*. Springer.
**Gatheral, J. (2006). *The volatility surface: A practitioner's guide*. Wiley Finance.
Hull, J. (2018). *Options, futures and other derivatives* (10th ed.). Prentice Hall Pearson Education.
Kolb, R. (2007). *Futures, options and swaps* (5th ed.). Blackwell.
Taleb, N. (1997). *Dynamic hedging: Managing vanilla and exotic options*. Wiley.

### Articles

Bakshi, G., Cao, C., & Chen, Z. (1997). Empirical performance of alternative option pricing models. *Journal of Finance, 52*(5), 2004–2049.
Bates, D. S. (2000). Post-'87 Crash fears in the S&P futures market. *Journal of Econometrics, 94*, 181–238.
Bergomi, L. (2004). Smile dynamics. *Risk*, 117–123.
Bergomi, L. (2005). Smile dynamics II. *Risk*, 67–73.

---

[28]More precisely, this last equation written for $x = S_T$ shows the payoff $\Psi(S_T)$ is obtained using a static portfolio created on date 0 and containing: a zero-coupon yielding $\Psi(F_{0,T}) - F_{0,T}\frac{\partial \psi}{\partial k}(F_{0,T})$ on $T$; $\frac{\partial \psi}{\partial k}(F_{0,T})$ units of the underlying; and a continuum $\frac{\partial^2 \psi}{\partial k^2}(k)dk$ of options (call or put) out of the money and with strike $k$.

# Suggestions for Further Reading

Bergomi, L. (2008). Smile dynamics III. *Risk*, 90–96.

Black, F. (1989). How to use the holes in Black-Scholes. *Journal of Applied Corporate Finance, 1*, 67–73.

Chaput, J. S., & Ederington, L. H. (2003). Option spread and combination trading. *Journal of Derivatives, 10*(4), 70–88.

Cookson, R. (1993). Moving in the right direction. *Risk*.

Daglish, T., Hull, J., & Suo, W. (2007). Volatility surfaces: Theory, rules of thumb, and empirical evidence. *Quantitative Finance, 7*(5), 507–524.

Derman, E. (1999). Regimes of volatility. *Risk*, 55–59.

Dumas, B., Flemming, J., & Whaley, R. (1998). Implied volatility functions: Empirical tests. *Journal of Finance, 53-6*, 2059–2106.

Figlewski, S. (1989). Options arbitrage in imperfect markets. *Journal of Finance, 44*, 1289–1311.

Harvey, C., & Whaley, R. (1991). S&P 100 Index option volatility. *Journal of Finance, 46*, 1551–1661.

Jackwerth, J., & Rubinstein, M. (1996). Recovering probability distributions from option prices. *Journal of Finance, 51*, 1611–1631.

Xu, X., & Taylor, S. J. (1994). The term structure of volatility implied by foreign exchange options. *Journal of Financial and Quantitative Analysis, 29*, 57–74.

# American Options and Numerical Methods 13

Standard (vanilla) European options have for them the simplicity in implementation, valuation, and replication. Compared with European options, Americans are generally a little more expensive because they can be exercised before expiration. This poses the problem of valuing the option to exercise early embedded in an American option. We showed in an earlier chapter that the rule according to which an American option is worth more that the European equivalent has one exception, namely that of a call written on a spot underlying asset (UA) not paying dividends or coupons (in this exceptional case the two options have the same value). In the other cases (in particular, for all puts) there are paths of the UA price for which it is optimal to exercise early the option. Contrary to European options whose payoff only depends on the price of the underlying upon expiry, the payoff of an American option is therefore *path-dependent*. Since the number of different cases to be considered is large and their technical complexity high, this chapter is not exhaustive, even if we provide many intuitive explanations, justifications, and proofs.

Section 13.1 of this chapter is devoted to conditions for early exercise of American options and to call-put "parity" relationships. Section 13.2 presents analytical methods of valuation which, except in few cases, do not yield exact analytic formulas. Numerical methods are flexible and powerful tools, adapted to a wide variety of financial problems for which no analytical solution is available. In Sect. 13.3, we explain the simplest numerical method for valuing American options, which is an adapted version of the binomial model. However, it is also relatively inaccurate except when numerous time steps are used, which may make it too slow. Three other types of numerical methods, finite differences, trinomial and multidimensional trees, are presented in Sect. 13.4. *Simple* Monte Carlo simulation is inappropriate for pricing American options and the method consisting of combining it with linear regressions to mitigate this inadequacy is relegated to Chap. 26 (Sect. 26.5).

© The Author(s), under exclusive license to Springer Nature Switzerland AG 2022
P. Poncet, R. Portait, *Capital Market Finance*, Springer Texts in Business and
Economics, https://doi.org/10.1007/978-3-030-84600-8_13

## 13.1 Early Exercise and Call-Put Parity for American Options

In all cases, except that of a call written on a spot asset not paying dividends, it may be optimal to exercise an in-the-money American option *before* its expiration date. We examine first the optimal conditions for early exercise for different types of options (Sect. 13.1.1), and then the parity relationships between American calls and puts (Sect. 13.1.2).

### 13.1.1 Early Exercise of American Options

We study first the optimal early exercise policy of American calls under different dividend distribution regimes (Sects. 13.1.1.1, 13.1.1.2, and 13.1.1.3) before considering the case of American puts (Sects. 13.1.1.4 and 13.1.1.5) and options on forward contracts (Sect. 13.1.1.6).

#### 13.1.1.1 A Refresher on the Early Exercise of a Call Written on a Dividend Paying Asset

Detaching a dividend or a coupon $D$ on date $t < T$ produces on that date a drop in the UA price by the same amount $D$. By exercising before detachment, the holder of the call can prevent the harmful consequences of this loss of value.

According to the call-put parity with dividend and because in the absence of arbitrage opportunity an American call $(C^a)$ can never be worth less than its European analogue $(C)$, we have on any date $t \leq t$:

$$C^a_t \geq C_t = P_t + \left(S_t - De^{-r(\tau-t)}\right) - Ke^{-r(T-t)}$$

$$\geq \left(S_t - De^{-r(\tau-t)}\right) - Ke^{-r(T-t)}, \tag{13.1}$$

where, to simplify the notation, we assume a flat yield curve and use a continuous interest rate.

Recall that if there is no dividend ($D = 0$) the final term is necessarily larger than $S_t - K$: the American call should therefore *never* be exercised early (that is "exchanged" for $S_t - K$) since its time (or speculative) value is always strictly positive, and its value is consequently equal to that of its European analogue (the American feature is worthless).

*On the other hand*, if $D$ is positive and sufficiently large, the last term in inequality (13.1) is not necessarily larger than $S_t - K$. Then the call-put parity with dividends does not exclude the case of a European call whose market price is less than its intrinsic value (its time value is negative). This implies that, under certain conditions, an American call (which under no-arbitrage cannot be worth less than its intrinsic value) should be exercised early so as to provide the gain $S_t - K$. These conditions will now be specified for calls and put written on different types of underlying assets.

## 13.1 Early Exercise and Call-Put Parity for American Options

### 13.1.1.2 Early Exercise of an American Call Written on a Spot Underlying Paying a Single Discrete Dividend

We now make explicit the precise conditions under which early exercise may take place for an American call on a spot asset which pays a single known dividend or coupon in the amount $D$, on a known date $\tau$ such that $\tau < T$.[1] We denote by $t^-$ an instant before $t$ (in practice, the day before the ex-dividend day referred to detachment or payment) and $t^+$ an instant after $t$; by $S_{t-}$ and $S_{t+}$ the values of the underlying just before and just after the dividend payment, respectively, $(S_{t-} = S_{t+} + D)$; and by $C^a_t$ and $P^a_t$ the American call and put premiums on any date $t$, and $C_t$ and $P_t$ those of their European counterparts (same underlying asset, strike $K$ and expiration date $T$). In several occasions it will be convenient to use the notation $z = T-t$, time to maturity instead of calendar time; in which case the reader will keep in mind that: $z = 0 <=> t=T$; $t = 0 <=> z=T$.

### Proposition 1

*When the underlying asset pays a discrete dividend, it is never optimal to exercise the call before the day $\tau^-$ preceding the detachment.*

### Proof

Consider an American call expiring at $T$ whose value is $C^a_t$ on date $t$ $(t < \tau^- < T)$. Consider also a second American call on the same UA with the same strike $K$ but with maturity $\tau^-$ (hence shorter than the first one and with no rights to dividends) with value $_{\tau^-}C^a_t$. We have already proved that the value of this second call is strictly greater than its intrinsic value: $C^a_t > S_t - K$ (this is the reason why it should not be early exercised). Since $C^a_t \geq {}_{\tau^-}C^a_t$ (the value of an American call increases with its duration), one may thus write: $C^a_t \geq {}_{\tau^-}C^a_t > S_t - K$. Therefore, on any date $t$ before $\tau$ an American call with maturity $T > \tau$ has a value strictly greater than its intrinsic value $S_t - K$; hence, it should never be exercised before the eve of the detachment day.

Intuitively, the reason one should wait until the day before detachment to (possibly) exercise the call is that an earlier date would uselessly lose some time value.

The following rules describe the behavior of the price of an American call around the detachment date ($\tau^-$ and $\tau^+$). They follow from the fact that exercise produces *the payoff* $(S_{\tau-} - K)$ and, if not exercised, the continuation value of the option will be $C^a_{\tau+}$ which is equal to $C_{\tau+}$ (the price of the European call on the following day, since there is no distribution between $\tau^+$ and $T$). Besides, in situations of no exercise, $C^a_{\tau-} = C^a_{\tau+} = C_{\tau+}$ *a continuity condition* that must prevail in absence of arbitrage (the knowledge of a drop of $C^a$ on $\tau$ would trigger sales on $\tau^-$ followed by immediate repurchases on $\tau^+$). For the same reason $C_{\tau-} = C_{\tau+} = C_\tau$ (continuity of the European

---

[1] It is possible to extend the analysis to the case of several dividends whose amounts and payment dates are known at the time the American call is valued.

premium across the distribution date in all cases): only for the sake of clarity we may distinguish $C_{\tau-}$ from $C_{\tau+}$ in the following argumentation.

- $C^a_{\tau-} = C^a_{\tau+} = C_{\tau+} = C_\tau > S_{\tau-} - K$ characterize the situation of **no exercise** of the American call on date $\tau^-$.
- $C^a_{\tau-} = S_{\tau-} - K > C_{\tau+} = C_\tau$ characterize the situation of **exercise** on date $\tau^-$. (when $C^a_{\tau-} = C^a_{\tau+} = C_\tau = S_{\tau-} - K$, the holder is indifferent to exercise or not)
- In all cases:

$$C^a_{\tau-} = \mathrm{Max}(S_{\tau-} - K, C_\tau) = \mathrm{Max}(S_{\tau+} - (K - D), C_\tau) \qquad (13.2)$$

Rule (i) states that, if the continuation value $C_{\tau+}$ is higher than the payoff, on the one hand, the option will not be exercised and, on the other hand, $C^a_{\tau-} = C^a_{\tau+} = C_\tau$ by continuity. The second rule simply states that the American call will be exercised if the payoff is higher than the continuation value $C_\tau$. Eq. (13.2) follows from the first two rules.

The following proposition provides an intuitive condition for early exercise that will also be useful in the following analysis.

### Proposition 2

*A necessary and sufficient condition for an optimal exercise of the call on $\tau^-$ is that the dividend D exceeds the time value of the European option on $\tau^+$:*

$$D > TV_{\tau+} = C_{\tau+} - (S_{\tau+} - K). \qquad (13.3)$$

### Proof

Since $S_{\tau-} = S_{\tau+} + D$, the necessary and sufficient condition for early exercise $S_{\tau-} - K > C_{\tau+}$ can also be written $S_{\tau+} + D - K > C_{\tau+}$, or $D > [C_{\tau+} - (S_{\tau+} - K)]$; the term in square brackets being the time value at $\tau^+$ (of the European as well as the American call), Proposition 2 obtains.

Intuitively, the reason for this result is that, by early exercising, one loses the time value but gains the dividend: the gain must outweigh the loss to justify the early exercise. Remark that $C_{\tau+}$ is the price of a European call and thus given by a BSM-type formula.

To go deeper into the analysis we specify the call's time value at $\tau^+$ by applying the European call-put parity:

$$TV_{\tau+} = C_{\tau+} - (S_{\tau+} - K) = P_{\tau+} + S_{\tau+} - Ke^{-r(T-\tau)} - (S_{\tau+} - K), \text{i.e.}$$

$$TV_{\tau+} = K(1 - e^{-r(T-\tau)}) + P_{\tau+} \qquad (13.4)$$

The time value of the call is thus the sum of two components: (*i*) $K(1 - e^{-r(T-\tau)})$, which is the financial gain on postponing the payment of the strike (time value of $K$). Moreover, $K(1 - e^{-r(T-\tau)})$ is *fixed and known since the issue of the call* whenever $\tau$ is

## 13.1 Early Exercise and Call-Put Parity for American Options

predetermined and $r$ assumed constant or known; (ii) $P_{\tau+} = P(S_{\tau+})$, the value of a European put, is a decreasing function of the underlying price $S_{\tau+}$ (and depends also on other, here fixed, parameters including time to maturity). It is an insurance against the risk of a future decrease on the underlying price, which mitigates the risk borne on postponing exercise; therefore, it is a component of the time value.

Since $P_{\tau+} > 0$, $K(1 - e^{-r(T-\tau)})$ is a lower bound for $TV_{\tau+}$.

The following proposition states that the American call *will never be exercised* before its maturity $T$ if the known dividend is smaller than a given threshold and that, if it exceeds this threshold, the call *will be* exercised *iff* the underlying price exceeds a minimum value called the early exercise price bound.

## Proposition 3

(i) *If $D < K(1 - e^{-r(T-\tau)})$, the American call should not be exercised whatever is the underlying asset price $S_\tau$. If this « dividend threshold » $K(1 - e^{-r(T-\tau)})$ is known on the call's issue date and is higher than the known dividend, $C^a = C$ at all times.*

(ii) *If $D > K(1 - e^{-r(T-\tau)})$, an "early exercise price bound" $S_\tau{}^{**}$ exists such that the call should be exercised (at $\tau^-$) iff the cum-dividend price $S_{\tau-} \geq S_\tau{}^{**}$. This early exercise price bound is the unique solution to the « value matching condition »:*

$$C_{\tau+}(S_\tau{}** - D) = S_\tau ** - K.$$

*It is equivalent, and often convenient, to define the early exercise price in reference to the ex-dividend price $S_{\tau+}$: this price threshold $S_\tau{}^* = S_\tau{}^{**} - D$ is such that the call should be exercised (at $\tau^-$) iff the ex-dividend price $S_{\tau+} \geq S_\tau{}^*$; it satisfies the value matching condition:*

$$C_{\tau+}(S_\tau*) = S_\tau * + D - K.$$

(iii) *The early exercise price bound $S_\tau{}^*$ increases with time to maturity $z \equiv T\text{-}t$ (decreases with calendar time t).*

## Proof

(*i*) is a direct consequence of Proposition 2 stating that the necessary and sufficient condition for exercise at $\tau^-$ is $D > TV_{\tau+}$. Or equivalently, from (13.4), $D > K(1 - e^{-r(T-\tau)}) + P_{\tau+}$. Since $P_{\tau+} > 0$, a necessary condition for a possible early exercise is $D > K(1 - e^{-r(T-\tau)})$. If $D < K(1 - e^{-r(T-\tau)})$ (a fixed, known threshold under our set of assumptions) the American call should *never* be exercised and the right of early exercise is worthless: $C^a = C$ at all times.

(*ii*) Assume now $D > K(1 - e^{-r(T-\tau)})$. Then the necessary and sufficient condition for exercising at $\tau^-$ (13.3) writes:

$D > K(1- e^{-r(T-\tau)}) + P_{\tau+}$ or, being more explicit on the put price: $P(S_{\tau+}) < D - K(1 - e^{-r(T-\tau)})$. Since the r.h.s. of this inequality is positive and $P(S_{\tau+})$ is a continuous decreasing function of $S_{\tau+}$ tending to 0 when $S_{\tau+}$ tends to infinity, there exists $S_\tau^*$ such that: $P(S_{\tau+}) < D - K(1- e^{-r(T-\tau)})$ for $S_{\tau+} > S_\tau^*$ (exercise); $P(S_\tau^*) = D - K(1- e^{-r(T-\tau)})$ for $S_{\tau+} = S_\tau^*$; and $P(S_{\tau+}) > D - K(1- e^{-r(T-\tau)})$ for $S_{\tau+} < S_\tau^*$ (continuation). $S_\tau^*$ is thus the solution to the equation $P(S_\tau^*) = D - K(1- e^{-r(T-\tau)})$, which rewrites, using the call-put parity: $C_{\tau+}( S_\tau^*) = S_\tau^* + D - K$.

However, the exercise decision being taken at time $t^-$, it may be more natural to express these exercise conditions as a function of $S_{t-}$. To this end, it suffices to let $S_t^{**} \equiv S_t^* + D$ (since $S_{t-} = S_{t+} + D$) and make the appropriate substitutions in all the previous results to obtain the whole Proposition 3(ii).

*(iii)* Let us prove finally that the early exercise price $S_\tau^*$ *increases with maturity* $T$-$\tau$. It is here more convenient to use $\mathbf{z} = T$-$\tau$ (rather than $\tau$) and to consider the set of $S^*(\mathbf{z})$ for different maturities $\mathbf{z}$. To avoid notational confusion, we use different fonts for $\mathbf{z}$ and a function of $\mathbf{z}$ and for a function of $t$ (or $\tau$). For instance, $S(\mathbf{z}) = S(T - t)$, which is different from $S(t) = S(T - \mathbf{z})$. It then should be kept in mind that: $\mathbf{z} = 0 \Leftrightarrow t = T; t = 0 \Leftrightarrow \mathbf{z} = T$; and $\partial(.)/\partial\mathbf{z} = -\partial(.)/\partial t$.

$S^*(\mathbf{z})$ satisfies the value matching condition for any $\mathbf{z} \equiv T - t$: $C(S^*(\mathbf{z}), \mathbf{z}) = S^*(\mathbf{z}) - D - K$ (this can be considered as an identity defining $S^*(\mathbf{z})$). Computing the total differential of this last relation, we obtain:

$$\frac{\partial C}{\partial S}\bigg]_{S=S*} dS * + \frac{\partial C}{\partial \mathbf{z}}\bigg]_{S=S*} d\mathbf{z} = dS *, \quad i.e.$$

$$\frac{dS^*}{d\mathbf{z}} = \frac{\frac{\partial C}{\partial \mathbf{z}}\big]_{S=S^*}}{(1 - \frac{\partial C}{\partial S}\big]_{S=S^*})},$$

which is the slope of $S^*(\mathbf{z})$. Since both the numerator and denominator of the r.h.s. of this expression are positive (C increases with time to maturity[2] and the delta $\partial C/\partial S$ < 1), $S^*(\mathbf{z})$ is upward sloping.

## Comments and Interpretations

Proposition 3(i) is an extension of a well-known result, presented in Chap. 10 (Proposition 4), stating that when no dividend (or coupon) is paid before its expiration date the American call on a spot asset should never be exercised. Not only must a dividend exist for the right to early exercise have some value, but *it must be large enough* to outweigh the financial opportunity cost $K(1 - e^{-r(T-\tau)})$ of paying $K$ prematurely. However, the statement that the American call should never be early exercised hence has no higher value than its European counterpart, relies here on the assumption that the dividend is smaller than a given and known threshold

---

[2] The European call price $C$ does not always increase with time to maturity while the American call price $C^a$ does (at least in the continuation region). Here, however, $C = C^a$ as there is no dividend between $t$ and $T$.

## 13.1 Early Exercise and Call-Put Parity for American Options

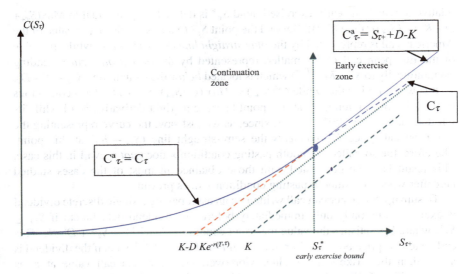

**Fig. 13.1** American and European calls at $\tau^-$ (just before discrete dividend detachment)

$K(1 - e^{-r(T-\tau)})$, which in turn presumes $D$, $\tau$ and $r$ known from the start. While assuming that the first two are known is often acceptable, assuming that $r$ (the interest rate prevailing at $\tau$) is known on date 0 is more questionable and weakens the practical importance of this statement.

Proposition 3(ii) implies that, if $D$ outweighs the opportunity cost $K(1 - e^{-r(T-\tau)})$, the American call *may be* exercised before expiration and *will be* exercised on $\tau^-$ iff the current underlying price is higher than a bound determined by the value matching condition. This condition is an equation involving the price of a European call given by a BSM-type formula which can be solved for $S_\tau^*$ (or $S_\tau^{**}$). Indeed, the value matching condition can be explicitly written: $BS(S_\tau^*, T - \tau) = S_\tau^* - D - K$. It means that when $S_{\tau+} = S_\tau^*$ (or $S_{\tau-} = S_\tau^{**}$), the investor is indifferent between exercising early or continuing to hold the option whose value is equal to that of its European counterpart given by $BS(S_{\tau+})$. Finally, the longer is the time to expiration $\mathbf{z}$ ($= T - \tau$), the larger is the opportunity cost $K(1 - e^{-r(T-\tau)})$ of paying the strike early, thus the smaller is the domain of optimal exercise and the higher is the early exercise price bound $S_\tau^*$. The upward-sloping curve representing $S_\tau^*$ as a function of $T-\tau$ (or, equivalently $S^*(\mathbf{z})$ as a function of $\mathbf{z}$) will be referred to as the *Early Exercise Frontier* (EEF).[3] But one must keep in mind that here (contrary to cases considered in the sequel) $T-\tau$ is *known*, hence the early exercise bound is a given point of the EEF given by the equation $BS(S_\tau^*, T - \tau) = S_\tau^* + D - K$.

The previous results yield the following graphic representation (Fig. 13.1) of the situation at the eve of ex-dividend day ($\tau^-$), if $D > K(1 - e^{-r(T-\tau)})$, which implies $K - D < Ke^{-r(T-\tau)}$. In Fig. 13.1, the abscissa represents the ex-dividend price $S_{\tau+}$,

---

[3] This frontier (which can be crossed) should not be confused with the Optimal Exercise Boundary defined below (which cannot be crossed).

relative to which the early exercise bound $S_\tau^*$ is defined. $C^a{}_{\tau-}$ is equal to $\text{Max}(S_{\tau+}+D-K, C_\tau(S_{\tau+}))$ (Eq. (13.2)). Beyond the point $S_\tau^*$ (exercise region), the value of the American call is represented by the *blue straight line* $(S_{\tau+}-(K-D))$, while the value of the European call $C_\tau$ is smaller, represented by the *dotted blue curve* tending asymptotically to $S_{\tau+}-Ke^{-r(T-\tau)})$ which is located *below* the straight line $S_{\tau+}-(K-D)$.

It is worth noting that while $C^a(S_{\tau+})=\text{Max}\,(C_\tau(S_{\tau+}),\,S_{\tau+}+D-K)$ is continuous in $S^*$, it is not differentiable at this point because its right derivative is $+1$ while its left derivative is $\partial C_\tau/\partial S|_{S=S^*}<1$, since, as we just saw, the curve representing the European call price $C_\tau$ intersects the semi-straight line $(S_{\tau-}-K)^+$ at this point. Therefore, the so-called « smooth pasting condition » does *not* prevail in this case. This result is to be contrasted with those obtained in most of the cases studied hereafter where a « smooth pasting condition » does prevail.

To sum up, the American call written on an asset paying a single discrete dividend is exercised optimally only immediately before the dividend detachment if $S_{\tau+}\geq S_\tau^*$, where $S_\tau^*$ satisfies the value matching condition $C_{\tau+}(S_\tau^*)=S_\tau^*+D-K$. This early exercise price bound $S_\tau^*$ exists only if $D>K(1-e^{-r(T-\tau)})$, i.e. if the dividend is larger than the strike's time value. Moreover, the American call value at $\tau^-$ is equal to: $\text{Max}\,(C_\tau(S_{\tau+}),\,S_{\tau+}+D-K)\,(=\text{Max}\,(C_\tau(S_{\tau-}-D),\,S_{\tau-}-K)$.

### 13.1.1.3 Early Exercise of an American Call on a Spot Asset Paying a Continuous Dividend

We now suppose that the underlying asset pays a continuous dividend $cS(t)dt$ in the intervals $(t,\,t+dt)$. As with a single discrete dividend, the disadvantage of foregoing the future dividend *stream* may outweigh the disadvantage of losing time value and induce an early exercise. However, in this more complex situation, there is no longer a fixed and known detachment date determining the unique optimal moment of exercise but, depending on the situation, *any date of the life of the American call option may be optimal for early exercise.*

- Consider first a European call for which, as shown in Chap. 10, the call-put parity writes at any date $t$: $C_t=P_t+S_t\,e^{-c(T-t)}-Ke^{-r(T-t)}$, implying a time value for the European call:

$$TV^e(t)=C_t-(S_t-K)=P_t+S_t\,e^{-c(T-t)}-Ke^{-r(T-t)}-(S_t-K)$$

$$=P_t+K\left(1-e^{-r(T-t)}\right)-S_t\left(1-e^{-c(T-t)}\right)$$

The third term, $-S_t\,(1-e^{-c(T-t)})\,(<0$ if $c>0)$, can be interpreted as the value of the stream of dividends foregone by postponing the exercise from $t$ to $T$. If this loss outweighs the other component of time value $[P_t+K\,(1-e^{-r(T-t)})]$, $TV^e$ becomes negative. Since the value of the foregone dividends increases to infinity when $S_t$ tends to infinity, for $S_t$ large enough $TV^e$ becomes negative. We thus obtain a result analogous to that of Proposition 3: there is a critical value $S^c$ such that $TV^e$

## 13.1 Early Exercise and Call-Put Parity for American Options

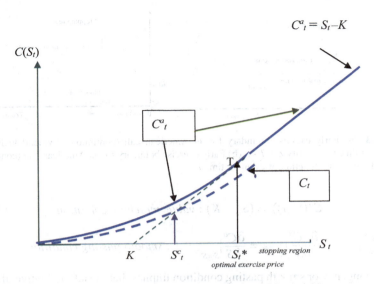

**Fig. 13.2** American and European calls at $\tau < T$ (continuous dividend yield)

$< 0$ if and only if $S_t > S^c$. This critical value is a function $S^c(t)$ of calendar time or $S^c(T-t)$ of time to maturity. Since the European call cannot be exercised before maturity, its time value becomes negative when it is deep enough in the money (it is worth less than its intrinsic value). The dotted curve in Fig. 13.2 represents this situation.

By contrast, the holder of an American call will not allow the value of his option to be smaller than its intrinsic value. However, his exercise threshold is higher than $S^c(t)$ since the time value of the American call is higher than that of its European counterpart (see Fig. 13.2). Nevertheless, at a sufficiently large underlying price $S^c(t)$, it becomes optimal to exercise the call before expiration, to prevent its price to drop below the intrinsic value. The continuous increasing function of time to maturity, $S^*(T - t)$ (decreasing $S^*(t)$), is called the *Optimal Exercise Boundary* (OEB hereafter). $S^*(T-t)$ and $S^*(t)$ will be respectively represented in Fig. 13.3a, b.

Figure 13.2 represents the values of the European and American calls as a function of the underlying price, at any given date $t$ before maturity.

In Fig. 13.2 the American call option price curve $C^a(S, t)$ is *tangent* to the dotted line representing the call's intrinsic value at a point T corresponding to the optimal exercise price $S^*(t)$. For $S \geq S^*(t)$, the American call price is equal to its intrinsic value $S - K$. The fact that the curve representing $C^a(S, t)$ is tangent to the dotted line for $S = S^*(t)$ implies two conditions:

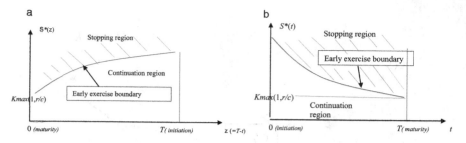

**Fig. 13.3** (a) Early exercise boundary for an American call (continuous dividend yield) as a function of time to maturity $z = T - t$. (b) Early exercise boundary for an American call (continuous dividend yield) as a function of calendar time $t$

$$C^a(S^*(t)) = (S^* - K) : value\ matching\ condition \quad (13.5a)$$

$$\frac{\partial^- C^a}{\partial S}\bigg|_{S=S^*} = \frac{\partial^+ C^a}{\partial S}\bigg|_{S=S^*} = \frac{\partial C^a}{\partial S}\bigg|_{S=S^*} = 1 : smooth\ pasting\ condition \quad (13.5b)$$

This tangency or smooth pasting condition implies that the left derivative of $C^a(S)$ with respect to $S$ at the optimal exercise price $S^*$ is equal to its right derivative (i.e. the delta of the call is continuous), and equal to the slope of the straight line $(S - K)$. Recall that this property does not prevail in the discrete dividend case where the path of the underlying asset price is discontinuous. In fact, the smooth pasting condition, in general, prevails only when the payoff path is continuous. A mathematical proof of the smooth pasting condition is presented in Appendix 1. The conceptual insights provided in this short but rather subtle proof are important and maybe explained as follows. Let again $z$ denote time to maturity $T - t$. An arbitrary and in general suboptimal exercise boundary b(z) could be imposed on a fictitious call $C^b$ otherwise identical to the considered American call $C^a$. This particular boundary b(z) imposes a stopping rule for $C^b$: « Exercise the call the first time S (z) reaches b(z) ». In which case $C^b(b(z)) = b(z) - K$ is a value matching condition associated to the boundary b(z). When b(z) is optimized (i.e., the value of $C^b$ is maximized), b(z) is *chosen* to be the *Optimal Exercise Boundary* $S^*(z)$. It is the optimization condition on b(z) that yields the smooth pasting condition. In brief, the value matching condition is an expression of the fact that $S^*(z)$ is *an* early exercise boundary while the smooth pasting condition expresses *the optimality of* $S^*(z)$.

The optimal exercise boundary is represented in Fig. 13.3a (using z) and 13.3b (using $t$). The American call option should not be exercised within the *continuation* region $\{(S, t): 0 \leq S < S^*(t), 0 < t \leq T\}$ and should be exercised as soon as the boundary is hit from below (*stopping* region).

Recall that the call's OEB increases with time to maturity (decreases with calendar time); we will show later on that $lim_{z \to 0}(S*(t)) = K \max(1; r/c)$.

In addition to the smooth pasting condition, two major differences make the analysis of an American call in case of continuous distribution more complex than that of discrete distribution:

## 13.1 Early Exercise and Call-Put Parity for American Options

- The time $t$ at which the boundary may be hit is *not* predetermined. While in the discrete case, the call holder waits till the eve of distribution day and exercises the option if the underlying price lies *inside* the stopping region (i.e., $S_{t-} \geq S^*$), in the continuous case the call holder will *never let the call penetrate the stopping region*.
- The equation of the *OEB* is *not known* since the pricing function $C^a(S^*, t)$ is not determined, while in the discrete dividend case $S^*$ is determined by the BS formula $(BS(S^*, t) = S^*+D-K)$. Therefore, in valuation models of American calls written on an underlying asset paying a continuous dividend, the OEB $S^*(t)$ *must be determined jointly with the pricing of the call*.
As we will see later on, the same situation prevails for options on forward/futures contracts.

This rather complex technical issue can be formulated as an *optimal stopping time* problem, as shown in Sect. 13.2.2.2

### 13.1.1.4 Early Exercise of American Puts Written on a Spot Asset Paying a Continuous Dividend

The analysis being analogous to that of the previous paragraph, the results will be more briefly justified. We assume that the underlying asset pays a continuous stream of dividends $cS(t)dt$ $(c \geq 0)$, in which case the call-put parity for European options writes $P_t = C_t - S_t e^{-c(T-t)} + Ke^{-r(T-t)}$. This situation encompasses the particular case of no dividend. The European put time value thus writes:

$$P_t(S_t) - (K - S_t) = \left[ C_t(S_t) + S_t \left( 1 - e^{-c(T-t)} \right) \right] - K \left( 1 - e^{-r(T-t)} \right).$$

Early exercising the put loses the time value associated with the implicit call plus the dividends foregone and gains the early payment time value of $K$. When $S_t$ tends to 0, the put time value decreases asymptotically to $-K(1 - e^{-r(T-t)})$, which is negative for any $r > 0$. There exists then a critical value $S_t^c$ such that: $P_t(S_t^c) = (K - S_t^c)$ and $P_t(S_t) > (K - S_t)$ iff $S_t > S_t^c$, as represented by the dotted curve in Fig. 13.4. The American put price $P^a_t$, higher than the European put price $P_t$, attains (from above) the critical zero time-value at the Optimal Exercise Boundary (OEB), for $S_t = S^*(t)$, as represented by the American Put price curve in Fig. 13.4. This curve representing $P^a_t$ is tangent to the straight half-line representing the intrinsic value for $S_t = S^*(t)$, which implies the two conditions:

$$P^a(S^*(t)) = (K - S^*) \qquad \text{: value matching condition} \qquad (13.6a)$$

$$\frac{\partial P^a}{\partial S} \bigg|_{S=S^*} = -1 \qquad \text{: } \textit{smooth pasting condition} \qquad (13.6b)$$

The OEB for American puts increases with time $t$ (decreases with time to maturity $z$) as represented in Fig. 13.5 for two increasing dividend yields $c_1 < c_2$. Unlike the

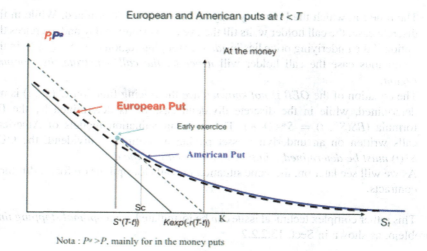

**Fig. 13.4** American and European puts before expiry

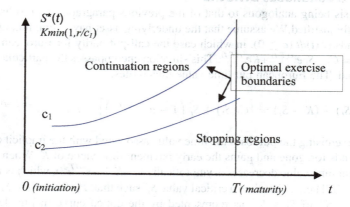

**Fig. 13.5** Early exercise boundaries for American puts (dividend yield $c_1 < c_2$)

call, a high dividend implies for the put an advantage and a higher likelihood of being exercised at expiry. However, the larger is the distribution yield $c$, the larger is the dividend foregone in case of early exercise of the put and the smaller is the threshold $S^*$ that induces this early exercise. The absence of distribution ($c = 0$) is a special case that does not change the thrust of the analysis. We will see later on that when the put nears maturity, $S^*$ gets close to $K \min(1, r/c)$: at time $T$ the exercise threshold is $K$ but if $c > r$ the put is kept alive as long as $cS > rK <=> S > Kr/c$. In particular, in absence of distribution, $\lim_{t->T} S^*(t) = K$. This implies that if $c < r$, *almost all exercised puts should be exercised (a bit) early!*

## 13.1 Early Exercise and Call-Put Parity for American Options

### 13.1.1.5 American Put on a One-Dividend Paying Asset

Consider now an American put whose expiration date is $T$, written on an underlying asset detaching a single dividend $D$ on a known date $\tau < T$. Note first that between $\tau^+$ and $T$ the behavior of the put is identical to that of American put on an asset with no dividend studied in Sect. 13.1.1.4 (no dividend is a particular case of a continuous dividend yield $c = 0$). We thus consider only the question of a possible exercise of the put *before* the ex-dividend date $\tau$. The early exercise policy of the put holder has a motivation opposite to that of the call holder. The earlier is the put exercised, the larger is the *gained* time value of $K$, but the dividend is *lost* if the exercise takes place before the detachment date: Waiting till the eve of detachment thus loses on both counts. Therefore, intuitively, if an early exercise takes place before $\tau$ it should be implemented early enough. The following analysis clarifies and specifies this idea. Its results are summarized in Proposition 4.

We use again the decomposition of $S_t$ into a zero-coupon bond valued $De^{-r(\tau-t)}$ intended for paying the dividend $D$ at date $\tau$ and a non-dividend paying stock whose value $S'_t$ follows a geometric Brownian motion with a risk-neutral expected return equal to $r$.

Let us assume that, at current date $t$ $(< \tau)$, $S_t < K$. When comparing the consequences of an immediate exercise to those of postponing the exercise until $\tau^+$ or after, the key factor appears to be $K[e^{r(\tau-t)} - 1] - D$, the difference between the strike time value gained and the dividend lost when exercising earlier than $\tau^+$, as stated in Proposition 4.

### Proposition 4

*(i) If $t < \tau$ and $K[e^{r(\tau-t)} - 1] \leq D \Leftrightarrow \tau - \ln(D/K + 1)/r \leq t < \tau$, early exercise is sub-optimal. Thus, there exists a no-exercise period prior to the detachment date $\tau$, beginning at $t_c = \tau - \ln(D/K + 1)/r$, during which the put should not be exercised under any circumstances.*

*(ii) If $K[e^{r(\tau-t)} - 1] > D$, before this no-exercise period (for $t < t_c$), early exercise becomes optimal if the UA price becomes small enough. An optimal exercise boundary $S'^*(t)$ thus exists with $\lim_{t \to t_c}(S'^*(t)) = 0$.*

### Proof of (i)

Let us assume first that $K[e^{r(\tau-t)} - 1] \leq D$ $(t_c < t < \tau)$. It suffices to find a particular (possibly sub-optimal) alternative that dominates immediate exercise to show that exercising at $t$ is sub-optimal. Deciding at $t$ to exercise at $\tau^+$ is such a strategy. Indeed, exercising at $t$ generates an immediate payoff $K - S_t = K - S'_t - De^{-r(\tau-t)}$ while postponing the exercise to date $\tau^+$ will yield a random payoff $K - S'_t$. The difference of the present values (at $t$) of these two payoffs writes:

$K - S'_t - De^{-r(\tau-t)} - (Ke^{-r(\tau-t)} - S'_t) = K[1 - e^{-r(\tau-t)}] - De^{-r(\tau-t)} = e^{-r(\tau-t)}[K(e^{r(\tau-t)} - 1) - D]$.

When compared to an exercise at $\tau^+$, immediate exercise is therefore a losing proposition iff $K[e^{-r(\tau-t)} - 1] < D$. We thus obtain a *sufficient* condition for discarding exercise at $t$ since we found one better alternative.

**Justification of (ii)**

Assume now that $K[e^{r(\tau-t)} - 1] > D \, (\Leftrightarrow t < t_c)$. An exercise at $t$ is then preferable to a pre-determined exercise at $\tau^+$ but since the latter is not the best alternative to the former, we cannot conclude that exercise at $t < \tau$ should be undertaken under this (sufficient) condition only. In fact, a more relevant alternative is to keep the put alive till $\tau^+$ and *then* decide whether to exercise it immediately or to keep it further alive. Therefore, the relevant comparison is between an immediate payoff (at $t$) and a right to an unexercised American put at $\tau$ whose value is $P^a(S'_t, \tau) = K - S'_t + TV^a(S'_t, \tau)$, where $TV^a$ stands for the put's positive time value.

The difference $d_t$ between the payoff at $t$ $(K - S'_t - De^{-r(\tau-t)})$ and the present value (at $t$) of $P^a(S'_t, \tau)$ is our comparison criterion between exercising at $t$ and waiting till $\tau^+$, which writes:

$d_t(S'_t) = K - S'_t - De^{-r(\tau-t)} - [Ke^{-r(\tau-t)} - S'_t + e^{-r(\tau-t)} E^Q_t(TV^a(S'_t, \tau))]$ where $E^Q_t$ stands for the risk-neutral expectation taken at $t$. This difference can be rewritten:

$$d_t(S'_t) = e^{-r(\tau-t)}\left[K\left(e^{r(\tau-t)} - 1\right) - D - E^Q_t[TV^a(S'_t, \tau)]\right], \text{ or more precisely :}$$

$$d_t(S'_t) = e^{-r(\tau-t)}\left[K\left(e^{r(\tau-t)} - 1\right) - D - E^Q_t[TV^a(S'_t \exp(r - 0.5\sigma^2(\tau - t) + \sigma(W\tau - W_t)))]\right],$$

where $W$ denotes a standard Brownian motion.

If $S'_t = 0$ at date $t$, $S'_t = 0$ surely at date $\tau$, hence $TV^a(0, \tau) = 0$. Therefore: $d_t(S'_t = 0) = e^{-r(\tau-t)}[K(e^{r(\tau-t)} - 1) - D]$, which is $> 0$, implying that immediate exercise is better than waiting till $\tau^+$. More generally, by continuity, for $S'_t$ small enough, $E^Q_t[TV^a(S'_t, \tau)]$ is small enough for $K(e^{r(\tau-t)} - 1) - D > E^Q_t[TV^a(S'_t, \tau)]$, in which case $d_t(S'_t) > 0$ and immediate exercise is preferable to leaving the put unexercised till $\tau^+$.

Equipped with the results of Proposition 4 and those of the previous Sect. 13.1.1.4, we obtain the OEB profile. In any $(t, T)$ period with $t < t_c < \tau < T$, three sub-periods can clearly be distinguished:

- $(t, t_c)$: if starting from a positive $S*(t)$, an OEB decreasing after some point towards 0 as $t \to t_c$
- $(t_c, \tau)$: a no exercise sub-period during which $S*(t) = 0$
- $(\tau, T)$: an upward-sloping OEB as described in Fig. 13.5

### 13.1.1.6 Early Exercise of American Options Written on a Forward Contract

It may also be optimal to exercise early an in-the-money American call or put written on a forward or a futures contract. The analysis follows the same paths as for options

## 13.1 Early Exercise and Call-Put Parity for American Options

written on spot assets. In fact, as the futures/forward price[4] behaves as the spot price of an asset paying a continuous dividend yield $c = r$ (see Eq. (9.4) of Chap. 9), our justifications are brief.

- For a European put, we have from call-put parity on a forward: $P_t = C_t + (K - F_t) e^{-r(T-t)}$.
  If the put is deep in-the-money ($K >> F_t$), then $C_t \# 0$ and $P_t \# (K - F_t) e^{-r(T-t)} < (K - F_t)$; however, $P^a{}_t > P_t$ but, possibly, for a small enough price $F_t$, $P^a{}_t = K - F_t$ and it is then optimal to exercise the American put.
- For a call, the justification is the same, if $F_t >> K$.

### 13.1.2 Call-Put "Parity" for American Options

We have just emphasized that the exact call-put parity ($C_t - P_t = (.)$) only holds for European options. The value of ($C^a{}_t - P^a{}_t$) lies in all cases between two bounds that it is easy to specify if the interest rates are constant or evolve in a deterministic manner.

The next proposition makes explicit the call-put "parity" relevant in the various possible situations. We provide the proof of only one of the results and leave the others to the reader as an exercise (Assume that interest rates remain constant for simplicity).

**Proposition 5**
*At date t, for a spot underlying (possibly distributing a discrete positive dividend D at time t with $t < t \leq T$), the call-put "parity" for American options writes:*

$$S_t - \frac{D}{(1+r)^{\tau - t}} - K \leq C^a{}_t - P^a{}_t \leq S_t - \frac{K}{(1+r)^{T-t}}. \qquad (13.7a)$$

In this formula, $r$ is a discrete zero-coupon rate.
If the dividend is issued continuously (for example, if the spot underlying is a stock market index involving a large number of stocks paying dividends at different dates, or a foreign interest rate $c$), then the "parity" can be written similarly as

$$S_t e^{-c(T-t)} - K \leq C^a{}_t - P^a{}_t \leq S_t - K e^{-r(T-t)}, \qquad (13.7b)$$

where the continuous domestic rate $r$ is used to preserve homogeneity in the formula.

---

[4]Recall that there is no theoretical difference between futures and forward prices whenever the interest rate is constant (or more generally deterministic). When the interest rate is stochastic, but remaining surely positive, the following results and argument hold but the $r$ present in the call-put parity is the zero-coupon rate prevailing at $t$ for a duration $T$-$t$.

516	13 American Options and Numerical Methods

**Table 13.1** A proof by arbitrage

|  | $t$ | $t < u < T$ | $T$ |
|---|---|---|---|
| Short call | $C^a_t$ | $-(S_u - K)$ | $-(S_T - K)^+$ |
| Long underlying | $-S_t$ | $S_u$ | $S_T$ |
| Long put | $-P^a_t$ | $P^a_u$ | $(K - S_T)^+$ |
| Borrowing | $Ke^{-r(T-t)}$ | $-Ke^{-r(T-u)}$ | $-K$ |
| Net cash flow or value of unwound position | $>0$ | $P^a_u + K(1 - e^{-r(T-u)}) > 0$ | $0$ |

## Proposition 6

*If the underlying is a forward or Futures contract, then the call-put "parity" for American options writes:*

$$F_t \, e^{-r(T-t)} - K \leq C^a_t - P^a_t \leq F_t - K \, e^{-r(T-t)}. \tag{13.8}$$

## Proof

We prove by contradiction that: $C^a_t - P^a_t \leq S_t - K \, e^{-r(T-t)}$.

Assume we observe that: $C^a_t - P^a_t - S_t + K \, e^{-r(T-t)} > 0$.

Table 13.1 presents the possible cash flows generated by selling the (relatively expensive) call, buying the UA and the (relatively cheap) put and borrowing $K \, e^{-r(T-t)}$ at current time $t$. Dealing with American options, one must consider the risk of an adverse early exercise on the short call at any time $u$ ($t < u < T$). This risk is accounted for in the third column of the table. The fourth column shows the net value of the positions on date $T$. Note that we do not need to consider the possibility of an early exercise of the (long) put, which can only be beneficial. Such a portfolio is an arbitrage since it generates an initial positive cash flow and yields positions that have a non-negative value in all situations, even in the case of an adverse early exercise of the call.

To conclude, let us remark that it might seem natural to try to exploit these call-put "parities" to prove the optimality of the early exercise of American options in some situations. This approach in fact turns out to be a deadlock because the no-arbitrage intervals for $C^a - P^a$ determined in Propositions 5 and 6 are not narrow enough. For instance, the inequality $C^a_t \leq S_t - K \, e^{-r(T-t)} + P^a_t$, which implies that the time value of the put $C^a_t - (S_t - K) \leq K(1 - e^{-r(T-t)}) + P^a_t$, *does not prove* that this time value becomes negative for high values of $S_t$ triggering early exercise, since $K(1 - e^{-r(T-t)}) > 0$.

## 13.2 Pricing American Options: Analytical Approaches

Pricing American options is a complicated problem for which, most often, there are no analytical formulas. So it is usually necessary to call upon numerical methods or/and more or less satisfactory approximations. We present in this section some

## 13.2 Pricing American Options: Analytical Approaches

mathematical approaches which lead to analytical solutions in few cases, analytical or quasi-analytical approximations in other cases, but, in any case, to interesting insights and concepts. Moreover, they provide the basis for some numerical methods.

In fact, analytical approaches yield closed form expressions of American options prices only for perpetual puts and calls and calls on an asset paying a single discrete dividend before maturity. Among those three cases, we only present the pricing of a call on a spot asset paying a single, discrete dividend (in Sect. 13.2.1). All other cases (puts, calls on assets paying a continuous dividend, options on forward contracts) are considered in Sect. 13.2.2 where the PDE approach and the stopping time formulation are presented and analytical approximations sketched. In Sect. 13.2.3, we present some order of magnitude of the error made when pricing American options with the standard Black–Scholes–Merton model suited to their European counterparts.

### 13.2.1 Pricing an American Call on a Spot Asset Paying a Single Discrete Dividend or Coupon

We consider here an American call of maturity $T$ whose UA pays a coupon or a dividend $D$ on date $t < T$. Both $D$ and t are known. The current date is $t = 0$. Recall that in this case the option of early exercise has a positive value hence the American option has a strictly higher value than its European counterpart (unless the coupon is very small, as stated in Proposition 3). We begin by a crude and simple approximation before presenting the approach leading to an analytical expression.

#### 13.2.1.1 Black's Approximation

We begin with the simplest way of pricing such an American call. Current time is $\tau = 0$. Fischer Black (1975) suggested a crude pricing method which consists in taking the maximum of two values: that of a *European* call with maturity $\tau \, (> 0)$ on a not-paying dividend UA priced $S_0$, denoted by $_tC(S_0)$; and that of a *European* call with maturity $T$ on an UA distributing $D$, which is equal to $_TC(S_0 - D^*)$:

$$C^a(S_0) = \text{Max}[_TC(S_0 - D*), \; _tC(S_0)],$$

where $D^* = De^{-\tau \, r\tau}$ is the present value of the dividend ($r_\tau$ is the $\tau$-zero-coupon rate prevailing at 0).

This method underestimates the value of an American call because Black's approximation amounts to assuming that the decision to exercise or not on $\tau^-$ is taken on date 0 (on the base of the relative value of two European options at date 0), and is applied regardless of the evolution of the UA price between 0 and $\tau$. In practice, this underestimation remains within limits that are often acceptable for non-professionals.[5]

## 13.2.1.2 Pricing the Call on a Spot Asset Detaching A Single Dividend with a Compound Option

First recall from Sect. 11.2.1.2 that it would not be consistent to assume that $S_t$ follows a GBM between $t$ and $\tau^-$ (just before the distribution date) because $S(\tau^-)$ would then be lognormal, implying that the event $S(\tau^-) < D$ has some positive probability, which is incompatible with the assumption that $D$ is certain. The solution consists in distinguishing two components in the underlying asset price $S(t)$: the first, $De^{-r(\tau-t)}$, is the (sure) value of the discounted dividend; the second, denoted $S'(t)$, is assumed to obey a GBM with constant volatility $\sigma$. The stock value then is $S(t) = S'(t) + De^{-r(\tau-t)}$. Therefore, we have:

$$S(t) = S'(t) + De^{-r(\tau-t)} \text{ for } t < \tau, \text{ and in particular } S(\tau^-) = S'(\tau^-) + D,$$

$$S(\tau^+) = S'(\tau^+), \text{ and}$$

$$S(t) = S'(t) \text{ for } t > \tau.$$

The value at time $\tau^-$ of the American call written on $S(t)$ with maturity $T$ is given by (13.2), and at time $t$ this call can be valued as a right to the payoff at $\tau$: Max $(C(S(\tau^+)), S(\tau^+) + D - K))$.

Provided $D > K(1 - e^{-r(T-\tau)})$, we have shown in Proposition 3 that there exists an early exercise price $S^*$ such that:

$$C^a(S(\tau^+), \tau) = S(\tau^+) + D - K \quad \text{if } S(\tau^+) \geq S^* \text{ (early exercise)}$$
$$= BS(S(\tau^+), T - \tau) \quad \text{if } S(t^+) < S^* \text{ (continuation)}$$

where $BS(.)$ is the value from the Black-Scholes formula (without dividends) for the remaining option life $(T - t)$ and $S^*$ the optimal *ex-dividend* exercise price of the UA for which the value of the unexercised option equals its value upon early exercise. $S^*$ then is the solution to the value matching condition:

$$C^a(S^*, K, \tau) = [BS(S^*, K, T - \tau) =] S^* + D - K.$$

We start by solving for $S^*$ this value matching condition by iteration. Then, from now on, $S^*$ is *given*. At time $t$ ($< \tau$), we have, using the ex-dividend price $S'(t)$:

---

[5]The underestimate is *on average* in the order of 2%, but unfortunately depends on $S/K$, $T$, $D$, $\tau$, and $\sigma$ (if it were roughly constant, it could be easily corrected).

## 13.2 Pricing American Options: Analytical Approaches

$$C^a(S', K, T - t)$$

(i)
$$= e^{-r(\tau-t)} \int\limits_{S'_\tau < S*} f(S'_\tau) BS(S'_\tau, K, T - \tau) dS'_\tau + e^{-r(\tau-t)} \int\limits_{S'_\tau \geq S*} f(S'_\tau)(S'_\tau + D - K) dS'_\tau.$$

where $f(S'_\tau)$, short notation for $f_t(S'_\tau)$, is the conditional risk-neutral density function of $S'_\tau$ at time $\tau$, *given* the current asset price $S'_t$, and where, since $S'_t$ follows a GBM, $S'_\tau$ is equal to: $S'_\tau = S'_t e^{\left(r - \frac{\sigma^2}{2}\right)(\tau-t) + \sigma\sqrt{\tau-t}\,U}$, with $U$ the standard normal distribution.

The first term on the r.h.s. of (*i*) corresponds to the call's value in the continuation region and the second term to its value in the stopping region.

Denote by $I_1$ and $I_2$, respectively, these two terms. Since $BS(S', K, T - t) < S* - K + D$ implies that $S' < S*$, we let $(S* - K + D)$ appear in $I_1$ as a strike price in order to recognize in $I_1$ the value of an option on $BS(.)$, and we rewrite $I_1$ and $I_2$ as:

$$I_1 = e^{-r(\tau-t)} \left[ \int\limits_{S'_\tau < S*} f(S'_\tau)[BS(S'_\tau, K, T - \tau) - (S* - K + D)] dS'_\tau \right.$$

$$\left. + (S* - K + D) \int\limits_{S'_\tau < S*} f(S'_\tau) dS'_\tau \right],$$

$$I_2 = e^{-r(\tau-t)} \int\limits_{S'_\tau \geq S*} f(S'_\tau)(S'_\tau - S*) dS'_\tau + e^{-r(\tau-t)}(S* - K + D)) \int\limits_{S'_\tau \geq S*} f(S'_\tau) dS'_\tau.$$

Now, since $e^{-r(\tau-t)} \int\limits_{S'_\tau < S*} (S* - K + D) f(S'_\tau) dS'_\tau + e^{-r(\tau-t)}(S* - K + D)$ $\int\limits_{S'_\tau \geq S*} f(S'_\tau) dS'_\tau = e^{-r(\tau-t)}(S* - K + D)$, summing up $I_1$ plus $I_2$ yields, after rearranging:

(ii) $C^a(S', K, T - t) = e^{-r(\tau-t)} \int\limits_{S'_\tau \geq S*} f(S'_\tau)(S'_\tau - S*) dS'_\tau$

$$-e^{-r(\tau-t)} \int\limits_{S'_\tau < S*} f(S'_\tau)[(S* - K + D) - BS(S'_\tau, K, T - \tau)] dS'_\tau + e^{-r(\tau-t)}(S* - K + D).$$

To visualize that (*ii*) is an explicit, closed-form solution, let us rewrite it less formally as:

$$C^a(S', K, T - t) = Call(S'_\tau, S^*, \tau - t)$$
$$- Put(Call(S'_t, K, T - t), S^* - K + D, \tau - t)$$
$$+ e^{-r(\tau-t)}(S^* - K + D). \tag{13.9}$$

The first term in (13.9) is a *vanilla European call*, written on $S'$, of maturity $\tau$ (not $T$) and strike $S^*$ (not $K$). The second is a *short* put (of strike $(S^*-K+D)$ and maturity $\tau$) written *on a call* (of strike $K$ and maturity $T$), i.e., a *compound option* whose explicit valuation formula is given in Sect. 14.1.4, Eq. (14.12). This formula involves the bivariate normal distribution function $N2(.)$ which can be calculated using standard numerical integration. The third is just a (riskless) *zero-coupon bond* whose principal is $(S^*-K+D)$.

We leave to the reader as an exercise to show that at $t = \tau$, solution (13.9) yields either $BS(S'_\tau, K, T - \tau)$ (if $S'_\tau \leq S^*$, because $S'_\tau \leq S^*$ implies $Call(S'_\tau, K, T - \tau) \leq S^* + D - K$) or $(S'_\tau - K + D)$ (if $S'_\tau > S^*$).

Whaley (1981) was the first to derive formula (13.9) written in a slightly different way.[6]

### 13.2.2 Pricing an American Option (Call and Put) on a Spot Asset Paying a Continuous Dividend or Coupon

When the UA pays a continuous dividend, American options are exercised when the UA price hits for the first time the Optimal Exercise Boundary (OEB), which, depending on the random path followed by the UA price, may occur at any time up to maturity. Recall that the OEB of an American option is not known and its determination is part of the pricing process.

We will briefly explore three formulations for the American option pricing problem: the first two are grounded on the Black–Scholes–Merton Partial Differential Equation (BSM PDE); the third on the notion of stopping time. We will then present briefly an analytical approximation to the price of an American put.

#### 13.2.2.1 The PDE Approach: The Free Boundary and the Linear Complementarity Formulations

(a) *The BSM (fundamental evaluation) PDE in the continuation and stopping regions.*

    (i) Consider an American option written on a UA paying a continuous dividend yield $c$ whose price $S$ is given by: $dS/S = (r-c)\, dt + \sigma\, dW$ in the risk-neutral world. The option price is $O^a$ and its given payoff is denoted by $\Psi(S)$ $(S-K$ for a call, and $K-S$ for a put).

---

[6]Geske (1979) analyses the problem of the valuation of a call (of strike $k$) written on a non-dividend paying stock $S$ which itself is analyzed as a call option written on the value of the issuing firm's assets $V$, the strike $K$ being the firm's total debt. So, clearly, the compound option involved is, unlike here, a call on a call. There is neither discrete nor continuous dividend in his model.

## 13.2 Pricing American Options: Analytical Approaches

Before reaching the OEB, *inside* the continuation region, the option price $O^a(S, t)$ satisfies, as any tradable derivative, the BSM (fundamental evaluation) PDE:

$$D^t O^a - r O^a = 0 \tag{13.10a}$$

where $D^t$ is the Dynkin operator (see Sect. 18.5.2.2)

$$D^t O^a \equiv \left[ \frac{\partial O^a}{\partial t}(.) + (r - c)\frac{\partial O^a}{\partial S}(.) + \frac{1}{2}\sigma^2 \frac{\partial^2 O^a}{\partial S^2}(.) \right].$$

This PDE holds *inside* the continuation region, where $O^a > \Psi(S)$; it simply states that the expected return on a traded asset is equal to the risk-free rate in an arbitrage-free risk-neutral world, the different assets being differentiated by their boundary conditions which involve $\Psi(S)$.

But only in the continuation region does the American option obey the BSM PDE (13.10a). As soon as the OEB is hit and inside the stopping region, the BSM PDE *does not hold anymore*: since it is sub-optimal to hold the option, its risk-neutral expected return $D^t O^a / O^a$ is *smaller* than $r$:

$$D^t O^a - r O^a < 0. \tag{13.10b}$$

This inequality prevails inside the stopping region, the OEB included (everywhere $O^a = \Psi(S)$).

We may thus characterize the stopping and the continuation regions in different alternative but equivalent ways:

$\{(S, t): O^a(S,t) > \Psi(S)\}$ for the continuation region ; $\{(S, t): O^a(S,t) = \Psi(S)\}$ for the stopping region;

$\{(S, t): D^t O^a - r O^a = 0\}$ for the continuation region; $\{(S, t): D^t O^a - r O^a < 0\}$ for the stopping region;

$\{(S, t): S < S_{call}{}^*(t)\}$ for the continuation region of a call; $S > S_{put}{}^*(t)$ for that of a put;

$\{(S, t): S \geq S_{call}{}^*(t)\}$ for the stopping region of a call; $S \leq S_{put}{}^*(t)$ for that of a put.

Let us clarify the situation prevailing *inside* the stopping region by considering, for instance, an investor holding (sub-optimally) a *call* at time $t$ while $S(t) > S^*(t)$. The value of such a position is $S(t) - K$ (the investor can exercise the call but cannot do better). The expected capital gain on such a position in the interval $(t, t+dt)$ is: $D^t(S(t) - K)\, dt = D^t S(t)\, dt = (r - c)S\, dt$ (the expected capital gain on $S$ is the only return component since the holder of an unexercised call has no right to the dividend).

The return on such a position will be smaller than $r$ iff:

$$(r - c) S(t) \, dt < r(S(t) - K) \, dt <=> cS(t) > rK <=> S > S^\circ = rK/c.$$

The condition $cS(t) > rK$ is now familiar: in the stopping region the dividend foregone by holding one more instant the unexercised call is larger than the gain obtained by delaying the payment $K$ during this instant. Whenever $c > 0$ (the relevant case), such a situation prevails iff $S$ is higher than the price threshold $S^\circ = rK/c$.

Symmetrically, holding inside the stopping region an American put whose value is $K - S$ implies: $(c - r) S(t) < r(K - S(t)) <=> cS(t) < rK <=> S < S^\circ = rK/c$.

These results are summarized in the following Proposition:

## Proposition 7

*The stopping region (OEB included) is characterized by risk-neutral returns on holding the option smaller than r.*

*In the call's stopping region the dividend foregone is higher than the* time value *of K: $cS > rK <=> S > S^\circ = rK/c$. The opposite prevails in the* put's *stopping region where $cS < rK <=> S < S^\circ$.*

*A word of caution:* The condition $cS > rK$ ($cS < rK$) is a *necessary but not a sufficient* condition for being in the call's (put's) stopping region.

In fact, the *continuation region contains* the threshold value $S^\circ = rK/c$. In the case of a call, we obtain the following figure for $t < T$, $0 < c < r$:

0 ————————————————$K$————————$S^\circ$————————————$S^*(t,c,r)$)////////////////

      *Continuation $cS < rK$*     *Continuation $cS > rK$*    *Stopping region*

In addition, $c = 0$ precludes early stopping ($S^\circ$ and $S_{call}{}^* \to \infty$). When $c \geq r$: $S^\circ < K$ (the positions of $K$ and $S^\circ$ in the diagram are just « swapped »).

The following configuration prevails for a put when $0 < c < r$:

0 ///////////////////////////$S^*$————————————————$K$————————$S^\circ$————————————

  *Stopping region*       *Continuation $cS < rK$*    *Continuation $cS > rK$*

When $c \geq r$: $S^\circ < K$ (the positions of $K$ and $S^\circ$ are also « swapped »).

## 13.2 Pricing American Options: Analytical Approaches

- We now state two additional results.

*Asymptotical behavior of the OEB* near maturity:

For a call: $lim_{t \to T}\left(S^*_{call}(t,c,r)\right) = \max\left(S^\circ, K\right) = K \max\left(r/c, 1\right)$.

For a put: $lim_{t \to T}\left(S^*_{put}(t,c,r)\right) = \min\left(S^\circ, K\right) = K \min\left(r/c, 1\right)$.

For a call close to its maturity $T$, the exercise threshold is indeed $K$ if the dividend is large ($c > r$) and outweighs the opportunity loss on paying $K$ early but, if it is small ($c < r$), the call is kept alive even when $S$ is somewhat larger than $K$ provided the time value on $K$ outweighs the dividend loss: $rK > cS$.

For a put, the same analysis leads to the opposite results (e.g., if $rK > cS$, the threshold is $K$ since receiving $K$ at once outweighs the fact that the price of the delivered asset includes a dividend).

*A property of the time value* of an American option *at the early exercise point*: $\frac{\partial O^a}{\partial t}\big]_{S=S^*_t} = 0$. This property means that, when the option is early exercised, the time left to expiration is irrelevant to its value. The proof of this property rests on the value matching and smooth pasting conditions. Indeed, on the optimal exercise boundary $S^*(z)$, we have:

$O^a(S^*(z), z) = \Psi(S^*(z))$. Total differentiation yields:

$$\frac{\partial O^a}{\partial S}\bigg]_{S=S^*(z)} \frac{dS^*}{dz} + \frac{\partial O^a}{\partial z}\bigg]_{S=S^*(z)} = \frac{\partial \Psi}{\partial S}\bigg]_{S=S^*(z)} \frac{dS^*}{dz}.$$

Therefore, by virtue of the smooth pasting condition $\frac{\partial O^a}{\partial S}\big]_{S=S^*(z)} = \frac{\partial \Psi}{\partial z}\big]_{S=S^*(z)}$, we obtain: $\frac{\partial O^a}{\partial z}\big]_{S=S^*(z)} = 0$, hence $\frac{\partial O^a}{\partial t}\big]_{S=S^*(t)} = 0$.

The previous analysis underpins two analytical pricing formulations briefly presented now: the free-boundary and the linear complementarity formulations.

(b) *The free-boundary formulation for American option pricing*

For an American call, for instance, the PDE-free-boundary formulation writes:

- $D^t C^a(S, t) - rC^a(S, t) = 0$ for $S < S^*(t)$;
- $C^a(S^*(t), t) = S^*(t) - K$ (value matching condition meaning that $S^*(t)$ is *an exercise boundary*);
- $\frac{\partial C^a}{\partial S}\big]_{S=S^*(t)} = 1$

(smooth pasting condition expressing the *optimality* of $S^*(t)$ as an exercise boundary);
- $C^a(S, T) = \text{Max}\,(S - K, 0)$ (terminal condition).

Similarly for a put:

- $D^t P^a(S, t) - rP^a(S, t) = 0$ for $S > S^*(t)$;
- $P^a(S^*(t), z) = K - S^*(t)$;

524      13 American Options and Numerical Methods

- $\frac{\partial P^a}{\partial S}\big|_{S=S^*(t)} = -1;$
- $P^a(S, T) = \text{Max}(K - S, 0).$

In general, for an American derivative whose price $O^a$ obeys PDE (13.10a) in the continuation region, and whose payoff is $Y(S, t)$, the value matching, smooth pasting and terminal conditions write respectively:

$$O^a(S^*(t), t) = \Psi(S^*(t), t),$$

$$\frac{dO^a(S, t)}{dS}\bigg|_{S=S^*(t)} = \frac{d\Psi(S, t)}{dS}\bigg|_{S=S^*(t)},$$

$$O^a(S, T) = \Psi(S, T).$$

This system is an example of a free-boundary value problem: the determination of the optimal exercise boundary $S^*(t)$ is part of the problem.

(c) *The linear complementarity formulation*

(13.10a) and (13.10b) are « quasi equivalent » to the so-called « linear complementarity formulation »:

$$(i)\ (D^t O^a - r O^a)(O^a(S, t) - \Psi(S)) = 0; \tag{13.11}$$

$$(ii)\ \ O^a(S, t) \geq \Psi(S);$$

$$(iii)\ \ D^t O^a - r O^a \leq 0;$$

And, in addition, the terminal value condition: $O^a(S, T) = \Psi(S)$.

No explicit mention is made to the smooth pasting condition or to the OEB $S^*(t)$ in the linear complementarity formulation. Again, American options differ by the boundary conditions associated with (13.11).

As already noted, there is no analytical solution to this free-boundary or linear complementarity problem, with the exception of perpetual options which are not examined in this book (except briefly at the end of Sect. 13.2.2.3). However, numerical pricing methods such as the finite difference method presented in Sect. 13.4 are based on a discretized version of the PDE approach.

### *13.2.2.2 Stopping Time Formulation

We now briefly introduce another modeling of American options based on the notion of stopping time, for the sake of some interesting concepts and insights. Pushing the modeling further than is intended in this book, it would be akin to the previous PDE formulations. Let $t$ be the current time. A stopping time t is a random variable taking here values in $[t, T]$ corresponding to the decision to exercise at time t. The condition t > $t$ means that the exercise must be delayed till a future random date t. The stopping decision at t can and should only be based on the information available at t, including the asset price path followed till time t $[S(u), t \leq u \leq t]$, but *not on future information*

## 13.2 Pricing American Options: Analytical Approaches

available only after t. Mathematically, this is expressed by the condition that the event $t \leq t$ (the call has been or should be exercised at $t$) is $F_t$-measurable, where $F_t$ is the information (filtration) available at $t$. It may be useful to consider a stopping time as resulting from a *stopping rule*. A viable stopping rule must be based on current and not future information. In our context for instance, an example of stopping rule could be derived from an early exercise frontier b(z) given *a priori* (which is in general sub-optimal). The decision to stop (exercise) is taken the first time $S(u)$ reaches the frontier and the corresponding stopping time is $t = Inf_u\{t \leq u \leq T: S(u) = b(T-u)\}$.

However, the exercise of an American option is associated with a more complex stopping time since the optimal exercise boundary $S^*(t)$ is not given *a priori* and is unknown.

Consider a fictitious American call exercised under a given, not necessarily optimal stopping rule R implying a stopping time t(R) that, will generate the payoff $Max(S(t(R)) - K, 0)$ at some time t(R) with $t \leq t(R) \leq T$. Its value at $t$ would be $V(t) = E^Q_t [e^{-r(t(R)-t)} Max(S(t(R)) - K, 0)]$ where Q is the risk-neutral probability measure and $E^Q_t$ is the RN expectation taken at $t$ ($E^Q[. | F_t]$). This valuation formula is generic and would be valid for any call supposed exercised at a given date t random or not random. In fact, since R can and will be optimally chosen so as to maximize the call value, the price of the « true » American call will be: $C(S,t) = Sup_R E^Q_t [e^{-r(t(R)-t)} Max(S(t(R)) - K, 0)]$ or more simply, identifying the stopping time to its corresponding stopping rule:

$$C(S,t) = Sup_{t \in T(t,T)} E^{Q_t} \left[ e^{-r(t-t)} Max(S(t) - K, 0) \right], \quad (13.12)$$

where the supremum is taken over all feasible stopping times taking values in $(t, T)$ (set $T(t,T)$) and is attained for the optimal stopping time (rule) $t^*$. Since $C(S, t)$ is strictly larger than the payoff before the stopping time, and equal to the payoff at the stopping time $t^*$, we can write: $t^* = Inf_u \{t \leq u \leq T: C(S(u),u) = Max(S(u) - K, 0)\}$. The *optimal stopping time* $t^*$ is attained the first time the expected discounted call value cannot be improved by waiting still more, i.e. the first time the call value is equal to its current intrinsic value (payoff).

### 13.2.2.3 An Approximate Analytical Solution (Barone-Adesi and Whaley)

Several analytical approximations for valuing American options have been proposed by researchers. Some of them are based on interpolating between an upper bound and a lower bound for the option value[7] and others on representing the exercise frontier.[8] The interested reader is referred to Fusai and Roncoroni (2008) for a presentation of these methods and additional references.

---

[7] See Broadie and Detemple (1996).

[8] See Omberg (1987), Chesnay and Lefoll (1996), and Carr et al. (1992).

One of the possible approaches consists in accepting an approximation to the BSM partial differential equation so that, with the appropriate boundary conditions, it has an analytical solution. This is the way chosen, for example, by Barone-Adesi and Whaley (1987). The American option is first decomposed into two parts. Its European part is obtained by a BSM formula. The American part is the value of the option to exercise early, thus equal to the difference in value between the American and the European options. It is obtained by an iterative procedure involving an approximation.[9] The fundamental idea is to find the critical value $S*$ (OEB) above which the call should be exercised early (or below which the put should be exercised). For $S = S*$, the American option must be worth its intrinsic value and its delta must equal $+1$ for a call or $-1$ for a put (value matching and smooth pasting conditions). We just sketch the solution without details to focus on ideas, not computations. Current time is $t = 0$. The option maturity is $T$. We consider first the call and then the put.

- *Pricing a call on a* spot *asset paying a continuous dividend.*

After some calculation using approximations, we obtain a first equation that holds for a call

$$S* - K = C(S*, T) + A_2 \quad \text{with} \quad A_2 \equiv \frac{S*}{\beta_2}[1 - e^{-cT}N(d_1(S*))], \qquad (13.13)$$

where $C$ is the value of the European call, $c$ the continuous dividend yield, $\beta_2$ a known parameter,[10] and $A_2$ is derived from the *smooth pasting* condition. This equation is the *value matching* condition. $S*$ $(>K)$ is the only unknown and is found by successive iterations (in practice two or three are usually sufficient for an efficient scheme).

The value of the American call then is given by

$$\begin{cases} C^a(S, T) = C(S, T) + A_2(S/S*)^{\beta_2} & \text{if } S < S* \\ C^a(S, T) = S - K & \text{if } S \geq S* \end{cases} \qquad (13.14)$$

**Remarks**
- $A_2(S/S*)^{\beta_2}$ is the approximate value of the early exercise option, simply added to the value of the vanilla European call.

---

[9]The approximation consists in neglecting a term in the four-term BSM pricing Partial Differential Equation (to which obeys the value of the early exercise option) to remove its dependence on time. The PDE then reduces to a second-order Ordinary Differential Equation, which has two known independent solutions, one of which being the option value sought after.

[10]$\beta_2 = \frac{1}{2}\left[-(n-1) + \sqrt{(n-1)^2 + 4m/g}\right]$, where $m \equiv 2r/\sigma^2$, $n \equiv 2(r-c)/\sigma^2$ and $g \equiv 1 - e^{-rT}$.

## 13.2 Pricing American Options: Analytical Approaches

- When the option's expiration date is near ($T$ tends to 0), $A_2$ in (13.13) tends to zero since $N(d_1(S^*))$ and $e^{-cT}$ tend to one. The value of an American call tends to its European analogue's.
- $A_2$ is positive if $e^{-cT} < 1$, that is $c > 0$. Recall that if the underlying does not pay a dividend ($c = 0$) or generates a negative "dividend" (e.g., the storage cost for a commodity), an American call has the same value as its European counterpart and the whole procedure is irrelevant.
- *Pricing a put on a* spot *asset*

The value of an *American* put, *similarly to that of a call, is given by*

$$\begin{cases} P^a(S,T) = P(S,T) + A_1(S/S^*)^{\beta_1} & \text{if } S > S^* \\ P^a(S,T) = K - S & \text{if } S \le S^* \end{cases} \tag{13.15}$$

where $S^*$, the critical value ($<K$) of the UA price, is the solution, computed iteratively, to

$$K - S^* = P(S^*,T) + A_1, \text{with } A_1 \equiv -\frac{S^*}{\beta_1}[1 - e^{-cT}N(-d_1(S^*))]$$

where $A_1$ is derived from the smooth pasting condition and $\beta_1$ is a known parameter.[11] The critical value $S^*$ is of course different from the $S^*$ relative to the call.

### Remarks
- Unlike the case of the call, the dividend yield $c$ may be zero without impact on the method.
- It is easy to deduce from this analysis the price of a *perpetual American put* (Merton, 1973):

$$P^a(S,\infty) = \frac{K}{1+m}\left[\frac{(1+m)S}{mK}\right]^{-m} \quad \text{with } m = 2r/\sigma^2. \tag{13.16}$$

Indeed, since the put's maturity is infinite, the partial differential equation of the Black–Scholes model for the put's price $P^a$ becomes the following ordinary differential equation (the time derivative disappears):

$$\frac{\sigma^2}{2}S^2\frac{\partial^2 P^a}{\partial S^2} + bS\frac{\partial P^a}{\partial S} - rP^a = 0$$

whose general solution is $P^a = aS^b$.

---

[11]$\beta_1 = \frac{1}{2}\left[-(n-1) - \sqrt{(n-1)^2 + 4m/g}\right]$, with the same parameters as in the previous footnote.

## General Remarks

One advantage of the Barone-Adesi and Whaley method is that it is not necessary to compute iteratively the threshold values $S^*$ for various American options (calls or puts) with the same maturity but different strike prices. In fact, since the option price is homogeneous in the strike price and the underlying asset price, it is easy to show that

$$\frac{S_i^*}{K_i} = \frac{S_1^*}{K_1} \quad \text{for } i = 2, \dots, N.$$

Therefore, it suffices to carry out the iterative computation only once (for each option type, call and put) with strike $K_1$, which saves much time if the number $N$ of different strikes to consider is large.

If the underlying is a *forward* contract, one replaces $S$ by $F$ and *the risk-neutral drift of the spot price process* $(r - c)$ by 0, since the net cost-of-carry of a forward is nil.

In practice, however, although the method is very fast, the approximation may be too crude for professional applications.

### 13.2.3 Prices of American and European Options: Orders of Magnitude

Table 13.2 compares the values of European and American calls and puts written on a notional long-term debt *forward* contract with coupon rate $(c)$ of 10%. The common values are: strike $= 100$, maturity $T = 0.5$ (year), short-term rate $r = 8\%$ (discrete), and UA volatility $s = 15\%$. The quoted price $F$ varies from 80 to 120 by steps of 10. Black's model is used for European options and Barone-Adesi and Whaley's for American ones.

As we can see, the difference becomes sizeable for in-the-money options. In particular, the two extreme European options (in bold) have a negative time value.

**Table 13.2** European (Black) and American (Barone-Adesi and Whaley) options

| $F$ | Calls | | Puts | |
|-----|-------|------|------|------|
|     | Black | BA-W | Black | BA-W |
| 80  | 0.0584 | 0.0603 | **19.3034** | 20.0000 |
| 90  | 0.8162 | 0.8264 | 10.4387 | 10.6117 |
| 100 | 4.0698 | 4.1139 | 4.0698 | 4.1139 |
| 110 | 10.6994 | 10.8664 | 1.0769 | 1.0897 |
| 120 | **19.4400** | 20.0031 | 0.1950 | 0.1992 |

## 13.3 Pricing American Options with the Binomial Model

We begin with the simplest numerical method based on an « adapted » binomial model. We start by constructing a recombining binary tree where the UA distributes a discrete dividend or coupon then a continuous one, before examining the valuation of an American option (a put, or a call for which the American feature is relevant) using the binomial model.

### 13.3.1 Binomial Dynamics of Price $S$: The Case of a Discrete Dividend

Most of the time, the dividend or coupon paid out by the underlying is discrete. Even if we assume known the amount *of the payment* $D$ and its date $\tau$ ($< T$, the option expiration date), using the Cox–Ross–Rubinstein (CRR) binomial model poses a problem: after the date of payment of the dividend $D$, the branches of the tree do not recombine and the tree "explodes" into "bushes." Indeed, if $d.\, u.\, S = u.\, d.\, S$ (a down followed by an up gives the same value as an up followed by a down, which permits the tree branches to rejoin), one does not get $d(u.S - D) = u(d\, S - D)$ since $d.D \neq u.D$. [12]

For *a single* coupon detachment *at step* $i$ ($0 < i < n$), we obtain at step $n$ no longer $(n+1)$ possible values for the underlying but $(i+1)(n - i +1)$. For example, for $n = 500$ and $i = 250$, there will be, not just 501 final values for the underlying, but 63,001! [13] If there are many discrete dividends, this number increases like a power and the algorithm takes too much time to carry out. This is the reason why, in practice, methods that retain the recombining character of the tree are used. We give three such below, for which the results are usually very close (Fig. 13.6).

- **The first method** is to assume *the dividend is proportional* to the UA price. Let $i$ ($0 < i < n$) be the number of the period in which the dividend date falls and $S_{i,j}$ ($0 \leq j \leq i$) the ($i + 1$) possible UA values on this date, $j$ being the number of price rises (the larger $j$ is, the larger $S_{i,j}$). On date $i$, instead of *reducing* the value of $S_{ij}$ by the discrete dividend $D$, one *multiplies* each $S_{i,j}$ by ($1 - c$), where $c$ is the coupon rate, defined by $c = D/S$. The tree branches thus recombine since, on date ($i + 1$), we have: $S_{i,j}(1 - c)u = S_{i,j+1}(1 - c)d = S_{i+1,j+1}$. The paths starting from nodes ($i, j$) and ($i, j+1$) therefore join up at ($i+1, j+1$) if the first encounters a price rise between $i$ and $i+1$ and the second a drop. Figure 13.6 illustrates this property when the dividend takes place at period $i =1$.

  This approach's advantage is its simplicity. Its drawback is that it makes the dollar value of the dividend depend on the value taken on by the UA price in the tree,

---

[12] A *sufficient* condition for a binary tree to recombine is that, for each of its moves $u$ and $d$, $S_{i+1}/S_i$ is constant along the path. This is not a *necessary* condition: we will see that if $S_{i+1}/S_i$ varies *in the same proportion in both moves (up and down) then the tree remains recombining*.

[13] Taking the derivative of the equation with respect to $i$ shows that the maximum number of final values is obtained for $i=n/2$. For $i = 0$ or $n$, we obviously recover the no-dividend case where the number of final values is $n+1$.

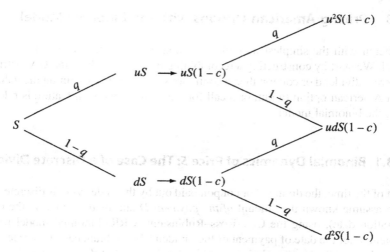

**Fig. 13.6** Proportional dividends and the binomial method. One period before expiration; Dividend proportional to $S$ ($D = c.S$)

which can be justified for long-term options (in the long run, the ratio between the market price of a share and its dividend is often rather constant), but much less so for short-term options, for which the dollar amount of the discrete dividend is known.

- **The second method** consists in *adjusting* the risk-neutral probability $q$, which one may recall is equal to $(1+r-d)/(u-d)$, and is used in calculating the values of the UA between dates $i$ (when the dividend is paid) and $(i+1)$. To compute the values of $S_{i+1,j+1}$ from $S_{i,j}$, we use the modified probability $(1+(r-c)-d)/(u-d)$, where $c$ is still defined as $D/S$. This amounts to reducing (increasing) the probability of a rise (fall) in the UA price, instead of reducing its price, which has the (desired) result of diminishing the value of a call and increasing that of a put.[14] Then one re-uses for subsequent steps in the CRR tree the usual probability $q$.

  The advantage of this approach is that it is easy to generalize it to the case when several dividends are distributed between now and the option expiration: We calculate beforehand the coupon rates $c_1 = D_1/S_0$, $c_2 = D_2/S_0$, etc., then, in the tree we modify $q$ each time a dividend is detached using $q_i = (1+(r-c_i)-d)/(u-d)$.

- **The third method**, in contrast to the first two, does not use the dividend rate $c$ but the absolute value $D$, which is assumed known. As we remarked at the start of this subsection, the recombining nature of the tree would be lost if one applied the

---

[14] More precisely, since the probability of a rise is reduced by $\frac{c}{u-d}$, the expectation of the rate of increase of $S$ between $i$ and $i+1$ is reduced by $\frac{c}{u-d}(u-d) = c$; therefore, $E(S_{i+1})$ is less by $cS_i = D$, which reduces the value of a call as when the price of the underlying goes down by $D$ in the two states $u$ and $d$ (and for any $j$).

### 13.3 Pricing American Options with the Binomial Model

binomial method directly. To retain the recombining feature, we distinguish, as in Sect. 13.2.1.2, two components in $S(t)$ which are assumed to evolve independently: the first, denoted by $D^*(t)$, represents the present value on date $t$ of the dividends expected from $t$ and $T$; the second, denoted by $S'(t)$, represents the remaining value of the stock which is assumed to be described by a binomial process. The total value of the UA is $S(t) = S'(t) + D^*(t)$. Starting from the date $\tau$ when the last coupon is distributed before $T$, $D^* = 0$ and $S = S'$. We already considered a decomposition such like this when we studied the adjustment to the Black-Scholes model that allowed for discrete dividends (Chap. 11, Sect. 11.2.1–11.2.2). We interpreted it by imagining that the firm has bought enough zero-coupon bonds to cover exactly the payment of the dividends, and has invested the rest of its funds in its usual physical assets. The binomial process only describes the risky part $S'$ and is represented by a standard recombining tree. The stock's total value $S'+D^*$ can also be represented by a recombinant tree derived from the previous one, with the two trees coinciding from the date of the last coupon distribution preceding $T$.

## 13.3.2 Binomial Dynamics of Price $S$: The Continuous Dividend Case

If there is a continuous dividend payment at rate $c$, the needed adjustment to the standard binomial tree is very simple: it is sufficient to modify its calibration so that the expected price growth rate is $r - c$ and not $r$. This adjustment is carried out on the probability $q$. Calibrations 1 and 2 described in Sect. 10.4.2 then become

$$
\text{Calibration 1 :} \quad
\begin{cases}
u_N = e^{\sigma\sqrt{\frac{T}{N}}}; \quad d_N = \dfrac{1}{u_N} = e^{-\sigma\sqrt{\frac{T}{N}}} \\[2ex]
q_N = \dfrac{e^{(\rho-c)\frac{T}{N}} - e^{-\sigma\sqrt{\frac{T}{N}}}}{e^{\sigma\sqrt{\frac{T}{N}}} - e^{-\sigma\sqrt{\frac{T}{N}}}}
\end{cases}
$$

$$
\text{Calibration 2 :} \quad
\begin{cases}
u_N = e^{\left(\rho-c-\frac{1}{2}\sigma^2\right)\frac{T}{N}+\sigma\sqrt{\frac{T}{N}}} \\[1.5ex]
d_N = e^{\left(\rho-c-\frac{1}{2}\sigma^2\right)\frac{T}{N}-\sigma\sqrt{\frac{T}{N}}} \\[1.5ex]
q_N = 0.5
\end{cases}
$$

Recall that modeling the dividend or coupon as continuous is appropriate for a large stock index ($c$ is an average dividend rate), an exchange rate ($c$ is the interest rate paid by the foreign currency), or a commodity ($c$ then is a *convenience yield*).

## 13.3.3 Pricing an American Option Using the Binomial Model

We saw in Chap. 10 how to use the CRR algorithm to price a European option. The analysis was based on the following recursive computation: $C_i = E^q\left[\frac{C_{i+1}}{1+r}\middle|S_i\right]$, i.e., at each node $(i,j)$ in the tree, $C_{i,j} = \frac{q\, C_{i+1,j+1}+(1-q)C_{i+1,j}}{1+r}$, where $C$ is the option value (call or put), $S$ the UA price (spot or forward), $i$ the number of periods that have elapsed, $j$ the number of price increases, and $q$ the risk-neutral probability. Starting from $i = N-1$ (for $i = N$, one takes the option's payoff) and back-tracking step by step until $i = 0$, we compute all the values of $C_{i,j}$, to finally obtain the desired price $C_0$ (see Diagram 10.2). This recursive method allows us to avoid the direct calculation of $C_0 = E^q\left[\frac{C_N}{(1+r)^N}\,|S_0\right]$.

To take (approximately) into account the American feature of the option, we replace the expression for $C_{i,j}$ above by

$$C_{i,j} = \text{Max}\left[\frac{q\, C_{i+1,j+1}+(1-q)C_{i+1,j}}{1+r}; IV_{i,j}\right]$$

where $IV_{i,j}$ denotes the option's intrinsic value, $(S_{i,j}-K)$ or $(K-S_{i,j})$. This ensures that the American option value is *at least* its intrinsic value in every circumstance. When $\frac{q\, C_{i+1,j+1}+(1-q)C_{i+1,j}}{1+r} < IV_{i,j}$ the call is thus implicitly exercised, but usually *inside* the stopping region (see previous Sects. 13.3.1 and 13.3.2), the Optimal Exercise Boundary being crossed. Only with a very large number of (very short) time steps will these sub-optimal early exercises be close enough from the OEB and the resulting under-estimation of the option acceptable. Figure 13.7 gives an example of pricing over 2 periods for an American call on a spot asset paying out a dividend on date 1. The binomial tree is constructed using the first method explained above (proportional dividends).

Table 13.3 offers a comparison of the values of European and American puts written on a spot asset that does not pay dividends. The common data are: $S_0 = 100$, $r = 4\%$, and $s = 30\%$. The strikes range from 80 to 130 by steps of 10, and the three maturities considered are 3, 6, and 12 months. The number of time steps $N$ always equals 30.

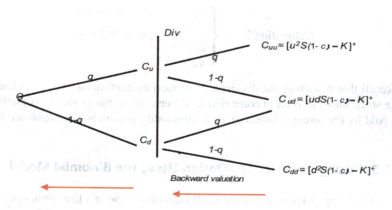

**Fig. 13.7** American call and binomial method (CRR)

## 13.4 Numerical Methods: Finite Differences, Trinomial and Three-Dimensional Trees

**Table 13.3** European and American puts (binomial model)

| Strike | 3 months European | 3 months American | 6 months European | 6 months American | 12 months European | 12 months American |
|---|---|---|---|---|---|---|
| 80 | 0.3310 | 0.3335 | 1.1688 | 1.1905 | 2.6970 | 2.7787 |
| 90 | 1.7853 | 1.8097 | 3.4597 | 3.5210 | 5.5860 | 5.7991 |
| 100 | 5.4314 | 5.5234 | 7.3685 | 7.5670 | 9.7724 | 10.2054 |
| 110 | 11.7184 | 11.9317 | 13.2878 | 13.6870 | 15.3475 | 16.0702 |
| 120 | **19.8476** | 20.3501 | 20.6212 | 21.3618 | 22.0117 | 23.1944 |
| 130 | **29.0454** | 30.0000 | **28.9846** | 30.242 | **29.4620** | 31.3157 |

The bold figures are those for European puts which are worth less than their respective intrinsic values. The value of the implicit option of early exercise can be relatively significant for *deep-in-the-money* puts.

Although it is not very precise, this model also allows taking into account the payment of one or more dividends on arbitrary dates, as we saw in Sect. 13.3.1. More generally, it is flexible enough to allow imposing any sort of constraint on any node of the binomial tree.

### 13.3.4 Improving the Procedure with a Control Variate

To improve the accuracy of the binomial model, Hull and White (1988) suggested a method using a control variate. We explore the control variate technique in more detail in Sect. 26.4.2 devoted to Monte Carlo simulations. In the context of option pricing using trees, the method is as follows. The BSM model is assumed to give the exact value of a European option, but it cannot be applied to an American option; the binomial model allows us to compute the value of an American option, but with an error. It is reasonable to presume that the binomial model suffers from practically the same error whether it prices an American or a European option using the same tree to represent the dynamics of the UA price. For a European option, it is enough to compare the BSM result with the binomial model's to obtain the error value $e$. The American option price that is to be used is therefore that given by the binomial model corrected by the error $e$ estimated from the European option.

## 13.4 Numerical Methods: Finite Differences, Trinomial and Three-Dimensional Trees

The valuation of any complex financial instrument, like an American option, as well as the solution to a wide variety of financial problems for which no analytical solution is available, rely on numerical methods such as binomial trees or Monte Carlo simulations which are powerful and flexible tools. We have already presented the recombining binomial trees for the dynamics of an asset price which are flexible enough to accommodate dividend distributions and we will devote the whole

534                                    13    American Options and Numerical Methods

Chap. 26 to Monte Carlo simulations. In this section, we briefly present three other numerical methods: finite difference methods for solving partial differential equations (Sect. 13.4.1), trinomial trees for representing the evolution of *one* process following a trinomial path (Sect. 13.4.2) and three-dimensional trees for representing the evolution of *two* processes following two binomial paths that may be correlated (Sect. 13.4.3).

## 13.4.1 Finite Difference Methods (*)

The price of any derivative asset obeys a partial differential equation (PDE) associated with a boundary condition that depends on the type of derivative product being considered, and, possibly, on other characteristics (American option, barriers, choosers, etc. See Chap. 14). Most often the PDE does not have any known analytic solution (BSM is a remarkable exception). However, one can resort to a numerical solution by discretizing the PDE and imposing boundary conditions to obtain a system of difference equations that can be solved iteratively.

In the sequel, the UA is a spot asset with price $S$, which distributes a continuous dividend $c$ up to the option maturity.[15] The option is an American *put* with value $V$, strike $K$, and maturity $T$, the current date being $t = 0$. The risk-free rate $r$ is assumed to be constant. The case of a *call* would be treated in a similar way, only boundary conditions being different. We are only illustrating the method here although it applies much more generally. Furthermore, we only describe the method called *implicit*, which is a little more difficult to implement than the rather similar method called *explicit*,[16] but which, unlike the latter, does not suffer from convergence problems (it is unconditionally stable, i.e., regardless of the respective sizes of the time step and the UA price step. The explicit method is stable only if the ratio time step size over UA price step size squared $(\Delta t/(\Delta S)^2)$ is smaller than or equal to $0.5$).[17,18]

We start by the standard implicit method which assumes that the boundaries are known and fixed, which leads to approximate but reasonable put values. Then we

---

[15] The case of no dividend (interesting only for a put) and the case of a forward asset (or a futures with deterministic interest rates) can be handled in the same way, with an adjustment in the second term of PDE (6) below in the text: one replaces $(r - c)$ with $r$ in the first case, and by zero (no net carrying cost) in the second. The case of a unique discrete dividend is slightly more complicated. The case of multiple discrete dividends is much more complex for the put, and the reader is referred to more specialized books or papers on the subject.

[16] It is interesting to note that the explicit scheme is the analog of constructing a trinomial tree (with states up, intermediate and down). See the trinomial model due to Hull and White (1990) in § 2.

[17] This, for example, implies that doubling the number of values of $n$ for $S$ (i.e., dividing $\Delta S$ by 2) requires at least quadrupling the number $m$ of dates (division of $\Delta t$ by 4).

[18] The Crank–Nicolson scheme is an « average » of these two methods. See, for example, Hull's book. Like the implicit method, it is stable. As a combination of the implicit and explicit methods, it is the most accurate of the three but requires more computation time than either one of them.

## 13.4 Numerical Methods: Finite Differences, Trinomial and Three-Dimensional Trees 535

briefly explain how to deal with the issue of an unknown optimal exercise boundary (OEB), the solution to which is much more involved but leads to a more accurate approximation.

### 13.4.1.1 The Standard Implicit Method

Since the UA price obeys, under the risk-neutral probability measure, the geometric Brownian motion (*see* Chap. 11, Eq. (20a))

$$\frac{dS}{S} = (r - c)dt + \sigma dW(t),$$

the PDE governing the put value $V$ writes:

$$V_t + (r - c)SV_S + \frac{1}{2}\sigma^2 S^2 V_{SS} \leq rV, \tag{13.17}$$

where the subscript on $V$ denotes a partial derivative (theta, delta, and gamma).

The discretization of this PDE is based on the construction of a two-dimensional grid representing different values of $t$ and $S$. To obtain our grid, we divide the duration $T$ into $m$ equal intervals of length $\Delta t = T/m$, so that there are $(m+1)$ dates. The time step $\Delta t$ can be chosen as small as desired, knowing that there is a tradeoff between accuracy and computation time. The specific dates are denoted by $j\Delta t$, for $j = 0, 1, 2, \ldots, m$. Note that the present corresponds to $j = 0$ and the option maturity to $j = m$ so that $m\Delta t = T$.

Then one chooses a maximum value $S^{max}$ for the UA price, so that the put price is practically zero, and also so that one of the prices obtained as will be described is the present price $S(0)$. We divide $S^{max}$ by $n$ to get a discretization step for the UA price (as small as desired) $\Delta S = S^{max}/n$. A typical UA price is denoted by $i\Delta S$, for $i = 1, 2, \ldots, n$, which gives $(n + 1)$ possible values for $S$, between 0 and $S^{max}$, the current price $S(0)$ being one of them.

In this way, we obtain a rectangular mesh of $(m+1)\times(n+1)$ points in which the $(m+1)$ dates are read from left to right and the $(n+1)$ possible prices of $S$ from bottom to top. Denote by $(i, j)$ the point corresponding to the price $i\Delta S$ and the date $j\Delta t$, and by $V_{i, j}$ the option value at this point. The basic idea underlying the implicit scheme is to compute the value $V_{i,j+1}$ from the three *unknown* values (except on the edges of the rectangle) $V_{i-1,j}$, $V_{i,j}$, and $V_{i+1,j}$.[19] Note that here we use PDE (13.17) with the inequality replaced by the equality, since in the standard scheme the stopping *region* (not the boundaries) is implicitly ignored so that the BSM PDE holds.

The discretization of $V_t$ (theta) yields

---

[19] In the explicit method, one computes $V_{i,j}$ from $V_{i-1,j+1}$ $V_{i,j+1}$ and $V_{i+1,j+1}$. Since then one calculates a value at time $j$ from (three) values at time $j+1$ and the valuation process, as in the binomial method, is done backwards starting from $j = m$ (expiry date $T$), this method is simpler and (more) "explicit."

$$\frac{\Delta V}{\Delta t} = \frac{V_{i,j+1} - V_{i,j}}{\Delta t}.$$

To discretize $V_S$ (delta), it is usual (for symmetry) to take the average of the right and left derivatives, which gives

$$\frac{\Delta V}{\Delta S} = \frac{V_{i+1,j} - V_{i-1,j}}{2\Delta S}.$$

The discretization of $V_{SS}$ (gamma), taking the difference between the right and left first derivatives, $((V_{i+1,j}-V_{i,j})-(V_{i,j}-V_{i-1,j}))/\Delta S$ yields

$$\frac{\Delta^2 V}{\Delta S^2} = \frac{V_{i+1,j} - 2V_{i,j} + V_{i-1,j}}{\Delta S^2}.$$

Putting these values back in PDE (13.17) expressed with equality, and using the fact that $S_i = i\Delta S$ and $V=V_{i,j}$, then multiplying by $\Delta t$ on both sides, and rearranging terms, we obtain

$$a_i V_{i-1,j} + b_i V_{i,j} + c_i V_{i+1,j} = V_{i,j+1}, \quad \text{with} \tag{13.18}$$

$$a_i = \frac{1}{2}\left((r-c)i - \sigma^2 i^2\right)\Delta t; \quad b_i = 1 + \left(r + \sigma^2 i^2\right)\Delta t; \quad \text{and} \quad c_i$$

$$= -\frac{1}{2}\left((r-c)i + \sigma^2 i^2\right)\Delta t.$$

Equations (13.18) hold for $i = 1, \ldots, (n-1)$, and for $j = 0, \ldots, (m-1)$. For $j = m$ *(expiry date)*, the values $V_{i,m}$ correspond to the put payoffs: $V_{i,m} = Max(K - S_{i,m;\ 0)}$.
For *a given j* $(<m)$, there are only $(n-1)$ equations for $(n+1)$ unknowns (the $V_{i,j}$, for $i = 0, \ldots, n$). The two missing equations are obtained by using the following *boundary conditions*:

- The equation for $i = 0$ (involving $S = 0$) writes:
  $V_{0,j} = K$; we label it equation (13.Aj-0).[20]
- The equation for $i = n$ (involving $S = S^{max}$) is derived from the fact that the put is deep out-of-the-money and thus has a delta of practically zero:
  $V_{n,j} - V_{n-1,j} = 0$ (equation (13.Aj-n)).

Using equation (13.Aj-0), we derive equation (13.Aj-1):

---

[20] Note that the equation would be $V_{0,j} = Ke^{-r(T-j\Delta t)}$ *for a European put as the payoff K could not be immediately cashed in (only at T).*

## 13.4 Numerical Methods: Finite Differences, Trinomial and Three-Dimensional Trees    537

$$b_1 V_{1,j} + c_1 V_{2,j} = V_{1,j+1} - a_1 K \equiv d_{1,j+1}. \qquad (13.\text{Aj-1})$$

For $2 \leq i \leq (n-1)$ we have the $(n-2)$ equations:

$$a_i V_{i-1,j} + b_i V_{i,j} + c_i V_{i+1,j} = V_{i,j+1} \equiv d_{i,j+1}, \qquad (13.\text{Aj-i})$$

and for $i = n$, we have

$$-V_{n-1,j} + V_{n,j} = 0 \equiv d_{n,j+1}. \qquad (13.\text{Aj-n})$$

For each $j < m$, this would make up a system of $n$ equations for $n$ unknowns if we knew the right-hand sides $(d_{i,j+1})$ of these equations. We can determine recursively the $d_{i,j+1}$ beginning by $j+1=m$ and going backward. Indeed, we do know the $d_{i,m}$, as was already noticed, since they are the different payoffs of the put according to the UA price $i\Delta S$ on date $T$: $d_{i,m} = V_{i,m} = \text{Max}(K - S_{i,m})$ for $i=0, 1, .., n$. The system can then be solved for $j+1 = m$ and yields the $V_{i,m-1}$. Using those, and *backtracking* in time as we do in a binomial tree, we can calculate the $V_{i,m-2}$. Then we obtain $V_{i,m-3}$ and so on.

Now, we have seen that the American nature of the put has been partly taken care of by using $V_{0,j} = K$ instead of $V_{0,j} = Ke^{-r(T-j\Delta t)}$ as equation (13.Aj-0). This is not enough as the value of the put at each node should be at least equal to its intrinsic value. A simple approximation to achieve this, akin to the one encountered in the binomial tree method, consists, *at each step* $k$ of the backtracking process previously described, in taking as the final value for $V_{i,m-k}$, $k=1, \ldots, m$ (in this order), the *maximum* between the intrinsic value $(K - S_{i,m-k})$ and the value $V_{i,m-k}$ resulting from the backward application of equation $(A_{(m-k)}$-i$)$. Each node $(i, j)$ where the put value is equal to the payoff is a point of the stopping region whose frontier corresponds to the OEB.

Doing this $m$ times, we finally get the $V_{i,0}$. The desired option value is the $V_{i,0}$ for which the value of $i$ *makes* $i\Delta S$ equal to the current price $S(0)$.

On *a priori* grounds, one would think that solving the system requires at each step $j$ (except the first when $j+1=m$) inverting an $(n \times n)$ matrix which can take time if $n$ is large. Fortunately, this is not the case. In fact, we see that, on the one hand, the middle $(n-2)$ rows of the matrix contain 3 terms only, and, on the other and most importantly, the first and the last rows have only 2. As a result, a very fast solution can be computed using a simple substitution that avoids inverting the matrix at all.

To do this, we use Eq. (13.Aj-1) to eliminate $V_{1,j}$ from Eq. (13.A$_j$-2) which then has only two unknowns, $V_{2,j}$ and $V_{3,j}$. Then we use this new Eq. (13.A$_j$-2) to get rid of $V_{2,j}$ in Eq. (A$_j$-3). We continue in this way, equation by equation, through all the equations that thus have each no more than 2 terms (involving $V_{i,j}$ and $V_{i+1,j}$) up to the last equation (13.Aj-n) which, since it only had 2 terms to start with, now has but a single one. Once this first-degree equation with the single unknown $(V_{n,j})$ is trivially solved, we back up the steps of the computation to obtain the $V_{n-1,j}, \ldots,$ and $V_{1,j}$ as solutions to first-degree equations in one unknown. This property makes

538 13 American Options and Numerical Methods

the overall computation time very small, even for small time steps and UA price increments.

**Remarks**
- The system is stable as the estimation error does not increase with the number of iterations.
- Numerical analysts have shown that it is more efficient to use log $S$ rather than $S$, since then the coefficients $a_i$, $b_i$ and $c_i$ in Eq. (13.18) no longer depend on $i$. Using the notation $X = \ln S$, we have to rewrite PDE (13.12) expressed with equality as follows, where the extra term involving the delta $V_x$ comes from Itô's Lemma applied to $dX$:

$$V_t + ((r - c) - \frac{\sigma^2}{2})V_X + \frac{1}{2}\sigma^2 V_{XX} = rV.$$

It is then recommended to impose $\Delta X = s\sqrt{3\Delta t}$ for maximum efficiency. As an alternative, one can use the following change of variables used by Black and Scholes themselves to reduce their four-term EDP to the two-term "heat transfer equation" of physics:

$$X = \ln S + (r - c - 0.5\sigma^2)(T - t),$$

$$\tau = 0.5\sigma^2 (T - t),$$

$$Z(X, \tau) = e^{-r(T - 2\tau/\sigma^2)} V(e^{X - (2r/\sigma^2 - 1)\tau}, T - 2\tau/\sigma^2),$$

and then obtain from Eq. (5) the simpler EDP: $Z_\tau = Z_{XX}$.

### *13.4.1.2 The Implicit Method with a Free Boundary

The standard finite difference methods are not very accurate since the PDE governing the option value is not the BSM PDE with a strict equality but Eq. (13.5) with the $<$ sign inside the stopping region. Among several approaches improving the method, we present briefly the ideas behind the linear complementarity (equivalent to the free-boundary) method that does not use the unknown $S^*$ explicitly associated with the so-called penalty method.

The idea is to have a unified treatment of PDE (13.5) which holds on the entire half strip $S \in [0, \infty[$, (and not only for $S > S^*$ for a put or $S < S^*$ for a call). Denoting by $\Psi(S,t)$ the payoff function at time $t$, and using the Dynkin operator for brevity, the free-boundary or linear complementarity problem (LCP) writes (see Eq. (13.11)):

$$(D^t V(S, t) - rV(S, t))(V(S, t) - \Psi(S, t)) = 0,$$

## 13.4 Numerical Methods: Finite Differences, Trinomial and Three-Dimensional Trees

$$V(S,t) - \Psi(S,t) \geq 0,$$

$$(D^t V(S,t) - rV(S,t)) \leq 0, (\text{PDE } (11)),$$

$$V(S,T) = \Psi(S,T),$$

$$Lim_{S \to +\infty} V(S,t) = Lim_{S \to +\infty} \Psi(S,t),$$

$$Lim_{S \to 0} V(S,t) = Lim_{S \to 0} \Psi(S,t).$$

Among many numerical methods which solve this problem, a simple and efficient one is the *penalty method* (see for instance Seydel 2017, Chap. 4) or Palczewski (2002, lecture 9)). The penalty function $p(V) \geq 0$ is introduced in PDE (5) of the LCP such that the following "penalty PDE" holds:

$$(*)(D^t V(S,t) - rV(S,t)) + p(V(S,t)) = 0,$$

where the penalty term $p(V)$ is *positive* in the stopping region and *zero* in the continuation region.

As the difference between $V(.)$ and the payoff $\Psi(.)$ is known, it can be used as a control. Then, the selected $p(V)$ function can be for instance given by:[21]

$p(V(S,t)) = q \max[\Psi(S,t) - V(S,t); 0]$ with $q$ a large positive number.

Note that the initial linear LCP thus is replaced by a non-linear LCP. Since the PDE $(*)$ is now written with an equality, the same iterative procedure as in Sect. 13.4.1 can be applied, with just an extra term in the $a_i$, $b_i$ and $c_i$ coefficients involving $q$ $[\Psi_{i,j} - V_{i,j}]$ when $\Psi_{i,j} > V_{i,j}$, $\forall i$, $j = 1, \ldots, m - 1$.

It can be shown that the convergence (stability) property is preserved and can be made faster by using advanced techniques such as Newton iterates or the SOR (successive over-relaxation) method.

### 13.4.2 Trinomial Trees

The trinomial recombining tree is a generalization of the binomial recombining tree. It too is an alternative to finite difference methods. It was introduced by Boyle (1986) and developed by Hull and White (1994, 2016) in several papers on interest rates and derivatives modeling. As its name suggests, a trinomial tree has three branches emanating from each node. The underlying asset price being $S_i$ on date $i$, it can take one of three following values on date $i+1$:

---

[21] Another typical choice is $p(V(S,t)) = \varepsilon/(V(S,t) - \Psi(S,t))$ with $\varepsilon$ a small positive number.

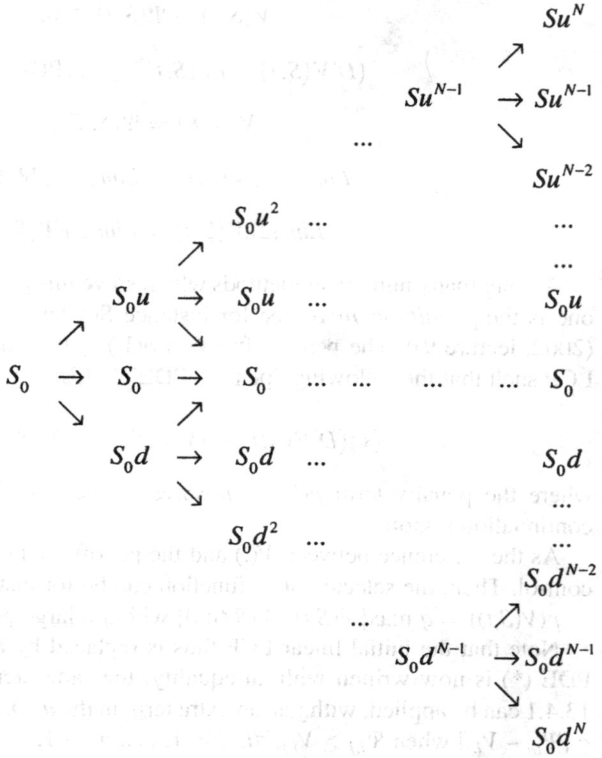

**Fig. 13.8** A trinomial recombining tree representing a price dynamics

$$Su^N$$
$$Su^{N-1} \rightarrow Su^{N-1}$$
$$\cdots$$
$$Su^{N-2}$$

$$S_0u^2 \quad \cdots$$
$$\cdots$$
$$\cdots$$
$$S_0u \rightarrow S_0u \quad \cdots \qquad S_0u$$
$$S_0 \rightarrow S_0 \rightarrow S_0 \quad \cdots \cdots \quad \cdots \cdots \quad S_0$$
$$S_0d \rightarrow S_0d \quad \cdots \qquad S_0d$$
$$\cdots$$
$$S_0d^2 \quad \cdots$$
$$\cdots$$
$$S_0d^{N-2}$$
$$S_0d^{N-1} \rightarrow S_0d^{N-1}$$
$$S_0d^N$$

$$S_{i+1} = \begin{cases} uS_i & \text{with probability } q_u \\ mS_i & \text{with probability } q_m = 1 - q_u - q_d \\ dS_i & \text{with probability } q_d \end{cases}$$

For a tree to be recombining we must have $ud = m^2$ (a rise in value followed by a fall leads to the same point as a fall followed by a rise or two average changes); so the final value of the asset price can take $2N+1$ values. Figure 13.8 shows such a tree with $m = 1$ and $d = 1/u$.

Along the lines of the binomial model, parameters need to be chosen for a trinomial tree. The following values are suggested by Hull and White (1990):

$$m = 1, u = e^{\sigma\sqrt{3\frac{T}{N}}}, d = 1/u,$$

$$q_d = \frac{1}{6} - \sqrt{\frac{T}{12N\sigma^2}}\left(r - c - \frac{\sigma^2}{2}\right), q_u = \frac{1}{6} + \sqrt{\frac{T}{12N\sigma^2}}\left(r - c - \frac{\sigma^2}{2}\right), \text{ and thus } q_m$$

$$= \frac{2}{3}.$$

## 13.4 Numerical Methods: Finite Differences, Trinomial and Three-Dimensional Trees 541

This leads to an expected risk-neutral return rate $r$ and a volatility $s$, for an asset paying a continuous coupon rate $c$, neglecting terms of order higher than the first.

There are other parameter calibrations that accelerate convergence. We present one that has two definite advantages over Hull and White's. On the one hand, it does not require any approximation and, on the other, it yields the exact values for the first *four* moments of the UA log-normal distribution. It consists in choosing the central value taken on by $S$ not equal to $m = 1$ (a zero rate of growth) but equal to its risk-neutral expected value: $m = \exp\left(r - c - \frac{\sigma^2}{2}\right)\frac{T}{N}$. Adopting the other parameters

$$q_u = q_d = \tfrac{1}{6} \quad, \quad q_m = \tfrac{2}{3} \quad, \quad u = \exp\left[\left(r - c - \frac{\sigma^2}{2}\right)\frac{T}{N} + \sigma\sqrt{\frac{3T}{N}}\right] \quad \text{and} \quad d =$$

$\exp\left[\left(r - c - \frac{\sigma^2}{2}\right)\frac{T}{N} - \sigma\sqrt{\frac{3T}{N}}\right]$ leads to the desired result for the first *four* moments.[22] One can check that with these parameters the tree is recombining ($ud = m^2$).

The option price is computed, as in a binomial tree, recursively. For an American option, on each date $i$, the value $O'_i$ is the maximum of its intrinsic and its continuing value; the latter equals its expected risk-neutral value on date $i+1$, discounted at the risk-free rate.

It can be shown that the trinomial tree method is equivalent to the finite differences method and that it can improve the tradeoff accuracy-computing time over the standard binomial model, especially when few steps are used.

### 13.4.3 Three-Dimensional Trees Representing Two Correlated Processes

Three-dimensional trees, not to be confused with the trinomial trees just presented, can efficiently represent *the evolution of the prices of two correlated assets* or, more generally, that of two correlated processes. This is particularly useful in the case of an option whose value depends on two processes (the UA price and the stochastic interest rate, for example, or an option written on two risky UAs).

The simple binomial and the trinomial trees are two-dimensional insofar as they represent $(t, S(t))$ on the plane. The tree we now describe combines two binomial trees, respectively for $S_1(t)$ and $S_2(t)$; it can be considered as three-dimensional insofar as it represents the triplet $(t, S_1(t), S_2(t))$. We explain first the construction of the three-dimensional tree under the assumption that $S_1(t)$ and $S_2(t)$ are independent before studying the case where they are correlated.[23]

---

[22] The system to be solved has for four unknowns: $q_u$, $q_d$, $u$ and $d$. The four equations needed are those for the first four moments of the UA log-return (Gaussian) distribution (e.g., the first moment is the $m$ given in the text and the third is nil).

[23] Here we use interchangeably the terns "uncorrelated" and "independent" since we work with Gaussian processes.

### 13.4.3.1 Construction of the Tree with Independent $S_1(t)$ and $S_2(t)$

In the interval $(i, i+1)$, $S_1$ can change from $S_1(i)$ to $u_1 S_1(i)$ or to $d_1 S_1(i)$ with respective risk-neutral probabilities $q_1$ and $1 - q_1$. During the same interval, $S_2$ can change from $S_2(i)$ to $u_2 S_2(i)$ or to $d_2 S_1(i)$ with respective risk-neutral probabilities $q_2$ and $1 - q_2$. Thus, in the interval $(i, i+1)$ there are four states of the world, corresponding to the evolutions of $S_1$ and $S_2$. The table below displays the probabilities of $(u_1, u_2)$, $(u_1, d_2)$, $(d_1, u_2)$, $(d_1, d_2)$:

|  | $S_2$ changes at rate $u_2$ | $S_2$ changes at rate $d_2$ |
|---|---|---|
| $S_1$ changes at rate $u_1$ | $q_1 \, q_2$ | $q_1(1 - q_2)$ |
| $S_1$ changes at rate $d_1$ | $(1 - q_1) \, q_2$ | $(1 - q_1)(1 - q_2)$ |

The three-dimensional tree thus leads to four branches from each node with the risk-neutral probabilities given in the table. This makes for $(N + 1)^2$ different final values for the pair $(S_1(N), S_2(N))$. Each node can be characterized by the triplet $(i, j, k)$ where $i$ denotes the date, $j$ the number of increases in $S_1$ and $k$ the number of increases in $S_2$. At the node $(i, j, k)$: $S_1(i) = S_1(0) \, u_1^{\,j} \, d_1^{\,(i-j)}$ and $S_2(i) = S_2(0) \, u_2^{\,k} \, d_2^{\,(i-k)}$. There are $(i+1)^2$ different nodes at date $i$. One chooses the calibration of each process independently and such that the desired moments (at least the first and second moments of each process must be recovered; see Appendix 2, Chap. 10) obtain.

### 13.4.3.2 Construction of the Tree with $S_1(t)$ and $S_2(t)$ Correlated

The case of two correlated processes $S_1$ and $S_2$ can be handled using different approaches. Hull and White (2016), for instance, suggest two methods.

- **Method 1.**[24] Construct orthogonal combinations of $\ln S_1$ and $\ln S_2$, and thus obtain uncorrelated variables $X$ and $Y$, as linear combinations of $\ln S_1$ and $\ln S_2$. Then build a first tree, as explained above, describing the dynamics of $X$ and $Y$. To each pair $(X, Y)$ at each node of this tree assign the corresponding pair $(S_1, S_2)$, which leads to a second tree representing the dynamics of $(S_1, S_2)$.
  Appendix 2 describes in greater detail how to do the orthogonalization and to build the three-dimensional tree representing the evolution of $S_1$ and $S_2$.
- **Method 2.**[25] Modify the probabilities so as to obtain the desired correlation.
  Start from calibration 2 given in Chap. 10 (Eq. (10.32)) applied separately to each variable $S_1$ and $S_2$. This calibration uses, for both $S_1$ and $S_2$, probabilities of a price increase ($u_1$ for $S_1$ and $u_2$ for $S_2$) equal to 0.5. These are marginal (or unconditional) probabilities since they depend on no assumption about the change in the other variable.

---

[24] Hull and White, Valuing Derivative Securities Using the Explicit Finite Difference Method, *Journal of Financial and Quantitative Analysis*, 25, 1990, pp. 87–100.

[25] Hull and White, « Numerical Procedures for Implementing Term Structure Models II: Two-Factor Models », *Journal of Derivatives*, Winter 1994, pp. 37–48.

## 13.5 Summary

If we assume the log-returns of $S_1$ and $S_2$ are linked by a correlation $r$, we associate variations in the pair $(S_1, S_2)$ to the joint probabilities as in the following table:

| | $S_2$ varies at rate $u_2$ | $S_2$ varies at rate $d_2$ | *Marginal probability* |
|---|---|---|---|
| $S_1$ varies at rate $u_1$ | 0.25(1 + r) | 0.25(1 − r) | 0.5 |
| $S_1$ varies at rate $d_1$ | 0.25(1 − r) | 0.25(1 + r) | 0.5 |

The three-dimensional tree (four branches at a node) will therefore be associated with the probabilities displayed in the table (rather than with probabilities of 0.25).

The reader can check that the correlation coefficient obtained from the joint probabilities given in the table is indeed $r$ for all values of $u_1, d_1, u_2, d_2$.

Computing the option price is carried out, as for a binomial tree, by recursion. For an American option, on date $i$, the value $O'_i$ is the maximum of its intrinsic value and its continuing value; the latter is equal to its risk-neutral expected value on date $i + 1$, discounted at the risk-free rate.

Three-dimensional trees are an interesting device for modeling two correlated processes in different contexts. They have, for instance, been applied to represent the joint evolution of the OIS and the LIBOR by Hull and White (2016).

## 13.5 Summary

- An American option is worth more than its European analogue, except for a call written on a spot underlying asset (UA) not paying dividends or coupons (or paying a very small dividend) before expiration, since, except in this last case, it *may* be optimal to exercise it before maturity.
- An American call on a spot asset which pays a fixed dividend $D$ on date $\tau$, should never be exercised before the eve of $\tau$; the necessary and sufficient condition for an optimal exercise of the call just before detachment is that $D$ exceeds the time value of the call just after detachment, which is the case if the UA price at $\tau$ is higher than or equal to an *early exercise price bound* $S_\tau^*$.
- In the case of puts, or calls on a UA paying a continuous dividend, there exists an Optimal Exercise boundary $S^*(t)$, with $t < T$, such that the option should be early exercised the first time the UA price $S(t)$ hits it from below for a call, or from above for a put. At this point of optimal early exercise, the option value $O(S,t)$ whose payoff is $\Psi(S)$ satisfies the *value matching* condition $O(S^*(t), t) = \Psi(S^*(t))$, and the *smooth pasting* condition $\frac{\partial O}{\partial S}\big]_{S=S^*} = \frac{\partial \Psi}{\partial S}\big]_{S=S^*}$. The last equation is a condition for the boundary $S^*(t)$ to be optimal and implies that the $O(S)$ curve is tangent to the $\Psi(S)$ curve for $S = S^*$.
- In the case of an American put on a spot asset that pays a fixed dividend $D$, there exists a no-exercise period prior to the detachment date during which the put should not be exercised under any circumstances.

- It may be optimal to exercise early an in-the-money American call or put written on a forward or futures contract whose price behaves as the price of a spot UA paying a continuous coupon yield $c$ equal to the continuous interest rate $r$.
- A call-put "parity" obtains in the form of an upper bound and a lower bound for $C^a - P^a$, these bounds depending on the type of UA and of its mode of distribution.
- A crude pricing method of an American call on a spot UA distributing $D$ on date $\tau$ consists in taking the maximum of two values: that of a *European* call of maturity $\tau$ on the not-paying dividend UA; and that of a *European* call of maturity $T$ on the UA paying $D$. Such an evaluation underestimates the American option value. It can be « exactly » valued as a put with expiry $\tau$ written on a call with expiry $T > \tau$ (a *compound option*). The formula obtained involves the bivariate normal distribution function.
- An American option on a spot UA paying a continuous dividend can be priced from BSM's PDE (which is satisfied only *inside* the continuation region) with the appropriate boundary conditions. This approach leads to two formulations, a *free boundary* valuation and a *linear complementarity* formulation, on which numerical methods can be grounded. American option pricing can also be formulated as an *optimal stopping time* problem.
- Analytical approximations of an American option price can also be obtained, for instance with the Barone-Adesi–Whaley method.
- An American option can be priced with the binomial model provided that the paths of a dividend paying spot UA recombine. One solution to obtain a recombining tree relies on assuming the dividends proportional to the current UA price. As in the European case, the solution is implemented backwards; at each node of the tree, the option value is the maximum of its intrinsic value and its continuation value, the latter being the risk-neutral expectation of its two next possible values (previously determined) discounted at the risk-free rate. Several calibrations (choice of parameters) of the binomial model are possible.
- The differences between American and European option values are *sizeable* for in-the-money options.
- The finite difference method (whether implicit, explicit, or Crank–Nicolson) yields a numerical solution to the pricing partial *differential* equation (PDE) of BSM type by discretizing it so as to obtain a system of *difference* equations that are solved iteratively. Since all financial derivatives obey such a PDE with different boundary conditions, this method applies to a wide scope of financial assets.
- The trinomial recombining tree, an alternative to finite difference methods, is a generalization of the binomial tree with three branches emanating from each node. From date $i$, the UA price can take one of three values on date $i+1$: $uS_i$ with probability $q_u$; $mS_i$ with probability $q_m = 1 - q_u - q_d$; *and* $dS_i$ with probability $q_d$. For a recombining tree, we must have $ud = m^2$. The option price is computed recursively, starting from its maturity date. On each date, the value of the option is the maximum of its intrinsic and its continuation value, the latter being equal to its

Appendix 1 545

expected risk-neutral value on the next date, discounted at the risk-free rate. As for a binomial tree, several calibrations are possible.

- Three-dimensional trees, which must not be confused with trinomial trees, can represent the evolution of *two correlated price processes* $S_1(t)$, $S_2(t)$, e.g., two asset prices, or an asset price and a stochastic interest rate. They involve *two correlated binomial processes*. They are called three-dimensional because they represent the possible paths of the triplet $(t, S_1(t), S_2(t))$.
- The construction of a three-dimensional tree requires building a first tree describing the dynamics of two independent processes $X(t)$ and $Y(t)$ by a couple of independent binomial trees. The desired correlation and parameters are obtained as linear combinations of $X$ and $Y$, leading to a second tree representing the dynamics of $(S_1, S_2)$.

## Appendix 1

### Proof of the Smooth Pasting (Tangency) Condition (13.5b)

Recall the true call price is denoted by $C^a(S, z)$, with $z = T\text{-}t$. Consider a fictitious call $C'$ of the exact same characteristics but to which a sub-optimal exercise boundary $b(z)$ is imposed. Its price is $C'(S, z, b(z)) \leq C^a(S, z)$, and by definition, we have $C'(S, z, S^*(z)) = C^a(S, z)$, where $S^*(z)$ is the Optimal Exercise Boundary (OEB): the true call is the particular $C'$ obtained for $b(z) = S^*(z)$.

We consider a *fixed* time to maturity $z$, but *different* $b(z)$, and thus can use without ambiguity the simplified notations $C^a(S, z) \equiv C^a(S)$ and $C'(S, z, b(z)) \equiv C'(S, b)$.

Since b is *an* exercise boundary, we have: $C'(b, b) = b - K$ (value matching condition at point b).

By totally differentiating this expression, we obtain:

$$\frac{\partial C'}{\partial S}(b, b) + \frac{\partial C'}{\partial b}(b, b) = 1. \tag{13.19}$$

*(13.19) holds for any exercise boundary, in particular for* $S^*(z)$, *which yields:*

$$\frac{\partial C'}{\partial S}(S^*, S^*) + \frac{\partial C'}{\partial b}(S^*, S^*) = 1. \tag{13.20}$$

*In addition, $C'$ is maximized for $S^*$. Hence, we have:*

$$\frac{\partial C'}{\partial b}(S^*, S^*) = 0, \tag{13.21}$$

*as the first-order condition at a maximum point.*
*Then combining (13.20) and (13.21) yields:*

$$\frac{\partial C'}{\partial S}(S^*, S^*) \equiv \frac{\partial C^a}{\partial S}(S^*) = 1,$$

*which is the smooth pasting condition (13.5b):*

$$\frac{\partial C^a}{\partial S}\bigg]_{S=S^*} = 1,$$

*where $C^a$ is expressed as a function of t. This equation is an optimality condition that* is satisfied because S*(z) is the optimal exercise boundary.

## Appendix 2

## Orthogonalization of the Processes ln $S_1$ and ln $S_2$ and Construction of a Three-Dimensional Tree

We start from the risk-neutral dynamics of ln $S_1$ and ln $S_2$ which can be written in terms of uncorrelated Wiener processes $W_1$ and $W_2$:

$$d \ln S_1 = \left(r - \frac{\sigma_1^2}{2}\right)dt + \sigma_1 dW_1$$

$$d \ln S_2 = \left(r - \frac{\sigma_2^2}{2}\right)dt + r\sigma_2 dW_1 + \sqrt{1 - \rho^2}\sigma_2 dW_2.$$

The reader can check that $\sigma_1$ and $\sigma_2$ represent the volatilities of $S_1$ and $S_2$, and $\rho$ the correlation of their returns ($\sigma_1$, $\sigma_2$, and $\rho$ are fixed at the desired levels). Then we posit:

$$X(t) = \ln S_1(t); \tag{13.22}$$

$$Y(t) = \ln S_2(t) - \rho(\sigma_2/\sigma_1)\ln S_1(t) \tag{13.22'}$$

which implies

$$dX = \left(r - \frac{\sigma_1^2}{2}\right)dt + \sigma_1 dW_1, \tag{13.23}$$

$$dY = \left[r(1 - \rho\frac{\sigma_2}{\sigma_1}) + \rho\frac{\sigma_2\sigma_1}{2} - \frac{\sigma_2^2}{2}\right]dt + \sqrt{1 - \rho^2}\sigma_2 dW_2. \tag{13.23'}$$

The two processes $X$ and $Y$ are independent since $W_1$ and $W_2$ are independent.

Thus we may construct, independently, two additive[26] trees for $X$ and $Y$ and combine them into a three-dimensional tree (as explained in the main text); calibration of the two trees $X$ and $Y$ is done independently of each other, so that the discrete parameters $u_X$, $d_X$, and $u_Y$, $d_Y$ are respectively adjusted to match the first two moments of the dynamics (13.23) and (13.23′) (*see* Appendix 2 of Chap. 10 for this type of calibration). By inverting formulas (13.22) and (13.22′), one deduces from this the tree which governs $S_1$ and $S_2$:

$$S_1(t) = \exp[X(t)]; \quad S_2(t) = \exp[Y(t) + \rho \frac{\sigma_2}{\sigma_1} X(t)].$$

In this way, to the node $(i, j, k)$ at which $X(i,j) = X(0) + ju_X + (i-j)d_X$ and $Y(i,k) = Y(0) + ku_Y + (i-k)d_Y$, we assign

$$S_1(i,j) = S_1(0) \exp(ju_X + (i-j)d_X) \text{ and}$$

$$S_2(i,j,k) = S_2(0)\exp[kuY + (i-k)dY + \rho \frac{\sigma_2}{\sigma_1}(juX + (i-j)dX)].$$

---

## Suggestion for Further Reading

### Books

*Fusai, G., & Roncoroni, A. (2008). *Implementing models in quantitative finance: Methods and cases*. Springer.

Hull, J. (2016). *Options, futures and other derivatives* (10th ed.). Prentice Hall Pearson Education.

**Jeanblanc, M., Yor, M., & Chesney, M. (2009). *Mathematical models for financial markets*. Springer.

***Karatzas, I., & Shreve, S. (1998). *Methods of mathematical finance*. Springer.

**Kwok, Y.-K. (2008). *Mathematical models of financial derivatives*. Springer.

**Musiela, M., & Rutkowski, M. (2007). *Martingale Methods in financial modelling*. Springer, Applications of Mathematics.

***Øksendal, B. (1998). *Stochastic differential equations*. Springer.

**Palczewski, A. (2002). *Mathematical finance in discrete time*. In Lecture Notes Summer School, Warsaw.

***Prigent, J. L. (2003). *Weak convergence of financial markets*. Springer.

**Seydel, R. (2017). *Tools for computational finance* (6th ed.). Springer.

*Wilmott, P. (1998). *Derivatives: The theory and practice of financial engineering*. Wiley.

*Wilmott, P., Dewynne, J., & Howison, S. (1993). *Option pricing*. Oxford Financial Press.

**Wilmott, P., Howison, S., & Dewynne, J. (1996). *The mathematics of financial derivatives* (2nd ed.). Cambridge University Press.

---

[26]The trees are multiplicative in $e^X$ and $e^Y$. At node $(i, j, k)$: $X(i,j) = X(0) + j u_X + (i-j) d_X$ ; $Y(i,k) = Y(0) + k u_Y + (i-k) d_Y$. The parameters $u_X$, $d_X$, $u_Y$ and $d_Y$ correspond to $\ln u$ or $\ln d$ of the usual multiplicative tree.

## Articles

Barone-Adesi, G., & Whaley, R. (1987). Efficient analytic approximation of American option values. *Journal of Finance, 42*, 301–320.

Barone-Adesi, G. (2005). The saga of the American put. *Journal of Banking and Finance, 29*, 2909–2918.

Black, F. (1975). Fact and fantasy in the use of options. *Financial Analyst Journal, 36–41*, 61–72.

Boyle, P. (1986). Option valuation using a three jump process. *International Options Journal, 3*, 5–12.

Broadie, M., & Detemple, J. (1996). American option valuation: new bounds, approximations and a comparison of existing methods. *Review of Financial Studies, 9*, 1211–1250.

Carr, P. (1998). Randomization and the American put. *Review of Financial Studies, 11*, 597–626.

Carr, P., Jarrow, R., & Myneni, R. (1992). Alternative characterizations of American puts. *Mathematical Finance, 2*, 87–106.

Chesnay, M., & Lefoll, J. (1996). Predicting premature exercise of an American put on stocks: Theory and empirical evidence. *European Journal of Finance, 2*, 21–39.

Cox, J., Ross, S., & Rubinstein, M. (1979). Option pricing, a simplified approach. *Journal of Financial Economics, 7*, 229–264.

Crank, P., & Nicolson, P. (1947). A practical method for numerical evaluation of solutions of PDE of the heat conduction types. *Proc Camb Philos Soc, 43*, 50–57.

Geske, R. (1979). The valuation of compound options. *Journal of Financial Economics, 7*, 63–81.

Geske, R., & Johnson, H. (1984). The American put valued analytically. *Journal of Finance, 39*, 1511–1524.

Hull, J., & White, A. (2016). Multi-curve modeling using trees. *Innovations in Derivative Markets*, Springer Proceedings in Mathematics and Statistics, pp. 2171–189.

Hull, J., & White, A. (1994). Numerical procedures for implementing term structure models I and II. *Journal of Derivatives*, 7–16 and 37–48.

Hull, J., & White, A. (1990). Pricing interest rate derivative securities. *Review of Financial Studies, 3*, 573–592.

Hull, J., & White, A. (1988). The use of the control variate technique in option pricing. *Journal of Financial and Quantitative Analysis, 23*, 237–251.

Johnson, H. E. (1983). An analytical approximation to the American put price. *Journal of Financial and Quantitative Analysis, 18*, 141–148.

Longstaff, F. A., & Schwartz, E. (2001). Valuing American options by simulation: a simple least squares approach. *Review of Financial Studies, 14*, 113–147.

MacMillan, L. W. (1986). An analytical approximation for the American put prices. *Advances in Futures and Options Research, 1*, 119–139.

Merton, R. (1973). The theory of rational option pricing. *Bell Journal of Economics and Management Science, 4*, 141–183.

Omberg, E. (1987). The valuation of American puts with exponential exercise policies. *Advances in Options and Futures Research, 2*, 117–142.

Roll, R. (1977). An analytical valuation formula for unprotected american call options on stocks with known dividends. *Journal of Financial Economics, 5*, 251–258.

Whaley, R. (1981). On the valuation of American call options with known dividends. *Journal of Financial Economics, 9*, 207–211.

Whaley, R. (1986). Valuation of American futures options: theory and empirical tests. *Journal of Finance, 41*, 127–150.

# *Exotic Options

# 14

This chapter is devoted to exotic options, and we will study their essential characteristics, use, and valuation. We cannot claim to provide an exhaustive treatment and will restrict ourselves to the most important options. Indeed, the variety of these instruments is extremely rich and involves all types of assets according to the specific needs of the investors and market participants (shares, market indices, interest and exchange rates, commodities, and even non-negotiable assets like the consumer price index, or levels of natural catastrophes or temperature). Offsetting this diversity is the weakness in liquidity that often is typical of such options, which makes the problem of valuing them more difficult in practice and sometimes even hedging them. Furthermore, some exotic characteristics are often present in various securities issued by enterprises to finance their investments (e.g., convertible bonds which are callable and/or putable under certain conditions).

There is no established classification for exotic options. What we adopt consists of distinguishing *path-independent* options whose payoffs only depend on the final value of the underlying asset and *path-dependent* options whose payoffs depend also on its trajectory.

Amongst the first type we describe, in Sect. 14.1, *forward start* options, *binary—* or *digital* and *double digital—*options, *multi-underlying* options, options on options or *compound* options, and *quantos* and *combos* (or *compos*) options.

Among the second we analyze, in Sect. 14.2, *barriers* (and *double barriers* and *Parisians*), *lookbacks*, options *on averages* or *Asians,* and *choosers.*

Closed formulas for valuations can be obtained in simple cases, but the solution of more complex problems may be difficult, approximate, or even non-existent. This explains why numerical methods are extensively used [trees (Chap. 13), Monte Carlo simulation (Chap. 26)]. Here we concentrate on mathematical models of valuation which lead to closed formulas, even if they are complicated.

Most of the proofs, except the simplest, are postponed to the mathematical appendices to this chapter and can be skipped on first reading.

© The Author(s), under exclusive license to Springer Nature Switzerland AG 2022
P. Poncet, R. Portait, *Capital Market Finance*, Springer Texts in Business and
Economics, https://doi.org/10.1007/978-3-030-84600-8_14

## 14.1 Path-Independent Options

### 14.1.1 The Forward Start Option (with Deferred Start)

This option is a vanilla European option except that its strike is not known at the time the buyer evaluates and pays for it, but only at some later time specified in the contract. Denote by $T_2$ the expiry date, $t$ the current date of valuation, and $T_1$ ($t < T_1 < T_2$) the date at which the strike is set. Typically this strike will be equal to the price of the underlying asset (UA) observed at $T_1$.

Such an option is of interest in several situations. The first is when the buyer wants to fix the volatility at its present level (for which she pays), for example, because she fears or expects that it will increase, while allowing some time before fixing the strike as she bets there will be a temporary reduction (for a call) or increase (for a put) in the UA price.

More often, these options are (extensively) used by *guaranteed capital funds*. Such funds are studied in Chap. 24. In its simplest form, the fund is made up of a risk-free non-defaultable zero-coupon bond that guarantees the capital contributed by the client plus a call written on some UA that allows, if it expires in-the-money, for a gain proportional to the relative increase in the UA price. However, these guaranteed funds, so as to be commercially attractive, need to offer their potential clients a relatively long subscription period, frequently of the order of several months. For example, the subscription period extends from 01/09/n to 31/12/n, and the performance of the UA is evaluated between 31/12/n and 31/12/n + 5. The manager of the guaranteed fund will buy, just before the subscription period, a *forward start call* with 31/12/n as the date for fixing the strike, and expiration date 31/12/n+5 (as well as a zero-coupon with the same maturity).

The increase in the number of guaranteed funds and their diversity has led to the creation of numerous products with a differed start, for which the performance may be calculated as an average, may be "capped" or "floored," may include ratchets, and so on.

In the classic context of the Black–Scholes–Merton (BSM) model, valuing forward start options turns out to be easy since the value of an option is homogeneous (of degree one) in the UA price and in the strike.

Let $S_t$ be the price at time $t$ of the UA and $C_t (= C(S_t, K, T_2 - t))$ the value at $t$ of a vanilla call with maturity $T_2$. At the time $T_1$ of fixing the strike, the call is worth:

$$C(S_{T1}, S_{T1}, \tau \equiv T_2 - T_1) = S_{T1} \, C(1, 1, \tau).$$

At time $T_1$, the option is, therefore, "equivalent" to $C(1, 1, \tau)$ *units* of the UA, where $C(1, 1, \tau)$ is a constant given by the BSM formula. Therefore, on date $t$, assuming the UA does not distribute between $t$ and $T_1$, the value of the forward start option is that of $C(1, 1, \tau)$ units of the UA, that is $S_t \, C(1, 1, \tau)$. Then, using the same homogeneity property in the opposite sense yields:

## 14.1 Path-Independent Options

$$S_t \ C(1, 1, \tau) = C(S_t, S_t, \tau).$$ (14.1)

*Therefore the value at t ($0 < t < T_1 < T_2$) of a forward start option is given by the BSM formula where: the strike is the current price of the UA and the duration of the option is not the variable $T_2 - t$, but the constant $\tau \equiv T_2 - T_1$.*

### 14.1.2 Digital and Double Digital Options

These options, also termed binary options, owe their name to the English word *digit*. They promise to their holder at maturity $T$, either a fixed sum (typically \$1), or nothing, according to whether the value $S(T)$ of the UA is greater than the strike (for a call) or less (for a put). Formally, the payoff may be written:

$$\begin{cases} 1 & \text{if } S(T) > K \\ & \quad \text{for the call,} \\ 0 & \text{otherwise} \\ 1 & \text{if } S(T) < K \\ & \quad \text{for the put.} \\ 0 & \text{otherwise} \end{cases}$$

The formula for the value of a digital option follows directly from the BSM formula. Recall that the risk-neutral probability that the call expires in-the-money is $N(d_2)$, where $d_2 \equiv \frac{\ln(S_0/K)}{\sigma\sqrt{T}} + \frac{\left(r - \delta - \sigma^2/2\right)T}{\sigma\sqrt{T}}$, where $\delta$ is the continuous dividend rate (often denoted by $c$ in the preceding chapters).

The digital call thus is worth \$1 discounted times this probability, i.e.,

$$C_d = e^{-rT} N(d_2).$$ (14.2)

In the same way, the digital put is worth

$$P_d = e^{-rT} \left[1 - N(d_2)\right] = e^{-rT} N(-d_2).$$ (14.3)

### Remarks
- The options called *clicks*, rather popular among individual investors, are digital options whose payoff is either $X$ (dollars) or 0. This product is in fact nothing other than a basket made up of $X$ identical digital options.
- The sum of a binary call and a binary put is worth \$1 discounted since the buyer is sure to have \$1 at maturity: $C_d + P_d = e^{-rT}$ [this is also apparent by summing Eqs. (14.2) and (14.3)].

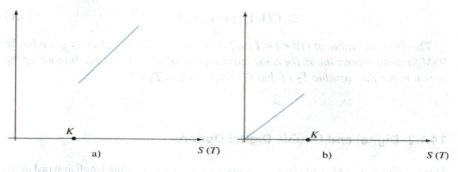

**Fig. 14.1** Asset-or-nothing options

Some professionals, by analogy with the fact that digital options deliver a certain amount of money or nothing (*cash-or-nothing*), classify *asset-or-nothing* options in this category. The payoff for these can be written:

$$\begin{cases} S(T) & \text{if } S(T) > K \\ & \text{for the call,} \\ 0 & \text{otherwise} \\ S(T) & \text{if } S(T) < K \\ & \text{for the put.} \\ 0 & \text{otherwise} \end{cases}$$

Notice that a long portfolio of call and put has a value equal to that of the UA (possibly discounted at the rate of continuous dividends).

Similarly, notice that options are involved here which, contrary to truly digital options (or to classic puts), have a payoff with slope +1 (with respect to the UA) as in Fig. 14.1.

We note that a long portfolio in a call and a put has a value equal to that of the underlying (possibly discounted at the continuous dividend rate $\delta$).

Also note that these options, in contrast with the true digital options (and to the classic put), have a payoff with slope +1 (with respect to the UA) as shown in Fig. 14.1.

The formula for valuing these asset-or-nothing options is likewise taken directly from the BSM formula. For a call, the strike paid upon exercise is 0. Hence the second term in the BSM formula vanishes and one obtains:

$$C_{aon} = S_0 \, e^{-\delta T} \, N(d_1) \qquad (14.4)$$

with

$$d_1 \equiv \frac{\ln(S_0/K)}{\sigma\sqrt{T}} + \frac{\left(r - \delta + \sigma^2/2\right)T}{\sigma\sqrt{T}}.$$

## 14.1 Path-Independent Options

For a put, one has:

$$P_{aon} = S_0 \ e^{-\delta T} \ N(-d_1) \ \text{ (since } S_0 \ e^{-\delta T} = C_{aon} + P_{aon}). \tag{14.5}$$

We note that, under no-arbitrage, the sum of these two options is not worth $S_0$ if the underlying pays a dividend, but $S_0 \ e^{-\delta T}$, because holding them, in contrast, to actually holding the underlying, gives no right to the dividend. If $\delta$ is zero, then we have: $C_{aor} + P_{aor} = S_0$.

A *gap* option is an option whose strike $K$ (on which exercising the option or not at maturity depends) differs from the amount $L$ used to calculate its payoff. The payoff writes:

$$\begin{cases} S(T) - L & \text{if } S(T) > K \\ & \text{for the call,} \\ 0 & \text{otherwise} \\ L - S(T) & \text{if } S(T) < K \\ & \text{for the put.} \\ 0 & \text{otherwise} \end{cases}$$

Typically, we have $L < K$ for a call and $L > K$ for a put, to make such options attractive, but the opposite does occur, which reduces the initial value of the premium paid by the buyer.

Valuing such a *gap* option is easy if one notices that it may be replicated from digital options and a *cash-or-nothing* option. For example, a *gap* call is made of $L$ binary calls sold and one cash-or-nothing call bought. From this, we get:

$$C_g = S_0 \ e^{-\delta T} \ N(d_1) - Le^{-rT} \ N(d_2) \tag{14.6}$$

where $K$ and not $L$ is involved in $d_1$ and $d_2$. One immediately recovers the BSM formula if $K = L$.

Similarly, a *gap* put is made of $L$ binary puts bought and one cash-or-nothing put *sold*. Consequently:

$$P_g = Le^{-rT} \ N(-d_2) - S_0 \ e^{-\delta T} \ N(-d_1) \tag{14.7}$$

with the same remarks as for $C_g$.

Furthermore, we recover a call-put parity, which writes:

$$C_g - P_g = S_0 \ e^{-\delta T} - Le^{-rT}.$$

Finally, *double digital options* (DDO) are elementary components from which one may structure numerous products. A DDO is composed of a long position in a digital option with strike $K_1$ and a short position in a digital option with strike $K_2$ (> $K_1$), where the two digital options have the same underlying and the same expiry date $T$. The payoff of this DDO thus is:

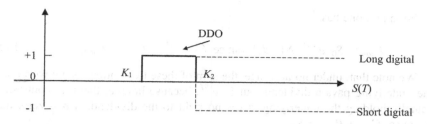

**Fig. 14.2** A double digital option

$$\begin{cases} 1 & \text{if } K_1 < S(T) > K_2 \\ 0 & \text{otherwise} \end{cases}$$

Such a payoff is represented by a "brick" of height 1 and length $(K_2 - K_1)$, as shown in Fig. 14.2.

### Example 1

As an illustration, we describe a fictional product that is inspired by products offered commercially by large banks, of *Corridor* or *Boost* type, which amounts to a bet on the stability of the price of the underlying around its trend, rather than its volatility.

Assume that the 12-month risk-free rate is 3.30% and that the OIS is 3.20%. The nominal value of the structured product, called "corridor," is $1 and its dollar payoff in exactly one year is given by the expression [1 + (4% × x/ 360)], where x is the number of calendar days between 01-01-n and 31-12-n (including the end-points of the interval) on which the OIS $r(t)$ remains in the interval (window) [2.90% - 3.50%].

For example, if $r(t)$ only remains a single day within this window, the yield to the investor will be 0.011%, but if it remains in the window all the time, the yield will be 4.056%. In other words, investors give up a certain yield of 3.30% over a year in the hope of getting, at best, 4.056%, and will have no interest in this product unless they expect a durable stability of interest rates.

The financial analysis of such a product is simple: it amounts to a portfolio of 365 DDO each lasting 1 day and each providing $0.04/360 if the OIS of that day is within the window [2.90–3.50%], and $0 otherwise. The first DDO expires at the end of the first day; the second (forward) begins at the start of the second day and expires at the end of it, etc. Valuing such a product thus is easy, knowing the valuation formula for digital options (calls).

It is a general practice, since expecting stability for an underlying is unusual, to allow the boundaries of the window $[K_1, K_2]$ to change according to a function specified in the contract. If the expectation of the market is a slight rise in rates, one could specify the following functions:

(continued)

## 14.1 Path-Independent Options

> Lower bound: 2.90% + [(3.30% − 2.90%)×D/365)]
> Upper bound: 3.50% + [(3.90% − 3.50%)×D/365)],
> where $D$ is the number of calendar days since 1-1-n. The "corridor" evolves linearly over time between the range [2.90–3.50%] ($D = 0$) and the range [3.30% − 3.90%] ($D = 365$).
> One would obviously adopt analogous formulas for an expected fall in market rates.

### 14.1.3 Multi-underlying (Rainbow) Options (*)

These options, sometimes called *rainbows*, depend on two or more underlying assets.

#### 14.1.3.1 Exchange Options

We have already described the option of exchanging one risky asset for another one, paying Max $(0, S_T − X_T)$ upon expiry, as studied by Margrabe and valued by a suitable change in the probability measure and in the numeraire (see Sect. 11.2.7). For the reader's convenience, we recall the formula in the case of constant variances and covariances:

$$C_t = S_t N(d_1) - X_t N(d_2) \text{ with } d_1 = \frac{\ln\frac{S_t}{X_t} + \frac{1}{2}v^2(T-t)}{v\sqrt{T-t}}, \quad d_2 = d_1 - v\sqrt{(T-t)} \quad (14.8)$$

and $v^2 = \sigma_S^2 + \sigma_X^2 - 2\sigma_{SX}$ is the variance of $S/X$.

We essentially study in what follows the case of two UAs, since understanding the generalization to $n$ UAs is easy. This is not so for valuing and hedging such options, and we devote a short part of the section to those issues.

#### 14.1.3.2 Best of or Worst of Options

In addition to exchange options, *best of* or *worst of* options are often used, especially when a portfolio is under delegated management.

The payoff of the option of receiving the *best of* two risky assets is equal to

$$\text{Max } (S_T, X_T).$$

Such an option, whose interest for the buyer is clear, naturally is not cheap.
The payoff of the option of receiving the *worst of* two risky shares is

$$\text{Min } (S_T, X_T).$$

The buyer of a *worst of* expects that *each of the two* underlyings will perform well and is attracted by the relative cheapness of such an option.

556                                                    14   *Exotic Options

Knowing Margrabe's formula, it is easy to value these two options. Indeed, we have

$$\text{Max } (S_T, X_T) = X_T + \text{Max } (0, S_T - X_T),$$

$$\text{Min } (S_T, X_T) = X_T - \text{Max } (0, X_T - S_T),$$

from which we see that replicating a *best of* or a *worst of* with one of the two underlyings and the appropriate exchange option is extremely simple. As a result, when the underlying does not distribute between $t$ and $T$, their value on date $t$ *is a function of $C_t$ given by* (14.8): $X_t + C_t$ for the *best of* ; $X_t - C_t$ for the *worst of.*

As an illustration, consider a bond manager (denote by X the value of his bond portfolio) who thinks that the stock market (denote by S the value of the basket representative of this market) will probably outperform the bond market. He buys an exchange option paying off Max $(0, S_T - X_T)$, and is thus assured to provide his clients with the better of the two, stock and bond, market performances, whichever occurs, less the cost of the initial option premium.

An interesting aspect of these options is that their prices [see (14.8) for example] reflect both the volatility of each underlying and their correlation. Knowing the volatility of each UA, estimated for example from liquid vanilla options, one may then infer their *implicit correlation*. This is why professionals often call these instruments *correlation products*. With the volatilities and correlation at hand, other more complex products involving the two UAs can be valued.

### 14.1.3.3 Options on the Minimum or on the Maximum

Calls and puts *on the minimum* or *on the maximum* of two assets are also correlation products that enjoy a relatively good liquidity.

Payoffs of options on the *minimum* write, respectively:

$$\text{Max } (0, \text{Min } (S_T, X_T) - K) \text{ for a call,}$$

$$\text{Max } (0, K - \text{Min } (S_T, X_T)) \text{ for a put,}$$

and the payoffs of options on the *maximum* write, respectively:

$$\text{Max } (0, \text{Max } (S_T, X_T) - K) \text{ for a call,}$$

$$\text{Max } (0, K - \text{Max } (S_T, X_T)) \text{ for a put.}$$

### Remarks
- The *best of* and *worst of* options are special cases of options on the minimum and on the maximum, respectively, obtained for $K = 0$.
- To avoid creating many valuation programs and to save computing time, one uses the *call-put parity* equations between different options. We have for instance:

# 14.1 Path-Independent Options

$$\text{Max } (0, K - \text{Min } (S_T, X_T)) = \text{Max } (0, \text{Min } (S_T, X_T) - K) - \text{Min } (S_T, X_T) + K,$$

$$\text{Max } (0, K - \text{Max } (S_T, X_T)) = \text{Max } (0, \text{Max } (S_T, X_T) - K) - \text{Max } (S_T, X_T) + K.$$

- There are also equations for *Min-Max parity*:

$$\text{Max } (0, \text{Min } (S_T, X_T) - K) = \text{Max } (0, S_T - K) + \text{Max } (0, X_T - K) \\ - \text{Max } (0, \text{Max } (S_T, X_T) - K),$$

$$\text{Max } (0, K - \text{Min } (S_T, X_T)) = \text{Max } (0, K - S_T) + \text{Max } (0, K - X_T) \\ - \text{Max } (0, K - \text{Max } (S_T, X_T)).$$

Closed-form formulas do exist for all these options. They involve the bivariate normal distribution function $N2$, where $N2[x; y; \rho]$ is the probability that the first normal variable is smaller than $x$ and the second smaller than $y$; $\rho$ denotes the correlation coefficient between the two components.

As an example, a *call on the minimum* has a payoff on date $T$

$$CMin_T = S_T\, 1_{S(T)<X(T)} 1_{S(T)>K} + X_T\, 1_{S(T)>X(T)} 1_{X(T)>K} - K\, 1_{S(T)>K} 1_{X(T)>K},$$

and consequently a value at $t = 0$ equal to

$$CMin_0 = E_Q\left[e^{-rT}\left[S_T\, 1_{S(T)<X(T)} 1_{S(T)>K} + X_T\, 1_{X(T)<S(T)} 1_{X(T)>K} - K\, 1_{S(T)>K} 1_{X(T)>K}\right]\right].$$

In the last term, involving $K$, the expectation of the product of the two indicator functions corresponds to the joint probability that the calls written on $S$ and on $X$ will both be in-the-money, which gives

$$- Ke^{-rT}\, N2[d_2(\sigma_S); d_2(\sigma_X); \rho]$$

where the $d_2$ comes from BSM and $\rho$ is the correlation between $S_T$ and $X_T$.

The first two terms are symmetric in $S$ *and* $X$. Therefore, it is sufficient to value the first. Since the first indicator function involves two random variables, we divide, as we did for exchange options, $S(T)$ and $X(T)$ by $S(T)$ to have only one random variable. Then we have

$$E_Q\left[e^{-rT} S_T\, 1_{X(T)/S(T)>1} 1_{S(T)>K}\right].$$

The second indicator function gives the usual probability $N(d_2(\sigma_S))$.

The first indicator function gives the probability $N(d_2(v))$, where n is defined in Eq. (14.8), and $d_2(v) = [[ln\, (X_0/S_0) - 0.5v^2 T]/v\sqrt{T}]$.

The joint probability of these two events is, therefore, $N2[d_2(\sigma_S); d_2(\nu); \rho']$ where $\rho'$, the correlation coefficient between $S_T$ and $X_T/S_T$, is equal to $(\rho \, \sigma_X - \sigma_S)/\nu$.[1]

Since the expectation involves $e^{-rT}S_T$, we recover, by using the value of $S_T$ and the usual change of variables, $S_0$ and $d_1(\sigma_S)$. Therefore, we have

$$E_Q[e^{-rT}S_T \; 1_{S(T)>K} \; 1_{X(T)/S(T)>1}] = S_0 \; N2[d_1(\sigma_S); d_2(\nu); \rho'].$$

Finally, applying the same reasoning to the term involving $X_T$, we have

$$CMin_0 = S_0 \; N2[d_1(\sigma_S); d_2(\nu); \rho'] + X_0 \; N2[d_1(\sigma_X); d'_2(\nu); \rho'']$$
$$- K \; e^{-rT} \; N2[d_2(\sigma_S); d_2(\sigma_X); \rho] \tag{14.9}$$

where, in addition to terms already defined, $d'_2(\nu) = [[\ln (S_0/X_0) - 0.5\nu^2 T]/\nu \sqrt{T}]$ and $\rho'' = (\rho \, \sigma_S - \sigma_X)/\nu$.

**Remarks**

- The function $N2[x; y; \rho]$ can be computed by numerical integration using the fact that

$$N2[x; y; \rho] = \int_{-\infty}^{x} \frac{\exp(-0.5u^2)}{\sqrt{2\pi}} N\left(\frac{y - \rho u}{\sqrt{1 - \rho^2}}\right) du$$

- When the UA pays a dividend, the preceding formulas are adjusted in the same way as Merton's model generalizes Black–Scholes.
- It is easy, starting from equations such as (14.9), to value the best of and worst of options *on two risky assets and the risk-free asset*. We have, for example

$$\text{Max} \; (S_T, X_T, K) = K + \text{Max} \; (0, \text{Max} \; (S_T, X_T) - K).$$

- Generalizing the equations above to $n$ risky assets is problematic for valuation and hedging since it is necessary to evaluate $n$-uple integrals with numerical methods, which is not practical. We then must use the Monte Carlo simulation (*see Chap.* 26). Furthermore, $n$ variances and $n(n-1)/2$ covariances must be estimated, a number that increases with $n$ squared, which brings about serious estimation errors. These products consequently often display large spreads between their bid and ask prices.

---

[1] The instantaneous variance (divided by $dt$) of $(dX/X)$ is $\sigma^2_X$. This is trivially also the product of $(dS/S)$ by $[(dX/X)/(dS/S)]$. By Itô's Lemma, the variance (divided by $dt$) of this product is equal to $\sigma^2_X = \sigma^2_S + \nu^2 + 2\sigma_S\nu\rho'$. In the same way, the variance (divided by $dt$) of the ratio $(dX/X)/(dS/S)$ equals $\nu^2 = \sigma_X^2 + \sigma_S^2 - 2\rho\sigma_X\sigma_S$ (see, for example, Proposition 15 of Chap. 11). Eliminating $\nu$ from these two equations and dividing each term by $2\sigma_S$ yields the result.

## 14.1.4 Options on Options or "Compounds"

An option (called the *mother*) on an option is the right but not the obligation, to buy or sell on some future date $T_1$ at a strike $K_1$, a call or a put (the *daughter*) with expiration date $T_2$ ($> T_1$) and strike $K_2$ (usually $> K_1$), written on an some underlying asset. Consequently, there are four types of compound options: a call on a call, a call on a put, a put on a call, and a put on a put.

An option on an option is cheaper (in dollars) than a vanilla option since its underlying is an option, but its relative cheapness cost obviously depends on the strikes of the two component options (that is, essentially on their relative *moneyness*).

As examples of the use of these instruments we can mention:

- Taking a position, on the rise (a mother call) or the fall (a mother put) in the future implicit volatility of an underlying, since a change in the volatility means a change in the same direction in the price of the daughter option.
- Taking a position on the direction in which the price of the underlying will vary at small or even very small cost, since the (double) leverage in a compound option can be huge.
- Anticipating a significant increase in the size of an underlying's price fluctuations without betting on their direction, by the double purchase of a call on a call and a call on a put, at low or very low cost relative to a vanilla *straddle* or *strangle*.
- Hedging, as is often done in practice, a conditional position. Suppose, for example, that a European industrial corporation participates in some large international tender in US dollars. If it wins the tender, then it will be long in this currency and is at risk if the dollar depreciates with respect to the euro. If not, it has no such position in dollars. A possible solution would be to buy a \$/€ put to cover the position if the tender is won. The risk is then of not winning the tender and having to resell the no longer useful put at a loss due to the passage of time (theta) and/or to appreciation of the dollar. Usually, a better solution is to buy a call on a \$/€ put, which will be cheaper than the put mentioned above. If the tender is lost, the potential loss is less, and if it is won and the dollar depreciates, that call on the put can be sold at a profit and the corporation covers its position by selling dollars on a fixed date.

*Valuing options on options is due to Geske (1977, 1979). Insofar as its underlying is itself an option (daughter), the solution involves the bivariate normal distribution $N2$, like for an option written on two underlyings. The solution is only pseudo-closed, like that of Barone-Adesi and Whaley for American options (see Chap. 13), in that one must find a threshold $S*$ for the underlying's price using numerical methods.[2] Denoting the dividend rate by $\delta$, we obtain the following formulas:

---

[2] However, in contrast to that of Barone-Adesi and Whaley, this solution is exact.

- For the *call on a call*:

$$CC_0 = S_0 \, e^{-\delta T_2} \, N2[a_1;b_1;\rho] - K_2 \, e^{-rT_2} \, N2[a_2;b_2;\rho] - K_1 \, e^{-rT_1} \, N(a_2) \quad (14.10)$$

$$\text{with} \quad a_1 = \frac{\ln(S_0/S^*) + (r - \delta + 0.5\sigma^2)T_1}{\sigma\sqrt{T_1}}, \quad a_2 = a_1 - \sigma\sqrt{T_1}$$

$$b_1 = \frac{\ln(S_0/K_2) + (r - \delta + 0.5\sigma^2)T_2}{\sigma\sqrt{T_2}}, \quad b_2 = b_1 - \sigma\sqrt{T_2}, \quad \rho = \sqrt{T_1/T_2},$$

and $S^*$ is the underlying price so that the value of the call $C(S^*, K_2, T_2 - T_1) = K_1$.

- For the *call on a put*:

$$CP_0 = -S_0 \, e^{-\delta T_2} \, N2[-a_1; -b_1;\rho] + K_2 \, e^{-rT_2} \, N2[-a_2; -b2;\rho]$$
$$- K_1 \, e^{-rT_1} \, N(-a_2) \quad (14.11)$$

where $S^*$ is the underlying price so that the value of the put $P(S^*, K_2, T_2 - T_1) = K_1$.

- For the *put on a call*:

$$PC_0 = -S_0 \, e^{-\delta T_2} \, N2[-a_1;b_1; -\rho] + K_2 \, e^{-rT_2} \, N2[-a_2;b_2; -\rho]$$
$$+ K_1 \, e^{-rT_1} \, N(-a_2) \quad (14.12)$$

where $S^*$ is the same as for (14.10).

- For the *put on a put*:

$$PP_0 = S_0 \, e^{-\delta T_2} \, N2[a_1; -b_1; -\rho] - K_2 \, e^{-rT_2} \, N2[a_2; -b_2; -\rho]$$
$$+ K_1 \, e^{-rT_1} \, N(a_2) \quad (14.13)$$

where $S^*$ is the same as for (14.11).

## 14.1.5 Quantos and Compos

One of the major risks for investors is the volatility in foreign exchange rates (Forex). Even for the major international currencies (US dollar, euro, sterling, yuan, and yen) fluctuations in exchange rates can be quite large and make a very profitable investment in foreign currency disastrous when its payoff is converted into domestic currency.

# 14.1 Path-Independent Options

The primary motivation for *quanto products* is, whether these are interest rate *swaps*[3] or options, to avoid exchange rate risk while betting on an asset denominated in foreign currency. The name *quanto* comes from the creation of these products as a reaction to the collapse of the Mexican peso at the end of the 1980s.

> **Example 2**
> A European investor wants to bet on a rise in the US stock market but fears a devaluation of the dollar with respect to the euro. The solution consists in buying a *quanto* call written on the Dow Jones Industrial Average. The underlying's price and the strike are denominated in US \$, but the call's payoff (if positive) is multiplied by a €/\$ exchange rate fixed in advance, and so denominated in euros. Let us suppose the exchange rate negotiated between the buyer and the seller (writer) will be the current spot rate, say 0.90 (€/\$). Assume the strike of the call written on the Dow is 21,400 (\$), and that upon expiry of the call the Dow is 21,800 (\$). The holder will receive \$(21,800 − 21,400) × 0.9 €/\$ = 360 €, whether the dollar has appreciated or depreciated relative to the euro.

**Definition** *More generally, a product is said to be quanto when its payoff is denominated in a currency different from the original currency of the product's fundamental components.*

Most often, the currency used in a transaction is the buyer's, which avoids any exchange risk to her. One may, however, consider the case where a *quanto* product is intended for a bet on both a foreign underlying and another foreign currency. In the preceding example, the payoff could have been given in yen, with a ¥/\$ exchange rate fixed in advance. The European investor would then have received yens, which exposes her not just to the fundamental risk of a bear market in the New York but also to a risk on the €/¥ exchange rate, profiting (losing) for appreciation (depreciation) of the Japanese currency relative to the euro.

Furthermore, even if the fixed exchange rate agreed in advance is usually the spot rate prevailing at the time of the contract, nothing prevents the two parties from agreeing upon a different rate, such as the current forward exchange rate or any other rate they find appropriate. Finally, in another type of option called a *compo (or combo)*, only the underlying's price is denominated in foreign currency, the strike being expressed in domestic currency.

In what follows, we will restrict ourselves to the case of calls, with puts being valued and hedged along the same lines.

---

[3] On the interest rate *swap* market, a *quanto swap* is called a *diff-swap* (differential *swap*). For example, a Japanese interest rate is exchanged for a US interest rate, the payments received and paid being all made in dollars and computed on a nominal amount denominated in dollars.

## *14.1.5.1 The Quanto Call

The payoff of this call is equal to:

$$CQ(T) = \widehat{X}\left(S^f(T) - K^f\right)^+$$

where the exponent $f$ indicates the *foreign* nature of the original currency in which the underlying and strikes are denominated, and $\widehat{X}$ denotes the predetermined exchange rate (domestic currency/foreign currency).

Let us value this call at the present date $t = 0$. We have:

$$CQ(0) = \widehat{X} E_Q\left[e^{-\int_0^T r(s)ds}\left(S^f(T) - K^f\right)^+\right] = \widehat{X} A(T) E_Q\left[\left(S^f(T) - K^f\right)^+\right]$$

where $r(t)$, the domestic interest rate, is assumed deterministic; the domestic discount factor $A(T) \equiv \exp\left(-\int_0^T r(s)\,ds\right)$ is therefore known. We use the risk-neutral probability $Q$.

At first sight, this problem may appear trivial: since $\widehat{X}$ and $A(T)$ are known, it is only necessary to apply BSM to the expectation under $Q$ of the payoff given in foreign currency, then multiply the BSM result by $\widehat{X} A(T)$ to obtain the price in domestic currency. This intuition, however, also suggests this solution is wrong, because it does not involve either the exchange rate $X(t)$ or its correlation with the underlying price $S^f(t)$. This hunch is vindicated when one considers the position of the call's writer, *who is exposed to an exchange risk in addition to that affecting the underlying*. In fact, the trivial solution is wrong since $Q$, the risk-neutral probability used by the domestic investor differs from $Q^f$, the risk-neutral probability used by the foreign investor. Under $Q$, the expected instantaneous returns are all equal to the domestic risk-free rate $r$, while under $Q^f$ they are equal to the foreign risk-free rate $r^f$.

More generally, $E_Q[\cdot]$, calculated from the domestic point of view, is *not* equal to $E_{Q^f}[\cdot]$ which would lead to the BSM formula. What is needed is to express the dynamics of the underlying price $S^f(t)$ and the exchange rate $X(t)$ under the *domestic* risk-neutral probability $Q$.

Assume that, under the historical probability, the exchange rate satisfies

$$\frac{dX(t)}{X(t)} = \mu_X(t)\,dt + \sigma_X(t)\,d\widehat{W}^x(t)$$

where $\sigma_X(t)$ is deterministic, and $\widehat{W}^x(t)$ is one-dimensional Brownian motion.

All Brownian motions used here are one-dimensional, but they are correlated (the generalization to multi-dimensional Brownian processes is easy). All variances and covariances are assumed deterministic, as well as the domestic and foreign riskless

# 14.1 Path-Independent Options

563

rates. Assume in addition that, under the same historical probability, the value $S^f(t)$ of the underlying satisfies[4]

$$\frac{dS^f(t)}{S^f(t)} = \mu_{S^f}(t)\,dt + \sigma_{S^f}(t)\,d\widehat{W}^{S^f}(t).$$

We will first prove two intermediate results.

**Lemma 1**

Under the domestic risk-neutral probability, the dynamics of the exchange rate $X(t)$ and the foreign underlying price $S^f(t)$ can be written, respectively:

$$\frac{dX(t)}{X(t)} = \left(r(t) - r^f(t)\right)dt + \sigma_X(t)\,dW^X(t) \tag{14.14}$$

$$\frac{dS^f(t)}{S^f(t)} = \left(r^f(t) - \rho(t)\sigma_X(t)\sigma_{S^f}(t)\right)dt + \sigma_{S^f}(t)\,dW^{S^f}(t) \tag{14.15}$$

where $r^f(t)$ is the foreign risk-free rate, $\rho(t)$ is the correlation between the relative variations in the exchange rate and in the UA price, and the two Brownian motions are under $Q$.

**Proof**

Let us define $Z(t) \equiv S^f(t)\,X(t)\,A(t)$ and $Y(t) \equiv [X(t)/A^f(t)]\,A(t)$, where

$$A(t) \equiv \exp - \int_0^t r(s)ds \quad \text{and} \quad A^f(t) \equiv \exp - \int_0^t r^f(s)ds.$$

$Z(t)$ thus is the price $S^f(t)X(t)$ of the foreign UA denominated in domestic currency discounted at the domestic risk-free rate $A(t)$, and is, therefore, a martingale under $Q$.

Similarly, $Y(t)$ is the value of the foreign money market account converted into domestic currency $\left[\frac{1}{A^f(t)}X(t)\right]$ and discounted with the domestic rate, and thus is also a $Q$-martingale. The drift terms of these two processes, therefore, vanish under $Q$. From this we infer, by Itô's Lemma, that

$$\mu_Y = r^f + \mu_X - r = 0 \Rightarrow \mu_X = r - r^f \quad \text{and}$$
$$\mu_Z = \mu_{S^f} + \mu_X + \rho\sigma_X\sigma_{S^f} - r = 0 \Rightarrow \mu_{S^f} = r - \mu_X - \rho\sigma_X\sigma_{S^f} = r^f - \rho\sigma_X\sigma_{S^f}.$$

---

[4]To simplify, we assume the underlying does not pay out any dividend. If it does, at the continuous rate $\delta^f(t)$, we only need to replace $S^f(0)$ by $S^f(0)\,D(0,T)$ in all the formulas, where $D(0,T) = \exp - \int_0^T \delta^f(s)\,ds$.

564            14   *Exotic Options

Besides, the volatility parameters are left unaffected by the change from the historical probability $P$ to the risk-neutral $Q$.

The *value of a quanto call* $\hat{X}A(T)E_Q\left[(S^f(T) - K^f)^+\right]$, computed using the underlying's price dynamics given in (14.15) is equal to:

$$CQ(0) = \hat{X} A(T) \left[\frac{S^f(0)}{A^f(T)} \gamma(T)N(d_1) - K^f N(d_2)\right] \qquad (14.16)$$

where $d_1 = \frac{\ln\left[S^f(0)\gamma(T)/A^f(T)K^f\right] + \frac{1}{2}Vol^2(S^f,T)T}{Vol(S^f,T)\sqrt{T}}$, $d_2 = d_1 - Vol(S^f,T)\sqrt{T}$, $\gamma(T) \equiv \exp\left(-\int_0^T \rho(s)\,\sigma_X(s)\,\sigma_{S^f}(s)\,ds\right)$, $Vol^2(S^f,T) = \frac{1}{T}\int_0^T \sigma_{S^f}^2(s)\,ds$, the last expression being the average, from 0 to $T$, of the UA price variance (volatility squared).

**Proof**
By integrating (14.15), we obtain

$$S^f(T) = S^f(0)\exp\left[\int_0^T \left(r^f(s) - \rho(s)\sigma_X(s)\sigma_{S^f}(s) - \frac{1}{2}\sigma_{S^f}^2(s)\right)ds + \int_0^T \sigma_{S^f}(s)\,dW^{S^f}(s)\right]$$

$$= \left[\frac{S^f(0)}{A^f(T)}\gamma(T)\right]\exp\left[-\frac{1}{2}\int_0^T \sigma_{S^f}^2(s)\,ds + \int_0^T \sigma_{S^f}(s)\,dW^{S^f}(s)\right].$$

The conditions for the BSM model (with non-constant volatility and rates) are then again satisfied, if one takes as the present value of the underlying, not $S^f(0)$, as the foreign investor will, but $\frac{S^f(0)}{A^f(T)}\gamma(T)$. Multiplying $E_Q[\bullet]$ by $\hat{X}A(T)$, we obtain the value of a *quanto* call as given in (14.16).

**Remarks**
- If $\hat{X} = 1$ and, for all $t$, $r(t) = r^f(t) = r$, $\rho(t) = 0$ and $\sigma_S^f(t) = \sigma_S^f$, we recover the BSM formula.
- The exchange rate volatility $\sigma_X$ plays *no* role, the volatility of the *quanto* product remaining that of the underlying's price.
- The randomness of the exchange rate is only accounted for through its correlation $\rho$ with the underlying. If it is nil, and the domestic and foreign interest rates are equal, the BSM formula can be applied directly (one just needs to multiply it by the constant $\hat{X}$).
- From the point of view of the writer of a *quanto call*, it is useful to rewrite solution (14.16) as

## 14.1 Path-Independent Options

$$CQ(0) = \frac{\widehat{X} \, X(0)}{F_X(0,T)} \left[ S^f(0) \gamma(T) N(d_1) - K^f A^f(T) N(d_2) \right] \qquad (14.17)$$

where $F_X(0,T) = X(0) A^f(T)/A(T)$ is, under no-arbitrage (*cash-and-carry relationship*), the forward exchange rate (for maturity $T$) prevailing at time $t = 0$.

We thus understand that the writer's hedging of such a product involves the foreign underlying price paid in domestic currency and a forward contract on the *inverse* of the exchange rate,[5] with such positions subject to continuous adjustment. These transactions in the forward contract allow the writer to dynamically manage the exchange risk to which he is exposed in lieu of the buyer of the quanto call. For example, if the foreign currency depreciates with respect to the domestic currency, the writer loses money on the call. He compensates for this loss by a gain on the foreign currency sold forward (which is equivalent to the domestic currency bought forward).

---

**Example 3**

Suppose we have an American underlying worth $S^f(0) = \$100$, and a standard call with strike $\$100$. For $T = 0.5$ year, $r^f = 4\%$ and $\sigma_{S\,f} = 40\%$, the BSM formula yields a call value of $\$12.15$, or $10.935$ € for $X(0) = 0.9$ (€/$). If the call is *quanto*, with $\widehat{X} = 0.8$, $r = 2.5\%$, and $\rho.\sigma_X.\sigma_{S\,f} = -0.02$, applying Eq. (14.16) gives $CQ(0) = 11.554$ €, compared with $10.935$ €, or a relative gain of $5.66\%$.

---

### *14.1.5.2  A Compo Call

The second type of option, the *compo* or *combo*, is intermediate between vanilla and *quanto*, in that the exchange rate risk affects the underlying price only, not the strike. Thus, the payoff is

$$CC(T) = \left( X(T) \, S^f(T) - K \right)^+$$

for a call, where $K$ is denominated in domestic currency.

Therefore, the investor is betting as much on a rise in the foreign underlying asset price as on an appreciation of the foreign currency relative to the domestic, while maintaining the strike free from exchange risk. We illustrate the difference in the payoff of a *compo* call from that of a *quanto* call with an example.

---

[5] As in the spot market, where $X(T) \left( \frac{1}{X(T)} \right) = 1$, we have on the futures market $F_X(t,T) F^f_{1/X}(t,T) = 1$, where $F^f_{1/X}(t,T)$ is the forward exchange rate (foreign currency/domestic currency) for $0 \leq t \leq T$.

## Example 4

Assume the data are: $S^f(0) = \$100$, $X(0) = 0.8 = \widehat{X}$, $K^f = \$100$ and $K = 80\ €$. Let us consider, upon expiry of the options, the four following scenarios: $S^f(T) = 120$ with $X(T) = 0.64$ or $X(T) = 0.96$, and $S^f(T) = 95$ with $X(T) = 0.64$ or $X(T) = 0.96$. The payoffs of the *quanto* are, respectively: $[16; 16; 0; 0]$, which reflects its complete immunity to exchange rate risk. The payoffs of the *compo* are, respectively: $[0; 35.2; 0; 11.2]$, reflecting a great sensitivity to exchange rate risk. We also see that the final values of these options are wide apart according to which scenario is realized.

Valuing a *compo* option is a bit more difficult because the underlying asset is now the product $X(T)S^f(T)$ whose dynamics under the $Q$-domestic probability is required. We give here immediately the result, the proof being relegated to the mathematical Appendix 1 to this chapter.

The *value of the compo call is equal to*

$$CC(0) = X(0)\,S^f(0)N(d_1) - K.A(T)N(d_2) \qquad (14.18)$$

where $d_1$, $d_2$ and $Vol^2\ (XS^f,\ T)$ are, respectively, defined by: $d_1 = d_2 + Vol\left(XS^f, T\right)\sqrt{T}$

$$d_2 = \frac{\ln\left(X(0)\,S^f(0)/K.A(T)\right) - 0,5\,Vol^2\left(XS^f, T\right)T}{Vol\left(XS^f, T\right)\sqrt{T}}$$

and

$$Vol^2\left(XS^f, T\right) \equiv \frac{1}{T}\int_0^T \left(\sigma_X^2(s) + \sigma_{S^f}^2(s) + 2\rho(s)\sigma_X(s)\sigma_{S^f}(s)\right)ds,$$

the last expression being the variance of $X(T)S^f(T)$.

## Remarks

- In contrast with the *quanto* case, it is in the volatility of the underlying $X(t)S^f(t)$ that the volatility of the exchange rate change $X(t)$ (which plays no role for *quantos*) and its correlation ($\rho(t)$) with the foreign underlying $S^f(t)$ come in.
- The initial value of the underlying, $X(0)S^f(0)$, is not modified by a factor taking into account the correlation ($\rho(t)$), in contrast again with the *quanto case*.
- Dynamic hedging of the compo call by its writer involves buying the support $S^f$ converted into domestic currency and the sale of a domestic risk-free zero-coupon. The foreign risk-free rate does not play any role, which explains why there is no need, unlike for *quantos*, to trade forward contracts on the foreign currency.

## 14.2 Path-Dependent Options

We discuss in order *barrier* options (among which are to be found *double barriers* and *Parisians*), and *digital barriers*, options with a maximum or minimum strike called *lookbacks*, options on *averages* or *Asian* options, and *chooser* options.

### 14.2.1 Barrier Options

Like a vanilla option, a *barrier* option's value upon expiration depends on the difference between the UA's price at expiry and the strike. However, the contract defining this type of option stipulates a condition that binds the exercise of the option to the UA's price crossing a specified (*barrier*, or *limit*) level, or on the contrary to not crossing that level. There are consequently two sorts of barrier options.

- A *knock-in* option (known more simply as an *In*) does not exist until the price of the underlying hits the barrier. If the UA's price never hits it, the option expires worthless at maturity. However, if the barrier is reached at any time, the option becomes vanilla and may expire at maturity in the money or not. Thus, the non-zero probability that the barrier is never hit reduces the initial value of such an option relative to its vanilla counterpart, and the more so the further the barrier is from the initial price of the underlying.
- A *knock-out* option (a.k.a. an *Out*) is the opposite of the *In*: the option is initially vanilla and stays so as long as the price of the underlying does not reach the barrier. As soon as the barrier is hit, the option simply becomes worthless, regardless of how the UA's price may thereafter evolve. Thus, the non-zero probability that the barrier is reached makes the initial value of the option less than its vanilla counterpart's, and the more so the closer the barrier is to the initial price of the underlying.

It is precisely their relative cheapness that makes these options so popular: Buyers attempt to benefit from an accurate enough forecast of the UA's price trajectory to reduce the cost of their investment, but by doing so they take on more risk than for vanilla options since their "*In*" options may never been activated (switched on) and their "*Out*" options may evaporate (be switched off).

*On a priori grounds*, one should have 8 ($= 2 \times 2 \times 2$) types of barrier options: There are calls and puts, *In* and *Out barriers*, and these may be reached by a rise in the underlying price, called an *Up option*, or a fall, called a *Down option*. The barrier has obviously to be higher than the initial UA price for *Up* options and lower for *Down* options. We will use a three-letter notation for these 8 options: CUI (*call up-and-in*), CDI (*call down-and-in*), CUO (*call up-and-out*), CDO (*call down-and-out*), PUI (*put up-and-in*), PDI (*put down-and-in*), PUO (*put up-and-out*), and PDO (*put down-and-out*). The price of any of these options will be denoted by the name of the option written in italic (e.g., *CUI* is the price of a CUI).

The value of the barrier, denoted by $L$ (for limit) in what follows, relative to the strike $K$ plays a crucial role in the economic interest and financial value of these options, which leads to distinguishing 16 ($= 8 \times 2$) options according to whether $L$ is larger than $K$ or smaller. However, it is easy to see that 2 "*In*" options and 2 "*Out*" options are not traded in the market, the first because they are in fact simple vanillas and the second because they are always worthless, since the barrier switches them off as soon as they become in-the-money. The reader should check, as an exercise in barrier options, that the cases just mentioned are, respectively a CUI with $L < K$, a PDI with $L > K$, a CUO with $L < K$ and a PDO with $L > K$. Consequently, there are only 12 barrier options actually traded, 6 "*In*" and 6 "*Out*."

Furthermore, it is often the case that the contract contains a clause according to which, if an "*In*" option is never switched on or an "*Out*" option is switched off, the holder does have a right to a cash *rebate* agreed upon in advance, that will be denoted by $R$. Naturally, options with *rebate* are more expensive than their counterparts with none. For an "*Out*" option, $R$ may be paid by the writer according to the contract either when the barrier is reached (*rebate at hit*) or at the maturity of the option (*rebate at expiry*). The first modality seems usually to prevail on the market. For an "*In*" option, $R$ can only be paid at maturity (*rebate at expiry*) since only then is it known that switching on has not occurred.

We first examine how to value barrier options without *rebate* then we compute the values of *rebates at expiry* and *at hit*.

### *14.2.1.1 Valuing Barrier Options

It is not necessary to have 12 closed-form formulas for the values of the 12 barrier options. Indeed, it is sufficient to know, in addition to the BSM formula, the prices of two of them and certain parity relationships among them all to obtain the other ten.

The first parity equations used here are as follows, where C and P denote, respectively, the standard European call and put:

(i)  $C = CDO + CDI = CUO + CUI$

(ii)  $P = PDO + PDI = PUO + PUI.$

These equations follow immediately from the definition of barrier options. For example, buying a CDO and a CDI, with the same underlying, expiration date, strike and barrier, is equivalent to a standard call since, if the barrier is hit (as a result of a large enough fall in the underlying's price), the CDO is worthless when it expires but the CDI is activated, and if it is not, then the CDO is alive but the CDI is not activated.

The parity equations of the second group are called the "inverses." We show in this chapter's mathematical Appendix 3 that knock-out calls and puts are linked, just like knock-in calls and puts:

## 14.2 Path-Dependent Options

$$
\text{(iii)} \begin{cases}
PUO(S,K,L,T,r,\delta) = S.K.CDO\ (\tfrac{1}{S},\tfrac{1}{K},\tfrac{1}{L},T,\delta,r) \\[2mm]
PDO(S,K,L,T,r,\delta) = S.K.CUO(\tfrac{1}{S},\tfrac{1}{K},\tfrac{1}{L},T,\delta,r) \\[2mm]
PUI(S,K,L,T,r,\delta) = S.K.CDI(\tfrac{1}{S},\tfrac{1}{K},\tfrac{1}{L},T,\delta,r) \\[2mm]
PDI(S,K,L,T,r,\delta) = S.K.CUI(\tfrac{1}{S},\tfrac{1}{K},\tfrac{1}{L},T,\delta,r)
\end{cases}
$$

where we note, in particular, the inversed roles played by the risk-free rate $r$ and the dividend rate $\delta$.

For example, suppose we know the price of a CDO ($L < K$) and of a CUO ($L > K$). The mathematical Appendix 4 provides the proof of the second result (the proof of the first one is similar). We can then obtain all the option prices using a double inference chain:

(iv) $CDO \rightarrow PUO$ [by "inverse" (iii)] $\rightarrow PUI$ [from (ii)]
  $\rightarrow CDI$ (from (iii), or (i) knowing $CDO$).

(v) $CUO \rightarrow CUI$ [from (i)] $\rightarrow PDI$ [from (iii)]
  $\rightarrow PDO$ (from (ii), or (iii) knowing $CUO$).

Table 14.1 presents a synopsis of the values of the various barrier options, in absence of *rebate*. The assumptions on the UA's price dynamics are those adopted by BSM. We use the following notation and results:[6]

$$
\mu \equiv \left(r - \delta - \frac{\sigma^2}{2}\right) \quad \varepsilon = \frac{\mu}{\sigma^2} + 1 \quad \phi, \eta = +1 \text{ or } -1
$$

$$
x = \frac{\ln(S/K)}{\sigma\sqrt{T}} + \varepsilon\sigma\sqrt{T} \quad x_1 = \frac{\ln(S/L)}{\sigma\sqrt{T}} + \varepsilon\sigma\sqrt{T}
$$

$$
y = \frac{\ln(L^2/S.K)}{\sigma\sqrt{T}} + \varepsilon\sigma\sqrt{T} \quad y_1 = \frac{\ln(L/S)}{\sigma\sqrt{T}} + \varepsilon\sigma\sqrt{T}
$$

$$
[1] \quad = \phi Se^{-\delta T}N(\phi x) - \phi Ke^{-rT}N\left(\phi x - \phi\sigma\sqrt{T}\right)
$$

$$
[2] \quad = \phi Se^{-\delta T}N(\phi x_1) - \phi Ke^{-rT}N\left(\phi x_1 - \phi\sigma\sqrt{T}\right)
$$

$$
[3] \quad = \phi Se^{-\delta T}\left(\frac{L}{S}\right)^{2\varepsilon}N(\eta y) - \phi Ke^{-rT}\left(\frac{L}{S}\right)^{2(\varepsilon-1)}N\left(\eta y - \eta\sigma\sqrt{T}\right)
$$

$$
[4] \quad = \phi Se^{-\delta T}\left(\frac{L}{S}\right)^{2\varepsilon}N(\eta y_1) - \phi Ke^{-rT}\left(\frac{L}{S}\right)^{2(\varepsilon-1)}N\left(\eta y_1 - \eta\sigma\sqrt{T}\right)
$$

---

[6]These results are due to Rubinstein and Reiner (1991).

**Table 14.1** Value of barrier options

|      | $K > L$ | $K < L$ | $\varphi$ | $\eta$ |
|------|---------|---------|-----------|--------|
| CDI  | [3] | [1]-[2]+[4] | 1 | 1 |
| PDI  | [2]-[3]+[4] | [1] | -1 | 1 |
| CUI  | [1] | [2]-[3]+[4] | 1 | -1 |
| PUI  | [1]-[2]+[4] | [3] | -1 | -1 |
| CDO  | [1]-[3] | [2]-[4] | 1 | 1 |
| PDO  | [1]-[2]+[3]-[4] | 0 | -1 | 1 |
| CUO  | 0 | [1]-[2]+[3]-[4] | 1 | -1 |
| PUO  | [2]-[4] | [1]-[3] | -1 | -1 |

**Remark**

We recover in this table the two types of parity {(i) + (ii)} and (iii) mentioned above, and the fact that 2 "*In*" options are vanilla and 2 "*Out*" options are worth nothing (since there is no rebate).

### *14.2.1.2 Value of *Rebates*

As indicated above, one has to distinguish *rebate at expiry*, which is easy to value, from *rebate at hit*, which is tricky to calculate. The first concerns "*In*" options (and very rarely "*Out*" ones, so we do not consider that case) and the second is relevant to "*Out*" options only.

The initial value of a *rebate at expiry*, which is added to what we read in Table 14.1, is easy to compute: It is equal to the discounted *rebate R* ($R\ e^{-rT}$) times the probability that the option is not knocked in. The computation, which follows closely that that led to the results in Table 14.1 (see Appendix 5 to this chapter), yields the following result:

$$[5]\quad R(0) = Re^{-rT}\left[N\left(\eta x_1 - \eta\sigma\sqrt{T}\right) - \left(\frac{L}{S}\right)^{2(\varepsilon-1)}N\left(\eta y_1 - \eta\sigma\sqrt{T}\right)\right]$$

We note that the value of the *rebate* does not depend on the relative values of the barrier and the strike, since the latter plays no role here.

Furthermore, only the direction of activation of the barrier option is relevant ($\eta = 1$ or -1 according to whether the option is "Down" or "Up"), the nature of the option (call or put) being irrelevant (f does not appear in the equation). Results are summarized in Table 14.2.

The initial value of a *rebate at hit* is much more difficult since it is (possibly) received at an unknown future date, denoted by $\tau$ (in mathematical terms, $\tau$ is a stopping time). As a result, $R\ e^{-r\tau}$ is random and we must compute $R.E^Q[e^{-r\tau}\ 1_{\{\tau<T\}}]$. We show in mathematical Appendix 5 that, using the density of the first time ($\tau$) that $S(t)$ attains the barrier $L$, we end up with

$$[6]\quad R(0) = R\left[\left(\frac{L}{S}\right)^{\varepsilon-1+b}N(\eta y_2) + \left(\frac{L}{S}\right)^{\varepsilon-1-b}N\left(\eta y_2 - 2\eta b\sigma\sqrt{T}\right)\right]$$

where $y_2 = \frac{\ln(L/S)}{\sigma\sqrt{T}} + b\sigma\sqrt{T}$   and   $b = \frac{\sqrt{\mu^2+2r\sigma^2}}{\sigma^2}$.

## 14.2 Path-Dependent Options

**Table 14.2** Value of
*rebates*

|  | Formula | η |
|---|---|---|
| *CDI* and *PDI* | [5] | 1 |
| *CUI* and *PUI* | [5] | -1 |
| *CDO* and *PDO* "at hit" | [6] | 1 |
| *CUO* and *PUO* "at hit" | [6] | -1 |

The preceding remarks about the absence of a role for the strike and the type (call or put) of the option also hold here. Table 14.2 summarizes the results about the value of *rebates*.

To obtain the value of a barrier option with *rebate*, one adds up the results of Tables 14.1 and 14.2.

**Remarks**
- Values for the various Greek parameters can easily be obtained (although some of the formulas are big) from the exact results above. The behavior of some of them, notably if the underlying price is close to the barrier, is particularly interesting but out of the scope of this book and we refer the reader to more specialized publications.
- We can nonetheless remark that generally speaking an *"In"* option is more sensitive to an increase in the volatility than a standard option of the same type (its vega is larger) as long as it has not been switched on and thus has not become vanilla. This is because a volatility increase implies a larger probability that the option will hit the knock-in barrier (from above or below) and so will be activated.
- Symmetrically, an *"Out"* option that has not been knocked out is less sensitive to an increase in volatility than a standard option of the same type (its vega is smaller). In fact, a volatility increase certainly increases the probability that the option expires in the money, as for a vanilla option, but also increases the probability that it is knocked out. The second effect offsets the first and the vega can even become negative as the UA's price comes close to the barrier.
- Furthermore, hedging these options is an acute problem for the writer. Only in some cases is a static duplication of the barrier option with vanilla options possible. In all other cases, dynamic hedging cannot be avoided, which in practice cannot be perfect.

### 14.2.1.3 Other Barriers
Among the numerous other options that constitute the "barrier" family, we only mention the main ones.

- **"Partial barriers"**. Such an option offers a specific time interval $[T_1, T_2]$, with $T_1 \geq 0$ and $T_2 \leq T$, during which the conditions of knock-in or knock-out are in effect. Their other properties are those of simple barrier options. Valuing such options involves the bivariate normal law $N2\ (x;\ y;\ \rho)$ just as for options on options.

- **"Double barriers"**. The contract specifies two barriers $L_H$ (high) and $L_L$ (low) located on each side of the UA's initial value. An "*In*" option is triggered if the underlying price hits $L_H$ in rising or $L_L$ in falling. An "*Out*" option is killed if one of the limits is hit or crossed.

  There are numerous variants. For example, a double "*In*" barrier could require that $L_L$ then $L_H$ are crossed in that order (or the reverse) to become alive. A double "*Out*" option barrier can be triggered if the two limits are crossed, in any order, or in a specific order agreed upon initially.

  Another option is the *successive touch*: the option is at the same time *Up-and-In* for the upper limit and then *Down-and-Out* for the lower barrier.

  Valuing such options is complicated and generally requires resorting to numerical methods at some stage in the process.

- **"Contingent barrier"**. Here, knocking in the option or knocking it out depends on crossing the barrier (or the double barrier) by the price of an asset *different* from the UA. The UA, for example, is a foreign stock index and the other asset is an interest rate, an exchange rate or some other stock index.

- **"Soft barrier"**. In this case, the payoff is partly reduced from what it would be for a vanilla option, proportionally to the distance to the two limits of the soft barrier attained by the UA price. Consider, for example, a CDO, with strike 100 and soft barrier range [60; 80]. If the minimum "min" reached by the underlying is between 60 and 80, the payoff is equal to $\left[1 - \frac{80 - \min}{80 - 60}\right]\%$ of the call's final value. For a minimum of 70, for instance, the payoff will be 50% of that of a vanilla call of strike 100.

- **"Roll option"**. This option is a double barrier, the two limits $L_H$ and $L_L$ being on the *same side* of the UA's initial value. So both are of the same type *Up* or *Down*. The closest limit switches the option on, the furthest switches it off. For example, consider a *roll down* call: the buyer expects a fall in the UA's price followed by a rise (or at least not a further fall). If she is right, her option will be switched on but not switched off later on.

- **"Barrier with strike reset"**. This is a roll option for which the strike changes when the knock-in barrier is hit.

- **"Ratchet option"**. Such an option has a lifetime divided into sub-periods, usually of same length (typically 5 or 10 years divided into years). The first strike is known. If at the end of the first period the UA price is lower than the strike (for a call), the strike remains unchanged and the holder's gain is nil. If, on the contrary, the UA price is higher than the strike, the holder does book a gain but the strike is increased to the level of the first barrier $L_{H_1}$ specified in the contract. One continues in the same way at the end of each sub-period, with the possible gains accumulating. This option insures investors, who bet on an increase in the UA price (for a call, a fall for a put) and are temporarily right, against a trend reversal which would make them lose everything with a long-lived vanilla option.

- **"Parisian options"**. A disadvantage that could be serious for the holder of a simple knock-out option and for the writer of a simple knock-in barrier is that the option dies (respectively, is activated) even if the underlying price hits the barrier $L$ only once and then gets away from it ever after. To mitigate this disadvantage, a Parisian option stipulates that the option is not switched on (*In*) or switched off

## 14.2  Path-Dependent Options

(*Out*) unless the underlying price remains on the same side of $L$, after crossing it, for a *consecutive period of time* called a "window" and denoted by $D$ (for duration). Consider the case of a CUI with expiry in one year, with $D = 2$ months. If, during the relevant year, the UA price remains above $L$ for any two consecutive months, the call is knocked in. If not, the call is not activated, even though the *total* amount of time the UA price spent above $L$ is, say, 8 months (there were too many oscillations about $L$).

Such an option can be useful if the buyer wishes to avoid price manipulations (intended to push the price beyond the barrier $L$ to kill an "*Out*" option, for example) during the last trading days before expiration, or a sudden but temporary increase in the UA's price because of an unsubstantiated rumor of a tender offer, or a large but temporary change in a currency exchange rate due for instance to central bank(s)' intervention, and so on.

Quasi-explicit formulas for the value of such options exist (see, e.g., Chesney et al. 1997) but they do require the use of numerical methods to calculate complicated integrals. However, there exist two types of parity, along the lines of those for simple barriers:

First, a portfolio of a Parisian DI and a Parisian DO is a vanilla option.

Second, the "inverses" parity continues to hold:

$$PDO_{Par} \ (S, K, L, D, T, r, \delta) = S.K.CUO_{Par} \ (1/S, 1/K, 1/L, D, T, \delta, r).$$

- **"Cumulative options"**. These options are variants of the preceding ones. The difference is that the time $D$ spent above or below the barrier $L$ is not counted consecutively but *cumulatively*: the duration counter is not reset to zero when the underlying price leaves the relevant range. A cumulative option "*In*" is, therefore, more expensive than its Parisian equivalent, and a cumulative "*Out*" is less expensive. Hugonnier (1999) proposed a general method for valuing options defined as a function of the time spent below or above a given barrier $L$, among which we find Parisian and cumulative options.

### 14.2.2  Digital Barriers

Like digital options, digital (or binary) barriers (hereafter DBs) can be *asset-or-nothing* or *cash-or-nothing options*. To avoid pedantic repetition or over-long lists, and because they are by far those most traded on the market, *we will only analyze true digitals with a binary payoff* (1 or 0).[7] Like a barrier option, a digital barrier is *path-dependent*. We will identify digitals with a subscript $d$.

---

[7] Whatever will be said about *cash-or-nothing* options also holds, *mutatis mutandis*, for *asset-or-nothing options*. For more details, see Rubinstein and Reiner (1991) who uncover 28 types of digital barriers and provide 44 formulas for valuing them (in some cases, the formula depends on whether $L$ *is larger or smaller than $K$*).

A first classification distinguishes the DB whose final payoff is received *at hit* (so these must be DB "*In*") from those whose final payoff is received *at expiry* (DB "*In*" or "*Out*"). The first ones are *Down-and-In* or *Up-and-In* and their payoff writes:

$DI_d$ *at hit* : 1 (on date t) if there exists $t \leq T$ such that $S(t) \leq L$; 0 otherwise.

$UI_d$ *at hit* : 1 (on date t) if there exists $t \leq T$ such that $S(t) \geq L$; 0 otherwise.

The other DB, whose possible payoff is received at expiry $T$, are of types DI, UI, DO or UO. A natural second classification then distinguishes the DB that depend only on the barrier $L$ from those that depend on both $L$ and strike $K$ (we must then differentiate calls from puts).

The respective payoffs of four DB involving only the barrier $L$ can be written (here we omit the phrasing "*at expiry*" for simplicity):

$$DI_d : 1 \quad (\text{at } T) \text{ if } \exists \ t \leq T \text{ such that } S(t) \leq L; \ 0 \text{ otherwise}$$

$$UI_d : 1 \quad (\text{at } T) \text{ if } \exists \ t \leq T \text{ such that } S(t) \geq L; \ 0 \text{ otherwise}$$

$$DO_d : 1 \quad (\text{at } T) \text{ if, for all } t \leq T, \ S(t) > L; \ 0 \text{ otherwise}$$

$$UO_d : 1 \quad (\text{at } T) \text{ if, for all } t \leq T, \ S(t) < L; \ 0 \text{ otherwise.}$$

These are the four DB most frequently traded.

The respective payoffs of the four "*In*" DB (2 calls and 2 puts) that involve $K$ in addition to $L$ write:

$$CDI_d : 1 \ (\text{at } T) \text{ if } S(T) > K \text{ and } \exists \ t \leq T \text{ such that } S(t) \leq L; \ 0 \text{ otherwise.}$$

$$PDI_d : 1 \ (\text{at } T) \text{ if } S(T) < K \text{ and } \exists \ t \leq T \text{ such that } S(t) \leq L; \ 0 \text{ otherwise}$$

$$CUI_d : 1 \ (\text{at } T) \text{ if } S(T) > K \text{ and } \exists \ t \leq T \text{ such that } S(t) \geq L; \ 0 \text{ otherwise}$$

$$PUI_d : 1 \ (\text{at } T) \text{ if } S(T) < K \text{ and } \exists \ t \leq T \text{ such that } S(t) \geq L; \ 0 \text{ otherwise.}$$

The respective payoffs of the four "*Out*" DB involving $L$ and $K$ are:

$$CDO_d : 1 \quad (\text{at } T) \text{ if } S(T) > K \text{ and, for all } t \leq T, \ S(t) > L; \ 0 \text{ otherwise}$$

$$PDO_d : 1 \quad (\text{at } T) \text{ if } S(T) < K \text{ and, for all } t \leq T, \ S(t) > L; \ 0 \text{ otherwise}$$

$$CUO_d : 1 \quad (\text{at } T) \text{ if } S(T) > K \text{ and, for all } t \leq T, \ S(t) < L; \ 0 \text{ otherwise}$$

$$PUO_d : 1 \quad (\text{at } T) \text{ if } S(T) < K \text{ and, for all } t \leq T, \ S(t) < L; \ 0 \text{ otherwise.}$$

There exists parity relations between these DB with the same barrier, maturity date and, if relevant, strike. For example, it is easy to show that on date $t = 0$:

## 14.2 Path-Dependent Options

$$DI_d + DO_d \;=\; UI_d + UO_d \;=\; e^{-rT} \qquad (14.19)$$

since the sum of the first two options brings \$1 with certainly upon expiry, just like the sum of the second two. In addition, we have the equations:

$$CDI_d + CUI_d = C_d = CDO_d + CUO_d \;\; ; \;\; PDI_d + PUI_d = P_d$$
$$= PDO_d + PUO_d \qquad (14.20)$$

where we recall that $C_d$ and $P_d$ denote, respectively, the digital call and put .

Finally, we have

$$
\begin{cases}
UOd = PUO(K = L+1;L) - PUO(K = L;L) \\
DOd = CDO\ (K = L-1;L) - CDO\ (K = L;L) \\
DId = PDI(K = L+1;L) - PDI(K = L;L) \\
UId = CUI(K = L-1;L) - CUI(K = L;L)
\end{cases}
\qquad (14.21)
$$

We will only prove the first of these formulas, since the others result from similar analyses. The DB $UO_d$ is worth 0 upon expiry if the UA's value hits the barrier $L$. The two relevant PUO, with the same barrier $L$, are knocked out in such a case and so are also worth 0. If the underlying value stays below $L$, the DB $UO_d$ is worth 1 upon expiry. In that case, the two puts are exercised upon expiry (since, necessarily, $S(T)$ $< L$) and the holder gains $\{L + 1 - S(T)\} - \{L - S(T)\} = 1$.

Finally, we give formulas for valuing, on date $t = 0$, the four DB for which the strike is not involved:

$$
\begin{cases}
DI_d = e^{-rT}\left[N\left(-x_1 + \sigma\sqrt{T}\right) + \left(\frac{L}{S}\right)^{2(\varepsilon-1)} N\left(y_1 - \sigma\sqrt{T}\right)\right] \\[2ex]
UI_d = e^{-rT}\left[N\left(x_1 - \sigma\sqrt{T}\right) + \left(\frac{L}{S}\right)^{2(\varepsilon-1)} N\left(-y_1 + \sigma\sqrt{T}\right)\right] \\[2ex]
DO_d = e^{-rT}\left[N\left(x_1 - \sigma\sqrt{T}\right) - \left(\frac{L}{S}\right)^{2(\varepsilon-1)} N\left(y_1 - \sigma\sqrt{T}\right)\right] \\[2ex]
UO_d = e^{-rT}\left[N\left(-x_1 + \sigma\sqrt{T}\right) - \left(\frac{L}{S}\right)^{2(\varepsilon-1)} N\left(-y_1 + \sigma\sqrt{T}\right)\right]
\end{cases}
\qquad (14.22)
$$

where $x_1$, $y_1$, and $\varepsilon$ are defined as in Sect. 14.2.1.1. The proof of these formulas is analogous to that given in the mathematical Appendix 4 to this chapter.

The DB, and mainly the four valued in (14.22), are important because they form the "building blocks" used in designing more complex instruments and tailor-made structured products.

## 14.2.3 Lookback Options (*)

A *lookback* option is a typical *path-dependent option*. A *lookback call* is a European option whose strike, unknown at inception, is the minimum UA price, $m_{T_0}^T$, observed between two dates $T_0$ (usually the option's creation date) and $T$, *its expiry date*. The payoff at $T$, therefore, writes: $Max\left[0, S_T - m_{T_0}^T\right]$, or simply $\left(S_T - m_{T_0}^T\right)$, since obviously $S_T \geq m_{T_0}^T$.

The holder thus is ensured to pay for the underlying the minimum price observed during the contract period, whence the frequent phrasing *"no regret call"*. The flip side is obviously that such an option is expensive.

Similarly, a *lookback* put is a claim on the payoff $\left(M_{T_0}^T - S_T\right)$ at $T$, where $M_{T_0}^T$ is the maximum UA price observed between $T_0$ and $T$.

We provide only a proof for the call, since that for the put is similar. The current date being by convention $t = 0$, we have in general $T_0 < 0$, except on the day the option is created ($T_0 = 0$). We use the following additional notation:

$$\mu \equiv r - \delta - \sigma^2/2; \quad \mu' \equiv r - \delta + \sigma^2/2; \quad \lambda \equiv \frac{\sigma^2}{2(r-\delta)}; \quad b \equiv \ln\left(S_0/m_{T_0}^0\right),$$

where $m_{T_0}^0$ is the smallest UA price (already) observed between $T_0$ and 0. Thus we have $b \geq 0$.

We must value, in the BSM world, the payoff $\left(S_T - m_{T_0}^T\right)$ received at $T$. On $t = 0$, and under the risk-neutral probability $Q$, we have

$$CLB_0 = e^{-rT}\left[E(S_T) - E\left[m_{T_0}^0 1_{m_0^T > m_{T_0}^0}\right] - E\left[m_0^T 1_{m_0^T < m_{T_0}^0}\right]\right]$$

$$CLB_0 = e^{-\delta T}S_0 - e^{-rT}m_{T_0}^0 E\left[1_{m_0^T > m_{T_0}^0}\right] - e^{-rT}E\left[m_0^T 1_{m_0^T < m_{T_0}^0}\right] \tag{14.23}$$

since there is a positive probability that the current minimum ($m_{T_0}^0$) remains the minimum at $T$ *and* a positive probability that a new minimum appears between 0 and $T\left(m_0^T\right)$.

The first expectation in (14.23) can be computed using Lemma (14.L4) (*see mathematical* Appendix 2) by noting that $y = -b = \ln\left(m_{T_0}^0/S_0\right) \leq 0$:

$$E\left[1_{m_0^T > m_{T_0}^0}\right] = N\left[\frac{b + \mu T}{\sigma\sqrt{T}}\right] - e^{-2\mu b\sigma^{-2}}N\left[\frac{-b + \mu T}{\sigma\sqrt{T}}\right] \tag{14.24}$$

The second expectation is more difficult to calculate and notably requires knowing the density of $m_0^T$. The computation, carried out in mathematical Appendix 6, leads to the following expression for the value of a *lookback call*:

## 14.2 Path-Dependent Options

$$CLB_0 = e^{-\delta T} S_0 N\left(\frac{b + \mu' T}{\sigma\sqrt{T}}\right)$$

$$- e^{-rT} m_{T_0}^0 \left[ N\left(\frac{b + \mu T}{\sigma\sqrt{T}}\right) - \frac{\lambda S_0}{m_{T_0}^0}\left(\frac{m_{T_0}^0}{S_0}\right)^{1/\lambda} N\left(\frac{-b + \mu T}{\sigma\sqrt{T}}\right) \right]$$

$$- e^{-\delta T} S_0 \lambda N\left(-\frac{b + \mu' T}{\sigma\sqrt{T}}\right) \tag{14.25}$$

Similarly, using the density of the maximum $M_0^T$, the value of a *lookback* put is given by:

$$PLB_0 = -e^{-\delta T} S_0 N\left(-\frac{d + \mu' T}{\sigma\sqrt{T}}\right)$$

$$+ e^{-rT} M_{T_0}^0 \left[ N\left(-\frac{d + \mu T}{\sigma\sqrt{T}}\right) - \frac{\lambda S_0}{M_{T_0}^0}\left(\frac{M_{T_0}^0}{S_0}\right)^{1/\lambda} N\left(\frac{d - \mu T}{\sigma\sqrt{T}}\right) \right]$$

$$+ e^{-\delta T} S_0 \lambda N\left(\frac{d + \mu' T}{\sigma\sqrt{T}}\right) \tag{14.26}$$

where $d \equiv \ln\left(S_0/M_{T_0}^0\right)$ and $M_{T_0}^0$ is the maximum UA price observed between $T_0$ and 0.

### Remarks

- The *initial value of a* lookback option is obtained by replacing in formulas (14.25) or (14.26) $m_{T_0}^0$ or $M_{T_0}^0$ by $S_0$ and therefore $b$ or $d$ by 0.
- Also traded are *partial* lookback options for which the strike is $\alpha m_{T_0}^0$ (with $\alpha > 1$) for the call and $\beta M_{T_0}^0$ (with $\beta < 1$) for the put, which makes them cheaper than the classic lookbacks. However, they are no longer *no regret* since they may expire worthless, which is all the more probable because $\alpha$ is larger and $\beta$ is smaller.
- Table 14.3 gives an idea of the relative orders of magnitude of lookback values. We have assumed $S_0 = K = 100$, $\sigma = 20\%$, $r = 10\%$, $\delta = 0\%$, $T_0 = 0$ and $\alpha = \beta = 1$.
- The lookback values in Table 14.3 are calculated with the exact formulas (14.25) and (14.26) obtained for continuous time. Actual market prices are, under the

**Table 14.3** Comparative values of lookback options

|  | Expiry | |
|---|---|---|
|  | 3 months | 6 months |
| Vanilla European call | 5.30 | 8.28 |
| Lookback call | 8.95 | 13.19 |
| Vanilla European put | 2.83 | 3.40 |
| Lookback put | 6.98 | 9.29 |

same conditions, strictly smaller than these theoretical values, as the frequency with which the UA price is observed is not infinite, nor even very large: For most contracts, the frequency is not daily, but weekly or even monthly for longer options. In all cases, it is usually the closing prices on the relevant days that are considered. The minimum (maximum) of the UA price effectively observed thus is strictly larger (smaller) than its theoretical continuous time value. For example, a lookback worth 17.4 with "continuous" observations is worth no more than 16.5 with daily observations and 15.1 for monthly monitoring. Valuation in discrete time requires an improved version of the Cox-Ross-Rubinstein binomial model in which the time step is taken equal to the time step used for monitoring UA price observations (it is therefore, $n$, the number of periods, that is adapted). The improvement comes from the fact that to explore all the possible paths in the CRR tree to determine each trajectory's minimum (or maximum) value is much too costly in computation time. The trick is to use as the numeraire the UA price itself (which is thus worth 1 on $t = 0$) and to construct the tree for the variable $Y(t) \equiv Max(t)/S(t)$ for the put or $Y(t) \equiv Min(t)/S(t)$ for the call.[8] This change in the numeraire in discrete time is analogous to the change in numeraire used in continuous time (see Sect. 19.5).

### 14.2.4 Options on Averages (Asians)

Options on averages were created on the foreign exchange market in Tokyo, whence their name Asians. Two types exist: either the UA price is an average and the strike is fixed, or it is the strike that is an average. The respective payoffs are therefore:

$$Max[0, M(T_1, T_2) - K] \quad \text{(call)} \quad \text{or} \quad Max[0, K - M(T_1, T_2)] \text{ (put)}$$

$$Max[0, S(T) - M(T_1, T_2)] \quad \text{(call)} \quad \text{or} \quad Max[0, M(T_1, T_2) - S(T)] \text{ (put)}$$

where $M(T_1, T_2)$ is the average (mean) price of the underlying between $T_1$ ($\geq 0$) and $T_2$ ($T_2 \leq T$, the expiry date of the option).

Such options are cheaper than their vanilla analogues, since the UA average price is by construction less volatile than the UA price. In addition, for an agent who carries out many transactions on the underlying (e.g., a currency in the case of an importer or exporter) the average price is more relevant than the spot price at any given date. These two reasons contribute to explaining their popularity.

Valuing these options faces two problems. The first is the discrete, non-continuous, nature of the UA prices observed. The second concerns the computation of the average. Closed-from solutions à la BSM exist only in the case, very rare in practice but helpful as a benchmark, of a *geometric average* (continuous or

---

[8] The relative value of the payoff of the lookback put, for example, is: $PLB(T)/S\,T) = (Max\,S(T) - S(T))/S(T) = Y(T) - 1$ ($\geq 0$), which allows dealing with a single process, $Y(t)$, just as in the binomial model.

## 14.2 Path-Dependent Options

discrete). This is because the geometric mean of a geometric Brownian motion is itself a geometric Brownian motion, which is not the case for the arithmetic mean.

Unfortunately, the average used for these options is almost always arithmetic. Therefore, we need to make an approximation which will be discussed later on. Since in most cases one uses all or part of the closed-form solution obtained for geometric averages, we provide the solution for some of these, first in continuous then in discrete time.[9]

### 14.2.4.1 Options on a Geometric Average Price

We first study the case where the average is computed as if monitoring the UA price was continuous, then when it is computed from discrete price observations.

*(a) **Continuous average***

Suppose that the risk-neutral dynamics of the UA price obeys the stochastic differential equation

$$\frac{dS}{S} = b\,dt + \sigma\,dW_t$$

where $b \equiv (r - \delta)$.

Let the current time be $t = 0$. The geometric average is computed between $T_1$ ($\geq 0$) *and* $T_2 = T$. Since the calculation is continuous, we have

$$M(T_1, T) = \exp\left[\frac{1}{T - T_1} \int_{T_1}^{T} \ln S(t)dt\right].$$

The computations carried out in Appendix 7 yield the following results:

- The value at date 0 of a *call on an average price* with payoff $[M(T_1, T) - K]^+$ is equal to

$$CAP(0) = e^{-\delta T} Z(0) N(d_1) - Ke^{-rT} N(d_2) \tag{14.27}$$

where

---

[9]There are numerous possible cases, since the average may be computed between any two dates $T_1$ and $T_2$, the first not necessarily coinciding with the date of writing the option and the second possibly being before the option expiry date. In addition, one can value the option before or after $T_1$, the starting date for computing the average. Thus, on the option writing date, we have $T_1 \geq 0$ but during the option lifetime we have $T_1 < 0$. In the second case, the solution involves the average already observed. We limit the analysis to the case where $T_2 = T$, by far the most popular. Besides, we derive explicitly the case $T_1 \geq 0$ but provide directly the solution for the case $T_1 < 0$.

$$d_1 = \frac{\ln(Z(0)/K) + (b + \Sigma^2/2)T}{\Sigma\sqrt{T}}, \quad d_2 = d_1 - \Sigma\sqrt{T}$$

$$\Sigma = \sigma\sqrt{\frac{T + 2T_1}{3T}} \quad \text{and} \quad Z(0) = S(0)\exp\left[-b\frac{T - T_1}{2} - \sigma^2\frac{T - T_1}{12}\right].$$

- The value on date 0 of a *put on an average price* with payoff $[K - M(T_1,T)]^+$ is equal to

$$PAP(0) = Ke^{-rT}N(-d_2) - e^{-\delta T}Z(0)N(-d_1). \qquad (14.28)$$

**Remarks**
- The procedure amounts to adjusting the drift and the volatility of the underlying price and applying BSM.
- The classical call-put parity holds: $CAP(0) - PAP(0) = e^{-\delta T}Z(0) - Ke^{-rT}$.
- If the mean's computation begins at the option's writing ($T_1 = 0$), the formula is simpler as $\Sigma$ is equal to $\sigma/\sqrt{3}$.
- If $T_1 < 0$, *the computation of the average having started before the valuation date* $(t = 0)$, the call price (the put price can be derived from the call-put parity) is equal to

$$CAP(0, T_1 < 0) = e^{-\delta T}\widehat{Z}(0)N(d_1) - Ke^{-rT}N(d_2) \qquad (14.29)$$

where:

$$d_1 = \frac{\ln\left(\widehat{Z}(0)/K\right) + \left(b + \widehat{\Sigma}^2/2\right)T}{\widehat{\Sigma}\sqrt{T}}, \quad d_2 = d_1 - \widehat{\Sigma}\sqrt{T}$$

$$\widehat{\Sigma} = \frac{\sigma}{\sqrt{3}}\frac{T}{T - T_1}$$

$$\widehat{Z}(0) = M_{T_1,0}^{-T_1/(T-T_1)}S(0)^{T/(T-T_1)}\exp\left[b\left(\frac{T^2}{2(T - T_1)} - T\right) + \frac{\sigma^2 T^2}{(T - T_1)^2}\frac{3T_1 - T}{12}\right]$$

and $M_{T_1,0}$ is the average between $T_1$ ($< 0$) and $0$.[10]

---

[10] The average $M(T_1,T)$ writes:

$$\exp\left[\frac{1}{T - T_1}\left\{\int_{T_1}^0 \ln S(t)\,dt + \int_0^T \ln S(t)dt\right\}\right] = \widehat{M}_{T_1,0}\exp\left[\frac{1}{T - T_1}\int_0^T \ln S(t)\,dt\right]$$

## 14.2 Path-Dependent Options

- Formula (14.29) is obviously identical to (14.27) if one sets $T_1 = 0$ in each formula.
- We have adopted the standard case $T_2 = T$. If $T_2 < T$, independently of whether $T_1$ is equal to or less than 0, it suffices to notice that the option payoff $[M(T_1, T_2) - K]^+$ is known at $T_2$, although paid at $T$. The option value thus is found by replacing $T$ by $T_2$ in formulas (14.27) to (14.29) and multiplying the result by $e^{-r(T-T_2)}$.

**(b) *Discrete Average***

In practice, the average is computed from discrete observations. The contract stipulates $n$ observation dates for the UA price: $\{iT/n\}_{i=1,\ldots,n}$. Under the risk-neutral probability, the UA price on date $iT/n$ is equal to

$$S_{\frac{iT}{n}} = S(0) \exp \left[ \left( b - \frac{\sigma^2}{2} \right) \frac{iT}{n} + \sigma W_{\frac{iT}{n}} \right]$$

The discrete geometric mean, computed between 0 (the date of writing the option) and $T$ (its expiration), is

$$M(0, T, n) = \left( \prod_{i=1}^n S_{\frac{iT}{n}} \right)^{1/n} = S(0) \exp \left[ \left( b - \frac{\sigma^2}{2} \right) \frac{(n+1)T}{2n} + \frac{\sigma}{n} \sum_{i=1}^n W_{\frac{iT}{n}} \right].$$

Remarking that $\sum_{i=1}^n W_{\frac{iT}{n}} = \sqrt{\frac{T}{n}} \sum_{i=1}^n W_i$ (the scaling property of the Brownian motion), that $\sum_{i=1}^n W_i = \sum_{i=1}^n W_i(n + 1 - i)(W_i - W_{i-1})$, and hence that the variance of $\sum_{i=1}^n W_i$ is equal to $\sum_{i=1}^n (n + 1 - i)^2 = [n(n + 1)(2n + 1)]/6$, we obtain that $\frac{\sigma}{n} \times \sum_{i=1}^n W_{\frac{iT}{n}}$ is a centered Gaussian with variance $\sigma^2 T(n+1)(2n+1)/6n^2$.

The calculations follow directly from BSM and the value of a *call on a discrete geometric average price* (CDAP) writes, as of $t = 0$:

$$CDAP(0) = ZD(0) N(d_1) - Ke^{-rT} N(d_2) \tag{14.30}$$

---

where $\widehat{M}_{T_{1,0}} \equiv \exp \left[ \frac{1}{T-T_1} \int_{T_1}^0 \ln S(t) \, dt \right]$ is known at $t = 0$. In addition, we have:

$$\widehat{M}_{T_{1,0}} = \exp \left[ \left( \frac{1}{-T_0} \int_{T_1}^0 \ln S(t) \, dt \right) \frac{-T_1}{T - T_1} \right] = M_{T_{1,0}}^{-T_1/(T-T_1)} \quad \text{where } M_{T_{1,0}}$$

$$\equiv \exp \left[ \frac{1}{-T_0} \int_{T_1}^0 \ln S(t) \, dt \right].$$

$$\text{with } ZD(0) = S(0) \exp\left[\left(b - \frac{\sigma^2}{2}\right)\frac{(n+1)T}{2n}\right] \exp\left[\frac{\Sigma^2 T}{2} - rT\right]$$

$$\Sigma^2 = \frac{\sigma^2(n+1)(2n+1)}{6n^2}$$

$$d_1 = \frac{1}{\Sigma\sqrt{T}} \ln\left(S(0) \exp\left[\left(b - \frac{\sigma^2}{2}\right)\frac{(n+1)}{2n}T\right]/K\right) + \Sigma\sqrt{T}$$

$$d_2 = d_1 - \Sigma\sqrt{T}.$$

The value for a *put on a discrete geometric average price* (PDAP) is obtained from call-put parity:

$$CDAP(0) - PDAP(0) = ZD(0) - Ke^{-rT}. \tag{14.31}$$

If the computation of the mean has *already started* at the time of valuation, and $\tau$ observations have already been made, the discrete geometric mean can be split into two terms the first of which (denoted by $\widehat{M}(0, \tau, n)$) is known:

$$M(0, T, n) = \left(\prod_{i=1}^{\tau} S_{\frac{iT}{n}}\right)^{1/n} \left(\prod_{i=\tau+1}^{n} S_{\frac{iT}{n}}\right)^{1/n} = \widehat{M}(0, \tau, n)\left(\prod_{i=\tau+1}^{n} S_{\frac{iT}{n}}\right)^{1/n}.$$

Following the same lines as above, the value of the call on date $t$ corresponding to the $t^{\text{th}}$ observation ($t = \tau T/n$) obtains:

$$CDAP(\tau) = \widehat{ZD}(\tau)N(d_1) - Ke^{-rT}N(d_2) \tag{14.32}$$

with $\widehat{ZD}(\tau) =$

$$\widehat{M}(0, \tau, n)S(t)^{\frac{n-\tau}{n}}\exp\left[(b - \frac{\sigma^2}{2})\frac{(n-\tau)(n+\tau+1)T}{2n^2}\right] \times \exp\left[\frac{\widehat{\Sigma}^2 T}{2} - rT\right]$$

$$\widehat{\Sigma}^2 = \frac{\sigma^2(n-\tau)(n-\tau+1)(2n-2\tau+1)}{6n^3} + \frac{\sigma^2(n-\tau)^2\tau}{n^3}$$

$$d_1 = \frac{1}{\widehat{\Sigma}\sqrt{T}}\ln(\widehat{M}(0, \tau, n)S(t)^{\frac{n-\tau}{n}}\exp\left[(b - \frac{\sigma^2}{2})\frac{(n-\tau)(n+\tau+1)T}{2n^2}\right]/K)$$

$$d_2 = d_1 - \widehat{\Sigma}\sqrt{T}.$$

The put value is obtained from the appropriate call-put parity.

## 14.2 Path-Dependent Options

583

### 14.2.4.2 Options with a Geometric Average Strike

For these options, it is the strike which is an average of the prices observed for UA. Valuing them is more complicated because two random variables are involved, $S(T)$ and $M(T_1, T)$. As BSM cannot be applied directly, we must use the change of numeraire technique. We provide here the results and a discussion, the main proof being relegated to the mathematical Appendix 8.

*(a) *Continuous average*

The values of options with a geometric average strike are given by the following formulas:

- The value of a *call with a geometric average strike with* payoff $[S(T) - M(T_1, T)]^+$ equals

$$CAS(0) = e^{-\delta T} S(0) \left[ N(d'_1) - Z'(0)N(d'_2) \right] \tag{14.33}$$

where:

$$d'_1 = \frac{\ln (1/Z'(0)) + \frac{1}{2}\Sigma'^2 T}{\Sigma'\sqrt{T}} \quad \text{and} \quad d'_2 = d'_1 - \Sigma'\sqrt{T},$$

$$\Sigma' = \sigma\sqrt{\frac{T - T_1}{3T}} \quad \text{and} \quad Z'(0) = \exp\left[\frac{1}{2}\left(b + \frac{\sigma^2}{2}\right)(T_1 - T) + \frac{1}{2}\Sigma'^2 T\right].$$

- The value of a *put with a geometric average strike* $M(T_1, T)$ is, similarly, equal to

$$PAS(0) = e^{-\delta T} S(0) \left[ Z'(0)N(-d'_2) - N(-d'_1) \right] \tag{14.34}$$

The proof relies on choosing the probability measure associated with the UA price itself as numeraire, which leads to martingale (relative to this numeraire) asset prices.

### Remarks
- If $T_1 = 0$, the computation of the average starting at the valuation date (e.g. on the date of writing the option), the formula is simpler since then $\Sigma' = \sigma\sqrt{3}$ and $Z'(0) = \exp\left[-\frac{bT}{2} - \frac{\sigma^2 T}{12}\right]$. Note also that then $\Sigma' = \Sigma$ and $S(0)Z'(0) = Z(0)$, which leads to formulas similar to those for options on average prices.
- If $T_1 < 0$, the computation of the mean *having already started* at valuation time, the value of the call (the put resulting again from a call-put parity) writes

$$CAS(0, T_1 < 0) = e^{-\delta T} S(0) \left[ N(d_1') - \hat{Z}'(0) N(d_2') \right] \qquad (14.35)$$

where

$$\hat{Z}'(0) \equiv M_{T_1,0}^{-T_1/(T-T_1)} \; S(0)^{T/(T-T_1)} \; \exp\; [\tfrac{1}{2}(b + \tfrac{\sigma^2}{2})(\tfrac{T^2}{T-T_1} - 2T) + \tfrac{1}{2} \hat{\Sigma}' \,^2 T],$$

$$\hat{\Sigma}' \equiv \frac{\sigma}{\sqrt{3T}} \sqrt{\frac{T_1^3}{(T-T_1)^2} + T - T_1},$$

$M_{T1,0}$ is the average observed between $T_1$ and 0, and $d_1'$ and $d_2'$ are defined as in (14.33) with $\hat{Z}\,(0)$ and $\hat{\Sigma}$ instead of $Z'(0)$ and $\Sigma'$, respectively.

### **(b) Discrete Average

Using the same conventions as for options on average prices, the valuation formula for the call, in the case $T_1 \geq 0$:

$$CDAS(0) = e^{-\delta T} S_0 [N(d_1) - ZD(0) N(d_2)] \qquad (14.36)$$

where:

$$ZD(0) = \exp \left[ -\frac{b}{2} \frac{(n-1)}{n} T - \frac{\sigma^2}{12} \frac{(n-1)(n+1)}{n^2} T \right]$$

$$\Sigma^2 = \sigma^2 \frac{(n-1)(2n-1)}{3n^2}$$

$$d_1 = \frac{\ln[1/ZD(0)]}{\Sigma\sqrt{T}} + \frac{1}{2}\Sigma\sqrt{T}$$

$$d_2 = d_1 - \Sigma\sqrt{T}.$$

The put value is obtained from the appropriate call-put parity.

### 14.2.4.3 Options on Arithmetic Means

In practice, contracts use arithmetic means to compute the option payoff. According to whether time is continuous or discrete, the average writes, for an arbitrary date $T_1$ ($< T$):

## 14.2 Path-Dependent Options

$$M(T_1 - T) = \frac{1}{T - T_1} \int_{T_1}^{T} S(t)dt \quad \text{or}$$

$$M(T_1 - T) = \frac{1}{n} \sum_{i=1}^{n} S\left(T_1 + \frac{i(T - T_1)}{n}\right).$$

As mentioned in the introduction to Sect. 14.2.4, an arithmetic mean of a geometric Brownian motion is not a geometric Brownian motion. This is why there is no closed-form solution for the Asian options that are actually traded on the market. Two main approaches then are used by traders.[11] The first accepts an *approximation* to find closed-form formulas similar to those obtained for geometric means. For example, the arithmetic mean, whose expectation is easily computed, is assumed to have the same variance as the geometric mean, which is of course erroneous. Another solution is to postulate that the sum of log-normal laws is log-normal, characterized by *ad hoc* expectation and variance, even though the law of such a sum is in fact unknown.

The second approach is Monte Carlo *simulation,* enhanced by using a *control variate,* a technique discussed in Sect. 26.4. The idea is the following.

Since an exact solution is known for an option on a geometric mean, instead of directly simulating the value of an option on an arithmetic mean, one simulates the *value gap* between the two options. One adds this gap to the known exact solution for the geometric mean ($OG$) to obtain the desired arithmetic mean ($OA$). Formally, we estimate by simulation

$$h(0) = e^{-rT} E[OA(T) - OG(T)],$$

then compute

$$OA(0) = OG(0) + h(0),$$

instead of simulating

$$OA(0) = e^{-rT} E[OA(T)].$$

Intuitively, the estimation error on a price difference is smaller than that on a price.

The advantage of this Monte Carlo technique over the preceding approximation is that it is well adapted to the problem, since averages are actually discrete. The adopted Monte Carlo simulation naturally uses a time step equal to the interval between observations specified in the contract defining the option.

---

[11] A third solution, which is exact but goes beyond the technical level intended for this book, is due to Geman and Yor (1993). It is not analytical and requires the use of numerical methods (e.g., to invert a Laplace transform). The procedure consists in transforming the geometric Brownian motion into a time-changed squared Bessel process.

586                      14   *Exotic Options

## 14.2.5 Chooser Options (*)

A *chooser* option allows its holder to choose, any time from the present (0) to the maturity $t$, between a call with strike $K_c$ and expiration date $T_c$ ($> t$) and a put with strike $K_p$ and expiration date $T_p$ ($> t$). This option is suitable when the buyer bets that there will be a large variation in the UA price between now and date $t$ without suspecting in what direction it will vary. Sometimes this option is called commercially "*as you like it.*" Its payoff on date $t$ is therefore

$$CH_t = \text{Max}\left[C_t\left(K_c, T_c\right), P_t\left(K_p, T_p\right)\right].$$

As will be seen later, valuing this exotic option involves a bivariate normal law $N2(.)$ as the joint distribution of the UA price at $t$ and at $T_c$ (or $T_p$) is needed. However, there is a simpler, but still interesting, case of chooser that only requires knowing the BSM formula: In the case $K_c = K_p = K$ and $T_c = T_p = T$, we have on date $t$

$$CH_t\left(t, K, T\right) = \text{Max}\left[C_t\left(K, T\right), P_t\left(K, T\right)\right]$$

$$= \text{Max}\left[C_t\left(.\right), C_t\left(.\right) - S_t + Ke^{-r(T-t)}\right], \text{from the call} - \text{put parity,}$$

$$= C_t\left(.\right) + \text{Max}\left[0, -S_t + Ke^{-r(T-t)}\right]$$

$$= C_t\left(K, T\right) + P_t\left(Ke^{-r(T-t)}, t\right)$$

that is the sum of the values on date $t$ of a call with expiry $T$ and strike $K$ and a put with expiry $t$ and strike $Ke^{-r(T-t)}$.

Therefore, on date $t = 0$, this *simple chooser is worth*

$$CH_0(t, K, T) = C_0(S_0, K, T) + P_0\left(S_0, Ke^{-r(T-t)}, t\right). \tag{14.37}$$

**Remarks**

- If we have $t = T$, then the buyer will only decide at the last moment, at $T$, and the chooser has exactly the same value as a *straddle* composed of a call and a put with same expiries and strikes, only one leg of which will *expire* in the money.
- We do remark from formula (14.37) that, if $t$ tends to $T$, the chooser's value tends to that of the straddle. The straddle's value, $C_0(S_0, K, T) + P_0(S_0, K, T)$, then is *an upper bound* for the chooser's price. The chooser is (doubly) cheaper than the straddle because the put included in (14.37) has a shorter maturity ($t < T$) *and a* lower strike ($Ke^{-r(T-t)} < K$) than the put included in the straddle.
- We could have used the call-put parity in the opposite way to substitute the value of a put with expiry $T$ for that of the call (rather than the other way around) to obtain

## 14.3 Summary

$$CH_0(t, K, T) = P_0(S_0, K, T) + C_0\left(S_0, Ke^{-r(T-t)}, t\right).$$

*In the general case, i.e. with $K_c \neq K_p$ and/or $T_c \neq T_p$, we still use the BSM framework but have to solve the equation

$$CH_0 = E_Q[e^{-rt}Max(C_t(\bullet), P_t(\bullet))]$$
$$= e^{-rt}\left[\int_{C_t \geq P_t} C_t(\bullet)\frac{e^{-u^2/2}}{\sqrt{2\pi}}\,du + \int_{C_t \leq P_t} P_t(\bullet)\frac{e^{-u^2/2}}{\sqrt{2\pi}}\,du\right]$$

To carry out the calculation we need first to find the particular value $A$ of the UA price on date $t$ such that $C_t = P_t$, that is such that

$$Ae^{-\delta(T_c-t)}N(z_1) - K_c e^{-r(T_c-t)}N(z_2) + Ae^{-\delta(T_p-t)}N(-z'_1)$$
$$- K_p e^{-r(T_p-t)}N(-z'_2) = 0 \tag{14.38}$$

where

$$z_1 = \frac{\ln(A/K_c) + \mu'(T_c - t)}{\sigma\sqrt{T_c - t}}, \quad z_2 = \frac{\ln(A/K_c) + \mu(T_c - t)}{\sigma\sqrt{T_c - t}}$$

with $\mu = r - \delta - \sigma^2/2$, $\mu' = r - \delta + \sigma^2/2$, and where $z'_1$ and $z'_2$ are defined like $z_1$ and $z_2$, respectively, but with $K_p$ and $T_p$.

Then we can show that the chooser's value is equal to

$$CH_0 = S_0 e^{-\delta T_c} N2\left(x_1, y_1, \sqrt{t/T_c}\right) - K_c e^{-rT_c} N2\left(x_2, y_2, \sqrt{t/T_c}\right)$$
$$- S_0 e^{-\delta T_p} N2\left(-x_1, -w_1, \sqrt{t/T_p}\right) + K_p e^{-rT_p} N2\left(-x_2, -w_2, \sqrt{t/T_p}\right) \tag{14.39}$$

with $A$ defined by (14.38), $x_1 = \frac{\ln(S_0/A)+\mu't}{\sigma\sqrt{t}}$, $x_2 = \frac{\ln(S_0/A)+\mu t}{\sigma\sqrt{t}}$, $y_1 = \frac{\ln(S_0/K_c)+\mu'T_c}{\sigma\sqrt{T_c}}$, $y_2 = \frac{\ln(S_0/K_c)+\mu T_c}{\sigma\sqrt{T_c}}$, $w_1 = \frac{\ln(S_0/K_p)+\mu'T_p}{\sigma\sqrt{T_p}}$ and $w_2 = \frac{\ln(S_0/K_p)+\mu T_p}{\sigma\sqrt{T_p}}$,

and where the different $N2(.)$ are computed by numerical integration.

---

## 14.3 Summary

- *Path-independent* options have payoffs that only depend on the final value of the underlying asset (UA), while *path-dependent* options have payoffs that depend also on its trajectory.

- A *forward start* option is a vanilla European option except that its strike is not known at the time the buyer evaluates and pays for it, but only at some later time specified in the contract. It admits a BSM closed-form solution.
- *Digital*, or binary, options, pay at maturity $T$, either a fixed sum (typically \$1) or nothing, according to whether the value of the UA is greater than the strike (for a call) or less (for a put).
- *Double digital options* are composed of a long position in a digital option with strike $K_1$ and a short position in a digital option with strike $K_2$ ($> K_1$).
- A *gap* option is an option whose strike $K$ (on which exercising the option or not at maturity depends) differs from the amount $L$ used to calculate its payoff.
- *Multi-asset* options, sometimes called *rainbows*, depend on two or more underlying assets. They include *exchange* options, *best of* or *worst of* options, and options on *maximum* or *minimum*. The former obey *call-put parity* or the latter *Min-Max parity*.
- A *compound* option is the right to buy or sell on some future date $T_1$ at a strike $K_1$, a call or a put with expiration date $T_2$ ($>T_1$) and strike $K_2$ (usually $> K_1$), written on some underlying asset. There are four types of compound options: a call on a call, a call on a put, a put on a call, and a put on a put.
- An instrument is said to be *quanto* when its payoff is denominated in a currency different from the original currency of the product's fundamental components.
- For instance, the payoff of a *quanto call* is equal to: $CQ(T) = \widehat{X}\left(S^f(T) - K^f\right)^+$ where the exponent $f$ indicates the *foreign* nature of the original currency in which the underlying and strikes are denominated, and $\widehat{X}$ denotes the fixed exchange rate (domestic currency/foreign currency). The investor thus bets on the value of the foreign UA without bearing the foreign currency risk.
- The *compo* or *combo option* is intermediate between vanilla and *quanto*, in that the exchange rate risk affects the underlying price only, not the strike. Thus, the payoff for a *combo call* is $CC(T) = (X(T) \, S^f(T) - K)^+$, where $K$ is denominated in domestic currency. The foreign currency risk here bears on the value of the UA, not the strike.
- The options above are all *path-independent*. All the options below are *path-dependent*.
- There are two kinds of *barrier* options. A *(Knock-)In* option does not exist until the price of the underlying hits the specified barrier. If the barrier is reached at any time, the option becomes vanilla and may expire at maturity in the money or not. A *(Knock-)Out* option is the opposite: the option is initially vanilla and stays so as long as the price of the UA does not reach the barrier. If the barrier is hit, the option becomes instantaneously worthless. These options are either *Up* or *Down*, depending on whether the barrier is set above or below the initial price of the UA.
- When an In option has not been activated or an Out option has been deactivated, the contract may provide the option holder a *rebate* so that the payoff is in fact not zero.

# Appendix 1

- Many variants of barrier options exist, such as *partial* barriers, *contingent* barriers, *double barriers*, *cumulative* and *Parisian* options. Another class is that of *digital and double digital barriers*.
- A *lookback call* is a European option whose strike, unknown at inception, is the minimum UA price, $m_{T_0}^T$, observed between two dates $T_0$ (usually the option's creation date) and $T$, its expiry date. The payoff at $T$ therefore is $\left(S_T - m_{T_0}^T\right)$. Likewise, a *lookback put* is a European option whose strike is the maximum UA price, $M_{T_0}^T$, observed between two dates $T_0$ (usually the option's creation date) and $T$, its expiry date. The payoff at $T$ thus is $\left(M_{T_0}^T - S_T\right)$. These options are said "no regret" as they always expire in the money.
- Two types of *average* or *Asian options* exist: either the UA price is an average and the strike is fixed, or the strike is an average. The respective payoffs are therefore: $Max[0, M(T_1,T_2) - K]$ for a call or $Max[0, K - M(T_1,T_2)]$ *for a* put, and $Max[0, S(T) - M(T_1,T_2)]$ for a call or $Max[0, M(T_1,T_2) - S(T)]$ for a put, where $M(T_1,T_2)$ is the average (mean) price of the underlying between $T_1$ ($\geq 0$) and $T_2$ ($T_2 \leq T$, the expiry date of the option).
- Closed-from solutions à la BSM exist only in the very rare case of a *geometric average* (continuous or discrete). This is because the geometric mean of a geometric Brownian motion is itself a geometric Brownian motion, which is not the case for the arithmetic mean. As the average used is almost always *arithmetic*, one needs to make an approximation or to have recourse to Monte Carlo *simulation* to value them.
- A *chooser* option allows its holder to choose, at any time up to maturity $t$, between a call with strike $K_c$ and expiry $T_c$ ($> t$) and a put with strike $K_p$ and expiry $T_p$ ($> t$). This option is suitable when the buyer expects that there will be a large variation in the UA price between now and date $t$ without suspecting in what direction.

---

# Appendix 1

## **Value of a Compo Call

Applying Itô's Lemma to the product $X(t)S^f(t)$, we obtain, taking into account Eqs. (14.14) in (14.15) the main text

$$\frac{d(X(t)S^f(t))}{X(t)S^f(t)} = [(r(t) - r^f(t)) + (r^f(t) - \rho(t)\sigma_X(t)\sigma_{S^f}(t)) + \rho(t)\sigma_X(t)\sigma_{S^f}(t)]dt$$

$$+ \sigma_X(t)dW^X(t) + \sigma_{S^f}(t)dW^{S^f}(t)$$

$$= r(t)dt + \sigma_X(t)dW^X(t) + \sigma_{S^f}(t)dW^{S^f}(t)$$

and we can check that the UA price (given in domestic currency) discounted at the domestic rate is a $Q$-martingale. We need to compute

(i) $CC(0) = A(T) E_Q[(X(T) S^f(T) - K)^+]$

$$= E_Q[X(T) S^f(T)A(T) 1_{\mathcal{A}}] - K.A(T) Q(\mathcal{A})$$

where $\mathcal{A} \equiv \{\omega \in \Omega : X(T)S^f(T) > K\}$ is the event "the option expires in the money". We denote the variance of $\ln[X(T)S^f(T)]$ by

(ii) $Vol^2\left(XS^f, T\right).T \equiv \int_0^T \left(\sigma_X^2(s) + \sigma_{S^f}^2(s) + 2\rho(s)\sigma_X(s)\sigma_{S^f}(s)\right) ds.$

Since $X(t)S^f(t)$ is the UA of this option and $X(t)S^f(t)A(t)$ is a $Q$-martingale, the term $K.A(T).Q(\mathcal{A})$ in expression (i) obtains immediately from applying the BSM formula, i.e.,

$$K.A(T).N(d_2)$$

where

(iii) $d_2 = \dfrac{\ln\left(X(0)\, S^f(0)/K.A(T)\right) - .5\, Vol^2\left(XS^f, T\right)T}{Vol\left(XS^f, T\right)\sqrt{T}}$

From BSM, we infer that the term $E_Q[.]$ in Eq. (i) is equal to

$$X(0)\, S^f(0)\, N(d_1)$$

where

(iv) $d_1 = d_2 + Vol\left(XS^f, T\right)\sqrt{T}.$

To prove this, one may either compute explicitly $E_Q[.]$ "by hand," or change the probability measure (and the numeraire) by defining (see Sect. 19.4)

$$\frac{dQ^{XS^f}}{dQ} = \frac{X(T)\, S^f(T)A(T)}{X(0)\, S^f(0)}$$

knowing that $A(0)$ equals 1.

That gives

$$E_Q[.] = X(0)S^f(0)E_{Q^{XS^f}}[1_{\mathcal{A}}] = X(0)S^f(0)N(d_1),$$

where the last equality follows from Girsanov's Theorem (see Sect. 19.4.2).

Finally, we obtain the value of a *compo* call:

(v) $CC(0) = X(0)\, S^f(0)N(d_1) - K.\, A(T)N(d_2)$

where $d_1$, $d_2$ and $Vol^2\left(XS^f, T\right)$ are defined, respectively, by (iv), (iii) and (ii).

# Appendix 2

## **Lemmas on Hitting Probabilities for a Drifted Brownian Motion

We want to compute the probabilities of a drifted Brownian motion attaining a particular value given *a priori* (such as a barrier). As the following lemmas involve a standard (arithmetic) Brownian motion and, in the BSM framework we work with geometric Brownian motions, we must show first how to get from the latter to the former.

Consider, under the risk-neutral measure $Q$, the following dynamics of the UA price:

$$\frac{dS_t}{S_t} = (r - \delta)\, dt + \sigma dW_t,$$

the integral solution to which is

$$S_t = S_0 \exp\left[\left(r - \delta - \frac{\sigma^2}{2}\right)t + \sigma W_t\right].$$

Define $X_t \equiv \ln(S_t/S_0) = \left(r - \delta - \frac{\sigma^2}{2}\right)t + \sigma W_t$, and use the notation $\mu \equiv \left(r - \delta - \frac{\sigma^2}{2}\right)$.

We then obtain a standard drifted Brownian motion under Q, $dX_t = \mu dt + \sigma dW_t$, and we will apply all the lemmas to an UA whose price writes: $X_t \equiv \ln(S_t/S_0)$.

Now define

$$M_t \equiv \text{Sup}\,(X_s, s \le t) \quad \text{and} \quad m_t \equiv \text{Inf}\,(X_s, s \le t).$$

We prove the following lemma

### Lemma 1
Let $(x, y)$ be such that $y \ge 0$ and $x \le y$. Then we have

$$P(X_t \le x; M_t \le y) = N\left(\frac{x - \mu t}{\sigma\sqrt{t}}\right) - e^{2\mu y/\sigma^2} N\left(\frac{x - 2y - \mu t}{\sigma\sqrt{t}}\right). \tag{14.L1}$$

### ***Proof
By complementarity, we have

$$P(X_t \le x; M_t \le y) = P(X_t \le x) - P(X_t \le x; M_t \ge y)$$

$$= N\left(\frac{x - \mu t}{\sigma\sqrt{t}}\right) - P(X_t \le x; M_t \ge y)$$

Denote by $\Delta$ the probability still to be computed and rewrite it as

$$\Delta = P(\gamma t + W_t \le x/\sigma; \ \mathrm{Sup}(\gamma s + W_s \ge y/\sigma)) \quad \text{where } \gamma \equiv \mu/\sigma.$$

Define the new probability $\widehat{Q}$ by

$$\frac{d\widehat{Q}}{dQ} \equiv L_t = \exp\left[-\gamma W_t - \frac{\gamma^2}{2}t\right].$$

We note that the UA price dynamics is now $\sigma \widehat{W}_t$ or $S_t = S_o e^{\sigma \widehat{W}_t}$, with

$$W_t = \widehat{W}_t - \gamma t.$$

This yields, using the indicator function of the desired event,

$$\Delta = E^Q\left[1_{\left[\gamma t + W_t \le \frac{x}{\sigma};\ \mathrm{Sup}\ (\gamma s + W_s) \ge \frac{y}{\sigma}\right]}\right] = E^{\widehat{Q}}\left[1_{\left[\widehat{W}_t \le \frac{x}{\sigma};\ \mathrm{Sup}\widehat{W}s \ge \frac{y}{\sigma}\right]} \cdot e^{\gamma W_t + \frac{\gamma^2}{2}t}\right]$$

(since $E^Q(Y) = E^{\widehat{Q}}(L_t^{-1}Y)$ for an $\mathcal{F}_t$−measurable $Y$)

$$= E^{\widehat{Q}}\left[1_{\left[\widehat{W}_t \le \frac{x}{\sigma};\ \mathrm{Sup}\widehat{W}_s \ge \frac{y}{\sigma}\right]} \cdot e^{\gamma \widehat{W}_t - \frac{\gamma^2}{2}t}\right].$$

Now consider the stopping time

$$\tau(y) = \inf\left\{s : \widehat{W}_s \ge y/\sigma\right\}, \ +\infty \quad \text{otherwise,}$$

and the new Brownian process $\overline{\widehat{W}}_s$ such that

$$\begin{cases} \overline{\widehat{W}}_s = \widehat{W}_s & \text{if } s \in [0, \tau(y) \wedge t] \\ \overline{\widehat{W}}_s = 2\dfrac{y}{\sigma} - \widehat{W}_s & \text{if } s \in [\tau(y) \wedge t, t] \end{cases}$$

where the symbol $\wedge$ indicates the smallest. Then we have

$$\Delta = E^{\widehat{Q}}\left[1_{\left[\widehat{W}_t \le \frac{x}{\sigma};\ \tau(y) \le t\right]}e^{\gamma \widehat{W}_t - \frac{\gamma^2}{2}t}\right] = E^{\widehat{Q}}\left[1_{\left[\frac{2y}{\sigma} - \overline{\widehat{W}}_t \le \frac{x}{\sigma};\ \tau(y) \le t\right]}e^{\frac{2\gamma y}{\sigma} - \gamma \overline{\widehat{W}}_t - \frac{\gamma^2}{2}t}\right]$$

$$= e^{\frac{2\gamma y}{\sigma}}E^{\widehat{Q}}\left[1_{\left[\overline{\widehat{W}}_t \ge \frac{2y-x}{\sigma}\right]}e^{-\gamma \overline{\widehat{W}}_t - \frac{\gamma^2}{2}t}\right]$$

where the simplification of the indicator function results from $\frac{2y-x}{\sigma} \ge \frac{y}{\sigma}$ since $y \ge x$.

In this simplification lies the benefit of using the symmetry property of the Brownian motion (the *reflection theorem*) through the definition of $\overline{\widehat{W}}_s$.

# Appendix 2

Using $\overline{W}_t = z\sqrt{t}$, where $z$ is $N(0, 1)$-distributed, and $\gamma = \mu/\sigma$, we have

$$\Delta = e^{\frac{2\mu y}{\sigma^2}}\left[\int_{\frac{2y-x}{\sigma\sqrt{t}}}^{\infty} e^{-\gamma\sqrt{t}z - \frac{\gamma^2}{2}t}\frac{1}{\sqrt{2\pi}}e^{-\frac{z^2}{2}}dz\right] = e^{\frac{2\mu y}{\sigma^2}}\left[\int_{\frac{2y-x}{\sigma\sqrt{t}}}^{\infty}\frac{1}{\sqrt{2\pi}}e^{\frac{-(z+\gamma\sqrt{t})^2}{2}}dz\right]$$

$$= e^{\frac{2\mu y}{\sigma^2}}N\left[-\left(\frac{2y-x}{\sigma\sqrt{t}} + \gamma\sqrt{t}\right)\right] = e^{\frac{2\mu y}{\sigma^2}}N\left(\frac{x-2y-\mu t}{\sigma\sqrt{t}}\right).$$

Lemma (14.L2) below is the symmetric of L1 and is proved in the same way. Lemmas (14.L3)–(14.L3$'$) and (14.L4)–(14.L4$'$) are special cases, respectively, of (14.L1) and (14.L2).

**Lemma 2**

Let $(x, y)$ be such that $y \leq 0$ and $y \leq x$. Then

$$P(X_t \geq x; m_t \geq y) = N\left(\frac{-x+\mu t}{\sigma\sqrt{t}}\right) - e^{\frac{2\mu y}{\sigma^2}}N\left(\frac{-x+2y+\mu t}{\sigma\sqrt{t}}\right) \tag{14.L2}$$

**Lemma 3**

Let $y \geq 0$ and $x = y$. Then

$$P(M_t \leq y) = N\left(\frac{y-\mu t}{\sigma\sqrt{t}}\right) - e^{\frac{2\mu y}{\sigma^2}}N\left(\frac{-y-\mu t}{\sigma\sqrt{t}}\right) \tag{14.L3}$$

and

$$P(M_t \geq y) = N\left(\frac{-y+\mu t}{\sigma\sqrt{t}}\right) + e^{\frac{2\mu y}{\sigma^2}}N\left(\frac{-y-\mu t}{\sigma\sqrt{t}}\right). \tag{14.L3$'$}$$

**Lemma 4**

Let $y \leq 0$ and $x = y$. Then

$$P(m_t \geq y) = N\left(\frac{-y+\mu t}{\sigma\sqrt{t}}\right) - e^{\frac{2\mu y}{\sigma^2}}N\left(\frac{y+\mu t}{\sigma\sqrt{t}}\right) \tag{14.L4}$$

and

$$P(m_t \leq y) = N\left(\frac{y-\mu t}{\sigma\sqrt{t}}\right) + e^{\frac{2\mu y}{\sigma^2}}N\left(\frac{y+\mu t}{\sigma\sqrt{t}}\right). \tag{14.L4$'$}$$

One can easily derive, from these lemmas and the complementarity property, other lemmas corresponding to other events, for example

$$P(X_t \geq x; M_t \geq y) = P(M_t \geq y) - P(X_t \leq x; M_t \geq y)$$
$$= (L3')\text{plus the second term in}(L1).$$

# Appendix 3

## **Proof of the "Inverses" Relation for Barrier Options

We will prove only the first of the four equations (iii) given in the main text (Sect. 14.2.1.1), relative to the value of a put up-and-out, since the other three can be obtained similarly:

$$PUO\ (S, K, L, r, \delta\ ) = S.K.CDO\left(\frac{1}{S}, \frac{1}{K}, \frac{1}{L}, T, \delta, r\right)$$

where $S = S_0$ to keep the notation simple.

On date $t = 0$, recalling that $\mu = (r - \delta - \sigma^2/2)$ and $\gamma = \mu/\sigma$, the PUO's value is

$$PUO(\bullet) = E^Q\left[e^{-rT}\left[K - Se^{\mu T + \sigma W_T}\right]^+ 1_{Sup(S_t) \leq L}\right]$$
$$= E^{\widehat{Q}}\left[e^{-rT}e^{\gamma \widehat{W}_T - \frac{\gamma^2}{2}T}\left[K - Se^{\sigma \widehat{W}_T}\right]^+ 1_{Sup(S_t) \leq L}\right]$$
$$= e^{-\left(r + \frac{\gamma^2}{2}\right)T}E^{\widehat{Q}}\left[S.K.e^{(\gamma + \sigma)\widehat{W}_T}\left[\frac{e^{-\sigma \widehat{W}_T}}{S} - \frac{1}{K}\right]^+ 1_{\widehat{M}_T \leq l}\right]$$

where the second equality is shown in Appendix 2, and where $l \equiv \frac{1}{\sigma} \ln (L/S)$ (> 0 since this is a PUO) and $\widehat{M}_T = Sup\left(\widehat{W}_s, s \leq T\right)$.

Let $\widehat{W}_T = -Z_T$. This gives

$$PUO(\bullet) = S.K.e^{-\left(r + \frac{\gamma^2}{2}\right)T}E^{\widehat{Q}}\left[e^{\lambda Z_T}\left[\frac{e^{\sigma Z_T}}{S} - \frac{1}{K}\right]^+ 1_{\widehat{m}_T \geq -l}\right]$$

where $-l \equiv \frac{1}{\sigma} \ln (S/L)$ (< 0), $\widehat{m}_T = Inf\ (Z_s, s \leq T)$ and $\lambda \equiv (\gamma + \sigma) = \frac{1}{\sigma}(\delta - r - \frac{\sigma^2}{2})$.

Notice the reversal of the roles of the two parameters $r$ and $\delta$. Now, we can write

$$PUO(\bullet) = S.K.e^{-\left(r + \frac{\gamma^2}{2}\right)T}e^{\left(\delta + \frac{\lambda^2}{2}\right)T}e^{-\left(\delta + \frac{\lambda^2}{2}\right)T}E^{\widehat{Q}}[\bullet]$$

From the definitions of $\gamma$ and $\lambda$, it is easy to check that $-\left(r + \frac{\gamma^2}{2}\right) + \left(\delta + \frac{\lambda^2}{2}\right) = 0$. Hence, we obtain

$$PUO(\bullet) = S.K\left[e^{-\left(\delta+\frac{\sigma^2}{2}\right)T}E^{\widehat{Q}}[\bullet]\right]$$

$$= S.K\left[e^{-\delta T}E^{\widehat{Q}}\left[e^{-\frac{\sigma^2}{2}T+\lambda Z_T}\left[\frac{e^{\sigma Z_T}}{S} - \frac{1}{K}\right]^+ 1_{\widehat{m_T \geq -l}}\right]\right]$$

$$= S.K\ CDO\left[\frac{1}{S}, \frac{1}{K}, \frac{1}{L}, T, \delta, r\right]$$

The last equality results from the fact that $r$ and $\delta$ have inverse roles and that $-l = \frac{1}{\sigma}\ln(S/L)$.

---

## Appendix 4

### **Valuing a Call Up-and-Out with L (Barrier) > K (Strike)

We have chosen this option because it is one of those whose valuation formula involves the most terms (8, with no *rebate*; see Table 14.1). Recall first that a CUO for which $L$ would be less than $K$ is worth nothing.

We have on date 0 (again, remark that $S_0 = S$)

$$CUO(S, K, L, T, r, \delta) = e^{-rT}E^{Q}\left[\left[Se^{\mu T+\sigma W_T} - K\right]1_{\{S_T>K\}\cap\{Sup(S_t)\leq L\}}\right]$$

$$= -Ke^{-rT}E^{Q}\left[1_{\{\mu T+\sigma W_T \geq \ln(K/S)\}\cap\{Sup(\mu t+\sigma W_t)\leq \ln(L/S)\}}\right]$$

$$+Se^{-\delta T}E^{Q}\left[e^{-\frac{\sigma^2}{2}T+\sigma W_T}1_{\{\mu T+\sigma W_T\geq \ln(K/S)\}\cap\{Sup(\mu t+\sigma W_t)\leq \ln(L/S)\}}\right]$$

From the lemmas in Appendix 2, it is easy to obtain the following result for $y \geq 0$ and $x \leq y$ (we let $t = T$):

$$E^{Q}\left[1_{\{\bullet\}\cap\{\bullet\}}\right] = P(X_t \geq x; M_t \leq y) = N\left(\frac{-x + \mu T}{\sigma\sqrt{T}}\right) - N\left(\frac{-y + \mu T}{\sigma\sqrt{T}}\right)$$

$$-e^{\frac{2\mu y}{\sigma^2}}N\left(\frac{-y - \mu T}{\sigma\sqrt{T}}\right) + e^{\frac{2\mu y}{\sigma^2}}N\left(\frac{x - 2y - \mu T}{\sigma\sqrt{T}}\right)$$

which has 4 terms. We first compute the 4 terms multiplied by $-Ke^{-rT}$, then the 4 terms multiplied by $Se^{-\delta T}$. Notice that $x = \ln(K/S)$ and $y = \ln(L/S)$.

- $N\left(\frac{-x+\mu T}{\sigma\sqrt{T}}\right) = N\left[\frac{\ln(S/K)+\mu T}{\sigma\sqrt{T}}\right] \equiv N(d_2)$
- $-N\left(\frac{-y+\mu T}{\sigma\sqrt{T}}\right) = -N\left[\frac{\ln(S/L)+\mu T}{\sigma\sqrt{T}}\right] \equiv -N(d_{2L})$

where $L$, in $d_{2L}$, means that one uses $\ln(S/L)$ instead of the usual $\ln(S/K)$.

- $-e^{\frac{2\mu y}{\sigma^2}} N\left(\frac{-y-\mu T}{\sigma\sqrt{T}}\right) = -\left(\frac{L}{S}\right)^{2(\varepsilon-1)} N\left(\frac{\ln(S/L)-\mu T}{\sigma\sqrt{T}}\right)$

where $(\varepsilon - 1) \equiv \mu/\sigma^2 = \frac{r-\delta}{\sigma^2} - \frac{1}{2}$.

- $e^{\frac{2\mu y}{\sigma^2}} N\left(\frac{x-2y-\mu T}{\sigma\sqrt{T}}\right) = \left(\frac{L}{S}\right)^{2(\varepsilon-1)} N\left(\frac{\ln(KS/L^2)-\mu T}{\sigma\sqrt{T}}\right)$.

To obtain the terms involving $Se^{-\delta T}$, notice that, as for vanilla options in the BSM framework, the indicator function is multiplied by $e^{-\frac{\sigma^2}{2}T+\sigma W_T}$. So we change the probability measure, adopting that associated with a new numeraire, namely the UA price, denoted by $Q*$ and defined by

$$\frac{dQ*}{dQ} = \exp\left[\sigma W_T - \frac{\sigma^2}{2} T\right].$$

From Girsanov's theorem, we have: $W_T^* = W_T - \sigma T$ or $\sigma W_T = \sigma W_T^* + \sigma^2 T$. As a result:

$$E^Q\left[e^{-\frac{\sigma^2}{2}T+\sigma W_T} 1_{\{\mu T+\sigma W_T \geq \ln(K/S)\} \cap \{Sup(\mu t+\sigma W_t) \leq \ln(L/S)\}}\right]$$

$$= E^{Q^*}\left[1_{\{\mu T+\sigma W_T \geq \ln(K/S)\} \cap \{Sup(\mu t+\sigma W_t) \leq \ln(L/S)\}}\right]$$

$$= E^{Q^*}\left[1_{\{\mu' T+\sigma W_T^* \geq \ln(K/S)\} \cap \{Sup(\mu' t+\sigma W_t^*) \leq \ln(L/S)\}}\right]$$

with $\mu' \equiv \mu + \sigma^2 \Rightarrow \mu'T = \left(r - \delta + \frac{\sigma^2}{2}\right)T$.

As in the BSM formula, we obtain the same 4 formulas as before, except we replace $\mu$ by $\mu'$. In particular, this implies that $d_1$ replaces $d_2$, $d_{1L}$ replaces $d_{2L}$ and $\varepsilon$ replaces $(\varepsilon - 1)$. Then finally we have:

$$CUO(S,K,L,T,r,\delta) =$$

$$Se^{-\delta T}\left[N(d_1) - N(d_{1L}) + \left(\frac{L}{S}\right)^{2\varepsilon}\left\{N\left(\frac{\ln(KS/L^2)-\mu'T}{\sigma\sqrt{T}}\right) - N\left(\frac{\ln(S/L)-\mu'T}{\sigma\sqrt{T}}\right)\right\}\right]$$

$$-Ke^{-rT}\left[N(d_2) - N(d_{2L}) + \left(\frac{L}{S}\right)^{2(\varepsilon-1)}\left\{N\left(\frac{\ln(KS/L^2)-\mu T}{\sigma\sqrt{T}}\right) - N\left(\frac{\ln(S/L)-\mu T}{\sigma\sqrt{T}}\right)\right\}\right].$$

We see that this solution is indeed that given in Table 14.1 (row 7) of the main text.

# Appendix 5

## **Valuing Rebates

We first value *rebates at expiry* which are typical of "*In*" options. We derive only one case, since the others can be obtained in the same way using the appropriate lemmas (see Appendix 2).

Consider the CDI $(S, K, L, T, r, \delta)$ or the PDI $(S, K, L, T, r, \delta)$ offering a rebate $R$ on date $T$ if the option has not been activated. The value of this rebate at $t = 0$ thus is

$$Re^{-rT}E_Q\left[1_{\{Inf(S_t)\geq L\}}\right] = Re^{-rT}E_Q\left[1_{\{m_T\geq\ell\}}\right]$$

where $\ell \equiv \ln(L/S)$ is negative since the option is a CDI or a PDI.

By Lemma (14.L4), we have (setting $y = \ell, \ell < 0$):

$$P[m_T \geq \ell] = N\left(\frac{-\ell + \mu T}{\sigma\sqrt{T}}\right) - e^{\frac{2\mu y}{\sigma^2}}N\left(\frac{\ell + \mu T}{\sigma\sqrt{T}}\right).$$

The first term, already met in Appendix 4, is equal to $N\left(\frac{\ln(S/L)+\mu T}{\sigma\sqrt{T}}\right) \equiv N(d_{2L})$.

The second term is equal to $-\left(\frac{L}{S}\right)^{2(\varepsilon-1)}N\left(\frac{\ln(L/S)+\mu T}{\sigma\sqrt{T}}\right)$, with $(\varepsilon - 1) \equiv \mu/\sigma^2$.

The current value of the rebate at expiry is therefore

$$Re^{-rT}\left[N(d_{2L}) - \left(\frac{L}{S}\right)^{2(\varepsilon-1)}N\left(\frac{\ln(L/S) + \mu T}{\sigma\sqrt{T}}\right)\right].$$

One can check that this solution is indeed that given in Table 14.2 (row 1) of the main text.

Now we value the *rebates at hit* typical of "*Out*" options. Here the problem is more complex since we must use the density of the first hitting time $(t)$, by a drifted Brownian, of a given value $\ell$.

We must compute: $R\int_0^T e^{-rt}h_\ell(t)dt$. We know how to do the calculation, since we know the density $h_\ell(t)$.[12] However, this requires a number of changes of variable some of which are difficult to find. A more direct method is to use the following probability result:

---

[12]For example, from the symmetry property of the standard Brownian, we have $P[M_t \geq \ell] = 2N(-\ell/\sqrt{t})$. Taking the derivative of this expression with respect to $t$ yields: $h_\ell(t) = \frac{\ell}{\sqrt{2\pi t^3}}e^{-\frac{\ell^2}{2t}}$ (for $\ell > 0$, otherwise one must use $|\ell|$).

(i) $E^Q[e^{-r\tau_\ell}] = \exp\left[\frac{\mu\ell}{\sigma^2} - |\ell|\frac{\sqrt{\mu^2+2r\sigma^2}}{\sigma^2}\right] \equiv \exp\left[\frac{\mu\ell}{\sigma^2} - |\ell|b\right]$

where $\tau_\ell$ is the first time $\ell$ (the barrier) is hit by the drifted Brownian $dX_t = \mu dt + \sigma dW_t$, and $b \equiv \sqrt{\mu^2 + 2r\sigma^2}/\sigma^2$.

Let us consider the case of a CDO or a PDO. The barrier $\ell \equiv \ln(L/S)$ is therefore negative. The value at $t = 0$ of the *rebate at hit* equals

(ii) $R.E^Q\left[e^{-r\tau_\ell}1_{\{\tau_\ell \leq T\}}\right] = R.E^Q\left[e^{-r\tau_\ell}1_{\{m_T \leq \ell\}}\right].$

Using the appropriate change in the probability measure (and the numeraire), we have

(iii) $E^Q\left[e^{-r\tau_\ell}1_{\{m_T \leq \ell\}}\right] = E^Q[e^{-r\tau_\ell}]E^{\widehat{Q}}\left[1_{\{\widehat{m} \leq \ell\}}\right]$

where we can show [as in the proof of (i)] that the drift of $dX_t$ under the measure $\widehat{Q}$ is no longer $\mu$ but $\widehat{\mu} = -\sqrt{\mu^2 + 2r\sigma^2} \equiv -b\sigma^2$.

Furthermore, by (14.L4$'$) in Appendix 2 with $t = T$ and $y = \ell \ (< 0)$, we have

(iv) $P[\widehat{m}_T \leq \ell] = E^{\widehat{Q}}\left[1_{\{\widehat{m}_T \leq \ell\}}\right] = N\left(\frac{\ell - \widehat{\mu}T}{\sigma\sqrt{T}}\right) + e^{\frac{2\widehat{\mu}\ell}{\sigma^2}}N\left(\frac{\ell + \widehat{\mu}T}{\sigma\sqrt{T}}\right).$

Using (i), (iii) and (iv), (ii) becomes (with $\ell < 0$)

$$RE^Q\left[e^{-r\tau_\ell}1_{\{m_T \leq \ell\}}\right] = R.e^{\frac{\mu\ell}{\sigma^2}+\ell b}\left[N\left(\frac{\ell - \widehat{\mu}T}{\sigma\sqrt{T}}\right) + e^{\frac{2\widehat{\mu}\ell}{\sigma^2}}N\left(\frac{\ell + \widehat{\mu}T}{\sigma\sqrt{T}}\right)\right],$$

which, using $\widehat{\mu} \equiv -b\sigma^2$, $\ell = \ln(L/S)$ and $(\varepsilon - 1) \equiv \mu/\sigma^2$, gives

$$R\left[\left(\frac{L}{S}\right)^{\varepsilon-1+b}N\left(\frac{\ln(L/S) + b\sigma^2 T}{\sigma\sqrt{T}}\right) + \left(\frac{L}{S}\right)^{\varepsilon-1-b}N\left(\frac{\ln(L/S) - b\sigma^2 T}{\sigma\sqrt{T}}\right)\right]$$

in accordance with row 3 of Table 14.2 in the main text.

---

# Appendix 6

## **Proof of the Price of a Lookback Call

We must compute the second expectation in Eq. (14.23) of the main text, $E\left[m_0^T 1_{m_0^T < m_{T_0}^0}\right]$. This requires knowing the density of $m_0^T$. Starting from (14.L4) (in Appendix 2), we have

# Appendix 6

599

$$P(m_0^T < y) = 1 - P(m_0^T \geq y) = N\left(\frac{y - \mu T}{\sigma\sqrt{T}}\right) + e^{2\mu y \sigma^{-2}} N\left(\frac{y + \mu T}{\sigma\sqrt{T}}\right).$$

The density of $m_0^T = y$ is therefore

$$\frac{1}{\sigma\sqrt{T}}\frac{1}{\sqrt{2\pi}}e^{-\frac{1}{2}\left(\frac{y-\mu T}{\sigma\sqrt{T}}\right)^2} + 2\mu\sigma^{-2}e^{2\mu y \sigma^{-2}} N\left(\frac{y + \mu T}{\sigma\sqrt{T}}\right) + e^{2\mu y \sigma^{-2}}\frac{1}{\sigma\sqrt{T}}\frac{1}{\sqrt{2\pi}}e^{-\frac{1}{2}\left(\frac{y+\mu T}{\sigma\sqrt{T}}\right)^2}.$$

Now, we have: $\dfrac{e^{2\mu y \sigma^{-2}}}{\sigma\sqrt{T}\sqrt{2\pi}}e^{-\frac{1}{2}\left(\frac{y+\mu T}{\sigma\sqrt{T}}\right)^2} = \dfrac{1}{\sigma\sqrt{T}\sqrt{2\pi}}e^{-\frac{1}{2}\left(\frac{y-\mu T}{\sigma\sqrt{T}}\right)^2}.$

Therefore, this density, denoted by $n_{min}(y)$, is equal to

$$n_{min}(y) = \frac{2}{\sigma\sqrt{T}\sqrt{2\pi}}e^{-\frac{1}{2}\left(\frac{y-\mu T}{\sigma\sqrt{T}}\right)^2} + 2\mu\sigma^{-2}e^{2\mu y \sigma^{-2}} N\left(\frac{y + \mu T}{\sigma\sqrt{T}}\right).$$

We have: $E\left[m_0^T 1_{m_0^T < m_{T_0}^0}\right] = E\left[S_0 e^{\min \ln (S_t/S_0)} 1_{\min_0^T(\ln (S_t/S_0)) < -b}\right]$, with $b \equiv \ln\left(S_0/m_{T_0}^0\right).$

Therefore, we obtain:

$$E\left[m_0^T 1_{m_0^T < m_{T_0}^0}\right] = \int_{-\infty}^{-b} S_0 e^y n_{min}(y)dy$$

(i)

$$= \int_{-\infty}^{-b} S_0 e^y \left[\frac{2}{\sigma\sqrt{T}\sqrt{2\pi}}e^{-\frac{1}{2}\left(\frac{y-\mu T}{\sigma\sqrt{T}}\right)^2} + 2\mu\sigma^{-2}e^{2\mu y \sigma^{-2}} N\left(\frac{y + \mu T}{\sigma\sqrt{T}}\right)\right] dy.$$

To compute the first integral, define $Z \equiv \frac{y-\mu T}{\sigma\sqrt{T}}$. We then have

$$2S_0 e^{\mu T}\int_{-\infty}^{\frac{-b-\mu T}{\sigma\sqrt{T}}} e^{Z\sigma\sqrt{T}}\frac{1}{\sqrt{2\pi}}e^{-\frac{Z^2}{2}}dZ.$$

Using a very general result,[13] this simplifies to

---

[13] $\displaystyle\int_{-\infty}^{x} e^{hu}\frac{e^{-u^2/2}}{\sqrt{2\pi}}\,du = \int_{-\infty}^{x} e^{-\frac{1}{2}[(u-h)^2-h^2]}\frac{du}{\sqrt{2\pi}} = e^{\frac{h^2}{2}}\int_{-\infty}^{x} e^{-\frac{1}{2}(u-h)^2}\frac{du}{\sqrt{2\pi}}$

$= (\textit{letting } t = u - h)\ e^{\frac{h^2}{2}}\int_{-\infty}^{x-h} e^{-\frac{t^2}{2}}\frac{dt}{\sqrt{2\pi}} = e^{\frac{h^2}{2}}N(x - h).$

$$2S_0 e^{\mu T} e^{\frac{\sigma^2 T}{2}} N\left(\frac{-b-\mu T}{\sigma\sqrt{T}} - \sigma\sqrt{T}\right) = 2S_0 e^{(r-c)T} N\left(\frac{-b-\mu'T}{\sigma\sqrt{T}}\right).$$

Let us compute the second integral using integration by parts:

$$\int_{-\infty}^{-b} 2S_0 e^y \mu\sigma^{-2} e^{2\mu y\sigma^{-2}} N\left(\frac{y+\mu T}{\sigma\sqrt{T}}\right) dy = \left[\frac{S_0 2\mu\sigma^{-2}}{2\mu\sigma^{-2}+1} e^{(2\mu\sigma^{-2}+1)y} N\left(\frac{y+\mu T}{\sigma\sqrt{T}}\right)\right]_{-\infty}^{-b}$$

$$- \int_{-\infty}^{-b} \frac{S_0 2\mu\sigma^{-2}}{2\mu\sigma^{-2}+1} e^{(2\mu\sigma^{-2}+1)y} \frac{1}{\sqrt{2\pi}\sigma\sqrt{T}} e^{-\frac{1}{2}\left(\frac{y+\mu T}{\sigma\sqrt{T}}\right)^2} dy.$$

Now, we have

$$\frac{2\mu\sigma^{-2}}{2\mu\sigma^{-2}+1} = \frac{[2(r-\delta)-\sigma^2]/\sigma^2}{[2(r-\delta)/\sigma^2]} = 1 - \frac{\sigma^2}{2(r-\delta)} = 1 - \lambda \quad \text{and}$$

$$2\mu\sigma^{-2}+1 = \frac{2(r-\delta)-\sigma^2}{\sigma^2} + 1 = \frac{2(r-\delta)}{\sigma^2} = \frac{1}{\lambda}.$$

Therefore, the integral is

$$S_0(1-\lambda) e^{-b/\lambda} N\left(\frac{-b+\mu T}{\sigma\sqrt{T}}\right) - 0 - S_0(1-\lambda)\int_{-\infty}^{-b} e^{y/\lambda} \frac{1}{\sigma\sqrt{T}\sqrt{2\pi}} e^{-\frac{1}{2}\left(\frac{y+\mu T}{\sigma\sqrt{T}}\right)^2} dy =$$

$$S_0(1-\lambda)\left(\frac{m_{T_0}^0}{S_0}\right)^{1/\lambda} N\left(\frac{-b+\mu T}{\sigma\sqrt{T}}\right) - S_0(1-\lambda) e^{(r-c)T} N\left(\frac{-b-\mu'T}{\sigma\sqrt{T}}\right)$$

where computing the second term is analogous to computing the first integral in (i). Equation (i) consequently rewrites

(ii) $2S_0 e^{(r-c)T} N\left(\frac{-b-\mu'T}{\sigma\sqrt{T}}\right) + S_0(1-\lambda)\left(\frac{m_{T_0}^0}{S_0}\right)^{1/\lambda} N\left(\frac{-b+\mu T}{\sigma\sqrt{T}}\right) -$

$$S_0(1-\lambda) e^{(r-c)T} N\left(\frac{-b-\mu'T}{\sigma\sqrt{T}}\right)$$

Using Eq. (14.24) in the text, multiplied by $-e^{-rT} m_{T_0}^0$, and result (ii), multiplied by $-e^{-rT}$, canceling some terms and rearranging, Eq. (14.23) giving the value of a *lookback* call becomes Eq. (14.25):

$$CLB_0 = e^{-cT} S_0 N\left(\frac{b + \mu'T}{\sigma\sqrt{T}}\right)$$

$$- e^{-rT} m_{T_0}^0 \left[ N\left(\frac{b + \mu T}{\sigma\sqrt{T}}\right) - \frac{\lambda S_0}{m_{T_0}^0}\left(\frac{m_{T_0}^0}{S_0}\right)^{1/\lambda} N\left(\frac{-b + \mu T}{\sigma\sqrt{T}}\right) \right]$$

$$- e^{-cT} S_0 \lambda N\left(-\frac{b + \mu'T}{\sigma\sqrt{T}}\right).$$

## Appendix 7

### **Options on an Average Price

The solution to the SDE describing the UA price $S(t)$ writes, for $t \geq T_1$,

$$S(t) = S(T_1) \exp\left[\left(b - \frac{\sigma^2}{2}\right)(t - T_1) + \sigma(W(t) - W(T_1))\right].$$

From this, using $\ln S(t)$, we obtain the geometric average computed between $T_1$ and $T$:

$$M(T_1, T) = \exp\left[\frac{1}{T - T_1}\int_{T_1}^{T} \ln S(t)\,dt\right]$$

$$= S(T_1) \exp\left[\frac{1}{2}\left(b - \frac{\sigma^2}{2}\right)(T - T_1) + \frac{\sigma}{T - T_1}\int_{T_1}^{T}(W(t) - W(T_1))\,dt\right]$$

The random variable $\int_{T_1}^{T}(W(t) - W(T_1))\,dt$ has a conditional (at $T_1$) Gaussian distribution

$$N\left(0; \frac{1}{3}(T - T_1)^2\right).$$

Indeed, $-\int_{T_1}^{T} W(T_1)\,dt = -W(T_1)(T - T_1)$ and $\int_{T_1}^{T} W(t)\,dt = -\int_{T_1}^{T} t\,dW(t) + [tW(t)]_{T_1}^{T}$ (integration by parts) $= -\int_{T_1}^{T} t\,dW(t) + TW(T) - T_1 W(T_1)$.

Whence $\int_{T_1}^{T}(W(t) - W(T_1))\,dt = -\int_{T_1}^{T} t\,dW(t) + TW(T) - TW(T_1)$

$$= \int_{T_1}^{T}(T - t)\,dW(t).$$

602                14   *Exotic Options

This Wiener integral has a Gaussian distribution with zero mean. In addition, its variance is equal to $\int_{T_1}^{T}(T-t)^2 dt = \frac{1}{3}(T-T_1)^3$.

Since $S(T_1) = S(0)\exp\left[\left(b - \frac{\sigma^2}{2}\right)T_1 + \sigma W(T_1)\right]$, $\quad M(T_1,T)$ rewrites

$$M(T_1,T) = S(0)\exp\left[\frac{1}{2}\left(b - \frac{\sigma^2}{2}\right)(T+T_1) + \sigma W(T_1) + \frac{\sigma}{T-T_1}\int_{T_1}^{T}(W(t)-W(T_1))dt\right].$$

Since Brownian increments are independent, the last two terms in the exponential are independent. Their sum is therefore Gaussian with mean zero and variance

$$\left[\sigma^2 T_1 + \frac{\sigma^2}{(T-T_1)^2}\frac{(T-T_1)^3}{3}\right] = \sigma^2\left(\frac{T+2T_1}{3}\right).$$

We can therefore write, with $z$ $N(0,1)$-distributed,

(i)   $M(T_1,T) = S(0)\exp\left[\frac{1}{2}\left(b - \frac{\sigma^2}{2}\right)(T+T_1) + \sigma\sqrt{\frac{T+2T_1}{3}}z\right]$

To recover the BSM framework, we need the following form for $M(T_1,T)$:

(ii)   $M(T_1,T) = Z(0)\exp\left[\left(b - \frac{\Sigma^2}{2}\right)T + \Sigma\sqrt{T}z\right]$

Identifying the diffusion parameters, then the drifts, of equations (i) and (ii), we obtain

$$\Sigma = \sigma\sqrt{\frac{T+2T_1}{3T}} \quad \text{and} \quad Z(0) = S(0)\exp\left[-b\frac{T-T_1}{2} - \sigma^2\frac{T-T_1}{12}\right].$$

The payoff of a call on an average price being $[M(T_1,T) - K]^+$, directly applying BSM leads to its value, at $t = 0$, given by Eq. (14.27). The proof for the put [Eq. (14.28)] is analogous.

---

## Appendix 8

### **Options with an Average Strike**

We value, for $T_1 \geq 0$, the call paying $Max\ [0,\ S(T) - M(T_1,\ T)]$ at expiry date $T$. Under the risk-neutral probability $Q$, we have, on date $t = 0$: $CAS(0) = e^{-rT}E_Q[[S(T) - M(T_1 - T)]^+] = e^{-rT}E_Q[S(T)[1 - M(T_1 - T)S^{-1}(T)]^+]$

As usual, we transform the expectation of a product into a product of expectations by changing the probability measure. Let $Q_S$, the measure associated with the UA price itself as the numeraire, be defined by

# Appendix 8

$$\frac{dQ_S}{dQ} \equiv \frac{S(T)}{S(0)} \cdot \frac{1}{e^{bT}} = \exp\left[-\frac{\sigma^2}{2}T + \sigma W_T\right]$$

The value $CAS(0)$ then becomes

$$CAS(0) = e^{-cT}S(0)E_{QS}\left[[1 - M(T_1, T)S^{-1}(T)]^+\right].$$

The problem thus reduces to computing the expectation $M(T_1, T)\, S^{-1}(T)$ under $Q_S$. We have

$$M(T_1, T) = S(0)\exp\left[\frac{1}{2}\left(b - \frac{\sigma^2}{2}\right)(T + T_1) + \sigma W(T_1) + \frac{\sigma}{T - T_1}\int_{T_1}^{T}(W(t) - W(T_1))\,dt\right]$$

$$= S(0)\exp\left[\frac{1}{2}\left(b + \frac{\sigma^2}{2}\right)(T + T_1) + \sigma W^S(T_1) + \frac{\sigma}{T - T_1}\int_{T_1}^{T}(W^S(t)\right.$$
$$\left. - W^S(T_1))\,dt\right]$$

since, by Girsanov's theorem, we have $W(t) = W^S(t) + \sigma t$.

As $S(T) = S(0)\exp\left[\left(b - \frac{\sigma^2}{2}\right)T + \sigma W(T)\right] = S(0)\exp\left[\left(b + \frac{\sigma^2}{2}\right)T + \sigma W^S(T)\right]$, we obtain $M(T_1, T)S^{-1}(T) = \exp\left[\frac{1}{2}\left(b + \frac{\sigma^2}{2}\right)(T_1 - T) + \sigma W^S(T_1) - \sigma W^S(T) + \frac{\sigma}{T - T_1}\int_{T_1}^{T}(W^S(t) - W^S(T_1))\,dt\right] = \exp\left[\frac{1}{2}\left(b + \frac{\sigma^2}{2}\right)(T_1 - T) - \sigma W^S(T) + \frac{\sigma}{T - T_1}\int_{T_1}^{T}W^S(t)\,dt\right]$.

Conditional on $\mathscr{F}_{T_1}$, the random variable $\left\{-\sigma W^S(T) + \frac{\sigma}{(T-T_1)}\int_{T_1}^{T}W^S(t)\,dt\right\}$ is, under the probability $Q_S$, a Gaussian $N\left(0, \frac{\sigma^2}{3}(T - T_1)\right)$. Indeed, by defining $f(t) \equiv -\sigma + \frac{\sigma(T-t)}{T-T_1}$, we can show that this random variable can be written $\int_{T_1}^{T}f(t)\,dW^S(t)$. The expectation of this Wiener integral, which is Gaussian, is zero. It variance is equal to

$$\int_{T_1}^{T}f^2(t)\,dt = \sigma^2\int_{T_1}^{T}\left(1 - \frac{2(T-t)}{T-T_1} + \frac{(T-t)^2}{(T-T_1)^2}\right)dt = \sigma^2\left(\frac{T-T_1}{3}\right).$$

We thus obtain

$$M(T_1, T)S^{-1}(T) = \exp\left[\frac{1}{2}\left(b + \frac{\sigma^2}{2}\right)(T_1 - T) + \int_{T_1}^{T}f(t)\,dW^S(t)\right].$$

Let us define

$$\Sigma' \equiv \sigma \sqrt{\frac{T - T_1}{3T}} \quad \text{and} \quad Z'(0) \equiv \exp\left[\frac{1}{2}\left(b + \frac{\sigma^2}{2}\right)(T_1 - T) + \frac{1}{2}\Sigma'^2 T\right].$$

Then we have

$$M(T_1, T)S^{-1}(T) = Z'(0)\exp\left(-\frac{1}{2}\Sigma'^2 T + \Sigma'\sqrt{T}z\right).$$

The value of a call with a geometric average strike is therefore [Eq. (14.33)]:

$$CAS(0) = e^{-\delta T}S(0)\left[N\left(d_1'\right) - Z'(0)N\left(d_2'\right)\right]$$

where $d_1' = \frac{\ln(1/Z'(0)) + \frac{1}{2}\Sigma'^2 T}{\Sigma'\sqrt{T}}$, $d_2' = d_1' - \Sigma'\sqrt{T}$ and $Z'(0)$ and $\Sigma'$ are defined as above.

Similarly, from the call-put parity, the value of a put with a geometric average strike is equal to [Eq. (14.34)]:

$$PAS(0) = e^{-\delta T}S(0)\left[Z'(0)N\left(-d_2'\right) - N\left(-d_1'\right)\right].$$

## Suggestions for Further Reading

### Books

Clewlow, L., & Strickland, C. (1997). *Exotic options, The state of the Art*. Thomson Business Press.
*Fusai, G., & Roncoroni, A. (2008). *Implementing models in quantitative finance: Methods and cases*. Springer.
Hull, J. (2020). *Options, futures and other derivatives* (11th ed.). Prentice Hall Pearson Education.
**Musiela, M., & Rutkowski, M. (2007). *Martingale methods in financial modeling*. Springer, Applications of Mathematics.
*Wilmott, P., Dewynne, J., & Howison, S. (1993). *Option pricing*. Oxford Financial Press.
Zang, P. G. (1998). *Exotic options: A guide to second generation options*. World scientific.

### Articles

Broadie, M., Glasserman, P., & Kou, S. G. (1998). Connecting discrete and continuous path-dependent options. *Finance and Stochastics, 2*, 1–28.
Broadie, M., Glasserman, P., & Kou, S. G. (1997). A continuity correction for discrete barrier options. *Mathematical Finance, 4*, 325–349.
Chesney, M., Jeanblanc-Piqué, M., & Yor, M. (1997). Brownian excursion and parisian barrier options. *Advances in Applied Probabilities, 29*, 165–184.
Conze, A., & Viswanathan, R. (1991). Path dependent options: The case of lookback options. *Journal of Finance, 46*(5), 1893–1907.
Derman, E., Ergener, D., & Kani, I. (1995). Static options replication. *Journal of Derivatives, 2*(4), 78–95.
Geman, H., & Yor, M. (1996). Pricing and hedging double-barrier options. *Mathematical Finance, 6*(4).

## Suggestions for Further Reading

Geman, H., & Yor, M. (1993). Bessel processes, Asian options, and perpetuities. *Mathematical Finance, 3*(4).

Geske, R. (1977). The valuation of corporate liabilities as compound options. *Journal of Financial and Quantitative Analysis, 12*, 541–552.

Geske, R. (1979). The valuation of compound options. *Journal of Financial Economics, 7*, 63–81.

Goldman, B., Sosin, H., & Gatto, M. A. (1979). Path dependent options: Buy at the low, sell at the high. *Journal of Finance, 34*, 1111–1127.

Hugonnier, J. N. (1999). The Feynman-Kac formula and pricing occupation time derivatives. *International Journal of Theoretical and Applied Finance, 2*(2), 153–178.

Kemna, A., & Vorst, A. (1990). A pricing method for options based on average asset values. *Journal of Banking and Finance, 14*, 113–129.

Margrabe, W. (1978). The value of an option to exchange one asset for another. *Journal of Finance, 33*, 177–185.

Ritchken, P. (1995). On pricing barrier options. *Journal of Derivatives, 3*(2), 19–28.

Rubinstein, M., & Reiner, E. (1991) Breaking down the barriers. *RISK*, September, 28–35.

Turnbull, S. M., & Wakeman, L. M. (1991). A quick algorithm for pricing european average options. *Journal of Financial and Quantitative Analysis, 26*, 377–389.

# Futures Markets (2): Contracts on Interest Rates

# 15

Forward and futures contracts on interest rates have already been presented in Chap. 9. However, given their importance, special nature and complexity, futures on interest rates deserve separate consideration. These contracts, which are very innovative, are very popular in the main financial markets. They originated in the United States in the mid-1960s and were introduced in Europe and Asia a few years later. Today the main markets in interest rate futures are the CME Group (for notional contracts and short-term rate contracts) in the U.S. and ICE Futures Europe (previously NYSE-EURONEXT-LIFFE) in Europe.

Interest rate futures are traded on exchange markets and thus subject to daily margin calls. The exchange organizing the market acts as a clearing-house and is intermediary and counterparty in all transactions. At any moment and for each contract the exchange sets the price so that its net position is zero. At this equilibrium price, the number of contracts bought is therefore equal to the number of contracts sold.

We have explained in Chap. 9 that a forward or futures contract obeys one of the two following principles, according to whether or not it gives rise to an actual delivery of the underlying asset. In the first case, if it is not settled before its maturity $T$, there is an actual delivery of a well pre-defined asset (*physical delivery*): The final settlement price $F_T$ then coincides with the spot price of the physical underlying asset delivered. In the second case (*cash settlement*), the price $F_T$ of the expired contract is defined unambiguously in the contract as a function of variables observable and measured at $T$ and the seller remits a cash amount $F_T$ in place of the underlying. In both cases, the price $F_t$ on any date $t$ before $T$ is set by supply and demand, and as $t$ nears $T$ and uncertainty over the value of $F_T$ diminishes, $F_t$ converges to $F_T$. Interest rate futures are governed by these two principles, many contracts on long- or medium-term underlying assets (2 years or more) leading to physical delivery and contracts on short-term assets usually being cash-settled.

This chapter provides an in-depth analysis of the *two principal types* of futures contracts on interest rates: The notional contracts on short, medium or long term securities in Sect. 15.1, and the Short Term Interest Rate contracts (STIRS) in Sect.

© The Author(s), under exclusive license to Springer Nature Switzerland AG 2022
P. Poncet, R. Portait, *Capital Market Finance*, Springer Texts in Business and
Economics, https://doi.org/10.1007/978-3-030-84600-8_15

15.2. Most interest rate futures traded in the main financial markets belong to one of these two types of contracts.

## 15.1 Notional Contracts

Notional contracts are futures whose underlying is a fixed-rate security (bond or note) issued by a government (a *sovereign bond*). If they are not settled before maturity, they cause delivery of a security by the seller and payment of a delivery price by the buyer. In contrast to the general case, these contracts are defined relative to an "abstract" reference (a "notional security") without physical existence and lead, at the will of the seller, to the delivery of one among different eligible physical securities considered similar enough to the notional security. This peculiar feature justifies a specific explanation. The essential properties of these "notional" contracts (notional security, basket of physical underlying bonds eligible for delivery) are described in Sect. 15.1.1. The Euro-bund contract is introduced as an example in Sect. 15.1.2. The delivery mechanism (delivery price, conversion factor, cheapest to deliver) is analyzed in Sects. 15.1.3 and 15.1.4. *Cash-and-carry* and *reverse* arbitrages as well as the spot-forward parity are presented in Sect. 15.1.5. The contract's sensitivity to rates is studied in Sect. 15.1.6. Hedging with futures is treated in Sect. 15.1.7. Finally, Sect. 15.1.8 exhibits an overview of different notional contracts on interest rates, with particular attention to *swap notes* since they are based on principles different from those of other notional contracts and do not permit physical delivery. In this whole section the reader will bear in mind the bond quotation practices: bonds are quoted in percent of par and when an interest is accrued since the last coupon during a fraction $\tau$ of the year, the *clean price* denoted by $C_\tau$ is, by definition, the full (dirty) price $C_\tau'$ reduced by the accrued interest: $C_\tau = C_\tau' - 100 \, k \, \tau$ ($k$ being the coupon rate), as explained in Chap. 4.

### 15.1.1 Basket of Deliverable Securities (DS) and Notional Security

For each contract, the Exchange handling the quotation defines a *basket of securities* eligible for delivery made up of 2–30 fixed-rate sovereign securities, as need be. The deliverable (or eligible) securities composing a given basket are rather homogeneous (they are of the bullet type, have a fixed coupon rate and a maturity lying within a rather narrow range). Any of the deliverable securities (DS) making up the basket can be delivered at the contract's maturity, and this on the seller's decision. The conversion factor system (explained in Sect. 15.1.3) sets the possibilities for settlement in accordance with the particular DS delivered by the seller. A bullet security (bond or note) often called *notional*, whose properties are defined by the Exchange precisely but arbitrarily (coupon rate and maturity) and this for the entire duration of the contract, serves as the benchmark for the settlement. The maturities of the DS are "close" to the maturity of the notional. This notional is "virtual" as it is neither issued nor quoted on the spot market. We stress the following principle:

## 15.1 Notional Contracts

*The notional security serves as a benchmark for the underlying of the futures contract. In case of physical delivery, any DS of the basket can be delivered, according to the seller's decision, but the exact price paid by the buyer depends on the particular security delivered.*

The properties defining a notional security are the following:

- Its maturity: For example, 2, 5, 10, or 30 years to be counted from the maturity date of the relevant futures contract. This maturity is a reference for the maturities of the corresponding DS.[1]
- Its coupon: It is (fictitiously) paid once a year in the case of most European contracts, twice a year in the case of American and UK contracts on the yearly (or semi-annual) anniversaries of the expiry dates of the futures contract, and is computed from the nominal interest rate,[2] which we denote by $k$ ($k = 6\%$ in the main US and European long-term futures contracts).
- Its nominal value, denoted by $M$: It specifies the amount of the contract. At expiry, the seller of a still open position must settle by providing securities in the amount of this nominal value.
- Its clean price: It is given as a percentage of the nominal value, like an ordinary bond. The price on date $t$ of a futures with an expiry date $T$ is denoted by $F_t^T$, or simply $F_t$ when there is no ambiguity about the expiry date.

Finally, recall that the DS is chosen from sovereign bonds or Treasury bonds or notes with maturities near that of the notional.

### 15.1.2 The Euro-Bund Contract

We take the Euro-bund contract, which is the most widely traded in Europe, as a representative example of a long-term notional futures contract.

#### 15.1.2.1 Contract Description

A Euro-bund contract is quoted on the EUREX. Its notional maturity is 10 years. The basket of DS is made up of annual bullet bonds issued by the German Republic (*bunds*), with maturities which, at the expiration of the contract, range from 8.50 to 10.50 years. The DS of a June 2005 contract, for example, were the following 9- or 10-year maturity bonds:

---

[1] This reference may be loose or even misleading: In the US, the maturities of the DS of the 30-year CME futures contract were restricted for some time to a range of 15–25 years! This contract is now called the classical T-bond, while the Ultra T-bond or new 30-year contract is settled with delivery of T-bonds whose maturity is longer than 25 years.

[2] The value of the face rate $k$ can be adjusted occasionally, but remains fixed for a given contract and this during its entire life. Remark, for example, that CME and EUREX has kept the face rate for notional bonds at 6% for several years, although the rates on the bond market are currently (2022) much below that figure.

| BUND 4.25%, 04- 01- 2014 |
| BUND 4, 25%, 04- 07- 2014 |
| BUND 3, 75%, 04- 01- 2015 |

The nominal worth of a Euro-bund contract is 100,000 €: a contract that is not settled before its expiry leads to a delivery of settlement securities in this amount.

The quoted expiries are the three next quarters (March, June, September, and December) plus the next 3 months (with a maximum of six contracts). For example, at the start of January contracts are quoted that expire in January, February, March, June, and September. The settlement date is fixed as the 10th of the expiry month if it is a business day, or, if not, the first business day thereafter; closing of transactions on a contract takes place two business days before the settlement date.

In the sequel, however, we mostly identify the closing day of transactions with the settlement date. We call it the "expiry date" and denote it by $T$. When a contract expires, a new contract, relative to a new quarter expiry, is opened (e.g., a contract expiring in March leads to opening a December contract). Notice that the quarter cycle (March, June, September, and December) is common to all notional contracts traded in large financial markets (the number of expiries quoted and the exact dates of expiry of course vary).

The closing price on date $j$ ($F_j^T$) (a price called the *daily settlement*—or *closing*—price) is an average of the prices of the five final transactions on date $j$ and the closing price upon expiry ($F_T^T$) (called the *final settlement price*) is the average of the six last transaction prices carried out on the day $T$ of closure of transactions.

The price of a Euro-bund contract is quoted as a percentage up to two decimal places and the allowed minimum variation in its price (the *tick*) is 0.01%, or 10 € per contract. Most transactions are wound up before expiry and so do not lead to physical transfers. This is why the *open interest*, i.e., the number of contracts open at a given moment, decreases abruptly in days preceding expiry.

### 15.1.2.2 Example (1) of Transactions for a Euro-Bund Contract Wound Up Before Its Expiry

On the morning of Monday 09-25-n, an agent buys ten contracts with December expiry.

The order is carried out at 107.40.

On the evening of 09-25, the closing price is 107.46. The agent's account is thus credited with:

$$10 \times (107.46 - 107.40) \times 100,000/100$$
$$= 600 \text{ € (or simply } 10 \times 6 \text{ ticks} \times 10 \text{ €).}$$

(continued)

## 15.1 Notional Contracts

On succeeding days the margins evolve as in the following table:

| Dates | Closing price | Margins (€) Calls (−) or repayments (+) |
|---|---|---|
| 09-25 | 107.46 | +600 |
| 09-26 | 107.50 | +400 |
| 09-27 | 107.42 | −800 |
| 09-28 | 107.38 | −400 |
| 09-29 | 107.44 | +600 |

On the evening of 09-29 the agent winds up their position by selling 10 December contracts (at 107.44): the account is credited with 600 €. In all, the investor has gained: $10 \times 4$ ticks $\times 10$ € $= 400$ € $= (600+400 - 800 - 400 + 600)$ €, without discounting the cash flows. Below, we adopt the Euro-bund contract as the archetypical notional contract and describe both the notional contract and its DS as "bonds." The principles underlying different notional contracts are, up to a few details, identical. It is useful to refer to the internet site *www.eurexchange.com* for information about interest rate Futures quoted on the EUREX in general, and the Euro-bund contract in particular.

### 15.1.3 Settlement and Conversion Factors

By contract, on the settlement date, the seller chooses the DS she delivers to the clearinghouse for a nominal amount equal to the nominal value of the futures contract (for instance $100,000 or 100,000 €). In exchange, she receives the *Delivery Price* ($DP_i$), which depends on the characteristics of the bond $i$ she chooses to deliver. The cash flows between buyer and seller are of course symmetric (the $DP_i$ is paid by the buyer). The clearinghouse gives to the buyers the bonds delivered by the sellers and collects and repays the resulting delivery prices to the sellers. Delivering DS $i$ "instead of" the notional bond makes necessary an adjustment to the delivery price. Without such an adjustment, the seller would simply select the bond in the basket with the lowest dirty price. Due to the relative homogeneity of the DS in terms of duration and to the fact that they are all issued by the same issuer, the dirty price is smaller if the nominal rate is lower or the coupon has accrued for a shorter period. Adjusting the delivery price $DP_i$ requires using a multiple called the *conversion factor* and involves the accrued interest; it does suppress much of the price bias by making each delivery price $DP_i$ more in line with the "true value" of DS $i$. As we will show, because the way the delivery price is calculated depends on the DS delivered, the bond with the lowest dirty price is *not* always the cheapest to deliver by the seller.

Recall that we have denoted by $F_t^T$, or more simply $F_t$, the price of a contact with maturity $T$ that holds on $t$. $F_T \equiv F_T^T$ is thus the last quote for this contract (*final*

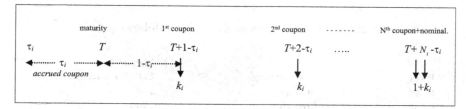

**Fig. 15.1** Diagram of cash flows for the DS $i$

settlement price). In exchange for the bond $i$ of nominal value 100, the buyer pays a *delivery price* equal to

$$DP_i = f_i F_T + ac_i(T), \qquad (15.1)$$

where $ac_i(T)$ is the accrued coupon for bond $i$ on date $T$, and $f_i$ is an adjustment factor associated with bond $i$ and calculated (for a nominal of 1) as explained below.[3] For a contract for a nominal amount $M$ of synonym $i$, $DP_i$ is simply multiplied by $\frac{M}{100}$.

Now suppose we are at the contract's expiration date $T$ and consider some $DS_i$, with coupon rate $k_i$ (which usually differs from the nominal rate $k$ *of the notional security*), and repayment of the capital in $N_i - \tau_i$ periods (on date $T + N_i - \tau_i$), which has to distribute $N_i$ coupons in $1 - \tau_i, 2 - \tau_i, \ldots, N_i - \tau_i$ periods (on $T$ the coupon has already accrued for $\tau_i$ periods), as in the following diagram for a nominal of 1 (Fig. 15.1):

The *conversion factor* $f_i$ is equal to the present value, obtained on delivery date $T$, by discounting (with the nominal rate $k$ of the notional bond) the cash flows of the $DS_i$ (coupon $k_i$ and capital of 1) net of the accrued coupon $k_i \tau_i$:

$$f_i = \sum_{j=1}^{N_i} \frac{k_i}{(1+k)^{j-\tau_i}} + \frac{1}{(1+k)^{N_i - \tau_i}} - k_i \tau_i. \qquad (15.2)$$

The duration $\tau_i$ of the accrued coupon is computed using the actual/actual convention.

The conversion factor $f_i$ can thus be interpreted as *the clean price at which $1 nominal of $i$ would trade on date $T$ if its yield to maturity (YTM) were equal to $k$*.

The conversion factor $f_i$, hence the delivery price $DP_i$, increases with the DS coupon $k_i$. Furthermore, the conversion factors, since computed by discounting fixed cash flows with the known fixed rate $k$, are *known at the start date of the futures contract and fixed for the duration of the contract*. They are made public at the start of the contract by the clearinghouse responsible for the trading and quotation.

---

[3] A conversion factor is particular to each DS and also depends on the maturity of the contract. The notation $f_i$ is therefore short for the more precise notation $f_{i,T}$.

## 15.1 Notional Contracts

**Example 2**

The Euro-bund contract expiring in June 2005 introduced at the outset of Sect. 15.1.2.1 publishes the following conversion factors for three DS (recall that the contract's face rate $k$ is 6%):

| | | |
|---|---|---|
| 4.25%,04- *Jan*- 2014 | 0.885049 |
| 4,25%,04- *Jul*- 2014 | 0.880217 |
| 3,75%,04- *Jan*- 2015 | 0.839315 |

**Remarks**

- The conversion factor is not a quoted price, unless the bond yields to maturity (YTM) are equal to $k$, the nominal rate of the notional bond, in which case it is equal to the clean price (on date $T$, for a nominal of 1). Let us consider, for one moment, such a situation (even though it is extremely improbable). The conversion factor table then does allow us to distinguish the different DS of the basket by their values. For example, a bond with a large coupon like the Bund 4.25%, 04-Jan-2014 has a higher value than a bond with a smaller coupon like the Bund 3.75%, 04-Jan-2015. Its conversion factor thus is larger (0.885049 compared to 0.839315). The adjustment in the delivery price as given in Eq. (15.1) therefore implies a larger payment for the Bund 4.25%.
- As the conversion factor is computed as a clean price, it allows comparing (clean) bond prices independently of the payment dates for their respective coupons. These are then comparable, or homogeneous, prices (see Chap. 4). For this reason, they are good conversion indicators. However, the delivery price in Eq. (15.1) is augmented by the accrued coupon, since the delivered bond actually bears an accrued coupon.

To sum up, the futures $F_T$ price is homogeneous to a clean price, as is the conversion factor (which is computed for a nominal of 1). The product $f_i F_T$ is therefore an adjusted clean price, to which is added the accrued coupon to obtain the delivery price, which is thus homogeneous to a dirty price.

## 15.1.4 Cheapest to Deliver and Quoting Futures at Expiration

We first examine the mechanism for choosing the cheapest bond to deliver, which allows determining the futures price at the expiration of the contract. We then offer a deeper analysis of what is the cheapest to deliver, according to various situations regarding bond yields.

### 15.1.4.1 Seller's Choice and Quotation at Expiry

As the contract's seller has the right, at expiry date, to select the bond to deliver among the $N$ DS, he, therefore, holds an option (*delivery* option or *quality option*),

614 15 Futures Markets (2): Contracts on Interest Rates

analyzed in Appendix 1. Therefore, he chooses the most advantageous one, *taking into account the cost he pays as well as the delivery price he receives.* We have already remarked that the *cheapest to deliver* (CTD) is in general different from that which is cheapest to buy (the one that has the smallest dirty price) because the delivery price depends on the DS selected. The gain $G_i$ that the seller realizes upon delivery of a $DS_i$ is equal to the difference between the delivery payment $DP_i$ he receives and the dirty price $C_i'(T)$ of the bond $i$ to be delivered:

$$G_i = DP_i - C_i'(T).$$

Since the full coupon rate $C_i'(T)$ is the sum of the clean price $C_i(T)$ and the accrued coupon, $C_i'(T) = C_i(T) + ac_i(T)$, using formula (15.1) for the invoice price we obtain:

$$G_i = f_i F_T - C_i(T), \tag{15.3}$$

The accrued coupon $cc_i(T)$ has vanished from this expression of the seller's P&L. As a result, the CTD (denoted by $m$ in what follows) is the bond that produces the maximal gain:

$$m = \arg \max {}_i G_i.$$

The final step in the argument leads to the determination of the futures price $F_T$: At expiry and under no-arbitrage, $F_T$ is set so that the seller's realized gain upon delivering bond $m$ vanishes:

$$G_m = 0.$$

Indeed, if $G_m > 0$, the arbitrage consisting in selling the contract at $T$ (at zero cost) and delivering $m$ immediately after produces a positive gain. If $G_m < 0$, the arbitrage consists in buying the contract, which produces the gain $-G_m > 0$.[4] The following proposition establishes the relationship between the cheapest to deliver and the futures price at expiration.

**Proposition 1**
(a) *The futures price at expiry, denoted $F_T$, represents the minimum ratio of the clean price of a deliverable bond to its conversion factor:*

$$F_T = \min_i \frac{C_i(T)}{f_i}. \tag{15.4}$$

---

[4]If the seller were to select a different DS than $m$ (which she obviously would never do) the gain would be even larger.

## 15.1 Notional Contracts

(b) *The cheapest to deliver is the DS for which the minimum is attained:*

$$m = \arg \min{}_i \frac{C_i(T)}{f_i}.$$

**Proof**

Since the gain $G_i$ is negative or zero for any $i = 1, \ldots, N$, we obtain: $f_i F_T - C_i(T) \leq 0$.

Therefore, for any $i = 1, \ldots, N$: $F_T \leq \frac{C_i(T)}{f_i}$.

When the gain vanishes (for $i = m$), we obtain the equality: $F_T = \frac{C_m(T)}{f_m}$.

---

**Example 3**

Let us resume the example of a Euro-bund contract (expiring on June 10, 2005).

Assume that on the expiration date the yields to maturity of the deliverable bonds are all 5%. The following table gives the details of the calculation leading to the cheapest to deliver and the quoted price $F_T$.

| Bond | Clean price | Clean price/fc | Gain |
|------|-------------|----------------|------|
| Bund 4.25%, 04-Jan-2014 | 94.85 | 107.17 | 0 |
| Bund 4.25%, 04-Jul-2014 | 94.63 | 107.51 | −0.30 |
| Bund 3.75%, 04-Jan-2015 | 90.65 | 107.99 | −0.69 |

We see that the Bund 4.25% 04-Jan-2014 minimizes the ratio clean price/ conversion factor. In the absence of arbitrage, the futures contract price $F_T$ is therefore equal to 107.17. The gain resulting to the seller for this bond thus is zero, and negative for the other bonds.

---

### 15.1.4.2 Detailed Examination of the Cheapest to Deliver

The goal here is to highlight a feature that allows distinguishing the CTD from the other DS. We show in particular that, according to whether the yield $y$ on the bond market is larger or smaller than the nominal rate of the notional bond, the CTD is the DS with the longest or shortest duration.

Rigorously speaking, this result only holds if the market yields of the different DS coincide. Although it is unlikely that the yields to maturity (YTM) of the deliverable bonds are all strictly equal, they are, in practice, very close since the DS originate from the same issuer and are rather homogeneous in terms of maturity. To be precise in the determination of the CTD, we compare the gains of the different alternatives and use again the argument presented in Sect. 15.1.4.1. We suppose here that the

YTM of the DS are all equal to $y$ at the expiry of the futures contract. Recall that Eq. (15.4) allows calculating the contract's price at expiry.

## (a) *The yields to maturity of the DS are all equal to the face rate of the notional*

We have already noted that, when the bond's yield to maturity $y$ equals the notional face rate $k$, the clean prices $C_i(T)$ equal 100 times[5] the conversion factor $f_i$. From this: $C_i(T)/f_i = 100$ for all $i$.

It also follows that the futures price is at par:

$$F_T = 100.$$

It is remarkable that, for $y = k$, the minimum occurring in Eq. (15.4) is realized for all $i = 1, \ldots, N$; all bonds in the deliverable basket thus realize the minimum (15.4): there is *indeterminacy in the cheapest to deliver*. This means that, if $y = k$, the adjustment in the delivery price using the conversion factor does exactly what is required and makes all the DS "equivalent."

Intuition suggests that, in the situation $y < k$, the delivery price $DP_i$ (calculated using the discount rate $k$ then overestimated) is undervalued for all deliverable bonds, and this undervaluation decreases as a DS has a smaller duration (since it is less sensitive to the discount rate): The seller trying to find *the least undervalued delivery price $DP_i$* then selects the bond with *the shortest duration*. On the contrary case $y > k$, we assume that the cheapest to deliver has *the longest duration*. We confirm this intuition with a proof grounded on a Taylor's series expansion of the DS prices around rate $k$.

## (b) *Taylor's expansion in the neighborhood of the notional contract face rate*

From the preceding intuition, we infer that the particular value of the market yield $y = k$ is pivotal for the determination of the CTD: If $y = k$, all DS are equivalent.

Consider the values of $y$ near $k$. By expanding the function $C_i(y)$ around $k$, we obtain for all $i = 1, \ldots, N$:

$$C_i(y) = C_i(k) - \frac{1}{1+k} D_i^{cp}(k) C_i(k)(y - k) + o(y - k),$$

where $o(y - k)$ is of the first order in $(y - k)$ and $D_i^{cp}(k)$ is the clean price duration of bond $i$ at expiry when its YTM is $k$ (Chap. 4). This duration is given by:

---

[5]The multiplier 100 comes from the fact that the price $C_i$ is quoted for a nominal amount of $100 while the conversion factor is calculated for an amount of 1.

## 15.1 Notional Contracts

$$D_i^{cp}(k) = \frac{100 \left[ \sum_{j=1}^{N_i} \frac{k_i}{(1+k)^{j-\tau_i}} (j - \tau_i) + \frac{1}{(1+k)^{N_i-\tau_i}} (N_i - \tau_i) \right]}{C_i(k)}. \qquad (15.5)$$

Using $C_i(k) = 100f_i$, the expansion becomes

$$C_i(y) = 100 \left[ f_i - \frac{1}{1+k} f_i D_i^{cp}(k)(y - k) \right] + o(y - k).$$

The ratio $\frac{C_i(T)}{f_i}$ thus may be written to the first order in $y - k$ as

$$\frac{C_i(T)}{f_i} = 100 \left[ 1 - \frac{1}{1+k} D_i^{cp}(y - k) \right] + o(y - k). \qquad (15.6)$$

This equation leads to the following conclusions as to the cheapest to deliver:

- When $y > k$, the ratio $\frac{C_i}{f_i}$ is strictly less than 100, and is minimum for the bond with the longest duration.
- When $y < k$, the ratio $\frac{C_i}{f_i}$ is strictly larger than 100 and is minimum for the bond with the shortest duration.

These results are summed up in the following proposition.

**Proposition 2**
*Assuming only one* yield to maturity *y on the expiration date T, the futures price $F_T$ can be found by the following method:*

- *Futures are quoted at par if $y = k$. All DS are equally costly to deliver.*
- *Futures are quoted above par if $y > k$. The bond which is cheapest to deliver is the one with the longest duration.*
- *Futures are quoted below par if $y < k$. The bond which is cheapest to deliver is the one with the shortest duration.*

---

**Example 4**
Let us take up again the example of a Euro-bund June 2005 contract, and let us compare the clean price durations at expiry:

| Bond | Clean price duration |
|---|---|
| Bund 4.25%, 04-Jan-2014 | 7.33 |
| Bund 4.25%, 04-Jul-2014 | 7.67 |
| Bund 3.75%, 04-Jan-2015 | 8.14 |

We showed in the preceding section that the Bund 4.25% 04-Jan-2014 is cheapest to deliver when $y = 5\%$. Recall that the face rate of the notional bond

(continued)

is $k = 6\%$. Therefore, we are in the case $k > y$. We again find that the CTD is the bond with the shortest clean price duration.

The change in CTD takes place in the neighborhood of rate $y = k = 6\%$, the CTD then being the Bund 4.25% 04-Jan-2014 when $y < 6\%$, and the Bund 3.75% 04-Jan-2015 when $y > 6\%$.

## 15.1.5 Arbitrage and Cash-Futures Relationship

The preceding subsection analyzed the quotation of a futures contract *at expiry*. This subsection discusses the arbitrage that can be carried out between the cash and the futures market prices and establishes their relationship *before expiry*, which leads to an arbitrage-free quote for a futures contract at any instant of time.

Here, we assume that the CTD is known, and does not change before the contract's expiration date. Appendix 1 provides a more complete analysis of the contract, including the risk of a change in the CTD over time which gives the *delivery option* (or quality option) held by the seller a positive value.

### 15.1.5.1 Cash and Carry

This arbitrage, whose general mechanism has been studied in Chap. 9, is carried out by investors who wish *to benefit from an overvaluation of a futures price*. The arbitrage is initiated on date 0, the contract's expiration date being $T$.

As in the general case (see Chap. 9), the idea is to sell the contract on date 0 and to buy simultaneously the spot (cash) underlying, this purchase being financed by a loan on the money market for a duration $T$. Note that, in the present case, the underlying is the presumed cheapest to deliver, $m$. On date $T$, the investor delivers $m$ to the contract's buyer and in return receives the delivery price in cash with which she can pay off the loan. If the ensuing net cash flow is strictly positive, the investor realizes an arbitrage profit at $T$.

We denote by $C_m(0)$ and $C'_m(0)$ the clean and dirty prices of CTD $m$ on date 0 and retain the notation defined in Fig. 15.1 (for $i = m$) above, which leads us to the following Fig. 15.2 *for a nominal* of 100 (we always use this nominal amount in what follows):

The arbitrage period $(0, T)$ is assumed to fall between two coupon disbursements (respectively on dates $T - \tau_m$ and $T + 1 - \tau_m$). We therefore assume that the CTD $m$ does not distribute a coupon between 0 et $T$ and we later indicate what adjustment is required when this is not the case. With these conventions:

## 15.1 Notional Contracts

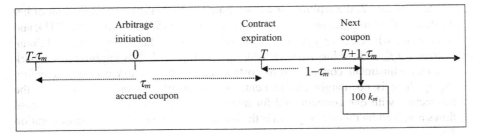

**Fig. 15.2** Cash and carry arbitrage involving the CTD $m$

**Table 15.1** Cash and carry arbitrage

| Transaction | Cash flows |
|---|---|
| Date 0 | |
| Sale of 100 $f_m$ contracts | 0 |
| Purchase of CTD | $-C'_m(0)$ |
| Borrowing in the money market | $-C'_m(0)$ |
| Intermediate dates $t \leq T$ | |
| Margin calls | $\sum_{t \leq T} f_m \left( F^T_{t-1} - F^T_t \right)$ |
| Date $T$ | |
| Adjustement of futures amount to 100 | 0 |
| Delivery of the CTD $m$ | 0 |
| Cashing in the delivery price | $f_m F_T + ac_m(T)$ |
| Repayment of the loan | $-C'_m(0)(1 + r_0 T)$ |

- The coupon on date $T$ has accrued for a period $\tau_m$ and has the value $ac_m(T) = 100 k_m \tau_m$.
- The accrued coupon *during the arbitrage period*, i.e., between 0 and $T$, is equal to

$$ac_m(T) - ac_m(0) = 100 k_m T.$$

Margin calls happen daily. Denoting by $F^T_t$ the quoted price for the contract on an intermediate date $t$, the margin on date $t$ on a short position is $F^T_{t-1} - F^T_t$.

The delivery price on final date $T$ is

$$DP_m = f_m F^T_T + ac_m(T).$$

Setting up the arbitrage involves adjusting the amount of each transaction, taking into account the special properties of the notional contract and the synonym $m$. In particular, *cash and carry* arbitrage consists in selling contracts for $100 f_m$ on date 0 so as to match the amounts of the margin calls with the delivery price. In this case, as can be seen from the summary Table 15.1, the margin calls amount to $f_m \left( F^T_{t-1} - F^T_t \right)$.

On final date $T$, the arbitrageur has to restore to $100 the nominal amount of her position in the futures contract. In this way, she delivers $100 worth of the CTD, and receives the whole delivery price $DP_m$. This is done by selling $100(1 - f_m)$ additional contracts if $f_m < 1$, or buying back $100(f_m - 1)$ if $f_m > 1$. It is important to remark that this adjustment comes at zero cost: since the transaction occurs just before expiry there is no margin requirement, and its purpose is only to terminate the transaction with one contract sold for one CTD held. Table 15.1 displays the cash flows involved by this arbitrage ($r_0$ is the interest rate on date 0 for a loan of duration $T$):

**Remarks**
- For simplicity, we assumed the CTD did not distribute any coupon between dates 0 and $T$. If that is not so, one may just add the first coupon to the delivery price received on date $T$, or else subtract its value from the price $C'_m(0)$.
- In addition, we simply added up the margins paid on the different intermediate dates $t$, neglecting the cost of financing them (and/or the gain on their placement) between $t$ and $T$, which amounts to considering the futures as a forward.

From Table 15.1, the sum of the cash flows involved in the arbitrage, or the gain of the *cash and carry*, denoted $G_{CC}$, is equal to

$$G_{CC} = \sum_{t \leq T} f_m \left( F_{t-1}^T - F_t^T \right) + f_m F_T^T + ac_m(T) - C'_m(0)(1 + r_0 T).$$

By adjusting the amount of the futures to be $100 f_m$, the first two terms in the equation simplify and we have

$$G_{CC} = f_m F_0^T + ac_m(T) - C'_m(0)(1 + r_0 T)$$

or again, remembering that $C'_m(0) = C_m(0) + ac_m(0)$,

$$G_{CC} = f_m F_0^T + ac_m(T) - C'_m(0) r_0 T - C_m(0) - ac_m(0).$$

Since $ac_m(T) - ac_m(0)$ is the coupon accrued between 0 and $T$, $ac_m(T) - ac_m(0) = 100 k_m T$, and we have

$$G_{CC} = f_m F_0^T - C_m(0) - \left( r_0 C'_m(0) - 100 k_m \right) T.$$
$$< \text{---basis---} >< \text{--net cost of carry--} > \tag{15.7}$$

The profit from this transaction is therefore a certainty on date 0, and only depends on market conditions on date 0 (the futures price $F_0^T$, the cash price $C_m(0)$, the interest rate $r_0$). It can be seen as the difference between the basis, here defined as $f_m F_0^T - C_m(0)$, and the net cost of carrying the CTD $m$ from 0 to $T$, i.e.,

# 15.1 Notional Contracts

$(r_0 C'_m(0) - 100k_m)T$. We note that the basis is defined as the difference between the futures price *adjusted* by the conversion factor and the clean price of $m$. If the gain $G_{CC}$ is strictly positive, the sequence of transactions described is an arbitrage. Consequently, under no-arbitrage, this gain can only vanish or be negative:

$$G_{CC} \leq 0 \Leftrightarrow f_m F_0^T - C_m(0) \leq (r_0 C'_m(0) - 100k_m)T. \tag{15.8}$$

## 15.1.5.2 Reverse Cash and Carry and Spot-Futures Parity

The second arbitrage is symmetrical to the preceding. It *profits from undervalued futures prices, by buying futures, selling the cheapest to deliver, and placing the proceeds of the sale on the money market*. The cash flow table is symmetrical to that describing the cash flows of *cash and carry* (the positions and the cash flows have opposite signs). The gain of this *reverse cash and carry* is equal to

$$G_{RCC} = (r_0 C'_m(0) - 100k_m)T - (f_m F_0^T - C_m(0)).$$

Under AAO, this gain cannot be strictly positive, so

$$G_{RCC} \leq 0 \Leftrightarrow f_m F_0^T - C_m(0) \geq (r_0 C'_m(0) - 100k_m)T. \tag{15.9}$$

Putting together the two conditions $G_{CC} \leq 0$ and $G_{RCC} \leq 0$ leads to the following proposition:

**Proposition 3 (Spot-Futures Parity for Notional Contracts)**
*In absence of arbitrage, the price on date 0 of a futures contract with expiry T is linked to the* spot *price of the CTD m by the relation*

$$f_m F_0^T - C_m(0) = (r_0 C'_m(0) - 100k_m)T. \tag{15.10a}$$

*where $f_m F_0^T - C_m(0)$ can be viewed an (adjusted) basis, $r_0$ is the money market rate on date 0 for maturity T and $(C'_m(0)r_0 - 100k_m)T$ is the net cost of carrying m. This relation thus reflects the equality of the (adjusted) basis and the net cost of carry. It can be written in an equivalent form which expresses the theoretical futures/forward price as a function of the price of the CTD:*

$$F_0^T = \frac{C_m(0) + (C'_m(0)r_0 - 100k_m)T}{f_m},$$

or:

$$F_0^T = \frac{C'_m(0)(1 + r_0 T) - cc_m(T)}{f_m} \tag{15.11}$$

*Reverse cash and carry transactions* deserve two additional comments.

The first concerns the cash sale of the CTD $m$: to sell a bond requires holding or borrowing it. In theory, it is sufficient that investors who hold $m$ could realize a profit through a *reverse cash and carry* for sales of $m$ and purchases of futures to exercise enough pressure on prices to prevent the situation $f_m F_0^T - C_m(0) < (C'_m(0)r_0 - 100k_m)T$.

However, most arbitrageurs do not actually hold the CTD in their portfolios. To implement the reverse, they have either to borrow it or to acquire it temporarily (during the period $(0, T)$) through a *repurchase agreement* (*repo*; see Chap. 3). In both cases, this involves a cost of several basis points. This cost is expressed as a spread, denoted by $s$, which decreases the interest rate $r_0$ obtained from the money market instrument. Note that a liquid market in sovereign debts usually makes for small spreads (e.g., 5 basis points annually). The spread nonetheless could rise sharply on less liquid debt markets, or briefly during a liquidity crisis.

The second comment concerns the uncertainty on date 0 about the identity of the CTD on $T$ and *the option offered to the seller to choose* on date $T$ the DS to be delivered. This *delivery option*, also called *quality option*, becomes a *wild card option* when the contract contains other optional clauses to the seller's benefit.[6] The quality option exposes an arbitrageur carrying out a *reverse cash and carry* to a risk. Indeed, if the seller delivers on date $T$ a bond $m'$ differing from the bond $m$ sold on the initial date by the arbitrageur (because, contrary to our assumption, $m'$ turns out on date $T$ to be cheaper to deliver than $m$), the result of the *reverse* is compromised. By symmetry, this risk favors the *cash and carry* arbitrage. Appendix 1 is devoted to valuing the quality option held by the seller of the notional contract.

Taking into account these two difficulties leads to revising the spot-futures parity which can be written for the *reverse* as follows:

$$f_m F_0^T - C_m(0) \geq \left((r_0 - s)C'_m(0) - 100k_m\right)T.$$

The spread $s$ represents the cost of borrowing the security (or the cost of the repo), possibly incremented by the cost of the delivery/wild card option.

As a result, the spot-futures relation is not expressed in the form of the exact equality (15.10a) but rather of a range of values:

$$\begin{aligned}\left((r_0 - s)C'_m(0) - 100k_m\right)T \leq & f_m F_0^T - C_m(0) \\ \leq & \left(r_0 C'_m(0) - 100k_m\right)T. \end{aligned} \tag{15.10b}$$

In the sequel, however, we ignore this asymmetry between *cash and carry* and its *reverse*. We set $s = 0$ and use the exact parity (15.10a), following the common practice.

---

[6] Some futures on bonds (such as the Chicago T-bond contract) give sellers a delay for choosing the exact moment of delivery. This is called a *timing option*. The *quality option* and the *timing option* together make up the *wild card option*.

## 15.1 Notional Contracts

**Table 15.2** Replicating a loan

| Transaction | Cash flows |
|---|---|
| Date 0 | |
| Sale of 100 $f_m$ contracts | 0 |
| Purchase of CTD | $-C'_m(0)$ |
| Intermediate dates $t \leq T$ | |
| Margin calls | $\sum_{t \leq T} f_m \left( F^T_{t-1} - F^T_t \right)$ |
| Date $T$ | |
| Adjustement of futures amount to 100 | 0 |
| Delivery of the CTDm | 0 |
| Cashing in the delvery price | $f_m F_T + ac_m(T)$ |

### Remarks

— The spot-futures parity is the result of the redundancy of three instruments: the futures contract, the underlying security ($m$), and a zero-coupon bond (borrowing/lending) of duration $T$. Indeed, risk-free lending (borrowing) for a duration $T$ can be replicated by a short (long) position in $f_m$ contracts and long (short) position in security $m$. Table 15.2 displays the case of lending (a *cash and carry* without the borrowing component).

By summing up the margins on date $T$, we obtain the characteristic sequence of a cash investment from 0 to $T$, $\left\{ -C'_m(0), \; f_m F_0 + cc_m(T) \right\}$, leading to the spot-futures parity. The return rate on this investment is

$$i_0 = \frac{f_m F_0 + ac_m(T) - C'_m(0)}{TC'_m(0)} = \frac{f_m F_0 + 100 k_m T - C_m(0)}{TC'_m(0)}.$$

Under no-arbitrage, the replicated rate is $r_0$, which implies Eq. (15.10a).

— The spot-futures Eq. (15.10a) does not take into account the impact of the margin call system. We already showed that the difference between forwards and futures vanishes when the risk-free interest rate evolves deterministically (see the Appendix to Chap. 9). However, such an assumption is not acceptable when analyzing products for which interest rate risk is an essential ingredient.

\* In that Appendix, we also presented the formula linking forward and futures prices under stochastic interest rates. In this chapter's Appendix 2, we apply that formula to forward/futures on interest rates (notional or short-term). *We also derive, in a particular but important case, the exact convexity adjustment to be made to the forward price when futures are used instead of forwards.* We denote by $F(t, T, T_B)$ and $\Phi(t, T, T_B)$, respectively, the price on date $t$ of a $T$-futures contract and a $T$-forward contract written on a bond $B(t, T_B)$ with maturity $T_B > T$. Using the one-factor Heath-Jarrow-Morton model explained in Chap. 17 (Sect. 2.1.2), we prove the following relationship:

$$\Phi(t, T, T_B) = F(t, T, T_B) \exp\left[\frac{\sigma^2}{2}(T_B - T)(T - t)^2\right] \tag{15.12}$$

which shows that the futures price is *smaller* than the forward price for contracts written on long-term bonds.

– The spot-futures equation and the futures prices (15.11) that results from it involve the proportional short-term money market rate with the same maturity $T$ as that of the futures. This can be also written using the compound version $y_0$ of the same rate using $(1 + y_0)^T = (1 + r_0 T)$ (see Chap. 2). We thus obtain the futures/forward price as a function of the dirty price of the CTD:

$$F_0^T = \frac{C_m'(0)(1 + y_0)^T - ac_m(T)}{f_m}. \tag{15.13a}$$

For reasons of computational convenience, we are sometimes led to use the continuous version of the money rate, i.e., $e^{z_0 T} = 1 + r_0 T$. The spot-futures equation then reads

$$F_0^T = \frac{C_m'(0)e^{z_0 T} - ac_m(T)}{f_m}. \tag{15.13b}$$

## 15.1.6 Interest Rate Sensitivity of Futures Prices

In this section, we analyze the sensitivity of futures prices to variations in interest rates. First, we examine the consequences of parallel shifts of the yield curve on the spot-futures parity (15.12). Then, as in Chap. 6, we extend the results to the general case of multifactor changes in the yield curve.

### 15.1.6.1 Parallel Shift of the Yield Curve

In this subsection, we keep the framework of simple sensitivity and duration analysis that we recall along with providing the notation. The zero-coupon rates are assumed to be all equal and subject to the same variations, as do the yields on the fixed-income risk-free securities like the DS in the basket. We use $y$ to denote this yield to maturity common to all these securities.

As in Chap. 5, we denote $S^F$ the sensitivity and $D^F$ the duration. By definition, applied to the futures price on date 0, we have,

$$S_0^F = -\frac{1}{F_0^T}\frac{\partial F_0^T}{\partial y} \text{ and } D_0^F = S_0^F(1 + y).$$

# 15.1 Notional Contracts

$(S^F_0 = D^F_0$ when $y$ is a continuous rate). We denote $C'_m(0) = C^m_0(y)$ to specify that the price for the bond $m$ is a function of the market rate $y$, or simply $C^m_0$. Recall that $C^m_0$ here is a dirty price.

Taking the derivative with respect to $y$ of the spot-futures parity (15.13a) we have

$$\frac{\partial F^T_0}{\partial y}(y) = \frac{C^m_0(y)(1+y)^{T-1}T}{f_m} + \frac{\frac{\partial C^m_0}{\partial y}(1+y)^T}{f_m},$$

where, by definition of the duration $D^m_0$ of the CTD on date 0, $\frac{\partial C^m_0}{\partial y} = -\frac{D^m_0}{1+y}C^m_0(y)$. Factorizing out $\frac{1}{1+y}$, we obtain

$$\frac{\partial F^T_0}{\partial y}(y) = \frac{1}{1+y}\left(\frac{C^m_0(y)(1+y)^T}{f_m}T - \frac{C^m_0(y)(1+y)^T}{f_m}D^m_0\right).$$

Factorizing out $\frac{C^m_0(y)(1+y)^T}{f_m}$ and rearranging now yields

$$\frac{\partial F^T_0}{\partial y} = -\frac{1}{1+y}\frac{(1+y)^T C^m_0}{f_m}(D^m_0 - T).$$

Therefore, the sensitivity $S^F_0$ is given by

$$S^F_0 = -\frac{1}{F^T_0}\frac{\partial F^T_0}{\partial y} = \frac{1}{1+y}\frac{(1+y)^T C^m_0}{(1+y)^T C^m_0 - ac_m(T)}(D^m_0 - T), \qquad (15.14)$$

and the duration $D^F_0 = (1+y)S^F_0$ by

$$D^F_0 = \frac{(1+y)^T C^m_0}{(1+y)^T C^m_0 - ac_m(T)}(D^m_0 - T).$$

We remark that the futures' duration equals that of the CTD $m$, reduced by the time to maturity $T$ of the contract, and adjusted by the ratio $A^F_0 = \frac{(1+y)^T C^m_0}{(1+y)^T C^m_0 - ac_m(T)}$. This adjustment coefficient, *which is slightly above one*, is linked to the presence on the date $T$ of an accrued coupon on the CTD which is missing on the futures (quoted at its clean price); it can be seen as the ratio of the dirty forward price of the CTD to its clean forward price. Indeed, on date 0, if no coupon has been detached by $m$ between 0 and $T$, the forward price $\Phi^m_0$ of the CTD equals

$$\Phi^m_0 = (1+y)^T C^m_0.$$

Additionally, note that the adjustment ratio $A^F_0$ can be written, from Eq. (15.13a), in terms of the futures price $F^T_0$: $A^F_0 = \frac{f_m F^T_0 + ac_m(T)}{f_m F^T_0}$. As a consequence

$$D_0^F = \frac{f_m F_0^T + ac_m(T)}{f_m F_0^T} \left(D_0^m - T\right). \tag{15.15}$$

These results are summarized in the following proposition.

**Proposition 4**

*The duration of the futures prices is equal to the duration of the cheapest to deliver reduced by the maturity $T$ of the contract, and corrected by the ratio of the dirty forward price of the cheapest to deliver to its clean forward price:*

$$D_0^F = A_0^F \left(D_0^m - T\right), \tag{15.16}$$

*where $A_0^F = \frac{(1+y)^T C_0^m}{(1+y)^T C_0^m - ac_m(T)} = \frac{f_m F_0^T + ac_m(T)}{f_m F_0^T}$.*

*In addition, the sensitivity $S_0^F$ is equal to $(1+y)D_0^F$ (with continuous rates $S_0^F = D_0^F$).*

**Remarks and Interpretations**

1. When the accrued coupon of the CTD on date $T$ is nil, the duration of the futures contract is exactly equal to $D_0^m - T$.
2. This result also allows recovering the interpretation of the forward contract's duration as the center of gravity (barycenter) of the payment dates of the cash flows with weights given by their discounted values. The forward contract's cash flows, seen from the buyer's point of view, are:
   (a) A negative cash flow on date $T$ equal to the delivery price.
   (b) Positive cash flows from the CTD corresponding to the coupon(s) and the capital on date $T$.

The duration of the CTD corresponds to the contribution to the barycenter of the positive cash flows (b), and the date $T$ contributes negatively to the barycenter because of the negative cash flow (a).

3. We have already remarked that the forward contract is equivalent to a portfolio made up of a long position in the underlying and a short position of the same amount in a zero-coupon with duration $T$. This portfolio's duration equals the difference in durations of the two components, that is $D_0^m - T$. The adjustment coefficient $A^F$ comes from the fact that the futures price is a clean price. This is also the reason why $D_0^m - T$ can be viewed as the forward duration $D_m^\Phi$ of the CTD $m$ on date 0. Furthermore, the durations of the futures contract prices and the forward price of $m$ are at any instant (in particular at 0) related by $D^F = A^F D_m^\Phi$.

### 15.1.6.2 Multifactor Deformations of the Rate Curve

Here we again take up the analysis in Sect. 5.2.1 of Chap. 5, using continuous rates in preference to the discrete yields to maturity, for mathematical simplicity. We denote by $(z_\theta)_{\theta \geq 0}$ the continuous zero-coupon rate curve on the valuation date

# 15.1 Notional Contracts

0, where $\theta$ represent the maturity. Any variation in the rate curve is described by a family of infinitesimal variations in the zero-coupon rates $(dz_\theta)_{\theta \geq 0}$.

We consider first the spot-futures equation (15.13b) and express it in terms of zero-coupon rates. We denote by $T_1, T_2, \ldots, T_m$ the coupon dates for the CTD *after* the expiry on date $T$ of the futures contract and by $\varphi_\theta$ the cash flow on date $\theta$ (coupon, or coupon plus nominal).[7] We have

$$F_0^T = \frac{e^{z_T T} \sum_{\theta=T_1}^{T_m} \varphi_\theta e^{-z_\theta \theta} - ac_m(T)}{f_m}.$$

Differentiating, we obtain

$$f_m dF_0^T = e^{z_T T} T dz_T \sum_{\theta=T_1}^{T_m} \varphi_\theta e^{-z_\theta \theta} - e^{z_T T} \sum_{\theta=T_1}^{T_m} \varphi_\theta e^{-z_\theta \theta} \theta dz_\theta. \tag{15.17}$$

Now let us choose a model for the way the rate curve changes. We saw in Chap. 5 that a model with $q$ factors $(b_1, \ldots, b_q)$ can be written on date $t$ as

$$z_\theta(t) = g\big(\theta, b_1(t), \ldots, b_q(t)\big),$$

which leads to the following description of deformation modes:

$$dz_\theta = \sum_{i=1}^{q} \frac{\partial g(\theta, \underline{b})}{\partial b_i} db_i.$$

The interest risk analysis of an instrument involves its sensitivity $b_i$ to each risk, with other factors being held constant:

$$dz_\theta = \frac{\partial g(\theta, \underline{b})}{\partial b_i} db_i. \tag{15.18}$$

- Replacing the terms $dz_\theta$ in Eq. (15.17) by their expressions from Eq. (15.18) yields

$$f_m dF_0^T = \left( e^{z_T T} \frac{\partial g(T, \underline{b})}{\partial b_i} T \sum_{\theta=T_1}^{T_m} \varphi_\theta e^{-z_\theta \theta} - e^{z_T T} \sum_{\theta=T_1}^{T_m} \varphi_\theta e^{-z_\theta \theta} \frac{\partial g(\theta, \underline{b})}{\partial b_i} \theta \right) db_i.$$

Dividing term by term by $C_0^m e^{z_T T}$, we get

---

[7] Recall that $\varphi_\theta = 100 k_m$ for the coupon payments on dates $(T_j = T + j - \tau_m)_{j=1,\ldots,N_m-1}$ and $100(1 + k_m)$ on the final date $T_m = T + N_m - \tau_m$.

$$\frac{f_m dF_0^T}{C_0^m e^{z_T T}} = \left( \frac{\partial g(T, \underline{b})}{\partial b_i} T - \frac{\sum_{\theta=T_1}^{T_m} \varphi_\theta e^{-z_\theta \theta} \frac{\partial g(\theta, \underline{b})}{\partial b_i} \theta}{C_0^m} \right) db_i.$$

Using the spot-futures parity $f_m F_0^T + ac_m = C_0^m e^{z_T T}$ gives

$$\frac{dF_0^T}{F_0^T} = \frac{f_m F_0^T + ac_m}{f_m F_0^T} \left( \frac{\partial g(T, \underline{b})}{\partial b_i} T - \frac{\sum_{\theta=T_1}^{T_m} \varphi_\theta e^{-z_\theta \theta} \frac{\partial g(\theta, \underline{b})}{\partial b_i} \theta}{C_0^m} \right) db_i.$$

The following proposition sums up these results.

**Proposition 5**

*The duration $D_{b_i}$ of the cheapest to deliver relative to the factor $b_i$ writes*

$$D_{b_i} = \frac{\sum_{\theta=T_1}^{T_m} \varphi_\theta e^{-z_\theta \theta} \frac{\partial g(\theta, \underline{b})}{\partial b_i} \theta}{\sum_{\theta=T_1}^{T_m} \varphi_\theta e^{-z_\theta \theta}}. \tag{15.19}$$

*The sensitivity of $F_0^T$ to factor $b_i$ is given by*

$$-\frac{1}{F_0^T} \frac{\partial F_0^T}{\partial b_i} = A_F \left( D_{b_i} - \frac{\partial g(T, \underline{b})}{\partial b_i} T \right) \tag{15.20}$$

where
$A_F = \frac{f_m F_0^T + cc_m}{f_m F_0^T}.$

It is useful to compare this result with that of the preceding proposition on the sensitivity of futures to a parallel shift of the rate curve. Comparing the two Eqs. (15.16) and (15.20), we see three similarities:

- The multiplicative coefficient $A_F = \frac{f_m F_0^T + cc_m}{f_m F_0^T}$ is the same in both cases and corresponds to the adjustment for the difference between clean and dirty prices.
- Since the computations involve continuous rates, duration and sensitivity coincide in this paragraph. The notion of duration/sensitivity, defined for a unique yield common to all maturities, is replaced by a more general notion which accounts for more complex changes in the rate curve than simple parallel shifts. The generalized duration can still be interpreted as a barycenter: The cash payment dates are just multiplied by the sensitivities of the rate curve to factors $b_i$. The weights remain unchanged and equal the discounted cash flows.
- In these two equations, we recover the fact that the duration of the CTD is decreased by the time to maturity $T$ of the futures contract, multiplied, in the case of a deformation governed by the factor $b_i$, by the sensitivity of the rate curve to this factor at maturity $T$.

## 15.1 Notional Contracts

**Remarks**

– Faced with a variation $(db_i)_{i=1,\ldots,q}$ in the $q$ risk factors, the relative variation of the futures price writes

$$-\frac{dF_0^T}{F_0^T} = A_F \sum_{i=1}^{q} \left( D_{b_i} - \frac{\partial g(T,\underline{b})}{\partial b_i} T \right) db_i$$

– Recall that these results have been obtained with continuous rates. A similar calculation can be carried out using discrete zero-coupon rates. Results are not much different, but the interpretation in terms of duration is not as easy, and the equations do not simplify in the same way as with continuous rates.

### 15.1.7 Hedging Interest Rate Risk Using Notional Bond Contracts

In this subsection, we handle the problem of hedging with notional futures an interest rate risk on an existing or an expected position. In the case of hedging daily an existing position, one protects the value of the position at a very short horizon (tomorrow), while for an expected position the hedging horizon can be arbitrarily distant.

#### 15.1.7.1 Hedging a Current Position

We consider here hedging an existing position in a financial security, and assume the problem is to compensate as much as possible for the daily variations in this position. In both cases, we assume that the financial security to be hedged is a debt obligation with sure cash flows, whose amounts and payment dates are known on the date of hedging. For example, this could be a Treasury bond or a (presumed default-free) corporate debt. Denote by $P_t$ the market value of the position to be hedged on date $t$, as a percentage of the nominal, and let $100w$ be the nominal amount of this position, so that its market value equals $wP_t$; $w$ is positive for a long position and *negative* for a short one. $S_t^P$ and $D_t^P$ denote, respectively, the sensitivity and the duration of the position to be hedged on the current date $t$.

(a) *Single-factor hedging (parallel shifts of the rate curve)*

As in the preceding Sect. 15.1.6, we first consider the case of parallel shifts of the rate curve. The duration quantifies the risk one wishes to eliminate, namely variations in a unique rate, denoted by $y$, that determines the value of both the position to be hedged and the futures contract used for hedging. The idea is to build a portfolio composed of the existing position and futures contracts whose global duration vanishes. We consider an *instantaneous* variation $\Delta y$ provoking value changes in all fixed-income securities. We suppose that the variation is sufficiently small that a Taylor series expansion in the neighborhood of 0 is reasonable. $F_t$ *is* the

630 15 Futures Markets (2): Contracts on Interest Rates

futures price on date $t$, quoted as a percentage of the nominal, and $S_t^F$ its sensitivity. Let $100x_t$ be the unknown nominal amount of the position in futures ($x_t > 0$ or $x_t < 0$ according to whether the position is long or short); $x_t \Delta F$ then represents the margin caused by a variation $\Delta F$ in the futures price. For a very small change $\Delta y$, the change in the position's value is, at the first order

$$w\Delta P = -wS_t^P P_t \Delta y. \tag{15.21}$$

A hedge at $t$ consists in selecting the amount $x_t$ of the futures position so that the margin call compensates for the change in the existing position's value. Since the margin call is $x_t \Delta F$, a first-order approximation is

$$x_t \Delta F = -x_t S_t^F F_t \Delta y.$$

Canceling the sum of the two variations, $w\Delta P + x_t \Delta F = 0$, gives $wS_t^P P_t \Delta y = -x_t S_t^F F_t \Delta y$.

Therefore, choosing $x_t/w = S_t^P P_t / S_t^F F_t$ eliminates interest rate risk up to the first order.

Now denote by $y_p$ the yield-to-maturity (YTM) of the position whose value is $P$ and by $y_m$ the YTM of the CDT (they may differ but are assumed subject to the same shift $\Delta y$). In terms of durations, the previous equation writes: $x_t/w = D_t^P P_t (1 + y_p) / D_t^F F_t (1 + y_m)$. When rates are continuous, or when $y_p$ is close enough to $y_m$, the last equation simplifies to: $x_t/w = D_t^P P_t / D_t^F F_t$.

The next proposition summarizes these results.

## Proposition 6

*Denote by $100w$ the nominal amount of the position to be hedged and by $100x$ that of the futures contracts.*

(i) *The hedge ratio $\frac{x_t}{w}$ for an immediate horizon is equal to the ratio (changed in sign) of the sensitivities multiplied by the prices:*

$$\frac{x_t}{w} = \frac{S_t^P P_t}{S_t^F F_t}, \tag{15.22a}$$

*and, equivalently, in terms of durations:*

$$\frac{x_t}{w} = \frac{D_t^P P_t}{D_t^F F_t} \frac{(1 + y_m)}{(1 + y_p)},$$

*or, approximately:*

## 15.1 Notional Contracts

$$\frac{x_t}{w} = \frac{D_t^P P_t}{D_t^F F_t} = \frac{D_t^P P_t}{A_t^F \left(D_t^m - (T - t)\right) F_t}. \tag{15.22b}$$

(ii) *The number $N_t$ of futures contracts with nominal $M'$ needed to hedge a position with a nominal $M$ is therefore:*

$$N_t = -\frac{M}{M'} \frac{S_t^P P_t}{S_t^F F_t} \simeq -\frac{M}{M'} \frac{D_t^P P_t}{D_t^F F_t} = -\frac{M}{M'} \frac{D_t^P P_t}{A_t^F \left(D_t^m - (T - t)\right) F_t}.$$

For example, for a Euro-bund contract whose nominal is $M' = 100{,}000$ €, a position in $N$ contracts corresponds to an $x = 1000$ € $N$. If the position's nominal to be hedged is $M = 100$ € $w$, the number of contracts bought ($N > 0$) or sold ($N < 0$) on date $t$ must be equal to:

$$N_t = -\frac{M}{100{,}000} \frac{S_t^P P_t}{S_t^F F_t} \simeq -\frac{M}{100{,}000} \frac{D_t^P P_t}{D_t^F F_t}.$$

### (b) *Extension to multifactor hedging*

The hedging just discussed protects against the risk of a parallel shift of the zero-coupon rate curve. Cancelling out a multifactor risk means one has to use a number of *different* futures contracts equal to the number of sources of risk (factors). We retain the notation from Sect. 15.1.6 above but, for the sake of simplicity, we do not make explicit the subscript $t$ indicating the current date. Let us consider the zero-coupon rate curve as a function of $q$ sources of risk $(b_i)_{i = 1, \dots, q}$:

$$z_\theta = g\left(\theta, b_1, \dots, b_q\right).$$

The $q$ hedging instruments are futures contracts (with different underlyings having different durations) whose current prices are denoted by $F_1, \dots, F_q$. A more complex set of hedging instruments would include short-term futures contracts of the type studied in the following Sect. 15.2. We denote by $S_P^1, \dots, S_P^q$ the sensitivities of the position $P$ to be hedged relative, respectively, to variations $\Delta b_1$, $\dots, \Delta b_q$ in the $q$ sources of risk. In the same way, we use $S_j^i$ for the sensitivities of the futures contract $j$ with respect to the variation in risk $b_i$, for $i, j = 1, \dots, q$. We need to determine the $q$ unknown values $x_1, \dots, x_q$ representing the amounts of the $q$ futures to be bought or sold. Repeating the approach above for a very small variation in each $b_i$, we obtain for any $i = 1, \dots, q$ :

$$wS^i_P P \Delta b_i = - \sum_{j=1}^{q} x_j S^i_j F_j \Delta b_i.$$

This result is the object of the next proposition.

## Proposition 7

*The quantities $x_1, \ldots, x_q$ representing the nominal amounts of futures contracts to be bought or sold to hedge an existing position are solutions to the following linear system of $q$ equations in $q$ unknowns:*

$$wS^1_P P = - \sum_{j=1}^{q} x_j S^1_j F_j$$

$$\cdots \qquad (15.23)$$

$$wS^q_P P = - \sum_{j=1}^{q} x_j S^q_j F_j.$$

Recall that sensitivities and durations are equivalent either when interest rates are continuous or, if discrete, when the yields $y_1, \ldots, y_q$ and $y_p$ respectively related to the CTD of contracts $1, \ldots, q$ and to the position $P$ can be considered equal.

---

### Example 5

We have to hedge a long position with a nominal amount of 100M € invested in a European long-term government bond (coupon rate 5%, maturity 12 years) whose price is denoted by $P$. We consider two cases of increasing complexity.

*First case: Cancelling the risk of variation in the yield to maturity (parallel shift)*

As in the previous examples, we assume that the only YTM on the market is 4.50%. The investor fears an increase in rates leading to a capital loss in her investment in fixed-rate bonds and hedges with the 10-year June 2005 contract previously introduced. The following table reports the intermediate calculations needed to establish the hedge ratio.

| Rate | 4.50% | 4.60% |
|------|-------|-------|
| $100w$ | 100M € | 100M € |
| $P$ | 104.56 | 103.63 |
| $D^P$ | 9.37 | |
| $F$ | 110.99 | 110.21 |
| $D^F$ | 7.35 | |

Futures prices are computed by taking as CTD the 4.25% 04-Jan-2014, in accordance with our previous examples. Note that for $y = 4.50\%$, the duration

(continued)

# 15.1 Notional Contracts

633

of the cheapest is equal to 7.47 years. The futures contracts expire in 0.25 year, and the adjustment factor $A^F$ is 1.018. With these values, we can compute the futures' duration using Eq. (15.16): $D^F = 1.018 \times 7.47 - 0.25 = 7.35$ (the figure appearing in the table).

Applying Eq. (15.22b), we find $100\, x = 120.09M$ €.

The investor minimizing her interest rate risk must sell 120.09M € worth of 10-year futures contracts, i.e., 1201 contracts. The approximation consisting in neglecting the adjustment factor $A^F$ would lead to selling 1223 contracts, a significant difference.

We can check the quality of the hedge by considering a scenario where the rate increases by 10 basis points, raising the YTM to 4.60%. From the table above we deduce that the loss is: $-104.56 + 103.63 = -0.93$. Margin calls on the $120.09M in futures amount to: $-120.09 \times (110.21 - 110.99) = 0.93$. The sum of these two quantities (delaying the rounding to the basis point) is zero. We thus have checked that hedging cancels to the first order the variations in the position's value. This result, however, does depend on the assumption of a uniform variation in all rates.

*Second case: cancelling a multifactor risk.*

We consider the example of a two-factor linear exponential model of the zero-coupon rate curve (derived from a model due to Vasicek with two variables):

$$\Delta z_\theta(t) = \frac{1 - e^{-a_1\theta}}{a_1\theta} \Delta b_1(t) + \frac{1 - e^{-a_2\theta}}{a_2\theta} \Delta b_2(t).$$

For the numerical calculation, we take $a_1 = 0.05$ and $a_2 = 0.2$.

Hedging against this type of rate curve deformation is achieved with two futures contracts. In addition to the 10-year June 2005 contract already introduced, we use a contract on a notional 5-year Euro-BOBL, 6%, expiring 10-06-2005. The corresponding basket of DS consists in two BOBL (medium-term debt issued by the German state) described in the table below:

| Security | Conversion factor |
|---|---|
| BOBL 5.375%, 04-Jan-2010 | 0.975265 |
| BOBL 5.25%, 04-Jul-2010 | 0.967955 |

The cheapest to deliver is the BOBL 5.375%– 04-Jan-2010 (the one with the smallest duration). We assume that on the date of hedging the rate curve is flat at the level 4.50%. In contrast, the deformations considered are of

(continued)

exponential type, and the rate curve does not remain horizontal after a variation in $b_1$ and/or $b_2$. The next table provides the values needed to calculate the hedge ratios.

| $y$ | $w$ | $P$ | $S_P^1$ | $S_P^2$ | $F_1$ | $S_1^1$ | $S_1^2$ | $F_2$ | $S_2^1$ | $S_2^2$ |
|------|-----|--------|------|------|--------|------|------|--------|------|------|
| 4.50% | 100 | 104.55 | 7.25 | 3.93 | 110.99 | 5.98 | 3.51 | 106.14 | 3.71 | 2.65 |

The linear system of two equations and two unknowns to be solved then writes

$$757.82 = -(6.639x_1 + 3.9403x_2)$$
$$410.43 = -(3.8958x_1 + 2.8132x_2)$$

and its solution is

$$x_1 = -154.74 \text{ and } x_2 = 68.39.$$

The two-factor hedge is therefore achieved by selling 154.74 M € of 10-year futures (1547 contracts) and buying 68.39 M € of 5-year futures (684 contracts).

### 15.1.7.2 Hedging an Expected, but Known, Position

When a position is expected *for sure* on a future date $H$ (horizon), e.g. a new investment or a new financing, the random variable to be canceled is the one that affects its value *on this date H*. Two possible strategies among several conceivable are:

- To hedge at every instant $t$ the position's value on $t + dt$. This is *dynamic* hedging, in practice implemented daily, which was studied in the preceding subsection. The hedge has to be adjusted (daily) so as to maintain the level $x_t$ given by Eqs. (15.22a) or (15.22b).
- To build, on date 0, a hedge that is not modified from 0 to $H$. This is *static* hedging to which we turn now.

A perfect static hedge consists simply in forward buying (or selling) the (sure) expected position whose value is to be guaranteed on date $H$. Since, in general, the expiry $T$ of the contract does not coincide with the horizon $H$ and the hedge's underlying (e.g., the Bund $m$) differs from the expected position, hedging is imperfect, since tainted by basis and correlation risks (see Chap. 9). We examine the same two cases as in the previous subsection.

## 15.1 Notional Contracts

### (a) *Parallel shift of the rate curve*

In this case, the interest rate risk can be perfectly hedged at the first order. Following the conventions introduced above, we denote by $P_H(y)$ the value on date $H$ of the expected position as a function of its yield to maturity $y$, and by $S_H^P(y)$ its sensitivity at $H$. $100w$ still is the nominal amount of the position to be hedged (long, $w > 0$, or short, $w < 0$). If the rate $y$ varies from $y_0$ to $y_0 + \Delta y$ between 0 and $H$, *the position's value at $H$ equals* $wP_H(y_0 + \Delta y)$. The random variation $\Delta y$ is the source of the risk which is to be hedged by a position in the notional contract. This hedge, *established on date 0 and not revised between 0 and $H$*, is made up of a nominal amount $100x$ of notional contracts with expiry $T \geq H$ which generates daily margin calls that accumulate from date 0 and are equal to $x(F_H^T - F_0^T)$ on date $H$. This hedge aims to offset, exactly if possible, the risk that affects $wP_H(y_0 + \Delta y)$ with a random term of opposite sign $x(F_H^T - F_0^T)$.

The value $P_H^*$ at $H$ *of the hedged* position writes, as a function of $\Delta y$:

$$P_H^*(y_0 + \Delta y) = wP_H(y_0 + \Delta y) + x[F_H^T(y_0 + \Delta y) - F_0^T(y_0)] = 0.$$

A Taylor series expansion at the first order gives

$$P_H^*(y_0 + \Delta y) - P_H^*(y_0) = [wS_H^P P_H(y_0) + xS_H^F F_H^T(y_0)]\Delta y,$$

where the sensitivities $S_H^P$ and $S_H^F$ are computed on date H with a yield $y_0$.

Choosing the futures position that, up to the first order, cancels the effect of $\Delta y$ on $P_H^*$, that is the special value $x^*$ that cancels the term in brackets in the last equation, leads to the next proposition whose result is used in practice.

### Proposition 8

– *The hedge ratio $\frac{x^*}{w}$ to be established on date 0 to hedge a position with nominal $100w$ expected on date $H$ is equal to*

$$\frac{x^*}{w} = -\frac{S_H^P P_H(y_0)}{S_H^F F_H^T(y_0)}. \tag{15.24}$$

*where the sensitivities are computed on date H.*

– *The number of futures contracts with nominal $M'$ needed to hedge an expected position with nominal $M$ thus equals*

$$N^* = -\frac{M}{M'}\frac{S_H^P P_H(y_0)}{S_H^F F_H^T(y_0)}.$$

This equation generalizes Eq. (15.22a) that holds (with $H = 0$) for instantaneous hedging. We emphasize the difference between Eqs. (15.24) and (15.22): For hedging on date $H$, the durations and prices present in the hedge ratio are calculated at $H$ and not at 0. Furthermore, this ratio is computed using the yield $y_0$ prevailing on date 0: it is therefore known on date 0, the date at which the hedge is established. We note that $D_H^P = D_0^P - H$ (if no coupon is distributed from 0 to $H$) and that $D_H^F = D_0^F - H$. Also worth noting is that, if there is a unique yield irrespective of the maturity, $y_0$ denotes both the spot rate and the forward rate on date 0. Therefore, the hedge ratio is the ratio of the forward prices multiplied by the forward durations.

---

**Example 6**

Resuming the preceding example of a long position in the nominal amount of 100 M € invested in a 12-year, 5%, bond hedged with a Euro-bund with expiry $(T)$ on date 10/July/2005. Let us consider the situation on 10/Feb/2005 (date 0) and suppose the position has to be hedged on 10/April/2005 (date $H$). We assume that on date 0 a unique YTM of 4.50% prevails on the market. Futures prices are computed by taking as the cheapest the bund 4.25% 04-Jan-2014, in accordance with the results previously given. There are 59 days between 10/Feb/2005 and 10/April/2005 and, for $y_0 = 4.50\%$, $D_H^P = D_0^P - H$

$$= 9.37 - 59/365 = 9.21, \ D_H^F = D_0^F - H = 7.35 - 59/365 = 7.19,$$

$P_H = 104.56 + 5 \times 59/365 = 105.37$ and the futures price is 111.02. Applying formula (15.24) then gives a hedge ratio of $-1.216$ (instead of $-1.2009$ in the previous example). It is thus necessary to sell 1216 contracts.

---

**Remark**

When the hedge is perfect, the futures position allows *locking in the yield (or the cost)* of the expected position $100w$ or, equivalently, locking in the unknown price $P_H$ of the position to be hedged *at the current forward price*.

### (b) *Deformation of the rate curve*

The preceding analysis assumes that the same rate $y$ governs $P_H(y)$ and $F_H^T(y)$. This condition can only prevail if the two following conditions are satisfied: (1) $H = T$; (2) the position to be hedged is identical or very close to the underlying $m$, in terms of maturity, coupon rate, and credit risk.

In effect, from Eq. (15.13a) written for the date $H$, at a distance $T$-$H$ from the contract's expiration $T$: $F_H^T = \frac{C_m'(H)(1+y_{T-H}(H))^{(T-H)} - ac_m(T)}{f_m}$ where $y_{T-H}(H)$ denotes the short-term rate (with maturity $T - H$) prevailing at $H$. As a result:

## 15.1 Notional Contracts

- If $T > H$, $F_H^T$ is a function of a short-term rate and a long-term rate $(F_H^T(y_{T-H}(H), y))$ and cannot be a perfect hedge of the long-term rate risk: Hedging with only notional futures introduces a *basis risk*;
- If $C_m'$ and $P$ are controlled by two different long-term rates,[8] the contract cannot be a perfect hedge for the interest rate risk affecting $P$, even if $H = T$: The hedge introduces a *correlation risk*.

A more precise hedge of the risk affecting $P$ must especially handle the risk of a variation in rates that differs according to their maturities (a deformation and not a simple translation of the rate curve) and involves several different contracts. For example, the basis risk that involves a short-term rate could be hedged by adding a futures contract on this rate to the notional contract on the long-term rate.

More generally, for a multifactor deformation of the rate curve, let us denote by $S_P^{\Phi(1)}, \ldots, S_P^{\Phi(q)}$ the forward sensitivities relative to the variations $\Delta b_1, \ldots, \Delta b_q$ in the sources of risk introduced in the previous subsection. By following the method used to determine an instantaneous hedge, we obtain the next proposition.

**Proposition 9**
*The hedges $x_1, \ldots, x_q$ giving the nominal amounts of futures contracts to buy or sell to hedge an expected position as described above are solutions to the linear system of q equations in q unknowns*

$$w S_P^{\Phi(1)} P_a^f = -\sum_{j=1}^{q} x_j S_j^{\Phi(1)} F_H^j$$

$$\cdots \tag{15.25}$$

$$w S_P^{\Phi(q)} P_a^f = -\sum_{j=1}^{q} x_j S_j^{\Phi(q)} F_H^j,$$

*where $P_a^f$ denotes the forward price of the expected position to be hedged.*
*The hedger thus can lock, on date H, the random value $P_H$ at its current forward value.*

### 15.1.8 The Main Notional Contracts

Our presentation of the main notional contracts worldwide is brief and the reader thus is invited to check the internet sites of the different markets for complementary material and updates.

---

[8] The gap between the two rates may be due to differences in maturities, coupon rates, liquidity or credit risk.

### 15.1.8.1 Brief Description of the Main Medium- and Long-Term Notional Contracts

In the United States, the major part of the activity on interest rate futures is handled by the CME Group which resulted from the merger in 2007 of the Chicago Board of Trade (rather specialized on notional contracts with long- and medium-term bonds issued by the US Treasury) and the Chicago Mercantile Exchange (which includes the International Money Market, very active in contracts with short underlying assets such as US Treasury notes or 3-month deposits in Eurodollars). The success of American contracts has emulated numerous other financial markets. London was the first European market (LIFFE) to introduce futures on interest rates. It was followed by Paris (MATIF), Amsterdam, Copenhagen, Barcelona, Zurich, Brussels, and Frankfurt. Following upon the different restructurings mentioned in previous chapters, the two main markets in Europe are presently the EUREX (in Frankfurt) and the ICE Futures Europe (ex EURONEXT-LIFFE).[9]

In these two markets, notional contracts are traded on medium or long-term notes and bonds, the main ones of which are the Euro-bund on a notional bond quoted on EUREX. Markets outside Europe have not remained inactive: interest rate futures are traded in Sidney, Singapore, Tokyo, Montreal, Rio de Janeiro, etc. In particular, contracts on Japanese State bonds are traded in the Tokyo and Singapore exchanges.

In terms of volume, the Euro-bund contract is the most important contract traded in Europe. The other important long-term notional futures are traded on ICE Futures Europe and CME Group in the US.

- On the CME, the main notional futures is the classical T-bond contract. Since its introduction in 1977, its success has never been denied. Its nominal is $100,000 and its basket of deliverable securities is made up of US Treasury bonds with 15–30 years to live from the delivery date. T-bond futures function according to the same principles as Euro-bund futures, the latter being in fact copied from the former. However, notably because of the differences in bond coupon payment frequency (twice a year in the US case), certain formulas used in computations must be changed, in particular those relating to the conversion factor. The nominal rate $k$ of the notional bond (with which the conversion factors are computed) is 6%.

The CME also trades other futures contracts on bonds and US Treasury Notes. Their main characteristics are summarized in the following table.

| CME Notional futures contracts on Treasury notes and bonds | | |
|---|---|---|
| Name of the contract | Remaining time to maturity of eligibles for delivery | Nominal amount |
| 2 year T-Note | Between 1 year 9 months and 2 years | $200,000 |
| 3 year T-Note | Between 2 years 9 months and 3 years | $200,000 |

(continued)

---

[9] See Chap. 9, Sect. 9.4.

## 15.1 Notional Contracts

CME Notional futures contracts on Treasury notes and bonds

| Name of the contract | Remaining time to maturity of eligibles for delivery | Nominal amount |
|---|---|---|
| 5 year T-Note | Between 4 years 2 months and 5 years 3 months | $100,000 |
| 10 year T-Note | Between 6 years 9 months and 10 years | $100,000 |
| Ultra 10 years T-Note | Between 9 years 5 months and 10 years | $100,000 |
| Classic T-Bond | Between 15 years and 25 years | $100,000 |
| Ultra T-Bond | More than 25 years | $100,000 |

– In Europe, the Euro-bund contract, which served as a reference in the analysis and numerical examples above, is the most important futures contract in terms of volume, as previously mentioned. The other main notional futures are traded on EUREX and ICE Futures Europe, EUREX being the most active. In addition to the Euro-bund, different contracts whose underlyings are fixed-rate securities issued by European governments are traded on EUREX. Their common features are: the face rate of the notional is 6% (with one exception; see the table below); the nominal amount is 100,000 € or CHF100,000; the expiration dates are the same as for a Euro-bund contract. They differ by the issuer, the maturity of the underlyings, and exceptionally by the face rate of the notional used for computing conversion factors. These characteristics are summarized in the following table.

EUREX notional contracts on short-, medium-, and long-term European Treasury securities

| Name of the contract | Government issuer | DS remaining term in years | Notional face rate $(k)$ |
|---|---|---|---|
| Euro-Schatz Futures | German | 1.75–2.25 | 6% |
| Euro-Bobl Futures | German | 4.5–5.5 | 6% |
| Euro-Bund Futures | German | 8.5–10.5 | 6% |
| Euro-Buxl® Futures | German | 24.0–35.0 | 4% |
| Short-Term Euro-BTP Futures | Italian | 2.0–3.25 | 6% |
| Mid-Term Euro-BTP Futures | Italian | 4.5–6.0 | 6% |
| Long-Term Euro-BTP Futures | Italian | 8.5–11.0 | 6% |
| Mid-Term Euro-OAT Futures | French | 4.5–5.5 | 6% |
| Euro-OAT Futures | French | 8.5–10.5 | 6% |
| Euro-BONO Futures | Spanish | 8.5–10.5 | 6% |
| CONF Futures (CHF) | Swiss | 8.0–13.0 | 6% |

– On the ICE Futures Europe, contracts in Sterling are traded on British state securities (Gilts) with very long (DS having a maturity of around 30 years), long (10 years), medium (5 years), and short (2 years) terms. These contracts are governed by the same principles (notional security, delivery of a DS) as their analogs on the EUREX or CME. By contrast, contracts on swap notes, briefly discussed below, are based on different principles.

### 15.1.8.2 Contracts on Swap Notes

If most notional futures are based on the principle of delivering a genuine security whose properties are near those of the notional (which may be understood as a loose reference), it is possible to construct contracts defined with respect to a very precise theoretical standard that does not give rise to physical delivery but to *cash settlement*. To achieve this, one must define precisely how the exact price $F_T$ upon expiry is calculated.

On its London platform, ICE offers such contracts, written on swap notes. The underlying notes have very long (30 years), long (10 years), medium (5 years), or short (2 years) terms, and they are denominated in different currencies (€, $ and £). We describe briefly the contracts in Euros, the others being similar. There are four futures defined on notional bullet instruments having exactly 2, 5, 10, or 30 years to run upon expiry of the contract, for a nominal 100,000 €, with a face rate fixed at 6%. As they do not give rise to delivery, the contract is defined by the procedure specified for computing the closing price at expiration (*settlement price*). This final price $F_T$ is computed by discounting the sequence of cash flows (coupons and principal) of the notional. Such a discounting is carried out using the zero-coupon rates read from the swap rate curve in Euros prevailing on the expiration date, whence the terminology *swap notes*.[10] The contract's underlying thus is to be considered as a medium-term note (for swap notes of 2 and 5 years) or as a bond (for a 10-year swap note) whose price is governed by the vanilla swap rates. Before and at expiry, the price resulting from supply and demand must equal the theoretical forward price of the underlying, i.e., the discounted value, on the date of valuing, of the cash flows it generates. All discounting has to be carried out using the zero-coupon rates extracted from the swap rate curve. Recall that swap rates are slightly higher than rates for government securities with the same maturity (the spread is some twenty basis points or so, but is variable) and slightly lower (by several basis points) than and highly correlated with rates for corporate securities. These spreads are mainly explained by the differences in credit and liquidity risks (see Chap. 7). The correlation between swap and corporate rates higher than between swap and sovereign rates indicates that swap notes are better hedging instruments than notional contracts on government securities, if the position to be hedged is made up of private securities.

---

[10]The zero-coupon rates, or equivalently the conversion factors, are read from the rate curve constructed by ICE, computed daily using a standard algorithm (see Chap. 7) and displayed on Reuters.

## 15.2 Short-Term Interest Rate Contracts (STIR) (3-Month Forward-Looking Rates and Backward-Looking Overnight Averages)

Generally speaking, Short-Term Interest Rate futures contracts (STIR) share the following features:

- They are referenced on a short-term underlying rate (1 month, 3 months, 1 or 3-month average of an overnight rate) but the contract maturity may be much longer, up to 10 years.
- They are *cash settled* (*no delivery* of an underlying asset).
- The price is quoted as 100 minus a reference rate, which is expressed in % and annualized.
- Expiry dates are usually (but not always) scheduled on International Monetary Market dates: the third Wednesday of March, June, September, and December.
- STIR futures are based on two different types of rate: either the reference rate is a forward-looking interest rate, usually a 3-month LIBOR, or it is a backward-looking average of overnight rates. The 3-month LIBOR-type futures are studied in Sect. 15.2.1, the contracts on 3-month and 1-month overnight averages in Sect. 15.2.2, and hedging interest rate risk with STIR in Sect. 15.2.3.

### 15.2.1 STIR 3-Month Contracts (LIBOR Type, Forward-Looking)

Futures contracts on short term interest rates (STIR) of the forward-looking (FL) type often reference a 3-month FL interbank rate: 3-month $-Libor for Eurodollars futures and 3-month T-bill on the CME Group, Euribor-3 months for Eurex and ICE Futures Europe contracts, etc. Sects. 15.2.1.1 and 15.2.1.2 describe the general properties of such a STIR-3-month contract and its underlying and Sect. 15.2.1.3 compares a FL STIR with a FRA. The main contracts on short-term rates (3-month LIBOR type) are briefly presented in Sect. 15.2.1.4.

In what follows, we will take the 3-month Eurodollar contract on the CME as an example that can be directly transposed to other contracts on 3-month FL rates (Euribor, Libor, T-bill futures contracts, etc.).

#### 15.2.1.1 Quotation, General Description and Margin Calls for the 3-Months STIR Contracts

Let us first note that, in this section, *a rate expressed as a percentage is denoted by a capital letter and its decimal version by a letter in lower-case* (e.g., $R = 5$ and $r = 0.05$ represent the same 5% rate). Forward rates are denoted by an $f$ in decimal form and either by $100f$ in their percentage version or by a generic $R$; for instance $100f_{T,3}(t)$ denotes the 3-month-forward for $(T, T + 3$ months$)$ prevailing at $t$ *expressed in percentage form, per annum.*

The quotation convention for the 3-month STIR contracts is that futures prices are equal to 100 *less an annual interest rate expressed as a percentage*, that we call the short-term rate implicit in the contract. This rate, conventionally denoted by a generic $R$, represents in fact here a 3-month *forward* rate, as we will soon show (it converges toward the 3-month *spot* rate $R_3(T)$ as $t$ tends to maturity $T$). As a result, the quoted futures "price" for maturity $T$ on date $t$ writes explicitly:

$$F_t^T = 100 - R_t^T. \tag{15.26}$$

The reader should avoid any confusion between this quotation and a genuine futures price. The $F$ resulting from an equation such as Eq. (15.26) is a quotation *convention different* from a "true" futures price of a real security. Moreover, like long-term notional contracts, the futures price is quoted as a percentage of the nominal amount of the contract (up to 3 decimal places), which for most STIR-3-month contracts is 1,000,000 in local currency.[11] On expiration date $T$, the closing price is fixed by the clearing house using the 3-month spot rate prevailing at $T$ (or at $T+2$ days),[12] denoted by $R_3(T)$, as given by the formula

$$F_T^T = 100 - R_3(T).$$

For the *Eurodollar* contract, for instance, $R_3(T)$ denotes the 3-month spot $-LIBOR (in %).

Before expiration, the margin call for day $j+1$ is computed with the following formula, for a nominal of $100,

$$MC^T{}_{j,j+1} = 0.25\left(F^T{}_{j+1} - F^T{}_j\right),$$

where $F^T{}_{j+1}$ and $F^T{}_j$ are, respectively, the closing quoted prices on days $j$ and $j+1$, and 0.25 is the duration of the quarter. Since most STIRs on 3-month LIBORs have a face value of 1,000,000 units of the local currency, a 0.01% move on $F$ (or $R$) entails a change of 25 units of local currency. In the case of the Eurodollar contract, for instance, one *tick* move (0.005 on the price $F$, i.e., 0.5 *basis point*) is thus equal to $12.5.

The cumulated gain (loss) of holding between $t$ and $T$ a long position on a nominal of 100 of such a contract thus is: $G(t,T) = 0.25\left(F_T^T - F_t^T\right) = 0.25(R_t^T - R_3(T))$ $[= 0.25(R_t^T - \text{LIBOR}_{3\text{month}}(T))$ in most contexts].

This contract thus allows to exchange on date $t < T$ the *future* spot $\text{LIBOR}_{3\text{month}}$ $(T)$ for the *current* rate $R_t^T$: with some provision, $R_t^T$ can thus be interpreted as a 3-month *forward* rate prevailing at $t$ for the period $(T, T+3$ months).

---

[11] A notable exception is Japanese contracts whose nominal is JPY100,000,000.

[12] We do not distinguish in the following analysis the closing date $T$ of transactions from the reference date for the final settlement price (often $T+2$ days). However, we make the distinction in the numerical example (7, item b).

## 15.2 Short-Term Interest Rate Contracts (STIR) (3-Month Forward-Looking...

**Example 7: Eurodollar Futures Transactions**

On the morning of Monday 09-25-n, an investor buys 20 Eurodollar contracts with a December expiry. The order is carried out at 98.400, implying that the implicit 3-month Eurodollar forward rate is 1.60%.

1. *Position wound up before expiry*

On the evening of 09-25 the closing price is 98.430 (the forward Eurodollar rate has thus decreased by 3 basis points (bps)), i.e., 6 ticks up per contract. The investor's account is thus credited with:

$$20 \times (98.430 - 98.400) \times 0.25 \times 1,000,000/100$$
$$= \$1500 \text{ (or simply } 20 \times 6 \text{ ticks} \times \$12.5).$$

On following days the margins on these 20 contracts evolve as in the table below:

| Dates | Daily closing Price | Margins (\$) Calls (−) or Repayments (+) |
|---|---|---|
| 09-25 | 98.430 | +1500 |
| 09-26 | 98.455 | +1250 |
| 09-27 | 98.410 | −2250 |
| 09-28 | 98.390 | −1000 |
| 09-29 | 98.420 | +1500 |

On the evening of 09-29 the agent closes the position by selling 20 contracts (at 98.420). The account is credited with \$1500 as indicated in the table. Since contracts were bought at 98.400 and sold at 98.420, the net gain is 4 ticks per contract so that the total net gain is: $20 \times 4 \text{ ticks} \times \$12.5 = \$1000$ (= +1500 + 1250 − 2250 − 1000 + 1500), without taking into account the discounting of the cash flows.

2. *Position held until expiry*

We assume now that the buyer of these 20 Eurodollar contracts at 98.400 holds his position until the expiry of the contract on Monday December 17. On Wednesday 19 (the third Wednesday of December), the 3-month spot \$LIBOR equals 1.505% entailing a final settlement price for the December Eurodollar contract of 98.495 and a gain for the buyer of $(98.495 - 98.400) = 9.5 \text{ bps} = 19$ ticks per contract, hence a total gain on the whole position of $20 \times 19 \times 12.5 = \$4750$.

(continued)

## 3. Remarks

(a) The buyer of these 20 Eurodollar futures may hedge against the risk of decreased earnings on \$20M to be placed for 3 months on mid-December. In that respect, the Eurodollar futures works like an FRA (Forward Rate Agreement) with slight differences that are indicated in the sequel. In both cases, the interest rate on this future investment is locked in at 1.60%.

(b) We chose Eurodollar contracts for this numerical application. In fact, all the STIR-3 month-futures being almost identical, the example is a relevant illustration for almost any such contract. One can, for instance, simply substitute "Euribor contracts" for "Eurodollar contracts" and € for \$ without any other change.

### 15.2.1.2 Alternative Formulation and Definition of the Underlying Security

Let us consider a second (fictitious) futures contract whose quoted price on date $t$ is computed using the previous rate $R_t^T$ in the formula

$$\Psi_t^T = 100 - 0.25 R_t^T, \tag{15.27}$$

with, at expiry $T$, $\Psi_T^T = 100 - 0.25 R_3(T)$ where $R_3(T)$ is the referenced 3-month spot rate prevailing at $T$, usually a 3-month LIBOR. As the true STIR, this fictitious contract is subject to margin calls which, for a nominal of 100, are equal to: $MC^T_{j,j+1} = \Psi^T_{j+1} - \Psi^T_j = 0.25\left(F^T_{j+1} - F^T_j\right)$.

The *price* $\Psi_t^T$ of this fictitious contract thus differs from the *posted price* $F_t^T$ of the true STIR contract previously defined, but *daily margin calls* are identical. *The two contracts thus are identical* except for a *quotation convention*.

The merit of the second quotation convention is that it can be considered as an approximately "true" forward/futures price of an identified *security* underlying the STIR contract. Indeed, the price given by Eq. (15.27) can be viewed as the forward price, set at $t$, of a zero-coupon with maturity $T+3$ months deliverable at $T$, and $R_t^T$ as the (approximate) forward/futures 3-month rate at $t$ for the period $(T, T+3)$.

Note also that $\widehat{F}_T^T = 100 - 0.25\widehat{R}_3(T)$ is the price at $T$ of a 3-month zero-coupon when $\widehat{R}_3(T)$ is the 3-month *bank discount rate* prevailing at $T$ (i.e., a rate set and *paid in advance* while a LIBOR, for instance, is set in advance and paid in arrears[13]). Ignoring the difference between a bank discount rate and a money market rate, hence

---

[13] See Chap. 2, Sect. 2.2.4 for the definition of a bank (or simple) discount rate, which is paid in advance, and its difference with a money market rate which is paid in arrears, even if set in advance.

## 15.2 Short-Term Interest Rate Contracts (STIR) (3-Month Forward-Looking...

accepting the approximation $R_3(T) = \widehat{R}_3(T)$, we have[14]: $\Psi_t^T = \widehat{F}_t^T$: The zero-coupon bond of nominal 100 and maturity $T+3$ months can therefore be (approximately) considered as the underlying asset of the STIR futures contracts.

These developments lead to the following Proposition.

### Proposition 10
*The underlying of a STIR futures contract on a 3-month forward-looking reference is the zero-coupon bond with maturity 3 months beyond the expiration date of the contract. Ignoring the difference between futures and forwards, the implicit rate of a futures contract is (the bank discount version of) the 3-month forward rate.*

### Remarks
Considering that the implicit rate of a STIR contract is the 3-month forward rate is an approximation for two reasons:

- The implicit rate $R$ of a STIR contract is a bank discount rate paid in advance while the LIBOR is a money market rate paid in arrears. This point will be developed in the following Sect. 15.2.1.3.
- The STIR contract is a futures and not a forward (because of daily margin calls) which implies a difference in prices. This difference increases with the expiration date of the contract (see Chap. 9). For this same reason, the STIR futures is not identical to a FRA, as also shown in Sect. 15.2.1.3. A convexity adjustment is often made by practitioners to account for the difference between forward and futures *rates*. We show at the end of Appendix 2 (part **4**) that a possible convexity adjustment, when the volatility of the instantaneous forward rate is assumed to be the constant $\sigma$, is given by:

$$Forward\ rate = Futures\ rate - 0.5\sigma^2 T(T + K) \qquad (15.28)$$

where $T$ is the contract maturity and $K$ the duration of the underlying short-term rate (typically 3 months, i.e., 0.25).

- By considering the futures price as defined in Eq. (15.27) and the zero-coupon of maturity $T+3$ months as the underlying security, one can build an (approximate) cash and carry arbitrage and establish the corresponding spot-futures relationship.
- The implicit rates extracted from the STIR futures prices for different maturities can be used to extend or fill an incomplete rate curve (a LIBOR zero curve for instance).

---

[14]This approximation is legitimate only for very low levels of short-term interest rates, as they currently (early 2021) are.

646 | 15 Futures Markets (2): Contracts on Interest Rates

| $t$ | $T$ | $T+3$ months |
|---|---|---|
| Forward at $t$ over $(T, T+3)$ | -1 | $(1+0.25f_{T,3}(t))$ |
| Spot at $T$ over $(T, T+3)$ | +1 | $-(1+0.25r_3(T))$ |
| **F1**: *Net cash flow* | 0 | $0.25(f_{T,3}(t)-r_3(T))$ |

| **F2**:FRA | $0.25(f'_{T,3}(t)-r_3(T))/(1+r_3(T))$ |
|---|---|

**F3**: 3-month STIR; margin calls $\longrightarrow$ total: $0.25(r_t^T - r_3(T))$ spread between $t$ and $T$

**Fig. 15.3** STIR vs. FRA

### 15.2.1.3 Forward-Looking STIR and FRA

It is also useful to compare an FL STIR futures to a Forward Rate Agreement (FRA) analyzed in Chap. 9. Recall first that a forward or an FRA involves three dates as illustrated in the first two panels of the following diagram (the nominal is 1 or 1M) (Fig. 15.3):

A \$1 standard forward investment at $t$ for the period $(T, T+3)$ associated with borrowing spot at $T$ yields a net cash flow $\mathbf{F1}= 0.25(f_{T,3}(t) - r_3(T))$ at $T+3$ months as shown in the first panel ($f_{T,3}$ is the forward rate). A cash flow $\mathbf{F2}= 0.25(f'_{T,3}(t) - r_3(T))/(1+r_3(T))$, is generated at $T$ by a FRA ($f'_{T,3}$ is the FRA rate). Since under no-arbitrage $\mathbf{F2}$ should be the present value of $\mathbf{F1}$, $f'_{T,3}$ and $f_{T,3}$ *should in theory be equal*. In fact, the forward transactions described in the first panel entail a higher credit risk than the FRA that involves only an exchange of interests. Consequently, the FRA rate is usually closer to a risk-free rate than is the forward rate implicit, for instance, in a LIBOR rate curve.

The third panel recaps the cash flows generated by a long position on a 3-month-LIBOR-type STIR contract bought on date $t$ at a price $F_t^T = 100 - R_t^T$ $=100 - 100r_t^T$. The margins amount to $0.25(r_t^T - r_3(T))$ but they are spread over $(t, T)$.[15] As far as credit risk is concerned, the STIR is closer to the FRA than to the forward transactions $\mathbf{F1}$. As to the timing and amount of cash flows, $\mathbf{F3}$ differs from $\mathbf{F2}$ and $\mathbf{F1}$ since the STIR contract is a futures and its margins are spread between $t$ and $T$. The assimilation of a STIR to a FRA and its implied rate $R$ to a FRA rate or a forward rate is tantamount to the usual approximation of a futures price by a forward price.

### 15.2.1.4 The Main STIR Futures on 3-Month Forward-Looking Rates

Most 3-month Short Term Interest Rate (STIR) futures on forward-looking short rates are virtually identical to the Eurodollar-3-month contract which we focused on and the reader can refer to the internet sites of the main markets for possible tiny differences.

---

[15] Between $t$ and $T+3$ for a STIR on a backward-looking rate (overnight rate average), as explained in Sect. 15.2.2 below.

## 15.2 Short-Term Interest Rate Contracts (STIR) (3-Month Forward-Looking...

In almost all cases, the contract's nominal is 1M units of local currency. The price (100 − the forward rate) is quoted up to three decimal places. The tick is either 0.01 bp, hence 25 units of local currency ($0.01\% \times 0.25 \times 1,000,000$) or 0.005 bp, hence 12.5 units of local currency, depending on the contract. Expiration takes place two business days before the third Wednesday of the contract month for the CME and ICE Futures Europe contracts. On the CME, up to 44 Eurodollar contracts are traded: the March, June, September, and December contracts for 10 years plus the contracts expiring the next 4 months. On the ICE Futures Europe, 28 Euribor contracts are traded.

As for notional futures, contracts on 3-month interest rates started in the US, and given their success, were created in various places on other continents.

The future of the contracts based on LIBOR, however, is uncertain. Most new contracts should shift to the backward-looking type (on overnight averages) studied in the next subsection and as of the end of 2020 trading venues plan to convert all existing Eurodollar contracts into 3-month SOFR futures.

- In the US, the main futures contract on forward-looking short-term rates is still (end of 2020) by far the Eurodollar futures previously described and traded on the International Monetary Market (IMM) of the CME Group.
- In Europe, the ICE Futures Europe is the most active market for STIR futures on FL rates, among which the Euribor contract is the most important. This contract is also traded on the EUREX, in a more or less identical form.
  In addition to the Euribor contract, and similar to it, the ICE Futures Europe still trades (end of 2020) STIR contracts on the 3-month £-Libor (Short Sterling), on the 3-month Eurodollar rate as well as on the 3-month CHF-Libor (Euro-Swiss).
- In Asia and Australia are traded the 3-month Euro-Yen on the 3 month-TIBOR (Tokyo) and the 90-day Bank Bill (Sydney), among others.

Futures markets also offer other STIRs, based on different principles, like the 3-month Swap Index and contracts on 1-month and 3-month overnight rate indexes, that are supposed to progressively replace most LIBOR-type contracts, and that we now study.

### 15.2.2 Futures Contracts on an Average Overnight Rate

STIR contracts of a second type are referenced on a *backward-looking* short-term rate index. This reference is usually an average of overnight rates over a month or a quarter. Like other STIR futures, they are quoted as $F = (100 - R)$ where $R$ is homogeneous to an annual rate expressed in % and are *cash settled* (no physical delivery). The final cash settlement coincides with the *end of the period over which the average of overnight rates is computed*, often referred to as the *reference* or *accrual period*. This reference period of duration $K$ is denoted by $(T, T+K)$ in the sequel (0 being the current date); it may have:

648                                   15 Futures Markets (2): Contracts on Interest Rates

- Either a *1-month duration* ($K$ is approximately 1/12), in which case the underlying is generally a simple arithmetic average of overnight rates over one month.
- Or a *3-month duration* ($K$ is approximately 0.25), in which case the overnight rate is business-day compounded over the reference quarter (the result is a kind of geometric average, expressed on a yearly basis), as for the OIS index described in Chap. 7.

The business-day compound calculation aims at faithfully mirroring an actual roll-over of overnight investments. We will focus on what follows on the 3-month contract on compound overnight rates since the 1-month contract on arithmetic average is broadly similar, except that the arithmetic average is an approximation for a compound average considered acceptable over a 1-month period. Again, the quotation $F = (100 - R)$, commonly called "price" for convenience, is in fact a conventional quotation. It must not be confused with a "true" futures or forward price which is more alike to $(100 - K.R)$.

After a general description of these contracts we examine the arbitrage mechanisms that link futures prices to spot interest rates. We finally describe the main contracts.

### 15.2.2.1 General Description of 3-Month Contracts on a Compound Average of Overnight Rates

Recall that the business-day compound overnight index (or OIS index) is computed as follows for a period $n$ (see Chap. 7, Sect. 7.1.2.3):

- We first calculate $c_n$, the compound interest on \$1 placed overnight between the beginning and the end of period $n$ (the reference quarter) capitalized on business days. When the interest convention is "actual/360," we obtain:

$$1 + c_n = \prod_{j=1}^{o_n} \left[ 1 + \left( n_j/360 \right) r_j \right]$$

where

$r_j$ = overnight rate fixed on business day $j$ (used to compute interest on $n_j$ calendar days), in decimal form
$o_n$ = number of overnight rate fixings used in quarter $n$ to capitalize interests
$n_j$ = number of calendar days to which the overnight rate $r_j$ applies

We will keep in mind that $1 + c_n$ can be obtained at the end of period $n$ by investing \$1 at the beginning of period $n$ and rolling over this investment overnight during the whole period. Here, period $n$ is the quarter $(T, T+K)$.

- Then, $c_n$, which is homogeneous to a quarterly rate in decimal form, may be expressed in an annualized-% -proportional-360 basis, which writes for quarter $n$:

## 15.2 Short-Term Interest Rate Contracts (STIR) (3-Month Forward-Looking...

$$I_n = 100\, c_n\, (360/N_n) = 100\, c_n/K,$$

where $N_n$ denotes the number of calendar days in quarter $n$. For the following analysis we often use the approximation $N_n/360 = K = 0.25$.

The result is a backward-looking rate accruing and progressively revealed during the reference quarter $(T, T+K)$ and fully known only at $T+K$.

When the convention is "actual/365," as in the UK, we just replace 360 by 365 in the formulas.

We will often select the 3-month SOFR of the CME as a prototypical example to which we occasionally compare the 3 month-forward-looking Eurodollar contract described in the previous section. The contract is referred to as the date $T$ of the *beginning* of the accrual period. Hence, $F_t^T = 100 - R_t^T$ represents the futures price of the $T$-contract whose reference period is $(T, T+K)$. As there is no physical delivery, the contract is defined by the method used for computing the *final settlement* price, at maturity $T+K$, hence by the calculation of the terminal rate $R_{T+K}^T$. For the $T$-contract with reference quarter $n = (T, T+K)$, we have *by construction* $R_{T+K}^T = I_n$ as previously defined. This implies a final settlement price, on date $T+K$, equal to $F_{T+K}^T = 100 - I_n$. Daily margins, as for the 3-month-forward-looking-LIBOR-type STIR, are equal to $0.25\left(F_j^T - F_{j-1}^T\right) = 0.25\left(R_{j-1}^T - R_j^T\right)$ for day $j$ and a nominal of \$100. At dates $t$ prior to $T+K$, prices $F_t^T$ are determined by economic forces involving the pseudo-arbitrages described in Sect. 15.2.2.2 below.

It is instructive to compare a 3-month contract on compound overnight rates to the corresponding 3-month LIBOR futures. For example, SOFR futures contract is tailored as the 3-month Eurodollar futures in terms of contract size (\$1,000,000, hence \$25 per basis point) and IMM calendar.

Consider, for instance, a SOFR contract whose final settlement is the third Wednesday of September (the settlement date $T+K$). The reference quarter thus starts on the third Wednesday of the preceding June, called the contract date. The corresponding 3-month Eurodollar contract is settled in June at a final settlement price based on the 3-month \$LIBOR prevailing on the third Wednesday[16] of June for the following 3-month period. The *reference quarter* $(T, T+K)$ *is thus the same* for the two contracts. Both are called June $(T)$ contracts. The difference is that the LIBOR is forward-looking, hence known at the beginning $T$ of the reference quarter allowing a final settlement at the *beginning* of the quarter (in June for the June Eurodollar contract of our example) while the overnight average is backward-looking implying a *final* settlement *postponed at the end* $T+K$ of the reference period (September for the June-SOFR contract of our example).

---

[16] Or Monday.

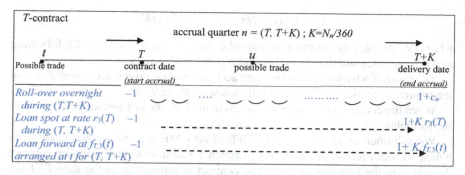

**Fig. 15.4** T-contract and three related transactions

Figure 15.4 illustrates the previous description of the *T*-contract (top black panel) as well as 3 other transactions involved in the analysis that follows (bottom blue panel). Trading on the *T*-contract starts before *T* (at dates like *t* in the diagram), continues during all the accrual period (at dates like *u*) and ends at *T+K*.

A long position on this contract, initiated at *t* and held up to *T+K*, yields daily margins whose cumulated value is $K\left(F_{T+K}^T - F_t^T\right) \simeq 0.25\left(F_{T+K}^T - F_t^T\right) = 0.25\left(R_t^T - R_{T+K}^T\right)$ per \$100 nominal. At expiry *T+K*, the final settlement price is perfectly defined by the formulas $R_{T+K}^T = I_n$ and $F_{T+K}^T = (100 - I_n)$. Cumulated margins over the whole period (*t, T+K*) thus amount to $0.25\left(R_t^T - I_n\right)$ per \$100.

Before expiry, prices $F_t^T$ are theoretically determined by market arbitrage to which we now turn.

### 15.2.2.2 Arbitrage and Prices of Overnight Rate Futures

1. To assess the no-arbitrage price of the 3-month overnight contract that should prevail at any time, we recall some concepts presented in Chap. 7 and then present more formal cash-and-carry and reverse arbitrages.

   Consider first the beginning of the reference quarter *n* (date *T*) and compare the value of two rights on cash flows available at *T+K* (represented in the blue panel of Fig. 15.4):

   (a) A right to $1+c_n$ obtained by a roll-over overnight during quarter *n* from \$1 invested at *T*; this cash flow is only known at *T+K*. In Fig. 15.4, it is represented by the first row of the blue panel.

   (b) A right to the terminal payoff of a \$1 loan made at *T*, at the 3-month risk-free rate $r_3(T)$ prevailing (thus *known*) at *T* (money-market spot riskless rate, in decimal form), and equal to $1+K\, r_3(T)$ (second blue row; note that here $r_3 = r_K$).

## 15.2 Short-Term Interest Rate Contracts (STIR) (3-Month Forward-Looking...

**Table 15.3** Cash and carry arbitrage involving a 3-month contract on compound overnight rate

| Nominal 100; $K$=0.25 | $t$ | $T$ | $T+K$ |
|---|---|---|---|
| Sell contract at date $t$ | 0 | Margins | $0.25\left(F^T_{T+K} - F^T_t\right) = 0.25\left(R^T_t - I_n\right)$ |
| Roll-over overnight in $(T, T+K)$ | | $-100$ | $100 + 0.25\,I_n$ |
| Borrow at $t$ forward for $(T,T+K)$ | 0 | 100 | $-100\,(1+0.25 f_{T,3}(t))$ |
| Net cash flow | | 0 | $0.25\left(R^T_t - 100 f_{T,3}(t)\right)$ |
| Decomposition of the forward: lending/$(t,T)$ + borrowing/$(t,T+K)$ | $-100/(1+(T-t)$ $r_{T-t}(t)) +100/$ $(1+(T-t) r_{T-t}(t))$ | 100 | $-100(1+(T + K - t)r_{T + K - t}(t))/$ $(1+(T-t)r_{T-t}(t))$ $= -100\,(1 + K f_{T,3}(t)) = -100$ $(1 + 0.25\, f_{T,3}(t))$ |

The rights on these two cash flows, which can be obtained with the same down payment of $1, should have in AAO the same value of $1 at $T$. It is therefore equivalent at $T$, and *at any* $t < T$ as well, to own a right to a (random) cash flow at $T+K$ equal to $1+c_n$ or to a flow $1+ Kr_3(T)$ known at $T$. Besides, viewed from $t$ ($< T$), a right to the random rate $Kr_3(T)$ available at $T+K$ is equivalent to the sure rate $K f_{T,3}(t)$, where $f_{T,3}(t)$ is the 3-month forward rate prevailing on date $t$ for the period $(T, T+K)$ annualized and in decimal form (third blue row of Fig. 15.4). It follows that $100 f_{T,3}(t)$, $R_3(t)$ and $I_n$ are "equivalent" since they are the % versions of $f_{T,3}(t)$, $r_3(T)$ and (annualized) $c_n$, respectively. Since the $T$-contract exchanges $R^T_t$ for $I_n$, we conclude that in AAO:

(a) At the beginning of quarter $n$ (date $T$) the contract should quote $100 - R_3(T)$;

(b) At any date $t$ prior to $T$ the contract should quote $100 - 100\, f_{T,3}(t)$, so that the no-arbitrage rule implies: $R^T_t = 100\, f_{T,3}(t)$ (this result is more formally established by the cash-and-carry arbitrage of Table 15.3).

We stress the importance of the condition that the *credit risks* entailed by these alternative investments are *identical*. In the case of a continuously refreshed overnight rate, especially if it is a secured rate such as the SOFR, credit risk can be considered as nil. Therefore, there will always be a *credit spread* when comparing the implicit rate of an overnight contract to a forward-looking market rate tainted with default risk but otherwise equivalent. In fact, the implicit rates of overnight contracts are comparable to the forward swap rates of overnight indexed swaps (OIS). When comparing the rates implicit in the prices of overnight futures to forward rates extracted from a LIBOR curve, an OIS-LIBOR spread emerges (see Chap. 7). For the same reason, the 3-month forward-looking and the 3-month overnight average prices for $t \le T$ (both theoretically equal to $100 - 100\, f_{T,3}(t)$) differ: For a Eurodollar contract, for instance, the forward rate $f$ implicit in a futures price is comparable to a LIBOR-curve forward-rate, while for the SOFR contract it derives from an OIS rate curve.

652                     15 Futures Markets (2): Contracts on Interest Rates

A second important comment is also in order. The interest $0.25 f_{T,3}(t)$ of a standard forward transaction is settled at the end of the reference period (on $T+K$) while the total margin induced by a futures contract is progressively settled between $t$ and $T+K$. As mentioned several times, this difference between futures and forward prices (or rates) justifies a *convexity adjustment* to make them (practically) equivalent.

Let us now determine the futures price $F_u^T$ and the corresponding implicit rate $R_u^T$ at a date $u$ between $T$ and $T+K$, *during* the accrual period. Consider first a roll-over overnight investment during $(u, T+K)$, starting from $(1 + c(T,u))$ on date $u$; $c(T,u)$ denotes the compound-overnight interests on a unit investment during $(T, u)$ which have already accrued and are known on date $u$. Such an investment yields $1 + c_n$ on $T+K$, because, since interests are compounded, we have: $1+c_n = 1 + c(T, T+K) = (1+c(T,u))(1+c(u, T+K))$. The same amount $(1+c(T,u))$ invested on date $u$ at the money market rate $r_D(u)$ yields $(1+c(T,u))(1+ D r_D(u))$ where $D$ denotes the duration of the period $(u, T+K)$ in fraction of a year $(T+K-u)$. The implicit rate $R_u^T$ should thus satisfy:

$$1 + K R_u^T / 100 = (1 + c(T, u))(1 + D r_D(u)). \tag{15.29}$$

The preceding argument was developed for the valuation of an OIS rate instrument in Chap. 7 and the virtual arbitrage underpinning this result is presented in item 3 below.

2. We now describe formally the cash-and-carry (pseudo-) arbitrage on date $t$ ($t < T$) linking the futures price $F_t^T$ and its implicit rate $R_t^T$ $\left(F_t^T = 100 - R_t^T\right)$ to the spot rate. Such an arbitrage is presented in Table 15.3 (margin calls are cumulated up to date $T+K$, and $K = 0.25$):

Focusing on the first three rows, the arbitrage combines a short position on the contract, a roll-over overnight investment between $T$ and $T+K$ financed by a forward borrowing set at date $t$. The net outcome of these transactions is a sure cash flow $= 0.25 \left(R_t^T - 100 f_{T,3}(t)\right)$ at date $T+K$, which in AAO is non-positive.

Since the reverse cash and carry generates the same cash flows with opposite signs (by buying a contract at $t$, borrowing by roll-over overnight over $(T, T+K)$ and setting at $t$ a forward risk-free loan over $(T, T+K)$), AAO implies $R_t^T = 100 f_{T,3}(t)$, as stated previously.

**Remarks**

(a) To fully recognize a "straight" cash-and-carry in the transactions described in Table 15.3, one should decompose the forward borrowing in its two spot components (see the last, blue panel): Borrowing $100/(1 +(T - t)r_{T-t}(t))$ at $t$ over $(t, T+K)$ and simultaneously investing the *same* amount $100/(1+(T - t)r_{T-t}(t))$ over the shorter period $(t, T)$. The underlying "security $L$" of the contract then identifies with a two-step investment composed by a loan of $100/(1+(T-t)r_{T-t}(t))$ between $t$ and $T$ whose repayment at $T$ is immediately reinvested in a roll-over overnight between $T$ and $T+K$. The cash-and-carry consists in the usual three-fold position built at date $t$: short in the contract, long in the underlying "security $L$"

## 15.2 Short-Term Interest Rate Contracts (STIR) (3-Month Forward-Looking... 653

**Table 15.4** Cash and carry arbitrage involving a 3-month contract at date $u$ ($T < u < T + K$)

| Nominal 100; $K=0.25$ | $u$ | $T+K$ |
|---|---|---|
| Sell contract at date $u$ | 0 | $0.25\left(F^T_{T+K} - F^T_u\right) = 0.25\left(R^T_u - I_n\right)$ in fact received all along $(u, T+K)$ |
| Roll-over overnight in $(u, T)$ | $-100\,(1 + c(T,u))$ | $100 + 0.25\,I_n$ |
| Borrow spot over $(u,T)$ | $100\,(1+c(T,u))$ | $-100\,(1+c(T,u))(1+Dr_D(u))$ |
| Net certain cashflow | 0 | $0.25R^T_u - 100\{(1 + c(T,u))(1 + Dr_D(u)) - 1\}$ |

financed by borrowing between $t$ and $T+K$. Or, equivalently, *the contract is synthesized by a long position in L financed by a short position in a zero-coupon of maturity T+K.*

(b) Let us emphasize again that the previous set of transactions is *not a strict arbitrage*. First, the forward rate $f_{T,3}$ replicated by two spot transactions is not in general free of credit risk. Second, the pseudo-replication of the contract is a forward, not a futures.

(c) Taking a casual account of these two remarks, we may write: $R^T_t = 100 f_{T,3}(t) + a_1 - a_2$, where $a_1$ stands for a positive convexity adjustment and $a_2$ for a risk spread. The former increases with the volatility of the interest rate and the maturity of the contract, while $a_2$ increases with the default risk affecting $f_{\tau,3}$.

3. On date $u$ ($T < u < T+K$) part of the accrued interest, equal to $1+c(T,u)$, is known, while the unknown part $1+c(u,T+K)$ is replicable by a roll-over overnight during the period $(u,T+K)$. Denoting $D = T+K-u$ (fraction of a year $< 0.25$) and $r_D(u)$ the $D$-period spot rate prevailing at $u$, the cash-and-carry arbitrage determining the futures price is described in Table 15.4.

The reverse arbitrage generates the opposite sure cash flow. Therefore, in AAO:

$$0.25R^T_u - 100\{(1 + c(T,u))(1 + Dr_D(u)) - 1\} = 0.$$

Hence, as stated previously, the implicit rate $R^T_u$ is given by:
$$1 + 0.25R^T_u\,/100 = (1+c(T,u))(1+Dr_D(u)) = (1+(0.25-D)I(T,u))(1+Dr_D(u)),$$
which is Eq. (15.29).

Under no arbitrage, the futures contract quotes: $F^T_u = 100 - R^T_u$.

Since $r_D(u)$ is usually affected by default risk and the contract is a futures (the margins are spread over the period $(u,T)$), the "no-arbitrage" relation is in fact approximate.

---

**Example 8: 3-Month SOFR Contracts (with Negative Interest Rates)**
On the morning of Monday 09-25-n, three investors $A$, $B$, and $C$ buy 20 three-month SOFR contracts with December 18-n+2 expiry (September year n+2

*(continued)*

contract). The order is carried out at 98.400, meaning that the implicit 3-month OIS forward rate is 1.60%.

On succeeding days the margins on these 20 contracts evolve as in the following table:

| Dates | Daily closing Price | Margins on the 20 contracts($) Calls (−) or Repayments (+) |
|---|---|---|
| 09/25 | 98.430 | +1500 |
| 09/26 | 98.455 | +1250 |
| 09/27 | 98.410 | −2250 |
| 09/28 | 98.390 | −1000 |
| 09/29 | 98.420 | +1500 |

- $A$ unwinds the contracts 5 days later, at the closing price of September 29-n: as in the case of Example 7 her total gain amounts to:
  Gain $A = 20\times(98.420 - 98.400) \times 0.25\times1,000,000/100 = \$1000$ (or 20 × 4 ticks × $12.5, or 20 × 2 bps × $25). This part of the example is comparable to Example 7.
- $B$ holds his position for 2 years and 2 months till Wednesday 10-18-n+2, 1 month after the beginning of the reference (accrual) quarter (09-17-n+2, 12-18-n+2). During this month (09-17 to 10-18 n+2) the compound SOFR (OIS index) has *decreased* at an annualized rate of −0.30% (the SOFR has become *negative*). The accrued (negative) compound interest per $100 of nominal during this month leads, using our notations, to: $100(1+c(09\text{-}17\text{-}n+2, 10\text{-}18\text{-}n+2)) = 100 - 0.30/12 = \$99.975$.
  $B$ closes his position at a price of 100.400, 2 months before the expiry of the accrual period, meaning that the implicit rate is $R = 100 - 100.400 = -0.40\%$.

  The 2-month spot rate $r_{2month}$ implicit in this futures price is given by Eq. (15.28):
  $100(1+c(09\text{-}17\text{-}n+2, 10\text{-}18\text{-}n+2))(1+r_{2month} \times 2/12) = 100 - 0.25R = 100 - 0.25 \times 0.40 \Rightarrow 1+r_{2month} \times 2/12 = 99.90/99.975 \Rightarrow r_{2month} = -0.45\%$ (negative).

  $B$'s *cumulated gain* during his 2-year operation is: Gain $B = 20\times (100.400 - 98.400) \times 0.25 \times 1,000,000/100 = \$100,000$ (or simply 20 × 200 bps × $25).
- $C$ holds her position till the final settlement date (December 18-n+2). Assume that the annualized OIS index has *decreased* to the *negative* value of −0.46% during the 3-month accrual period. The final settlement price then is 100.460. Since $C$ bought her contracts at 98.400, her gain (cumulated margins) obtained throughout the 2 years and 3 months holding period is:
  Gain $C = 20 \times 206$ bps × $25 = \$103,000$.

## 15.2 Short-Term Interest Rate Contracts (STIR) (3-Month Forward-Looking...

### 15.2.2.3 The Case of a Reference Period of Duration $K$ Different from 0.25

The previous analysis and results are valid, *mutatis mutandis*, in the case of STIR contracts referenced on overnight averages over periods of duration different than a quarter. For 1-month overnight averages, for instance, all the previous results hold substituting 1/12 for $K$ and for 0.25, and $f_{T,1}(t)$, the forward rate prevailing at $t$ for the period $(T, T+1$ month), for $f_{T,3}(t)$. For any accrual duration $K$:

– The implicit rate $R_t^T$ in the futures price $F_t^T = 100 - R_t^T$ is (approximately) equal, in AAO, to $100 f_{T,K}(t)$.
– The margin call is $K(F_j^T - F_{j-1}^T) = K(R_{j-1}^T - R_j^T)$ for day $j$ and a nominal of \$100.

In fact, all the previous analyses are transposable only approximately in the case of a 1-month contract ($K = 1/12$) because, in this case, the reference is an *arithmetic* monthly average of the overnight rate instead of the geometric average obtained by business-day compounding, hence is imperfectly replicated by an overnight roll-over. But for short reference periods (as is 1 month) and small rates, this approximation is acceptable by most standards.

### 15.2.2.4 The Main Futures Contracts on Overnight Rates Averages

The contracts on overnight rate averages should progressively replace the LIBOR referenced STIRS.

1. The 3- and 1-month referenced STIRS are structured differently.
   (a) As previously mentioned, the 3-month STIR is a geometric average of overnight rates during the 3-month reference period (the overnight rate is business-day compounded), follows the IMM quarterly calendar (beginning the third Wednesday of March, June September, and December), and has a face value of 1M in local currency (LC) which implies a margin of $1M \times 0.25 \times 0.0001 = LC25$ per bp (0.01%) fluctuation. Depending on the contract, the tick can be 1bp (LC25), 0.5 bp (LC12.5), or 0.25 bp (LC6.25), the last, smallest tick usually concerning the contracts expiring first.
   (b) The 1-month reference is computed as a simple arithmetic average between the beginning and the end of the full calendar month and thus does not follow the IMM calendar. Its face value may be \$2.5 M (in the US) or LC3M (in Europe), implying respectively margins of \$22.83 ($= 2.5 \times 0.0001/12$) or LC25 ($= 3 \times 0.0001/12$) per basis point.
2. We briefly cite the main contracts on overnight rates averages traded in the US and in Europe (for details, see the CME and ICE websites):
   (a) The CME trades the 1-month SOFR, the 3-month SOFR and the 30 day Fed Fund rate. In addition, it offers an MPC SONIA contract (on £-overnight

# 656        15   Futures Markets (2): Contracts on Interest Rates

average rate) which follows a specific British calendar,[17] as well as a quarterly IMM SONIA obeying the conventional IMM calendar.

(b) The ICE futures Europe trades the 1-month EONIA/€ster (new contracts on €ster since October 2019), a 1-month SONIA, a 1-month SOFR, a global 3-month SONIA Futures (with IMM dates), a 3-month SOFR, a 3-month SARON (on CHF-overnight average rate).

Besides, the EUREX trades also contracts on EONIA/€ster and SARON averages.

## 15.2.3 Hedging Interest Rate Risk with STIR Contracts

After a presentation of the simple and extended durations for STIR contracts, we briefly analyze how to use them for hedging purposes.

### 15.2.3.1 Simple and Extended Durations

– First, we study the sensitivity of a STIR futures price to a "small" parallel shift of the yield curve and obtain the equivalent of a simple duration. We consider a forward-looking or backward-looking STIR, referenced to a rate of duration $K$ (usually 0.25 for a quarter or 1/12 for a month). The current date is $t$, the reference period is $(T, T + K)$ and in all cases, the current price writes $F_t^T = 100 - 100 f_{T,K}(t)$. A common variation $dr$ of all rates implies a variation $df_{T,K} = dr$, triggering, for a nominal 100, a margin call $d(MC) = -100K dr$. For example, for a 3-month STIR contract, a 1% increase in the implicit rate generates a margin of \$0.25 per \$100 of nominal (\$25 per bp for a contract of nominal \$1M).

**Proposition 11**

*It is convenient to define the sensitivity of a STIR contract price as a proportion of its nominal value. We thus obtain: $D^F = -(1/100)d(MC)/dr$. Since $d(MC) = -100K dr$, it follows that:*

$$D^F = K. \qquad (15.30)$$

*where $K$ is the duration of the reference period (for a 3-month contract: $D^F = 0.25$). This result applies to both forward-looking and backward-looking STIRs.*

In the sequel, $D^F$ or $K$ is indifferently called sensitivity or duration of the contract.[18]

---

[17] The dates are fixed by the Monetary Policy Committee and announced by the Bank of England.

[18] Alternatively, and more in line with standard definitions of sensitivity/duration, one can define the STIR sensitivity using the price $\Psi$ (and not the nominal value): $-d\Psi / \Psi \, dr = K/(1 - Kr)$, where $\Psi$ is the fictitious quotation $\Psi = 100 - KR = 100 - 100Kr$ introduced above in Sect. 15.2.1.2. This entails a small difference with the simpler definition (which leads to $- dF/100dr = K$), but an appropriate use of each risk measure yields identical results.

## 15.2 Short-Term Interest Rate Contracts (STIR) (3-Month Forward-Looking...

- Let us now consider the more complex risk of a rate curve deformation stemming from $q$ sources (factors) of risk $(b_1, \ldots, b_q) = \underline{b}$. For mathematical simplicity, we assume that the current date is $t=0$ and use continuous rates. Recall that $z_\theta$ is the continuous zero-coupon rate for maturity $\theta$, and that:

$$z_\theta = g(\theta, \underline{b}).$$

Denoting by $f_{T,K}$ the continuous forward rate on date 0 for the period $(T, T+K)$, we have:
$Kf_{T,K} = (T+K)z_{T+K} - Tz_T.$ [19] Taking the derivatives of this equation with respect to the risk sources $b_i$ for $i = 1, \ldots, q$, leads to the following proposition.

**Proposition 12**
*For a continuous zero-coupon rate curve given by a q-factor model $z_\theta = g(\theta, \underline{b})$, the durations $D_{b_i}^F$ with respect to factor $b_i$ is equal to*

$$D_{b_i}^F = (T+K)\frac{\partial g}{\partial b_i}(T+K, \underline{b}) - T\frac{\partial g}{\partial b_i}(T, \underline{b}), \quad i = 1, \ldots, q. \tag{15.31}$$

### 15.2.3.2 Hedge Ratios
As with notional contracts, we consider two sorts of hedge: hedging a current, existing position or an expected, but known, position.

- **Hedging a current position**

As in Sect. 15.1.7, we denote by $P$ the market value of the position to be hedged, as a percentage of its nominal value, $S_P$ its sensitivity, and by $100w$ the nominal amount to be hedged, so that its market value is equal to $wP$; $w > 0$ means a long position and $w < 0$ a short position. Here, we drop the time subscript to simplify the notation.

- We first assume a parallel shift $dy$ of the rate curve. We want to find the nominal amount $100x$ of contracts that cancels interest rate risk up to the first order. Setting the margin call $xd(MC) = -x100K\,dy$ equal to the change $wdP = -wS^P P\,dy$ in the position value, we obtain the next proposition which is equivalent to Proposition 6 for notional contracts (Eq. 15.22a).

**Proposition 13**
*The hedge ratio x/w for a current position w is:*

---

[19] This is because $\exp(Kf_{T,K}) = \exp((T+K)r_{T+K})/\exp(Tr_T)$ which implies $Kf_{T,K} = (T+K)\,r_{T+K} - Tr_T$.

$$x/w = S^P P/100K \tag{15.32a}$$

($K=0.25$ or $1/12$ for standard contracts), and the number $N$ of contracts whose nominal is $M'$ required for hedging a position of nominal $M$ is:

$$N = S^P PM/100KM' \tag{15.32b}$$

- We can also compute the extended duration/sensitivity to factors deforming the rate curve (see Eq. 15.31). The hedge involves $q$ contracts, including short-term ones, and can be determined using a system of $q$ equations in $q$ unknowns (see system 15.25). If one of the futures contracts, for example, the $j$th, is identified as the short-term contract, its durations $D_j^1, \ldots, D_j^q$ are given by Eqs. (15.30) or (15.31).

- *Hedging an expected, known position*

The analysis conducted in Sect. 15.1.7.2, remains valid for futures contracts on short-term rates. Just as for a current position, it is sufficient to identify one of the futures contracts as the short-term contract in Eqs. (15.24) and (15.25), using the appropriate durations as given by Eqs. (15.30) or (15.31).

---

**Example 9: Hedging Future Borrowing**

At the beginning of year $n$ a firm plans to borrow \$30 M for the last 6 months of year $n$ (from 07-01-$n$ till 01-01-n+1). To hedge against the risk of an increase in the 6-month interest rate the firm decides to *sell N* 3-month September SOFR contracts whose reference period is (09-18-$n$, 12-17-$n$). An upward parallel shift $dr$ of the short-term (1–6 months) rates will increase its financing costs by $dr \times 30M \times 6/12 = \$15Mdr$, while the gain $G$ on a short position on $N$ SOFR contracts (whose nominal is \$1M) will be:

$G = NKdr = \$0.25MNdr$. This gain compensates the additional financing cost provided that $0.25N = 15 \Rightarrow N = 60$ (contracts).

This hedge is imperfect (subject to basis and/or correlation risks) for various reasons:

- The futures price on the 07-01, date on which the contracts may be unwound, depends on a 3-month forward rate, while the financing cost depends on the 6-month spot rate (for instance a 6-month LIBOR as of the date 07-01). In the case of a non-parallel shift of the yield curve, the increment on the cost of financing may be higher than the profit on the hedge.
- The cost of financing depends on the LIBOR while the rates governing the SOFR futures prices are of the OIS-rate type. The hedge is thus subject to the risk that the OIS-LIBOR spread widens.

## 15.3 Summary

Interest rate futures are traded on official Exchanges and thus subject to daily margin calls. The two main types of futures are: (1) the notional contracts on sovereign short, medium notes and long-term bonds; (2) the Short Term Interest Rate contracts (STIRs).

1. *Notional contracts*

   In the case of a notional contract, a *basket of deliverable securities* (DS) is defined. Any DS in the basket can be delivered at the contract's maturity, on the seller's decision. A *conversion factor* sets the settlement price in accordance with the particular DS delivered.

   (a) The classical US T-bond and the Euro-bund contracts are similar, widely traded, and representative examples of long-term notional futures. In both cases, the DS have approximately 10 years of remaining life.

   (b) On expiry date $T$, the seller of an unwound contract chooses the DS to be delivered for a nominal amount equal to the nominal value of the futures contract (\$100,000 or 100,000 € for most contracts). In exchange, he receives the *Delivery Price* ($DP_i$), which depends on the delivered bond $i$ since $DP_i = f_i F_T + ac_i$, where $F_T$ is the final settlement price, $ac_i$ is the accrued interest on $i$ at $T$, and $f_i$ is the $i$-conversion factor. $f_i$ is interpreted as the clean price at which \$1 nominal of $i$ would trade if its yield to maturity (YTM) were equal to *a fixed given rate k* (6% for most US and European contracts).

   (c) The seller chooses the most advantageous DS, taking into account the cost she pays as well as the delivery price she receives (she holds a *delivery option*). The chosen DS, called the *Cheapest To Deliver* (CTD, denoted by $m$), is the DS that maximizes her gain $G_i$ equal to the difference between the delivery price $DP_i$ and the dirty price $C_i'(T)$ of the bond $i$ she delivers:
   $G_i = DP_i - C_i'(T) = f_i F_T - C_i(T)$, where $C_i(T)$ is the clean price. It is *thus the DS i that minimizes the ratio $C_i(T)/f_i$*.

   (d) Since the DS composing a given basket are close in terms of time to maturity and are issued by the same government, they yield (almost) the *same YTM* denoted by $y$.

      I. If $y > k$, Futures are quoted above par and the CTD is the one with the longest duration.

      II. If $y < k$, Futures are quoted below par and the CTD is the one with the shortest duration.

   (e) In absence of arbitrage, the price on date 0 of a futures contract with expiry $T$ is related to the spot price of the CTD $m$ by the cash-and-carry relation
   $f_m F_0^T - C_m(0) = \left( r_0 C_m'(0) - 100k_m \right) T$.
   where $f_m F_0^T - C_m(0)$ is an (adjusted) basis, $r_0$ is the spot money market rate on date 0 for maturity $T$ and $\left( r_0 C_m'(0) - 100k_m \right) T$ is the net cost of carrying $m$.

660    15  Futures Markets (2): Contracts on Interest Rates

(f)  The relation above is approximate since it is established as if the futures was a forward, the implementation of the reverse cash-and-carry may be difficult and costly, and the seller of the contract holds a *delivery option*.

(g)  The duration $D^F$ of the futures is approximately equal to the duration of the CTD reduced by the time to maturity $T$ of the contract, and its sensitivity $S^F = -dF^T/F^T$ is equal to $(1+y)D^F$.
A generalized sensitivity/duration can be defined in association with a multi-factor model of the yield curve.

(h)  An existing position of nominal amount $100w$ composed of a security of price $P$ and sensitivity $S^P$ can be hedged against the risk of a parallel shift of the yield curve by taking an opposite position on futures contracts in nominal amount $100x$. The hedge ratio is: $x/w = S^P P/S^F F$.
The number $N$ of contracts with nominal $M'$ hedging a position with a nominal $M$ is therefore: $N = -MS^P P/M'S^F F$.

(i)  A hedge against more complex movements of the yield curve can be constructed using the method of generalized duration associated with a $q$-factor model of the yield curve; the hedge component involves $q$ *different* futures contracts.

(j)  In addition to the 10-year classical T-bond and the Euro-bund contracts, the CME trades futures on 2, 5, and 30 years Treasury notes and bonds and the EUREX trades several notional futures on different government bonds and notes of different maturities.

2. *Short Term Interest Rate contracts (STIR)*

(a)  STIR futures share the following features. They are referenced on a short underlying rate (1 month, 3 months, 1 or 3-month average of an overnight rate). They are cash settled. Their price is quoted as 100 minus a reference interest rate expressed in % and annualized so that it writes on date $t$ for an expiration $T$: $F_t^T = 100 - R_t^T$. Their expiry dates are usually scheduled on International Monetary Market dates: the 3rd Wednesday of March, June, September, and December.

(b)  The STIR reference is either a forward-looking interest rate, often a 3-month LIBOR, or a backward-looking average of overnight rates.

(c)  *For a forward-looking 3-month reference*, the final settlement price (at $T$) writes: $F_t^T = 100 - R_3(T)$ where $R_3(T)$ is the 3-month spot rate (in %) prevailing at $T$ for the period $(T, T+3$ months$)$.

(d)  Before expiration, the margin call for day $j$ is, for a nominal of 100: $0.25(F_j^T - F_{j-1}^T) = 0.25(R_{j-1}^T - R_j^T)$. The total gain on a long position held between $t$ and $T$ thus is $0.25(F_T^T - F_t^T) = 0.25(R_t^T - R_3(T))$. The contracts allows to "exchange" at $t$ the 3-month spot rate that will prevail at $T$ for $R_t^T$, In absence of arbitrage, $R_t^T$ is thus (approximately) equal to the 3-month *forward* (or FRA) rate.

(e)  The main contracts are the 3-month Eurodollar of the CME (US), and the 3-month Euribor contract of the ICE Futures Europe. Some are to be

# 1 Valuation of the Delivery Option

661

replaced by the corresponding contracts on average overnight rates references.

(f) *The contracts on a backward-looking underlying are referenced on an average of overnight rates.* The average is computed either over a month or a quarter $(T, T+K)$ called the *reference period or accrual period* $(K = 1/12$ or $0.25)$. Quotations are also in the form $F_t^T = 100 - R_t^T$, where $R_t^T$ is an annual rate expressed in % and the margin of day $j$ is $K(F_j^T - F_{j-1}^T) = K(R_{j-1}^T - R_j^T)$.

(g) For the contract of maturity $T$, the final settlement price on $T+K$ is $F_{T+K}^T$
= $100 - C(T, T+K)$, where $C(T, T+K)$ is either a business-day-compounded overnight rate (as for an OIS) for $K = 0.25$ or a simple arithmetic average for $K = 1$ month. This produces a backward-looking rate accruing and progressively revealed during the reference period $(T, T+K)$, fully known only on date $T+K$.

(h) At $t < T+K$, the futures price $100 - R_t^T$ is determined by the no-arbitrage condition. At $t < T$, $R_t^T$ is (approximately) equal to the forward rate for $(T, T+K)$ implicit on the $t$-OIS swap-rate curve.

(i) The main contracts on overnight averages are the 1- and 3-month SOFR, 30-day Fed-Fund futures of the CME (US), and different €STER and SONIA contracts of the ICE Futures Europe.

(j) For STIR contracts, a change $dR$ of the implied forward rate entails a margin $-KdR$ for 100 of nominal. Their sensitivity then is simply $K$, although a multi-factor sensitivity can also be defined. STIR contracts thus are important elements of the interest rate hedging toolkit.

## Appendices

These appendices provide two improvements in the techniques for valuing futures contracts. The first concerns the option to change the delivered DS held by the seller of a notional contract. The second allows taking into account the margin call system.

## 1 Valuation of the Delivery Option

To obtain a closed formula for the delivery option included in a contract on a notional security, we make the following simplifying assumption which is justified by the uniformity of the basket of DS:

### Assumption
*The forward yield curve is flat for all DS. We denote by $R_T^f(t)$ the unique forward rate prevailing on date $t \geq 0$ for the maturity $T$.*

Applying the result from Sect. 15.1.4.2, the two candidates for CTD on the expiration date $T$ are the bond with the longest duration using the clean price (indexed by the letter $l$) and the bond with the shortest duration (indexed by the letter $s$). The choice of one or the other will depend on the actual bond yield on date $T$, denoted by $R(T)$.

We showed in Sect. 15.1.4.1, that the futures price on date $T$ is

$$F_T = \min\left(\frac{C_l(T)}{f_l}, \frac{C_s(T)}{f_s}\right).$$

Consider, for example, the case where the most probable CTD is the DS with longest duration $l$ (i.e. $R_T^f(t) > k$). The value at the expiry of the futures contract can be decomposed into two parts, involving an option of exchange between the two bonds $s$ and $l$:

$$F_T = \frac{C_l(T)}{f_l} - \max\left(\frac{C_l(T)}{f_l} - \frac{C_s(T)}{f_s}; 0\right). \tag{15.33}$$

Written this way, it is clear that the exchange option, with payoff $\max\left(\frac{C_l(T)}{f_l} - \frac{C_s(T)}{f_s}; 0\right)$, reduces the expiry price of the futures contract as computed in Sect. 15.1.4.1. This exchange option can be valued using *Margrabe's formula* (see Chap. 11, Sect. 11.2.7).

Suppose, for example, that the annual standard deviation of the bond rate is 1%. If the option is in the money ($R_T^f(t) = k$), applying the formula gives a futures price of 99.80%, which can be seen as the theoretical price of the futures without the delivery option (100% of the nominal amount) reduced by the time value of the in-the-money option: 0.20%. This effect is mitigated (the delivery option value is low) when the market yield goes noticeably above or below $k$ *since then the likelihood of a change in CTD is very low.*

## 2 Relationship Between Forward and Futures Prices

We examine here the impact of the margin call system on the futures price which, in particular, makes it different from the forward price.

1. We first recall two important results (*see Appendix 1 to Chap. 9*):
   (a) The *futures* price, in contrast to the forward price, is a *martingale* under the risk-neutral probability $Q$ associated with the standard (riskless asset) numeraire $\beta_t = \exp\left(\int_0^t r(s)ds\right)$.
   (b) The price $F_0^T$ on date 0 of the futures contract with underlying $S$ equals the forward price $\Phi_0^T$ corrected by a covariance term (see Eq. (9.A-3) in Appendix 1 to Chap. 9):

## 2 Relationship Between Forward and Futures Prices

$$F_0^T = \Phi_0^T \exp\left[\mathrm{cov}\left(\int_0^T r(t)dt,\ \ln S_T\right)\right].$$ (15.34)

The role of covariance is explained as follows: if there is a *negative* covariance between the short-term rate and the return on the underlying, then an increase in $S_t$ (thus in $F_t^T$) goes along with a decrease in the short-term rate. For the buyer of the futures, the positive margin received is invested at a low interest rate. If, on the contrary, $S_t$ (thus $F_t^T$) falls, the buyer must finance the negative margin at a higher interest rate. Therefore, the system of margin calls works against the buyer's interests, implying *a price of the futures smaller than that of the forward*. Symmetrically, a positive covariance would play in favor of the buyer, implying a futures price higher than that of the forward. Since rates and prices are negatively correlated, the *futures prices of financial assets are usually smaller than the corresponding forward prices (while futures rates are higher)*. Relation (15.34) is *general* and applies to contracts on both short-term and long-term securities.

2. Let us apply Eq. (15.34) to contracts on interest rates beginning with a *contract on a short-term rate*.

We start with the approximate result in Proposition 10 according to which the price of a short-term futures is equal to the forward price with expiration date $T$ of a zero-coupon with maturity $T + K$ (usually $K = 0.25$). This result does not take into account the impact of margin calls that affects futures. Noticing that the zero-coupon $B_{T + K}(t)$ of maturity $T + K$ (duration $K$ on date $T$) plays the role of $S_t$ in result (b) above, we deduce the short-term (fictitious) futures price *defined by Eq. (15.27) with $K$ instead of $0.25$*[20]:

The short-term (fictitious) futures price $\Psi_0^T$ *is linked to the short-term forward price* $\Phi_0^T$ *by:*

$$\Psi_0^T = \Phi_0^T \exp\left[-\mathrm{cov}\left(\int_0^T r(t)dt,\ K \cdot z_K(T)\right)\right],$$ (15.35)

*where $z_K(T)$ is the continuous zero-coupon spot rate of duration $K$ prevailing on expiration date $T$.*

**Proof**
Applying Eq. (15.34) in the case where $S_T = B_{T + K}(T)$ gives

---

[20]We must use this fictitious price since the quoted "price" $F_0^T$ *is not a meaningful price, as argued in the text.*

$$\Psi_0^T = \Phi_0^T \exp\left( \operatorname{cov}\left( \int_0^T r(t)dt; \ \ln\left(B_{T+K}(T)\right) \right) \right).$$

We obtain the desired result by using the fact that, by definition

$$B_{T+K}(T) = e^{-K.Z_K(T)}.$$

Again, note that the covariance in Eq. (15.35) is that of two processes strongly *positively* correlated. Indeed, the two terms depend on the evolution of short-term rates between 0 and $T$: $\int_0^T r(s)ds$ depends on the evolution of the instantaneous rate over the period $[0, T]$, and $z_K(T)$ is the rate of duration $K$ that will prevail at $T$. Since this covariance is multiplied by $(-1)$, the futures price is *smaller* than the corresponding forward price.

3. An analogous result holds for a *contract on a long-term, notional rate and for a contract on a bond price.*

Let $F_0^{i,T}$ be the price on date 0 of a fictitious futures contract written on DS $i$ and of maturity $T$. By applying Eq. (15.34), *the futures and forward clean prices are related by*

$$F_0^{i,T} = \Phi_0^{i,T} \exp\left( \operatorname{cov}\left( \int_0^T r(s)ds; \ \ln C_i(T) \right) \right),$$

*where $C_i$ is the clean spot price of DS $i$.*

The relationship between the futures price of a notional contract and its forward price is obtained by applying the last equation to the CTD $m$.

More precise relations can be obtained by combining this last result with an interest-rate model such as those developed in Chap. 17. Using the notations of that chapter, we denote by $F(t, T, T_B)$ and $\Phi(t, T, T_B)$, respectively, the price on date $t$ of a $T$-futures contract and a $T$-forward contract written on a bond $B(t, T_B)$ with maturity $T_B > T$. Using the one-factor Heath–Jarrow–Morton model (Chap. 17, Sect. 17.2. 1.4), according to which the instantaneous volatility of the instantaneous forward rate is the constant $\sigma$, we derive an explicit formula for the relationship between $F(t, T, T_B)$ and $\Phi(t, T, T_B)$ (Eq. 17.24 in Chap. 17):

$$F(t, T, T_B) = \Phi(t, T, T_B) \exp\left[ -\frac{\sigma^2}{2}(T_B - T)(T - t)^2 \right], \tag{15.36}$$

which is Eq. (15.12) in the text providing the *convexity adjustment between the futures and forward clean prices of a zero-coupon bond due to margin calls.*

4. Using the same one-factor Heath–Jarrow–Morton model, which is equivalent to the Ho and Lee (1986) model, we can derive a closed-form relation for the convexity adjustment relevant to contracts on *short-term rates* (note that, by contrast, Eq. (15.35) is general and relative to *prices*).

## 2 Relationship Between Forward and Futures Prices

Assuming that the instantaneous spot interest rate obeys:

$$dr = \theta(t)dt + \sigma dW(t), \tag{15.37}$$

where $W(t)$ is a Wiener process, the price $P(t,T)$ of the $T$-maturity zero-coupon bond is, in the risk-neutral world, equal to:

$$P(t,T) = A(t,T)e^{-r(T-t)}, \tag{15.38}$$

where $A(t, T)$ is a deterministic function of time.

Define the current (continuously compounded) forward rate $f(t, T_1, T_2)$ for lending/borrowing between $T_1$ and $T_2$ as:

$$f(t, T_1, T_2) = \frac{LnP(t, T_1) - LnP(t, T_2)}{T_2 - T_1}.$$

Note that in the main text relative to STIR contracts, $T_1$ is denoted by $T$ and $T_2$ by $T+K$.

Applying Ito's lemma to $f(t, T_1, T_2)$ using Eqs. (15.37) and (15.38) yields:

$$\begin{aligned} df(t, T_1, T_2) &= \frac{1}{T_2 - T_1}\left[(T_2 - T_1)\sigma dW + \frac{1}{2}\sigma^2\left[(T_2 - t)^2 - (T_1 - t)^2\right]dt\right] \\ &= \sigma dW + \frac{(T_2 - t)^2 - (T_1 - t)^2}{2(T_2 - T_1)}\sigma^2 dt \end{aligned}$$

$$\tag{15.39}$$

The risk-neutral conditional expectation of $df(.)$ is the second, deterministic, term on the r.h.s. of Eq. (15.39). By integrating this term between 0 and $T_1$, we obtain the risk-neutral expected change in the *forward* rate between the current date (0) and date $T_1$:

$$E_t[f(T_1) - f(t)] = \int_0^{T_1} \frac{(T_2 - t)^2 - (T_1 - t)^2}{2(T_2 - T_1)}\sigma^2 dt = \frac{1}{2}\sigma^2 T_1 T_2.$$

Now, we know that under the risk-neutral probability, the *futures* rate is a martingale (see the Appendix to Chap. 9) so that the conditional expectation of its change between any two dates is 0. Therefore, the *convexity adjustment* between the forward and futures rates is conform to that given in Eq. (15.28) in the main text, i.e., $0.5\sigma^2 T_1 T_2$.

## Suggestions for Further Reading

### Books

Duffie, D. (1989). *Futures markets*. Prentice-Hall.
Hull, J. (2018). *Options, futures and other derivatives* (10th ed.). Prentice Hall Pearson.
\* Lioui, A., & Poncet, P. (2005). *Dynamic asset allocation with forwards and futures*. Springer

### Articles

Adler, M., & Detemple, J. (1988). On the optimal hedge of a non-traded cash position. *Journal of Finance, 43*, 143–153.
Anderson, R., & Danthine, J. P. (1983). The time pattern of hedging and the volatility of futures prices. *Review of Economic Studies, 50*, 249–266.
Breeden, D. (1984). Futures markets and commodity options hedging and optimality in incomplete markets. *Journal of Economic Theory, 32*, 275–300.
Cox, J., Ingersoll, J. E., & Ross, S. A. (1981). The relation between forward and futures prices. *Journal of Financial Economics, 9*, 321–346.
Duffie, D., & Stanton, R. (1992). Pricing continuously resettled contingent claims. *Journal of Economic Dynamics and Control, 16*, 561–573.
Ho, T. S. Y., & Lee, S. B. (1986). Term structure movements and pricing interest rate contingent claims. *Journal of Finance, 41*(5), 1011–1029.
Kane, A., & Marcus, A. (1986). Valuation and optimal exercise of the wild card option in the treasury bond futures market. *Journal of Finance, 41*(1), 195–207.
Lioui, A., & Poncet, P. (1996). Optimal hedging in a dynamic futures market with a non-negativity constraint on wealth. *Journal of Economic Dynamics and Control, 20*, 1101–1113.
Lioui, A., & Poncet, P. (2000). The minimum variance hedge ratio under stochastic interest rates. *Management Science, 46*, 658–668.
Schroder, M. (1999). Changes of numeraire for pricing futures, forwards, and options. *Review of Financial Studies, 12*, 1143–1163.
Siddique, A. (2003). Common asset pricing factors in volatilities and returns in futures markets. *Journal of Banking and Finance, 27*, 2347–2368.

### Internet Sites

Australian market: www.asx.com.au
Brazilian market: www.bmf.com.br
CME Group: www.cmegroup.com
EUREX: www.eurexchange.com
ICE: www.theice.com
Japanese market (Tokyo): www.tfx.co.jp/en
NYMEX: www.nymex.com
Professional association of Futures markets: www.Futuresindustry.org
Singapore market: www.simex.com.sg

# Interest Rate Instruments: Valuation with the BSM Model, Hybrids, and Structured Products

# 16

Chapter 7 analyzed floating rate instruments and vanilla interest rate swaps. This chapter presents more complex products, because either they are not standard, or they contain optional clauses, or they are combinations of standard or nonstandard instruments. Valuation of the products with optional provisions is carried out using the Black-Scholes-Merton (BSM) model adapted to the context of stochastic interest rates, introduced in Chap. 11. This chapter does not consider explicit models of interest rate dynamics (such as those of Vasicek, Hull and White, Cox-Ingersoll-Ross, Heath-Jarrow-Morton, the Libor Market Model, or the Swap Market Model), which are postponed to Chap. 17.

Section 16.1 presents the principles of option valuation adapted to the context of stochastic rates (using the forward-neutral expectation of the payoff) as well as several versions of the generalized BSM model. Section 16.2 is devoted to swaps and swaptions. It resumes briefly the analysis of swaps given in Chap. 7 and deals with products that were just briefly mentioned there. Section 16.3 presents caps and floors and Sect. 16.4 combinations, hybrid and structured products, and the financial engineering needed for these various instruments. Section 16.5 is devoted to hybrid products embedding options or optional clauses, such as convertible bonds.

## 16.1  Valuation of Interest Rate Instruments Using Standard Models

The analyses and results developed in this section are based on valuation with the forward-neutral probability measure, which leads to generalizations of the standard Black-Scholes-Merton tailored to different stochastic rate contexts (Sect. 16.1.1).

We apply a version of the model, called the BSM-price model, to valuing bond options (Sect. 16.1.2). Then we study instruments that generate a payoff that is defined as a function of a rate, and not a price, such as a caplet that is valued using a BSM-rate model (Sect. 16.1.3). Finally, we present the convexity adjustment, that allows approximate valuation of some non-vanilla cash flows (Sect. 16.1.4).

© The Author(s), under exclusive license to Springer Nature Switzerland AG 2022
P. Poncet, R. Portait, *Capital Market Finance*, Springer Texts in Business and
Economics, https://doi.org/10.1007/978-3-030-84600-8_16

### 16.1.1 Principles of Valuation and the Black-Scholes-Merton Model Generalized to Stochastic Interest Rates

We recall first some principles underlying the valuation of payoffs that may depend on rates, then the generalized BSM model presented in Chap. 11. Several customizations and additions useful to the analysis and valuation of interest rate instruments lead to two versions, the BSM-rate and the BSM-price models.

#### 16.1.1.1 Valuation Principles

Generally, determining the value $V$ of a security that provides a random payoff $\Psi(T)$ on date $T$ requires computing its appropriately discounted expectation under the convenient probability measure. Its value on date 0 thus can be written

$$V(0) = E_0^* \{A(T)\Psi(T)\}.$$

The operator $E_0^*\{.\}$ is a risk-neutral (RN) or forward-neutral (FN) expectation computed on date 0 and $A(T)$ is the discounting factor. We distinguish the RN and FN probability distributions, leading to $E_0^Q\{.\}$ and $E_0^Q{}_T\{.\}$, respectively, the asterisk in $E_0^*\{.\}$ covering both cases.

– Under the RN probability distribution (of measure $Q$), the discounting factor writes

$$A(T) = e^{\int_0^T -r(t)dt}$$

where $r(t)$ is the risk-free instantaneous rate, and $\dfrac{V(t)}{e^{\int_0^t r(u)du}}$ is a $Q$-martingale. Hence, we have $V(0) = E_0^Q \left\{ e^{\int_0^T -r(t)dt} V(T) \right\}$, and since $V(T) = \Psi(T)$, this becomes

$$V(0) = E_0^Q \left\{ e^{\int_0^T -r(t)dt} \Psi(T) \right\}. \tag{16.1}$$

Remark that usually $r(t)$ is a stochastic process, so $A(T) = e^{\int_0^T -r(t)dt}$ is a random variable and $E_0^Q \left\{ e^{\int_0^T -r(t)dt} \Psi(T) \right\} = E_0^Q \left\{ e^{\int_0^T -r(t)dt} \right\} \cdot E_0^Q \{\Psi(T)\} + cov_0^Q \left\{ e^{\int_0^T -r(t)dt}, \Psi(T) \right\}$.

The valuation Eq. (16.1) allows, in particular, writing $B_T(0)$, the price on date 0 of a zero-coupon bond giving a sure payoff $\Psi(T) = \$1$ on date $T$, as

$$B_T(0) = E_0^Q \left\{ e^{\int_0^T -r(t)dt} \right\}. \tag{16.2}$$

Recall that the price of the zero-coupon equals the value of the single cash flow ($\$1$) it generates discounted at the zero-coupon rate, that is, $B_T(0) = e^{-r_T(0)T}$ if $r_T$ is

## 16.1 Valuation of Interest Rate Instruments Using Standard Models

the continuous zero-coupon rate of maturity $T$, or $B_T(0) = \frac{1}{1+r_T(0)T}$ if $r_T$ is a proportional rate.

- Under the FN probability measure denoted by $Q_T$, asset prices relative to the price of a zero-coupon expiring at $T$ (i.e., the forward-$T$ *price*) are martingales. The discounting factor equals $B_T(0)$ and is known at 0, which simplifies the valuation Eq. (16.1) rewritten[1]

$$V(0) = B_T(0)\, E_0^{Q_T}\{\Psi(T)\} \tag{16.3}$$

Equation (16.3) is fundamental and applies systematically in this chapter. It means that *the right to a payoff on a future date $T$ is obtained by simple discounting of the FN expectation with the maturity-$T$ zero-coupon rate.*

Recall that there is a FN probability distribution for each expiration date $T$; we use the notation $Q_T$ or FN-$T$ for those measures which make the maturity-$T$ forward prices $\Phi_T(t)$ $Q_T$-*martingales*, i.e., the spot prices using the maturity-$T$ zero-coupon bond price as numeraire $\left(\frac{S(t)}{B_T(t)}\right)$.

We stress the difference between the valuation equation (16.1), which holds under RN probability measures and involves a stochastic discount factor $\left(e^{\int_0^T -r(t)dt}\right)$, with the simpler equation (16.3) which holds under FN probability measures and for which the discount factor $(B_T(0))$ is known on date 0.

In fact, these two equations are equivalent under one of the following two assumptions:

(i) $r(t)$ is deterministic, in which case the measures FN and RN coincide, $e^{\int_0^T -r(t)dt}$ is not random and Eq. (16.1) simplifies to $V(0) = e^{\int_0^T -r(t)dt} E_0^Q\{\Psi(T)\} = B_T(0)E_0^{Q_T}\{\Psi(T)\}$.

(ii) $e^{\int_0^T -r(t)dt}$ is independent of $\Psi(T)$ (a less restrictive assumption than the preceding). Then Eq. (16.1) simplifies to $V(0) = E_0^Q\left\{e^{\int_0^T -r(t)dt}\right\}E_0^Q\{\Psi(T)\}$, from which, using Eq. (16.2), we have:

$$V(0) = B_T(0)\, E_0^Q\{\Psi(T)\} = B_T(0)E_0^{Q_T}\{\Psi(T)\}.$$

Unfortunately, neither of the two assumptions is acceptable if one has to value an interest rate instrument, since the random nature of interest rates and their correlation

---

[1] $\frac{V(t)}{B_T(t)}$ being a martingale, we have : $\frac{V(0)}{B_T(0)} = E_0^{Q_T}\left\{\frac{V(T)}{B_T(T)}\right\}$; since $B_T(T) = 1$ and: $\psi(T) = V(T)$, we obtain:
$V(0) = B_T(0)\, E_0^{Q_T}\{\psi(T)\}$.

with the payoff $\Psi(T)$ then are crucial. *It is thus more convenient to apply valuation equation* (16.3) (since the discount factor is not a random variable), *and so to use the FN probability measure rather than the RN one*. This approach allows us for instance, as we saw in Chap. 11, to generalize Black's model to price standard European options under stochastic interest rates.

### 16.1.1.2 Revisiting the Generalized BSM or Gaussian Model

The generalized BSM model (the Gaussian model) studied in Chap. 11 (Proposition 14) applies to a payoff that writes $\Psi(T) = Max\,(X(T) - K, 0)$ for a call, or $\Psi(T) = Max\,(K - X(T), 0)$ for a put when $X(T)$ is a log-normal variable depending on interest rates. Usually, $X(T)$ is either the price of the underlying asset of the option with expiration date $T$ or a rate when it is itself the option's underlying. The first interpretation leads to the BSM-price model and the second to the BSM-rate model discussed later on.

We consider here $X(T)$ to be the spot price of the underlying on date $T$. Note that this price is the same as the forward price $\Phi_T(T)$ at the expiry of the forward contract.

As is shown in Chap. 11, Eq. (11.54), the log-normality of $\Phi_T(T)$ implies we have

$$X(T) = \Phi_T(T) = \Phi_0 e^{-0.5[\sigma(\ln X)]^2 + \sigma(\ln X)U},$$

where $U$ is $N(0,1)$, $\Phi_0 = E_0^{Q_T}\{\,\Phi_T(T)\,\}$ and $\sigma(\ln X)$ is the standard deviation of ln $(X(T)) = \ln(\Phi_T(T))$.

At this stage, $\Phi_0$ is only a simple forward-neutral expectation, *unobservable* and as such *useless*. It is here that the martingale property of the forward price $\Phi_T(t)$ under $Q_T$ is useful, implying

$$\Phi_T(0) = E_0^{Q_T}\{\,\Phi_T(T)\,\} \equiv \Phi_0.$$

Therefore, we can identify $\Phi_0$ with the underlying's forward price on date 0 which can be *observed* or *computed* from the spot price using the spot-forward parity.

Finally, we could set $\sigma(\ln X) = \sigma\sqrt{T}$. The parameter $\sigma$ can be viewed as average volatility of $X(t)$ between 0 and $T$, as the instantaneous volatility can vary between these two dates. Since the change in probability does not change the volatility (only the drift of the process is affected), $\sigma$ is also the volatility for the historical probability (it can thus be estimated from market data). We therefore write

$$X(T) = \Phi_0 e^{-0.5\sigma^2 T + \sigma\sqrt{T}U}.$$

As a result, by applying Eq. (16.3) with the assumed log-normality of $X(T)$ we obtain the price of the European call and put written on $X(T)$. On date 0, they are equal, respectively, to

# 16.1 Valuation of Interest Rate Instruments Using Standard Models

$$C_0 = B_T(0)[\Phi_T(0)\, N(d_1) - K\, N(d_2)], \tag{16.4a}$$

$$P_0 = B_T(0)[K\, N(-d_2) - \Phi_T(0)\, N(-d_1)], \text{ with}: \tag{16.4b}$$

$$d_1 = \frac{\ln\frac{\Phi_T(0)}{K} + \frac{1}{2}\sigma^2 T}{\sigma\sqrt{T}} \left(\equiv \frac{\ln\frac{\Phi_T(0)}{K} + \frac{1}{2}\sigma^2(\ln X)}{\sigma(\ln X)}\right);$$

$$d_2 = d_1 - \sigma\sqrt{T}(\equiv d_1 - \sigma(\ln X)). \tag{16.4c}$$

All three Eqs. (16.4a–c) are identical to Eqs. (11.52) in Proposition 4 of Chap. 11 (written at $t = 0$ with $X(t) = S(t)$) and constitute what is called the *BSM-price model*, in contrast with the BSM-rate model where $X(t)$ denotes a rate. That model is analyzed in Sect. 16.1.3 below when we value a caplet.

The reader will recall that formulas (16.4) can be applied in the same way in the case of coupons distributed by the underlying asset since the forward price occurring in the equations takes into account such distributions.

## 16.1.2 Valuation of a Bond Option Using the BSM-Price Model

In the simplest case of valuation of an option on a fixed-rate instrument, the BSM-price model can be applied almost directly.

Let $S(t)$ be the cash, dirty, price of a fixed-income bond and $\Phi_t(T)$ its forward price for delivery at $T$. To value a European option written on it with strike $K$ and expiring on date $T$, the simplest method is to assume that its price $S(T)$ is log-normal and the standard deviation of $\ln(S(T))$ equal to $\sigma\sqrt{T}$. The forward and spot prices $\Phi_T(T)$ and $S(T)$ coincide on the contracts' expiration date $T$ (for both the option and forward contracts), so the BSM-price model (Eqs. 16.4) can be used *as such* (with $X(t) = S(t)$).

Recall that, in Eq. (16.4), $\Phi_T(0)$ is the forward price of the bond on valuation date 0 and is given as a function of the spot (full coupon) price $S_0$ using spot-forward parity written, for example, as $\Phi_T(0) = \frac{1}{B_T(0)}(S_0 - D^*)$, where $D^*$ is the discounted value on date 0 of the coupons paid by the bond between 0 and $T$.

Furthermore, it is often useful to express the bond price's volatility parameter $\sigma$ as a function of the volatility $\sigma_y$ of the bond yield $y(t)$. Recall that the approximation involving the changes $\Delta S$ in the price and $\Delta y$ in the yield between 0 and $T$ writes:

$$\frac{\Delta S}{S_0} \approx (\text{bond sensitivity at } T) \times y(0)\left(\frac{\Delta y}{y(0)}\right).$$

This implies the following approximate relationship between the annualized average volatilities $\sigma$ of the price and $\sigma_y$ of the yield:

$$\sigma\sqrt{T} = (\text{bond sensitivity at } T) \times y(0) \times \sigma_y\sqrt{T},$$

or else

$$\sigma_{bond\,price} = (\text{bond sensitivity at } T) \times y(0) \times \sigma_y$$

This formula is often used to estimate the volatility of the price from that of the yield.

The same BSM-price model also applies *as such* to an option written on a forward contract on a notional bond such as a Euro-bund.

---

**Example 1**

Consider a bond with 9.25 years to maturity, which distributes a 5% annual coupon in 3 months (0.25 year) and whose yield to maturity is 6% and full coupon price is 94 for a nominal of 100. The 3-month rate is 4% and the 9-month rate is 4.5% (these are continuous actual/actual Libor rates). The forward price of this bond to be delivered in 9 months is, therefore, $\Phi_T(0) = e^{0.045 \times 0.75}(94 - e^{-0.04 \times 0.25} \times 5) = 92.11$. In addition, the price of a zero-coupon providing \$1 in 9 months is $B_T(0) = e^{-0.045 \times 0.75} = 0.9668$.

We have to value a European put on this bond, which expires in 9 months (0.75 year) and whose strike is 100. The annual volatility $\sigma_y$ of the bond rate is estimated to be 25%.

In 9 months, at expiration of the option, the bond has a maturity of 8.5 years and the standard bond calculation shows that its sensitivity is 6.5 (for the same yield of 6%). The annualized volatility of this bond's forward price in 9 months can then be estimated to be $\sigma = 6.5 \times 0.06 \times 0.25 = 9.75\%$.

The formulas in Eq. (16.4) with $\sigma = 9.75\%$, $T = 0.75$, $\Phi_T(0) = 92.11$, $B_T(0) = 0.9668$, and $K = 100$ immediately give a put value equal to $P = 8.32$.

---

## 16.1.3 Valuation of the Right to a Cash Flow Expressed as a Function of a Rate and the BSM-Rate Model

In numerous situations, one has to value the right to a cash flow $\Psi(r(T))$ given as a function of a market rate $r(T)$ on date $T$ (rather than a function of a price $S(T)$). The payoff is, as the case may be, paid at $T$ or on a date after $T$. In addition, it could have an optional feature, for example, $\Psi(r(T)) = \text{Max}[r(T) - k, 0]]$. It could then be convenient to consider the rate as the underlying of an option and to apply the BSM-rate model that we now describe.

## 16.1 Valuation of Interest Rate Instruments Using Standard Models

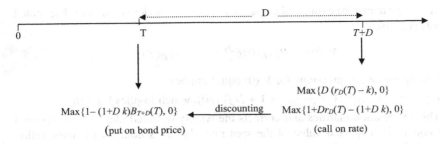

**Fig. 16.1** The analysis of a caplet as a call on a rate or a put on a bond price

### 16.1.3.1 Analysis of a Vanilla Cash Flow: Forward Rate and FN Expectation of a Spot Rate

We consider three dates: 0 (today); $T$ and $T+D$, two future dates (see Fig. 16.1). First, we examine the cash flow $D\, r_D(T)$ paid on date $T+D$ where $r_D(T)$ is the (proportional) spot rate for $D$ periods which holds on date $T$ and $D$ is a fraction of a year.[2] This is a cash flow with a "vanilla forward-looking (predetermined) rate" that is replicable by a spot investment of \$1 at $T$ for a duration $D$. Instruments with a vanilla floating rate as well as variable legs of vanilla swaps are made up of such components which have been studied in detail in Chap. 7.

Recall the properties of such a cash flow. We showed in Chap. 7 that, seen from date 0, the certainty-equivalent of the random cash flow $D\, r_D(T)$ paid on date $T+D$ is $D f_{T,D}(0)$ where $f_{T,D}(0)$ is the forward (proportional) rate for $D$ periods in $T$ periods: For the purpose of valuation on date 0 the random cash flow $D\, r_D(T)$ can be replaced by the sure cash flow $D f_{T,D}(0)$ and discounted at the risk-free rate. In addition, $1+D\, r_D(T)$ can be interpreted as the payoff of a security $x$ of duration $D$, issued at $T$ and with nominal value \$1, whose market value is also \$1 on date $T$. Its value on date 0 is, therefore, $B_T(0)$. We will recover this result by another method which depends on an important result expressed in the next proposition.

### Proposition 1
*The forward rate $f_{T,D}(0)$ equals the expectation of the spot rate $r_D(T)$ calculated with the probability measure $Q_{T+D}$ (the one that makes martingales the prices expressed in zero-coupon bonds with maturity $T + D$, i.e., the forward prices with maturity $T + D$):*

$$f_{T,D}(0) = E_0^{Q_{T+D}}(r_D(T)). \tag{16.5}$$

### Proof
Let us consider the security $x$ previously described, whose payoff on $T+D$ is $1+D\, r_D(T)$, the value at $T$ is $V_x(T) = 1$, and therefore the value at 0 is $V_x(0) = B_T(0)$. The

---
[2] Generally computed on an actual/360 basis.

latter value can also be expressed using the expectation under $Q_{T+D}$ of its discounted payoff (Eq. 16.3):

$$V_x(0) = B_{T+D}(0)E_0^{Q_{T+D}}(1 + Dr_D(T)).$$

Setting the two expressions for $V_x(0)$ equal implies
$E_0^{Q_{T+D}}(1 + Dr_D(T)) = \frac{B_T(0)}{B_{T+D}(0)} = 1 + D f_{T,D}(0)$, which implies Eq. (16.5).

This proposition clarifies and corrects the widely believed idea that the forward rate equals the expected value of the spot rate: this does hold for the expectation computed under $Q_{T+D}$ but is wrong under the historical probability. In addition, it is only valid for the vanilla case of a rate $r_D(T)$ paid in arrears, on date $T+D$; if it were paid in advance (on date $T$), the previous result would have to be modified (see the case studied in Sect. 16.1.4.2 below).

### 16.1.3.2 Valuation of a Caplet or a Floorlet: the BSM-Rate Model

We now value a caplet using a BSM model. A caplet for nominal $1 is the right to the payment Max$\{D(r_D(T) - k), 0\}$ paid on $T+D$, where $k$ is a rate fixed when the caplet is issued (on date 0, today).

In a similar way, the payoff of a floorlet is Max$\{D (k - r_D(T)), 0\}$.

We will see in Sect. 16.3 of this chapter that caps are made up of caplets and floors of floorlets. A caplet can be considered either as a call on a rate, or as a put on a price, the reverse being true for a floorlet. We justify and analyze in turn these two points of view and their consequences.

The payoff of a caplet allows one to consider it directly as a call on the rate $r_D(T)$, with strike $k$ and expiring on $T+D$ (a floorlet is put on the same rate). From this vantage, *the rate is the underlying* of the option. If the rate $r_D(T)$ is assumed to be log-normal, it can be written under the probability $Q_{T+D}$ as

$$r_D(T) = f_{T,D}(0)e^{-0.5\sigma_r^2 T + \sigma_r \sqrt{T} U},$$

where $f_{T,D}(0)$ is the forward rate on date 0 relative to the period $(T, T+D)$, $U$ is $N(0,1)$ and $\sigma_r$ interpreted as the average volatility of the rate $r_D(t)$ between 0 and $T$. As a result, the generalized BSM model whose validity depends only on the log-normality of the underlying, is applicable in a modified form called the BSM-rate model; it leads to equations similar to Eq. (16.4) which we write in the next proposition.

**Proposition 2. Valuation of a Caplet and a Floorlet in the BSM-Rate Model**
*Let us assume that the rate $r_D(T)$ is log-normal.*

– *The value $C_0$ on date 0 of the caplet (with payoff $MD [r_D(T) - k]^+$ on $T+D$) writes*

$$C_0 = M B_{T+D}(0)\big[ f_{T,D}(0) N(d_1) - k N(d_2)\big]D, \tag{16.6}$$

## 16.1 Valuation of Interest Rate Instruments Using Standard Models 675

– *The value $P_0$ of a floorlet (with payoff MD $[k - r_D(T)]^+$ on T+D) writes*

$$P_0 = M \, B_{T+D}(0) \left[ k \, N(-d_2) - f_{T,D}(0) \, N(-d_1) \right] D, \qquad (16.6')$$

$$\text{with} : d_1 = \frac{\ln \frac{f_{T,D}(0)}{k} + \frac{1}{2}\sigma_r^2 T}{\sigma_r \sqrt{T}} ; d_2 = d_1 - \sigma_r \sqrt{T}$$

*where $f_{T,D}(t)$ is the forward rate at t relative to the period $(T, T + D)$, $\sigma_r$ the average volatility of the rate $r_D(t)$ from 0 to T and M the nominal of the caplet and floorlet.*

This proposition is proved exactly like Eqs. (16.4a, 16.4b and 16.4c) ($r_D$ plays the role of X and $f_{T,D}$ that of $\Phi$).

As we will see in Chap. 17, the BSM-rate model (a Gaussian model) can be combined with other rate models (particularly those of Vasicek, Hull, and White, and the Libor Market Model) some of which are amenable to parameter calibration. Propositions 1 and 2 lie in fact at the core of the Libor Market Model discussed in more detail in Chap. 17.

---

### Example 2

Suppose we have to value a caplet of nominal $100 whose payoff is Max$\{D (r_D(T) - k), 0\}$ on date T+D, where k is the strike rate of 4.2%, $r_D$ is the 3-month Euribor, D, therefore, is 0.25 and T is the caplet's expiration in 6 months (T= 0.5). The 6-month Euribor is presently 4.0% and the 9-month Euribor is 4.1%. The annualized average volatility of the 3-month Euribor is estimated to be 20%.

To apply formula (16.6), one must first compute the discounting factor $B_{T+D}(0) = B_{0.75}(0)$ and the forward rate $f_{T,D}(0) = f_{0.5,0.25}(0)$. The first calculation gives $1/(1.041)^{0.75} = 0.97031$. The second gives $((1.041)^{0.75}/(1.04)^{0.5})^4 - 1 = 0.043 = 4.30\%$. Knowing that the nominal M is $100 and D equals 0.25, formula (16.6) gives the caplet's value as $0.0711.

The floorlet with the same parameters would be worth $0.0469.

---

Alternatively and equivalently, one may consider a caplet as a *put P on a zero-coupon bond providing* $(1 + Dk)$ on date T +D (with a duration D on expiration date T), the call's expiration date being T and the strike 1 (see Fig. 16.1). To see this, it suffices to:

(i) Start from the identity giving the caplet's payoff when it is not zero ($r_D(T) > k$):

$$D(r_D(T) - k) = (1 + D \, r_D(T)) - (1 + D \, k).$$

(ii) Note that the cash flow available at $T+D$ is known since instant $T$, on which date its value is:

$$1 - \frac{1+Dk}{1+Dr_D(T)} = 1 - (1+Dk)B_{T+D}(T)$$

When $r_D(T) > k$, we do have the payoff of a put $P$ (if $r_D(T) \leq k$ both the caplet and the put $P$ have zero payoff).

For a Gaussian price $B_{T+D}(T)$, we could have applied the BSM-price model to value the caplet.

The same analysis allows considering a floorlet, whose payoff on $T+D$ is Max $\{D(k - r_D(T)), 0\}$, as a call on $(1+Dk)B_{T+D}(T)$, with strike 1, and expiration date $T$.

### 16.1.3.3 Digital Option on a Rate

Along the lines of digital options on asset prices studied in Chap. 14, there exist digital options on rates, which may be components of more complex financial instruments discussed in Sect. 16.4.

- A digital call on the reference rate $r(t)$, with nominal \$1, strike $k$, and exercise date $T$, yields \$1 if $r(T) \geq k$ and 0 otherwise, or in another notation $1_{r(T) \geq k}$ (the indicator function of the event $r(T) \geq k$). Computing the expectation under $Q_T$ means we can write the digital call's price as $B_T(0) \, E_0^{Q_T} \{1_{r(T) \geq k}\} = B_T(0) \, Q_T(r(T) \geq k)$. Since the probability of exercise, $Q_T(r(T) \geq k)$, is equal to $N(d_2)$, the BSM value of the digital call is

$$B_T(0)N(d_2) \text{ where } d_2 = \frac{\ln \frac{f_T(0)}{k} - \frac{1}{2}\sigma_r^2 T}{\sigma_r \sqrt{T}}, \qquad (16.7)$$

and $f_T(0)$ is the forward rate relative to $r$ and $\sigma_r$ the mean volatility of $r$ between 0 and $T$.

- The digital put yields $1_{r(T) < k}$ and its BSM value is $B_T(0)(1 - N(d_2))$.

## 16.1.4 Convexity Adjustments for Non-vanilla Cash Flows (*)

In this subsection, we give an adjustment that takes into account the convexity of the payoff and so allows us to value some non-vanilla interest rate products.

### 16.1.4.1 Adjustment for Convexity

Let us consider a right $x$ to a cash flow $\Psi(r(T))$ paid on $T$ whose value depends on a spot rate $r(T)$ (a zero-coupon or a bullet bond yield rate, compound or proportional, the fixed rate of a swap, etc.) prevailing on date $T$. $\Psi(r(T))$ could be:

# 16.1 Valuation of Interest Rate Instruments Using Standard Models 677

- Some payoff (e.g., the settlement of a swap or a non-vanilla caplet).
- The value of a fixed-income security, such as a bond or the fixed leg of a swap, generating future cash flows after $T$, whose yield is $r(T)$ on date $T$: $\Psi(r(T))$ is, therefore, the present value of those cash flows, discounted at $T$ at the rate $r(T)$.

In both cases, the value on date $T$ of the right $x$ is $V(T) = \Psi(r(T))$. We are interested in the value of $x$ at $t < T$ and more specifically at $t = 0$.

On date 0, the spot price of $x$ is $V(0)$ and its forward price for delivery or payment at $T$ (called forward-$T$ in this subsection) is $\Phi_T(0)$. On date $T$, the cash and forward prices for $x$ coincide: $V(T) = \Phi_T(T) = \Psi(r(T))$.

Now let us define the forward rate as the rate "deduced" from the forward price, so that it is equivalent to fixing on date 0 the forward rate or the forward price used at $T$:

**Definition** *On date 0, the forward-$T$ rate of the instrument $x$ is the rate $\phi_T(0)$ such that $\Phi_T(0) = \Psi(\phi_T(0))$.*

This definition, which generalizes the classical definition of a forward rate, coincides with it in the standard case. For example, if $\Psi(r_D(T))$ is the value at $T$ of a zero-coupon yielding \$ 1 at $T+D$, we can write $\Psi(r_D(T)) = \frac{1}{1+Dr_D(T)}$ ; the forward rate $\phi_T(0)$, that allows fixing on date 0 the price to be paid on $T$ at $\frac{1}{1+D\phi_T(0)}$, coincides with the forward rate $f_{T,D}(0)$ for $D$ periods in $T$ periods. The definition given in this chapter can be applied in more complex situations (the forward yield on a bond, the forward rate for a swap, etc.).

In contrast with the vanilla case, the forward rate $\phi_T(0)$ is different, for nonstandard cases, from the forward-neutral expectation of the spot rate.

Indeed, we know that the forward-$T$ price today does equal the FN-$T$ expectation of the forward (or spot) price on date $T$, that is, $\Phi_T(0) = E^{Q_T}(\Psi(r(T)))$. Putting this into the expression defining the forward rate, $\Phi_T(0) = \Psi(\phi_T(0))$, we obtain

$$\Psi(\phi_T(0)) = E^{Q_T}(\Psi(r(T))).$$

$\Psi(r)$ is often some convex function, for example, a present value, and calculating the expectation is not easy (there is no closed-form expression). Since in that case we have

$E^{Q_T}[\Psi(r(T))] > \Psi(E^{Q_T}[r(T)])$, it follows that $\Psi(\phi_T(0)) > \Psi(E^{Q_T}[r(T)])$.[3]

Consequently, the forward rate $\phi_T(0)$ is not equal to the FN-$T$ expectation of the spot rate $E^{Q_T}[r(T)]$. In particular, if $\Psi$ is a present value, which is convex and decreasing with the discount rate, we have $\phi_T(0) < E^{Q_T}[r(T)]$. An approximate relationship between the forward rate and the FN-$T$ expectation of the spot rate can

---

[3] The conclusion comes from Jensen's inequality that can be understood intuitively with a graph, and that turns up in numerous financial contexts. See, for example, Chap. 21, Sect. 21.1.3, the characterization of risk aversion by a concave utility function. There, we show the symmetrical result $U(E(X)) > E(U(X))$ for $U$ a concave function.

nevertheless be obtained from a Taylor expansion of $\Psi(r(T))$; it involves the first and second derivatives $\Psi'$ and $\Psi''$, the second being related to the convexity of $\Psi$. We write this as a proposition:

**Proposition 3**

*In the notation previously defined, the FN-T expectation of the spot rate and the forward-T rate are linked by the approximate relationship:*

$$E^{Q_T}(r(T)) \; \phi_T(0) - \frac{1}{2}\frac{\Psi''(\phi_T(0))}{\Psi'(\phi_T(0))}\sigma_\phi^2\phi_T^2(0)T, \tag{16.8}$$

*where $\sigma_\phi$ denotes the volatility of the forward rate $\phi_T(0)$.*

**Proof**

Let us proceed by expanding $\Psi(r(T))$ to the second order about $\phi_T(0)$:

$$\Psi(r(T)) \; \Psi(\phi_T(0)) + (r(T) - \phi_T(0))\Psi'(\phi_T(0))$$
$$+0.5\left[r(T) - \phi_T(0)\right]^2 \Psi''(\phi_T(0)).$$

By taking the FN-$T$ expectation of the two members of this equation we have:

$$E^{Q_T}\left[\Psi(r(T)\right] \; \Psi(\phi_T(0)) + (E^{Q_T}\left[r(T)\right] - \phi_T(0))\Psi'(\phi_T(0))$$
$$+0.5\,E^{Q_T}\left[(r(T) - \phi_T(0))^2\right]\Psi''(\phi_T(0)).$$

Since $E^{Q_T}[\Psi(r(T)] = \Psi(\phi_T(0))\,(= \Phi_T(0))$, this becomes

$$(E^{Q_T}\left[r(T)\right] - \phi_T(0))\,\Psi'(\phi_T(0)) + 0.5E^{Q}{}_T\left[(r(T) - \phi_T(0))^2\right]\Psi''(\phi_T(0))\; 0, \text{ or}$$

$$E^{Q_T}(r(T)) \; \phi_T(0) - \frac{1}{2}\frac{\Psi''(\phi_T(0))E^{Q_T}[r(T) - \phi_T(0))^2]}{\Psi'(\phi_T(0))}.$$

We obtain Eq. (16.8) by writing $E^{Q_T}\left[(r(T) - \phi_T(0))^2\right]$ in the form $(\phi_T(0)\sigma_\phi)^2\,T$, where $\sigma_\varphi$ is the volatility of the forward rate $\phi_T(0)$.

We note that the difference between the forward rate and the expectation under $Q_T$ of the corresponding spot rate equals $\frac{1}{2}\frac{\Psi''(\phi_T(0))E^{Q_T}[r(T)-\phi_T(0))^2]}{\Psi'(\phi_T(0))}$. This discrepancy is linked to the nonlinearity (and, in general, convexity) of $\Psi$: if $\Psi'' = 0$, it vanishes. If $\Psi$ is convex and decreasing, $\Psi''$ is positive, $\Psi'$ negative and, as we have seen, $E^{Q_T}$ $(r(T)) > \phi_T(0)$.

## 16.1.4.2 Application: Accounting for Time Lags

Let us consider, as in Sect. 16.1.4.1, a cash flow of $D\,r_D(T)$, where $r_D(T)$ is the proportional spot rate $T$ for transactions with duration $D$; but, in contrast to the vanilla case Sect. 16.1.3.1, *this cash is to be paid on date $T$* (and not on $T+D$), that is

## 16.1 Valuation of Interest Rate Instruments Using Standard Models 679

*in advance*. As a result, this is not the vanilla situation (it can no longer be replicated by a spot placement of \$1 realized on $T$) and Proposition 1 can no longer be applied. In particular, $E^{Q_T}[r_D(T)]$ is different from $f_{T,D}(0)$ (which equals $E^{Q_{T+D}}[r_D(T)]$). Although the difficulty does not seem to come from the convexity of the payoff (it is, in fact, linear), the adjustment for convexity described in Proposition 3 allows valuing such a payoff as stated in the next proposition.

### Proposition 4
With the notation and within the framework previously introduced:

- *The FN-T expectation of the rate $r_D(T)$ and the corresponding forward rate $f_{T,D}(0)$ are linked by an approximate relationship involving the volatility $\sigma_f$ of the forward rate (remark the difference between $\sigma_f$ and the $\sigma_\phi$ appearing in Proposition 3) :*

$$E^{Q_T}(r_D(T)) \# f_{T,D}(0) + \frac{1}{1+Df_{T,D}(0)}\sigma_f^2[f_{T,D}(0)]^2 \, D \, T. \qquad (16.9)$$

- *The value $V(0)$ on date 0 of the cash flow $D \, r_D(T)$ available on $T$ is approximately*

$$V(0) = B_T(0) \, D \left[ f_{T,D}(0) + \frac{1}{1+Df_{T,D}(0)}\sigma_f^2 D \, T \, (f_{T,D}(0))^2 \right]. \qquad (16.10)$$

### Proof
Recall that a zero-coupon bond yielding \$ 1 on $T+D$ is worth $\frac{1}{1+Dr_D(T)}$ on date $T$. As a result, the right to \$1 on $T+D$ has the same value as a bond $x$ yielding the right to the payoff $\Psi(r_D(T)) = \frac{1}{1+Dr_D(T)}$ paid on $T$. The analysis in this subsection can be applied to the bond $x$ whose forward-$T$ rate on date 0 is $\phi_T(0) = f_{T,D}(0)$ (the forward rate for $D$ periods in $T$ periods applied today to fix the price of $x$ paid on $T$). From Eq. (16.8), we have

$$E^{Q_T}(r_D(T)) \, f_{T,D}(0) - \frac{1}{2}\frac{\Psi''(f_{T,D}(0))}{\Psi'(f_{T,D}(0))}\sigma_f^2[f_{T,D}(0)]^2 \, T.$$

Since $\frac{1}{2}\frac{\Psi''(f_{T,D}(0))}{\Psi'(f_{T,D}(0))} = -\frac{D}{1+Df_{T,D}(0)}$ (taking the first and second derivatives of $\Psi(.)$), we obtain

$E^{Q_T}(r_D(T)) \# f_{T,D}(0) + \frac{1}{1+Df_{T,D}(0)}\sigma_f^2[f_{T,D}(0)]^2 \, D \, T$, which is Eq. (16.9).

Now, the value $V(0)$ on date 0 of the cash flow $D \, r_D(T)$ paid on $T$ equals its FN-T expectation discounted at the zero-coupon rate for a maturity $T$:

$V(0) = B_T(0) \, D \, E^{Q_T}(r_D(T))$, which implies, using Eq. (16.8):

$V(0) = B_T(0) \, D \, [f_{T,D}(0) + \frac{1}{1+Df_{T,D}(0)}\sigma_f^2 D \, T \, (f_{T,D}(0))^2]$, which is Eq. (16.10).

## 16.2 Nonstandard Swaps and Swaptions

In Chap. 7 we studied *plain vanilla swaps*, both forward- and backward-looking. After a brief review of the notation and key points in their analysis (Sect. 16.2.1), we discuss different non-standard swaps and their valuation (Sect. 16.2.2), then swap options (swaptions) and their valuation (Sect. 16.2.3).

### 16.2.1 Review of Swaps and Notation

A *swap* for a fixed rate $k$ against a floating reference $r(t)$ with nominal value $1 is denoted by $S_{r-k}$ (*fixed-rate payer* or *floating-rate receiver*). The counterparty's position, *fixed-rate receiver* or *floating-rate payer*, is denoted by $S_{k-r}$. Hence: $S_{r-k} = -S_{k-r}$.

Settlements occur on the dates $T_1, T_2, \ldots, T_N$. The *settlement on $T_i$ is determined by $r(T_{i-1})$* if the rate is *forward-looking* (pre-determined), which we assume in this subsection.

For a standard swap, the initiation date 0 coincides with the date $T_0$ on which the swap starts to run and the first reset date of the floating leg (see Fig. 16.2 with $T_0 = 0$). This implies that the first cash flow (on date $T_1$) is known on date 0. Denote by $D_i$ the duration of the $i$th period between settlements (*tenor*): $D_i = T_i - T_{i-1}$. The basis is usually actual/360 ($D_i = Nj(T_{i-1}, T_i)/360$). The $i$th settlement (cash flow) of a $k$ payer vanilla swap, therefore, writes $M D_i [r(T_{i-1}) - k]$ where $M$ is the notional or nominal. In the sequel, to simplify we most often use a nominal $M$ of $ 1, as for $S_{r-k}$ and $S_{k-r}$.

In the vanilla case, the frequency of settlements on the floating leg, and so the tenors $D_i$, are tied up with the maturity of the floating reference (four settlements per year, so $D_i \# 0.25$, for the 3-month Euribor; two annual settlements for the 6-month Libor, etc.).

The settlement dates for the fixed leg do not necessarily coincide with those for the floating leg. A simple example would be one annual settlement for the fixed leg

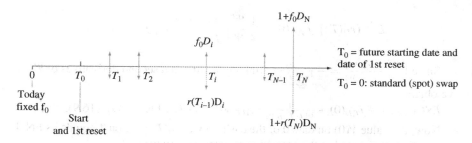

**Fig. 16.2** A forward swap

## 16.2 Nonstandard Swaps and Swaptions

and four settlements per year for the leg referencing the 3-month Euribor. For the sake of generality, we write the settlements of the floating leg as $M D_i \, r(T_{i-1})$ on the dates $T_1, T_2, \ldots, T_N$ and those of the fixed leg as $M D'_i \, k$ on the dates $T'_1, T'_2, \ldots, T'_N$.

An asset of nominal \$1, that yields a fixed rate $k$ and whose coupons $D'_i \, k$ are paid on dates $T'_1, T'_2, \ldots, T'_N$ is denoted by $A_k$, and a security with a floating rate $r(t)$ paying coupons $D_i \, r(T_{i-1})$ on dates $T_1, T_2, \ldots, T_N$ is denoted by $A_r$. The nominal values of these two securities are paid *in fine*, respectively on $T'_N$ and $T_N$ (usually, $T'_N = T_N$).

The *swap* $S_{r-k}$ can be thought of as a position long in $A_r$ and short in $A_k$. We thus write:

$$S_{r-k} = A_r - A_k. \tag{16.11a}$$

Similarly, the *swap* $S_{k-r}$ is short in $A_r$ and long in $A_k$:

$$S_{k-r} = A_k - A_r. \tag{16.11b}$$

From this point of view, the nominals $M$ of the fixed and floating components are fictitiously exchanged on the final dates, the net cash flow being zero.

Alternatively, one may consider the series of coupons $a_k$ and $a_r$ alone, without capital reimbursements, called abusively "annuities." For a nominal $M = 1$, the fixed annuity $a_k$ provides the sure cash flow $k \, D_i$ on $T'_i$ and the variable annuity $a_r$ the random cash flow $D_i \, r(T_{i-1})$ on $T_i$, for $i = 1, \ldots, N$. Unlike assets $A_k$ and $A_r$, these annuities do *not* pay the nominal \$1 at maturity. The annuities $a_k$ and $a_r$ are, respectively, the fixed and variable legs of the swap and we have:

$$S_{r-k} = a_r - a_k \quad \text{and} \quad S_{k-r} = a_k - a_r. \tag{16.12}$$

To avoid misunderstandings we use the wording "fixed and floating *components*" of the swap for $A_k$ and $A_r$ and "fixed and variable *legs*" for $a_k$ and $a_r$.

Equations (16.11a and 16.11b) and (16.12) imply that, under no arbitrage, the market values of swaps write $V(S_{k-r}) = -V(S_{r-k}) = V(A_k) - V(A_r) = V(a_k) - V(a_r)$, which grounds the valuation methods called *stripping* (see Chap. 7). For vanilla swaps, $V(A_k)$ (or $V(a_k)$) is obtained by simple discounting using the zero-coupon swap rates, and $V(A_r)$ (or $V(a_r)$) by one of the following two equivalent methods:

- Noting that $A_r$ is at par on the next coupon date $T_1$ and discounting this full-coupon value $(1 + r(0))$ on date $0$,[4] we get

$$V_0(A_r) = B_{T_1}(0)(1 + r(0)) \Leftrightarrow V_0(a_r) = B_{T1}(0)(1 + r(0)) - B_{T_N}(0). \tag{16.13}$$

---

[4] The next coupon is known and equal to $r(0)$, from which it follows that $A_r(T_1^-) = 1 + r(0)$.

- Replacing the vanilla forward-looking rates with the corresponding forward rates (which are their certainty equivalents) and discounting the sure cash flows thus obtained by the zero-coupon swap rates, we find

$$V_0(a_r) = B_{T_1}(0)\, r(0) + \sum_{i=2}^{N} B_{T_i}(0) D_i f_{T_{i-1},D_i}(0).\qquad (16.14)$$

## 16.2.2 Some Nonstandard Swaps

Some swaps analyzed below were mentioned in Chap. 7. Here we proceed with their valuation.

### 16.2.2.1 Forward Swaps (*Forward Start*)

*Forward swaps* are contracts signed on date 0 for a swap that starts running on a future date $T_0 > 0$ on the basis *of a fixed rate specified on date 0*. On Figure 16.2 that illustrates the components of such a swap with nominal \$1, the fixed and floating cash flows are assumed synchronous ($T_i = T'_i$).

The rate specified at 0 for this forward swap can be interpreted as the forward rate of the swap and is denoted by $f_{T_0}(0)$ or simply $f_0$.

The first reset date of the variable leg is $T_0$ ($> 0$). Therefore, even with a predetermined reference, one does not know on date 0 the next variable settlement amount $M D_1 r(T_0)$, in contrast with the spot swap case for which $T_0 = 0$.[5] However, one does know that the floating component $A_r$ is at parity on the reset date $T_0$, which implies that, for a nominal of \$1, its value at 0 is $B_{T_0}(0)$.

The forward rate $f_0$ is fixed at a level so that the value of the swap is zero on date 0, that is, so that the value of the fixed component $A_{f_0}$ equals $B_{T_0}(0)$: $f_0 \sum_{i=1}^{N} D'_i B_{T'_i}(0) + B_{T_N}(0) = B_{T_0}(0)$. In this way we obtain $f_0 \equiv f_{T_0}(0)$, the forward-$T_0$ rate of the swap:

$$f_{T_0}(0) = \frac{B_{T_0}(0) - B_{T'_N}(0)}{\sum_{i=1}^{N} D'_i B_{T'_i}(0)}.\qquad (16.15)$$

---

[5] A spot swap thus is a special, degenerate case of a forward swap, like any spot asset is a degenerate forward asset.

## 16.2 Nonstandard Swaps and Swaptions

**Example 3**

We assume the following curve for the zero-coupon rates $r_\theta$ extracted from the *swap rate curve*.

| $r - 3$ months | $r - 9$ months | $r - 15$ months | $r - 21$ months | $r - 27$ months |
|---|---|---|---|---|
| 4% | 4.2% | 4.4% | 4.5% | 4.5% |

This implies the following prices $B_\theta(0) = \frac{1}{(1+r_\theta)^\theta}$ for the zero-coupon bonds:

| $B - 3$ months | $B - 9$ months | $B - 15$ months | $B - 21$ months | $B - 27$ months |
|---|---|---|---|---|
| 0.9902 | 0.9696 | 0.9476 | 0.9259 | 0.9057 |

Let us consider a swap of 6-month Libor for a fixed rate, with nominal 100, starting on 3 months, and lasting 2 years. Looking at the situation from 01/01/$n$ ($t = 0$), it starts on 01/04/$n$ ($T_0 = 3$ months) and yields four variable payments on 01/10/$n$ ($T_1=9$ months), on 01/04/$n$+1 ($T_2 =15$ months), on 01/10/$n$+1 ($T_3=21$ months) and on 01/04/$n$+2 ($T_4=27$ months). We assume the fixed settlements are annual and occur on 01/04/$n$+1 ($T'_1= T_2 =15$ months) and on 01/04/$n$+2 ($T'_2 = T_4 = 27$ months).

The floating component will be at par in 3 months and therefore its value today is $100 \times V_0(A_r) = 99.02$.

For a fixed rate $k$, the fixed component has the value $100 \times V_0(A_k) = k(94.76 + 90.57) + 90.57$.

The value of the forward swap is nil on date 0 for $k = f_0$ such that $f_0 = \frac{99.02-90.57}{94.76+90.57} = 4.56\%$.

### 16.2.2.2 Step-Down (Amortization) Swaps

A *step-down (or amortization) swap* generates cash flows computed from nominal values $M_0, M_1, \ldots, M_{N-1}$ that decrease according to a preset schedule. The swap is assumed to be issued on date 0 and the settlements take place on dates $T_1, T_2, \ldots, T_N$, so that the $i$th cash flow of the fixed leg (on date $T_i$) is $M_{i-1}D_i k$ and that on the variable leg is $M_{i-1} D_i r(T_{i-1})$.

The fixed-rate payer swap is analyzed as a combination of borrowing at a fixed rate with amortization of the principal and borrowing at a floating rate, with the same reimbursements of principal equal to $M_{i-1} - M_i$ on date $T_i$ for both components. The settlement payments can therefore be written as $M_{i-1}D_i k + (M_{i-1} - M_i)$ for the fixed leg and $M_{i-1}D_i r(T_{i-1}) + (M_{i-1} - M_i)$ for the variable leg (the reimbursement for the fixed and floating components are fictional, since the net cash flow is zero).

Note that the floating component is at par on the settlement date, that is $A_r(T_i^+) = M_i$ just after the $i$th payment (and $A_r(T_i^-) = M_{i-1} + M_{i-1}D_i r(T_{i-1})$ just before). This remark allows valuing the floating component on any date along the same lines as a vanilla floating rate instrument. The fixed component is valued using standard discounting.

---

**Example 4**

We want to value a step-down swap with the following properties:

- Remaining time to run: 3 years exactly.
- The nominal principals are, respectively: $M_0 = 100$, $M_1 = 75$ and $M_2 = 50$.
- Cash payments (fixed and floating) are annual; the variable leg pays off at the end of year $i$ an amount computed with the 1-year rate, $r_1(i-1)$, observed at the end of year $i-1$. More precisely, incorporating the fictional reimbursements of principal, the series of floating cash flows reads: $r_1(0)\times 100 + 25$, $r_1(1)\times 75 + 25$, and $r_1(2)\times 50 + 50$, paid on dates 1, 2, and 3, respectively. The fixed rate is 4% which implies that the fixed cash flows are, respectively, $25 + 4$, $25 + 3$, and $50 + 2$ on the same dates.

If we assume the swap was initiated on some past date, the value of the swap usually differs from 0 and can be calculated as follows:

Value of the fixed component $= V(A_k) = \frac{29}{1+r_1(0)} + \frac{28}{(1+r_2(0))^2} + \frac{52}{(1+r_3(0))^3}$; the rates $r_1(0)$, $r_2(0)$, $r_3(0)$ are, respectively, the zero-coupon rates for maturities 1, 2, and 3 years extracted from the vanilla swap rate curve.[6] For example, for a flat rate curve $r_1(0) = r_2(0) = r_3(0) = 5\%$, we obtain $V(A_k) = 97.94$ and the swap value is $+ 2.06$ or $- 2.06$ according to whether it is a fixed-rate payer or receiver.

The floating component is at par since it can be decomposed into three different *bullet* bonds at par: the first with maturity 3 years and value 50, the second maturing in 2 years and value 25, and the third in 1 year and value 25. Therefore, $V(A_r) = 100$.

The swap value follows: $V(A_r) - V(A_k)$ or $V(A_k) - V(A_r)$.

At the swap's initiation date, its value is zero and the fixed-rate $k$ is chosen so that $V(A_k) = 100$, that is, $\frac{25+100k}{1+r_1(0)} + \frac{25+75k}{(1+r_2(0))^2} + \frac{50+50k}{(1+r_3(0))^3} = 100$.

For example, with a flat rate curve $r_1(0) = r_2(0) = r_3(0) = x$, we have $k = x$.

---

### 16.2.2.3 In Arrears Swaps

In an *in arrears* swap, the variable payment on date $T_i$ is given by the formula $r(T_i)D_i$ rather than $r(T_{i-1})D_i$. Thus, this swap is not vanilla since it cannot be replicated by a

---

[6]The careful reader will note that, unduly but for the sake of simplicity, $r_1(0)$ is both the 1-year proportional actual/360 rate (in the formula for computing the floating coupon) and a zero-coupon yield in the discounting formula.

## 16.2 Nonstandard Swaps and Swaptions

spot investment on $T_{i-1}$. As we have shown in Sect. 16.1.4, this time lag can be compensated for by using a convexity adjustment consisting in adding to the forward rate $f_{T_i,D_i}(0)$ the correction term $\frac{1}{1+D_i f_{T_i,D_i}(0)} \sigma_i^2 T_i D_i (f_{T_i,D_i}(0))^2$ (Eq. (16.9)).[7] For valuation purposes, the variable cash flows $r(T_i) D_i$ can then be replaced by the certainty equivalents $[f_{T_i,D_i}(0) + \frac{1}{1+D_i f_{T_i,D_i}(0)} \sigma_i^2 T_i D_i (f_{T_i,D_i}(0))^2]D_i$ ; the value of the variable leg obtains by discounting:

$$V(a_r) = \sum_{i=1}^{N} B_{T_i}(0) \left[ f_{T_i,D_i}(0) \, \frac{1}{1 + D_i f_{T_i,D_i}(0)} \sigma_i^2 + T_i D_i \left( f_{T_i,D_i}(0) \right)^2 \right] D_i.$$

---

**Example 5**

Let us consider the following in arrears swap: the nominal is \$100 million, the fixed payer rate is 4.8%, the floating receiver rate is the 6-month Libor (delayed by one period), the time to run is 1 year, and the cash flows are half-yearly ($D_i = D = 0.5$ (for the sake of simplicity)). The yield curve is flat, for maturities less than 1 year, at 4.8%. The forward rates relevant to this swap are then obviously all 4.8%, which would make it of *zero value* if it were vanilla.

The volatility of the forward rates is estimated at 25% (annualized). The floating cash flows $r(T_i) D$ can thus be replaced by their certainty equivalents $\left[ f_{T_i,D}(0) + \frac{1}{1+D f_{T_i,D}(0)} \sigma_i^2 T_i D( f_{T_i,D}(0))^2 \right] D$, and therefore the convexity adjustments are here:

$$\left[ \frac{1}{1 + 0.5 \times 0.048} 0.25^2 \times 0.5 \times (0.048)^2 \times T_i \right] \times 0.5 = [0.00007 \, T_i] \times 0.5.$$

i.e., for $T_i = 1$: 0.000035, and for $T_i = 2$: 0.000070.

With a nominal of \$100 M, the payment adjustments are \$3500 and \$7000 so that the value of the swap is positive (floating cash flows are received) and equal to the sum of the cash flows discounted at the half-year proportional rate of 2.4%, i.e., 3500/(1.024) + 7000/(1.048) = \$10,097.

---

### 16.2.2.4 Constant Maturity Swaps

In a *constant maturity swap* (CMS), the reference rate $i(t)$ on which the computation of the floating payment is based is a *K-maturity yield* which is not related to the duration of the period between settlements. Typically, $i(t)$ is the fixed rate of a vanilla

---

[7]The volatility $\sigma_i$ of the forward rate $f_{T_i,D_i}(0)$ can be obtained from the implicit volatility of the corresponding caplet.

swap, or a bond rate (e.g., a bullet bond yield), with a given medium or long duration $K$. For example, the floating settlements take place every 3 months, computed with the 10-year swap rate, denoted $s_{10}(t)$, which means the floating payment on $T_i$ is $D_i s_{10}(T_i)$ (or $D_i s_{10}(T_{i-1})$) if there is a time lag).

*If there is no time lag*, the approximate valuation of the variable leg of the CMS involves a convexity adjustment. For each expiration $T_i$, one then replaces the bond rate $i(T_i)$ by the forward-$T_i$ rate, denoted $f_{T_i}(0)$, corrected by the factor $-\frac{1}{2} \frac{f_{T_i}(0)^2 \sigma_i^2}{V'} \frac{V''}{} T_i$, where $\sigma_i^2$ is the volatility of the forward rate, $V$ represents the value of a $K$-maturity bond, $V'$ and $V''$ are the first and second derivatives of $V$ with respect to the yield, and $V$, $V'$, and $V''$ are calculated at the point $f_{T_i}(0)$.

Recall that the forward rate $f_{T_i}(0)$ is such that $V(f_{T_i}(0)) = F_{T_i}(0) = \text{forward-}T_i$ price of the bond with maturity $K$; it can therefore be computed using the equation $V(f_{T_i}(0)) = V(0)/B_{T_i}(0)$.

To obtain the variable leg's value, it suffices to discount the series of the sure cash flows $D_i(f_{T_i}(0) - \frac{1}{2} \frac{f_{T_i}(0)^2 \sigma_i^2}{V'} \frac{V''}{} T_i)$. The value of the fixed leg is obtained by standard discounting.

*If there is a time lag*, in fact, a frequent situation (the floating cash flow on date $T_i$ depends on a yield or a swap rate prevailing on $T_{i-1}$, for example), it is necessary to make a *second* adjustment, analogous to that of a *swap in arrears*.

### 16.2.3 Swap Options (or Swaptions)

By definition, a European swaption gives its holder the right of entering into a swap contract on the exercise date $T$ on the basis of a fixed rate $k$ determined on date 0 (the initiation date). In contrast with a forward swap, it is a conditional asset, i.e., a right but not an obligation, which involves the payment of a premium up front. One distinguishes options on a $k$ payer swap from options on a $k$ receiver swap. We consider a European swaption on a vanilla swap, either payer or receiver, with a notional value of \$1 for which the $N$ settlements take place on dates $T_1, T_2, \ldots, T_N$. We denote by $D_i = T_i - T_{i-1}$ the duration of the $i$th period (*tenor*), expressed as a fraction of a year.

Two valuation methods are conceivable. The first applies the BSM-price model and the second the BSM-rate model. The first is simpler but is mentioned only as an example since the second is more precise and is favored by practitioners.

- The first method, for a fixed-rate receiver, consists in analyzing the swap as a long position in the fixed component (fixed-rate bond) associated with a short position in the floating component. Since the last component is at par (\$1), the option on a *k receiver* swap is a *call* of strike \$1, exercise date $T$, on a bond with fixed-rate $k$ and nominal \$1. The *payer* swaption is a put on the same bond. If the price of the bond is log-normal, the swaption can be valued using the BSM-price model.

## 16.2 Nonstandard Swaps and Swaptions

– The second, more accurate method is based on the BSM-rate model. It has the advantage of being appropriate for using and calibrating the *Swap Market Model*, as we will see in Chap. 17. In this chapter, we just present the basic valuation principles and ignore all calibration issues.

Let the fixed rate of a vanilla swap on date $T$ be $s(T)$. This is the fixed rate that makes the market value of the option's underlying swap on date $T$ nil (*par yield*). Seen from date 0, this market rate $s(T)$ is a random variable. The option on a $k$ payer swap is exercised on date $T$ if and only if $s(T) > k$. Note that on date $T$, the swap can be settled by a fixed-rate receiver swap with the same reference rate, duration, and amount, at the fixed rate $s(T)$. The position thus resulting from exercising the option generates the periodic cash flows $(s(T) - k)D_i$ on dates $T_1, T_2, \ldots, T_N$ ($T_i$ is the $i$th settlement date and $D_i = T_i - T_{i-1}$ the tenor of the $i$th period between payments). We remark that all the cash flows are known from date $T$ *onward*. An option on a fixed-rate $k$ *payer* swap can thus be analyzed as a basket of $N$ calls $C_i$ on the same *cash flow* $s(T)$, with the same strike $k$, and differing in their respective expiration dates $T_1$, $T_2, \ldots, T_N$.[8] The payoff of the call $C_i$ on $T_i$ writes $D_i Max(s(T) - k, 0)$. Note that, if $s(T) > k$, *all calls $C_i$ will be exercised* (and yield *approximately the same payoff D Max(s(T) – k, 0) on different dates $T_1, T_2, \ldots, T_N$). In the opposite case, $s(T) \leq k$, none of them will be exercised.* In this respect, these $N$ calls *cannot* be assimilated to the caplets $D_i Max(r(T_i) - k, 0)$ that compose a cap as these are exercised independently of each other. To value $C_i$ using the BSM-rate model discussed in Sect. 16.1 (Eq. 16.6), we assume the underlying rate $s(T)$ to be log-normally distributed, so that

$$s(T) = f_0 e^{-0.5\sigma_s^2 T + \sigma_s \sqrt{T} U}, \tag{16.16}$$

where $U$ is $N(0,1)$ and $\sigma_s$ is interpreted as an average volatility between dates 0 and $T$.

Besides, recall that $f_T(t)$ denotes the forward-$T$ swap rate holding on date $t$ and that $f_T(T) = s(T)$. Equation (16.16) can then be written in the equivalent form

$$f_T(T) = f_0 e^{-0.5\sigma_s^2 T + \sigma_s \sqrt{T} U}. \tag{16.16'}$$

From Eqs. (16.16) or (16.16'), it follows that $f_0$ is the expected value of $s(T)$ or of $f_T(T)$; this expectation depends on the adopted probability measure. It is convenient to use the probability $Q_a$ that makes the rate $f_T(t)$ of the forward swap a martingale, so that $E^{Q_a}[f_T(T)] = f_T(0) \equiv f_0$. *One can thus identify the rate $f_0$ in Eqs. (16.16) and (16.16') with the rate at 0 of a forward start-T swap.* The martingale probability $Q_a$ is characterized in the Appendix, along with details for this analysis.

The assumption that $s(T) = f_T(T)$ is log-normal then allows applying the BSM-rate model to the forward rate of the swap that underlies the call $C_i$ and ending up

---

[8] It could, alternatively, be analyzed as an option exchanging the variable annuity $a_r$ for the fixed annuity $a_k$. It is this point of view that explains the appropriate choice of the martingale numeraire (see the Appendix).

688        16 Interest Rate Instruments: Valuation with the BSM Model, Hybrids,...

with a formula analogous to Eq. (16.6); we can therefore state the following proposition:

**Proposition 5**

*Under the assumptions and with the notation specified previously:*

- *The call $C_i$, with strike $k$ and expiration $T_i$, is worth:*

$$C_i(0) = B_{T_i}(0)D_i \left[ f_T(0) N(d_1) - k N(d_2) \right], \text{ with } d_1 = \frac{\ln \frac{f_T(0)}{k} + \frac{1}{2}\sigma_s^2 T}{\sigma_s \sqrt{T}};$$

$$d_2 = d_1 - \sigma_s \sqrt{T},$$

*(the same values of $d_1$ and $d_2$ for every $i$).*

- *The value $OS_{r-k}$ of the option on this $k$ payer swap can be obtained by summing the $N$ values $C_i(0)$:*

$$OS_{r-k}(0) = \sum_{i=1}^{N} C_i(0) = \left[ f_T(0) N(d_1) - k N(d_2) \right] \sum_{i=1}^{N} B_{T_i}(0)D_i. \qquad (16.17)$$

- *The option on a $k$ receiver swap can be analyzed as a basket of $N$ puts $P_i$ on the same cash flow $s(T)$ but with different expirations; it is worth*

$$OS_{k-r}(0) = \sum_{i=1}^{N} P_i(0) = \left[ k N(-d_2) - f_T(0) N(-d_1) \right] \sum_{i=1}^{N} B_{T_i}(0)D_i. \qquad (16.18)$$

As we show in the following Chap. 17, this valuation using the BSM-rate (Gaussian) model can be combined with other interest rate models (Libor Market Model, Swap Market Model), which in particular make calibrating the volatility easier.

---

## 16.3   Caps and Floors

Caps and floors are over-the-counter (OTC) contracts for exchanging rates, with option properties, defined by the following parameters:

- An amount (nominal or notional), denoted $M$ in what follows.
- A floating reference rate such as a Libor or a Euribor, denoted $r(t)$.
- A fixed rate, also called guaranteed rate or strike, denoted $k$.

## 16.3 Caps and Floors

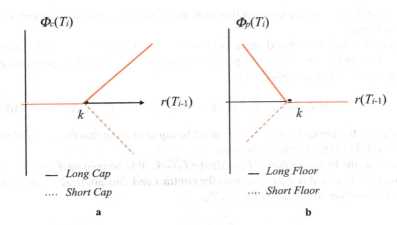

**Fig. 16.3** Payoff profiles of a cap (**a**), and a floor (**b**), as a function of the reference rate

- A schedule $T_0, T_1, T_2, \ldots, T_N$ of settlement payments. As for swaps, we denote by $D_i$ the *tenor* of the $i$th period: $D_i = T_i - T_{i-1}$ usually expressed with basis actual/360.

The frequency of settlements, and so the tenors $D_i$, are linked to the maturity of the reference chosen so that the underlying cash flows are vanilla (e.g., four annual payments, so $D_i \# 0.25$, for a 3-month Euribor).

Caps and floors can be viewed as chains of options on interest rates (caplets or floorlets studied in Sects. 16.3.1 and 16.3.2 below). They are instruments for managing interest rate risk used both by firms and financial institutions. Their trading volume is substantial. We examine caps and floors in turn.

### 16.3.1 Vanilla Caps

Caps take their name from the fact that they are often (but not exclusively) used to cap the cost of borrowing at a floating rate, or that of repeated short-term borrowing.

#### 16.3.1.1 Definition and Description

The holder or buyer of a vanilla cap in the notional amount $M$ receives from the seller on settlement date $T_i$ a payoff $\Phi_c(T_i)$ given by the formula

$$\Phi_c(T_i) = M\, D_i\, Max[r(T_{i-1}) - k, 0] \quad \text{for } i = 1, \ldots, N. \tag{16.19}$$

Figure 16.3a shows such a payment for the buyer (full line) and for the seller (dashed line) as a function of the floater $r(t)$ on date $T_{i-1}$.

To benefit from such a position (which can only lead to cash flows $\geq 0$), the buyer pays a premium, which is usually given as a rate. This premium is paid either up-front at the start of the contract (the most common case) or in installments ($\pi_c$ on each settlement).

Equation (16.19) is based on assuming an *up-front* premium (which therefore does not influence the settlement payments). Assuming a premium payment $\pi_c$ at each settlement, the $i$th cash flow is

$$\Phi_c(T_i) = M\, D_i\, Max[r(T_{i-1}) - k, 0] - \pi_c \quad \text{for } i = 1, \ldots, N, \tag{16.19'}$$

Note that the present value of all $\pi_c$ must be equal to $\pi_0$ so that the two settlement models (16.19) (16.19') are equivalent.

Often, as the first cash flow, $M\, D_1\, Max[r(T_0) - k, 0]$ to be paid on $T_1$, is known on the initiation date 0, it is excluded from the contract and thus not paid; so settlements are in fact relative to the dates $T_2, \ldots, T_N$.

---

**Example 6**

Consider a 3-month Libor cap for \$10M, of 3-year duration, with guaranteed rate = 5%, and 1% premium paid at the start. The next settlement being known, it is excluded from the contract and, as a result, the cap yields a sequence of 11 quarterly payments approximately equal to
$$\Phi_c(t) = \$2.5 \times Max[L_3(t - 3\ months) - 0.05,\ 0]\ M \text{ on } t = 2, \ldots, 12,$$ where $t$ is a quarterly maturity and the duration $D_t$ of the quarter is taken as 0.25 years. This is, for example, \$25,000 if Libor($t - 3\ months$) = 6%, or 0 if Libor ($t - 3\ months$) $\leq$ 5%. The buyer pays a premium to the seller of \$100,000 up-front on the cap's initiation date.

---

The following remarks end this general description:

- The premium $\pi_0$ is in principle equal to the value of the cap on the initiation date.
- The cap is only profitable in situations where rates are going up and the spreads $r(t) - k$ are large enough to cover the payment of the premium ($r(t) - k \geq \pi_c$ if the premium is paid in installments); thus, it can be used as an optional instrument to hedge against rate increases.
- As such a hedge, it allows capping the cost of borrowing at a floating rate. In effect, for a borrowing rate $r(t)$ in excess of the strike rate $k$ of the cap, the payoff of the cap hedges the spread $r(t) - k$, which caps the cost of the hedged loan. More precisely, a nominal $M$ loan at a vanilla floating rate generates a $i$th coupon of amount $M\, D_i\, r(T_{i-1})$ which, associated with the payoff of the cap $M\, D_i\, Max[r(T_{i-1}) - k, 0] - \pi_c$ leads to a net cost of the hedged borrowing equal to

$$M\, D_i\, \{r(T_{i-1}) - Max[r(T_{i-1}) - k, 0] + \pi_c\} = M\, D_i\, \{Min[r(T_{i-1}), k] + \pi_c\}.$$

The cost of financing is therefore capped at $k + \pi_c$.

## 16.3 Caps and Floors

691

**Example 7**

Suppose we have borrowed $10 M for 3 years at the 3-month Libor, reset every quarter. The risk of a large rate increase may be eliminated by buying a 3-month Libor cap, at strike 5%, with an *up-front* premium of 1%, and 3 years to run (11 settlements on the dates $T_2, \ldots, T_{12}$), as given in the preceding example.

Let us start by transforming the up-front premium into a periodic (quarterly) premium. We compute $\pi_c$ such that the present value of the sequence of 11 periodic premiums equals $\pi_0$, that is, $\pi_0 = \pi_c \times \sum_{t=T_2}^{T_{12}} B_t(0)$ where $B_t(0)$ is the price on date 0 of a zero-coupon yielding 1$ on date $t$, with the dates $t$ spaced out in quarters.

Assume that the calculation gives $\pi_c = \left(1 / \sum_{t=T_2}^{T_{12}} B_t(0)\right) \times 1\% = 0.10\%$.

The $t$th payoff of the cap is therefore equivalent to $2.5 \times [Max\{L_3(t - 3\ months) - 5\%, 0\} - 0.10\%]$.

This payoff is deducted from the $t$th variable coupon to compute the net financial costs, that is

$$\$2.5 \times \left[L_3(t - 3\ \text{months})\right.$$
$$\left. - Max\{L_3(t - 3\ \text{months}) - 5\%0\} + 0.10\%\right]M, \text{or alternatively,}$$

$$\$2.5 \times [L_3(t - 3\ \text{months}) + 0.10\%]\ M \text{ if } L_3(t - 3\ \text{months}) \le 5\%;$$

$$\$2.5 \times [5\% + 0.10\%]M \text{ if } L_3(t - 3\ \text{months}) > 5\%.$$

The borrowing cost thus is capped at 5.10% (i.e., $k + \pi_c$).

### 16.3.1.2 Valuation of a Vanilla Cap

The cap's $i$th settlement as given by Eq. (16.19) is a caplet, whose valuation with the BSM model has been studied in Sect. 16.1.3. This valuation can be based on an interest rate model of Vasicek-Hull-White type, CIR (Cox-Ingersoll-Ross), or the Libor Market Model, as we show in Chap. 17. The cap is analyzed as a chain of caplets, i.e., a portfolio of calls on rates or puts on bond prices. The decomposition of the cap into $N$ different caplets is called *stripping* the cap. The value $c_i$ of the $i$th caplet can, for example, be obtained for any date using Eq. (16.6) and the cap's value $C$ is just simply the sum of the values of the caplets $c_i$: $C = \sum_{i=1}^{N} c_i$. In using these

692        16   Interest Rate Instruments: Valuation with the BSM Model, Hybrids,...

formulas, the problem lies in choosing the volatility $\sigma_i$, on which $c_{i+1}$ depends.[9] Traders use volatility curves. Actually, two theoretically equivalent methods are used:

- A single volatility for all the caplets involved in the same cap (called *flat* volatility) but which depends on the cap's maturity.
- A different volatility for each caplet, depending on its maturity (called *spot* volatility).

Just as one can derive zero-coupon rates from bullet bond yields, one can extract the volatilities of caplets from the flat volatility curve applied to caps with different durations. In both cases, one uses a term structure of volatilities. All the notions, methods, and practices of option trading are applicable to caps and caplets (the quotation is that of an implicit volatility, Greeks are calculated, the smile or skew is taken into account, etc.). The Libor Market Model is suitable for such applications, as explained in Chap. 17 where numerical examples are also provided.

### 16.3.2 A Vanilla Floor

A floor is an instrument primarily (but not exclusively) used to transform a floating rate investment (or a repeated short-term investment) into a floating rate investment ensuring a minimum yield or floor. Therefore, it allows investors to profit from rate increases while guaranteeing a minimum yield. Analysis of floors being similar to that of caps, our exposition is brief.

#### 16.3.2.1 Definition and Description

A vanilla floor leads to a series of settlements that can be written, according to whether the premium $\pi_0$ is paid *up-front* or in periodic fractions $\pi_f$:

$$\Phi_f(T_i) = M \, D_i \, \text{Max}[k - r(T_{i-1}), 0] \text{ for } i$$
$$= 1, \dots, N \ (up\text{-}front \text{ premium}) \tag{16.20}$$

$$\Phi_f(T_i) = M \, D_i \, Max[k - r(T_{i-1}), 0] - \pi_f \text{ for } i$$
$$= 1, \dots, N \text{ (periodic premia).} \tag{16.20'}$$

Figure 16.3b shows such a settlement in the *up-front* case (Eq. 16.20), for the buyer (solid line) and the seller (dotted line).

Remark that a floor is profitable when interest rates fall and that, as a result, it can be used as an instrument hedging such a risk. For example, a Libor floor allows fixing a minimum yield for an investment indexed on the Libor. We leave it up to the

---

[9]Recall that this model assumes the log-normality of $r(T_i)$ and that one sets: $\sigma(\ln(r(T_i))) = \sigma_i \sqrt{T_i}$.

reader to check that a Libor floor with strike $k$ and periodic premium $\pi_f$ ensures that a coupon initially set at the floating rate $r(t)$ never falls below the floor $k - \pi_f$.

### 16.3.2.2 Valuation of a Floor

The $i$th payoff, $M\, D_i\, Max[k - r(T_{i-1}), 0]$, is that of a floorlet. The floor is thus made up of a chain (portfolio) of floorlets (see Eq. (16.6′) for their valuation using the BSM-rate model). Its value is the sum of the floorlet values. As for a cap, one uses a term structure of volatilities (volatility as a function of the maturity) to value the different floorlets.

---

## 16.4 Static Replications and Combinations; Structured Contracts

We previously pointed out that a cap associated with a floating rate loan provides a ceiling for its cost, and that a floor on top of a floating rate investment ensures a minimum yield. More generally, the various investment or borrowing floating rate instruments (swaps, caps, and floors) can be combined to make up positions that fit the market participants' various motivations. Such combinations are examined in this section.

### 16.4.1 Basic Instruments: Notation and General Remarks

We consider different instruments with *unit nominal value* and use a notation allowing us to develop an "algebra" of their combinations which is as simple, general, and unambiguous as possible. The most parsimonious system of instruments from which one may synthesize the different financial products presented in the sequel involves four families: bullet bonds with fixed rates, those with floating rates, caps, and digital (or binary) calls. Although payer and receiver swaps, floors, and digital puts can be replicated using the preceding ones, they are also discussed as basic instruments in what follows.[10]

### 16.4.1.1 Fundamental Instruments: Definitions and Notation

We adopt here a notation that differs slightly from the preceding one for the sake of both generality and simplicity: $i(t)$ is the floating reference rate that determines the payment on *date t*. Thus, in the vanilla case, $i(t)$ implicitly equal $r_D(t - D)$ if it is forward-looking, or is an index if it is backward-looking. In fact, we do not rigorously distinguish these two situations, and rather use the generic words "floating" or "variable" rate. In addition, most of the relationships that we establish still

---

[10] These instruments are actually not the most elementary, as zero-coupon bonds or forward rate agreements are, but they are "intermediate products" which are used in the combinations examined below.

hold in non-vanilla contexts (the duration of the reference rate differing from the tenor, payments *in-arrears*, etc.).

There are eight fundamental building blocks, of which the first six have already been discussed.

- The asset that pays a fixed rate $k$, denoted $A_k$, and the asset with floating rate $i(t)$, denoted $A_i$.
- The payer swap $S_{i-k}$ of fixed-rate $k$ and floating reference rate $i(t)$, and the receiver swap $S_{k-i}$.

With a slight abuse of language and for simplicity, we call *payoff* the simple exchange of rates (e.g., $i(t) - k$ for $S_{i-k}$, and not the effective settlement payment $M$ $D[i(t) - k]$).

- The cap with guaranteed rate $k$ and reference rate $i(t)$, denoted by $C_k$. If this involves payment of a premium $\pi_c^0$ at the start of the contract, its payoff writes $Max[i(t) - k, 0]$. Typically, the floating rate is forward-looking, so that $i(t) = r_D(t - D)$.

If the total premium is fractioned, each periodic premium is denoted by $\pi_c$, the cap's payoff writes $[i(t) - k, 0] - \pi_c$, and the cap is denoted by $C_k^{\pi_c}$.

- The floor denoted by $F_k$ yielding the payoff $Max[k - i(t), 0]$.

The premium is denoted by $\pi_f^0$ if it paid up-front and by $\pi_f$ if it is fractioned (the floor is then denoted by $F_k^{\pi_f}$).

- The digital call on the reference rate $i(t)$, with nominal \$ 1 and strike $k$, which gives \$1 if $i(t) \geq k$ and 0 otherwise, that is, $1_{i(t) \geq k}$. Its BSM-price is $B_t(0)N(d_2)$ (see Eq. (16.7)).
- The digital put yields $1_{i(t) < k}$, can be replicated by purchasing $A_k$ and selling the digital call $1_{i(t) \geq k}$, and its BSM price is $B_t(0)N(-d_2)$.

The minus sign "–" in front of any instrument denotes a short position or borrowing. All instruments are written on the same reference $i(t)$, with the same durations, and with synchronous payoffs. The market values are denoted by $V$ (e.g., $V(A_k)$ is the market value of a fixed-rate $k$ bond *of* \$1 *nominal*).

These eight basic instruments can be used to synthesize a vast class of financial instruments, like a toy construction set, which implies that the different prices, rates, premiums, strikes, gearing, and so on, are all connected by AAO relationships. The eight basic blocks are themselves redundant (only four of them are linearly independent), as we recall for some of them and prove for others.

## 16.4.1.2 Redundancy Between a Swap, a Fixed-Rate Asset, and a Floating-Rate Asset

We have seen that the swap $S_{i-k}$ can be considered as a long position in $A_i$ and a short one in $A_k$, with the opposite holding for $S_{k-i}$. This result, linked to the redundancy of the three instruments $A_i$, $A_k$, and $S$, writes as in Eq. (16.11) repeated here for convenience:

$$S_{i-k} = A_i - A_k.$$

This algebraic equation expresses the fact that each of these instruments can be synthesized from the other two: For example, we can write:

$A_i = S_{i-k} + A_k$ (a payer swap plus a fixed-rate amounts to a floating rate).

$A_k = A_i + S_{k-i}$ (a floating rate plus a receiver swap is a fixed rate).

$- S_{i-k} = S_{k-i} = A_k - A_i$ (a *swap* $S_{k-i}$ of a variable rate for a fixed rate is equivalent to a long position in the fixed leg $A_k$ and a short one in the variable leg $A_i$).

## 16.4.1.3 Redundancy Between a Cap, a Floor, and a Swap

A cap, a floor, and a swap with the same reference $i(t)$, the same fixed-rate $k$, and yielding synchronous payments of the same duration, also constitute a redundant set.

More precisely, the three instruments $C_k$, $F_k$, and $S_{i-k}$ (or $S_{k-i}$) are redundant and one may write:

$$C_k - F_k = S_{i-k}, \tag{16.21}$$

or similar equations obtained by simple algebra:

$$C_k = F_k + S_{i-k}, \quad F_k - C_k = S_{k-i}, \text{ etc.}$$

For instance, a portfolio long cap and short floor replicate a payer swap, and a receiver swap replicates a long floor and a short cap.

Justifying Eq. (16.21) amounts to check that the exchange in rates resulting from a long position in the cap and a short in the floor equals the cash flow generated by a payer swap: $Max\ (i(t) - k, 0) - Max\ (k - i(t), 0) = i(t) - k$. In AAO, the value of a payer swap is equal to the value of the cap minus that of the floor. On the initial date, the swap's fixed-rate $k$ is chosen so as to make zero the value of the swap, which implies that the cap's premium equals the floor's if their strikes $k$ are equal to the swap's fixed rate prevailing on the market at initiation.

We leave it to the reader to check that for fractional premiums $\pi_c$ and $\pi_f$, we have

$$C_k^{\pi_c} - F_k^{\pi_f} = S_{k'-i}, \text{with } k' = k + \pi_f - \pi_c,$$

as well as all the possible algebraic variations of this equation.

## 16.4.2 Replication of a Capped or Floored Floating-Rate Instrument Using a Standard Asset Associated with a Cap or a Floor

We have seen how we can use a cap to guarantee a minimum return on an investment or a floor to limit the cost of financing. Here, we develop and provide details for this idea. In addition to swaps and assets with fixed and floating rates, we consider an asset with a floating rate $i(t)$ and a floor $p$ or a ceiling (cap) $P$.

The asset with a floor rate $p$ and margin $m$ yields a coupon equal to $i(t) - m$ if $i(t) > p + m$ and to $p$ if $i(t) \leq p + m$, that is, $Max[i(t) - m, p\,]$. We denote it by $A^+(i(t) - m, p)$ or simply by $A^+$. Its coupon can be written in alternative ways:

$$Max[i(t) - m, p\,] \equiv p + Max[i(t) - p - m, 0] \equiv Max[i(t), p + m] - m.$$

We can interpret $m$ as a margin (to the benefit of the borrower, therefore deducted from the coupon). $p$ is the floor payment. Figure 16.4 (the solid red line) represents such a payoff for $p = 4.6\%$ and $m = 0.50\%$.

The asset with a cap rate $P$ and margin $m$, denoted $A^-(i(t) + m, P)$ or more simply $A^-$, provides a coupon $= Min[i(t) + m, P\,] \equiv Min[i(t), P - m] + m$ (the margin $m$ is here to the profit of the investor).

### 16.4.2.1 Replication of a Floored Floating-Rate Instrument

Let us first consider a position made up of a fixed-rate asset $A_k$ and a cap $C_{k'}$ with strike $k' > k$. The payoff of such a position writes

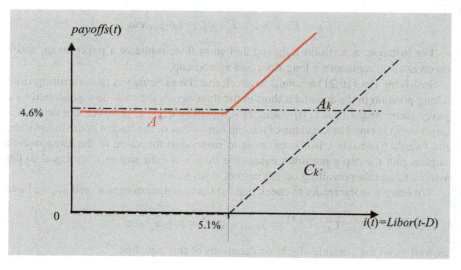

**Fig. 16.4** A floating-rate asset with floor $A^+$: payoff and replication with $A_k$ and $C_{k'}$

## 16.4  Static Replications and Combinations; Structured Contracts    697

$$k + Max[i(t) - k', 0] = Max[i(t) + k - k', k] \equiv A^+ [i(t) - (k' - k), k].$$

Thus, it is a coupon paid by a floating rate $i(t)$ with floor $k$ and margin $k' - k$. Using the notation and conventions previously defined, we can write

$$A_k + C_{k'} = A^+ [i(t) - (k' - k), k] \tag{16.22}$$

Equation (16.22) means that the fixed-rate $k$ asset and the cap allow us to replicate a floating rate asset guaranteeing a floor rate $k$. It translates the redundancy of $A_k$, $C_{k'}$ and $A^+$, and admits numerous variants:

$$C_{k'} = A^+ - A_k; \quad A_k - A^+ = -C_{k'}; \; \dots$$

---

**Example 8**

We consider a bullet note $A$ with fixed rate 4.6% and 5-year duration, and a cap $C$ on the 1-year Libor, also of 5-year duration, with guaranteed rate 5.1%.

A position with \$ 1 nominal in $A$ and $C$ yields the annual payments:

$$A + (t) = \underbrace{4.6\%}_{(A_k)} + \underbrace{Max(L_1(t-1) - 5.1\%, 0)}_{(C_{k'})} = \underset{\text{(minimum)}}{4.6\%} + \underset{\text{(surplus)}}{Max(L_1(t-1) - 5.1\%, 0)}$$

The $t$th payment varies as a function of the Libor as shown by the red line in Fig. 16.4. It is the result of the payoffs from the assets $A_k$ and $C_{k'}$ given in the dotted curve ($k = 4.6\%$ and $k' = 5.1\%$). It equals the coupon of a floating-rate asset with floor $p = 4.6\%$ and margin $m = 0.5\%$.

---

Now, let us consider a position made up of fixed-rate asset $A_k$ and a cap with strike $k'$, $C_{k'}{}^{\pi_c}$, assuming the premium $\pi_c$ is paid on each expiration date.[11] The payoff for such a position is

$$k + \underbrace{Max(i(t) - k', 0)}_{(C_{k'})} - \pi_c,$$
$$\phantom{k + }(A)$$

or else, $Max(i(t) + k - k' - \pi_c, k - \pi_c)$. Thus, we have a coupon paid by an asset with a variable rate $i(t)$, with a floor $k - \pi_c$ and margin $\pi_c + k' - k$.

One may also replicate a floating-rate asset $A^+$ with floor $k$, using a simple floating-rate asset $A_i$ and a floor $F_{k+\pi_f}{}^{\pi_f}$. Indeed, we have

---

[11] This convention is adopted without loss of generality since one can always divide the up-front premium $\pi_0$ over the cap's expiration dates $T_i$ such that the present value of the series of periodic premiums $\pi_c$ is equal to $\pi_0$.

$$i(t) \ + \ Max\big[k + \pi_f - i(t), 0\big] - \pi_f \ = \ Max\big[k, i(t) - \pi_f\big]$$
$$(A_i) \qquad\qquad (F) \qquad\qquad\qquad \big(\text{asset with floor } k, \text{margin } \pi_f\big)$$

Thus, we may write

$$A_i + F_{k+\pi_f}{}^{\pi_f} = A^+\big(i(t) - \pi_f, k\big), \tag{16.23}$$

as well as all the algebraic variations of this equation.

Equation (16.23) expresses a well-known result: investing in a floating rate guarantees a minimum yield by purchasing a floor. More AAO results follow.

## 16.4.2.2 Replication of a Capped Floating-Rate Instrument

As the following results are similar to Eqs. (16.22) and (16.23), their derivation is left to the reader as an exercise.

A capped floating-rate asset $A^-$ is replicated by a floating-rate asset $A_i$ and a short cap, or by a fixed-rate asset $A_k$ and a short floor:

$$A_i - C_k{}^{\pi_c} = A^-\big(i(t) + \pi_c, k + \pi_c\big), \tag{16.24}$$

$$A_k - F_k{}^{\pi_f} = A^-\big(i(t) + \pi_f, k + \pi_f\big), \tag{16.25}$$

All the algebraic combinations of these two equations also obtain.

Equation (16.24) means that a position made up of a floating-rate asset and a sold cap replicates a capped floating-rate instrument. It is obtained from the payoff of such a position:

$$i(t) - Max[i(t) - k, 0] + \pi_c = Min[i(t) + \pi_c, k + \pi_c].$$

Notice that by changing the signs of the left- and right-hand terms in Eq. (16.24) we obtain

$$C_k{}^{\pi_c} - A_i = -A^-\big(i(t) + \pi_c, k + \pi_c\big).$$

This equation is a classic result: borrowing at a floating rate and purchasing a cap is equivalent to borrowing at a floating rate with a ceiling.

Equation (16.25) means that a fixed-rate asset and a sold floor replicate a capped floating-rate instrument. It is obtained from the identity:

$$k - Max[k - i(t), 0] + \pi_f = Min[k, i(t)] + \pi_f.$$

Finally, note that a capped *and* floored floating-rate instrument can be replicated:

- Either by a floating-rate asset plus a long cap plus a short cap.
- Or by a fixed-rate asset plus a long floor plus a short floor.

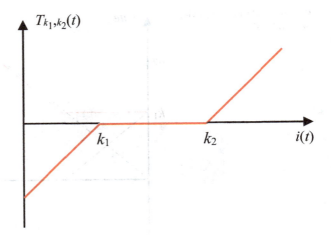

**Fig. 16.5** Payoff of a borrower collar $T$

### 16.4.3 Collars

We have seen that the combination of buying a cap and selling a floor, with the same strikes $k$, reference rates $i(t)$, nominal amounts, and durations, replicates a payer swap while the receiver swap is obtained by buying a floor and selling a cap with the same characteristics. If the strikes of the cap and the floor differ, one has, by definition a *collar* or *reverse collar*.

#### 16.4.3.1 The Collar

A borrower collar results from a short position in a floor with strike $k_1$ and a long one in a cap with strike $k_2$ such that $k_2 > k_1$; we use the notation $T_{k_1,k_2}$ and write

$$T_{k_1,k_2} = C_{k_2} - F_{k_1}. \tag{16.26}$$

The premium of this collar equals the cap's premium minus the floor's and could be nil. If not, it is often paid on the initial date. The payoff of $T_{k_1,k_2}$ is as shown in Fig. 16.5 and writes

$$T_{k_1,k_2}(t) = Max[i(t) - k_2, 0] - Max[k_1 - i(t), 0].$$

This instrument is sometimes called a borrower collar (in spite of its bayonet-like shape) because it is often associated with borrowing at a floating rate $i(t)$, which imposes a maximum and a minimum to the cost of borrowing. In effect, the *net interests paid* by the borrower, illustrated in Fig. 16.6, are equal to

$$i(t) - Max[i(t) - k_2, 0] + Max[k_1 - i(t), 0].$$

We see in Fig. 16.6 that, regardless of the value of $i(t)$, the net cost always lies between the floor $k_1$ and the cap $k_2$ (hence the phrasing "collar" or "tunnel" in some

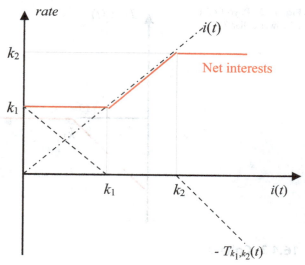

**Fig. 16.6** Cost of borrowing hedged by $T$

countries). The floating-rate borrower hedges the risk of an increase in the interest rate above the ceiling but gives up the benefit of the rate's falling below the floor. This sacrifice, due to selling the floor, is compensated by the reduction of the cost of hedging, i.e., the collar's premium $\pi_c^0 - \pi_f^0$. Sometimes, the premiums of the cap and the floor are periodic. Then the collar's payoff writes

$$T_{k_1,k_2}(t) = Max[i(t) - k_2, 0] - Max[k_1 - i(t), 0] + \pi_f - \pi_c.$$

The net cost of borrowing if hedged in this way then fluctuates between the floor $k_1 + \pi_c - \pi_f$ and the ceiling $k_2 + \pi_c - \pi_f$. Between these two boundaries it is equal to $i(t) + \pi_c - \pi_f$.

---

**Example 9**

Company XY has repeated overdrafts, for which its bank charges the 3-month Libor rate + 0.60%. The rate is set at the end of each quarter and applied to the average amount outstanding in the past quarter. The risk for XY is that the 3-month Libor goes up drastically. It is presently 4% (which implies an overdraft rate of 4.60% for the company), and XY thinks it could rise well above 4.50%, but that it is very unlikely that it falls below 3.50%. The price difference between a caplet with strike 4.50% and a floorlet with strike 3.50%, both extending over a single quarter, is 0.10%. The cost to XY of the overdraft associated with a borrower collar (purchased caplet + sold floorlet) during the next quarter is between a maximum of 5.20% [= 4.50% + 0.60% + 0.10%] and a minimum of 4.20% [= 3.50% + 0.60% + 0.10%].

## 16.4.3.2 The Reverse Collar

A reverse collar is made up of a long position in a floor with strike $k_1$ and a short position in a cap with strike $k_2$ such that $k_2 > k_1$; we use the notation $T'_{k_1,k_2}$ and write

$$T'_{k_1,k_2} = F_{k_1} - C_{k_2} \qquad (16.27)$$

Remark that $T'_{k_1,k_2} = -T_{k_1,k_2}$: a long position in a reverse collar is equivalent to a short position in its collar analogue.

The premium for a reverse collar equals the floor premium minus the cap premium. It is often paid on the date of inception and could be zero. The payoff $T'_{k_1,k_2}$ is illustrated in Fig. 16.7 and writes

$$T'_{k_1,k_2}(t) = Max[k_1 - i(t), 0] - Max[i(t) - k_2, 0].$$

This instrument is sometimes called a lender collar because it is often associated with lending at a floating rate $i(t)$ with net received interests restricted to the interval $(k_1, k_2)$.

As the reverse collar payoff is added algebraically to the interests on the floating-rate *loan*, the net interests derived from such a position write

$$i(t) + Max[k_1 - i(t), 0] - Max[i(t) - k_2, 0].$$

These interests, illustrated in Fig. 16.8, fluctuate between the floor $k_1$ and the ceiling $k_2$. With the reverse collar, the floating-rate lender hedges against rates falling below the floor $k_1$, but renounces the benefit of a rise in $i(t)$ above the ceiling $k_2$.

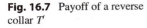

**Fig. 16.7** Payoff of a reverse collar $T'$

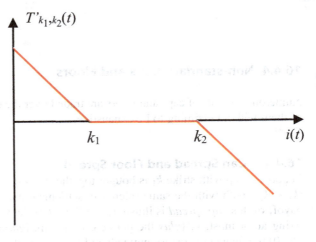

**Fig. 16.8** Interests on a loan hedged by $T'$

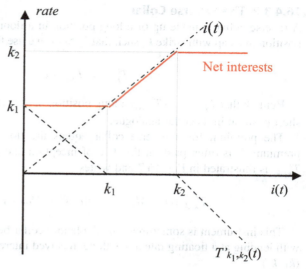

**Fig. 16.9** Cap spread payoff for a $1 notional: Max($i(t) - k_1$, 0) − Max($i(t) - k_2$, 0)

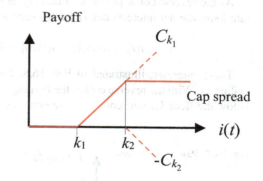

### 16.4.4 Non-standard Caps and Floors

Numerous variants of caps and floors are traded over the counter. We describe some of them without claiming to be exhaustive.

#### 16.4.4.1 Cap Spread and Floor Spread

Suppose a cap with strike $k_1$ is bought together with the sale of a cap with strike $k_2$ ($k_1 < k_2$), both with the same reference, settlement date, and notional amount; the payoff of this *cap spread* is illustrated in Fig. 16.9. Its value is $C_{k_1} - C_{k_2}$. It allows fixing at, at most, $k_1$ (*plus* the periodic premium, *ignored* in both Figs. 16.9 and 16.10) the financing cost for any value of $i(t) \leq k_2$ and reducing increases in excess of $k_2$ by the fixed amount $k_2 - k_1$ (see Fig. 16.10).

### 16.4 Static Replications and Combinations; Structured Contracts

**Fig 16.10** Net cost of floating-rate borrowing "hedged" by a cap spread

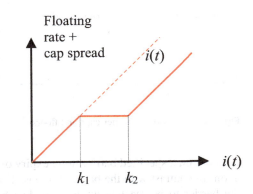

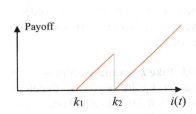

**Fig. 16.11** Cap with steps

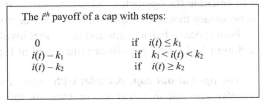

The $i^{th}$ payoff of a cap with steps:

| | |
|---|---|
| 0 | if $i(t) \leq k_1$ |
| $i(t) - k_1$ | if $k_1 < i(t) < k_2$ |
| $i(t) - k_2$ | if $i(t) \geq k_2$ |

There are also *floor spreads* composed of a long floor $k_2$ and a short floor $k_1$ ($k_1 < k_2$). This guarantees that a ($i(t)$) floating-rate loan will yield at least $k_2$ (*less* the periodic premium) for all values of $i(t) \geq k_1$.

#### 16.4.4.2 Caps and Floors with Steps

A cap with steps includes two strikes denoted, respectively, $k_1$ and $k_2$ ($k_1 < k_2$). Its payoff is shown in Fig. 16.11.

It can be replicated by buying a cap with strike $k_1$ and selling a digital call ($k_2 - k_1$) $1_{i(t) \geq k_2}$.

The formula for its valuation results from this replication.

Associated with borrowing at a floating rate $i(t)$, such a cap provides a ceiling at $k_1$ to the cost $i(t)$ (*plus* the periodic premium) as long as $i(t)$ is less than $k_2$, and at $k_2$ otherwise.

The premium for such a cap is, therefore, less than the premium of the vanilla cap with strike $k_1$ but larger than that of the cap with strike $k_2$.

There also exist floors associated with two different strikes $k_1$ and $k_2$.

#### 16.4.4.3 Caps and Floors with Barriers

Like barrier options, European caps and floors with barriers are made of caplets (or floorlets) activated or deactivated according to whether the reference rate does or

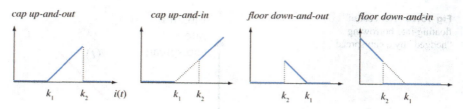

**Fig. 16.12** Payoffs of barrier caps and floors

does not hit a specified barrier upon expiry of the caplet (or floorlet) under consideration. In contrast with the barrier options studied in Chap. 14, it is not sufficient for the barrier to be reached upon some date before expiry, even if, as a result of a movement in the opposite direction the rate goes below the barrier (or above), but it is necessary that *upon expiry* of the caplet or floorlet the barrier is hit or exceeded.

Four types of barrier caps and floors that involve a strike $k_1$ and a barrier $k_2$ can be distinguished. Their payoffs are illustrated in Fig. 16.12.

- The **up-and-out cap**. A caplet with expiry $t$ and strike $k_1$ expire worthless if the reference rate $i(t)$ reaches or exceeds the barrier $k_2$; its payoff is therefore consistent with Fig. 16.12 and writes $Max(i(t) - k_1, 0)$ if $i(t) < k_2$ and 0 otherwise, i.e. $Max(i(t) - k_1, 0)1_{i(t) < k_2}$, where $1_{i(t) < k_2}$ denotes the indicator function. The up-and-out cap can be replicated by a long cap with strike $k_1$, a short cap with strike $k_2$ $(> k_1)$ and the sale of a chain of digital calls $(k_2 - k_1) \, 1_{i(t) < k_2}$ (one digital for each caplet).
 The borrower at a floating rate who buys such a product is protected only against rate increases between $k_1$ and $k_2$, but pays a smaller premium as compared with the standard cap with strike $k_1$.
- The **up-and-in cap**. The caplet is activated as soon as the reference rate upon expiry of the caplet reaches or exceeds the barrier $k_2$ $(> k_1)$: the payoff is as in Fig. 16.12 and writes $Max(i(t) - k_1, 0)1_{i(t) \geq k_2}$. The up-and-in cap can be replicated by a long cap with strike $k_2$ and a chain of digital calls $(k_2 - k_1)1_{i(t) \geq k_2}$ (one digital for each caplet).
- The **down-and-out floor**. The floorlet is deactivated if the reference rate is less than or equal to the barrier $k_2$ $(< k_1)$ at expiration. The *down-and-out floor* can be replicated by a long floor with strike $k_1$, a short floor with strike $k_2$ and the sale of a chain of digital puts $(k_1 - k_2)1_{i(t) \leq k_2}$.
- The **down-and-in floor**. The floorlet is activated if the reference rate is *smaller than or equal to* a barrier $k_2$ $(< k_1)$ at expiration. The *down-and-in floor* is equivalent to a long floor with strike $k_1$ and a chain of digital puts $(k_1 - k_2)1_{i(t) \leq k_2}$.

The valuation of barrier caps (floors) results from this replication: each caplet (floorlet) is valued using Eq. (16.6) (or the *Libor Market Model* presented in Chap. 17) and digital options using Eq. (16.7). The valuation of the cap (floor) is obtained by summing the values of the different caplets (floorlets).

## 16.4 Static Replications and Combinations; Structured Contracts

### 16.4.4.4 Cap and Floor with a Contingent Premium

The cap is a standard one with strike $k_1$, but its premium is only paid if the cap is exercised, or more generally if the reference exceeds an activating barrier $k_2$ above $k_1$.

There are also floors with contingent premium.

We leave to the reader the determination of the combinations of caps, floors, and digitals that allow the replication and valuation of these instruments.

### 16.4.4.5 Other Non-standard Caps and Floors

All the products presented above are static combinations of the basic instruments. Those that follow, and the list is not exhaustive, are obtained by more complicated strategies that we do not make explicit.

- The *N-cap* is a variant of the *up-and-out* cap. If the barrier is reached, the cap is replaced by another cap with a larger strike.
  The price of an *N*-cap is larger to that of an *up-and-out* cap and smaller than that of a vanilla cap since the level of protection that it provides is in-between that of these two instruments.
  In the case of an *N-floor*, if the barrier is hit, the floor is replaced by another floor with a lower strike.
- *Ratchet caps are made up of caplets* $C_i$ whose strike $k_i$ depends on the level reached by the reference rate at stage $i - 1$: $k_i = i(T_{i-1}) + spread$ (beware that $i(T_{i-1}) = r_D(T_{i-2})$ in the case of a vanilla forward-looking rate). For a *ratchet floor*: $k_i = i(T_{i-1}) - spread$.
- *Sticky caps are* characterized by caplets whose strike $k_i$ *equals* $\text{Min}(i(T_{i-1}), k') + spread$, where $k'$ is pre-specified. In the case of sticky floors, $k_i$ equals $\text{Min}(i(T_{i-1}), k') - spread$.
- *Flexi-caps (flexi-floors)* differ from standard caps (floors) in that the maximum number of caplets (floorlets) that can be exercised is bounded by contract.

### 16.4.5 Other Static Combinations; Structured Products; Contracts on Interest Rates with Profit-Sharing

#### 16.4.5.1 Generalities

The different possible combinations of the instruments $A$, $C$, $F$, and $S$ lead to very varied payoffs that share the property of being piecewise affine linear functions of the reference $i(t)$ and are described in professional circles as structured products.[12] The similarity of payoffs thus structured with those resulting from static strategies using options and their underlying assets (see Chap. 12, Sect. 12.1) is noteworthy. These static strategies can thus be applied to structured products.

---

[12] Furthermore, are considered structured products the credit derivatives, as analyzed in Chap. 30, as well as some products with optional covenants linked to stocks, indices, and exchange rates.

## 706    16 Interest Rate Instruments: Valuation with the BSM Model, Hybrids,...

After a brief description of some of these products, we analyze a contract that allows an investor to profit from favorable changes in interest rates while providing a guarantee in the opposite case.

### 16.4.5.2 Structured Products on Interest Rates

Among the numerous structured rate products traded on the market, let us mention, with no claim to completeness:

- *Reverse floaters*, which provide a coupon equal to $k - i(t)$ with a minimum of 0%, $k$ a pre-specified rate and $i(t)$ a reference rate.
- *Callable set ups, which* yield a coupon that increases or decreases in a predetermined way; for example, the $n$th coupon equals $k_1 + n\ k_2$, with the possibility of early reimbursement by the writer ($k_1$ and $k_2$ are pre-specified).
- *Range notes*, whose coupon accrues only when the reference rate stays within a predefined *range*; for instance, over a period of $N$ days, if the reference has remained within the interval $(l_1, l_2)$ during $m$ days, the coupon is $(k/360) \times (m/N)$.
- *Snowballs*, to which is applied the formula: coupon $(n) =$ coupon $(n -1) + (k_n - i(n))$ where $k_n$ is predefined and may be decreasing or increasing with $n$. Often snowballs include a covenant allowing for early repayment to the profit of the writer (*callable snowballs*).
- *Thunderballs*, the coupons of which are computed according to the formula: coupon $(n) = k_n \times$ coupon $(n -1) - i(n)$.

We leave to the reader as an exercise the problem of replicating these instruments. We just analyze, as an example, the following structured product.

### 16.4.5.3 Example of an Interest Rate Contract with Profit-Sharing

Consider an investor who currently invests liquid assets at a floating rate $i(t)$ (e.g., Libor($t - 1$)), and who wants to hedge against the risk of a fall in interest rates while retaining, at least in part, the profit to be derived from a rise.

Buying a floor $F_k$ meets their needs, since it guarantees a return of $Max(i(t), k)$, at least equal to the floor $k$.

However, the floor brings with it either the payment of a premium which the investor thinks too high or a floor rate $k$ deemed too low.

The investor could consider the deal of a bank which proposes, at no cost, a payoff $R$ equal to

$$R(t) = k - i(t) + \alpha \operatorname{Max}[i(t) - k', 0], \text{ with } 0 < \alpha < 1.$$

$R(t)$ added to the investment that gives $i(t)$ leads to a structured product that yields $\phi(t) = k + \alpha \operatorname{Max}[i(t) - k', 0]$, i.e., a guaranteed rate to which is added a fraction $\alpha$ of a possible increase in the reference rate above the threshold $k'$.

Such a contract $R$ is called a *receiver swap with profit-sharing*.

Remark that the payment $R(t)$ can be obtained by a swap of \$1, $i(t)$ for a fixed rate $k + \alpha \pi_c$, i.e., $S_{i(t)-(k+\alpha\pi_c)}$, plus a cap $C_{k'}{}^{\pi_c}$, of amount \$$\alpha$ with strike $k'$, and periodic

## 16.5 Bonds with Optional Features and Hybrid Products

premium $\alpha\pi_c$ (as $\alpha$ is less than 1, the premium is smaller than that of a cap with notional $1).

The parameters $(k, \alpha, k')$ in the contract may be *calibrated*: for example, for a given $k$, $\alpha$ increases with $k'$; for a given $k'$, $\alpha$ decreases with $k$.

---

**Example 10**

Corporation ABC can invest its liquid assets at the floating rate Libor($t$) – 0.10%, i.e., currently 1.90%. It fears a fall in interest rates, but wants to profit partially from their possible rise.

ABC negotiates with its banker the following *receiver swap with profit-sharing*:

$$1.70\% - \text{Libor}(t) + 0.8 \times \text{Max}[\text{Libor}(t) - 2\%, 0]$$

and, furthermore, continues to invest at Libor($t$) – 0.10%. The return rate obtained is consequently 1.60% + 0.8×Max[Libor($t$) – 2%, 0]. If the Libor remains at 2% or falls, ABC has ensured a minimum rate of 1.60%; if the Libor rises, for example, to 2.40% (resp. 3.00%), it receives 1.92% (resp. 2.40%).

---

Replicating this contract allows the banker (1) to offer conditions $(k, \alpha, k')$ consistent with the market but permitting a fee compensating for brokering and financial services, and (2) to hedge the contract perfectly so as to reap a sure profit margin.

We note that, since $S = C - F$, the same contract can be alternatively replicated using a floor and a cap rather than a receiver swap.

Finally, remark that, in the same spirit, one may structure a contract that allows the borrower to cap its financing cost while benefiting in part from a possible fall in rates. Such a contract, called a *payer swap with profit-sharing*, leads to a payoff of the form

$$R'(t) = i(t) - k + \alpha\, Max[k' - i(t), 0] \text{ with } \alpha < 1.$$

The contract can be replicated with a payer swap and a floor for $\alpha$.

---

# 16.5 Bonds with Optional Features and Hybrid Products

Mixing options with standard securities provides a great flexibility to financial engineering and the combinations we discussed in the previous section are well-known examples. The securities we analyze in this section are bond and optional products that also have a stock component, and thus are often called hybrids. After convertible bonds, we consider other bonds with optional features.

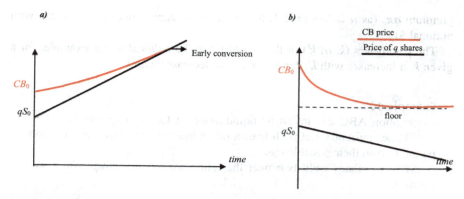

**Fig. 16.13** Evolution of the CB price as a function of the price of $q$ shares. (**a**) Rising share price. (**b**) Falling share price

### 16.5.1 Convertible Bonds

#### 16.5.1.1 General Description and Qualitative Analysis

As its name suggests, a convertible bond is a fixed-income corporate debt security that can also be converted into a predetermined number of common stock or equity shares:

- As a bond, it pays coupons and reimbursement of principal, denoted by $V_r$, at maturity.
- As a convertible asset, either at any time, or during a predetermined period, it gives the holder the right of converting it into a predetermined number $q$ of shares ($q$ is the *conversion rate*).

It can be analyzed, as a first approximation, as an option to buy (a call) $q$ shares with strike $V_r$, the reimbursement bond value. Thus, it is an American option to exchange one bond for $q$ shares.

Issuing convertible bonds (CB in what follows) has to be authorized by the existing shareholders under conditions that vary according to the legal framework in force. The shareholders actually forego subscription rights to buy the shares that may be created as a result of conversions.[13]

A CB is therefore a "hybrid," intermediate between a stock and a bond. Investors hope to profit from a hike in the stock price by converting the bond. They are also partly protected from the risk of a fall in the stock price by not converting the bond and so receiving coupons and then reimbursement of the principal. Figure 16.13 sketches the evolution of the CB price, when the market value of the $q$ shares increases regularly (case $a$) and when it decreases (case $b$).

---

[13] However, they still have the right to buy the convertible bonds, unless they explicitly renounce it.

## 16.5 Bonds with Optional Features and Hybrid Products

Figure 16.13b implicitly assumes a constant yield-to-maturity: the value of the standard bond, when any hope of conversion disappears (the embedded option value tends to zero), then remains approximately constant.[14] This value is a *floor* below which the CB price cannot fall and to which it tends when the probability of conversion decreases. Actually, this floor varies over time in an opposite direction to market interest rates and to the issuer's credit spread.

Holders of CBs may benefit from exercising their conversion option at or before the bond maturity:

- **In the first case (conversion at maturity)**, the bonds are called for reimbursement and the market value $qS$ of the shares is larger than the bond's reimbursement price. The rule "$qS > V_r$" is a necessary and sufficient condition for converting the bonds at maturity.
- **In the second case (early conversion)**, the bonds have not yet been called for reimbursement but the market value of the shares in which they can be converted reaches the market value of the CBs. This case can only occur if the dividend distributed by the relevant $q$ shares exceeds the coupon; the rule "dividend > coupon" is a necessary but *not* sufficient condition for early conversion.[15] In fact, the conditions for early conversion are analogous to those for early exercise of an American call: it happens on the day before a dividend payment if this exceeds the time value of the option (see Chap. 13, and Chap. 28 for a deeper analysis).

It is important to note that, to protect the CB from drops in the share's market value caused by possible new equity issue(s), the conversion right applies to *adjusted shares* (see Chap. 8 for this important notion).

In the same spirit, to avoid an overgenerous distribution of dividends compromising the rise of the share's market price (and hence the probability of profitable conversion), the contract usually specifies the company's dividend policy which it commits to follow in the future (e.g., the contract stipulates that the company cannot change the percentage of the profits to be distributed, or must adjust (increase) the conversion rate following dividend distributions). Finally, contracts often include an early reimbursement option to the issuer's benefit (*issuer call*). Mostly this issuer call has a time limit (it can only be exercised after a certain date, or on certain dates), and/or is restricted to a given fraction of the total bond issue and/or is possibly

---

[14] Actually, if the yield remains constant, the value of the bond (without the conversion option value), which is constantly below par (because the coupon rate offered on a CB is less than that on an ordinary bond as a result of the conversion option), increases with the passage of time to reach par on the reimbursement date. The floor thus is slightly upward sloping. Another phenomenon, however, makes it downward sloping: in general, if the stock price drops, the issuer's credit risk perceived by the market increases, which lowers the floor. It may happen that the two effects approximately cancel one another such that the floor is roughly horizontal.

[15] As long as the coupon is larger than the dividends on the $q$ shares, rational bond holders prefer the coupon and wait before converting (their conversion option has a positive time or speculative value).

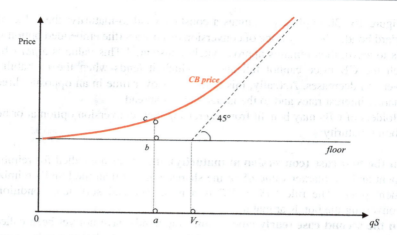

**Fig. 16.14** CB price as a function of the shares' value $qS$

contingent on the price of the $q$ shares being above the reimbursement value $V_r$. This last case amounts to allowing the issuer to elicit an early conversion. Finally, the contract may make the conversion rate $q$ decrease over time in a pre-specified way.

The advantages to a company of financing its investments through CBs are the following:

- The coupon rate offered is less than for classic bonds with the same profile (as the counterparty to the conversion option).
- The dilution of capital, if it occurs, is delayed and reduced, for the same amount of new cash, relative to an equity issue, since the price of the $q$ shares is initially less than that of the CB (see Fig. 16.13a: $qS_0 < CB_0$).

The analysis of the CB value is based on disentangling its two components: the floor (the security's value as a classic bond) and the value of the call option. Figure 16.14 illustrates this decomposition; it shows the CB price as a function of the value $qS$ of the shares on which it is a claim (the red curve).

The difference between the CB price and the floor is the value of the call option (*conversion premium*).

For example, for a value $qS$ of the shares equal to 0a on Fig. 16.14, the floor is equal to $ab$, the value of the conversion premium to $bc$ and the CB price to $ac$.

The floor is valued as the discounted value (at the rate prevailing on the bond market, taking into account the risk of the issuer's default) of the bond's remaining cash flows (coupons + reimbursement). The conversion premium is calculated using an appropriate option valuation model. However, applying a BSM model does bring a number of problems. First, the CB is an American option. Second, the strike of the option is actually floating since it is the market price of the bond, which fluctuates with the interest rate and the issuer's credit spread (in fact, this is an American exchange option, as already mentioned). Third, the equity capital is diluted as soon

## 16.5 Bonds with Optional Features and Hybrid Products

as the CBs are converted, which may justify adjusting the pricing model. Fourth, the CB contract most often includes an issuer call to the firm's benefit. Finally, the CB maturity is generally very far off (10–20 years) and standard option models are ill suited to value options of this duration. The following subsection very briefly takes up these questions, a deeper analysis being postponed to Chap. 28.

### 16.5.1.2 Quantitative Analysis and Valuation

According to the precision required, different methods are used to value a CB.

One method consists in using a BSM model, or Margrabe's model (exchange of one bond for $q$ shares; see Chap. 11, Sect. 11.2.7) to value the conversion option. The CB value is obtained by adding the option value $C$ to the value $B$ of the simple bond (floor): $CB = C + B$. The floor is valued by discounting the bond's future cash flows as just mentioned above.

This method, which has the merit of simplicity, leaves much to be desired, notably because it does not account for the American character of the conversion option, nor for the (pervasive) issuer call covenant, nor for the credit and default risk that taints the bonds. These problems can be solved by using numerical methods, in particular, those associated with modeling credit risk (see Chap. 28, Sect. 28.2.3).

### 16.5.2 Other Bonds with Optional Features

This subsection presents briefly debt securities incorporating one or more options, either in the form of the right inherent in the security, or in the form of an option attached to the security at issuance, later to be detached. These options attached to the asset are called *subscription warrants* and are types of *warrants* (which are not necessarily associated with another asset). We first describe such options before considering some securities which associate them with bonds or with shares.

### 16.5.2.1 Subscription Warrants for Shares and Warrants

Options on shares are not in general written (or sold) by the company that issues the shares. However, for certain transactions such as issuing *shares with subscription warrants or bonds with share subscription warrants*, a "package" security + option is offered to investors; in this context, the option is a *subscription warrant*. In addition, companies can also issue *warrants* which are (generally long term) options that then constitute a source of cash for the issuing company.

A share subscription warrant (which should not be confused with the right to subscribe to new shares involved in an equity issue) is an American call on a specified number of shares. Subscription warrants can possibly be issued independently but are mostly attached to bonds or shares. In this case, the subscription warrant, attached to the asset at issue, is then detachable and tradable as such.

Warrants are options (calls or puts) whose duration is, in general, longer than that of standard negotiable options (they can reach 20 years). The difference between a warrant and a subscription warrant is that the second is associated (initially) with another security while the first is not. Since they have a long duration, warrants are generally protected against dilution as they are written on *adjusted* shares. Warrants can be issued by the company whose shares are the underlying asset or by a third party (usually a bank). In the first case (*uncovered warrant*), it is a financing tool for the company, and the underlying shares are issued only if the warrants are exercised. In the second (*covered warrant*), it is a financial product offered by a bank to its clients, and the underlying shares already exist when the warrant is issued. Valuation of an *uncovered warrant* is like that of an option, although a dilution effect could, in certain situations, justify an adjustment.[16]

## 16.5.2.2 Bonds with Share Subscription Warrants

These are bonds to which *share subscription warrants* are initially attached. Like CBs, these bonds make investors participate to the issuing company's stock performance. In contrast to CBs, the holders do not have to choose between being shareholders or bondholders (they can opt to exercise the warrant and sell the bond or sell the warrant while retaining the bond).

The value of such a bond is the value of the standard bond plus that of the share subscription warrant. The latter can be valued using a BSM-type model.

> **Example 11**
> Company Z is considering issuing some bonds with share subscription warrants. The subscription warrant attached to a bond is valued at $40 and the yield for a classical bond that Z recently issued is 6%. The bond would have a nominal of $200. Two extreme solutions and all the intermediate ones may be considered: to choose a bond coupon rate of 6% (the bond then is at par and worth $200), in which case the issue price is $240; or to price the issue at $200, in which case the coupon rate is less than 6% and such that the market value of the standard bond equals $160.

---

[16]Exercising an *uncovered warrant* implies issuing new shares whose value is larger than the exercise price paid to the company. If, at initiation, the warrant is issued at market value, the value of the shares should not be affected by the issue and, in the future, this value should take into account the probability of exercise. Then the warrant is correctly valued by an appropriate option model *without adjustment*. Such is not the case for warrants to be issued below their market value, under the assumption that the future issue is not yet known by the market. Then, an adjustment compensates for the potential dilution effect which decreases the price of the underlying shares. For example, for an unexpected distribution of stock options for free, the option's value should be reduced by a factor $N/(N + n)$ where $N$ denotes the current number of shares and $n$ the number of warrants issued.

### 16.5.2.3 Bonds with Optional Features Disconnected from the Stock's Performance

There exist other bonds with optional covenants not associated with the issuing firm's stock performance but which protect either the investor or the issuer against interest rate fluctuations. They are not hybrids but more or less sophisticated debt securities. We sort out the two most popular.

- **Bonds reimbursable early (callable and putable)**
  These give the issuer (*callable* bonds) or the investor (*putable bonds*) the option of being redeemed *early at a contractually fixed price, under certain conditions and at* certain moments in the life of a bond. Thus, they partially protect the beneficiary of the early reimbursement option from interest rate risk. They are valued as a standard bond minus a call on the bond (callable bonds) or plus a put on the bond (putable bonds).
- **Bonds with bond subscription warrants**
  This product is a standard fixed-rate bond plus a *subscription warrant for another bond at a predetermined fixed rate.* It allows investors to benefit twice from a fall in market rates, due to the capital gain on the fixed-rate bond and also the option to subscribe to a new bond at profitable conditions.

### 16.5.2.4 Other Types of Convertible Bonds

- An *exchangeable bond*, unlike a CB, gives the issuer, not the holder, the right to decide if the bonds are redeemed in cash or converted into shares.
- A *reverse convertible bond* is a high-coupon bond that can be converted to equity, debt, or cash at the issuer's discretion at a predetermined date. At maturity, the issuer also has the option to redeem the bonds or convert them into a set number of shares.
- A *mandatory convertible* has a required (non-optional) conversion feature. Either on or before the contractual conversion date, the holder must convert into the pre-specified number of the company's shares. As a compensation for not benefiting from an optional feature, this bond has a higher coupon than the analogous CB.

---

## 16.6 Summary

- To value at date 0 a claim on a payoff $\Psi(T)$ available at $T$, one computes the expectation of the appropriately discounted cash payments: $V(0) = E_0^* \{A(T)\Psi(T)\}$.

  $E_0^* \{.\}$ is either the risk-neutral (RN) or forward-neutral (FN) expectation (the asterisk on $E_0^* \{.\}$ covers both cases) and $A(T)$ is the discounting factor. Under the FN probability $Q_T$, the forward prices $\Phi_T(t) = V(t)/B_T(t)$ are martingales. $B_T(t)$, the price of a zero-coupon yielding \$1 on $T$, is the martingale numeraire.

- Under the RN, the discounting factor $A(T) = \exp(-\int_0^T r(t)dt)$ is random if $r(t)$ is random, while under $Q_T$, the discounting factor $B_T(0)$ is *known* on 0, which simplifies the FN valuation : $V(0) = B_T(0) \, E_0^{Q_T} \{\Psi(T)\}$.
- The generalized BSM model (see Chap. 11) applies to the payoff $Max\,(X(T) - K, 0)$ for a call, or $Max\,(K - X(T), 0)$ for a put, whenever $X(T)$ *is a log-normal variable that depends or not on interest rates*. $X(T)$ is either the underlying *price* (*BSM-price* model) or the underlying *rate* (*BSM-rate* model).
- The option value $O$ writes $O = \text{BSM}(\Phi_T(0), B_T(0), \sigma, T, k)$, where $\sigma$ is the "average volatility" of $X(t)$ (price or rate) between 0 and $T$ and $\Phi_T(t)$ is either a *forward price* or a *forward rate*.
- BSM-rate models value vanilla caplets or floorlets. The payoff of a *caplet*, $D\text{Max}\{r_D(T) - k, 0\}$, is paid on $T+D$, where $r_D(T)$ is the spot rate (proportional) for $D$ periods prevailing at $T$ ($D$ is a fraction of a year). A *floorlet* yields $D\text{Max}\{k - r_D(T), 0\}$. Caps and floors are made up of such elements. The BSM-rate model can be combined with other rate models (Libor Market Model, Swap Market Model).
- The buyer of a *vanilla* cap with notional amount $M$ receives from the seller a stream of payments, the $i$th settlement on date $T_i$ being a caplet payoff $\Phi_c(T_i) = M D_i Max[r(T_{i-1}) - k, 0]$. The *vanilla floor* yields a stream of floorlet payoffs. The premium can be paid upfront or spread over the different settlements.
- A cap associated with a floating rate loan provides a *ceiling* to its *cost*, and a floor associated with a floating rate investment provides a *minimum yield*.
- The various investment or borrowing instruments with fixed and floating rates, swaps, caps, and floors, are *redundant* and can be combined to make up various payoffs.
- A cap, a floor and a swap with the same reference $r$, the same fixed-rate $k$ and synchronous payments make up a *redundant set of three instruments*, e.g., $C_k - F_k = S_{r-k}$, meaning that the combination of a long position on a cap and a short one on a floor replicates a floating-receiver swap.
- If the strikes of the cap and the floor differ, their combination replicates a *collar*. A *borrower collar* results from a short position on a floor with strike $k_1$ and a long one on a cap with strike $k_2 > k_1$: $C_{k_2} - F_{k_1}$; it imposes a floor $k_1$ and a ceiling $k_2$ to the cost of a floating rate borrowing. The *reverse* (or *lender*) *collar* $F_{k_1} - C_{k_2}$, when associated with lending at a floating rate, restricts interest rate received within the interval $(k_1, k_2)$.
- Caps and floors with barriers are made of caplets (or floorlets) activated or deactivated according to whether the reference rate does or does not hit or exceed a specified barrier *upon expiry* of each element.
- Many different combinations of fixed and floating rate instruments, swaps, caps, and floors, called structured products, lead to payoffs which are *piecewise affine linear* functions of the reference price or rate.
- A convertible bond (CB) has a *dual nature*. As a bond, it confers the *right to* coupons *and reimbursement* at maturity. As a convertible asset, either at any time or during a predetermined period, it confers the *right to convert it into a*

*predetermined number of shares*. Thus, as a first approximation, it is an American call on the shares whose exercise price is the bond's reimbursement value. However, a frequent covenant allows for an early reimbursement to the benefit of the issuer.

- Investing in a CB benefits from a hike in the share price by converting it, and in addition is partly protected from a fall in the share price by keeping it as a bond.
- Financing through a CB issue has two advantages: the coupon offered is smaller than for classic bonds (as the counter-part to the conversion option), and capital dilution, upon conversion, is delayed and reduced.
- The CB value is obtained by adding the option value $C$ to the pure bond value $B$ (actuarial floor). $B$ is obtained by standard discounting. Valuing $C$ requires either an option pricing model such as Margrabe's or the implementation of numerical methods. The latter are especially useful when additional complexities must be accounted for (option of early reimbursement to the issuer's benefit, credit risk, etc.)
- There is a wide variety of *debt securities embedding options*. The options are either rights inherent in the asset or options initially attached to the asset, later maybe to be detached: *subscription warrants* for shares, bonds *exchangeable* for shares (the *issuer*—not the holder—*decides* if the bonds are redeemed in cash or converted into shares), bonds refundable early giving the possibility of being redeemed *at the discretion of the issuer* (*callable* bonds) or *of the holder* (*putable* bonds), at a given price, conditions, and dates.

---

# Appendix

## The $Q_a$-Martingale Measure

We consider the asset that generates the sure payoffs $D_1$ at $T_1$, $D_2$ at $T_2$, ... $D_N$ at $T_N$. We called it, by slightly abusing the language, an "annuity," and used the notation $a_1$, simplified here as $a$. Its value on $t < T_1$ is

$$a(t) = \sum_{i=1}^{N} D_i B_{T_i}(t)$$

where $B_{T_i}(t)$ denotes, as usual, the value at $t$ of a zero coupon yielding \$1 on date $T_i$.

Asset $a$ is a viable numeraire (since it is a self-financed asset with values surely positive; see Chap. 19, Sect. 19.5). We denote by $Q_a$ the probability measure that makes martingales the prices of self-financing assets denominated in numeraire $a$, so that, for any self-financing asset with value $V(t)$:

$$\frac{V(0)}{a(0)} = E^{Q_a}\left(\frac{V(t)}{a(t)}\right) \text{ for } 0 \leq t \leq T < T_1.$$

We show that $Q_a$ *also makes martingale the rate* $f_T(t)$ *of the forward swap starting on date $T$* (the forward-$T$ rate of the *swap*).

Let us consider a vanilla swap producing payoffs on $T_1, T_2, \ldots, T_N$ ($T_1 > T$) whose variable and fixed legs are here denoted by $VL$ and $FL$. Recall that for a fixed rate $k$, $V_{FL} = ka$.

Start from date $t$ when the forward-$T$ market rate of the swap is $f_T(t)$. At this rate:

- The fixed-rate $f_T(t)$ swap starting at $T$ has zero value, so that $V_{FL}(t) = V_{VL}(t)$;
- The value of its fixed leg is $V_{FL}(t) = a(t)f_T(t)$.

It follows that $f_T(t) = V_{VL}(t)/a(t)$. As a result, since $V_{VL}(t)/a(t)$ is a $Q_a$-martingale,[17] $f_T(t)$ is also a $Q_a$-martingale and we can write $f_T(0) = E^{Q_a}(f_T(T))$.

In addition, the payoff of an option on a fixed-rate payer swap, of strike $k$, can be expressed as $a(T)Max[s(T) - k, 0] \equiv a(T)Max[f_T(T) - k, 0]$. The value of this swaption on its expiry date $T$, given in the numeraire $a$ writes $Max[f_T(T) - k, 0]$. Its value on date 0 is

$$OS_{r-k}(0)/a(0) = E^{Q_a} Max[f_T(T) - k, 0], \text{i.e.},$$

$$OS_{r-k}(0) = \left( \sum_{i=1}^{N} B_{T_i}(0)D_i \right) E^{Q_a} Max[f_T(T) - k, 0].$$

The log-normality of $f_T(T)$ allows using the BSM-rate model and leads to Eq. (16.17). Equation (16.18) for an option on a fixed-rate receiver swap can be obtained similarly.

---

## Suggestions for Further Reading

### Books

** Brigo, D., & Mercurio, F. (2007). *Interest rate models: Theory and practice* (2nd ed.). Springer.
Das, S. (2001). *Structured products and hybrid securities* (2nd ed.). Wiley.
Fabozzi, F. J. (Ed). (2000). *The handbook of fixed income securities* (6th ed.). McGraw-Hill.
Hull, J. (2018). *Options, futures and other derivatives* (10th ed.). Prentice Hall Pearson.
Martellini, L., Priaulet, P., & Priaulet, S. (2003). *Fixed income securities*. Wiley Finance.

### Articles

Ammann, A., Kind, A., & Wilde, C. (2003). Are convertible bonds underpriced? An analysis of the French market. *Journal of Banking and Finance, 27*(4), 635–753.
Black, F. (1976). The pricing of commodity contracts. *Journal of Financial Economics, 3*, 167–179.

---

[17] We consider $VL$ as a self-financing asset whose price expressed in numeraire $a$ is a $Q_a$-martingale.

## Suggestions for Further Reading

Black, F., & Scholes, M. (1973). The pricing of options and corporate liabilities. *Journal of Political Economy, 81*, 637–659.

Brennan, M. J., & Schwartz, E. S. (1980). Analyzing convertible bonds. *Journal of Financial and Quantitative Analysis, 15*(4), 907–929.

Jamshidian, F. (1989). An exact bond pricing formula. *Journal of Finance, 44*(1), 205–209.

Merton, R. (1973). The theory of rational option pricing. *Bell Journal of Economics & Management Science, 4*, 141–183.

# Modeling Interest Rates and Options on Interest Rates

# 17

Chapters 3 and 7 were devoted to basic rate instruments such as fixed-rate and floating-rate products and vanilla swaps. The analysis of these instruments is relatively simple since it is based on discounting or arguments about static arbitrage which do not require explicitly taking uncertainty into account. This chapter follows up Chap. 16 and is devoted to the study of interest rate models and the valuation of interest rate products that depend on the techniques, models, and probabilistic theories that were particularly studied in Chaps. 10 and 11 (options) and whose mathematical foundations are presented in Chaps. 18, 19, and 20. Here, we analyze models based on the dynamics of the spot rate (Vasicek, Hull and White, and Cox, Ingersoll and Ross) and on the dynamics of the forward rate (Heath-Jarrow-Morton, Libor and Swap Market Models), as well as options on bonds and caps and floors, whose valuation rests on interest rate models.

We can distinguish two types of probabilistic models for valuing interest rate instruments:

- Models of the first type, discussed in the preceding chapter, are not based on any explicit modeling of the dynamics of interest rates although they are assumed to be Gaussian. The price of a vanilla option is obtained from a BSM-type formula.
- Models of the second type rely on modeling the yield curve and are based on making explicit the stochastic process assumed to control the evolution of rates in the absence of arbitrage. These models, which lead to lognormal prices and rates, are compatible with the BSM model and can be associated with it, in particular for calibration purposes.

The main models, based on specific representations of the stochastic process governing the evolution of the spot rate, such as those of Vasicek, Hull and White, and Brennan and Schwartz, are the subject of Sect. 17.1. In Sect. 17.2, we describe the models due to Heath, Jarrow and Morton, the Libor Market Model due, among others, to Brace, Gatarek and Musiela, and the Swap Market Model due to

---

© The Author(s), under exclusive license to Springer Nature Switzerland AG 2022
P. Poncet, R. Portait, *Capital Market Finance*, Springer Texts in Business and
Economics, https://doi.org/10.1007/978-3-030-84600-8_17

# 17 Modeling Interest Rates and Options on Interest Rates

Jamshidian. These models are applied to the valuation of caps, floors, swaptions and are calibrated against the prices of the latter.

## 17.1 Models Based on the Dynamics of Spot Rates

The valuation models given in this chapter assume that the proportional rate of maturity $\theta$ in force on a future date $T$ is lognormal (the corresponding continuous rate is thus normal). Such an approach, which has the merit of simplicity, has nevertheless the following limitations:

- It is based on a relatively simplistic assumption about the distribution of the future rate $r_\theta(T)$.
- It makes no choice as to the relationship between an interest rate and its maturity ($r_\theta(T)$ as a function of $\theta$), i.e., the term structure of interest rates.
- It does not consider the path followed by rates between 0 and $T$.

These limitations make the model ill-adapted to the analysis and valuation of some complex products, in particular those whose payoffs depend on the path followed by the underlying asset price or rate and not just on its final value. Furthermore, as they are, they are insufficiently suited to calibrating parameters from the observed prices of very liquid instruments, a calibration used to value other less liquid assets, or more complex or exotic ones. More precise modeling of the yield curve and its evolution is thus useful. This section is devoted to models of the term structure that depend on the dynamics of the short-term spot rate $r(t)$ (Vasicek, Hull–White, Cox–Ingersoll–Ross). Those based on the dynamics of *forward* rates (Heath–Jarrow–Morton, Libor Market Model) are studied in Sect. 17.2.

The simplest model using the dynamics of the short-term rate is due to Ho and Lee (1986). Originally it was a discrete-time model but there is a continuous-time version. We discuss the latter as a special case of the Heath–Jarrow–Morton model in Sect. 17.2, and will not deal with the discrete-time Ho and Lee model.

Sticking with our usual conventions, we denote by $B_T(t)$ the price at $t$ of the default-free zero-coupon (ZC) asset that pays \$1 on a fixed date $T$ ($> t$). Its duration is therefore $(T - t)$ and its price is linked to the corresponding spot ZC rate $r_{T-t}(t)$ by the equation $B_T(t) = e^{-(T-t)r_{T-t}(t)}$. Models of the yield curve express the zero-coupon prices $B_T$ (or equivalently, the ZC rates $r_{T-t}$) as a function of one or several state variables or factors.

Some of these models involve forward rates. We denote by $f_{T,D}(t)$ the forward rate for the period $(T, T + D)$ prevailing on date $t$. With *continuous* rates, which are *the only ones used in the sequel* of this chapter, we write

$$\exp\left[Df_{T,D}(t)\right] = \frac{B_T(t)}{B_{T+D}(t)} \quad \Leftrightarrow \quad f_{T,D}(t) = \frac{1}{D}\left[\ln B_T(t) - \ln B_{T+D}(t)\right].$$

We also use the notion of an *instantaneous* forward rate.

## 17.1 Models Based on the Dynamics of Spot Rates

**Definition** *The instantaneous forward rate relative to the future date T which is implicit in the yield (or price) curve of zero-coupon bonds on date t is defined by:*
$$f_T(t) \equiv \lim_{D \to 0} f_{T,D}(t) = -\frac{\partial \ln B_T(t)}{\partial T}.$$

The instantaneous forward rate $f_T(t)$ is the rate that can be assured on date $t$ for a loan transaction in the infinitesimal future interval $(T, T + dT)$. It is given graphically by the slope, at point $T$ and with a minus sign, of the current curve of the log prices (ln $B_T$, a curve decreasing as a function of the maturity date $T$).

First, we present two one-factor models that do not necessarily fit the currently observed yield curve (Vasicek, and Cox, Ingersoll and Ross), then two extended versions of these models: the first allows fitting the observed initial yield curve (Hull and White's model); the second allows taking into account several risk factors. The analysis of models based on the dynamics of spot rates will be revisited in Chap. 20 using a different approach.

### 17.1.1 One-Factor Models (Vasicek, and Cox, Ingersoll and Ross)

We adopt a probabilistic approach and start from the risk-neutral (RN) dynamics of the short-term rate, while in Chap. 20 we will revisit the same models specified by partial differential equations and use the real dynamics of rates and the market price of risk. Studying these two approaches and comparing them will deepen the reader's understanding of these models.

#### 17.1.1.1 General Presentation and Analysis of One-Factor Models

Let us consider all zero-coupon financial assets with different expiration dates, all without default risk, but subject to interest rate risk.

As a first approximation, we can consider that all these prices (thus the spot rates for the different maturities) are perfectly correlated and therefore assume that they depend on a single, common risk factor (or state variable), for example, the very short-term rate $r(t)$.

The current value of $r(t)$ on date $t$ is denoted by $r$, or possibly $r_t$ to avoid ambiguity. For example, we write the zero-coupon price curve as $\{B_T(t,r)\}_{T \in (t, T_{max})}$.

First, we characterize the different functions $B_T(.)$, so that the relationship between prices $B_T(.)$ and interest rates then allows us to determine the yield curve.[1]

Prices $B_T(t, r)$ can be obtained from Eq. (17.2) which we rewrite in this context:

---

[1] It is worth noting that theories of the yield curve only determine *relative* rates, i.e. calculated with respect to a pivot rate, typically the nominal overnight interest rate (OIS), which is itself strongly influenced by the leading rate(s) of the relevant Central Bank (see Chap. 3). This pivot rate is exogenous unless the model includes a dynamic stochastic general equilibrium (DSGE) of the macro-economy.

$$B_T(t, r) = E_t \left\{ \exp \left( - \int_t^T r(u) du \right) \right\}.$$

Recall that $E_t(.)$ is the RN expectation taken at $t$, in particular knowing that $r(t) = r$:

$$B_T(t, r) = E \left\{ \exp \left( - \int_t^T r(u) du \, | r(t) = r \right) \right\}. \tag{17.1}$$

The process that $r(t)$ follows, which by assumption is the only state variable of the considered financial system, determines the prices of all zero-coupons and thus the entire yield curve.

The short-term rate $r$ is assumed to obey a diffusion process that we write directly under the RN probability measure:

$$dr = m(t, r) \, dt + \sigma(t, r) dW. \tag{17.2}$$

The analytic form of the function $B_T(.)$ obviously depends on the explicit form of the process for $r$, i.e., on the form of the functions $m(t, r)$ and $\sigma(t, r)$.

The models differ by the process $r(t)$ they assume. We examine two basic models due to Vasicek (1977) and to Cox et al. (1985a, 1985b).

### 17.1.1.2 The Vasicek Model (1977)

This model is the prototype of a Gaussian model and we describe it in detail. The proofs that do not appear in the main text are all provided in Appendix 1.

(a) *The model's assumption*

The Vasicek model assumes the short-term rate $r(t)$ is an Ornstein–Uhlenbeck process (see Chap. 18), for which the drift and the volatility are respectively equal to

$$m(t, r) = a(k - r) \quad \text{and} \quad \sigma(t, r) = \sigma.$$

The coefficients $a$, $k$, and $\sigma$ are positive constants; $k$ can be interpreted as the long-term "normal rate." The drift $a(k - r)$ implies that a mean-reverting force draws the rate $r$ toward its normal value $k$: if $r < k$ (too low) the drift is positive and if $r > k$ (too high) it is negative. This tension increases with $a$.

Under such a specification, Eq. (17.2) then means that the following RN dynamics for the short-term rate:

$$dr = a(k - r) \, dt + \sigma \, dW(t). \tag{17.3a}$$

## 17.1 Models Based on the Dynamics of Spot Rates

The solution to this stochastic differential equation is well known (see Chap. 18, Sect. 18.4.2.2) and can be written on a date $u > t$, starting from a date $t$ on which $r(t) = r$:

$$r(u) = e^{-a(u-t)}[r - k] + k + \sigma e^{-au}\sigma e^{-au}\int_t^u e^{az}dW(z). \qquad (17.3b)$$

As a sum of independent normal variables $e^{az}dW(z)$, $r(u)$ is *normal*.[2]

Let us emphasize that Eqs. (17.3a and 17.3b) are written in the RN universe; in the real world, the dynamics involves the market price of risk (which here is implicitly constant and incorporated in the value of $k$), as one may see from Chap. 20, Sect. 20.5.

### (b) *Zero-coupon prices and yield curve in the Vasicek model.*

The computation of zero-coupon prices amounts to evaluating the integral

$$B_T(t, r) = E\left\{ \exp\left( -\int_t^T r(u)du \right) | r(t) = r \right\},$$

where $r(u)$ is given by Eq. (17.3b). The computation is carried out in Appendix 1 and leads to Vasicek's formula:

$$B_T(t, r) \equiv b_\theta(t, r) = A(\theta) \exp[-(\sigma_\theta/\sigma)\, r] \qquad (17.4)$$

where $\theta$ denotes the zero-coupon's time to maturity $(T - t)$, $b_\theta$ is the discounting factor, and

$$\sigma_\theta = \sigma\frac{1 - e^{-a\theta}}{a}; \quad A(\theta) = \exp\left[ \left(\frac{\sigma_\theta}{\sigma} - \theta\right)\left(k - \frac{\sigma^2}{2a^2}\right) - \frac{\sigma_\theta^2}{4a} \right]. \qquad (17.5)$$

These results imply that:

- The price of a zero-coupon is lognormal and only a function of the variable $r$ for a given $\theta$.
- The zero-coupon rate is affine linear in $r$: $r_\theta(t, r) = -\frac{1}{\theta}\ln[b_\theta(t, r)] = \frac{1}{\theta}[(\sigma_\theta/\sigma)r - \ln A(\theta)]$.

---

[2] The expectation and variance that characterize its distribution are, respectively: $E[r(u) | r(t) = r] = e^{-a(u-t)}[r - k] + k$ ; $var[r(u) | r(t) = r] = \sigma^2 e^{-2au}\int_t^u e^{2az}dz = \sigma^2(1 - e^{-2a(u-t)})/2a$.

This last equation implies that $\lim_{\theta \to \infty} r_\theta = k - \frac{\sigma^2}{2a^2}$ (the term in $\sigma_\theta r/\sigma\theta$ tends to zero and $\ln A(\theta)/\theta$ is of order $\left(\theta\left(k - \frac{\sigma^2}{2a^2}\right)/\theta\right)$), which means that the zero-coupon rate "at infinity" is a constant.

- $\sigma_\theta$ is the instantaneous volatility of a zero-coupon with duration $\theta$ and $\sigma_\theta/\sigma$ is its sensitivity to $r$.[3]

We can see that these volatility and sensitivity decrease with the passage of time and tend to 0 as $t$ tends to $T$ (since $\theta = (T - t)$ goes to 0).

The RN dynamics[4] of the price of a zero-coupon that expires on date $T$ thus can be written

$$dB_T/B_T(t) = r(t)\, dt + \sigma_{T-t}\, dW(t).$$

The Vasicek model leads to yield curves that can be increasing, decreasing or with a hump.

### (c) *Valuation of options on bonds using the Vasicek model*

First, the lognormality of the zero-coupon price means one can apply the BSM price model (the Gaussian model given by Eqs. (16.4a, 16.4b and 16.4c) in Chap. 16) to value an option on a zero-coupon bond.

For a call with strike $K$ and expiration $T$ written on a zero-coupon expiring on $T + D > T$, of nominal \$1 and price $B_{T+D}(t)$, we obtain the following price on date 0:

$$C(0) = B_{T+D}(0)N(d_1) - K\, B_T(0)N(d_2); \quad d_1$$

$$= \frac{1}{\sigma_{\ln B}} \left( \ln \frac{B_{T+D}(0)}{B_T(0)K} + \frac{\sigma_{\ln B}^2}{2} \right); \quad d_2 = d_1 \sigma_{\ln B} \tag{17.6}$$

where $\sigma_{\ln B} = \sigma \frac{1-e^{-aD}}{a} \sqrt{\frac{1-e^{-2aT}}{2a}}$ is the standard deviation of $\ln(B_{T+D}(T))$ calculated on date 0 (see Appendix 1 for this calculation).

The put's price is obtained from the parity relation: $P(0) = C(0) + K\, B_T(0) - B_{T+D}(0)$.

Second, the Vasicek model also allows valuing vanilla options on *coupon bonds* or vanilla options on *any fixed-income asset*.

---

[3] This sensitivity writes $-\frac{1}{B_{T-t}} \frac{\partial B_{T-t}}{\partial r}$; it is sometimes called *stochastic* to distinguish it from the standard sensitivity calculated with respect the compound rate $r_{T-t}$ (i.e. $-\frac{1}{B_{T-t}} \frac{\partial B_{T-t}}{\partial r_{T-t}} = T - t$).

[4] Under the real probability, the dynamics writes $dB_T/B_T(t) = (r(t)+\lambda\sigma_{T-t})\, dt + \sigma_{T-t}\, dW(t)$ where $\lambda$ is the (assumed constant) market price of risk (see Chap. 20).

## 17.1 Models Based on the Dynamics of Spot Rates

Let us consider an asset $x$ that generates the sure cash flow sequence $c_{\theta|\theta=T_1,T_2,...,T_n}$ and a call with strike $K$ and expiration $T < T_1$ written on $x$. We denote by $X(t)$ the price of $x$ on date $t$; in particular,

$$X(T) = \sum_{i=1}^{n} B_{T_i}(T, r(T)) c_{T_i}.$$

We denote by $r_k$ the rate $r(T)$ for which $X(T) = K$ and $K_i \equiv B_{T_i}(T, r_k) c_{T_i}$, from which

$$K = \sum_{i=1}^{n} B_{T_i}(T, r_k) c_{T_i} = \sum_{i=1}^{n} K_i.$$

The condition for exercise on date $T$ can be written either as $X(T) > K$ or as $r(T) < r_k$. Therefore, the call's payoff writes

$$C(T) = \left[ \sum_{i=1}^{n} B_{T_i}(T, r(T)) c_{T_i} - \sum_{i=1}^{n} K_i \right] \mathbf{1}_{r(T)<rk}$$

$$= \sum_{i=1}^{n} [B_{T_i}(T, r(T)) c_{T_i} - K_i] \mathbf{1}_{r(T)<rk} = \sum_{i=1}^{n} \max [B_{T_i}(T, r(T)) c_{T_i} - K_i, 0].$$

The final equality comes from the fact that, in a one-factor model, all the zero-coupon prices $B_{Ti}$ are decreasing functions of $r$. Therefore, we can consider the option on a coupon bond as a basket of $n$ independent options, the $i$th of which is written on a zero-coupon yielding $c_{Ti}$ on date $T_i$ and of strike $K_i$ (the exercise condition is the same for each of the $n$ options, i.e., $r(T) < r_k$, which is essential for the argument). The value of the option on an asset with coupons can then be derived by the simple addition of the values of the $n$ options in the basket.

### 17.1.1.3 The Cox–Ingersoll–Ross Model (1985)

Within a general equilibrium model, Cox, Ingersoll, and Ross (CIR) derive a dynamics describing $r(t)$ (a square root process), slightly different from the Orstein–Uhlenbeck process, which we write, under the RN probability measure, as

$$dr = a[k - r] \, dt + \sigma\sqrt{r} \, dW(t) \tag{17.7}$$

where $a$, $k$, and $\sigma$ are positive constants and $r$, as we recall, is the value of $r(t)$ on $t$. The instantaneous variance of the rate is not constant but equals $\sigma^2 r$, and thus is proportional to the interest rate.

The CIR specification presents on a priori grounds one disadvantage and one advantage relative to the Vasicek model, which is connected to the model's non-Gaussian feature:

- The disadvantage is its relative complexity (since we leave the Gaussian framework).
- The advantage is that it yields more realistic results, excluding in particular negative interest rates.[5]

This last property deserves an explanation. If, in fact, at a given moment $r(t)$ takes on the value $r = 0$, the variance of $dr$ $(= \sigma^2 r \, dt)$ vanishes and $dr$ is with certainty equal to its expected value $akdt$ which is positive. Therefore, the interest rate (surely) increases if it hits the value zero; one says that zero is a reflecting barrier for the stochastic process governing $r$.

CIR derived the zero-coupon prices resulting from such a process. The solution writes:

$$B_{T-t}(r) = b_\theta(r) = \Phi(\theta) \exp\left[-r\Gamma(\theta)\right], \quad \text{with } \theta = (T - t) \tag{17.8}$$

and:

$$\Gamma(\theta) = \frac{2(e^{\gamma\theta} - 1)}{(\gamma + a)(e^{\gamma\theta} - 1) + 2\gamma}; \quad \Phi(\theta) = \left(\frac{2\gamma e(a + \gamma)^{\theta/2}}{(\gamma + a)(e^{\gamma\theta} - 1) + 2\gamma}\right)^{2ak/\sigma^2};$$

$$\gamma = \sqrt{a^2 + 2\sigma^2}.$$

Like in the Vasicek model, the rate $r_\theta$ is an affine function of the single variable $r$.

## 17.1.2 Fitting the Initial Yield Curve; the Hull and White Model

Generalizations of the preceding models allow fitting the yield curve observed on the initial date. Among these, we discuss that of Hull and White (1990) which generalizes the Vasicek model.

### 17.1.2.1 Fitting the Initial Yield Curve

The Vasicek and CIR models have one drawback which makes them unsuitable at empirical and operational levels: the initial yield curve is an output and *does not fit automatically to that actually observed on the current date*, even if the parameters are chosen with care (they do not have enough degrees of freedom). Other models, for which the yield curve is an input, are, by construction, compatible with all observed rates; such models are sometimes called "no-arbitrage models" while models of the Vasicek and CIR types are called "equilibrium models."[6]

---

[5]Negative rates do exist in some countries for some periods, but are generally not available to private borrowers. See, however, Footnote 9 below.

[6]The CIR is explicitly a general equilibrium model. For a literature review, see for example Rebonato (2004).

## 17.1 Models Based on the Dynamics of Spot Rates

However, the compatibility of some equilibrium models with the initial yield curve can be achieved by making the drift of the short-term rate depend on time (this flexibility allows the desired adjustment).

The table below shows the RN dynamics of the short-term rate $r(t)$ underlying the principal one-factor models that can be made compatible with the initial yield curve.

| General form | $dr = (k(t) - a f(r)) \, dt + \sigma g(r) dW(t)$ |
|---|---|
| Black and Karasinski (1991) | $d\ln(r) = (k(t) - a(t)\ln(r)) \, dt + \sigma(t) \, dW(t)$ |
| Ho and Lee (continuous time version) (1986) | $dr = k(t) \, dt + \sigma dW(t)$ |
| Hull and White – Vasicek (1990) | $dr = (k(t) - a \, r_t) \, dt + \sigma \, dW(t)$ |
| Hull and White – CIR (2000) | $dr = (k(t) - a \, r_t) \, dt + \sigma \, \sqrt{r} \, dW(t)$ |

We will only discuss the Hull and White – Vasicek model.

### 17.1.2.2 The Hull and White Model (1990)

(a) *The model*

Hull and White (1990) generalized the Vasicek model to make it fit the current yield curve by assuming the following random process for the short-term rate:

$$dr = (k(t) - a \, r_t) \, dt + \sigma \, dW(t). \qquad (17.9)$$

Note that here we use the notation $r(t) = r_t$ and not $r$, and that by setting $k(t) = ak$, we obtain Eq. (17.3a), the Vasicek model.

One can show that the model adapts to the presently observed yield curve (on date 0) if one chooses $k(t)$ in a way consistent with the forward-yield curve

$$k(t) = a \, f_t(0) + f_t'(0) + \frac{\sigma^2}{2a} \left(1 - e^{-2at}\right)$$

where $f_t(0)$ denotes the instantaneous forward rate for date $t$ that is implicit in the yield curve at $t = 0$ and $f_t'(0) = \frac{\partial f_t(0)}{\partial t}$ is the slope of the instantaneous forward yield curve (estimated numerically by discretization).

The term $\frac{\sigma^2}{2a} \left(1 - e^{-2a}\right)$ is in practice small and the drift term for the rate $r$ can be approximated by $a(f_t(0) - r_t) + f_t'(0)$. This result means that $r(t)$ is attracted toward the instantaneous forward rate $f_t(0)$ observed at 0 which is supposed to be the "normal rate" on date $t$. Indeed, if $r_t$ equals the forward rate $(f_t(0) - r_t = 0)$, it tends to follow the slope $f_t'(0)$ of the forward yield curve (thus to stay equal to the forward rate), and if it departs from it, $(f_t(0) - r_t \neq 0)$, there is the additional mean-reverting force of intensity $a$ that reduces the discrepancy.

The Hull and White model allows us, in this framework, to express the price $B_T(t)$ of a zero-coupon (ZC) at a future date $t$ $(t < T)$ and the price of an option on such a ZC or on a coupon bond as a function of the random variable $r(t)$.

728      17   Modeling Interest Rates and Options on Interest Rates

### (b) *Price of a zero-coupon bond*

The model gives the price on date $t$ of a ZC yielding \$1 on date $T$. Setting $\theta = (T - t)$, we have

$$B_T(t, r(t)) \equiv b_\theta(t, r(t)) = A(\theta) \exp\left[-(\sigma_\theta/\sigma)r(t)\right], \qquad (17.10)$$

with, as the only difference from formula (17.4) in the standard Vasicek model, the expression for $A(\theta)$:

$$A(\theta) = \frac{B_T(0)}{B_t(0)} \exp\left(\frac{\sigma_\theta}{\sigma} f_t(0) - \frac{\sigma^2}{4a^3}\left(e^{2at} - 1\right)\left(e^{-aT} - e^{-at}\right)\right). \qquad (17.11)$$

The last expression is computed from today's price (or yield) curve for ZCs (the instantaneous forward rate $f_t(0) = -\frac{\partial \ln B_t(0)}{\partial t}$ can be estimated by discretization: $-\frac{\partial \ln B_t(0)}{\partial t} \approx \frac{1}{\Delta t}\left(\ln B_t - \ln B_{t+\Delta t}\right) = f_{t,\Delta t}$, by choosing the maturity $(t + \Delta t)$ the closest possible to $t$ available in the market.

The volatility of the price of a ZC of duration $\theta$ is, as in the Vasicek model, $\sigma_\theta = \sigma \frac{1-e^{-a\theta}}{a}$, as is the standard deviation of $\ln(B_{T+D}(T))$, $\sigma_{\ln B} = \sigma \frac{1-e^{-aD}}{a}\sqrt{\frac{1-e^{-2aT}}{2a}}$.

### (c) *The value of options on bond*

We return to the problem of valuing an option with expiration date $T$ on a ZC with maturity $T+D$, and price $B_{T+D}$. The Hull and White model leads to exactly the same formulas for valuing these options as Vasicek's. Indeed, the model remains Gaussian and, as both the instantaneous volatility $\sigma_{T+D-t}$ of the ZC and the standard deviation of the log of its price $B_{T+D}(T)$ on date $T$ are identical in both models, the difference between them can only be the drift term which plays no part in valuing options.

Equation (17.6) for valuing options on a ZC is therefore applicable as such. Also, the decomposition of an option on a coupon bond into a basket of options on ZCs allows its valuation by simple addition of the prices in the basket. Finally, the Hull and White model can be used to value caps and floors (see Chap. 16) since it treats them as a collection of options on ZCs.

## 17.1.3 Multifactor Structures

A second possible generalization of the Vasicek and CIR rate models consists in letting the short-term rate dynamics depend on several processes (called state variables, or risk factors) and not just one. Here we limit the exposition to a short introduction, as multifactor models are discussed at length in Chap. 20 presenting state variable models.

## 17.2 Models Grounded on the Dynamics of Forward Rates

Let us consider $n$ "state variables"[7] forming a vector $\underline{X}(t) = (X_1(t), X_2(t), \ldots, X_n(t))$, which are assumed to be described by a multidimensional diffusion (thus Markovian) process, that we will not further specify here (see Chap. 20 for more details). The RN dynamics of the short-term rate writes:

$$dr = m(t, \underline{X}_t)\, dt + \sigma(t, \underline{X}_t)dW_r.$$

The ZC prices $B_T(t)$ are then equal to:

$$B_T(t, \underline{X}_t) = E\left\{ \exp\left( -\int_t^T r(u)du \right) | \underline{X}(t) = \underline{X}_t \right\}.$$

Itô's Lemma implies $dB_T = m_T(t, \underline{X})\, dt + \sum_{i=1}^{n} \frac{\partial B_T}{\partial X_i} dX_i$. Since the $X_i$ are imperfectly correlated, $dB_T$ and $dB_{T'}$ are imperfectly correlated for $T \neq T'$.

The following list of multifactor models is far from exhaustive. The oldest model is due to Langetieg (1980) who introduces a multifactor structure generalizing the Vasicek model. In his model, the state variables, among them the short-term rate, are governed by a generalized multidimensional Ornstein–Uhlenbeck process. This assumption retains the model's Gaussian character and leads to closed-form formulas for the prices of ZC bonds and options on interest rates. The model of Brennan and Schwartz (1982) involves two factors: the short-term rate and a long-term rate (that of a perpetuity). Longstaff and Schwartz (1992) generalize the CIR model to two factors. Brennan and Schwartz's and Langetieg's models are discussed in Chap. 20.

## 17.2 Models Grounded on the Dynamics of Forward Rates

Among the oldest models using the observed initial yield (or price) curve as exogenous input, aside from that of Hull and White, one finds the class of models à la Heath et al. (1992) who model the instantaneous forward rate dynamics (Sect. 17.2.1). Then, interest rate *market models* have been developed by Brace et al. (1997), Jamshidian (1997), Miltersen et al. (1997), and Musiela and Rutkowski (1997). The novelty of these models lies in their modeling the *discrete* forward rates directly observable on the market, whence their name, in contrast with the theoretical instantaneous rates used previously. They are the forward Libor (London Interbank Offered Rate) in the Libor market model, and the forward swap rate in the Swap market model (Sect. 17.2.2).

---

[7] The name comes from the fact that they characterize the state of the financial system. In Markovian models, the financial variables (rates, asset prices, etc.) are written as functions of these state variables (see Chap. 20).

## 17.2.1 The Heath–Jarrow–Morton Model (1992)

Rather than modeling the dynamics of ZC prices or the instantaneous short-term rate, Heath et al. (1992), hereafter HJM, model the instantaneous forward rate dynamics within a very general framework. They show what *restrictions* on the *drifts* of the RN processes governing these rates must be imposed so that the ZC rate curve is free of arbitrage. Furthermore, their model does recover, *by construction*, the exact initial prices of all ZC bonds. As an illustrative example, we apply a simple version of the HJM model where the volatility of forward rates is constant, which leads to closed-form solutions for the prices of bonds, bond futures, and options written on bond forward and futures contracts.

### 17.2.1.1 Representation of the Yield Curve

We defined in Sect. 17.2.1 the forward rate prevailing at $t$ and relative to the period $(T, T + D)$

$$f_{T,D}(t) = \frac{\ln B_T(t) - \ln B_{T+D}(t)}{D},$$

and the instantaneous forward rate

$$f_T(t) \equiv \lim_{D \to 0} f_{T,D}(t) = -\frac{\partial \ln B_T(t)}{\partial T}.$$

The curve for instantaneous forward rates $f_T(t)^{T \in (t, T_{\max})}$ can therefore be deduced, theoretically, from the ZC price curve, thus from the ZC spot rate curve, and conversely the instantaneous forward rate curve determines that of the ZC rates (and prices) by the inverse relation:

$$B_T(t) = \exp\left(-\int_t^T f_s(t)ds\right), \quad \text{for } T \in (t, T_{\max}). \tag{17.12}$$

Thus it is *equivalent* to use the forward rate, the ZC price or the ZC rate, as far as the initial curves and their dynamics are concerned (each actually determining the other two).

### 17.2.1.2 General Dynamics of Forward Rates and ZC Bond Prices

HJM assume that the instantaneous forward rate $f_T(t)$ obeys a diffusion process whose evolution can be explicitly written *under the RN probability measure*:

$$f_T(t) = f_T(0) + \int_0^t \mu(s, T, \underline{X}(s))ds + \int_0^t \sigma(s, T, \underline{X}(s))dW(s) \tag{17.13}$$

where the drifts $\mu(s, T, \underline{X}(s))$, denoted later simply by $\mu(s, T, .)$, and the volatilities $\sigma(s, T, \underline{X}(s))$, denoted by $\sigma(s, T, .)$, are for the moment very general (apart from certain

## 17.2 Models Grounded on the Dynamics of Forward Rates

technical conditions to ensure that the integrals involved are well defined) and depend on a vector $\underline{X}$ of state variables that may include interest rates, in particular the spot rate $r(t)$.[8]

We deduce immediately that the instantaneous spot rate $r(t) \equiv f_t(t)$ obeys the RN dynamics

$$r(t) = f_t(0) + \int_0^t \mu(s, t, .)ds + \int_0^t \sigma(s, t, .)dW(s). \tag{17.14}$$

The fundamental idea of the HJM model is that the RN drift of the forward rate has, under no arbitrage, to satisfy a constraint. The latter will be identified thanks to Eq. (17.12) (which binds the forward rate to the ZC price $B_T(t)$), by asserting that the RN yield of $B_T$ equals $r(t)$. This approach leads to the following proposition:

**Proposition 1**
*In the HJM model, the dynamics of the instantaneous forward $f_T(t)$ is, under the risk-neutral probability measure Q, given by*

$$df_T(t) = \mu(t, T, .)dt + \sigma(t, T, .)dW(t) = \left(\sigma(t, T, .)\int_t^T \sigma(t, s, .)ds\right)dt + \sigma(t, T, .)dW(t)$$

$$\tag{17.15}$$

*where $W(t)$ is a Brownian process under Q.*

**Proof**
Under the RN probability, we have $\frac{dB_T(t)}{B_T(t)} = r(t)dt + \sigma_B(t, T, .)\,dW(t)$ and by Itô's Lemma we obtain $d\ln B_T(t) = dB_T(t)/B_T(t) - 0.5(\sigma_B(t, T, .))^2\,dt$, where $\sigma_B(t, T, .)$ denotes the volatility of $B_T(t)$. From this we get

(i) $d \ln B_T(t) = [r(t) - 0.5(\sigma_B(t, T, .))^2]\,dt + \sigma_B(t, T, .)\,dW(t).$

Furthermore, the forward rate prevailing at $t$ for the period $(T, T+D)$ writes

$$f_{T,D}(t) = \frac{\ln B_T(t) - \ln B_{T+D}(t)}{D}.$$

Differentiating with respect to $t$, we have

---

[8] Remark, in particular, that the Brownian motion $\underline{W}(t)$ can be multidimensional, in which case $\sigma(s, T, .)$ is a row vector of the same dimension as the column vector $d\underline{W}(t)$. We then write, to simplify, $\sigma d\underline{W}$ as $\sigma_B dW$ using the change of variable: $dW = \frac{\sigma d\underline{W}}{|\underline{\sigma}|}$ and $\sigma_B = |\underline{\sigma}|$.

$$df_{T,D}(t) = \frac{d\ln B_T(t) - d\ln B_{T+D}(t)}{D},$$

and applying ($i$) yields

$$df_{T,D}(t)$$
$$= \frac{1}{D}\left\{0.5\left[(\sigma_B(t,T+D,.))^2 - (\sigma_B(t,T,.))^2\right]dt + [\sigma_B(t,T,.) - \sigma_B(t,T+D,.)]dW(t)\right\}.$$

Now, let $D$ go to 0 using the fact that $\lim_{D\to 0} f_{T,D}(t) = f_T(t)$ (the instantaneous forward rate) and focusing on the derivatives with respect to $T$ of the last term in the r.h.s. of the last equation,

$$df_T(t) = \frac{1}{2}\frac{\partial \sigma_B^2(t,T,.)}{\partial T}dt - \frac{\partial \sigma_B(t,T,.)}{\partial T}dW(t),$$

or else,

$$df_T(t) = \sigma_B(t,T,.)\frac{\partial \sigma_B(t,T,.)}{\partial T}dt - \frac{\partial \sigma_B(t,T,.)}{\partial T}dW(t). \tag{17.16}$$

We notice that the instantaneous standard deviation of $df_T$, $\sigma(t, T,.)$, is equal to $-\frac{\partial \sigma_B(t,T,.)}{\partial T}$ and, conversely, $\sigma_B(t, T,.) = -\int_t^T \sigma(t, s, .)ds$.

Equation (17.16) can then be written $df_T(t) = \left(\sigma(t,T,.)\int_t^T \sigma(t,s,.)ds\right)dt + \sigma(t, T, .)dW(t)$, which is Eq. (17.15).

HJM's fundamental result is that, *under the RN measure, the drift and the volatility of the forward rate $f_T(t)$ must be linked, under no arbitrage, by the relation*

$$\mu(t,T,.) = \sigma(t,T,.)\int_t^T \sigma(t,s,.)ds. \tag{17.17}$$

The no-arbitrage condition (17.17) is very general and all interest rate models must satisfy it. In this sense, all interest rate models based on the no-arbitrage rule can be viewed as special cases of HJM. For example,

- The simplest model, $\sigma(t,T,.) = \sigma = $ constant, is the continuous-time version of the model due to Ho and Lee (1986). We develop and use this simple model in what follows.
- The Vasicek/Hull and White model ($dr = a(k(t) - r)dt + \sigma\,dW$) can be obtained by setting $\sigma(t,T,.) = \sigma\,e^{-a(T-t)}$.

To provide an example enjoying an explicit solution, we specialize the model to the very simple case where the Brownian motion is one-dimensional and the volatility of the instantaneous forward rate, $\sigma$, is constant (the continuous-time

## 17.2 Models Grounded on the Dynamics of Forward Rates

version of the Ho and Lee model).[9] We then easily obtain the (Gaussian) dynamics of the forward rate as well as the (lognormal) dynamics of the ZC price:

$$f_T(t) = f_T(0) + \sigma^2 t\left(T - \frac{t}{2}\right) + \sigma W(t) \tag{17.18}$$

$$B_T(t) = \frac{B_T(0)}{B_t(0)} \exp\left[-\frac{\sigma^2}{2} tT(T - t) - \sigma(T - t)W(t)\right]. \tag{17.19}$$

Indeed, Eq. (17.15) with $\sigma$ constant simplifies to $df_T(t) = \sigma^2(T - t)dt + \sigma dW(t)$, which yields Eq. (17.18) by integration. As for Eq. (17.19), we have

$$
\begin{aligned}
B_T(t) &= \exp\left(-\int_t^T f_s(t)ds\right) = \exp\left[-\int_t^T \left(f_s(0) + \sigma^2 t\left(s - \frac{t}{2}\right) + \sigma W(t)\right)ds\right] \\
&= \exp\left[\int_0^t f_s(0)ds - \int_0^T f_s(0)ds\right] \exp\left[-\int_t^T \left(\sigma^2 t\left(s - \frac{t}{2}\right) + \sigma W(t)\right)ds\right] \\
&= \frac{B_T(0)}{B_t(0)} \exp\left[-\sigma^2 t\left(\frac{T^2}{2} - \frac{t^2}{2}\right) + \sigma^2 \frac{t^2}{2}(T - t) - \sigma(T - t)W(t)\right],
\end{aligned}
$$

which, after simplifying, is Eq. (17.19). Note that from Eq. (17.19) we do recover $B_T(T) = 1$ at maturity $T$.

### 17.2.1.3 Application 1: Valuation of Options on Bonds and on Bond Forwards

Using the one-factor model with constant volatility for the forward rate, we first value options on forward ZC bonds and then options on spot ZC bonds, the second case being a special case of the first. As to the notation, $t$ denotes the current date, $T$ the expiration of the option, $T_F$ the maturity of the forward contract written on a ZC bond with maturity $T_B$, whence $t < T < T_F < T_B$.

Since the underlying asset prices are all Gaussian, the BSM framework applies.

#### (a) *Options on forward bond contracts*

Denote by $F_{T_F}(t, T_B)$ the price of a forward contract maturing on $T_F$ written for a ZC bond of maturity $T_B$. Under no-arbitrage, we have $F_{T_F}(t, T_B) = \frac{B_{T_B}(t)}{B_{T_F}(t)}$. The payoff on date $T$ of the European call $C_F$ written on this forward contract is: $C_F(T) = B_{T_F}(T)[F_{T_F}(T, T_B) - K]^+$, with the discount factor $B_{T_F}(T)$ coming from the

---

[9] HJM give, in addition, a two-factor model that can still be easily handled and that avoids the drawback of having perfectly correlated instantaneous forward rates. It should be noted that, in these two special versions of HJM, rates can be negative with a positive probability (see Eq. 17.18). Calibration should ensure that such a probability is very small. The current trend among practitioners is to use models that do not exclude (slightly) negative rates,

requirement that the (possible) gain will only be realized by the call's holder upon maturity at $T_F$ of the forward contract and not on the option's expiry date $T$.

The price on date $t$ of this call is given by the following formulas (*BSM-price*):

$$C_F(t) = B_{T_F}(t)$$
$$\times (F_{T_F}(t, T_B)N(d_1) - KN(d_2)) \quad (\text{as a function of the forward price } F_{T_F}) \tag{17.20}$$

$$C_F(t) = B_{T_B}(t)N(d_1)$$
$$- KB_{T_F}(t)N(d_2) \quad (\text{as a function of the spot price } B_{T_B}) \tag{17.21}$$

where $d_1 = \dfrac{\ln\left(F_{T_F}(t, T_B)/K\right)}{\sigma_F\sqrt{T-t}} + \frac{1}{2}\sigma_F\sqrt{T-t} = \dfrac{\ln\left(B_{T_B}(t)/KB_{T_F}(t)\right)}{\sigma_F\sqrt{T-t}} + \frac{1}{2}\sigma_F\sqrt{T-t}$, $d_2 = d_1 - \sigma_F\sqrt{T-t}$; $\quad \sigma_F = \sigma(T_B - T_F)$.

### Remarks

- Equation (17.21) can be derived immediately from Eq. (17.20) by using the value of $F_{T_F}(t, T_B)$ and the no-arbitrage rule.
- The value of the call on the forward contract (Black's formula) is equal to the value of the call on the spot bond of maturity $T_B$ (from BSM).
- The discount factor in Eq. (17.20) is not $B_T(t)$ as in Black's original model (in which the bond maturity coincides with that of the forward contract) but $B_{T_F}(t)$.
- The volatility $\sigma_F$ of the option's underlying (the forward price) is proportional to the duration between the maturity of the bond and that of the forward contract.
- The values of the corresponding European puts are obtained from the call-put parity, for example,

$$C_F(t) - P_F(t) = B_{T_F}(t)(F_{T_F}(t, T_B) - K). \tag{17.22}$$

### (b) *Options on* spot *bonds*

The payoff on date $T$ of the European call $C_B$ written on a bond with maturity $T_B$ is equal to $C_B(T) = [B_{T_B}(T) - K]^+$. Using Eq. (17.21) with $T_F = T$, (the forward contract is expiring and now is a spot instrument), we immediately obtain this call's value on date $t$ as

$$C_B(t) = B_{T_B}(t)N(d_1) - KB_T(t)N(d_2) \quad (\text{BSM-price}) \tag{17.23}$$

where

$$d_1 = \frac{\ln\left(B_{T_B}(t)/KB_T(t)\right)}{\sigma_B\sqrt{T-t}} + \frac{1}{2}\sigma_B\sqrt{T-t},$$

$$d_2 = d_1 - \sigma_B\sqrt{T-t}; \quad \sigma_B = \sigma(T_B - T)$$

The corresponding European put is obtained from the call-put parity: $C_B(t) - P_B(t) = B_{T_B}(t) - KB_T(t)$.

## 17.2 Models Grounded on the Dynamics of Forward Rates

### 17.2.1.4 Application 2: Forward-Futures Relationship and Options on Bond Futures Contracts

From Chap. 11, the futures price of an underlying is a martingale under the RN measure. From the Appendix to Chap. 9, when interest rates are stochastic, the forward price (a martingale under the FN measure) and the futures price differ by a covariance term. We recover this relationship by differentiating the exact equation relating the two prices, in the one-factor HJM model with constant forward rate volatility. We then provide the price of options written on bond futures.

### (a) *Relationship between forward and futures prices*

We denote by $H_{T_F}(t, T_B)$ the price of a futures contract with maturity $T_F$ written on a ZC bond with maturity $T_B$. The relationship between $H_{T_F}(t, T_B)$ and $F_{T_F}(t, T_B)$ is given in Proposition 2.

### Proposition 2
*In the one-factor HJM model with constant volatility for the instantaneous forward rate, the relationship under no-arbitrage between futures and forward prices writes*

$$H_{T_F}(t, T_B) = F_{T_F}(t, T_B) \exp\left(-\frac{\sigma^2}{2}(T_B - T_F)(T_F - t)^2\right). \qquad (17.24)$$

### Proof
On the futures maturity date $T_F$, under no-arbitrage, we must have $H_{T_F}(T_F, T_B) = B_{T_F}(T_B)$. We know that, under the RN measure, the futures price is a martingale, so we can write

(i) $H_{T_F}(t, T_B) = E_Q[H_{T_F}(T_F, T_B) \mid \Im_t] = E_Q[B_{T_F}(T_B) \mid \Im_t].$

Using Eq. (17.19), we have, on the one hand,

$$B_{T_F}(T_B) = \frac{B_{T_B}(0)}{B_{T_F}(0)} \exp\left[-\frac{\sigma^2}{2} T_B T_F(T_B - T_F) - \sigma(T_B - T_F)W(T_F)\right],$$

and, on the other,

$$F_{T_F}(t, T_B) = \frac{B_{T_B}(t)}{B_{T_F}(t)}$$

$$= \frac{B_{T_B}(0)}{B_{T_F}(0)} \exp\left[-\frac{\sigma^2}{2} t(T_B^2 - T_F^2 - t(T_B - T_F)) - \sigma(T_B - T_F)W(t)\right]$$

Putting these two equations together and rearranging, we obtain

$$B_{T_F}(T_B)$$

$$= F_{T_F}(t, T_B) \exp\left[-\frac{\sigma^2}{2}(T_B - T_F)(T_F - t)(T_B - t) - \sigma(T_B - T_F)[W(T_F) - W(t)]\right]$$

Therefore, equation ($i$) becomes

$$H_{T_F}(t, T_B)$$

$$= F_{T_F}(t, T_B)E_Q\left[\exp\left[-\frac{\sigma^2}{2}(T_B - T_F)(T_F - t)(T_B - t) - \sigma(T_B - T_F)[W(T_F) - W(t)]\right] /\Im_t\right]$$

Recalling that, for $X$ Gaussian, $E[\exp(aX)] = \exp[aE(X) + 0.5a^2 Var(X)]$, and that $[W(T_F) - W(t)]$ is a Gaussian variate with zero mean and variance $(T_F - t)$, computing the expectation above leads, after simplification and regrouping, to $\exp\left(-\frac{\sigma^2}{2}(T_B - T_F)(T_F - t)^2\right)$.

**Remarks**
- One recovers from Eq. (17.24) the equality of the forward and futures prices if the rates are deterministic since then $\sigma = 0$.
- Remark that here the futures price $H(.)$ is *smaller than* the forward price $F(.)$. More generally, irrespective of the asset underlying the contracts, this is true if variations in the spot price of the ZC bond with maturity $T_F$ are positively correlated with the forward price; here the correlation is perfect (and positive) since there is only one factor and the forward contract's underlying is a rate instrument.

### (b) *Options on bond futures*

The payoff on date $T$ of the European call $C_H$ written on futures is

$$C_H(T) = [H_{T_F}(T, T_B) - K]^+.$$

It is rather long to derive the value of such a call and we provide only a sketch. It is based on the device of defining the *forward* price at $t$, for a maturity $T$ corresponding to that of the option, of the *futures* contract, that is, the price at $t$ of a forward contract with maturity $T$ allowing the buyer to take a position at $T$ on a futures contract (written on a bond $T_B$) with maturity $T_F$, say $\Phi_T(t, T_F, T_B)$. This forward price is given by

$$\Phi_T(t, T_F, T_B) = \frac{E_Q\left[\exp\left(-\int_t^T r(s)ds\right) H_{T_F}(T, T_B) /\Im_t\right]}{B_T(t)},$$

## 17.2 Models Grounded on the Dynamics of Forward Rates

where the numerator is the RN expectation of the futures price at $T$ discounted to the current date $t$, and the denominator allows this expectation to be valued back on date $T$.

By first finding the relationship at $t$ between $\Phi_T(t,T_F,T_B)$ and $H_{T_F}(t,T_B)$,[10] then changing from the RN probability to the $Q_T$-FN probability, so that $\Phi_T(t,T_F,T_B)$ is a $Q_T$-martingale, we recover formally the case of an option on a forward contract, whose solution is Eq. (17.20):

**Proposition 3**

*In the one-factor HJM model with constant volatility for the instantaneous forward rate, the value of a European call on a bond futures is given by*

$$C_H(t) = B_T(t)(D(t,T)H_{T_F}(t,T_B)N(d_1) - KN(d_2)) \qquad (17.25)$$

*where:*

$$d_1 = \frac{\ln\left(D(t,T)H_{T_F}(t,T_B)/K\right)}{\sigma_H\sqrt{T-t}} + \frac{1}{2}\sigma_H\sqrt{T-t},$$

$$d_2 = d_1 - \sigma_H\sqrt{T-t}; \qquad \sigma_H = \sigma(T_B - T_F);$$

$$D(t,T) = \exp\left(\frac{\sigma^2}{2}(T_B - T_F)(T-t)^2\right).$$

**Remarks**

- The comparison of Eqs. (17.20) for $C_F(t)$ and (17.25) for $C_H(t)$ highlights two differences: the presence of a correction term $D(t,T)$ on the one hand, and of $B_T(t)$ instead of $B_{T_F}(t)$ on the other. The first is due to the fact that futures are assumed to be continuously *marked-to-market*, the second from the fact that possible gain is paid at $T$ with futures but at $T_F (> T)$ only with forwards.
- If interest rates are deterministic (the standard BSM assumption that can be justified as a first approximation if the contract's underlying is not an interest rate instrument), the term $D(t,T)$ equals one (and $H(.) = F(.)$) and only the second difference remains.

## 17.2.2 The Libor (LMM) and Swap (SMM) Market Models

One main disadvantage of HJM and HJM-like models is that they use unobservable instantaneous interest rates. The so-called "market models" were designed in the late 1990s to cope with this issue.[11] This new approach, using the *observable discrete market rates*, the forward Libor (London Interbank Offered Rate) in the Libor Market Model (hereafter LMM) framework, and the forward Swap rate for the Swap Market Model (SMM) has two merits: it avoids resorting to the estimation methods needed

---

[10]The ratio of the two prices equals the term $D(t, T)$ defined in Eq. (17.25).

[11]See Brace et al. (1997), Miltersen et al. (1997), Jamshidian (1997) and Musiela and Rutkowski (1997).

738  17 Modeling Interest Rates and Options on Interest Rates

to obtain the instantaneous rate; it allows direct valuation of the most liquid of options, caps and swaptions, and the calibration of the model from their observed prices. Market models provide a rigorous theoretical foundation for professionals' practices in valuing such products, since they assume, as the market does, that rates are lognormal. In this connection, the BSM-rate model (presented in the previous chapter) and the LMM are essentially the same and one may consider the first to be the core of the second.

We assume in what follows that the relevant *Brownian motion is one-dimensional* (i.e., we discuss one-factor models). If the generalization to a multi-dimensional Brownian process (multifactor models) does not pose additional theoretical problems, it does lead to considerable complications that are outside the reach of this book. These difficulties occur particularly in creating and deploying algorithms to calibrate the models, in estimating functions of the volatility and co-variation of the relevant discrete rates, and taking into account the dynamics of the volatility smile or skew.[12] What follows is therefore just an introduction to the practical valuation of interest rate derivatives.

### 17.2.2.1 The Libor Market Model (LMM)

(a) *Dynamics of the forward Libor rate*

Starting from the derivation of this LMM (sometimes also called BGM model, after Brace et al. (1997)) under the appropriate forward-neutral probability measure, we first show how to obtain easily closed-form solutions for cap prices. Then we show why using this model to value swaptions leads to inconsistencies from which one can only escape by some, more or less satisfactory, approximations, which justifies the existence of the Swap Market Model SMM presented in Sect. 17.2.2.2.

We assume throughout this Sect. 17.2.2 the existence of a series of maturities, each called a *tenor*, $\{T_0, T_1,...,T_N\}$. The present date is denoted $t$ $(0 \le t \le T_0)$, the date an option contract has been negotiated being $t = 0$. The set of all $N$ periods $(T_0, T_1),..., (T_{N-1}, T_N)$ is called a *tenor structure*. We also use the notation, for reasons of legibility, $B(t,T_i)$ [instead of $B_{T_i}(t,r)$] for the price on date $t \le T_i$ of a ZC bond with expiration $T_i$. The forward Libor rates are linked to the ZC bond prices by the relationship:

$$1 + DL(t, T_i, T_{i+1}) = \frac{B(t, T_i)}{B(t, T_{i+1})}, \qquad (17.26)$$

where $L(t, T_i, T_{i+1})$ is a forward Libor rate observed on date $t$ and $D$ is the length of the period delimited by the $T_i$ and $T_{i+1}$, which is assumed to be constant, for the sake of simplicity, for all $i$.[13] We will also write for simplicity $L(t, T_i, T_{i+1}) \equiv L(t, T_i)$.

---

[12] The reader will find these developments in the books cited with an asterisk or in some articles at the end of this chapter.

[13] One could have, without further complications other than notational ones, assumed different durations $D_i$.

## 17.2  Models Grounded on the Dynamics of Forward Rates

In the LMM, the rates $L(t, T_i)$ are assumed lognormal. This assumption is consistent with market practice which uses the BSM-rate formula to value caps, floors and swaptions (see Chap. 16, Sect. 16.1.3.2 for the analysis of the BSM-rate model).

In accordance with the exposition given in Sect. 17.2.1, the measure under which it is convenient to carry out computations is the forward neutral measure $Q_T$. Indeed, as we showed in Chap. 16 (Proposition 1, Sect. 16.1.3.1), each forward rate, in particular the forward Libor rate $L(t,T_i)$, is a martingale under its own forward measure, specifically the measure associated to the numeraire $B(t,T_{i+1})$. Calling $\gamma_i(t)$ the instantaneous volatility of the forward Libor rate $L(t,T_i)$, we immediately have the following dynamics for $L(t,T_i)$ under its forward-neutral (FN) measure $Q^{T_{i+1}} \equiv Q^{i+1}$:

$$dL(t, T_i) = L(t, T_i)\gamma_i(t)\,dW^{i+1}(t),\qquad (17.27)$$

with $W^{i+1}$ a standard Brownian motion under $Q^{i+1}$. Let us emphasize that $L(t,T_i)$ is only a martingale under $Q^{i+1}$; In particular, the expression for the dynamics of this rate under the forward terminal measure $Q^n$ to which the numeraire $B(t,T_n)$ is associated is not a martingale and is given by the following equation (proved in Appendix 2, Sect. 1):

$$dL(t, T_i) = L(t, T_i)\gamma_i(t)\left[dW^n(t) - \sum_{j=i+1}^{n-1} \frac{D\gamma_j(t)L(t, T_j)}{1 + DL(t, T_j)}\,dt\right].\qquad (17.28)$$

**Remarks**
- The forward Libor rates are not $Q^n$-martingales, except the final rate $L(t, T_{n-1})$.
- The dynamics of the forward Libor rate $L(t,T_i)$ can also be obtained, in a similar way to Eq. (17.28), under any forward probability measure $Q^k$ where $k \neq i + 1$.
- Often one chooses, to reduce the problems of calibration connected with valuing numerous rate products, another measure, called the *spot measure*, to which another numeraire is associated and whose value depends on the current date $t$. The latter corresponds to $\$1$ on the initial date ($t = 0$) invested in a number $1/B(0, T_0)$ of ZC bonds with maturity $T_0$ each worth $B(0, T_0)$. On date $T_0$, the portfolio numeraire is, therefore, worth $\$ 1/B(0, T_0)$, and this sum is reinvested (a *roll-over* strategy) in $1/(B(0, T_0)B(T_0, T_1))$ ZC bonds with maturity $T_1$. One continues in this way until the date $T_k$ in the tenor structure that just precedes the present date $t$. The value of the spot numeraire is therefore the inverse of the product of the prices of successive ZC bonds of the same duration $D$.

## 740          17   Modeling Interest Rates and Options on Interest Rates

### (b) *Valuation of a caplet*

Let us consider a vanilla caplet with maturity $T_{i+1}$, strike $k$ and payoff $D(L(T_i, T_i) - k)^+$ on date $T_{i+1}$ (recall that $L(T_i, T_i)$ is the spot Libor on $T_i$). The price of this caplet is given, under the forward probability $Q^{i+1}$ for which $L(t,T_i)$ is a martingale, as

$$C_i(0) = E^{i+1}\left[DB(0, T_{i+1})\left(L(T_i, T_i) - k\right)^+\right]. \tag{17.29}$$

Since the forward Libor rate is assumed lognormal, we can apply the *BSM-rate* solution to Eq. (17.29) and obtain the following result, already derived in Chap. 16 (Proposition 2, Sect. 16.1.3.2) but rewritten in this chapter's notation:

The price $C_i(0)$ of the caplet covering the period $(T_i, T_{i+1})$ (expiration at $T_{i+1}$) writes:

$$C_i(0) = DB(0, T_{i+1})\, Black(L(0, T_i), k, \sigma(i)) \tag{17.30}$$

*with*    $Black(L(0, T_i), k, \sigma(i)) \quad = \quad L(0, T_i)N(d_1) \quad - \quad kN(d_2)$   *where*   $d_1 = \frac{\ln(L(0,T_i)/k) + \frac{1}{2}\sigma^2(i)T_i}{\sigma(i)\sqrt{T_i}}$,   $d_2 = d_1 - \sigma(i)\sqrt{T_i}$, and $\sigma^2(i) = \frac{1}{T_i}\int_0^{T_i}\gamma_i(s)^2\,ds$.

The expression in Eq. (17.30) is Black's formula applied to an underlying rate (BSM-rate), rather than a price as in Sect. 17.2.1.2 of this section. The volatility $\sigma(i)$ is relative to the forward Libor $L(t,T_i)$, its square is the mean of the instantaneous variances of that rate between 0 and $T_i$. Attention must be paid to the notation: $C_i$ is the price of the caplet with expiration $T_{i+1}$ that starts at $T_i$.

In practice, however, only the implicit volatilities of the *caps*, and not of the *caplets*, are quoted on the market. Nevertheless, given that a cap is the sum of caplets, one may extract the volatilities of the caplets from the volatilities of the caps.

### (c) *Volatilities implicit in Libor caps*

First notice that, if the market deems the volatility of interest rates to be constant throughout the duration of the cap (a flat volatility term structure), then trivially the volatility of each caplet is equal to that of the cap and to that of any cap of duration smaller than or equal to that of the said cap. Since, in fact, this is not the case, a volatility term structure must be estimated. One may actually define two curves: one for the *flat volatilities* and one for the *spot* volatilities, and both are used in practice. Their precise definitions, the way to compute them and to go from one to the other are the subject of this subsection.

To avoid confusions in understanding what follows, the notation needs to be defined carefully. Recall that the $N+1$ dates, $\{T_0, T_1, ..., T_N\}$, define the structure of $N$ tenors $\{(T_0, T_1), (T_1, T_2), ..., (T_{N-1}, T_N)\}$ ( tenor structure), where $T_N$ denotes the maturity of the longest liquid cap.

Here we consider, to simplify, that the present date is $t = 0$, and $T_0 > 0$ denotes the first reset date.

## 17.2 Models Grounded on the Dynamics of Forward Rates

Any cap with expiration $T_n \leq T_N$ is thus composed of $n$ caplets with start dates $T_i$, $\{i = 0, \ldots, n - 1\}$ and respective prices $C_0, \ldots, C_{n-1}$. As a caplet is defined by its start date, we identify the cap by *the start date of its final caplet* ($T_n$ for an expiration date $T_{n+1}$). As a result, cap $n$ is relevant for the period $T_0, T_{n+1}$ and includes the $n + 1$ caplets $(T_0, T_1), \ldots, (T_n, T_{n+1})$ denoted by 0, 1,.., $n$. The market value of a cap on date $t$, denoted $\Lambda_n(0)$, is obtained by applying Black's formula (17.30) to each component of the cap. Let us denote by $C_i(0, \sigma)$ the price at 0 of the caplet with start date $T_i$ calculated with Eq. (17.30) and a volatility $\sigma$. Since the value of the cap $n$ is a sum of $n + 1$ values of caplets, on can compute, from the observed value $\Lambda_n(0)$ of the cap, a unique implicit volatility $\sigma^c_{flat}(n)$ such that

$$\Lambda_n(0) = \sum_{i=0}^{n} C_i\left(0, \sigma^c_{flat}(n)\right). \tag{17.31}$$

It is important to note that the value of the $n+1$ caplets is here computed with the *same volatility* $\sigma^c_{flat}(n)$, which explains why it is called *flat volatility* by practitioners. The bijection between $\Lambda_n$ and $\sigma^c_{flat}(n)$ allows quoting caps by their volatilities (called quoted volatilities). There is a flat volatility for *each* cap maturity; as the $N$ prices $\Lambda_n(0)$ ($n = 0, 1, \ldots, N-1$) are observed one may compute $N$ flat volatilities $\sigma^c_{flat}(n)$.

Alternatively, one can determine, from the $N$ quoted market prices $\Lambda_n(0)$ ($n = 0, \ldots, N-1$) of $N$ different caps, the $N$ volatilities $\sigma_{spot}(i)$ of the caplets $C_i$ ($i = 0, \ldots, N-1$), called *spot* volatilities. The curve for $\sigma_{spot}(i)$ is compatible with that for $\sigma^c_{flat}(i)$ and one may deduce, as we will show, one from the other. The method is similar to that used to extract ZC rates from the bullet bond yield curve.

The quoted flat volatilities are in fact at-the-money forward, so that the cap's strike $k$ is the forward swap rate (see the paragraph d) below), here denoted by $s^{n+1}(0)$, for the swap negotiated on date 0 for a start at $T_0$ and a final expiration at $T_{n+1}$.

On the one hand, applying Eq. (17.30) and using $\sigma^c_{flat}(n)$ allows writing for the cap $n$

$$\Lambda_n(0) = \sum_{i=0}^{n} C_i\left(0, \sigma^c_{flat}(n)\right)$$
$$= \sum_{i=0}^{n} DB(0, T_{i+1}) Black\left(L(0, T_i), s^{n+1}(0), \sigma^c_{flat}(n)\right). \tag{17.32}$$

On the other hand, Eq. (17.30), used with volatility $\sigma_{spot}(i)$ and $s^{i+1}(0)$ as the given strike, gives for the caplet $i$, regardless of which cap it is part of,

$$C_i\left(0, \sigma_{spot}(i)\right) = C_i = DB(0, T_{i+1}) Black\left(L(0, T_i), s^{i+1}(0), \sigma_{spot}(i)\right). \tag{17.33}$$

The $N$ volatilities $\sigma_{spot}(i)$ we seek are those that make Eq. (17.33) compatible with the $N$ observed cap values $\Lambda_n(0)$.

742                    17 Modeling Interest Rates and Options on Interest Rates

A very simple algorithm consists in proceeding in two steps.

– Step 1: solve the $N$ equations in $N$ unknowns $C_0, C_1, \ldots, C_{N-1}$:

$$\Lambda_n(0) = \sum_{i=0}^{n} C_i \text{ for } n = 0, \ldots, N-1. \tag{17.34}$$

The solution to this system is obvious, and we can check that

$$C_n = \Lambda_n(0) - \Lambda_{n-1}(0), \text{ for } n = 1, \ldots, N-1 \text{ and } C_0 = \Lambda_0(0). \tag{17.35}$$

– Step 2: with the $N$ values $C_0, \ldots, C_{N-1}$, we invert the $N$ equations

$$C_i = DB(0, T_{i+1}) Black\big(L(0, T_i), s^{i+1}(0), \sigma_{spot}(i)\big),$$

to obtain the $\sigma_{spot}(i)$ for $i = 0, \ldots, N-1$.

---

### Example

Consider, to simplify, date $t = T_0 = 0$ (the caps start today) and three caps of respective maturities $T_n$ of 2, 3, and 4 quarters, all made up of synchronous quarterly caplets. The first cap thus contains 2 caplets (0 and 1), the second 3 (0, 1, and 2) and the last 4 (0, 1, 2, and 3). The volatility of the caplet 0 included in *any* of these caps is zero, its value being known on the present date $t = 0$ since one knows the relevant forward Libor rate on this date. We thus work with the values $\Lambda_n(0)$ of these caps *amputated* of their first caplet (caplet 0). The volatilities $\sigma_{spot}(i)$ of the three caplets $\{i = 1, 2 \text{ and } 3\}$ are computed successively, in increasing order of $i$, as follows (here, $D = 0.25$):

– The desired volatility $\sigma_{spot}(1)$ is such that

$$C_1\big(0, \sigma_{spot}(1)\big) = 0, 25\, B(0, 2)\, Black\big(L(0, 1), s^2(0), \sigma_{spot}(1)\big) =$$
$$C_1\big(0, \sigma^c_{flat}(1)\big) = 0, 25\, B(0, 2)\, Black\big(L(0, 1), s^2(0), \sigma^c_{flat}(1)\big) = \Lambda_1(0)\overset{.}{\phantom{,}}$$

This gives immediately $\sigma_{spot}(1) = \sigma^c_{flat}(1)$.

– The volatility $\sigma_{spot}(2)$ is computed from the quoted prices $\Lambda_1(0)$ and $\Lambda_2(0)$ of the first and second caps. It must be such that

*(continued)*

## 17.2 Models Grounded on the Dynamics of Forward Rates

$$\Lambda_2(0) = C_1(0) + C_2(0) = \Lambda_1(0) + C_2(0),$$

which gives $C_2(0) = \Lambda_2(0) - \Lambda_1(0)$ and therefore, inverting equation (17.33) with $i = 2$, the desired value $\sigma_{spot}(2)$.

- The volatility $\sigma_{spot}(3)$ is computed from the quoted prices $\Lambda_2(0)$ and $\Lambda_3(0)$ of the second and third caps:

$$\Lambda_3(0) = C_1(0) + C_2(0) + C_3(0) = \Lambda_2(0) + C_3(0),$$

which yields $C_3(0) = \Lambda_3(0) - \Lambda_2(0)$ and thus, by inverting equation (17.33) with $i = 3$, the volatility $\sigma_{spot}(3)$.

In this way we obtain successively the different spot volatilities for the caplets, using the quoted prices of caps with increasing maturities.

It can be useful or even compulsory, for example for valuing more complex instruments in this framework, to estimate the functions $\gamma_i(s)$ of the instantaneous volatilities entering the calculation of the spot volatilities $\sigma(i)$: $\sigma_{spot}^2(i) = \frac{1}{T_i}\int_0^{T_i}\gamma_i(s)^2 ds$ (see Eq. 17.30). Assume, for example, that the $\gamma_i(s)$ are piecewise constant between two reset dates. We then have

$$\sigma_{spot}^2(i)T_i = \int_0^{T_i}\gamma_i(s)^2 ds = \int_0^{T_0}\gamma_i(s)^2 ds + \sum_{k=1}^{i}\int_{T_{k-1}}^{T_k}\gamma_i(s)^2 ds$$

$$= T_0\lambda_0^2 + \sum_{k=1}^{i}D\lambda_k^2, \tag{17.36}$$

where $\lambda_k$ denotes the constant spot volatility (annualized) of the forward Libor rate $L(t, T_i)$ between the two reset dates $T_{k-1}$ and $T_k$. We then continue by recurrence.

For example, let the $\sigma_{spot}(i)$ be inferred from market data as previously explained, $i = 0, 1, \ldots, N - 1$. We first obtain $\lambda_0 = \sigma_{spot}(0)$, then $\lambda_1$ from $\sigma_{spot}^2(1)T_1 = T_0\lambda_0^2 + D\lambda_1^2$, then $\lambda_2$ from $\sigma_{spot}^2(2)T_2 = T_0\lambda_0^2 + D(\lambda_1^2 + \lambda_2^2)$, and so on.

### (d) *Valuation of a swaption: problems with using the LMM model*

In this chapter we use the notation $s(t,T_i,T_n)$ or simply $s^{i,n}(t)$, for the forward swap rate on date $t$ ($0 \le t \le T_1$) for the period $(T_i, T_n)$, $\{i = 1, , n-1\}$. From this, $s(T_i, T_i, T_n) \equiv s^{i,n}(T_i)$ or simply $s(T_i)$ denotes the spot swap rate on $T_i$ (the forward rate is the spot rate at expiration $T_i$). These *swap rates thus have different start dates but a common maturity* $T_n$. We continue to assume, for notational simplicity, that the accrual periods $D$ are of equal lengths: $D = T_{i+1} - T_i$. These rates are then the underlyings for the swaptions described in Chap. 16, whose market is extremely liquid and whose mechanism we briefly recall.

Swaptions allow their holders to enter into a swap by paying the fixed rate $k$, the strike of the swaption, and receiving a floating rate (for a *payer* swaption; the opposite is true for a *receiver* swaption).

If upon the expiration date $T_i$ of a *payer* swaption its holder exercises it, she will have to pay cash amounts computed at the fixed rate $k$, and receive payments computed using the floating rate until the swap matures on date $T_n$. So she can enter into a *receiver* swap on the spot market which allows her to cancel the floating leg.[14] This transaction thus results in paying, from date $T_i$, fixed amounts at the known fixed rate $k$ and receiving fixed amounts at the new fixed-rate (obviously unknown at the start) $s^{i,n}(T_i) = s(T_i)$, which are certainly seen from $T_i$. The cash flow series thus generated is, therefore: $D(s(T_i) - k)$ on dates $T_{i+1}, \cdots, T_n$. Exercising the swaption is profitable only if $s(T_i) > k$ and this series' value on date $T_i$ writes

$$(s(T_i)k) \sum_{j=i+1}^{n} DB(T_i, T_j). \tag{17.37}$$

As seen in Chap. 16 (Sect. 16.2.3), the market values swaptions with the BSM-rate formula, which assumes that the underlying, the forward swap rate, is lognormal. Unfortunately, this assumption is *incompatible* with supposing that the forward Libor rate is lognormal. Indeed, the forward swap rate $s^{i,n}(t)$ is fixed[15] so as to equalize the values of the fixed leg $\left( s^{i,n}(t) \sum_{j=i+1}^{n} DB(t, T_j) \right)$ and the floating leg $\left( \sum_{j=i+1}^{n} DL(t, T_{j-1}) B(t, T_j) \right)$, whence

$$s^{i,n}(t) = \frac{\sum_{j=i+1}^{n} DL(t, T_{j-1}) B(t, T_j)}{\sum_{j=i+1}^{n} DB(t, T_j)} \equiv \sum_{j=i+1}^{n} \alpha_j(t) L(t, T_{j-1}), \quad \text{for } t$$

$$\in [0, T_i] \text{ and } i \leq (n-1). \tag{17.38}$$

Thus, the forward swap rate is a weighted *sum* of forward Libor rates. As a result, if the latter are lognormal, the former cannot be.[16] To use nevertheless the LMM model for swaptions, together with caps and floors, several authors (such as Rebonato (1999), Hull and White (2000), Jäckel and Rebonato (2003), and Kawai

---

[14] She pays the floating rate and receives the forward swap rate quoted on the market at that instant.
[15] See Sect. 17.2.2.2 below for more details.
[16] It is the *product* of lognormal variables that is lognormal.

## 17.2 Models Grounded on the Dynamics of Forward Rates

(2002)) have suggested approximations, some of which are presented in Appendix 2, Sect. 2.

It is better, however, to associate the BSM formula for valuing swaptions with the Swap Market Model which we analyze in Sect. 17.2.2.2.

### (e) *Difficulties in implementing the LMM* [17]

To use the model, the following inputs are required:

– The prices of many liquid bonds with different maturities.
– All the forward Libor rates.
– The instantaneous variances and covariances of the forward rates.

It is obviously the third inputs that are difficult to obtain. This is the whole point of *calibration*. The relevant variances and covariances *should be estimated or chosen* so that LMM provides prices for the most liquid vanilla rate instruments that differ as little as possible from actual market quotes. To achieve this, the prices of the most liquid caplets and swaptions are used.

As we saw in Sect. 17.2.2.1, item c, this calibration is easy enough in the one-factor case, since then the different forward rates are perfectly correlated locally (which eliminates the problem of estimating covariances) and it is sufficient to assume volatilities piecewise constant between reset dates (or to give them some convenient parametric form) to implement a very simple computational algorithm.

However, since practitioners use multi-factor models (which justifies using the LMM framework), calibration is much more complicated and lies outside the scope of this book. On the one hand, covariances, which must be estimated, turn up in valuations formulas; on the other, the variances and volatilities need to be decomposed into as many elements as there are sources of risk assumed (three-factor models are often proposed, but this is not necessarily optimal) or estimated. For example, one can use principal component analysis (PCA) to determine both the factors and the sensitivities (*factor loadings*) of each forward Libor rate to these factors, which then allows to express the forward rate dynamics and to compute their variances and covariances. Another solution consists in finding parameters that minimize the sum of the squares of deviations of the market prices of the liquid instruments from the prices computed using the parameters. Many alternate methods have been suggested and are in use.

### 17.2.2.2 The Swap Market Model (SMM)

First, we derive the SMM market model under different equivalent probability measures. We then link it to the BSM model, which leads to the closed-form solution

---

[17] What has been asserted here can be applied, mutatis mutandis, to the Swap market model (SMM) analyzed below. We will only touch upon the subject, letting the interested reader consult the references cited at the end of the chapter for technical details.

746                    17  Modeling Interest Rates and Options on Interest Rates

for swaption prices already given in Chap. 16. Finally, we point out how the use of this model for valuing caps leads to inconsistency.

## (a) *The dynamics of the forward swap rate*

Jamshidian (1997) has developed a model of the swap market in which the forward swap rates are assumed to be lognormal. This assumption, as we have seen, is consistent with market practice which uses Black's model to value European swaptions.

We continue, as in Sect. 17.2.2.1, item c, to study the forward swap rates $s(t, T_i, T_n) = s^{i,n}(t)]_{i = 1, .., n - 1}$, with different start dates $T_i$ and a common maturity $T_n$.[18]

The forward swap rate, that gives the swap an initial value of zero, satisfies, for $t \in [0, T_i]$, Eq. (17.15) from the previous Chap. 16, that we reproduce here using the notation of this chapter:

$$s^{i,n}(t) = \frac{B(t, T_i) - B(t, T_n)}{\sum\limits_{j=i+1}^{n} DB(t, T_j)} \qquad i \le n - 1, \tag{17.39}$$

where $D$ is the length of the period (tenor) delimited by $T_i$ and $T_{i+1}$, which is assumed to be constant for all $i$. Indeed, the swap rate (spot or forward) allows us to equalize the values of the fixed leg $\left[ s^{i,n}(t) \sum\limits_{j=i+1}^{n} DB(t, T_j) \right]$ and the floating leg $[B(t, T_i) - B(t, T_n)]$.[19]

In what follows (and in Appendix 2, Sect. 3), we show how to obtain the dynamics of the forward swap rate under different probability measures.

First, let us denote by $a_{i,n}(t)$ the process for the swap's fixed leg

$$a_{i,n}(t) = D \sum\limits_{j=i+1}^{n} B(t, T_j). \tag{17.40}$$

We need the price on date $t$ of an asset that delivers a constant cash flow of \$1 on the dates $T_{i+1}, \ldots, T_n$ which we have called, in Chap. 16 with an abuse of language, an "annuity" and used as the numeraire. Let us consider the probability measure $Q^{i, n}$ which makes prices expressed in the numeraire $a_{i,n}(t)$ martingales (it is the *forward swap measure*). We mentioned in Chap. 16, and showed in that chapter's

---

[18] Other families of forward swap rates can be considered. In the first, the forward swaps have different maturities but same initial date. In the second, more generally, the forward swaps $s^{i,i+h}(t)$ have both different initial and maturity dates. It should be noted that the forward LIBOR rate $L(t, T_i)$ is a special case where $h = 1$.

[19] The floating leg can be valued equivalently by $\sum\limits_{j=i+1}^{n} DL(t, T_{j-1}) B(t, T_j)$ (see Chaps. 7 and 16).

## 17.2 Models Grounded on the Dynamics of Forward Rates

Appendix,[20] that the forward swap rate process $s^{i,n}(t)$ is a martingale under $Q^{i,n}$ whose dynamics can be written

$$ds^{i,n}(t) = s^{i,n}(t)\nu_{i,n}(t)dW^{i,n}(t) \tag{17.41}$$

where $\nu_{i,n}(t)$ is the rate's instantaneous volatility.

Under the measure $Q^{n-1,n}$, a particular measure of $Q^{i,n}$, for which $DB(t,T_n)$ is the associated numeraire, the dynamics of $s^{n-1,n}(t)$ becomes

$$ds^{n-1,n}(t) = s^{n-1,n}(t)\nu_{n-1,n}(t)dW^{n-1,n}(t). \tag{17.42}$$

Note that the forward swap measure $Q^{n-1,n}$ is homologous to the terminal forward measure $Q^n$ (for which $B(t,T_n)$ is the chosen numeraire).

### (b) The possible probability measures

Implementing the SMM, like the LMM, may require for the valuation of certain interest rate derivatives, the specification of different rate processes *under a single probability measure*. A number of choices are possible, at the user's whim: either the *terminal forward* measure $Q^n$, or a *general forward* swap measure (that is, not specific to a given forward swap rate) $Q^{q,n}$, or even the *risk-neutral* measure $Q$. To allow the reader to make an efficient choice and benefit from the flexibility it brings with it, we introduce these three measures briefly in Appendix 2, Sect. 3.

### (c) Valuation of a swaption

The SMM is compatible with the BSM-rate model since the *underlying* $s(T_i)$ is assumed to be lognormal. We thus take up again the closed-form solution for the value of a swaption given in Proposition 5 of Chap. 16, using the notation of the SMM.

The price on date $t$ of a swaption with maturity $T_i$ is given in the SMM framework by

$$OS(t) = a_{i,n}(t)\left[s^{i,n}(t)N(d_1) - kN(d_2)\right] \tag{17.43}$$

$$\text{with}: \quad d_{1,2} = \frac{\ln\left(s^{i,n}(t)/k\right) \pm \frac{1}{2}\left(\sigma_{i,n}(t)\right)^2}{\sigma_{i,n}(t)} \quad \text{and} \quad (\sigma_{i,n})^2(t) = \int_t^{T_i}\left(\nu_{i,n}(s)\right)^2 ds,$$

$s^{i,n}(t)$ being given by Eq. (17.39).

Note that these formulas are closer to the standard results of BSM if we express them as functions of $\sigma_{i,n}^{average}$ by letting $\sigma_{i,n}(t) = \sigma_{i,n}^{average}\sqrt{T_i - t}$.

---

[20] The probability $Q^{i,n}$ and its numeraire $a_{i,n}$ were defined and analyzed in the Appendix to Chap. 16 with the simplified notation $Q^a$ and $a$, respectively.

## (d) *Valuation of a caplet*

As the LMM model, the SMM model can be used to value caplets. Obviously, this leads to the same type of inconsistency as that encountered in valuing a swaption in the LMM framework. This is because the Libor forward rate is not lognormal as it was in the LMM. Indeed, by Eq. (17.38), the forward swap rate is a weighted mean of forward Libor rates, and conversely, a forward Libor rate can be written as a linear combination of two forward swap rates. If the forward swap rates are lognormal, the forward Libor rates cannot be.

However, the inconsistency mentioned can also be circumvented by an approximation in the same way as in Sect. 17.2.2.1, item d.[21]

### 17.2.2.3 Numerical Estimates and Extension of the Basic Models

The LMM and SMM market models both assume that all the underlying rates follow lognormal laws. Since the Libor and forward swap rates are determined from zero-coupon prices, it is standard to use those prices to check lognormality.

Empirical tests, among them the Jarque–Bera test, indicate that the distribution's tails are in fact thick, a characteristic of leptokurtic distributions, and also are asymmetric on the left, for all maturities. The geometric Browning motion can thus be considered as a partly inappropriate process for describing the evolution of the yields of zero-coupon bonds, and thus of the Libor and forward swap rates. Besides, as for options on stocks, or on stock-market indices, options on interest rates also exhibit a volatility *smile* (*skew* in fact) reflecting the dependence of the implicit volatility on the strike. One finds additionally a term structure effect, according to which the implied volatility depends on the option's maturity, for any given strike. Taking into account both these effects, strike and time to maturity, allows constructing, or observing, *volatility surfaces*. For example, it is common to observe that the higher is the ratio strike/underlying rate (the inverse of the option's *moneyness*) the lower is the implicit volatility of the cap (the *skew* property). In addition, the further off is the option maturity, the smaller is the volatility.[22]

It, therefore, seems desirable, by modeling in a different way the dynamics of interest rates, to refine the valuation of caps, floors, and swaptions to better reproduce the observed market prices. Many alternatives have been pursued. Some have specified a particular function for the instantaneous volatility, for example, a variance with constant elasticity (Andersen and Andreasen (2000)). Andersen and Brotherton-Ratcliffe (2005) and Wu and Zhang (2006) modeled interest rate volatility as a stochastic process, Glasserman and Kou (2003) and Glasserman and Merener (2003) have proposed an extended version of the SMM incorporating jumps in the

---

[21] Thus, one may generalize Rebonato's approximation (1999) as used in Galluccio et al. (2007).

[22] This characteristic, in contrast with the skew, depends on the current average value of the volatility; if it is rather high, the implicit volatility tends to decrease with the option maturity, but if it is rather small, the implicit volatility tends to increase with the option maturity. This reveals a mean-reverting process on the part of the volatility.

## 17.3 Summary

dynamics of interest rates. These models, or a combination of them, however, have experienced only a rather limited success (see Jarrow et al., 2007).[23] To be able to generate Libor rates that are negative (but bounded), as well as a volatility smile, the LMM has been generalized to the DDLMM (Displaced Diffusion LMM) where the dynamics of Libor rates displaced by a displacement factor still is lognormal (e.g., Brigo & Mercurio, 2001).[24] This model has been generalized to a stochastic volatility by Joshi and Rebonato (2003). Alternatively, we can give up the framework of the LMM and SMM to use a class of models originally developed by Hunt et al. (1998). These models, called Markov functional models, assume that interest rates are determined by a small number of factors obeying a specific Markov process. Finally, we indicated at the end of Chap. 11 (Sect. 11.2.8.3) that the SABR model with stochastic volatility due to Hagan et al. (2002), is well suited to the evolution of forward prices of rate instruments. More specifically to the swaps and swaptions markets, following Bergomi's stochastic volatility model (2005, 2016), Oya (2018a) uses variance swap contracts on interest rates to obtain a two-factor model of swap rates and forward variance curve under the appropriate "annuity" measure that seems quite efficient in pricing and hedging vanilla and Bermuda swaptions.

## 17.3  Summary

- Interest rate models based on the dynamics of the spot rate (Vasicek, Hull and White, and Cox, Ingersoll and Ross) can be distinguished from those based on the dynamics of the forward rate (Heath-Jarrow-Morton, Libor and Swap Market Models).
- One-factor models of the spot rate $r(t)$ are grounded on its RN dynamics: $dr = m(t, r)\, dt + \sigma(t, r)dW$. They differ by the assumed functions $m(t, r)$ and $\sigma(t, r)$.

  ZC bond prices then are given by $B_T(t,r) = E\left\{ \exp\left( -\int_t^T r(u)du\, |r(t) = r \right) \right\}$,

  where $Et\{.\}$ is the RN expectation taken at $t$, knowing that $r(t) = r$.
- The Vasicek model assumes that $r(t)$ is an Ornstein-Uhlenbeck (mean-reverting) process: $m(t, r) = a(k - r)$ and $\sigma(t, r) = \sigma$, where $k$ is the long-term "normal rate".
- This model implies a Gaussian $r$ (that can thus become negative), a ZC rate affine linear in r also normal, hence ZC prices log-normal functions of r and maturity. The BSM model can thus be applied to interest rate options and yields closed-form solutions.

---

[23] For example, swaptions having generally long maturities and jumps affecting essentially short-term rates, modeling the underlying swap rate dynamics as a jump-diffusion process may not be really appropriate for these options.

[24] The idea is to consider that the displaced (by a constant factor called shift) process of each forward Libor rate is lognormal under its own martingale measure.

- The more complex Cox–Ingersoll–Ross (CIR) model rests on a non-Gaussian process:

  $dr = a[k - r] \, dt + \sigma \sqrt{r} dW(t)$,

  where the instantaneous variance of $r(t)$ thus is not constant but equals $\sigma^2 r$.
- Multifactor models of the spot rate are grounded on the following RN dynamics: $dr = m(t, \underline{X}t) \, dt + \sigma(t, \underline{X}t)dWr$, where $\underline{X}$ is a multi-dimensional factor, $r(t)$ being usually one of the factors.

  ZC prices BT$(t)$ are then functions of $\underline{X}t$ and their movements (and those of their yields $rT$) are all locally imperfectly correlated (in contrast with those of one-factor models).
- Adaptation of the previous one-factor models allows fitting the observed initial yield curve. Making the drift of the short-term rate time-varying provides the desired adjustment.
- For instance, Hull and White generalize the Vasicek model by assuming: $dr = (k(t) - a \, r) \, dt + \sigma \, dW(t)$. The model conforms to the initial yield curve if $k(t)$ is consistent with the current instantaneous forward-yield curve $f_t(0)$ and its slope $f_t'(0)$: $k(t) = a \, f_t(0) + f_t'(0) + \frac{\sigma^2}{2a}\left(1 - e^{-2at}\right)$, often approximated by $a(\, f_t(0) - r_t) + f_t'(0)$. Pricing models of ZC bonds and of options on ZC bonds calibrated with the current yield curve can then be obtained.
- Heath, Jarrow, and Morton (HJM) model the instantaneous forward rate dynamics within a very general framework. They show that restrictions on the drifts of the RN processes governing these forward rates must be imposed so that the ZC rate curve is free of arbitrage. Furthermore, their model recovers the exact initial prices of all ZC bonds.
- Assuming that the instantaneous forward rate $f_T(t)$ obeys the RN diffusion process $f_T(t) = f_T(0) + \int_0^t \mu(s, T, \underline{X}(s))ds + \int_0^t \sigma(s, T, \underline{X}(s))dW(s)$,

  their fundamental result is that the drift and the volatility of the forward rate $f_T(t)$ are linked, under no-arbitrage, by the relation $\mu(t, T, \underline{X}(t)) = \sigma(t, T, \underline{X}(t)) \int_t^T \sigma(t, s, \underline{X}(t))ds$.
- This very general no-arbitrage condition must be satisfied by all arbitrage-free interest rate models, which then can be seen as special cases of HJM. For example, the Vasicek/Hull-White model $(dr = a(k(t) - r)dt + \sigma dW)$ obtains by setting $\sigma(t,T, \underline{X}(t)) = \sigma \, e^{-a(T - t)}$.
- The HJM model generates *closed-form BSM-price solutions* for options on spot bonds and on forward and futures bond contracts as well as for the relationship between futures and forward bond prices.
- "Market models" using the *observable discrete* market rates, the forward Libor and the forward Swap rates, avoid resorting to the estimation methods needed to obtain the instantaneous rate and allows direct valuation of interest rate derivatives.
- In the *Libor Market Model*, the forward Libor rates are linked to the ZC bond prices by: $1 + DL(t, T_i, T_{i+1}) = \frac{B(t, T_i)}{B(t, T_{i+1})}$, where $D$ is the length of the period between $T_i$ and $T_{i + 1}$.

# Appendix 1

- As $L(t, T_i, T_{i+1})$ is assumed lognormal and is a martingale under the *forward-neutral* measure $Q^{i+1}$, *caplet and floorlet* prices obtain from *BSM-rate-like formulas.*
- One obtains successively the different *spot* volatilities for the caplets (or floorlets), using the quoted prices of caps (or floors) with increasing maturities, in a way similar to that of the extraction of ZC rates from the yield curve.
- Assuming the forward Libor rate lognormal is *incompatible* with assuming the forward swap rate also lognormal. Swaptions thus cannot be valued correctly with the Libor Market Model. An adjustment is needed to allow for an approximate value.
- The valuation of swaptions is correct, however, if one assumes that the forward swap rate is lognormal, which leads to the *Swap Market Model*. The forward swap rate, which gives the swap an initial value of zero, satisfies, for $t \in [0, T_i]$,

$$s^{i,n}(t) = \frac{B(t, T_i) - B(t, T_n)}{\sum_{j=i+1}^{n} DB(t, T_j)} \qquad i \leq n - 1.$$

- As $s^{i,n}(t)$ is a martingale under the *forward swap* measure $Q^{i,n}$, *swaption* prices obtain from *BSM-rate-like formulas.* By symmetry, the model needs an adjustment to produce approximate values for caplets and floorlets.
- The smile (or skew) effect and the term structure of interest rate volatilities produce implicit *volatility surfaces*. To account for these, researchers have developed various extensions to the original "market models". They assume for instance a variance with constant elasticity, a stochastic volatility, jumps in the dynamics of interest rates, or displaced lognormal distributions. Alternatively, the framework of the "market models" can be replaced by Markov functional models where interest rates are determined by factors obeying specific Markov processes.

---

## Appendix 1

### *The Vasicek Model

We first formulate a technical lemma about the distribution for the short-term rate $r(t)$ when it obeys an Ornstein–Uhlenbeck process, then we prove Vasicek's formulas (17.4) and (17.5).

**Lemma 1**
Consider the Ornstein-Uhlenbeck process given by

$$dr = a(k - r)dt - \sigma dW \Leftrightarrow r(s)$$

$$= (r_0 - k) \exp[-as] + k - \sigma \exp(-as) \int_0^s \exp(au)dW(u).$$

(a) $r(s)$ is normally distributed and its expectation and variance are, respectively:

$$E\left[r(u)/r(t)=r\right]=e^{-a(u-t)}\left[r-k\right]+k; \qquad (17.44)$$

$$\text{Variance}\left[r(u)\mid r(t)=r\right]=\sigma^2 e^{-2au}\sigma^2 e^{-2au}\int_t^u e^{2az}dz$$

$$=\sigma^2(1-e^{-2a(u-t)})/2a. \qquad (17.45)$$

(b) $\int_t^T r(s)ds$ is normal and we can write:

$$\int_t^T r(s)ds=\alpha(t)-\int_t^T \sigma_{T-s}(s)dW(s), \text{with}: \sigma_{T-t}=\frac{\sigma\left(1-e^{-a(T-t)}\right)}{a}. \qquad (17.46)$$

$$\int_t^T r(s)ds \sim N\left[\alpha(t),\ \eta^2(t)\right], \text{with}: \qquad (17.47)$$

$$\alpha(t)=k(T-t)+(r(t)-k)\frac{\sigma_{T-t}}{\sigma}$$

$$\eta^2(t)=\frac{\sigma^2}{a^2}\left[(T-t)-2\frac{\sigma_{T-t}}{\sigma}+\frac{\sigma_{2(T-t)}}{2\sigma}\right]$$

$$=\frac{\sigma^2}{a^2}\left[(T-t)-\frac{\sigma_{T-t}}{\sigma}-a\frac{\sigma_{T-t}^2}{2\sigma^2}\right].$$

**Proof**

(a) The solution to an Ornstein–Uhlenbeck process is well known (see Chap. 18, Sect. 18.4.2) and implies that, for $s>t$, $r(s)$ is given by

$$r(s)=(r(t)-k)\exp\left[-a(s-t)\right]+k-\sigma\exp\left(-as\right)\int_t^s \exp\left(au\right)dW(u),$$

from which we deduce Eq. (17.44).

In addition, since as a sum of independent Gaussians $r(s)$ is normally distributed, we have

# Appendix 1

$$\text{Variance}[r(u) \mid r(t) = r] = \sigma^2 e^{-2au} \int_t^u e^{2az} dz = \sigma^2 (1 - e^{-2a(u-t)})/2a, \text{ which proves}$$

Eq. (17.45).

(b) From (a) it results that

$$\int_t^T r(s)ds = (r(t) - k) \int_t^T \exp[-a(s-t)]ds + k(T-t) - \sigma \int_t^T \exp(-as) \int_t^s \exp(au)dW(u)ds$$

$$= (r(t) - k)\frac{1 - \exp[-a(T-t)]}{a} + k(T-t) - \sigma \int_t^T \int_t^s \exp[a(u-s)]dW(u)ds$$

$$= (r(t) - k)\frac{1 - \exp[-a(T-t)]}{a} + k(T-t) - \sigma \int_t^T \frac{1 - \exp[-a(T-u)]}{a} dW(u)$$

This implies Eq. (17.46) since $(1-\exp(-a(T-t))/a = \sigma_{T-t}/\sigma$.

Furthermore, $\sigma \int_t^T \frac{1-\exp[-a(T-u)]}{a} dW(u)$ is normally distributed with mean zero and variance equal to

$$\sigma^2 \int_t^T \left(\frac{1 - \exp[-a(T-u)]}{a}\right)^2 du = \frac{\sigma^2}{a^2}\left[(T-t) - 2\frac{\sigma_{T-t}}{\sigma} + \frac{\sigma_{2(T-t)}}{2\sigma}\right] = \eta^2(t)$$

which is the first expression for $\eta^2(t)$ presented in Eq. (17.47); to obtain the second, we first check that

$$\sigma_{T-t}^2 = \frac{\sigma^2 (1 - e^{-a(T-t)})^2}{a^2} = \frac{\sigma}{a}\left(2\sigma_{T-t} - \sigma_{2(T-t)}\right), \text{ or } \sigma_{2(T-t)} = -\frac{a}{\sigma}\sigma_{T-t}^2 + 2\sigma_{T-t};$$

We then substitute this value for $\sigma_{2(T-t)}$ into the first expression for $\eta^2(t)$ to obtain $\eta^2(t) = \frac{\sigma^2}{a^2}\left[(T-t) - \frac{\sigma_{T-t}}{\sigma} - a\frac{\sigma_{T-t}^2}{2\sigma^2}\right]$, which is the second expression.

The price of a ZC bond follows immediately from this lemma. Indeed, we have under the RN probability

$$B_T(t,r) = E\left\{\exp\left(-\int_t^T r(s)ds\right)\right\} \text{ or } \int_t^T r(s)ds \sim N[\alpha(t), \eta^2(t)].$$

Then, from the generating function of the normal law, we obtain

$$B_T(t,r) = \exp\left[-\alpha(t) + \frac{1}{2}\eta^2(t)\right].$$

Substituting in this equation the expressions for $\alpha(t)$ and $\eta(t)$ given in Eq. (17.47) yield the results (17.4) and (17.5) in the main text.

This lemma also allows valuing options on ZC bonds with a BSM model since the ZC price is lognormal. To calculate the value on date 0 of an option with expiration date $T$ on a ZC with maturity $T+D$ simply requires computing the standard deviation of $\ln B_{T+D}(T) = \ln\{A(D) \exp[-(\sigma_D/\sigma)r(T)]\} = -(\sigma_D/\sigma)\, r(T) + \ln A(D)$. By Eq. (17.45), this standard deviation, seen from date 0, is

$$\sigma_{\ln B} = \sigma_D \sqrt{\frac{1 - e^{-2aT}}{2a}} = \sigma \frac{1 - e^{-aD}}{a} \sqrt{\frac{1 - e^{-2aT}}{2a}}.$$

---

## Appendix 2

## *The LMM and SMM Models

### 1 Proof of Eq. (17.28), the Dynamics of $L(t,T_i)$ Under the Final Forward Measure $Q^n$

Let us start with the differentiated form of Eq. (17.26): $1 + D dL(t, T_i) = d\left(\frac{B(t,T_i)}{B(t,T_{i+1})}\right)$.

The right-hand side of the equality can be expanded as follows:

$$d\frac{B(t,T_i)}{B(t,T_{i+1})} = \frac{1}{B(t,T_{i+1})} dB(t,T_i) + B(t,T_i)d\left(\frac{1}{B(t,T_{i+1})}\right)$$

$$+ d\left\langle B(t,T_i), \frac{1}{B(t,T_{i+1})}\right\rangle$$

Under the probability $Q$, we have $\frac{dB(t,T_i)}{B(t,T_i)} = r(t)dt + \sigma(t,T_i)\,dW(t)$.

Thus, we obtain

(i)
$$dL(t,T_i) = \frac{1}{D}\frac{B(t,T_i)}{B(t,T_{i+1})}[\sigma(t,T_{i+1})\{\sigma(t,T_{i+1}) - \sigma(t,T_i)\}\,dt - \{\sigma(t,T_{i+1}) - \sigma(t,T_i)\}\,dW(t)].$$

Since $\gamma_i(t)$ is the volatility of the rate $L(t,T_i)$, its dynamics under $Q$ writes

(ii) $dL(t,T_i) = [.]\, dt + \gamma_i(t)L(t,T_i)dW(t)$.

The two diffusion terms in (i) and (ii) are equal, so we have

(iii) $-\frac{1}{D}\frac{B(t,T_i)}{B(t,T_{i+1})}\left(\sigma(t,T_{i+1}) - \sigma(t,T_i)\right) = \gamma_i(t)L(t,T_i)$

from which, using Eq. (17.26) $\left[\frac{B(t,T_i)}{B(t,T_{i+1})} = 1 + DL(t,T_i)\right]$, we have

# Appendix 2

(iv) $\sigma(t, T_{i+1}) - \sigma(t, T_i) = -\frac{D\gamma_i(t)L(t, T_i)}{1+DL(t, T_i)}$.

Iterating this expression yields

(v) $\sigma(t, T_{i+1}) = \sigma(t, T_1) - \sum_{j=1}^{i} \frac{D\gamma_j(t)L(t, T_j)}{1+DL(t, T_j)}$.

The measure $Q^n$ is defined as the Radon–Nikodym derivative given by the ratio of the numeraires (see Proposition 11 of Chap. 19):

$$\frac{dQ^n}{dQ} = \frac{B(t, T_n)}{B(0, T_n) \exp\left(\int_0^t r(s)ds\right)} = \exp\left(\int_0^t \sigma(s, T_n)dW(s) - \frac{1}{2}\int_0^t \sigma^2(s, T_n)ds\right).$$

Applying Girsanov's theorem gives: $dW^n(t) = dW(t) - \sigma(t, T_n)\, dt$.
Using (ii), the dynamics of $L(t, T_i)$ becomes

$$dL(t, T_i) = L(t, T_i)\gamma_i(t)[dW^n(t) + (\sigma(t, T_n) - \sigma(t, T_{i+1}))dt]$$

Combining this with (v), we obtain result Eq. (17.28). Also notice that, by adopting the measure $Q^{i+1}$ instead of $Q^n$, Eq. (17.27) obtains directly from Eq. (17.28).

## 2 Valuation of Swaptions in the LMM Framework

To use the LMM model for valuing swaptions, together with valuing caps, several authors (among them Rebonato (1999), Hull and White (2000), Jäckel and Rebonato (2003), and Kawai (2002)) have proposed approximations for the instantaneous volatility of the forward swap rate as a function of the forward Libor rates and their instantaneous volatilities and correlations. From Eq. (17.38), if the forward swap rate can well be written as a linear combination of forward Libor rates, the difficulty comes from the fact that the weights $\alpha_j(t)$ are random variables.

One may then, for example, assume with Rebonato (1999) that the time variations in the different $\alpha_j$ are negligible in comparison with those of Libor rates, which is what empirical data suggest. This translates into freezing the $\alpha_j$ at their initial values, which greatly simplifies differentiating (17.38). Calculating the quadratic variation of $s^{i,n}$, we obtain[25]

$$d\langle s^{i,n}\rangle_t \approx \sum_{l,j=i}^{n-1} \alpha_{j+1}(0)\gamma_j(t)L(t, T_j)\alpha_{l+1}(0)\gamma_l(t)\rho_{j,l}L(t, T_l)\, dt.$$

---

[25] Recall that the quadratic variation of $X$, denoted by $d<X>_t$, is equal to $\sigma_X^2 dt$.

# 17 Modeling Interest Rates and Options on Interest Rates

Dividing both sides by $(s^{i,n}(t))^2$ and accepting a second approximation by considering the forward Libor rates on initial date $t = 0,$[26] we obtain[27]

$$\frac{d\langle s^{i,n}\rangle_t}{(s^{i,n}(t))^2} \approx \frac{\sum\limits_{l,j=i}^{n-1} \alpha_{j+1}(0)\gamma_j(t)L(0,T_j)\alpha_{l+1}(0)\gamma_l(t)L(0,T_l)}{(s^{i,n}(0))^2} dt.$$

Knowing that the forward swap rates are assumed lognormal, and denoting by $v_{i,n}(t)$ the instantaneous volatility of the rate $s^{i,n}(t)$, the variance of the forward swap rate $(\sigma_{i,n})^2$ writes [28]:

$$(\sigma_{i,n})^2(0) = \int_0^{T_i} (v_{i,n}(u))^2 du$$

$$= \frac{\sum\limits_{l,j=i}^{n-1} \alpha_{j+1}(0)L(0,T_j)\alpha_{l+1}(0)L(0,T_l)}{(s^{i,n}(0))^2} \int_0^{T_i} \gamma_l(u)\gamma_j(u)du.$$

Brigo and Mercurio (2001, chap. 8) test this approximation and that developed by Hull and White (2000), by comparing them with the implicit volatility they obtain by inverting the Monte Carlo price of a swaption where the dynamics of the forward swap rate used is given by our Eq. (17.41). The discrepancy between the approximate volatilities and the Monte Carlo volatility is, in general, less than 2%.

## *3 Three Probability Measures for the SMM Model

We show how to obtain the dynamics of the forward swap rate under the terminal forward swap measure $Q^n$, then using the general forward swap measure $Q^{q,n}$ and then the risk-neutral measure $Q$.

### (a) *The terminal forward measure $Q^n$*

– Defining

$$\text{(i)} \quad \tau_i^n(t) \equiv D \sum_{j=i}^{n-1} \prod_{k=i+1}^{j} \left(1 + Ds^{k,n}(t)\right);$$

with $s^{k,\,n}(t) = 0$ for $k > j$, we show that

---

[26] This translates in obtaining $(s^{i,n}(0))^2$ in the denominator of the r.h.s. of the equation.

[27] In the numerator of the right-hand term of the expression obtained, and under the sum sign, we have, in multifactor models, a multiplicative term $\rho_{jl}$ that takes into account the instantaneous correlation of the relevant Libor rates. Here, all the $\rho_{jl}$ are equal to one since the model has a single factor, which implies a perfect correlation between all the Libor rates. It is especially the presence of such correlations that makes the practical implementation of the models difficult.

[28] Indeed, we have: $\int_0^{T_i} (v_{i,n}(u))^2 du = \int_0^{T_i} (d\ln(s^{i,n}(u)))^2 = \int_0^{T_i} \frac{d\langle s^{i,n}\rangle_t}{(s^{i,n}(u))^2}.$

# Appendix 2

(ii) $\frac{B(t,T_i)}{B(t,T_n)} = 1 + \tau_i^n(t)s^{i,n}(t).$

Indeed, from Eq. (17.39), we can write

$$1 = \widetilde{B}(t,T_i) - s^{i,n}(t)\sum_{j=i+1}^{n} D\widetilde{B}(t,T_j)$$

where $\widetilde{B}(t,T_i) \equiv \frac{B(t,T_i)}{B(t,T_n)}$. In matrix form, we have

$$
\begin{pmatrix}
1 & -Ds^{1,n} & -Ds^{1,n} & \cdots & -Ds^{1,n} \\
0 & 1 & -Ds^{2,n} & \cdots & -Ds^{2,n} \\
\cdots & \cdots & \cdots & \cdots & \cdots \\
0 & \cdots & \cdots & 1 & -Ds^{n-1,n} \\
\cdots & \cdots & \cdots & \cdots & 1
\end{pmatrix}
\begin{pmatrix}
\widetilde{B}(t,T_1) \\
\widetilde{B}(t,T_2) \\
\cdots \\
\widetilde{B}(t,T_{n-1}) \\
\widetilde{B}(t,T_n)
\end{pmatrix}
=
\begin{pmatrix}
1 \\
1 \\
\cdots \\
1 \\
1
\end{pmatrix}.
$$

By inverting this matrix we obtain

$$
\begin{pmatrix}
\widetilde{B}(t,T_1) \\
\widetilde{B}(t,T_2) \\
\cdots \\
\widetilde{B}(t,T_{n-1}) \\
\widetilde{B}(t,T_n)
\end{pmatrix}
=
\begin{pmatrix}
1 & Ds^{1,n} & s^{1,n}(D(1+Ds^{2,n})) & \cdots & s^{1,n}\left(D\prod_{k=2}^{n-1}(1+Ds^{k,n})\right) \\
0 & 1 & Ds^{2,n} & \cdots & s^{2,n}\left(D\prod_{k=3}^{n-1}(1+Ds^{k,n})\right) \\
\cdots & \cdots & \cdots & \cdots & \cdots \\
0 & 0 & & 1 & Ds^{n-1,n} \\
\cdots & \cdots & \cdots & \cdots & 1
\end{pmatrix}
$$

$$
\times
\begin{pmatrix}
1 \\
1 \\
\cdots \\
1 \\
1
\end{pmatrix}.
$$

For the $i$th row, we obtain

$$\widetilde{B}(t,T_i) \equiv \frac{B(t,T_i)}{B(t,T_n)} = 1 + s^{i,n}(t)\underbrace{\sum_{j=i}^{n-1} D\prod_{k=i+1}^{j}(1+Ds^{k,n}(t))}_{\equiv \tau_i^n(t)},$$

which is the desired result (ii).

From (ii) and the definition of $a_{i,n}(t)$ given by Eq. (17.40), we deduce

$$\frac{a_{i,n}(t)}{B(t, T_n)} = \tau_i^n(t).$$  (17.48)

- Besides, by Eqs. (17.39) and (17.40) we see that when the forward swap rate is multiplied by its corresponding "fixed-leg" process, $a_{i,n}(t)$, it becomes a traded or tradable asset. Consequently, $\frac{s^{i,n}(t) a_{i,n}(t)}{B(t, T_n)}$ ($= s^{i,n}(t) \tau_i^n(t)$ by Eq. 17.48) is a $Q^n$-martingale. Applying Itô's lemma then gives

$$d\left(s^{i,n}(t) \tau_i^n(t)\right) = \tau_i^n(t) ds^{i,n}(t) + s^{i,n}(t) d\tau_i^n(t) + d\left\langle \tau_i^n(t), s^{i,n}\right\rangle_t.$$

Since $s^{i,n}(t) \tau_i^n(t)$ is a martingale under $Q^n$, its drift vanishes. Denoting by $\mu_{i,n}(t)$ the drift in the dynamics of the forward swap rate $s^{i,n}$, and noticing that the drift for the process $\tau^{i,n}$ is zero (since it is a $Q^n$-martingale), we have

(iii) $\mu_{i,n}(t) = -\frac{1}{\tau_i^n(t)} d\left\langle \tau_i^n(t), s^{i,n}(t)\right\rangle_t$

where

(iv) $d\left\langle \tau_i^n(t), s^{i,n}(t)\right\rangle_t = \sum_{j=i+1}^{n-1} \frac{Ds^{j,n}(t) \sigma_{j,n}(t)}{1 + Ds^{j,n}(t)} \tau_{ij}^n(t)$

with

$$\tau_{ij}^n(t) \equiv \sum_{k=j}^{n-1} D \prod_{h=i+1}^{k} \left(1 + Ds^{h,n}(t)\right) \qquad i < j \le n - 1.$$

Substituting (iv) in (iii) and using definition (i), we obtain the following lemma:

**Lemma 2**
Under the terminal forward probability $Q^n$, the dynamics of the forward swap rate satisfies the following stochastic differential equation (SDE):

$$ds^{i,n}(t) = s^{i,n}(t) \sigma_{i,n}(t) \left[ dW^n(t) - \sum_{j=i+1}^{n-1} \frac{Ds^{j,n}(t) \sigma_{j,n}(t)}{1 + Ds^{j,n}(t)} \frac{\tau_{ij}^n(t)}{\tau_i^n(t)} dt \right] \qquad (17.49)$$

where $\tau_{ij}^n(t)$ is given by (iv).

# Appendix 2

759

## (b) *The general forward swap measure $Q^{q,n}$*

Noticing that $\frac{s^{i,n}(t)a_{i,n}(t)}{a_{q,n}(t)}$ is a $Q^{q,n}$-martingale (see the Appendix to Chap. 16), we immediately obtain Lemma 3 by applying the same procedure as for Lemma 2:

**Lemma 3**
Under any forward swap measure $Q^{q,n}$, the dynamics of the forward swap rate satisfies the following SDE:

$$ds^{i,n}(t) = s^{i,n}(t)\sigma_{i,n}(t)$$

$$\times \left[ dW^{q,n} + \left( \sum_{j=q+1}^{n-1} \frac{Ds^{j,n}(t)\sigma_{j,n}(t)}{1+Ds^{j,n}(t)} \frac{\tau_{qj}^n(t)}{\tau_q^n(t)} - \sum_{j=i+1}^{n-1} \frac{Ds^{j,n}(t)\sigma_{j,n}(t)}{1+Ds^{j,n}(t)} \frac{\tau_{qj}^n(t)}{\tau_i^n(t)} \right) dt \right]$$

$$(17.50)$$

## (c) *The risk-neutral measure $Q$*

– The preceding measures are not appropriate when certain extensions of the standard SMM model are considered, such as those where the volatility itself is a stochastic process that exhibits a correlation with the rate process. Also, sometimes it can be useful to write the dynamics of the forward swap rate under the risk-neutral measure $Q$. We recover this from the forward swap measure $Q^{i,n}$ by taking a Radon–Nikodym derivative

$$\frac{dQ^{i,n}}{dQ} = \frac{a_{i,n}(t)}{a_{i,n}(0)\exp\left(-\int_0^{T_i} r(s)ds\right)} = \zeta_t$$

with

$$d\zeta_t = \zeta_t \sum_{j=i+1}^{n} D\frac{B(t,T_j)}{a_{i,n}(t)}\sigma(t,T_j)dW(t)$$

By Girsanov's theorem, we have

$$dW^{i,n}(t) = dW(t) - \sum_{j=i+1}^{n} D\frac{B(t,T_j)}{a_{i,n}(t)}\sigma(t,T_j)dt.$$

Thus, from the definition of the fixed-leg process of the swap $a_{i,n}(t)$, we have under $Q$ the following result:

## Lemma 4

Under the risk-neutral measure $Q$, the dynamics of the forward swap rate satisfies the following SDE:

$$ds^{i,n}(t) = s^{i,n}(t)\sigma_{i,n}(t)\left[dW(t) - \sum_{j=i+1}^{n} D\frac{B(t,T_j)}{a_{i,n}(t)}\sigma(t,T_j)\,dt\right]. \tag{17.51}$$

- Moreover, from the dynamics of $s^{i,\,n}(t)$ under the terminal forward measure $Q^n$, it is easy to show, using the same kind of analysis conducted from Eq. (17.48), that we also have, under $Q$:

$$ds^{i,n}(t) = s^{i,n}(t)\,\sigma_{i,n}(t)$$
$$\times\left[dW(t) - \left(\sigma(t,T_n) + \sum_{j=i+1}^{n-1}\frac{Ds^{j,n}(t)\sigma_{j,n}(t)\,\tau_{ij}^n(t)}{1+Ds^{j,n}(t)}\frac{\tau_{ij}^n(t)}{\tau_i^j(t)}\right)dt\right]. \tag{17.52}$$

- Finally, it is instructive to note the relation between the volatility of ZC bonds and that of forward swap rates. Setting the drifts in Eqs. (17.51) and (17.52) equal, and using Eq. (17.48), we obtain

$$\sigma(t,T_n) + \sum_{j=i+1}^{n-1}\frac{Ds^{j,n}(t)\sigma_{j,n}(t)}{1+Ds^{j,n}(t)}\frac{\tau_{ij}^n(t)}{\tau_i^n(t)}$$

$$= \sum_{j=i+1}^{n}\frac{D\left(1+\tau_j^n(t)s^{j,n}(t)\right)}{\tau_i^n(t)}\sigma(t,T_j). \tag{17.53}$$

This equation shows that the volatility of a ZC bond cannot be expressed solely in terms of forward swap instantaneous volatilities (and rates), in contrast with the situation for the LMM (see equation (v) in Sect. 1 of this Appendix). In fact, the two types of volatility are *interdependent*.

Equation (17.53) allows us to prove the following intuitive result: Given that the forward swap rate $s^{i,\,i+1}(t)$ is nothing else than the forward Libor rate $L(t,T_i)$, *the instantaneous volatility of the first must be equal to that of the second*. Formally, we obtain

$$\gamma_i(t) = \sigma_{i,i+1}(t). \tag{17.54}$$

## Proof

Generally, we may write Eq. (17.53) as

# Appendix 2    761

$$\sigma(t, T_{u+h}) + \sum_{j=u+1}^{u+h-1} \frac{Ds^{j,u+h}(t)\sigma^{j,u+h}(t)}{1 + Ds^{j,u+h}(t)} \frac{\tau_{uj}^n(t)}{\tau_u^n(t)}$$

$$= \sum_{j=u+1}^{u+h} \frac{D\left(1 + \tau_j^n(t)s^{j,u+h}(t)\right)}{\tau_u^n(t)} \sigma(t, T_j)$$

By considering the special case $h = 2$ and $u = i-1$, we obtain

$$\sigma(t, T_{i+1}) + \frac{Ds^{i,i+1}(t)\sigma_{i,i+1}(t)}{1 + Ds^{i,i+1}(t)} \frac{\tau_{i-1i}^n(t)}{\tau_{i-1}^n(t)} = \frac{D\left(1 + \tau_i^n(t)s^{i,i+1}(t)\right)}{\tau_{i-1}^n(t)} \sigma(t, T_i)$$

$$+ \frac{D\left(1 + \tau_{i+1}^n(t)s^{i+1,i+1}(t)\right)}{\tau_{i-1}^n(t)} \sigma(t, T_{i+1}).$$

Rearranging, this equation becomes

(v) $\frac{s^{i,i+1}(t)}{1+Ds^{i,i+1}(t)} \sigma_{i,i+1}(t)\tau_{i-1j}^n(t) = \frac{B(t,T_i)}{B(t,T_{i+1})} (\sigma(t, T_i) - \sigma(t, T_{i+1}))$.

But, the instantaneous variance of the forward Libor rate, $\gamma_i^2(t) = d <$ $\ln L(t, T_i) >$, is

$$\gamma_i^2(t) = \frac{1}{L^2(t, T_i)} \left[\frac{1}{D} \frac{B(t, T_i)}{B(t, T_{i+1})} \left(\sigma(t, T_i) - \sigma(t, T_{i+1})\right)\right]^2.$$

Making use of (v), we obtain: $\gamma_i^2(t) = \left[\frac{1}{D} \frac{\tau_{i-1i}^n(t)}{1+Ds^{i,i+1}(t)} \sigma_{i,i+1}(t)\right]^2$.

Since we know that, by (iv), $\tau_{i-1,i}^n(t)$ is equal to $D(1 + Ds^{i,\,i+1}(t))$, we obtain Eq. (17.54).

## Suggestions for Further Reading

### Books

** Bergomi, L. (2016). *Stochastic volatility modeling.* Chapman & Hall / CRC financial mathematics series.

** Brigo, D., & Mercurio, F. (2007). *Interest rate models: Theory and practice* (2nd ed.). Springer.

* Fusai, G., & Roncoroni, A. (2008). *Implementing models in quantitative finance: Methods and cases,* Springer.

* Gatarek, D., Bachert, P., & Maksymiuk, R. (2006). *The LIBOR market model in practice* (Finance series). Wiley.

Hull, J. (2018). *Options, futures and other derivatives* (10th ed.). Prentice Hall Pearson.

Hunt, P., & Kennedy, J. (2000). *Financial derivatives in theory and practice.* Wiley.

Martellini, L., Priaulet, P., & Priaulet, S. (2003). *Fixed income securities.* Wiley Finance.

** Musiela, M., & Rutkowski, M. (2007). *Martingale methods in financial modelling* (Applications of mathematics). Springer.

* Rebonato, R. (1998). *Interest rate option models* (2nd ed.). Wiley.

** Rebonato, R. (2002). *Modern pricing of interest rate derivatives: The LIBOR market model and beyond* (2nd ed.). Princeton University Press.

### Articles

Andersen, L., & Andreasen, J. (2000). Volatility skews and extensions of the libor market model. *Applied Mathematical Finance, 7*(1), 1–32.

Andersen, L., & Brotherton-Ratcliffe, R. (2005). Extended libor market models with stochastic volatility. *Journal of Computational Finance, 9.*

Bergomi L. (2005, October). Smile dynamics ii. *Risk Magazine.*

Black, F. (1976). The pricing of commodity contracts. *Journal of Financial Economics, 3,* 167–179.

Black, F., Derman, E., & Toy, W. (1990). A one-factor model of interest rates and its application to treasury bond options. *Financial Analysts Journal, 46*(1), 33–39.

Black, F., & Karasinski, P. (1991). Bond and option pricing when short-term rates are lognormal. *Financial Analysts Journal, 47*(4), 52–59.

Brace, A., Gatarek, D., & Musiela, M. (1997). The market model of interest rate dynamics. *Mathematical Finance, 7*(2), 127–155.

Brennan, M., & Schwartz, E. (1982). An equilibrium model of bond pricing and a test of market efficiency. *Journal of Financial and Quantitative Analysis, 17*(3), 301–329.

Brigo, D., & Mercurio, F. (2001). Displaced and mixture diffusions for analytically-tractable smile models. In *Mathematical Finance-Bachelier Congress 2000.* Springer Finance.

Cox, J., Ingersoll, J., & Ross, S. (1985a). An intertemporal general equilibrium model of asset prices. *Econometrica, 53,* 363–384.

Cox, J., Ingersoll, J., & Ross, S. (1985b). A theory of the term structure of interest rates. *Econometrica, 53,* 385–408.

Galluccio, S., Huang, Z., Ly, J.-M., & Scaillet, O. (2007). Theory and calibration of swap market models. *Mathematical Finance, 17*(1), 111–141.

Glasserman, P., & Kou, S. (2003). The term structure of simple forward rates with jump risk. *Mathematical Finance, 13*(3), 383–410.

Glasserman, P., & Merener, N. (2003). Cap and swaption approximations in libor market models with jumps. *Journal of Computational Finance, 7*(1).

Hagan, P., Kumar, D., Lesniewski, A., & Woodward, D. (2002). Managing smile risk. *Wilmott Magazine,* 84–108.

## Suggestions for Further Reading

Heath, D., Jarrow, R., & Morton, A. (1992). Bond pricing and the term structure of interest rates: A new methodology for contingent claims valuation. *Econometrica, 60*, 77–105.

Ho, T. S. Y., & Lee, S. B. (1986). Term structure movements and the pricing of interest rate contingent claims. *The Journal of Finance, 41*, 1011–1029.

Hull, J., & White, A. (1990). Pricing interest-rate derivative securities. *Review of Financial Studies, 3*, 573–592.

Hull, J., & White, A. (2000). Forward rate volatilities, swap rate volatilities, and implementation of the libor market model. *Journal of Fixed Income*, 46–62.

Hunt, P., Kennedy, J., & Pelsser, A. (1998). Fit and run. *Risk, 11*, 65–67.

Jäckel, P., & Rebonato, R. (2003). The link between caplet and swaption volatilities in a Brace–Gatarek–Musiela/Jamshidian framework: Approximate solutions and empirical evidence. *Journal of Computational Finance, 6*(4), 35–45.

Jamshidian, F. (1995). A simple class of square root interest rate models. *Applied Mathematical Finance, 2*, 61–72.

Jamshidian, F. (1997). Libor and swap market models and measures. *Finance and Stochastics, 1*, 293–330.

Jarrow, R., Li, H., & Zhao, F. (2007). Interest rate caps smile too! But can the libor market models capture it? *Journal of Finance, 62*, 345–382.

Joshi, M., & Rebonato, R. (2003). A stochastic-volatility, displaced-diffusion extension of the Libor market model. *Quantitaive Finance, 3*(6), 458–469.

Kawai, A. (2002). Analytical and Monte Carlo swaption pricing under the forward swap measure. *Journal of Computational Finance, 6*(1), 101–111.

Langetieg, T. C. (1980). A multivariate model of the term structure. *Journal of Finance, 35*, 71–97.

Longstaff, F., & Schwartz, E. (1992). Interest rate volatility and the term structure: A two-factor general equilibrium model. *Journal of Finance, 47*(4), 1259–1282.

Miltersen, K., Sandmann, K., & Sondermann, D. (1997). Closed form solutions for term structure derivatives with lognormal interest rates. *Journal of Finance, 52*(1), 409–430.

Musiela, M., & Rutkowski, M. (1997). Continuous-time term structure models: Forward measure approach. *Finance and Stochastics, 1*, 261–291.

Oya, K., (2018a, July) The swap market model with local stochastic volatility. *Risk Magazine*

Oya, K. (2018b). The co-terminal swap market model with Bergomi stochastic volatility.

Rebonato, R. (1999). On the pricing implications of the joint lognormal assumption for the swaption and cap markets. *Journal of Computational Finance, 2*(3), 57–76.

Rebonato, R. (2004). Interest-rate term-structure pricing models: A review. *Proceedings of the Royal Society A: Mathematical, Physical and Engineering Sciences, 460*, 667–728.

Vasicek, O. (1977). An equilibrium characterization of the term structure. *Journal of Financial Economics, 5*(2), 177–188.

Wu, L., & Zhang, F. (2006). Libor market model with stochastic volatility. *Journal of Industrial and Management Optimization, 2*, 199–227.

# Elements of Stochastic Calculus

# 18

The use of the stochastic calculus that applies to the continuous-time Brownian motion, and more generally to random processes whose variations can be expressed as functions of such Brownian motions, is particularly fruitful in the analysis of financial asset prices.

These processes were introduced by the French mathematician Louis Bachelier, in 1900, to represent the movements of stock prices. The mathematics for such processes were later developed, in particular by Einstein, Wiener, and Itô, to solve certain problems in physics, and then once more applied to finance. It turns out that the differential and integral calculus for functions of stochastic processes goes significantly beyond the ordinary calculus that holds for functions on $\mathbb{R}^n$.

This chapter presents the elements of calculus relevant to Itô processes, which are extensively used in the theory and techniques of modern finance. Rigorous proofs (notably employing measure theory) are not provided in this chapter whose focus is on computational tools and their meaning since these tools are needed to understand many chapters.[1] More detailed mathematical exposition (only useful for understanding the more technical parts of the book) is to be found in footnotes and in the following chapter. In addition to Itô processes, Poisson processes are briefly discussed. While the former is continuous, the latter represent jumps occurring at random times and are used particularly for modeling credit risk.

After providing several general definitions about stochastic processes of various types (Sect. 18.1), we examine Brownian motions (Sect. 18.2), one-dimensional Itô and diffusion processes (Sect. 18.3), properties of functions of stochastic processes and the rules of Itô differential and integral calculus (Sects. 18.4 and 18.5) and, finally, jump processes (Sect. 18.6).

---

[1] The tools given in the first four sections of this chapter are used in about 15 other chapters. Sections 18.5 and 18.6 are less crucial and can be skipped on a first reading.

© The Author(s), under exclusive license to Springer Nature Switzerland AG 2022
P. Poncet, R. Portait, *Capital Market Finance*, Springer Texts in Business and Economics, https://doi.org/10.1007/978-3-030-84600-8_18

765

# 18.1 Definitions, Notation, and General Considerations About Stochastic Processes

A stochastic process (or a random process) is a one- or multi-dimensional variable that depends on chance and time. There are infinitely many examples of such processes. Just sticking to finance, we can mention the price of a share, an interest rate, a stock market index, or a set of variables including several interest rates, prices, indices, exchange rates, etc.

## 18.1.1 Notation

Random variables and processes will be denoted by capital letters and nonrandom elements of $\mathbb{R}$ or $\mathbb{R}^n$ will *usually* be written in small letters.

Underlining denotes a vector (examples: $\underline{X}$ is a vector $(X_1, \ldots, X_n)$ of random variables; and $\underline{x}$ is a nonrandom element of $\mathbb{R}^n$).

A matrix will be written with a bold capital letter such as $\mathbf{\Sigma}$.

## 18.1.2 Stochastic Processes: Definitions, Notation, and General Framework

### 18.1.2.1 Probability Framework (Simplified)

Just like a random variable, a stochastic process depends on the state of the world denoted by $\omega$ (the set of all states will be written $\Omega$); it also depends on time. A complete notation indicates this double dependence; thus we denote by $X(t,\omega)$ the value taken on by the process $X$ at a time $t$ with the state of the world $\omega$. We consider the case of processes defined and observed at particular instants $0, t_1, t_2, \ldots, t_n = T$ as well as the case of processes defined for all $t$ in a time interval $(0, T)$.[2] In the first case one says the process is "in discrete time" and in the second that it is "in continuous time." The date 0 often (but not always) denotes the present date. *The state $\omega$ should be understood as the complete history of the system studied between 0 and T.* For a given state $\omega$ every realization of the process $X(t,\omega)$ is therefore known. As a result, for a given $\omega$, we have that $X(t,\omega) \equiv x_\omega(t)$ is a simple function of time (with no randomness) which gives a specific realization of the process under the state $\omega$. This function of time $x_\omega(t)$ for a given $\omega$ is called a path or a trajectory. A path is therefore a discrete or continuous sequence of realizations of $X$. For a continuous-time process

---

[2] Formally it is a measurable mapping $X$ of $[0, T] \times \Omega$ into $\mathbb{R}$ (or $\mathbb{R}^n$), where $\Omega$ is the fundamental probability space equipped with a $\sigma$-algebra and a probability measure. The $\sigma$-algebra $\mathbb{F}_t$ represents the information set available at $t$; the enrichment of the information set over time is represented by a series $\mathbb{F}_t$ of growing $\sigma$-algebras ($\mathbb{F}_0 \subseteq \mathbb{F}_1 \subseteq \ldots \subseteq \mathbb{F}_t \mathbb{Z} \ldots \subseteq \mathbb{F}_T$), called a filtration. We will only consider processes that are said to be adapted or $\mathbb{F}_t$-measurable, i.e. such that, for any $t$, the random variable $X(t, .)$ is "known" at $t$, i.e., is measurable with respect to $\mathbb{F}_t$.

# 18.1 Definitions, Notation, and General Considerations About Stochastic Processes 767

it can be represented by a simple curve (in the plane, if the process is one-dimensional, or in space with $n + 1$ dimensions if it is $n$-dimensional).

For a given $t$, $X(t,\omega) \equiv X_t(\omega)$ is a simple random variable whose value is only known at $t$.

Using probabilistic wording, a property of $X(t,\omega)$ is said to be true "almost surely" (a.s.) if it holds for every $\omega$ except possibly in a subset of $\Omega$ with probability zero. One could, therefore, interpret "a.s." as a synonym for "with (almost complete) certainty" or "with probability 1."

The notation $X(t,\omega)$ is fully explicit in denoting a random process although we shall often simply denote it by $X(t)$. The ambiguity is partially alleviated by our convention that capital letters denote random quantities (small letters usually being used for deterministic elements).

The stochastic processes that we consider are one-dimensional variables or multi-dimensional vectors, that is, are random and functions of time, taking their values in $\mathbb{R}^n$. A random vector process will, for example, be written $\underline{X}(t)$.

Sometimes we will consider two instants $s$ and $t$ and always assume that $\underline{X}(s)$ is known at $t$ for any $s \leq t$. This latter condition characterizes processes that are adapted (the values are known without delay, "in real time").[3]

Insofar as more information accrues with the passage of time, uncertainty is reduced. The amount of information available at $t$ is given by $\mathbb{F}_t$.[4] It is from $\mathbb{F}_t$ that, at $t$, probabilities about the future, which are described as *conditional* since they depend on the then-available information, are assigned to observable events. We use the notation $\mid \mathbb{F}_t$ for such a conditioning; for example, $E(X(T) \mid \mathbb{F}_t)$ denotes the expectation of the final value $X(T)$ assigned at $t$, using the information $\mathbb{F}_t$ available at that time. Similarly, we denote by $E(X(T) \mid X(t) = x)$ "the expected value of $X(T)$ knowing that $X(t) = x$."

## 18.1.2.2 Processes Without Memory: Markov Processes

Let us begin by considering a one-dimensional process $X(t)$, an arbitrary sequence of successive times $t_0, t_1, ..., t_{m-1}$ within $[0, T)$, arbitrary real numbers $x_0, x_1, ..., x_{m-1}$ and $x$, and the probability of $X(t_m)$ being less than or equal to $x$, knowing that $X(t_0) = x_0$, $X(t_1) = x_1, ..., X(t_{m-1}) = x_{m-1}$; we can write this conditional probability as

$$\text{Prob}\{X(t_m) \leq x \mid X(t_0) = x_0, X(t_1) = x_1, ..., X(t_{m-1}) = x_{m-1}\}.$$

A process without memory, is, by definition, characterized by conditional probabilities for $X(t_m)$ that, whatever $x$ and the sequence $(t_0, ..., t_m)$, only depend on variables $(x_{m-1}, t_{m-1})$ that characterize the last known state; the preceding conditional probability thus simplifies to $\text{Prob}\{X(t_m) \leq x \mid X(t_{m-1}) = x_{m-1}\} \equiv F(x_{m-1}, t_{m-1}; x, t_m)$. This probability of passing from the state $x_{m-1}$ at $t_{m-1}$ to a state $\leq x$ (in fact, an event) at $t_m$ is called the transition probability.

---

[3] See Footnote 2.
[4] See Footnote 2.

To sum this up, a memoryless process is such that its most recent realization contains all the relevant information needed to assess probabilities about the future: "The past only influences the future through the intermediation of the present."

Therefore, to assume that stock prices follow a Markov process is in contradiction to chartist and technical analysis but in agreement with the market efficiency hypothesis (see Chap. 1).

The definitions we have just presented can easily be generalized to multivariate processes: $X$ and $x$ simply become $\underline{X}$ and $\underline{x}$, which are random vectors.

**Definition** *A Markov process is a process without memory whose transition probabilities obey certain technical conditions.*[5]

### 18.1.2.3 Processes with Continuous Paths

The preceding definitions cover two cases that we have distinguished: where the process $\underline{X}(t)$ is only defined for discrete times; where it can be defined for continuous time, that is, for every instant in a time interval $[0, T]$.

Continuous-time processes can be divided into two categories: continuous processes and jump processes.

Continuous processes, also called processes with continuous paths, are those for which the trajectories are, with certainty, continuous while those of jump processes show discontinuities. As a result, a continuous process can only change its value by an infinitesimal amount in an infinitesimal time interval $dt$. Continuous processes were particularly studied by the Japanese mathematician Kiyoshi Itô.

Finally, among the processes in continuous or discrete time we are particularly interested in stationary processes, that it is those for which the transition probabilities Prob $\{\underline{X}(t) \leq \underline{x}_t \mid \underline{X}(s) = \underline{x}_s\}$ can be written, for all $s$ and $t \in [0, T]$ with $s < t$, as functions of $(\underline{x}_s, \underline{x}_t, t - s)$.[6]

---

[5] A memoryless process is Markovian if the following two conditions are met:
- $F(\underline{x}_s, s\,;.,\,t)$ is a measurable function with respect to the $\sigma$-algebra $\mathbb{F}_t$ and the Borel sets of $\mathbb{R}^n$.
- The Chapman-Kolmogorov equation is satisfied for $r < s < t$: $F(\underline{x}, r\,;\,\underline{z},\,t) = \int_y F(\underline{y}, s\,;\,\underline{z},\,t)\,dF(\underline{x}, r\,;\,\underline{y},\,s)$ (the integration is done on $\underline{y}$).

This equation simply means that the probability for $X$ to go from $(r, \underline{x})$ to $(t, \leq \underline{z})$ is equal to the sum of the probabilities of all the trajectories passing at time $s$ by all the possible intermediary states $\underline{y}$ and leading at time $t$ to a value $\leq \underline{z}$.

[6] Stationary processes are not defined only on the special class of Markovian processes; generically, a process $X(t)$ is stationary if its distribution function is such that:

$\forall h$, Proba $(\underline{X}(t_m) \leq \underline{x}_m, \underline{X}(t_{m-1}) \leq \underline{x}_{m-1}, ..., \underline{X}(t_0) \leq \underline{x}_0)$ = Proba $(\underline{X}(t_m + h) \leq \underline{x}_m, ..., \underline{X}(t_0 + h) \leq \underline{x}_0)$,

i.e., it is invariant for any time translation. This general definition of course implies the particular one we have given for Markovian processes.

## 18.2 Brownian Motion

The notion of Brownian motion was introduced in 1828 by the botanist Robert Brown to describe the movement of pollen particles suspended in water. Seventy-two years later, the French mathematician Louis Bachelier, in his 1900 thesis titled "Théorie de la spéculation" (Theory of Speculation) studying the behavior of prices quoted on the Paris Stock Market, developed and formalized the theory and used the continuous-time Brownian motion to represent the variations of stock prices. The work of Bachelier preceded that of the physicist A. Einstein, and of the mathematicians N. Wiener, P. Lévy, and K. Itô, who used Brownian motion to describe physical phenomena and develop the theory of continuous stochastic processes. Seventy years after Bachelier's work, and its rediscovery by the economist Paul Samuelson, Robert Merton, in a series of pioneering articles, laid down an essential part of the current theory of financial markets relying on the mathematics of stochastic processes derived from the Brownian motion.

We start by studying the one-dimensional Brownian motion in Sect. 18.2.1, then provide useful calculus rules relevant to such processes in Sect. 18.2.2, and finally discuss the multi-dimensional Brownian motion in Sect. 18.2.3.

### 18.2.1 The One-Dimensional Brownian Motion

In this subsection, we study one-dimensional Brownian motions which have their values in $\mathbb{R}$.

#### 18.2.1.1 Introduction: Discrete Time

To introduce several concepts essential for our needs, we first consider a one-dimensional Markov process observed at regularly spaced instants, with successive values $X(0), X(1), ..., X(t), ...$, with **independent increments $X(t) - X(t-1)$, normally and identically distributed** with mean $\mu$ and standard deviation $\sigma$.

Let us set $U(t-1) \equiv \frac{X(t+1)-X(t)-\mu}{\sigma}$

From the properties of processes $X$ as described above, the $U(t)$ is reduced normal, centered, and independently distributed variables, and the equation governing the motion of $X$ can be written

$$X(t+1) - X(t) = \mu + \sigma\, U(t+1). \qquad (18.1)$$

Writing $X(t)$ as the sum of $X(0)$ and the $t$ increments between 0 and $t$, we have $X(t) - X(0) = \sum_{i=1}^{t} X(i) - X(i-1)$; the increments $X(i) - X(i-1)$ are independent Gaussians, so we have $E(X(t)) - X(0) = \mu\, t$; $\mathrm{var}((X(t)) = \sigma^2\, t$ and $(X(t) - X(0))$ is $N(\mu t, \sigma^2 t)$-distributed.

This very simple, first-order auto-regressive, process is routinely used in econometrics. The Brownian motion can be considered as a limit of this process as one goes from discrete time to continuous.

770          18 Elements of Stochastic Calculus

### 18.2.1.2 Continuous Time

We generalize this approach by considering a process $X(t)$ defined for continuous time, that is, for any $t \in (0, T)$ that we define as a "non-standard arithmetic[7] Brownian motion" or simply as "a Brownian motion."

Among the different possible definitions of the Brownian motion we choose the following:

### Definition of the Arithmetic Brownian Motion

$X(t)$ is an arithmetic Brownian motion if:

(i) $X(t) - X(0)$ is Gaussian with mean $\mu t$ and variance $\sigma^2 t$ (i.e., with the law $N(\mu t, \sigma^2 t)$);

(ii) Its increments are independent (if computed for disjoint periods); as result, if $t_1$, $t_2$, $t_3$, $t_4$ are four successive times with $0 \leq t_1 < t_2 < t_3 < t_4 \leq T$: $X(t_4) - X(t_3)$ is independent of $X(t_2) - X(t_1)$.

More generally, the increment of a Brownian motion for any period $(s, t)$ is independent of all events that have occurred up to and including the date $s$.

Let us consider any two successive dates $s$ and $t$.

Writing $X(t) - X(s) = (X(t) - X(0)) - (X(s) - X(0))$, we obtain $E(X(t) - X(s)) = \mu (t - s)$.

Writing $X(t) - X(0) = (X(t) - X(s)) + (X(s) - X(0))$, we have

$$Var(X(t) - X(0)) = Var(X(t) - X(s)) + Var(X(s) - X(0)),$$

from which follows

$$Var(X(t) - X(s)) = \sigma^2 (t - s).$$

One may also write (with $s < t$):

$$
\begin{aligned}
Cov(X(s), X(t)) &= Cov(X(s), X(s) + (X(t) - X(s))) \\
&= Var(X(s)) + Cov(X(s), X(t) - X(s)) \\
&= Var(X(s)) = \sigma^2 s.
\end{aligned}
$$

Furthermore, the increments $X(t) - X(s)$ do not depend on events before $s$, so the process is Markovian. In addition, $X$ is stationary since the first two moments of $X(t) - X(s)$ (which determine all probabilities, since $X$ is Gaussian) only depend on $t - s$.

Because of this, if we consider two times $t$ and $t + \Delta t$, then $X(t + \Delta t) - X(t)$ is distributed according to the law $N (\mu \Delta t, \sigma^2 \Delta t)$, and we conclude

---

[7] Simple Brownian motions are sometimes dubbed "arithmetic" to be distinguished from Geometric Brownian motions (see Sect. 18.3.2, Example 2).

## 18.2 Brownian Motion

$$X(t + \Delta t) - X(t) \equiv \Delta X = \mu \Delta t + \sigma \sqrt{\Delta t}\, U(t), \qquad (18.2)$$

where $U(t)$ represents a standard Gaussian variate.

Equation (18.2) also shows the process is continuous, since letting $\Delta t$ go to 0 we find $\Delta X$ tends to 0 (in a certain sense better defined below). Indeed, the normal and i.i.d. character of the increments is enough to make the process continuous, as we shall show below.

The properties of Brownian motion identified above are summarized in the form of a proposition.

**Proposition 1. First Series of Properties of the Arithmetic Brownian Motion**
- *The increments of a Brownian motion have a mean and a variance proportional to the length of the interval over which they are calculated: $X(t) - X(s)$ is distributed according to the law $N(\mu\,(t - s),\ \sigma^2\,(t - s))$. Two increments computed for periods of the same length are identically distributed. Two increments computed for non-overlapping periods are independent.*
- *For $s < t$: $Cov(X(s), X(t)) = \sigma^2 s$.*
- *Trajectories (paths) of the Brownian motion are continuous.*
- *The Brownian motion is a Markovian and stationary process.*

Among arithmetic Brownian motions, we distinguish those whose increments have mean zero ($\mu = 0$) and whose variance per unit time equals one ($\sigma^2 = 1$).

**Definition of the Standard Brownian Motion (or Wiener Process)** *An arithmetic Brownian motion whose increments have mean $\mu = 0$ and variance per unit time $\sigma^2 = 1$ is called a standard Brownian motion, or a Wiener process (from the name of the mathematician who first studied its properties); we denote it by $W(t)$. By definition, $W(0) = 0$, i.e. the Wiener process starts at the origin.*

From this, we immediately deduce the following properties:

**Second Series of Properties of a Wiener Process (Standard Brownian Motion)**
- *The Wiener process $W(t)$ is distributed according to the law $N(0, t)$;*
- *The increment between $t$ and $t + \Delta t$, $\Delta W \equiv W(t + \Delta t) - W(t)$, is $N(0, \Delta t)$;*
- *The conditional expectation is $E(W(t) \mid W(s)) = W(s)$; i.e., $W$ is a martingale.*[8]

The standard Brownian process plays, with respect to a Brownian motion, a role analogous to that played by the reduced centered normal random variable for normal random variables with arbitrary means and variances. In particular, it is possible to write any arithmetic Brownian motion $X(t)$ with parameters $\mu$ and $\sigma$ as a function of a Wiener process $W(t)$. It suffices to check that $W(t) = \frac{X(t) - \mu t}{\sigma}$ is a Wiener process [$X(t)$

---

[8] More generally, the increment $W(t) - W(s)$ does not depend on any event prior to date $s$, *so that we can write*: $E(W(t) - W(s) \mid \mathbb{F}_s) = 0$, hence $E(W(t) \mid \mathbb{F}_s) = W(s)$, where $\mathbb{F}_s$ is the set of all the information available at date $s$ represented mathematically by a $\sigma$-algebra.

772                                                                18  Elements of Stochastic Calculus

provides its Gaussian property as well as the independence of its increments, and, by construction, $W(t)$ has mean zero and variance equal to $\text{var}(X(t)/\sigma^2) = t]$. Thus, we write

$$X(t)\mu t + \sigma W(t). \tag{18.3}$$

For increments, this implies

$$\Delta X = \mu \Delta t + \sigma \Delta W. \tag{18.4}$$

Since $\Delta W = W(t + \Delta t) - W(t)$ is $N(0, \Delta t)$-distributed, this increment can be expressed as a function of a standard normal variable $U$

$$\Delta W = \sqrt{\Delta t}U. \tag{18.5}$$

Note that Eq. (18.5) is a special case of Eq. (18.2) (with $\mu = 0$ and $\sigma = 1$).

We have already remarked that, as intuition suggests, a Brownian motion's trajectories are continuous everywhere. Let us take up this important point more rigorously. At any point $(t, W(t))$ of an arbitrary trajectory of the Wiener process, consider the random increment $\Delta W$ that will accrue in the period $(t, t + \Delta t)$.

$Var(\Delta W) = E\left[(\Delta W)^2\right] = \Delta t$ goes to 0 as $\Delta t$ tends to 0, which shows that any trajectory is continuous in the quadratic norm. It is also continuous in the sense that, for any $\varepsilon > 0$, $\lim_{\Delta t \to 0} \text{Prob}(|\Delta W| > \varepsilon) = 0$, since $\Delta W$ is a normal random variable centered on 0 whose variance goes to 0 as $\Delta t$ tends to 0; this makes the probability of any jump vanish.

Furthermore, the continuity property of trajectories that we just showed for the Wiener process continues to hold for nonstandard Brownian processes as a result of Eq. (18.3).

As processes $X(t)$ are $W(t)$ are defined in continuous time, we can write Eq. (18.4) for an infinitesimal interval $(t, t + dt)$ using differentials

$$dX(t) = \underbrace{\mu\,dt}_{\substack{non-random \\ = E(dX)}} + \underbrace{\sigma\,dW}_{\substack{centered\ Gaussian\ random \\ variable\ with\ variance\ =\ \sigma^2 dt}} \tag{18.6}$$

($dW$ is a normal variable with mean zero and variance $dt$).

Equation (18.6) is a stochastic differential equation (SDE). It says that the variation $dX$ has a deterministic part $\mu dt$ equal to its mean and a stochastic component $\sigma dW$ with mean zero and variance $\sigma^2 dt$; $\mu$ is called the *drift*, $\sigma$ the *diffusion parameter* and $\sigma^2$ the instantaneous variance.

SDE (18.6) has an integral form (or solution):

## 18.2 Brownian Motion

$$X(t) - X(0) = \int_0^t \mu \, ds + \int_0^t \sigma dW(s) = \mu t + \sigma W(t).$$

SDE (18.6) is the generalization of the simpler ordinary differential equation (ODE) (without a stochastic term): $dX = \mu \, dt \Leftrightarrow X(t) - X(0) = \int_0^t \mu \, ds = \mu t$.

Notice that $\int_0^t \sigma dW(s) = \sigma \int_0^t dW(s) = \sigma W(t)$ since $\sigma$ is a constant and that $\int_0^t dW(s)$ is the stochastic integral simply giving the sum of all increments of $W$ between 0 and $t$: this result, and even the definition of the stochastic integral, would be less obvious if $\sigma$ was not constant, as in some of the more general processes we study later.

The Brownian motion can be used when the evolution of a system results from a constant force that is represented by the drift term ($\mu \, dt$) perturbed by successive continual time-independent random shocks that characterize erratic Gaussian innovations (expressed by the term $\sigma \, dW$). This modeling has been especially useful to describe, following Brown's observations, the motion of a particle suspended in a fluid.[9]

Now Eq. (18.5) allows us to write: $\frac{\Delta W}{\Delta t} = \frac{\sigma}{\sqrt{\Delta t}} U$.

Since this expression does not have a limit as $\Delta t$ tends to 0, Brownian motions (both $W$ and $X$) do not have a derivative although they are continuous.

The trajectories of Brownian motions are therefore continuous at every point although they are not differentiable anywhere (each point is "needle-like," which is difficult to imagine).

This double property of continuity and non-differentiability is only one of the aspects of Brownian motions that may appear strange. Let us mention, among others.

**Third Property of a Brownian Motion**

Whatever the timescale at which the Brownian motion is observed (i.e., whatever the "zoom" level), the trajectory shows the "similarity in form" that is characteristic of a fractal structure.

Let us be more precise about this fractal structure of the Brownian motion. Generally speaking, a fractal has a special form of invariance or symmetry relating its whole to its parts: the whole may be split into smaller parts, each of which seems

---

[9] $X(t)$ is one of the coordinates of the location of the particle (a tri-dimensional Brownian would represent its location in the three-dimensional space). The particle is continuously shocked by the endlessly moving molecules of the fluid. These shocks induce numerous random and independent moves ("walks") of the particle which are represented by the Gaussian term s$dW$ (*the Gaussian feature is due to the Central Limit Theorem*). In addition, a drift $\mu dt$ is generated by a "stream" that flows inside the fluid and gives the particle its general direction. In financial markets, the shocks are in particular due to the arrival of new information.

to be an "echo" of the whole. It is, in this sense, scale invariant. Consider a Wiener process. During a period of length $\Delta t$ this standard Brownian motion has variation $\sqrt{\Delta t}\, U$ where $U$ is a standard Gaussian. The interval may be further decomposed into $n$ subintervals of length $\Delta t/n$ and the variation in the Brownian motion then is the sum of $n$ normal variables $\sqrt{\frac{\Delta t}{n}} U_i$ where the $U_i$ are standard Gaussian. In the same way, on one of the subintervals of length $\Delta t/n$, the variation can, in turn, be split up into $n$ subintervals of length $\Delta t/n^2$ on each of which the Brownian variation is $\sqrt{\frac{\Delta t}{n^2}} U_j$; and so on for ever finer decompositions as diagrammed below.

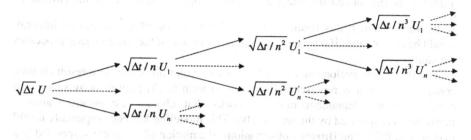

The standard Gaussian $U$ is the common pattern, the generator of the whole weft, whatever the scale at which it is observed.

However, the random variables $U_i$ take on different values, which explains why the same exact forms are not to be found in the nested images corresponding to different zooms of a single trajectory.

Figure 18.1 illustrates, from the infinity of different possible realizations, two trajectories of a Brownian motion $X(t)$.

Figure 18.1a shows two simulated trajectories of a Brownian motion with drift $\mu = 0.04\%$ (i.e., $0.004 \times 252 \# 10\%$ for a year of 252 business days) and diffusion parameter $\sigma = 0.02$ per day (i.e., $0.02 \times \sqrt{252} \# 32\%$ annualized). The figure shows the evolution of $X(t)$ over 4000 days. As we will see later, $X(t)$ could be the logarithm of a stock price with a volatility of 32%.

Notice these trajectories are continuous but everywhere rough (without jumps but changing direction at every point) due to the non-differentiability of the Brownian motion.

Figure 18.1b is zoomed in on one of the trajectories (a magnification that looks closer at the path from day 1200 to day 2200). The same roughness as in the 4000-day trajectory is observed: it could represent, on a priori grounds, a Brownian trajectory at any scale whatsoever, in accordance with the characteristic scale invariance of a fractal structure.

## 18.2 Brownian Motion

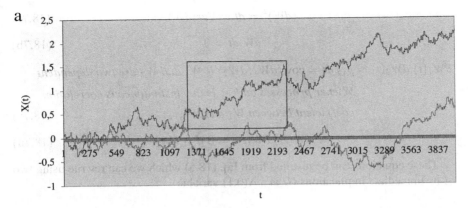

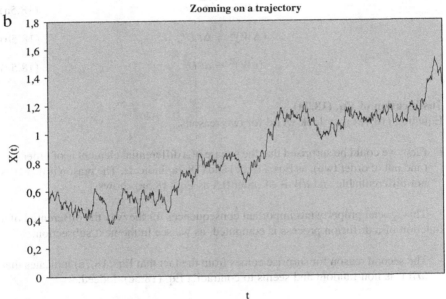

**Fig. 18.1** Trajectories of a Brownian motion

### 18.2.2 Calculus Rules Relative to Brownian Motions

We now establish a number of heuristic calculus rules for arithmetic Brownian motions that are the basis of stochastic calculus.

Differential calculus for Wiener processes satisfies the following rules:

$$(dW)^2 = dt = Var(dW) \tag{18.7a}$$

$$dW.dt = 0 \tag{18.7b}$$

$dW_1(t).dW_2(t) = \rho_{12}\,dt = \text{cov}\,(dW_1, dW_2)$ *if* $W_1$ *and* $W_2$ *are two dependent*

*Wiener processes and* $\rho_{12}$*is the instantaneous correlation*

*coefficient between* $W_1$ *and* $W_2$. $\tag{18.7c}$

$$dW(t_1).dW(t_2) = 0 \text{ if } t_1 \neq t_2 \tag{18.7d}$$

These equations can be justified from Eq. (18.5) which we can rewrite using two of its immediate implications ($U$ is a $N(0,1)$ variable):

$$\Delta W = \sqrt{\Delta t}\, U, \tag{18.5a}$$

$$(\Delta W)^2 = \Delta t\, U^2, \tag{18.5b}$$

$$(dW)^2 = dt\, U^2. \tag{18.5c}$$

**Justification of Eq. (18.7a)**

Equation (18.7a) may look weird for two reasons.

- First, we could be surprised that the square of a differential element is of order one (and not of order two), as Eqs. (18.5c) and (18.7a) indicate. The reason is that $W$ is not differentiable and $dW$ is of order 0.5 as Eq. (18.5a) shows.

This special property has important consequences for the way the differential of a function of a diffusion process is computed, as we see in the next subsection.

- The second reason for surprise comes from the fact that Eq. (18.7a) indicates that $(dW)^2$ is non-random and seems to contradict Eq. (18.5c). Indeed:

$$\text{E}\left[(dW)^2\right] = dt$$

and Eq. (18.5c) implies $Var\,[(dW)^2] = (dt)^2\,Var[U^2] = 2(dt)^2$ (since $Var[U^2] = 2$).[10]

---

[10]To recover this well-known result, let us compute: $Var[U^2] = E[U^4] - (E[U^2])^2$. We have $E[U^2] = Var\,(U) = 1$ and $E[U^4] = \text{kurtosis}[U]$ which is equal to 3 since $U$ is standard Gaussian, from which $Var[U^2] = 3 - 1 = 2$.

## 18.2 Brownian Motion

The variable $(dW)^2 - dt$, therefore, has mean zero and variance $2(dt)^2$, of order two: the variance can thus be considered to be zero,[11] so that $(dW)^2$ can be assimilated to its expected value $dt$.

### Justification of Eqs. (18.7b) and (18.7c)

Rules (18.7b) and (18.7c) obey the same principles as Eq. (18.7a).

First, $Var\{dt.dW\} = (dt)^{3/2}$, so $dt\,dW$ can also be assimilated to its expectation, which is zero, from which we have $dW\,dt = 0$.

Then we have $dW_1\,dW_2 = U_1\,U_2\,dt$, where $U_1$ and $U_2$ are two standard normal variables with correlation coefficient $\rho_{12}$ (also their covariance). As a result:

$$E\{dW_1\,dW_2\} = dt\,E\{U_1\,U_2\} = \rho_{12}dt;$$

$$Var\{dW_1\,dW_2\} = (dt)^2\,Var\{U_1.U_2\},$$

which is therefore of order two. The product can then be assimilated to its expectation, so that:

$$dW_1\,dW_2 = \rho_{12}dt.$$

### Justification of Eq. (18.7d)

Rule (18.7d) is an immediate consequence of Eq. (18.7c) and the fact that the increments $dW$ are independent.

These rules apply to a nonstandard arithmetic Brownian process $X$.

*"Heuristic" rules applicable to non-standard Brownian processes with parameters $\mu$ and $\sigma$:*

$$dX^2 = \sigma^2 dt; \tag{18.8a}$$

$$dX\,dt = 0; \tag{18.8b}$$

$$dX_1(t)\,dX_2(t) = \text{cov}(dX_1, dX_2) = \sigma_{12}\,dt = \rho_{12}\sigma_1\sigma_2 dt. \tag{18.8c}$$

We only show Eq. (18.8a), since the other equations have similar justifications. From Eq. (18.6), $(dX)^2 = (\mu\,dt + \sigma\,dW)^2 = (\mu\,dt)^2 + \sigma^2(dW)^2 + 2\mu\sigma dt\,dW$, and applying rules (18.7a to 18.7d), i.e., keeping only first-order terms, we obtain $dX^2 = \sigma^2 dt$.

## 18.2.3 Multi-dimensional Arithmetic Brownian Motions

The Brownian motion studied above is one-dimensional since its values lie in $\mathbb{R}$, but it is possible to generalize all previous results to the multi-dimensional case.

---

[11] The space of centered, finite variance, random variables can be equipped with a scalar product, the covariance, and consequently with a norm, the variance. An element with zero norm is the null element.

We start by defining a multi-dimensional Wiener process $\underline{W}(t)$ as a process with $m$ standard Brownian components that are mutually independent: for $j \neq k$, $W_j$ is a Wiener process, independent of $W_k$, so that we may write, using Eq. (18.7c):

$$dW_j \, dW_k = 0.$$

Therefore, we may define the non-standard multi-dimensional arithmetic Brownian motion from $\underline{W}$ as a random vector in $m$ dimensions, $\underline{X}(t)$, which satisfies:

$$\underline{X}(t) - \underline{X}(0) = \underline{\mu} \, t + \Sigma \, \underline{W}(t), \tag{18.9}$$

where $\mu$ is a constant vector in $\mathbb{R}^n$ and $\Sigma$ an $n \times m$ matrix of constants called the *diffusion matrix*. The latter equation is analogous to Eq. (18.3). $X(t)$ thus satisfies the following SDE (18.10) analogous to Eq. (18.6):

$$d\underline{X} = \underline{\mu} \, dt + \Sigma \, d\underline{W}, \tag{18.10}$$

where $\underline{W}$ is a multi-dimensional Wiener process with $m$ independent components.[12]

The correlation structure of the components of $d\underline{X}$ is characterized by the diffusion matrix $\Sigma$. Indeed, looking at the $i$th component of Eq. (18.10), we can write

$$dX_i = \mu_i dt + \sum_{j=1}^{m} \sigma_{ij} dW_j.$$

This implies that

$$cov(dX_i, dX_k) = cov\left(\sum_{j=1}^{m} \sigma_{ij} dW_j, \sum_{l=1}^{m} \sigma_{kl} dW_l, \right) = \sum_{j=1}^{m} \sigma_{ij}\sigma_{kj} var\left(dW_j\right)$$

$$= \sum_{j=1}^{m} \sigma_{ij}\sigma_{kj} dt$$

i.e., the instantaneous variance-covariance matrix of the components of the vector $d\underline{X}$ is the matrix product $\Sigma \, \Sigma'$ (where $\Sigma'$ is the transpose of $\Sigma$). When $\Sigma$ is diagonal the components of $d\underline{X}$ are independent.

---

[12] Alternatively, one may consider a Wiener process $\underline{Z}$ with dependent components $(dZ_i dZ_j = \rho_{ij} dt)$ and write the SDE relative to the $i$th component as: $dX_i = \mu_i + \sigma_i \, dZ_i$.

## 18.3 More General Processes Derived from the Brownian Motion; One-Dimensional Itô and Diffusion Processes

In spite of its many successful applications, the Brownian motion is, in fact, a very special case, notably because its drift $\mu$ and diffusion parameter $\sigma$ are assumed to be constant. A generalization leads to processes whose drifts and diffusion coefficients are not constant. We discuss univariate processes here, multivariate processes being the subject of Sect. 18.5.

### 18.3.1 One-Dimensional Itô Processes

We wish to describe the dynamics of a univariate process more general than the Brownian motion whose drift $\mu(t)$ and diffusion parameter $\sigma(t)$ are themselves stochastic.

The variation in $X$ between $t$ and $t + dt$ thus satisfies the SDE

$$dX = \mu(t)dt + \sigma(t)dW. \tag{18.11a}$$

It should be remarked that the notations for the random elements $\mu(t)$ and $\sigma(t)$ depart exceptionally from our convention of using capitals for random quantities. We assume that the random processes $\mu(t)$ and $\sigma(t)$ are known at $t$: recall that, as a result, they are said to be "adapted."

Equation (18.11a) can be analyzed like Eq. (18.6) for the Brownian motion: seen from the instant $t$, the change $dX$ from $t$ to $t+dt$ is the sum of the expected value $\mu(t)dt$ and a random contribution $\sigma(t)dW$ with variance $\sigma^2(t)\,dt$.

The change in $X$ between 0 and $t$, as a sum of infinitesimals between 0 and $t$, can be written

$$X(t) - X(0) = \int_0^t \mu(s)ds + \int_0^t \sigma(s)dW(s). \tag{18.11b}$$

To have Eq. (18.11b), as a solution to SDE (18.11a), well defined depends on the definitions of the two integrals on the right-hand side. These two integrals are themselves adapted stochastic processes; their definition, which will be briefly discussed in Sect. 18.3.3 below, requires that the processes $\mu(t)$ and $\sigma(t)$ satisfy certain technical conditions. If these conditions are satisfied, SDE (18.11a) has the solution (18.11b) which, for given initial conditions, is unique in the two following, equivalent senses:

780           18 Elements of Stochastic Calculus

- All processes satisfying SDE (18.11a) and the same initial condition have the same realization (the same trajectory): if two processes $X(t)$ and $Y(t)$ obey Eq. (18.11a) and $X(0) = Y(0)$ then $X(t) = Y(t)$, a.s.,[13] and for any time $t \in (0, T)$.
- To any value $X(0)$ and any trajectory of the Brownian motion $W(t)]_{t \in (0, T)}$ corresponds a.s. a single trajectory $X(t)]_{t \in (0, T)}$ given by Eq. (18.11b).

**Definition** *A stochastic process obeying Eqs. (18.11a) and (18.11b) ($\mu(t)$ and $\sigma(t)$ satisfying certain technical conditions) is called an Itô process.*[14]

Like all the Brownian motions from which they are derived, **Itô processes are everywhere continuous and nowhere differentiable**. In contrast to Brownian motions, **Itô processes are not necessarily Markovian** (although some are), since the coefficients $\mu(t)$ and $\sigma(t)$ generally depends on the whole history preceding date $t$ and, consequently, are not necessarily memory-less processes.

The calculus rules for Brownian motions generalize to Itô processes.

*Heuristic rules applicable to Itô processes with parameters $\mu(t)$ and $\sigma(t)$ :*

$$dX^2 = \sigma(t)^2 dt; \quad dX \, dt = 0; \quad dX_1(t) \, dX_2(t) = \mathrm{cov}(dX_1, dX_2)$$
$$= \sigma_{12}(t) \, dt. \tag{18.12}$$

The justification for Eq. (18.12) is similar to that for the rules (18.9) relevant to Brownian motions.

### 18.3.2 One-Dimensional Diffusion Processes

Among Itô processes, which make up a vast class of continuous processes, we consider those whose drift and diffusion terms are deterministic functions of time and of the current value of the process. Denoting, respectively, by $\mu(t, X(t))$ and $\sigma(t, X(t))$ the values of the drift and diffusion terms, such a process is then described by the equations

$$dX = \mu(t, X(t)) \, dt + \sigma(t, X(t)) \, dW. \tag{18.13a}$$

$$X(t) - X(0) = \int_0^t \mu(s, X(s)) ds + \int_0^t \sigma(s, X(s)) dW(s). \tag{18.13b}$$

As in the more general case of Itô processes, the two integrals on the right-hand side of Eq. (18.13b) are well defined and provide "unique" trajectories, if the functions $\mu(t, X)$ and $\sigma(t, X)$ satisfy certain technical conditions (see Sect. 18.3.3 below).

---

[13] Almost surely, i.e., for any w *except possibly in a subset of* $\Omega$ of zero probability.

[14] From the Japanese mathematician who was the first to define and study them (just after World War II).

In addition, note that the coefficients $\mu(t, X(t))$ and $\sigma(t, X(t))$ obey stochastic processes, although they have been calculated as ordinary functions of $\mathbb{R}^2$ into $\mathbb{R}$ called $\mu(.)$ and $\sigma(.)$, since the argument $X(t)$ is stochastic.

**Definition** *Stochastic processes obeying Eqs. (18.13a) and (18.13b) are called diffusion processes.*

Just like the Brownian motions from which they are derived and the Itô processes of which they are special cases, **diffusion processes are everywhere continuous and nowhere differentiable**. But like Brownian motions, and in contrast to some Itô processes, **diffusion processes are Markovian** since the coefficients $\mu(t, X(t))$ and $\sigma(t, X(t))$, and hence the changes in $X$ after the date $t$, only depend on the final state $(t, X(t))$.

---

**Example 1. The Geometric Brownian Motion (GBM)**

A one-dimensional GBM is a process $X$ satisfying the SDE: $dX = \mu X \, dt + \sigma X \, dW$.

The coefficients $\mu$ and $\sigma$ are constant. This SDE is a special case of Eq. (18.13a) with $\mu(t, X(t)) = \mu X(t)$ and $\sigma(t, X(t)) = \sigma X(t)$. GBM is therefore a diffusion process. It is often used to describe the evolution of a market price. In this case, the initial price $X(0)$ is positive and $\mu$ is the expected instantaneous return, which is typically positive.

At this stage, we do not yet have mathematical tools adequate to solve the SDE governing GBM, that is to provide a formula for $X(t) - X(0)$. Simply note that, for $X(t) = 0$, the SDE governing $X$ does imply that $dX = 0$, and so recursively $X(t') = 0$ for all $t' > t$. As a result, if the process ever hits the value zero, it remains "stuck" to this value and the process is said to have an absorbing barrier at $X = 0$. Furthermore, if $X(0) > 0$, since $X(t)$ is continuous, its value cannot "jump" below zero, and so $X(t) \geq 0$ for all $t > 0$. We take up again later the study of GBM which is a process widely used to describe market price evolutions.

---

**Example 2. The Ornstein-Uhlenbeck Process**

This process, sometimes used to describe the evolution of short-term interest rates, obeys the SDE written $dX(t) = a \, [b - X(t)] \, dt + \sigma \, dW(t)$. The coefficients $a$, $b$ and $\sigma$ are constant with $a > 0$.

So, we have here a diffusion process whose drift $\mu(t, X(t)) = a \, [b - X(t)]$ is proportional to the gap between a value $b$ (a "normal" value) and the current value $X(t)$. A mean-reverting force toward $b$, with strength depending on the value of $a$, tends to push $X(t)$ back to its normal value; on top of this a random shock $\sigma \, dW$ is added whose intensity depends on the value of $\sigma$.

Later we revisit this process when we have more powerful technical tools.

## 18.3.3 Stochastic Integrals (*)

We now describe what the integrals appearing in Eqs. (18.11b), (18.13b), (18.14b), and (18.15b) mean.

### 18.3.3.1 The Itô Process Case

Recall Eq. (18.11b): $X(t) - X(0) = \int_0^t \mu(s)ds + \int_0^t \sigma(s)dW(s)$.

Call $I_1$ the first integral on the right-hand side, and rewrite it so as to emphasize the random aspect: $I_1(\omega, t) = \int_0^t \mu(\omega, s)ds$.

For a given $\omega$, $\mu(\omega, s)$ is a deterministic function of time describing a particular trajectory of the process $\mu$, and $I_1$ becomes an ordinary integral which can be defined as an ordinary Riemann integral, for example. So, for any process $\mu(s)$ for which all the trajectories are integrable,[15] $I_1(\omega, t) = \int_0^t \mu(\omega, s)ds$ can then simply be defined point by point, that is for each trajectory (for each $\omega$).

The second integral in Eq. (18.11b), $I_2 \equiv \int_0^t \sigma(s)dW(s)$, is less easy to define.

In more general terms, we have to define the stochastic integral $I \equiv \int_0^t Y(s)dW(s)$, where $Y(t)|_{t\in(0,T)}$ is a stochastic process, and to specify conditions on $Y(t)$ so that the definition makes sense. One can think of defining $I$ like a Stieltjes integral, using the standard method that consists in, starting from a partition $[t_0 = 0, t_1, ..., t_{k-1}, t_k, ..., t_m = t]$ of the interval $[0, t]$, considering the Riemann sums:

$$S_m = \sum_{k=1}^m Y(t_k^*)(W(t_k) - W(t_{k-1})) \text{ where } t_k^* \in (t_{k-1}, t_k),$$

and defining $I$ as the limit of the $S_m$ (almost surely or in quadratic mean[16]) as the partition becomes finer and finer ($m$ tends to infinity and $\text{Sup}_k(t_k - t_{k-1})$ goes to 0).

One can show that the sums $S_m$ converge (almost surely or in quadratic mean) if the process $Y(t)$ satisfies certain conditions,[17] but that the limit depends on the position of $t_k^*$ in the interval $[t_{k-1}, t_k]$.

---

[15] We then will assume $\int_0^t |\mu(\omega, s)| \, ds < \infty$ a.s.

[16] $S_m$ tends to $I$ in quadratic mean if $\lim_{m\to\infty} E[(S_m - I)^2] = 0$.

[17] A limit in quadratic mean of the sums $S_m$ can be obtained if $Y(t)$ is square integrable, i.e., if: $E\left[\int_0^t Y^2(s)ds\right] < \infty$. A limit a.s. can be obtained if $\int_0^t Y^2(s)ds < \infty$ a.s. Recall that the limit a.s. implies the limit in quadratic mean but that the converse is not always true. The integrability condition $E\left[\int_0^t Y^2(s)ds\right] < \infty$ is easier to verify and most often adopted.

## 18.3 More General Processes Derived from the Brownian Motion; One-Dimensional... 783

It is common to choose the integral corresponding to the choice $t^*_k = t_{k-1}$, and thus to define $I$ as the limit of the sums $S_m = \sum_{k=1}^{m} Y(t_{k-1})[W(t_k) - W(t_{k-1})]$.

Defined in this way, we obtain the Itô integral which is the only one used in finance.

To sum up, the Itô process, $X(t) - X(0) = \int_0^t \mu(s)ds + \int_0^t \sigma(s)dW(s)$, with the integrals involved, can be unambiguously defined if $\mu(t)$ and $\sigma(t)$ are two adapted processes satisfying integrability conditions for non-stochastic integrals that are most often written as:

$$(\text{CI}) \qquad E\left(\int_0^t |\mu(s)|ds\right) < \infty \quad \text{and} \quad E\left(\int_0^t \sigma^2(s)ds\right) < \infty.$$

The most important properties of the Itô integral are presented as a proposition.

**Proposition 2. Some Properties of the Itô Stochastic Integral**

$$E\left[\int_s^t Y(u)dW(u)|\mathbb{F}_s\right] = 0. \tag{18.14a}$$

*Define* $I(t) = \left[\int_0^t Y(u)dW(u)\right]$ *for* $t \in (0, T)$. *$I(t)$ is a martingale, that is,*

$$E\left[I(t) \mid \mathbb{F}_s\right] = E\left[I(t) \mid I(s)\right] = I(s) \tag{18.14b}$$

Furthermore,

$$E\left[\left(\int_s^t Y(u)dW(u)\right)^2 |\mathbb{F}_s\right] = \text{var}\left[\int_s^t Y(u)dW(u)|\mathbb{F}_s\right]$$

$$= \left[\int_s^t E(Y^2(u))du|\mathbb{F}_s\right]; \tag{18.14c}$$

$$E\left[\int_s^t Y(u)dW(u) \int_s^t Z(v)dW(v)|\mathbb{F}_s\right] = E\left[\int_s^t Y(u)Z(u)du|\mathbb{F}_s\right]. \tag{18.14d}$$

784　　　　　　　　　　　　　　　　　　　　18　Elements of Stochastic Calculus

Recall that $\mathbb{F}_s$ describes the information available on date $s$ and that $E(.\mid\mathbb{F}_s)$ is a conditional expectation.[18]

The following arguments make use of the heuristic rules of the stochastic calculus and do not constitute a rigorous proof of the proposition.

**Justification of Eq. (18.14a)**

$$E\left[\int_s^t Y(u)dW(u)\mid\mathbb{F}_s\right] = \left[\int_s^t E(Y(u)dW(u)\mid\mathbb{F}_s)\right]$$

$$= \left[\int_s^t E(Y(u)\mid\mathbb{F}_s)E(dW(u)\mid\mathbb{F}_s)\right] = 0$$

(using the fact that $dW(u) = W(u+du) - W(u)$ is independent of $Y(u)$ and that $E(dW(u)\mid\mathbb{F}_s) = 0$).

**Justification of Eq. (18.14b)**

$I(t) = \int_0^t Y(u)dW(u) = \int_0^s Y(u)dW(u) + \int_s^t Y(v)dW(v)$, from which we have

$I(t) = I(s) + \int_s^t Y(v)dW(v)$, and therefore, by Eq. (18.14a), since $I(s)$ is $\mathbb{F}_s$ - measurable ($I(s)$ is known on date $s$): $E[I(t)\mid I(s)] = I(s)$.

**Justification of Eq. (18.14c)**

$$E\left[\left(\int_s^t Y(u)dW(u)\right)^2\mid\mathbb{F}_s\right] = \mathrm{var}\left[\int_s^t Y(u)dW(u)\mid\mathbb{F}_s\right]$$

$$+\left(E\left[\int_s^t Y(u)dW(u)\mid\mathbb{F}_s\right]\right)^2 = \mathrm{var}\left[\int_s^t Y(u)dW(u)\mid\mathbb{F}_s\right]$$

by Eq. (18.14a);
    Furthermore,

$$\left[\left(\int_s^t Y(u)dW(u)\right)^2\right] = \int_s^t Y(u)dW(u)\int_s^t Y(v)dW(v)$$

$$= \iint Y(u)Y(v)dW(u)dW(v) = \int_s^t Y^2(u)du$$

---

[18] To be understood in concrete terms as an expectation formed at date $s$ using the available information $\mathbb{F}_s$.

18.4 Differentiation of a Function of an Itô Process: Itô's Lemma

since, by Eq. (18.7), $dW(u)\, dW(v) = 0$ if $u \neq v$ and $dW(u)\, dW(v) = du$ if $v = u$; the final equality in Eq. (18.14c) follows if we take into account the conditions imposed on the process $Y(t)$.

**Justification of Eq. (18.14d)**
The justification of Eq. (18.14d) is similar to that of Eq. (18.14c).

### 18.3.3.2 The Case of Diffusion Processes

All the results for Itô processes, in particular those given in the previous subsection, can be applied to the special case of diffusion processes. However, for the diffusion case, the integrability conditions can be given directly for the two-variable functions $\mu(t, x)$ and $\sigma(t, x)$: these have to be Lipschitz, that is, there are two constants $c$ and $k$ such that for all $x$, $y$ and $t \in (0, T)$,

$$\mid \mu\ (t,x) \mid \le\ c(1 + |x|); \ \mid \sigma\ (t,x) \mid \le\ c(1 + |x|);$$

$$\mid \mu\ (t,x) - \mu\ (t,y) \mid \le\ k \mid x - y \mid; \ \mid \sigma\ (t,x) - \sigma\ (t,y) \mid \le\ k \mid x - y \mid.$$

## 18.4   Differentiation of a Function of an Itô Process: Itô's Lemma

Recall that, for an ordinary differentiable function $f(t, X)$ of two deterministic variables $t$ and $X$, the differential of $f$ can be written as $df = \frac{\partial f}{\partial t} dt + \frac{\partial f}{\partial X} dX$.

*This rule does not hold when X is a stochastic process.* Therefore, we study the rule for differentiation which must be used when $X$ is an Itô or a diffusion process. We start with the case of a one-dimensional $X$ process before dealing with a vector process $\underline{X}$ in Sect. 18.5.

### 18.4.1 Itô's Lemma

Consider the following:

– A one-dimensional Itô process $X(t)$ obeying the SDE

$$dX = \mu(t)\ dt + \sigma(t)\ dW$$

where $\mu(t)$ and $\sigma(t)$ satisfy the integrability conditions and $W$ is a Wiener process.

– A function $f$ from $\mathbb{R}^2$ to $\mathbb{R}$, with values denoted by $f(t, x)$, that is once continuously differentiable with respect to $t$ and twice continuously differentiable with respect to the second component $x$. The partial derivatives are denoted by $\partial f/\partial t$, $\partial f/\partial x$ and $\partial^2 f/\partial x^2$.

We examine the process $f(t, X(t))$ which is stochastic since $X(t)$ also is. For example, $f(t, X(t))$ may describe the value of an option and $X(t)$ that of its underlying.

We consider the interval $(t, t+dt)$ during which $X$ varies by $dX = X(t+dt)) - X(t)$.

Using the notation $df = f(t + dt, X(t + dt)) - f(t, X(t))$ the variation in $f(t, X)$ is induced by the double infinitesimal variation $(dt, dX)$. To simplify the writing and keep some equations to a single line, (.) will be short for $(t, X(t))$.

The following proposition characterizes the process $f(t, X(t))$.

**Proposition 3. Itô's Lemma, One-Dimensional Case**
*With the notation and under the assumptions given above:*

(i) $f(t, X(t))$ is an Itô process.
(ii) *The differential df can be written in the following three ways:*

$$df = \frac{\partial f}{\partial t}(.)dt + \frac{\partial f}{\partial x}(.)dX + \frac{1}{2}\frac{\partial^2 f}{\partial x^2}(.)(dX)^2; \tag{18.15a}$$

*which leads to*

$$df = \left(\frac{\partial f}{\partial t}(.) + \frac{1}{2}\frac{\partial^2 f}{\partial x^2}(.)\sigma^2(t)\right)dt + \frac{\partial f}{\partial x}(.)dX; \tag{18.15b}$$

or:

$$df = \left(\frac{\partial f}{\partial t}(.) + \frac{\partial f}{\partial x}(.)\mu(t) + \frac{1}{2}\frac{\partial^2 f}{\partial x^2}(.)\sigma^2(t)\right)dt + \frac{\partial f}{\partial x}(.)\sigma(t)dW. \tag{18.15c}$$

*In integral form*, Eq. (18.15c) *reads*

$$f(t, X(t)) - f(0, X(0)) = \int_0^t \left(\frac{\partial f}{\partial s}(.) + \frac{\partial f}{\partial x}(.)\mu(s) + \frac{1}{2}\frac{\partial^2 f}{\partial x^2}(.)\sigma^2(s)\right)ds + \int_0^t \frac{\partial f}{\partial x}$$
$$\times (.)\sigma(s)dW.$$

Equations (18.15a to 18.15c) are of great use and can be applied to computing the differential of any function of an Itô or diffusion process.

**Justification of Eq. (18.15)**
First, remark that rules (18.15) are different from the corresponding rules for classical differentials. For example, Eq. (18.15a) includes the added term $\frac{1}{2}\frac{\partial^2 f}{\partial x^2} \times (.)(dX)^2$ that is of first order in stochastic calculus but is absent, since it would be of second order, in ordinary differential calculus.

## 18.4 Differentiation of a Function of an Itô Process: Itô's Lemma

To understand the origin of this additional term, and sticking to simple arguments, let us consider the variation $\Delta f = f(t+\Delta t, X(t + \Delta t)) - f(t, X(t))$ calculated over a short, but not infinitesimal, interval $\Delta t$. We set $\Delta X \equiv X(t + \Delta t) - X(t)$ and denote by $\varepsilon$ the terms of order higher than $(\Delta t)^2$ or $(\Delta X)^2$. The expansion of $f(t + \Delta t, X(t + \Delta t))$ to second order around $(t, X(t))$ gives

$$\Delta f = \frac{\partial f}{\partial t}(.)\Delta t + \frac{\partial f}{\partial x}(.)\Delta X + \frac{1}{2}\frac{\partial^2 f}{\partial x^2}(.)(\Delta X)^2 + \frac{1}{2}\frac{\partial^2 f}{\partial t^2}(.)(\Delta t)^2 + \frac{\partial^2 f}{\partial x \partial t}(.)\Delta X \Delta t$$
$$+ \varepsilon$$

Let us look at the behavior of $\Delta f$ as $\Delta t$ goes to zero by replacing $\Delta t$ with the differential $dt$, $\Delta X$ by $dX$, and removing terms of higher than first order. Since $(dX)^2$ is of order 1, we arrive at Eq. (18.15a).

Equation (18.15a) leads directly to Eq. (18.15b) since $(dX)^2 = \sigma^2(t)\,dt$ using rule (18.8a). Finally, Eq. (18.15b) yields Eq. (18.15c) directly since $dX = \mu(t)\,dt + \sigma(t)\,dW$.

### 18.4.2 Examples of Application

To illustrate the stochastic calculus just given, and since its rules have given rise to numerous applications in quantitative finance, we study two special processes (see Examples 1 and 2): the Geometric Brownian motion and the Ornstein-Uhlenbeck process.

### 18.4.2.1 Geometric Brownian Motion

We have already introduced Geometric Brownian motion (GBM) as a process obeying the SDE

$$dX = \mu X\,dt + \sigma X\,dW \tag{18.16a}$$

where the coefficients $\mu$ and $\sigma$ are two constants.

We have already noted that if $X(0)$ is positive GBM is necessarily positive for all time, which allows writing the relative variations of $X$ in the equivalent form

$$dX/X = \mu\,dt + \sigma\,dW. \tag{18.16b}$$

The diffusion parameter $\sigma$ is called the volatility.

We characterize the GBM with the following proposition.

**Proposition 4**

If $X(t)$ is a GBM satisfying Eqs. (18.16a or 18.16b) (*with drift $\mu$ and volatility $\sigma$*):

(i) $\ln X(t)$ is governed by an arithmetic Brownian motion whose drift is $(\mu - \sigma^2/2)$ and instantaneous variance is $\sigma^2$;

(ii) $X(t)$ is a log-normal variable that can be written as:

$$X(t) = X(0)e^{(\mu-\sigma^2/2)t+\sigma W(t)} \tag{18.16c}$$

Equation (18.16c) is the solution to the SDE (18.16a) or (18.16b);

(iii) The expected value of $X(t)$ increases exponentially at rate $\mu$.

**Proof**

To show (i), apply Itô's Lemma to $\ln(X)$ (Eq. (18.15a)):

$$d \ln (X) = \frac{dX}{X} - \frac{1}{2} \frac{(dX)^2}{X^2}.$$

Since $\frac{dX}{X} = \mu dt + \sigma dW$ and $\left(\frac{dX}{X}\right)^2 = \sigma^2 dt$, we can write

$$d \ln(X) = (\mu - \sigma^2/2) \; dt + \sigma \; dW.$$

Notice that this last result could be obtained directly from Eq. (18.15c). This does give us the first result (i): $\ln(X)$ is an arithmetic Brownian motion; in addition,

$$\ln(X(t)) - \ln(X(0)) = \ln(\frac{X(t)}{X(0)}) = (\mu - \sigma^2/2)t + \sigma W(t).$$

$X(t)$ is therefore log-normal and $X(t) = X(0)e^{(\mu-\sigma^2/2)t+\sigma W(t)}$, which proves (ii).

Writing $W(t) = U\sqrt{t}$ (where $U$ is $N(0,1)$) we have: $E(X(t)) = X(0)E\left(e^{(\mu-\sigma^2/2)t+\sigma\sqrt{t}U}\right) = X(0)e^{\mu t}$, which proves (iii) .[19]

The GBM has often been used to describe variations in stock prices. The log-normality of the price is compatible with one important property: A stock price cannot be negative (from the limited liability of the shareholders). This property does not hold for other distribution laws, notably the normal law.

### 18.4.2.2 The Ornstein-Uhlenbeck Process

This diffusion process, sometimes used to describe the evolution of a short-term interest rate, satisfies, as seen in Example 2, a SDE written

---

[19] Recall that if $X$ follows a normal distribution $N(\mu, \sigma^2)$, then we have: $E(exp(aX)) = exp(a\mu + 0.5a^2\sigma^2)$ (moment generating function).

## 18.5 Multi-dimensional Itô and Diffusion Processes (*)

$$dX(t) = a\,[b - X(t)]\,dt + \sigma\,dW(t). \tag{18.17}$$

To integrate this SDE, let

$$Y(t) = [X(t) - b]e^{at},$$

whose differential is

$$dY(t) = e^{at}\,dX(t) + [\,X(t) - b\,]\,a\,e^{at}\,dt$$

or else, using Eq. (18.17), $dY(t) = \sigma\,e^{at}\,dW(t)$.

From this it follows, by definition of $Y(t)$ and integrating over $dY$:

$$Y(t) - Y(s) = [X(t) - b]e^{at} - [X(s) - b]e^{as} = \sigma \int_s^t e^{au}dW(u)$$

or

$$X(t) = e^{-a(t-s)}[X(s) - b] + b + \sigma e^{-at\int_s^t e^{au}dW(u)}$$

As a "sum" of normal variables $e^{au}dW(u)$, $\int_s^t e^{au}dW(u)$ is also *normal*, just as $X(t)$ is.

Denoting by $x_s$ the realization at time $s$ of $X(s)$, we can compute the first two moments which completely specify its distribution, since it is normal:

$$E\,[X(t)\mid X(s) = x_s] = e^{a(ts)}\,[x_s - b] + b \equiv \varphi(x_s, (ts))$$

$$Var[X(t)|X(s) = x_s] = \sigma^2 e^{-2at}\int_s^t e^{2au}\,du = \sigma^2(1 - e^{-2a(t-s)})/2a \equiv \psi(ts).$$

These last results follow directly from Proposition 2 and show that the Ornstein-Uhlenbeck is stationary.

## 18.5 Multi-dimensional Itô and Diffusion Processes (*)

We introduced in Sect. 18.2.3 of Sect. 18.2 multivariate Brownian processes and we keep the notation there given for multivariate Itô processes. Furthermore, we extend the definitions and results in Sects. 18.3 and 18.4 for one-dimensional Itô and diffusion processes to the multivariate case.

## 18.5.1 Multivariate Itô and Diffusion Processes

– An $n$-dimensional Itô process, associated to a standard Brownian process $\underline{W}(t)$ with $m$ independent components, is a stochastic process $\underline{X}(t)$ with $n$ components governed by the following equations:

$$d\underline{X} = \underline{\mu}(t)dt + \Sigma(t)d\underline{W} \tag{18.18a}$$

$$\underline{X}(t) - \underline{X}(0) = \int_0^t \underline{\mu}(s)ds + \int_0^t \Sigma(s)d\underline{W}(s). \tag{18.18b}$$

The drift vector $\underline{\mu}(t)$ is an adapted stochastic process with $n$ components (satisfying certain technical conditions) and the diffusion (or co-volatility) matrix $\Sigma(t)$ is an $n \times m$ matrix each of whose $nm$ components $\sigma_{ij}(t)$ is an adapted process, satisfying technical conditions that ensure the existence and uniqueness of Eq. (18.18b).

The $i$th component of Eq. (18.18a) can be written $dX_i = \mu_i(t)dt + \sum_{j=1}^m \sigma_{ij}(t)dW_j$.

The variance-covariance matrix for the random vector $d\underline{X}(t)$ (with the generic term $\text{cov}(dX_i, dX_j)$) is, at time $t$, equal to $\Sigma(t) \Sigma'(t)$.

The multi-dimensional Itô process, like its analog in one dimension, is continuous, non-differentiable and not necessarily Markov.

– *An n-dimensional diffusion process* is a special Itô process whose dynamics can be written:

$$d\underline{X} = \underline{\mu}(t, \underline{X}(t)) \ dt + \Sigma(t, \underline{X}(t)) \ d\underline{W} \tag{18.19a}$$

$$\underline{X}(t) - \underline{X}(0) = \int_0^t \underline{\mu}(s, \underline{X}(s))ds + \int_0^t \Sigma(s, \underline{X}(s))d\underline{W}(s). \tag{18.19b}$$

The terms $\mu_i(t, \underline{X})$ and $\sigma_{ij}(t, \underline{X})$ are functions from $\mathbb{R}^{n+1}$ to $\mathbb{R}$ satisfying Lipschitz conditions to ensure the existence and uniqueness of the solution to Eq. (18.19b).

A multi-dimensional diffusion process, like its analog in one dimension, is continuous, non-differentiable, and Markov.

## 18.5 Multi-dimensional Itô and Diffusion Processes (*)

### 18.5.2 Itô's Lemma (Differentiation of a Function of an n-Dimensional Itô Process)

#### 18.5.2.1 Itô's Lemma for a Multivariate Process $\underline{X}$

Itô's lemma as formulated in Proposition 3 and applicable to univariate processes can be generalized to multivariate processes as follows.

**Proposition 5. Itô's Lemma, Multidimensional Case**

*Let $\underline{X}(t) = (X_1(t), \ldots, X_n(t))$ be a multivariate Itô process and $f(t, \underline{x})$ a function from $\mathbb{R}^{n+1}$ to $\mathbb{R}$ such that the partial derivatives $\partial f/\partial t$, $\partial f/\partial x_i$ and $\partial^2 f/\partial x_i \partial x_j$ exist and are continuous for all $(i, j)$.*

*Recalling that $(.)$ stands for $(t, \underline{X}(t))$, we can write*

$$df = \frac{\partial f}{\partial t}(.)dt + \sum_{i=1}^{n} \frac{\partial f}{\partial x_i}(.)dX_i + \sum_{i=1}^{n}\sum_{j=1}^{n} \frac{1}{2} \frac{\partial^2 f}{\partial x_i \partial x_j}(.)dX_i dX_j. \tag{18.20}$$

*This implies*

$$df = \left[ \frac{\partial f}{\partial t}(.) + \frac{1}{2}\sum_{i=1}^{n}\sum_{j=1}^{n} \frac{\partial^2 f}{\partial x_i \partial x_j} v_{ij}(.) + \sum_{i=1}^{n} \mu_i(.) \frac{\partial f}{\partial x_i}(.) \right] dt + \sum_{i=1}^{n}\sum_{j=1}^{m}$$

$$\times \frac{\partial f}{\partial x_i}(.)\sigma_{ij}(.)dW_j \tag{18.21}$$

*where $v_{ij}$ denotes a generic entry of the matrix $\Sigma \Sigma'$, i.e. $v_{ij} = \sum_{k=1}^{n} \sigma_{ik}\sigma_{jk}$.*

*In its integral form, Eq. (18.21) can be written*

$$f(t, \underline{X}(t)) - f(s, \underline{X}(s)) = \int_{s}^{t} \left( \frac{\partial f}{\partial u}(.) + \frac{1}{2}\sum_{i}^{n}\sum_{j}^{n} \frac{\partial f^2}{\partial x_i \partial x_j} v_{ij}(.) + \sum_{i}^{n} \mu_i(.) \frac{\partial f}{\partial x_i}(.) \right) du$$

$$+ \sum_{j=1}^{m} \int_{s}^{t} \sum_{i=1}^{n} \frac{\partial f}{\partial x_i}(.)\sigma_{ij}(.)dW_j$$

$$\tag{18.22}$$

Justification of these equations can be carried out in a similar way to that for Eqs. (18.15), i.e., by a series expansion of $f(t + dt, \underline{X} + d\underline{X})$ around $(t, \underline{X})$ and dropping terms of order higher than one.

#### 18.5.2.2 The Dynkin Operator

Writing down stochastic differential equations and Itô's Lemma can be greatly simplified by using the Dynkin operator (or more simply the Dynkin).

As a simplification, denote by $E_t(.)$ the conditional expectation $E( . | \mathbb{F}_t )$ and examine $E_t(df) = \underset{t}{E}\{ f(t+dt, \underline{X}(t+dt)) - f(t, \underline{X}(t)) \}$. In the one-dimensional case, this expectation can, by Eq. (18.15c), be written as

$$\underset{t}{E}\, df = \left[ \frac{\partial f}{\partial t}(.) + \mu(.)\frac{\partial f}{\partial x}(.) + \frac{1}{2}\sigma^2(.)\frac{\partial^2 f}{\partial x^2}(.) \right] dt \qquad (18.23a)$$

and, in the multi-dimensional case, using Eq. (18.22) as

$$\underset{t}{E}\, df = \left[ \frac{\partial f}{\partial t}(.) + \sum_{i=1}^{n} \mu_i(.)\frac{\partial f}{\partial x_i}(.) + \frac{1}{2}\sum_{i=1}^{n}\sum_{j=1}^{n} v_{ij}(.)\frac{\partial^2 f}{\partial x_i \partial x_j}(.) \right] dt. \qquad (18.23b)$$

Then one can write the instantaneous conditional expectation of the variation in $f$ in the form

$$\frac{\underset{t}{E}df}{dt} \equiv D^t f = \frac{\partial f}{\partial t}(.) + \sum_{i=1}^{n}\mu_i(.)\frac{\partial f}{\partial x_i}(.) + \frac{1}{2}\sum_{i=1}^{n}\sum_{j=1}^{n}\frac{\partial^2}{\partial x_i \partial x_j}v_{ij}(.). \qquad (18.24)$$

The operator $D^t$ defined in this way is called the Dynkin operator, or the differential generator, and is useful in simplifying the notation. For example, Eq. (18.22) can be rewritten as

$$df = D^t f\ dt + \sum_{i=1}^{n}\sum_{j=1}^{m}\frac{\partial f}{\partial x_i}(.)\sigma_{ij}(.)dW_j \iff$$

$$f(t, \underline{X}(t)) - f(s, \underline{X}(s)) = \int_s^t D^u f du + \sum_{j=1}^{m}\int_s^t [\sum_{i=1}^{n}\frac{\partial f}{\partial x_i}(.)\sigma_{ij}(.)dW_j]$$

which implies, if $\underline{X}(s) = \underline{x}$,

$$\underset{s}{E}\{f(t, \underline{X}(t))|\ \underline{X}(s) = \underline{x}\} = f(s, \underline{x}) + \underset{s}{E}\{\int_s^t D^u f du\}.$$

## 18.6 Jump Processes

We briefly examine the second type of process whose trajectories are characterized by randomly occurring discontinuities. These processes have been applied to different aspects of finance, notably in the area of credit risk. The reader will find a deeper study of jump processes in Chap. 28, specifically concerning credit risk.

### 18.6.1 Description of Jump Processes

In contrast to Ito processes, for which only infinitesimal changes are possible in infinitely small time intervals,[20] processes with jumps, often described by Poisson processes, are characterized by discrete changes taking place at random times.

Consider a process $X(t)$ whose value, during a given time interval, either remains stable or undergoes a finite and discrete modification; for a Poisson process the probability of such a jump is proportional to the length of the interval.

In a time interval of duration $dt$ the jump process $X(t)$ will, therefore, display *a finite change with an infinitesimal probability* while a continuous process in the same interval of duration $dt$ displays an almost certain but infinitesimal change.

### 18.6.2 Modeling Jump Processes

The *continuous time Poisson process* is the fundamental model for the description of jump processes.

A continuous time Poisson process, describing a certain event (e.g., a jump), is defined by the two following properties:

- Property 1
  In any time interval of duration $\Delta t$, the probability that the event occurs:
  - Never, equals $1 - \lambda \Delta t + o_1(\Delta t)$,
  - Once only, equals $\lambda \Delta t + o_2(\Delta t)$,
  - More than once, equals $o_3(\Delta t)$. $\quad o(\Delta t)$ denotes, as usual, a value such that $\lim_{\Delta t \to 0} \frac{o(\Delta t)}{\Delta t} = 0$ and $\lambda$ is a positive constant that describes the average number of occurrences in unit time.

- Property 2
  Events taking place in two disjoint time intervals are independent.

---

[20] Their paths, as previously stated, are almost surely continuous.

There is a great deal of latitude in defining the event to which a Poison process refers; for example, it could be the arrival of income at random times, or something that changes the value of a stock (bankruptcy, major fire, lawsuit, etc.), or a devaluation, etc. In the context of finance, such an event mostly leads to a discontinuity (a jump) in income or wealth.

Denote by $N(t)$ the number of events occurring between 0 and $t$. In continuous time $dN = N(t + dt) - N(t)$ is the number of events in the interval $(t, t + dt)$ and equals 1 with probability $\lambda dt$ and zero with probability $1 - \lambda dt$.

More generally, one may consider in continuous time a process $X(t)$ whose variation has a continuous evolution as well as discrete jumps of amplitude $J(t, X(t))$. The SDE governing the one-dimensional process $X$ then can be written

$$dX = \mu(t,X) \ dt + J(t,X) \ dN. \tag{18.25}$$

Recall that $dN$ can take on two values: 1 with probability $\lambda dt$, if there is a discontinuity (jump) between $t$ and $t+dt$; zero with probability $1 - \lambda dt$.

Now consider a real function of two variables, $f(t, x)$ and the process $f(t, X(t))$ where $X(t)$ is a jump process satisfying Eq. (18.25).

The variation's instantaneous expected value, in the interval $(t, t + dt)$, for an $f(t, X)$ that has continuous partial derivatives, is at the point $(t, X)$ given by

$$E\left(\frac{df}{dt}\right) \equiv D'f = \frac{\partial f}{\partial t} + \frac{\partial f}{\partial x}\mu(t,X) + E[\lambda(f(t,X + J(t,X)) - f(t,X) \mid \mathbb{F}_t].$$

The last equation is similar to the one used to define the Dynkin for continuous processes.

To conclude, let us remark that the generalization to multi-dimensional processes works here in a similar way as for diffusion processes.

## 18.7 Summary

- A stochastic process is a one- or multi-dimensional random variable depending on the state of the world $\omega$ (the set of all states is $\Omega$) and on time. It is denoted $X(t,\omega)$ or, less explicitly, $X(t)$.
- $X(t)$ may be defined or observed at particular instants $0, t_1, t_2, \ldots, t_n = T$ (discrete time), or for all t in an interval $(0, T)$.
- Continuous processes have trajectories (paths) surely continuous while jump processes show discontinuous ones.
- $X(t)$ is an arithmetic Brownian motion if its increments $X(t + \Delta t) - X(t)$ are independent if non-overlapping, and distributed according to the normal law $N(\mu\Delta t, \sigma^2\Delta t)$ with drift $\mu$ and diffusion coefficient $\sigma$.
- In the special case where $\mu = \sigma = 1$, it is a standard Brownian motion (or Wiener process) and is denoted $W(t)$.

## 18.7 Summary

- Any nonstandard Brownian increment can be written as a function of a standard Brownian:
$X(t + \Delta t) - X(t) \equiv \Delta X = \mu \, \Delta t + \sigma \, \Delta W$.
- Over an infinitesimal interval $(t, t + dt)$ this increment writes as a Stochastic Differential Equation (SDE): $dX = \mu \, dt + \sigma \, dW$, the "integral" of which is $X(t) - X(0) = \mu \, t + \sigma \, (W(t) - W(0))$.
- Differential calculus for Brownian motions satisfies specific heuristic rules such as: $(dW)^2 = dt$ ; $dWdt = 0$, hence $(dX)^2 = \sigma^2 \, dt$.
- Itô processes are much more general than the Brownian motion since their drift $\mu(t)$ and diffusion parameter $\sigma(t)$ are themselves stochastic. The variation in $X$ between t and $t + dt$ satisfies the Stochastic Differential Equation (SDE):
- $dX = \mu(t) \, dt + \sigma(t) \, dW$. Itô processes make up a vast class of continuous processes (their paths are continuous).
- A diffusion process is a special Itô process whose drift and diffusion terms are deterministic functions of time and of their current values and whose dynamic is either described by the following SDE or by a stochastic integral:
- $dX = \mu(t, X(t))dt + \sigma(t, X(t))dW <==> X(t) - X(0) = \int_0^t \mu(s, X(s))ds$

$$+ \int_0^t \sigma(s, X(s))dW(s)$$

- A stochastic integral of the form $I(t) \equiv \int_0^t Y(s)dW(s)$ can be defined in the Itô sense as "a limit" of the sums $S_m = \sum_{k=1}^m Y(t_{k-1})(W(t_k) - W(t_{k-1}))$. It is therefore a martingale process. This definition is subject to some technical integrability conditions.
- A Geometric Brownian motion (GBM) is a particular diffusion process satisfying the SDE: $dX = \mu X \, dt + \sigma X \, dW$, with $\mu$ and $\sigma$ constants. GBM is often used to describe the evolution of a market price. In this case $\mu$ is the expected instantaneous rate of return, and $\sigma$ the volatility.
- The Ornstein-Uhlenbeck process, another example of diffusion process, is sometimes used to describe the evolution of short-term interest rates. It obeys the SDE: $dX(t) = a \, [b - X(t)] \, dt + \sigma \, dW(t)$. The coefficients $a$, $b$ and $\sigma$ are constant with $a > 0$; a mean-reverting force tends to push $X(t)$ toward a "normal" value $b$.
- When considering an ordinary twice differentiable function $f(t, X)$, Itô's lemma states the rule for differentiation which must be used when $X$ is an Itô or a diffusion process obeying the SDE: $dX = \mu(t, X(t)) \, dt + \sigma(t, X(t)) \, dW$.
- Itô's lemma yields: $df = \left[ \frac{\partial f}{\partial t}(.) + \mu(t) \frac{\partial f}{\partial x}(.) + \frac{1}{2} \sigma^2(t) \frac{\partial^2 f}{\partial x^2}(.) \right] dt + \frac{\partial f}{\partial x} \times (.)\sigma(t)dW(t)$.
- Itô's lemma is the main tool of stochastic calculus and obtains by expanding $f(t + \Delta t, X + \Delta X)$ up to the second order and applying the rules $(dW)^2 = dt$ and $dW . dt = 0$.

- The expression for $df$ can be simplified by using the Dynkin operator (differential generator) $D^t$:

$$D^t f \equiv E_t df/dt = \left[ \frac{\partial f}{\partial t}(.) + \mu(.) \frac{\partial f}{\partial x}(.) + \frac{1}{2}\sigma^2(.) \frac{\partial^2 f}{\partial x^2}(.) \right].$$

- When $X(t)$ follows a GBM, Itô's lemma implies that $\ln(X)$ is an arithmetic Brownian motion obeying $d\ln X = (\mu - \sigma^2/2)\, dt + \sigma\, dW$. Hence $X(t)$ is log-normal and: $X(t) = X(0)e^{\left(\mu - \sigma^2/2\right)t + \sigma W(t)}$.
- It is possible to generalize the previous developments and define multi-dimensional Itô processes, in which case $\underline{X}$, $\mu(t)$, and $\underline{W}$ are vectors, $\Sigma(t)$ is a matrix and $\underline{X}(t)$ satisfies the multi-dimensional SDE : $d\underline{X} = \mu(t)\, dt + \Sigma(t)\, d\underline{W}$,
- A generalized version of Itô's lemma applies to multivariate processes.
- In contrast to continuous processes like Itô processes, for which only infinitesimal changes are possible in infinitely small time intervals, processes with jumps, often described by Poisson processes, are characterized by discrete changes taking place at random times.

## Suggestions for Further Reading

### Books

Baxter, M., & Rennie, A. (1999). *Financial calculus*. Cambridge University Press.
** Dumas, B., & Luciano, E. (2017). *The economics of continuous-time finance*. MIT Press.
* Elliot, R., & Kopp, E. (2004). *Mathematics of financial markets* (2nd ed.). Springer.
* Fusai, G., & Roncoroni, A. (2008). *Implementing models in quantitative finance: methods and cases*. Springer.
*** Karatzas, I., & Shreve, S. (1991). *Brownian motions and stochastic calculus*. Springer.
** Lamberton, D., & Lapeyre, B. (2007). *Introduction to stochastic calculus applied to finance* (2nd ed.). CRC Press.
* Merton, R. (1999). *Continuous time finance*. Basil Blackwell.
** Musiela, M., & Rutkowski, M. (2007). *Martingale methods in financial modelling* (Applications of mathematics). Springer.
*** Revuz, D., & Yor, M. (1994). *Continuous martingales and brownian motion* (2nd ed.). Springer.
** Shreve, S. (2008). *Stochastic calculus for finance II: continuous-time models*. Springer.

# *The Mathematical Framework of Financial Markets Theory

# 19

In the parts of this book devoted to valuation of derivative products and to portfolio management in continuous time, different notions were introduced and computation methods employed with a minimum of justification, raising in fact questions concerning, among other things:

— Changing probabilities. For example, how does one define the probability $Q$ in the risk-neutral world, and what precisely is its relationship to the true probability $P$? Why is the volatility the same in both the risk-neutral and the real worlds? What is the connection between the market price of risk (from which risk premiums depend) and the probability used?
— Changing the numeraire. For example, what is the precise definition of a numeraire, and can one always choose a probability to make prices martingales for an arbitrary given numeraire? What is the relation between a numeraire and its martingale probability, if there is one?
— The notion of complete markets. For example, what is the precise definition of a complete market? How does one recognize that a market is effectively complete? When can a price be unambiguously determined with the simple hypothesis of Absence of Arbitrage Opportunities (AAO)? What is the connection between the market price of risk, AAO and completeness of markets?

The modern theory of financial markets provides very precise answers to all these questions. It was developed in the early 1980s based on the work of Harrisson, Kreps, and Pliska,[1] is founded on the theory of probability and stochastic processes, and provides a very coherent analytical framework, to which we devote this chapter. Reading this chapter requires some basic knowledge of probability theory

---

[1] Harrison and Kreps, "Martingales and arbitrage in multiperiod securities markets", Journal of Economic Theory 20, 1979, pp. 381–408. Harrison and Pliska, "Martingales and stochastic integrals in the theory of continuous trading", Stochastic Processes and Their Applications 11, 1981, pp. 215–260.

© The Author(s), under exclusive license to Springer Nature Switzerland AG 2022
P. Poncet, R. Portait, *Capital Market Finance*, Springer Texts in Business and
Economics, https://doi.org/10.1007/978-3-030-84600-8_19

(probability spaces, tribes (σ-algebras), filtration, conditional expectation, etc.), that we will recapitulate, albeit briefly, and of stochastic processes (the preceding chapter is sufficient in this respect). The development that follows does not go beyond the analysis and algebra courses provided in a full undergraduate program in Maths, Physics, or Engineering or a math-oriented graduate program in Economics. Readers desiring more mathematical rigor can consult specialized works in Mathematical Finance.[2] Furthermore, we introduce some notions regarding portfolios (description using weights, redundant securities, APT). It could, however, be useful to refer to Chaps. 21 and 23 to clarify certain parts of the exposition that follows.

The general framework is given in Sect. 1; then a more constrained framework in which asset prices obey Itô processes is presented in Sect. 2; the risk-neutral universe and transforming prices into martingales are described in Sect. 3; changing probabilities is studied in Sect. 4 and changing numeraires associated with martingale probabilities in Sect. 5; Section 6 is devoted to the logarithmic portfolio which is the numeraire that leads to martingale prices under the true probability $P$; finally, Sect. 7 discusses incomplete markets.

## 19.1 General Framework and Basic Concepts

The general framework (probabilistic framework for defining random variables and an information system) is described in §1; the financial market, portfolio strategies, the notions of AAO and of complete markets are presented in §2; the conditions needed for the existence and uniqueness of a price system compatible with AAO are analyzed in §3.

### 19.1.1 The Probabilistic Framework

Consider a time interval $(0, T)$; an observer can have as a vantage point any date in this interval, but, unless otherwise specified, will be placed at the date 0.

We consider a probability space $(\Omega, \mathbb{F}_T, P)$; $\Omega$ is the space of states and an event is a subset of $\Omega$; $\mathbb{F}_T$ is a σ-algebra on $\Omega$ giving the events observable (measurable) on date $T$ to which probabilities can be given on date 0 (unconditional probabilities) and on any date $t$ ($0 < t < T$, conditional probabilities); $P$ is an unconditional probability measure.

We recall briefly the definitions of a σ-algebra (or tribe) and a probability measure.

---

[2] Among the numerous excellent works on this topic, we mention, in order of increasing difficulty: M. Baxter and A. Rennie, 1998 (very pedagogical, with the same mathematical level as the present book); M. Musiela and M. Rutkowski, 2007; M. Jeanblanc, M. Yor and M. Chesney, 2009; and I. Karatzas and S. Shreve, 1999 (a sequel to the volume published in the same collection "Brownian Motions and Stochastic Calculus", 1991, very thorough and rigorous).

## 19.1 General Framework and Basic Concepts

- A $\sigma$-algebra $\mathbb{F}_T$ is a collection of subsets of $\Omega$ such that: $(i)$ $\Omega$ and $\varnothing$ belong to $\mathbb{F}_T$; $(ii)$ if $A$ and $B$ are in $\mathbb{F}_T$, $A \cup B$ ($A$ or $B$) and $A \cap B$ ($A$ and $B$) are in $\mathbb{F}_T$.
- A probability measure $P$ is a mapping of $\mathbb{F}_T$ on $[0,1]$ such that:

$(i)$ $P(\Omega) = 1$ and $P(\varnothing) = 0$; $(ii)$ if $A$ and $B$ belong to $\mathbb{F}_T$, $P(A \cup B) = P(A) + P(B) - P(A \cap B)$.

The information available at some time $t$ between 0 and $T$ is described by the $\sigma$-algebra $\mathbb{F}_t$ comprising the events observable on $t$: So on date $t$ it is known if an event in $\mathbb{F}_t$ has taken place or not. The number of observable events increases with the passage of time and thus $\mathbb{F}_t$ is supplemented at each instant by the arrival of new information. This dynamics of the information between 0 and $T$ is described by a continuous succession of $\sigma$-algebras $\mathbb{F}_t$, called a filtration and denoted by $\{\mathbb{F}_t\}_{t \in [0,T]}$. A filtration satisfies the following three conditions:

(i) $\mathbb{F}_0$ is made up of events with probabilities 0 or 1; this condition trivially means that it is known, from date 0, if an event observable on this date has occurred (its probability is then 1) or has not (its probability is then 0).

(ii) $s < t \Leftrightarrow \mathbb{F}_s \subset \mathbb{F}_t$; this condition means that the observer retains a memory of all previous observations while accumulating new information and that, as a result, the information collection $\mathbb{F}_t$ is larger than $\mathbb{F}_s$ if and only if $t > s$.

(iii) $\mathbb{F}_s = \underset{t>s}{\cap} \mathbb{F}_t$ this technical condition means that information does not arrive discretely but in a continuous, progressive way.

Recall that a stochastic process $X$ is said to be $\mathbb{F}_t$ -measurable, or adapted, if its value $X(t)$ on date $t$ is known on date $t$.[3]

When the sources of uncertainty are given by a vector of Brownian motions $\underline{W}(t)$, the information assumed to be known at time $t$ is often the trajectory $\{\underline{W}(s)\}_{s \in [0,t]}$ of this random vector between 0 and $t$. Information $\mathbb{F}_t$ allows "reconstructing" the trajectory between 0 and $t$ of any Itô process governed by $\underline{W}(t)$. One then says the filtration is generated by $\underline{W}$.[4]

We often consider an event that happens "a.s." meaning "almost surely," i.e., except in circumstances with probability zero.

## 19.1.2 The Market, Securities, and Portfolio Strategies

In mathematical finance, the standard model of the market describes the securities that are traded there, the portfolio strategies which are dynamic combinations of traded securities, contingent claims which are actually or virtually traded (and, in the

---

[3] Formally, the event $X(t) \leq a \in \mathbb{F}_t$, for any real number $a$.

[4] More precisely, a filtration $\{\mathbb{F}_t\}_{t \in [0,T]}$ is generated by the $m$-dimensional Brownian vector $\underline{W}(t)$ if $\mathbb{F}_t$ comprises all the events $\underline{W}(s) \leq \underline{a}_s$ for all $s \leq t$ and all the vectors $\underline{a}_s$ of $R^m$.

800           19 *The Mathematical Framework of Financial Markets Theory

second case, which can be replicated or not by a dynamic strategy) and the prices of contingent claims that are compatible with AAO.

### 19.1.2.1 Primitive Securities

The market is assumed continuously open and the transactions that take place there are exempt from friction or transaction costs.

Certain financial securities that are traded on the market are called *primitive securities* (also sometimes *pure securities*). There are $n + 1$ of them, denoted by $0, 1, \ldots, n$ and they constitute a set which will be called the *pricing base*, or simply the *base*.[5] Unless otherwise mentioned, these are *unconditional* securities, which, in particular, excludes options. The security numbered 0 is risk-free on an infinitesimal horizon (in practice, this means an overnight money market security) while securities $1, \ldots, n$ bear risks. We assume that these securities do not distribute dividends or coupons.[6] $S_i(t)$ is the price at $t$ of security $i$ and rate of returns (returns in short) during the interval $(t, t + dt)$ write

$$\frac{dS_0}{S_0} = r(t)dt; \quad \frac{dS_i}{S_i} \equiv dR_i \quad i = 1, \ldots, n \qquad (19.1)$$

where $r(t)$ is the short-term rate (that varies randomly) and $dR_i$ the return on security $i$.[7]

The vector of $n$ risky returns on the primitive securities is denoted $d\underline{R}$ (the underlining indicates a vector). Remark that, in absence of dividends or coupons, the returns are only due to capital gains (or losses).

For the moment we do not assume any particular probability distribution for $d\underline{R}$ apart from supposing that the first and second moments of the distribution of prices $S_i(t)$ exist.

Note that non-primitive securities can be traded or created. We consider them below in Subsection 2.3.

## 19.1.3 Portfolio Strategies

Continuous and discrete-time formulations are related and, in this respect, reading this chapter may be made easier by first looking at Chap. 21.

At time $t$, and up to a scaling factor, a portfolio combining the $n + 1$ primitive securities can be defined by the weights $x_1(t), \ldots, x_n(t)$ of the $n$ risky securities. Since

---

[5] With no relation with the *basis* of a vector space.

[6] This is without loss of generality since, if asset $i$ *does pay out dividends or coupons*, one can consider a capitalization fund that re-invests them to acquire additional assets $i$. $S_i$ then represents the value of one share of this fund.

[7] For $dS_i/S_i$ to be the return on $i$, the possible dividend must be re-invested in $i$ and included in price $S_i$ (see previous footnote). Besides, for $dS_i/S_i$ to be well defined, $S_i(t)$ must be positive a.s. (it is less problematic in this respect to write $dS_i = S_i \, dR_i$).

## 19.1 General Framework and Basic Concepts

the weights of all the assets sum to one, the weight $x_0(t)$ of the risk-free asset equals $x_0(t) = 1 - \sum_{i=1}^{n} x_i(t)$, or in vector notation, $x_0(t) = 1 - \underline{1}'\underline{x}(t)$ ($\underline{1}$ is the unit vector in $\mathbb{R}^n$, the prime denotes a transpose and $\underline{1}'\underline{x}$ is a scalar product).

To a portfolio strategy $\underline{x}(t)_{(t\in[0,T])}$ and an initial investment $X(0)$ is associated with the portfolio's value $X(t)$. The strategy is denoted by $(\underline{x}, X)$ to make explicit the weights and the value.

**Definition of Self-financing and Admissible Strategies (SA)** *Recall that by definition, a self-financing portfolio has no cash inflows or outflows between 0 and T.*

*Furthermore, a self-financing strategy is said admissible if its final value X(T) is a random variable with a well-defined second moment $E\left[(X(T))^2\right] < \infty$.*[8]

*We denote by SA the double property of being Self-financing and Admissible.*

The relative increase $dX/X$ of the value of a self-financing portfolio equals its return (no dividend); it is also equal to the weighted average of the returns on its $n + 1$ elements.[9] If a portfolio is self-financing we can, therefore, write

$$\frac{dX}{X} = x_0(t)r(t)dt + \underline{x}'(t)d\underline{R}. \tag{19.2}$$

Over the period $(0, T)$ a self-financing strategy generates two cash flows $\{-X(0), X(T)\}$ on dates 0 and $T$ so its total return is $\frac{X(T)-X(0)}{X(0)}$.

In the rest of this subsection, we only consider self-financing strategies; we carry out a more general analysis in §3.3 below.

### Definition of an Arbitrage
*A self-financing and admissible strategy generating $\{-X(0), X(T)\}$ is an arbitrage if $X(T) \geq 0$ a.s. and either $E[X(T)] > 0$ and $X(0) \geq 0$ (requires no investment), or $X(0) < 0$ (the initial investment is negative) and $E[X(T)] \geq 0$.*

### Important Remark
We will sometimes skip the notation "a.s." (almost surely, that is with probability 1). All equalities and inequalities involving random variables are to be understood as holding a.s., even if a.s. does not appear explicitly.

---

[8] Recall that, if a random variable is in $\mathbb{L}^2$ (its second moment is defined), it is also in $\mathbb{L}^1$ (its expectation is defined).

[9] See chapter 21 (Section 2, Subsection 4) for this rather obvious property.

## 19.1.4 Contingent Claims, AAO, and Complete Markets

We have already described the primitive securities that are effectively traded and quoted, whose returns are assumed to satisfy Eqs. (19.1, 19.2). These primitive securities form a pricing base relative to which other securities or claims can, if required, be evaluated by arbitrage. Some of these other securities or claims could possibly be actually traded and others not. Generally, a security or a claim is defined as a right to future, in general random, cash flows and for this reason is termed *contingent* (to the state of the world that will materialize). The following definitions make precise several notions relative to these contingent claims.

**Definition of a Simple Contingent Claim**  *A simple contingent claim is a right to a single cash flow $C_T$ on T where $C_T$ is an $\mathbb{F}_T$-measurable random variable whose second moment is defined: $E\left(C_T^2\right) < \infty$.*

We tend to use the word "securities" for existing financial assets (in the base or not) and the word "claims," which is a *broader* category of assets including the first, for assets that exist or may potentially exist and may be simply "virtual." Recall that the $\mathbb{F}_T$-measurable random variables with defined second moments form a vector space called $\mathbb{L}^2$ for which the second moment provides a norm.[10] We thus call $\mathbb{L}^2$ the space of simple contingent claims.

In the more general framework considered in Subsection 3.3, we also examine *complex contingent securities and claims* which generate discrete or continuous cash flow sequences over $(0, T)$. Unless otherwise specified, we use the term contingent claim to mean a simple contingent claim, and denote it by its final payoff (cash flow, value); thus, for example, we use $C_T$ for a contingent claim that yields the terminal cash flow $C_T$.

Some examples of simple contingent claims:

– Any primitive security $i$ in the base ($i = 0, 1, \ldots, n$) can be considered a contingent claim as it provides a right to the final cash flow $S_i(T)$.
– Any admissible and self-financing strategy $(x, X)$ can also be considered a contingent claim as it is a right to its terminal cash flow $X(T)$.
– A derivative product such as a European option written on a primitive asset (or a portfolio or index combining such securities) is also a contingent claim. For example, a European call with strike $K$ and expiry $T$ written on the primitive asset $i$ is a right to $(S_i(T) - K)^+$ and, consequently, is a contingent claim.

---

[10]Formally, $\mathbb{L}^2$ is a Hilbert space equipped with the scalar product $E(XY)$.

## 19.1 General Framework and Basic Concepts

**Definition of an Attainable (or Replicable) Contingent Claim**
*A simple contingent claim $C_T$ is attainable if there is at least one SA strategy (a dynamic combination of the n primitive securities) whose final value equals the cash flow $C_T$: $(x, X)$ attains $C_T$ if and only if $X(T) = C_T$ a.s.*
  *$X(0)$ thus is the investment required to attain $C_T$ using the strategy $(x, X)$.*
  *The set of all attainable claims is denoted by $\mathbb{A}$.*

For example, in the Black-Scholes universe, a European option is attainable (or replicable) with a dynamic portfolio containing $\delta$ underlying securities and an adequate number of risk-free assets.

Generally speaking, $\mathbb{L}^2$, the set of contingent claims, includes both elements that are attainable and elements that are not. Furthermore, a contingent claim $C_T$ can, in certain cases, be replicated using several different SA strategies. Finally, as already noted, some contingent claims (whether attainable or not) are effectively traded, while others (attainable or not) are not traded and are virtual.

The following three subsets of the space $\mathbb{L}^2$ of contingent claims thus respect the following inclusions:

{base} $\subset$ {static portfolios (an infinite number of them) combining base securities without the possibility of reallocation between 0 and $T$} $\subset$ {contingent claims attainable by SA strategies combining primitive securities} $\subseteq$ {all contingent claims = rights to all payoffs $\in \mathbb{L}^2$.}

*A condition for absence of arbitrage opportunity (AAO):*
If two different SA strategies denoted $(x, X)$ and $(y, Y)$ attain $C_T$ (so $X(T) = Y(T) = C_T$), then $X(0) = Y(0)$ under AAO.

For example, if $X(0) < Y(0)$, an arbitrage consists in combining the strategy $(x, X)$ with the strategy $(-y, -Y)$; such a combination generates zero cash flow on date $T$ (since $X(T) - Y(T) = 0$) and a positive cash flow $Y(0) - X(0)$ on date 0.

**Definition of a Complete Market** *The market is complete if all contingent claims are attainable.*

As a result, in a complete market, any random variable in $\mathbb{L}^2$ is the final value of a dynamic SA portfolio combining the $n + 1$ primitive securities.

Note that the notion of completeness is doubly relative:

- $\mathbb{L}^2$ will contain more elements as the filtration $\{\mathbb{F}_t\}_{t \in [0,T]}$ becomes larger (so that $\mathbb{F}_T$ contains more observable events) and the completeness condition will then be more difficult to satisfy.
- There will be more attainable claims (and completeness will be all the more likely) as the number $n$ of primitive securities is higher since SA strategies are built from these securities.

In general, one includes in the base as many quoted nonredundant securities as possible (see Section 2, §3.2). Furthermore, we often consider a filtration generated by Brownian motions. This is the framework used in this chapter from Sect. 2 where we assume that the filtration $\{\mathbb{F}_t\}_{t\in[0,T]}$ is generated by $m$ Brownian processes; increasing $m$ makes the filtration larger. We show that the market is incomplete if $n < m$ but complete if $n = m$ (under the condition, satisfied by definition, that the primitive securities are not redundant).

### 19.1.5 Price Systems

Generally, a price system $\pi$ defined on the space $\mathbb{L}^2$ of contingent claims is a mapping from $\mathbb{L}^2$ into $\mathbb{R}$. Among the infinite number of possible mappings, only some are considered *viable* for their coherence properties and compatibility with AAO.

#### 19.1.5.1 Viable Price Systems
**Definition**

*A viable price system satisfies the three following conditions:*
   *(i) $C_T \geq 0$ a.s. and $E(C_T) > 0$ implies $\pi(C_T) > 0$; $C_T = 0$ a.s. implies $\pi(C_T) = 0$.*
   *(ii) If $C_T$ is attainable, $\pi(C_T) = X(0)$ for every $(x, X)$ strategy attaining $C_T$.*
   *(iii) $\pi$ is a linear mapping of $\mathbb{L}^2$ into $\mathbb{R}$.*

($i$) and ($ii$) are AAO conditions we have met before.

($iii$) Is simply an additional necessary AAO condition. Assume, for example, that it is not satisfied, which means that there exist three conditional assets $X_T, Y_T, Z_T$ such that $Z_T = \lambda X_T + \mu Y_T$ ($\lambda$ and $\mu$ are scalars) and $\pi(Z_T) \neq \lambda\, \pi(X_T) + \mu\, \pi(Y_T)$. A portfolio with $\lambda$ units of $X_T$ and $\mu$ units of $Y_T$ replicates $Z_T$; therefore, if, for example, $\pi(Z_T) > \lambda\, \pi(X_T) + \mu\, \pi(Y_T)$, the relevant arbitrage consists in selling $Z_T$ while simultaneously purchasing the replicating portfolio.[11]

#### 19.1.5.2 Existence and Uniqueness of a Viable Price System
– By definition, a price system provides a price for all contingent claims, whether they are attainable or not. With AAO, there is a viable price, and only one, for an *attainable* contingent claim ($\in \mathbb{A}$), namely the initial investment $X(0)$ required to attain it (condition ($ii$)), which we know to be unique under AAO. By contrast, ($ii$) does not provide a price for a non-attainable $C_T$ and an infinity of *prices can be considered for it*. Meanwhile, in a complete market, all contingent claims are attainable and there is one single viable price only for every contingent claim ($\in \mathbb{L}^2$).
– By definition, if the AAO rule does not hold, ($ii$) is violated and *a viable price system cannot exist.*

---

[11] Possibly one or several of these contingent assets are not actually *traded although they are tradable*: In this case, such an arbitrage is potential (it could be made if the assets were traded and priced with the system p).

In what follows, the space of viable price systems will be denoted by $\Pi$.
These conclusions are summarized in the following proposition.

**Proposition 1**
- *The existence of at least one viable price system is a synonym for AAO; there can exist many viable price systems but the price of an attainable claim is the same for all viable price systems, so the systems differ only in the prices assigned to non-attainable contingent claims.*
- *Uniqueness. Under AAO, the necessary and sufficient condition for the existence of only one viable price for every contingent claim is that the market is complete. The price of any contingent claim is then equal to the initial investment required to attain it with a SA strategy.*

### 19.1.5.3 Generalization to Non-self-Financing Strategies and Contingent Securities
The preceding analysis can be generalized to non-self-financing strategies and contingent claims. We sketch briefly such a generalization.

- The dynamics of the value $X(t)$ of a portfolio distributing $c(t)dt$ in the interval $(t, t + dt)$ writes

$$dX = X(t)\left[x_0(t)r(t)dt + \underline{x}'(t)d\underline{R}\right] - c(t)dt. \tag{19.3}$$

Equation (19.3) generalizes (19.2). Since (19.2) can be written $dX = X(t)[x_0(t)r(t) dt + \underline{x}'(t)d\underline{R}]$, a distribution of $c(t)dt$ reduces by that amount the increment $dX$, which yields (19.3).

- A contingent claim (not necessarily self-financing) is the right to the random density of coupons $\{c_t\}_{t\in(0,T)}$ to which should be added the payoff or value $C_T$.
- The non-self-financing strategy $(\{c(t)\}_{t\in(0,T)}, X)$ attains or replicates the contingent claim $(\{c_t\}_{t\in(0,T)}, C_T)$ if $c(t) = c_t$ a.s. for any $t$ and if $X(T) = C_T$ a.s.
- Under AAO, the price $C_0$ of a replicable contingent claim equals the initial investment $X(0)$ of any strategy that replicates it, and this price is unique.
- By definition, in a complete market, all claims are replicable. As a result, under AAO and in a complete market, the price of each contingent claim is given uniquely by the initial investment required to replicate it.

---

## 19.2 Price Dynamics as Itô Processes, Arbitrage Pricing Theory and the Market Price of Risk

The analysis provided in Sect. 1 is very general as it does not rely on special assumptions about the processes describing asset price dynamics (we only assumed that the payoffs of the securities and strategies are in $\mathbb{L}^2$). In this section and the rest

# 19 *The Mathematical Framework of Financial Markets Theory

of the chapter, we describe the sources of uncertainty as multidimensional Brownian motions and we assume that security prices are governed by Itô processes.

## 19.2.1 Price Dynamics as Itô Processes

We assume from now on that uncertainty stems from $m$ sources of risk such that the filtration $\{F_t\}_{t \in [0,T]}$ is generated by $m$ standard Brownian motions $\underline{W}(t)$ (the components $W_i$ being mutually independent) and that security prices are governed by Itô processes depending on these $m$ Brownians. In particular, the $n$-dimensional vector of primitive security returns writes

$$dR = \underline{\mu}(t)dt + \Sigma(t)d\underline{W}. \tag{19.4}$$

The $n$-vector $\underline{\mu}(t)$ and the $n \times m$ matrix of co-volatilities $\Sigma(t)$ (whose entries are denoted by $\sigma_{ik}(t)$) are $F_t$-measurable and obey integrability conditions.[12]

Additionally, we assume that the diffusion matrix $\Sigma(t)$ has rank $n$, which means, as we see in Subsection 3, that the $n$ primitive securities form a nonredundant set of securities. We return to this characterization of the base in Subsection 3.

More generally, the return on any security (primitive or not) or claim $z$ in the interval $(t, t + dt)$ can be written as

$$dR_z = \mu_z(t)dt + \sum_{k=1}^{m} \sigma_{zk}(t)dW_k. \tag{19.5}$$

In this context, from (19.2, 19.4), the value $X(t)$ of any *self-financing* portfolio satisfies

$$\frac{dX}{X} = \left[x_0(t)r(t) + \underline{x}'(t)\underline{\mu}(t)\right]dt + \underline{x}'(t)\Sigma(t)d\underline{W}. \tag{19.6}$$

We will occasionally set $\underline{\sigma}_X(t) = \underline{x}'(t)\Sigma(t)$; $\underline{\sigma}_X$ is called the "volatility vector."

## 19.2.2 Arbitrage Pricing Theory in Continuous Time

Equation (19.5) expresses the sensitivity $\sigma_{zk}(t)$ of $R_z$ to the fluctuation in the random factor $W_k$, which we will call the risk $W_k$ affecting security $z$.

Note that the structure of Eq. (19.5) is analogous to that of the generating process for returns which led to the Arbitrage Pricing Theory (APT) in discrete time

---

[12] See *in the previous chapter (Section 3, Subsection 3) the* integrability conditions (IC): $E\left(\int_0^T |\mu_i(t)|dt\right) < \infty; E\left(\int_0^T \sigma_{ik}^2(t)dt\right) < \infty.$

## 19.2 Price Dynamics as Itô Processes, Arbitrage Pricing Theory and the...

$(R_z = \mu_z + \sum_{k=1}^{m} b_{ik} F_k$; see Chap. 23); here the factors $\underline{F}$ are given by the $m$ Brownian motions.

In this way, the same arguments that lead to the APT in discrete time, based on linear algebra (Chap. 23, Sect. 2, §2.2 which is useful to read before starting on the remainder of this section), also lead here to the APT in continuous time.

**Proposition 2. APT in Continuous Time.**
*If the security returns depend on $m$ Brownian risk factors $W_1, \ldots, W_m$ as in Eq. (19.5), in AAO there exist $m + 1$ coefficients $r(t)$, $\lambda_1(t)$, $\ldots$, $\lambda_m(t)$ the same for all securities (and the first of which is the risk-free rate) such that, at any date $t$, the instantaneous expected return on any security $z$ can be written*

$$\mu_z(t) = r(t) + \sum_{k=1}^{m} \lambda_k(t)\, \sigma_{zk}(t). \tag{19.7}$$

The term $r(t)$ in Eq. (19.7) can be identified as the risk-free interest rate. Indeed, in accordance with Eq. (19.5), the return $dR_z$, whose $\sigma_{zk}(t)$ all vanish, is not random and its instantaneous expectation is $r(t)$; this then must, under AAO, equal the return on the risk-free asset with horizon $dt$.

Equation (19.7) means that the expected return on any security or portfolio equals, in AAO, the risk-free rate $r$ augmented with $m$ risk premia $\lambda_1 \sigma_{z1}, \ldots, \lambda_k \sigma_{zk}, \ldots, \lambda_m \sigma_{zm}$. The premium $\lambda_k \sigma_{zk}$ for risk $W_k$ is the product of the risk $W_k$ affecting $z$ whose intensity is $\sigma_{zk}$ with the coefficient $\lambda_k$, common to all securities; $\lambda_k$ is thus interpreted as *the market price of risk $W_k$* (risk premium for a unit $\sigma_{zk}$). The $m$-vector $\lambda(t)$ is called the MPR (market price of risk). Remark that Proposition 2 asserts that such a vector exists under AAO but *does not state that it is unique*. We see, in Section 3, §3, that the uniqueness of this MPR vector is the characteristic of a complete market.

The space of MPR vectors compatible with AAO is denoted by $\Lambda_P$.

The APT of type (19.7) can be applied in the same way to primitive and non-primitive assets. For the $n$ primitive securities, the APT writes

$$\mu_i(t) = r(t) + \sum_{k=1}^{m} \lambda_k(t)\, \sigma_{ik}(t) \text{for } i = 1, \ldots, n; \tag{19.8-a}$$

or, in matrix notation

$$\underline{\mu}(t) = r(t)\underline{1} + \Sigma(t)\, \underline{\lambda}(t).$$

This implies

$$dR = [r(t)\underline{1} + \Sigma(t)\underline{\lambda}(t)]dt + \Sigma(t)d\underline{W}. \tag{19.8-b}$$

For a *non-primitive* $z$ with return $dR_z$, APT (19.7) implies

$$dR_z(t) = \left[ r(t) + \sum_{k=1}^{m} \lambda_k(t)\, \sigma_{zk}(t) \right] dt + \sum_{k=1}^{m} \sigma_{zk}(t)\, dW_k(t).$$

The MPR $\lambda(t)$ is an $m$-dimensional $\mathbb{F}_t$-measurable stochastic process; we assume it satisfies the integrability condition:

(IC') $E\left\{ exp \int_0^T \|\underline{\lambda}'(t)\|^2 dt \right\} < \infty,$

which ensures that the stochastic integrals we use are well defined.

## 19.2.3 Redundant Securities and Characterizing the Base of Primitive Securities

### 19.2.3.1 Redundant Securities

In Chap. 21, Sect. 3.1, we study redundancy and the conditions for its existence in discrete time. This analysis can be extended to continuous time as we briefly sketch.

**Definition** *The $p$ risky securities $i = 1, \ldots, p$ are redundant between dates $t$ and $t + dt$ if there exist $(p + 1)$ non-zero $\mathbb{F}_t$-measurable scalars $a_1(t), \ldots, a_p(t), k(t)$ such that*

$$\sum_{i=1}^{p} a_i(t)dR_i = k(t)dt.$$

The redundancy condition simply says that there is a risk-free linear combination of random variables $dR_i$.

Without loss of generality, we assume that $\sum_{i=1}^{p} a_i(t) = 1$ (we can always recover this case by dividing $k(t)$ by $\sum_{i=1}^{p} a_i(t)$); then $\sum_{i=1}^{p} a_i(t)dR_i$ can be interpreted as the return on a portfolio $\underline{a}$ allocating on date $t$ the weight $a_i(t)$ to security $i$ ($i = 1, \ldots, p$). If $\sum_{i=1}^{p} a_i(t)dR_i = k(t)dt$, the portfolio $\underline{a}$ is risk-free from $t$ to $t + dt$ and the $p$ risky securities can replicate a risk-free security. We note that, in this case and under AAO, the return $k(t)$ is equal to the risk-free rate $r(t)$.

The condition that the securities are not redundant can be written in terms of the variance-covariance matrix of their returns. Let us denote by $\mathbf{V}_p$ the variance-covariance matrix of $p$ returns $dR_i$, with general entry $cov(dR_i, dR_j)/dt$.

## 19.3 The Risk-Neutral Universe and Transforming Prices into Martingales

We show, in Chap. 21 Sect. 3-1.3, that a necessary and sufficient condition for the $p$ securities not to be redundant is that $\mathbf{V}_p$ be positive definite, or equivalently, invertible.

Furthermore, if $d\underline{R} = \underline{\mu}(t)dt + \Sigma_p(t)d\underline{W}$ (where $\Sigma_p$ is the diffusion matrix with $p$ rows and $m$ columns), $\mathbf{V}_p = \Sigma_p\Sigma_p{}'$; a result known from linear algebra means that $\mathbf{V}_p$ is invertible if and only if $\Sigma_p$ is of rank $p$ (that is, its $p$ rows are linearly independent). Also, in order that $\Sigma_p$ be of rank $p$, we need $p \leq m$, which states that the collection is necessarily redundant if the number of securities exceeds the number of Brownian (this is a necessary condition but not a sufficient one).

To sum up, we have the following criteria for non-redundancy:

*The $p$ securities are not redundant s* $\Leftrightarrow$ $\mathbf{V}_p$ *is invertible* $\Leftrightarrow$ $\Sigma_p$ *has rank* $p \Rightarrow p \leq m$.

### 19.2.3.2 More on Primitive assets and Conditions for Pricing by Arbitrage

Recall that the $n$ primitive securities from a nonredundant set of securities called the base.[13] This non-redundancy condition implies that, in Eq. (19.4), the diffusion matrix $\sum(t)$ has rank $n$.

In addition, and unless otherwise indicated, *the base is made up of the greatest possible number of nonredundant quoted securities*. The value of a security or claim $z$ that does not belong to the base is a function of the prices of these securities. As mentioned earlier, this valuation is possible if security $z$ can be replicated by a dynamic portfolio involving securities in the base, i.e., when the set $\{z, base\}$ is redundant. Under AAO, the price of $z$ is simply equal to the value of the portfolio that replicates it.

---

## 19.3 The Risk-Neutral Universe and Transforming Prices into Martingales

We give some useful results about martingales (§1), show how to transform prices into martingales in the risk-neutral universe (§2) and characterize a complete market (§3).

### 19.3.1 Martingales, Driftless Processes, and Exponential Martingales

In the rest of this chapter, we transform price processes into martingales by changing the numeraire and the overall probability distribution. In this subsection, we examine some characteristics of martingale processes.

---

[13] Any redundant asset would simply be withdrawn from the base.

## 19 *The Mathematical Framework of Financial Markets Theory

### 19.3.1.1 Definition and an Example

Recall that $E(X(T)|F_t)$, i.e., the conditional expectation of $X(T)$ on date $t$, is a random variable (viewed from an instant $s < t$), and $F_t$-measurable (that is, its price is known from instant $t$).

**Definition** *An adapted process $X(t)$ is a martingale if $E[|X(t)|] < \infty$ and $E[X(T)/F_t] = X(t)$, for all $t \in [0, T]$.*

It follows from the definition and the properties of iterated conditional expectations that $X(t)$ is a martingale if and only if $E[|X(t)|] < \infty$, and.
  $E[X(t)|F_s] = X(s)$, for all $s$ and $t$ such that $0 \le s \le t \le T$.
Indeed, these conditions imply those required by our definition of a martingale (just set $t = T$). Conversely, by definition, if $X(t)$ is a martingale, we have for all $t$ and $s$.

(i)
$$E(X(T)|F_t) = X(t);$$
  (ii)
$$E(X(T)|F_s) = X(s).$$
Furthermore, again from iterated conditional expectations, for $s < t$.
  $E(X(T)|F_s) = E[E(X(T)|F_t)|F_s];$
Substituting from (i) and (ii) into the latter expression, we have $X(s) = E[X(t)|F_s]$.
The condition $X(s) = E[X(t)|F_s]$ for all $t \in [s, T]$, that seems stronger than what is needed by the definition itself, $X(s) = E[X(T)|F_s]$, is then actually equivalent to it, and constitutes an alternative definition of a martingale. To put it simply, a martingale is a process whose expected future value is always equal to its present value.

### 19.3.1.2 Representing a martingale as a Driftless Itô Process

We assume that $\{F_t\}_{t\in[0,T]}$ is the filtration generated by the $m$-dimensional Brownian motion $\{\underline{W}(t)\}_{t\in[0,T]}$. We consider some $m$-dimensional processes $\left\{\underline{\phi}(t)\right\}_{t\in[0,T]}$, that are adapted (that is, $F_t$- measurable) and satisfy the integrability conditions (IC):
$$E\left[\int_0^t \left\|\underline{\phi}(s)\right\|^2 ds\right] < \infty.$$
IC is imposed so that the stochastic integrals built from $\underline{\phi}(t)$ are well-defined; $\|\underline{\phi}\|$ is the Euclidean norm and $\underline{\phi}'d\underline{W}$ the scalar product $\sum_{k=1}^{m} \phi_k dW_k$.

**Proposition 3. Describing a martingale as a Driftless Process.**

(a) *Any process $M(t)$ that obeys a driftless stochastic differential equation (SDE), of the type $dM(t) = \underline{\phi}'(t) d\underline{W}$.*

## 19.3 The Risk-Neutral Universe and Transforming Prices into Martingales

$\Leftrightarrow M(t) = M(0) + \int_0^t \underline{\phi}'(s)d\underline{W}$, where $\underline{\phi}'$ is adapted and satisfies the integrability condition IC, is a martingale.

(b) Conversely, if $M(t)$ is a continuous $\mathbb{F}_t$-measurable martingale, there is a unique $m$-dimensional process $\{\phi(t)\}_{t \in [0, T]}$, that is adapted and satisfies IC, such that

$$dM(t) = \underline{\phi}'(t)\, d\underline{W}, or, equivalently, M(t) = M(0) + \int_0^t \underline{\phi}'(s)d\underline{W}.$$

### Proof

We have already seen in the preceding chapter that a driftless process is a martingale. We just remark here that if $X(t)$ is an Itô process obeying the SDE $dX(t) = \mu(t)dt + \underline{\phi}'(t)\, d\underline{W}$ with $\underline{\phi}'$ adapted and satisfying IC, the necessary and sufficient condition for it to be a martingale is that the drift term $\mu(t)$ vanishes for all $t$. Indeed:

- If on date $t$ $\mu(t) \neq 0$: $E[X(t + dt)] = X(t) + \mu(t)dt \neq X(t)$ and the process is not a martingale.
- If $\mu(s) = 0$ for all $s \leq t$, $X(t) = X(s) + \int_s^t \underline{\phi}'(u)d\underline{W}(u)$.

And, since $E\left\{\int_s^t \underline{\phi}'(u)d\underline{W}(u)\Big|\mathbb{F}_s\right\} = 0$, it follows that $E\{X(t)|\mathbb{F}_s\} = X(s)$, proving that $X(t)$ is a martingale.

The conditions "$\underline{\phi}'$ adapted and satisfying IC" ensure that the stochastic integral, and thus the SDE, is well defined (see the preceding chapter).

### 19.3.1.3 Return Dynamics and Exponential Martingales

The dynamics of securities or portfolios is often described using the notion of return. Let us consider a process $X(t)$ obeying

$$dX = X(t)\left[\mu(t)dt + \underline{\sigma}'(t)d\underline{W}\right].$$

If $X(t)$ is the value of a self-financing portfolio, its return is $\mu(t)dt + \underline{\sigma}'(t)\, d\underline{W}$. Here we write an integrability condition for $\mu(t)$ and $\underline{\sigma}(t)$[14]:

(IC')

$$E\left\{exp\int_0^T|\mu(t)|dt\right\} < \infty \text{ and } E\left\{exp\tfrac{1}{2}\int_0^T\|\underline{\sigma}(t)\|^2dt\right\} < \infty.$$

The instantaneous variance of $dX/X$ equals $\|\underline{\sigma}\|^2$[15] and Itô's Lemma applied to the function $\ln(X)$ yields

---

[14] Rather than for the drift and diffusion terms of $X(t)$, equal to $X(t)\mu(t)$ and $X(t)\sigma(t)$, respectively.

[15] Indeed we have: $\frac{1}{dt}var(dX/X) = \frac{1}{dt}var(\underline{\sigma}'\, d\underline{W}) \equiv \frac{1}{dt}var\left(\sum_{k=1}^{m}\sigma_k dW_k\right) = \frac{1}{dt}\sum_{k=1}^{m}\sigma_k{}^2 dt = \sum_{k=1}^{m}\sigma_k{}^2 = \|\underline{\sigma}\|^2$.

Global volatility thus is $\|\underline{\sigma}\|$.

$$d \ln (X) = \frac{dX}{X} - \frac{1}{2}\left(\frac{dX}{X}\right)^2 = \frac{dX}{X} - \frac{1}{2}\|\underline{\sigma}\|^2 dt = \left(\mu(t) - \frac{1}{2}\|\underline{\sigma}(t)\|^2\right)dt + \underline{\sigma}'(t)dW,$$

or

$$X(t) = X(0) \exp\left(\int_0^t \left(\mu(u) - \frac{1}{2}\|\underline{\sigma}(u)\|^2\right)du + \int_0^t \underline{\sigma}'(u)d\underline{W}(u)\right).$$

The integrability condition CI' means the two integrals in the last expression do actually exist.

As a result, we can assert the equivalence of the SDE and its solution:

$$\frac{dX}{X} = \mu(t)dt + \underline{\sigma}'(t)dW \Leftrightarrow X(t) = X(0)e^{\int_0^t \left(\mu(u) - \frac{1}{2}\|\underline{\sigma}(u)\|^2\right)du + \int_0^t \underline{\sigma}'(u)d\underline{W}(u)}. \tag{19.9}$$

Three important special cases of (19.9) are worth mentioning:

– *Geometric Brownian motion*: $\underline{\sigma}(t)$ are $\mu(t)$ constants denoted by $\underline{\sigma}$ and $\mu$.

In this case (19.9) can be written $\frac{dX}{X} = \mu\, dt + \underline{\sigma}'dW \Leftrightarrow X(t) = X(0)e^{\left(\mu - \frac{1}{2}\|\underline{\sigma}\|^2\right)t + \underline{\sigma}'\underline{W}(t)}$.

Another, even more special, case is that of a Geometric Brownian motion depending on a Brownian $W(t)$ with one-dimensional $\sigma$.

– *Locally risk-free process* (also called "with finite variation"): $\underline{\sigma}(t) = 0$ for all $t$.

This is notably the case for the price $S_0(t)$ of a risk-free asset satisfying the following SDE:

$$\frac{dS_0}{S_0} = r(t)dt \Leftrightarrow S_0(t) = S_0(0)\, exp\int_0^t r(u)du. \tag{19.10-a}$$

Without loss of generality, we can choose the unit of asset 0 so that its initial price equals one ($S_0(0) = 1$) and write

$$S_0(t) = e^{\int_0^t r(u)du}. \tag{19.10-b}$$

– *Driftless processes $X(t)$ and exponential martingales*: $\mu(t) = 0$ for all $t$.

$X(t)$ is then a martingale (since $E(dX) = \mu(t)X(t)dt = 0$); it is called an exponential martingale and is characterized by its dynamics:

## 19.3 The Risk-Neutral Universe and Transforming Prices into Martingales 813

$$\frac{dX}{X} = \underline{\sigma}'(t)d\underline{W} \Leftrightarrow X(t)$$

$$= X(0)\,exp\left(\int_0^t -\frac{1}{2}\|\underline{\sigma}(u)\|^2 du + \int_0^t \underline{\sigma}'(u)d\underline{W}(u)\right). \qquad (19.11)$$

In the following, we encounter different processes that obey SDEs like (19.11) and thus are exponential martingales.

In particular, we show in the next subsection that the prices of securities and values of self-financing portfolios denominated in the numeraire $S_0$ obey, in the risk-neutral universe, processes of type (19.11) characterizing exponential martingales.

Finally, note that, if $X(t)$ is an Itô process satisfying Eq. (19.9):

$$X(t) = X(0)e^{\int_0^t \left(\mu(u)-\frac{1}{2}\|\underline{\sigma}(u)\|^2\right)du + \int_0^t \underline{\sigma}'(u)d\underline{W}(u)}, \text{ the process } Y(t) = X(t)e^{-\int_0^t \mu(u)du} \text{ is}$$

a martingale.

### 19.3.2 Price and Return Dynamics in the Risk-Neutral Universe, Transforming Prices into martingales and Pricing Contingent Claims

In the real world, asset returns obey Eq. (19.4). In the risk-neutral (RN) universe, we have $d\underline{R} = r(t).\underline{1}\,dt + \Sigma(t)d\underline{W}^*$ (where $\underline{W}^*(t)$ is a standard Brownian vector in this universe) so that the expected returns of all assets equal the risk-free rate $r(t)$. *The probability measure giving these expectations is different from the probability measure P prevailing in the real world.* In this Section $Q$ denotes the RN probability, to distinguish it from the real or historical probability P; consequently, $E^Q$ is the expectation operator under $Q$ (and is different from the expectation $E^P$ calculated using probability $P$). Furthermore, in the rest of this chapter, SDEs written under $Q$ will involve the Brownian vector $\underline{W}^*$ to distinguish them from SDEs under $P$ involving the Brownian vector $\underline{W}$.

More generally, in the RN universe, on date $t$ the instantaneous expected return on any portfolio $(\underline{x}, X)$, self-financing or not, is equal to $r(t)$.

In addition, if the portfolio is self-financing, its return stems only from a capital gain (no dividend) and equals the relative variation of the portfolio value, which writes

$$\frac{dX}{X} = r(t)dt + \underline{x}'(t)\Sigma(t)d\underline{W}^* = r(t)dt + \underline{\sigma}_x'(t)d\underline{W}^*,$$

or else,

$$X(t) = X(0)e^{\int_0^t \left(r(u)-\frac{1}{2}\|\underline{\sigma}_x(u)\|^2\right)du + \int_0^t \underline{\sigma}_x'(u)d\underline{W}^*(u)} \qquad (19.12)$$

where $\underline{\sigma}_x' = \underline{x}'(t)\Sigma(t)$ is the volatility vector for the portfolio considered.

# 814                                    19 *The Mathematical Framework of Financial Markets Theory

The price $X$ is implicitly denominated in current monetary units (euros, dollars, etc.) but it is possible to give the price using another numeraire. If the risk-free security numbered 0 is chosen as the numeraire, the portfolio's value on date $t$ writes $\frac{X(t)}{S_0(t)}$. This value gives the number of units of asset 0 that must be paid in exchange for portfolio $x$ (i.e., it is a relative price). It can also be seen as a Present Value. Recall that, by Eq. (19.10-b), $S_0(t) = \exp \int_0^t r(u) du$, therefore $\frac{X(t)}{S_0(t)} = \frac{X(t)}{\exp \int_0^t r(u) du}$, and we recognize here the expression of a Present Value.

From now on we denote by $\widehat{X}(t)$ this value denominated in numeraire 0:

$$\widehat{X}(t) \equiv \frac{X(t)}{\exp \int_0^t r(u) du}.$$

Using (19.12) and noticing that $X(0) = \widehat{X}(0)$, we have.

$$\widehat{X}(t) = \widehat{X}(0) e^{\int_0^t -\frac{1}{2} \|\underline{\sigma}(u)\|^2 du + \int_0^t \underline{\sigma}'(u) d\underline{W}^*(u)}.$$

In this dynamics for $\widehat{X}(t)$ we recognize an exponential martingale (i.e. an SDE of type (19.11)). To transform price processes into martingales, it is sufficient to "deflate" the RN prices using the price $S_0(t)$ of the risk-free security (i.e., use security 0 as numeraire), and we can state the following proposition.

## Proposition 4

(i) *In the RN universe and with the numeraire $S_0(t)$, the price $\widehat{X}(t) \left( \equiv \frac{X(t)}{S_0(t)} \right)$ of any self-financing portfolio is a martingale.*

(ii) *The price on date $s$ of a right to a random cash flow $X_t$ available on date $t > s$ equals*

$$X(s) = S_0(s) E^Q \left\{ \frac{X_t}{S_0(t)} | \mathbb{F}_s \right\}, or \ else, \tag{19.13-a}$$

$$X(s) = E^Q \left\{ X_t e^{-\int_s^t r(u) du} | \mathbb{F}_s \right\}. \tag{19.13-b}$$

## Proof
Part (*i*) of the proposition results from what just precedes.

## 19.3 The Risk-Neutral Universe and Transforming Prices into Martingales

To justify part $(ii)$, consider the right to a single $X_t$ as a self-financing asset between $s$ and $t$, which is worth $X(t) = X_t$ on date $t$: part $(ii)$ of the proposition is a consequence of the martingale character of $\frac{X(s)}{S_0(s)}$ that implies $\frac{X(s)}{S_0(s)} = E^Q\left\{\frac{X_t}{S_0(t)} | \mathbb{F}_s\right\}$, i.e. (19.13-a), or again,

$$X(s) = E^Q\left\{\frac{S_0(s)X_t}{S_0(t)} | \mathbb{F}_s\right\} = E^Q\left\{X_t \frac{e^{\int_0^s r(u)du}}{e^{\int_0^t r(u)du}} | \mathbb{F}_s\right\} = E^Q\left\{X_t e^{-\int_s^t r(u)du} | \mathbb{F}_s\right\},$$

which proves (19.13-b).

Let us consider a right to a stochastic cash flow such as $X_t$ as a *contingent* security or claim by granting this term a more general meaning than in Sect. 1 (where a simple contingent claim was defined as the right to a random cash flow $X_T$). Equations (19.13-a) and (19.13-b) allow us to price such contingent claims, provided that one can calculate expectations under the RN probability.

---

### Example. The Standard Black-Scholes (BS) Model

Under BS assumptions ($r$, $\mu$ and $\sigma$ are constants and no dividend is paid out by the underlying spot asset), in the RN universe:

- The price of the underlying spot asset satisfies $S(t) = S(0)e^{(r-0.5\sigma^2)t + \sigma W^*(t)}$.
- A risk-free asset satisfies $S_0(t) = e^{rt}$.
- The price $\hat{S}(t) = S(t)/S_0(t)$ of the underlying relative to numeraire 0 satisfies $\hat{S}(t) = S(0)e^{-0.5\sigma^2 t + \sigma W^*(t)}$;
- the price on date 0 of the call of strike $K$ and expiry $T$ is,

$$C(0) = E^Q\{(C(T)e^{-rT}\} = E^Q\left\{(S(T) - K)^+ e^{-rT}\right\} =$$

$$E^Q\left\{\left(S(0)e^{(r-\frac{1}{2}\sigma^2)T + \sigma W*(T)} - K\right)^+ e^{-rT}\right\}.$$

Since $W^*(T)$ is distributed according to $N(0,T)$ under the probability $Q$, one can write.

$$C(0) = E\left\{\left(S(0)e^{(r-\frac{1}{2}\sigma^2)T + \sigma\sqrt{T}\, U} - K\right)^+ e^{-rT}\right\}, \quad \text{with} \quad U \text{ distributed}$$

according to $N(0,1)$, which is one way of stating the BS formula (see Chap. 11).

---

Equation (19.13-b) also allows to compute the present value of a cash flow sequence in the RN world, by summing (or integrating in case of a density) the present values of the different flows in the sequence. In this way is obtained an expression for the value $v(s)$ on date $s$ of an asset that pays out dividends $[c(t)dt]_{t \in [s,}$ $_{T]}$ as well as a final payoff $l(T)$:

$$v(s) = E^Q \left[ \int_s^T c(t) e^{-\int_s^t r(z)dz} dt + l(T) e^{-\int_s^T r(z)dz} | \mathbb{F}_s \right]. \tag{19.14}$$

### 19.3.3 Characterizing a Complete market and Market Prices of Risk

We have already remarked that completeness of a market is a relative notion: it depends on the filtration defining the observable events as well as on the base of primitive assets that is used. In this subsection, we show that the necessary and sufficient condition for a market to be complete, relatively to the filtration generated by $m$ Brownian motions and to a base made up of $n$ nonredundant primitive assets, is simply that $n = m$, and that this condition is also necessary and sufficient for the MPR vector $\underline{\lambda}(t)$ to be uniquely derived from the dynamics of the primitive assets.

Let us consider an arbitrary contingent claim $X_T$ $(X_T \in \mathbb{L}^2)$ whose payoff given in numéraire 0 is $\widehat{X}_T \equiv \frac{X_T}{S_0(T)}$. We have seen in Proposition 4-($i$) that $\widehat{X}_t \equiv E^Q\left[\widehat{X}_T | F_t\right]$ is a martingale (under $Q$) and in proposition 3 (on the representation of a martingale by a stochastic integral) that there exists an $m$-dimensional adapted process $\underline{\phi}(t)$ such that

(i) $\widehat{X}_t = \widehat{X}_0 + \int_0^t \underline{\phi}'(u)d\underline{W}^* \Leftrightarrow d\widehat{X}_t = \underline{\phi}'(t)d\underline{W}^*$, where $\underline{W}^*(t)$ is a Brownian vector under $Q$.

Furthermore, if the strategy $(\underline{x}, X)$ is self-financing, we have

(ii) $d\widehat{X} = \widehat{X}(t)\underline{x}'(t) \Sigma(t) d\underline{W}^*$.

This strategy $(\underline{x}, X)$ attains $X_T$ iff

$$\widehat{X}(t) \equiv \frac{X(t)}{S_0(t)} = E^Q\left[\widehat{X}_T | F_t\right] \text{ for all } t \in [0, T].$$

If this condition is satisfied, $\widehat{X}(t) = \widehat{X}_t$, and $d\widehat{X}$ also obeys ($i$).

Using the fact that $\widehat{X}(t) = E^Q\left[\widehat{X}_T | F_t\right]$, comparing ($i$) and ($ii$) shows that the weights $\underline{x}(t)$ have to satisfy

## 19.3 The Risk-Neutral Universe and Transforming Prices into Martingales

$$E^Q\left[\widehat{X}_T|\mathbb{F}_t\right]\underline{x}'(t)\Sigma(t) = \underline{\phi}'(t). \tag{19.15}$$

A priori we know that $E^Q\left[\widehat{X}_T|\mathbb{F}_t\right]$ and $\underline{\phi}$ exist (they depend on $\widehat{X}_T$); however, we do not know if there exists an $\underline{x}(t)$ satisfying (19.15), that is if $X_T$ is attainable. Equation (19.15) provides conditions for the existence of such an $\underline{x}(t)$. Recall that $\underline{x}$ is an $n$-dimensional vector, $\Sigma$ an $n \times m$ matrix of rank $n$ ($n \leq m$) and $\underline{\phi}$ is an $m$-dimensional vector. Equation (19.15) can thus be seen as a system of $m$ equations in $n$ unknowns (the $x_i(t)$) for which a solution exists for any $\underline{\phi}(t)$ if and only if $n = m$, in which case the co-volatility matrix $\Sigma(t)$ is invertible and

$$\underline{x}'(t) = \left[E^Q\left(\widehat{X}_T|\mathbb{F}_t\right)\right]^{-1}\Sigma^{-1}(t)\underline{\phi}'(t).$$

Note that the condition $n = m$ is the same condition as for the uniqueness of an MPR vector $\underline{\lambda}(t)$ compatible with the dynamics of the primitive assets. Indeed, we have seen that under AAO there is *at least* one vector $\underline{\lambda}(t)$ such that $\Sigma(t)\underline{\lambda}(t) = \underline{\mu} - r(t)\underline{1}$ (see Eq. (19.8-a)). This system of $n$ equations in $m$ unknowns (the $\lambda_k(t)$) has a unique solution if and only if $n = m$, that is if and only if $\Sigma(t)$ is invertible. The solution then writes $\underline{\lambda}(t) = \Sigma^{-1}(t)\left[\underline{\mu}(t) - r(t)\underline{1}\right]$.

For $n < m$, there are infinitely many $\underline{\lambda}(t)$ satisfying $\Sigma(t)\underline{\lambda}(t) = \underline{\mu}(t) - r(t)\underline{1}$.

The MPR satisfying this equation are called viable or compatible with AAO, and form a set that we denote by $\Lambda_P$:

$$\Lambda_P = \left\{\underline{\lambda}|\Sigma(t)\underline{\lambda}(t) = \underline{\mu}(t) - r(t)\underline{1} \text{ for all } t \in [0, T]\right\}.$$

We summarize the main conclusions of the preceding analysis in a proposition.

### Proposition 5

- *The following four statements are equivalent:*
  - (i) *The market is complete, relative to the filtration generated by the $m$-dimensional Brownian process $\underline{W}(t)$ and the nonredundant base formed by $n$ primitive assets;*
  - (ii) *The base includes at any date $n = m$ nonredundant assets;*
  - (iii) *The set $\Lambda_P$ of MPR compatible with AAO has a single element;*
  - (iv) *$\Sigma(t)$ is a.s. invertible and for any $t$.*
- *Furthermore, if the market is complete, we can write*

$$\underline{x}'(t) = \left[E^Q\left(\widehat{X}_T|\mathbb{F}_t\right)\right]^{-1}\Sigma^{-1}(t)\underline{\phi}'(t) \tag{(19.16)$^c$}$$

$$\underline{\lambda}(t) = \Sigma^{-1}(t)\left[\underline{\mu}(t) - r(t)\underline{1}\right]. \tag{(19.17)$^c$}$$

**Notational Convention** Some equations only hold for complete markets. When these equations are given a number, we add a superscript "c" to the equation number to distinguish them from more general relations that can also hold for incomplete markets. Eqs. $(19.16)^c$ and $(19.17)^c$ in Proposition 5 thus only hold in complete markets.

## 19.4 Change of Probability Measure, Radon-Nikodym derivative and Girsanov's Theorem

We have emphasized that the probability measure $Q$ which prevails in the RN universe differs from the historical probability measure $P$ without formally defining $Q$. We also presumed without further justification that a change from $P$ to $Q$ just amounts to a simple change in the drifts (under $Q$, all expected returns equal $r$) without affecting the volatility (diffusion) parameters.

In this Section, we justify these presumptions, define the underlying probability model more rigorously and provide more precise tools for changing a probability measure and for obtaining the risk-neutral measure.

Changing probabilities and the Radon-Nikodym derivative are presented in §1, Girsanov's Theorem in §2, the formal definition of risk-neutral measure in §3, and the correspondence between systems of viable prices and risk-neutral probabilities in §4.

### 19.4.1 Changing Probabilities and the Radon-Nikodym Derivative

*By definition, two probability measures $P'$ and $P$ are equivalent (denoted $P'{\sim}P$) if the events with probability zero are the same for both measures; more formally, $P'{\sim}P$ if and only if, for all $A \in \mathbb{F}_T$, $P'(A) = 0 \Leftrightarrow P(A) = 0$.*

Note that the sure events (those with probability 1) are also the same for two equivalent probabilities $P$ and $P'$. Indeed, if $P(B) = 1$, $P(\Omega - B) = 0 = P'(\Omega - B)$ (since $P'{\sim}P$), therefore $P'(B) = 1$. The events true a.s. thus are the same under both measures.

From now on, $E^{P'}$ and $E^P$ respectively denote the expectations under $P'$ and under $P$.

#### Proposition 6 and Definition of the Radon-Nikodym Derivative

(i) *If $P'{\sim}P$, there is a random variable $\xi$ that is $\mathbb{F}_T$-measurable with $E^P[\xi] = 1$ and $\xi > 0$ a.s. such that for all $A \in \mathbb{F}_T$:*

## 19.4 Change of Probability Measure, Radon-Nikodym derivative and Girsanov's...

$$P'(A) = \int_A \xi(\omega)dP(\omega). \tag{19.18}$$

$\xi$ is called the Radon-Nikodym derivative of $P'$ with respect to $P$ and we write

$$\xi \equiv \frac{dP'}{dP}.$$

(ii) Any random variable $\xi$ that is $\mathbb{F}_T$-measurable with $E^P[\xi] = 1$ and $\xi > 0$ a.s. is a "valid" Radon-Nikodym derivative, meaning that it unambiguously defines a probability measure $P'$ equivalent to $P$ by $P'(A) = \int_A \xi(\omega)dP(\omega)$ for all $A \in \mathbb{F}_T$ (or more simply by $dP' = \xi\, dP$).

(iii) If $\xi = \frac{dP'}{dP}$, for any $\mathbb{F}_T$-measurable variable $X$ whose expectation is defined, we have

$$EP'(X) = E^P(\xi\, X). \tag{19.19}$$

**Proof**

We only provide a justification for (*ii*) and (*iii*).

– First, we check that $P'$ defined by (19.18) is a probability measure. We do have

$$P'(\Omega) = \int_\Omega \xi(\omega)dP(\omega) = E^P[\xi] = 1.$$

In addition,

$$P'(A \cup B) = \int_{A \cup B} \xi(\omega)dP(\omega) = \int_A \xi(\omega)dP(\omega) + \int_B \xi(\omega)dP(\omega) - \int_{A \cap B} \xi(\omega)dP(\omega)$$
$$= P'(A) + P'(B) - P'(A \cap B).$$

– Then we check $P' \sim P$. Since $\xi > 0$ and $dP(\omega) \geq 0$, we have.

$$P'(A) = \int_A \xi(\omega)dP(\omega) = 0\text{ if and only if} P(A) = \int_A dP(\omega) = 0.$$

## Remark

A change in probability means intuitively a change in density (or mass); the density $dP(\omega)$ around $\omega$ is multiplied by the positive factor $\xi(\omega)$ (as a result $dP'(\omega) = \xi(\omega)\, dP(\omega)$. The condition $E^P[\xi] = 1$ ensures that $P'(\Omega) = 1$.

Part (*iii*) of the proposition results immediately from this interpretation:

$$
E^{P'}[X] = \int_\Omega X dP' = \int_\Omega X\xi dP = E^P[\xi X].
$$

### 19.4.2 Changing Probabilities and Brownian Motions: Girsanov's Theorem

We consider $\underline{W}$ a standard $m$-dimensional Brownian motion under the historical probability $P$, the filtration $\{\mathbb{F}_t\}_{t\in[0,T]}$ generated by $\underline{W}$ and an $m$-dimensional process $\left\{\underline{\gamma}(t)\right\}_{t\in[0,T]}$ such that $E\left\{\exp \frac{1}{2}\int_0^T \left\|\underline{\gamma}(t)\right\|^2 dt\right\}$ is defined (the second element in the condition IC', known as *Novikov's condition*). We define

$$
\xi(t) \equiv \exp\left[-\frac{1}{2}\int_0^t \left\|\underline{\gamma}(s)\right\|^2 ds - \int_0^t \underline{\gamma}'(s)d\underline{W}\right]. \tag{19.20}
$$

We recall that

(i) $\xi(t)$ is a martingale under $P$, since (19.20) conforms with Eq. (19.11) characterizing an exponential martingale;

(ii) Since $\xi(t)$ is a martingale, $E^P[\xi(T)] = \xi(0) = 1$; also $\xi(T)$ is $>0$ and $\mathbb{F}_T$-measurable. As a result, $\xi(T)$ satisfies the three conditions required from a Radon-Nikodym derivative: So we may define a new probability measure $P' \sim P$ by:

$$
\frac{dP'}{dP} = \xi(T) = e^{-\frac{1}{2}\int_0^T \left\|\underline{\gamma}(t)\right\|^2 dt - \int_0^T \underline{\gamma}'(t)\, d\underline{W}}
$$

Furthermore, choosing some arbitrary instant $t$ as the origin for time, we see that the two conditional probabilities $P'_t \equiv P'(.|\mathbb{F}_t)$ and $P_t \equiv P(.|\mathbb{F}_t)$ are connected by the Radon-Nikodym derivative:

## 19.4 Change of Probability Measure, Radon-Nikodym derivative and Girsanov's... 821

$$\frac{dP'_t}{dP_t} = e^{-\frac{1}{2}\int_t^T \|\underline{\gamma}(t)\|^2 dt - \int_t^T \underline{\gamma}'(t)d\underline{W}} = \frac{\xi(T)}{\xi(t)}$$

where $\xi(t)$ is defined by (19.20).

An important consequence is that if $M(t)$ is a martingale under $P'$, $\xi(t) M(t)$ is a martingale under $P$ and vice versa; indeed, if $M(t)$ is a $P'$-martingale

$$M(t) = E^{P'}_t[M(T)] = E^{P'}_t\left[\frac{\xi(T)}{\xi(t)}M(T)\right] = \frac{1}{\xi(t)}E^{P}_t[\xi(T)M(T)],$$

therefore $\xi(t) M(t) = E^P_t[\xi(T) M(T)]$, so that $\xi(t) M(t)$ is a $P$-martingale. The converse results from the "symmetry" of the change in probability measure from $P$ to $P'$.

We can now state an important theorem that we will often use.

### Proposition 7. Girsanov's Theorem.

\With the definitions and notation and under the conditions just introduced:

$\underline{\widehat{W}}(t) \equiv \underline{W}(t) + \int_0^t \underline{\gamma}(s)ds$ is a standard m-dimensional Brownian motion under the probability measure $P'$.

### Proof

We know that $\xi(t) \equiv \exp\left[-\frac{1}{2}\int_0^t \|\underline{\lambda}(s)\|^2 ds - \int_0^t \underline{\lambda}'(s) \ d\underline{W}\right]$ is a $P$-martingale for any vector process $\underline{\lambda}(t.)$ satisfying IC'. To assert that $\underline{\widehat{W}}(t)$ is a standard Brownian process under $P'$, it is sufficient to check that $\widehat{\xi}(t) \equiv \exp\left[-\frac{1}{2}\int_0^t \|\underline{\gamma}(s)\|^2 ds - \int_0^t \underline{\gamma}'(s) \ d\underline{\widehat{W}}\right]$ is a $P'$-martingale. This holds if $\xi(t) \ \widehat{\xi}(t)$ is a $P$-martingale, as a previous result states. Replacing $\xi(t)$ and $\widehat{\xi}(t)$ by their values and $\underline{\widehat{W}}(t)$ by $\underline{W}(t) + \int_0^t \underline{\gamma}(s) \ ds$, we have

$$\xi(t)\widehat{\xi}(t) = \exp\left[-\frac{1}{2}\int_0^t \left(\|\underline{\gamma}(s)\|^2 + \|\underline{\lambda}(s)\|^2 + 2\underline{\gamma}'(s)\underline{\lambda}(s)\right)ds - \int_0^t \left(\underline{\gamma}'(s) + \underline{\lambda}'(s)\right)d\underline{W}\right].$$

Defining $\underline{v}(t) \equiv \underline{\gamma}(t) + \underline{\lambda}(t)$, we obtain

$$\xi(t)\widehat{\xi}(t) = \exp\left[-\frac{1}{2}\int_0^t \|\underline{v}(s)\|^2 ds - \int_0^t \underline{v}'(s)d\underline{W}\right],$$

which is indeed a $P$-martingale.

Girsanov's theorem allows us to write down the dynamics of a process $X(t)$ under a probability $P'$ defined by $\frac{dP'}{dP} = e^{-\frac{1}{2}\int_0^T \|\underline{\gamma}(t)\|^2 dt - \int_0^T \underline{\gamma}(t) d\underline{W}}$, starting from its SDE under $P$: It suffices to replace the term $d\underline{W}$ in the latter by $d\widehat{\underline{W}} - \underline{\gamma}(t)dt$ to obtain the SDE governing $X(t)$ under $P'$. So we obtain the following rule:

$$d\underline{X} = \underline{\mu}(t)dt + \sum(t) \ d\underline{W} \text{ under } P \Leftrightarrow d\underline{X}$$
$$= \left[\underline{\mu}(t) - \sum(t)\underline{\gamma}(t)\right]dt + \sum(t) \ d\widehat{\underline{W}} \text{ under } P'.$$

(the fact that the implication holds in both directions comes again from the "symmetry" implicit in changing from one probability measure to another).

It is crucial to remember that a change in the probability leads to a modification in the vector of drifts $\underline{\mu}(t)$ but does not affect the co-volatility matrix $\sum(t)$.

### 19.4.3 Formal Definition of RN Probabilities

Let us take up again the description of the financial market, as described in Sect. 2, in which $n$ risky primitive assets obeyed Itô processes depending on $m$ Brownian processes so that with AAO, Eq. (8) holds in the historical or real universe; we recall (19.8-b)

$$d \ \underline{R} = [r(t) \cdot \underline{1} + \Sigma(t)\underline{\lambda}(t)]dt + \Sigma(t)d\underline{W} \tag{19.21}$$

Remember that $\underline{\lambda}(t)$ is a vector of market prices of risk market (MPR). Generally, Eq. (19.21) holds for all $\underline{\lambda}(t) \in \Lambda_P$, where $\Lambda_P$ is the set of all MPR compatible with AAO, meaning they satisfy $\Sigma(t)\underline{\lambda}(t) = \underline{\mu}(t) - r(t)\underline{1}$ for all $t$. We recall that $\Lambda_P$ is not empty under AAO and only contains a single element if the market is complete (the number of nonredundant assets equals $m$).

In the RN universe, the drift term in the return vector $d\underline{R}$ is equal to $r(t)\underline{1}$. Girsanov's theorem allows us to characterize the RN probabilities that are compatible with such a universe.

### Proposition 8
*For any MRP $\underline{\lambda}(t) \in \Lambda_P$ one may define a risk-neutral probability $Q$ from the historical probability $P$ using the Radon-Nikodym derivative:*

$$\frac{dQ}{dP} = e^{-\frac{1}{2}\int_0^T \|\underline{\lambda}(s)\|^2 ds - \int_0^T \underline{\lambda}'(s) \ d\underline{W}}. \tag{19.22}$$

*The existence of a risk-neutral probability is therefore dependent on AAO and its uniqueness on whether the market is complete or not.*

## 19.4 Change of Probability Measure, Radon-Nikodym derivative and Girsanov's...

**Proof**

By Girsanov's theorem, the process defined by $d\underline{W}^* = d\underline{W} + \underline{\lambda}(t)dt$ is Brownian under $Q$. Replacing $d\underline{W}$ by $d\underline{W}^* - \underline{\lambda}(t)dt$ in Eq. (19.21) yields

$$d\underline{R} = [r(t) \cdot \underline{1} + \Sigma(t)\underline{\lambda}(t)] \, dt + \Sigma(t)(d\underline{W}^* - \underline{\lambda}(t)dt), \text{ or } d\underline{R} = r(t)\underline{1}dt + \Sigma(t)d\underline{W}^*.$$

Since $\underline{W}^*$ is a Brownian under $Q$, the drift term, under this probability, is indeed equal to $r(t)$ for all risky assets, which is characteristic of the RN universe.

We note that the expected return on any portfolio that is a combination of risky and risk-free assets also equals $r(t)$.

We check an assertion made but not justified previously: Under a RN probability, the volatilities $\Sigma(t)$ are the same as under the historical probability, as only the drift terms are modified.

The Radon-Nikodym derivative $dQ/dP$ leads, from Eq. (19.18), to a pricing formula that involves the historical probability $P$. Indeed, consider a contingent claim on a single payoff $X_T$ available on date $T$. We look for its price $X(0)$ on date 0. Eqs. (19.13-a), (19.19) and (19.22) imply we can write.

$$X(0) = E^Q\left\{\frac{X_T}{S_0(T)}\right\} = E^P\left\{\frac{dQ}{dP}X_T e^{-\int_0^T r(t)dt}\right\}, \text{ or}$$

$$X(0) = E^P\left\{X_T e^{-\int_0^T \left(r(t)+\frac{1}{2}\|\underline{\lambda}(t)\|^2\right)dt - \int_0^T \underline{\lambda}'(t)d\underline{W}}\right\}.$$

(19.23)

The term $e^{-\int_0^T \left(r(t)+\frac{1}{2}\|\underline{\lambda}(t)\|^2\right)dt - \int_0^T \underline{\lambda}'(t)d\underline{W}}$ can be interpreted as a *stochastic discount factor*.

Finally, we reformulate Proposition 8 in set theory terms. Let us denote by $\mathbb{Q}$ the set of RN probabilities. Equation (19.22) defines a mapping from $\Lambda_P$ to $\mathbb{Q}$: to each $\underline{\lambda}(t) \in \Lambda_P$ it assigns a $Q_{\underline{\lambda}} \in \mathbb{Q}$ such that $\frac{dQ_{\underline{\lambda}}}{dP} = e^{-\frac{1}{2}\int_0^T \|\underline{\lambda}(s)\|^2 ds - \int_0^T \underline{\lambda}'(s)d\underline{W}}$.

If the market is not complete, $\Lambda_P$ and $\mathbb{Q}$ contain an infinite number of elements; if it is complete, each set has only one element.

### 19.4.4 Relations between Viable Price Systems, RN Probabilities, and MPR

We first consider self-financing strategies providing a single cash flow $X_T$ on date $T$.

The case with non-self-financing assets or portfolios is then briefly mentioned (in §4.3).

By definition, a probability measure $Q{\sim}P$ is RN if the value $\widehat{X}(t)\left(\equiv \frac{X(t)}{S_0(t)}\right)$ of any self-financing portfolio is a $Q$-martingale. Recall that $\mathbb{Q}$ denotes the set of RN probabilities and may a priori be empty or contain one or more elements.

We have characterized a system $\pi$ of viable prices (see Sect. 1) and we denoted $\Pi$ the set of viable price systems. From the analysis in Sect. 1, $\Pi$ is non-empty in AAO and only contains one element if the market is complete.

### 19.4.4.1 Relationship between $\Pi$ and $\mathbb{Q}$

The following proposition states there is a one-to-one correspondance (bijection) between $\Pi$ and $\mathbb{Q}$.

**Proposition 9**

(i) *The three following propositions are equivalent:*
   $\mathbb{Q}$ *is non-empty;* $\Pi$ *is non-empty; the market is free of arbitrage opportunities.*
(ii) *If there are no arbitrage opportunities, there is a bijection between the sets* $\mathbb{Q}$ *and.* $\Pi$
(iii) *The following three propositions are equivalent:*
   $\mathbb{Q}$ *has a single element;* $\Pi$ *has a single element; the market is complete.*

**Proof**

Assume first $\mathbb{Q}$ and $\Pi$ non-empty. We show in the appendix that the mapping defined below is indeed a bijection of $\mathbb{Q}$ onto $\Pi$.

$\mathbb{Q} \rightarrow \Pi$; to each $Q \in \mathbb{Q}$ assign $\pi_Q \in \Pi$ such that $\pi_Q(X_T) \equiv E^Q\left(\widehat{X}_T\right)$ for all $X_T \in \mathbb{L}^2$;

$\Pi \rightarrow \mathbb{Q}$ to each $\pi \in \Pi$ assign $Q_\pi \in \mathbb{Q}$ such that $Q_\pi(A) \equiv \pi(1_A S_0(T))$ for all $A \in \mathbb{F}_T$ ($1_A$ is the indicator function of the event $A$ and so is a random variable with $1_A(\omega) = 1$ if $\omega \in A$ and $1_A(\omega) = 0$ if $\omega \notin A$).

Since, by Proposition 1, $\Pi$ is not empty if and only if there are no arbitrage opportunities, we have shown (*i*) and (*ii*). Furthermore, by Proposition 1, $\Pi$ is not empty if and only if the market is complete; the bijection relating $\mathbb{Q}$ and $\Pi$ then implies that $\mathbb{Q}$ only has a single element if and only if the market is complete, which proves (*iii*).

### 19.4.4.2 Relationship between $\Pi$, $\Lambda_P$, and $\mathbb{Q}$ when Asset Prices Obey Itô Processes

Notice that Proposition 9 is very general; indeed, since its derivation is not based on a special assumption regarding the underlying processes, it holds in the general framework of Sect. 1. This applies, for example, for price trajectories containing some discontinuities (corresponding to possible stock market crashes). In the special Brownian framework of Sects. 2 and 3, the MPR $\underline{\lambda}(t)$ can be defined under AAO; they form the set $\Lambda_P = \left\{\underline{\lambda}|\Sigma(t)\underline{\lambda}(t) = \underline{\mu}(t) - r(t)\underline{1}\right\}$ for all $t$} and the criteria for completeness stated in Proposition 5 imply that the following equivalences hold:

# 19.5 Changing the Numeraire

$\mathbb{Q}$ has a single element $\Leftrightarrow \Pi$ has a single element $\Leftrightarrow$ the market is complete $\Leftrightarrow n = m \Leftrightarrow \Lambda_P$ has a single element $\Leftrightarrow \Sigma(t)$ is invertible $\Leftrightarrow V(t) \equiv \Sigma(t)\Sigma'(t)$ is invertible.

In the more general context of possibly incomplete markets, recall that Eq. (19.22) defines a one-to-one correspondence between $\Lambda_P$ and $\mathbb{Q}$ (see Sect. 7 below). The three sets $\Lambda_P$, $\Pi$ and $\mathbb{Q}$ therefore in correspondence: in particular, to each MPR compatible with AAO there corresponds a RN probability and a viable price system.

### 19.4.4.3 The Case of Non-self-Financing Securities and Portfolios

The theory can be extended to non-self-financing portfolios by considering the values $X(t)$ of (self-financing) capitalization funds and defining $\mathbb{Q}$ as the set of probability measures making values $\widehat{X}(t)$ martingales. In such a context, one ends up with the pricing formula (19.14).

## 19.5 Changing the Numeraire

In the preceding sections, we show how the choice of the risk-free asset (asset numbered 0) associated with a RN probability leads to martingale prices. In fact, it is possible to have different numeraires for the risk-free asset. Under AAO, to each numeraire corresponds one (or more) probability measure(s) equivalent to the historical probability $P$ that turn(s) prices into martingales. Conversely, to each probability measure equivalent to $P$ there is a corresponding numeraire that leads to martingale prices. Pricing a particular contingent security is often easier when one uses the appropriate pair of probability measure and numeraire.

We define viable numeraires in §1, then we study the martingale probabilities associated with a given numeraire in §2.

### 19.5.1 Numeraires

#### 19.5.1.1 Definition of a Numeraire

**Definition** *A numeraire is an admissible and self-financing portfolio whose value is almost surely (a.s) positive at any date.*

We denote by $(\underline{n}, N)$, or more simply by $N$, an arbitrary numeraire and by $\mathbb{N}$ the set of all numeraires.

Recall the two conditions required of a viable numeraire:

- It must be possible to construct it using a SA strategy that combines primitive securities: Under no circumstances can any stochastic process, even a positive one, be a numeraire if it cannot be replicated by some SA strategy. The condition

826                                   19  *The Mathematical Framework of Financial Markets Theory

that it is self-financing excludes portfolios that "lose value" as a result of payment (s) of dividends or coupons.

The first condition implies that $\mathbb{N}$ is a subset of all SA strategies.

– The value $N(t)$ of a numeraire must always be strictly positive a.s., which in particular ensures that the value $\frac{X(t)}{N(t)}$ of any portfolio denominated in the numeraire $N$ is well defined.

### 19.5.1.2 Examples of Numeraires

– The risk-free asset of value $S_0(t) = \exp \int_0^t r(u) du$ (we set $S_0(0) = 1$ without loss of generality) is a numeraire we have already met many times;
– A zero-coupon bond yielding \$1 at its maturity $\theta$ and priced $B_\theta(t)$ is a numeraire between dates 0 and $\theta$;
– Any of the $n$ primitive risky assets is a numeraire, insofar as it is self-financing and that its value is strictly positive a.s. on any date $t$.

### 19.5.1.3 Properties of Numeraires

One important, and obvious, property of numeraires is that *any self-financing portfolio remains self-financing after a change of numeraire.* The reason for this is that since the portfolio is self-financing, whether its value is expressed relative to that of one security (or portfolio) rather than another one cannot involve addition or subtraction of funds. We leave to the reader the task of a rigorous proof, which follows easily from the linearity of the covariance operator.[16]

## 19.5.2 Numeraires and Probabilities that yield martingale Prices

We establish a correspondence between the set $\mathbb{N}$ of viable numeraires and the set $\mathbb{P}$ of probabilities equivalent to the historical probability $P$.

### 19.5.2.1 Correspondence $\mathbb{N} \to \mathbb{P}$ and Characterization of the Probabilities that Make Prices Denominated in numeraire $N$ martingales ($N$-Martingale Probabilities)

We generalize Proposition 9 to arbitrary numeraires.

**Definition of martingale Probabilities Associated to a numeraire $N$** *For any numeraire N, we define $\mathbb{Q}_N \in \mathbb{P}$ as the set of probabilities equivalent to P which make the price $\frac{X(t)}{N(t)}$ of any SA portfolio $(\underline{x}, X)$ martingale.*

*We say that $\mathbb{Q}_N$ is the set of martingale probabilities associated with N or simply the set of N-martingale probabilities.*

---

[16]The covariance between a sum of weighted random variables $x_i Y_i$ and a random variable $Z$ is equal to the weighted sum of all the individual covariances between $Y_i$ and $Z$: $\mathrm{cov}(\Sigma x_i Y_i, Z) = \Sigma x_i \, \mathrm{cov}(Y_i, Z)$.

## 19.5 Changing the Numeraire

Note that the set $\mathbb{Q}_0$ of martingale probabilities associated to the numeraire $0$ is the same as the set $\mathbb{Q}$ of RN probabilities defined in the preceding Sect. 4, which we know to be non-empty under AAO and a singleton in perfect markets.

*A priori* we do not know that such $N$-martingale probabilities exist, i.e. if $\mathbb{Q}_N$ is non-void. The following proposition states that a bijection may be constructed between $\mathbb{Q}$ and $\mathbb{Q}_N$, which implies that the conditions for existence and uniqueness are the same for both sets.

### Proposition 10
*Consider an arbitrary numeraire $N$ and the set $\mathbb{Q}_N$ of martingale probabilities associated with it.*

*(i) In AAO, there is a bijection between $\mathbb{Q}$ (set of RN probabilities)and$\mathbb{Q}_N$ that may be defined as follows: the mapping $\begin{smallmatrix} f_N \\ \mathbb{Q} \to \mathbb{Q}_N \end{smallmatrix}$ assigns to any $Q \in \mathbb{Q}$ a probability $Q_N \in \mathbb{Q}_N$ such that $\frac{dQ_N}{dQ} = \frac{N(T)}{N(0)S_0(T)}$; its inverse $f_N^{-1}$ maps any $Q_N \in \mathbb{Q}_N$ onto a probability $Q \in \mathbb{Q}$ such that $\frac{dQ}{dQ_N} = \frac{N(0)S_0(T)}{N(T)}$.*

*(ii) $\mathbb{Q}_N$ is not empty if an only if the market is free of arbitrage and is a singleton if the market is complete.*

### Proof
Denote by $N$ some numeraire, $Q$ a RN probability and $\mathbb{Q}$ the collection of all RN probabilities.

We use several times the results that $X(0) = E^Q\left[\frac{X(T)}{S_0(T)}\right]$ for all SA $(\underline{x}, X)$ and $E^Q_N(X) = E^Q\left(\frac{dQ_N}{dQ} X\right)$ as well as the convention $S_0(0) = 1$. We assume first that the market is free of arbitrage so that $\mathbb{Q}$ is not empty.

We proceed in two steps.

**first step.** We check that $\xi_N \equiv \frac{dQ_N}{dQ} \equiv \frac{N(T)}{N(0)S_0(T)}$ satisfies the three conditions required of a Radon-Nikodym derivative:

(i) $\xi_N$ is $\mathbb{F}_T$-measurable, since $N(T)$ and $S_0(T)$ are $\mathbb{F}_T$-measurable;
(ii) $\xi_N > 0$ a.s., since $N(T)$ and $S_0(T)$ are $> 0$ a.s.;
(iii) $E^Q(\xi_N) = 1$, since $\frac{N(t)}{S_0(t)}$ is a $Q$-martingale (because $(\underline{n}, N)$ is SA), and therefore:

$$E^Q\left[\frac{N(T)}{S_0(T)}\right] = N(0), \text{ from which we have } E^Q\left[\frac{N(T)}{N(0)S_0(T)}\right] = 1.$$

**second step.** We construct a bijective mapping $f_N$ from $\mathbb{Q}$ to $\mathbb{Q}_N$.

We assign to each $Q \in \mathbb{Q}$ the probability $Q_N$ that is equivalent to $Q$ (and thus to $P$) defined by $\frac{dQ_N}{dQ} = \frac{N(T)}{N(0)S_0(T)}$ and check that the prices $\frac{X(t)}{N(t)}$ are $Q_N$-martingales for all SA $X(t)$, i.e. that $Q_N$ does belong to $\mathbb{Q}_N$. Indeed:

$$E^{Q_N}\left[\frac{X(T)}{N(T)}\right] = E^Q\left[\frac{dQ_N}{dQ}\frac{X(T)}{N(T)}\right] = E^Q\left[\frac{N(T)}{N(0)S_0(T)}\frac{X(T)}{N(T)}\right] = \frac{1}{N(0)}E^Q\left[\frac{X(T)}{S_0(T)}\right]$$
$$= \frac{X(0)}{N(0)}.$$

Furthermore, the inverse $f_N^{-1}$ of the mapping defined in this way assigns to any $Q_N$ in $\mathbb{Q}_N$ the probability $Q$ in $\mathbb{Q}$ defined by $\frac{dQ}{dQ_N} = \frac{N(0)S_0(T)}{N(T)}$.

By Proposition 9, $\mathbb{Q}$ is non-empty if and only if the market is free of arbitrage and has only one element if the market is complete: As a result, the bijection between $\mathbb{Q}$ and $\mathbb{Q}_N$ ensures that AAO is also a necessary and sufficient condition for $\mathbb{Q}_N$ to be non-empty and have a single element if the market is complete.

Note that the preceding proposition can easily be generalized:

**Proposition 11**

*Consider two arbitrary numeraires $N$ and $N'$. The following mapping $f$ is a bijection between $\mathbb{Q}_N$ and $\mathbb{Q}_{N'}$:*

- *to each probability $Q_N$ in $\mathbb{Q}_N$, $f$ assigns the probability $Q_{N'} = f(Q_N)$ defined by:*

$$\frac{dQ_{N'}}{dQ_N} = \frac{N(0)N'(T)}{N'(0)N(T)} \qquad (19.24)$$

- *to each $Q_{N'}$ in $\mathbb{Q}_{N'}$, $f^{-1}$ assigns $Q_N$ such that $\frac{dQ_N}{dQ_{N'}} = \frac{N'(0)N(T)}{N(0)N'(T)}$.*

The proof is similar to that given for the preceding proposition.

We have in this way constructed a correspondence that associates to any $N \in \mathbb{N}$ a subset $\mathbb{Q}_N$ of $\mathbb{P}$ and we have shown that the different $\mathbb{Q}_N$ are in a mutual one-to-one correspondence, as illustrated in the figure below:

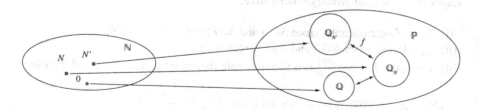

In complete markets, the subsets $\mathbb{Q}_N$ each have a single element, therefore to each $N$ corresponds a single martingale probability $Q_N$, thus defining a map from $\mathbb{N}$ to $\mathbb{P}$ that, as we will see in the following subsection, is bijective.

# 19.5 Changing the Numeraire

Finally, note that Eq. (19.24) allows us to evaluate the Radon-Nikodym derivative $\frac{dQ_{N'}}{dQ_N}$ for any pair $(N, N')$ of viable numeraires for which we know the final values $N(T)$ and $N'(T)$.

> **Example of using Eq. (19.24).**
> The *forward neutral* probability, that we denote by $Q_T$ and corresponds to a zero-coupon with expiry $T$ priced at $B_T(t)$ as numeraire, is such that
>
> $$\frac{dQ_T}{dQ} = \frac{B_T(T)}{B_T(0)S_0(T)} = \frac{1}{B_T(0)S_0(T)}$$
>
> This probability is of great use in pricing interest rate instruments and is employed in many chapters of this book (notably 16, 17, 28 and 29). Later we will encounter several other examples of application of Eq. (19.24).

### 19.5.2.2 The Mapping $\mathbb{P} \to \mathbb{N}$ and the Characterization of Numeraires
**Proposition 12**
*Let $\Phi$ be an arbitrary probability equivalent to P. Under AAO, there exists a single numeraire that makes the prices of SA portfolios martingales under $\Phi$.*

*The uniqueness of this "martingale numeraire" or "$\Phi$-numeraire" holds even in incomplete markets.*

This proposition, that we prove in Sect. 7, amounts to stating that the different $Q_{NS}$ for all $N \in \mathbb{N}$ provide a partition of $\mathbb{P}$, that is, $\underset{N\in\mathbb{N}}{\cup} Q_N = \mathbb{P}$ (to every probability corresponds a martingale numeraire) and $Q_N \cap Q_{N'} = \varnothing$ if $N \neq N'$ (the same probability cannot be a martingale probability for two different numeraires).

In complete markets, where the different $Q_N$ are all singletons, the two sets $\mathbb{N}$ and $\mathbb{P}$ are bound together by a bijection.

Remark that the historical probability $P$ belongs to $\mathbb{P}$ since it is trivially equivalent to itself. Proposition 12 therefore implies there exists a unique numeraire which makes martingales the values of all SA strategies (portfolios) *under the true probability P*. We characterize this P-numeraire in Sect. 6.

### 19.5.2.3 Volatility of numeraires and Market Prices of Risk in Complete Markets
In this subsection we assume the market is free of arbitrage and complete. As a result, to each numeraire N corresponds a unique martingale probability $Q_N$ and a unique MPR vector $\lambda_N(t)$. We thus can write, for any SA portfolio of value $X(t)$ and volatility vector $\underline{\sigma}_x(t)$, the following SDE under $Q_N$:

$$\frac{dX}{X} = \left[ r(t) + \underline{\sigma}_X{}'(t)\underline{\lambda}_N(t) \right] dt + \underline{\sigma}_X{}'(t) d\underline{W}^N$$

where $\underline{W}^N$ is a Brownian process under $Q_N$.

We characterize the MPR $\underline{\lambda}_N(t)$ by showing that it is equal to the volatility vector $\underline{\sigma}_N(t)$ of numeraire $N$.

## Proposition 13

*The MPR vector $\underline{\lambda}_N(t)$ associated with the martingale probability $Q_N$ is equal to the volatility vector $\underline{\sigma}_N(t)$ of numeraire $N$: $\underline{\lambda}_N(t) = \underline{\sigma}_N(t)$.*

*It follows that, for any SA portfolio with value $X(t)$ and volatility vector $\underline{\sigma}_X(t)$, we have under $Q_N$:*

$$\frac{dX}{X} = \left[ r(t) + \underline{\sigma}_X{}'(t)\underline{\sigma}_N(t) \right] dt + \underline{\sigma}_X{}'(t)\ d\underline{W}^N \qquad ((19.25-\mathrm{a})^c)$$

where $W^N$ is a Brownian vector under $Q_N$.

*In particular, we have:*

$$\frac{dN}{N} = \left[ r(t) + \|\underline{\sigma}_N(t)\|^2 \right] dt + \underline{\sigma}_N{}'(t)\ d\underline{W}^N$$
$$\Leftrightarrow N(t) = N(0)e^{\int_0^t \left( r(u)+\frac{1}{2}\|\sigma_N(u)\|^2 \right)du + \int_0^t \underline{\sigma}_N{}'(u)d\underline{W}^N(u)} \qquad ((19.25-\mathrm{b})^c)$$

*In addition[17]:*

$$\frac{dQ_N}{dQ} = e^{\int_0^T \frac{1}{2}\|\underline{\sigma}_N(u)\|^2 du + \int_0^T \underline{\sigma}_N{}'(u)dW^N(u)}. \qquad (19.26)$$

In Sect. 7 we will see that Eqs. $(19.25\text{-a})^c$ and $(19.25\text{-b})^c$, valid only in complete markets, can be generalized to incomplete markets.

## Proof

Since $\underline{\lambda}_N(t)$ is the MPR vector under $Q_N$, for all SA $X(t)$, we have under $Q_N$

$$\frac{dX}{X} = \left[ r(t) + \underline{\sigma}_X{}'(t)\underline{\lambda}_N(t) \right] dt + \underline{\sigma}_X{}'(t)d\underline{W}^N$$

and, in particular,

---

[17] Note that the Radon-Nikodym derivative given in (19.26) is not an exponential martingale under $Q$. One must use Girsanov's theorem and the Brownian vector $\underline{W}^*$ to obtain the exponential martingale under $Q$: $e^{\int_0^T -\frac{1}{2}\|\underline{\sigma}_N(u)\|^2 du + \int_0^T \underline{\sigma}_N{}'(u)d\underline{W}_*(u)}$

## 19.5 Changing the Numeraire

$$\frac{dN}{N} = \left[r(t) + \underline{\sigma_N}'(t)\underline{\lambda_N}(t)\right]dt + \underline{\sigma_N}'(t)d\underline{W}^N.$$

We need to show $\underline{\lambda_N}(t) = \underline{\sigma_N}(t)$. Itô's lemma implies that

$$d\ln\left(\frac{X}{N}\right) = d(\ln X) - d(\ln N) = \frac{dX}{X} - \frac{dN}{N} - \frac{1}{2}\|\underline{\sigma_X}\|^2 dt + \frac{1}{2}\|\underline{\sigma_N}\|^2 dt,$$

from which follows

(i) $d\ln\left(\frac{X}{N}\right) = \left[(\underline{\sigma_X}'(t) - \underline{\sigma_N}'(t))\underline{\lambda_N}(t) - \frac{1}{2}\|\underline{\sigma_X}\|^2 + \frac{1}{2}\|\underline{\sigma_N}\|^2\right]dt + \left[\underline{\sigma_X}'(t) - \underline{\sigma_N}'(t)\right]d\underline{W}^N.$

Furthermore,

$$\frac{d(X_{/N})}{X_{/N}} = d\ln\frac{X}{N} + \frac{1}{2}\|\underline{\sigma_X} - \underline{\sigma_N}\|^2 dt,$$

therefore, using (i):

$$\frac{d(X_{/N})}{X_{/N}} = \left[(\underline{\sigma_X}'(t) - \underline{\sigma_N}'(t))\underline{\lambda_N}(t) + (\underline{\sigma_N}'(t) - \underline{\sigma_X}'(t))\underline{\sigma_N}(t)\right]dt$$
$$+ \left[\underline{\sigma_X}'(t) - \underline{\sigma_N}'(t)\right]d\underline{W}^N$$

(ii) $\frac{d(X_{/N})}{X_{/N}} = \left[\underline{\sigma_X}'(t) - \underline{\sigma_N}'(t)\right]\left[(\underline{\lambda_N}(t) - \underline{\sigma_N}(t)]dt + \left[\underline{\sigma_X}'(t) - \underline{\sigma_N}'(t)\right]d\underline{W}^N.$

For $N$ to be a martingale numeraire, $X/N$ must be a martingale: the drift in equation *(ii)* must therefore vanish for all $\underline{\sigma_x}(t)$, which can only be true if $\underline{\lambda_N}(t) = \underline{\sigma_N}(t)$: the MPR is then a.s. equal to the volatility vector of the martingale numeraire and Eqs. $(19.25\text{-a})^c$ and $(19.25\text{-b})^c$ hold.

The preceding results and Eq. (19.24) (applied to the pair of numeraires $(N, S_0)$) imply

$$\frac{dQ_N}{dQ} = \frac{N(T)}{S_0(T)N(0)} = \frac{e^{\int_0^T \left(r(u)+\frac{1}{2}\|\underline{\sigma_N}(u)\|^2\right)du} + \int_0^T \underline{\sigma_N}'(u)d\underline{W}^N(u)}{e^{\int_0^T (r(u))du}}$$

$$= e^{\int_0^T \frac{1}{2}\|\underline{\sigma_N}(u)\|^2 du + \int_0^T \underline{\sigma_N}'(u)d\underline{W}^N(u)},$$

which proves (19.26).

832                    19 *The Mathematical Framework of Financial Markets Theory

**Corollary**
*Let X be the price of a self-financing asset with a volatility vector $\underline{\sigma}_{x'}(t)$. Changing from one probability $Q_{N1}$ to another probability $Q_{N2}$ therefore increases the drift of dX/X by $\underline{\sigma}'x(t)(\underline{\sigma}_{N2}(t) - \underline{\sigma}_{N1}(t))$.*

The corollary is an immediate consequence of $(19.25\text{-a})^c$.

## 19.6 The P-Numeraire (Optimal Growth or Logarithmic Portfolio)

Since the historical probability is trivially equivalent to itself, Proposition 12 implies that there exists a unique numeraire making martingales the values of all SA strategies *under the true probability P*.

This portfolio, denoted $(\underline{h}, H)$, is analyzed in Chap. 24 where it is defined as the portfolio that, among all SA portfolios with a given initial price, maximizes the expectation of the log of the final price. Here, adopting a different perspective, we define $(\underline{h}, H)$ as the P-numeraire and deduce that it maximizes the expected log of the final price. We then recover some of its other properties and its exact composition when financial asset prices are governed by Itô processes.

### 19.6.1 Definition of the Portfolio ($\underline{h}$, H) as the P-Numeraire

**Definition** *Call $(\underline{h}, H)$ the numeraire portfolio that makes prices martingales under the historical probability P (the P-numeraire portfolio). By definition, it is a SA portfolio with positive values such that, for any SA strategy $(\underline{x}, X)$, $\frac{X(t)}{H(t)}$ is a P-martingale, that is, $\frac{X(t)}{H(t)} = E^P{}_t\left(\frac{X(T)}{H(T)}\right)$ where $E^P{}_t(.)$ denotes the conditional expectation $E^P(.|\mathbb{F}_t)$.*
*Without loss of generality, we set $H(0) = 1$.*

Proposition 12 implies that the P-numeraire portfolio thus defined exists and is unique in an arbitrage-free, even incomplete, market.

From the definition of the portfolio $(\underline{h}, H)$, the price $X(t)$ on date $t \leq T$ of any contingent asset leading to a payoff $X_T$ and attained by the SA strategy $(\underline{x}, X)$ writes

$$X(t) = E^P_t\left(\frac{H(t)}{H(T)}X_T\right)$$

(19.27)

$\frac{H(t)}{H(T)}$ is, therefore, the *stochastic discounting factor* that in theory allows pricing random future cash flows using a simple expectation under the true probability $P$.

Since $H(0) = 1$, on date 0 we have $X(0) = E^P\left(\frac{1}{H(T)}X_T\right)$.

## 19.6 The P-Numeraire (Optimal Growth or Logarithmic Portfolio)

The numeraire portfolio $(\underline{h}, H)$ is also called the *logarithmic* or *optimal growth portfolio* for reasons given below.

### 19.6.2 Characterization and Composition of the P-Numeraire Portfolio $(\underline{h}, H)$

The portfolio $(\underline{h}, H)$ has a number of remarkable properties.

#### 19.6.2.1 The Portfolio $(\underline{h}, H)$ Maximizes the Expectation of Logarithmic Utility

**Proposition 14**
*The strategy $(\underline{h}, H)$ maximizes the expectation of the logarithm of the portfolio's final value. More precisely, the portfolio $(\underline{h}, H)$ solves the optimization program:*

$$\underset{\underline{x}, X}{MaxE}[\ln (X(T))] \text{ for a } SA(\underline{x}, X) \text{ and a given } X(0).$$

In economic terms, $(\underline{h}, H)$ is the optimal strategy for the investor exhibiting a logarithmic utility function (log investor).

**Proof**
We show that, whatever the SA portfolio $(\underline{x}, X)$, for the same initial investment $X(0) = H(0)$, the log investor prefers the payoff $H(T)$ to $X(T)$. Indeed:

– by Jensen's inequality[18] that applies to the concave logarithm function.

(i) $E\left[\ln\left(\frac{X(T)}{H(T)}\right)\right] < \ln\left[E\left(\frac{X(T)}{H(T)}\right)\right]$ if $X(T) \neq H(T)$.

and since $H$ is the $P$-numeraire:

(ii) $E\left[\ln\left(\frac{X(T)}{H(T)}\right)\right] = \frac{X(0)}{H(0)} = 1$.

(i) and (ii) imply $E\left[\ln\left(\frac{X(T)}{H(T)}\right)\right] < \ln(1) = 0$, and therefore $E[\ln(X(T))] < E[\ln (H(T))]$ for any SA payoff $X(T)$, which means that the portfolio $(\underline{h}, H)$ maximizes the expectation of the logarithm of the final payoff.

Generally, the portfolio $(\underline{h}, H)$ is the solution to the following optimization program:

---

[18] Jensen's inequality states that $E(f(X)) < f(E(X))$ for a strictly concave function $f$ and $X$ a random variable.

$$MaxE\{\ln[X(T)]\}$$
$$x, X$$

$$s.a. \quad \frac{dX}{X} = (1 - \underline{x}'\underline{1})rdt + \underline{x}' \ d\underline{R} \quad \text{and} \quad X(0) = 1$$

As explained in Chap. 24, solving this program is made easier by the *myopia* that characterizes the log utility function: $\underline{h}(t)$ simply solves

$$MaxE_t[d\ln(X(t))] \quad s.t. \quad \frac{dX}{X} = (1 - \underline{x}'\underline{1})rdt + \underline{x}'d\underline{R} \quad \text{and} \quad X(0) = 1.$$
$$\underline{x}(t)$$

The name *logarithmic portfolio* inherits from this property of the portfolio $(\underline{h}, H)$.

Note that the solution to the problem of maximizing $E[\ln(X(T))]$ implies the uniqueness of the $P$-numeraire, whether markets are complete or not.[19]

### 19.6.2.2 Other Properties of the P-Numeraire Portfolio

Let us mention two of the numerous properties that are given and proved in Chap. 24:

- portfolio $(\underline{h}, H)$ is efficient for a horizon $t + dt$;
- at each time $t$, $(\underline{h}, H)$ maximizes the expected growth rate $G$ of the portfolio's value, for any horizon $T$. In effect, as $G$ is equal to $\frac{1}{T-t}[\ln(X(T)) - \ln(X(t))]$, the strategy that maximizes $E_t\{\ln(X(T))\}$, also maximizes $E_t(G)$.

The second property of the log portfolio gives it its other name *optimal growth portfolio*.

### 19.6.2.3 Composition, Volatility, and Dynamics of the Logarithmic Portfolio

For prices governed by Itô processes, we can find closed-form expressions for the weights $\underline{h}(t)$, as shown in Chap. 24, in accordance with the following proposition:

**Proposition 15**
- *If the prices of primitive assets are governed by Itô processes and their returns satisfy $d\underline{R} = \underline{\mu}(t)dt + \Sigma(t)d\underline{W}$ with the variance-covariance matrix $\mathbf{V} = \Sigma\Sigma'$, the weights $\underline{h}(t)$ of risky assets in the logarithmic portfolio write.*

$$\underline{h}(t) = \mathbf{V}(t)^{-1}\left[\underline{\mu}(t) - r(t)\underline{1}\right]. \tag{19.28}$$

---

[19]The solution is unique since ln is a strictly concave function. One can indeed check that $X_1(T) \neq X_2(T)$ cannot be both a solution as they would be dominated by e.g. $0.5[X_1(T) + X_2(T)]$ (attainable with the same initial investment).

## 19.6 The P-Numeraire (Optimal Growth or Logarithmic Portfolio)   835

- *In addition, if the markets are complete, $\underline{h}(t)$ can be written as a function of the MPR $\underline{\lambda}(t)$:*

$$\underline{h}(t) = \Sigma'(t)^{-1}\underline{\lambda}(t) \qquad ((19.29)^c)$$

**Proof**
First, we give, briefly, the proof of (19.28) as carried out in Chap. 24.

Itô's lemma implies that $d(\ln(X)) = \frac{dX}{X} - \frac{1}{2}\left(\frac{dX}{X}\right)^2$ and the optimization program writes

$$\underset{\underline{x}(t)}{MaxE}\left\{\frac{dX}{X} - \frac{1}{2}\left(\frac{dX}{X}\right)^2\right\} \Leftrightarrow \underset{\underline{x}(t)}{MaxE}\left\{(1 - \underline{x}'\underline{1})rdt + \underline{x}'d\underline{R} - \frac{1}{2}((1 - \underline{x}'\underline{1})rdt + \underline{x}'d\underline{R})^2\right\}$$

Recalling that $\mathbf{V} = \Sigma\,\Sigma'$ denotes the matrix of instantaneous variances-covariances of returns, the program writes

$$\underset{\underline{x}(t)}{Max}\left\{\underline{x}'\left(\mu - r\underline{1}\right) - \frac{1}{2}\underline{x}'\,\mathbf{V}\underline{x}\right\};$$

The first-order condition is

$$\frac{d}{dx}\left[\underline{x}'\left(\mu - r\underline{1}\right) - \frac{1}{2}\underline{x}'\,\mathbf{V}\underline{x}\right] = 0$$

whose solution is $\underline{x} = \underline{h}$ given by (19.28).

Note that Eq. (19.28) holds for complete and incomplete markets since it is based only on the invertibility of the matrix $\mathbf{V} = \Sigma\,\Sigma'$ (ensured by avoiding redundancy in the primitive securities).

Recall that Eq. (8) (APT applied to the primitive assets) allows writing the premia $\mu(t) - r(t)\underline{1}$ as functions of the MPR: $\mu(t) - r(t)\underline{1} = \Sigma(t)\underline{\lambda}(t)$. Therefore, to each MPR $\underline{\lambda}$ corresponds a vector of premia $\mu - r\underline{1}$, but the converse is only true if the markets are complete.

Assume that the markets are complete, which ensures that the matrices $\Sigma$, $\Sigma'$ (and $\mathbf{V}$) are invertible so that $\mathbf{V}(t)^{-1} = \Sigma'(t)^{-1}\Sigma(t)^{-1}$. Equation (19.28) then implies: $\underline{h}(t) = \Sigma'(t)^{-1}\Sigma(t)^{-1}\Sigma(t)\underline{\lambda}(t) = \Sigma'(t)^{-1}\underline{\lambda}(t)$, which proves $(29)^c$.

Note that the vector $\underline{h}'\Sigma$ of volatilities for the P-numeraire $(\underline{h}, H)$ equals the MPR $\underline{\lambda}$ holding under P, in accordance with Proposition 13. The result is that, from Eq. (19.25-b)c), the dynamics of $H(t)$ can be written, under the probability $P$, as a function of the MPR $\underline{\lambda}(t)$:

$$\frac{dH}{H} = \left[r(t)dt + \|\underline{\lambda}(t)\|^2\right]dt + \underline{\lambda}'(t)d\underline{W} \qquad ((19.30 - a)^c)$$

$$H(t) = e^{\int\limits_0^t \left(r(u) + \frac{\|\underline{\lambda}(u)\|^2}{2}\right)du + \int\limits_0^t \underline{\lambda}'(u)d\underline{W}} \qquad ((19.30 - b)^c)$$

Equations (19.30-a)c) and (19.30-b)c) are special cases of (19.25-b)c) (N $=$ H, $P_N = P$, $\underline{\lambda}_N = \underline{\lambda} = \underline{\sigma}_H$).

### 19.6.2.4 P-Numeraire Portfolio and Radon-Nikodym Derivatives

Here we assume complete markets. Consider a numeraire $N$ and its associated $N$-martingale probability $Q_N$ along with the historical probability $P$ and the numeraire $H$ associated with it.

By Proposition 10, we have

$$\frac{dQ_N}{dP} = \frac{N(T)}{H(T)N(0)} = \frac{N(T)}{N(0)} e^{-\int\limits_0^t \left(r(u) + \frac{\|\underline{\lambda}(u)\|^2}{2}\right)du - \int\limits_0^t \underline{\lambda}'(u)d\underline{W}} . \qquad ((19.31)^c)$$

Equation (19.31)c)$^c$ expresses the Radon-Nikodym derivative $\frac{dQ_N}{dP}$ corresponding to any viable numeraire $N$ for which the expression of the final value $N(T)$ is known. It can be generalized to the case of incomplete markets (see Sect. 7).

---

**Example**
The RN probability $Q$ corresponding to the numeraire numbered 0 (the risk-free asset) is such that.

$$\frac{dQ}{dP} = \frac{S_0(T)}{H(T)} = \frac{e^{\int_0^T r(u)du}}{e^{\int_0^T \left\{r(u) + \frac{1}{2}\|\underline{\lambda}(u)\|^2\right\}du + \int_0^T \underline{\lambda}'(u)^2 d\underline{W}(u)}} , \text{i.e.} \quad \text{which is Eq. (19.22)}$$

$$\frac{dQ}{dP} = e^{-\int_0^T \frac{1}{2}\|\underline{\lambda}(u)\|^2 du - \int_0^T \underline{\lambda}'(u)d\underline{W}(u)},$$

of Proposition 8 obtained there from Girsanov's theorem and here from Eq. (19.31)c)$^c$.

---

## 19.7 ** Incomplete Markets

We characterize the different martingale probabilities and price systems compatible with AAO in the context of incomplete markets. In §1, we describe the MPR compatibles with AAO using the kernel of the diffusion matrix $\Sigma(t)$. The set of deflators (or pricing kernels) is defined and characterized in §2.

## 19.7.1 MPR and the Kernel of the Diffusion Matrix $\Sigma(t)$

After recalling some results of linear algebra, we describe the set of MPR using the kernel of the diffusion matrix $\Sigma$. This then allows us to express the Radon-Nikodym derivatives $dQ/dP$ for different RN probabilities $Q$, to decompose risks into two orthogonal components, one of which can be replicated and the other not, and to characterize the $P$-numeraire using a particular MPR.

### 19.7.1.1 Several Useful Results from Linear Algebra

If markets are incomplete, $n < m$, $\Sigma(t)$ is not invertible. We denote by $\mathbb{K}(t)$ the *kernel* of the matrix $\Sigma(t)$ : $\mathbb{K}(t) = \{\kappa(t)|\Sigma(t)\kappa(t) = 0\}$, and by $\mathbb{K}(t)^{\perp}$ the vector subspace orthogonal to $\mathbb{K}(t)$.

We put together below, in a lemma, some elements of linear algebra used in this section; to simplify the formulas our notation omits the explicit dependence on time $(t)$.

Lemma. Results from linear algebra

(i) *If the $(n \times m)$ matrix $\Sigma$ is of rank $n$ ($< m$), its kernel $\mathbb{K}$ is a subspace of $\mathbb{R}^m$ with dimension $m - n$ and its orthogonal $\mathbb{K}^{\perp}$ is a subspace of dimension $n$;*

(ii) *Every $\underline{\lambda} \in \mathbb{R}^m$ can be uniquely decomposed as the sum of its projections onto $\mathbb{K}^{\perp}$ and $\mathbb{K}$: $\underline{\lambda} = \underline{\lambda}_1 + \underline{\lambda}_2$, where $\lambda_1 \in \mathbb{K}^{\perp}$ and $\underline{\lambda}_2 \in \mathbb{K}$;*

(iii) *To each $\underline{y} \in \mathbb{K}^{\perp}$ is associated a single $\underline{x} \in \mathbb{R}^n$ such that $\underline{x}'\Sigma = \underline{y}$, and conversely.*

(iv) *For any $\underline{q} \in \mathbb{R}^m$, the $m$-dimensional vectors solutions to the equation $\Sigma\underline{\lambda} = \underline{q}$ can be written as $\underline{\lambda} = \underline{\lambda}^* + \kappa$, where $\underline{\lambda}^*$ is the unique vector in $\mathbb{K}^{\perp}$ which satisfies the equation $\Sigma\underline{\lambda}^* = \underline{q}$ and is equal to $\Sigma'(\Sigma\Sigma')^{-1}\underline{q}$, and $\underline{\kappa}$ is an arbitrary vector in $\mathbb{K}$: all these solutions, therefore, have the same projection $\underline{\lambda}^*$ onto $\mathbb{K}^{\perp}$ and differ in their projections $\underline{\kappa}$ onto $\mathbb{K}$.*

### Proof

Assertions *(i)* and *(ii)* are formulated and proved in any basic linear algebra textbook.

To prove *(iii)*, consider an arbitrary $\underline{x}$ in $\mathbb{R}^n$ : $\underline{x}'\Sigma \in \mathbb{K}^{\perp}$ because, for any $\underline{\kappa} \in \mathbb{K}$, $\Sigma\underline{\kappa} = \underline{0}$ and so $\underline{x}'\Sigma\kappa = 0$. It follows that $\underline{x} \to \underline{x}'\Sigma$ is a bijective linear mapping of $\mathbb{R}^n$ onto $\mathbb{K}^{\perp}$ since $\mathbb{K}^{\perp}$ has dimension $n$ and $\Sigma$ has rank $n$, which proves *(iii)*.

To prove *(iv)*, consider any two solutions $\underline{\lambda}$ and $\underline{\lambda}'$ of the equation $\Sigma\underline{\lambda} = \underline{q}$. By *(ii)* they can, respectively, be decomposed as $\underline{\lambda} = \underline{\lambda}^* + \kappa$ and $\underline{\lambda}' = \underline{\lambda}^{*'} + \kappa'$, with $\underline{\lambda}^*$ and $\underline{\lambda}^{*'} \in \mathbb{K}^{\perp}$ and $\underline{\kappa}$ and $\underline{\kappa}' \in \mathbb{K}$.

Furthermore, $\Sigma(\underline{\lambda} - \underline{\lambda}') = \underline{0}$ and $\Sigma(\underline{\kappa} - \underline{\kappa}') = \underline{0}$; therefore $\Sigma(\underline{\lambda}^* - \underline{\lambda}^{*'}) = \underline{0}$, which implies that $\underline{\lambda}^* - \underline{\lambda}^{*'} \in \mathbb{K}$; since $\underline{\lambda}^* - \underline{\lambda}^{*'} \in \mathbb{K}^{\perp}$ and two orthogonal subspaces have only the zero vector in common, we have $\underline{\lambda}^* = \underline{\lambda}^{*'}$.

We thus have shown that all solutions of $\Sigma\underline{\lambda} = \underline{q}$ can be written as $\underline{\lambda} = \underline{\lambda}^* + \underline{\kappa}$, where the projection $\underline{\lambda}^*$ onto $\mathbb{K}^\perp$ is the same for all these solutions, which then only differ by the projections $\underline{\kappa}$ onto $\mathbb{K}$, as claimed by $(iv)$.

Finally, we can check that $\underline{\lambda}^* = \Sigma'(\Sigma\Sigma')^{-1}\underline{q}$ since it does satisfy the equation $\Sigma\underline{\lambda}^* = \underline{q}$ and is orthogonal to every $\underline{\kappa} \in \mathbb{K}$ (since $\underline{\lambda}^*{}'\underline{\kappa} = \underline{q}'(\Sigma\Sigma')^{-1}\Sigma\underline{\kappa} = 0$).

In the context of financial markets, these results from linear algebra lead to an immediate characterization of the MPRs compatible with AAO.

### 19.7.1.2 Characterization of the Set $\Lambda_P$ of MPRs Compatible with AAO

First recall that $\Lambda_P$ denotes the set of MPRs compatible with AAO, i.e. $\Lambda_P = \left\{\underline{\lambda} | \Sigma(t)\,\underline{\lambda}(t) = \underline{\mu}(t) - r(t)\underline{1} \text{ for all } t \in [0,T]\right\}$.

By Lemma $(iv)$, the set $\Lambda_P$ of MPR is, therefore, an affine subspace of dimension $m - n$: $\quad \Lambda_P = \{\underline{\lambda} | \underline{\lambda}^* + \underline{\kappa}, \text{for } \underline{\kappa} \in \mathbb{K}\}$ . In this characterization $\underline{\lambda}^* = \Sigma'(\Sigma\Sigma')^{-1}\left[\underline{\mu} - r\underline{1}\right]$ is simultaneously a specific MPR and a common component of every MPR; also, to each $\underline{\kappa} \in \mathbb{K}$ corresponds an MPR in $\Lambda_P$ and conversely.

### 19.7.1.3 Radon-Nikodym Derivatives

Recall the bijection between the set $\mathbb{Q}$ of RN probabilities and $\Lambda_P$, so that with each RN probability $Q_i$ is associated the MPR $\underline{\lambda}^{(i)} \in_P \Lambda_P$ such that

$$\frac{dQ_i}{dP} = e^{-\int_0^T \frac{1}{2}\|\underline{\lambda}^{(i)}(u)\|^2\,du - \int_0^T \underline{\lambda}^{(i)}{}'(u)d\underline{W}(u)}.$$

By the preceding arguments, $\underline{\lambda}^{(i)}$ can be split into $\underline{\lambda}^* + \underline{\kappa}_i$, and these two components are orthogonal, we have $\|\underline{\lambda}^* + \underline{\kappa}_i\|^2 = \|\underline{\lambda}*\|^2 + \|\underline{\kappa}_i\|^2$ (by Pythagoras) and we can write

$$\frac{dQ_i}{dP} = e^{-\int_0^T \frac{1}{2}\|\underline{\lambda}^*(u)\|^2\,du - \int_0^T \frac{1}{2}\|\underline{\kappa}_i(u)\|^2\,du - \int_0^T \underline{\lambda}^*{}'(u)d\underline{W}(u) - \int_0^T \underline{\kappa}_i(u)d\underline{W}(u)}$$

$$\frac{dQ_i}{dP} = e^{-\int_0^T \frac{1}{2}\|\underline{\lambda}^*(u)\|^2\,du - \int_0^T \underline{\lambda}*(u)d\underline{W}(u)} \; e^{-\int_0^T \frac{1}{2}\|\underline{\kappa}_i(u)\|^2\,du - \int_0^T \underline{\kappa}_i(u)d\underline{W}(u)}$$

$$(19.32)$$

The first exponential in (19.32) is the same for all RN probabilities, while the second is specific for each one of them.

Since $\frac{X(t)}{S_0(t)}$ is a $Q_i$-martingale, we have $X(0) = E^{Q_i}\left[\frac{X_T}{S_0(T)}\right] = E^P\left[X_T\left(\frac{1}{S_0(T)}\frac{dQ_i}{dP}\right)\right]$.

This equation leads to pricing equations such as.

## 19.7 ** Incomplete Markets

$$X(0) = E^P \left[ X_T \left( e^{-\int_0^T \left( r(u) + \frac{1}{2} \| \underline{\lambda}^{(i)}(u) \|^2 \right) - \int_0^T \underline{\lambda}^{(i)'}(u) d\underline{W}(u)} \right) \right]$$

$$X(0) = E^P \left[ X_T \left( e^{-\int_0^T \left( r(u) + \frac{1}{2} \| \underline{\lambda}^*(u) \|^2 \right) du - \int_0^T \underline{\lambda}^{*'}(u) d\underline{W}(u)} \ e^{-\int_0^T \frac{1}{2} \| \underline{\kappa}_i(u) \|^2 du - \int_0^T \underline{\kappa}_i'(u) d\underline{W}(u)} \right) \right].$$

If we are on date $t$, and no longer on date 0, we obtain

$$X(t) = E^P \left[ X_T e^{-\int_t^T \left[ r(u) + \frac{1}{2} \| \underline{\lambda}^{(i)}(u) \|^2 \right] du - \int_t^T \underline{\lambda}^{(i)'}(u) d\underline{W}(u)} \right]. \tag{19.33}$$

### 19.7.1.4 Decomposition of Random Variables; Replicable and Non-replicable Orthogonal Elements

Assume we are under a RN probability $Q$ and consider the variation $dZ = \underline{z}'(t)d\underline{W}^*$ of some $Q$-martingale that, by (ii) and (iii), can be split up as

$$dZ = \underline{x}'(t)\Sigma(t)d\underline{W}^* + \underline{y}'(t)d\underline{W}^*, \tag{19.34}$$

where $\underline{x}'(t)\Sigma(t)$ is the projection of $\underline{z}(t)$ onto $\mathbb{K}(t)^\perp$ and $\underline{y}(t)$ its projection onto $\mathbb{K}(t)$.

We can thus decompose a random quantity into a replicable part and a non-replicable part orthogonal to the first: Since the variation in the price of a SA portfolio $(\underline{u}, X)$, denominated in the numeraire $S_0$, is $dX = X(t) \underline{u}'(t)\Sigma(t)d\underline{W}^*$, the first term in (19.34) is replicable with a portfolio with weights $\underline{u}(t) = \underline{x}(t)/X(t)$, while $y'(t)d\underline{W}^*$ is not replicable. In particular, we recover that every $Q$-martingale can be thought of as the value, denominated in the numeraire $S_0$, of a SA portfolio if and only if the kernel of $\Sigma(t)$ is the vector $\underline{0}$ (hence $\Sigma(t)$ is invertible), which is a necessary and sufficient condition for the market to be complete.

We can also interpret the MPR $\underline{\lambda}^*$ as the one that leads to *zero* premia on *non-replicable* risks (of the form $\underline{y}'d\underline{W}^*$ with $\underline{y} \in \mathbb{K}$).

### 19.7.1.5 P-Numeraires

We pointed out that, for both complete and incomplete markets, the $P$-numeraire is unambiguously defined by its weights on risky assets $\underline{h}(t) = \mathbf{V}(t)^{-1} \left[ \underline{\mu}(t) - r(t)\underline{1} \right]$.

The reason is that, since the $n$ primitive assets are not redundant, $\Sigma$ is an $n \times m$ matrix of rank $n$ and the variance-covariance matrix of returns $\mathbf{V} = \Sigma \Sigma'$ is $n \times n$ invertible (even in incomplete markets where $n < m$). Since the MPR $\underline{\lambda}^*$ defined above equals $\Sigma'(\Sigma\Sigma')^{-1} \left( \underline{\mu} - r\underline{1} \right)$, it follows that

$$\underline{\lambda}^*(t) = \Sigma(t)'\underline{h}(t). \tag{19.35}$$

This equation generalizes (19.29)c) that holds only in a complete market (characterized by a square and invertible matrix $\Sigma(t)$). It allows us to write, in incomplete markets, the dynamics under $P$ and the value $H(t)$ of the logarithmic portfolio as a function of the MPR $\underline{\lambda}^*(t)$:

$$\frac{dH}{H} = \left[ r(t) + \underline{h}'(t)\Sigma(t)\underline{\lambda}^*(t) \right] dt + \underline{h}'(t)\Sigma(t)d\underline{W}, \text{i.e.}$$
$$\frac{dH}{H} = \left[ r(t) + \|\underline{\lambda}^*(t)\|^2 \right] dt + \underline{\lambda}^{*'}(t)d\underline{W} \tag{19.36-a}$$

$$H(t) = e^{\displaystyle\int_0^t \left( r(u) + \frac{\|\underline{\lambda}^*(u)\|^2}{2} \right) du + \int_0^t \underline{\lambda} * I(u)d\underline{W}}. \tag{19.36-b}$$

Equations (19.36-a), generalizing (19.30-a)c) which holds for complete markets only, means that the volatility vector for the $P$-numeraire is equal to the particular MPR $\underline{\lambda}^*$ that belongs to $\mathbb{K}^\perp$ and whose premia are nil for non-replicable risks.

### 19.7.2 Deflators

First, we define the set of deflators before establishing the relationships between the set of MPR, Radon-Nikodym derivatives and the $P$-numeraire.

#### 19.7.2.1 Deflators and the Pricing Kernel
**Definition** *A deflator is a stochastic process* $\pi(t)$, *a.s. positive, such that, for any SA strategy* $(\underline{x}, X)$, $X(t)\pi(t) - E[X(T)\pi(T)|F_t]$, *making* $X(t)\pi(t)$ *a martingale.*

The term $\pi(T)/\pi(t)$ is sometimes called a *stochastic discount factor* (since $X(t) = E\left[\frac{\pi(T)}{\pi(t)}X(T)|F_t\right]$. The set $\mathbb{D}$ of deflators is called the *pricing kernel.*

All deflators lead to the same price for any attainable security or SA portfolio, but to different prices for non-attainable ones.

It is important not to mix up deflators and numeraires. The inverse $1/H(t)$ of the $P$-numeraire is a particular deflator, and besides, is the only one in complete markets. But, in an incomplete market, $\mathbb{D}$ contains an infinity of elements, not necessarily attainable. As a result, while the inverse of the $P$-numeraire portfolio is indeed a deflator, the inverse $1/\pi(t)$ of some arbitrary deflator is not necessarily a numeraire since it is not necessarily replicable by some SA strategy. These differences and the relationship between deflators and the $P$-numeraire are made precise in the following subsection.

### 19.7 ** Incomplete Markets

#### 19.7.2.2 Deflators, MPR, Radon-Nikodym Derivatives, and the P-Numeraire

The Radon-Nikodym derivatives $dQ_i/dP$, where $Q_i$ is one of the RN probabilities in $\mathbb{Q}$, play a role analogous to that of a deflator; for any SA strategy attaining $X_T$.

$$X(0) = E^P\left[X_T\left(\tfrac{1}{S_0(T)}\tfrac{dQ_i}{dP}\right)\right]; \text{ from this, } \tfrac{1}{S_0(T)}\tfrac{dQ_i}{dP} \text{ "plays the role" of } \pi(T).$$

Recall also that each RN probability $Q_i$ is associated with an MPR $\underline{\lambda}^{(i)} \in \Lambda_P$ such that $\frac{dQ_i}{dP} = e^{-\int_0^T \frac{1}{2}\left\|\underline{\lambda}^{(i)}(u)\right\|^2 du - \int_0^T \underline{\lambda}^{(i)\prime}(u)d\underline{W}(u)}$, leading to (19.33) that we repeat here:

$$X(t) = E^P\left[X_T e^{-\int_t^T \left[r(u)+\frac{1}{2}\left\|\underline{\lambda}^{(i)}(u)\right\|^2\right]du - \int_t^T \underline{\lambda}^{(i)\prime}(u)d\underline{W}(u)}\right].$$

We define the process $\pi_i(t)$ by

$$\pi_i(t) \equiv e^{-\int_0^t \left(r(u)+\frac{1}{2}\left\|\underline{\lambda}^*(u)\right\|^2\right)du - \int_0^t \underline{\lambda}^{(i)\prime}(u)d\underline{W}(u)}. \tag{19.37-a}$$

From (19.33) and (19.37-a), we obtain $X(t)\pi_i(t) = E^P[X_T\pi_i(T)/\mathbb{F}_t]$.

Thus, the process $\pi_i(t)$ defined by (19.37-a) is a deflator. Moreover, the same Eq. (19.37-a) defines a bijection between the set of MPR and that of deflators.

Breaking up $\underline{\lambda}^{(i)}$ into its two projections $\underline{\lambda}^*$ onto $\mathbb{K}^\perp$ and $\underline{\kappa}_i$ onto $\mathbb{K}$ allows writing (19.37-a) as

$$\pi_i(t) = e^{-\int_0^t \left(r(u)+\frac{1}{2}\left\|\underline{\lambda}^*(u)\right\|^2\right)du - \int_0^t \underline{\lambda}^{*\prime}(u)d\underline{W}(u)} e^{-\int_0^t \frac{1}{2}\left\|\underline{\kappa}_i(u)\right\|^2 du - \int_0^t \underline{\kappa}_i(u)d\underline{W}(u)}. \tag{19.37-b}$$

We recognize the inverse $1/H(t)$ of the $P$-numeraire in the first exponential term on the r.h.s. of (19.37-b). Also, let us set

$$m_i(t) = e^{-\int_0^t \frac{1}{2}\left\|\underline{\kappa}_i(u)\right\|^2 du - \int_0^t \underline{\kappa}_i(u)d\underline{W}(u)} \Leftrightarrow \frac{dm_i}{m_i} = \kappa_i'(t)dW. \tag{19.38}$$

We can thus express any deflator in the form

$$\pi_i(t) = \frac{m_i(t)}{H(t)}. \tag{19.37-c}$$

The preceding results are gathered in the following proposition:

**Proposition 16**

*Denote by $\mathbb{M}$ the set of all P-martingales $m_i(t)$ satisfying.*

$$m_i(t) = e^{-\int_0^t \frac{1}{2}\|\underline{\kappa}_i(u)\|^2 du - \int_0^t \underline{\kappa}_i(u)d\underline{W}(u)}, \text{ i.e. } \frac{dm_i}{m_i} = \underline{\kappa}_i'(t)d\underline{W} \text{ with } \underline{\kappa}_i(t) \in \mathbb{K}(t).$$

- *Any deflator $\pi_{i\,i}$ can be written as $\pi_i = \frac{m_i}{H}$ for some $m_i \in \mathbb{M}$.*
- *Conversely, any $\pi_i$ satisfying this last equation is a deflator.*
- *It follows that any $m_i \in \mathbb{M}$ leads to a possible AAO price for any contingent asset $X_T$:*

$$X_0^{(i)} = E\left[\frac{X_T m_i(T)}{H(T)}\right].$$

- *If $X_T$ is attainable, the last equation leads to the same price $X_0$ irrespective of which martingale $m_i \in \mathbb{M}$ is used.*
- *The six sets $\Pi$, $\Lambda_P$, $\mathbb{Q}$, $\mathbb{D}$, $\mathbb{K}$, and $\mathbb{M}$ are in bijective correspondences with one another.*

In the academic literature on contingent asset pricing, authors emphasize, according to the case, changes in probabilities and numeraires, deflators or stochastic discount factors of the pricing kernel. The choice really is just a matter of presentation.

Finally, consider the inverse $\gamma_i$ of some deflator $\pi_i : \gamma_i(t) \equiv 1/\pi_i(t)$.

Equation (19.37-b) implies that

$$\frac{d\gamma_i}{\gamma_i} = \left[r(t) + \|\underline{\lambda}^*(t)\|^2\right]dt + \underline{\lambda}^*(t)d\underline{W} + \|\underline{\kappa}_i(t)\|^2 dt + \underline{\kappa}_i(t)d\underline{W}$$

$$= \frac{dH}{H} + \|\underline{\kappa}_i(t)\|^2 dt + \underline{\kappa}_i(t)d\underline{W}.$$

Since the component $\underline{\kappa}_i(t)\,d\underline{W}$ is not replicable, the only deflator whose inverse $\gamma_i$ is replicable with a SA strategy, hence the only $\gamma_i$ that is a numeraire, is the one for which $\underline{\kappa}_i$ equals 0, that is H.

## 19.8   Summary

- Mathematical Finance is grounded on the standard probabilistic framework ($\Omega$, $\mathbb{F}_T$, $P$), the revelation of the information described by a filtration $\mathbb{F}_t$, often generated by a vector of Brownian motions $\underline{W}(t)$. "Admissible" means $\mathbb{F}_t$-measurable with defined second moment.
- The standard mathematical model describes the securities that are traded, the portfolio strategies which are dynamic combinations of traded securities, contingent claims (traded or not) and the prices compatible with AAO.

## 19.8 Summary

- The *primitive securities,* in number $n + 1$, are traded, non-redundant and as such form together the *base.* Asset 0 yields a risk-free rate $r(t)$ while securities $i = 1, \ldots, n$ are risky. Their prices $S(t)$ and returns over $(t, t + dt)$ write: $dS_0/S_0 = r(t)dt$; $dS_i/S_i = dR_i$ with $d\underline{R} = \underline{\mu}(t)dt + \Sigma(t)d\underline{W}$. $\Sigma(t)$ is an $(n \times m)$ diffusion matrix, and $r(t)$ may be a stochastic process.
- A *portfolio strategy* is defined by the $n$-dimensional vector of $\mathbb{F}_t$-measurable weights $\underline{x}(t)$ on the $n$ risky securities to which is associated the portfolio's value $X(t)$ observed between 0 and $T$.
- A *self-financing* strategy yields only two cash-flows $[-X(0), X(T)]$ and its return is governed by $dX/X = x_0(t)r(t) + \underline{x}'(t)d\underline{R}$.
- A self-financing "admissible" (SA) strategy is an *arbitrage* if cash flows $-X(0)$ and $X(T)$ *cannot be negative,* and at least one of them *may* be positive.
- A *contingent claim* is a right to an "admissible" cash flow $C_T$ on date $T$. The market is *complete* if all contingent claims are *attainable* (*replicable*) by (at least) one SA (i.e. $X(T) = C_T$, a.s.).
- A viable price system $\pi$ is a linear mapping of contingent claims into $\mathbb{R}$ *consistent* with AAO. In particular, if $C_T$ is attainable, $\pi(C_T) = X(0)$ for every SA $(\underline{x}, X)$ attaining $C_T$.
- There is thus a *single viable price,* for any *attainable* contingent claim, which is the initial investment $X(0)$ required to attain it; for non-attainable $C_T$ an infinite number of *prices are possible.* In an *arbitrage-free and complete market,* all contingent claims have *a single* viable price since all of them are attainable. When arbitrage opportunities exist, there is no viable price system.
- *Continuous time* APT: when the return of the $n$ risky primitive assets are driven by an $m$-dimensional Brownian motion $\underline{W}$, in AAO there exists an $m$-dimensional vector $\underline{\lambda}(t)$ such that the instantaneous expected returns can be written: $\underline{\mu}(t) = r(t)\mathbf{1} + \Sigma(t)\underline{\lambda}(t)$. The vector of *market prices of risk* (MPR) $\underline{\lambda}(t)$, common for all securities, is *unique iff markets are complete.*
- The following statements are equivalent: (i) The market is complete; (ii) The number of non-redundant traded assets is $n = m$; (iii) There is only one MPR vector $\underline{\lambda}(t) = \Sigma^{-1}(t)\left[\underline{\mu}(t) - r(t)\underline{1}\right]$.
- Any driftless Itô process $M(t)$, of the type $dM(t) = \underline{\phi}'(t)d\underline{W}$, is a martingale and conversely.
- In the risk neutral (RN) world, the drifts of all returns are equal to $r(t)$, and using the *numeraire* $S_0(t)$, the value $X(t)/S_0(t)$ of *any self-financing portfolio* or asset is a *martingale.*
- A change in probability means intuitively that the density $dP(\omega)$ around $\omega$ is multiplied by a positive random factor $\xi(\omega) : dP'(\omega) = \xi(\omega) dP(\omega))$. A « valid » probability P′ is obtained iff $\xi$ is a.s. positive and $E^P[\xi] = 1$; $\xi = dP'/dP$ is a Radon-Nikodym derivative.
- Girsanov's theorem allows *changing the drifts* of Brownian motions by an appropriate change in probability, the volatilities remaining unchanged. In

particular the *risk-neutral probability measure Q corresponding to any viable MPR* $\underline{\lambda}(t)$ can be defined by: $\frac{dQ}{dP} = e^{-\frac{1}{2}\int_0^T \|\underline{\lambda}(s)\|^2 ds - \int_0^T \underline{\lambda}'(s)d\underline{W}^*}$ .

- Under Q, all expected returns are equal to $r(t)$ and asset-0-denominated prices are martingales, which implies the pricing formula: $X(0) = E^Q\left\{\frac{X_T}{S_0(T)}\right\} = E^P\left\{\frac{dQ}{dP}X_T e^{-\int_0^T r(t)dt}\right\}$.

- When they exist, MPRs, viable price systems, and risk-neutral probabilities are in a one-to-one relationship; *their existence relies on AAO and their uniqueness on market completeness.*

- Any admissible self-financing asset or portfolio of value a.s. positive is eligible as a *numeraire*, such as the risk-free asset, a zero-coupon bond or the log-optimal portfolio. In AAO, *to any numeraire N may correspond probabilities $Q_N$ yielding $Q_N$-martingale prices* $X(t)/N(t)$ for all SA portfolios. Their existence *relies on AAO and their uniqueness on market completeness.*

- The P-numeraire that makes values of all SA strategies martingales *under the historical ("true") probability P* is the portfolio $(\underline{h}, H)$ maximizing the *expectation of the logarithm* of the final value. The following *pricing formula* obtains: $X(0)/H(0) = E^P(X(T)/H(T))$, which involves the "true" probability $P$ and a stochastic discount factor $H(0)/H(T)$.

- The weights characterizing the P-numeraire are: $\left[\underline{h}(t) = \mathbf{V}(t)^{-1}\left[\underline{\mu}(t) - r(t)\underline{1}\right]\right]$, where $\mathbf{V} = \Sigma\,\Sigma'$ is the variance-covariance matrix of the returns on the (non-redundant) primitive securities. This remarkable log-optimal portfolio is studied in chap. 24.

- Under *incomplete markets*, the *sets* of viable MPRs and of the corresponding RN probabilities can be characterized in terms of the kernel of the matrix $\Sigma$. Any random variable can be decomposed into two orthogonal components, one of which can be replicated and the other not.

---

## Appendix

### Construction of a One-to-one Correspondence between $\mathbb{Q}$ and $\Pi$

First recall that $1_A$ is the indicator function of the event $A \subseteq \Omega$ i.e. $1_A(\omega) = 1$ if $\omega \in A$ and $1_A(\omega) = 0$ if $\omega \notin A$. As a result, for any $A \in \mathbb{F}_T$, $1_A$ is an $\mathbb{F}_T$-measurable random variable with $Q(A) = E^Q(19.1_A)$. $1_A$ may be considered as a particular contingent claim, like an Arrow-Debreu security.

Let us assume $\mathbb{Q}$ and $\Pi$ non-empty and construct a one-to-one correspondence between $\mathbb{Q}$ and $\Pi$.

We start with $\mathbb{Q} \xrightarrow{f} \Pi$: to any $Q \in \mathbb{Q}$ assign $f(Q) \equiv \pi_Q$ such that $\pi_Q(X_T) \equiv E^Q(X_T)$ for all $X_T \in \mathbb{L}^2$. We must check that, so defined, $\pi_Q$ does satisfy the conditions required of a system of viable prices, and thus belongs to $\Pi$. Indeed:

$-X_T \geq 0$ a.s. and $\mathbb{E}(X_T) > 0$ implies $\mathbb{E}\left[\frac{dQ}{dP}\frac{X_T}{S_0(T)}\right] = \mathbb{E}^Q(X_T) = \pi_Q(X_T) > 0$ (since $\mathbb{E}\left[\frac{dQ}{dP}\frac{1}{S_0(T)}\right] > 0$ a.s.). Also, $X_T = 0$ a.s. does imply that $\pi_Q(X_T) = 0$;

– Suppose $X_T$ can be attained by a self-financing strategy $(\underline{x}, X)$ (therefore $X(T) = X_T$);

$Q \in \mathbb{Q}$ implies that $X(t)$ is a Q-martingale, from which follows $X(0) = \mathbb{E}^Q(X_T) = (X_T)$.

– Finally we check that $f$ is a *linear* mapping from $\mathbb{L}^2$ to $\mathbb{R}$.

Now construct $\mathbf{\Pi} \xrightarrow{g} \mathbb{Q}$ : For each $\pi \in \mathbf{\Pi}$ define $g(\pi) \equiv Q_\pi$ such that $Q_\pi(A) \equiv \pi(1_A S_0(T))$ for all $A \in \mathbb{F}_T$.

We must check that $Q_\pi$ satisfies the different axioms required for a probability. Indeed:

– $Q_\pi(\Omega) \equiv \pi(1_\Omega S_0(T)) = \pi(S_0(T)) = 1$;

– Also, for any $A \in \mathbb{F}_T$, $0 \leq Q_\pi(A) \leq Q_\pi(\Omega) = 1$;

– Finally, if $A$ and $B$ are two events in $\mathbb{F}_T$, $Q_\pi(A \cup B) \equiv \pi(1_{A \cup B} S_0(T))$.

But $1_{A \cup B} = 1_A + 1_B - 1_{A \cap B}$, therefore

$$Q_\pi(A \cup B) = \pi(1_A S_0(T)) + \pi(1_A S_0(T)) - \pi(1_{A \cap B} S_0(T))$$
$$= Q_\pi(A) + Q_\pi(B) - Q_\pi(A \cap B).$$

Finally, we show that $g^{-1} = f$ by verifying $g \circ f = $ Identity, i.e. starting from $Q \in \mathbb{Q} : g(f(Q)) \equiv Q_\pi = Q$. Indeed, for any $A \in \mathbb{F}_T : Q_\pi(A) = \pi_Q(S_0(T)1_A) = \mathbb{E}^Q(1_A) = Q(A)$.

---

## Suggestions for further reading

### Books

\*\*Back K. (2017). (2nd ed.), *Asset pricing and portfolio choice theory*, Oxford University Press.

\*\* Björk T. (2020). (4th ed.), *Arbitrage theory in continuous time*, Oxford University Press.

Cochrane, J. H. (2005). *Asset Pricing*. Princeton University Press.

\*\* Duffie D. (1988). *Security markets: Stochastic models*, Academic Press.

\*\* Duffie D. (2001). (3rd ed.), *Dynamic asset pricing theory*, Princeton University Press.

Elliot R., & Kopp E. (2005). (2nd ed.), *Mathematics of financial markets*, Springer.

\*\* Jeanblanc M., Yor M., & Chesney M. (2009). *Mathematical methods for financial markets*. Springer.

Joshi, M. (2003). *The Concepts and Practice of Mathematical Finance*. Cambridge University Press.

\*\*\* Karatzas I., & Shreve S. (2017). *Methods of mathematical finance*, Springer.

\*\* Musiela M., & Rutkowski M. (2007) (3rd ed.), *Martingale methods in financial Modelling*, Springer, Applications of Mathematics.

\*\* Prigent J.L., (2003). *Weak convergence of financial markets*. Springer.

## Articles

Bajeux-Besnainou I., & Portait R. (1997, December) *"The Numeraire Portfolio: a New Approach to Financial Theory"*. *European Journal of Finance*.

Cox, J., & Huang, C. F. (1989). Optimal consumption and portfolio policies when asset prices follow a diffusion process. *Journal of Economic Theory, 49*, 33–83.

Cox, J., & Huang, C. F. (1991). A variational problem arising in financial economics. *Journal of Mathematical Economics, 20*, 465–487.

Cox J., Ingersoll J.E., and S.A. Ross, "An Intertemporal General Equilibrium Model of Asset Prices", *Econometrica*, 1985, 53, 363-384.

Dybvig, P. H., & Huang, C. F. (1988). Non-Negative Wealth, Absence of Arbitrage and Viable Consumption Plans. *Review of Financial Studies, 1*, 377–401.

Geman, H., El Karoui, N., & Rochet, J. C. (1995). Changes of Numeraire, Changes of Probability Measures and Pricing of Options. *Journal of Applied Probability, 32*, 443–458.

Harrison, J. M., & Kreps, D. M. (1979). Martingales and Arbitrage in Multiperiod Securities Markets. *Journal of Economic Theory, 20*, 381–408.

*Harrison J.M., and S. Pliska, Martingales and Stochastic Integrals in the Theory of Continuous Trading, *Stochastic Processes and their Applications*, 1981, 11, 215-260.

He, H., & Pearson, N. D. (1991). Consumption and portfolio policies with incomplete markets and short sale constraints: The infinite dimensional case. *Journal of Economic Theory, 54*, 259–304.

He, H., & Pearson, N. D. (1991). Consumption and portfolio policies with incomplete markets and short sale constraints: The finite dimensional case. *Mathematical Finance, 1*(3), 1–10.

Karatzas, I., Lehoczky, J. P., Shreve, S. E., & Xu, G. L. (1991). Martingale and duality methods for utility maximization in an incomplete market. *SIAM Journal of Control and Optimization, 29*, 702–730.

Long, J. B. (1990). The Numeraire Portfolio. *Journal of Financial Economics, 26*, 29–69.

# The State Variables Model and the Valuation Partial Differential Equation

# 20

In the preceding chapter, we studied the case of assets whose prices obey Itô processes. In this chapter, prices are assumed to be governed by diffusion processes that depend on state variables. These variables, which are intended to represent the state of the economy, also behave as diffusion processes. The diffusion processes being special cases of Itô processes, the analysis and results from the preceding chapter can be applied. However, these particular processes lead to additional results.

After introducing the framework and notation for our analysis (Sect. 20.1), we provide two ways of defining the factors governing returns (Sect. 20.2), two versions of the *Arbitrage Pricing Theory* (*APT*) that can be applied in this context (Sect. 20.3), the partial differential equation governing asset prices (Sect. 20.4), some applications in modeling interest rates (Sect. 20.5), the pricing of assets in the risk-neutral universe (Sect. 20.6) and discounting under uncertainty using Feynman-Kac formula (Sect. 20.7).

## 20.1 Analytical Framework and Notation

We express the dynamics of state variables as diffusion processes (§1) and then formalize the problem of valuing financial assets in this context (§2).

## 20.1.1 Dynamics of State Variables

Assume that the state of the financial market depends on a vector $\underline{X}(t)$ of $q$ variables called state variables, or factors, such as the interest rate, exchange rate, stock prices, market indices, etc.

These state variables are taken to be described by the $q$-dimensional diffusion process obeying the stochastic differential equation (SDE)

© The Author(s), under exclusive license to Springer Nature Switzerland AG 2022
P. Poncet, R. Portait, *Capital Market Finance*, Springer Texts in Business and Economics, https://doi.org/10.1007/978-3-030-84600-8_20

847

$$dX(t) = \underline{a}(t, \underline{X}(t))dt + \Omega(t, \underline{X}(t))d\underline{W}. \tag{20.1-a}$$

where $\underline{W}$ is a standard $m$-dimensional Brownian motion, $\underline{a}(t, \underline{X})$ is a vector in $\mathbb{R}^q$, $\Omega(t, \underline{X})$ is the $q \times m$ matrix of the diffusion, the components $a_i(.)$ of the vector of drifts and $\omega_{ij}(.)$ of the diffusion matrix are functions from $\mathbb{R}^{q+1}$ to $\mathbb{R}$ that satisfy Lipschitz conditions so that the SDE (20.1-a) has a solution. We call $\Gamma$ the $q \times q$ variance-covariance matrix $\Omega\Omega'$ whose typical term is

$$\left(dX_i dX_j\right)/dt = \gamma_{ij} = \sum_{k=1}^{m} \omega_{ik}\omega_{jk}.$$

The $j^{th}$ row of the vector Eq. (20.1-a) can then be written[1]

$$dX_j = a_j(t, X(t))dt + \sum_{k=1}^{m} \omega_{jk}\, dW_k, \quad \text{for } j = 1, \ldots, q \tag{20.1-b}$$

## 20.1.2 The Asset Pricing Problem

Let us assume that the state variables $\underline{X}(t)$ determine three items:

- The value $v$ of any financial asset that can be written, at time $t \in (0, T)$, as $v(t, \underline{X}(t))$; this function from $(0, T) \times \mathbb{R}^q$ to $\mathbb{R}$ is not known but it is assumed to be once differentiable with respect to the first component $t$ and twice with respect to $\underline{x}$;
- The dividend (or coupon) distributed by this asset in the interval $(t, t + dt)$ that can be written $c(t, \underline{X}(t))dt$, with the function $c(t, \underline{x})$ assumed known;
- The value of this asset on date $T$, which we denote by $l(\underline{X}(T))$, with $l(\underline{x})$ assumed given.

We need to determine the pricing function $v(t, \underline{x})$.

A very simple example of this situation is a standard option in the Black-Scholes-Merton universe: the only state variable is the price of the underlying, (thus $q = m = 1$; and $X(t) = S(t)$); the underlying distributes a continuous dividend at the constant rate $c$ ($c(t, S(t)) = c\, S(t)$); the call has a final value $l(S(T)) = (S(T)-K)^+$ and the put $(K-S(T))^+$. The BSM problem is to find the value $v(t, S)$ of the option.

In the general case, the framework for the analysis involves any number of state variables $\underline{X}$, any dividend function $c(t, \underline{x})$, and a limit condition $l(\underline{x})$.

The holder of an asset whose price is $v(t, \underline{X}(t))$ receives, in the interval $(t, t + dt)$, a coupon $c(t, \underline{X}(t))dt$ and realizes a random capital gain or loss equal to $dv(t, \underline{X}(t))$. Denoting $(t, \underline{x})$ by $(.)$, the total return on the investment in the interval $(t, t + dt)$ is the sum of these two components and writes

---

[1] Recall that the $m$ Brownian motions that compose $\underline{W}$ are independent.

## 20.2 Factor Decomposition of Returns

$$dR(.) = \frac{dv(.) + c(.)\,dt}{v(.)}$$

The expectation of the capital gain equals $E_t(dv) = D_t v\,dt$, where $E_t$ is the conditional expectation $E(.|F_t)$ and $D_t$ the Dynkin operator computed on date $t$ and relative to $\underline{X}(t)$. Recall that if $\underline{X}(t)$ satisfies (20.1), the Dynkin operator is defined by

$$D_t v \equiv \frac{\partial v}{\partial t}(t, \underline{x}) + \sum_{i=1}^{q} a_j(t, \underline{x})\frac{\partial v}{\partial x_j}(t, \underline{x}) + \frac{1}{2}\sum_{i=1}^{q}\sum_{j=1}^{q}\gamma_{ij}(t, \underline{x})\frac{\partial^2 v}{\partial x_i \partial x_j}(t, \underline{x})$$

where $\gamma_{ij}$ is the generic element of the matrix $\Gamma = \Omega\Omega'$.

The expected return is, therefore, $\dfrac{E_t(dR)}{dt} = \dfrac{1}{dt}\dfrac{E_t(dv) + c(.)dt}{v(.)} \equiv \dfrac{D_t v + c(.)}{v(.)}$.

## 20.2 Factor Decomposition of Returns

The return $dR$ of the considered asset results from random factors $d\underline{X}$ (which are themselves dependent on the underlying sources of risk $d\underline{W}$ according to Eq. (20.1)). In this section, we discuss two factorial models explaining asset returns that are the continuous time analogs of those analyzed in the following Chapters 21 and 23 and relevant to discrete time.

Itô's lemma allows writing the return $dR$ in two *equivalent* forms: the first uses its sensitivity to the $X_j$ factors and the second its sensitivity to the Brownian motions $W_k$.

### 20.2.1 Expressing the Return $dR$ as a Function of the $dX_j$

Itô's lemma implies

$$dR = \frac{1}{v(.)}\left(\frac{\partial v}{\partial t}(.) + \frac{1}{2}\sum_{i=1}^{q}\sum_{j=1}^{q}\gamma_{ij}(.)\frac{\partial^2 v}{\partial x_i \partial x_j}(.) + c(.)\right)dt + \sum_{j=1}^{q}\frac{1}{v}\frac{\partial v}{\partial x_j}(.)\,dX_j$$

or, upon "centering' the $dX_j$,

$$dR = \frac{1}{v(.)}\left(\frac{\partial v}{\partial t}(.) + \frac{1}{2}\sum_{i=1}^{q}\sum_{j=1}^{q}\gamma_{ij}(.)\frac{\partial^2 v}{\partial x_i \partial x_j}(.) + \sum_{j=1}^{q}\frac{\partial v}{\partial x_j}a_j(.) + c(.)\right)dt + \sum_{j=1}^{q}$$

$$\times \frac{1}{v}\frac{\partial v}{\partial x_j}(.)\,\left(dX_j - a_j(.)dt\right).$$

The drift ($[.]\,dt$) is nothing other than $\dfrac{D_t v + c(.)}{v(.)}$, and the $\dfrac{1}{v}\dfrac{\partial v}{\partial x_j}(.)$ terms represent the sensitivities $\beta_j$ of the asset return to the factors $x_j$. Thus we can express the return

850                    20  The State Variables Model and the Valuation Partial Differential Equation

$dR$ more simply, using the Dynkin operator and the sensitivities, as a standard multi-beta factorial model (see Chap. 23):

$$dR = \frac{D_t v + c(.)}{v(.)} dt + \sum_{j=1}^{q} \beta_j (dX_j - \alpha_j(.)dt). \qquad (20.2)$$

The random return of an asset is therefore equal to its expected value plus a correction of arbitrary sign but mean zero that depends on the sensitivity of the asset to the deviations of the state variables from their (also zero) means. Recall that a risk-free asset has zero sensitivity $\beta_j$ to each and every $j$.

### 20.2.2  Expressing the Return $dR$ as a Function of the $dW_k$

Replacing $dX_j$ by (20.1-b) in (20.2), we see immediately that

$$dR = \frac{D_t v + c(.)}{v(.)} dt + \sum_{k=1}^{m} \left( \sum_{j=1}^{q} \frac{1}{v} \frac{\partial v}{\partial x_j} (.)\omega_{jk}(.) \right) dW_k$$

This equation involves the $m$ sensitivities $\sigma_k \equiv \sum_{j=1}^{q} \frac{1}{v} \frac{\partial v}{\partial x_j}(.)\omega_{jk}(.)$ (or co-volatilities) of the asset return to the Brownian motions $W_k$. It can be rewritten more simply as:

$$dR = \frac{D_t v + c(.)}{v(.)} dt + \sum_{k=1}^{m} \sigma_k dW_k \qquad (20.3)$$

Equation (20.3) thus plainly shows the influence of variations in the Brownian motions. The volatilities $\sigma_k$ reflect the sensitivities of the return to the $W_k$ and can also be interpreted here as the betas of a multifactor model. According to which problem has to be solved, it is more ready to use Eq. (20.2) or its analog (20.3).

## 20.3  Expected Asset Returns and Arbitrage Pricing Theory (APT) in Continuous Time

We gave in Sect. 19.2 (Proposition 2) of the preceding chapter a version of APT valid for Itô processes. This version also works in this chapter's state variable model, but additional results, as well as a second version of APT, can be obtained in this more restricted framework.

## 20.3 Expected Asset Returns and Arbitrage Pricing Theory (APT) in Continuous Time

### 20.3.1 First Formula for Expected Returns

The expectation of the continuous return required by the market on the asset under consideration (and of assets with "comparable" durations and risks) is $\mu(t, \underline{X}(t))$. In AAO, the expected return on the asset must conform to this required market rate. Consequently, for $\underline{X}(t) = \underline{x}$, the following equation must hold:

$$\frac{D_t v(t, \underline{x}) + c(t, \underline{x})}{v(t, \underline{x})} = \mu(t, \underline{x}) \tag{20.4-a}$$

The APT gives an expression of this required return $\mu(.)$ which is assumed known for the moment. Making the Dynkin operator explicit yields the stochastic partial differential equation (PDE) that governs the pricing function $v(t, x)$:

$$\frac{\partial v}{\partial t}(.) + \sum_{i=1}^{q} \alpha_j(.) \frac{\partial v}{\partial x_j}(.) + \frac{1}{2} \sum_{i=1}^{q} \sum_{j=1}^{q} \gamma_{ij}(.) \frac{\partial^2 v}{\partial x_i \partial x_j}(.) = m(.)v(.) - c(.) \tag{20.4-b}$$

The functions $\alpha_j(t, \underline{x})$, $\gamma_{ij}(t, \underline{x})$ and $c(t, \underline{x})$ being assumed to be known, PDE (20.4-b) would theoretically allow us to find $v(t, \underline{x})$ if $\mu(t, \underline{x})$ were known. However, without a theoretical model for the returns required at equilibrium providing an explicit form for $\mu(t, x)$, Eqs. (20.4-a) or (20.4-b) are incompletely defined. Such a theory does exist: It is the continuous-time APT that was presented in Proposition 2 of the previous chapter. However, this formulation of APT has to be adapted to the state variables model discussed in this chapter. This version of APT, together with Eq. (20.4), leads to a general valuation method that will be provided in Sect. 20.4.

### 20.3.2 Continuous Time APT in a State variables Model

We derive the APT in continuous time in the present framework where a $q$-dimensional diffusion $\underline{X}(t)$ captures the state of the financial system. The dynamics of these state variables obey the diffusion process (Eq. 20.1) that depends on $m$ Brownian motions $\underline{W}$. We consider a financial market made up of $\eta$ different assets of which there are $n$ non-redundant primitive securities or assets. We assume $\eta > m \geq q$. Most often, one may reduce to the case $m = q$ by redefining the Brownian motions or the state variables, but we consider for the present the general case where $m$ and $q$ may be different. The asset $z$, with price $v_z(t, \underline{X}(t))$, distributes the continuous dividend $c_z(t, \underline{X}(t))$. The functions $v_z(t, \underline{x})$ and $c_z(t, \underline{x})$ are functions of $\mathbb{R}^{q+1}$ into $\mathbb{R}$ and assumed to be once differentiable with respect to time $t$ and twice with respect to $\underline{x}$ which denotes a special value taken on by $\underline{X}(t)$ at $t$. The analysis carried out previously, and especially Eqs. (20.2, 20.3), can be individually applied to each asset $z$ ($z = 1, \ldots, \eta$):

$$dR_z = \mu_z(.)dt + \sum_{k=1}^{m} \sigma_{zk}dW_k \tag{20.5}$$

$$dR_z = \mu_z(.)dt + \sum_{j=1}^{q} \beta_{zj}(dX_j - \alpha_j(.)dt) \tag{20.6}$$

As before, (.) stands for the arguments $(t, \underline{x})$.

We also assume that, for any date $t$, there exists a risk-free asset in the interval $(t, t + dt)$. Generally, the locally risk-free rate, $r(t, \underline{X}(t))$, depends both on time and the state of the economy $\underline{X}(t)$, but most often it is treated as one of the state variables.

We notice that (20.5) or (20.6) is not the most general formulation for a return-generating process leading to APT, for there may exist additional terms corresponding to risks specific to $dR_z.$[2]

In a state variables model, the continuous time APT is given by the following proposition:

### Proposition 1. APT in a State variables Model

*According to whether the Brownian motions $\underline{W}$ or the state variables $\underline{X}$ are considered as the sources of risk, the APT can be formulated in one of two equivalent ways.*
- *Let us consider the Brownian motions $\underline{W}$ as the sources of risk (the returns on different financial assets obey Eq. (20.5)). Under AAO, there exist $m + 1$ coefficients $r(t, \underline{x})$, $\lambda_1(t, \underline{x})$, $\ldots$, $\lambda_m(t, \underline{x})$, common to all assets, (and the first of which is the risk-free rate) such that the different instantaneous expected returns can be written as*

$$\mu_z(t,x) = r(t,x) + \sum_{k=1}^{m} \lambda_k(t,\underline{x})\sigma_{zk}(t,\underline{x}), \text{ for } z = 1, \ldots, \eta; \tag{20.7}$$

- *Let us consider the state variables $\underline{X}$ as the risk factors. The returns on different financial assets obey equations (20.6) and, under AAO, there exist $(q + 1)$ coefficients $r(t,\underline{x})$, $\kappa_1(t,\underline{x})$, $\ldots$, $\kappa_q(t,\underline{x})$, common to all assets, such that the instantaneous expected returns can be written as*

---

[2]Equation (20.5) may be written in a more general way, by introducing additional risks $S_i$:

$$dR_z = \mu_z(.) dt + \sum_{k=1}^{m} \sigma_{zk}dW_k + dS_z$$

where $dS_i$ is not correlated with $dS_z$ and $dW_k$ for $z \neq i$ and $k = 1, \ldots, m$. $dS_z$ represents the risk specific to firm $z$ in contrast with the (systematic) risks, common to all firms, that are related to $dW$ (GDP, inflation,...). In all well diversified portfolios, specific risks vanish (asymptotically) due to the law of large numbers and, therefore, APT still holds (asymptotically).

$$\mu_z(t,\underline{x}) = r(t,\underline{x}) + \sum_{j=1}^{q} \kappa_j(t,\underline{x})\beta_{zj}(t,\underline{x}), \quad \text{for } z = 1, \ldots, \eta \qquad (20.8)$$

The proof and the interpretation of Proposition 1 are the same as for Proposition 2 in Sect. 19.2 of the previous chapter. Eq. (20.7) means that the expected return on an arbitrary asset $z$ is equal, in AAO, to the return $r$ of the risk-free asset increased by the $m$ risk premia $\lambda_1 \sigma_{z1}, \ldots, \lambda_k \sigma_{zk}, \ldots, \lambda_m \sigma_{zm}$ relative to the corresponding risks $W_k$. Eq. (20.8) means that the expected return on an arbitrary asset $z$ is equal to the return $r$ of the risk-free asset increased by the $q$ risk premia $\kappa_1 \beta_{z1}, \ldots, \kappa_j \beta_{zj}, \ldots, \kappa_m \beta_{zq}$ corresponding to the risks $X_j$. $\kappa_j$ can be seen as the *market price of risk (MPR)* $X_j$ and $\beta_{zj}$ as the *intensity* of the risk $X_j$ borne by $z$.

The APT just presented does impose a restriction on the expected returns of the $\eta$ securities that all are linear affine combinations of $m$ MPR, or $q$ of them, according to which version of APT is adopted. The constraint is obviously stronger with fewer factors and a priori it seems advantageous to use whichever equation involves the smallest number of factors ($m$ or $q$).[3]

## 20.4   The General valuation PDE

The APT of the preceding section (version (20.7) or (20.8)), in spite of its name, is not a complete pricing model. It is necessary to add the PDE (20.4) given in §1 of the previous section. *Adding up* these two models yields a valuation PDE that we now derive.

### 20.4.1 Derivation of the General valuation PDE

Assume again that the state of the economy is given by the $q$-dimensional process $\underline{X}(t)$ that obeys Eq. (20.1). Consider a financial asset $z$ that could be one of the $n$ primitive securities or an asset derived from these securities. Asset $z$ distributes a continuous coupon $c_z(t, \underline{X}(T))$ and on date $T$ is worth $l_z(\underline{X}(T))$. The *functions* $c_z(t, \underline{x})$ and $l_z(\underline{x})$ are known, and we wish to find the pricing function $v_z(t, \underline{x})$.

The general method consists in getting the expected return $\mu_z(t, \underline{x})$ from APT (Eq. (20.7) or (20.8)) and writing that this return must equal $\frac{D_t v_z(t,\underline{x}) + c_z(t,\underline{x})}{v_z(t,\underline{x})}$. This

---

[3] As already mentioned, one can in general recover the case $m = q$ by redefining the Brownians or the state variables: if $m > q$, the number of Brownians can be reduced without affecting the joint probability distribution of returns that depends only on the $q$ variables $d\underline{X}$ (equation (20.6)), and if $q > m$ the variations $dX_j$ in the state variables are linearly dependent and we can redefine these variables such that they are reduced to a system of $m$ independent factors.

854      20 The State Variables Model and the Valuation Partial Differential Equation

leads to a partial differential equation (PDE) which the function $v_z(t, \underline{x})$ we seek must satisfy.

**Proposition 2 (valuation PDE)**
*Equation (20.7) leads to a first PDE for the price of any security $z$:*

$$D_t v_z - \sum_{k=1}^{m} \lambda_k(t, \underline{x}) \sigma_{zk}(t, \underline{x}) = r(t, \underline{x}) v_z(t, \underline{x}) - c_z(t, \underline{x}) \qquad (20.9)$$

*Alternatively, Eq. (20.8) leads to a second PDE:*

$$D_t v_z - \sum_{j=1}^{q} \kappa_j(t, \underline{x}) \frac{\partial v_z}{\partial x_j}(t, \underline{x}) = r(t, \underline{x}) v_z(t, \underline{x}) - c_z(t, \underline{x}) \qquad (20.10)$$

with, in both cases, the boundary condition:

$$v_z(T, \underline{x}) = l_z(\underline{x}). \qquad (20.11)$$

Recall that $\sigma_{zk} = \sum_{j=1}^{q} \frac{1}{v_z} \frac{\partial v_z}{\partial x_j}(.) \omega_{jk}(.)$ is the sensitivity of $z$ to the Brownian motion $k$.

All the financial assets obey the same Eq. (20.9) or (20.10) but differ by their terminal condition (20.11) which is specific to each one.

If *the market prices of risk ($\underline{\lambda}$ or $\underline{\kappa}$) are known*, PDE (20.9) or (20.10) with condition (20.11) can be solved, according to the case:

- By classical methods (change of variables, Laplace transforms...);
- By the Feynman-Kac formula, discussed in Sect. 20.7 below, that allows giving the solution $v_z(t, x)$ in the form of a conditional expectation;
- Bby numerical methods, such as finite differences (presented in Sect. 13.4 of Chap. 13), if analytic methods are insufficient.

We use this general methodology to build several models of the term structure of interest rates in Sect. 20.5.

## 20.4.2 Market Prices of Risk and Risk Premia

Recall that the $m$ market prices of risk (MPR) $\underline{\lambda}$ are relative to the risks $\underline{W}$ and the $q$ MPR $\underline{\kappa}$ to the risks $\underline{X}$. Consider the case where the number of Brownian motions is

## 20.4 The General valuation PDE

equal to the number of state variables, $q = m$, to which it is, in general, possible to reduce.[4]

Let us also assume that the market is complete, i.e. $m = q \leq \eta$.

Among the $\eta$ securities we can choose, therefore, $m$ non-redundant primitive assets, denoted $i = 1, \ldots, m$, as the *base* (the set of primitive securities) that serves as a reference for pricing all securities that do not belong to this set.

We rewrite Eqs. (20.7, 20.8) restricting them to the $m$ base securities:

$$\mu_i = r(t,\underline{x}) + \sum_{k=1}^{m} \lambda_k(t,\underline{x})\sigma_{ik}(t,\underline{x}) \text{ or } \mu_i = r + \sum_{j=1}^{q} \kappa_j(t,\underline{x}) \frac{1}{v_i} \frac{\partial v_i}{\partial X_j}(t,\underline{x}), \quad \text{for } i$$

$$= 1, \ldots, m$$

i.e., in matrix notation

$$\underline{\mu} = r\underline{1} + \Sigma\,\underline{\lambda}, \text{ or} \tag{20.12-a}$$

$$\underline{\mu} = r\underline{1} + \mathbf{B}\,\underline{\kappa}. \tag{20.12-b}$$

In these equations, $\underline{1}$ is the unit vector in $\mathbb{R}^m$, $\Sigma$ the $m \times m$ matrix of co-volatilities with generic term $\sigma_{ik}$ (volatility of security $i$ in relation to $W_k$) and $\mathbf{B}$ the $m \times q$ ($= m \times m$) matrix of sensitivities (betas) to the factors $X_j$. The general term of $\mathbf{B}$ is $\beta_{ij} = \frac{1}{v_i} \frac{\partial v_i}{\partial X_j}(t,\underline{x})$ and we recall that $\sigma_{ik} = \Sigma\,\beta_{ij}\,\omega_{ij}(.)$

This last equation in matrix form is $\Sigma = \mathbf{B}\Omega$, where $\Omega$ is is the $m \times m$ diffusion matrix for the state variables $\underline{X}$ (recall that $m = q$); by substitution in (20.12-a) and comparison with (20.12-b), we obtain $\mathbf{B}\Omega\,\underline{\lambda} = \mathbf{B}\underline{\kappa}$.

A necessary and sufficient condition for the $m$ securities in the base to be linearly independent, i.e., not redundant, is that the matrices $\mathbf{B}$ and $\Sigma$ are invertible. As a result, the MPR $\underline{\kappa}$ and $\underline{\lambda}$ are linked by the equation.

$$\underline{\kappa} = \Omega\,\underline{\lambda}. \tag{20.13}$$

### 20.4.3 The Relation between MPR and Excess Returns on Primitive Securities and the Condition for Market Completeness

Let us call $\pi_i$ the total premium, or excess return, on security $i$ required by the market. For each primitive security $i = 1, \ldots, m$, we write

$$\mu_i(t,\underline{x}) = r(t,\underline{x}) + \pi_i(t,\underline{x}).$$

From APT, this total premium $\pi_i$ equals the sum of the premia for the different risks $W_k$ or $X_j$; for example, (Eq. 20.12) leads to

---

[4] See Footnote 3.

$$\pi_i(t,\underline{X}) = \sum_{k=1}^{m} \lambda_k(t,\underline{x})\sigma_{ik}(t,\underline{x}), \quad \text{or} \quad \pi_i(t,\underline{X}) = \sum_{j=1}^{m} \kappa_j(t,\underline{x})\beta_{ij}(t,\underline{x})$$

or, in matrix notation,

$$\underline{\pi}(t,\underline{x}) = \Sigma(t,\underline{x})\,\underline{\lambda}(t,\underline{x}), \text{or} \qquad (20.14\text{-a})$$

$$\underline{\pi}(t,\underline{x}) = B(t,\underline{x})\,\underline{\kappa}(t,\underline{x}). \qquad (20.14\text{-b})$$

Since $\Sigma$ and $B$ are invertible (the $m$ primitive securities are not redundant), the MPR are given, uniquely, as a function of the excess return $\underline{\pi}$ by the relations: $\underline{\lambda} = \Sigma^{-1}\underline{\pi}$ and $\underline{\kappa} = B^{-1}\underline{\pi}$.

Often, $\underline{\pi}$ is assumed known as well as the $\beta_{ij}$, as a result, for example, of a multifactor regression on a sufficiently long history of returns; in this way, one may infer the MPR $\underline{\lambda}$ and $\underline{\kappa}$ from the dynamics of the base securities' returns, and price any security which is not in the base by using the valuation PDE. It should be pointed out that this presumption rests on the assumption that there exist at least $m$ non-redundant securities and that the market is therefore complete. In such a market a security not in the base is redundant and can thus be priced using arbitrage. If the market is incomplete, the MPR cannot be uniquely inferred from the dynamics of the returns on the primitive securities ($\Sigma$ and B are not invertible) and, the MPRs being unknown, the valuation PDEs (20.9) or (20.10) are indeterminate.

---

### Example of the Black–Scholes Model

The Black–Scholes (BS) model is based on the most restrictive set of assumptions and illustrates the preceding analysis well. In the BS universe, the underlying asset price $S$ is the only state variable; $S$ plays the double role of price of the (only one) primitive security and of state variable. $S(t)$, furthermore, does not distribute any dividend ($c(.) = 0$), and is governed by the following process with constant volatility:

$$dS = S\mu dt + S\sigma dW.$$

$\mu$ denotes the expected return on $S$, and the APT applied to $S$ implies that $\mu - r = \lambda\sigma$.

In addition, the diffusion coefficient of $S(t)$ is $\omega = S\sigma$, and consequently $\kappa = \omega\lambda = S\sigma\lambda = S(\mu - r)$.

We want to find the pricing function $C(t, S)$ of a call whose expected return can, because of Eq. (20.10), be written.

$D_t C - \kappa \frac{\partial C}{\partial S} = r\,C$, or else :

$\frac{\partial C}{\partial t} + \frac{\partial C}{\partial S}S\mu + \frac{1}{2}\frac{\partial^2 C}{\partial S^2}S^2\sigma^2 - S(\mu - r)\frac{\partial C}{\partial S} = r\,C$, i.e.

$\frac{\partial C}{\partial t} + \frac{\partial C}{\partial S}Sr + \frac{1}{2}\frac{\partial^2 C}{\partial S^2}S^2\sigma^2 = r\,C$, which is the PDE leading to the BS formula.

## 20.5 Applications to the Term Structure of Interest Rates

As an application of the method described in the preceding section, we again take up the model of the interest rate curve discussed in Chap. 17 that involves one risk factor, then, more succinctly, two-factor models and some extensions.

The analysis we adopt is an alternative approach to that in Chap. 17. The reader will usefully compare the two approaches.

We note $B_z(t)$ the price at $t$ of the zero-coupon free of credit risk that pays 1\$ on a fixed date $z$ ($> t$); the duration of this zero-coupon is consequently $(z-t)$ and its price is linked to the corresponding spot rate $r_{z-t}(t)$ by the equation $B_z(t) = e^{-(z-t)r_{z-t}(t)}$. Rate curve models give the prices $B_z$ of zero-coupons (or, equivalently, the rates $r_{z-t}$) as a function of one or more state variables or factors.

### 20.5.1 Models with One State Variable

The simplest models consider that all prices $B_z$ (and therefore the spot rates for different maturities) depend on a single state variable, for example, the very short-term (instantaneous) risk-free rate $r_0(t)$ that we denote $r(t)$ for simplicity.

The actual value of $r(t)$ at $t$ will be denoted by $r$. As a result, we write the zero-coupon price curve $\{B_z(t, r)\}_{z \in (t, t+T)}$.

The short-term rate $r$, which is by assumption the only state variable for the financial system under consideration, is required to obey a diffusion process that we write:

$$dr = \mu(t, r)dt + \sigma(t, r)dW.$$

The reader will note that this dynamics, written under the historical probability, differs in its drift ($\mu \neq m$) from the RN dynamics of Chap. 17 (Eq. 17.2).

With this notation, the diffusion coefficient for the state variable $r$ is denoted by $\sigma$ (it was denoted by $\omega$ in the preceding sections). Itô's lemma applied to $B_z(t, r)$ implies

$$\frac{dB_z}{B_z} = \frac{D_t B_z}{B_z} + \frac{1}{B_z} \frac{\partial B_z}{\partial r} \sigma(t, r)dW$$

We note that the problem thus posed is a special case of that treated in the previous section and comes under APT with a single state variable. Equation (20.9) thus can be applied and implies that the instantaneous expected return on each security, $\frac{D_t B_z}{B_z}$, equals the sum of the risk-free rate $r$ and a risk premium, $\lambda \frac{1}{B_z} \frac{\partial B_z}{\partial r} \sigma$ that is,

$$\frac{D_t B_z}{B_z} = r + \lambda(t, r) \frac{1}{B_z} \frac{\partial B_z}{\partial r} \sigma(t, r)$$

where $\lambda$ is the market price of the risk $W$ common to all securities, $\frac{1}{B_z} \frac{\partial B_z}{\partial r}$ the sensitivity of security $z$ to the interest rate $r$ and $\sigma$ its volatility.[5]

The product $\frac{1}{B_z} \frac{\partial B_z}{\partial r}$ can, therefore, be seen as the sensitivity of $B_z$ to fluctuations of $W$ and measures the only risk affecting $B_z$. Making $D_t$ explicit, we have

$$\frac{\partial B_z}{\partial t}(.) + \frac{\partial B_z}{\partial r}(.)[\mu(.) - \lambda(.)\sigma(.)] + \frac{1}{2} \frac{\partial^2 B_z}{\partial r^2}(.)\sigma^2(.) = rB_z(.). \qquad (20.15)$$

The function $B_z(.)$ whose form we want to find is thus the solution to PDE (20.15) with the boundary condition $B_z(z, r) = 1$. This condition constrains the value of the zero-coupon bond to be \$1 at maturity.

Remark that PDE (20.15) governs the value of an arbitrary contingent security, such as an interest-rate option, with only the boundary condition being specific to the security.

The formula for the function $B_z(.)$ obviously depends on the explicit form of the process assumed for $r(t)$, i.e. the form of the functions $\mu(t, r)$ and $\sigma(t, r)$ as well as the function $\lambda(t, r)$ representing the market price of risk.

The different models proposed differ in the $r(t)$ processes postulated. We examine two models already presented in Chap. 17, the Vasicek (1977) and Cox, Ingersoll, and Ross (1985 b) models.

### 20.5.1.1 The Vasicek Model

The model supposes that the short-term rate $r(t)$ obeys an Ornstein-Uhlenbeck process, as presented in Chapters 17 and 18, for which $\mu(t, r) = a(b-r)$; the volatility $\sigma(t, r) = \sigma$ and the price of risk $\lambda$ are assumed constant.[6]

The coefficients $a$, $b$, $\sigma$ are positive constants and $b$ is the "normal" long-term rate. The trend $a(b - r)$ implies that a mean-reverting force pulls the rate $r$ toward its normal value $b$.[7]

These assumptions imply that PDE (20.15) governing $B_z(t, r)$ writes

---

[5]Let us remark that this is *not* the standard definition of a bond sensitivity, which writes $\frac{1}{B_z} \frac{\partial B_z}{\partial r_{z-t}}$ where $r_{z-t}$ is the zero-coupon rate of duration $z$-$t$.

[6]Assuming a constant MPR is rather restrictive. Yet, as the MPR can be shown to be a function of the investors' average risk aversion, the assumption boils down to supposing this aversion stable over time (at least at short or medium horizons).

[7]Like $\mu$ differs from $m$, $b$ differs from $k$ present in the risk-neutral dynamics of the interest rate process (see Chap. 17). The relationships between the two sets of parameters are: $m = \mu$ - $\lambda\sigma$ and $k = b + \lambda\sigma/a$.

## 20.5 Applications to the Term Structure of Interest Rates

$$\frac{\partial B_z}{\partial t}(.) + \frac{\partial B_z}{\partial r}(.)[a(b-r) - \lambda\sigma] + \frac{1}{2}\frac{\partial^2 B_z}{\partial r^2}\sigma^2 = rB_z$$

with $B_z(z, r) = 1$ (the boundary condition).

Vasicek showed that the solution to this PDE reads:

$$B_z(t, r) = \exp[\alpha_\theta(1-r) - \theta r_\infty - \sigma^2\alpha_\theta^2/4a]$$

where $\theta$ denotes the duration ($z$–$t$) of the zero-coupon and $r_\infty$ is the interest rate at infinity:

$$r_\infty \equiv \lim_{\theta\to\infty} r_\theta(t, r) = b + \frac{\sigma\lambda}{a} - \frac{1}{2}\frac{\sigma^2}{a^2} \quad \text{and} \quad \alpha_\theta = \frac{1 - e^{-a\theta}}{a}.$$

This result means that the return on a zero-coupon tends toward a constant when its duration tends to infinity.

Furthermore, the equation $B_z(t, r) = e^{-\theta r_\theta(t)}$ implies that $r_\theta(t, r) = r_\infty - \frac{\alpha_\theta}{\theta} \times (1 - r) - \frac{\sigma^2}{4a\theta}\alpha_\theta^2$, a quantity that does tend to $r_\infty$ as $\theta$ tends to infinity.

These results can be written in the equivalent form of Eqs. (17.4, 17.5) in Chapter (17) (see footnote 7 for the relationship between the two formulations).

### 20.5.1.2 The One-Factor Cox–Ingersoll–Ross Model

Recall (see Chap. 17) that the interest rate model due to Cox, Ingersoll, and Ross (CIR, 1985 b) rests on a $r(t)$ process, different from Ornstein-Uhlenbeck, that writes under the historical probability

$$dr = a[b - r]dt + c\sqrt{r}dW \tag{20.16}$$

where $a$, $b$, and $c$ are positive constants.[8] The instantaneous variance of the rate is not constant but is equal to $c^2 r$. Note that negative interest rates are excluded.[9]

Equation (20.16) implies that $\mu(.) = a(b - r)$ and $\sigma(.) = c\sqrt{r}$ which, by substitution in (20.15), leads to a PDE that CIR solved. The solution is given in Chap. 17.

### 20.5.2 Multi-Factor models and valuation of Fixed-Income Securities

The Vasicek and CIR models present some disadvantages from an empirical viewpoint: on the one hand, the asymptotic end of the rate curve is fixed even if the rest of the curve can change, and on the other hand, the variations in rates of different

---

[8] Again, note that $b$ differs from $k$ present in the risk-neutral dynamics of chap. 17 (see equation (17. 3-a)).

[9] If on some date $t$ $r(t)$ hits zero, the variance of $dr$ (= $c^2 r dt$) is nil and $dr$ is surely equal to its expectation $ab\,dt$ which is positive. The interest rate then certainly increases from zero: zero is a *reflecting barrier* for the process (20.16) which $r$ obeys.

860    20 The State Variables Model and the Valuation Partial Differential Equation

durations are perfectly correlated (for a period of duration $dt$) and determined by $dr$. Models with several state variables make these disadvantages disappear.

### 20.5.2.1 Models with Two State Variables; the Brennan and Schwartz Model (1979, 1982)

For these models, the price $B_T$ depends on the current value $r$ of the short-term rate and on a second variable. Brennan and Schwartz choose as the second variable the long-term rate, denoted by $l$, which has an influence upon the normal rate toward which the short-term rate is attracted.[10] Therefore we denote the zero-coupon price at $t$, for a maturity $z$, by $B_z(t, r, l)$. Furthermore, the two-state variables $r$ and $l$ are assumed to follow a two-dimensional diffusion process.

As above, two-factor models are special cases of APT. As a result, one knows that the instantaneous expectation of the returns $\frac{D_t B_z}{B_z}$ of every zero-coupon $z$ is equal to the sum of the risk-free rate, $r$, and two risk premia. This leads to the following PDE for $B_z(t, r, l)$:

$$D_t B_z = r B_z + \kappa_1 \frac{\partial B_z}{\partial r} + \kappa_2 \frac{\partial B_z}{\partial l}. \tag{20.17}$$

Such a PDE, with its boundary conditions, $B_z(z, r, l) = 1$, can be solved numerically, for example by a finite-difference method such as the one given in Sect. 13.4 of Chap. 13, or (with difficulty) in exact form for certain special $(r, l)$ processes.

### 20.5.2.2 Multi-Factor Models; the APT Approach

Generally, when the rate curve depends on several factors $X_1, X_2, \ldots, X_q$, governed by an Itô process, the price of a zero-coupon $b_z(t, X_1, X_2, \ldots, X_q)$ obeys the APT:

$$D_t B_z = r B_z + \sum_{j=1}^{q} \kappa_j \frac{\partial B_z}{\partial X_j} \tag{20.18}$$

with the boundary condition $B_z(z, X_1, X_2, \ldots, X_q) = 1$.

Note that the models we present can be used to price securities whose values depend on interest rates, notably options. Indeed, any instrument whose value only depends on interest rates satisfies a PDE of type (15, 17), or (20.18) according to whether the rate curve is driven by one or more factors.[11] These instruments only differ in their boundary conditions that determine the precise form of the solution to the PDE.

---

[10] Choosing the second variable may pose theoretical problems. In particular, Dybvig, Ingersoll, and Ross (1996) have shown that the zero-coupon rate at infinity cannot decrease under AAO. In what follows, $l$ refers to the yield of the default-free perpetuity.

[11] Empirically, three factors generally suffice to produce a good enough approximation of the rate curve behavior. They are interpreted as the level, slope, and convexity of the curve, respectively.

20.5 Applications to the Term Structure of Interest Rates

### 20.5.2.3 Langetieg's Multi-factor Model (1980)

Langetieg's model involves several state variables gathered in a vector $\underline{X}(t) = (X_1(t), X_2(t), \ldots, X_q(t))$. We will just sketch it. To apply this interesting (and relatively unknown) model, the reader will supplement our discussion with some of the references suggested at the end of the chapter. The model rests on two assumptions:

(i) The vector $\underline{X}(t)$ of state variables obeys a multi-dimensional Ornstein-Uhlenbeck process written

$$d\underline{X}(t) = [-\mathbf{A}\underline{X}(t) + \underline{h}]dt + \mathbf{\Omega}d\underline{W}(t).$$

where $\underline{W}$ is standard $m$-dimensional Brownian motion, and $\mathbf{A}$ and $\mathbf{\Omega}$ are constant matrices; the process at work thus is mean-reverting.[12]

(ii) The spot rate $r(t)$ is an affine combination of state variables:

$$r(t) = \mu'\underline{X}(t) + \mu_0; \quad \mu \text{ and } \mu_0 \text{ are constants.}$$

This linearity greatly facilitates the econometric approach (using regressions).

The Vasicek model is a special case of Langetieg's ($\mu_0 = 0$; $\mu_1 = 1$ and $\mu_2 = \ldots = \mu_n = 0$), which encompasses much more general cases such as: $X_1(t) = $ real rate, $X_2(t) = $ inflation rate and $r(t) = X_1(t) + X_2(t)$.

A closed-form solution to the multi-dimensional PDE governing $X(t)$, that we do not give explicitly here, is a generalization of the one-dimensional Ornstein-Uhlenbeck SDE. The model is much simplified if $\mathbf{A}$ is diagonalizable.

The logarithm of the price $B_T(t)$ of zero-coupon instruments and the corresponding zero-coupon rate $r_{T-t}(t)$ can be written as an affine combination of state variables[13]:

$$r_\theta(t) = \underline{a}'_\theta[\underline{X}(t) - \underline{X}_0] + r_\theta(0)$$

The vector $\underline{a}_\theta$ is a function of $\mathbf{A}$, $\underline{h}$, $\mathbf{\Omega}$ and $\mu$. As in other multifactor models, the rate curve is not entirely fixed by the short-term rate but is influenced by other factors, conferring flexibility and realism to the model. Finally, remark that, since the model is Gaussian, options can be priced by a BSM-type model.

---

[12] Indeed the $i^{th}$ row of the SDE writes:

$$dX_i = \left(-\sum_{j \neq i}a_{ij}X_j - a_{ii}X_i + h_i\right)dt + \sum_{k=1}^{m}\omega_{ik}dW_k = a_{ii}\left(\frac{h_i}{a_{ii}} - \sum_{j \neq i}\frac{a_{ij}}{a_{ii}}X_j - X_i\right)dt + \sum_{k=1}^{m}\omega_{ik}dW_k,$$

and $X_i$ is attracted to $\frac{h_i}{a_{ii}} - \sum_{j \neq i}\frac{a_{ij}}{a_{ii}}X_j$ which varies randomly with the $X_j$.

[13] As we have seen, the models of Vasicek, Hull, and White, and CIR also are affine in their unique state variable, $r$. In this respect, all these models belong with an important "affine structure" class.

# 20.6 Pricing in the Risk-Neutral Universe

In this section, we recover the general valuation PDE simply from the dynamics of the relevant processes under the risk-neutral (RN) measure.

## 20.6.1 Dynamics of Returns, of Brownian Motions and of State Variables in the Risk-Neutral Universe

In an RN universe, the risk premiums vanish, the expectations of the returns of all risky assets equal the risk-free rate $r$ and, in the state-variables model, the returns of the basic securities can be written

$$dR = r(t,\underline{x})\underline{1}\, dt + \mathbf{\Sigma}(t,\underline{x})d\underline{W}*  \tag{20.19}$$

where $d\underline{W}^*$ is a standard Brownian motion in the RN universe.

In the real (or historical) universe, this dynamics writes

$$d\underline{R} = (r(.)\underline{1} + \underline{\pi}(.))dt + \mathbf{\Sigma}(.)d\underline{W} = (r(.)\underline{1} + \mathbf{\Sigma}(.)\underline{\lambda}(.))dt + \mathbf{\Sigma}(.)d\underline{W}, \text{ or}$$

$$d\underline{R} = r(.)\underline{1}dt + \mathbf{\Sigma}(.)\,[\underline{\lambda}(.)dt + d\underline{W}]  \tag{20.20}$$

where $d\underline{W}$ is a standard Brownian motion in the real universe and $(.)$ represents the set $(t, \underline{x})$.

We showed in the preceding chapter that moving from the real to the RN world increases the drift of processes without changing their diffusion or volatility parameters.

Comparing the two expressions for the returns in the RN and real universes shows that the dynamics of the RN Brownian $\underline{W}^*$ is connected to that of the historical Brownian by the equation

$$d\underline{W}^* = \underline{\lambda}(.)dt + d\underline{W}.$$

*The RN dynamics of any process is therefore easily obtained from its real-world dynamics by replacing* $d\underline{W}$ *by* $d\underline{W}^* - \underline{\lambda}(.)dt$. This simple rule is the thrust of the Girsanov theorem presented in the previous chapter which usually suffices to go back and forth from the real to the RN universes.

Let us apply this method, for example, to state variables whose evolution obeys Eq. (20.1):

$d\underline{X}(t) = \underline{a}(.)\, dt + \mathbf{\Omega}(.)\, d\underline{W}$, where $\underline{W}$ is a Brownian vector.

In the RN universe, their dynamics differ and their value, denoted by $\underline{X}^*$, obeys

## 20.6 Pricing in the Risk-Neutral Universe

$$dX^*(t) = [\underline{\alpha}(.) - \underline{\Omega}(.)\underline{\lambda}(.)]dt + \Omega(.)dW^*, \qquad (20.21\text{-}a)$$

or, alternatively,

$$dX^*(t) = [\underline{\alpha}(.) - \underline{\kappa}(.)]dt + \Sigma(.)dW^* \qquad (20.21\text{-}b)$$

where $\underline{W}^*$ is a Brownian vector in the RN universe.

### 20.6.2 The Valuation PDE

Let us take up again the valuation PDE, for example in the form (20.10), and spell out the Dynkin operator:

$$\frac{\partial v}{\partial t}(t,\underline{x}) + \sum_{j=1}^{q} \alpha_j(t,\underline{x})\frac{\partial v}{\partial x_j}(t,\underline{x}) + \frac{1}{2}\sum_{i=1}^{q}\sum_{j=1}^{q}\gamma_{ij}(t,\underline{x})\frac{\partial^2 v}{\partial x_i \partial x_j}(t,\underline{x}) - \sum_{j=1}^{q}\kappa_j(t,\underline{x})\frac{\partial v}{\partial x_j}$$

$$\times (t,\underline{x})$$

$$= r(t,\underline{x})v(t,\underline{x}) - c(t,\underline{x}).$$

Denoting $(t, \underline{x})$ by $(.)$ and regrouping the terms, we get the following pricing PDE followed by the value $v(.)$ of any asset, under AAO:

$$\frac{\partial v}{\partial t}(.) + \sum_{i=1}^{q}\left(a_j(.) - \kappa_j(.)\right)\frac{\partial v}{\partial x_j}(.) + \frac{1}{2}\sum_{i=1}^{q}\sum_{j=1}^{q}\gamma_{ij}(.)\frac{\partial^2 v}{\partial x_i \partial x_j}(.)$$

$$= r(.)v(.) - c(.) \qquad (20.22\text{-}a)$$

This PDE is completed by the boundary condition $v(T, \underline{x}) = l(\underline{x})$.

Taking (21) into account, the left-hand term of the PDE (20.22-a) is the Dynkin operator applied to $v$ but *relative to the dynamics of* state variables *in the RN world.* In this context, the drift of the state variables $\underline{X}^*$ is $a_j(t, \underline{x}) - \kappa(t, \underline{x})$ $(= a_j - \sum_{k=1}^{m}\omega_{jk}\lambda_k)$ and not $a_j(t, \underline{x})$. Denoting this "RN Dynkin operator" by $D^*_t$, we can write (20.22-a) in the equivalent and more compact form.

$$D^*_t v = r(.)v(.) - c(.) \qquad (20.22\text{-}b)$$

The reader will recall that PDE (20.22) is strictly equivalent to PDE (20.10), since it is obtained by a simple algebraic rearrangement of the terms of the latter. Pricing in the RN world consequently yields the same result as pricing in the real world, if care is taken to adjust the dynamics of the state variables or, equivalently, to adjust the Brownian motions.

864    20 The State Variables Model and the Valuation Partial Differential Equation

> **Examples**
> - Let us consider the BS model again. In the RN world, the dynamics of $S(t)$ writes $dS = Srdt + S\sigma\, dW^*$. With $D^*_t$ made explicit, Eq. (20.22-b) is BS's PDE.
> - Taking another look at a single-factor rate, the RN rate dynamics writes:
>
> $$dr^* = [\mu(t,r) - \lambda(t,r)\sigma(t,r)]dt + \sigma(t,r)dW^*.$$
>
> Equation (20.22-b) applied to the zero-coupon $b_z$ is PDE (20.15).

## 20.7 Discounting under Uncertainty and the Feynman–Kac Theorem

The Feynman-Kac theorem gives the solution to PDE (20.22) under the boundary condition in the form of a mathematical expectation (§1). It allows calculating the value of a financial instrument granting the right to a random cash flow, and thus constitutes the generalization to the uncertainty of the standard discounting technique under certainty (§2). The Appendix provides a proof in an important special case.

### 20.7.1 The Cauchy-Dirichlet PDE and the Feynman-Kac Theorem

The Cauchy-Dirichlet problem involves a PDE encountered in physics, which initially does not have a stochastic character.

Let $v$ be a function from $[0, T] \times \mathbb{R}^q$ to $\mathbb{R}$, whose values, written $v(t, \underline{x})$, satisfy the following three conditions:

- $v$ is once continuously differentiable w.r.t. $t$ in the interval $(0, T)$ and twice continuously differentiable w.r.t. $\underline{x}$ in an open set $Q$ of $\mathbb{R}^q$;
- $v$ obeys the following PDE:

$$\frac{\partial v}{\partial t}(t,\underline{x}) + \sum_{j=1}^{q} \eta_j(t,\underline{x})\frac{\partial v}{\partial x_j}(t,\underline{x}) + \frac{1}{2}\sum_{i=1}^{q}\sum_{j=1}^{q}\gamma_{ij}(t,\underline{x})\frac{\partial^2 v}{\partial x_i \partial x_j}(t,\underline{x})$$

$$= \psi(t,\underline{x})v(t,\underline{x}) - c(t,\underline{x}) \tag{20.23-a}$$

- $v$ satisfies the boundary condition $v(T, \underline{x}) = l(\underline{x})$.

The functions $\psi(.)$, $\eta_j(.)$ and $\gamma_{ij}(.)$ are given, continuous and bounded in $(0,T) \times \overline{Q}$ where $\overline{Q} = Q \cup \partial Q$ is the boundary of $Q$; $l(\underline{x})$ is given, continuous and bounded in $Q$; the matrix $\Gamma(.)$ with general term $\gamma_{ij}(t, \underline{x})$ is symmetric and positive definite for all $(t, \underline{x})$ in $(0,T) \times \overline{Q}$.

## 20.7 Discounting under Uncertainty and the Feynman–Kac Theorem

The problem is to characterize the solution $v(t, \underline{x})$ to the Cauchy–Dirichlet PDE (20.23-a).

The brilliant idea underlying the Feynman–Kac theorem consists in reformulating the problem in a probabilistic framework that a priori has nothing to do with it.

Denote by $\Omega(t, \underline{x})$ the invertible matrix such that $\Omega\Omega' = \Gamma$; such a matrix $\Omega$ exists and is unique since $\Gamma$ is symmetric and positive definite. Then we consider the $q$-dimensional random process $\underline{X}(t)$, which is to be seen as a vector of state variables obeying the SDE:

$$d\underline{X}(t) = \underline{\eta}(t, \underline{x})dt + \Omega(t, \underline{x})d\underline{W}.$$

Noticing that the left-hand term of (20.23-a) is the Dynkin operator applied to $v$, calculated at $t$ and with respect to the state variables $\underline{X}$, the PDE rewrites in the more compact equivalent form.

$$D_t v = \psi(.)v(.) - c(.). \tag{20.23-b}$$

The Feynman–Kac theorem gives a solution to this PDE as a mathematical expectation!

**Feynman-Kac Theorem** *Under the assumptions above, the solution to PDE (20.23-a) or (20.23-b) with the boundary condition $v(T, \underline{x}) = l(\underline{x})$ writes*

$$v(t, x) = E\left[\int_t^T c(u, \underline{X}(u))e^{-\int_t^u \psi(z, \underline{X}(z))dz} du + l(X(T))\, e^{-\int_t^T \psi(u, \underline{X}(u))du} \,\bigg|\, X(t) = x\right]$$
$$\tag{20.24}$$

The right-hand side of (20.24) is an expectation over the random variables $\underline{X}(u)$ for $u > t$ and conditional on the information available at time $t$ entirely contained in the value $x$ taken on by $\underline{X}(t)$ on date $t$. The Feynman–Kac theorem expresses the solution to the PDE (with no stochastic character) as an appropriately defined conditional expectation!

We will see that this theorem has extremely fruitful financial interpretations. The Appendix provides a proof in a special but important case that suits the valuation of standard options.

### 20.7.2 Financial Interpretation of the Feynman–Kac Theorem and Discounting under Uncertainty

Let us interpret the function $v(.)$ involved in the Cauchy–Dirichlet problem as the value of a financial asset that has to be found, $c(.)$ as the coupon rate it distributes, $\psi(.)$ as its instantaneous expected return at equilibrium, and $l(.)$ as its final value on

# 866    20 The State Variables Model and the Valuation Partial Differential Equation

date $T$. PDE (20.23) is then nothing else than Eq. (20.4) (with $\eta(.) = \alpha(.)$ and $\psi(.) = \mu(.)$), which states that its expected return under the historical measure equals $\mu$, or PDE (20.22) (with $\eta(.) = \alpha(.) - \kappa(.)$ and $\psi(.) = r(.)$) according to which its expected return equals $r$ in the RN universe.

The Feynman–Kac Theorem (20.24) asserts that, under certain conditions that are usually assumed to be satisfied in finance, the value of an asset equals the expectation of the present value (on date $t$) of the cash flows $c(.)$ that it pays from $t$ to $T$ and of its final value $l(.)$. Discounting is done either using the return rate $\mu(\cdot)$ in the historical universe, or the risk-free rate $r(\cdot)$ of the RN universe.

Applying eq. (20.24) is one of the most efficient ways to obtain an asset value. It emphasizes the equivalence between solving the PDE, which is this section's subject, and adopting the probabilistic approach using a conditional expectation discussed in the preceding chapter.

## 20.8    Summary

- In this chapter, prices are diffusion processes depending on a vector $\underline{X}(t)$ of $q$ state variables, which are themselves diffusion processes driven by a $m$-dimensional Brownian vector $\underline{W}(t)$.
- We consider any asset, with price $v(t, \underline{X}(t))$, value $l(\underline{X}(T))$ on terminal date $T$ and continuous payment stream $c(t, \underline{X}(t))dt$. The *pricing function* $v(t, \underline{X}(t))$ *must be determined* knowing the functions $l(.)$ and $c(.)$.
- Itô's lemma allows writing the return $dR$ in two *equivalent* forms: the first brings forward the sensitivity to the $X_j$ factors and the second the sensitivity to the Brownian motions $W_k$:

$$dR = \frac{D_t v + c(.)}{v(.)} dt + \sum_{j=1}^{q} \beta_j (dX_j - \alpha_j(.)dt) \text{ and } dR = \frac{D_t v + c(.)}{v(.)} dt + \sum_{k=1}^{m} \sigma_k dW_k$$

where $D_t$ is the Dynkin operator computed at $t$ relative to $\underline{X}(t)$ and involves partial derivatives of $v(t, \underline{X}(t))$.

- Then, continuous time APT returns obey two alternative non-arbitrage conditions:

There exist $m + 1$ *coefficients* $r(.), \lambda_1(.), ..., \lambda_m(.)$, the *same* for all assets, (the first being the risk-free rate, the $\lambda$'s being the $m$ *market prices of risk*) such that *all instantaneous expected returns* write as

## 20.8 Summary

(a) $\mu(t,\underline{X}(t)) = \dfrac{D_t v(t,\underline{X}(t)) + c(t,\underline{X}(t))}{v(t,\underline{X}(t))} = r(t,\underline{X}(t)) + \sum\limits_{k=1}^{m} \lambda_k(t,\underline{X}(t))\sigma_k(t,\underline{X}(t));$

There exist $q + 1$ *coefficients* $r(.), \kappa_1(.), ..., \kappa_q(.)$ (the $\kappa$'s are $q$ MPRs), the *same* for all assets, such that *all instantaneous expected returns* write as.

(b) $\mu(t,\underline{X}(t)) = \dfrac{D_t v(t,\underline{X}(t)) + c(t,\underline{X}(t))}{v(t,\underline{X}(t))} = r(t,\underline{X}(t)) + \sum\limits_{j=1}^{q} \kappa_j(t,\underline{X}(t))\beta_j(t,\underline{X}(t))$ .

a) and b) are 2 *partial differential equations* (PDE) with the *same boundary condition*: $v(T, \underline{X}(t)) = l(\underline{X}(T))$. Their solution is the pricing function $v(t, \underline{X}(t))$ sought after.

- The Black–Scholes–Merton model (one state variable $S$ and one Brownian), one-factor models of the return curve (one state variable $r$ and one Brownian) such as those of Vasicek and Cox–Ingersoll–Ross, and multi-factor models (e.g., Langetieg) are special cases of the previous general pricing model.
- The two general pricing PDEs can be established in the risk-neutral (RN) universe simply by adjusting the drifts of the relevant processes. The RN dynamics of any process is easily obtained from its real-world dynamics by replacing $d\underline{W}$ by $d\underline{W}*- \underline{\lambda}(.)dt$ where $d\underline{W}*$ is a standard Brownian motion in the RN universe and $\underline{\lambda}(\cdot)$ is the MPR present in PDE a).
- The RN *dynamics of the* state variables $\underline{X}$ thus can be obtained by this drift adjustment leading to a "RN Dynkin operator" $D^*_t$, and to the first *PDE* that rewrites:
- $D^*_t v = r(.) \, v(.) - c(.)$, called the *RN valuation PDE*.
- It expresses that the instantaneous expected return of any asset is equal to $r(.)$.
- The Feynman-Kac theorem gives the solution of the *valuation PDE* under a boundary condition in the form of a *mathematical expectation*. It allows calculating the value of a financial instrument returning random cash-flows, and is thus the generalization *to uncertainty* of the standard discounting:

$$v(t,X(t)) = E\left[ \int_t^T c(u,\underline{X}(u)) e^{-\int_t^u \mu(z,\underline{X}(z))dz} \, du + l(\underline{X}(T)) e^{-\int_t^T \mu(u,\underline{X}(u))du} \, \Big| X(t) = x \right]$$

- This result asserts that the value of an asset equals the expectation of the present value (at $t$) of the cash-flows $c(.)$ that it pays from $t$ to $T$ and of its final value $l(.)$. Discounting is done either using the return rate $\mu(\cdot)$ (i.e. $r(\cdot)$ plus a risk premium) in the historical universe, or the risk-free rate $r(\cdot)$ in the RN universe.

## Appendix

Proof of the Feynman–Kac theorem in a special case.

We are proving the Feynman–Kac formula in the special case of the PDE $D_t v(t, \underline{x}) = \psi(t, \underline{x}) v(t, \underline{x})$, with the boundary condition $v(T, \underline{x}) = l(\underline{x})$. Compared to the general problem (20.23-b) and its solution (20.24), the $c(t, \underline{x})$ term here vanishes, and we have to show:

$$\text{(i)} \quad v(t, x) = E[l(X(T))\exp - \int_t^T \psi(s, \underline{X}(s)ds | X(t) = x]$$

where $\underline{X}(t)$ is a vector of state variables obeying the diffusion $d\underline{X} = \underline{\mu}(t, \underline{X})dt + \underline{\Sigma}(t, \underline{X}) d\underline{W}$.

In finance, the problem is to price contingent security whose payoff (or final value) is $l(\underline{X}(T))$ on date $T$ and which pays no dividend $c(.)$ before $T$.

– Let us start by specializing the problem once more by assuming $\psi(.) = r = $ constant, which gives:

$$\text{(ii)} \quad D_t v(t, \underline{x}) = r \, v(t, \underline{x}), \text{ with the boundary condition } v(T, \underline{x}) = l(\underline{x}).$$

We could think of the value of a standard European option written on an underlying asset with price $X$, that pays no dividend between $t$ and $T$, in the Black-Scholes framework (RN universe, constant risk-free $r$, and $l(X(T)) = (X(T)-K)^+$ for a call and $(K-X(T))^+$ for a put.)

The Feynman-Kac formula $(i)$ we need to prove then simplifies to:

$$\text{(iii)} \quad v(t, \underline{x}) = E \, [l(\underline{X}(T)) \, e^{-r(T-t)} \, | \, \underline{X}(t) = \underline{x}].$$

Define $h(t, \underline{x}) = v(t, \underline{x}) \exp(-rt)$. Differentiating gives:
$D_t h(t) = D_t[v(t, \underline{x}) \exp(-rt)] = -r \exp(-rt) \, v(t, \underline{x}) + \exp(-rt) \, D_t v(t, \underline{x}) = 0$ by $(ii)$.
The process $h(.)$ has no drift, so is a martingale, whence.

$$E[v(T, X(T)) \, \exp(-rT) \, | X(t) = x] = v(t, x) \, \exp(-rt).$$

Replacing $v(T, X(T))$ with its value $l(X(T))$ we obtain $(iii)$.

Notice that, in the special case when $X(T)$ is log-normal, explicit computation of $(iii)$ gives the Black-Scholes formula.

– In the most general form of the PDE, $D_t v(t, \underline{x}) = \psi t, \underline{x}) v(t, \underline{x})$, we consider the process $h(t, x) = v(t, x) \exp - \int_0^t \psi(s, \underline{X}(s)ds$ which can be shown to be a martingale (since $D_t h(t) = 0$), which leads to: $v(t, \underline{x}) = E\left[l(X(T))\exp -\int_t^T \psi\left(s\underline{X}(s)dsX(t) = x\right)\right]$, which is $(i)$.

# Suggestions for Further Reading

## Books

Baxter, M., & Rennie, A. (1999). *Financial Calculus*. Press.
**Björk T., (2004). (second ed.), *Arbitrage theory in continuous time*, Oxford University Press.
**Duffie D. (1992). *Dynamic asset pricing models*, Princeton University Press.
*Elliot R., & Kopp, E. (2004). (second ed.), *Mathematics of financial markets*, Springer.
***Karatzas I., & S. Shreve, (1998). *Methods of mathematical finance*. Springer.
*Merton R., (1999). *Continuous time finance*. Basil Blackwell.
Wilmott, P., Howison, S., & Dewynne, J. (1995). *The mathematics of financial derivatives, a student introduction*. Press.

## Articles

Black, F. (1973). And M. Scholes, "the pricing of options and corporate liabilities". *Journal of Political Economy, 81*, 637–659.
Brennan, M. (1979). And E. Schwartz, "a continuous time approach to the pricing of bonds". *Journal of Banking and Finance, 3*, 133–155.
Brennan, M. (1982). And E. Schwartz, "an equilibrium model of bond pricing and a test of market efficiency". *Journal of Financial and Quantitative Analysis, 17*(3), 301–329.
Cox, J., & Ingersoll, J. (1985). And S. Ross (a), "an intertemporal general equilibrium model of asset prices". *Econometrica, 53*, 363–384.
Cox, J., & Ingersoll, J. (1985). And S. Ross (b), "a theory of the term structure of interest rates". *Econometrica, 53*, 385–408.
Dybvig, P. H., & Ingersoll, J. E. (1996). And S.a. Ross, "long forward and zero-coupon rates can never fall". *Journal of Business, 69*(1), 1–25.
Langetieg, T. C. (1980). A multivariate model of the term structure. *Journal of Finance, 35*, 71–97.
Merton, R. C. (1971). Optimum consumption and portfolio rules in a continuous time model. *Journal of Economic Theory, 3*, 373–413.
Merton, R. C. (1973). An intertemporal capital asset pricing model. *Econometrica, 41*, 867–888.
Vasicek, O. (1977). An equilibrium characterization of the term structure. *Journal of Financial Economics, 5*(2), 177–188.